Intracellular Delivery

FUNDAMENTAL BIOMEDICAL TECHNOLOGIES

Series Editor:
Mauro Ferrari, Ph.D. Houston, TX.

Volume 1: Martin, Donal. Nanobiotechnology of Biomimetic Membranes. 978-0-387-37738-4. 2006
Volume 2: Cristini, V., M. Ferrari, and P. Decuzzi. Nanoparticulate Delivery to Cancerous Lesions. 978-0-387-29085-0. Forthcoming
Volume 3: Bulte, J.W., and Mike Modo. Nanoparticules in Biomedical Imaging. 978-0-387-72026-5. 2007
Volume 4: Vladimir Torchilin. Multifunctional Pharmaceutical Nanocarriers. 978-0-387-76551-8. 2008
Volume 5: Aleš Prokop. Intracellular Delivery: Fundamentals and Applications. 978-94-007-1247-8. 2011

For further volumes:
http://www.springer.com/series/7045

Aleš Prokop
Editor

Intracellular Delivery

Fundamentals and Applications

Springer

Editor
Aleš Prokop
Chemical and Biological Engineering
Vanderbilt University
VU Station B 351604
37235 Nashville, Tennessee
USA
ales.prokop@Vanderbilt.Edu

ISSN 1559-7083
ISBN 978-94-007-1247-8   e-ISBN 978-94-007-1248-5
DOI 10.1007/978-94-007-1248-5
Springer Dordrecht Heidelberg London New York

Library of Congress Control Number: 2011929631

Printed on acid-free paper

Springer is part of Springer Science+Business Media (www.springer.com)

Contents

Part II Nanocarrier Formulation

Part III Medical Applications

Contributors

J. H. Adair Department of Materials Science and Engineering, Penn State University, University Park, PA 16802, USA

Maite Agüeros Department of Pharmaceutics and Pharmaceutical Technology, School of Pharmacy, University of Navarra, Pamplona 31080, Spain

Jenolyn F. Alexander Department of Nanomedicine, The Methodist Hospital Research Institute, 6670 Bertner Ave, MS R7-414, Houston, TX 77030, USA

Isabel D. Alves Laboratory of Biomolecule, Universite Pierre et Marie Curie, Paris, 75252 Cedex 05, France

Sándor Balogh Department of Information Technology, Kaposvár University, 40 Guba S, Kaposvár 7400, Hungary

Muriel Barberi-Heyob Centre de Recherche en Automatique de Nancy (CRAN), UMR 7039, Nancy-Université, CNRS, Centre Alexis Vautrin Brabois, Avenue de Bourgogne, 54511 Vandoeuvre-les-Nancy Cedex, France

Gillian Barratt CNRS UMR 8612, Faculty of Pharmacy, Université Paris-Sud XI, 5 rue J.B. Clément, Chatenay-Malabry F 92296, France

Nikhil Bhargava Department of Nanomedicine and Biomedical Engineering, University of Texas Graduate School of Biomedical Sciences at Houston, Houston, TX 77030, USA

Kawthar Bouchemal Physico-chimie, Pharmacotechnie, Biopharmacie, UMR CNRS 8612, Univ Paris-Sud, F-92296 Chatenay-Malabry, France

Ben J. Boyd Monash Institute of Pharmaceutical Sciences, Monash University, Melbourne, Australia

Jerry Chan UQ Centre for Clinical Research, University of Queensland, Brisbane, QLD 4029, Australia
and
Experimental Fetal Medicine Group, Department of Obstetrics and Gynecology, Yong Loo Lin School of Medicine, National University of Singapore, 5 Lower Kent Ridge Road, Singapore 119074, Singapore
and
Department of Reproductive Medicine, KK Women's and Children's Hospital, 100 Bukit Timah Road, Singapore 229899, Singapore

Stefano Chiesa Departamento Inteligencia Artificial, Facultad de Informatica, Biomedical Informatics Group, Universidad Politécnica de Madrid, Campus de Montegancedo, Madrid, Spain

Jihye Choi Department of Chemical and Biomolecular Engineering, Yonsei University, Seoul 120-752, Korea

Thibaud Coradin CNRS, Laboratoire de Chimie de la Matière Condensée de Paris, Collège de France, UPMC Univ Paris 06, 11 Place Marcelin Berthelot, F-75005 Paris, France

José Crespo Departamento Inteligencia Artificial, Facultad de Informatica, Biomedical Informatics Group, Universidad Politécnica de Madrid, Campus de Montegancedo, Madrid, Spain

Sophie Cribier Laboratory of Biomolecule, Universite Pierre et Marie Curie, Paris, 75252 Cedex 05, France

Béla Csukás Department of Information Technology, Kaposvár University, 40 Guba S, Kaposvár 7400, Hungary

Gerard G.M. D'Souza Department of Pharmaceutical Sciences, Massachusetts College of Pharmacy and Health Sciences, 179 Longwood Ave, Boston, MA 02115, USA

Jean-Olivier Durand Equipe Chimie Moléculaire et Organisation du Solide, Institut Charles Gerhardt Montpellier, UMR 5253 CNRS-UM2-ENSCM-UM1, CC1701 Place Eugène Bataillon 34095, Montpellier Cedex 05, France

Tarek El-Bialy Department of Biomedical Engineering, Faculty of Medicine & Dentistry and Engineering, University of Alberta, Alberta, Canada
and
Department of Dentistry, Faculty of Medicine & Dentistry, University of Alberta, Alberta, Canada

Irene Esparza Department of Pharmaceutics and Pharmaceutical Technology, School of Pharmacy, University of Navarra, Pamplona 31080, Spain

Socorro Espuelas Department of Pharmaceutics and Pharmaceutical Technology, School of Pharmacy, University of Navarra, Pamplona 31080, Spain

Mauro Ferrari Department of Nanomedicine, The Methodist Hospital Research Institute, 6670 Bertner Ave, MS R7-414, Houston, TX 77030, USA
and
Department of Bioengineering, Rice University, Houston, TX 77005, USA

Silvia Ferrati Department of Nanomedicine and Biomedical Engineering, University of Texas Graduate School of Biomedical Sciences at Houston, Houston, TX 77030, USA
and
Department of Nanomedicine, The Methodist Hospital Research Institute, 6670 Bertner Ave, MS R7-414, Houston, TX 77030, USA

Nicholas M. Fisk UQ Centre for Clinical Research, University of Queensland, Brisbane, QLD 4029, Australia

Ross Fitzsimmons Department of Biomedical Engineering, Faculty of Medicine & Dentistry and Engineering, University of Alberta, Alberta, Canada

Martin Fritts Nanotechnology Characterization Laboratory and SAIC-Frederick of Computer Science, National Cancer Institute, US National Institutes of Health, Baltimore, MD, USA

Céline Frochot Laboratoire Réactions et Génie des Procédés, UPR 3349, Nancy-Université, 1 rue Grandville, BP20451, 54001 Nancy Cedex, France

Miguel García-Remesal Departamento Inteligencia Artificial, Facultad de Informatica, Biomedical Informatics Group, Universidad Politécnica de Madrid, Campus de Montegancedo, Madrid, Spain

Biana Godin Department of Nanomedicine, The Methodist Hospital Research Institute, 6670 Bertner Ave, MS R7-414, Houston, TX 77030, USA

T.M. Goff Department of Materials Science and Engineering, Penn State University, University Park, PA 16802, USA

Sven Gohla Department of Pharmaceutics, Biopharmaceutics and NutriCosmetics, Institute of Pharmacy/Pharmaceutical Technology, Freie Universität Berlin, Kelchstr. 31, 12169 Berlin, Germany

Seungjoo Haam Department of Chemical and Biomolecular Engineering, Yonsei University, Seoul 120-752, Korea

Kateřina Horská Department of Nanobiotechnology, Institute of Nanobiology and Structural Biology of GCRC, Academy of Sciences of the Czech Republic, Na Sádkách 7370 05, České Budějovice, Czech Republic

Yong-Min Huh Department of Radiology, College of Medicine, Yonsei University, Seoul 120-752, Korea

Diana de la Iglesia Departamento Inteligencia Artificial, Facultad de Informatica, Biomedical Informatics Group, Universidad Politécnica de Madrid, Campus de Montegancedo, Madrid, Spain

Juan M. Irache Department of Pharmaceutics and Pharmaceutical Technology, School of Pharmacy, University of Navarra, Pamplona 31080, Spain

Yi Jia Department of Pharmaceutics, College of Pharmacy, Third Military Medical University, 30 Gaotanyan Zhengjie, Shapingba District, Chongqing 400038, China

Zhe Jin Company 13, Brigade of Cadets, Military Medical University, 30 Gaotanyan Zhengjie, Shapingba District, Chongqing 400038, China

Maria Jose Morilla Programa de Nanomedicinas, Departamento de Ciencia y Tecnologia, Universidad Nacional de Quilmes, Roque Saenz Peña 180, Bernal 1876, Buenos Aires, Argentina

Lisa M. Kaminskas Monash Institute of Pharmaceutical Sciences, Monash University, Melbourne, Australia

Cornelia M. Keck Department of Pharmaceutics, Biopharmaceutics and NutriCosmetics is a single word. Do not split. Institute of Pharmacy/Pharmaceutical Technology, Freie Universität Berlin, Kelchstr. 31, 12169 Berlin, Germany

Gabriela Kratošová Nanotechnology Centre, VŠB-Technical University in Ostrava, 17. listopadu 15, 708 33 Ostrava, Czech Republic

Casimir A. Kulikowski Department of Computer Science, Rutgers, The State University of New Jersey, Newark, NJ, USA

Afsaneh Lavasanifar Faculty of Pharmacy and Pharmaceutical Sciences, University of Alberta, Edmonton, AB T6G 2N8, Canada

Eddy S.M. Lee Richard M. Lucas Center for Imaging, Radiology Department, Stanford University, 1201 Welch Road, Stanford, CA 94305–5488, USA
and
UQ Centre for Clinical Research, University of Queensland, Brisbane, QLD 4029, Australia

Min-Hui Li Institut Curie, Centre de recherche; CNRS, UMR 168; Université Pierre et Marie Curie, 26 rue d'Ulm, F-75248 Paris, France

Xiaohui Li Department of Pharmaceutics, College of Pharmacy, Third Military Medical University, 30 Gaotanyan Zhengjie, Shapingba District, Chongqing 400038, China

Xiaodong Li Department of Oral and Maxillofacial Surgery, The Affiliated Stomatology Hospital, College of Medicine, Zhejiang University, Hangzhou 310006, Zhejiang, China

Eder Lilia Romero Programa de Nanomedicinas, Departamento de Ciencia y Tecnologia, Universidad Nacional de Quilmes, Roque Saenz Peña 180, Bernal 1876, Buenos Aires, Argentina

Jing Liu Department of Biomedical Engineering, School of Medicine, Tsinghua University, Beijing 100084, P. R. China

Y. Liu Department of Materials Science and Engineering, Penn State University, University Park, PA 16802, USA

Liping Lou Polytechnic Institute of NYU, 6 Metro Tech Center, Brooklyn, NY 11201, USA

Rainer H. Müller Department of Pharmaceutics, Biopharmaceutics and NutriCosmetics, Institute of Pharmacy/Pharmaceutical Technology, Freie Universität Berlin, Kelchstr. 31, 12169 Berlin, Germany

Victor Maojo Departamento Inteligencia Artificial, Facultad de Informatica, Biomedical Informatics Group, Universidad Politécnica de Madrid, Campus de Montegancedo, Madrid, Spain

Sylvie Masse CNRS, Laboratoire de Chimie de la Matière Condensée de Paris, Collège de France, UPMC Univ Paris 06, 11 Place Marcelin Berthelot, F-75005 Paris, France

Julien Nicolas Physico-chimie, Pharmacotechnie, Biopharmacie, Univ Paris-Sud, UMR 8612, Chatenay-Malabry F-92296, France
and
CNRS, Chatenay-Malabry F-92296, France

Seung Jae Oh Department of Radiology, College of Medicine, Yonsei University, Seoul 120-752, Korea

David Pérez-Rey Departamento Inteligencia Artificial, Facultad de Informatica, Biomedical Informatics Group, Universidad Politécnica de Madrid, Campus de Montegancedo, Madrid, Spain

Andrew M. Peters Department of Nanomedicine and Biomedical Engineering, University of Texas Graduate School of Biomedical Sciences at Houston, Houston, TX 77030, USA

O.A. Pinto Department of Materials Science and Engineering, Penn State University, University Park, PA 16802, USA

Aleš Prokop Nano Delivery International, Břeclav 69141, Czech Republic
and
Chemical and Biological Engineering, Vanderbilt University, Nashville, TN 37235, USA

Sandrine Quignard CNRS, Laboratoire de Chimie de la Matière Condensée de Paris, Collège de France, UPMC Univ Paris 06, 11 Place Marcelin Berthelot, F-75005 Paris, France

Laurence Raehm Equipe Chimie Moléculaire et Organisation du Solide, Institut Charles Gerhardt Montpellier, UMR 5253 CNRS-UM2-ENSCM-UM1, CC1701 Place Eugène Bataillon, 34095, Montpellier Cedex 05, France

Sébastien Richeter Equipe Chimie Moléculaire et Organisation du Solide, Institut Charles Gerhardt Montpellier, UMR 5253 CNRS-UM2-ENSCM-UM1, CC1701 Place Eugène Bataillon, 34095, Montpellier Cedex 05, France

Nicolas Rodriguez Laboratory of Biomolecule, Universite Pierre et Marie Curie, Paris, 75252 Cedex 05, France

Laura Rose Department of Biomedical Engineering, Faculty of Medicine & Dentistry and Engineering, University of Alberta, Alberta, Canada

John J. Rossi Irell and Manella Graduate School of Biological Sciences, Beckman Research Institute of City of Hope, City of Hope, 1500 East Duarte Rd, Duarte, CA 91010, USA

Brian K. Rutt Richard M. Lucas Center for Imaging, Radiology Department, Stanford University, 1201 Welch Road, Room PS064, Stanford, CA 94305–5488, USA

Sandrine Sagan Laboratory of Biomolecule, Universite Pierre et Marie Curie, Paris, 75252 Cedex 05, France

Rupa R. Sawant Department of Pharmaceutical Sciences and Center for Pharmaceutical Biotechnology and Nanomedicine, Northeastern University, Mugar Building, Room 312, 360 Huntington Avenue, Boston, MA 02115, USA

Adam Schröfel Nanotechnology Centre, VŠB-Technical University in Ostrava, 17. listopadu 15, 708 33 Ostrava, Czech Republic

Rita E. Serda Department of Nanomedicine, The Methodist Hospital Research Institute, 6670 Bertner Ave, MS R7-414, Houston, TX 77030, USA

Ranjita Shegokar Department of Pharmaceutics, Biopharmaceutics and NutriCosmetics is a single word. Do not split. Institute of Pharmacy/Pharmaceutical Technology, Freie Universität Berlin, Kelchstr. 31, 12169 Berlin, Germany

Melani Solomon Department of Pharmaceutical Sciences, Massachusetts College of Pharmacy and Health Sciences, 179 Longwood Ave, Boston, MA 02115, USA

Joo-Hiuk Son Department of Physics, University of Seoul, Seoul 130-743, Korea

Nelly E. Song Department of Nanomedicine and Biomedical Engineering, University of Texas Graduate School of Biomedical Sciences at Houston, Houston, TX 77030, USA

Srimeenakshi Srinivasan Department of Nanomedicine and Biomedical Engineering, University of Texas Graduate School of Biomedical Sciences at Houston, Houston, TX 77030, USA
and
Department of Nanomedicine, The Methodist Hospital Research Institute, 6670 Bertner Ave, MS R7-414, Houston, TX 77030, USA

Agathe K. Streiff Department of Nanomedicine and Biomedical Engineering, University of Texas Graduate School of Biomedical Sciences at Houston, Houston, TX 77030, USA

Jin-Suck Suh Department of Radiology, College of Medicine, Yonsei University, Seoul 120-752, Korea

Ivo Šafařík Department of Nanobiotechnology, Institute of Nanobiology and Structural Biology of GCRC, Academy of Sciences of the Czech Republic, Na Sádkách 7, 370 05 České Budějovice,Czech Republic
and
Regional Centre of Advanced Technologies and Materials, Palacký University, Šlechtitelů 11, 783 71, Olomouc, Czech Republic

Mirka Šafaříková Department of Nanobiotechnology, Institute of Nanobiology and Structural Biology of GCRC, Academy of Sciences of the Czech Republic, Na Sádkách 7, 370 05 České Budějovice, Czech Republic

A. Tabaković Department of Materials Science and Engineering, Penn State University, University Park, PA 16802, USA

Ennio Tasciotti Department of Nanomedicine, The Methodist Hospital Research Institute, 6670 Bertner Ave, MS R7-414, Houston, TX 77030, USA

Vladimir P. Torchilin Department of Pharmaceutical Sciences and Center for Pharmaceutical Biotechnology and Nanomedicine, Northeastern University, Mugar Building, Room 312, 360 Huntington Avenue, Boston, MA 02115, USA

Hasan Uludağ Department of Biomedical Engineering, Faculty of Medicine & Dentistry and Engineering, University of Alberta, Alberta, Canada
and
Department of Chemical & Materials Engineering, Faculty of Engineering, University of Alberta, Alberta, Canada
and
Faculty of Pharmacy & Pharmaceutical Sciences, University of Alberta, Alberta, Canada

Régis Vanderesse Laboratoire de Chimie-Physique Macromoléculaire, UMR CNRS-INPL 7568, Nancy Universite, 1 rue Grandville, BP20451, 54001 Nancy Cedex, France

Mónika Varga Department of Information Technology, Kaposvár University, 40 Guba S, Kaposvár 7400, Hungary

Christine Vauthier Physico-chimie, Pharmacotechnie, Biopharmacie, Univ Paris-Sud, UMR 8612, Chatenay-Malabry F-92296, France
and
CNRS, Chatenay-Malabry F-92296, France

Shih-Chang Wang Discipline of Medical Imaging, University of Sydney, Sydney, NSW 2006, Australia
and
Department of Radiology, Westmead Hospital, Westworthville, NSW 2145, Australia

Qian Wang Department of Biomedical Engineering, School of Medicine, Tsinghua University, Beijing 100084, P. R. China

Xiao-Bing Xiong Faculty of Pharmacy and Pharmaceutical Sciences, University of Alberta, Edmonton, AB T6G 2N8, Canada

Jianxiang Zhang Department of Pharmaceutics, College of Pharmacy, Third Military Medical University, 30 Gaotanyan Zhengjie, Shapingba District, Chongqing 400038, China

Jiehua Zhou Division of Molecular and Cellular Biology, Beckman Research Institute of City of Hope, City of Hope, 1500 East Duarte Rd, Duarte, CA 91010, USA

Yuxuan Zhu Department of Pharmaceutics, College of Pharmacy, Third Military Medical University, 30 Gaotanyan Zhengjie, Shapingba District, Chongqing 400038, China

Editorial and Introduction

Intracellular delivery: Fundamentals and applications A special volume of the series – Fundamental Biological Technologies (http://www.springer.com/series/7045), Professor M. Ferrari, Editor.

This book features a special subsection of Nanomedicine, an application of nanotechnology to achieve breakthroughs in healthcare. It exploits the improved and often novel physical, chemical and biological properties of materials only existent at the nanometer scale. As a consequence of small scale, nanosystems in most cases are efficiently uptaken by cells and appear to act at the intracellular level. Nanotechnology has the potential to improve diagnosis, treatment and follow-up of diseases, and includes targeted drug delivery and regenerative medicine; it creates new tools and methods that impact significantly existing conservative practices. This book more specifically targets nanotechnology in the area of drug delivery, i.e. the application of various nanoparticulates based on natural or synthetic, organic or inorgarnic materials as drug carriers, first of all to deliver drugs inside cells.

During the last decade, intracellular drug delivery has become an emerging area of research in the medical and pharmaceutical field. Many therapeutic agents can be delivered to a particular compartment of a cell to achieve better activity. It is very prolific field as it appears that the pace of discovery of intracellular drugs (to cellular or organ compartments) is proceeding faster than conventional ones. The lipidic nature of biological membranes is the major obstacle to the intracellular delivery of macromolecular and ionic drugs. Additionally, after endocytosis, the lysosome, the major degradation compartment, needs to be avoided for better activity. Various carriers have been investigated for efficient intracellular delivery, either by direct entry to cytoplasm or by escaping the endosomal compartment. These include cell penetrating peptides, and carrier systems such as liposomes, cationic lipids and polymers, polymeric nanoparticles, etc. Various properties of these carriers, including size, surface charge, composition and the presence of cell specific ligands, alter their efficacy and specificity towards particular cells. This book summarizes various aspects of targeted intracellular delivery of therapeutics including pathways, mechanisms, and approaches. Several case studies featuring a high success at an industrial scale are reviewed.

This volume is a collection of authoritative reviews. We first cover fundamental routes of nanodelivery devices cellular uptake, types of delivery devices, particularly in terms of localized cellular delivery, both for small drug molecules, macromolecular drugs and genes; all at academic and applied levels. Following is dedicated to enhancing delivery via special targeting motifs. Second, we introduce different types of intracellular nanodelivery devices (chemistry, although the coverage is far from complete) and ways of producing these different devices. Third, we put special emphasis on particular disease states and on other biomedical applications. Diagnostic and sensing is also included.

This is a very pregnant topic which will stir great interest. Intracellular delivery enables much more efficient drug delivery since the impact (on different organelles and sites) is intracellular as the drug is not supplied externally. There is a great potential for targeted delivery to certain cells or even to certain intracellular compartments with improved localized delivery and efficacy.

The Part I of this Volume deals with nanoparticle uptake and targeting and functionalization tools available to facilitate and enhance their internalization. The passive uptake and entry of nanoparticles into the subcellular space and its organelles could be enhanced with a targeting motifs via an active entry. Targeting, localized and intracellular delivery present still a key challenge to effective delivery. To establish an effective fight against diseases, we have to have the ability to selectively attack specific cells, while saving the normal tissue from excessive burdens of drug toxicity. However, because many drugs are designed to simply kill specified cells, in a semi-specific fashion, the distribution of drugs in healthy organs or tissues is especially undesirable due to the potential for severe side effects. Consequently, systemic application of these drugs often causes severe side effects in other tissues (e.g. bone marrow suppression, cardiomyopathy, neurotoxicity), which greatly limits the maximal allowable dose of the drug. In addition, rapid elimination and widespread distribution into non-targeted organs and tissues requires the administration of a drug (in a suitable carrier) in large quantities, which is often not economical and sometimes complicated due to non-specific toxicity. This vicious cycle of large doses and the concurrent toxicity is a major limitation of many current therapies. In cancer treatment, in many instances, it has been observed that the patient succumbs to the ill effects of the drug toxicity far earlier than the tumor burden (for example). All above calls for focused attention to mechanisms of entry and targeting and studies how to avoid reticulo-endothelial system (RES) capture (sometimes such affinity is employed to our advantage – see lymphatic targeting). Futhermore, while the nanoparticles entry is always associated with particle repulsion, exocytosis is also covered, although it deserves greater atention as many carriers/drugs could be eliminated from inside the cell even they enter the intracellular millieau. Some special peptide-based targeting mechanisms are covered at the end of this part. Such peptide motifs are often employed to facilitate proper cellular entry once associated with nanocarriers.

For a single, even very sizable book, some important topics unfortunately are not covered here, such as dual targeting, combination therapy, multivalency, transport across the blood-brain barrier and some aspects of toxicity. Other neglected area is that of chemical vectors for non-viral gene delivery that mimic some viral functions and ultrasound-enhanced uptake.

Nanoparticle preparation is the subject of Part II of the Volume. Although not all possible chemistries are being covered, still the available information presents this important area in good details. Some chapters deal with employment of natural polymers, others with synthetic or semi-synthetic ones. Inorganic systems are also described. A special emphasis is on the synthesis of magnetic nanoparticles and metalic-based, biosynthetically synthesized nanoparticles, which are present in a bionanocomposite (enmesshed) form with microroganisms. This Part is closed by articles on processing and scale-up, the very significant consideration often limiting the commerical application in general. Some special formulations are also covered in Part III of this Volume (e.g., inorganic – calcium phosphate nanoparticles).

The Part III deals with specific medicinal applications. In doing so, we were unable to cover all related topics because of Volume space limitation. Among those covered, some are quite new, such as new imaging modes and stem cell tracking. In our opinion, a special attention should be paid to coverage of bioinformatics (and systems biology) as an important tool, which is going to shape the whole medicine in a near future.

We strongly believe that the intracellular delivery/therapy is a very pregnant topic, which will stir great interest. Intracellular delivery enables much more efficient drug delivery since the impact (on different organelles and sites) is intracellular as the drug is not supplied externally within the blood stream.

As was said, the topic of intracellular delivery is so broad that not all aspects could be covered. Great emphasis is on the state-of-the art topics. Unfortunately, we could not avoid some overlap in covering individual topics. Thus, while there is a considerable attention (and some overlap) devoted to entry mechanisms and targeting, we would like to emphasize its importance towards the rational nanocarrier design.

The Editor would like to profoundly thank all Contributors to this Volume for their cooperation and enthusiasm. It is also noted that the authorship coverage is truly international, in spite of apparent "concentration" of nanoactivity in USA.

Aleš Prokop, Editor and
Vladimir P Torchilin
Northeastern University, Boston, USA

Part I
Mechanisms of Uptake and Targeting

Mass Transport via Cellular Barriers and Endocytosis

Silvia Ferrati, Agathe K. Streiff, Srimeenakshi Srinivasan, Jenolyn F. Alexander, Nikhil Bhargava, Andrew M. Peters, Nelly E. Song, Ennio Tasciotti, Biana Godin, Mauro Ferrari, and Rita E. Serda

Abstract Mass transport within body compartments and across biological barriers negatively impacts drug delivery but also presents opportunities to optimally design drug carriers that benefit from novel differentials presented in pathological tissue. As an example, cancer presents unique alterations in vascular permeability, osmotic pressure, cellular zip-codes, and numerous other physical parameters that can be used to achieve preferential accumulation of imaging and therapeutic agents at the cancer lesion. This chapter describes the journey of drug delivery from the site of administration to the appropriate subcellular compartment within the target cell. Design parameters for optimal fabrication of nanoparticle-based carriers, including size, shape, elemental composition, surface staging, and hierarchical ordering of multi-particle complexes are presented. The overall objective of this chapter is to enhance our understanding of mass transport in order to facilitate the development of carriers for therapy and diagnostics of various pathological conditions.

Keywords Mass transport • Cellular targeting • Drug delivery • Endocytosis • Nanoparticle • Cellular zip codes • Nanoparticle carriers • Multi-particle complexes • Nanoparticle parameter tuning • Particle delivery • Intracellular transportation • Intracellular delivery • Paracellular transport

---

S. Ferrati, A.K. Streiff, S. Srinivasan, N. Bhargava, A.M. Peters, and N.E. Song
Department of Nanomedicine and Biomedical Engineering, University of Texas Graduate School of Biomedical Sciences at Houston, Houston, TX 77030, USA

S. Ferrati, S. Srinivasan, J.F. Alexander, E. Tasciotti, B. Godin, and M. Ferrari
Department of Nanomedicine, The Methodist Hospital Research Institute, 6670 Bertner Ave, MS R7-414, Houston, TX 77030, USA

M. Ferrari
Department of Bioengineering, Rice University, Houston, TX 77005, USA

R.E. Serda (✉)
Department of Nanomedicine, The Methodist Hospital Research Institute, 6670 Bertner Ave, MS R7-414, Houston, TX 77030, USA
e-mail: reserda@tmhs.org

A. Prokop (ed.), *Intracellular Delivery: Fundamentals and Applications*, Fundamental Biomedical Technologies 5, DOI 10.1007/978-94-007-1248-5_1,

Abbreviations

RES	reticulo-endothelial system
SLC	solute carriers
OATP	organic anion transporting polypeptides
PEPT-1 and SLC15A1	Oligopeptide Transporter 1
MCT1 and SLC16A1	mono-carboxylic acid transporters
gp60	cell surface glycoprotein receptor
ECM	extracellular matrix
VEGF	vascular endothelial growth factor
EPR	enhanced permeability and retention
TGF-β1	transforming growth factor beta one
CendR	C end Rule
iRGD	internalizing Arginine-Glycine-Aspartic acid
HGC	hydrophobically modified glycol chitosan
LAMP-1	Lysosomal-associated membrane protein 1
SWNT	single walled carbon nanotubes
I-CAM	intercellular adhesion molecule one
GRP78	78 kDa glucose-regulated protein
DTPA	Diethylene triamine pentaacetic acid
EC20	Endocyte 20
EC17	Endocyte 17
EC145	Endocyte 145
LHRH	leutinizing hormone releasing hormone
PLGA	Poly(lactide-co-glycolide)
LDL	low density lipoproteins
PSMA	Prostate Specific Membrane Antigen
PEG	polyethylene glycol
siRNA	small interfering RNA
gp120	HIV-1_{BAL} envelop glycoprotein
PK2	doxorubicin carrying polymer conjugated to galactose
ADEPT	antibody directed enzyme prodrug therapy
NV	nanovehicular
QD	quantum dots
PEI	polyethylenimine
NLS	nuclear localization signal
HIV-TAT	human immunodeficiency virus Trans-Activator of Transcription
ER	endoplasmic reticulum
V+ATPase	vacuolar proton transporter
EMMA	Endosomolysis by Masking of a Membrane-Active Agent
RER	rough endoplasmic reticulum
SER	smooth endoplasmic reticulum
PDI	protein disulfide isomerase
BiP	binding immunoglobulin protein

KDEL	Lys-Asp-Glu-Leu
KDEL-R	Lys-Asp-Glu-Leu receptor
GFP	green fluorescent protein
PERK	proline-rich, extensin-like receptor kinase
STX-B	Shiga toxin
mtDNA	mitochondrial DNA
OMM	outer mitochondrial membrane
IMS	intermembrane space
IMM	inner mitochondrial membrane
TOM	translocases of the outer membrane
TIM	translocases of the inner membrane
ANT	adenine nucleotide translocase
BODIPY	boron-dipyrromethene
TPP	triphenylphosphonium
HPMA	hexamethylphosphoramide
iSur-pDNA	suppressor of metastatic and resistance-related protein survivin
DQAsomes	mitochondria-specific nanocarrier system prepared from the amphiphilic quinolinium derivative dequalinium chloride
MITO-porter	liposome-based carrier with a high density of fusogenic R8
R8	eight arginines
Mito-8	mitochondria selective peptide
NPC	nuclear pore complex
NLS	nuclear localization signal
Ran	RAs-related Nuclear protein
K	lysine
R	arginine
GFP	green fluorescent protein
CPP	cell penetrating peptide
LEV	logic-embedded vector
VEGFR-2	vascular endothelial growth factor receptor two
PECAM	Platelet endothelial cell adhesion molecule

1 Introduction

Depending on the mode of administration, the nature of the drugs, and their formulation, molecules or particles have to bypass multiple biological barriers before reaching the target site. Among these barriers are sequestration by the reticulo-endothelial system (RES); transport across cellular (vascular endothelium and gastrointestinal epithelial lining) and stromal (tissue interstitium) barriers; and crossing cell and subcellular organelle membranes. These processes involve mass transport at multiple levels bridging biology, chemistry, physics and mathematics with the goal of understanding the mechanisms that govern transport across bio-barriers in order to develop more effective therapeutics with fewer side effects.

In this chapter we will review the main mechanisms of drug uptake for the most common routes of administration starting from the macro-level of absorption (across intestinal epithelium or vascular endothelium), then moving to cellular uptake, and finally arriving at the level of intracellular trafficking and subcellular targeting. The objective of this chapter is to enhance our understanding of mass transport to facilitate the development of carriers for cancer therapy and diagnostics. Unique characteristics that originate from the tumor microenvironment will be emphasized, as well as the ability to use these traits to achieve preferential accumulation of imaging and therapeutic agents at the cancer lesion. Nanoparticles, which are being explored as drug carriers, are discussed with an emphasis on the ability to tune their chemical and physical properties to achieve preferential accumulation at the lesion.

2 Epithelial/Endothelial/Stromal Barriers: Mechanisms to Cross Barriers

2.1 *Gastrointestinal Administration: Epithelial Barriers*

When given orally, a drug must cross barriers such as the mucus layer and the unstirred water layer, or the epithelia cell membrane of the intestine and the underlying tissues before reaching the systemic circulation and arriving at the site of action. In general, biological membranes are composed primarily of lipid bilayers, containing cholesterol and phospholipids which provide stability and determine their permeability characteristics. Proteins that are involved in receptor-mediated transport and the regulation of cellular metabolism are also embedded in these membranes. Since epithelial cells are polarized, the membrane facing the two sides, that is, the apical (facing the intestinal lumen) and the basolateral (facing the internal milieu and the blood supply) have different lipid and protein compositions and thus different permeability properties. As shown in Fig. 1, transepithelial transport of drugs can be achieved either across the cells (via transcellular or intracellular pathways) or through the junctions between cells (paracellular or intercellular pathways).

Transcellular transport involves drugs crossing the barrier through exploitation of one of the following mechanisms: passive diffusion, facilitated passive diffusion, active transport or pinocytosis (Fig. 1) (Stenberg et al. 2000; Gad 2008; Siccardi et al. 2005).

2.1.1 Passive Transcellular Diffusion

Most drugs are transported across the gastrointestinal epithelium by passive diffusion moving from a compartment at higher concentration, the intestinal lumen, to one of lower concentration, the blood circulation, where the drug is more diluted

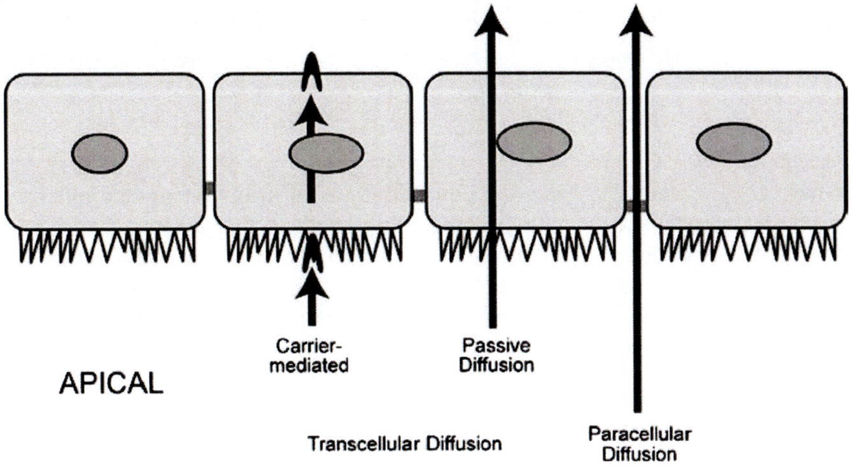

Fig. 1 Schematic illustrating several mechanisms of drug transport across the epithelium (Reprinted from Gad (2008). Courtesy of Wiley Publishing Group)

and rapidly distributed to tissues. The kinetics of transport are therefore linear or first order and the drug absorption can highly depend on the dissolution of the dosage form (e.g. tablet, capsule), which can be the rate limiting step (Muranishi 1990). Transport is affected by the physico-chemical characteristics of the drug, such as lipophilicity, size, and ionization, the latter depending on the pKa of the drug. Since the molecules have to cross the lipid bilayer, small hydrophobic molecules are favored. Moreover, uncharged molecules display less repulsion for charged groups on the membrane surface and are therefore assimilated more easily. Drugs that are weakly acidic, such as aspirin, are neutral in acidic environments and are therefore more easily assimilated in the stomach, while weak bases, such as most benzodiazepines, are better absorbed in more alkaline environments such as the duodenum. However, it is important to emphasize that the main site of absorption in the gastrointestinal tract is the small intestine and, in particular, jejunum. Physiologically, this segment of the gastrointestinal system is designed to absorb the nutrients from ingested food, while the large intestine's role is to absorb water and ions. The small intestine is lined with simple columnar epithelium covered with folds, called plicae circulars. These folds consist of finger-like projections, or villi, characteristic of epithelial cells, and each cell also has hair-like projections, called microvilli. Each villus is fed with a network of blood and lymphatic vessels. This very specific structure has the important function of increasing the surface area available for absorption of the exogenous substances.

The fundamental equation describing passive drug transport through membranes is based on Fick's first law, where the constant of proportionality between the drug flux through a membrane and the gradient of concentration define the effective permeability coefficient of the drug through the membrane (P_{eff}) (Pang 2003).

P_{eff} can be estimated experimentally by measuring the drug loss across the lumen at any given segment:

$$P_{eff} = \frac{Q_{lumen}}{2\pi rl} ln\left(\frac{C_{out,lumen}}{C_{in,lumen}}\right)$$

Where $C_{out,lumen}$ and $C_{in,lumen}$ are the concentrations of drug leaving and entering the lumen respectively, r is the radius, l the length of the lumen segment and Q_{lumen} is the experimentally determined luminal flow rate. P_{eff} is dependent on the physico-chemical properties of the drug which ultimately determine drug partitioning between the hydrophilic exterior of the membrane (P_{aq}) and the lipophilic bilayer (P_m):

$$\frac{1}{P_{eff}} = \frac{1}{P_{aq}} + \frac{1}{P_m}$$

2.1.2 Facilitated Passive Diffusion and Active Transport

Some of the endogenous substances or nutrients (e.g. vitamins, ions, sugars, amino acids) are absorbed through the processes such as facilitated passive diffusion and active transport. Both involve the drug reversibly binding to a carrier protein at the cell membrane exterior, followed by the carrier-substrate complex being transported across the lipid layers by proteins embedded in the membrane (Tsuji and Tamai 1996). Carrier-mediated diffusion is characterized by selectivity, saturability and regional variability (Tanaka et al. 1998). The carrier transports only substrates with a relatively specific molecular configuration and the process is limited by the availability of carriers. Saturation of carriers results in a non-linear dose-response relationship. Facilitated passive diffusion does not require energy expenditure by the cell and therefore transport against a concentration gradient does not occur through this mechanism.

Transport against a concentration barrier can be achieved by active transport, which requires both associated carrier proteins and energy expenditure by the cell. Membrane transporters can be divided in two major super-families: ATP-binding cassette and solute carriers (SLC). These carrier proteins are expressed in different degrees in different regions and of particular importance are those present in the intestinal epithelia, hepatocytes, and proximal tubules of the kidney and in the endothelium of the blood-brain barrier, as shown in Fig. 2. Among the specialized small-intestine transport systems are organic anion transporting polypeptides (OATP), Oligopeptide Transporter 1 (PEPT-1 and SLC15A1), and mono-carboxylic acid transporters (MCT1 and SLC16A1) (Giacomini et al. 2010). In contrast to these specialized drug transport mechanisms, there are processes that obstruct drug penetration by expelling therapeutics from cells. These include membrane efflux pumps. An example is cell surface p-glycoprotein. P-glycoprotein actively opposes drug absorption by transporting many drugs from the intracellular environment to the intestinal lumen (Hunter and Hirst 1997).

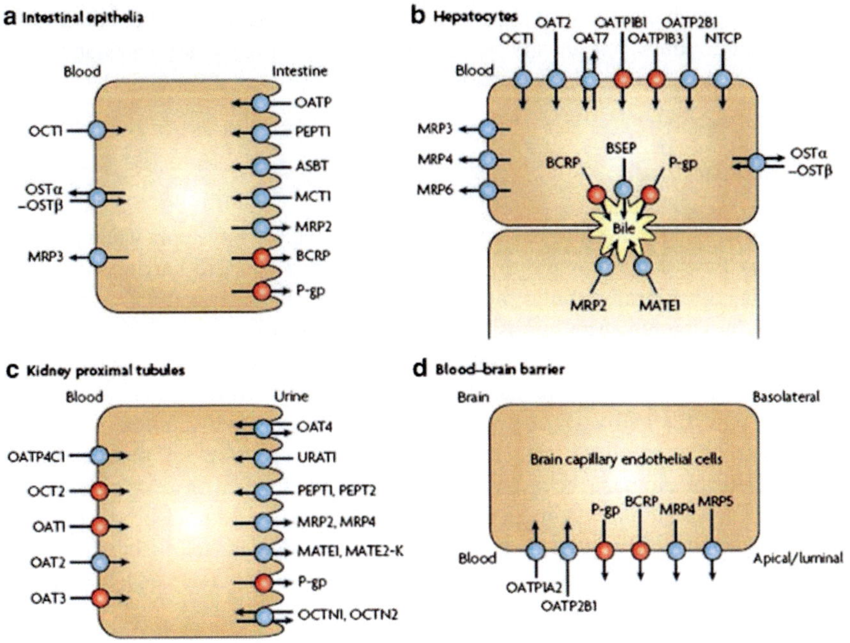

Fig. 2 Specialized transport systems for drugs and endogenous substances (Reprinted from Giacomini et al. (2010). Courtesy of Nature Publishing Group)

2.1.3 Paracellular Transport

Paracellular transport is defined as the pathway along the water-filled tight junctions between adjacent cells. Tight junctions normally regulate the trafficking of nutrients, fluids and small molecules, with a cut-off of 500 Da (Siccardi et al. 2005; Gad 2008). This type of transport occurs through diffusion and based on its aqueous nature, it is most relevant for small hydrophilic drugs that cannot penetrate cell membranes (Nellans 1991; Karlsson et al. 1999).

2.2 Intravenous Administration: Endothelial Barriers

The delivery of therapeutic agents to a target site after systemic administration involves three processes which all require bypassing biological barriers: distribution through vascular compartments, extravasation and transport in the interstitium (Au et al. 2001; Ferrari 2010). These processes are influenced by the physicochemical properties of the drug, including size, surface charge, surface flexibility, and binding with extracellular and intracellular macromolecules, as well as the physiological properties of the targeted site, which, in the case of cancer, includes abnormal vasculature, altered extracellular matrix components, interstitial fluid pressure, and tumor cell density. All of these aspects affect drug distribution and thereby the cytotoxicity of the treatment.

2.2.1 Transcytosis

Transcytosis is equivalent to the small-intestine transporter systems where macromolecules pass the endothelium barrier bound to a specific carrier through a receptor-mediated transcellular mechanism. A well characterized pathway is that mediated by albumin. Albumin is the most abundant protein found in plasma and can function as a cargo chaperone binding to many substrates in the plasma, facilitating their delivery across the endothelial barrier. The albumin-molecule complex interacts with the cell-surface glycoprotein (gp60) receptor which subsequently binds to protein caveolin-1 triggering the formation of transcytotic vesicles (caveolae). These phenomena have an important clinical impact on drug uptake and efficacy. One example is Abraxane, a Cremophor® free albumin-bound 130-nm particle formulation of the chemotherapeutic agent paclitaxel which has been shown to improve paclitaxel tumor cell penetration with reduced side effects (Desai et al. 2006).

2.2.2 Vasculature Permeability and Models for Passive Drug Transport Across the Endothelium

The vascular endothelium is a semi-permeable barrier which controls the passage of macromolecules and fluids in the interstitial space assuring tissue fluid homeostasis. Interactions between endothelial cells and pericytes, or integrin-mediated interactions with extracellular matrix (ECM) proteins, such as collagen IV, fibronectin and laminin, activates signaling pathways, keeping the endothelium quiescent and leading to stable vessels with controlled junctions and permeability. The passive extravasation of molecules across the endothelium occurs by diffusion and convection through two primary mechanisms: paracellular and transcellular routes. In the paracellular route, small molecules (<3 nm), such as urea and glucose, can pass the endothelium barrier through junctions between cells, moving down a concentration gradient. The transcellular pathway instead mediates the transport of large molecules (>3 nm), such as plasma proteins (Mehta and Malik 2006). For example, albumin exploits the receptor-mediated caveolae pathway as well as transiently formed transcellular channels. This architecture can be altered during disease states in which inflammatory mediators, such as VEGF (vascular endothelial growth factor) and thrombin, alter the organization of intercellular junctions and interactions between cells and ECM, creating intercellular gaps which allow the uncontrolled flow of plasma protein and liquids across the endothelium, causing swelling and immunological responses.

These two mechanisms of transport can be described by fluidic models (Jain 1987, 1990, 1999) where the diffusive components of the transport are proportional to the exchange vessel's surface area, S (cm^2) and the difference between the plasma and interstitial concentration, $C_p - C_i$ (g/ml):

$$J_s \propto S * (C_p - C_i)$$

The constant of proportionality is, as per the epithelial transport, the permeability coefficient, P (cm/s). According to Starling's hypothesis the convection component, J_f (ml/s), is instead proportional to the rate of fluid leakage, from the vessels:

$$J_f \propto S * [(P_v - P_i) - \sigma * (\pi_V - \pi_i)]$$

where S is the vessel's surface area, $P_v - P_i$ (mmHg) is the difference between the vascular and interstitial hydrostatic pressure, σ is the osmotic reflection coefficient (which is close to 1 for macromolecules and almost zero for small molecule) and $\pi_V - \pi_i$ (mmHg) is the difference between the vascular and interstitial osmotic pressure. The constant of proportionality is the hydraulic conductivity L_p (cm/mm Hg*s).

In the presence of both diffusion and convection we have the following equation:

$$J_s = S * (C_p - C_i) + J_f (1 - \sigma_F) \Delta C_{lm}$$

Where σ_F is the solvent-drag reflection coefficient, $1 - \sigma_F$ is a measurement of the coupling between fluid and solute transport and ΔC_{lm} is the mean concentration within the pore across the membrane.

2.2.3 Tumor Vasculature and the EPR Effect

In addition to the physico-chemical properties of the drugs, the characteristics of the target site play a crucial role in controlling the diffusion of drugs in tissues. Trans-vascular transport in tumors is characterized by unique properties (Di Paolo and Bocci 2007; Jain 1990; Minchinton and Tannock 2006). Tumors present an abnormal vasculature in terms of structure and functionality. The vessels often have an incomplete endothelial layer and lack intact pericyte coverage and basement membrane which result in the opening of fenestrations in the vasculature walls that make the vessels more permeable than normal tissues (Cairns et al. 2006; Tredan et al. 2007). The increase in vascular permeability and hydraulic conductivity in the tumor makes diffusion and convection through the large gaps formed in the endothelium the primary pathways of drug transport across the tumor microvascular walls. Since the vascular cut-off across the endothelium is bigger compared to normal tissue ranging from 100 to 1,200 nm in diameter (Hobbs et al. 1998), particle-based drug carriers, such as liposomes or polymeric nanoparticles, tend to passively accumulate preferentially in the tumor. Moreover, since the cancer lymphatic drainage is defective, there is a decreased clearance of the macromolecules and particles from tumor interstitium, resulting in an enhanced permeability and retention (EPR) effect that makes nanoparticles accumulate and be retained longer at the tumor site (Iyer et al. 2006; Maeda et al. 2000). These tumor characteristics have been successfully exploited in clinic for the delivery of doxorubicin-encapsulated in liposomes, showing reduced cardiotoxicity while maintaining antitumor efficacy (Batist et al. 2001). However, given the heterogeneous nature of cancer, vessel permeability may vary significantly even within the same tumor, which therefore leads to heterogeneous extravasation of therapeutic agents. An additional result of the lack of permeability selectivity of the vasculature is the accumulation of fluid and osmotic proteins from the blood in the interstitial space resulting in increased osmotic pressure within the tumor interstitium (Cairns et al. 2006; Tredan et al. 2007). Increased fluid pressure is high and relatively uniform in the center of the tumor, but it drops to normal levels at the periphery and in the surrounding normal tissues (Boucher et al. 1990). Consequently, transvasculature flow and convective transport of therapeutic molecules in the tumor is greatly non-uniform, resulting in low distribution of drugs at the tumor center. Poor vascular organization also leads to hypoxia regions within the tumor which trigger the activation of cell survival pathways and alternative energy sources, such as glycolytic energy production. This generates the build-up of metabolic products that lower the extracellular pH affecting protonation of drugs and therefore cellular uptake (Tredan et al. 2007). All of these characteristics have to be taken into account in order to successfully engineer novel therapeutics able to bypass the multiple barriers that oppose extravasation of drugs into the tumor site.

2.3 *Interstitial Transport in Solid Tumor: Stromal Barriers*

Following extravasation, therapeutic agents still have to be transported across the interstitial space by diffusion and convection in order to reach the targeted cells, often far from the blood vessels. The ratio of non vascularized to well-perfused regions, and tumor specific characteristics such as composition and structure of the extracellular matrix, as well as cell density, are potential causes for the limited delivery of therapeutics within the tissue (Tannock et al. 2002).

Various methods are available to characterize these barriers and to predict and improve therapeutic agent penetration across tissues. Among the most common methods are multi-cellular spheroids (Minchinton and Tannock 2006) and multi-layered *in vitro* cell culture models (Tannock et al. 2002; Minchinton and Tannock 2006), as well as *in vivo* analysis through magnetic resonance imaging (MRI), positron emission tomography (PET), and fluorescent microscopy (Reddy et al. 2006).

One barrier to drug transport in tumors is elevated interstitial fluid pressure, which affects the convection flow of the extravasated molecules, slowing their movement and causing them to move radially outward (Jain 1987). This leads to diffusion-dominated interstitial transport, which is very slow for particles and macromolecules (Jain 1987), as well as to a possible clearance of the drugs.

Another barrier is the extracellular matrix (Netti et al. 2000). The major components of extracellular matrix are fibrous proteins (collagen and fibronectin) and polysaccharides (hyaluronan and proteoglycans). *In vitro* models used to study transport include gels with immobilized cells and collagen (Ramanujan et al. 2002). As an example, these systems are used to study the impact of cell density and collagen concentration on drug movement. Studies have shown that drug transport is inhibited by high cell density and over-expression of ECM components (Netti et al. 2000; Jain 1987).

Due to these barriers, only tumor cells close to the vasculature are exposed to an effective drug concentration. In addition to these physical barriers, the physicochemical properties of molecules and particles such as size, shape, charge and aqueous solubility affect the rate of diffusion through tissues (Minchinton and Tannock 2006). For instance, small molecules diffuse readily due to their low steric hindrance, and globular proteins diffuse slower than their linear counterparts.

2.4 Approaches to Enhance Uptake and Penetration of Intravenously-Administered Therapeutic Agents

A wide range of studies have been done on chemical permeation enhancers for oral delivery for both paracellular and trancellular routes (Muranishi 1990). In the first case the enhancers alter the cell membrane structure while in the second case they make the tight junctions between the cells more open. Included in this group of molecules are melittin, surfactants such as sodium cholate, and palmitoyl carnitine (Whitehead and Mitragotri 2008; Godin 2006). Further in this chapter we will focus on the cellular barriers present mostly when the substance is administered intravenously. In this context, many approaches have been studied to overcome transport barriers and promote the delivery of therapeutic agents to the target lesion. Some of these are focused on directly improving drugs in terms of structure, lipophilicity, and formulation. Other methods target the environment, modifying, for instance, the blood flow or enhancing intestinal and vascular extravasation as well as interstitial penetration (Cairns et al. 2006). A promising strategy is the use of carriers such as micro- and nano-particles to enhance the delivery of already employed drugs, exploiting, in the case of cancer, the EPR effect (Iyer et al. 2006) and taking advantage of the large variety of possible chemistries involved in the assembly of these particles (Ferrari 2010). Moreover these carriers can be decorated with molecules such as antibodies, aptamers and peptides which selectively recognize markers on the surface of the cells at the site of action. Several studies have also evaluated the use of pH-sensitive and temperature-sensitive carriers for the triggered release of the payload in the tumor microenvironment (Hafez et al. 2000; Lee et al. 2008). Among the approaches that target the environment, enhancing, for instance, the diffusion and/or convection of therapeutic agents across the vessel, are anti-angiogenic therapies (Ferrara and Kerbel 2005), employed in cancer therapy as well as other chronic diseases such as psoriasis (Buckland), rheumatoid arthritis (Longo et al. 2002) or diabetic retinopathy (Avery et al. 2006). Various compounds have been developed to block VEGF pathways such as monoclonal antibodies (Avastin), small molecules tyrosine kinases inhibitors such as lapatinib and sunitinib. These molecules have been shown to enhance the effect of chemotherapeutics normalizing the architecture of the tumor vasculature and reducing interstitial fluid pressure, hence increasing the pressure gradient across the tumor vessels, which leads to greater extravasation and deeper penetration of macromolecules into tumors.

As stated previously, high cellular and collagen densities have also been associated with increased interstitial fluid pressure and various strategies have also been employed to reduce both factors before administering the therapy. Possible methods include pretreatment with anti-adhesive agents or bacterial collagenases (Minchinton and Tannock 2006; McKee et al. 2006), although there are concerns about the possible effect on metastasis. For some tumors, it is possible to identify the pathway responsible for the overexpression of ECM components. For instance there is an abundance of evidence that indicates that advanced-stage pancreatic tumors overproduce transforming growth factor (TGF)-β1 which is involved in the functional regulation of tumor interstitium (Roberts and Wakefield 2003; Pasche 2001). In light of these findings, recent research has focused on strategies that antagonize the TGF-β1 pathway as a possible treatment modality for tumors with a fibrotic microenvironment (Biswas et al. 2006). Strategies include the use of monoclonal TGF-β1–neutralizing antibodies (Arteaga et al. 1993), antisense technology to reduce the translational efficacy of TGF-β1 ligands (Spearman et al. 1994), and small molecule inhibitors that antagonize the TGF-β1 receptor I/II kinase function (Peng et al. 2005).

Other approaches to enhance drug extravasation and tissue penetration rely on the conjugation or co-administration of peptides containing a tissue penetration motif, such as the CendR (C end Rule) motif of the cyclic tumor penetrating peptide iRGD (internalizing Arginine-Glycine-Aspartic acid) (Ruoslahti et al.; Sugahara et al.). iRGD targets tumors by binding to integrins expressed on the tumor vasculature. The peptide is then proteolytically cleaved, generating a peptide fragment with the CendR motif exposed which interacts with the neuropilin-1 receptor, promoting drug penetration into tissues. iRGD has been shown to improve penetration of different types of molecules and particles in four different cancer models.

Local hyperthermia treatment, which is usually employed to improve radiation therapy, also increases blood flow and tumor microvascular permeability, thereby enhancing the extravasation and interstitial diffusion of therapeutic agents such as liposomes or antibodies (Kong et al. 2000).

All of the above mentioned mechanisms for enhancing the permeation to the target tissue should be taken into consideration when the delivery to the specific population of the cells or to the specific cell organelles is desired. The following sections will deal with the mechanisms of cellular uptake and targeting the therapeutic entities to their intended site of action intracellularly.

3 Mechanisms of Cellular Uptake

3.1 *Overview of Endocytosis Pathways*

3.1.1 Endocytosis

Endocytosis is the process in which a cell internalizes exogenous substances such as extracellular fluid, small molecules, proteins, debris, and even entire cells such as bacteria and apoptotic cells. There are two main categories of endocytosis, phagocytosis (cell eating) and pinocytosis (cell drinking). Pinocytosis occurs through four mechansims: clathrin-dependent endocytosis, caveolin-mediated endocytosis, macropinocytosis, and clathrin/dynamin-independent endocytosis (Liu and Shapiro 2003). While virtually all cells are capable of pinocytosis, phagocytosis may only occur in certain cells, including neutrophils, macrophages, monocytes, and endothelial cells (Serda et al. 2009a). In the context of nanoparticles, understanding which pathways and cellular compartments are involved in the delivery of a drug helps to determine its bioavailability and pharmacological activity.

During pinocytosis, the cell membrane invaginates and then fuses, nonspecifically entrapping fluid and particles in a vesicle that pinches off and moves to the cell interior. Receptor-mediated endocytosis occurs when the solute binds to a receptor on the cell membrane and is selectively internalized (Swaan 1998). Phagocytosis, on the other hand, involves outward extension of membrane processes known as pseudopodia. Pseudopodia wrap around the target in a process that is dependent on actin polymerization (Yacobi et al. 2009). Once surrounded by these extensions, the target is pulled into the cell into a membrane-bound vesicle known as a phagosome. During maturation, phagosomes fuse with lysosomes and become phagolysosomes (Hartwig et al. 1977). This fusion exposes the phagocytosed contents to proteolytic enzymes and an acidic pH, which aids in digestion of the contents.

Similar to phagocytosis, macropinocytosis is dependent on actin polymerization. The process involves cell surface ruffling and leads to the formation of large macropinosomes (Swanson and Watts 1995). This route is believed to be less selective that phagocytosis and results in endocytosis of solute macromolecules. Agents such as cytochalasin D and latrunculin B can be used to study internalization by this pathway (Yacobi et al. 2009). Other studies use amiloride to selectively inhibit macropinocytosis by blocking the Na+/H+ exchange pump (Park et al. 2010).

Clathrin-mediated endocytosis is an energy-dependent pathway that involves cargo being recognized by adapter protein complexes. These complexes then recruit clathrin coats using other effector proteins. Invaginations, known as clathrin-coated pits, pinch off into clathrin-coated vesicles holding the selected cargo. Once internalized, the clathrin is uncoated and the vesicles fuse together to form early endosomes (Yacobi et al. 2009). Subsequently, late endosomes are formed. Cargo in late endosomes may be shipped to various organelles, including the lysosome. Lysosomal pathways will be discussed in greater detail, but in the context of endosomal compartments, further distinctions between early endosomes and late endosomes exist (Park et al.). For some nanoparticles, it has been determined that clathrin-mediated internalization is the predominant pathway by using chlorpromazine to inhibit internalization (Park et al.). Chlorpromazine causes dissociation of clathrin from the cell membrane, effectively blocking clathrin-mediated endocytosis. Dansylcadaverine has also been used as an effective inhibitor of clathrin-mediated endocytosis, but the exact mechanism of this drug is not known (Yacobi et al. 2009). Clathrin-mediated endocytic pathways can lead to the lysosomal degradation, reducing the effectiveness of a drug (Park et al.).

Similar to clathrin-mediated endocytosis, caveolae-mediated endocytosis creates vesicles 50–70 nm in diameter, called caveolae, through invaginations of the cell membrane (Brown and London 1998). Caveolae have been demonstrated to be localized to lipid rafts, areas of plasma membrane highly concentrated in sphingolipids and cholesterol (Brown and London 1998). It is this property of caveolae that allows their function to be disrupted in the presence of sterol-binding drugs that sequester cholesterol, such as methyl-beta-cyclodextrin (Yacobi et al. 2009) and filipin. Park et al. (2010b) used filipin to demonstrate the involvement of the caveolae-mediated pathway in the endocytosis of hydrophobically modified glycol chitosan (HGC). Confirmation that the caveolae pathway is the mechanism for cellular uptake has also been demonstrated through colocalization of labeled substrates, such as nanoparticles, with caveolin-1, the major protein in caveolae (Yacobi et al. 2009). Caveolin-1 is also found in trans-Golgi and post-Golgi vesicles.

Pinocytic pathways that do not fall under the above categories are referred to as clathrin- and caveolae-independent endocytosis (Park et al. 2010b). These pathways are a challenge to study and the extent of involvement of these pathways is hard to quantify due to the lack of know inhibitors. Thus, inhibition studies of defined pathways are performed, and this mechanism will often be named by default.

3.1.2 Targeting Specific Endocytic Pathways

A cell uses various pathways in order to internalize molecules and particles. What remains an unanswered question is what characteristics of a substance render it more favorable for one pathway over another. However, this is obviously not the case for cargo known to bind to specific cell surface receptors, and for molecules that enter cells by diffusion, whose characteristics are well defined. For novel therapeutics, such as nanoparticle drug carriers and exogenous drugs, future studies will be required to determine the physiological pathways involved during their endocytosis by target cells. Understanding which properties of nanoparticles dictate which pathways are utilized will benefit the development of particles and will aid in designing drug carriers that traffic along desired pathways to reach the targeted intracellular locations.

3.2 *Pathways Exploited by Nanoparticles*

Many nanoparticles, especially those used in therapeutic drug delivery, are intended to deposit a drug at a specific intracellular location, such as the cytosol. The transit time for the cell to internalize the nanoparticle and deliver it to the cytosol, as well as the steady-state level of the nanoparticle and its cargo, influenced by degradation and dose, will ultimately determine the overall efficacy of the drug delivery system. This section of the chapter will explore which endocytic pathways are candidates for nanoparticle internalization.

In many instances, nanoparticles may enter cells via multiple endocytic pathways. Park et al.'s (Hyung Park et al. 2006) investigation of hydrophobically modified glycol chitosan (HGC) is one such example. HGC is a promising nanoparticle-based drug carrier used for transport of anti-cancer drugs, genes, and peptides. Hence, understanding the endocytic pathways used in its uptake and subsequently its intracellular localization and bioavailability, is of clinical interest. The advantages of using HGC are high circulation time, rapid cellular uptake, low toxicity and small size of the particle. The inhibition studies conducted on this type of particle employed chlorpromazine, filipin III, and amiloride to inhibit clathrin-mediated endocytosis, caveolae-mediated endocytosis, and macropinocytosis, respectively (Schnitzer et al. 1994). These studies demonstrated that while HGC used all three endocytic pathways, their primary method of entry into the cell is macropinocytosis.

Nanoparticles also enter cells using non-endocytic pathways. For example, Taylor et al. (2010) demonstrated that incubation of gold nanoparticles with cells at 4°C failed to inhibit nanoparticle uptake, suggesting an energy-independent pathway for entry into the cell, such as diffusion. The internalized gold nanoparticles also failed to co-localize with Rab5a or LAMP-1 (Lysosomal-Associated Membrane Protein 1), markers for early endosomes and lysosomal membranes, respectively.

Although targeting to lysosomal pathways is generally undesirable and actively avoided in drug delivery due to degradation of the drug, some nanoparticle drug delivery systems take advantage of this pathway in order to increase delivery. For instance, Sahay et al. (2009) have described the delivery of the cytotoxic drug, doxorubicin, via cross-linked-micelles, or polymeric micelles with cross-linked ionic cores made of poly(methacrylic acid) and nonionic shells made of poly(ethylene oxide). The cross-linked-micelles are not internalized in normal epithelial cells, but are instead taken up via a caveolae-mediated endocytic pathway in cancer cells. Subsequently, the cross-linked-micelles bypass the early endosomes and arrive at the lysosomes, where the pH-sensitive hydrazone bonds of doxorubicin are cleaved. This event leads to the accumulation of doxorubicin in the nucleus of the cancer cell 5 h after administration. Hence, the method of uptake and intracellular destination in this instance confers the selective toxicity of this nanoparticle drug delivery method. Many other investigators also have documented the use of pH sensitive carriers for lysosomal release of therapeutics.

The above examples demonstrate the importance of endocytic pathways involved in a nanoparticle's uptake. Determining which pathway is taken in target and non-target cells will allow a greater understanding of the efficacy and specificity of the nanoparticle, and hence, permit greater knowledge of the systemic clinical effects of a drug from the cellular level. Future work in this direction will help to elucidate which pathways are favored by emerging nanoparticle carrier technologies, and what determines this selection.

3.3 *Impact of Size and Shape on Uptake*

The geometry of nanocarriers has been recently recognized as an important design parameter for sculpturing the interactions of the system with biological substances in addition to elemental composition (Roh et al. 2005; Akagi et al. 2010; Kennedy et al. 2009; Godin and Touitou 2004), surface chemistry (Shi et al. 2002; Park et al. 2010a; Satija et al. 2007), attachment of targeting ligands (Rawat et al. 2007; Torchilin 2006) and mechanical properties (Yum et al. 2009; Cho et al. 2009). Various studies point towards the importance of particle geometry in cellular functions such as endocytosis, vesiculation, phagocytic internalization, transport in the vasculature and adhesion to the target receptors (Mitragotri 2009; Lee et al. 2010a; Decuzzi and Ferrari 2007, 2008b; Serda et al. 2009a, b; Doshi et al. 2010; Champion and Mitragotri 2009; Ferrati et al. 2010).

In vitro studies using phagocytic cells have shown that macrophages internalize IgG-coated polystyrene spherical particles (200 nm–2 μm) through different intracellular delivery pathways. Nanospheres are internalized by clathrin-mediated endocytosis, while polystyrene microspheres undergo classical phagocytosis, trafficking more rapidly to the lysosomes (Koval et al. 1998; Rejman et al. 2004). In a study (Jin et al. 2009) on the intracellular uptake rates of length-fractionated single-walled carbon nanotubes (SWNT) 130–660 nm in diameter, the authors assessed single particle tracking using their intrinsic photoluminescence. It was suggested that nanoparticles aggregate on the cell membrane to form a cluster size-sufficient to generate a large enough enthalpic contribution for overcoming the elastic and entropic energy barriers associated with membrane vesicle formation. The endocytosis rate for nanotubes was 1,000 times higher than for spherical gold nanoparticles (Jin et al. 2009).

Data from a study on Inter-Cellular Adhesion Molecule 1 (I-CAM-1) targeted spherical and elliptical shaped polymeric nanoparticles, 100 nm–10 μm in size, have shown the effect of carrier geometry on the rate of endocytosis and lysosomal transport in endothelial cells. Discoidal particles had higher targeting specificity; and larger micron-size particles had an extended residency in prelysosomal compartments. Submicron carriers trafficked to lysosomes more readily (Muro et al. 2008). In another work, spheres, elongated and flat particles with an effective diameter of 500 nm–1 μm and different zeta potentials were found to differently affect endothelial cells. Needle shaped particles significantly impaired spreading and motility of the cells through induced disruption of the cell membrane (lasting for up to 48 h) while spherical and elliptical disc-shaped particles did not have an impact on these properties (Doshi and Mitragotri 2010).

Recent studies focused on dynamic manipulation of particle geometry as a tool to precisely control particle-cell interactions (Caldorera-Moore et al. 2010). As an example, Yoo and Mitragotri have designed polymeric particles able to switch shape in a stimulus-responsive manner (Yoo and Mitragotri 2010). The shape-switching behavior was a result of a fine balance between polymer viscosity and interfacial tension and could be tuned based on external stimuli (temperature, pH, or chemicals in the medium). Shape changes altered phagocytosis of elliptical particles that were previously not internalized by the cells (Yoo and Mitragotri 2010). These results clearly emphasize the importance of size, shape and surface physico-chemical properties in controlling the rate of uptake of nanovectors.

A number of mathematical models and design maps were proposed to explain the effect of particles geometry on the intracellular uptake kinetics and to enable rational design of the nanocarriers, respectively. Based on the above experimental results, the rate of uptake can be described through a first order kinetic law where the intracellular concentration $C_i(t)$ increases with time following the relationship (1)

$$\frac{d\tilde{C}_i(t)}{dt} = k_{\text{int}}\left[\chi - \tilde{C}_i(t)\right] \quad k_{\text{int}} = \tau_w^{-1} \tag{1}$$

where τ_w is the characteristic time for the nanovector to be wrapped by the cell membrane, which can be related to the nanovector geometry (size, shape) and surface chemistry (zeta potential, specific ligands). Mathematical modeling for receptor-mediated internalization based on an energetic analysis has shown that there is a minimal threshold particle radius to enable intracellular uptake. Below this point the internalization is energetically unfavorable. Similar analysis shows that the surface physico-chemical properties of the nanovector related to that of the cell membrane can dramatically increase or decrease the uptake rate (Decuzzi and Ferrari 2007, 2008b). A mathematical model developed to predict the rate of uptake for ellipsoidal particles as a function of their aspect ratio indicate that spherical or oval particles can be more rapidly internalized by cells compared to elongated particles.

The effect of geometry on the intracellular delivery properties of nanovectors can further be integrated together with other geometry-affected biological processes (e.g. margination, vascular transport, adhesion to vessel walls, etc.) to generate generalized design maps which recapitulate the performances of nanovectors in terms of transport, specific recognition and adhesion, and uptake as a function of the design parameters and physiological/biophysical conditions. As an example, design maps have been generated in the simpler case of spherical particles as a function of the non-specific interaction factor, which accounts for the steric and electrostatic surface interactions between the particle and a cell; and of the ratio between the number of ligand molecules distributed over the particle surface and the number of receptor molecules expressed over the cell membrane. As a function of these parameters, design maps allow for estimating the propensity of a circulating nanovector to adhere and be internalized by cells (Decuzzi and Ferrari 2008a).

Influence of carrier geometry on intracellular delivery can be explained in part by different effects that particles with variable sizes and shapes may have on membrane remodeling. This process, important in endocytosis, vesiculation and vesicle transport through endoplasmic reticulum and Golgi apparatus, protein sorting and other vital cellular functions, is highly dependent on membrane curvature, naturally affected by some proteins through curvature-mediated attractive interactions (Reynwar et al. 2007). The formation of a coordinated actin cup is crucial to initializing phagocytosis and probably follows the local geometry of the particle. Because actin remodeling is a metabolically intensive process, it may provide the basis for the fact that particles requiring only gradual expansion of actin rings are phagocytosed more effectively (Champion and Mitragotri 2006).

Recent developments in micro and nano-fabrication of shape and size-specific vectors, following rational design, show great promise in overcoming one of the chief constrains in drug delivery – the specific delivery into the diseased cells. Further incorporation of stimuli-responsive biomaterials into the rationally designed vectors will enable the development of conceptually novel drug delivery systems.

3.4 Receptor-Mediated Uptake of Nanoparticles

One of the major problems encountered with existing therapeutics is their harmful effects on healthy tissues. To overcome these systemic effects, nanocarriers are intended to concentrate the drug selectively at the lesion. This is achieved through targeted delivery of therapeutics via encapsulation in nanocarriers that are surface-coated with lesion-specific ligands.

Targeting moieties are, predominately, focused on receptors or antigens that are overexpressed on the plasma membrane of diseased cells or adjacent vasculature (Fuchs and Bachran 2009; Cho et al. 2008). Interaction of these targeting moieties or ligands with their respective receptors aids in the internalization of the nanocarriers into the cells through various processes such as receptor-mediated endocytosis, phagocytosis, macropinocytosis, etc. Several targeting ligands have been examined and tested for the delivery of nanocarriers into cells.

Molecular signatures on the cell surface can be identified and used as a target. One such technique to reveal new and unique markers on cell surface is *in vivo* phage display. Using this technology, various tumor blood vessel endothelial markers such as Aminopeptidase N (Pasqualini et al. 2000), 78 kDa Glucose-Regulated Protein (GRP78) (Wood et al. 2008), integrins including $\alpha_v\beta_3$ (Pasqualini et al. 1997) and $\alpha_v\beta_5$ (Nie et al. 2008), and CRKL (Mintz et al. 2009) have been exposed. Other markers, specific for lung vasculature (targeted by CGSPGWVRC peptide) (Giordano et al. 2008) and white adipose tissue (prohibitin) (Kolonin et al. 2004) have also been described. These markers can be exploited for inducing targeted apoptosis when conjugated with proapoptotic peptides, such as D-$(KLAKLAK)_2$ (Ko et al. 2009).

Some of the main membrane targets and related delivery systems are summarized in Table 1 and discussed below. Folate (Vitamin B9) is an essential vitamin for

Table 1 Examples of receptors used for internalization of nanoparticles

Receptor	Ligand	Target cells	Nanovector	Mechanism	References
Cancer therapeutics					
$\alpha_v\beta_3$ Integrin	RGD-4C	Tumor cells	Hybrid viral particles called AAVP	Receptor mediated endocytosis	Pasqualini et al. (1997) and Hajitou et al. (2006)
LHRH receptor	LHRH peptide	Ovarian, breast and prostate tumor cells	Campothecin-PEG-LHRH		Dharap et al. (2005)
Folate receptor	Folic acid	Tumor cells	Doxorubicin/antisense oligonucleotides against EGFR loaded liposomes	Receptor mediated endocytosis	Guo et al. (2000), Leamon and Low (1991), Pan et al. (2003), and Wang et al. (1995)
LDL receptor	Poloxamer 188	Brain blood vessel endothelial cells	Doxorubicin loaded PLGA nanoparticles	Phagocytosis	Gelperina et al. (2010)
Aminopeptidase N	NGR peptide	Tumor endothelial cells	Liposomes, phage		Pasqualini et al. (2000) and Loi et al. (2010)
Transferrin receptor	Transferrin	Gastric and colon cancer cells	Cisplatin containing PEG-liposomes	Receptor mediated endocytosis	Ishida et al. (2001) and Iinuma et al. (2002)
Insulin receptor	Insulin	Fibroblasts	Superparamagnetic iron oxide nanoparticles	Prevention of endocytosis	Gupta et al. (2003)
Vascular Endothelial Growth Factor receptor (FLK1)	Anti-Vascular Endothelial Growth Factor Receptor	Tumor endothelial cells	Doxil®		ElBayoumi and Torchilin (2009)
ERBB-2 receptor	Anti-ERBB2 (Herceptin)	Breast and ovarian cancer cells	Anti-ERBB2-Hygromycin conjugated phage particles	Receptor mediated endocytosis	Bar et al. (2008)
SMA	Aptamer	Prostate tumor cells	Pt(IV) prodrug-PLGA-PEG nanoparticles	Receptor mediated endocytosis	Dhar et al. (2008)

(continued)

Table 1 (continued)

Receptor	Ligand	Target cells	Nanovector	Mechanism	References
GRP-78	WIFPWIQL	Breast and prostate tumor cells	PEG-PAMAM	Receptor mediated endocytosis	Wood et al. (2008)
	CAYHRLRRC	Leukemia and Lymphoma cells		Macropinocytosis	Nishimura et al. (2008)
Vascular (atherosclerosis, atherothrombosis, etc.)					
VCAM-1	VCAM-1 Antibody	Endothelial cells (cerebral)	Microparticles of iron oxide		McAteer et al. (2007)
P-selectin + VCAM-1	P-selectin and VCAM-1 antibodies	Atherosclerotic plaque endothelium	Microparticles of iron oxide		McAteer et al. (2008) and Ye et al. (2008)
GPIIb/IIIa receptor	LIBS antibody	Activated platelets	Microparticles of iron oxide		Stoll et al. (2007) Schwarz et al. (2006), and Schwarz et al. (2004)
HIV					
Con A	Mannose	Macrophage like cells (HIV targeting)	Liposomes, gelatin nanoparticles, dendrimers		Garg et al. (2006), Jain et al. (2008), and Dutta and Jain (2007)
HIV-1_{BAL} gp120	Aptamer targeting gp120	HIV, T-cells and PBMC	Aptamer-siRNA	Receptor mediate endocytosis	Zhou et al. (2009)
Lung					
	CGSPGWVRC	Lung endothelial cells	Peptide – $(KLAKLAK)_D$ conjugate		Giordano et al. (2008)
Fat					
Prohibitin	CKGGRAKDC	Adipose tissue endothelial cells	Peptide – $(KLAKLAK)_D$ conjugate		Kolonin et al. (2004)

AAVP adeno-associated virus/phage, *LHRH* leutinizing hormone releasing hormone, *PEG* poly ethylene glycol, *LDL* low density lipoprotein, *PLGA* poly (lactide-co-glycolide), *PSMA* prostate specific membrane antigen, *GRP-78* glucose regulated protein 78, *VCAM* vascular cell adhesion molecule, *GP* glycoprotein, *HIV* human immunodeficiency virus, *PBMC* peripheral blood mononuclear cells various cell functions including nucleotide biosynthesis. The concentration of folate receptors on the tumor cells has been shown to increase as the tumor progresses. This, along with the inaccessibility of the receptor on normal epithelial cells has made it an attractive target for designing tumor-specific drug delivery systems (Lu and Low 2002a). Guo et al showed that liposomes conjugated with folic acid and non-conjugated liposomes accumulated at tumor sites similarly *in vivo*, although the targeted liposomes were internalized by tumor cells more efficiently than their untargeted counterparts (Guo et al. 2000). This indicates that the targeting moieties may not directly affect the organ distribution of the nanocarriers, but rather enhance the partitioning of the drug within the tumor tissue between cancer and adjacent normal cells. The suggested mechanism of internalization for macromolecules conjugated to folic acid involves a nondestructive (non-lysosomal) endocytic pathway (Leamon and Low 1991). A number of therapeutic and imaging agents conjugated to folate were developed and are currently in clinical trials. Among these are the imaging agents ^{111}In-DTPA (diethylene triamine pentaacetic acid)-folate (Siegel et al. 2003) and EC20 (Endocyte20) (Fisher et al. 2008), as well as the therapeutic agents EC17 (Lu and Low 2002b) and EC145 (Vlahov et al. 2006).

Leutinizing hormone releasing hormone (LHRH) is a peptide that was used as a targeting moiety by Dharap and Minko to deliver campothecin to various tumor cells in vitro and human ovarian carcinoma xenograft tumors in vivo (Dharap et al. 2005). The receptors for LHRH are overexpressed in multiple hormone-dependent tumors such as ovarian, breast and prostate cancers. The nanocarriers decorated with the LHRH peptides were shown to accumulate in the tumors preferentially with minimal to no accumulation in other organs and showed significant reduction in tumor size when injected in human ovarian carcinoma xenograft containing mice (Dharap et al. 2005).

Poloxamer 188 coated Poly(lactide-co-glycolide) (PLGA) nanoparticles use their similarity to low density lipoproteins (LDL) and bind to LDL receptors on endothelial cells, thereby inducing a receptor mediated endocytosis of these nanoparticles into the brain blood vessel endothelial cells and across the blood brain barrier. Intravenous injection of doxorubicin-loaded poloxamer 188 PLGA nanoparticles showed a long-term remission (<100 days without tumor) in 40% of rats with intracranially transplanted glioblastomas, in comparison to non-treated control mice that died in less than 20 days post tumor implantation (Gelperina et al. 2010).

Aptamers, nucleic acid ligands selected for target specificity, are promising agents for homing drug delivery vehicles to the disease site. Aptamers have been used both as targeting agents and as therapeutic agents. For example, aptamers for Prostate Specific Membrane Antigen (PSMA) have been used to label PLGA-PEG (polyethylene glycol) nanoparticles that are loaded with Cisplatin, a platinum-based drug. The labeled nanoparticles were efficient in targeting and eliminating prostate tumor cells, as compared to non-targeted nanoparticles (Dhar et al. 2008). Another example is the conjugation of aptamers to siRNA (small interfering RNA). Aptamers targeted to gp120 (HIV-1_{BAL} envelop glycoprotein) in conjugation with dicer substrate siRNA demonstrated specific inhibition of HIV replication and infectivity to T-cells *in vitro* (Zhou et al. 2009).

Table 2 Some ligand targeted therapeutic agents in clinics and clinical trials (Modified from Davis et al. (2008))

Name	Targeting agent	Therapeutic agent	Status	Nanovector	References
Gemtuzumab ozogamicin (Mylotarg; UCB/Wyeth)	Humanized anti-CD33 antibody	Calicheamicin	Approved	Antibody–drug conjugate	Bross et al. (2001)
Denileukin diftitox (Ontak; Ligand Pharmaceuticals/ Eisai)	Interleukin 2	Diphtheria toxin fragment	Approved	Fusion protein of targeting agent and therapeutic protein	Kawakami et al. (2006)
Ibritumomab tiuxetan (Zevalin; Cell Therapeutics)	Mouse anti-CD20 antibody	90Yttrium	Approved	Antibody–radioactive element conjugate	Allen (2002)
Tositumomab (Bexxar; GlaxoSmithKline)	Mouse anti-CD20 antibody	131Iodine	Approved	Antibody–radioactive element conjugate	Allen (2002)
Abraxane	Albumin	Paclitaxel	Approved	Albumin bound nanoparticles	Iglesias (2009)
Ad5Delta24RGD	RGD-4C	Herpes Simplex Virus (HSV) thymidine kinase (TK) suicide gene and the somatostatin receptor type 2	Phase I completed	Adenovirus	Matthews et al. (2009)
FCE28069 (PK2)	Galactose	Doxorubicin	Phase I (stopped)	Small-molecule targeting agent conjugated to polymer nanoparticle	Duncan (2006)
MCC-465	F(ab′)2 fragment of human antibody GAH	Doxorubicin	Phase I	Liposome nanoparticle containing antibody fragment targeting agent	Matusumura et al. (2004)

MBP-426	Transferrin	Oxaliplatin	Phase I	Liposome nanoparticle containing human transferrin protein targeting agent	Phan (2007)
SGT-53	Antibody fragment to transferrin receptor	Plasmid DNA with p53 gene	Phase I	Liposome nanoparticle containing antibody fragment targeting agent	Clinicaltrials.gov – NCT00470613 (2008)
CALAA-01	Transferrin	Small interfering RNA	Phase I	Polymer-based nanoparticle containing human transferrin protein targeting agent	Davis et al. (2010)
^{111}In-DTPA-folate	Folic acid	^{111}In-diethylenetriamine-pentaacetic acid	Phase I/II	Radiopharmaceutical	Siegel et al. (2003)
^{99m}Tc-EC-20	Folic acid		Phase I/II	Radiopharm.	Fisher et al. (2008)

Targeted therapeutics currently approved by the FDA for the treatment of different types of cancer are mostly antibody conjugates. Recently, several ligand-directed nanoparticulate therapeutics have entered/completed different phases of clinical trials (Table 2) (Phan 2007; Matsumura et al. 2004; Duncan 2006; Siegel et al. 2003; Fisher et al. 2008; Matthews et al. 2009). MCC-465 (an immunoliposomal formulation of doxorubicin conjugated to F(ab)2 fragment of human mAb GAH) was the first "active" targeted nanoparticle that reached clinical trials for the treatment of metastatic stomach cancer (Wang et al 2008).

Internalization of drug delivery systems following binding of the targeting moiety may or may not be a desirable feature. It is important that internalization occurs when delivery of immunoliposomes are concerned, however, in the case of radiolabeled antibodies or antibody directed enzyme prodrug therapy (ADEPT), internalization might be immaterial or even undesirable. (Allen 2002). Further, it is important to remember that macromolecules and nanoparticles that enter through receptor-mediated endocytosis, may invariably end up in lysosomes, where active degradation of the drug and/or nanoparticles can take place by action of the lysosomal enzymes. This leads to inefficient drug delivery into the cell cytoplasm. Thus, ways to escape the endosomal pathway are being designed to enhance the intracellular delivery of drugs. For example, the use of pH sensitive liposomes, (Straubinger et al. 1985) or cell penetrating peptides (Torchilin 2008) can either aid in escaping lysosomal degradation or bypassing the endosomal route entirely as will be discussed later in this chapter.

4 Intracellular Targeting

4.1 Introduction

In order for nanoparticles to be internalized by a cell, and have a therapeutic effect, they must be capable of crossing the cellular membrane and steering to the proper subcellular compartment. The combination of nanoparticle attributes and cellular properties determine whether a nanoparticle will be internalized, by which endocytic pathway, and what is the ultimate subcellular fate of the nanoparticle (Xia et al. 2008; Sahay et al. 2010a). A major barrier to successful drug delivery is entrapment of drug carriers in endosomal vesicles. Two mechanisms for endosomal avoidance exist: (1) the use of select pathways for nanoparticle internalization, i.e. clathrin-independent mechanisms; and (2) escape from the endosome into the cytosol, perhaps via disruption of the vesicular membrane. The endocytic process utilized by a cell to internalize nanoparticles has a significant impact on the pathway and intracellular trafficking of a nanoparticle, and these pathways depend on extracellular signaling, cell type, and the properties of nanoparticle, including size, shape, charge, and surface properties (Xia et al. 2008; Sahay et al. 2010a).

4.2 Cellular Zip-Codes

Blobel and Dobberstein hypothesized in 1975 that "zip-codes" were responsible for targeting proteins toward the endoplasmic reticulum (Blobel and Dobberstein 1975). Blobel was awarded the Nobel Prize for Physiology or Medicine in 1999 (Heemels 1999), when the signal hypothesis based on "zip-codes" held true for not only transporting the proteins towards the endoplasmic reticulum but for other intracellular organelles as well. By maneuvering the "zip-code" of a protein, the sub-cellular destination of the protein could be changed. Since various diseases have different subcellular origination, an opportunity is presented to use zipcodes to drive intracellular trafficking in a disease-specific manner. Therefore it is crucial to analyze and understand the multiple signaling elements and their role in "zip-code" change to target an individual organelle and the specific disease (Davis et al. 2007). For example, targeting lysosomes has been applied in treatment of lysosomal storage diseases such as TaySach's disease, Lesh-Nyhan-syndrome and adenosine deaminase insufficiency. Targeting caveolae has been proposed to influence cancer, atherosclerosis, Alzheimer's disease, and muscular dystrophy (Prokop and Davidson 2008).

A number of intracellular DNA delivery studies for gene delivery have been directed towards the nuclear genome, or more recently, to the mitochondria (Torchilin et al. 2002). With mitochondria being a suitable target of proapoptotic drugs for cancer therapy, and lysosomes being the target for lysosomal storage diseases, the need to create efficient protocols for targeting such intracellular organelles is augmented (Torchilin 2006).

Trafficking substances inside cells is an intrinsic process of prime importance. As an example, it has been demonstrated that an oncogene can be transformed into an apoptotic factor by manipulating the intracellular location of the protein, as in the case of Bcr-Abl protein in chronic myelogenous leukemia (Kanwal et al. 2004). Thus an externally introduced ligand can alter the location of an intracellular protein modifying its functionality (Kakar et al. 2007). Nanotechnology being in the interface of a variety of disciplines on the microscopic and molecular scales provides great opportunities to manipulate trafficking of the substances specifically to the target organelles. A number of nanocarriers have been evaluated for the ability to deliver the substances to different subcellular compartments. These include lipid based systems (liposomes, ethosomes), magnetic nanoparticles, polymeric carriers (drug conjugates, micelles, nanoparticles). The different systems, their form factors, and targeting mechanism are summarized in Table 3.

Prokop and Davidson, have discussed the emergence of nanovehicular (NV) intracellular delivery systems which differ from microdelivery and commercially employed nanotechnology. NVs are created *to deliver cargo to particular intracellular sites, and if possible, exert a local action* that may range over a wide scale of nanosizes (Prokop and Davidson 2008). Intracellular targeting such as endo-lysosomal, cytoplasmic, nuclear and mitochondrial trafficking can be achieved using nanovectors with specific physical characteristics as carriers.

Table 3 Summary on nanovehicles: likely uptake mechanism examples of drug carrier systems for intracellular and organelle-specific drug delivery (Based on Torchilin 2006; Prokop and Davidson 2008)

Vehicle type	Details	Purpose	Targeting mechanism
Logic-embedded vectors (LEV) (Serda et al. 2010a)	Multiple nano-particulate entities – porous silicon loaded with chitosan or amine – SPIONS	Targeting late endosome, cytoplasm and perinuclear vesicles.	Receptor mediated endocytosis
Liposomes and solid lipid nanoparticles (SLN)	Positively charged (Lipoplexes, DOTAP)	Intracellular DNA delivery	Endocytosis-independent (lipid rafts)
	pH-sensitive (DOPF, oleyl alcohol, morpholino-lipids)	Cytosolic drug delivery by endosomal escape	
	Cytoskeleton-specific immunoliposomes	Transfection into reversibly hypoxic cells	
	Cell-penetrating peptide-modified liposomes	Intracytoplasmic delivery of drugs and DNA	
	pH-sensitive polymer-modified liposomes (NIPAM)	Cytosolic drug delivery by endosomal escape	
	DQAsomes	Delivery of drug and DNA to mitochondria	
Magnetic Nanoparticles	Cell-penetrating peptide-modified particles	Intracellular penetration for imaging purposes	Macropinocytosis or lipid-raft mediated
Adenoviral particle delivery	Adenoviral particles or vectors (AdV)	Intracellular siRNA delivery	Endocytosis, receptor mediated
Cell permeating peptides (CPP)	Amphipathic helical peptides – Transportan	Nuclear or cytosol delivery	Various endocytic mechanisms including lipid rafts
	Arginine rich peptides – Trans-Activating Transcriptional Activator (TAT)		
Dendrimers	Poly amidoamine (PAMAM) dendrimers	Gene delivery	Lipid raft-mediated or combination of lipid-raft and caveolar endocytosis for PAMAM dendrimers; surface modified with ligand-HER2 mAb via receptor-mediated endocytosis

Ethosomal carriers (Godin and Touitou 2004)	Vesicular Carriers	Intracellular and transmembrane delivery of antibiotics	Penetration enhancement
Gene vehicles	Cationically charged gene vehicles	Intracellular DNA delivery	Non specific adsorptive endocytosis or fluid-phase endocytosis
	Lipid-based gene vehicles		Speculated to be macropinocytosis or lipid-raft mediated unless functionalized (endocytosis or receptor-mediated)
Micelles	pH-sensitive (NIPAM) micelles Positively charged micelles	Cytosolic drug delivery by endosomal escape	Endocytosis-dependent, clathrin mediated
	Amphiphilic Polymers (Micelles)		Macropinocytosis or lipid-raft mediated, (Speculated, mechanism not yet determined)
Multifunctional envelope-type nano device (MEND) (Kogure et al. 2008)	Non-viral gene delivery system	Nuclear targeting of plasmid DNA Cytosolic delivery of siRNA and oligonucleotides	Vesicular transport-independent pathway
Nanosuspension	Nano-sized solid crystals	Intracellular drug delivery	Macropinocytosis or lipid-raft mediated due to hydrophobic character
Polymers	Polyalkylcyanoacrylate (PACA) Polyelectrolyte Complex (PEC) Polylactate/glycolate (PLGA) Positively charged PEI-based polyplexes Cell-penetrating peptide-modified polymers Polymeric drug conjugates	Intracellular DNA Delivery	Non specific adsorptive endocytosis or fluid-phase endocytosis, or depending on the chemical modification of the nanovehicle periphery, adsorptive endocytosis and clathrin-coated pathway or receptor mediated endocytosis for targeted drugs
SMART drug delivery systems (Sawant et al. 2006)	PEGylated Drug Delivery Systems (DDS) – micelles and liposomes	Double-targeted pH responsive pharmaceutical nanocarriers	Target accumulation and intracellular penetration by TAT peptide mediated interaction

In the majority of cases the intracellular internalization of nanocarriers involves enodocytosis (Prokop and Davidson 2008). Endosomes are naturally designed to start the degradation process of the substances invading the cell membrane, and as a result have much lower pH than that of the cytoplasm. Thus, endosomal escape is one of the most important processes to prevent the deactivation of pharmaceutics and enable their efficient action. A variety of strategies to enhance endosomal escape are discussed in the literature, including pH-sensitive carriers (Park et al. 2010a; Kakimoto et al. 2010; Torchilin 2006), photochemical activation (Selbo et al. 2010) and the use of endosome-disruptive agents (Nakase et al. 2010). As an example, surface charge of nanoparticles plays a vital role in determining their destination. Transition from pH 7 to an endosomal acidic pH resulted in the release of the particles from the endosome into the cytoplasm (Panyam et al. 2003; Panyam and Labhasetwar 2004). Cationic quantum dots (QDs) conjugated with polyethylenimine (PEI) have been demonstrated to be suitable siRNA carriers to silence specific target genes and also to study intracellular trafficking pathways (Lee et al. 2010b). Cell Penetrating Peptides have been used to deliver therapeutic agents to intracellular sites using nanoparticles. Mitochondria are characteristic of possessing a high membrane potential and so attract positively charged particles like oligoguanidines organelle (Rothbard et al. 2005). Nanocarriers modified with mitochondriotropic triphenylphosphonium have shown effective delivery of a model lipophilic molecule, ceramide, to mammalian mitochondria, increasing its therapeutic efficiency both *in vitro* and *in vivo* (Sarathi et al. 2008). Localization signals specific to organelles need to be incorporated on the nanovector surface to be directed to specific organelles (Prokop and Davidson 2008). NLS (nuclear localization signal) peptides interact with cytosolic agents and facilitate targeted delivery to the nucleus. Low molecular weight protamine, which is structurally similar to HIV-TAT (Human Immunodeficiency Virus Trans-Activator of Transcription) peptide, has been coupled to plasmid DNA, permitting nuclear specific delivery (Park et al. 2003). Viral proteins and combinations of organelle targeting peptides and cell penetrating peptides are being investigated for intracellular targeting (Tkachenko et al. 2003; Vasir and Labhasetwar 2007). Current research on targeting small peptides and peptide-conjugating molecules which are specific to the targeting site and of therapeutic potential, are being conducted (Mendoza et al. 2005).

The following sections summarize various mechanisms used to traffic peptides, proteins, and drugs within vesicles to four significant organelles: lysosomes (and late-endosomes), endoplasmic reticulum (ER), mitochondria, and nucleus.

4.3 Endosomes, Lysosomes and Cytoplasmic Targeting

The treatment for a variety of genetic, metabolic, and oncologic diseases are focused on targets located within the cytoplasm. Thus, mechanisms to release nanoparticles into the cytoplasm (either through extra lysosomal trafficking or through direct vesicle escape) are an important focus of nano-medical research. Recent research has shown that these objectives can be accomplished. For example, nanoparticles have been utilized to deliver siRNA (Chen et al. 2009), RNase A (Bale et al. 2010), Cyt C (Slowing et al. 2007; Park et al. 2010a) and plasmids into the cytoplasm, from where nucleic core complexes direct the plasmids into the nucleus (Tros de Ilarduya et al. 2010). With the exception of a select few particles which directly penetrate the cellular membrane in order to reach the cytoplasm, a nanoparticle must be capable of escaping the degradative environment of the late-endosomes and lysosomes which serve three main objectives: (1) to digest and store metabolic material for the cell, (2) to digest and recycle aged proteins and organelles, and (3) to protect the cell from engulfed external agents. In order for these organelles to accomplish these three objectives, they must be able to digest their internalized material and selectively transport desired agents across their intracellular membranes. Though these two functions are distinct, they are closely linked.

To perform a digestive function, the internal environment of organelles must be closely maintained by their specialized membranes at a low pH. The low pH of these organelles is optimal for many proteolytic, glycanolytic, lipolytic and nuclease enzymes. The abundance of protons also serves as a catalyst for degradative organic reactions (Holtzman 1989, Storrie B; Ivanov 2008). The pH decreases constantly as material travels from the cellular membrane (in early endosomes) to the lysosome, and a particle can escape the vesicle at any point along the pathway; however the two organelles are intrinsically different. In fact, lysosomes are known to exist as separate entities before merging with endosomes and subsequently engulfing their contents. In this way, lysosomes are able to conserve energy and recycle their enzymes rather than expending energy by creating new zymogens, enzymes, membrane proteins, and lysosomal membrane phospholipids (Holtzman 1989; Storrie B; Ivanov 2008).

A variety of membrane transporters for products such as carbohydrates, amino acids, nucleic acids, enzymes, and ions serve to regulate the hostile environment inside of late-endosomes and lysosomes. However, two families of membrane markers and one membrane transporter are of particular interest. The Rab-family membrane markers distinguish between endosomes (specific species distinguish between different types) and lysosomes (the lysosomal associated membrane proteins (LAMP)). The membrane transporter of note is the vacuolar proton transporter V+ATPase. This transporter is considered to be the dominant transporter in the acidification of endosomes to lysosomes.

Since the 1990's, at least four different mechanisms have been proposed for endosomal or lysosomal escape: the use of pH-sensitive fusogenic peptides, "fusogenic" lipoplexes, dynamic polyconjugates and proton-sponge polymers or materials (Paillard et al. 2010; Sasaki et al. 2008; Zelphati and Szoka 1996; Rozema et al. 2007; Akinc et al. 2005). Of these four, the latter three utilize the low pH of the organelle for their escape while the first capitalizes on the properties of the phospholipid bilayer, which is predominantly zwitterionic on the inside and anionic on the outside. The "fusogenic" lipoplexes are composed of cationic lipids complexed around an oligonucleotide core. The cationic lipids not only protect the

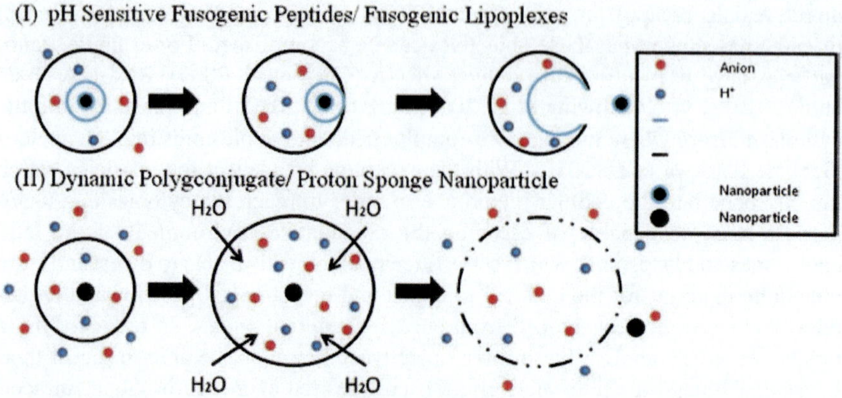

Fig. 3 The figures display the two principles behind the four endosomal and lysosomal escape methods discussed. (I) is a simplified version depicting how pH sensitive fusogenic peptides and fusogenic lipoplexes escape, and (II) is a simplified depiction of how dynamic polyconjugates and nanoparticles which act as proton sponges escape from an endosome or lysosome. In the case of pH sensitive fusogenic peptides and fusogenic lipoplexes, the particles fuse with the endosomal or lysosomal membrane and release their contents into the cytoplasm. Nanoparticles that act as proton sponges create an osmotic gradient that eventually causes the endosome or lysosome to burst releasing the contents. Meanwhile, dynamic polyconjugates disrupt and destabilize the endosomal or lysosomal membranes leading to the release of the nanoparticle's payload

oligonucleotide payload from the degradative enzymes of the extracellular and lysosomal compartments, but when the particle comes into contact with the predominantly zwitterionic interior surface of the endosomal membrane, it attracts the anionic phospholipid particles on the outer surface causing some of them to flip. Once flipped the attraction between the two phospholipids creates an ionic pair, which then leads to membrane destabilization and causes the two bilayers to merge. With the bilayers merged, the oligonucleotide complex is delivered to the cytoplasm unharmed (Fig. 3) (Zelphati and Szoka 1996).

Another two methods focus on protonation of the nanoparticle or of a complex bound to its surface. The first and simplest of these concepts was originally explored with polyethylenimine (PEI) in the application of DNA transfection. First described by Bousiff et al. in 1995 (Boussif et al. 1995) and verified a decade later by Akinc et al. (2005), this concept explores molecules capable of buffering the internal lysosomal environment. In this proton sponge mechanism, PEI, a molecule with a high cationic charge-density potential, absorbs the protons pumped into the lysosome. The internal environment of the lysosome is buffered, preventing lysosomal enzymes from degrading its payload. As the internal environment's pH remains elevated, the V+ATPase continues to pump protons into the lumen, which is followed by anions such as Cl^- (Sonawane et al. 2002). Eventually, the ions within the lysosome create an ionic gradient that causes the lysosome to swell and eventually burst, releasing its contents into the cell (Fig. 3-II). This application, while applied originally to PEI, has also been applied to PEI-conjugated particles and nanoparticles composed of other material, such a silica (Slowing et al. 2006). Yet, while it has been shown to be effective, it has been postulated that the rupturing of too many lysosomes is cytotoxic to the cell; hence the application of too many of these particles to the cell can be cytotoxic (Xia et al. 2008; Terman et al. 2006; Reiners et al. 2002).

In 2003, Wakefield et al. introduced another mechanism for endosomal escape employing the use of a dynamic polyconjugate known as EMMA (Endosomolysis by Masking of a Membrane-Active Agent). This macromolecule is designed to transition from a nonreactive entity at normal pH to a reactive membrane-active agent at an acidic pH (Fig. 3-II). The membrane agent, known as melittin, contains a reactive region that is masked using a mealic anhydride. Once protonated it unmasks itself, exposing the membrane active domain of the molecule (Wakefield et al. 2005). It has been suggested that the design of EMMA could be improved by increasing the number of hydrophobic regions on the peptide, thereby increasing its transfection capability (Rozema et al. 2003).

The last of the four nanoparticle escape mechanisms utilizes a viral model. Beginning in 2004, Harashima et al. has been working on transferrin modified liposomes. To improve the nanoparticle's ability to escape the late-endosome or lysosome, a pH-sensitive fusogenic peptide was added to the particle. This 30-amino acid fusogenic peptide, composed of the nonpolar tryptophan, alanine, and leucine and the polar acidic and basic glutamate and histidine, respectively, is known to make a conformational change from a random coil at a pH of 7.4 to a more hydrophobic alpha-helix at a pH of 5.5 (Simoes et al. 1998). This change increases its interaction with the endosomal membrane. A cholesterol moiety and a PEGylated peptide improve the interaction and penetration of the fusogenic peptide with the endosomal membrane (Kakudo et al. 2004). These actions then induce fusion of the liposomal and endosomal membrane releasing the nanoparticle's contents into the cytoplasm (Fig. 3-I) (Sasaki et al. 2008).

4.4 Endoplasmic Reticulum Targeting

Targeting the endoplasmic reticulum (ER) is of great importance because there are several degenerative processes which involve the ER, such as degenerative diseases (e.g. cystic fibrosis and cholera) and the drug-efflux mechanism in multi-drug resistance. Some drugs aggravate the ER stress response to induce apoptosis in cancer cells. However, since the ER is a complex organelle, consisting in a diffuse network of vesicles, and it is involved in multiple cellular functions, it cannot be directly targeted by certain drugs, and therefore treatments have not been developed for these diseases (Lai et al. 2008). Furthermore, intracellular trafficking dynamics are not entirely understood. Among the most complicated variables to control are the variations in trafficking dynamics which can be modified by attaching ligands to the nanoparticles. Understanding how to favor one transport pathway or how to optimize the efficiency of transport from one structure to another (e.g. endosome to lysosome or endosome to Golgi) is essential for successful intracellular targeting.

The ER is a network of tubules, vesicles and sacs that are interconnected. It is continuous with the outer membrane of the nucleus and is subdivided into the rough ER (RER) and smooth ER (SER) based on the association of ribosomes with the ER membrane. The ER is more voluminous in cells involved in drug metabolism or secretion of proteins or lipid. The ER is involved in (1) secretory proteins, membrane proteins and lipid synthesis and exchange, (2) protein modification, folding and quality control, (3) calcium sequestration and (4) drug detoxification (Dancourt and Barlowe 2010; Hu et al. 2010; Michalak and Opas 2009; Neve and Ingelman-Sundberg 2010). Protein synthesis begins on the ribosome complex, which may be transported to the ER surface via a signal-sequence-recognition mechanism, after which translation resumes. The ER is also responsible for glycosylation, the enzyme-mediated placement of an oligosaccharide tag that marks the state of protein folding. Correct protein folding is mediated by several ER-resident, chaperone proteins, such as Protein Disulfide Isomerase (PDI), and Binding immunoglobulin Protein (BiP), which bind the glucose moiety on improperly folded proteins. The chaperone proteins bind misfolded or unassembled proteins, preventing aggregation and directing the protein for degradation or proper folding. Proteins, such as lysosomal enzymes, are sorted and sequestered from the other luminal contents due to their potential degradative or aggregative effects. Only properly folded, nascent polypeptides are released from the chaperones and transported from the rough ER to the Golgi. Shuttling between the ER and Golgi is mediated by signal peptides and oligosaccharide chains. Coat proteins and KDEL (Lys-Asp-Glu-Leu)-receptors (KDEL-R) facilitate directed transport between organelles (Fig. 4). KDEL-R has different affinity for KDEL in Golgi (slightly acidic pH) and ER (neutral pH). Thus the receptor goes between the Golgi and ER to retrieve more KDEL-bearing proteins (Capitani and Sallese 2009). Several studies exploit this property and use KDEL-GFP as a fluorescent ER marker (Watson et al. 2005).

Other ER functions include lipid elongation and desaturation, calcium sequestration, drug detoxification and cell-protective mechanisms. Lipid synthesis is enzyme-mediated, utilizing fatty acid synthase complexes and desaturases. Deviations in this process correlate with diabetes, obesity, cardiovascular disease, and cancer. The ER also contains enzymes for drug detoxification, as seen in the hepatocyte. Cytochrome p450 enzymes metabolize water insoluble drugs or metabolites to facilitate excretion. Another enzyme, protein disulfide isomerase, oxidizes sulfhydryl groups. The lumen of the ER is thus an oxidizing environment, which contrasts with the reducing environment of the cytosol. The ER Stress Response is an important function to adjust the cell's protein content. An accumulation of unfolded proteins causes the ER to attenuate translation, arrest the cell cycle by the PERK (proline-rich, extensin-like receptor kinase) receptor, and upregulate the chaperone proteins to assist in protein folding. If the accumulation is severe or prolonged, apoptosis is induced.

Several methods have been used to target the ER. Peptides exploit natural trafficking mechanisms to the ER. In particular, bacterial and viral peptides navigate the intracellular compartments to mediate their toxic effects. Cholera toxin is one of the peptides known for its ability to traffic to the ER and escape to the cytosol to

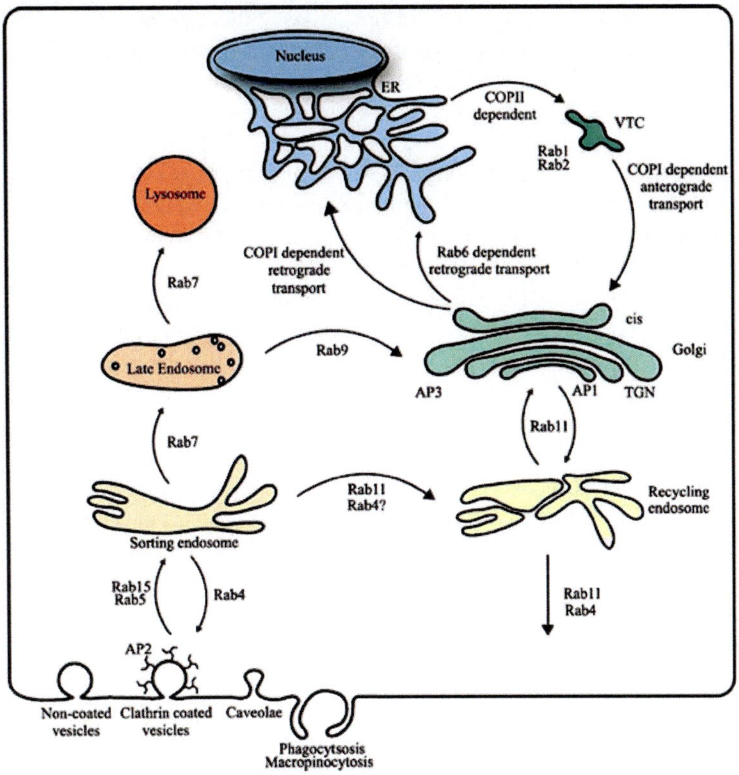

Fig. 4 A summary of membrane-protein-mediated trafficking (Reprinted from Watson et al. (2005). Courtesy of Elsevier Publishing Group)

alter adenylyl cyclase activity. Other proteins, such as Shiga toxin, Pseudomonas exotoxin and ricin toxin, utilize a similar route: the peptides are internalized via caveolin or clathrin-mediated processes and trafficked to the ER, whereupon the peptides exploit an unfolded-protein degradation/escape mechanism to enter the cytosol and cause their respective cellular dysfunctions (Sandvig et al. 2010).

Intracellular membrane trafficking mechanisms vary based on cargo. These different mechanisms allow for specific localization and different interactions en route (Fig. 4). Clathrin-mediated endocytosis traffics to endosomes where cargo can be either recycled (Rab4), sorted for degradation (lysosomes) or delivered to the Golgi (Rab 11 perinuclear sorting) (Rajendran et al. 2010; Fantini et al. 2002). To enter cells, viruses and bacterial toxins utilize lipid rafts, which are enriched in cholesterol and sphingolipids. Clustering of these raft components induces budding, and these vesicles follow a pathway that bypasses early endosomes and traffics to the Golgi (Fantini et al. 2002).

Studies to specifically target nanoparticles to the ER are rare. Quantum dots (QD) have been coupled to numerous ligands, such as Shiga toxin and ricin, to study internalization and localization within organelles. Although the toxins alone pass through the Golgi/ER, the toxin-QD conjugates tend to accumulate in endosomes (Tekle et al. 2008; Sandvig et al. 2010). Pluronic or poloxamer utilizes similar pathways as pathogens to gain entry into the cell. Sahay notes that the polymer bypasses endosomes/lysosomes and enters a transient co-localization with the ER before the polymer localizes to the mitochondria (Sahay et al. 2010b). Ma et al. study the internalization and localization of a recombinant peptide containing the KDEL motif, which appears to localize to the ER (Motosugi et al. 2009). Others use fluorescent protein-conjugates with –KDEL to fluorescently tag the organelle. Most non-modified nanoparticles have shown little to no localization with the Golgi/ER. Lai et al. show that negatively charged 43 nm polymer nanoparticles can enter HeLa cells through clathrin-mediated endocytosis and reside in the endo-lysosomal compartment. Smaller particles, approximately 24 nm in size, utilize a clathrin- and caveolin-independent pathway, bypass lysosomes and localize near the perinuclear space. In all cases, whether the particles localize within the lumen of the ER or on the outside membrane, has yet to be clearly and definitively shown (Lai et al. 2008).

Some drugs attempt to aggravate the ER stress response in cancer cells to induce apoptosis. However, as with most chemotherapeutic agents, global toxicity is observed. Modification of nanoparticles to target tumor cells and then the target organelle within, can better direct a chemotherapeutic drug to the specific location necessary. The mechanism of Cholera and Shiga toxins to target the ER is well studied (Rajendran et al. 2010). They are sorted from the plasma membrane to the Golgi complex through the pentameric B-subunit of the toxin. From there, the toxins are retrograde sorted to the ER, in which the A subunit of the toxin is released into the cytosol (Sandvig et al. 2010). The B-domain of Shiga toxin (STX-B) has been used to transport a prodrug against colon cancer to the ER, where it is released and transported to the cytosol (Johannes and Romer 2010). Targeting non-cancer diseases via the ER is a field that has hardly been breached due to the relative inability until now to specifically target the ER. Potential drugs may be used in targeting the ER stress response, downregulating the exocytosis of proteins or drugs (responsible for multi-drug resistance), or interacting with cholesterol (and lipid) synthesis and modification. Future research is needed to develop drugs that include novel targeting methods.

4.5 Mitochondria Targeting

Mitochondria are known as the “power plants” of the cells because they supply most of the cellular ATP. In addition, they are involved in cellular signaling, differentiation and in the control of the cell cycle. More specifically, mitochondria are involved in the (1) storage of calcium, (2) regulation of the membrane potential, (3) apoptosis-programmed cell death (4) calcium signaling (including calcium-evoked apoptosis), (5) cellular proliferation regulation, (6) regulation of cellular metabolism, (7) certain heme synthesis reactions, (8) steroid synthesis, and (9) heat production. Given their many functions, mitochondria have been associated with several human neuromuscular (e.g. mitochondrial DNA (mtDNA)) and metabolic diseases, as well as with cancer progression (Holt et al. 1988; Harding 1991; de Moura et al. 2010; Breunig et al. 2008; Andre et al. 2006). Schaefer et al. approximates that around 1 in 10,000 people have clinically manifest mtDNA disease, making this one of the most common inherited neuromuscular disorders. In addition, 16.5 in 100,000 people younger than 65 are at risk for development of mtDNA disease (Schaefer et al. 2008). These estimates confirm that mtDNA diseases can represent a common cause of chronic morbidity that is more prevalent than what previously estimated.

The mitochondrion is composed of an outer mitochondrial membrane (OMM), intermembrane space (IMS), inner mitochondrial membrane (IMM), and matrix. The mitochondrion also contains special translocases of the outer membrane (TOM) and translocases of the inner mitochondrial membrane (TIM) which are involved in mitochondrial trafficking. Mitochondria contain several unique proteins, many of which are coded by the nucleus and must be imported from the cytosol into the organelle. Mitochondrial precursor proteins are delivered to the organelle by virtue of specific mitochondrial targeting signals. These signals are relatively divergent in nature, but are all recognized by the specialized receptor proteins, which are located in the OMM and expose their receptor domains to the cytosol. The receptors deliver the precursor proteins to the translocation channel of the TOM complex, through which the pre-proteins cross the OMM. The TOM complex itself is sufficient for translocation of a small subset of OMM proteins and some IMS proteins. For translocation of all other mitochondrial pre-proteins, the TOM complex cooperates with other mitochondrial translocases. According to Mokranjac et al., six complex molecular machines, i.e. protein translocases, mediate this process (Mokranjac and Neupert 2005). Studies involving purified TOM40 of *Neurospora crassa* suggest a pore diameter of 2.5 nm (Ahting et al. 2001). The TIM23 complex is a major translocase in the IMM. It is an energy-dependent complex that facilitates pre-protein translocation across the IMM and insertion into the IMM (Fig. 5).

Although there are currently no drug therapies that target the mitochondria, there are some anticancer agents (paclitaxel, vinblastine, lonidamine, etoposide, and arsenic trioxide) already in clinical use that permeabilize mitochondria (Breunig et al. 2008; Andre et al. 2006). Overall, these drugs aim to activate the cell-death machinery in cancer cells by inhibiting metabolic tumor-specific alterations or by stimulating mitochondrial membrane permeabilization. In clinical trials, therapeutics like Lonidamine, Gadolinium Texaphyrin, Mangafodipir show few side effects, whereas drugs like 2-deoxy-D-glucose and ANT1 (adenine nucleotide translocase 1) or ANT3 show significant side effects, such as compromised glycolytic metabolism of the brain and of the heart (Fulda et al. 2010). These anticancer agents are toxic due to nonspecific targeting and thus hinder therapeutic efficacy. To avoid severe side effects and improve efficacy, drugs should not only be delivered to the correct cell, but ideally also to the correct subcellular organelle.

Despite some challenges in mitochondrial targeting following systemic administration, which include: (1) immune clearance of nanoparticles (2) traversing the

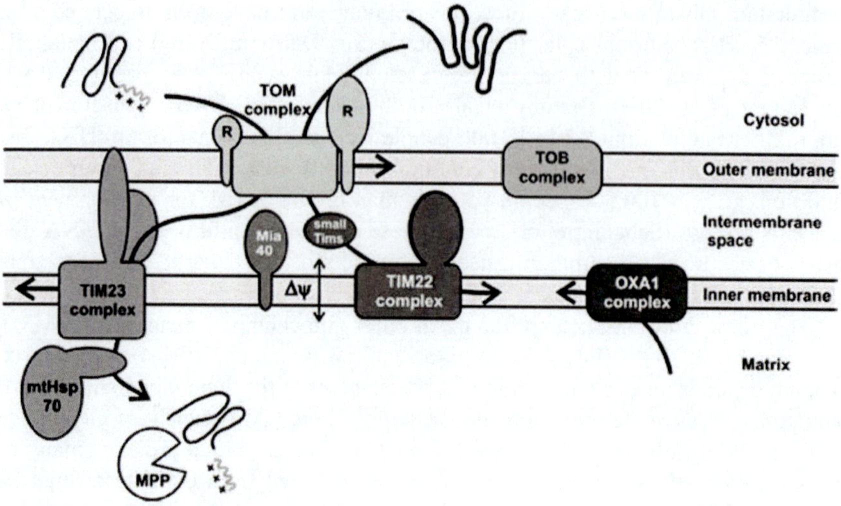

Fig. 5 Schematic of mitochondrial translocases (Reprinted from Mokranjac and Neupert (2005). Courtesy of Portland Press Limited Publishing Innovation)

lipophilic cell membrane (3) multidrug resistance (4) lysosomal accumulation and inactivation of nanoparticles and (5) cationic polymer- or lipid-associated plasmid DNA accumulation around the nucleus in the absence of targeting moieties (Holt et al. 1988); targeting the mitochondria is possible through both "passive" and "active" strategies. While some scientists suggest that only lower molecular weight (<5 kDa) BODIPY (boron-dipyrromethene) FL-labeled TPP (triphenylphosphonium)-semitelechelic HPMA (hexamethylphosphoramide) copolymers exhibit significant organelle localization or uptake (Callahan and Kopecek 2006), folate-modified multifunctional nanoassemblies, approximately 200 nm in size loaded with both docetaxel and iSur-pDNA (suppressor of metastatic and resistance-related protein survivin) for hepatocellular carcinoma therapy, showed mitochondrial accumulation (Biasutto et al. 2010; Xu et al. 2010). A proposed mechanism for mitochondrial targeting takes advantage of the negative membrane potential that builds up in the mitochondria (130–150 mV) (Xu et al. 2010). Lipophilic cations such as tetra-guanidinium oligomers or rhodamine-123 accumulate selectively within mitochondria due to their membrane potential. "DQAsomes" (mitochondria-specific nanocarrier system prepared from the amphiphilic quinolinium derivative dequalinium chloride) loaded with mitochondrial leader sequence-DNA conjugates and paclitaxel have accumulated in mitochondria and resulted in reduced growth of a human colon tumor mass in nude mice (Weissig 2003; Weissig et al. 2004, 2006). Alternatively, conventional liposomes can be rendered mitochondria-specific via the attachment of known mitochondriotropic residues to the liposomal surface (Boddapati et al. 2005, 2008). MITO-Porter, a liposome-based carrier with a high density of fusogenic R8 (octaArg) peptides conjugated to its surface has facilitated membrane fusion, and displays mitochondriotropic features (Biasutto et al. 2010; Fujiwara et al. 2010; Yamada et al. 2007; Rapoport and Lorberboum-Galski 2009). Similarly, Mito-8, a mitochondria-selective peptide fused to QDs was able to induce a strong mitochondrial localization in living cells (Hoshino et al. 2004a). Finally, Pluronic® block copolymers exhibit the ability to reach the mitochondria, where they exert unique pharmacological activities, such as ATP depletion in multidrug resistant cancer cells (Sahay et al. 2010b; Gupta et al. 2010).

4.6 Nuclear Targeting

The nucleus is a membrane-enclosed organelle that houses the cell's genetic material. The outer nuclear membrane is continuous with the ER. Nuclear pores, approximately 9 nm in diameter, allow certain molecules to enter the nucleus, making the nucleoplasm topologically equivalent to the cytosol. The nucleolus is a dense structure of highly packed heterochromatin that resides within the nucleus. The nucleus is involved in three major functions: nucleic acid compartmentalization and storing of the genetic material; DNA replication, transcription and cell replication; and processing of pre-mRNA (e.g. splicing).

The Nuclear Pore Complex (NPC) is a transport channel in the nuclear membrane (a double lipid bilayer) composed of nucleoporins. The NPC regulates trafficking of water-soluble macromolecules, such as proteins, carbohydrates, lipids and ribosomes between the nucleus and the cytoplasm in a signal-dependent manner. It has an inner (functional) diameter of 9 nm and an outer diameter of 120 nm. The NPC can dilate to facilitate the bidirectional translocation of a wide size range of protein complexes. Small particles (<30 kDa) are able to pass through the NPC by passive diffusion. Larger particles require a Nuclear Localization Signal (NLS) to cross. Aside from the standard cellular barriers to entry as mentioned earlier, internalization through the nuclear pore represents a unique barrier to nuclear targeting. Nature provides some solutions to circumventing the barriers. For example, viruses utilize various proteins to navigate intracellular trafficking mechanisms and localize to the nucleus. These viral models serve as the basis for the design of some targeted, functionalized nanoparticles (Mao et al. 2010).

Targeting the nucleus is of great importance due to its key resident: DNA. Multiple diseases, including cancer and genetic disease, arise from nuclear malfunction, DNA damage and mutation. Many cancer drugs target the nucleus to inhibit or interfere with cell replication. Nanoparticles containing certain cancer therapeutics target the cytoplasm and rely on diffusion of the drug through the nuclear pore for therapeutic action. For drugs that cannot cross through the pores, direct nuclear targeting of nanoparticles may be able to increase transport into the nucleus. The most extensively exploited way to target nanoparticles to the nucleus is through surface conjugation to a NLS. NLSs consist of one or more short sequences of positively charged lysine (K) or arginine (R) residues. A protein translated with a NLS will bind strongly to importin alpha/beta, and this complex will move through the NPC (Goldfarb et al. 2004). Once the complex is inside the nucleus, Ran (RAs-related Nuclear protein)-GTP binds and causes the NPC to lose affinity and release the NLS-containing cargo. The dissociated carrier complex exits the nucleus, where the GTP is hydrolyzed, and Ran dissociates from importin. Ran-GDP is recycled to the nucleoplasm, where it exchanges its GDP for GTP to facilitate further importation (Gilchrist et al. 2002; Gilchrist and Rexach 2003). Gold colloids (Gao et al. 2002; Hines and Guyot-Sionnest 1996), gold nanoparticles (about 15 nm in diameter) (Nativo et al. 2008), glucose-derived carbon nanospheres (Selvi et al. 2008), and other nanoparticles as large as 39 nm (Hoshino et al. 2004b) have been shown to avoid the lysosome and successfully target the nucleus, as confirmed by transmission electron microscopy. PEG-quantum dots are generally found in the cytoplasm, but the conjugation of NLS to their surface alters their localization toward the nucleus, suggesting the NLS is sufficient to alter trafficking. Given the restriction of inner nuclear pore diameter, only substances 9 nm or less in diameter can enter the nucleus via passive transport, yielding a possible explanation of why certain particles have been noted around the nucleus, but not within; however, particles larger than 9 nm have been observed within the nucleus through a mechanism that is still not fully understood (Hanaki et al. 2003; Aglipay et al. 2003). While several studies have confirmed nuclear targeting through the attachment of NLS to nanoparticles, others suggest that an additional peptide is required to escape endosomes, such as a receptor-mediated endocytosis peptide (or a cell-penetrating peptide) (Xie et al. 2009; Tkachenko et al. 2003). Optimal efficiency is observed when these peptides are short and separate, rather than combined as a single, long peptide (Tkachenko et al. 2003).

CPPs (cell penetrating peptides) such as the HIV Tat protein transduction domain facilitate nanoparticle entry into the cell and may facilitate endosomal escape such that particles can be targeted to the nucleus via the cytosol (Shiraishi and Nielsen 2006). Once in the cytosol, particle content can dissociate and diffuse to the nucleus; however certain payloads are susceptible to degradative enzymes in the cytosol. Tat-mediated delivery may occur via macropinocytosis, and others may enter cells via electrostatic interactions in non-energy dependent mechanisms (Breunig et al. 2008). The conjugation of Tat to a fusogenic peptide facilitates endosomal escape and enhances nuclear transport after internalization via macropinocytosis (Torchilin 2008). The fusogenic peptide is the N-terminus domain of the influenza virus hemagglutinin protein HA2, and it triggers the release of the Tat-fusion protein from endosomes and thereby enhances its nuclear transport as first described in 1999 (Josephson et al. 1999). Tat peptide-mediated translocation of gold nanoparticles into the cell nucleus has been heavily investigated (de la Fuente and Berry 2005). This functionalization allows the nanoparticles to penetrate the cell membrane and target the nucleus.

Nuclear targeting research has many potential applications in cell imaging, but also extends to wider biomedical applications, such as gene therapy and drug delivery. Tumor targeting and nuclear imaging have been validated using nanoparticles directed to the nucleus (Rogach and Ogris 2010). Additionally, cancer therapeutics that inhibit cell replication, such as 5-fluorouracil, paclitaxel and gemcitabine, have been delivered successfully to the nucleus. Several nanoparticles are already in clinical trials for metastatic and solid tumors, though delivery of therapeutics to the cytosol is just as effective, though dependent on the properties of the therapeutic itself. Gene therapy is a novel and upcoming field as nanoparticle payloads can include DNA, RNA, RNAi and siRNA, all of which mediate specific therapeutic effects depending on their individual properties (Hart 2010; Lares et al. 2010).

5 Multi-Site Targeting

Targeting therapeutics toward a specific organelle is in vogue. One step further is manipulating the numerous signaling elements of available proteins to design novel carriers capable of targeting multiple organelles (Lim 2007). With the discovery of specific intracellular targets for advanced therapeutics we can envision systems programmed for simultaneously delivery of agents to various subcellular locations, providing a paradigm shift in our approach to disease management. Bearing this in mind, we have recently introduced a concept of advanced nanovectors called Logic-Embedded Vectors (LEVs) (Ferrari 2010), which describe a nanoassembly responsible for the the co-delivery of multiple nanoparticles that cooperate with biomolecules and intracellular organelles to target multiple subcelluar locations simultaneously or sequentially. These LEVs are designed to overcome biological barriers to transport and distribute therapeutic agents to unique sites. A prominent example of LEV is the multistage delivery systems comprised of stage-one porous silicon microparticles loaded with second stage nanoparticles (Serda et al. 2010b). Each stage of the system is formulated to overcome a biological barrier such as enzyme degradation, transport through the vascular endothelium and molecular efflux pumps. As an example, stage one components target afflicted endothelia surface moieties through attachment of ligands, such as antibodies, peptides and aptamers, to the surface of the particles. Second-stage components, concentrated within the stage one vector, target at a second level, leading to delivery of a large quantity of therapeutic and/or imaging agents. LEVs that incorporate superparamagnetic iron oxide nanoparticles into porous silicon have recently been demonstrated (Ferrati et al. 2010; Serda et al. 2010b). Aminosilylation of the nanoparticles creates functional sites for conjugation of targeting ligands, such as antibodies specific for VEGFR-2 and PECAM (Platelet Endothelial Cell Adhesion Molecule) which facilitate endothelial association. Intracelluar trafficking of the LEVs have been demonstrated in transmission electron microscopy images using J774 mouse macrophages (Fig. 6; Serda et al. 2010a, b).

We have demonstrated that intracellular trafficking of LEVs can be designed to target several organelles simultaneously as a function of surface modifications of each of the components (Serda et al. 2010b). LEVs loaded surface-modified iron oxide nanoparticles have been shown to deliver nanoparticles to specific organelles, specifically the endosome and cytoplasm in one case, and in another case, endosomes and novel membrane-bound vesicles that are candidates for exocytosis. Silicon microparticles were transported along the endosomal pathway while

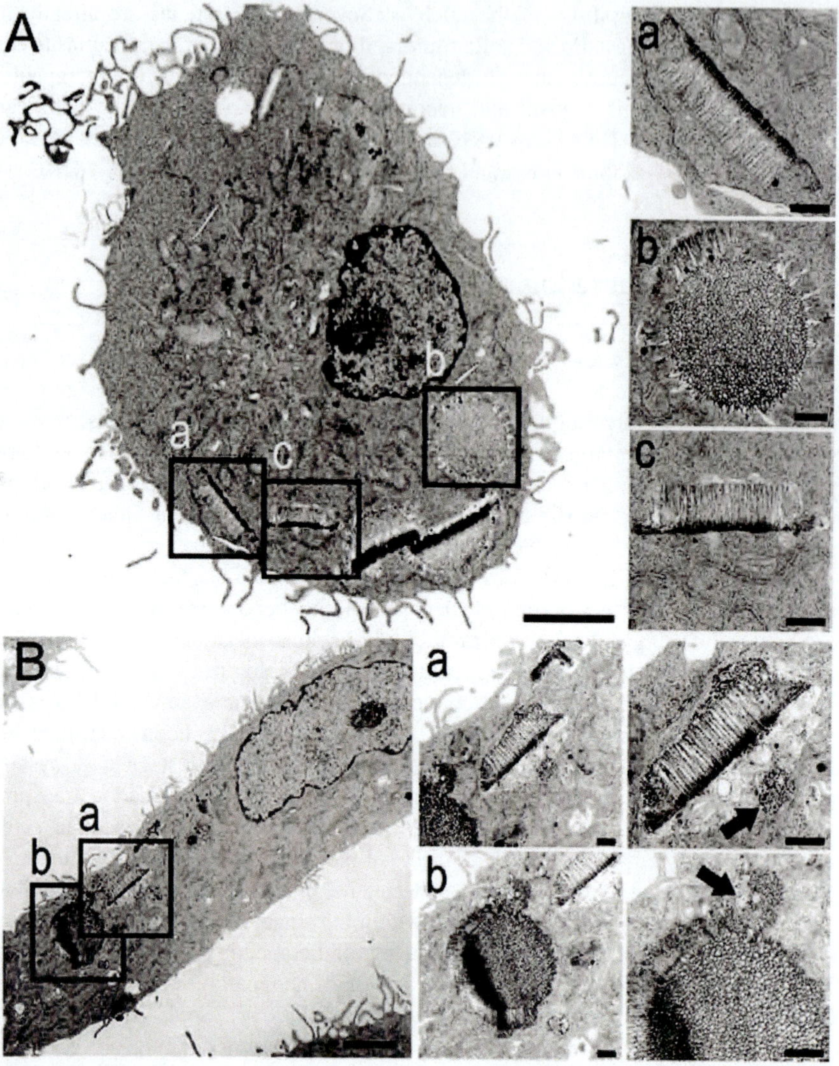

Fig. 6 Intracellular Trafficking of the LEVs: transmission electron micrographs of J774 cells 24 h after introduction of control (**A**) or iron oxide-loaded (**B**) porous silicon particles (ratio of 1:5). Cells on the left are at a 6k magnification (bar: 3 μm) and to the right at 25k and 50k magnification (bars: 500 nm). Internalized LEVs are depicted in different orientations, with black arrows indicating clusters of iron oxide nanoparticles (Reprinted from Serda et al. (2010a). Courtesy of Wiley Publishing Group)

PEGylated amine iron oxide nanoparticles were delivered to novel secretory vesicles. Chitosan-coated nanoparticles, on the other hand, were released from endosomes into the cytosol. LEVs are introduced into a cell in membrane bound units, creating endosomes or phagosomes, which in period of time, distribute their contents into specific organelles (e.g. endosomal sorting). The vesicular secretion may occur at any point of the endo-lysosomal pathway. Exosomal secretion is often associated with transport of protein, mRNA and microRNA content between cells, giving rise to specific cell signals. Therefore, LEVs can be used for both intracellular and extracellular targeting simultaneously.

In summary, targeting therapeutic nanovectors to unique or multiple intracellular sites can efficiently combat a particular disease at the site of origin of the disease. Moreover, the design and physical properties of the nanovectors or nanovehicles can be formulated based on the type of target desired to achieve the therapeutic goal.

6 Conclusions

Mass transport of drug carriers begins with the route of administration and moves across epithelial and endothelial barriers, stromal barriers, and cellular barriers, as well as an abundance of additional physical and chemical barriers that challenge arrival at the target site and drug integrity. Unique characteristics that originate from the tumor microenvironment have been summarized with an emphasis on utilizing these traits to achieve preferential accumulation of imaging and therapeutic agents at the cancer lesion. Nanoparticle carrier properties including size, shape, density and surface chemistry dominate convective transport in the blood stream, margination, cell adhesion, selective cellular uptake, sub-cellular trafficking and localization. The understanding of transport differentials in cancer promises to enhance the development of lesion-specific delivery carriers that exploit these transport differentials to achieve greater therapeutic efficacy and reduced side effects.

References

Aglipay, J. A., Lee, S. W., Okada, S., Fujiuchi, N., Ohtsuka, T., Kwak, J. C., Wang, Y., Johnstone, R. W., Deng, C., Qin, J. and Ouchi, T. (2003) 'A member of the Pyrin family, IFI16, is a novel BRCA1-associated protein involved in the p53-mediated apoptosis pathway', *Oncogene,* 22(55), 8931–8.

Ahting, U., Thieffry, M., Engelhardt, H., Hegerl, R., Neupert, W. and Nussberger, S. (2001) 'Tom40, the pore-forming component of the protein-conducting TOM channel in the outer membrane of mitochondria', *J Cell Biol,* 153(6), 1151–60.

Akagi, T., Kim, H. and Akashi, M. (2010) 'pH-dependent disruption of erythrocyte membrane by amphiphilic poly(amino acid) nanoparticles', *J Biomater Sci Polym Ed,* 21(3), 315–28.

Akinc, A., Thomas, M., Klibanov, A. M. and Langer, R. (2005) 'Exploring polyethylenimine-mediated DNA transfection and the proton sponge hypothesis', *J Gene Med,* 7(5), 657–63.

Allen, T. M. (2002) 'Ligand-targeted therapeutics in anticancer therapy', *Nat Rev Cancer,* 2(10), 750–63.

Andre, N., Rome, A. and Carre, M. (2006) '[Antimitochondrial agents: a new class of anticancer agents]', *Arch Pediatr,* 13(1), 69–75.

Arteaga, C. L., Hurd, S. D., Winnier, A. R., Johnson, M. D., Fendly, B. M. and Forbes, J. T. (1993) 'Anti-transforming growth factor (TGF)-beta antibodies inhibit breast cancer cell tumorigenicity and increase mouse spleen natural killer cell activity. Implications for a possible role of tumor cell/host TGF-beta interactions in human breast cancer progression', *Journal of Clinical Investigation,* 92(6), 2569.

Au, J. L., Jang, S. H., Zheng, J., Chen, C. T., Song, S., Hu, L. and Wientjes, M. G. (2001) 'Determinants of drug delivery and transport to solid tumors', *J Control Release,* 74(1–3), 31–46.

Avery, R. L., Pearlman, J., Pieramici, D. J., Rabena, M. D., Castellarin, A. A., Nasir, M. A., Giust, M. J., Wendel, R. and Patel, A. (2006) 'Intravitreal bevacizumab (Avastin) in the treatment of proliferative diabetic retinopathy', *Ophthalmology,* 113(10), 1695.

Bale, S. S., Kwon, S. J., Shah, D. A., Banerjee, A., Dordick, J. S. and Kane, R. S. (2010) 'Nanoparticle-mediated cytoplasmic delivery of proteins to target cellular machinery', *ACS Nano,* 4(3), 1493–500.

Bar, H., Yacoby, I. and Benhar, I. (2008) 'Killing cancer cells by targeted drug-carrying phage nanomedicines', *BMC Biotechnol,* 8, 37.

Batist, G., Ramakrishnan, G., Rao, C. S., Chandrasekharan, A., Gutheil, J., Guthrie, T., Shah, P., Khojasteh, A., Nair, M. K. and Hoelzer, K. (2001) 'Reduced cardiotoxicity and preserved antitumor efficacy of liposome-encapsulated doxorubicin and cyclophosphamide compared with conventional doxorubicin and cyclophosphamide in a randomized, multicenter trial of metastatic breast cancer', *Journal of Clinical Oncology,* 19(5), 1444.

Biasutto, L., Dong, L. F., Zoratti, M. and Neuzil, J. (2010) 'Mitochondrially targeted anti-cancer agents', *Mitochondrion.*

Biswas, S., Criswell, T. L., Wang, S. E. and Arteaga, C. L. (2006) 'Inhibition of transforming growth factor- signaling in human cancer: targeting a tumor suppressor network as a therapeutic strategy', *Clinical Cancer Research,* 12(14), 4142.

Blobel, G. and Dobberstein, B. (1975) 'Transfer of proteins across membranes. I. Presence of proteolytically processed and unprocessed nascent immunoglobulin light chains on membrane-bound ribosomes of murine myeloma', *J Cell Biol,* 67(3), 835–51.

Boddapati, S. V., D'Souza, G. G., Erdogan, S., Torchilin, V. P. and Weissig, V. (2008) 'Organelle-targeted nanocarriers: specific delivery of liposomal ceramide to mitochondria enhances its cytotoxicity in vitro and in vivo', *Nano Lett,* 8(8), 2559–63.

Boddapati, S. V., Tongcharoensirikul, P., Hanson, R. N., D'Souza, G. G., Torchilin, V. P. and Weissig, V. (2005) 'Mitochondriotropic liposomes', *J Liposome Res,* 15(1–2), 49–58.

Boucher, Y., Baxter, L. T. and Jain, R. K. (1990) 'Interstitial pressure gradients in tissue-isolated and subcutaneous tumors: implications for therapy', *Cancer research,* 50(15), 4478.

Boussif, O., Lezoualc'h, F., Zanta, M. A., Mergny, M. D., Scherman, D., Demeneix, B. and Behr, J. P. (1995) 'A versatile vector for gene and oligonucleotide transfer into cells in culture and in vivo: polyethylenimine', *Proc Natl Acad Sci USA,* 92(16), 7297–301.

Breunig, M., Bauer, S. and Goepferich, A. (2008) 'Polymers and nanoparticles: intelligent tools for intracellular targeting?', *Eur J Pharm Biopharm,* 68(1), 112–28.

Bross, P. F., Beitz, J., Chen, G., Chen, X. H., Duffy, E., Kieffer, L., Roy, S., Sridhara, R., Rahman, A., Williams, G. and Pazdur, R. (2001) 'Approval summary: gemtuzumab ozogamicin in relapsed acute myeloid leukemia', *Clin Cancer Res,* 7(6), 1490–6.

Brown, D. A. and London, E. (1998) 'Functions of lipid rafts in biological membranes', *Annual Review of Cell and Developmental Biology,* 14(1), 111–136.

Buckland, J. 'Psoriasis: Anti-VEGF antibody therapy for psoriasis?', *Nature Reviews Rheumatology,* 6(3), 119.

Cairns, R., Papandreou, I. and Denko, N. (2006) 'Overcoming physiologic barriers to cancer treatment by molecularly targeting the tumor microenvironment', *Molecular Cancer Research,* 4(2), 61.

Caldorera-Moore, M., Guimard, N., Shi, L. and Roy, K. (2010) 'Designer nanoparticles: incorporating size, shape and triggered release into nanoscale drug carriers', *Expert Opin Drug Deliv,* 7(4), 479–95.

Callahan, J. and Kopecek, J. (2006) 'Semitelechelic HPMA copolymers functionalized with triphenylphosphonium as drug carriers for membrane transduction and mitochondrial localization', *Biomacromolecules,* 7(8), 2347–56.

Capitani, M. and Sallese, M. (2009) 'The KDEL receptor: new functions for an old protein', *FEBS Lett,* 583(23), 3863–71.
Champion, J. A. and Mitragotri, S. (2006) 'Role of target geometry in phagocytosis', *Proc Natl Acad Sci USA,* 103(13), 4930–4.
Champion, J. A. and Mitragotri, S. (2009) 'Shape induced inhibition of phagocytosis of polymer particles', *Pharm Res,* 26(1), 244–9.
Chen, A. M., Zhang, M., Wei, D., Stueber, D., Taratula, O., Minko, T. and He, H. (2009) 'Co-delivery of Doxorubicin and Bcl-2 siRNA by Mesoporous Silica Nanoparticles Enhances the Efficacy of Chemotherapy in Multidrug-Resistant Cancer Cells', *Small.*
Cho, E. C., Shim, J., Lee, K. E., Kim, J. W. and Han, S. S. (2009) 'Flexible magnetic microtubules structured by lipids and magnetic nanoparticles', *ACS Appl Mater Interfaces,* 1(6), 1159–62.
Cho, K., Wang, X., Nie, S., Chen, Z. G. and Shin, D. M. (2008) 'Therapeutic nanoparticles for drug delivery in cancer', *Clin Cancer Res,* 14(5), 1310–6.
Clinicaltrials.gov – NCT00470613 (2008) 'Safety study of infusion of SGT-53 to treat solid tumors', *Clinicaltrials.gov.*
Dancourt, J. and Barlowe, C. (2010) 'Protein sorting receptors in the early secretory pathway', *Annu Rev Biochem,* 79, 777–802.
Davis, J. R., Kakar, M. and Lim, C. S. (2007) 'Controlling protein compartmentalization to overcome disease', *Pharm Res,* 24(1), 17–27.
Davis, M. E., Chen, Z. G. and Shin, D. M. (2008) 'Nanoparticle therapeutics: an emerging treatment modality for cancer', *Nat Rev Drug Discov,* 7(9), 771–82.
Davis, M. E., Zuckerman, J. E., Choi, C. H., Seligson, D., Tolcher, A., Alabi, C. A., Yen, Y., Heidel, J. D. and Ribas, A. (2010) 'Evidence of RNAi in humans from systemically administered siRNA via targeted nanoparticles', *Nature,* 464(7291), 1067–70.
de la Fuente, J. M. and Berry, C. C. (2005) 'Tat peptide as an efficient molecule to translocate gold nanoparticles into the cell nucleus', *Bioconjug Chem,* 16(5), 1176–80.
de Moura, M. B., dos Santos, L. S. and Van Houten, B. (2010) 'Mitochondrial dysfunction in neurodegenerative diseases and cancer', *Environ Mol Mutagen,* 51(5), 391–405.
Decuzzi, P. and Ferrari, M. (2007) 'The role of specific and non-specific interactions in receptor-mediated endocytosis of nanoparticles', *Biomaterials,* 28(18), 2915–22.
Decuzzi, P. and Ferrari, M. (2008a) 'Design maps for nanoparticles targeting the diseased microvasculature', *Biomaterials,* 29(3), 377–84.
Decuzzi, P. and Ferrari, M. (2008b) 'The receptor-mediated endocytosis of nonspherical particles', *Biophys J,* 94(10), 3790–7.
Desai, N., Trieu, V., Yao, Z., Louie, L., Ci, S., Yang, A., Tao, C., De, T., Beals, B. and Dykes, D. (2006) 'Increased antitumor activity, intratumor paclitaxel concentrations, and endothelial cell transport of cremophor-free, albumin-bound paclitaxel, ABI-007, compared with cremophor-based paclitaxel', *Clinical cancer research,* 12(4), 1317.
Dhar, S., Gu, F. X., Langer, R., Farokhzad, O. C. and Lippard, S. J. (2008) 'Targeted delivery of cisplatin to prostate cancer cells by aptamer functionalized Pt(IV) prodrug-PLGA-PEG nanoparticles', *Proc Natl Acad Sci USA,* 105(45), 17356–61.
Dharap, S. S., Wang, Y., Chandna, P., Khandare, J. J., Qiu, B., Gunaseelan, S., Sinko, P. J., Stein, S., Farmanfarmaian, A. and Minko, T. (2005) 'Tumor-specific targeting of an anticancer drug delivery system by LHRH peptide', *Proc Natl Acad Sci USA,* 102(36), 12962–7.
Di Paolo, A. and Bocci, G. (2007) 'Drug distribution in tumors: mechanisms, role in drug resistance, and methods for modification', *Current Oncology Reports,* 9(2), 109–114.
Doshi, N. and Mitragotri, S. (2010) 'Needle-shaped polymeric particles induce transient disruption of cell membranes', *J R Soc Interface,* 7 Suppl 4, S403-10.
Doshi, N., Prabhakarpandian, B., Rea-Ramsey, A., Pant, K., Sundaram, S. and Mitragotri, S. (2010) 'Flow and adhesion of drug carriers in blood vessels depend on their shape: a study using model synthetic microvascular networks', *J Control Release,* 146(2), 196–200.
Duncan, R. (2006) 'Polymer conjugates as anticancer nanomedicines', *Nat Rev Cancer,* 6(9), 688–701.
Dutta, T. and Jain, N. K. (2007) 'Targeting potential and anti-HIV activity of lamivudine loaded mannosylated poly (propyleneimine) dendrimer', *Biochim Biophys Acta,* 1770(4), 681–6.

ElBayoumi, T. A. and Torchilin, V. P. (2009) 'Tumor-targeted nanomedicines: enhanced antitumor efficacy in vivo of doxorubicin-loaded, long-circulating liposomes modified with cancer-specific monoclonal antibody', *Clin Cancer Res,* 15(6), 1973–80.

Fantini, J., Garmy, N., Mahfoud, R. and Yahi, N. (2002) 'Lipid rafts: structure, function and role in HIV, Alzheimer's and prion diseases', *Expert Rev Mol Med,* 4(27), 1–22.

Ferrara, N. and Kerbel, R. S. (2005) 'Angiogenesis as a therapeutic target', *Nature,* 438(7070), 967–974.

Ferrari, M. (2010) 'Frontiers in cancer nanomedicine: directing mass transport through biological barriers', *Trends Biotechnol,* 28(4), 181–8.

Ferrati, S., Mack, A., Chiappini, C., Liu, X., Bean, A. J., Ferrari, M. and Serda, R. E. (2010) 'Intracellular trafficking of silicon particles and logic-embedded vectors', *Nanoscale,* 2(8), 1512–20.

Fisher, R. E., Siegel, B. A., Edell, S. L., Oyesiku, N. M., Morgenstern, D. E., Messmann, R. A. and Amato, R. J. (2008) 'Exploratory study of 99mTc-EC20 imaging for identifying patients with folate receptor-positive solid tumors', *J Nucl Med,* 49(6), 899–906.

Fuchs, H. and Bachran, C. (2009) 'Targeted tumor therapies at a glance', *Curr Drug Targets,* 10(2), 89–93.

Fujiwara, T., Akita, H. and Harashima, H. (2010) 'Intracellular fate of octaarginine-modified liposomes in polarized MDCK cells', *Int J Pharm,* 386(1–2), 122–30.

Fulda, S., Galluzzi, L. and Kroemer, G. (2010) 'Targeting mitochondria for cancer therapy', *Nat Rev Drug Discov,* 9(6), 447–64.

Gad, S. C. (2008) 'Preclinical Development Handbook: ADME and Biopharmaceutical Properties'.

Gao, X., Chan, W. C. and Nie, S. (2002) 'Quantum-dot nanocrystals for ultrasensitive biological labeling and multicolor optical encoding', *J Biomed Opt,* 7(4), 532–7.

Garg, M., Asthana, A., Agashe, H. B., Agrawal, G. P. and Jain, N. K. (2006) 'Stavudine-loaded mannosylated liposomes: in-vitro anti-HIV-I activity, tissue distribution and pharmacokinetics', *J Pharm Pharmacol,* 58(5), 605–16.

Gelperina, S., Maksimenko, O., Khalansky, A., Vanchugova, L., Shipulo, E., Abbasova, K., Berdiev, R., Wohlfart, S., Chepurnova, N. and Kreuter, J. (2010) 'Drug delivery to the brain using surfactant-coated poly(lactide-co-glycolide) nanoparticles: influence of the formulation parameters', *Eur J Pharm Biopharm,* 74(2), 157–63.

Giacomini, K. M., Huang, S. M., Tweedie, D. J., Benet, L. Z., Brouwer, K. L., Chu, X., Dahlin, A., Evers, R., Fischer, V., Hillgren, K. M., Hoffmaster, K. A., Ishikawa, T., Keppler, D., Kim, R. B., Lee, C. A., Niemi, M., Polli, J. W., Sugiyama, Y., Swaan, P. W., Ware, J. A., Wright, S. H., Yee, S. W., Zamek-Gliszczynski, M. J. and Zhang, L. (2010) 'Membrane transporters in drug development', *Nat Rev Drug Discov,* 9(3), 215–36.

Gilchrist, D., Mykytka, B. and Rexach, M. (2002) 'Accelerating the rate of disassembly of karyopherin.cargo complexes', *J Biol Chem,* 277(20), 18161–72.

Gilchrist, D. and Rexach, M. (2003) 'Molecular basis for the rapid dissociation of nuclear localization signals from karyopherin alpha in the nucleoplasm', *J Biol Chem,* 278(51), 51937–49.

Giordano, R. J., Lahdenranta, J., Zhen, L., Chukwueke, U., Petrache, I., Langley, R. R., Fidler, I. J., Pasqualini, R., Tuder, R. M. and Arap, W. (2008) 'Targeted induction of lung endothelial cell apoptosis causes emphysema-like changes in the mouse', *J Biol Chem,* 283(43), 29447–60.

Godin B, T. E. (2006) 'Surfactants in enhanced enteral delivery of macromolecules. In Touitou E, Barry BW. (Eds) Enhancement in drug delivery, CRC press, Taylor & Francis Group, Boca Raton-London-New York', 37–56.

Godin, B. and Touitou, E. (2004) 'Mechanism of bacitracin permeation enhancement through the skin and cellular membranes from an ethosomal carrier', *J Control Release,* 94(2–3), 365–79.

Goldfarb, D. S., Corbett, A. H., Mason, D. A., Harreman, M. T. and Adam, S. A. (2004) 'Importin alpha: a multipurpose nuclear-transport receptor', *Trends Cell Biol,* 14(9), 505–14.

Gradishar, W. J., Tjulandin, S., Davidson, N., Shaw, H., Desai, N., Bhar, P., Hawkins, M. and O'Shaughnessy, J. (2005) 'Phase III trial of nanoparticle albumin-bound paclitaxel compared with polyethylated castor oil-based paclitaxel in women with breast cancer', *J Clin Oncol,* 23(31), 7794–803.

Gupta, A. K., Berry, C., Gupta, M. and Curtis, A. (2003) 'Receptor-mediated targeting of magnetic nanoparticles using insulin as a surface ligand to prevent endocytosis', *IEEE Trans Nanobioscience,* 2(4), 255–61.

Gupta, N. P., Nayyar, R. and Sahay, S. C. (2010) 'Periureteric venous ring with renal calculi and transitional cell carcinoma: report of a rare case', *Surg Radiol Anat,* 32(4), 405–7.

Hafez, I. M., Ansell, S. and Cullis, P. R. (2000) 'Tunable pH-sensitive liposomes composed of mixtures of cationic and anionic lipids', *Biophysical Journal,* 79(3), 1438–1446.

Hajitou, A., Trepel, M., Lilley, C. E., Soghomonyan, S., Alauddin, M. M., Marini, F. C., 3rd, Restel, B. H., Ozawa, M. G., Moya, C. A., Rangel, R., Sun, Y., Zaoui, K., Schmidt, M., von Kalle, C., Weitzman, M. D., Gelovani, J. G., Pasqualini, R. and Arap, W. (2006) 'A hybrid vector for ligand-directed tumor targeting and molecular imaging', *Cell,* 125(2), 385–98.

Hanaki, K., Momo, A., Oku, T., Komoto, A., Maenosono, S., Yamaguchi, Y. and Yamamoto, K. (2003) 'Semiconductor quantum dot/albumin complex is a long-life and highly photostable endosome marker', *Biochem Biophys Res Commun,* 302(3), 496–501.

Harding, A. E. (1991) 'Neurological disease and mitochondrial genes', *Trends Neurosci,* 14(4), 132–8.

Hart, S. L. (2010) 'Multifunctional nanocomplexes for gene transfer and gene therapy', *Cell Biol Toxicol,* 26(1), 69–81.

Hartwig, J. H., Davies, W. A. and Stossel, T. P. (1977) 'Evidence for contractile protein translocation in macrophage spreading, phagocytosis, and phagolysosome formation', *Journal of Cell Biology,* 75(3), 956.

Heemels, M. T. (1999) 'Medicine Nobel goes to pioneer of protein guidance mechanisms', *Nature,* 401(6754), 625.

Hines, M. A. and Guyot-Sionnest, P. (1996) 'Synthesis and Characterization of Strongly Luminescing ZnS-Capped CdSe Nanocrystals', *The Journal of Physical Chemistry,* 100(2), 468–471.

Hobbs, S. K., Monsky, W. L., Yuan, F., Roberts, W. G., Griffith, L., Torchilin, V. P. and Jain, R. K. (1998) 'Regulation of transport pathways in tumor vessels: role of tumor type and microenvironment', *Proceedings of the National Academy of Sciences of the United States of America,* 95(8), 4607.

Holt, I. J., Harding, A. E. and Morgan-Hughes, J. A. (1988) 'Deletions of muscle mitochondrial DNA in patients with mitochondrial myopathies', *Nature,* 331(6158), 717–9.

Holtzman, E. (1989) *Lysosomes,* New York, NY: Plenum Press.

Hoshino, A., Fujioka, K., Oku, T., Nakamura, S., Suga, M., Yamaguchi, Y., Suzuki, K., Yasuhara, M. and Yamamoto, K. (2004a) 'Quantum dots targeted to the assigned organelle in living cells', *Microbiol Immunol,* 48(12), 985–94.

Hoshino, A., Fujioka, K., Oku, T., Suga, M., Sasaki, Y. F., Ohta, T., Yasuhara, M., Suzuki, K. and Yamamoto, K. (2004b) 'Physicochemical Properties and Cellular Toxicity of Nanocrystal Quantum Dots Depend on Their Surface Modification', *Nano Letters,* 4(11), 2163–2169.

Hu, J., Zhang, Z., Shen, W. J. and Azhar, S. (2010) 'Cellular cholesterol delivery, intracellular processing and utilization for biosynthesis of steroid hormones', *Nutr Metab (Lond),* 7, 47.

Hunter, J. and Hirst, B. H. (1997) 'Intestinal secretion of drugs. The role of P-glycoprotein and related drug efflux systems in limiting oral drug absorption', *Advanced drug delivery reviews,* 25(2–3), 129–157.

Hyung Park, J., Kwon, S., Lee, M., Chung, H., Kim, J. H., Kim, Y. S., Park, R. W., Kim, I. S., Bong Seo, S. and Kwon, I. C. (2006) 'Self-assembled nanoparticles based on glycol chitosan bearing hydrophobic moieties as carriers for doxorubicin: in vivo biodistribution and anti-tumor activity', *Biomaterials,* 27(1), 119–126.

Iglesias, J. (2009) 'nab-Paclitaxel (Abraxane(R)): an albumin-bound cytotoxic exploiting natural delivery mechanisms into tumors', *Breast Cancer Res,* 11 Suppl 1, S21.

Iinuma, H., Maruyama, K., Okinaga, K., Sasaki, K., Sekine, T., Ishida, O., Ogiwara, N., Johkura, K. and Yonemura, Y. (2002) 'Intracellular targeting therapy of cisplatin-encapsulated transferrin-polyethylene glycol liposome on peritoneal dissemination of gastric cancer', *Int J Cancer,* 99(1), 130–7.

Ishida, O., Maruyama, K., Tanahashi, H., Iwatsuru, M., Sasaki, K., Eriguchi, M. and Yanagie, H. (2001) 'Liposomes bearing polyethyleneglycol-coupled transferrin with intracellular targeting property to the solid tumors in vivo', *Pharm Res,* 18(7), 1042–8.

Ivanov, A. I. (2008) *Exocytosis and Endocytosis: Methods in Molecular Biology,* Totowa, NJ: Humana Press.

Iyer, A. K., Khaled, G., Fang, J. and Maeda, H. (2006) 'Exploiting the enhanced permeability and retention effect for tumor targeting', *Drug discovery today,* 11(17–18), 812–818.

Jain, R. K. (1987) 'Transport of molecules across tumor vasculature', *Cancer and Metastasis Reviews,* 6(4), 559–593.

Jain, R. K. (1990) 'Vascular and interstitial barriers to delivery of therapeutic agents in tumors', *Cancer and Metastasis Reviews,* 9(3), 253–266.

Jain, R. K. (1999) 'Transport of molecules, particles, and cells in solid tumors', *Annual Review of Biomedical Engineering,* 1(1), 241–263.

Jain, S. K., Gupta, Y., Jain, A., Saxena, A. R. and Khare, P. (2008) 'Mannosylated gelatin nanoparticles bearing an anti-HIV drug didanosine for site-specific delivery', *Nanomedicine,* 4(1), 41–8.

Jin, H., Heller, D. A., Sharma, R. and Strano, M. S. (2009) 'Size-dependent cellular uptake and expulsion of single-walled carbon nanotubes: single particle tracking and a generic uptake model for nanoparticles', *ACS Nano,* 3(1), 149–58.

Johannes, L. and Romer, W. (2010) 'Shiga toxins--from cell biology to biomedical applications', *Nat Rev Microbiol,* 8(2), 105–16.

Josephson, L., Tung, C. H., Moore, A. and Weissleder, R. (1999) 'High-efficiency intracellular magnetic labeling with novel superparamagnetic-Tat peptide conjugates', *Bioconjug Chem,* 10(2), 186–91.

Kakar, M., Davis, J. R., Kern, S. E. and Lim, C. S. (2007) 'Optimizing the protein switch: altering nuclear import and export signals, and ligand binding domain', *J Control Release,* 120(3), 220–32.

Kakimoto, S., Tanabe, T., Azuma, H. and Nagasaki, T. (2010) 'Enhanced internalization and endosomal escape of dual-functionalized poly(ethyleneimine)s polyplex with diphtheria toxin T and R domains', *Biomed Pharmacother,* 64(4), 296–301.

Kakudo, T., Chaki, S., Futaki, S., Nakase, I., Akaji, K., Kawakami, T., Maruyama, K., Kamiya, H. and Harashima, H. (2004) 'Transferrin-modified liposomes equipped with a pH-sensitive fusogenic peptide: an artificial viral-like delivery system', *Biochemistry,* 43(19), 5618–28.

Kanwal, C., Mu, S., Kern, S. E. and Lim, C. S. (2004) 'Bidirectional on/off switch for controlled targeting of proteins to subcellular compartments', *Journal of Controlled Release,* 98(3), 379–393.

Karlsson, J., Ungell, A. L., Gråsjö, J. and Artursson, P. (1999) 'Paracellular drug transport across intestinal epithelia: influence of charge and induced water flux', *European journal of pharmaceutical sciences,* 9(1), 47–56.

Kawakami, K., Nakajima, O., Morishita, R. and Nagai, R. (2006) 'Targeted anticancer immunotoxins and cytotoxic agents with direct killing moieties', *ScientificWorldJournal,* 6, 781–90.

Kennedy, I. M., Wilson, D. and Barakat, A. I. (2009) 'Uptake and inflammatory effects of nanoparticles in a human vascular endothelial cell line', *Res Rep Health Eff Inst,* (136), 3–32.

Ko, Y. T., Falcao, C. and Torchilin, V. P. (2009) 'Cationic liposomes loaded with proapoptotic peptide D-(KLAKLAK)(2) and Bcl-2 antisense oligodeoxynucleotide G3139 for enhanced anticancer therapy', *Mol Pharm,* 6(3), 971–7.

Kogure, K., Akita, H., Yamada, Y. and Harashima, H. (2008) 'Multifunctional envelope-type nano device (MEND) as a non-viral gene delivery system', *Adv Drug Deliv Rev,* 60(4–5), 559–71.

Kolonin, M. G., Saha, P. K., Chan, L., Pasqualini, R. and Arap, W. (2004) 'Reversal of obesity by targeted ablation of adipose tissue', *Nat Med,* 10(6), 625–32.

Kong, G., Braun, R. D. and Dewhirst, M. W. (2000) 'Hyperthermia enables tumor-specific nanoparticle delivery: effect of particle size', *Cancer research,* 60(16), 4440.

Koval, M., Preiter, K., Adles, C., Stahl, P. D. and Steinberg, T. H. (1998) 'Size of IgG-opsonized particles determines macrophage response during internalization', *Exp Cell Res,* 242(1), 265–73.

Lai, S. K., Hida, K., Chen, C. and Hanes, J. (2008) 'Characterization of the intracellular dynamics of a non-degradative pathway accessed by polymer nanoparticles', *J Control Release,* 125(2), 107–11.

Lares, M. R., Rossi, J. J. and Ouellet, D. L. (2010) 'RNAi and small interfering RNAs in human disease therapeutic applications', *Trends Biotechnol.*

Leamon, C. P. and Low, P. S. (1991) 'Delivery of macromolecules into living cells: a method that exploits folate receptor endocytosis', *Proc Natl Acad Sci USA,* 88(13), 5572–6.

Lee, E. S., Gao, Z. and Bae, Y. H. (2008) 'Recent progress in tumor pH targeting nanotechnology', *Journal of Controlled Release,* 132(3), 164–170.

Lee, H., Fonge, H., Hoang, B., Reilly, R. M. and Allen, C. (2010a) 'The Effects of Particle Size and Molecular Targeting on the Intratumoral and Subcellular Distribution of Polymeric Nanoparticles', *Mol Pharm.*

Lee, H., Kim, I. K. and Park, T. G. (2010b) 'Intracellular trafficking and unpacking of siRNA/quantum dot-PEI complexes modified with and without cell penetrating peptide: confocal and flow cytometric FRET analysis', *Bioconjug Chem,* 21(2), 289–95.

Lim, C. S. (2007) 'Organelle-specific targeting in drug delivery and design', *Advanced Drug Delivery Reviews,* 697–697.

Liu, J. and Shapiro, J. I. (2003) 'Endocytosis and signal transduction: basic science update', *Biol Res Nurs,* 5(2), 117–28.

Loi, M., Marchio, S., Becherini, P., Di Paolo, D., Soster, M., Curnis, F., Brignole, C., Pagnan, G., Perri, P., Caffa, I., Longhi, R., Nico, B., Bussolino, F., Gambini, C., Ribatti, D., Cilli, M., Arap, W., Pasqualini, R., Allen, T. M., Corti, A., Ponzoni, M. and Pastorino, F. (2010) 'Combined targeting of perivascular and endothelial tumor cells enhances anti-tumor efficacy of liposomal chemotherapy in neuroblastoma', *J Control Release,* 145(1), 66–73.

Longo, R., Sarmiento, R., Fanelli, M., Capaccetti, B., Gattuso, D. and Gasparini, G. (2002) 'Anti-angiogenic therapy: rationale, challenges and clinical studies', *Angiogenesis,* 5(4), 237–256.

Lu, Y. and Low, P. S. (2002a) 'Folate-mediated delivery of macromolecular anticancer therapeutic agents', *Adv Drug Deliv Rev,* 54(5), 675–93.

Lu, Y. and Low, P. S. (2002b) 'Folate targeting of haptens to cancer cell surfaces mediates immunotherapy of syngeneic murine tumors', *Cancer Immunol Immunother,* 51(3), 153–62.

Maeda, H., Wu, J., Sawa, T., Matsumura, Y. and Hori, K. (2000) 'Tumor vascular permeability and the EPR effect in macromolecular therapeutics: a review', *Journal of Controlled Release,* 65(1–2), 271–284.

Mao, Z., Wan, L., Hu, L., Ma, L. and Gao, C. (2010) 'Tat peptide mediated cellular uptake of SiO2 submicron particles', *Colloids Surf B Biointerfaces,* 75(2), 432–40.

Matthews, K., Noker, P. E., Tian, B., Grimes, S. D., Fulton, R., Schweikart, K., Harris, R., Aurigemma, R., Wang, M., Barnes, M. N., Siegal, G. P., Hemminki, A., Zinn, K., Curiel, D. T. and Alvarez, R. D. (2009) 'Identifying the safety profile of Ad5.SSTR/TK.RGD, a novel infectivity-enhanced bicistronic adenovirus, in anticipation of a phase I clinical trial in patients with recurrent ovarian cancer', *Clin Cancer Res,* 15(12), 4131–7.

Matsumura, Y., Gotoh, M., Muro, K., Yamada, Y., Shirao, K., Shimada, Y., Okuwa, M., Matsumoto, S., Miyata, Y., Ohkura, H., Chin, K., Baba, S., Yamao, T., Kannami, A., Takamatsu, Y., Ito, K. and Takahashi, K. (2004) 'Phase I and pharmacokinetic study of MCC-465, a doxorubicin (DXR) encapsulated in PEG immunoliposome, in patients with metastatic stomach cancer', *Annals of Oncology,* 15, 517–525.

McAteer, M. A., Schneider, J. E., Ali, Z. A., Warrick, N., Bursill, C. A., von zur Muhlen, C., Greaves, D. R., Neubauer, S., Channon, K. M. and Choudhury, R. P. (2008) 'Magnetic resonance imaging of endothelial adhesion molecules in mouse atherosclerosis using dual-targeted microparticles of iron oxide', *Arterioscler Thromb Vasc Biol,* 28(1), 77–83.

McAteer, M. A., Sibson, N. R., von Zur Muhlen, C., Schneider, J. E., Lowe, A. S., Warrick, N., Channon, K. M., Anthony, D. C. and Choudhury, R. P. (2007) 'In vivo magnetic resonance imaging of acute brain inflammation using microparticles of iron oxide', *Nat Med,* 13(10), 1253–8.

McKee, T. D., Grandi, P., Mok, W., Alexandrakis, G., Insin, N., Zimmer, J. P., Bawendi, M. G., Boucher, Y., Breakefield, X. O. and Jain, R. K. (2006) 'Degradation of fibrillar collagen in a human melanoma xenograft improves the efficacy of an oncolytic herpes simplex virus vector', *Cancer research,* 66(5), 2509.

Mehta, D. and Malik, A. B. (2006) 'Signaling mechanisms regulating endothelial permeability', *Physiological reviews,* 86(1), 279.

Mendoza, F. J., Espino, P. S., Cann, K. L., Bristow, N., McCrea, K. and Los, M. (2005) 'Anti-tumor chemotherapy utilizing peptide-based approaches--apoptotic pathways, kinases, and proteasome as targets', *Arch Immunol Ther Exp (Warsz),* 53(1), 47–60.

Michalak, M. and Opas, M. (2009) 'Endoplasmic and sarcoplasmic reticulum in the heart', *Trends Cell Biol,* 19(6), 253–9.

Minchinton, A. I. and Tannock, I. F. (2006) 'Drug penetration in solid tumours', *Nature Reviews Cancer,* 6(8), 583–592.

Mintz, P. J., Cardo-Vila, M., Ozawa, M. G., Hajitou, A., Rangel, R., Guzman-Rojas, L., Christianson, D. R., Arap, M. A., Giordano, R. J., Souza, G. R., Easley, J., Salameh, A., Oliviero, S., Brentani, R. R., Koivunen, E., Arap, W. and Pasqualini, R. (2009) 'An unrecognized extracellular function for an intracellular adapter protein released from the cytoplasm into the tumor microenvironment', *Proc Natl Acad Sci USA,* 106(7), 2182–7.

Mitragotri, S. (2009) 'In drug delivery, shape does matter', *Pharm Res,* 26(1), 232–4.

Mokranjac, D. and Neupert, W. (2005) 'Protein import into mitochondria', *Biochem Soc Trans,* 33(Pt 5), 1019–23.

Motosugi, U., Ichikawa, T., Nakajima, H., Sou, H., Sano, M., Sano, K., Araki, T., Iino, H., Fujii, H. and Nakazawa, T. (2009) 'Imaging of small hepatic metastases of colorectal carcinoma: how to use superparamagnetic iron oxide-enhanced magnetic resonance imaging in the multi-detector-row computed tomography age?', *J Comput Assist Tomogr,* 33(2), 266–72.

Muranishi, S. (1990) 'Absorption enhancers', *Critical reviews in therapeutic drug carrier systems,* 7(1), 1.

Muro, S., Garnacho, C., Champion, J. A., Leferovich, J., Gajewski, C., Schuchman, E. H., Mitragotri, S. and Muzykantov, V. R. (2008) 'Control of endothelial targeting and intracellular delivery of therapeutic enzymes by modulating the size and shape of ICAM-1-targeted carriers', *Mol Ther,* 16(8), 1450–8.

Nakase, I., Kobayashi, S. and Futaki, S. (2010) 'Endosome-disruptive peptides for improving cytosolic delivery of bioactive macromolecules', *Biopolymers.*

Nativo, P., Prior, I. A. and Brust, M. (2008) 'Uptake and intracellular fate of surface-modified gold nanoparticles', *ACS Nano,* 2(8), 1639–44.

Nellans, H. N. (1991) '(B) Mechanisms of peptide and protein absorption::(1) Paracellular intestinal transport: modulation of absorption', *Advanced drug delivery reviews,* 7(3), 339–364.

Netti, P. A., Berk, D. A., Swartz, M. A., Grodzinsky, A. J. and Jain, R. K. (2000) 'Role of extracellular matrix assembly in interstitial transport in solid tumors', *Cancer research,* 60(9), 2497.

Neve, E. P. and Ingelman-Sundberg, M. (2010) 'Cytochrome P450 proteins: retention and distribution from the endoplasmic reticulum', *Curr Opin Drug Discov Devel,* 13(1), 78–85.

Nie, J., Chang, B., Traktuev, D. O., Sun, J., March, K., Chan, L., Sage, E. H., Pasqualini, R., Arap, W. and Kolonin, M. G. (2008) 'IFATS collection: Combinatorial peptides identify alpha5beta1 integrin as a receptor for the matricellular protein SPARC on adipose stromal cells', *Stem Cells,* 26(10), 2735–45.

Nishimura, S., Takahashi, S., Kamikatahira, H., Kuroki, Y., Jaalouk, D. E., O'Brien, S., Koivunen, E., Arap, W., Pasqualini, R., Nakayama, H. and Kuniyasu, A. (2008) 'Combinatorial targeting of the macropinocytotic pathway in leukemia and lymphoma cells', *J Biol Chem,* 283(17), 11752–62.

Paillard, A., Hindre, F., Vignes-Colombeix, C., Benoit, J. P. and Garcion, E. (2010) 'The importance of endo-lysosomal escape with lipid nanocapsules for drug subcellular bioavailability', *Biomaterials,* 31(29), 7542–54.

Pan, X. Q., Wang, H. and Lee, R. J. (2003) 'Antitumor activity of folate receptor-targeted liposomal doxorubicin in a KB oral carcinoma murine xenograft model', *Pharm Res,* 20(3), 417–22.

Pang, K. S. (2003) 'Modeling of intestinal drug absorption: roles of transporters and metabolic enzymes (for the Gillette Review Series)', *Drug Metabolism and Disposition,* 31(12), 1507.

Panyam, J. and Labhasetwar, V. (2004) 'Targeting intracellular targets', *Curr Drug Deliv,* 1(3), 235–47.

Panyam, J., Sahoo, S. K., Prabha, S., Bargar, T. and Labhasetwar, V. (2003) 'Fluorescence and electron microscopy probes for cellular and tissue uptake of poly(D,L-lactide-co-glycolide) nanoparticles', *Int J Pharm,* 262(1–2), 1–11.

Park, H. S., Kim, C. W., Lee, H. J., Choi, J. H., Lee, S. G., Yun, Y. P., Kwon, I. C., Lee, S. J., Jeong, S. Y. and Lee, S. C. (2010a) 'A mesoporous silica nanoparticle with charge-convertible pore walls for efficient intracellular protein delivery', *Nanotechnology,* 21(22), 225101.

Park, S., Lee, S. J., Chung, H., Her, S., Choi, Y., Kim, K., Choi, K. and Kwon, I. C (2010). 'Cellular uptake pathway and drug release characteristics of drug-encapsulated glycol chitosan nanoparticles in live cells', *Microscopy Research and Technique,* 73, 857–865.

Park, S., Lee, S. J., Chung, H., Her, S., Choi, Y., Kim, K., Choi, K. and Kwon, I. C. (2010b) 'Cellular uptake pathway and drug release characteristics of drug-encapsulated glycol chitosan nanoparticles in live cells', *Microsc Res Tech,* 73(9), 857–65.

Park, Y. J., Liang, J. F., Ko, K. S., Kim, S. W. and Yang, V. C. (2003) 'Low molecular weight protamine as an efficient and nontoxic gene carrier: in vitro study', *J Gene Med,* 5(8), 700–11.

Pasche, B. (2001) 'Role of transforming growth factor beta in cancer', *Journal of Cellular Physiology,* 186(2), 153–168.

Pasqualini, R., Koivunen, E., Kain, R., Lahdenranta, J., Sakamoto, M., Stryhn, A., Ashmun, R. A., Shapiro, L. H., Arap, W. and Ruoslahti, E. (2000) 'Aminopeptidase N is a receptor for tumor-homing peptides and a target for inhibiting angiogenesis', *Cancer Res,* 60(3), 722–7.

Pasqualini, R., Koivunen, E. and Ruoslahti, E. (1997) 'Alpha v integrins as receptors for tumor targeting by circulating ligands', *Nat Biotechnol,* 15(6), 542–6.

Peng, S. B., Yan, L., Xia, X., Watkins, S. A., Brooks, H. B., Beight, D., Herron, D. K., Jones, M. L., Lampe, J. W. and McMillen, W. T. (2005) 'Kinetic Characterization of Novel Pyrazole TGF- Receptor I Kinase Inhibitors and Their Blockade of the Epithelial- Mesenchymal Transition', *Biochemistry,* 44(7), 2293–2304.

Phan, A. E. A. (2007) 'Open label phase I study of MBP-426, a novel formulation of oxalipatin, in patients with advanced or metastatic solid tumors', *Proceedings of the international conference on molecular targets cancer therapeutics, discovery, development and validation.*

Prokop, A. and Davidson, J. M. (2008) 'Nanovehicular intracellular delivery systems', *Journal of Pharmaceutical Sciences,* 97(9), 3518–3590.

Rajendran, L., Knolker, H. J. and Simons, K. (2010) 'Subcellular targeting strategies for drug design and delivery', *Nat Rev Drug Discov,* 9(1), 29–42.

Ramanujan, S., Pluen, A., McKee, T. D., Brown, E. B., Boucher, Y. and Jain, R. K. (2002) 'Diffusion and convection in collagen gels: implications for transport in the tumor interstitium', *Biophysical Journal,* 83(3), 1650–1660.

Rapoport, M. and Lorberboum-Galski, H. (2009) 'TAT-based drug delivery system--new directions in protein delivery for new hopes?', *Expert Opin Drug Deliv,* 6(5), 453–63.

Rawat, A., Vaidya, B., Khatri, K., Goyal, A. K., Gupta, P. N., Mahor, S., Paliwal, R., Rai, S. and Vyas, S. P. (2007) 'Targeted intracellular delivery of therapeutics: an overview', *Pharmazie,* 62(9), 643–58.

Reddy, S. T., Berk, D. A., Jain, R. K. and Swartz, M. A. (2006) 'A sensitive in vivo model for quantifying interstitial convective transport of injected macromolecules and nanoparticles', *Journal of applied physiology,* 101(4), 1162.

Reiners, J. J., Jr., Caruso, J. A., Mathieu, P., Chelladurai, B., Yin, X. M. and Kessel, D. (2002) 'Release of cytochrome c and activation of pro-caspase-9 following lysosomal photodamage involves Bid cleavage', *Cell Death Differ,* 9(9), 934–44.

Rejman, J., Oberle, V., Zuhorn, I. S. and Hoekstra, D. (2004) 'Size-dependent internalization of particles via the pathways of clathrin- and caveolae-mediated endocytosis', *Biochem J,* 377(Pt 1), 159–69.

Reynwar, B. J., Illya, G., Harmandaris, V. A., Muller, M. M., Kremer, K. and Deserno, M. (2007) 'Aggregation and vesiculation of membrane proteins by curvature-mediated interactions', *Nature,* 447(7143), 461–4.

Roberts, A. B. and Wakefield, L. M. (2003) 'The two faces of transforming growth factor in carcinogenesis', *Proceedings of the National Academy of Sciences of the United States of America,* 100(15), 8621.

Rogach, A. L. and Ogris, M. (2010) 'Near-infrared-emitting semiconductor quantum dots for tumor imaging and targeting', *Curr Opin Mol Ther,* 12(3), 331–9.

Roh, K. H., Martin, D. C. and Lahann, J. (2005) 'Biphasic Janus particles with nanoscale anisotropy', *Nat Mater,* 4(10), 759–63.

Rothbard, J. B., Jessop, T. C. and Wender, P. A. (2005) 'Adaptive translocation: the role of hydrogen bonding and membrane potential in the uptake of guanidinium-rich transporters into cells', *Adv Drug Deliv Rev,* 57(4), 495–504.

Rozema, D. B., Ekena, K., Lewis, D. L., Loomis, A. G. and Wolff, J. A. (2003) 'Endosomolysis by masking of a membrane-active agent (EMMA) for cytoplasmic release of macromolecules', *Bioconjug Chem,* 14(1), 51–7.

Rozema, D. B., Lewis, D. L., Wakefield, D. H., Wong, S. C., Klein, J. J., Roesch, P. L., Bertin, S. L., Reppen, T. W., Chu, Q., Blokhin, A. V., Hagstrom, J. E. and Wolff, J. A. (2007) 'Dynamic PolyConjugates for targeted in vivo delivery of siRNA to hepatocytes', *Proc Natl Acad Sci U S A,* 104(32), 12982–7.

Ruoslahti, E., Bhatia, S. N. and Sailor, M. J. 'Targeting of drugs and nanoparticles to tumors', *The Journal of cell biology,* 188(6), 759.

Sahay, G., Alakhova, D. Y. and Kabanov, A. V. (2010a) 'Endocytosis of nanomedicines', *J Control Release,* 145(3), 182–95.

Sahay, G., Gautam, V., Luxenhofer, R. and Kabanov, A. V. (2010b) 'The utilization of pathogen-like cellular trafficking by single chain block copolymer', *Biomaterials,* 31(7), 1757–64.

Sahay, G., Kim, J. O., Kabanov, A. V. and Bronich, T. K. (2009) 'The exploitation of differential endocytic pathways in normal and tumor cells in the selective targeting of nanoparticulate chemotherapeutic agents', *Biomaterials.*

Sandvig, K., Torgersen, M. L., Engedal, N., Skotland, T. and Iversen, T. G. (2010) 'Protein toxins from plants and bacteria: probes for intracellular transport and tools in medicine', *FEBS Lett,* 584(12), 2626–34.

Sarathi, V. B., Gerard, G. M. D. S., Suna, E., Vladimir, P. T. and Volkmar, W. (2008) 'Organelle-Targeted Nanocarriers: Specific Delivery of Liposomal Ceramide to Mitochondria Enhances Its Cytotoxicity in Vitro and in Vivo', *Nano Letters,* 8(8), 2559–2563.

Sasaki, K., Kogure, K., Chaki, S., Nakamura, Y., Moriguchi, R., Hamada, H., Danev, R., Nagayama, K., Futaki, S. and Harashima, H. (2008) 'An artificial virus-like nano carrier system: enhanced endosomal escape of nanoparticles via synergistic action of pH-sensitive fusogenic peptide derivatives', *Anal Bioanal Chem,* 391(8), 2717–27.

Satija, J., Gupta, U. and Jain, N. K. (2007) 'Pharmaceutical and biomedical potential of surface engineered dendrimers', *Crit Rev Ther Drug Carrier Syst,* 24(3), 257–306.

Sawant, R. M., Hurley, J. P., Salmaso, S., Kale, A., Tolcheva, E., Levchenko, T. S. and Torchilin, V. P. (2006) '"SMART" drug delivery systems: double-targeted pH-responsive pharmaceutical nanocarriers', *Bioconjug Chem,* 17(4), 943–9.

Schaefer, A. M., McFarland, R., Blakely, E. L., He, L., Whittaker, R. G., Taylor, R. W., Chinnery, P. F. and Turnbull, D. M. (2008) 'Prevalence of mitochondrial DNA disease in adults', *Ann Neurol,* 63(1), 35–9.

Schnitzer, J. E., Oh, P., Pinney, E. and Allard, J. (1994) 'Filipin-sensitive caveolae-mediated transport in endothelium: reduced transcytosis, scavenger endocytosis, and capillary permeability of select macromolecules', *The Journal of cell biology,* 127(5), 1217.

Schwarz, M., Meade, G., Stoll, P., Ylanne, J., Bassler, N., Chen, Y. C., Hagemeyer, C. E., Ahrens, I., Moran, N., Kenny, D., Fitzgerald, D., Bode, C. and Peter, K. (2006) 'Conformation-specific blockade of the integrin GPIIb/IIIa: a novel antiplatelet strategy that selectively targets activated platelets', *Circ Res,* 99(1), 25–33.

Schwarz, M., Rottgen, P., Takada, Y., Le Gall, F., Knackmuss, S., Bassler, N., Buttner, C., Little, M., Bode, C. and Peter, K. (2004) 'Single-chain antibodies for the conformation-specific blockade of activated platelet integrin alphaIIbbeta3 designed by subtractive selection from naive human phage libraries', *FASEB J,* 18(14), 1704–6.

Selbo, P. K., Weyergang, A., Hogset, A., Norum, O. J., Berstad, M. B., Vikdal, M. and Berg, K. (2010) 'Photochemical internalization provides time- and space-controlled endolysosomal escape of therapeutic molecules', *J Control Release.*

Selvi, B. R., Jagadeesan, D., Suma, B. S., Nagashankar, G., Arif, M., Balasubramanyam, K., Eswaramoorthy, M. and Kundu, T. K. (2008) 'Intrinsically fluorescent carbon nanospheres as a nuclear targeting vector: delivery of membrane-impermeable molecule to modulate gene expression in vivo', *Nano Lett,* 8(10), 3182–8.

Serda, R. E., Gu, J., Bhavane, R. C., Liu, X., Chiappini, C., Decuzzi, P. and Ferrari, M. (2009a) 'The association of silicon microparticles with endothelial cells in drug delivery to the vasculature', *Biomaterials,* 30(13), 2440–8.

Serda, R. E., Gu, J., Burks, J. K., Ferrari, K., Ferrari, C. and Ferrari, M. (2009b) 'Quantitative mechanics of endothelial phagocytosis of silicon microparticles', *Cytometry A,* 75(9), 752–60.

Serda, R. E., Mack, A., Pulikkathara, M., Zaske, A. M., Chiappini, C., Fakhoury, J., Webb, D., Godin, B., Conyers, J. L., Liu, X. W., Bankson, J. A. and Ferrari, M. (2010a) 'Cellular Association and Assembly of a Multistage Delivery System', *Small* 6(12), 1329–40.

Serda R. E., M. A., van de Ven A., Ferrati S., Godin B., Chiappini C., Dunner Jr. K., Landry M., and Brousseau L. L. X., Bean A. J., Ferrari M. (2010b) 'Logic-embedded vectors for intracellular partitioning and exocytosis of nanoparticles', *Small,* 6(23), 2691–700.

Shi, G., Guo, W., Stephenson, S. M. and Lee, R. J. (2002) 'Efficient intracellular drug and gene delivery using folate receptor-targeted pH-sensitive liposomes composed of cationic/anionic lipid combinations', *J Control Release,* 80(1–3), 309–19.

Shiraishi, T. and Nielsen, P. E. (2006) 'Enhanced delivery of cell-penetrating peptide-peptide nucleic acid conjugates by endosomal disruption', *Nat Protoc,* 1(2), 633–6.

Siccardi, D., Turner, J. R. and Mrsny, R. J. (2005) 'Regulation of intestinal epithelial function: a link between opportunities for macromolecular drug delivery and inflammatory bowel disease', *Advanced drug delivery reviews,* 57(2), 219–235.

Siegel, B. A., Dehdashti, F., Mutch, D. G., Podoloff, D. A., Wendt, R., Sutton, G. P., Burt, R. W., Ellis, P. R., Mathias, C. J., Green, M. A. and Gershenson, D. M. (2003) 'Evaluation of 111In-DTPA-folate as a receptor-targeted diagnostic agent for ovarian cancer: initial clinical results', *J Nucl Med,* 44(5), 700–7.

Simoes, S., Slepushkin, V., Gaspar, R., de Lima, M. C. and Duzgunes, N. (1998) 'Gene delivery by negatively charged ternary complexes of DNA, cationic liposomes and transferrin or fusigenic peptides', *Gene Ther,* 5(7), 955–64.

Slowing, II, Trewyn, B. G. and Lin, V. S. (2007) 'Mesoporous silica nanoparticles for intracellular delivery of membrane-impermeable proteins', *J Am Chem Soc,* 129(28), 8845–9.

Slowing, I., Trewyn, B. G. and Lin, V. S. (2006) 'Effect of surface functionalization of MCM-41-type mesoporous silica nanoparticles on the endocytosis by human cancer cells', *J Am Chem Soc,* 128(46), 14792–3.

Sonawane, N. D., Thiagarajah, J. R. and Verkman, A. S. (2002) 'Chloride concentration in endosomes measured using a ratioable fluorescent Cl- indicator: evidence for chloride accumulation during acidification', *J Biol Chem,* 277(7), 5506–13.

Spearman, M., Taylor, W. R., Greenberg, A. H. and Wright, J. A. (1994) 'Antisense oligodeoxyribonucleotide inhibition of TGF-[beta] 1 gene expression and alterations in the growth and malignant properties of mouse fibrosarcoma cells', *Gene,* 149(1), 25–29.

Stenberg, P., Luthman, K. and Artursson, P. (2000) 'Virtual screening of intestinal drug permeability', *Journal of Controlled Release,* 65(1–2), 231–243.

Stoll, P., Bassler, N., Hagemeyer, C. E., Eisenhardt, S. U., Chen, Y. C., Schmidt, R., Schwarz, M., Ahrens, I., Katagiri, Y., Pannen, B., Bode, C. and Peter, K. (2007) 'Targeting ligand-induced binding sites on GPIIb/IIIa via single-chain antibody allows effective anticoagulation without bleeding time prolongation', *Arterioscler Thromb Vasc Biol,* 27(5), 1206–12.

Storrie B., M., R.F. *Endosomes and lysosomes: a dynamic relationship,* Greenwich, CT; London.

Straubinger, R. M., Duzgunes, N. and Papahadjopoulos, D. (1985) 'pH-sensitive liposomes mediate cytoplasmic delivery of encapsulated macromolecules', *FEBS Lett,* 179(1), 148–54.

Sugahara, K. N., Teesalu, T., Karmali, P. P., Kotamraju, V. R., Agemy, L., Greenwald, D. R. and Ruoslahti, E. 'Coadministration of a Tumor-Penetrating Peptide Enhances the Efficacy of Cancer Drugs', *Science,* 328(5981), 1031–5.

Swaan, P. W. (1998) 'Recent advances in intestinal macromolecular drug delivery via receptor-mediated transport pathways', *Pharmaceutical research,* 15(6), 826–834.

Swanson, J. A. and Watts, C. (1995) 'Macropinocytosis', *Trends Cell Biol,* 5(11), 424–8.

Tanaka, H., Miyamoto, K., Morita, K., Haga, H., Segawa, H., Shiraga, T., Fujioka, A., Kouda, T., Taketani, Y. and Hisano, S. (1998) 'Regulation of the PepT1 peptide transporter in the rat small intestine in response to 5-fluorouracil-induced injury', *Gastroenterology,* 114(4), 714–723.

Tannock, I. F., Lee, C. M., Tunggal, J. K., Cowan, D. S. M. and Egorin, M. J. (2002) 'Limited Penetration of Anticancer Drugs through Tumor Tissue', *Clinical cancer research,* 8(3), 878.

Taylor, U., Klein, S., Petersen, S., Kues, W., Barcikowski, S. and Rath, D. (2010) 'Nonendosomal cellular uptake of ligand-free, positively charged gold nanoparticles', *Cytometry A,* 77(5), 439–46.

Tekle, C., Deurs, B., Sandvig, K. and Iversen, T. G. (2008) 'Cellular trafficking of quantum dot-ligand bioconjugates and their induction of changes in normal routing of unconjugated ligands', *Nano Lett,* 8(7), 1858–65.

Terman, A., Kurz, T., Gustafsson, B. and Brunk, U. T. (2006) 'Lysosomal labilization', *IUBMB Life,* 58(9), 531–9.

Tkachenko, A. G., Xie, H., Coleman, D., Glomm, W., Ryan, J., Anderson, M. F., Franzen, S. and Feldheim, D. L. (2003) 'Multifunctional gold nanoparticle-peptide complexes for nuclear targeting', *J Am Chem Soc,* 125(16), 4700–1.

Torchilin, V. P. (2006) 'Recent approaches to intracellular delivery of drugs and DNA and organelle targeting', *Annu Rev Biomed Eng,* 8, 343–75.

Torchilin, V. P. (2008) 'Tat peptide-mediated intracellular delivery of pharmaceutical nanocarriers', *Adv Drug Deliv Rev,* 60(4–5), 548–58.

Torchilin, V. P., Khaw, B.-A. and Weissig, V. (2002) 'Intracellular Targets for DNA Delivery: Nuclei and Mitochondria', *Somatic Cell and Molecular Genetics,* 27(1), 49–64.

Tredan, O., Galmarini, C. M., Patel, K. and Tannock, I. F. (2007) 'Drug resistance and the solid tumor microenvironment', *JNCI Journal of the National Cancer Institute,* 99(19), 1441.

Tros de Ilarduya, C., Sun, Y. and Duzgunes, N. (2010) 'Gene delivery by lipoplexes and polyplexes', *Eur J Pharm Sci,* 40(3), 159–70.

Tsuji, A. and Tamai, I. (1996) 'Carrier-mediated intestinal transport of drugs', *Pharmaceutical research,* 13(7), 963–977.

Vasir, J. K. and Labhasetwar, V. (2007) 'Biodegradable nanoparticles for cytosolic delivery of therapeutics', *Adv Drug Deliv Rev,* 59(8), 718–28.

Vlahov, I. R., Santhapuram, H. K., Kleindl, P. J., Howard, S. J., Stanford, K. M. and Leamon, C. P. (2006) 'Design and regioselective synthesis of a new generation of targeted chemotherapeutics. Part 1: EC145, a folic acid conjugate of desacetylvinblastine monohydrazide', *Bioorg Med Chem Lett,* 16(19), 5093–6.

Wakefield, D. H., Klein, J. J., Wolff, J. A. and Rozema, D. B. (2005) 'Membrane activity and transfection ability of amphipathic polycations as a function of alkyl group size', *Bioconjug Chem,* 16(5), 1204–8.

Wang, A. Z., Gu, F., Zhang, L., Chan, J. M., Radovic-Moreno, A., Shaikh, M. R., Farokhzad, O. C. (2008) 'Biofunctionalized targeted nanoparticles for therapeutic applications', *Expert Opin Biol Ther,* 8(8), 1063–70.

Wang, S., Lee, R. J., Cauchon, G., Gorenstein, D. G. and Low, P. S. (1995) 'Delivery of antisense oligodeoxyribonucleotides against the human epidermal growth factor receptor into cultured KB cells with liposomes conjugated to folate via polyethylene glycol', *Proc Natl Acad Sci USA,* 92(8), 3318–22.

Watson, P., Jones, A. T. and Stephens, D. J. (2005) 'Intracellular trafficking pathways and drug delivery: fluorescence imaging of living and fixed cells', *Adv Drug Deliv Rev,* 57(1), 43–61.

Weissig, V. (2003) 'Mitochondrial-targeted drug and DNA delivery', *Crit Rev Ther Drug Carrier Syst,* 20(1), 1–62.

Weissig, V., Boddapati, S. V., Cheng, S. M. and D'Souza, G. G. (2006) 'Liposomes and liposome-like vesicles for drug and DNA delivery to mitochondria', *J Liposome Res,* 16(3), 249–64.

Weissig, V., Cheng, S. M. and D'Souza, G. G. (2004) 'Mitochondrial pharmaceutics', *Mitochondrion,* 3(4), 229–44.

Wenjin Guo, T. L., Jennifer Sudimack, Robert J. Lee (2000) 'Receptor-Specific Delivery of Liposomes Via Folate-Peg-Chol', *Journal of Liposome Research,* 10(2&3), 179–195.

Whitehead, K. and Mitragotri, S. (2008) 'Mechanistic analysis of chemical permeation enhancers for oral drug delivery', *Pharmaceutical research,* 25(6), 1412–1419.

Wood, K. C., Azarin, S. M., Arap, W., Pasqualini, R., Langer, R. and Hammond, P. T. (2008) 'Tumor-targeted gene delivery using molecularly engineered hybrid polymers functionalized with a tumor-homing peptide', *Bioconjug Chem,* 19(2), 403–5.

Xia, T., Kovochich, M., Liong, M., Zink, J. I. and Nel, A. E. (2008) 'Cationic polystyrene nanosphere toxicity depends on cell-specific endocytic and mitochondrial injury pathways', *ACS Nano,* 2(1), 85–96.

Xie, W., Wang, L., Zhang, Y., Su, L., Shen, A., Tan, J. and Hu, J. (2009) 'Nuclear targeted nanoprobe for single living cell detection by surface-enhanced Raman scattering', *Bioconjug Chem,* 20(4), 768–73.

Xu, Z., Zhang, Z., Chen, Y., Chen, L., Lin, L. and Li, Y. (2010) 'The characteristics and performance of a multifunctional nanoassembly system for the co-delivery of docetaxel and iSur-pDNA in a mouse hepatocellular carcinoma model', *Biomaterials,* 31(5), 916–22.

Yacobi, N. R., Malmstadt, N., Fazlollahi, F., DeMaio, L., Marchelletta, R., Hamm-Alvarez, S. F., Borok, Z., Kim, K. J. and Crandall, E. D. (2009) 'Mechanisms of alveolar epithelial translocation of a defined population of nanoparticles', *American journal of respiratory cell and molecular biology.*

Yamada, Y., Akita, H., Kogure, K., Kamiya, H. and Harashima, H. (2007) 'Mitochondrial drug delivery and mitochondrial disease therapy--an approach to liposome-based delivery targeted to mitochondria', *Mitochondrion,* 7(1–2), 63–71.

Ye, Q., Wu, Y. L., Foley, L. M., Hitchens, T. K., Eytan, D. F., Shirwan, H. and Ho, C. (2008) 'Longitudinal tracking of recipient macrophages in a rat chronic cardiac allograft rejection model with noninvasive magnetic resonance imaging using micrometer-sized paramagnetic iron oxide particles', *Circulation,* 118(2), 149–56.

Yoo, J. W. and Mitragotri, S. (2010) 'Polymer particles that switch shape in response to a stimulus', *Proc Natl Acad Sci USA,* 107(25), 11205–10.

Yum, K., Na, S., Xiang, Y., Wang, N. and Yu, M. F. (2009) 'Mechanochemical delivery and dynamic tracking of fluorescent quantum dots in the cytoplasm and nucleus of living cells', *Nano Lett,* 9(5), 2193–8.

Zelphati, O. and Szoka, F. C., Jr. (1996) 'Mechanism of oligonucleotide release from cationic liposomes', *Proc Natl Acad Sci U S A,* 93(21), 11493–8.

Zhou, J., Swiderski, P., Li, H., Zhang, J., Neff, C. P., Akkina, R. and Rossi, J. J. (2009) 'Selection, characterization and application of new RNA HIV gp 120 aptamers for facile delivery of Dicer substrate siRNAs into HIV infected cells', *Nucleic Acids Res,* 37(9), 3094–109.

Approaches to Achieving Sub-cellular Targeting of Bioactives Using Pharmaceutical Nanocarriers

Melani Solomon and Gerard G.M. D'Souza

Abstract It is well accepted that the ability of a biologically molecule to selectively find its target influences its potential as a successful therapeutic drug. For many molecules the molecular target is located inside sub-cellular structures. Molecules with such sub cellular targets and the inability to specifically accumulate at the location of the target can potentially be made more active by targeting strategies that improve their accumulation at the target. Pharmaceutical nanocarriers form the basis of several such targeting strategies. This chapter deals with the rational approach underlying the current uses of nanocarriers to deliver bioactive molecules to sub cellular compartments.

Keywords Nanocarrier • Targeting • Sub-cellular • Organelle-specific

1 Introduction

Drug therapy at the most fundamental level is based on the interaction of two molecules. An exogenous molecule administered to a patient and the molecule in the patient that the administered molecule interacts with to initiate a physiological response. In an ideal scenario, the administered molecule interacts with only one physiological molecule and produces a physiological response that improves a patient's condition. In this context it is clear that the term target may be applied to the physiological molecule and the administered molecule is a drug. The concept of targeting has multiple definitions. From a drug discovery perspective, targeting is very often described in terms of the drug molecule's ability to interact only with the target. This concept is more appropriately described by the use of the term

M. Solomon and G.G.M. D'Souza (✉)
Department of Pharmaceutical Sciences, Massachusetts College of Pharmacy
and Health Sciences, 179 Longwood Ave, Boston, MA 02115, USA
e-mail: gerard.dsouza@mcphs.edu

A. Prokop (ed.), *Intracellular Delivery: Fundamentals and Applications*, Fundamental Biomedical Technologies 5, DOI 10.1007/978-94-007-1248-5_2,

selectivity and is very different from the concept of targeting from a military perspective where the term arguably first originated. Consider the firing of a bullet from a gun as an example. The object the bullet is intended to hit is the target, while targeting is associated with the act of aiming the gun so the bullet hits the target. The action that the bullet produces is destruction of the target. This action is indiscriminate in that if the bullet hits an object other than the target, that object will be destroyed as well. Using the gunshot analogy to illustrate the drug discovery perspective on targeting would involve firing bullets that only destroy the target but leave the non-targets unharmed. Most approaches to disease therapy have followed such an argument: finding such selective molecules has been relatively easy when there were significant differences between the disease causing process and normal human biochemical pathways. Not surprisingly, infectious diseases are relatively easier to treat than inherent disorders. The selectivity is however dose dependent and most drugs that are considered to be selectively toxic to invading pathogens are in fact toxic to human cells as well, but at higher doses.

The current challenges in drug therapy lie in the treatment of diseases associated with malfunctions of normal human biochemical pathways in certain tissues. More often that not, even dose dependent selectivity is hard to achieve. Therefore the concept of targeting is becoming more and more associated with selective delivery. The term 'targeting' should ideally imply that the molecule is in some way able to selectively accumulate at an intended site of action and that the selective accumulation is associated with its selective action. This distinction is particularly important in developing targeted therapy for a disease like cancer. Unless unique molecular targets found exclusively (or at sufficiently higher levels) in the diseased state and not in normal state are discovered, selective accumulation at the disease site is crucial to the improvement of therapy. In summary, it can be said that there appear to be two distinct approaches to targeting in the context of drug therapy. The first involves selective action on the target while the second involves selective accumulation at the target. Most if not all examples of targeting seem to end up being the combination of some degree of selective action on the target and some degree of selective accumulation at the site of the target. Improving the degree of selective accumulation has the added advantage, even for molecules with high target selective action, of reducing the required dose and hence should be a major focus of all targeting approaches.

In the context of drug molecules the properties of selective accumulation are associated with the concept of bioavailability and biodistribution that are related to the physico-chemical properties of the molecule. To overcome the limitations that a compound's physico-chemical properties can impose on its potential pharmaceutical application, the process of large-scale screening of chemical libraries has been extended beyond identifying desired bioactivity. Screening approaches routinely incorporate selection for physico-chemical properties that are known to confer high bioavailability as well. Unfortunately, this approach often leads to many potent molecules being excluded from further development. These molecules often have a potent pharmacological action at a desired molecular target but aren't able to find their way exclusively to that target. It is almost certain that there is a growing list of such molecules that are in essence potential drugs if only a delivery strategy can be devised to get them to their molecular target in the human body. Delivery strategies aimed at mediating the selective accumulation of a biologically active molecule fall into two broad classes. The first involves direct chemical modification of the molecule and includes the traditional approaches of designing chemical analogs of the molecule as well as more recent approaches of conjugation to macromolecules or ligands capable of directing site-specific binding of the bioactive molecule. The second approach is the use of pharmaceutical carriers in particular those approaches that involve physical entrapment of the bioactive in the carrier thus offering what might be viewed as a non-chemical approach to modify the disposition of drug molecules. All chemistry can be performed on the components of the nanocarrier system that can then be loaded with the drug to afford targeted delivery.

Generally speaking, the role of nanocarriers in drug targeting to tissues is well accepted. For example, nanocarriers are able to improve the tumor specific accumulation of anticancer drugs by virtue of the Enhanced Permeability and Retention (EPR) effect (Maeda et al. 2000; Greish 2010; Fang et al. 2003). Such targeting approaches have been applied to anticancer drugs like doxorubicin (Working et al. 1999; Northfelt et al. 1996; Gabizon et al. 1989a, b), daunorubicin (Gill et al. 1996; Cervetti et al. 2003), paclitaxel (Yang et al. 2007; Sharma et al. 1996, 1997), and cisplatin (Stathopoulos 2010) with commercial success in some cases (eg Doxil, Daunosome). Current approaches also routinely focus on improving cell specific accumulation of the nanocarriers as well. However, nanocarrier targeting at a sub-cellular level has until recently not been as widely pursued perhaps due to technological limitations or the argument that once a drug gets inside a cell it will eventually find its way to the subcellular target. There often seems to be an assumption that mediating cell cytosolic internalization is adequate to ensure the interaction of the drug molecule with its final sub-cellular target by virtue of simple diffusion of the drug molecule and random interaction with various sub-cellular structures in the cell. However, it has become increasingly evident during the last decade that such an assumption cannot always be made (Duvvuri and Krise 2005; Breunig et al. 2008; Torchilin 2006; Li and Huang 2008; Kaufmann and Krise 2007; Panyam and Labhasetwar 2003; Weissig 2003, 2005; Weissig et al. 2004, 2007). The potential role that nanocarriers play in controlling the subcellular disposition of bioactives is, therefore, certainly worth examining.

2 Potential Roles of the Nanocarrier in Controlling Sub-cellular Disposition of Bioactives

Bioactive molecules can potentially act in various sub cellular locations that can be roughly divided into general cytosolic locations, the surface of organelles and specific regions inside organelles. The nature of the molecule and its potential sites of action together influence the design of the nanocarrier delivery strategy. Cytosolic access, transport through the cytosol and organelle entry thus become the major barriers that need to be overcome to mediate the efficient delivery of the bioactive to its intended site of action.

2.1 Cytosolic Access

In the case of nanocarrier based intracellular delivery strategies, access to the cytosol is not a trivial matter. Based on current understanding it is accepted that all nanocarrier systems are subject to some form of endocytosis either receptor mediated or more commonly non specific endocytosis. Nanocarriers may be surface functionalized with endocytic targeting moieties such as transferrin, folic acid, low-density lipoprotein, cholera toxin, riboflavin, nicotinic acid and the tripeptide RGD, which lead to internalization by either clathrin-dependent receptor mediated endocytosis, caveolin-assisted endocytosis, lipid-raft assisted endocytosis or macropinocytosis (Bareford and Swaan 2007; Rajendran 2010). Strictly speaking a nanocarrier and the associated drug that is inside an endosome is not in the cytosol and therefore much research has been devoted to mediating the release of the bioactive cargo from the endosome into the cytosol (Rajendran 2010). Endosomal release may be mediated by a fusogenic mechanism where lipid-containing nanocarriers fuse with the endosomal membrane increasing the fluidity of the membrane and therefore the release of the cargo (Martin and Rice 2007). Alternatively endosomolytic agents that can disrupt the endosomal membrane by the so called proton sponge effect leading to the release of endosomal contents (Yessine and Leroux 2004) or photochemical internalization (PCI) using photosensitizer molecules that can rupture the endosomal membrane can also be utilized (Shiraishi and Nielsen 2006). Such approaches are arguably necessary in the case of agents such as siRNA since the target of the drug molecule is in the cytoplasm itself (Patil and Panyam 2009; Yuan et al. 2006; Tahara 2010; Song 2010). On the other hand retention in the endosome followed by subsequent lysosomal fusion is a desirable outcome in the case of bioactives requiring delivery into the lysosome. Recombinant human acid sphingomyelinase for example has been delivered intracellularly by nanocarriers surface functionalized with the anti-ICAM antibody that mediates endosomal uptake and subsequent lysosomal association of the enzyme. This approach has shown promise in the treatment of Nieman Pick's disease type A and B (Muro et al. 2006). Most viruses utilize a cell penetrating peptide (CPP) that aids in the internalization of the virus in the cell. Some of these peptides can be anchored with the therapeutic drug molecule to facilitate its internalization into the cell. HIV's Tat peptide is one such CPP, which has aided in the delivery of large proteins and DNA into the cell either when it is directly conjugated to the molecule (Fittipaldi and Giacca 2005; Schwarze et al. 1999; Fawell et al. 1994) or conjugated to a nanocarrier such as a liposome (Torchilin et al. 2001). Other CPPs, D-penetratin and Syn-B, have been used to enhance the intracellular delivery of doxorubicin in the brain (Rousselle et al. 2000).

2.2 *Transport Through the Cytoplasm*

The cytoplasm is a highly viscous space (Seksek et al. 1997; Goins et al. 2008), which has a high concentration of dissolved solutes (Ellis and Minton 2003; Minton 2006) and as such can present a significant barrier to the movement of many potential bioactive molecules especially those with high molecular weight. However, in the eukaryotic cells, transport of biomolecules such as proteins and lipids is governed by the actin and microtubule cytoskeletons (Vale 1987). The extent to which this transport system can be utilized for nanocarrier mediated intracellular transport remains to be explored. Interestingly it has recently been reported that PEGylation of nanoparticles improves their cytoplasmic transport presumably by allowing nanoparticles to evade non specific interactions in the highly crowded environment of the cytoplasm (Suh et al. 2007). It is also interesting to consider in this context the fact that endosomes are routinely trafficked through the cytosol in their normal progression to lysosomes. If the endosomal vesicle can somehow be rerouted to afford association with other membrane bound organelles like mitochondria or the nucleus, there may well be no need to have the biologically active molecule enter the cytosol. If for example, the nanocarrier components were to undergo a redistribution to become part of the endosome and the targeting ligand was able to redistribute to the surface of the endosomal vesicle, it might be possible then that the vesicle would have an altered sub-cellular fate that could involve transport to and association with a target compartment other than the lysosome. While there is emerging evidence to suggest that in fact cells actively traffic nanocarriers in cell membrane-derived vesicles (Ruan et al. 2007), the concept remains speculative until more is understood about the mechanisms of intracellular molecular versus vesicular transport. For now, the trend towards mediating endosomal release of internalized nanocarrier and associated bioactive is arguably based on insights gleaned from intracellular dynamics of viral particles that are in essence naturally occurring nanocarriers. Viral particles are endocytosed and then are able to mediate endosomal escape and subsequent nucleus specific delivery of their DNA. Based on the premise that to efficiently deliver DNA to the nucleus, a delivery system must penetrate through the plasma membrane and the nuclear envelope, prior to DNA release in the nucleus, a strategy that involved step-wise membrane fusion was devised. Using a multi-layered nanoparticle called a Tetra-lamellar Multi-functional Envelope-type Nano Device (T-MEND) consisting of a DNA-polycation condensed core coated with two nuclear membrane-fusogenic inner envelopes and two endosome-fusogenic outer envelopes, which are shed in stepwise fashion transgene expression in non-dividing cells was reported to be dramatically increased (Akita et al. 2009). A similar approach in designing a mitochondria specific delivery sytem has been reported as well. Liposomal carriers called MITO-Porters which carry octaarginine surface modifications to stimulate their entry into cells as intact vesicles (via macropinocytosis) were prepared with lipid compositions that were identified in various experiments to promote both fusion with the mitochondrial membrane and the release of liposomal cargo to the intra-mitochondrial compartment in living cells. Using GFP protein as a model cargo it was shown that MITOporter liposomes are able to selectively deliver their cargo to mitochondria (Yamada and Harashima 2008; Yamada et al. 2008). It is also possible that conferring upon a nanocarrier the properties of the membrane of a particular subcellular compartment might direct transport through the cytosol to the organelle of interest. This concept has been explored with a liposomal formulation of a crude mitochondrial fraction (Inoki et al. 2000). These so calle proteoliposomes were reported to colocalize with endogenous mitochondria when microinjected into pre implantation embryos.

In addition to particle composition, it has also been observed that changes in nanoparticle architecture affect sub-cellular disposition (Xu et al. 2008). Fluorescein isothiocyanate labeled layered double hydroxide (LDH) nanoparticles were prepared from Mg_2Al under conditions that yielded either hexagonal sheets (5–150 nm wide and 10–20 nm thick) or nanorods (30–60 nm wide and 100–200 nm long). A comparison of the sub-cellular distribution of these two types of preparations revealed that the nanorods trafficked to the nucleus but the hexagonal sheets remained in the cytoplasm (Xu et al. 2008). An active microtubule mediated transport process is hypothesized to be responsible for the observed rapid nuclear accumulation of the nanorods (Xu et al. 2008).

2.3 *Organelle Entry*

As with the previous two, barriers hitchhiking existing biological processes extend to mediating organellar entry in most delivery approaches. Proteins synthesized in the cytosol bear short peptide sequences within the protein as 'molecular zip codes' that determine the preferential transport of a protein into a membrane bound subcellular organelle (Walter et al. 1982). These leader sequences or localization signals are typically recognized by specialized import proteins that are associated with the organelle (Allen et al. 2000; Stoffler et al. 1999). Localization sequences for the nucleus and the mitochondria are well known and have been explored for mediating the entry of a variety of molecular cargos as well as nanocarriers into these organelles. The potential of leader sequence peptides for overcoming intracellular barriers to DNA delivery was first demonstrated a decade ago (Zanta et al. 1999). To facilitate the import of exogenous DNA into the nucleus, a capped 3.3-kbp CMVLuciferase-NLS gene containing a single nuclear localization signal peptide (PKKKRKVEDPYC) was synthesized. The resulting transfection enhancement due to the nuclear leader peptide was about 10-fold to 1,000-fold irrespective of the cationic vector or the cell type used. At that time the authors hypothesized that the 3 nm wide DNA present in the cytoplasm was initially docked to and translocated through a nuclear pore by the nuclear import machinery and as DNA enters the nucleus, it was quickly condensed into a chromatin-like structure, which provided a mechanism for threading the remaining worm-like molecule through the pore (Zanta et al. 1999). In this context an interesting question arises as to whether localization signals should be directly conjugated to the bioactive molecule or the nano-carrier. Besides protein import machinery, other properties of organellar compartments like membrane lipid composition (Fernandez-Carneado et al. 2005), membrane potential (D'Souza et al. 2008) and even intra-organellar pH (Torchilin et al. 2009) have been explored for achieving selective delivery to organelles.

3 Prevalent Nanocarrier Design Approaches

It currently appears that nanocarrier design for subcellular targeting is based on the fractal symmetry between the case of drug delivery to a cell and drug delivery to a molecular target inside a sub-cellular compartment. The cell could be viewed as being a small, slightly simpler but nonetheless highly organized "body" with "organs" (organelles) and "cells" (defined structures and molecular arrangements) within these organs. Therefore principles that have been explored for organ and cell specific targeting are being applied at the subcellular level. Nanocarriers are either being modified with sub cellular targeting ligands or are being prepared from materials that have inherent subcellular accumulation characteristics. As alluded to in the previous section, much of this line of thinking is based on current understanding of viral particles. Viruses could be considered naturally occurring nanocarriers with the ability to selectively deliver their DNA cargo to a sub-cellular target (the nucleus). It is perhaps safe to say that much of what we know about the cellular interaction and sub-cellular disposition of nanocarriers has some how been associated with investigations into mimicking the DNA delivery capability of viruses using artificial nanocarriers.

3.1 Nanocarriers Modified with Sub-cellular Targeting Ligands

Most nanocarriers are believed to enter the cell by endocytic mechanisms and could therefore be considered as having a predisposition for accumulation in endosomes and potentially lysosomes as well. This predisposition of particulate systems is particularly useful as pathological conditions associated with endosomes and lysosomes could potentially benefit from therapies targeting these pathways (Bareford and Swaan 2007; Gregoriadis and Ryman 1971; Castino et al. 2003; Tate and Mathews 2006). The fate of the nanocarrier is dependent on the mechanism of vesicular internalization (Bareford and Swaan 2007). For example, nanoscale drug carrier systems taken up by clathrin-dependent receptor-mediated endocytosis (RME) are most likely to undergo lysosomal degradation, while clathrin-independent RME may lead to endosomal accumulation (Bareford and Swaan 2007). Consequently, the type of targeting moiety displayed by the nanocarrier system determines whether the carrier delivers its cargo to either endosomes or lysosomes. Several endocytic targeting moieties have been studied and include folic acid, low-density lipoprotein, cholera toxin B, mannose-6-phosphate, transferrin, riboflavin, the tripeptide RGD, ICAM-1 antibody and nicotinic acid (Bareford and Swaan 2007). Perhaps the most widely used endocytic targeting ligands for the functionalizing of nanoscale drug delivery systems are transferrins (comprehensively reviewed in (Qian et al. 2002), a family of large nonheme iron-binding glycoproteins. The efficient cellular uptake of transferrins (Tf) has been and still is being explored for the intracellular delivery of anticancer agents, and also proteins and therapeutic genes. Iron-loaded transferrin binds to a specific cell-surface receptor (TfR1) and upon endocytosis via clathrin-coated pits the transferrin-receptor complex is routed into the endosomal compartment avoiding lysosomal digestion. This is an important feature of TfR1 for drug delivery, since normally glycoproteins taken up via receptor-mediated endocytosis are destined to eventually fuse with lysosomes. Such intracellular sorting of endocytosed transferrin from other endocytosed asialoglycoprotein has been found to occur immediately after cell internalization (Stoorvogel et al. 1987). Following loss of the clathrin coat, the endosome containing the Tf-TfR1 complex then starts taking up protons which causes the quick acidification of the lysosomal lumen to a pH of around 5.5. Recently a homologue to TfR1 was cloned, called TfR2 (Trinder and Baker 2003). Of importance for anticancer drug delivery, TfR2 was found to be frequently expressed in human cancer cell lines (Calzolari et al. 2007). Encapsulation of doxorubicin into liposomes bearing transferrin on the distal end of liposomal polyethylene glycol (PEG) chains resulted in significantly increased doxorubicin uptake into glioma cells, which are known to overexpress the transferrin receptor with the extent of overexpression correlated to the severity of the tumor (Eavarone et al. 2000). Transferrin modification of Doxorubicin-loaded palmitoylated glycol chitosan (GCP) vesicles resulted in higher uptake and increased cytotoxicity as compared to GCP Doxorubicin alone (Dufes et al. 2004). Tf vesicles were taken up rapidly with a plateau after 1–2 h and Doxorubicin reached the nucleus after 60–90 min.

Low-density lipoprotein (LDL) represents another endocytic targeting ligand. Furthermore, LDL itself actually provides a highly versatile natural nanoplatform for the delivery of diagnostic and therapeutic agents to normal and neoplastic cells that over express LDL receptors (LDLR) (Glickson et al. 2008, 2009). LDL-loading of contrast or therapeutic agents has been achieved by covalent attachment to protein side chains, intercalation into the phospholipid monolayer and extraction and reconstitution of the triglyceride/cholesterol ester core (Zheng et al. 2005). Glickson and coworkers have constructed a semi-synthetic nanoparticle by coating magnetite iron oxide nanoparticles with carboxylated cholesterol and overlaying a monolayer of phospholipid to which Apo A1, Apo E or synthetic amphoteric alpha-helical polypeptides were adsorbed for targeting HDL, LDL or folate receptors, respectively (Zheng et al. 2002). These semisynthetic particles have potential utility for the in situ loading of magnetite into cells for magnetic resonance imaging (MRI) monitored cell tracking or gene therapy (Zheng et al. 2005). In addition to the surface ligand, carrier geometry also might play a role in the endocytic process (Muro et al. 2008). Disks were found to display longer half-lives in circulation and higher targeting specificity in mice, whereas spheres underwent a more rapid endocytosis. Most interestingly from the aspect of intracellular drug delivery it was also found that the size of the carrier might determine its intracellular fate. While micron-size carriers had prolonged residency in prelysosomal compartments, submicron carriers trafficked more readily to lysosomes (Muro et al. 2008).

In addition to endolysosomal targeting there is a growing body of work that suggests the feasibility of modifying nanocarriers to redirect delivery of their cargo to other sub-cellular compartments. A fusigenic viral liposome fused with the Hemagglutinating virus of Japan (HVJ) envelope protein was used to efficiently encapsulate and deliver DNA to the cytoplasm through fusion of the liposome with the plasma membrane (Dzau et al. 1996). In another study, pH sensitive liposomes composed of dioleylphosphatidylethanolamine and cholestryl hemisuccinate have been reported to be efficient carriers of N-butylnojirimycin to the endoplasmic reticulum (ER) (Costin et al. 2002); this can potentially be applied to the treatment of melanoma. Surface functionalization with ligands like cholera toxin, shiga toxin due to which the nanocarrier will be internalized into the cell by caveolae or lipid raft mediated endocytosis surpasses the lysosome and delivers the moiety to the ER or golgi complex (Le and Nabi 2003; Tarrago-Trani and Storrie 2007). Such ER liposomes may be used for the delivery of anti virals (Pollock et al. 2010). On the other hand, presence of an ER retrieval sequence on a ligand may confer to it ER targeting abilities. Acyl coenzyme-A binding protein (ACBP) is a ligand consisting of a potential ER retrieval signal, a dilysine (KK) motif near its C-terminus, which was found, by live cell imaging and indirect immunohistochemistry, to preferentially accumulate in the endoplasmic reticulum and the golgi complex (Hansen et al. 2008). Liposomes modified with mitochondriotropic ligands have been shown to improve the efficacy of an anticancer drug both *in vitro* and *in vivo* (Boddapati et al. 2008). To render liposomes mitochondria-specific, the liposomal surface was modified with triphenyl phosphonium (TPP) cations (Boddapati et al. 2005). Methyltriphenylphosphonium cations (MTPP) are rapidly taken up by mitochondria in living cells (Liberman et al. 1969) and have been extensively explored for the delivery of biological active molecules to and into mitochondria (Murphy 2008; Murphy and Smith 2007; Ross et al. 2008; Smith et al. 2003). The replacement of the methyl group in MTPP with a stearyl residue was shown to facilitate the attachment of TPP cations to the surface of liposomes (Boddapati et al. 2005).

Solid nanoparticles prepared from polymers or colloidal metals also fall under the umbrella of pharmaceutical nanocarriers. Gold nanoparticles (AuNPs) are a flexible nanoscale platform for the conjugation of a variety of targeting ligands based on the affinity of thiol and amino groups for the gold surface. Of particular interest here is the report of the conjugation of the triphenyl phosphonium mitochondriotropic ligand (Horobin et al. 2007) to the surface of AuNPs (Ju-Nam et al. 2006). Triphenyl-phosphonioalkylthiosulfate and potassium tetrachloroaurate were dissolved in dichloromethane followed by drop-wise addition into an aqueous solution of sodium borohydride to generate 5–10 nm sized AuNPs with surface-attached triphenylphosphonium residues. While data describing the intracellular localization of these potentially mitochondriotropic AuNP's have not yet been made available, AuNPs have already been targeted to the nucleus using the adenoviral nuclear localization signal (NLS) and integrin binding domain (Tkachenko et al. 2004).

Such an approach has been reported to be useful in the development of probes for cell tracking by surface enhanced raman scattering. Gold nanoparticles were surface functionalized with the SV40 NLS, which led to accumulation of the nanoparticles in the nucleus of HeLa cells (Xie et al. 2009). Quantum dots conjugated to the amino group of the mitochondrial targeting sequence, Mito-8 were reported to be colocalized in the mitochondria (Hoshino et al. 2004). Chlorotoxin conjugated iron oxide nanoparticles coated with a copolymer of chitosan-PEG-PEI were found to specifically translocate to the nucleus of cancer cells permitting a dual targeting approach to a specific cell type and an organelle within the cell type; this approach was successful *in vivo*, both for imaging and therapeutic purposes Kievit (2010).

Modification with a leader sequence peptide has also been applied to creating delivery systems for mitochondria. A mitochondrial leader peptide (MLP), derived from the nucleo-cytosol expressed but mitochondria localized ornithine transcarbamylase was recently used to render polyethylenimine (PEI) mitochondriotropic (Lee et al. 2007). PEI had been developed in the mid 1990s as a versatile vector for gene and oligonucleotide transfer into cells in culture and *in vivo* (Boussif et al. 1995; Demeneix and Behr 2005). Lee et al. (2007) conjugated the mitochondrial leader peptide to PEI via a disulfide bond and confirmed the complex formation of PEI-MLP with DNA by a gel retardation assay. In vitro delivery tests of rhodamine-labeled DNA into living cells demonstrated that PEI-MLP/DNA complexes were localized at mitochondrial sites in contrast to controls carried out with PEI-DNA complexes lacking MLP. The author's data suggest that PEI-MLP can deliver DNA to the mitochondrial sites and may be useful for the development of direct mitochondrial gene therapy, a strategy for the cure of mitochondrial DNA diseases proposed earlier (Seibel et al. 1995; Weissig and Torchilin 2000, 2001a, b) as an alternative to allotropic expression (Ellouze et al. 2008; Oca-Cossio et al. 2003; Zullo 2001; Gray et al. 1996).

3.2 *Nanocarriers Prepared from Self-assembling Molecules with Known Sub-cellular Accumulation*

All the examples discussed in the previous section share a common assumption that unless a targeting ligand is incorporated into the design, the nanocarriers would remain in the endolysosomal compartment. However it is interesting to also consider the disposition of a nanocarrier made exclusively of a molecule with a predisposition for a sub-cellular compartment. A good example of a molecule that has a strong affinity for a sub-cellular compartment and that is also capable of self assembling to form a potential carrier system is the mitochondriotropic amphiphile dequalinium chloride. Vesicles prepared exclusively from dequalinium chloride (called DQAsomes) have been explored for the delivery of several bioactive molecules to mitochondria (D'Souza et al. 2003, 2005, 2008; Weissig et al. 2000, 2001). Most recently, the antitumor efficiency of DQAsomal encapsulated paclitaxel was enhanced by modifying the DQAsomal surface with folic acid (FA) (Vaidya et al. 2009). However, DQAsomes currently serve as a preliminary proof of concept and are still far from being the perfect delivery system. In order to design similar carriers for other sub-cellular compartments it is necessary to find suitable self-assembling molecules with an affinity for the intended sub-cellular compartment. To this end, recent work on the sub-cellular distribution of micelle forming agents offers some interesting insights (Bae et al. 2005; Maysinger et al. 2007; Savic et al. 2006, 2009; Xiong et al. 2008).

Imaging studies based on the use of a variety of organelle-specific dyes, gold and fluorescent polymers have provided detailed insight into the sub-cellular distribution of block copolymer micelles (Maysinger et al. 2007; Savic et al. 2003). Both imaging techniques, i.e. confocal fluorescence microscopy (to detect the fluorophore-labeled copolymers) and transmission electron microscopy (to detect the gold-labeled copolymers) demonstrate that poly(caprolactone)-b-poly(ethylene oxide) micelles (PCL-b-PEO micelles) do not enter the nucleus. With respect to the cytosolic distribution of PCL-b-PEO micelles, however, the two different imaging techniques used in these studies suggest quite a different sub-cellular disposition. TEM images show most of the gold labeled micelles to be localized in endosomes/lysosomes and a few of them were seen at or in mitochondria (Maysinger et al. 2007). Confocal fluorescence microscopic images, on the other hand, show fluorescent PCL-b-PEO micelles almost evenly distributed throughout the cytosol (Savic et al. 2003). Therefore, not surprisingly, cell staining with organelle-specific dyes and overlaying the corresponding confocal fluorescence images reveal partial colocalization of PCL-b-PEO micelles with lysosomes, with the Golgi apparatus and the Endoplasmic Reticulum, with the mitochondria and the Endoplasmic Reticulum and with mitochondria alone. Considering the nature of the micelle corona, which is entirely made up of non-functionalized polyethylene oxide, a highly hydrophilic polymer, any specific interaction with or any specific affinity for any of the cell organelles could not be expected per se. It would be very interesting to see to what extent modifying the micelle corona with organelle-specific ligands would alter the intracellular distribution of such micelles, which then potentially could become nanocontainers that distribute cargo to defined cytoplasmic organelles. However, the distinctive distribution of nonfunctionalized PCL-b-PEO micelles throughout the cytosol makes them highly suitable for multiple cytoplasmic targeting (Savic et al. 2003), which has most recently been proven to be relevant for the delivery of effector molecules of the cell signaling pathways that are activated in the cytosol (Savic et al. 2009). This study suggests that micelle-based intracellular delivery of potent, poorly water-soluble, cell-death-pathway inhibitors may represent a useful addition to established delivery of cytocidal block-copolymer micelle-incorporated bioactives (Savic et al. 2009).

4 Conclusion and Perspectives

From the examples discussed so far it would seem that nanocarrier systems could be designed to achieve true molecular level targeting inside cells. However to say that these systems will be available soon is perhaps premature given what little is currently known about the sub-cellular dynamics associated with nanocarrier trafficking. There are several unanswered questions. For example, do all nanocarriers remain intact upon cell entry and subsequent disposition? Are there differences in the disposition of vesicles in comparison to solid particles? What is the true influence of size on the intracellular disposition of various nanocarriers? Most important is however the question of the mechanism by which the nanocarrier is able to achieve selective uptake and delivery into the sub-cellular compartment. Despite several reports of altered or improved sub-cellular accumulation associated with an improvement in activity, it is still unclear how exactly this happens. It will take much active research to answer these and other questions before sub-cellular targeted nanocarrier-based therapies become commonplace.

References

Akita, H., et al., *Multi-layered nanoparticles for penetrating the endosome and nuclear membrane via a step-wise membrane fusion process.* Biomaterials, 2009. **30**(15): p. 2940–9.

Allen, T.D., et al., *The nuclear pore complex: mediator of translocation between nucleus and cytoplasm.* J Cell Sci, 2000. **113 (Pt 10)**: p. 1651–9.

Bae, Y., et al., *Preparation and biological characterization of polymeric micelle drug carriers with intracellular pH-triggered drug release property: tumor permeability, controlled subcellular drug distribution, and enhanced in vivo antitumor efficacy.* Bioconjug Chem, 2005. **16**(1): p. 122–30.

Bareford, L.M. and P.W. Swaan, *Endocytic mechanisms for targeted drug delivery.* Adv Drug Deliv Rev, 2007. **59**(8): p. 748–58.

Boddapati, S.V., et al., *Mitochondriotropic liposomes.* J Liposome Res, 2005. **15**(1–2): p. 49–58.

Boddapati, S.V., et al., *Organelle-targeted nanocarriers: specific delivery of liposomal ceramide to mitochondria enhances its cytotoxicity in vitro and in vivo.* Nano Lett, 2008. **8**(8): p. 2559–63.

Boussif, O., et al., *A versatile vector for gene and oligonucleotide transfer into cells in culture and in vivo: polyethylenimine.* Proc Natl Acad Sci U S A, 1995. **92**(16): p. 7297–301.

Breunig, M., S. Bauer, and A. Goepferich, *Polymers and nanoparticles: intelligent tools for intracellular targeting?* Eur J Pharm Biopharm, 2008. **68**(1): p. 112–28.

Calzolari, A., et al., *Transferrin receptor 2 is frequently expressed in human cancer cell lines.* Blood Cells Mol Dis, 2007. **39**(1): p. 82–91.

Castino, R., M. Demoz, and C. Isidoro, *Destination 'lysosome': a target organelle for tumour cell killing?* J Mol Recognit, 2003. **16**(5): p. 337–48.

Cervetti, G., et al., *Efficacy and toxicity of liposomal daunorubicin included in PVABEC regimen for aggressive NHL of the elderly.* Leuk Lymphoma, 2003. **44**(3): p. 465–9.

Costin, G.E., et al., *pH-sensitive liposomes are efficient carriers for endoplasmic reticulum-targeted drugs in mouse melanoma cells.* Biochem Biophys Res Commun, 2002. **293**(3): p. 918–23.

Demeneix, B. and J.P. Behr, *Polyethylenimine (PEI).* Adv Genet, 2005. **53**: p. 217–30.

D'Souza, G.G., et al., *DQAsome-mediated delivery of plasmid DNA toward mitochondria in living cells.* J Control Release, 2003. **92**(1–2): p. 189–97.

D'Souza, G.G., S.V. Boddapati, and V. Weissig, *Mitochondrial leader sequence--plasmid DNA conjugates delivered into mammalian cells by DQAsomes co-localize with mitochondria.* Mitochondrion, 2005. **5**(5): p. 352–8.

D'Souza, G.G., et al., *Nanocarrier-assisted sub-cellular targeting to the site of mitochondria improves the pro-apoptotic activity of paclitaxel.* J Drug Target, 2008. **16**(7): p. 578–85.

Dufes, C., et al., *Anticancer drug delivery with transferrin targeted polymeric chitosan vesicles.* Pharm Res, 2004. **21**(1): p. 101–7.

Duvvuri, M. and J.P. Krise, *Intracellular drug sequestration events associated with the emergence of multidrug resistance: a mechanistic review.* Front Biosci, 2005. **10**: p. 1499–509.
Dzau, V.J., et al., *Fusigenic viral liposome for gene therapy in cardiovascular diseases.* Proc Natl Acad Sci U S A, 1996. **93**(21): p. 11421–5.
Eavarone, D.A., X. Yu, and R.V. Bellamkonda, *Targeted drug delivery to C6 glioma by transferrin-coupled liposomes.* J Biomed Mater Res, 2000. **51**(1): p. 10–4.
Ellis, R.J. and A.P. Minton, *Cell biology: join the crowd.* Nature, 2003. **425**(6953): p. 27–8.
Ellouze, S., et al., *Optimized allotopic expression of the human mitochondrial ND4 prevents blindness in a rat model of mitochondrial dysfunction.* Am J Hum Genet, 2008. **83**(3): p. 373–87.
Fang, J., T. Sawa, and H. Maeda, *Factors and mechanism of "EPR" effect and the enhanced antitumor effects of macromolecular drugs including SMANCS.* Adv Exp Med Biol, 2003. **519**: p. 29–49.
Fawell, S., et al., *Tat-mediated delivery of heterologous proteins into cells.* Proc Natl Acad Sci U S A, 1994. **91**(2): p. 664–8.
Fernandez-Carneado, J., et al., *Highly efficient, nonpeptidic oligoguanidinium vectors that selectively internalize into mitochondria.* J Am Chem Soc, 2005. **127**(3): p. 869–74.
Fittipaldi, A. and M. Giacca, *Transcellular protein transduction using the Tat protein of HIV-1.* Adv Drug Deliv Rev, 2005. **57**(4): p. 597–608.
Gabizon, A., et al., *Systemic administration of doxorubicin-containing liposomes in cancer patients: a phase I study.* Eur J Cancer Clin Oncol, 1989. **25**(12): p. 1795–803.
Gabizon, A., R. Shiota, and D. Papahadjopoulos, *Pharmacokinetics and tissue distribution of doxorubicin encapsulated in stable liposomes with long circulation times.* J Natl Cancer Inst, 1989. **81**(19): p. 1484–8.
Gill, P.S., et al., *Randomized phase III trial of liposomal daunorubicin versus doxorubicin, bleomycin, and vincristine in AIDS-related Kaposi's sarcoma.* J Clin Oncol, 1996. **14**(8): p. 2353–64.
Glickson, J.D., et al., *Lipoprotein nanoplatform for targeted delivery of diagnostic and therapeutic agents.* Mol Imaging, 2008. **7**(2): p. 101–10.
Glickson, J.D., et al., *Lipoprotein nanoplatform for targeted delivery of diagnostic and therapeutic agents.* Adv Exp Med Biol, 2009. **645**: p. 227–39.
Goins, A.B., H. Sanabria, and M.N. Waxham, *Macromolecular crowding and size effects on probe microviscosity.* Biophys J, 2008. **95**(11): p. 5362–73.
Gray, R.E., et al., *Allotopic expression of mitochondrial ATP synthase genes in nucleus of Saccharomyces cerevisiae.* Methods Enzymol, 1996. **264**: p. 369–89.
Gregoriadis, G. and B.E. Ryman, *Liposomes as carriers of enzymes or drugs: a new approach to the treatment of storage diseases.* Biochem J, 1971. **124**(5): p. 58P.
Greish, K., *Enhanced permeability and retention (EPR) effect for anticancer nanomedicine drug targeting.* Methods Mol Biol, 2010. **624**: p. 25–37.
Hansen, J.S., et al., *Acyl-CoA-binding protein (ACBP) localizes to the endoplasmic reticulum and Golgi in a ligand-dependent manner in mammalian cells.* Biochem J, 2008. **410**(3): p. 463–72.
Horobin, R.W., S. Trapp, and V. Weissig, *Mitochondriotropics: a review of their mode of action, and their applications for drug and DNA delivery to mammalian mitochondria.* J Control Release, 2007. **121**(3): p. 125–36.
Hoshino, A., et al., *Quantum dots targeted to the assigned organelle in living cells.* Microbiol Immunol, 2004. **48**(12): p. 985–94.
Inoki, Y., et al., *Proteoliposomes colocalized with endogenous mitochondria in mouse fertilized egg.* Biochem Biophys Res Commun, 2000. **278**(1): p. 183–91.
Ju-Nam, Y., et al., *Phosphonioalkylthiosulfate zwitterions--new masked thiol ligands for the formation of cationic functionalised gold nanoparticles.* Org Biomol Chem, 2006. **4**(23): p. 4345–51.
Kaufmann, A.M. and J.P. Krise, *Lysosomal sequestration of amine-containing drugs: analysis and therapeutic implications.* J Pharm Sci, 2007. **96**(4): p. 729–46.
Kievit, F.M., et al., *Chlorotoxin labeled magnetic nanovectors for targeted gene delivery to glioma.* ACS Nano, 2010. **4**(8): p. 4587–94.

Le, P.U. and I.R. Nabi, *Distinct caveolae-mediated endocytic pathways target the Golgi apparatus and the endoplasmic reticulum.* J Cell Sci, 2003. **116**(Pt 6): p. 1059–71.

Lee, M., et al., *DNA delivery to the mitochondria sites using mitochondrial leader peptide conjugated polyethylenimine.* J Drug Target, 2007. **15**(2): p. 115–22.

Li, S.D. and L. Huang, *Pharmacokinetics and biodistribution of nanoparticles.* Mol Pharm, 2008. **5**(4): p. 496–504.

Liberman, E.A., et al., *Mechanism of coupling of oxidative phosphorylation and the membrane potential of mitochondria.* Nature, 1969. **222**(5198): p. 1076–8.

Maeda, H., et al., *Tumor vascular permeability and the EPR effect in macromolecular therapeutics: a review.* J Control Release, 2000. **65**(1–2): p. 271–84.

Martin, M.E. and K.G. Rice, *Peptide-guided gene delivery.* AAPS J, 2007. **9**(1): p. E18–29.

Maysinger, D., et al., *Fate of micelles and quantum dots in cells.* Eur J Pharm Biopharm, 2007. **65**(3): p. 270–81.

Minton, A.P., *How can biochemical reactions within cells differ from those in test tubes?* J Cell Sci, 2006. **119**(Pt 14): p. 2863–9.

Muro, S., E.H. Schuchman, and V.R. Muzykantov, *Lysosomal enzyme delivery by ICAM-1-targeted nanocarriers bypassing glycosylation- and clathrin-dependent endocytosis.* Mol Ther, 2006. **13**(1): p. 135–41.

Muro, S., et al., *Control of endothelial targeting and intracellular delivery of therapeutic enzymes by modulating the size and shape of ICAM-1-targeted carriers.* Mol Ther, 2008. **16**(8): p. 1450–8.

Murphy, M.P., *Targeting lipophilic cations to mitochondria.* Biochim Biophys Acta, 2008. **1777**(7–8): p. 1028–31.

Murphy, M.P. and R.A. Smith, *Targeting antioxidants to mitochondria by conjugation to lipophilic cations.* Annu Rev Pharmacol Toxicol, 2007. **47**: p. 629–56.

Northfelt, D.W., et al., *Doxorubicin encapsulated in liposomes containing surface-bound polyethylene glycol: pharmacokinetics, tumor localization, and safety in patients with AIDS-related Kaposi's sarcoma.* J Clin Pharmacol, 1996. **36**(1): p. 55–63.

Oca-Cossio, J., et al., *Limitations of allotopic expression of mitochondrial genes in mammalian cells.* Genetics, 2003. **165**(2): p. 707–20.

Panyam, J. and V. Labhasetwar, *Biodegradable nanoparticles for drug and gene delivery to cells and tissue.* Adv Drug Deliv Rev, 2003. **55**(3): p. 329–47.

Patil, Y. and J. Panyam, *Polymeric nanoparticles for siRNA delivery and gene silencing.* Int J Pharm, 2009. **367**(1–2): p. 195–203.

Pollock, S., et al., *Uptake and trafficking of liposomes to the endoplasmic reticulum.* FASEB J, 2010. **24**(6): p. 1866–78.

Qian, Z.M., et al., *Targeted drug delivery via the transferrin receptor-mediated endocytosis pathway.* Pharmacol Rev, 2002. **54**(4): p. 561–87.

Rajendran, L., et al., *Subcellular targeting strategies for drug design and delivery.* Nat Rev Drug Discovery, 2010. **9**(1): p. 29–42.

Ross, M.F., et al., *Rapid and extensive uptake and activation of hydrophobic triphenylphosphonium cations within cells.* Biochem J, 2008. **411**(3): p. 633–45.

Rousselle, C., et al., *New advances in the transport of doxorubicin through the blood-brain barrier by a peptide vector-mediated strategy.* Mol Pharmacol, 2000. **57**(4): p. 679–86.

Ruan, G., et al., *Imaging and tracking of tat peptide-conjugated quantum dots in living cells: new insights into nanoparticle uptake, intracellular transport, and vesicle shedding.* J Am Chem Soc, 2007. **129**(47): p. 14759–66.

Savic, R., et al., *Micellar nanocontainers distribute to defined cytoplasmic organelles.* Science, 2003. **300**(5619): p. 615–8.

Savic, R., et al., *Assessment of the integrity of poly(caprolactone)-b-poly(ethylene oxide) micelles under biological conditions: a fluorogenic-based approach.* Langmuir, 2006. **22**(8): p. 3570–8.

Savic, R., et al., *Block-copolymer micelles as carriers of cell signaling modulators for the inhibition of JNK in human islets of Langerhans.* Biomaterials, 2009.

Schwarze, S.R., et al., *In vivo protein transduction: delivery of a biologically active protein into the mouse.* Science, 1999. **285**(5433): p. 1569–72.

Seibel, P., et al., *Transfection of mitochondria: strategy towards a gene therapy of mitochondrial DNA diseases.* Nucleic Acids Res, 1995. **23**(1): p. 10–7.

Seksek, O., J. Biwersi, and A.S. Verkman, *Translational diffusion of macromolecule-sized solutes in cytoplasm and nucleus.* J Cell Biol, 1997. **138**(1): p. 131–42.

Sharma, A., U.S. Sharma, and R.M. Straubinger, *Paclitaxel-liposomes for intracavitary therapy of intraperitoneal P388 leukemia.* Cancer Lett, 1996. **107**(2): p. 265–72.

Sharma, A., et al., *Activity of paclitaxel liposome formulations against human ovarian tumor xenografts.* Int J Cancer, 1997. **71**(1): p. 103–7.

Shiraishi, T. and P.E. Nielsen, *Enhanced delivery of cell-penetrating peptide-peptide nucleic acid conjugates by endosomal disruption.* Nat Protoc, 2006. **1**(2): p. 633–6.

Smith, R.A., et al., *Delivery of bioactive molecules to mitochondria in vivo.* Proc Natl Acad Sci U S A, 2003. **100**(9): p. 5407–12.

Song, W.J., et al., *Gold nanoparticles capped with polyethyleneimine for enhanced siRNA delivery.* Small, 2010. **6**(2): 239–46.

Stathopoulos, G.P., *Liposomal Cisplatin: a new cisplatin formulation.* Anticancer Drugs, 2010. **21**(8): p. 732–6.

Stoffler, D., B. Fahrenkrog, and U. Aebi, *The nuclear pore complex: from molecular architecture to functional dynamics.* Curr Opin Cell Biol, 1999. **11**(3): p. 391–401.

Stoorvogel, W., H.J. Geuze, and G.J. Strous, *Sorting of endocytosed transferrin and asialoglycoprotein occurs immediately after internalization in HepG2 cells.* J Cell Biol, 1987. **104**(5): p. 1261–8.

Suh, J., et al., *PEGylation of nanoparticles improves their cytoplasmic transport.* Int J Nanomedicine, 2007. **2**(4): p. 735–41.

Tahara, K., et al., *Chitosan-modified poly(D,L-lactide-co-glycolide) nanospheres for improving siRNA delivery and gene-silencing effects.* Eur J Pharm Biopharm, 2010. **74**(3): p. 421–6.

Tarrago-Trani, M.T. and B. Storrie, *Alternate routes for drug delivery to the cell interior: pathways to the Golgi apparatus and endoplasmic reticulum.* Adv Drug Deliv Rev, 2007. **59**(8): p. 782–97.

Tate, B.A. and P.M. Mathews, *Targeting the role of the endosome in the pathophysiology of Alzheimer's disease: a strategy for treatment.* Sci Aging Knowledge Environ, 2006. **2006**(10): p. re2.

Tkachenko, A.G., et al., *Cellular trajectories of peptide-modified gold particle complexes: comparison of nuclear localization signals and peptide transduction domains.* Bioconjug Chem, 2004. **15**(3): p. 482–90.

Torchilin, V.P., *Recent approaches to intracellular delivery of drugs and DNA and organelle targeting.* Annu Rev Biomed Eng, 2006. **8**: p. 343–75.

Torchilin, V.P., et al., *TAT peptide on the surface of liposomes affords their efficient intracellular delivery even at low temperature and in the presence of metabolic inhibitors.* Proc Natl Acad Sci U S A, 2001. **98**(15): p. 8786–91.

Torchilin, V.P., *Nanotechnology for Intracellular Delivery and Targeting*, in *Nanotechnology in Drug Delivery*, M.M. de Villiers, G.S. Kwon, and P. Aramwit, Editors. 2009, Springer Publications: New York. p. 313–348.

Trinder, D. and E. Baker, *Transferrin receptor 2: a new molecule in iron metabolism.* Int J Biochem Cell Biol, 2003. **35**(3): p. 292–6.

Vaidya, B.P., R., Rai, S., Khatri, K., Goyal, A.K., Mishra, N., Vyas, S.P., *Cell-selective mitochondrial targeting: A new approach for cancer therapy.* Cancer Therapy, 2009. **7**: p. 141–148.

Vale, R.D., *Intracellular transport using microtubule-based motors.* Annu Rev Cell Biol, 1987. **3**: p. 347–78.

Walter, P., et al., *The protein translocation machinery of the endoplasmic reticulum.* Philos Trans R Soc Lond B Biol Sci, 1982. **300**(1099): p. 225–8.

Weissig, V., *Mitochondrial-targeted drug and DNA delivery.* Crit Rev Ther Drug Carrier Syst, 2003. **20**(1): p. 1–62.

Weissig, V., *Targeted drug delivery to mammalian mitochondria in living cells.* Expert Opin Drug Deliv, 2005. **2**(1): p. 89–102.

Weissig, V. and V.P. Torchilin, *Mitochondriotropic cationic vesicles: a strategy towards mitochondrial gene therapy.* Curr Pharm Biotechnol, 2000. **1**(4): p. 325–46.

Weissig, V. and V.P. Torchilin, *Towards mitochondrial gene therapy: DQAsomes as a strategy.* J Drug Target, 2001. **9**(1): p. 1–13.

Weissig, V. and V.P. Torchilin, *Cationic bolasomes with delocalized charge centers as mitochondria-specific DNA delivery systems.* Adv Drug Deliv Rev, 2001. **49**(1–2): p. 127–49.

Weissig, V., C. Lizano, and V.P. Torchilin, *Selective DNA release from DQAsome/DNA complexes at mitochondria-like membranes.* Drug Deliv, 2000. **7**(1): p. 1–5.

Weissig, V., G.G. D'Souza, and V.P. Torchilin, *DQAsome/DNA complexes release DNA upon contact with isolated mouse liver mitochondria.* J Control Release, 2001. **75**(3): p. 401–8.

Weissig, V., S.M. Cheng, and G.G. D'Souza, *Mitochondrial pharmaceutics.* Mitochondrion, 2004. **3**(4): p. 229–44.

Weissig, V., et al., *Mitochondria-specific nanotechnology.* Nanomed, 2007. **2**(3): p. 275–85.

Working, P.K., et al., *Reduction of the cardiotoxicity of doxorubicin in rabbits and dogs by encapsulation in long-circulating, pegylated liposomes.* J Pharmacol Exp Ther, 1999. **289**(2): p. 1128–33.

Xie, W., et al., *Nuclear targeted nanoprobe for single living cell detection by surface-enhanced Raman scattering.* Bioconjug Chem, 2009. **20**(4): p. 768–73.

Xiong, X.B., et al., *Multifunctional polymeric micelles for enhanced intracellular delivery of doxorubicin to metastatic cancer cells.* Pharm Res, 2008. **25**(11): p. 2555–66.

Xu, Z.P., et al., *Subcellular compartment targeting of layered double hydroxide nanoparticles.* J Control Release, 2008. **130**(1): p. 86–94.

Yamada, Y. and H. Harashima, *Mitochondrial drug delivery systems for macromolecule and their therapeutic application to mitochondrial diseases.* Adv Drug Deliv Rev, 2008. **60**(13–14): p. 1439–62.

Yamada, Y., et al., *MITO-Porter: A liposome-based carrier system for delivery of macromolecules into mitochondria via membrane fusion.* Biochim Biophys Acta, 2008. **1778**(2): p. 423–32.

Yang, T., et al., *Enhanced solubility and stability of PEGylated liposomal paclitaxel: in vitro and in vivo evaluation.* Int J Pharm, 2007. **338**(1–2): p. 317–26.

Yessine, M.A. and J.C. Leroux, *Membrane-destabilizing polyanions: interaction with lipid bilayers and endosomal escape of biomacromolecules.* Adv Drug Deliv Rev, 2004. **56**(7): p. 999–1021.

Yuan, X., et al., *SiRNA drug delivery by biodegradable polymeric nanoparticles.* J Nanosci Nanotechnol, 2006. **6**(9–10): p. 2821–8.

Zanta, M.A., P. Belguise-Valladier, and J.P. Behr, *Gene delivery: a single nuclear localization signal peptide is sufficient to carry DNA to the cell nucleus.* Proc Natl Acad Sci USA, 1999. **96**(1): p. 91–6.

Zheng, G., et al., *Low-density lipoprotein reconstituted by pyropheophorbide cholesteryl oleate as target-specific photosensitizer.* Bioconjug Chem, 2002. **13**(3): p. 392–6.

Zheng, G., et al., *Rerouting lipoprotein nanoparticles to selected alternate receptors for the targeted delivery of cancer diagnostic and therapeutic agents.* Proc Natl Acad Sci USA, 2005. **102**(49): p. 17757–62.

Zullo, S.J., *Gene therapy of mitochondrial DNA mutations: a brief, biased history of allotopic expression in mammalian cells.* Semin Neurol, 2001. **21**(3): p. 327–35.

Delivery to Intracellular Targets by Nanosized Particles

Gillian Barratt

Abstract Nanosized drug carrier systems, including liposomes and nanoparticles, have the potential of delivering their contents to the interior of cells. However, engineering of the particle size and surface properties is necessary to achieve targeting to particular cell types. Conventional particles with hydrophobic surfaces are rapidly engulfed by phagocytic cells. Modification of the surface with hydrophilic polymers yields so-called "Stealth" particles which avoid phagocytosis and remain in the circulation longer after intravenous injection. The addition of specific ligands to the surface of these particles can confer more specific targeting to particular cell types. Nanoparticles and liposomes are normally taken up by endocytosis in non phagocytic cells, leading to their delivery to the lysosomal compartment. In order for the cargo to reach other cell compartments, a mechanism of endosomal escape is necessary. Examples are given of drug delivery in two particular applications: delivery to macrophages for immunomodulating and anti-infectious functions, and delivery of antisense oligonucleotides and small interfering RNA to cells.

Keywords Liposome • Nanoparticle • Endocytosis • Macrophage • Nucleic acids

Abbreviations

AS-ODN	antisense oligo deoxynucleotide
CHEMS	cholesteryl hemisuccinate
DC-Chol	3ß-[N-(N′,N′-dimethylaminoethane)-carbamoyl]cholesterol
DNA	deoxyribonucleic acid
DOGS	dioctadecylamidoglycylspermine
DOPE	dioleoylphosphatidylethanolamine

G. Barratt (✉)
CNRS UMR 8612, Faculty of Pharmacy, Université Paris-Sud XI, 5 rue J.B. Clément, Chatenay-Malabry F92296, France
e-mail: gillian.barratt@u-psud.fr

A. Prokop (ed.), *Intracellular Delivery: Fundamentals and Applications*, Fundamental Biomedical Technologies 5, DOI 10.1007/978-94-007-1248-5_3,

DOTAP	1,2-dioleoyl-3-trimethylammonium-propane
DOTMA	1,2-di-O-octadecenyl-3-trimethylammonium propane
EPR	enhanced permeation and retention
GAPDH	glyceraldehyde 3-phosphate dehydrogenase
GFP	green fluorescent protein
GPI	glycosylphosphatidylinositol
HUVEC	human umbilical vein endothelial cells
IgG	immunoglobulin G
LDL	low density lipoprotein
LHRH	luteinizing-hormone-releasing hormone
MDP	muramyl dipeptide
MTP-Chol	muramyl tripeptide cholesterol
MTP-PE	muramyl tripeptide phosphatidylethanolamine
NC	nanocapsules
PACA	poly (alkylcyanoacrylate)
PAMAM	poly (amido amine)
PEG	poly (ethylene glycol)
PEG-PMMA	poly (ethylene glycol) – poly (methyl methacrylate) copolymer
PEI	poly (ethyleneimine)
PIBCA	poly (isobutylcyanoacrylate)
PIHCA	poly (isohexylcyanoacrylate)
PLA	poly (D, L-lactide)
PLGA	poly (lactide *–co-*glycolide)
RGD	Arg-Gly-Asp tripeptide
SAINT-2	N-methyl-4(dioleyl)methylpyridinium chloride
SiRNA	small interfering ribonucleic acid
VEGF-R2	vascular endothelial growth factor receptor 2

1 General Remarks About Drug Carriers

The use of drug carrier systems to improve the efficacy of therapeutic molecules has been attracting attention over the last five decades. Among the different constructions that can be made to modify the distribution of an active molecule, colloidal or nanosized particles present a number of advantages (Barratt 2003). They are large enough to transport a large number of guest molecules but small enough (typically between 20 and 200 nm) to pass through some biological barriers and to be taken up by cells.

The first form of colloidal particle to be considered as a potential drug delivery system was the liposome (Ryman and Tyrell 1980). These particles, which were first proposed as models of cell membranes, consist of one or more phospholipid bilayers surrounding aqueous compartments. Water-soluble drugs can be entrapped in the aqueous phase and lipophilic or amphiphilic ones can be inserted into the lipid bilayers. According to the application, the particle diameter can be adjusted from a few microns down to about 25 nm by the choice of preparation technique and the phospholipid composition can also be chosen to provide adequate stability (Szoka 1990). Other supramolecular assemblies based on amphiphilic molecules have also be proposed for drug delivery. For example, oil-in- water micelles can be used to solubilize lipophilic drugs for delivery. Of particular interest are micelles formed from amphiphilic copolymers which are very stable to dilution (Kakizawa and Kataoka 2002).

Emulsified systems have either an oil phase dispersed in an aqueous phase (o/w) or an aqueous phase dispersed in an oil phase (w/o) with the aid of surfactants; o/w systems are more suitable for biological applications. Despite the name, the so-called microemulsions that are attracting much attention as delivery systems for water-insoluble drugs have droplet sizes in the sub-micronic range. They are thermodynamically stable and form with a minimum of energy input, because they contain a high proportion of surfactant and often a co-surfactant. On the other hand, nanoemulsions, with a similar droplet size but a different internal structure are only kinetically stabilized and require a large energy input to generate their small droplet size (Heuschkel et al. 2008). Self micro-emulsifying drug delivery systems (SMEDDS) have been developed recently in attempt to improve the oral bioavailability of some water-insoluble drugs. A mixture of drug, oil, surfactant and co-surfactant is administered and the micro-emulsion forms on dilution in the intestinal fluid (Kyatanwar et al. 2010). Apolar lipids that are solid at physiological temperatures can be formed into nanoparticulate systems with the aid of surfactant and an energy input; these are called solid lipid nanoparticles (Wissing et al. 2004).

Nanoparticulate systems can also be prepared from macromolecules. Proteins (e.g. albumin), poly (amino acids), polysaccharides (dextran, chitosan) and synthetic polymers have all been used (Vauthier and Bouchemal 2009). Obviously for drug delivery applications, the polymer chosen to prepare nanoparticles should be biodegradable to non toxic products. Therefore, the polyesters poly (D,L-lactide) (PLA) and poly (glycolide-*co*-lactide) (PLGA) are very often chosen as the basis for nanoparticles. A particulate system can be formed from a single highly branched polymer molecule: a dendrimer. Guest molecules can be attached by absorption or covalent linkage. The most usual type of nanoparticle is a matrix of entangled polymer chains. Depending on the affinity of the drug for the polymer, it can be included in the matrix or adsorbed on the surface. These nanoparticles are sometimes referred to nanospheres to distinguish them from another type of organization, the nanocapsule (Couvreur et al. 2002). This is a reservoir form consisting of an oily or aqueous core surrounded by a polymer shell. An appropriate drug molecule can be dissolved in the core liquid.

The major role of a drug carrier is to modify the distribution of the drug, re-routing it away from sites of toxicity and delivering more to the site of action. The carrier can also protect a fragile molecule from degradation in physiological fluids. It follows that the biodistribution of the carrier is primordial in determining its range of application (Gregoriadis and Senior 1982). Simple colloidal particles are "recognized" by the immune system in the same way as other foreign bodies such as bacteria. That is, when they are introduced into the blood stream some components of the complement system, in particular C3b, and other proteins known as opsonins, adsorb onto the surface of the particles. This makes the particles susceptible to phagocytosis by macrophages, especially those of the liver and spleen (Szebeni 1998). Thus intracellular delivery occurs, but only in a particular cell type. As described below, there are some therapeutic scenarios in which intracellular delivery to macrophages is a useful strategy. However, it is often desirable to deliver drugs to other cell types; for example cancer cells. Opsonization and interaction with complement proteins can be reduced by decorating the surface of the drug delivery system with end-attached hydrophilic polymer chains (Jeon et al. 1991). Poly (ethylene glycol) (PEG) is the most commonly used hydrophilic polymer (Woodle 1998). The resulting liposomes or nanoparticles persist in the circulation after intravenous injection and are often referred to as "Stealth" particles. They are then able to carry drug to other cell types. In particular, they are able to extravasate into solid tumors by the "EPR" effect (Jain 1987) and also penetrate into infected or inflamed tissue (Oyen et al. 1996).

Cell-specific delivery with colloidal drug carriers can be achieved by attaching a ligand to the surface (ideally at the far end of a PEG chain). Antibodies and fragments thereof have often been used for this purpose, although a smaller ligand has some advantages. Small molecules that have been used for targeted drug delivery systems include folic acid and RGD-containing peptides. Carrier systems that bind to cell-surface receptors in this way will be internalized if receptor ligation normally leads to internalization and if the particle size is consistent with the size of the endocytic vesicle formed (in non phagocytic cells generally 150–200 nm). However, uptake by endocytosis or phagocytosis results in the carrier system being sequestered in endosomes or phagosomes, which are then acidified and fuse with lysosomes containing hydrolytic enzymes. Labile, hydrophilic drugs which cannot escape from this compartment may be destroyed without reaching their target. In response to this, pH-sensitive carrier systems which destabilize the endosome membrane and allow the drug to reach the cytoplasm have been developed. These are discussed in the section dealing the intracellular delivery of nucleic acids.

2 Interactions Between Colloidal Drug Carriers and Cells

As explained above, colloidal drug carriers have the potential to deliver their cargo to the interior of cells under certain conditions. Whether intracellular delivery occurs and to which compartment depends on the cell type and the composition of the carrier. The interactions of nano-sized carriers with cells have recently been reviewed by Hillaireau and Couvreur (2009). Whatever the mechanism of uptake, it must be preceded by contact and binding between the particle and the cell surface. Both hydrophobic and electrostatic interactions can occur, and increased binding is seen with charged particles, whether they are negatively or positively charged. As explained in the section above, binding to the cell surface is enhanced by the presence of proteins adsorbed on the cell surface. The presence of PEG or other hydrophilic polymers at the surface of the particle can reduce protein absorption and in this way hinder binding to the cell surface. This is a useful property when the aim is to reduce clearance by phagocytic cells, but will also reduce binding to target cells. Specific binding to the target can be achieved by the attachment of a ligand to the particle surface. Many different types of molecule have been proposed as targeting ligands for liposomes and nanoparticles. Antibodies are obvious candidates because of their extreme specificity; however saccharides and other small molecules may also be employed. The efficiency of ligand-mediated binding will depend on the orientation of the ligand on the surface. For example, when liposomes are covered with PEG, the ligand should be attached to the distal end of the PEG chain, to allow unhindered interaction with the receptor (Mercadal et al. 1999). It should also be noted that the binding of a ligand to a receptor will not necessarily lead to internalization; this depends on the nature of the receptor and also on the size of the particle carrying the ligand (Allen and Moase 1996).

When the particle is a liposome, there are several theoretical possibilities after binding to the cell surface: internalization of the intact particle by endocytotic processes; fusion of the liposome membrane with the cell membrane, leading to delivery of its contents directly to the cytoplasm; exchange of lipids between the liposome membrane and the plasma membrane. A very early study by Pagano and Huang (1975) using Chinese hamster V79 cells and liposomes prepared from dioleylphosphatidylcholine and cholesterol gave evidence for the last two mechanisms. However, subsequent studies have shown that liposome-cell fusion is quite rare unless specific fusogenic peptides are included (see for example Fattal et al. 1994; Pecheur et al. 1997). Koning et al. (2002) exploited lipid exchange between liposomes immobilized at the surface of colon carcinoma cells by means of a specific monoclonal antibody to deliver a lipophilic prodrug of 5-fluorodeoxyuridine.

In the case of nanoparticles, fusion with the cell membrane is not possible. Transfer of lipophilic cargo to the cell can occur after binding to the cell surface, as shown by Mosqueira et al. (2001) with a fluorescent marker within the oily phase of non PEGylated nanocapsules. However, most interactions take place by endocytotic mechanisms, including phagocytosis, clathrin- or caveolin-mediated endocytosis and macropinocytosis.

Phagocytosis ("cell-eating") is restricted to specialized cells of the immune system: monocytes and neutrophils in the blood and macrophages and dendritic cells in the tissues. The physiological role of phagocytosis is to clear foreign bodies and to present antigens from them to other actors in the immune response. The size of particles that can be engulfed by macrophages reaches several microns; senescent red blood cells, for example. Binding to the phagocyte surface is greatly increased by opsonization with plasma proteins, among which immunoglobulins, fibronectin and complement component C3b play important roles. The phagocyte membrane carries receptors for these proteins, as well as receptors for saccharides (mannose/fucose and galactose), apolipoproteins and the less specific scavenger receptors. Pseudopodia grow out around the particle and a "zip fastener" effect of binding to a number of receptors on the same particle encloses it within a vacuole known as a phagosome. This vacuole is subsequently acidified and fuses with lysosomes (containing a panoply of hydrolytic enzymes), while receptors are recycled to the cell surface. It follows that a drug carrier that is internalized in this way, and its contents, will come into contact with an environment in which it can be degraded. This is an important feature for polymeric delivery systems because it allows drug release and prevents the accumulation of polymer within the cells but can inactivate or sequester the active agent.

Factors which influence the uptake of drug carriers by macrophages include size, surface charge, surface coating and the presence of specific ligands. *In-vitro* studies have often observed more rapid uptake of larger particles, although this may be a consequence of more rapid sedimentation of larger objects in unstirred culture medium (Barratt et al. 1986; Tabata and Ikada 1988). However, more rapid clearance of larger particles is also observed in vivo (Senior and Gregoriadis 1982) and complement consumption also increases with particle size (Vonarbourg et al. 2006), suggesting that the surface of larger particles is more susceptible to opsonization than that of particles with a smaller radius of curvature. Nanoparticles and liposomes with positively or negatively charged surfaces are taken up more rapidly than neutral ones (Tabata and Ikada 1988; Heath et al. 1985) and those with hydrophobic surfaces are phagocytosed more readily than those with a hydrophilic surface. In particular, the presence of end-on hydrophilic chains such as PEG (Woodle 1998; Gref et al. 2000) or dextran (Jaulin et al. 2000) on the surface reduces uptake by phagocytes. Analysis of the kinetics of binding and internalization for various particle types suggests that the rate-limiting step is binding to the particle surface and that once they are bound PEG-covered particles are internalized at the same rate as non PEGylated ones (Mosqueira et al. 2001; Martina et al. 2007).

Some specific ligands can increase particle uptake by macrophages and this phenomenon can be exploited to deliver biologically active material to these cells. Liposomes containing phosphatidylserine are preferentially taken up by macrophages (Schroit and Fidler 1982) by means of a receptor whose primary purpose is to clear senescent erythrocytes and fragments of cells after apoptosis (Fadok et al. 2000). Another receptor which can be utilized to promote capture of drug carriers by macrophages is the mannose/fucose receptor, which allows these cells to capture and destroy a number of microorganisms. One example is the delivery of an immunomodulator in mannose-grafted liposomes, in order to stimulate the antitumoral properties of macrophages (Barratt et al. 1987). More recently, this strategy has been applied to the delivery of Amphotericin B (Vyas et al. 2000; Nahar et al. 2010) and another fungally derived antibiotic (Mitra et al. 2005) to macrophages for treatment of leishmaniasis. Another targeting ligand which has been used in a similar application is the tetrapeptide tuftsin (Thr-Lys-Pro-Arg, Agrawal et al. 2002). This peptide has the advantage of being both a targeting element and a macrophage activator. The anti-leishmanial activity of the drug is thus reinforced by macrophage-mediated effects.

Most other cell types are capable of internalizing carrier systems by endocytic mechanisms. The best documented pathway is that of clathrin-dependent endocytosis, involving so-called "coated pits". These are invaginations of the plasma membrane enriched in the protein clathrin, to which the intracellular portion of some membrane-bound receptors can attach. Binding of ligand to these receptors triggers assembly of further clathrin molecules to form a vesicle around the particle which is pinched off to become completely internalized. The clathrin coat is lost and the vacuole becomes an early endosome. As in phagocytosis, acidification, receptor recycling and fusion with endosomes occur. Another vesicular endocytotic pathway has been described more recently: the caveolae pathway. Caveolae are flask-shaped invaginations in the membrane coated with the protein caveolin. The membrane composition is different from the bulk composition, rich in cholesterol and resembling that of lipid rafts; some receptors are particularly associated with these areas, particularly those which have a GPI anchor (Anderson 1998). The vesicles formed after pinching off of the caveolae by the protein dynamin are not acidified and do not fuse with lysosomes. Finally, macropinocytosis involves the actin-driven formation of membrane ruffles which collapse and fuse with the plasma membrane to enclose a vesicle, internalizing a droplet of the extracellular medium without any specific receptor. Unlike clathrin- and caveolin-coated vesicles which are about 200 nm in diameter, vesicles formed by macropinocytosis can be as large as 5 μm (Swanson and Watts 1995), and thus presents a mechanism of uptake for larger-sized particles.

The endocytic pathway taken by drug delivery systems is usually determined by the use of specific inhibitors, such as cytochalasin B for clathrin-mediated endocytosis, filipin for caveolae-mediated endocytosis and amiloride for macropinocytosis. Cholesterol depletion is also used to detect caveolae-mediated processes. Thus, Rejman et al. (2004) were able to reveal the influence of size on the mechanism of uptake of fluorescent polystyrene nanoparticles by non phagocytic B16 cells. While particles smaller than 200 nm were internalized by clathrin-coated pits, larger particles from 200 to 500 nm in diameter were preferentially taken up by caveolae.

A variant of the endocytic pathway is transcytosis. In this pathway, vesicles formed by endocytosis do not fuse with lysosomes but cross the cell, fuse with the plasma membrane in another region of the cell and release their contents into the extracellular medium. In particular, receptor-mediated transcytosis is a mechanism of carrying macromolecules across endothelial cells. For example, transcytosis of insulin, IgG, LDL and iron bound to the transport protein transferrin cross the endothelial cells of the blood-brain barrier by transcytosis. This pathway can be exploited for drug delivery to the brain. The group of Pardridge has reported results using monoclonal antibodies targeting either the transferrin receptor or the insulin receptor conjugated to long-circulating liposomes (Pardridge 2010a). Jallouli et al. (2007) observed that neutral or cationic polysaccharide nanoparticles of 60 nm in diameter without any surface modification underwent transcytosis by the caveolae pathway across a model of the blood-brain barrier. PLGA nanoparticles coated with transferrin followed the same route in this model (Chang et al. 2009). On the other hand, PEGylated poly (alkylcyanoacrylate)-based nanoparticles were taken up by rat brain endothelial cells by a clathrin-dependent pathway (Kim et al. 2007). This uptake was found to be via LDL receptors, since Apolipoprotein E is adsorbed onto PEGylated nanoparticles. These results illustrate the complexity of uptake mechanisms for drug delivery systems.

In the second part of this chapter, two aspects of intracellular drug delivery with colloidal drug carriers will be discussed in more detail: intracellular delivery to macrophages and intracellular delivery of nucleic acids.

3 Intracellular Delivery to Macrophages

The accumulation of colloidal drug carriers within phagocytic cells can be exploited in some drug delivery applications. For example, muramyldipeptide and analogues which stimulate the antimicrobial and antitumoral activity of macrophages can be delivered more efficiently as carrier-associated molecules. Muramyldipeptide (MDP) is a low-molecular-weight, soluble, synthetic compound derived from the structure of peptidoglycan from mycobacteria. Such compounds would be generated within macrophages after the ingestion of bacteria; therefore they act on intracellular receptors but, because of their hydrophilicity, they penetrate poorly into the cells and are eliminated rapidly after i.v. administration. Therefore, muramyl peptides have been associated with both liposomes and nanocapsules. The first studies using using soluble MDP within liposomes showed activity against pulmonary (Fidler et al. 1981) and liver (Daemen et al. 1990) metastases in mice. However, the low molecular weight and water-solubility means that this compound was poorly encapsulated and leaked easily from liposomes. In response to this, lipophilic derivatives such as muramyltripeptide-cholesterol (MTP-Chol; Barratt et al. 1989) and muramyltripeptide-phosphatidylethanolamine (MTP-PE; Asano and Kleinerman 1993) were developed. These systems promoted increased intracellular penetration of muramyl peptides into macrophages in vitro. *In-vitro* studies of nanocapsules loaded with MTP-Chol indicated that nanocapsules were taken up by phagocytosis and that a soluble derivative was released in the lysosomes (Seyler et al. 1999; Mehri et al. 1996). Thereafter, a number of effector mechanisms are induced in the macrophages, such as the production of nitric oxide (Morin et al. 1994), cytokines and arachadonic acid derivatives (Seyler et al. 1997).

Nanocapsules were also active against hepatic metastases in mice; however, the treatment was only curative when the tumour burden was low (Barratt et al. 1994). Similar observations were made with liposomes containing MTP-PE (Asano and Kleinerman 1993). Nevertheless, these liposomes have been proposed for the treatment of osteosarcoma (Mori et al. 2008).

Similar liposomes were also able to activate macrophages to control bacterial infections, for example, *Klebsiella pneumoniae* (Melissen et al. 1994). As well as activating the non specific defence mechanisms of macrophages, muramyl peptides can act as adjuvants facilitate the development of a specific immune response to an antigen and in this respect as well liposomal encapsulation increases their efficiency (Turanek et al. 2006). The potential of liposomes as immunological adjuvants (reviewed by Kersten and Crommelin 2003) was recognized as early as 1974. Both liposomes and nanoparticles can be used to deliver immunological adjuvants or antigens or combinations of the two to antigen-presenting cells (macrophages and dendritic cells) (Peek et al. 2008). This approach can be applied to both protein antigens (van Broekhoven et al. 2004) and DNA vaccines (Greenland and Letvin 2007).

A number of microorganisms – bacteria, viruses and parasites – are able to live within macrophages. Colloidal drug carriers loaded with antibiotic drugs can be used to reach these infections (Pinto-Alphandary et al. 2000). Poly (isohexylcyanoacrylate) nanospheres loaded with Ampicillin allowed a large increase in efficacy compared with free antibiotic in mice infected with *Salmonella typhimurium* and *Listeria monocytogenes* (Fattal et al. 1989; Youssef et al. 1988). Studies using electron microscopy and confocal fluorescence microscopy with labeled *S. typhimurium* and nanospheres revealed the carrier system and the bacteria in the same intracellular compartment (Pinto-Alphandary et al. 1994).

A fluoroquinolone antibiotic, ciprofloxacin, was encapsulated within poly (isobutylcyanoacrylate) (PIBCA) and PIHCA nanospheres in an attempt to kill both dividing and non-dividing bacteria; however, the formulation was not effective against persistent Salmonella (Page-Clisson et al. 1998). More recently, the same antibiotic was encapsulated in PLGA nanospheres (Jeong et al. 2008). PLGA nanospheres are also able to deliver gentamicin to the liver and spleen of *Brucella melitensis*-infected mice (Lecaroz et al. 2007). Nanospheres prepared from poly (D,L-lactide) containing the antiparasitic drug primaquine were also efficient at delivering this drug to the liver (Rodrigues et al. 1994). Co-localization of nanospheres and *Leishmania donovani* parasites in Kupffer cells was observed.

Another infectious disease in which the organism responsible is to be found in macrophages is visceral leishmaniasis. One of the effective drugs against this disease is Amphotericin B, a polyene antibiotic. This amphiphilic molecule has been associated with several lipid-based drug delivery systems (Barratt and Bretagne 2007). Although the main advantage brought by nanoencapsulation is the reduction of Amphotericin B's dose-limiting toxicity, the use of colloidal carriers also means that the drug can reach the same intracellular compartment as the parasite.

4 Intracellular Delivery of Nucleic Acids

A second application in which colloidal drug carriers can provide intracellular delivery is that of the administration of nucleic acids. Progress in molecular biology has led to the availability of therapeutic genes and shorter nucleic acid sequences, in particular anti-sense oligonucleotides (AS-ODN) and small interfering RNA (siRNA). However, these large, negatively charged molecules cannot penetrate cell membranes and are also susceptible to degradation by nucleases, particularly the single-stranded AS-ODN. The stability problem for AS-ODN can be overcome by chemical modification, yielding structures such as phosphorothionate, methylphosphonate and boranophosphonate analogues that are resistant to enzymes but retain the capacity to bind to messenger RNA. However, the barrier of intracellular penetration remains. Furthermore, passage of the cell plasma membrane is not the last barrier to nucleic acid delivery. Therapeutic genes and oligonucleotides which form triple helices with DNA must be transported into the nucleus, while AS-ODN and siRNA act in the cytoplasm at the level of protein synthesis on the ribosomes. However, if the formulation containing the nucleic acid is internalized by phagocytosis or endocytosis, it will be sequestered in a membrane-bound vacuole, which is subsequently acidified and fused with a lysosome containing acid hydrolases. Therefore, some mechanism must be included to allow the nucleic acid to escape from the endosome before it is degraded.

Viral vectors have been developed for gene therapy, but have serious drawbacks in terms of toxicity, immunogenicity and the size of the plasmid which can be inserted. A number of other strategies have been adopted: physical methods such as the "gene gun" and electroporation, complexes with cationic polymers such as chitosan, poly (lysine) and poly (ethyleneimine) (PEI) and complexes with cationic lipids, known as lipoplexes. This section will concentrate on recent developments in intracellular delivery systems for AS-ODN and siRNA.

4.1 Lipid-Based Systems

The concept of "lipofection" was advanced by Felgner et al. (1987). Small liposomes, formed from dioleoylphosphatidylethanolamine (DOPE) and a cationic lipid, DOTMA, were mixed with DNA and the resulting "lipoplexes" were able to introduce reporter genes into various cell lines. The cationic lipid is able to condense the linear DNA into a complex, but the original lipid vesicle morphology is not conserved. DOTMA is available commercially as Lipofectamine®. Many other cationic lipids have been developed, including DOTAP, DOGS, DC-cholesterol and SAINT-2. These lipids are usually mixed with a "helper" lipid, such as DOPE or cholesterol which improves their stability and may aid cellular penetration. However, these complexes show low transfection efficiency compared with viral systems and in particular do not perform well in vivo. The net positive charge of the complexes is probably responsible for their high toxicity and also promotes the adsorption of plasma protein which leads to their rapid elimination. As a result, attempts have been made to modify the surface of these complexes. The surface charge can be modified by the addition of anionic lipids (Lee and Huang 1996) or by the inclusion of PEGylated lipids (Fenske et al. 2001).

Li and Szoka (2007) have reviewed the development of lipid-based colloidal particles with a diameter of less than 100 nm, which would be better adapted to *in-vivo* nucleic acid delivery. In particular, they describe a detergent dialysis method which allows the different components to be associated in a controlled fashion. They also present a model for the interaction of these lipid articles with cells. It was originally assumed that cationic lipoplexes were able to fuse directly with the plasma membrane, which has a negative charge and deliver their cargo directly to the cytoplasm. However, it is now accepted that the cationic complexes are taken up by endocytosis after electrostatic interaction with the plasma membrane. Within the endosome, endogenous lipids in the endosomal membrane are transferred into the particle to form ion pairs with the cationic lipid. This destabilizes the endosomal membrane and the nucleic acid is able to escape into the cytoplasm (Xu and Szoka, 1996).

Another, similar, strategy for cytoplasmic delivery is pH-sensitive liposomes (reviewed by Fattal et al. 2004). In this case the nucleic acid is encapsulated in neutral or anionic liposomes including lipids in their composition which undergo a change in organization at the pH of the endosomes, destabilize the endosomal membrane and release the nucleic acid into the cytoplasm. DOPE is one of the lipids used in this context, because it adopts a hexagonal phase at low pH. Oleic acid or cholesteryl hemisuccinate (CHEMS) are often added to the formulation. Some results have been obtained in vitro showing that these liposomes increase nucleic acid delivery to cells; for example, the replication of Friend virus in NIH 3 T3 cells was inhibited by an AS-ODN in pH-sensitive cells (Ropert et al. 1996). Interestingly, this work suggested that virally infected cells preferentially take up particulate carrier systems by reflex endocytosis during virus budding. However, observations in vivo with pH-sensitive liposomes have shown limited efficacy, because interactions with plasma proteins reduce the pH-sensitivity (De Oliveira et al. 2000). Recently, more sophisticated systems have been developed. Thus, Mudhakir et al. (2008) coupled an arginine-rich peptide onto the surface of lipid systems, to direct them to the macropinocytosis uptake pathway. When siRNA was incorporated into this type of particle, which also contained a pH-sensitive lipid combination DOPE/CHEMS, gene expression was effectively inhibited. Another pH-sensitive system, developed by the laboratory of Robert Langer, used a coating of PEG-polycation copolymer on pH-sensitive liposomes encapsulating siRNA (Auguste et al. 2008). At the pH of the endosomes, the protective polymer is desorbed and the nucleic acid released. Efficient knockdown of GFP in transfected HeLa cells and GAPDH in HUVEC cells by these particles has been demonstrated. This is interesting in the light of results reported by Remaut et al., (2007) showing that PEGylation of liposomes fails to protect the nucleic acid cargo from degragation in the lysosomes. On the other hand, an assembly of cationic lipid and AS-ODN conjugated to PEG was able to promote rapid delivery to the nucleus in KB cells (Jeong et al. 2006). Experimental results from Pakunlu et al. (2006) showed that PEGylated liposomes containing both AS-ODN to drug resistance genes and doxorubicin could be taken up by cancer cells and in this way the efficacy of the cytotoxic drug was enhanced.

A alternative strategy for intracellular delivery is the use of so-called "fusogenic" liposomes in which a viral peptide included in the formulation allows cytoplasmic delivery. An example is the promotion of cytoplasmic delivery of DNA oligonucleotides with a system using inactivated Sendai virus (Kunisawa et al. 2005).

The intracellular delivery of antisense oligonucleotides has been reported in a large number of publications. Delivery has been evidenced by detection of fluorescent oligonucleotides within the cells and by an antisense effect on the target gene. Thus, Ruozi et al. (2005) showed the role of DOTAP in liposomes for the intracellular delivery of AS-ODN to COS 1 and HaCaT cells.

In the same way, uptake of siRNA loaded lipid complexes has been demonstrated in many different cells types. For example, Santel et al. (2006) showed uptake of lipid complexes containing siRNA against CD31 in mouse vascular endothelial cells and also observed inhibition of tumor growth by an antiangiogenic effect. Lavigne and Thierry (2007) were able to quantify the delivery of siRNA directed against cyclin D1 associated with a lipoplex to specific cell compartments in MCF-7 cells. Yadava et al. (2007) compared the performance of lipoplexes with poly(ethyleneimine) (PEI, see below) and saw similar cellular uptake of a model siRNA from the two systems, but a better silencing activity from the lipoplexes, which they attributed to a more rapid dissociation of the lipid complexes.

Recent developments include the use of specific targeting ligands to promote uptake of the nucleic acid complexes by specific cells types. Cardoso et al. (2007) used lipoplexes targeted by transferrin to deliver siRNA silencing marker genes to glioma, hepatocarcinoma and HT-22 cells. Transferrin was also used as a ligand to prepare "Trojan horse" liposomes to carry nucleic acids and other therapeutic agents across the blood-brain barrier, taking advantage of the presence of transferrin receptors promoting transcytosis across endothelial cells (Pardridge 2010a). The insulin receptor has been used for the same purpose (Pardridge 2010b). The folate receptor, overexpressed on many tumor cells, is another target which has been exploited for intracellular nucleic acid delivery (Yu et al. 2009). For example, Bcl2 downregulation in KB cells was shown to be greatly enhanced by the presence of folate acid on the surface of cationic liposomes (Chiu et al. 2006). The ligand is attached to the extremity of PEG chains themselves coupled to phosphatidylethanolamine in the liposome membrane. Delivery of anti HER-2 AS-ODN to head and neck cancer cells was also improved by folate-liposomes and increased the sensitivity of the cells to conventional chemotherapy (Rait et al. 2003).

4.2 Polymer-Based Systems

As far as polymer-based systems are concerned, three strategies have been adopted for the delivery of nucleic acids: complexation with cationic molecules, adsorption onto preformed nanoparticles and encapsulation within nanoparticles.

The most commonly used macromolecule for complexation of nucleic acids is PEI, which provides a high density of positive charge. Complexes can be formed easily by mixing the polymer and the nucleic acid; the resulting assemblies retain a net positive charge, which leads to electrostatic interactions with the cell membrane and uptake by endocytosis. Within the acidic endosome, the amino groups on the polymer are protonated, leading to osmotic swelling and rupture of the endosome. This so-called proton sponge effect ensures the release of the nucleic acid into the cytoplasm. However, a simple PEI/nucleic acid formulation presents drawbacks in terms of toxicity and an unfavorable biodistribution. PEI, particularly branched or high-molecular weight chains is toxic because of its ability to bind to cell membranes and cause necrotic cell death (Roques et al. 2007). The net positive charge of the complexes also provokes rapid adsorption of plasma proteins after intravenous administration, leading to particle aggregation. These aggregates are physically trapped in small capillaries, particularly in the lung, or taken up by phagocytosis by macrophages (Chemin et al. 1998). Despite these drawbacks, some success has been obtained in cell culture, including in non dividing cells (Boussif et al. 1995; Dheur et al. 1999; Gomes dos Santos et al. 2006). Some positive results have also been obtained in vivo: Grzelinski et al. (2006) were able to suppress the expression of the growth factor pleiotrophin and Urban-Klein et al. (2005) achieved an anti-tumor effect with a PEI complex of siRNA targeting the HER-2 receptor.

The biocompatibility of PEI-based formulations can be improved by coating them with PEG (Mao et al. 2006) or by glycosylation of the surface (Leclercq et al. 2000). Specific targeting ligands have also been attached. One such ligand is the RGD (Arg-Gly-Asp) peptide, which was used by Schiffelers et al., (2004) to block vascular endothelial growth factor receptor-2 (VEGF-R2) expression in vascular endothelial cells and thereby prevent tumor angiogenesis.

Dendrimers composed of poly (amidoamine) (PAMAM) carry a high density of positive charge and can therefore be used to bind nucleic acids on their surface (Zhou et al. 2006; Shen et al. 2007). Helin et al. (1999) studied the intracellular distribution of a fluorescent ODN adsorbed onto dendrimers and found that this depended on the stage of the cell cycle. Fluorescence was found in the nuclei during G2/M phase while in other phases it seemed to be concentrated in acidic vesicles. Inhibition of target gene expression was observed with these systems. Other examples of inhibition of gene expression by AS-ODN attached to dendrimers are given by Bielinska et al., (1996), Yoo et al. (1999) and Li and Morcos (2008). The subject has been reviewed recently by Raviña et al. (2010). Specific targeting of dendrimers carrying AS-ODN against the Epidermal Growth Factor receptor to glioma cells was achieved by coupling folic acid to their surface (Kang et al. 2010).

SiRNA has also been immobilized on dendrimers (Kang et al. 2005; Zhou et al. 2006). These systems have been rendered more biocompatible by acetylation (Waite et al. 2009; Patil et al. 2008) and they have also been targeted: with the Tat peptide (Kang et al. 2005) and with the peptide hormone LHRH (Minko et al. 2010). Other, more water-soluble, polymers have been used to prepare dendrimers for nucleic acid delivery. Among these are carbosilane (Gras et al. 2009) and poly (L-lysine) (Eom et al. 2007).

Nucleic acids can also be complexed with the cationic arginine-rich polypeptide protamine. Junghans et al. (2001) observed uptake of protamine-AS-ODN particles by endocytosis and delivery to the cytoplasm in Vero cells. More stable particles were prepared using both protamine and human serum albumin (Weyermann et al. 2005). These showed both uptake and an anti-sense effect in mouse fibroblasts. Attempts were made to target these particles to the blood-brain barrier, by coating them with apolipoprotein A-1 to promote transcytosis (Kratzer et al. 2007). More sophisticated delivery systems can be obtained by mixing protamine, lipids and nucleic acids. Thus, Li et al. (2008) were able to deliver siRNA to the cytoplasm of NCI-H460 tumor cells, after modifying the surface with PEG and anisamide to target a tumor receptor. Similar particles have been used to downregulate Bcl-2 expression in cells with AS-ODN, using transferrin as a targeting ligand (Yang et al. 2009).

A positively charged polysaccharide, chitosan, has also been used to complex nucleic acids. Gene silencing in H1299 human lung carcinoma cells has been achieved (Liu et al. 2007). Cyclodextrins, molecular cages of glucose units, are not positively charged but can be coupled to a cationic polymer to form effective systems for complexing nucleic acids. When modified with PEG and a targeting moiety (again transferrin) these particles show gene knock-down in vitro and inhibition of the EWS-Fli-1 gene in Ewing's sarcoma in vivo (Bartlett and Davis 2007; Hu-Lieskovan et al. 2005).

Micellar systems have also been used to deliver nucleic acids to intracellular compartments. Examples are PEG-poly (aspartic acid) copolymers (Kakizawa et al. 2004) which can exert a "proton sponge" effect in endosomes, lactosylated PEG which can form complexes with poly (lysine) and AS-ODNs and allow targeting to galactose receptors on liver cells (Oishi et al. 2005; Oishi et al. 2007) and PEG-PMMA block copolymers (Kakizawa et al. 2006) which can be used to produce hybrid organic-inorganic nanoparticles.

The adsorption of nucleic acids onto preformed nanoparticles requires a positively charged surface. The first systems described used cationic surfactant on the surface of poly (alkylcyanoacrylates) (PACA) (Fattal et al. 1998) or poly (lactide-co-glycolide) (Singh et al. 2003; Oster et al. 2005). AS-ODN adsorbed onto PACA nanoparticles are protected from nucleases (Lambert et al. 1998) and can be delivered intracellularly (Chavany et al. 1994). These formulations were found to be able to inhibit the proliferation of mutated Ha-ras-transformed cells proliferation and reduce tumorigenicity in nude mice (Schwab et al. 1994). More recently, cationic nanoparticles made up of a biodegradable core of poly (isobutylcyanoacrylate) surrounded by a shell of chitosan were prepared by de Martimprey et al. (2008). A siRNA directed against the Ret/PTC1 junction oncogene was adsorbed onto these particles and was able to silence the gene in papillary thyroid carcinoma cells and exert an anti-tumor effect in mice.

PLGA-based nanoparticles have also been modified to provide a cationic surface. Notably, Nafee et al. (2007) coated them with chitosan and were able increase intracellular uptake of AS-ODN into A549 lung cells. Cationic nanoparticles have also been prepared by co-polymerization of methylmethacrylate and aminoalkylmethacrylate monomers (Zobel et al. 2000). These nanoparticles can also promote intracellular accumulation of AS-ODN (Tondelli et al. 1998; Zobel et al. 2000).

Despite the fact that adsorption onto the nanoparticle surface has produced some results with the intracellular delivery of nucleic acids, better protection from nucleases would be obtained if the nucleic acids were incorporated within the interior of the particles. However, the hydrophilic character of the nucleic acids is not compatible with the hydrophobic polymers; Thus, Lambert et al. (2000) developed a method of producing nanocapsules (NC) with an aqueous core and a PIBCA shell which could encapsulate hydrophilic molecules such as nucleic acids. These nanocapsules were able to protect AS-ODN from nuclease degradation and showed evidence of intracellular uptake by their ability to transfect cells, including vascular smooth muscle cells, which are notoriously difficult to transfect (Toub et al. 2005). The same system was also able to deliver siRNA to cells (Toub et al. 2006). Fluorescent labeling showed a punctate pattern of siRNA within the cells, suggesting that it was delivered to endosomes. In vivo, these NC containing AS-ODN targeted to the EWS-Fli-1 spliced gene showed a true antisense effect against Ewing's sarcoma in mice (Maksimenko et al. 2003).

Nanoparticles encapsulating nucleic acids have also been prepared from the well known PLGA polymer. Thus, the uptake of AS-ODN directed against VEGF mRNA by ARPE-19 human retinal pigment epithelial cells was increased 4-fold by encapsulation and protein expression was thereby inhibited (Aukunuru et al. 2003). More recently, intracellular delivery of siRNA was achieved by PLGA nanoparticles, silencing the GFP gene in 293 T cells (Yuan et al. 2006).

Coating of PLGA nanoparticles with the positively charged polysaccharide chitosan promoted their uptake by A549 lung cancer cells (Beisner et al. 2010). In this way, 2′-O-methyl RNA was able to inhibit telomerase activity in a sequence-specific way in these cells. Chitosan alone has also been used to form particles for siRNA delivery, by ionic gelation with tripolyphosphate (Katas and Alpar 2006). These particles were efficient at silencing the marker gene luciferase expression in CHO K1 and HEK 293 cells and showed better activity than simple chitosan-RNA complexes. Later, the same group combined PLGA and PEI in nanoparticles. These were able to adsorb siRNA and allow a good gene silencing effect in CHO K1 cells, showing better activity and lower toxicity than PEI alone (Katas et al. 2008).

Finally, methacrylic polymers, such as those in the Eudragit series, are attractive as nucleic acid delivery systems because of their positive charge. Yessine et al. (2006) showed that complexes between these polymers and AS-ODN could deliver the nucleic acid to the cytoplasm, probably because of an endosome-perturbing effect. Recently, nanoparticles for gene delivery have been formulated from Eudragit polymers and shown to be able to transfect some tumor cells lines (Gargouri et al. 2009).

5 Conclusion

Particulate nanocarriers have shown their potential for intracellular delivery of biologically interesting substances in numerous studies over the past decades. However, a number of unresolved issues remain, such as the choice of truly specific ligands for *in-vivo* applications and the formulation of systems which allow the cargo to escape from the lysosomes while remaining non toxic and stable in biological fluids. Therefore, this is likely to remain an active research field in the years to come.

References

Agrawal A J, Agrawal A, Pal A, Guru P Y and Gupta C M (2002), 'Superior chemotherapeutic efficacy of Amphotericin B in tuftsin-bearing liposomes against *Leishmania donovani* infection in hamsters', *J Drug Target*, 10, 41–45.

Allen T M and Moase E H (1996), 'Therapeutic opportunities for targeted liposomal drug delivery', *Adv Drug Deliv Rev*, 21, 117–133.

Anderson R G (1998) 'The caveolae membrane system', *Ann Rev Biochem*, 67, 199–225.

Asano T and Kleinerman E S (1993), 'Liposome-entrapped MTP-PE, a novel biologic agent for cancer therapy', *J Immunother*, 14, 286–292.
Aukunuru J V, Ayalasomayajula S P, Kompella U B (2003), 'Nanoparticle formulation enhances the delivery and activity of a vascular endothelial growth factor antisense oligonucleotide in human retinal pigment epithelial cells' *J Pharm Pharmacol*, 55, 1199–1206.
Auguste D T, Furman K, Wong A, Fuller J, Armes S P, Deming T J, Langer R (2008), 'Triggered release of siRNA from poly(ethylene glycol)-protected, pH-dependent liposomes', *J Control Rel*, 130, 266–274.
Barratt G (2003), 'Colloidal drug carriers: achievements and perspectives', *Cell Mol Life Sci*, 60, 21–37.
Barratt G and Bretagne S (2007), 'Optimizing efficacy of Amphotericin B through nanomodification', *Int J Nanomedicine*, 2, 301–313.
Barratt G M, Tenu J P, Yapo A and Petit J F (1986), 'Preparation and characterization of liposomes containing mannosylated phospholipids capable of targeting drugs to macrophages', Biochim Biophys Acta, 862, 153–164.
Barratt G M, Nolibé D, Yapo A, Petit J-F and Tenu J-P (1987), 'Use of mannosylated liposomes for *in vivo* targeting of a macrophage activator and control of artificial pulmonary metastases', *Ann Inst Pasteur (Immunol)*, 138, 437–450.
Barratt G M, Yu W P, Fessi H, Devissaguet J Ph, Petit J F, Tenu J P, Israel L, Morère J F and Puisieux F (1989), 'Delivery of MDP-L-alanyl-cholesterol to macrophages: comparison of liposomes and nanocapsules', *Cancer J*, 2, 439–443.
Barratt G M, Puisieux F, Yu W-P, Foucher C, Fessi H and Devissaguet J-Ph (1994), 'Anti-metastatic activity of MDP-L-alanyl-cholesterol incorporated into various types of nanocapsules', *Int J Immunopharmacol*, 16, 457–461.
Bartlett D W and Davis M E (2007), 'Physicochemical and biological characterization of targeted, nucleic acid-containing nanoparticles', *Bioconjug Chem*, 18, 456–468.
Beisner J, Dong M, Taetz S, Nafee N, Griese E U, Schaefer U, Lehr C M, Klotz U and Mürdter T E (2010), 'Nanoparticle mediated delivery of 2'-O-methyl-RNA leads to efficient telomerase inhibition and telomere shortening in human lung cancer cells', *Lung Cancer*, 68, 346–354.
Bielinska A, Kukowska-Latallo J F, Johnson J, Tomalia D A and Baker J R Jr (1996), 'Regulation of *in vitro* gene expression using antisense oligonucleotides or antisense expression plasmids transfected using starburst PAMAM dendrimers', *Nucleic Acids Res*, 24, 2176–2182.
Boussif O, Lezoualc'h F, Zanta M A, Mergny M D, Scherman D, Demeneix B and Behr J P (1995), 'A versatile vector for gene and oligonucleotide transfer into cells in culture and *in vivo*: polyethylenimine', *Proc Natl Acad Sci USA*, 92, 7297–7301.
Cardoso A L, Simões S, de Almeida L P, Pelisek J, Culmsee C, Wagner E and Pedroso de Lima M C (2007), 'siRNA delivery by a transferrin-associated lipid-based vector: a non-viral strategy to mediate gene silencing', *J Gene Med*, 9, 170–183.
Chang J, Jallouli Y, Kroubi M, Yuan X B, Feng W, Kang C S, Pu P Y and Betbeder D, (2009), 'Characterization of endocytosis of transferrin-coated PLGA nanoparticles by the blood-brain barrier', *Int J Pharm*, 379, 285–292.
Chavany C, Saison-Behmoaras T, Le Doan T, Puisieux F, Couvreur P and Helene C (1994), 'Adsorption of oligonucleotides onto polyisohexylcyanoacrylate nanoparticles protects them against nucleases and increases their cellular uptake', *Pharm Res*, 11, 1370–1378.
Chemin I, Moradpour D, Wieland S, Offensperger W B, Walter E, Behr J P and Blum H E (1998), 'Liver-directed gene transfer: a linear polyethlenimine derivative mediates highly efficient DNA delivery to primary hepatocytes *in vitro* and *in vivo*', *J Viral Hepat*, 5, 369–375.
Chiu S J, Marcucci G and Lee R J (2006), 'Efficient delivery of an antisense oligodeoxyribonucleotide formulated in folate receptor-targeted liposomes', *Anticancer Res*, 26, 1049–1056.
Couvreur P, Barratt G, Fattal E, Legrand P and Vauthier C (2002), 'Nanocapsule technology: a review', *Crit Rev Drug Del Sys*, 19, 101–136.
Daemen T, Dontje B H, Veninga A, Scherphof G L and Oosterhuis W L (1990), 'Therapy of murine liver metastases by administration of MDP encapsulated in liposomes', *Select Cancer Ther*, 6, 63–71.

De Martimprey H, Bertrand JR, Fusco A, Santoro M, Couvreur P, Vauthier C and Malvy C (2008), 'siRNA nanoformulation against the ret/PTC1 junction oncogene is efficient in an *in vivo* model of papillary thyroid carcinoma,' *Nucleic Acids Res* 36, e2.

De Oliveira MC, Boutet V, Fattal E, Boquet D, Grognet JM, Couvreur P and Deverre JR (2000), 'Improvement of in vivo stability of phosphodiester oligonucleotide using anionic liposomes in mice', *Life Sci*, 67, 1625–1637.

Dheur S, Dias N, van Aerschot A, Herdewijn P, Bettinger T, Remy J S, Hélène C and Saison-Behmoaras E T (1999), 'Polyethylenimine but not cationic lipid improves antisense activity of 3'-capped phosphodiester oligonucleotides,' *Antisense Nucleic Acid Drug Dev*, 9, 515–525.

Eom K D, Park S M, Tran H D, Kim M S, Yu R N and Yoo H (2007), 'Dendritic alpha,epsilon-poly(L-lysine)s as delivery agents for antisense oligonucleotides', *Pharm Res,*, 24, 1581–1589.

Fadok V A, Bratton D L, Rose D M, Pearson A, Ezekewitz R A and Henson P M (2000), 'A receptor for phosphatidylserine-specific clearance of apoptotic cells', *Nature*, 405, 85–90.

Fattal E, Vauthier C, Aynie I, Nakada Y, Lambert G, Malvy C and Couvreur P (1998), 'Biodegradable polyalkylcyanoacrylate nanoparticles for the delivery of oligonucleotides', *J Control Rel*, 53, 137–143.

Fattal E, Youssef M, Couvreur P and Andremont A (1989), 'Treatment of experimental salmonellosis in mice with ampicillin-bound nanoparticles', *Antimicrob Agents Chemother*, 33, 1540–1543.

Fattal E, Nir S, Parente R A and Szoka F C Jr (1994), 'Pore-forming peptides induce rapid phospholipid flip-flop in membranes', *Biochemistry*, 31, 6721–6731.

Fattal E, De Rosa G and Bochot A (2004), 'Gel and solid matrix systems for the controlled delivery of drug-carrier associated nucleic acids', *Int J Pharm* 277, 25–30.

Felgner P L, Gadek T R, Holm M, Roman R, Chan H W, Wenz M, Northrop J P, Ringold G M and Danielsen M (1987), 'Lipofection: a highly efficient, lipid-mediated DNA-transfection procedure', *Proc Natl Acad Sci U S A*, 84, 7413–7417.

Fenske D B, Palmer L R, Chen T, Wong K F and Cullis P R (2001), 'Cationic poly(ethyleneglycol) lipids incorporated into pre-formed vesicles enhance binding and uptake to BHK cells', *Biochim Biophys Acta*, 1512, 259–272.

Fidler I J, Sone S, Fogler W E and Barnes Z L (1981), 'Eradication of spontaneous metastases and activation of alveolar macrophages by intravenous injection of liposomes containing muramyl dipeptide', *Proc Natl Acad Sci USA*, 78, 1680–1684.

Gargouri M, Sapin A, Bouli S, Becuwe P, Merlin J L and Maincent P (2009), 'Optimization of a new non-viral vector for transfection: Eudragit nanoparticles for the delivery of a DNA plasmid', *Technol Cancer Res Treat*, 8, 433–444.

Gomes dos Santos A L, Bochot A, Tsapis N, Artzner F, Bejjani R A, Thillaye-Goldenberg B, de Kozak Y, Fattal E and Behar-Cohen F (2006), 'Oligonucleotidepolyethylenimine complexes targeting retinal cells: structural analysis and application to anti-TGFbeta-2 therapy', *Pharm Res*, 23, 770–781.

Gras R, Almonacid L, Ortega P, Serramia MJ, Gomez R, de la Mata F J, Lopez-Fernandez L A and Muñoz-Fernandez M A (2009), 'Changes in gene expression pattern of human primary macrophages induced by carbosilane dendrimer 2 G-NN16', *Pharm Res*, 26, 577–586.

Greenland J R and Letvin N L (2007), 'Chemical adjuvants for plasmid DNA vaccines', *Vaccine*, 25, 3731–3741.

Gref R, Lück M, Quellec P, Marchand M, Dellacherie E, Harnisch S, Blunk T and Müller R H (2000), "Stealth' corona-core nanoparticles surface modified by polyethylene glycol (PEG): influences of the corona (PEG chain length and surface density) and of the core composition on phagocytic uptake and plasma protein adsorption', *Colloids Surf B Biointerfaces*, 18, 301–313.

Gregoriadis G and Senior J (1982), 'Control of fate and behaviour of liposomes *in vivo*', *Prog Clin Biol Res*, 102 pt A, 263–279

Grzelinski M, Urban-Klein B, Martens T, Lamszus K, Bakowsky U, Hobel S, Czubayko F and Aigner A (2006), 'RNA interference-mediated gene silencing of pleiotrophin through polyethylenimine-complexed small interfering RNAs *in vivo* exerts antitumoral effects in glioblastoma xenografts', *Hum Gene Ther*, 17, 751–766.

Heath T D, Lopez N G and Papahadjopoulos D (1985), 'The effects of liposome size and surface charge on liposome-mediated delivery of methotrexate-gamma-aspartate to cells *in vitro*', *Biochim Biophys Acta*, 820, 74–84.

Helin V, Gottikh M, Mishal Z, Subra F, Malvy C and Lavignon M (1999), 'Cell cycle-dependent distribution and specific inhibitory effect of vectorized antisense oligonucleotides in cell culture', *Biochem Pharmacol*, 58, 95–107.

Heuschkel S, Goebel A and Neubert R H (2008), 'Microemulsions - modern colloidal carrier for dermal and transdermal drug delivery', *J Pharm Sci*, 97, 603–631.

Hillaireau H and Couvreur P (2009), 'Nanocarriers' entry into the cell: relevance to drug delivery', *Cell Mol Life Sci*, 66, 2873–2896.

Hu-Lieskovan S, Heidel J D, Bartlett D W, Davis M E and Triche T J (2005), 'Sequence-specific knockdown of EWS-FLI1 by targeted, nonviral delivery of small interfering RNA inhibits tumor growth in a murine model of metastatic Ewing's sarcoma', *Cancer Res*, 65, 8984–8992.

Jain R K (1987), 'Transport of molecules across tumor vasculature', *Cancer Metastasis Rev*, 6, 559–593.

Jallouli Y, Paillard A, Chang J, Sevin E and Betbeder D (2007), 'Influence of surface charge and inner composition of porous nanoparticles to cross blood-brain barrier *in vitro*', *Int J Pharm*, 344, 103–109.

Jaulin N, Appel M, Passirani C, Barratt G and Labarre D (2000), 'Reduction of the uptake by a macrophagic cell line of nanoparticles bearing heparin or dextran covalently bound to poly(methyl methacrylate)', *J Drug Target*, 8, 165–172.

Jeon S I, Lee J H, Andrade J D and De Gennes P G (1991), 'Protein-surface interactions in the presence of polyethylene oxide; 1. Simplified theory' *J Colloid Interf Sci*, 142, 149–166.

Jeong J H, Kim S H, Kim S W, Park T G (2006), 'Intracellular delivery of poly(ethylene glycol) conjugated antisense oligonucleotide using cationic lipids by formation of self-assembled polyelectrolyte complex micelles', *J Nanosci Nanotechnol* 6, 2790–2795.

Jeong Y I, Na H S, Seo D H, Kim D G, Lee H C, Jang M K, Na S K, Roh S H, Kim S I and Nah J W (2008), 'Ciprofloxacin-encapsulated poly(DL-lactide-co-glycolide) nanoparticles and its antibacterial activity', *Int J Pharm*, 352, 317–323.

Junghans M, Kreuter J and Zimmer A (2001), 'Phosphodiester and phosphorothioateoligonucleotide condensation and preparation of antisense nanoparticles', *Biochim Biophys Acta*, 1544, 177–188.

Kakizawa Y and Kataoka K (2002), 'Block copolymer micelles for delivery of gene and related compounds', *Adv Drug Deliv Rev*, 54, 203–222.

Kakizawa Y, Furukawa S and Kataoka K (2004), 'Block copolymer-coated calcium phosphate nanoparticles sensing intracellular environment for oligodeoxynucleotide and siRNA delivery', *J Control Rel*, 97, 345–356.

Kakizawa Y, Furukawa S, Ishii A and Kataoka K (2006), 'Organic-inorganic hybrid-nanocarrier of siRNA constructing through the self-assembly of calcium phosphate and PEG-based block aniomer', *J Control Rel*, 111, 368–370.

Kang H, DeLong R, Fisher M H and Juliano R L (2005), 'Tat-conjugated PAMAM dendrimers as delivery agents for antisense and siRNA oligonucleotides', *Pharm Res*, 22, 2099–2106.

Kang C, Yuan X, Li F, Pu P, Yu S, Shen C, Zhang Z and Zhang Y (2010), 'Evaluation of folate-PAMAM for the delivery of antisense oligonucleotides to rat C6 glioma cells *in vitro* and *in vivo*', J Biomed Mater Res A, 93, 585–594.

Katas H and Alpar H O (2006), 'Development and characterisation of chitosan nanoparticles for siRNA delivery', *J Control Rel*, 115, 216–225.

Katas H, Chen S, Osamuyimen A A, Cevher E and Oya Alpar H (2008), 'Effect of preparative variables on small interfering RNA loaded Poly(D,L-lactide-co-glycolide)-chitosan submicron particles prepared by emulsification diffusion method', *J Microencapsul*, 25, 541–548.

Kersten G F A and Crommelin D J A (2003), 'Liposomes and ISCOMs', *Vaccine*, 21, 915–920.

Kim H R, Gil S, Andrieux K, Nicolas V, Appel M, Chacun H, Desmaële D, Taran F, Georgin D and Couvreur P (2007), 'Low-density lipoprotein receptor-mediated endocytosis of PEGylated nanoparticles in rat brain endothelial cells', *Cell Mol Life Sci*, 64, 356–364.

Koning G A, Kamps J A and Scherphof G L (2002), 'Efficent intracellular delivery of 5-fluorodeoxyuridine into colon cancer cells by targeted immunoliposomes', *Cancer Detect Prevent*, 26, 299–307.

Kratzer I, Wernig K, Panzenboeck U, Bernhart E, Reicher H, Wronski R, Windisch M, Hammer A, Malle E, Zimmer A and Sattler W (2007), 'Apolipoprotein A-I coating of protamineoligonucleotide nanoparticles increases particle uptake and transcytosis in an *in vitro* model of the blood-brain barrier', *J Control Rel*, 117, 301–311.

Kunisawa J, Masuda T, Katayama K, Yoshikawa T, Tsutsumi Y, Akashi M, Mayumi T and Nakagawa S (2005), 'Fusogenic liposome delivers encapsulated nanoparticles for cytosolic controlled gene release', *J Control Rel*, 105, 344–353.

Kyatanwar A U, Jadhav K R and Kadam V J (2010)' 'Self micro-emulsifying drug delivery system (SMEDDS) : Review', *J Pharm Res*, 3, 75–83

Lambert G, Fattal E, Brehier A, Feger J and Couvreur P (1998), 'Effect of polyisobutylcyanoacrylate nanoparticles and lipofectin loaded with oligonucleotides on cell viability and PKC alpha neosynthesis in HepG2 cells', *Biochimie* 80, 969–976.

Lambert G, Fattal E, Pinto-Alphandary H, Gulik A and Couvreur P (2000), 'Polyisobutylcyanoacrylate nanocapsules containing an aqueous core as a novel colloidal carrier for the delivery of oligonucleotides', *Pharm Res*, 17, 707–714.

Lavigne C and Thierry A R (2007), 'Specific subcellular localization of siRNAs delivered by lipoplex in MCF-7 breast cancer cells', *Biochimie*, 89, 1245–1251.

Lecaroz M C, Blanco-Prieto M J, Campanero M A, Salman H and Gamazo C (2007), 'Poly(D,L-lactide-coglycolide) particles containing gentamicin: pharmacokinetics and pharmacodynamics in *Brucella melitensis*-infected mice', *Antimicrob Agents Chemother*, 51, 1185–1190.

Leclercq F, Dubertret C, Pitard B, Scherman D and Herscovici J (2000), 'Synthesis of glycosylated polyethylenimine with reduced toxicity and high transfecting efficiency', *Bioorg Med Chem Lett*, 10, 1233–1235.

Lee R J and Huang L (1996), 'Folate-targeted, anionic liposome-entrapped polylysine-condensed DNA for tumor cell-specific gene transfer', *J Biol Chem*, 271, 8481–8487.

Li Y F and Morcos P A (2008), 'Design and synthesis of dendritic molecular transporter that achieves efficient *in vivo* delivery of morpholino antisense oligo', *Bioconjug Chem*, 19, 1464–1470.

Li W and Szoka F C Jr (2007), 'Lipid-based nanoparticles for nucleic acid delivery', *Pharm Res*, 24, 438–449.

Li S D, Chen Y C, Hackett M J and Huang L (2008), 'Tumor-targeted delivery of siRNA by self-assembled nanoparticles', *Mol Ther*, 16, 163–169.

Liu X, Howard K A, Dong M, Andersen M Ø, Rahbek U L, Johnsen M G, Hansen O C, Besenbacher F and Kjems J (2007), 'The influence of polymeric properties on chitosan/siRNA nanoparticle formulation and gene silencing', *Biomaterials*, 28, 1280–1288.

Maksimenko A, Malvy C, Lambert G, Bertrand J R, Fattal E, Maccario J and Couvreur P (2003), 'Oligonucleotides targeted against a junction oncogene are made efficient by nanotechnologies', *Pharm Res*, 20, 1565–1567.

Mao S, Neu M, Germershaus O, Merkel O, Sitterberg J, Bakowsky U and Kissel T (2006), 'Influence of polyethylene glycol chain length on the physicochemical and biological properties of poly(ethylene imine)- graft-poly(ethylene glycol) block copolymer/SiRNA polyplexes', *Bioconjug Chem*, 17, 1209–1218.

Martina M S, Nicolas V, Wilhem C, Ménager C, Barratt G and Lesieur S (2007), 'The *in vitro* kinetics of the interactions between PEG-ylated magnetic-fluid-loaded liposomes and macrophages', *Biomaterials*, 28, 4143–4153

Mehri G, Coleman AW, Devissaguet J Ph and Barratt G M (1996), 'Synthesis and immunostimulating properties of lipophilic ester and ether muramyl peptide derivatives', *J Med Chem*, 39, 4483–4488.

Melissen P M B, van Vianen W and Bakker-Woudenberg, I A J M (1994), 'Treatment of *Klebsiella pneumoniae* septicemia in normal and leukopenic mice by liposome-encapsulated muramyl tripeptide phosphatidylethanolamine', *Antimicrob Agents Chemother*, 38, 147–150.

Mercadal M, Domingo J C, Petriz J C, Garcia J and De Madariaga M M (1999), 'A novel strategy affords high-yield coupling of antibody to extremities of liposomal surface grafted PEG chains', *Biochim Biophys Acta*, 1418, 232–238.

Minko T, Patil M L, Zhang M, Khandare J J, Saad M, Chandna P and Taratula O (2010), 'LHRH-targeted nanoparticles for cancer therapeutics', *Methods Mol Biol*, 624, 281–294.

Mitra M, Mandal A K, Chatterjee T K and Das N (2005), 'Targeting of mannosylated liposome incorporated benzyl derivative of *Penicillium nigricans* derived compound MT81 to reticuloendothelial systems for the treatment of visceral leishmanias', *J Drug Target*, 13, 285–293.

Mori K, Ando K and Heymann D (2008), 'Liposomal muramyl tripeptide phosphatidyl ethanolamine: a safe and effective agent against osteosarcoma pulmonary metastases', Expert Rev Anticancer Ther, 8, 151–159.

Morin C, Barratt G M, Fessi H, Devissaguet J-Ph and Puisieux F (1994), 'Improved intracellular delivery of a muramyldipeptide analog by means of nanocapsules', *Int J Immunopharmacol*, 16, 451–456.

Mosqueira VCF, Legrand P, Gulik A, Bourdon O, Gref R, Labarre D and Barratt G (2001), 'Relationship between complement activation, cellular uptake and surface physicochemical aspects of novel PEG-modified nanocapsules', *Biomaterials*, 22, 2969–2979.

Mudhakir D, Akita H, Tan E and Harashima H (2008), 'A novel IRQ ligand-modified nanocarrier targeted to a unique pathway of caveolar endocytic pathway', *J Control Rel*, 125, 164–173.

Nafee N, Taetz S, Schneider M, Schaefer U F and Lehr C M (2007), 'Chitosan-coated PLGA nanoparticles for DNA/RNA delivery: effect of the formulation parameters on complexation and transfection of antisense oligonucleotides', *Nanomedicine,* 3, 173–183.

Nahar M, Dubey V, Mishra D, Mishra P K, Dube A and Jain N K (2010), '*In vitro* evaluation of surface functionalized gelatin nanoparticles for macrophage targeting in the therapy of visceral leishmaniasis', *J Drug Target*, 18, 93–105.

Oishi M, Nagasaki Y, Itaka K, Nishiyama N and Kataoka K (2005), 'Lactosylated poly(ethylene glycol)-siRNA conjugate through acid-labile beta-thiopropionate linkage to construct pH-sensitive polyion complex micelles achieving enhanced gene silencing in hepatoma cells', *J Am Chem Soc,* 127, 1624–1625.

Oishi M, Nagasaki Y, Nishiyama N, Itaka K, Takagi M, Shimamoto A, Furuichi Y and Kataoka K (2007), 'Enhanced growth inhibition of hepatic multicellular tumor spheroids by lactosylated poly(ethylene glycol)-siRNA conjugate formulated in PEGylated polyplexes', *Chem Med Chem,* 2, 1290–1297.

Oster C G, Kim N, Grode L, Barbu-Tudoran L, Schaper A K, Kaufmann S H and Kissel T (2005), 'Cationic microparticles consisting of poly(lactide co- glycolide) and polyethylenimine as carriers systems for parental DNA vaccination', *J Control Rel,* 104, 359–377.

Oyen W J, Boerman O C, Storm G, van Bloois L, Koenders E B, Crommelin D J, van der Meer J W and Corstens F H (1996), 'Labeled Stealth® liposomes in experimental infection: an alternative to leukocyte scintigraphy?', *Nucl Med Commun*, 17, 742–748.

Pagano R E and Huang L (1975), 'Interaction of phospholipid vesicles with cultured mammalian cells. II. Studies of mechanism', *J Cell Biol*, 67, 49–60.

Page-Clisson M E, Pinto-Alphandary H, Chachaty E, Couvreur P and Andremont A (1998), 'Drug targeting by polyalkylcyanoacrylate nanoparticles is not efficient against persistent *Salmonella*', *Pharm Res*, 15, 544–549.

Pakunlu RI, Wang Y, Saad M, Khandare J J, Starovoytov V and Minko T (2006), '*In vitro* and *in vivo* intracellular liposomal delivery of antisense oligonucleotides and anticancer drug', *J Control Rel*, 114, 153–162.

Pardridge WM (2010a), 'Preparation of Trojan horse liposomes (THLs) for gene transfer across the blood-brain barrier', *Cold Spring Harb Protoc*, 2010, pdb.prot5407.
Pardridge WM (2010b), 'Biopharmaceutical drug targeting to the brain', *J Drug Target*, 18, 157–167.
Patil M L, Zhang M, Betigeri S, Taratula O, He H and Minko T (2008), 'Surface-modified and internally cationic polyamidoamine dendrimers for efficient siRNA delivery', *Bioconjug Chem*, 19, 1396–1403.
Pecheur E I, Hoekstra D, Sainte-Marie J, Maurin L, Bienvenüe A and Philippot J R (1997), 'Membrane anchorage brings about fusogenic properties in a short synthetic peptide', *Biochemistry*, 36, 3773–3781.
Peek L J, Middaugh C R and Berkland C (2008), 'Nanotechnology in vaccine delivery', *Adv Drug Deliv Rev*, 60, 915–928.
Pinto-Alphandary H, Balland O, Laurent M, Andremont A, Puisieux F and Couvreur P (1994), 'Intracellular visualization of ampicillin-loaded nanoparticles in peritoneal macrophages infected *in vitro* with *Salmonella typhimurium*', *Pharm Res*, 11, 38–46.
Pinto-Alphandary H, Andremont A and Couvreur P (2000), 'Targeted delivery of antibiotics using liposomes and nanoparticles: research and applications', *Int J Antimicrob Agents* 13, 155–168.
Rait A S, Pirollo K F, Ulick D, Cullen K and Chang E H (2003), 'HER-2-targeted antisense oligonucleotide results in sensitization of head and neck cancer cells to chemotherapeutic agents', *Ann N Y Acad Sci*, 1002, 78–89.
Raviña M, Paolicelli P, Seijo B and Sanchez A (2010), 'Knocking down gene expression with dendritic vectors', *Mini Rev Med Chem*, 10, 73–86.
Rejman J, Oberle V, Zuhorn I S and Hoekstra D (2004), 'Size-dependent internalization of particles via the pathways of clathrin- and caveolae-mediated endocytosis', *Biochem J*, 377(Pt 1), 159–169.
Remaut K, Lucas B, Braeckmans K, Demeester J and De Smedt S C. (2007), 'Pegylation of liposomes favours the endosomal degradation of the delivered phosphodiester oligonucleotides', *J Control Rel*, 117, 256–266.
Rodrigues J M Jr, Croft S L, Fessi H, Bories C and Devissaguet J P (1994), 'The activity and ultrastructural localization of primaquine-loaded poly (d,l-lactide) nanoparticles in *Leishmania donovani* infected mice', *Trop Med Parasitol*, 45, 223–228.
Ropert C, Mishal Z Jr, Rodrigues J M, Malvy C and Couvreur P (1996), 'Retrovirus budding may constitute a port of entry for drug carriers', *Biochim Biophys Acta*, 1310, 53–59.
Roques C, Salmon A, Fiszman M Y, Fattal E and Fromes Y (2007), 'Intrapericardial administration of novel DNA formulations based on thermosensitive Poloxamer 407 gel', *Int J Pharm*, 331, 220–223.
Ruozi B, Battini R, Tosi G, Forni F, Vandelli M A (2005), 'Liposome-oligonucleotides interaction for in vitro uptake by COS I and HaCaT cells', J Drug Target, 13, 295–304.
Ryman B E and Tyrrell D A (1980), 'Liposomes - bags of potential', *Essays Biochem*, 16, 49–98.
Santel A, Aleku M, Keil O, Endruschat J, Esche V, Durieux B, Löffler K, Fechtner M, Röhl T, Fisch G, Dames S, Arnold W, Giese K, Klippel A and Kaufmann J (2006), 'RNA interference in the mouse vascular endothelium by systemic administration of siRNA-lipoplexes for cancer therapy', *Gene Ther*, 13, 1360–1370.
Schiffelers R M, Ansari A, Xu J, Zhou Q, Tang Q, Storm G Molema G, Lu P Y, Scaria P V and Woodle M C (2004), 'Cancer siRNA therapy by tumor selective delivery with ligand-targeted sterically stabilized nanoparticle', *Nucleic Acids Res*, 32, e149.
Schwab G, Chavany C, Duroux I, Goubin G, Lebeau J, Helene C and Saison-Behmoaras T (1994), 'Antisense oligonucleotides adsorbed to polyalkylcyanoacrylate nanoparticles specifically inhibit mutated Ha-ras-mediated cell proliferation and tumorigenicity in nude mice', *Proc Natl Acad Sci USA*, 91, 10460–10464.
Schroit A J and Fidler I J (1982), 'Effects of liposome structure and lipid composition on the activation of the tumoricidal properties of macrophages by liposomes containing muramyl dipeptide', *Cancer Res*, 42, 161–167.
Senior J and Gregoriadis G (1982), 'Stability of small unilamellar liposomes in serum and clearance from the circulation: the effect of the phospholipid and cholesterol components', *Life Sci*, 30, 2123–2136.

Seyler I, Appel M, Devissaguet J Ph, Legrand P and Barratt G (1997), 'Modulation of nitric oxide production in RAW 264.7 cells by Transforming Growth Factor-β and Interleukin-10', *J Leukoc Biol*, 62, 374–380.

Seyler I, Appel M, Devissaguet J Ph, Legrand P and Barratt G (1999), 'Macrophage activation by a lipophilic derivative of muramyldipeptide within nanocapsules: investigation of the mechanism of drug delivery', *J Nanoparticle Res*, 1, 91–97.

Shen X C, Zhou J, Liu X, Wu J, Qu F, Zhang Z L Pang D W, Quélever G, Zhang C C and Peng L (2007), 'Importance of size-to-charge ratio in construction of stable and uniform nanoscale RNA/dendrimer complexes', *Org Biomol Chem,* 5, 3674–3681.

Singh M, Ugozzoli M, Briones M, Kazzaz J, Soenawan E and O'Hagan D T (2003), 'The effect of CTAB concentration in cationic PLG microparticles on DNA adsorption and *in vivo* performance', *Pharm Res,* 20, 247–251.

Swanson J A and Watts C (1995), 'Macropinocytosis', *Trends Cell Biol*, 5, 424–428.

Szebeni J (1998), 'The interaction of liposomes with the complement system', *Crit Rev Ther Drug Carrier Syst*, 15, 57–88.

Szoka FC Jr (1990), 'The future of liposomal drug delivery', *Biotechnol Appl Biochem*, 12, 496–500.

Tabata Y and Ikada Y (1988), 'Effect of the size and surface charge of polymer microspheres on their phagocytosis by macrophage', *Biomaterials*, 9, 356–362.

Tondelli L, Ricca A, Laus M, Lelli M and Citro G (1998), 'Highly efficient cellular uptake of c-myb antisense oligonucleotides through specifically designed polymeric nanospheres', *Nucleic Acids Res,* 26, 5425–5431.

Toub N, Angiari C, Eboue D, Fattal E, Tenu J P, Le Doan T and Couvreur P (2005), 'Cellular fate of oligonucleotides when delivered by nanocapsules of poly(isobutylcyanoacrylate)', *J Control Rel,* 106, 209–213.

Toub N, Bertrand J R, Malvy C, Fattal E and Couvreur P (2006), 'Antisense oligonucleotide nanocapsules efficiently inhibit EWS-Fli1 expression in a Ewing's sarcoma model', *Oligonucleotides,* 16, 158–168.

Turánek J, Ledvina M, Kasná A, Vacek A, Hríbalova V, Krejcí J and Miller A D (2006), 'Liposomal preparations of muramyl glycopeptides as immunomodulators and adjuvants', *Vaccine*, 24 Suppl 2:S90–S91

Urban-Klein B, Werth S, Abuharbeid S, Czubayko F and Aigner A (2005), 'RNAi-mediated gene-targeting through systemic application of polyethylenimine (PEI)-complexed siRNA *in vivo*', *Gene Ther,* 12, 461–466.

van Broekhoven C L, Parish C R, Demangel C, Britton W J and Altin J G (2004), 'Targeting dendritic cells with antigen-containing liposomes: a highly effective procedure for induction of antitumor immunity and for tumor immunotherapy', *Cancer Res*, 64, 4357–4365.

Vauthier C and Bouchemal K (2009), 'Methods for the preparation and manufacture of polymeric nanoparticles', *Pharm Res*, 26, 1025–1058.

Vonarbourg A, Passirani C, Saulnier P, Simard P, Leroux J C and Benoit J P (2006), 'Evaluation of pegylated lipid nanocapsules versus complement system activation and macrophage uptake', *J Biomed Mater Res A*, 78, 620–628.

Vyas, S P, Katare Y K, Mishra V and Sihorkar V (2000), 'Ligand directed macrophage targeting of amphotericin B loaded liposomes', *Int J Pharm*, 210, 1–14.

Waite C L, Sparks S M, Uhrich K E and Roth C M (2009), 'Acetylation of PAMAM dendrimers for cellular delivery of siRNA', *BMC Biotechnol*, 9, 38.

Weyermann J, Lochmann D, Georgens C and Zimmer A (2005), 'Albumin-protamine- oligonucleotide-nanoparticles as a new antisense delivery system. Part 2: cellular uptake and effect', *Eur J Pharm Biopharm,* 59, 431–438.

Wissing S A, Kayser O and Müller R H (2004), 'Solid lipid nanoparticles for parenteral drug delivery', *Adv Drug Deliv Rev*, 56, 1257–1272.

Woodle M C (1998), 'Controlling liposome blood clearance by surface-grafted polymers', *Adv Drug Deliv Rev,* 32, 139–152.

Xu Y and Szoka FC Jr (1996), 'Mechanism of DNA release from cationic liposome/DNA complexes used in cell transfection', *Biochemistry*, 35, 5616–5623.

Yadava P, Roura D and Hughes J A (2007), 'Evaluation of two cationic delivery systems for siRNA', *Oligonucleotides*, 17, 213–222.

Yang X, Koh CG, Liu S, Pan X, Santhanam R, Yu B, Peng Y, Pang J, Golan S, Talmon Y, Jin Y, Muthusamy N, Byrd J C, Chan K K, Lee L J, Marcucci G and Lee R J (2009), 'Transferrin receptor-targeted lipid nanoparticles for delivery of an antisense oligodeoxyribonucleotide against Bcl-2', *Mol Pharm*, 6, 221–230.

Yessine M A, Meier C, Petereit H U and Leroux J C (2006), 'On the role of methacrylic acid copolymers in the intracellular delivery of antisense oligonucleotides', *Eur J Pharm Biopharm*, 63, 1–10.

Yoo H, Sazani P and Juliano R L (1999), 'PAMAM dendrimers as delivery agents for antisense oligonucleotides', *Pharm Res*, 16, 1799–1804

Youssef M, Fattal E, Alonso M J, Roblot-Treupel L, Sauzières J, Tancrède C, Omnès A, Couvreur P and Andremont A (1988), 'Effectiveness of nanoparticle-bound ampicillin in the treatment of *Listeria monocytogenes* infection in athymic nude mice', *Antimicrob Agents Chemother*, 32, 1204–1207.

Yu B, Zhao X, Lee L J and Lee R J (2009), 'Targeted delivery systems for oligonucleotide therapeutics', *AAPS J*, 11, 195–203.

Yuan X, Li L, Rathinavelu A, Hao J, Narasimhan M, He M, Heitlage V, Tam L, Viqar S, Salehi M (2006), 'SiRNA drug delivery by biodegradable polymeric nanoparticles', *J Nanosci Nanotechnol* 6, 2821–2828.

Zhou J, Wu J, Hafdi N, Behr JP, Erbacher P and Peng L (2006), 'PAMAM dendrimers for efficient siRNA delivery and potent gene silencing', *Chem Commun (Camb),* 22, 2362–2364.

Zobel H P, Junghans M, Maienschein V, Werner D, Gilbert M, Zimmermann H, Noe C, Kreuter J and Zimmer A (2000), 'Enhanced antisense efficacy of oligonucleotides adsorbed to monomethylaminoethylmethacrylate methylmethacrylate copolymer nanoparticles', *Eur J Pharm Biopharm*, 49, 203–210.

The Potential and Current Progress of Internalizing Molecules in Targeted Drug Delivery

Jiehua Zhou and John J. Rossi

Abstract Safe and efficient drug delivery in the treatment of cancer and infectious diseases constitute a major challenge. Development of targeted intracellular delivery approaches to specific cell populations or tissues for different therapeutics is highly desirable to maximize potency and minimize toxicity. Extensive efforts to improve drug safety and effectiveness have resulted in numerous target-specific delivery strategies. Currently, various internalizing molecules, including antibodies, proteins, peptides, folate, carbohydrates, aptamers and other ligands, have been successfully adopted as active recognition moieties to target a distinct disease or tissue in a cell-type-specific manner. The use of direct covalent conjugation or non-covalent assembly of targeting ligands with the drug or delivery vehicle can be used to get drugs to their intended targets, thereby reducing the dosing requirements as well as minimizing unwanted toxicities. In this article we focus mainly on the potential and current progress of internalizing molecules in targeted drug delivery.

Keywords Cell-type specific delivery • Internalizing molecules • Targeted drug delivery • Targeted therapy

Abbreviations

AML	acute myeloid leukemia
Antp	antennapedia
Apo A-I	apolipoprotein A-I

J. Zhou
Division of Molecular and Cellular Biology, Beckman Research Institute of City of Hope, City of Hope, 1500 East Duarte Rd, Duarte, CA 91010, USA
e-mail: Jzhou@coh.org

J.J. Rossi (✉)
Irell and Manella Graduate School of Biological Sciences, Beckman Research Institute of City of Hope, City of Hope, 1500 East Duarte Rd, Duarte, CA 91010, USA
e-mail: Jrossi@coh.org

A. Prokop (ed.), *Intracellular Delivery: Fundamentals and Applications*, Fundamental Biomedical Technologies 5, DOI 10.1007/978-94-007-1248-5_4,

Ara-C	1-β-D-arabinofuranosylcytosine
AZT	azidothymidine
CCP	cell-penetrating peptide
CD4	cluster of differentiation 4
CD7	cluster of differentiation 7
CPT	camptothecin
DRBD	double stranded RNA-binding domain
CVB3	coxsackievirus B3
d4T	stavudine
Dox	doxorubicin
Dtxl	docetaxel
ECM	extracellular matrix
EFV	efavirenz
EGF	epidermal growth factor
EGFR	epidermal growth factor receptor
EPR	enhanced permeability and retention
HDL	high-density lipoprotein
GnRH	gonadotropin-releasing hormone
HA	hyaluronic acid
HBV	hepatitis B virus
HCV	hepatitis C virus
HER-2	human epidermal growth factor receptor-2
HIV gp120	HIV glycoprotein 120
LHRH	luteinizing-hormone-releasing hormone
LN	lymph nodes
mAb	monoclonal antibody
MAP	mitogen-activated protein
MR	magnetic resonance
MTX	methotrexate
MUC1	mucin protein
NCL	nucleolin
NP	nanoparticles
NR	nanorod
NRP1	neuropilin-1
ON	oligonucleotide
PAMAM	polyamidoamine
PEI	polyethylenimine
PHA	polyhydroxyalkanoates
PL	phospholipid
PLL	poly-L-lysine
PPI	poly(propyleneimine)
pRNA	packaging RNA
PSMA	prostate-specific membrane antigen
PTD	peptide transduction domain
PTK7	protein tyrosine kinase 7
QD	quantum dot

RNAi	RNA interference
scFv	single-chain Fv antibody
SELEX	systematic evolution of ligands by exponential enrichment
siRNA	small interfering RNA
SR-BI	scavenger receptor BI
sTRAIL	tumor necrosis factor-related apoptosis-inducing ligand
SWNT	single-wall carbon nanotube
TAT	trans-activating transcriptional activator
3TC	lamuivudine
Tf	transferrin
TfR	transferrin receptor
TN-C	tenasin-C
TNPO3	Transportin-3

1 Introduction

Since the "magic bullet" dream was first postulated by Dr. Paul Ehrlich in 1906 (Strebhardt and Ullrich 2008), "targeted therapy" has become a long-standing goal for human disease treatment but still remains a major challenge for clinical applications (Zhukov and Tjulandin 2008; Markman 2008; Katzel et al. 2009). Most therapeutic agents currently in use, such as conventional chemotherapy (Joensuu 2008), radiotherapy (Harrison et al. 2002), immunotherapy (Shapira et al. 2010) or gene therapy (Whitehead et al. 2009), generally are not functionalized to selectively target the site of disease. When these therapeutics agents are systemically administrated, they nonspecifically distribute throughout the body, thereby significantly reducing the therapeutic efficacy and leading to harmful side-effects associated with distribution to non-targeted sites (Langer 1998).

With the intent and enthusiasm for developing targeted therapeutic drug delivery strategies, many efforts have been made to develop targeted drug delivery strategies capable of selectively transporting drug to a specific site of disease (Yu et al. 2010; Zhou and Rossi 2009; Kim et al. 2009c; Levy-Nissenbaum et al. 2008; Allen 2002). The strategy mainly relies on specific interaction between the targeting ligand and its receptors expressed on the diseased cells or tissues. Such ligand-directed recognization events consequently increase the local concentration of the drug in the targeted cells or tissues. Furthermore, after the ligand-receptor binding, the cellular receptors are readily internalized and rapidly re-expressed on the cell surface to allow repeated targeting and internalization. Depending on the types of therapeutic agents or drug formulation, it is worth noting that internalization may be dispensable in some therapies (Allen 2002). For example, for antibody-directed enzyme-prodrug anticancer therapy (Senter and Springer 2001), the enzyme-prodrug is first selectively accumulated on the cell surface *via* antibody-receptor interaction; subsequently, it is activated into an anticancer drug by the enzyme that only functionalizes at the cell surface. Additionally, in the radioimmunotherapy (Ercan and Caglar 2000; Goldenberg 2002), antibody-conjugated beta- or alpha- radionuclides can directly functionalize at the cell surface once the conjugates are selectively localized on the target cells or tissues. However, internalization might still be beneficial for some alpha-emitting radionuclides with a very short range of alpha-particle emission and lack of penetrability. In this regard, an internalizing ligand therefore might more be efficient and helpful than non-internalizing ligands. Indeed, a specific internalizing molecule with high specificity and affinity to a cellular surface receptor is therefore a major factor in establishing a targeted drug delivery system.

Over the past few decades, a wide variety of internalizing molecules such as antibodies, proteins, peptides, folate, carbohydrates, aptamers and other small molecule ligands have been successfully adapted for the targeted delivery of active drug substances both *in vitro* as well as *in vivo* (Russ and Wagner 2007; Ciavarella et al. 2010; Yan and Levy 2009; Higuchi et al. 2010). Recent investigations have described targeted delivery of various therapeutic agents (e.g. anticancer, antifungal, antiviral agents, antibiotics, protein, peptide, genes, oligonucleotides and small interfering RNAs) using different strategies. Through precisely engineering the internalizing molecules with the drug or delivery vehicle, the recognition and internalization of the therapeutic compounds by the target tissue can be dramatically improved. In particular, the advent of nanotechnology has greatly accelerated the development of drug delivery, providing a large variety of nanocarriers for disease therapy including liposomes, polymers, carbon nanotubes and quantum dots etc. (Mok and Park 2009; Gullotti and Yeo 2009; Kim et al. 2009a; Wiradharma et al. 2009; Liu et al. 2007b; Zrazhevskiy and Gao 2009). Meanwhile, powerful multifunctional nanomedicines that combine several desirable functions such as therapeutics, targeting and imaging in one nanoscale carrier, have been developed to significantly enhance drug efficacy (Torchilin 2006; Zrazhevskiy and Gao 2009; Gao et al. 2010; Muthu and Wilson 2010; Hart 2010). In this review, we discuss recent advances in targeting strategies for tumors and infectious diseases, with a particular emphasis on cell internalizing molecules as a targeting strategy.

2 Targeting Strategies for Targeted Delivery

The key for targeted therapy is to accurately identify molecular targets that distinguish diseased cells from healthy ones. Ideally, a targeted drug delivery should enable selective accumulation of the drug in the specific target cells with minimal adverse side-effects. Generally, targeted delivery can be realized through two strategies (Torchilin 2010): **(1)** passive targeting and **(2)** active targeting.

2.1 Passive Targeting

In the first strategy, macromolecules and nanocarriers with sizes below 400 nm in diameter can travel through the bloodstream and accumulate in tumor *via* passive leakage, thereby increasing the concentration of drugs in tumors and enhancing the therapeutic index (Yuan et al. 1995; Moghimi et al. 2001). This passive mechanism is called the enhanced permeability and retention (EPR) effect (Matsumura and Maeda 1986; Greish 2007). Since passive targeting relies on a size-flow-organ/tissue filtration that is generally limited to tumors and lymph nodes (LNs), drug particles/carriers can theoretically be designed and engineered with appropriate sizes (e.g. 100–400 nm) and surfaces (e.g. PEGylated surface) (Maeda 2010). Currently, many studies have taken advantage of the EPR effect to achieve passive targeting of drug nanocarriers to most human tumors. And some of them have been approved for clinical use such as Doxil (Doxorubicin encapsulated by PEGylated liposome) (Gabizon 2001). In the treatment of HIV, LNs are important sites for targeting HIV-1 replication and they have been exploited as a promising passive targeting site (Gupta and Jain 2010; Gunaseelan et al. 2010).

2.2 *Active Targeting*

In the second strategy, an actively recognized moiety specific to the target sites of interest is an essential component. These actively recognized molecules can be precisely decorated on the therapeutic agents or their delivery vehicles, functioning as cell or tissue-specific homing agents (Peer et al. 2007; Gullotti and Yeo 2009). Typically, by utilizing biologically specific interactions such as antigen-antibody binding or ligand-receptor interactions, these targeting molecules facilitate the cellular uptake of therapeutic agents *via* receptor-mediated endocytosis or cellular membrane permeation, thereby increasing the local concentration of the drug in the targeted cells or tissues, thereby improving the therapeutic efficacy at lower doses.

2.3 *Combinatorial Targeting*

Despite numerous successful examples using either one of the above strategies, a combination of passive and active targeting also has been exploited to provide the most efficient targeted delivery system (Torchilin 2010). In this regard, a combinatorial targeting results in twice the enrichment at the cellular and tissue/organ levels (Akbulut et al. 2009). For example, for a targeted nanocarrier system, an appropriate nano-scale size can allow preferential accumulation in the tumor tissue/organ in the passive mode. Once the nanocarrier is concentrated at the tumor site, a cell type-specific affinity internalizing molecule such as an antibody, peptide, folate or aptamer engineered on the nanocarrier surface would further facilitate selective internalization into the targeted tumor cells. In comparison with traditional small molecule drugs or non-targeted nanocarriers, targeted nanocarriers that combine the above two desirable properties could not only prolong circulation time and improve drug biodistribution *via* passive targeting, but would also provide higher specificity *via* the active targeting component (Park et al. 2009).

3 Specific Internalizing Molecules for Targeted Delivery

Currently, the development of the internalizing molecules specifically targeting membrane receptors and their adaptation for active drug targeting has drawn attention in the field of targeted therapy. In the following section internalizing molecule-mediated active targeting will be discussed in greater detail.

3.1 Antibody-Mediated Drug Delivery

Antibodies (Abs, also known as immunoglobulins) that are used by the immune system to identify and neutralize foreign objects can specifically identify and bind their unique antigens. Due to the specificity of antigen-antibody binding, different internalizing antibodies and their genetically engineered fragments have been widely investigated as active targeting molecules (Wu and Senter 2005; Schrama et al. 2006). These include anti-CD33 monoclonal antibodies (mAb) (Simard and Leroux 2010), anti-HER2 mAb (anti-CD90 mAb) (Park et al. 2002; Chiu et al. 2004a), anti-CD7 mAb (Kumar et al. 2008), anti-HIV envelop gp120 mAb (Song et al. 2005), anti-β7 integrin mAb (Peer et al. 2008), anti-JL1 mAb (Suh et al. 2001), anti-EGF receptor Ab (Wu et al. 2004; Mamot et al. 2005), anti-CD31 (Li et al. 2000), anti-GAD Ab (Jeong et al. 2005), anti-D4.2 mAb (Ho et al. 1987), anti-H-$2K^k$ mAb (Connor and Huang 1986), anti-HLA-DR Ab (Gagne et al. 2002), anti-DC-SIGN mAb (Tacken et al. 2005; Dakappagari et al. 2006) and so on. Four representative antibodies targeting cancer cells or HIV infected cells are described below.

3.1.1 Anti-CD33 Antibody

The CD33 receptor is a 67 kDa glycoprotein expressed on the surface of leukemia cells in most patients with acute myeloid leukemia (AML) (Griffin et al. 1984; Scheinberg et al. 1989). Previous studies have demonstrated that CD33 can be rapidly internalized into leukemia cells after antibody binding (Simard and Leroux 2009). Therefore, anti-CD33 mAbs have been conjugated with cytotoxic agents or nanocarriers for targeted delivery. For example, Mylotag (Gemtuzumab Ozogamicin), a recombinant, humanized anti-CD33 antibody which was approved by the US Food and Drug Administration in 2000 as a single-agent therapy for CD33-positive AML, was linked with an anticancer agent (calicheamicin) and the resulting conjugate showed excellent clinical promise (Larson et al. 2005). Recently, an anti-CD33 mAb or its Fab' fragment were also chemically coupled to PEGylated liposomes (LPs) (Simard and Leroux 2010). Such a liposomal formulation has been demonstrated to specifically recognize the CD33 cell surface antigen and promote the efficient intracellular delivery of 1-β-D-arabinofuranosylcytosine (ara-C, anti-AML agent) to human myeloid leukemia cells.

3.1.2 Anti-HER2 Antibody

Another popular antibody for tumor targeting is the anti-HER-2 (human epidermal growth factor receptor-2) antibody (Mamot et al. 2005; Wu et al. 2004). Since HER-2 is highly-expressed on the surface of many human pancreatic cancer cell lines, it can be used as a therapeutic target. Therapeutic antibodies such as Trastuzumab and Herceptin have been routinely applied in the clinical treatment of breast cancer and leukemia (Chiu et al. 2004a). A number of reports have shown that the anti-HER2 antibody can be decorated on a nano-particle surface *via* avidin-biotin or covalently conjugated to polymers (e.g. polyethylenimine PEI) through different crosslinkers. Moreover, anti-HER2 Ab-mediated delivery of oligonucleotides encased in lipid nanoparticles was also shown to be capable of targeting mammary carcinoma cells (Kirpotin et al. 2006). These modified nanocarriers were effectively endocytosed by HER2-overexpressing cells. In addition, a multifunctional anti-HER2 Ab conjugated magnetic gold nano-shell particle showed the potential for targeted MR (magnetic resonance) photo-thermal therapy and tumor imaging (Kim et al. 2006).

3.1.3 Anti-CD7 Antibody

CD7 (Cluster of Differentiation 7), a human transmembrane protein, is found on thymocytes and the majority of human T cells. It has been documented that CD7 is rapidly internalized after antibody binding. Several research groups have exploited an anti-CD7 mAb and its recombinant single-chain Fv (scFvCD7) fragment for targeted delivery. For example, a single-chain Fv fragment was genetically linked to a truncated Pseudomonas exotoxin A fragment, thereby conferring restricted specificity for CD7 positive cells (Peipp et al. 2002). The recombinant immunotoxin promoted apoptosis in two CD7-positive cell lines and provided a potent inducer of apoptosis in acute leukemic T cells. In a similar manner, scFvCD7 was genetically fused with sTRAIL (tumor necrosis factor-related apoptosis-inducing ligand) (Bremer et al. 2005). As expected, treatment with scFvCD7 : sTRAIL specifically induced potent CD7-restricted apoptosis in a series of malignant T-cell lines, with low toxicity for normal human blood and endothelial cells.

In addition to the targeted delivery of a toxin, the scFvCD7 Ab was recently used for siRNA delivery *in vitro* and *in vivo*. Kumar et al. conjugated a Cys-modified scFvCD7 with an oligo-9-arginine peptide to obtain a cell type specific siRNA delivery system (scFvCD7-9R) (Kumar et al. 2008). A combination of siRNAs against the cellular CCR5 co-receptor and two conserved HIV-1 genes (*vif* and *tat*) was loaded into scFvCD7-9R system, subsequently was systemically administered to HIV-1 infected animal model. The weekly injection resulted in suppression of HIV-1 infection and protection of CD4+ T cell depletion in humanized mice.

3.1.4 Anti-HIV gp120 Antibody

The interaction of the HIV envelope glycoprotein gp120 with the cellular CD4 receptor is a crucial step in the entry of HIV into T-cells (Ugolini et al. 1999; Sattentau and Moore 1993). HIV gp120 is exposed on the surface of virus particles and the plasma membrane of HIV-1 infected cells. Therefore, it represents an excellent marker to distinguish HIV-1 infected host cells from non-infected cells (Kwong et al. 1998). F105, an anti-HIV gp120 monoclonal antibody, has been demonstrated to bind with high affinity to its ligand gp120 expressed on the surface of a wide range of HIV-1 laboratory strains and primary isolates (Posner et al. 1991). Importantly, its selective binding to HIV-1 infected cells triggers rapid cellular internalization. Thus, F105 has the potential to be an excellent internalizing molecule for selective drug delivery to HIV-1 infected cells (Clayton et al. 2007). Song et al. have successfully applied a heavy chain fragment of F105 to increase receptor-specific uptake to cells expressing the HIV envelope protein (Song et al. 2005). In this system, the F105 antibody fragment – protamine fusion protein was able to specifically bind to and deliver siRNAs to HIV-infected primary T cells or HIV envelope-expressing cells. Moreover, their results demonstrated that the siRNA targeting the HIV-1 *gag* capsid gene delivered by such a fusion protein inhibited HIV-1 replication only in cells expressing the HIV-1 envelope. Recently, PEGylated liposomes coated with a F105 Fab' fragment have been developed and evaluated for targeted delivery of a novel HIV-1 protease inhibitor PI1 (Clayton et al. 2009). The anti-HIV immunoliposomes were selectively taken up by HIV-1 infected cells. When compared with free drug or non-targeted drug, the targeted PI1 delivery showed a greater inhibitory effect on viral replication.

3.2 Protein-Mediated Drug Delivery

3.2.1 Transferrin and Lactoferrin

Transferrin (Tf), an 80 kDa glycoprotein that transports iron into cells, is one of most popular tumor targeting ligands, because its receptors (TfRs) are over-expressed in many types of cancer cells and TfR expression is correlated with the proliferation and malignancy of the tumor cells (Daniels et al. 2006). After interaction of Tf-TfR, Tf is internalized into cells through an endocytosis pathway (Qian et al. 2002). To date, numerous investigations have demonstrated that Tf-conjugated nanocarriers loaded with various drugs (e.g. anticancer agents, oligonucleotides, therapeutic genes, siRNAs) could facilitate specific cellular uptake and enhance the therapeutic efficacy of the drugs in tumor cells (Heidel et al. 2007; Davis 2009). One group developed a Tf-conjugated cyclodextrin-containing polycation delivery system for siRNAs, which resulted in 50% greater gene silencing in the tumor versus a non-targeted delivery system in an immunodeficient mouse model (Bartlett et al. 2007). Most recently, the same group conducted the first in-human phase I clinical trial using a Tf-targeted nanoparticle delivery system and their work provided direct evidence of intercellular delivery of the siRNA which triggered the RNA interference mechanism (Davis et al. 2010).

Since the brain capillaries are specifically known to express TfR, Tf-conjugated nanocarriers were also successfully applied for brain targeting (Chang et al. 2009). For example, Tf chemically conjugated to polyamidoamine (PAMAM) dendrimers resulted in an increase in the brain uptake of the therapeutic DNA (Huang et al. 2007). In a similar study, Tf-coated PEG nanoparticles delivered an antiretroviral agent (azidothymidine, AZT) across the blood-brain barrier to the central nervous system (Mishra et al. 2006). Additionally, another glycoprotein lactoferrin, a globular iron-binding protein, also could function as a lung and brain targeting molecule (Elfinger et al. 2007). It is known that lactoferrin receptors are predominantly expressed in brain capillaries and bronchial epithelial cells, thereby suggesting a new target for brain delivery as well (Huang et al. 2008a).

3.2.2 Epidermal Growth Factor (EGF)

The epidermal growth factor receptor (EGFR), the cell-surface receptor for members of the EGF-family of extracellular protein ligands, is over-expressed in the majority of human cancers (Muslimov 2008). EGF acts by high affinity binding to EGFR on the cell surface thereby stimulating the intrinsic protein-tyrosine kinase activity of the receptor (Agarwal et al. 2008). Through the formation of dimers EGF is endocytosed into the cells after receptor-binding. Moreover, because EGF only has 53 amino acids and a molecular weight of ~6 kDa, it has been widely employed as an internalizing ligand for tumor targeting. A number of such examples have been reported in the literatures (Lee et al. 2003; Lee and Park 2002). For example, biotin-modified EGF molecules were installed onto a Streptavidin-modified PEI complexed plasmid DNA *via* the interaction of biotin-Streptavidin (Lee et al. 2002). The resulting EGF-targeted PEI delivery system resulted in a significantly higher transfection efficiency than non-targeted system in a human epidermal carcinoma cell line. Recently, EGF was attached to single-wall carbon nanotubes (SWNTs) loaded with cisplatin or quantum dots (QDs) and used for targeted therapy and tumor imaging (Bhirde et al. 2009).

3.2.3 Apolipoprotein A-I

Apolipoprotein A-I (apo A-I) is a protein component of the high-density lipoprotein (HDL) (von Eckardstein et al. 2001). It has been demonstrated that apo A-I can predominantly be internalized to the liver *via* the cell-surface receptor scavenger receptor BI (SR-BI) (Kim et al. 2007). As a typical serum protein, apo A-I functions as a beneficial factor to decrease cholesterol levels and therefore should not trigger immunological side effects in clinical applications. Thus, apo A-I can be used as an internalizing moiety for liver targeting. A proof-of-concept study has shown the potential of apo A-I for the systemic delivery of nucleic acids to the liver using real-time *in vivo* imaging (Kim et al. 2007). Apo A-I was formulated onto the surface of the lipid bilayer of a representative cationic liposome and facilitated the hepatocyte-specific siRNA delivery *via* receptor-mediated endocytosis. The Apo A-I-coating liposome carrying an anti-hepatitis B virus (HBV) siRNA showed a prolonged inhibition of the target protein expression at low doses in a mouse model. A subsequent study further evaluated the efficacy of an apo A-I – mediated liver-specific delivery of an siRNA against hepatitis C virus (HCV) in a transient HCV model (Kim et al. 2009b). The results of this study showed Apo-liposome nanoparticles systemically delivered an siRNA into mouse hepatocytes expressing HCV resulting in siRNA mediated inhibition of the targeted HCV protein expression.

3.3 Peptide-Mediated Drug Delivery

3.3.1 RGD Peptide

Peptides based on the three-amino-acid sequence arginine-glycine-aspartate, known as RGD peptides have been used extensively as tumor cell-recognizable ligands in tumor diagnostics and therapeutics (Ruoslahti 2003). The RGD sequence, originally identified as a cell binding site in the extracellular matrix (ECM) protein (fibronectin and vitronectin), can bind to the integrin receptor $\alpha_v\beta_3$ with high affinity (Cheresh 1987). Since the integrins are specifically over-expressed at the surface of tumor cells and angiogenic endothelial cells at the tumor site, RDG-mediated drug delivery generally leads to high levels of accumulation in tumor tissues compared with free drug or non-targeted drug delivery systems. Both linear RGD and cyclic RGD have been popularly applied for targeted delivery of traditional drugs, genes and polymers. For example, a dimeric RGD peptide was conjugated to an oligonucleotide (ON), thereby increasing cellular uptake and the biological function of the ON (Alam et al. 2010; Alam et al. 2008). Additionally, Dai et al. chemically conjugated a cyclic RGD peptide to the terminal group in a phospholipid – single-walled carbon nanotube (PL-SWNT), which resulted in enhanced and specific delivery of doxorubicin (Dox) to RGD positive U87 cancer cells (Liu et al. 2007a).

Recently, Ruoslahti and colleagues identified a cyclic peptide iRGD (CRGDKGPDC and its related variants), which can target tumor cells by binding α_v integrins that are specifically expressed in the tumor vessel endothelium (Sugahara et al. 2009). They combined the tumor-homing RGD sequence with another peptide ligand (the CendR motif) for neuropilin-1 (NRP1, a transmembrane receptor) that mediates penetration into tissue and cells. The resulting peptide-mediated delivery system (iRGD) was first recruited to tumor tissue through the

RGD-integrin interaction and subsequently was proteolytically processed to generate a CendR motif with cell-penetrating activities, further facilitating the penetration of drug into the tumor. Therefore, drugs chemically conjugated to the iRGD could be selectively transported deep into the tumor parenchyma, and significantly improve the sensitively of tumor-imaging agents and the activity of an antitumor drug. Recent studies further demonstrated that co-administration of drugs and the iRGD peptide resulted in penetration of extravascular tumor tissue (Sugahara et al. 2010). They demonstrated in murine tumor models that systemic injection with iRGD can improve the therapeutic index of various therapeutic agents, such as a small molecule drugs (doxorubicin), nanoparticles (nab-paclitaxel and doxorubicin liposomes), and a monoclonal antibody (trastuzumab). Although the mechanism of action still requires further investigation, this system may represent a valuable strategy to enhance the efficacy of anticancer agents and to accelerate clinical translation of modification-intolerable drugs.

3.3.2 LHRH Peptide

Luteinizing-hormone-releasing hormone (LHRH), also known as luliberin and Gonadotropin-releasing hormone (GnRH), are known to be internalized and are good agents for tumor targeting (Dharap et al. 2003, 2005; Kim et al. 2008). Although LHRH receptors are present on the surface of most healthy human cells, they are over-expressed in ovarian, prostate and breast cancer cells. Through direct conjugation, the LHRH peptide or its analogues have been applied for the delivery of anticancer drugs and siRNAs. For example, Dharap et al. have shown that a LHRH-PEG-camptothecin (CPT), a cytotoxic quinoline alkaloid which inhibits the DNA enzyme topoisomerase I, significantly enhanced the drugs cytotoxicity against cancer cells (Dharap et al. 2003). Recently, a LHRH peptide analogue was appended to an siRNA *via* a PEG linker to target ovarian cancer cells (Kim et al. 2008). Once effective cellular internalization of the LHRH-targeted siRNA conjugates takes place the disulfide bond between PEG and siRNA is reductively cleaved to release the siRNA into the cytosol.

3.3.3 Peptide Transduction Domains (PTDs)

PTDs (also called cell-penetrating peptides), are short cationic peptide sequences with a maximum of 30 amino acids that mediate translocation across cell membranes (El-Sayed et al. 2009; Pangburn et al. 2009). Despite the different theories on the mechanism of PTD-mediated uptake, it is now widely accepted that PTDs bind the anionic cell surface through electrostatic interaction and then are rapidly internalized into cells by macropinocytosis, a specialized type of fluid phase endocytosis that all cells appear to perform (Nakase et al. 2004; Wadia et al. 2004). This feature therefore enhances the cellular uptake in a non-cytotoxic manner, making PTDs attractive candidates for intracellular drug delivery. PTDs such as trans-activating transcriptional activator (TAT), polyarginine, antennapedia (Antp), penetratin, transportan, and mitogen-activated protein (MAP) have been shown to deliver a wide variety of therapeutic loads to target cells, and some of the most well characterized PTDs are currently being tested in preclinical and clinical trials (Gump and Dowdy 2007).

The most commonly used PTDs is the HIV TAT peptide, a small polypeptide of 86 amino acids with a cysteine-rich region derived from the HIV TAT protein (Rao et al. 2009). This peptide possesses an arginine-rich sequence and is highly positively charged enable permeation of the cell membrane in a receptor- and transporter-independent fashion (Vives et al. 1997). Strong experimental results have validated the effectiveness of TAT-directed drug delivery. The TAT peptide directly conjugated with various agents, including horseradish peroxidase, β-galactosidase and nucleic acids, or decorated on the surfaces of nanocarriers such as liposomes and PEI, is able to deliver these molecules to various cells and tissues in the mouse, displaying high accumulation in the heart, lung and spleen (Pangburn et al. 2009). Chiu et al. covalently linked a TAT peptide to the 3′-terminus of the antisense strand of an siRNA (Chiu et al. 2004b). The Tat-siRNA conjugate showed rapid internalization into cells and specific siRNA mediated knockdown of the target gene.

Despite of these advances in PTD-mediated siRNA delivery, direct conjugation of a cationic PTD to an anionic siRNA often results in insoluble complexes thereby reducing the delivery efficiency and inducing some cytotoxicities (Turner et al. 2007; Meade and Dowdy 2008). In order to overcome this shortcoming, an efficient "mask" PTD-directed siRNA delivery approach that uses a TAT peptide transduction domain-double stranded RNA-binding domain (PTD-DRBD) fusion protein has been developed (Eguchi and Dowdy 2009, 2010; Eguchi et al. 2009). The DRBD is known to bind siRNA with high avidity, which functions as a "mask" to shield the negative charge of the siRNA allowing the PTD to efficiently deliver the siRNA into cells.

3.4 *Folate-Mediated Drug Delivery*

Folate is also known as folic acid or vitamin B_9, which is able to specifically bind to the folate receptor with nanomolar binding affinity (Weitman et al. 1992; Wang et al. 1996). It has been known that folate receptors are over-expressed in many human cancer cells including ovarian, breast, pharyngeal and liver cancer, while their distribution in normal tissues is minimal. In particular, it was found that expression of folate receptors are elevated on epithelial tumors of various organs such as colon, lung, prostate, ovaries, mammary lands and brain. To date, folate is well-characterized and shows many benefits, including a low-molecular weight (MW 441 Da), good stability, no immunogenicity and well-defined conjugation chemistry. Therefore, it has become one of the most popular targeting moieties for tumor specific drug delivery (Sudimack and Lee 2000). Numerous studies have already demonstrated that attachment of folate to various molecules allows rapid internalization by endocytosis and high accumulation in tumor cells.

Folate has been successfully decorated on polymer, nucleic acids, liposomes, dendrimers, quantum dots and polymeric micelles. For example, a folate-PEG-PEI nanoconjugate delivered plasmid DNA or oligonucleotides into tumor cells in a target specific manner (Cho et al. 2005). Polyhydroxyalkanoates (PHAs) were modified with folate *via* covalent bonds for targeted Dox delivery (Zhang et al. 2010). Similarly, folate-targeted liposomes specifically transported daunorubicin (Ni et al. 2002) and Dox (Goren et al. 2000) into various tumor cells and increased the cytoxic killing of these cells. Folate-linked packaging RNA (pRNA) was shown useful for selectively delivering siRNAs against coxsackievirus B3 (CVB3) thereby effectively inhibiting virus replication (Zhang et al. 2009). In addition, a folate analog, methotrexate (MTX), has been conjugated with magnetic nanoparticles for targeted drug delivery and specific imaging (Duthie 2001).

3.5 *Carbohydrate-Mediated Drug Delivery*

Lectin-carbohydrate interaction is another consideration for active targeting. It is known that the cell surface is rich in carbohydrate moieties attached to both membrane glycolipids and glycoproteins (Boraston et al. 2004; Kannagi et al. 2004). Therefore, the carbohydrate moieties represent a potential recognition ligand for targeted therapeutic delivery.

3.5.1 Galactose and Lactose

The asialoglycoprotein receptors are expressed on the surface of hepatocytes and are able to mediate endocytosis and subsequent internalization into hepatoma cells. The oligosaccharides, such as galactose (a monosaccharide) and lactose (a disaccharide) have been used as targeting molecules for hepatocyte-specific delivery through asialoglycoprotein receptor-mediated endocytosis (Cook et al. 2005; Lim et al. 2000). For example, a galactosylated liposome was reported to deliver Dox to the hepatocytes (Wang et al. 2006). The attachment of galactose to the cyclodextrin-containing polymer *via* a PEG-adamantine linker allowed for targeted nucleic acid delivery (Bartlett and Davis 2007). Oishi et al. also developed a lactosylated PEG-siRNA conjugate through an acid-labile linkage, which could efficiently release free siRNA once this conjugate entered acidic endosomal compartment (Oishi et al. 2005, 2007).

3.5.2 Mannose

It has been known that mannose receptors are highly expressed in macrophages, dendritic cells and other hematopoietic cells important for innate immune responses. Mannose is recognized by a mannose receptor and is internalized into cells (Ferkol et al. 1996). The mannose-coating nanoparticles have been widely used for targeted delivery of anti-HIV drugs. For example, Jain et al. have used mannose-conjugated, fifth generation poly(propyleneimine) (MPPI) dendrimers to deliver lamuivudine (3TC) to HIV-1 infected MT-2 cells. Significant increases in the cellular uptake of 3TC were observed compared with free drug and non-targeted PPI dendrimers (Dutta and Jain 2007). In another study, MPPI nanocarriers also delivered efavirenz (EFV) to human monocytes/macrophages. Similarly, stavudine (d4T, brand name Zerit) also has been loaded in mannosylated liposomes and has been evaluated *in vitro* for anti-HIV activity (Dutta et al. 2007).

3.6 Aptamer-Mediated Drug Delivery

Aptamers are evolved, single-stranded nucleic acids that can specifically recognize and bind their cognate targets by means of well-defined stable, three-dimensional structures (Mayer 2009; Famulok et al. 2007). Over the past 20 years, numerous nucleic-acid aptamers have been raised against a wide variety of targets (Nimjee et al. 2005). Moreover, their many favorable characteristics, such as the low nanomolar dissociation constants and exquisite specificity, small size, stability, lack of immunogenicity, and the ability to be chemically synthesized with various base and backbone modifications make them versatile tools for diagnostics, *in vivo* imaging, and targeted therapeutics. Currently, the advances in the development of DNA or RNA aptamers specifically targeting membrane receptors to deliver and enhance the efficacy of other therapeutics agents have drawn attention for exploiting specific cell-internalizing aptamers as targeted drug delivery carriers (Zhou and Rossi 2009; Yan and Levy 2009). An increasing number of cell-type specific RNA and DNA aptamers targeting PSMA (prostate-specific membrane antigen) (Lupold et al. 2002); HIV-1 gp120 (Cohen et al. 2008; Dey et al. 2005); CD4 (cluster of differentiation 4) (Kraus et al. 1998); TN-C (tenasin-C) (Hicke et al. 2001), EGFR (epidermal growth factor receptor) (Li et al. 2010), PTK7 (protein tyrosine kinase 7) (Xiao et al. 2008), TfR (transferrin receptor) (Chen et al. 2008), NCL (nucleolin) (Shieh et al. 2010) and MUC1 (mucin protein) (Ferreira et al. 2009), have been adopted for targeted delivery of the various molecules of interest. Four representative examples are discussed below.

3.6.1 Anti-PSMA RNA Aptamer

PSMA, a trans-membrane protein, is highly expressed in human prostate cancer and the vascular endothelium. It has been known that PSMA can be constitutively endocytosed into PSMA-positive LNCap cells and be continually recycled from the plasma membrane, thereby allowing for intracellular drug transport (Tasch et al. 2001; Liu et al. 1998). Since several 2′-Fluoro modified anti-PSMA RNA aptamers with nanomolar binding affinity were isolated by Lupold et al. (Lupold et al. 2002), they already have become popular cell-specific internalizing molecules for targeted drug delivery (Zhou and Rossi 2009).

Three independent studies have demonstrated anti-PSMA aptamer-mediated small interfering RNA (siRNA) delivery to PSMA-positive cells or human cancer cells-transplanted into nude mice. Through Streptavidin-biotin interactions, Chu et al. non-covalently assembled two biotinylated anti-PSMA aptamers and two biotinylated siRNAs into a Streptavidin connector (Chu et al. 2006b). The multivalent construct was effectively internalized into targeted cells and triggered specific gene silencing. Using a somewhat different approach, Giangrande and coworkers have developed a simple covalent aptamer-siRNA chimeric RNA which allowed effective PSMA mediated cell uptake along with siRNA-mediated gene silencing in athymic mice following both intra-tumoral delivery and systemic administration (McNamara et al. 2006; Dassie et al. 2009).

By taking advantage of the targeting properties of aptamers, a gelonin toxin has been successfully conjugated to an anti-PSMA aptamer (Chu et al. 2006a). The resulting conjugates dramatically increased cellular uptake and therapeutic efficacy in PSMA-positive cells. Reports from Farokhzad and colleagues also demonstrated the potential of anti-PSMA aptamers to mediate nanoparticles delivery (Peer et al. 2007). For example, they have constructed an anti-PSMA aptamer – poly (lactic acid)-block-PEG copolymer nanoparticle conjugate encapsulating the chemotherapeutic drug Docetaxel (Dtxl-NP-aptamer) (Farokhzad et al. 2004). Such Dtxl-NP-aptamer conjugates showed remarkable efficacy and completely suppressed tumor growth in a xenograft nude mouse model with a single intra-tumoral injection (Farokhzad et al. 2006).

A simple physical conjugation also was employed to assemble an aptamer with drugs. Anthracycline drugs (e.g. Dox) non-covalently intercalated into double-strand regions of an anti-PSMA aptamer *via* the flat aromatic ring, and formed a stable physical conjugate capable of effectively targeting PSMA-positive cells (Bagalkot et al. 2006). In another example, Bagalkot et al. functionalized a quantum dot (QD) with anti-PSMA aptamers to achieve a multifunctional QD-aptamer conjugate, serving both as a fluorescence imager and a drug delivery carrier (Bagalkot et al. 2007). Dox was loaded into QD-aptamer conjugate *via* physical intercalation within the aptamer strand. Once internalized inside cancer cells, the QD-aptamer-Dox system gradually released Dox (for targeting therapy) and recovered the fluorescence of the QD (for synchronous cancer imaging).

3.6.2 Anti-HIV-1 gp120 RNA Aptamer

As described in sect. 3.1.4, the HIV-1 gp120 envelope protein represents an attractive molecular target for receptor-mediated drug delivery. Several 2′-F modified RNA aptamers have been isolated against HIV-1 gp120, which can specifically bind and be rapidly internalized into HIV-1 infected cells (Zhou et al. 2008, 2009). Using an afore-mentioned strategy described by Giangrande, a dual-action anti-gp120 aptamer-siRNA chimeric RNA was developed (Zhou et al. 2008; Neff et al. 2011).

This dual function chimera blocked HIV-1 infectivity and also selectively delivered siRNAs into HIV-1 infected cells which suppressed HIV-1 replication. Continued efforts to improve the aptamer delivery resulted in an aptamer-mediated combinatorial multi-targeting RNAi therapeutic, in which a single anti-gp120 aptamer was tightly tethered with three different siRNAs targeting HIV-1 *tat/rev* transcripts and HIV-1 host dependency factors (CD4 and TNPO3) through a "sticky" bridge (Zhou et al. 2009). The resulting aptamer-stick-cocktail siRNA conjugates suppressed viral loads in cell culture and *in vivo* in a humanized mouse model for HIV-1 infection (Zhou et al., unpublished).

3.6.3 Anti-PTK7 DNA Aptamer

Protein tyrosine kinase-7 (PTK7) is a member of the receptor tyrosine kinase family, which is highly expressed during thymocyte development in both mice and humans. Receptor protein tyrosine kinases transduce extracellular signals across the cell membrane. Tan and colleagues have isolated a panel of DNA aptamers against CCRF-CEM cells using a whole cell-SELEX (systematic evolution of ligands by exponential enrichment) procedure (Shangguan et al. 2006). One of the selected DNA aptamers, sgc8 identified to target PTK7, has been demonstrated to be specifically taken up into lymphoblastic leukemia T-cells *via* a receptor-mediated endocytosis (Shangguan et al. 2007). Recent studies have exploited the anti-PTK7 aptamer as a promising targeting ligand for targeted drug delivery. For example, an anti-PTK7 aptamer with a thiol group was covalently attached to Dox *via* an acid-labile linkage (Taghdisi et al. 2009). Once the aptamer-Dox conjugates specifically internalized into target cancer cells, Dox was easily released inside the acidic endosomal environment, selectively killing cancer cells. Recently, an anti-PTK7 aptamer-modified liposome was also used to facilitate targeted Dextran delivery (Kang et al. 2010). Additionally, an aptamer-conjugated Au-Ag nanorod (NR) was designed as a selective photo-thermal agent to efficiently kill tumor cells (Huang et al. 2008b).

3.6.4 Anti-NCL DNA Aptamer

Human nucleolin (NCL), a multifunctional protein involved in the synthesis and maturation of ribosomes, is over-expressed on the plasma membrane of several cancer cells such as breast, prostate, and lymphocytic leukemia. Moreover, NCL can transfer molecules between the cell surface and the nucleus. Therefore, it provides a potential target for the higher NCL-expressing cancer cell specific therapy. An anti-NCL DNA aptamer AS1411 (also known as AGRO100), a 26-mer guanine-rich oligonucleotide with high affinity and specificity to NCL, has been found to internalize into several cancer cell lines, including breast cancer cells (Bates et al. 1999; Soundararajan et al. 2008). A reversible AS1411 aptamer-liposome bioconjugate was developed to effectively deliver cisplatin to cancer cells and to improve therapeutic efficacy (Cao et al. 2009). Importantly, when a complementary antidote was used to disrupt the aptamer's active structure, the cellular uptake of the aptamer-liposome was inhibited, thereby suggesting a novel controllable targeted drug delivery system. Most recently, Shieh et al. physically conjugated AS1411 with six molecules of a porphyrin derivative (TMPyP4), a broadly used photodynamic therapeutic agent (Shieh et al. 2010). By NCL-mediated internalization, the aptamer-TMP complex exhibited a higher TMPyP4 accumulation and specific photo-damage in MCF7 breast cancer cells.

3.7 Other Internalizing Molecule-Mediated Drug Delivery

In addition to the afore-mentioned more popular internalizing molecules, other specific ligands have been applied for targeted drug delivery.

Anisamide has served as a targeting ligand for tumor cells expressing the sigma receptor, a transmembrane protein that plays a role in regulating ion channels. It was found that sigma receptors are over-expressed in a diverse set of human and rodent tumor cell lines (Vilner et al. 1995). Small molecules such as asanisamide, haloperidol and opipramol have been reported as sigma receptor ligands and have been developed as radio-imaging agents for tumors (Maurice and Su 2009; Cobos et al. 2008; John et al. 1999). These receptor-specific ligands can be linked to various nanocarriers, or be directly conjugated to the oligonucleotide itself. For example, Huang and colleagues have successfully conjugated the high-affinity sigma receptor ligand asanisamide to lipopolyplex nanocarriers for specifically delivering doxorubicin or siRNAs to tumors in animals (Li and Huang 2006; Banerjee et al. 2004). Their subsequent studies confirmed that the cellular uptake was mediated *via* a sigma receptor dependent pathway. Similarly, haloperidol-modified lipoplexes were shown to mediate tenfold greater delivery of DNA to breast carcinoma cells compared with the control lipoplexes (Mukherjee et al. 2005). Most recently, anisamine also was directly conjugated to antisense oligonucleotides (ONs) (Nakagawa et al. 2010). The trivalent anisamide-ONs conjugate significantly enhanced receptor-specific cell uptake and biological activity.

Hyaluronan, also known as hyaluronic acid (HA) or sodium hyaluronate, is a natural anionic polysaccharide which can be efficiently taken up into cells by HA receptor mediated endocytosis (Stern et al. 2006; Prevo et al. 2001). Therefore, HA and its derivatives have been widely used as novel targeting ligands as well as target specific, long acting drug carriers of various therapeutic agents, including Dox, paclitaxel, proteins, peptides and nucleotide therapeutics (Oh et al. 2010). For example, the HA-modified long-circulating liposomes actively targeted tumors over-expressing the HA receptor (Peer and Margalit 2004a, b). The HA-poly-L-lysine (PLL) conjugate was shown to target sinusoidal epithelial cells in the liver (Jiang et al. 2009). The delivery of siRNA with HA modified PEI significantly improved gene silencing in HA receptor over-expressing cells. Most recently, HA derivative-quantum dot (HA-QD) conjugates facilitated targeted delivery and accumulated more efficiently in the cirrhotic liver than the normal liver after tail-vein injection (Kim et al. 2010).

4 Conclusion and Perspectives

Although Paul Ehrlich's dream – "magic bullets" – did not stand the test of time one century ago, the creative concept of "targeted therapy" that drugs are capable of going straight to their intended targets has inspired generations of scientists to continually pursue targeted based delivery approaches to selectively treat human diseases. In addition to passive targeting, active targeting that is based on the avid and specific interaction between the targeting ligands and the molecular targets has drawn much attention. Today, in view of the tremendous efforts in developing targeted therapy, an increasing number of specific internalizing molecules have been adapted for active targeting of tumors or infectious diseases. Moreover, a better understanding of the molecular events associated with human diseases and the emergence of new technologies (such as nanotechnology, RNAi, Cell-SELEX) have greatly accelerated the drug discovery process over the years.

Despite these advances bringing these targeted therapies to the clinic is currently a slow process. For example, one important concern for using cell surface receptors for active targeting is to engineer the targeting ligands with therapeutic agents or their carriers without disrupting the receptor-binding ability or reducing the therapeutic efficacies of the drugs. In the regard, a precisely engineered nanocarrier with multiple functions may be considered a promising therapeutic system. Ideally, an intelligent multifunctional nanocarrier with combinatorial targeting ability can efficiently encapsulate multiple drugs, prolong circulation time, accurately accumulate and release drugs in the targeted cells/tissues, eventually achieving maximal therapeutic efficacy and providing custom/tailored treatments. Meanwhile, such multifunctional nanocarriers can be used to visualize their location in the body for real-time monitoring of therapeutic responses.

Acknowledgements This work was supported by grants from the National Institutes of Health AI29329, AI42552 and HL07470 awarded to J.J.R.

Competing interests: The authors declare that they have no competing financial interests.

Authors' contributions: JZ drafted the manuscript. JR revised it and gave final approval of the version to be published. All authors read and approved the final manuscript.

References

Agarwal, A., Saraf, S., Asthana, A., Gupta, U., Gajbhiye, V. & Jain, N. K. (2008) Ligand based dendritic systems for tumor targeting. *Int J Pharm*, 350, 3–13.

Akbulut, M., D'Addio, S. M., Gindy, M. E. & Prud'homme, R. K. (2009) Novel methods of targeted drug delivery: the potential of multifunctional nanoparticles. *Expert Review of Clinical Pharmacology*, 2, 265–282.

Alam, M. R., Dixit, V., Kang, H., Li, Z. B., Chen, X., Trejo, J., Fisher, M. & Juliano, R. L. (2008) Intracellular delivery of an anionic antisense oligonucleotide via receptor-mediated endocytosis. *Nucleic Acids Res*, 36, 2764–76.

Alam, M. R., Ming, X., Dixit, V., Fisher, M., Chen, X. & Juliano, R. L. (2010) The biological effect of an antisense oligonucleotide depends on its route of endocytosis and trafficking. *Oligonucleotides*, 20, 103–9.

Allen, T. M. (2002) Ligand-targeted therapeutics in anticancer therapy. *Nat Rev Cancer*, 2, 750–63.

Bagalkot, V., Farokhzad, O. C., Langer, R. & Jon, S. (2006) An aptamer-doxorubicin physical conjugate as a novel targeted drug-delivery platform. *Angew Chem Int Ed Engl*, 45, 8149–52.

Bagalkot, V., Zhang, L., Levy-Nissenbaum, E., Jon, S., Kantoff, P. W., Langer, R. & Farokhzad, O. C. (2007) Quantum dot-aptamer conjugates for synchronous cancer imaging, therapy, and sensing of drug delivery based on bi-fluorescence resonance energy transfer. *Nano Lett*, 7, 3065–70.

Banerjee, R., Tyagi, P., Li, S. & Huang, L. (2004) Anisamide-targeted stealth liposomes: a potent carrier for targeting doxorubicin to human prostate cancer cells. *Int J Cancer*, 112, 693–700.

Bartlett, D. W. & Davis, M. E. (2007) Physicochemical and biological characterization of targeted, nucleic acid-containing nanoparticles. *Bioconjug Chem*, 18, 456–68.

Bartlett, D. W., Su, H., Hildebrandt, I. J., Weber, W. A. & Davis, M. E. (2007) Impact of tumor-specific targeting on the biodistribution and efficacy of siRNA nanoparticles measured by multimodality in vivo imaging. *Proc Natl Acad Sci U S A*, 104, 15549–54.

Bates, P. J., Kahlon, J. B., Thomas, S. D., Trent, J. O. & Miller, D. M. (1999) Antiproliferative activity of G-rich oligonucleotides correlates with protein binding. *J Biol Chem*, 274, 26369–77.

Bhirde, A. A., Patel, V., Gavard, J., Zhang, G., Sousa, A. A., Masedunskas, A., Leapman, R. D., Weigert, R., Gutkind, J. S. & Rusling, J. F. (2009) Targeted killing of cancer cells in vivo and in vitro with EGF-directed carbon nanotube-based drug delivery. *ACS Nano*, 3, 307–16.

Boraston, A. B., Bolam, D. N., Gilbert, H. J. & Davies, G. J. (2004) Carbohydrate-binding modules: fine-tuning polysaccharide recognition. *Biochem J*, 382, 769–81.

Bremer, E., Samplonius, D. F., Peipp, M., van Genne, L., Kroesen, B. J., Fey, G. H., Gramatzki, M., de Leij, L. F. & Helfrich, W. (2005) Target cell-restricted apoptosis induction of acute leukemic T cells by a recombinant tumor necrosis factor-related apoptosis-inducing ligand fusion protein with specificity for human CD7. *Cancer Res*, 65, 3380–8.

Cao, Z., Tong, R., Mishra, A., Xu, W., Wong, G. C., Cheng, J. & Lu, Y. (2009) Reversible cell-specific drug delivery with aptamer-functionalized liposomes. *Angew Chem Int Ed Engl*, 48, 6494–8.

Chang, J., Jallouli, Y., Kroubi, M., Yuan, X. B., Feng, W., Kang, C. S., Pu, P. Y. & Betbeder, D. (2009) Characterization of endocytosis of transferrin-coated PLGA nanoparticles by the blood-brain barrier. *Int J Pharm*, 379, 285–92.

Chen, C. H., Dellamaggiore, K. R., Ouellette, C. P., Sedano, C. D., Lizadjohry, M., Chernis, G. A., Gonzales, M., Baltasar, F. E., Fan, A. L., Myerowitz, R. & Neufeld, E. F. (2008) Aptamer-based endocytosis of a lysosomal enzyme. *Proc Natl Acad Sci U S A*, 105, 15908–13.

Cheresh, D. A. (1987) Human endothelial cells synthesize and express an Arg-Gly-Asp-directed adhesion receptor involved in attachment to fibrinogen and von Willebrand factor. *Proc Natl Acad Sci U S A*, 84, 6471–5.

Chiu, S. J., Ueno, N. T. & Lee, R. J. (2004a) Tumor-targeted gene delivery via anti-HER2 antibody (trastuzumab, Herceptin) conjugated polyethylenimine. *J Control Release*, 97, 357–69.

Chiu, Y. L., Ali, A., Chu, C. Y., Cao, H. & Rana, T. M. (2004b) Visualizing a correlation between siRNA localization, cellular uptake, and RNAi in living cells. *Chem Biol*, 11, 1165–75.

Cho, K. C., Kim, S. H., Jeong, J. H. & Park, T. G. (2005) Folate receptor-mediated gene delivery using folate-poly(ethylene glycol)-poly(L-lysine) conjugate. *Macromol Biosci*, 5, 512–9.

Chu, T. C., Marks, J. W., 3 rd, Lavery, L. A., Faulkner, S., Rosenblum, M. G., Ellington, A. D. & Levy, M. (2006a) Aptamer: toxin conjugates that specifically target prostate tumor cells. *Cancer Res*, 66, 5989–92.

Chu, T. C., Twu, K. Y., Ellington, A. D. & Levy, M. (2006b) Aptamer mediated siRNA delivery. *Nucleic Acids Res*, 34, e73.

Ciavarella, S., Milano, A., Dammacco, F. & Silvestris, F. (2010) Targeted therapies in cancer. *BioDrugs*, 24, 77–88.

Clayton, R., Ohagen, A., Goethals, O., Smets, A., Van Loock, M., Michiels, L., Kennedy-Johnston, E., Cunningham, M., Jiang, H., Bola, S., Gutshall, L., Gunn, G., Del Vecchio, A., Sarisky, R., Hallenberger, S. & Hertogs, K. (2007) Binding kinetics, uptake and intracellular accumulation of F105, an anti-gp120 human IgG1kappa monoclonal antibody, in HIV-1 infected cells. *J Virol Methods*, 139, 17–23.

Clayton, R., Ohagen, A., Nicol, F., Del Vecchio, A. M., Jonckers, T. H., Goethals, O., Van Loock, M., Michiels, L., Grigsby, J., Xu, Z., Zhang, Y. P., Gutshall, L. L., Cunningham, M., Jiang, H., Bola, S., Sarisky, R. T. & Hertogs, K. (2009) Sustained and specific in vitro inhibition of HIV-1 replication by a protease inhibitor encapsulated in gp120-targeted liposomes. *Antiviral Res*, 84, 142–9.

Cobos, E. J., Entrena, J. M., Nieto, F. R., Cendan, C. M. & Del Pozo, E. (2008) Pharmacology and therapeutic potential of sigma(1) receptor ligands. *Curr Neuropharmacol*, 6, 344–66.

Cohen, C., Forzan, M., Sproat, B., Pantophlet, R., McGowan, I., Burton, D. & James, W. (2008) An aptamer that neutralizes R5 strains of HIV-1 binds to core residues of gp120 in the CCR5 binding site. *Virology*, 381, 46–54.

Connor, J. & Huang, L. (1986) pH-sensitive immunoliposomes as an efficient and target-specific carrier for antitumor drugs. *Cancer Res*, 46, 3431–5.

Cook, S. E., Park, I. K., Kim, E. M., Jeong, H. J., Park, T. G., Choi, Y. J., Akaike, T. & Cho, C. S. (2005) Galactosylated polyethylenimine-graft-poly(vinyl pyrrolidone) as a hepatocyte-targeting gene carrier. *J Control Release*, 105, 151–63.

Dakappagari, N., Maruyama, T., Renshaw, M., Tacken, P., Figdor, C., Torensma, R., Wild, M. A., Wu, D., Bowdish, K. & Kretz-Rommel, A. (2006) Internalizing antibodies to the C-type lectins, L-SIGN and DC-SIGN, inhibit viral glycoprotein binding and deliver antigen to human dendritic cells for the induction of T cell responses. *J Immunol*, 176, 426–40.

Daniels, T. R., Delgado, T., Helguera, G. & Penichet, M. L. (2006) The transferrin receptor part II: targeted delivery of therapeutic agents into cancer cells. *Clin Immunol*, 121, 159–76.

Dassie, J. P., Liu, X. Y., Thomas, G. S., Whitaker, R. M., Thiel, K. W., Stockdale, K. R., Meyerholz, D. K., McCaffrey, A. P., McNamara, J. O., 2nd & Giangrande, P. H. (2009) Systemic administration of optimized aptamer-siRNA chimeras promotes regression of PSMA-expressing tumors. *Nat Biotechnol*, 27, 839–49.

Davis, M. E. (2009) The first targeted delivery of siRNA in humans via a self-assembling, cyclodextrin polymer-based nanoparticle: from concept to clinic. *Mol Pharm*, 6, 659–68.

Davis, M. E., Zuckerman, J. E., Choi, C. H., Seligson, D., Tolcher, A., Alabi, C. A., Yen, Y., Heidel, J. D. & Ribas, A. (2010) Evidence of RNAi in humans from systemically administered siRNA via targeted nanoparticles. *Nature*, 464, 1067–70.

Dey, A. K., Khati, M., Tang, M., Wyatt, R., Lea, S. M. & James, W. (2005) An aptamer that neutralizes R5 strains of human immunodeficiency virus type 1 blocks gp120-CCR5 interaction. *J Virol*, 79, 13806–10.

Dharap, S. S., Qiu, B., Williams, G. C., Sinko, P., Stein, S. & Minko, T. (2003) Molecular targeting of drug delivery systems to ovarian cancer by BH3 and LHRH peptides. *J Control Release*, 91, 61–73.

Dharap, S. S., Wang, Y., Chandna, P., Khandare, J. J., Qiu, B., Gunaseelan, S., Sinko, P. J., Stein, S., Farmanfarmaian, A. & Minko, T. (2005) Tumor-specific targeting of an anticancer drug delivery system by LHRH peptide. *Proc Natl Acad Sci U S A*, 102, 12962–7.

Duthie, S. J. (2001) Folic-acid-mediated inhibition of human colon-cancer cell growth. *Nutrition*, 17, 736–7.

Dutta, T., Agashe, H. B., Garg, M., Balakrishnan, P., Kabra, M. & Jain, N. K. (2007) Poly (propyleneimine) dendrimer based nanocontainers for targeting of efavirenz to human monocytes/macrophages in vitro. *J Drug Target*, 15, 89–98.

Dutta, T. & Jain, N. K. (2007) Targeting potential and anti-HIV activity of lamivudine loaded mannosylated poly (propyleneimine) dendrimer. *Biochim Biophys Acta*, 1770, 681–6.

Eguchi, A. & Dowdy, S. F. (2009) siRNA delivery using peptide transduction domains. *Trends Pharmacol Sci*, 30, 341–5.
Eguchi, A. & Dowdy, S. F. (2010) Efficient siRNA delivery by novel PTD-DRBD fusion proteins. *Cell Cycle*, 9, 424–5.
Eguchi, A., Meade, B. R., Chang, Y. C., Fredrickson, C. T., Willert, K., Puri, N. & Dowdy, S. F. (2009) Efficient siRNA delivery into primary cells by a peptide transduction domain-dsRNA binding domain fusion protein. *Nat Biotechnol*, 27, 567–71.
El-Sayed, A., Futaki, S. & Harashima, H. (2009) Delivery of macromolecules using arginine-rich cell-penetrating peptides: ways to overcome endosomal entrapment. *Aaps J*, 11, 13–22.
Elfinger, M., Maucksch, C. & Rudolph, C. (2007) Characterization of lactoferrin as a targeting ligand for nonviral gene delivery to airway epithelial cells. *Biomaterials*, 28, 3448–55.
Ercan, M. T. & Caglar, M. (2000) Therapeutic radiopharmaceuticals. *Curr Pharm Des*, 6, 1085–121.
Famulok, M., Hartig, J. S. & Mayer, G. (2007) Functional aptamers and aptazymes in biotechnology, diagnostics, and therapy. *Chem Rev*, 107, 3715–43.
Farokhzad, O. C., Cheng, J., Teply, B. A., Sherifi, I., Jon, S., Kantoff, P. W., Richie, J. P. & Langer, R. (2006) Targeted nanoparticle-aptamer bioconjugates for cancer chemotherapy in vivo. *Proc Natl Acad Sci U S A*, 103, 6315–20.
Farokhzad, O. C., Jon, S., Khademhosseini, A., Tran, T. N., Lavan, D. A. & Langer, R. (2004) Nanoparticle-aptamer bioconjugates: a new approach for targeting prostate cancer cells. *Cancer Res*, 64, 7668–72.
Ferkol, T., Perales, J. C., Mularo, F. & Hanson, R. W. (1996) Receptor-mediated gene transfer into macrophages. *Proc Natl Acad Sci U S A*, 93, 101–5.
Ferreira, C. S., Cheung, M. C., Missailidis, S., Bisland, S. & Gariepy, J. (2009) Phototoxic aptamers selectively enter and kill epithelial cancer cells. *Nucleic Acids Res*, 37, 866–76.
Gabizon, A. A. (2001) Pegylated liposomal doxorubicin: metamorphosis of an old drug into a new form of chemotherapy. *Cancer Invest*, 19, 424–36.
Gagne, J. F., Desormeaux, A., Perron, S., Tremblay, M. J. & Bergeron, M. G. (2002) Targeted delivery of indinavir to HIV-1 primary reservoirs with immunoliposomes. *Biochim Biophys Acta*, 1558, 198–210.
Gao, W., Xiao, Z., Radovic-Moreno, A., Shi, J., Langer, R. & Farokhzad, O. C. (2010) Progress in siRNA delivery using multifunctional nanoparticles. *Methods Mol Biol*, 629, 53–67.
Goldenberg, D. M. (2002) Targeted therapy of cancer with radiolabeled antibodies. *J Nucl Med*, 43, 693–713.
Goren, D., Horowitz, A. T., Tzemach, D., Tarshish, M., Zalipsky, S. & Gabizon, A. (2000) Nuclear delivery of doxorubicin via folate-targeted liposomes with bypass of multidrug-resistance efflux pump. *Clin Cancer Res*, 6, 1949–57.
Greish, K. (2007) Enhanced permeability and retention of macromolecular drugs in solid tumors: a royal gate for targeted anticancer nanomedicines. *J Drug Target*, 15, 457–64.
Griffin, J. D., Linch, D., Sabbath, K., Larcom, P. & Schlossman, S. F. (1984) A monoclonal antibody reactive with normal and leukemic human myeloid progenitor cells. *Leuk Res*, 8, 521–34.
Gullotti, E. & Yeo, Y. (2009) Extracellularly activated nanocarriers: a new paradigm of tumor targeted drug delivery. *Mol Pharm*, 6, 1041–51.
Gump, J. M. & Dowdy, S. F. (2007) TAT transduction: the molecular mechanism and therapeutic prospects. *Trends Mol Med*, 13, 443–8.
Gunaseelan, S., Gunaseelan, K., Deshmukh, M., Zhang, X. & Sinko, P. J. (2010) Surface modifications of nanocarriers for effective intracellular delivery of anti-HIV drugs. *Adv Drug Deliv Rev*, 62, 518–31.
Gupta, U. & Jain, N. K. (2010) Non-polymeric nano-carriers in HIV/AIDS drug delivery and targeting. *Adv Drug Deliv Rev*, 62, 478–90.
Harrison, L. B., Chadha, M., Hill, R. J., Hu, K. & Shasha, D. (2002) Impact of tumor hypoxia and anemia on radiation therapy outcomes. *Oncologist*, 7, 492–508.

Hart, S. L. (2010) Multifunctional nanocomplexes for gene transfer and gene therapy. *Cell Biol Toxicol*, 26, 69–81.

Heidel, J. D., Yu, Z., Liu, J. Y., Rele, S. M., Liang, Y., Zeidan, R. K., Kornbrust, D. J. & Davis, M. E. (2007) Administration in non-human primates of escalating intravenous doses of targeted nanoparticles containing ribonucleotide reductase subunit M2 siRNA. *Proc Natl Acad Sci U S A*, 104, 5715–21.

Hicke, B. J., Marion, C., Chang, Y. F., Gould, T., Lynott, C. K., Parma, D., Schmidt, P. G. & Warren, S. (2001) Tenascin-C aptamers are generated using tumor cells and purified protein. *J Biol Chem*, 276, 48644–54.

Higuchi, Y., Kawakami, S. & Hashida, M. (2010) Strategies for in vivo delivery of siRNAs: recent progress. *BioDrugs*, 24, 195–205.

Ho, R. J., Rouse, B. T. & Huang, L. (1987) Target-sensitive immunoliposomes as an efficient drug carrier for antiviral activity. *J Biol Chem*, 262, 13973–8.

Huang, R., Ke, W., Liu, Y., Jiang, C. & Pei, Y. (2008a) The use of lactoferrin as a ligand for targeting the polyamidoamine-based gene delivery system to the brain. *Biomaterials*, 29, 238–46.

Huang, R. Q., Qu, Y. H., Ke, W. L., Zhu, J. H., Pei, Y. Y. & Jiang, C. (2007) Efficient gene delivery targeted to the brain using a transferrin-conjugated polyethyleneglycol-modified polyamidoamine dendrimer. *Faseb J*, 21, 1117–25.

Huang, Y. F., Sefah, K., Bamrungsap, S., Chang, H. T. & Tan, W. (2008b) Selective photothermal therapy for mixed cancer cells using aptamer-conjugated nanorods. *Langmuir*, 24, 11860–5.

Jeong, J. H., Lee, M., Kim, W. J., Yockman, J. W., Park, T. G., Kim, Y. H. & Kim, S. W. (2005) Anti-GAD antibody targeted non-viral gene delivery to islet beta cells. *J Control Release*, 107, 562–70.

Jiang, G., Park, K., Kim, J., Kim, K. S. & Hahn, S. K. (2009) Target specific intracellular delivery of siRNA/PEI-HA complex by receptor mediated endocytosis. *Mol Pharm*, 6, 727–37.

Joensuu, H. (2008) Systemic chemotherapy for cancer: from weapon to treatment. *Lancet Oncol*, 9, 304.

John, C. S., Vilner, B. J., Geyer, B. C., Moody, T. & Bowen, W. D. (1999) Targeting sigma receptor-binding benzamides as in vivo diagnostic and therapeutic agents for human prostate tumors. *Cancer Res*, 59, 4578–83.

Kang, H., O'Donoghue, M. B., Liu, H. & Tan, W. (2010) A liposome-based nanostructure for aptamer directed delivery. *Chem Commun (Camb)*, 46, 249–51.

Kannagi, R., Izawa, M., Koike, T., Miyazaki, K. & Kimura, N. (2004) Carbohydrate-mediated cell adhesion in cancer metastasis and angiogenesis. *Cancer Sci*, 95, 377–84.

Katzel, J. A., Fanucchi, M. P. & Li, Z. (2009) Recent advances of novel targeted therapy in non-small cell lung cancer. *J Hematol Oncol*, 2, 2.

Kim, J., Park, S., Lee, J. E., Jin, S. M., Lee, J. H., Lee, I. S., Yang, I., Kim, J. S., Kim, S. K., Cho, M. H. & Hyeon, T. (2006) Designed fabrication of multifunctional magnetic gold nanoshells and their application to magnetic resonance imaging and photothermal therapy. *Angew Chem Int Ed Engl*, 45, 7754–8.

Kim, K. S., Hur, W., Park, S. J., Hong, S. W., Choi, J. E., Goh, E. J., Yoon, S. K. & Hahn, S. K. (2010) Bioimaging for targeted delivery of hyaluronic Acid derivatives to the livers in cirrhotic mice using quantum dots. *ACS Nano*, 4, 3005–14.

Kim, S., Kim, J. H., Jeon, O., Kwon, I. C. & Park, K. (2009a) Engineered polymers for advanced drug delivery. *Eur J Pharm Biopharm*, 71, 420–30.

Kim, S. H., Jeong, J. H., Lee, S. H., Kim, S. W. & Park, T. G. (2008) LHRH receptor-mediated delivery of siRNA using polyelectrolyte complex micelles self-assembled from siRNA-PEG-LHRH conjugate and PEI. *Bioconjug Chem*, 19, 2156–62.

Kim, S. I., Shin, D., Choi, T. H., Lee, J. C., Cheon, G. J., Kim, K. Y., Park, M. & Kim, M. (2007) Systemic and specific delivery of small interfering RNAs to the liver mediated by apolipoprotein A-I. *Mol Ther*, 15, 1145–52.

Kim, S. I., Shin, D., Lee, H., Ahn, B. Y., Yoon, Y. & Kim, M. (2009b) Targeted delivery of siRNA against hepatitis C virus by apolipoprotein A-I-bound cationic liposomes. *J Hepatol*, 50, 479–88.

Kim, S. S., Garg, H., Joshi, A. & Manjunath, N. (2009c) Strategies for targeted nonviral delivery of siRNAs in vivo. *Trends Mol Med*, 15, 491–500.
Kirpotin, D. B., Drummond, D. C., Shao, Y., Shalaby, M. R., Hong, K., Nielsen, U. B., Marks, J. D., Benz, C. C. & Park, J. W. (2006) Antibody targeting of long-circulating lipidic nanoparticles does not increase tumor localization but does increase internalization in animal models. *Cancer Res*, 66, 6732–40.
Kraus, E., James, W. & Barclay, A. N. (1998) Cutting edge: novel RNA ligands able to bind CD4 antigen and inhibit CD4+ T lymphocyte function. *J Immunol*, 160, 5209–12.
Kumar, P., Ban, H. S., Kim, S. S., Wu, H., Pearson, T., Greiner, D. L., Laouar, A., Yao, J., Haridas, V., Habiro, K., Yang, Y. G., Jeong, J. H., Lee, K. Y., Kim, Y. H., Kim, S. W., Peipp, M., Fey, G. H., Manjunath, N., Shultz, L. D., Lee, S. K. & Shankar, P. (2008) T cell-specific siRNA delivery suppresses HIV-1 infection in humanized mice. *Cell*, 134, 577–86.
Kwong, P. D., Wyatt, R., Robinson, J., Sweet, R. W., Sodroski, J. & Hendrickson, W. A. (1998) Structure of an HIV gp120 envelope glycoprotein in complex with the CD4 receptor and a neutralizing human antibody. *Nature*, 393, 648–59.
Langer, R. (1998) Drug delivery and targeting. *Nature*, 392, 5–10.
Larson, R. A., Sievers, E. L., Stadtmauer, E. A., Lowenberg, B., Estey, E. H., Dombret, H., Theobald, M., Voliotis, D., Bennett, J. M., Richie, M., Leopold, L. H., Berger, M. S., Sherman, M. L., Loken, M. R., van Dongen, J. J., Bernstein, I. D. & Appelbaum, F. R. (2005) Final report of the efficacy and safety of gemtuzumab ozogamicin (Mylotarg) in patients with CD33-positive acute myeloid leukemia in first recurrence. *Cancer*, 104, 1442–52.
Lee, H., Jang, I. H., Ryu, S. H. & Park, T. G. (2003) N-terminal site-specific mono-PEGylation of epidermal growth factor. *Pharm Res*, 20, 818–25.
Lee, H., Kim, T. H. & Park, T. G. (2002) A receptor-mediated gene delivery system using streptavidin and biotin-derivatized, pegylated epidermal growth factor. *J Control Release*, 83, 109–19.
Lee, H. & Park, T. G. (2002) Preparation and characterization of mono-PEGylated epidermal growth factor: evaluation of in vitro biologic activity. *Pharm Res*, 19, 845–51.
Levy-Nissenbaum, E., Radovic-Moreno, A. F., Wang, A. Z., Langer, R. & Farokhzad, O. C. (2008) Nanotechnology and aptamers: applications in drug delivery. *Trends Biotechnol*, 26, 442–9.
Li, N., Larson, T., Nguyen, H. H., Sokolov, K. V. & Ellington, A. D. (2010) Directed evolution of gold nanoparticle delivery to cells. *Chem Commun (Camb)*, 46, 392–4.
Li, S., Tan, Y., Viroonchatapan, E., Pitt, B. R. & Huang, L. (2000) Targeted gene delivery to pulmonary endothelium by anti-PECAM antibody. *Am J Physiol Lung Cell Mol Physiol*, 278, L504–11.
Li, S. D. & Huang, L. (2006) Targeted delivery of antisense oligodeoxynucleotide and small interference RNA into lung cancer cells. *Mol Pharm*, 3, 579–88.
Lim, D. W., Yeom, Y. I. & Park, T. G. (2000) Poly(DMAEMA-NVP)-b-PEG-galactose as gene delivery vector for hepatocytes. *Bioconjug Chem*, 11, 688–95.
Liu, H., Rajasekaran, A. K., Moy, P., Xia, Y., Kim, S., Navarro, V., Rahmati, R. & Bander, N. H. (1998) Constitutive and antibody-induced internalization of prostate-specific membrane antigen. *Cancer Res*, 58, 4055–60.
Liu, Z., Sun, X., Nakayama-Ratchford, N. & Dai, H. (2007a) Supramolecular chemistry on water-soluble carbon nanotubes for drug loading and delivery. *ACS Nano*, 1, 50–6.
Liu, Z., Winters, M., Holodniy, M. & Dai, H. (2007b) siRNA delivery into human T cells and primary cells with carbon-nanotube transporters. *Angew Chem Int Ed Engl*, 46, 2023–7.
Lupold, S. E., Hicke, B. J., Lin, Y. & Coffey, D. S. (2002) Identification and characterization of nuclease-stabilized RNA molecules that bind human prostate cancer cells via the prostate-specific membrane antigen. *Cancer Res*, 62, 4029–33.
Maeda, H. (2010) Tumor-selective delivery of macromolecular drugs via the EPR effect: background and future prospects. *Bioconjug Chem*, 21, 797–802.
Mamot, C., Drummond, D. C., Noble, C. O., Kallab, V., Guo, Z., Hong, K., Kirpotin, D. B. & Park, J. W. (2005) Epidermal growth factor receptor-targeted immunoliposomes significantly enhance the efficacy of multiple anticancer drugs in vivo. *Cancer Res*, 65, 11631–8.

Markman, M. (2008) The promise and perils of 'targeted therapy' of advanced ovarian cancer. *Oncology*, 74, 1–6.

Matsumura, Y. & Maeda, H. (1986) A new concept for macromolecular therapeutics in cancer chemotherapy: mechanism of tumoritropic accumulation of proteins and the antitumor agent smancs. *Cancer Res*, 46, 6387–92.

Maurice, T. & Su, T. P. (2009) The pharmacology of sigma-1 receptors. *Pharmacol Ther*, 124, 195–206.

Mayer, G. (2009) The chemical biology of aptamers. *Angew Chem Int Ed Engl*, 48, 2672–89.

McNamara, J. O., 2nd, Andrechek, E. R., Wang, Y., Viles, K. D., Rempel, R. E., Gilboa, E., Sullenger, B. A. & Giangrande, P. H. (2006) Cell type-specific delivery of siRNAs with aptamer-siRNA chimeras. *Nat Biotechnol*, 24, 1005–15.

Meade, B. R. & Dowdy, S. F. (2008) Enhancing the cellular uptake of siRNA duplexes following noncovalent packaging with protein transduction domain peptides. *Adv Drug Deliv Rev*, 60, 530–6.

Mishra, V., Mahor, S., Rawat, A., Gupta, P. N., Dubey, P., Khatri, K. & Vyas, S. P. (2006) Targeted brain delivery of AZT via transferrin anchored pegylated albumin nanoparticles. *J Drug Target*, 14, 45–53.

Moghimi, S. M., Hunter, A. C. & Murray, J. C. (2001) Long-circulating and target-specific nanoparticles: theory to practice. *Pharmacol Rev*, 53, 283–318.

Mok, H. & Park, T. G. (2009) Functional polymers for targeted delivery of nucleic acid drugs. *Macromol Biosci*, 9, 731–43.

Mukherjee, A., Prasad, T. K., Rao, N. M. & Banerjee, R. (2005) Haloperidol-associated stealth liposomes: a potent carrier for delivering genes to human breast cancer cells. *J Biol Chem*, 280, 15619–27.

Muslimov, G. F. (2008) Role of epidermal growth factor gene in the development of pancreatic cancer and efficiency of inhibitors of this gene in the treatment of pancreatic carcinoma. *Bull Exp Biol Med*, 145, 535–8.

Muthu, M. S. & Wilson, B. (2010) Multifunctional radionanomedicine: a novel nanoplatform for cancer imaging and therapy. *Nanomedicine (Lond)*, 5, 169–71.

Nakagawa, O., Ming, X., Huang, L. & Juliano, R. L. (2010) Targeted intracellular delivery of antisense oligonucleotides via conjugation with small-molecule ligands. *J Am Chem Soc*, 132, 8848–9.

Nakase, I., Niwa, M., Takeuchi, T., Sonomura, K., Kawabata, N., Koike, Y., Takehashi, M., Tanaka, S., Ueda, K., Simpson, J. C., Jones, A. T., Sugiura, Y. & Futaki, S. (2004) Cellular uptake of arginine-rich peptides: roles for macropinocytosis and actin rearrangement. *Mol Ther*, 10, 1011–22.

Neff, C. P., Zhou, J. H., Remling, L., Zhang, J., Li, H. T., Swiderski, P., Rossi J. J. & Akkina, R. (2011) An aptamer-siRNA chimera suppresses HIV-1 viral loads and protects from helper CD4+ T cells in humanized mice. *Science Translational Medicine*, 3(66), 66–6.

Ni, S., Stephenson, S. M. & Lee, R. J. (2002) Folate receptor targeted delivery of liposomal daunorubicin into tumor cells. *Anticancer Res*, 22, 2131–5.

Nimjee, S. M., Rusconi, C. P. & Sullenger, B. A. (2005) Aptamers: an emerging class of therapeutics. *Annu Rev Med*, 56, 555–83.

Oh, E. J., Park, K., Kim, K. S., Kim, J., Yang, J. A., Kong, J. H., Lee, M. Y., Hoffman, A. S. & Hahn, S. K. (2010) Target specific and long-acting delivery of protein, peptide, and nucleotide therapeutics using hyaluronic acid derivatives. *J Control Release*, 141, 2–12.

Oishi, M., Nagasaki, Y., Itaka, K., Nishiyama, N. & Kataoka, K. (2005) Lactosylated poly(ethylene glycol)-siRNA conjugate through acid-labile beta-thiopropionate linkage to construct pH-sensitive polyion complex micelles achieving enhanced gene silencing in hepatoma cells. *J Am Chem Soc*, 127, 1624–5.

Oishi, M., Nagasaki, Y., Nishiyama, N., Itaka, K., Takagi, M., Shimamoto, A., Furuichi, Y. & Kataoka, K. (2007) Enhanced growth inhibition of hepatic multicellular tumor spheroids by lactosylated poly(ethylene glycol)-siRNA conjugate formulated in PEGylated polyplexes. *ChemMedChem*, 2, 1290–7.

Pangburn, T. O., Petersen, M. A., Waybrant, B., Adil, M. M. & Kokkoli, E. (2009) Peptide- and aptamer-functionalized nanovectors for targeted delivery of therapeutics. *J Biomech Eng*, 131, 074005.

Park, J. W., Hong, K., Kirpotin, D. B., Colbern, G., Shalaby, R., Baselga, J., Shao, Y., Nielsen, U. B., Marks, J. D., Moore, D., Papahadjopoulos, D. & Benz, C. C. (2002) Anti-HER2 immunoliposomes: enhanced efficacy attributable to targeted delivery. *Clin Cancer Res*, 8, 1172–81.

Park, K., Lee, S., Kang, E., Kim, K., Choi, K. & Kwon, I. C. (2009) New Generation of Multifunctional Nanoparticles for Cancer Imaging and Therapy. *Advanced functional materials* 19, 1553–1566.

Peer, D., Karp, J. M., Hong, S., Farokhzad, O. C., Margalit, R. & Langer, R. (2007) Nanocarriers as an emerging platform for cancer therapy. *Nat Nanotechnol*, 2, 751–60.

Peer, D. & Margalit, R. (2004a) Loading mitomycin C inside long circulating hyaluronan targeted nano-liposomes increases its antitumor activity in three mice tumor models. *Int J Cancer*, 108, 780–9.

Peer, D. & Margalit, R. (2004b) Tumor-targeted hyaluronan nanoliposomes increase the antitumor activity of liposomal Doxorubicin in syngeneic and human xenograft mouse tumor models. *Neoplasia*, 6, 343–53.

Peer, D., Park, E. J., Morishita, Y., Carman, C. V. & Shimaoka, M. (2008) Systemic leukocyte-directed siRNA delivery revealing cyclin D1 as an anti-inflammatory target. *Science*, 319, 627–30.

Peipp, M., Kupers, H., Saul, D., Schlierf, B., Greil, J., Zunino, S. J., Gramatzki, M. & Fey, G. H. (2002) A recombinant CD7-specific single-chain immunotoxin is a potent inducer of apoptosis in acute leukemic T cells. *Cancer Res*, 62, 2848–55.

Posner, M. R., Hideshima, T., Cannon, T., Mukherjee, M., Mayer, K. H. & Byrn, R. A. (1991) An IgG human monoclonal antibody that reacts with HIV-1/GP120, inhibits virus binding to cells, and neutralizes infection. *J Immunol*, 146, 4325–32.

Prevo, R., Banerji, S., Ferguson, D. J., Clasper, S. & Jackson, D. G. (2001) Mouse LYVE-1 is an endocytic receptor for hyaluronan in lymphatic endothelium. *J Biol Chem*, 276, 19420–30.

Qian, Z. M., Li, H., Sun, H. & Ho, K. (2002) Targeted drug delivery via the transferrin receptor-mediated endocytosis pathway. *Pharmacol Rev*, 54, 561–87.

Rao, K. S., Ghorpade, A. & Labhasetwar, V. (2009) Targeting anti-HIV drugs to the CNS. *Expert Opin Drug Deliv*, 6, 771–84.

Ruoslahti, E. (2003) The RGD story: a personal account. *Matrix Biol*, 22, 459–65.

Russ, V. & Wagner, E. (2007) Cell and tissue targeting of nucleic acids for cancer gene therapy. *Pharm Res*, 24, 1047–57.

Sattentau, Q. J. & Moore, J. P. (1993) The role of CD4 in HIV binding and entry. *Philos Trans R Soc Lond B Biol Sci*, 342, 59–66.

Scheinberg, D. A., Tanimoto, M., McKenzie, S., Strife, A., Old, L. J. & Clarkson, B. D. (1989) Monoclonal antibody M195: a diagnostic marker for acute myelogenous leukemia. *Leukemia*, 3, 440–5.

Schrama, D., Reisfeld, R. A. & Becker, J. C. (2006) Antibody targeted drugs as cancer therapeutics. *Nat Rev Drug Discov*, 5, 147–59.

Senter, P. D. & Springer, C. J. (2001) Selective activation of anticancer prodrugs by monoclonal antibody-enzyme conjugates. *Adv Drug Deliv Rev*, 53, 247–64.

Shangguan, D., Cao, Z. C., Li, Y. & Tan, W. (2007) Aptamers evolved from cultured cancer cells reveal molecular differences of cancer cells in patient samples. *Clin Chem*, 53, 1153–5.

Shangguan, D., Li, Y., Tang, Z., Cao, Z. C., Chen, H. W., Mallikaratchy, P., Sefah, K., Yang, C. J. & Tan, W. (2006) Aptamers evolved from live cells as effective molecular probes for cancer study. *Proc Natl Acad Sci U S A*, 103, 11838–43.

Shapira, S., Lisiansky, V., Arber, N. & Kraus, S. (2010) Targeted immunotherapy for colorectal cancer: monoclonal antibodies and immunotoxins. *Expert Opin Investig Drugs*, 19 Suppl 1, S67–77.

Shieh, Y. A., Yang, S. J., Wei, M. F. & Shieh, M. J. (2010) Aptamer-based tumor-targeted drug delivery for photodynamic therapy. *ACS Nano*, 4, 1433–42.

Simard, P. & Leroux, J. C. (2009) pH-sensitive immunoliposomes specific to the CD33 cell surface antigen of leukemic cells. *Int J Pharm*, 381, 86–96.

Simard, P. & Leroux, J. C. (2010) In vivo evaluation of pH-sensitive polymer-based immunoliposomes targeting the CD33 antigen. *Mol Pharm*, 7, 1098–107.

Song, E., Zhu, P., Lee, S. K., Chowdhury, D., Kussman, S., Dykxhoorn, D. M., Feng, Y., Palliser, D., Weiner, D. B., Shankar, P., Marasco, W. A. & Lieberman, J. (2005) Antibody mediated in vivo delivery of small interfering RNAs via cell-surface receptors. *Nat Biotechnol*, 23, 709–17.

Soundararajan, S., Chen, W., Spicer, E. K., Courtenay-Luck, N. & Fernandes, D. J. (2008) The nucleolin targeting aptamer AS1411 destabilizes Bcl-2 messenger RNA in human breast cancer cells. *Cancer Res*, 68, 2358–65.

Stern, R., Asari, A. A. & Sugahara, K. N. (2006) Hyaluronan fragments: an information-rich system. *Eur J Cell Biol*, 85, 699–715.

Strebhardt, K. & Ullrich, A. (2008) Paul Ehrlich's magic bullet concept: 100 years of progress. *Nat Rev Cancer*, 8, 473–80.

Sudimack, J. & Lee, R. J. (2000) Targeted drug delivery via the folate receptor. *Adv Drug Deliv Rev*, 41, 147–62.

Sugahara, K. N., Teesalu, T., Karmali, P. P., Kotamraju, V. R., Agemy, L., Girard, O. M., Hanahan, D., Mattrey, R. F. & Ruoslahti, E. (2009) Tissue-penetrating delivery of compounds and nanoparticles into tumors. *Cancer Cell*, 16, 510–20.

Sugahara, K. N., Teesalu, T., Karmali, P. P., Kotamraju, V. R., Agemy, L., Greenwald, D. R. & Ruoslahti, E. (2010) Coadministration of a tumor-penetrating peptide enhances the efficacy of cancer drugs. *Science*, 328, 1031–5.

Suh, W., Chung, J. K., Park, S. H. & Kim, S. W. (2001) Anti-JL1 antibody-conjugated poly (L-lysine) for targeted gene delivery to leukemia T cells. *J Control Release*, 72, 171–8.

Tacken, P. J., de Vries, I. J., Gijzen, K., Joosten, B., Wu, D., Rother, R. P., Faas, S. J., Punt, C. J., Torensma, R., Adema, G. J. & Figdor, C. G. (2005) Effective induction of naive and recall T-cell responses by targeting antigen to human dendritic cells via a humanized anti-DC-SIGN antibody. *Blood*, 106, 1278–85.

Taghdisi, S. M., Abnous, K., Mosaffa, F. & Behravan, J. (2009) Targeted delivery of daunorubicin to T-cell acute lymphoblastic leukemia by aptamer. *J Drug Target*.

Tasch, J., Gong, M., Sadelain, M. & Heston, W. D. (2001) A unique folate hydrolase, prostate-specific membrane antigen (PSMA): a target for immunotherapy? *Crit Rev Immunol*, 21, 249–61.

Torchilin, V. P. (2006) Multifunctional nanocarriers. *Adv Drug Deliv Rev*, 58, 1532–55.

Torchilin, V. P. (2010) Passive and active drug targeting: drug delivery to tumors as an example. *Handb Exp Pharmacol*, 3–53.

Turner, J. J., Jones, S., Fabani, M. M., Ivanova, G., Arzumanov, A. A. & Gait, M. J. (2007) RNA targeting with peptide conjugates of oligonucleotides, siRNA and PNA. *Blood Cells Mol Dis*, 38, 1–7.

Ugolini, S., Mondor, I. & Sattentau, Q. J. (1999) HIV-1 attachment: another look. *Trends Microbiol*, 7, 144–9.

Vilner, B. J., John, C. S. & Bowen, W. D. (1995) Sigma-1 and sigma-2 receptors are expressed in a wide variety of human and rodent tumor cell lines. *Cancer Res*, 55, 408–13.

Vives, E., Brodin, P. & Lebleu, B. (1997) A truncated HIV-1 Tat protein basic domain rapidly translocates through the plasma membrane and accumulates in the cell nucleus. *J Biol Chem*, 272, 16010–7.

von Eckardstein, A., Nofer, J. R. & Assmann, G. (2001) High density lipoproteins and arteriosclerosis. Role of cholesterol efflux and reverse cholesterol transport. *Arterioscler Thromb Vasc Biol*, 21, 13–27.

Wadia, J. S., Stan, R. V. & Dowdy, S. F. (2004) Transducible TAT-HA fusogenic peptide enhances escape of TAT-fusion proteins after lipid raft macropinocytosis. *Nat Med*, 10, 310–5.

Wang, S., Lee, R. J., Mathias, C. J., Green, M. A. & Low, P. S. (1996) Synthesis, purification, and tumor cell uptake of 67Ga-deferoxamine–folate, a potential radiopharmaceutical for tumor imaging. *Bioconjug Chem*, 7, 56–62.

Wang, S. N., Deng, Y. H., Xu, H., Wu, H. B., Qiu, Y. K. & Chen, D. W. (2006) Synthesis of a novel galactosylated lipid and its application to the hepatocyte-selective targeting of liposomal doxorubicin. *Eur J Pharm Biopharm*, 62, 32–8.

Weitman, S. D., Weinberg, A. G., Coney, L. R., Zurawski, V. R., Jennings, D. S. & Kamen, B. A. (1992) Cellular localization of the folate receptor: potential role in drug toxicity and folate homeostasis. *Cancer Res*, 52, 6708–11.

Whitehead, K. A., Langer, R. & Anderson, D. G. (2009) Knocking down barriers: advances in siRNA delivery. *Nat Rev Drug Discov*, 8, 129–38.

Wiradharma, N., Tong, Y. W. & Yang, Y. Y. (2009) Self-assembled oligopeptide nanostructures for co-delivery of drug and gene with synergistic therapeutic effect. *Biomaterials*, 30, 3100–9.

Wu, A. M. & Senter, P. D. (2005) Arming antibodies: prospects and challenges for immunoconjugates. *Nat Biotechnol*, 23, 1137–46.

Wu, G., Barth, R. F., Yang, W., Chatterjee, M., Tjarks, W., Ciesielski, M. J. & Fenstermaker, R. A. (2004) Site-specific conjugation of boron-containing dendrimers to anti-EGF receptor monoclonal antibody cetuximab (IMC-C225) and its evaluation as a potential delivery agent for neutron capture therapy. *Bioconjug Chem*, 15, 185–94.

Xiao, Z., Shangguan, D., Cao, Z., Fang, X. & Tan, W. (2008) Cell-specific internalization study of an aptamer from whole cell selection. *Chemistry*, 14, 1769–75.

Yan, A. C. & Levy, M. (2009) Aptamers and aptamer targeted delivery. *RNA Biol*, 6, 316–20.

Yu, X., Zhang, Y., Chen, C., Yao, Q. & Li, M. (2010) Targeted drug delivery in pancreatic cancer. *Biochim Biophys Acta*, 1805, 97–104.

Yuan, F., Dellian, M., Fukumura, D., Leunig, M., Berk, D. A., Torchilin, V. P. & Jain, R. K. (1995) Vascular permeability in a human tumor xenograft: molecular size dependence and cutoff size. *Cancer Res*, 55, 3752–6.

Zhang, C., Zhao, L., Dong, Y., Zhang, X., Lin, J. & Chen, Z. (2010) Folate-mediated poly(3-hydroxybutyrate-co-3-hydroxyoctanoate) nanoparticles for targeting drug delivery. *Eur J Pharm Biopharm*, 76, 10–16.

Zhang, H. M., Su, Y., Guo, S., Yuan, J., Lim, T., Liu, J., Guo, P. & Yang, D. (2009) Targeted delivery of anti-coxsackievirus siRNAs using ligand-conjugated packaging RNAs. *Antiviral Res*, 83, 307–16.

Zhou, J., Li, H., Li, S., Zaia, J. & Rossi, J. J. (2008) Novel dual inhibitory function aptamer-siRNA delivery system for HIV-1 therapy. *Mol Ther*, 16, 1481–9.

Zhou, J. & Rossi, J. J. (2009) The therapeutic potential of cell-internalizing aptamers. *Curr Top Med Chem*, 9, 1144–57.

Zhou, J., Swiderski, P., Li, H., Zhang, J., Neff, C. P., Akkina, R. & Rossi, J. J. (2009) Selection, characterization and application of new RNA HIV gp 120 aptamers for facile delivery of Dicer substrate siRNAs into HIV infected cells. *Nucleic Acids Res.*

Zhukov, N. V. & Tjulandin, S. A. (2008) Targeted therapy in the treatment of solid tumors: practice contradicts theory. *Biochemistry (Mosc)*, 73, 605–18.

Zrazhevskiy, P. & Gao, X. (2009) Multifunctional Quantum Dots for Personalized Medicine. *Nano Today*, 4, 414–428.

Simulation Based Analysis of Nanocarrier Internalization: Exciting Challenges with a New Computational Tool

Béla Csukás, Mónika Varga, Aleš Prokop, and Sándor Balogh

Abstract A new computational tool was developed, for model-based analysis of the endocytosis and exocytosis mechanisms involved in nanocarrier delivery. This was a hypothetical study because current data are insufficient to identify the underlying process model.

The detailed case studies represent appropriate examples for useful applications of a quite hypothetical model. It allows the study of the individual mechanisms, as well as the synergistic and antagonistic effects of their combinations. It helps to understand how the possible limiting transportations and transformations, as well as the inherent degradations, limit the utilization of the injected drug. In addition, by switching on and off the respective processes, we can evaluate the beneficial or harmful effects of some pathways, like nanocarrier or drug-containing nanocarrier exocytosis.

The applied methodology was flexible and effective, permitting one to describe and run complex process models without any mathematical and computational assistance. The question is whether constructive applications can compensate for the possible malfunctions, caused by the limiting assumptions. Our answer is yes.

The development of this computational model is a valid example for the qualitative identification and validation, organized by the dialogue between field and model experts with the computational model. As such, it can help to organize new round of experiments.

We can state that the future of the computer assisted biosystem related, biotechnological and biomedical studies depends highly on this kind of interactive collaborations.

B. Csukás (✉), M. Varga, and S. Balogh
Department of Information Technology, Kaposvár University, 40 Guba S, Kaposvár 7400, Hungary
e-mail: csukas.bela@ke.hu; varga.monika@ke.hu; balogh.sandor@ke.hu

A. Prokop
NanoDelivery International, Břeclav 69141, Czech Republic
and
Chemical and Biological Engineering, Vanderbilt University, Nashville, TN 37235, USA
e-mail: ales.prokop@vanderbilt.edu

A. Prokop (ed.), *Intracellular Delivery: Fundamentals and Applications*, Fundamental Biomedical Technologies 5, DOI 10.1007/978-94-007-1248-5_5,

Direct Computer Mapping (DCM) seems to be an effective tool for the organization of these collaborations.

Looking at the Chapters of this Volume it is obvious that the majority of the contributions have very sophisticated and deep experimental background, related to various specific details. From an engineering point of views, they are well-elaborated pieces of a puzzle. However, our simplified, but systematic computational model emphasizes the hypothetical big picture. Similarly to bridge building, in computational systems biology we have to construct the bridge from both sides and, hopefully, to meet in the middle.

Keywords Endocytosis • Exocytosis • Internalization and trafficking • Direct Computer Mapping • Biosystem process modeling • Simulation tool

Abbreviations

NV	nanovehicle
L	ligand, making possible the association of the drug with nanovehicle
D	drug (or gene)
R	receptor
Lip	lipid, supporting the specific transport through the lipid raft
Targ	target component in the cytoplasm
Degr	fictitious terminating component for the degradation of NV, L, R and D containing components
Mot	motor proteins, contributing to the exocytosis of nanovehicles and of the drug
FPM	fluid phase macropinocytosis
RME	receptor-mediated endocytosis
NAE	nonspecific adsorptive endocytosis
LRME	lipid-rafts mediated endocytosis
DCM	Direct Computer Mapping
R1-R40	elementary reactions

1 Overview of the Problem to be Modeled

The cellular uptake of extracellular ligands, soluble molecules, proteins and lipids from the extracellular surface occurs by endocytosis, a process where localized regions of the plasma membrane invaginate and pinch off to form endocytic vesicles. Many of the endocytosed molecules and particles travel through the lysosomes and degrade. Endocytosis occurs both constitutively or in a triggered response to extracellular signals. Many cell types undergo endocytosis continuously where a large fraction of the plasma membrane is internalized every hour.

To make this possible, most of the internalized plasma membrane components (proteins and lipids) are continually returned to the cell surface by exocytosis. This large-scale endocytic-exocytic cycle is mediated largely by clathrin-coated pits and vesicles. Major trafficking pathways consist of an inward flux of endocytic vesicles from the plasma membrane and an outward flux of exocytic vesicles to the plasma membrane. Virtually all eucaryotic cells continually ingest their plasma membrane in the form of small pinocytic (endocytic) vesicles, which later return to the cell surface. The same amount of membrane that is being removed by endocytosis is being added to the cell surface by exocytosis. In this sense, endocytosis and exocytosis are linked processes that can be considered to constitute an endocytic-exocytic cycle (Alberts et al. 2002). In this paper, we discuss both arms of the cycle.

Kinetically, three modes of endocytosis can be defined: fluid-phase, adsorptive, and receptor-or lipid raft mediated endocytosis (Khalil et al. 2006): (1) Fluid-phase endocytosis is a low efficiency, nonspecific process that involves the bulk uptake. (2) Absorptive endocytosis molecules are bound to the cell surface and concentrated before internalization, with the molecules interacting with generic complementary binding sites, such as lectin or charged interaction. (3) Receptor-mediated endocytosis also involves concentration of the molecules, with certain ligands binding to receptors on the cell surface and becoming concentrated before internalization. Typically, clathrin-coated pits are involved. In addition, we consider lipid rafts as a separate uptake process.

Specific proteins, including receptors, are removed from early endosomes and recycled to their original plasma membrane domains; some proceed to a different domain of the plasma membrane, thereby mediating a process called transcytosis; and some progress to lysosomes, where they are degraded, while both the receptor and the ligand end up being degraded in lysosomes, resulting in receptor down-regulation. In other cases, both are transferred to a different plasma membrane domain, and the ligand is thereby released by exocytosis at a surface of the cell different from that where it originated (transcytosis). In the case of transcytosis, nanoparticles specifically fuse with the basolateral domain of the plasma membrane. In our model, we consider transcytosis as one of the key processes.

A central question is whether exocytosis occurs for nanoparticles. This is a little appreciated area, although it should be separated from endocytosis (see Alberts et al. above). Experimentally, exocytosis is measured by repeated incubations and wash (void of external nanoparticles) cycles at various time intervals after the initial exposure (Panyam and Labhasetwar 2003). There are several limitations of this approach, particularly in capturing nanoparticle dynamics and quantifying endocytosis and exocytosis rates simultaneously, in a steady-state fashion.

Likewise, the real magnitude and impact of spatiotemporal dynamics can be understood only by carrying out mathematical modeling (Kholodenko et al. 2010). Our recently developed model allows the incorporation of exocytosis in a virtual situation, providing more informative nanoparticle dynamics.

We describe the above events phenomenologically, without going into the deepest details of reaction steps. For simplicity, only the minimal routes are considered in this example. Of course, a number of uptake (endocytic) routes can be considered and explored. In our specific case, we only consider the following three routes: FPM/NAE (fluid phase macropinocytosis, which we lumped together with nonspecific adsorptive endocytosis), RME (receptor-mediated endocytosis) and LRME (lipid-rafts mediated endocytosis) (Prokop and Davidson 2008; Doherty and McMahon 2009) as the most predominant mechanisms. We also consider a molecular motor in our scheme, as cellular motor proteins are responsible for moving structural elements to specific cellular locations and trafficking of endocytic and exocytic vesicles (Holzbaur and Goldman 2010; Robert et al. 2010). We set the exocytosis rate by one order of magnitude less of the endocytosis rate as found experimentally by (Jin et al. 2009). Note, sorting and recycling is not properly delineated as it may occur from different endosomal stages (early endosomes, endosomal compartment and recycling endosome; Grant and Donaldson 2009). Motor proteins may be considered as a recycling vehicle at this simplified stage. There is very limited literature on simulation of endocytosis (see Wattis et al. 2008, on LDL uptake by hepatocytes).

We consider multiple entry mechanisms simultaneously as there are several means of internalization that operate concurrently at the cell surface. Various data suggests that the cellular internalization of nanoparticles occurs through a multifaceted internalization mechanism primarily involving caveolae, yet clathrin-coated vesicles and macropinosomes are also involved to a lesser degree. Various cells may use alternative entry mechanisms to a different degree (Kumari et al. 2010).

Endocytosis also samples the extracellular milieu and serves to regulate various processes initiated at the cell surface. These include nutrient uptake, signaling from cell-surface receptors, and many other processes essential for cell and tissue function, emphasizing why the knowledge and extent of use different routes is of prime importance. The ability to understand and predict the cellular uptake (and exocytosis) of nanoparticles quantitatively should find utility in designing nano-systems with controlled toxicity, efficacy and functionality.

2 Model Design

2.1 *Simplified Scheme of the Model*

Model design usually starts from the qualitative description of the biological processes and the involved components. In this case, we try to develop a preliminary model of the drug delivery related endocytosis and exocytosis. Modeling is limited by the fact that we do not have experimental data for the identification and validation of the model. Rather we use the model to gain a better understanding of the simultaneous trafficking and recycling processes. In addition, a preliminary model can help to plan the appropriate set of possible measurements that supports the identification and validation. Afterwards, with the knowledge gained from the preliminary model, the experimental and computational investigations support each other mutually in the deeper understanding of the investigated problem. Finally, the model can also participate in the further design of the effective drug delivery therapies.

Considering the overview in Chapter 1, the objective of modeling is to describe the essential transport and transformation mechanisms associated with the drug (or gene) delivery from the external phase to the various targets (in cytoplasm, nucleus or mitochondria). Also, exocytosis of the nanoparticle components is considered, e.g. a recirculation of the undelivered drug (or gene) components.

Accordingly, our model has to describe the following processes:

- Fluid phase macropinocytosis (FPM) is a low efficiency, nonspecific process that involves the bulk uptake of solutes in exact proportion to their concentration in the extracellular fluid;
- In nonspecific adsorptive endocytosis (NAE) the molecules are bound to the cell surface and concentrated before internalization, i.e. the molecules interact preferentially with generic complementary binding sites, such as lectin or charged surface elements;
- Receptor mediated endocytosis (RME) also involves concentration of the molecules, with certain ligands bound to receptors on the cell surface and becoming concentrated before internalization. The cytoplasm membrane folds inward to form coated pits, such as clathrin-coated pits. These inward budding vesicles bud to form cytoplasmic vesicles;
- Lipid rafts mediated endocytosis (LRME) is another kind of specific endocytosis, where the transport is supported by special lipids of the membrane channels;
- Exocytosis of the nanovehicles is enhanced by the motor proteins.

In the following model, nonspecific NAE will be taken into consideration in nonspecific FPM (called FPM/NAE).

2.2 *Compartments and Components*

In the cellular compartmentalization of nanoparticles, the specific role of the so-called early and late endosomes, and lysosomes have to be taken into consideration. These specific subsystems of the cell are involved in the trafficking and degradation of the receptor mediated components.

The cellular compartments with estimated percentages are the followings (Alberts et al. 2002, see Table 12-1, p. 661; note, this composition is reported for hepatocytes): cytosol (cyt): 54%; plasma membrane (pm): 0.5% (ER function in endo/exocytosis is not considered in this paper); mitochondria (mit): 22%; nucleus (nuc): 6%; early endosomes (ee): 1%; late endosomes (le): 1%; and lysosomes (lys): 1%. The capacity of the external phase is taken into account by an external/cytoplasm ratio of approximately 1:1.

The primary building elements of the various biological compounds are the followings: NV: nanovehicle; L: ligand, permitting the association of the drug (D) with the nanovehicle (NV); D: drug (or gene) to be transported from the external phase to the prescribed target (to the cytoplasm, to the mitochondria or e.g. in case of genes to the nucleus); R: receptor, supporting the specific transport through the plasma membrane; Lip: lipid supporting the specific transport through the lipid rafts; Targ: target component in the cytoplasm; Degr: fictitious terminating component for the degradation of NV, L, R, and D in e.g. ubiquitin containing components; Mot: motor proteins, contributing to the exocytosis of nanovehicles and of the drug.

2.3 *Transformations and Transportations*

For the better understanding, the "biologically inspired" scheme of our simplified model is illustrated in Fig. 1. The Figure shows a cell, surrounded by a plasma membrane in the external phase. The above-mentioned compartments are designated by colored fields.

Black dots symbolize the various primary components and their derived complex intermediates. There are 34 individual components in the respective phases. According to the Figure, in the notation

$$comp\_X\_Y\_\ldots\_Z$$

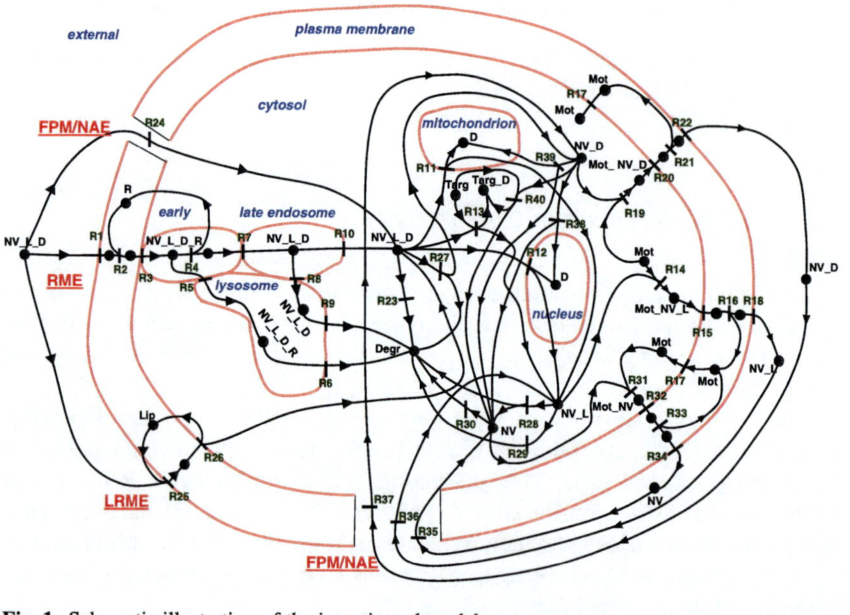

Fig. 1 Schematic illustration of the investigated model

designates a complex, consisting of primary components of X, Y and Z in the compartment called "comp". The actual names of the compartments and primary components correspond to the ones defined in Chapter 2.2.

There are two main classes of the elementary processes in the model, namely

- transformations, determining the composition and decomposition of the complexes, built from the primary components in a given compartment; as well as
- transportations for trafficking of the given component or complex between and/or along the compartments.

The 40 elementary processes, illustrated by small bars in Fig. 1, are collected in Table 1. It is to be noted that the set of compartments, components, transportations and transformations can be modified by the field expert. The objective of the present case study is rather to illustrate how one can build a hypothetical model with roughly estimated data, than to explain sophisticated model. The implementation and the case studies demonstrate only the possible power of consulting with a computational tool in the discovery, design or control of the underlying processes.

Table 1 Elementary processes of the model

Process number	Elementary processes	Description of the elementary process
R1	ext_NV_L_D→pm_NV_L_D	Transport of drug (or gene) containing NV_L_D complex from the external phase to the plasma membrane
R2	pm_NV_L_D+pm_R→pm_NV_L_D_R	Binding of drug (or gene) containing NV_L_D complex to the receptor in the plasma membrane
R3	pm_NV_L_D_R→ee_NV_L_D_R	Transport of drug (or gene) containing NV_L_D_R from the plasma membrane to the early endosomes
R4	ee_NV_L_D_R→ee_NV_L_D+pm_R	Transformation and transport of drug (or gene) containing NV_L_D_R complex in the early endosomes, and the dissociation of receptor, returning into the plasma membrane
R5	ee_NV_L_D_R→lys_NV_L_D_R	Transport of drug (or gene) containing NV_L_D_R complex from the early endosomes to the lysosomes
R6	lys_NV_L_D_R→cyt_Degr	Decomposition of drug (or gene) containing NV_L_D_R complex in the lysosomes
R7	ee_NV_L_D→le_NV_L_D	Transport of drug (or gene) containing NV_L_D complex from the early endosomes to the late endosomes
R8	le_NV_L_D→lys_NV_L_D	Transport of drug (or gene) containing NV_L_D complex from the late endosomes to the lysosomes
R9	lys_NV_L_D→cyt_Degr	Decomposition of drug (or gene) containing NV_L_D complex in the lysosomes

(continued)

Table 1 (continued)

Process number	Elementary processes	Description of the elementary process
R10	le_NV_L_D→cyt_NV_L_D	Transport of drug (or gene) containing NV_L_D complex from the late endosomes to the cytosol
R11	cyt_NV_L_D→cyt_NV_L+mit_D	Transfer of the drug to the mitochondria
R12	cyt_NV_L_D→cyt_NV_L+nuc_D	Transfer of the gene (drug) to the nucleus
R13	cyt_NV_L_D+cyt_Targ→cyt_NV_L+cyt_Targ_D	Delivery of the drug to the target in the cytosol
R14	cyt_NV_L+cyt_Mot→cyt_Mot_NV_L	Binding of free nanovehicle NV_L to the motor protein in the cytosol
R15	cyt_Mot_NV_L→pm_Mot_NV_L	Transport of motor protein guided NV_L complex from the cytosol to the plasma membrane
R16	pm_Mot_NV_L→pm_NV_L +pm_Mot	Decomposition of motor protein guided NV_L complex in the plasma membrane
R17	pm_Mot→cyt_Mot	Transport of motor protein from the plasma membrane to the cytosol
R18	pm_NV_L→ext_NV_L	Exocytosis of nanovehicle complex NV_L from the plasma membrane to the external fluid
R19	cyt_NV_D+cyt_Mot→cyt_Mot_NV_D	Binding of drug (gene) transporting nanovehicle complex NV_D to the motor protein in the cytosol
R20	cyt_Mot_NV_D→pm_Mot_NV_D	Transport of motor protein guided NV_D complex from the cytosol to the plasma membrane
R21	pm_Mot_NV_D→pm_NV_D+pm_Mot	Decomposition of motor protein guided NV_D complex in the plasma membrane
R22	pm_NV_D→ext_NV_D	Exocytosis of drug (gene) transporting nanovehicle complex NV_D from the plasma membrane to the external fluid
R23	cyt_NV_L_D→cyt_Degr	Decomposition of drug (or gene) containing NV_L_D complex in the cytosol
R24	ext_NV_L_D→cyt_NV_L_D	Fluid phase macropinocytosis of the complex NV_L_D from the external phase to the cytosol
R25	ext_NV_L_D+pm_Lip→pm_Lip_NV_L_D	Lipid raft mediated endocytosis of drug (or gene) containing NV_L_D complex to the plasma membrane
R26	pm_Lip_NV_L_D→cyt_NV_L_D+pm_Lip	Lipid raft mediated internalization of NV_L_D complex to the cytosol with proliferating lipid in the plasma membrane
R27	cyt_NV_L_D→cyt_NV_D+cyt_Degr	Degradation of NV_L_D in cytosol
R28	cyt_NV_L→cyt_NV+cyt_Degr	Decomposition of NV_L with degradation of L in cytosol
R29	cyt_NV_L→cyt_Degr	Degradation of NV_L in cytosol

(continued)

Table 1 (continued)

Process number	Elementary processes	Description of the elementary process
R30	cyt_NV → cyt_Degr	Degradation of NV in cytosol
R31	cyt_NV + cyt_Mot → cyt_Mot_NV	Binding of free nanovehicle NV to the motor protein in the cytosol
R32	cyt_Mot_NV → pm_Mot_NV	Transport of motor protein guided NV from the cytosol to the plasma membrane
R33	pm_Mot_NV → pm_NV + pm_Mot	Decomposition of motor protein guided NV in the plasma membrane
R34	pm_NV → ext_NV	Exocytosis of nanovehicle NV from the plasma membrane to the external fluid
R35	ext_NV → cyt_NV	Endocytosis of nanovehicle NV from the external fluid to the cytosol
R36	ext_NV_L → cyt_NV_L	Endocytosis of NV_L complex from the external fluid to the cytosol
R37	ext_NV_D → cyt_NV_D	Endocytosis of NV_D complex from the external fluid to the cytosol
R38	cyt_NV_D → cyt_NV + nuc_D	Delivery of NV_D containing drug to the nucleus
R39	cyt_NV_D → cyt_NV + mit_D	Delivery of NV_D containing drug to the mitochondrion
R40	cyt_NV _D + cyt_Targ → cyt_NV + cyt_Targ_D	Delivery of NV_D containing drug to the cytosol target

3 Implementation of the Computational Model

3.1 *Direct Computer Mapping (DCM) of Process Models*

The computer modeling of the functional and structural features of various processes evolved in two, more or less separated ways.

The functional models can usually be described by sets of algebraic, differential and/or integral equations (IPDAE) (Marquardt 1996). The *'a priori'* (white box) models can be derived from simple first principles, and then they are transformed into various sophisticated mathematical constructs. The *'a posteriori'* (black box) models are based on various flexible mathematical constructs with "artificial" parameters to be identified from the available data. Accordingly, they differ from the *'a priori'* models only in the origin and content of the equations, but the numerical solution of the mathematical equations is similar. The general formal model of these process systems developed before powerful computers appeared (Kalman et al. 1969).

Nowadays, the sophisticated mathematical constructs address the unified solution of the continuous and discrete problems.

The structural features of the process model are usually represented by network structures. Originally, they were designed for the illustration of the static structure, as well as for the steady state or dynamically changing fluxes along the network routes. However, starting in the early 1960s, the very innovative Petri Net (Petri 1962) and the subsequent General Net Theory (e.g. Brauer 1980) proposed a net model for the description of discrete process structures that inspired other avenues for representation of dynamic processes. In the past decade, extensive efforts have been devoted towards the implementation of the quantitative functionalities in the structural model, e.g. in the form of the Quantitative Petri Nets (Chen and Hofestaedt 1999; Peleg et al. 2005).

Nevertheless, the execution of the hybrid, discrete/continuous and optionally multiscale models is a difficult question, because the usual integrators do not tolerate the discrete events, while the usual representation of the continuous processes cannot be embedded into the discrete models conveniently (Meier-Schellersheim et al. 2009). Another challenge is the effective combination of quantitative models with rule-based qualitative knowledge. Moreover, a multiscale approach is necessary to treat the hierarchy of interactions ranging from molecular (nm, ns) to signaling (μm, ms) length and time scales, which necessitates the development and application of specialized modeling tools (Shih et al. 2008).

Having recognized the problem discussed above, in Direct Computer Mapping (Csukas and Perez Uriza 1995; Csukas et al. 1996; Csukas and Balogh 1998), the natural building blocks of the balance based (conservational) and rule based (informational) processes are mapped onto the state and transition elements of an executable program. The discrete or continuous, as well as quantitative or qualitative functioning are described by expert defined program prototypes. These brief programs are associated with the structural elements and executed by a general kernel (Csukas 2001; Csukas et al. 2005). The idea of DCM is illustrated in Fig. 2.

Related to the above approaches, the new methodology can be interpreted as a "naturally structured" Kalman model or a "very high order" Petri net of prototyped programs. As illustrated in Fig. 3, the methodology describes the essential structure of the processes, as well as the structured distribution and optionally parallel execution of the distributed elementary functionalities.

The generic structure of the process models can be characterized by a feedback structure of the "passive" state elements (P) and of the "active" elementary transitions (A).

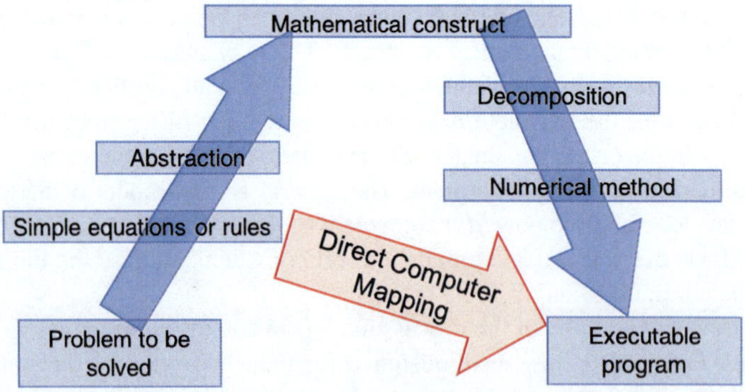

Fig. 2 Idea of direct computer mapping

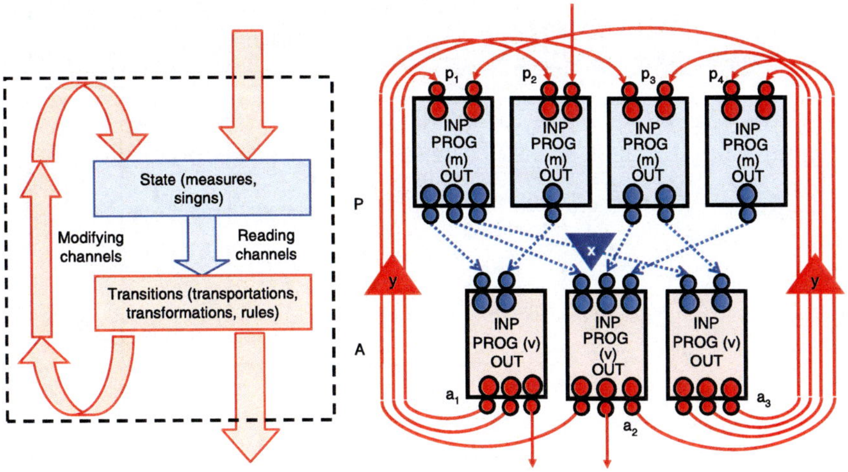

Fig. 3 General architecture of the process models in sense of direct computer mapping

The discrete or continuous and the quantitative or qualitative functioning of the structural modules can be determined by the expert-defined brief programs (m or v). The passive output→active input and the active output→passive input connections are described by the reading (X) and modifying (Y) connections, respectively. Considering our example model, we describe "passive" state elements for the components in the compartments, as well as "active" transition elements for the transportations and transformations summarized in Table 1. The calculations are described by the brief programs, associated with these elements, represented by boxes. The various edges (lines) correspond to the information flow between the elements. Accordingly the system determines and makes possible the execution of a dynamic model without the usual mathematical formalisms and without any special numerical solver (that is embedded in the general kernel, see in Fig. 4).

The method can be applied for the computer modeling, dynamic simulation and simulation based problem solving of various process systems (Csukas et al. 1999, 2000; Nagy et al. 2001; Temesvari et al. 2004; Varga et al. 2010).

The schematic architecture of the computer implementation is illustrated in Fig. 4. The brief programs determine the calculation of the active (v) and passive (m) functionalities. The respective expressions are declared by the expert. The structure and the parameters of the actual problem are described by the user.

The execution consists of four cyclically repeated consecutive steps, as follows:

1. active elements read the content of the associated passive elements through to the reading channels;
2. brief programs, associated with active elements calculate the changes;
3. passive elements are modified according to the changes carried via modifying channels;
4. brief programs, associated with passive elements calculate the new state.

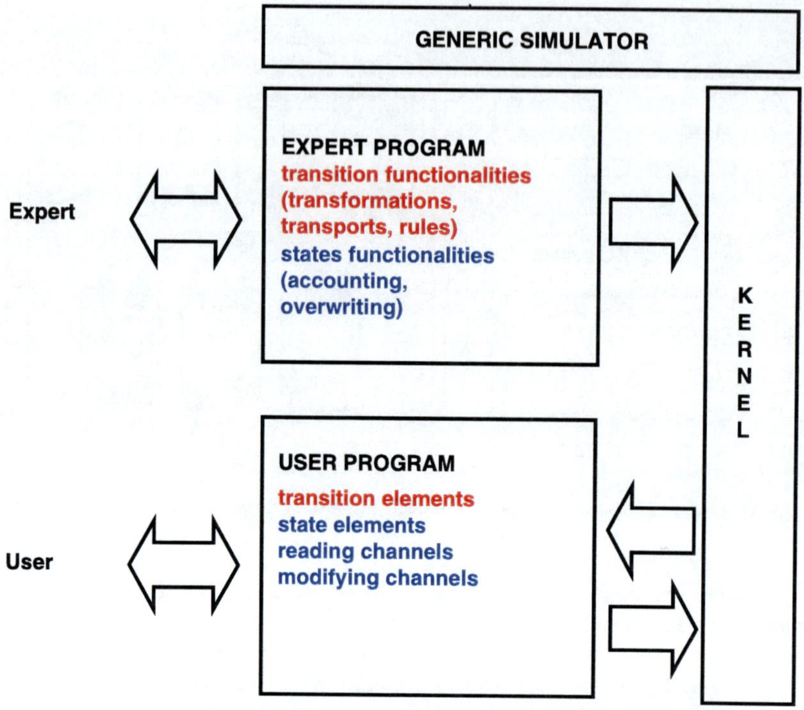

Fig. 4 Computer implementation of direct computer mapping

The recent implementation of the above architecture is written in GNU-Prolog (http://www.gprolog.org/), because the free unification of the high-level structures, contained in dynamic partitions, supports the methodology. Temporarily, we use a GraphViz (http://www.graphviz.org/) based configurable input/output user interface (Varga 2009). This is a platform-independent and partly open software tool, consisting of a general kernel, communicating with expert and user interfaces.

The software implementation can be applied for the simulation of multi-scale hybrid (discrete/continuous, quantitative/qualitative, deterministic/stochastic) processes. In hybrid models, the communication of the different parts can be supplied with the necessary transforming operators.

3.2 *DCM Description of the Internalization Problem*

In the temporarily used software implementation of Direct Computer Mapping, first we edit the structure of the model by the open source code GraphViz tool that generates a script. The graphically edited structure of the discussed model (see Fig. 1) is shown in Fig. 5. In this graphical representation

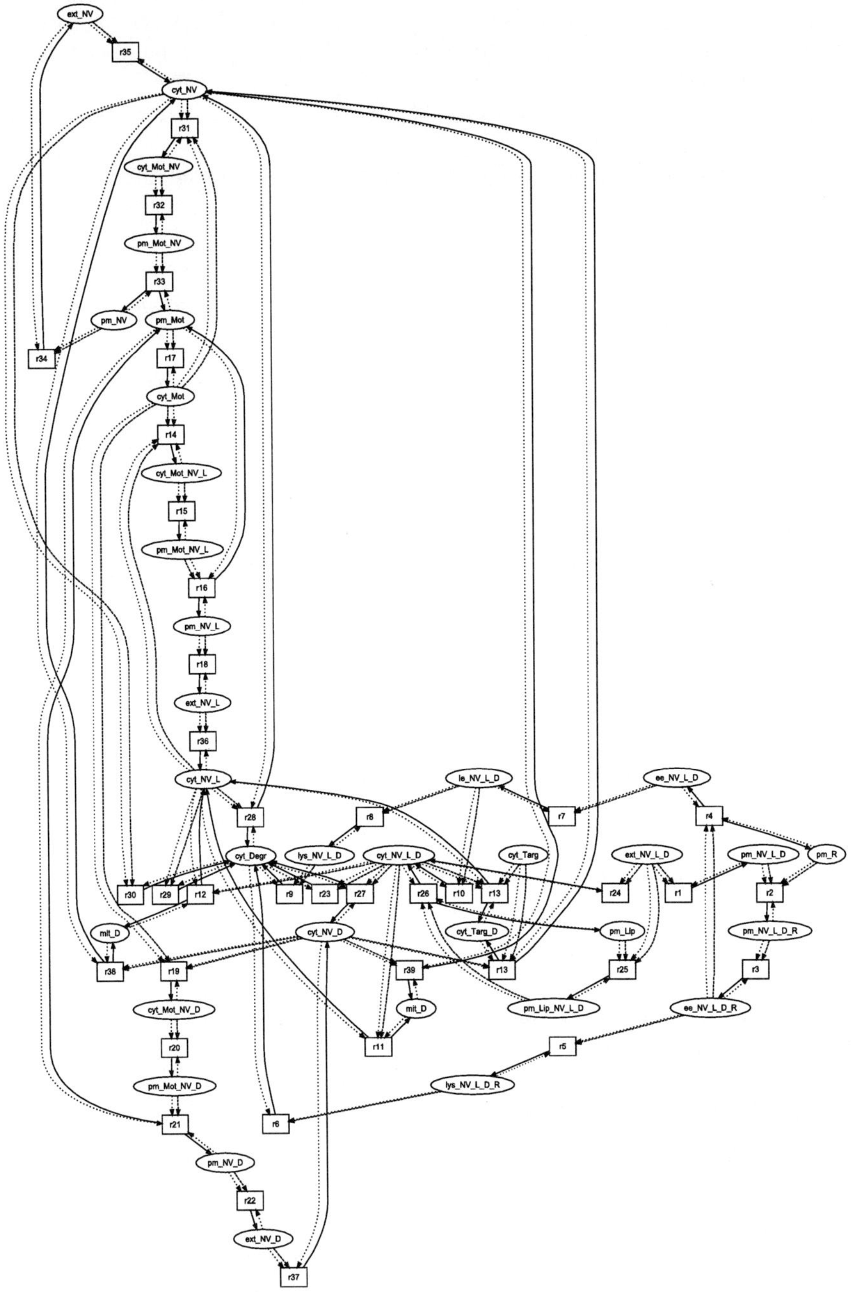

Fig. 5 GraphViz representation of the model of Fig. 1

- ellipses correspond to the state elements (primary and complex components);
- boxes represent the elementary processes (transformations and transportations);
- dotted lines declare the reading connections that forward the intensive parameters (e.g. concentrations) from the output slots of the state elements to the input slots of the transition elements;
- dashed lines define the modifying connections that forward the decreasing changes of the extensive characteristics (e.g. moles of components) from the output slots of the transition elements to the input slots of the state elements;
- solid lines define the modifying connections that forward the increasing changes of the extensive characteristics (e.g. moles of components) from the output slots of the transition elements to the input slots of the state elements.

Description of the model means that the automatically generated. DOT file is extended with the detailed definition of the process model. As an example, we use a small part of the extended script, related to the decomposition of ee_NV_L_D_R (only the characteristic parts will be shown).

The ee_NV_L_D_R complex refers to NV_L_D_R being in the inner side of the plasma membrane, in compartment of early endosome. The respective state element is defined by the declaration

```
'ee_NV_L_D_R' [
        shape='ellipse'
        model='component'
        data='[]'
        inp='[i(comp,dl,[d(ee,[0.01],nd),d(quant,[0],nd)])]'
        out='[]' ]
```

where model of ‘component’ refers to the respective prototype, declared at another state element, as

```
newprototype='component'
program='m(y,component,[],[i(comp,dl,Extenziv)],[o(conc,dl,Intenziv)])
:-calculate_intensive(Extenziv,Intenziv),!.'
```

while the calculate_intensive() predicate calls for the general clause, determining the concentrations of the components in the compartments.

The list “inp” contains the initial values for the mass of the compartment ee (= 0.01) and for the respective component NV_L_D_R (= 0).

The decomposition of this complex, accompanied by “returning” of the receptor to the plasma membrane is declared, as follows:

```
'r4' [shape='box'
      newprototype='decomp&transport'
      data='[c(par,dl,[d(kin,[0.1],nd),d(eq,[100000],nd)])]'
      program='v(y,
'decomp&transport,[c(par,dl,[d(kin,[Kin],nd),d(eq,[Eq],nd)])],
       [i(inp_1,dl,[d(Compartment1,[Comp1],nd),d(quant,[C1],nd)]),
       i(inp_2,dl,[d(Compartment2,[Comp2],nd),d(quant,[C2],nd)]),
       i(inp_3,dl,[d(Compartment1,[w],nd),d(quant,[C3],nd)])],
```

```
[o(minus_1,dl,[d(Compartment1,[DComp1],nd),d(quant,[DM1],nd)]),
o(plus_1,dl,[d(Compartment2,[DComp2],nd),d(quant,[DM2],nd)]),
o(plus_2,dl,[d(Compartment1,[DComp1],nd),d(quant,[DM3],nd)])]) :-
g(dt,DT), DComp1 is 0.0, DComp2 is 0.0,
Rate is Kin*(C1-(1/Eq)*C2*C3)*Comp1*DT,
DM1 is (-1)*Rate, DM2 is Rate, DM3 is Rate.' ]
```

It is a transition prototype with two (kinetic and equilibrium) parameters, as well as three input and three output slots, corresponding to the formulation of the "1–2" kinetic (mass action) expression. The three input slots correspond to the reading of concentration of the components, while the three output slots refer to the decrease of ee_NV_L_D_R, as well as to the increase of ee_NV_L_D and pm_R. The method makes possible also the change of the compartments' mass, but right now it is neglected (DComp1 = DComp2 = 0).

The rate of the process prototype is calculated by the bold italics signed part of the program, declared in the given Prolog clause. These brief program parts can be edited freely through a general interface. Accordingly, the open program code can be adapted to the various special hybrid and multiscale problems, easily.

An example for a reading connection is the following:

```
'ee_NV_L_D_R'→'r4' [
        style='dotted'
        sendop='read'
        sendslot='conc'
        receiveop='write'
        receiveslot='inp_1' ]
```

This declares reading of ee_NV_L_D_R concentration for the above discussed elementary process. The shown, significant definitions determine the sending slot and operator, as well as the receiving slot and operator, respectively. When running, the connection will carry the actual sign from the sending to the receiving element. As another example the modifying connection

```
'r4'→'ee_NV_L_D' [
        style='solid'
        sendop='read'
        sendslot='plus_2'
        receiveop='add'
        receiveslot='comp' ]
```

declares increasing of ee_NV_L_D concentration caused by the given transition. The shown significant definitions determine the respective sending and receiving slots and operators. When running, the connection will carry the actual molar change from the sending to the receiving slots.

Having started from this temporary input file, the general DCM process simulator first interprets the script and prepares the GNU-Prolog declarations of the expert and user modules. In the user's module, there are four types (partitions) of predicates, describing the state and transition elements, as well as the reading and modifying (increasing, decreasing or overwriting) connections.

The user module contains altogether 34 p() states, 40 a() transitions, 96 x() reading and 96 y() modifying connections. Its smaller part, generated by the above example script is the following:

```
p(y,ee_NV_L_D_R,[],component,[],[i(comp,dl,[d(ee,[0.01],nd),d(quant,[0],
nd)])],[],[],[],[]).
a(y,r4,[],decomp&transport, [c(par,dl,[d(kin,[0.1],nd),d(eq,[100000],nd)])],
      [i(inp_1,dl,[d(ee,[],nd),d(quant,[],nd)]),
      i(inp_2,dl,[d(pm,[],nd),d(quant,[],nd)]),
      i(inp_3,dl,[d(ee,[],nd),d(quant,[],nd)])],[],[],[],[]).
x(y,read,ee_NV_L_D,[],conc,write,r4,[],inp_3,dl,[],nd,[],[],[]).
y(y,read,r4,[],plus_2,addinc,ee_NV_L_D,[],comp,dl,[],nd,[],[],[]).
```

In the expert's module there are altogether 1 m() state functioning and 8 v() transition clauses for the declaration of functioning of the state and transition elements, respectively. Two clauses, generated by the above example script are the followings:

```
m(y,component,[],[i(comp,dl,Extenziv)],[o(conc,dl,Intenziv)]):-
ujintenziv(Extenziv,Intenziv),!.
v(y, decomp&transport,[c(par,dl,[d(kin,[Kin],nd),d(eq,[Eq],nd)])],[
       i(inp_1,dl,[d(Compartment1,[Comp1],nd),d(quant,[C1],nd)]),
       i(inp_2,dl,[d(Compartment2,[Comp2],nd),d(quant,[C2],nd)]),
       i(inp_3,dl,[d(Compartment1,[Comp1],nd),d(quant,[C3],nd)])],[
       o(minus_1,dl,[d(Compartment1,[DComp1],nd),d(quant,[DM1],nd)]),
       o(plus_1,dl,[d(Compartment2,[DComp2],nd),d(quant,[DM2],nd)]),
       o(plus_2,dl,[d(Compartment1,[DComp1],nd),d(quant,[DM3],nd)])]) :-
       g(dt,DT), DComp1 is 0.0, DComp2 is 0.0,
       Rate is Kin*(C1-(1/Eq)*C2*C3)*Comp1*DT,
       DM1 is (-1)*Rate, DM2 is Rate, DM3 is Rate.
```

During simulation, the general purpose DCM kernel consults the expert's and user's files, and organizes the simulation of the model. The prescribed output is automatically saved in a file that can be shown in runtime or afterwards.

3.3 *Example for the Simulated Results*

First, we will show an example for the input data and output results of a typical simulation. The state elements and the respective initial concentrations are summarized in Table 2. In Table 2, the columns refer to the various compartments with their reference measures (which express the estimated ratios of the compartments). The components are listed in the rows, and the blue colored cells represent the possible compartmental components. As indicated in Table 2, there are only five non-zero initial concentrations, as follows:

ext_NV_L_D

that determines the components describing the injected external component for the delivery, and

pm_R, pm_Lip, cyt_Targ, cyt_Mot

that describe the components contained in the cell. We emphasize that the given simplified set of the components can be extended in a more realistic case study easily. The quite arbitrary compartmental volumes and concentrations can also be modified.

Table 3 shows the transition elements. The columns correspond to the various characteristics of the transitions, as follows:

process ID, that corresponds to the same in Fig. 1;
roughly estimated process rate (kinetic parameter) in a second time basis;
equilibrium parameter (the large numbers mean that the given process is almost unidirectional);
name(s) of the rate determining components;
name(s) of the (decreasing) input components,
name(s) of the (increasing) output components.

The architecture of the rows illustrates the eight prototypes of the elementary processes, which are the followings:

Formation: two components associate forming the product;
Formation with transport: two components from two compartments associate forming the product;
Transport: one component moves from one compartment to the other;
Decomposition: one component dissociates to two products;
Degradation with transport: one component disappears and the amount of the waste components increases in another compartment;
Decomposition with transport: similar to the decomposition, however one produced component moves to another compartment;
2:2 transformation: a component moves from one complex to another one and *vice versa*;
Degradation: one component disappears and the amount of the waste components increases in the same compartment.

The eight prototypes correspond to the eight v() clauses of the expert module.

It is to be noted that the modeled set of the transitions can be extended in a more realistic investigation, easily. For example one can add the so-called "sorting endosomes", can take into account the independent transport of the drug, released from the nanocarrier. User can also take into consideration the more sophisticated model of the ubiquitin-proteosome system in the protein degradation, etc. The roughly estimated kinetic, equilibrium and other parameters can also be refined. Moreover, our simulation model, combined with a so-called genetic algorithm (a collaborating computational tool) can be applied for the identification of the structure and parameters of the model with the knowledge of the experimentally measured data, by itself. In addition,

Table 2 State elements and initial concentrations

	Compartment	ext (External)	pm (Plasma membrane)	cyt (Cytosol)	ee (Early endosome)	le (Late endosome)	lys (Lysosome)	mit (mitochondrial)	nuc (nucleus)
	Reference measures of compartments	5	0.005	5.4	0.01	0.01	0.01	2.2	0.6
Components, SU	**NV**	0	0	0					
	NV_L	0	0	0					
	D							0	0
	NV_L_D	0.3	0	0	0	0	0		
	NV_D	0	0	0					
	R		10						
	NV_L_D_R		0		0		0		
	Degr			0					
	Targ			0.2					
	Targ_D			0					
	Mot		0	0.3					
	Mot_NV_L		0	0					
	Mot_NV		0	0					
	Mot_NV_D		5	0					
	Lip		0						
	Lip_NV_L_D		0						

Table 3 Transition elements and initial concentrations of the model

Number of process	Process rate	Equilibrium	Rate determining components	Decreasing components	Increasing components
Formation					
R2	0.05	100000	pm_NV_L_D, pm_R	pm_NV_L_D, pm_R	pm_NV_L_D_R
R14	0.05	100000	cyt_NV_L, cyt_Mot	cyt_NV_L, cyt_Mot	cyt_Mot_NV_L
R19	0.01	100000	cyt_NV_D, cyt_Mot	cyt_NV_D, cyt_Mot	cyt_Mot_NV_D
R31	0.05	100000	cyt_NV, cyt_Mot	cyt_NV, cyt_Mot	cyt_Mot_NV
Formation with transport					
R25	0.0005	100000	ext_NV_L_D, pm_Lip	ext_NV_L_D, pm_Lip	pm_Lip NV_L_D
Transport					
R1	0.005	100000	ext_NV_L_D	ext_NV_L_D	pm_NV_L_D_R
R3	0.05	100000	pm_ NV_L_D_R	pm_ NV_L_D_R	ee_ NV_L_D_R
R5	0.01	100000	ee_ NV_L_D_R	ee_ NV_L_D_R	lys_NV_L_D_R
R7	0.1	100000	ee_ NV_L_D	ee_ NV_L_D	le_ NV_L_D
R8	0.01	100000	le_ NV_L_D	le_ NV_L_D	lys_ NV_L_D
R10	0.1	100000	le_ NV_L_D	le_ NV_L_D	cyt_ NV_L_D
R15	0.05	100000	cyt_ Mot_NV_L	cyt_Mot NV_L	pm_Mot_NV_L
R17	0.05	100000	pm_Mot	pm_Mot	cyt_Mot
R18	0.05	100000	pm_NV_L	pm_NV_L	ext_NV_L
R20	0.01	100000	cyt_Mot_NV_D	cyt_Mot_NV_D	pm_Mot_NV_D
R22	0.05	100000	pm_NV_D	pm_NV_D	ext_NV_D
R24	0.0001	1000	ext_NV_L_D	ext_NV_L_D	cyt_NV_L_D
R32	0.01	100000	cyt_Mot_NV	cyt_Mot_NV	pm_Mot_NV
R34	0.05	100000	pm_NV	pm_NV	ext_NV
R35	0.0001	1000	ext_NV	ext_NV	cyt_NV
R36	0.0001	1000	ext_NV_L	ext_NV_L	cyt_NV_L
R37	0.0001	1000	ext_NV_D	ext_NV_D	cyt_NV_D

(continued)

Table 3 (continued)

Number of process	Process rate	Equilibrium	Rate determining components	Decreasing components	Increasing components
Decomposition					
R16	0.05	100000	pm_Mot_NV_L	pm_Mot_NV_L	pm_Mot, pm_NV_L
R21	0.01	100000	pm_Mot_NV_D	pm_Mot_NV_D	pm_Mot, pm_NV_D
R27	0.01	100000	cyt_ NV_L_D	cyt_ NV_L_D	cyt_NV_D, cyt_Degr
R28	0.01	100000	cyt_NV_L	cyt_ NV_L	cyt_NV_D, cyt_Degr
R33	0.01	100000	pm_Mot_NV	pm_ Mot_NV	pm_NV, pm_Mot
Degradation with transport					
R6	0.001	100000	lys_NV_L_D_R	lys_NV_L_D_R	cyt_Degr
R9	0.001	100000	lys_NV_L_D	lys_NV_L_D	cyt_Degr
Decomposition with transport					
R4	0.1	100000	ee_NV_L_D_R	ee_NV_L_D_R	pm_R, ee_NV_L_D
R11	0.025	100000	cyt_NV_L_D	cyt_NV_L_D	mit_D, cyt_NV_L
R12	0.01	100000	cyt_NV_L_D	cyt_NV_L_D	nuc_D, cyt_NV_L
R26	0.005	100000	pm_Lip_NV_L_D	pm_Lip_NV_L_D	cyt_ NV_L_D, pm_Lip
R38	0.01	100000	cyt_NV_D	cyt_NV_D	nuc_D, cyt_NV
R39	0.025	100000	cyt_NV_D	cyt_NV_D	mit_D, cyt_NV
Transformation (2:2)					
R13	0.03	100000	cyt_NV_L_D, cyt_Targ	cyt_NV_L_D, cyt_Targ	cyt_Targ_D, cyt_NV_L
R40	0.03	100000	cyt_NV_D, cyt_Targ	cyt_NV _D, cyt_Targ	cyt_Targ_D, cyt_NV
Degradation					
R23	0.001	100000	cyt_NV_L_D	cyt_NV_L_D	cyt_Degr
R29	0.001	100000	cyt_NV_L	cyt_NV_L	cyt_Degr
R30	0.001	100000	cyt_NV	cyt_NV	cyt_Degr

the hypothetical simulations can also help to find the most effective possible measurements from the identification point of views.

The simulated results can be seen in Figs. 6 and 7. Because of the different volume of the various compartments, we show the calculated extensive amounts instead of concentrations, for the easier comparison. In the following Figures you will see the changes in the concentration of the various components, calculated by the hypothetical, simplified model, with the roughly estimated parameters, summarized in Tables 2 and 3. The results come from the automatic solution of the automatically generated model, prepared from the appropriate description language.

Figure 6 gives an overview of initial, external and terminal components. The Figure shows an explicit picture, outlining the utilization of the drug. The external drug level continuously decreases, and it is consumed within 10 min. Afterwards, the cell utilizes (or decomposes) the drug molecules, present in the various complexes. It is to be noted that we illustrated all of the targets. However, in real cases it might be very rare. Usually drugs travel to the cytosol or to mitochondria, while genes enter the nucleus. In our hypothetical example, the mitochondria target is preferred, followed by the nucleus target. Actually, there is a considerable overall degradation rate. Please pay attention to the fact that everything is controlled by the fictitious, hypothetical parameters, tuned according to some qualitative statements of the field expert.

Figure 7 illustrates the changing amount of transient components. The Figure depicts the dynamics of the subsequent processes. The injected drug is consumed within 10 min, and then all of the drug (or gene) containing components converge to zero. Regarding the exocytosis related components, it is clear that in the first

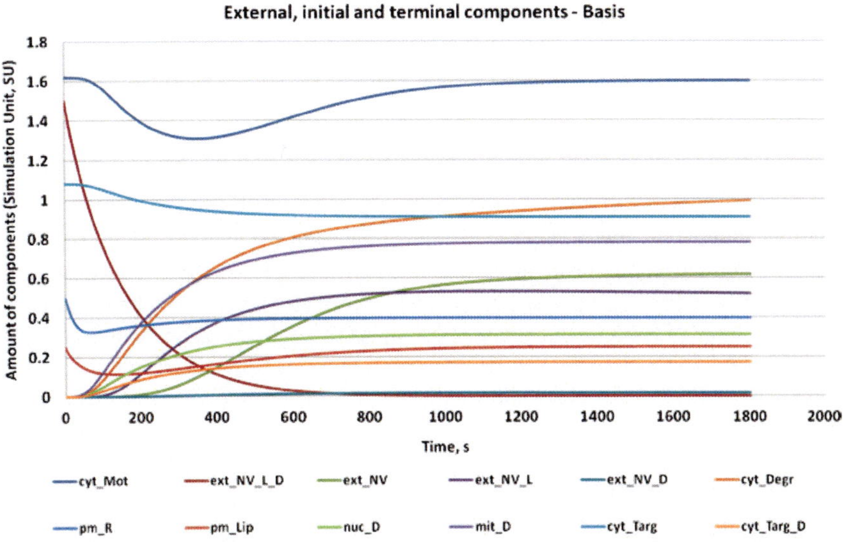

Fig. 6 Initial, external and terminal components

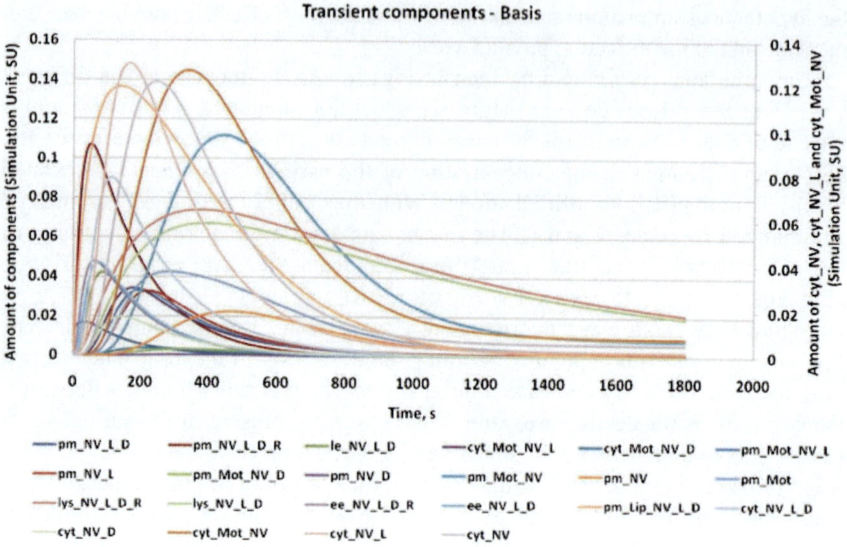

Fig. 7 Transient components

period the motor proteins will bind various complexes, but later on all of them are either delivered, or degraded, and finally the motor proteins will be free again. The receptor containing components converge to zero. The external concentration of NV and NV_L is greater than the same for NV_D, because it can deliver D after recycling.

4 Analysis of Hypothetical Case Studies

4.1 Comparison of the Individual Mechanisms

It is an interesting question how the three mechanisms (FPM/NAE, RME and LRME) collectively contribute to drug delivery. It provides the investigator a chance to tune the model with the knowledge of the expert's considerations (say, the percentage of receptor mediated endocytosis is estimated to be 70%, etc.).

DCM fully supports the discrete, structural changes, directly. Accordingly, we can switch off two mechanisms, only by changing the status flag of two elementary processes from 'y' to 'n'.

In the following two series of Figures we compare the dynamically interesting periods of four models, as follows:

complex model: upper left diagrams,
mechanism RME alone: upper right diagram,

mechanism FPM/NAE alone: lower left diagram,
mechanism LRME alone: lower right diagram.

Please consider that the amount (ordinate) scale, is different for the four diagrams about the transient components.

Figure 8 shows that the dominant mechanism is the RME. We see clearly that the major portion of delivery is carried with RME and LRME, while the percentage of FPM/NAE is less (in this order). However, it is very interesting that the three mechanisms together do not transport significantly more drug, than the receptor mediated mechanism alone. It means that the transporting capability is limited (inside the cell).

Figure 9 shows, that in the case of FPM/NAE or LRME, only one or two complexes take place in the delivery with a higher amount. The amount of exocytosis-related components is almost the same for the complex and for the RME cases. LRME and FPM/NAE mechanisms alone needs less amount of exocytosis related components, of course. Change of the nanovehicle components is in accordance with the above diagnosis. The receptor containing components are utilized only in the RME and complex cases. The very low concentrations, appearing in the FPM/NAE and LRME schemes are caused by the equilibrium character of the reactions and transports, which can produce a small amount of these complexes "backwards" (similar situation might appear in various biosystems). It is interesting that LRME alone can utilize more lipid rafts related resources, than the three mechanisms together.

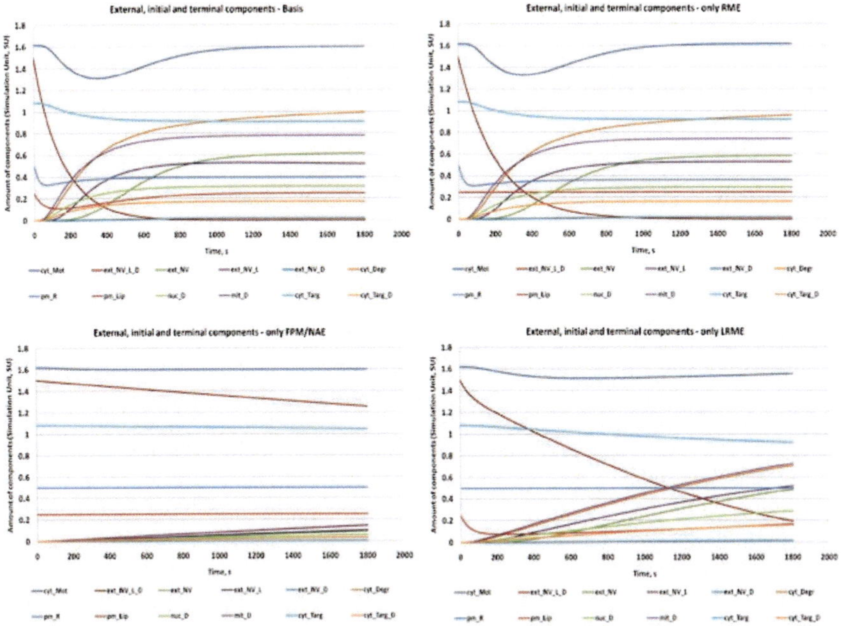

Fig. 8 External, initial and terminal components in the complete model and in the individual mechanisms

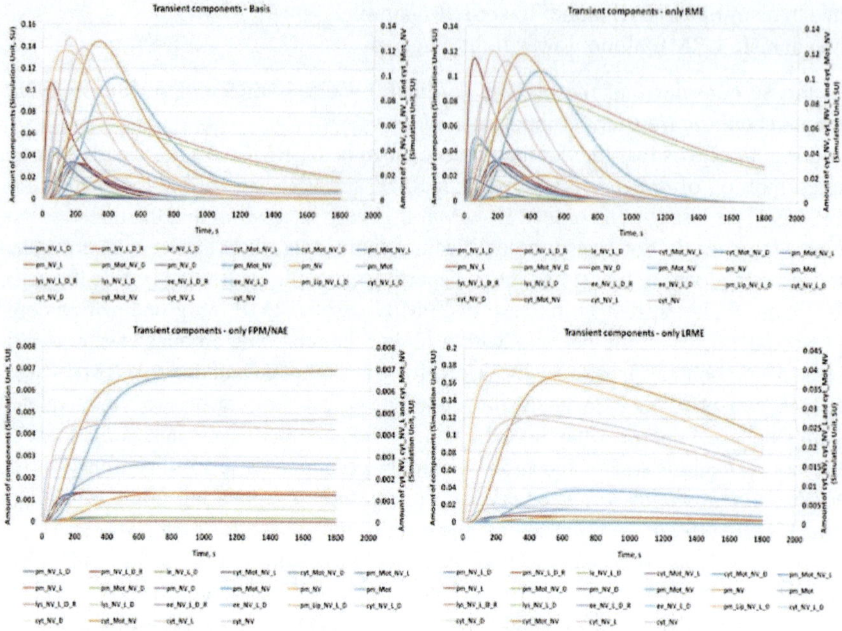

Fig. 9 Transient components in the complete model and in the individual mechanisms

4.2 *Effect of Pulse Size on the Drug Delivery*

The amount of the drug, i.e. the size of the injection is an important parameter to be considered. The knowledge of the feasible amount of drug will help to plan the medication strategy in another spatial and time scale.

In the following two series of Figures we compare the dynamically interesting periods of three models, as follows:

baseline model: upper diagrams,
model with less drug injection: lower left diagram,
model with more drug injection: lower right diagram.

Please note that there are magnitude differences in the amount (ordinate) scales of the four diagrams.

In Fig. 10 we see, that all injections are delivered almost with the same percentage, only the delivery time changes asymmetrically, because of the simultaneous change of driving forces.

Figure 11 shows similar non-linearity. Having increased the injected drug, slow degradation causes longer tailing of the curves.

The change of the nanovehicle components is in good accordance with the injected amounts.

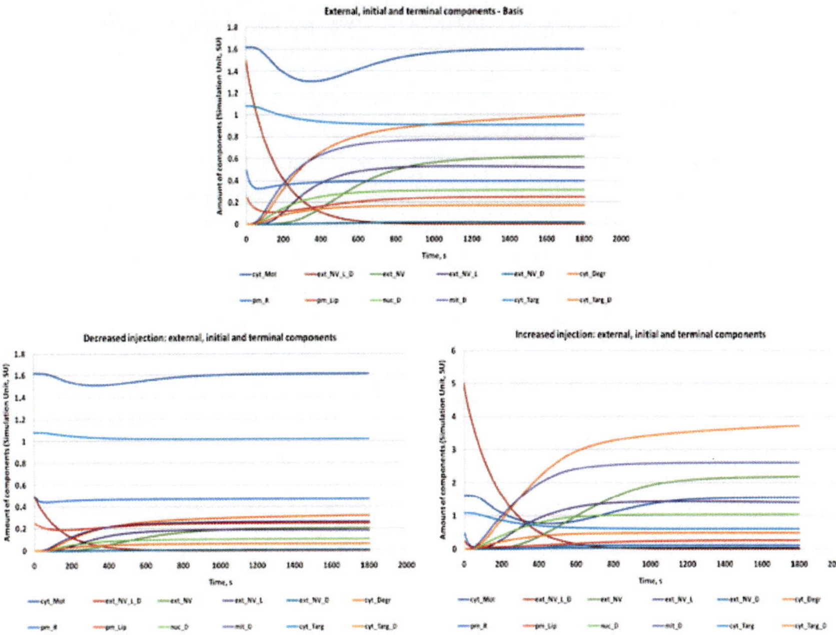

Fig. 10 External, initial and terminal components in case of three different drug injections

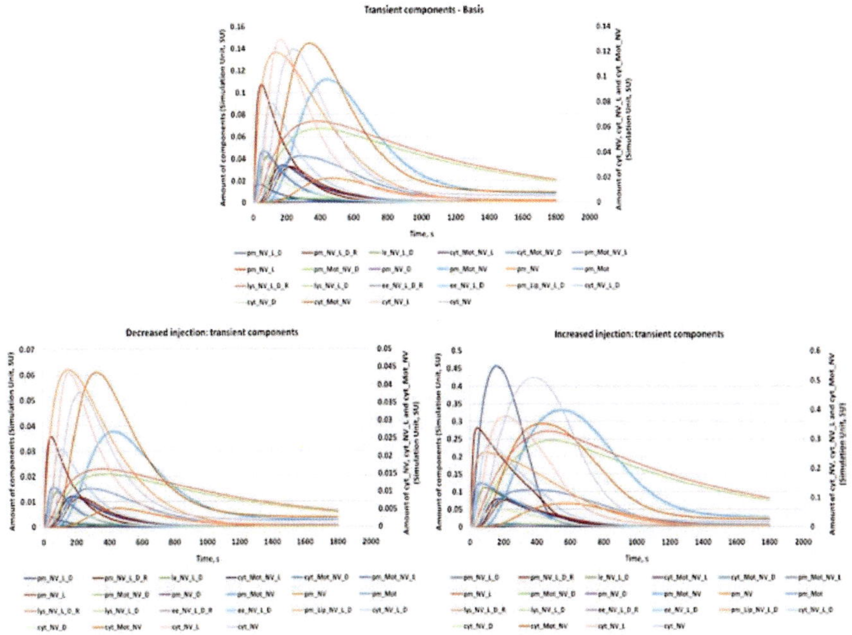

Fig. 11 Transient components in case of three different drug injections

4.3 *Role of Exocytosis Based Recycling in the Drug Delivery*

With the knowledge of the previous results we can conclude that amount of delivered drug is affected also by the exocytosis.

The useful exocytic processes transport drug containing NV_D vehicle through the plasma membrane back to the external phase. Accordingly, this recycling helps to deliver more drug. The useless exocytosis transports a given part of the released NV and NV_L complex through the plasma membrane back to the external phase. This causes a delayed degradation only.

In computational analysis, we can derive hypothetical studies that cannot be realized experimentally. In the following model, we switched off the exocytosis, fully (note that various inhibitors can only do this partially). The results, illustrated in Figs. 12 and 13 are compared with same diagrams obtained with switched on exocytosis (in Figs. 6 and 7), with same time scale.

Figure 12 illustrates the amount of the initial, external and terminal components. Having reached the limit values only the degradation follows slowly.

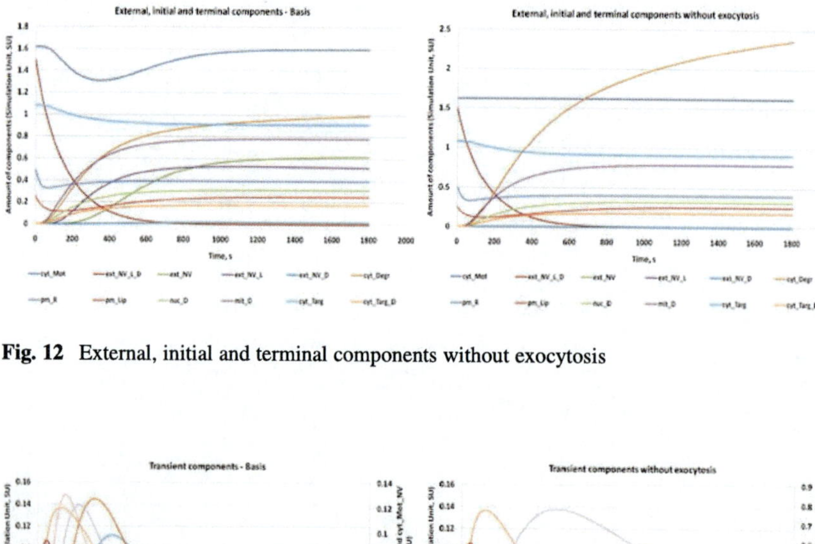

Fig. 12 External, initial and terminal components without exocytosis

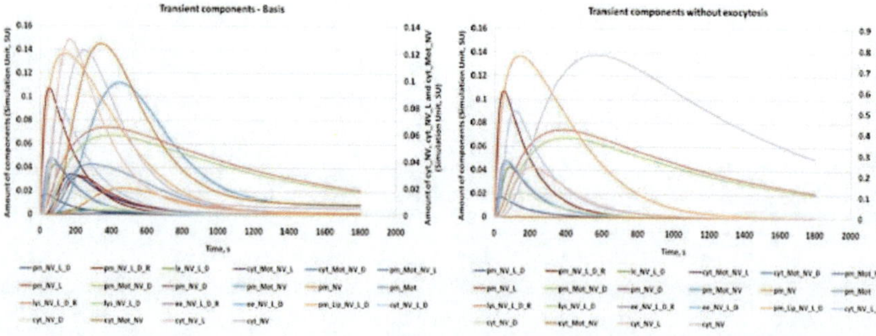

Fig. 13 Transient components without exocytosis

Figure 13 illustrates the change in the drug transient components, involved in the various delivery routes. It demonstrates the lack of exocytosis. The change of the nanovehicle containing components is in accordance with the slow degradation inside the cell. The receptor containing components converge either to the equilibrium value or to zero. The slowest processes are the decomposition of cyt_NV and lys_NV_L_D.

5 Conclusions

A new computational tool has been employed for the model-based analysis of the endocytosis and exocytosis mechanisms involved in nanocarrier delivery. It is to be emphasized that we made a quite hypothetical study, because at this moment the available data are not sufficient for a sophisticated identification of the underlying process model.

We demonstrated a hypothetical computation in the analysis of a significant, practical problem. Computational models may be a useful interacting partner for the field expert. The case studies allow studying the individual mechanisms, as well as the synergistic and antagonistic effects of their combinations. It helps to understand, the possible limiting transportations and transformations, as well as how inherent degradation limits the utilization of the injected drug. In addition, by switching on and off the respective processes, we can evaluate the effects of some deleting certain pathways, like exocytosis of the nanocarrier or of the drug containing nanocarrier.

The applied methodology proved to be flexible and effective enough to describe and to run complex process models without any mathematical and computational assistance.

The most interesting general conclusions can be summarized, as follows.

5.1 Limiting Assumptions vs. Constructive Usefulness

All of the computer simulations, especially the completely hypothetical ones, can be evaluated by the balance between assumptions and usefulness. The question is whether constructive applications can compensate for the possible malfunctions, caused by the limiting assumptions.

First, we describe the explicit and implicit assumptions behind our model. We assumed the following:

1. The modeled components, transportations and transformations cover all of the necessary and sufficient elements determining the states and transitions of the studied reality.
2. There are not hidden, un-modeled components, affecting the modeled transitions.

3. There are not hidden, un-modeled transportations and transformations, affecting the modeled components.
4. The applied elementary models are feasible.
5. The estimated (kinetic, equilibrium, etc.) parameters are feasible.
6. The estimated initial concentrations and boundary values (e.g. injection size) are feasible.

The usefulness of this model can be summarized, as follows:

(i) The computational model helps to gain deeper insight into the various endocytosis mechanisms and supports the major contribution of the specific pathways.
(ii) The case studies help to analyze the inherent non-linearities and limitations of the drug uptake (i.e. the relation between the injection size and the delivered drug).
(iii) Based on the hypothetical simulation, the functioning of exocytosis can be analyzed clearly.

Considering the assumptions, we can state that the installed model seems to be closed enough that helps to fulfill the conditions, covered by (1)–(3), relatively well. We summarize that assumptions (4)–(6) can be fulfilled by the feasible estimations of the field and model experts. Probably the constructive applications (i)–(iii) compensate for the uncertainties of (1)–(6).

It is to be emphasized that there are no inputs, except of the drug and ligand containing nanovehicle, so the model may be considered as a closed system.

5.2 *Qualitative Identification and Validation by Dialog Between the Field Expert and the Computational Model*

The development of this computational model is a good example for the qualitative identification and validation, organized by the dialogue between field expert and computational model. We started from immature descriptions and models with roughly estimated parameters (not seen in the present paper). Afterwards, utilizing the abilities of DCM, we attempted to build a model, consisting of the necessary and enough species of state and transition elements. This was controlled via a feedback from the field expert.

Some examples for the important feedbacks are the followings:

- The contribution of specific mechanisms to the endocytosis is cca. 70%;
- Consider motor protein enhanced exocytosis for NV, NV_L and NV_D.
- Various kinds of degradation exist, and they are commensurable with the delivery; etc.

We can state that the future of the computer-assisted biosystem-related, biotechnological and biomedical studies depends highly on this kind of collaboration, based on the mutual trust and understanding between the field expert and simulator.

In addition, DCM seems to be an effective methodology for the organization of these collaborations.

5.3 *Bi-Directional Bridge Building Between Experiments and Computations*

Looking at the other Chapters of this Volume, it is obvious that the majority of the contributions are of very sophisticated nature and exhibited deep experimental insight, related to various specific details. From an engineering point of views, they are well-elaborated pieces of a puzzle. On the contrary, the simplified, but systematic computational model attempted to describe a hypothetical, but comprehensive big picture. Similar to bridge building, in computational systems biology, we have to start from the two sides of the bridge and we have to meet in the middle, hopefully.

The hypothetical computations might help in the more conscious design of experiments that supports the effective identification and validation of the models. For example, nanotechnological measurement techniques make possible to select cellular compartments and to measure the drug or nanocarrier content of them, at least at some important points. On the other hand, the identified models could replace some exhaustive measurements.

For researchers and experimentalists who are interested, this program can be made available in an interactive format for exploration.

References

Alberts, B., Johnson, A., Lewis, J., Raff, M., Roberts, K. and Walter, P. (2002), Molecular biology of the cell, 4th edition, New York: Garland Science.

Brauer, W. (ed.) (1980) Net theory and applications, *Springer Lecture Notes in Computer Science*, 84.

Chen, M. and Hofestaedt, R. (1999) Quantitative petri net model of gene regulated metabolic networks in the cell, *In Silico. Biol (on-Line)*, 3, 0030.

Csukas, B. (2001) Direct Computer Mapping of conservational and informational processes, Manuscript of DSc Theses (in Hungarian).

Csukas, B. and Balogh, S. (1998) Combining genetic programming with generic simulation models in evolutionary synthesis, *Computers in Industry*, 36, 181–197.

Csukas, B. and Perez Uriza, S. (1995) Discrete modelling by Direct Mapping of the conservational processes, *Hung J Ind Chem*, 23(4), 277–287.

Csukas, B., Balogh, S. and Bankuti, Gy. (2005) Generic Bi-layered Net Model – General software for simulation of hybrid processes, In: Li, D., Wang, B. (eds.) *Artificial Intelligence Applications and Innovations II.*, Springer, USA, 701–710.

Csukas, B., Balogh, S., Kovats, S., Aranyi, A., Kocsis, Z. and Bartha, L. (1999) Process design by controlled simulation of the executable structural models, *Comput Chem Engng*, 23, 569–572.

Csukas, B., Rossiter, G. and Wooley, R. (2000) Model based comparison of various methods for three component separation in simulated moving bed, *Ion Exchange at the Millennium, Proceedings of IEX 2000*, Cambridge, 219–229.

Csukas, B., Varga, K. and Lakner, R. (1996) Dynamic simulation by algorithmic generated structural model I., Principles of the Model Generator, *Hung J Ind Chem*, 24(2), 107–130.

Diaz, D. The GNU Prolog web site, http://www.gprolog.org/

Doherty, G. J., McMahon, H. T. (2009) Mechanisms of endocytosis, Annu Rev Biochem 78, 857–902.

Graphviz – Graph visualization software tool, http://www.graphviz.org/

Grant, B. D., Donaldson, J.G. (2009) Pathways and mechanisms of endocytic recycling, Nat Rev Mol Cell Biol. 10(9), 597–608.

Holzbaur, E. L. and Goldman, Y. E. (2010) Coordination of molecular motors: from in vitro assays to intracellular dynamics, *Curr Opin Cell Biol*, 22(1), 4–13.

Jin, H., Heller, D. A., Sharma, R. and Strano, M. S. (2009) Size-dependent cellular uptake and expulsion of single-walled carbon nanotubes: single particle tracking and a generic uptake model for nanoparticles, *ACS Nano*, 3(1),149–58.

Kalman, R., Falb, P. and Arbib, M. (1969) Topics in mathematical system theory. McGraw Hill.

Khalil, I. A., Kogure, K., Akita, H. and Harashima, H. (2006) Uptake pathways and subsequent intracellular trafficking in nonviral gene delivery, *Pharmacol Rev,* 58(1), 32–45.

Kholodenko, B. N., Hancock, J. F. and Kolch, W. (2010) Signalling ballet in space and time, *Nat Rev Mol Cell Biol*, 11(6), 414–26.

Kumari, S., Mg, S. and Mayor S. (2010) Endocytosis unplugged: multiple ways to enter the cell, *Cell Res*, 20(3), 255–75.

Marquardt, W. (1996) Trends in computer-aided process modeling, *Computers chem. Engng.* 20(6/7), 591–609.

Meier-Schellersheim, M., Fraser, D. C. and Klauschen, F. (2009) Multiscale modeling for biologists, *John Wiley & Sons, Inc. WIREs Syst Biol Med* 1, 4–14.

Nagy, K., Csukas, B., Kiss, G., Bartha, L. and Balogh, S. (2001) Study on the preparation and properties of olefin-maleic-anhydride copolymers, 40th International Petroleum Conference, Bratislava, Slovak Republic, Paper L-E-1.

Panyam, J. and Labhasetwar, V., (2003) Dynamics of endocytois and exocytosis of poly(D,L-lactide-co-glycolide) nanoparticles in vascular smooth muscle cells, *Pharm Re*, 20(2), 212–20.

Peleg, M., Rubin, D. and Altman, R. B. (2005) Using petri tools to study properties and dynamics of biological systems, *Journal of the American Medical Informatics Association*, 12(2), 181–199.

Petri, C. A. (1962) Kommunikation mit automaten (Communication with automatons). *Schriften des Institut für Instrumentelle Mathematik*, Nr. 2, Bonn.

Prokop, A., and Davidson, J. M. (2008) Nanovehicular intracellular delivery systems, *J Pharm Sci*, 97(9), 3518–90.

Robert, D., Nguyen, T. H., Gallet, F. and Wilhelm, C. (2010) In vivo determination of fluctuating forces during endosome trafficking using a combination of active and passive microrheology, *PLoS One* 6;5(4):e10046.

Shih A. J., Purvis J and Radhakrishnan R. (2008) Modeling systems biology of ErbB1 signaling: bridging the gap through multiscale modeling and high-performance computing, *Mol Biosyst* 4(12), 1151–1159.

Temesvari, K., Aranyi, A., Csukas, B. and Balogh, S. (2004) Simulated moving bed chromatographic separation of a two-component steroid mixture, *Chromatographia (on Line)*, 60, S189-S199.

Varga, M., (2009) Economic optimization of sustainable complex processes, *PhD Theses* (in Hungarian)

Varga, M., Balogh, S. and Csukas, B. (2010) Sector spanning agrifood process transparency with Direct Computer Mapping, *Agricultural Informatics* 1(2), 73–83.

Wattis, J. A., O'Malley, B., Blackburn, H., Pickersgill, L., Panovska, J, Byrne, H.M., Jackson, K.G. (2008) Mathematical model for low density lipoprotein (LDL) endocytosis by hepatocytes, Bull Math Biol. 70(8), 2303–33

Nanosized Drug Delivery Vectors and the Reticuloendothelial System

Lisa M. Kaminskas and Ben J. Boyd

Abstract Nanomaterials have potential as drug delivery vectors that can improve the chemical stability and pharmacokinetic profile of small molecule drugs or improve drug uptake into solid tumours. However, one consequence of the use of nanosized drug delivery vectors is their potential recognition by tissue macrophages and accumulation in organs of the reticuloendothelial system (RES). While in some instances the uptake of drug loaded nanomaterials or 'nanomedicines' into organs of the RES is favoured, in most instances uptake into the RES can limit systemic exposure of the nanomedicine and limit therapeutic utility. Hence, this section discusses ways in which the RES uptake of nanomedicines can either be promoted or inhibited. Specifically, the effect of various physicochemical properties and presence or absence of key RES 'recognition ligands' on RES uptake are examined.

Keywords Drug delivery • Macrophage • Nanomedicine • Opsonisation • Reticuloendothelial

Abbreviations

RES	Reticuloendothelial system
PAMAM	Polyamidoamine
PEG	Polyethylene glycol
DPPC	Dipalmitoyl phosphatidylglycerol
Succinate	Succinic acid
Ph-sulphonate	Benzene sulphonate
Ph-disulphonate	Benzene disulphonate
TIM	T cell Ig domain and mucin domain

L.M. Kaminskas and B.J. Boyd (✉)
Monash Institute of Pharmaceutical Sciences, Monash University, Melbourne, Australia
e-mail: ben.boyd@monash.edu

A. Prokop (ed.), *Intracellular Delivery: Fundamentals and Applications*, Fundamental Biomedical Technologies 5, DOI 10.1007/978-94-007-1248-5_6,

NPC	Non-parenchymal cell
HPI	Hydrogenated phosphatidylinositol
DPPG	1,2-dipalmitoyl-sn-glycero-3-phospho-rac-(1-glycerol)
DAB	Diaminobutane
Gd-DAB	Gadolinium conjugated diaminobutane dendrimer
HLA-DR	Human leukocyte antigen DR
BAI	Brain specific angiogenesis inhibitor
GM1	Monosialoganglioside GM1
DPPC	1,2-dipalmitoyl-sn-glycero-3-phosphocholine
DSPC	L-alpha-phosphatidylcholine distearoyl
DSPG	L-alpha-phosphatidyl-DL-glycerol-distearoyl

1 Introduction

A number of attributes of nanosized drug delivery particles are driving their application in a wide range of uses in medicine and as drug delivery vectors. Their circulation behaviour in plasma can be extended by controlling size to be on the one hand sufficiently small to permit intravenous administration, while on the other hand being sufficiently large to hinder passage through endothelial gap junctions, leading to extended circulation time. Drugs can be incorporated into their structures either via entrapment or surface conjugation. Particles can be engineered to permit entrapped drugs to gradually leak out of the structures down a concentration gradient, to enable an extended release-type delivery of drug to systemic circulation, or to liberate drug specifically under physiological conditions encountered at the desired site of drug action. They can be engineered for tissue-specific drug delivery either passively (for instance via the enhanced permeation and retention effect in solid tumours) or actively via the surface presentation of ligands to cell surface receptors. In addition, association of drugs with nanosized delivery vectors allows delivery into the cells in the form of a nano-drug formulation bypassing the efflux mechanism giving rise to resistance. This is particularly relevant in the treatment of cancer, where cytotoxic drugs often develop multidrug resistance phenotypes.

A wide range of potential structures based on lipids, polymers, dendrimers, solid drug particles, surfactants and other matrix materials have been described for use as particulate or macromolecular drug delivery systems for intravenous application. A range of these structures are illustrated below in Fig. 1. The two classes of particulate nanomaterials, i.e. the macromolecules and matrix-based particles associated with drug molecules, are collectively called ‘nanomedicines’ for the purpose of this contribution.

Although nanomedicines have shown a number of beneficial attributes for their application in drug delivery, they are also prone to recognition and removal via the reticuloendothelial system (RES), which can severely limit their application. Specifically, recognition of nanomedicines by the RES can result in very rapid

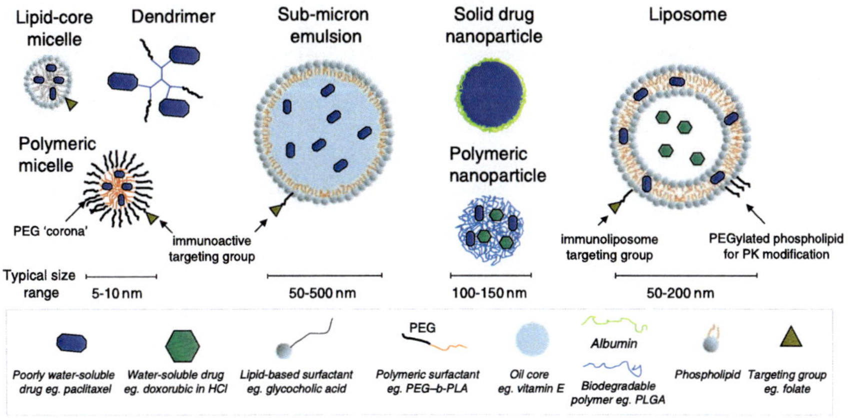

Fig. 1 Various commonly used nanostructures in drug delivery

removal of the particle from plasma within only minutes to a few hours, thus limiting the therapeutic usefulness of the nano-drug formulation where extended circulation time is important for performance. This chapter is therefore dedicated to explaining the reticuloendothelial system, its functions, pathophysiology and mechanisms of nanomedicine removal and detailing ways in which uptake via the system can either be avoided or promoted.

2 Pathophysiology of the Reticuloendothelial System

The reticuloendothelial system (commonly referred to as the RES) is comprised of a collection of organs (Fig. 2) that contain high proportions of the cells that make up the body's defence system against particulate pathogens. Specifically, it is a collection of monophagocytic cells that are largely manufactured by the bone marrow and transported throughout the body to aid in the removal of foreign organisms, damaged cells and products of cellular degradation. Immature macrophages in the systemic circulation are monocytes that migrate into tissues once matured. Mature, tissue fixed macrophages make up the Kupffer cells in the liver and the reticular cells in the spleen, lungs, lymph nodes and bone marrow. These cells remove foreign material via rapid phagocytosis. In some instances, initial priming via lymphocytes and the internalization of material into endosomes is required. The endosomes eventually fuse with lysosomes that release enzymes into the endosome that digest the engulfed material. For material that is resistant to degradation via lysosomal enzymes, the material remains trapped within the macrophage until the cell eventually dies. The largest concentration of macrophages in the body are located in the liver and lungs, however macrophages in different locations exhibit very different biochemical and functional activity that can influence their ability to scavenge various nanomedicines.

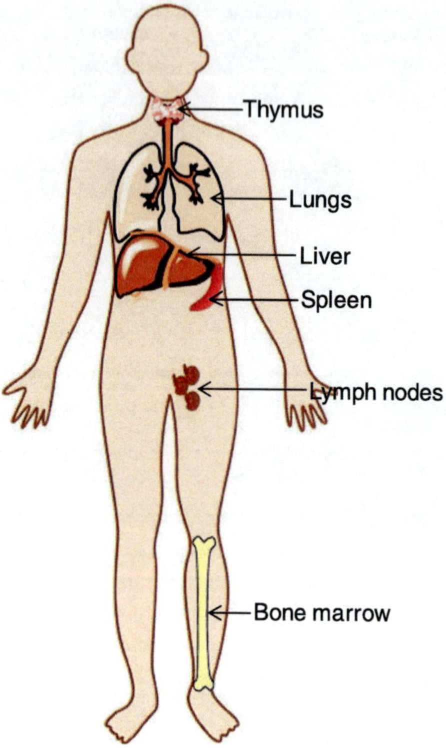

Fig. 2 Diagram of the organs that make up the reticuloendothelial system

The RES is therefore largely responsible for the removal of nanomedicines from the systemic circulation. The uptake of particles is largely size dependent. Compared to circulating biological proteins, drug delivery particles are usually large in size (Fig. 1), making them appear as foreign materials to the monophagocytic system. Opsonisation of drug delivery particles by plasma proteins and lipoproteins accelerates and enhances the recognition process. The pathophysiology of several major organs of the RES (liver, spleen, lymph nodes and lungs) and how this relates to particle uptake is therefore described in the following section.

2.1 *The Liver*

The liver is one of the most important metabolic organs in the body and is a part of the digestive system. The visceral surface of the liver contains the porta hepatis, the site at which vessels and ducts enter and leave the liver, including the portal vein, the hepatic artery and the common hepatic, cystic and bile ducts. At any one time, approximately 20% of the bodies blood supply resides in this organ and the liver is a major site for the removal of drugs absorbed into the blood from the gastrointestinal tract.

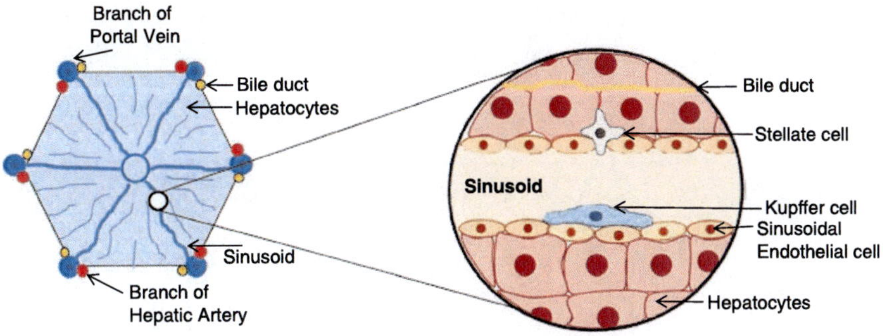

Fig. 3 Diagram of the basic lobular structure of the liver (*left*) and of the microstructure surrounding the sinusoids (*right*)

The major microstructural unit of the liver is the lobule (Fig. 3), a hexagonal structure that contains a central vein with blood tracts that protrude from the main structure (called the sinusoids that are lined with a thin layer of sinusoidal endothelial cells) and branches of the portal vein, hepatic artery and bile ducts at each corner of the hexagonal lobule. The major cell type that resides between the sinusoids are the hepatocytes. These are the major metabolic cells of the liver and are storage sites for a number of essential nutrients that occur in excess in the blood, such as iron and glycogen. The hepatocytes occur in a single layer between the liver sinusoids and bile cannaliculi and are responsible for absorbing drugs and materials from the blood and processing them for secretion into the bile ducts. Thus, the excretion of nanomedicines via the bile and subsequently via the feces requires initial uptake via the hepatocytes. Hepatic stellate cells also lie in between clusters of hepatocytes. These cells are the fat storage cells of the liver. Finally, Kupffer cells (the major phagocytic cells of the liver that account for approximately 2% of the liver volume) lie within the sinusoids. Their residence at this site optimizes their exposure to blood pathogens, foreign material and more importantly to this chapter, nanomedicines. Blood entering the liver from the hepatic artery and the portal vein flow towards the lobule via their branches, through the sinusoids where the blood is 'cleaned' and into the central vein that pools detoxified or cleaned blood into the hepatic vein. There are also two distinct types of Kupffer cells in the liver. The smaller, more immature cells are mainly located in the centrilobular region. The larger, more mature cells that play a more prominent role in phagocytosis, and hence are of more importance in the uptake of particles from the blood by the liver, are located in the periportal region.

2.2 *The Spleen*

The spleen is important in generating blood cells during fetal development, but it is not a vital organ in adults. It does, however, function in addition to the liver to destroy and remove damaged or aged red blood cells from the systemic circulation.

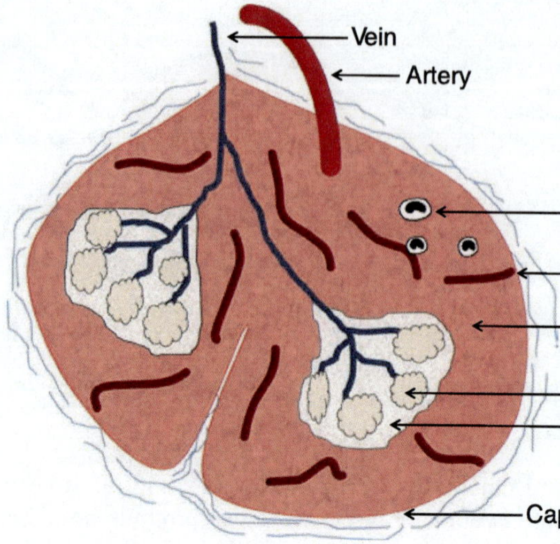

Fig. 4 Representation of the basic structure of the spleen containing red and white pulp

This function takes place in the red pulp of the spleen that comprises the majority of the organ's structural surface area (Fig. 4). The spleen is the largest organ within the lymphatic network and functions to filter blood in a similar manner to the way lymph nodes filter lymph. Blood enters the spleen via the splenic artery that extends into the white pulp and branches into the venous sinuses that remove damaged red cells. Splenic cords extend between venous sinuses that contain reticular connective tissue, macrophages and lymphocytes that act to remove dead cells and foreign material from the blood. The spleen also acts as a reservoir for red cells that can be added to the blood via splenic contractions when needed.

The spleen also has clusters of white pulp that are responsible for producing, storing and exchanging lymphocytes with the blood. Therefore, nanomedicines circulating in the blood can be removed either via macrophages and lymphocytes in both the red and white pulp of the spleen. As an example, the biodistribution of polylysine dendrimers conjugated with both polyethylene glycol and methotrexate display more avid uptake into the spleen than in the liver of athymic nu/nu rats that have a reduced number of T cells (Kaminskas et al. 2009a). Macrophages in the spleen likely act in this instance to compensate for the loss of T cell activity by becoming more phagocytically active. Also, the relative biodistribution of liposomes into liver and spleen varies according to composition, where liposomes composed of egg phosphatidylcholine show roughly equivalent uptake into liver and spleen, whereas liposomes containing phosphatidylserine or dipalmitoylphosphatidylglycerol (DPPC) are more avidly taken up into the spleen (Oussoren and Storm 1997).

2.3 *Lymph Nodes*

The lymphatic system is comprised of a series of capillaries, ducts, nodes and organs including the spleen and thymus, and is involved in the production and release of lymphocytes into the blood. Lymphatic capillaries lie in interstitial spaces along with blood capillaries. They collect excess interstitial fluid, extravasated protein and lipids absorbed from the digestive tract to form a creamy white fluid that flows from peripheral capillaries into lymphatic ducts that finally drain into the thoracic lymph duct and the right lymph duct at the base of the neck. From here, lymph fluid drains directly into the systemic circulation.

The main difference between blood and lymph capillaries is that while blood capillaries have tight junctions between endothelial cells and a fibrous basement membrane that limits the passage of macromolecules and particulate nanomedicines, lymphatic capillaries have wide intercellular junctions and a loose basement membrane. Blood flow is approximately 100–1000 times faster than that of lymph, so uptake of small molecules from the interstitium is preferentially via the blood, whereas the uptake of macromolecules is via the lymph. Prior to emptying into the systemic circulation, lymph fluid filters through lymph nodes placed at strategic positions to remove foreign particles preventing their access to the systemic circulation. The lymph nodes are therefore the main sites of deposition for nanomedicines that are absorbed from the interstitium, either after subcutaneous administration or after extravasation from the blood stream.

Lymph nodes are located at the end of afferent lymph capillaries that enable lymph flow in one direction via valves that are positioned along the entire length of the capillary system (Fig. 5). Lymph nodes contain an outer subcapsular sinus with

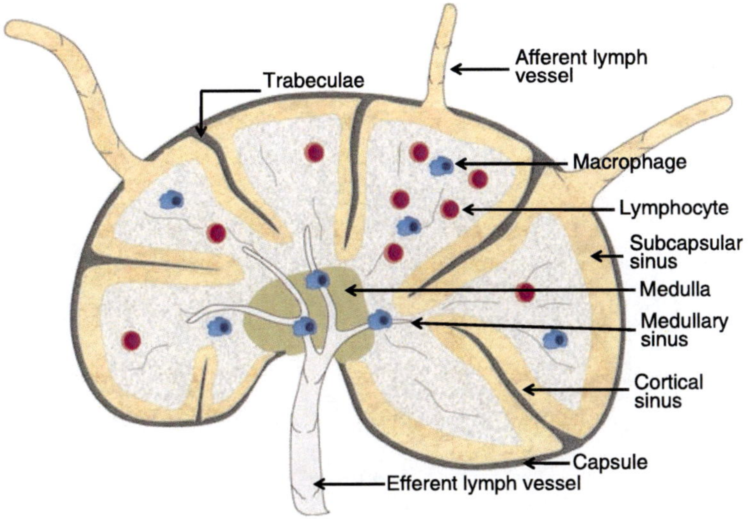

Fig. 5 Diagram of the microstructure of a lymph node

collagenised trabeculae extending deep into the node. Lymph flows from afferent lymph vessels into the subcapsular sinus where it is delivered into deep cortical sinuses that contain a network of reticular fibers, macrophages and lymphocytes that filter pathogens and foreign material from the lymph. Filtered lymph fluid then drains towards the efferent lymph vessel via medullary sinusoids that contain a high density of macrophages that represent the final filtration barrier for pathogens and particulate nanomedicines before lymph is delivered into the systemic circulation. When examining the lymph node localization of PEGylated PAMAM dendrimers after subcutaneous administration the greatest lymph node retention of the dose is in the subcapsular sinus and the medulla (Mounzer et al. 2007).

2.4 *The Lungs*

As the major internal organ exposed to the air, the lungs are densely populated with macrophages that act to prevent systemic exposure to airborne pathogens and particles. The airways are constructed such that larger airways (bronchi) branch out and thin down to smaller bronchioles. The bronchioles supply air to thin alveolar sacs that are composed of a single layer of alveolar endothelial cells, and a mucosal membrane that lines the inside of the alveolar sacs to prevent the alveolar membranes sticking to each other (Fig. 6). Lymphatic and blood capillaries lie on the outside of the alveoli encased by a membranous cover (the pleura) that encloses the whole organ.

There are two distinct populations of macrophages in the lungs that are located in the alveolar membrane and protrude out into the alveolus and interstitial macrophages that account for approximately half of the macrophages in the lungs. Macrophages resident in the alveolar mucosa act to phagocytose pathogens and

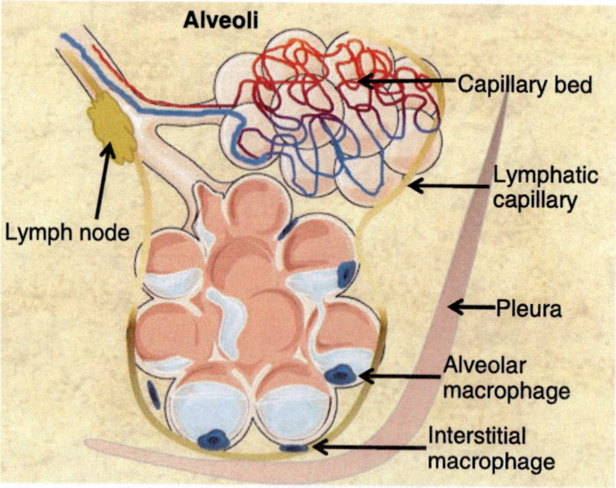

Fig. 6 Diagram of the alveoli and localisation of macrophages

particles that enter via the airways. Alveolar macrophages that trap pathogens then enter into the lymphatic capillaries where they drain into the pulmonary network of lymph nodes (both in the pleura and in the upper airways). Here they are presented to T lymphocytes that facilitate an immune response against the pathogen. The lungs therefore represent an alternate route for the delivery of vaccines that avoids the use of needles since the induction of an immune response in one mucosal membrane transfers immunity to other mucosal membranes. It is also this dense distribution of macrophages that makes the lungs a potential target for the uptake of particulate nanomedicines from the blood. This can be achieved to a certain degree by conjugating surface targeting ligands that make nanoparticles more attractive to alveolar macrophages such as mannose. For instance, the uptake of rifabutin via the lungs after IV administration can be improved by associating the drug with mannosylated lipid nanoparticles that display equivalent biodistribution to the liver and lungs (Nimje et al. 2009).

3 Particle Properties That Mediate Reticuloendothelial Uptake

3.1 Nanoparticles and Colloids That Target the Reticuloendothelial System and the Role of Opsonisation

Since the RES is designed to protect the body against invading pathogens and other foreign materials, any properties of an administered nanomedicine that indicates that the particle or macromolecule is foreign will make it susceptible for RES uptake. Nanomedicines, depending on definition, range in size from 1 to 1,000 nm in diameter. In comparison, the diameter of a virus particle can range from 25 nm to approximately 120 nm and the diameter of a bacterium can range from 200 to 250 μm. Thus, particulate nanomedicines typically have a size that mimics that of a pathogen. This alone does not determine the ultimate fate of a nanomedicine, since particulate nanomedicines have been administered that demonstrate very good biocompatibility and very limited uptake by organs of the RES. The surface characteristics of the material also strongly influence the RES targeting capacity of a nanomedicine, either by promoting opsonisation that essentially tags the particle for removal by macrophages, or by acting as ligands to receptors expressed on the macrophage surface. Receptor mediated processes will be discussed in a later section of this chapter.

It is important, however, to have an understanding of the opsonisation process. The surface of a macrophage is anionic as are the surfaces of many pathogens. This dictates that binding of a pathogen with the macrophage membrane would be unfavourable on purely electrostatic grounds. Thus, prior modification of the pathogen surface is required to enable phagocytosis of these species which would otherwise be repelled from the macrophage surface. Opsonisation is a process whereby a plasma component coats the surface of a pathogen or target particle and presents it to macrophage cells via receptor mediated recognition of the bound component. Classical plasma opsonins include complement factors, immunoglobulins, mannose binding lectin, fibronectin and β2-glycoprotein 1. Macrophage receptors to these factors include CR1, Fc and mannose receptors. This process is highly species specific. However, in some cases the binding of plasma components does not result in RES targeting, but rather protects the particle from macrophage uptake. Although the identity of these factors is largely unknown, several plasma components have been suggested as potential dysopsonins, including IgA, α1-acidic glycoprotein and albumin. As an example, polystyrene microparticles of approximately 1 μm diameter display avid uptake into dendritic cells in culture in the absence of serum, however uptake into dendritic cells is significantly reduced by preadsorption of human serum albumin (Thiele et al. 2003).

3.2 *Effect of Size*

In general, the plasma pharmacokinetics of nanomaterials after intravenous administration is largely dependent on their renal clearance. Particles smaller than approximately 20 kDa (or approximately 8 nm in diameter) generally pass through glomerular filtration slits relatively unhindered, enabling ready removal from the body via the urine. Particles that are too large for effective renal clearance are then subjected to alternative clearance processes including metabolism (for biodegradable particles), biliary excretion (after uptake via hepatocytes) and biodistribution into tissues. Very large particles have the capacity to circulate in the bloodstream for extended periods of time on account of their limited capacity for extravasation via fenestrated capillaries or compromised vasculature. The limited extravasation and biodistribution results in their increased presentation to macrophages.

All other physicochemical factors aside, smaller particles have a greater tendency to avoid uptake via the RES, whereas larger particles represent better substrates. This has been demonstrated in detail when comparing the uptake of liposomes or dendrimers into draining lymph nodes after subcutaneous administration. For example, increasing the size of small (<20 nm) dendrimers, results in both increased uptake via the lymph, and increased uptake into draining lymph nodes (Kaminskas et al. 2009b). However in the case of larger liposomes (40–400 nm), increasing size results in slower clearance of the dose from the injection site into the lymphatic system, but an increase in the proportion of the absorbed dose recovered in lymph nodes (Oussoren and Storm 2001). The complexity of examining RES uptake in this way is that retention of nanomedicines in lymph nodes is dependent both on uptake by macrophages as well as physical filtration, both of which have been demonstrated to play a role in the retention of relatively biologically inert particles.

For particles that show relatively limited renal clearance and that do not contain targeting groups intended to direct particles to particular organs, increasing particle size results in long term accumulation in the liver and spleen. This has been demonstrated for PEGylated dendrimers, although the cellular targets have not been specifically identified (Kaminskas et al. 2008). Thus it cannot be stated with certainty that targeting towards the liver and spleen over time is a result of cumulative uptake via macrophages. Given that increasing dendrimer size beyond approximately generation 7 (>100 kDa) results in a dramatic decrease in blood retention and an increase in uptake via the liver and spleen 15 min after an IV dose (Kobayashi et al. 2001a), it can be suggested that very large particles are recognized more avidly by RES macrophages than smaller ones. For dendrimers, uptake of large particles is preferentially via the liver, whereas for large liposomes uptake of particles larger than approximately 70 nm is preferentially via the spleen (on a percentage of injected dose per gram of tissue basis) (Litzinger et al. 1994). For liposomes based on phosphatidylcholine, although smaller particles show less uptake via the liver and spleen, uptake into these organs is almost entirely via the macrophages (Gabizon et al. 1990). Uptake via the lungs, however, is decreased, likely reflecting either the different function of spleen, liver and lung macrophages or the relative difficulty in accessing macrophages in the interstitium and alveolus from the vascular side.

Increasing the size of polystyrene nanospheres from 60 to 225 nm also resulted in a change in biodistribution from preferential absorption by liver Kupffer cells (no uptake via the hepatocytes or sinusoidal cells was evident) to uptake via macrophages in the red pulp of the spleen (Moghimi 2002).

3.3 *Effect of Surface Charge*

As discussed above, macrophages have an overall intrinsic anionic charge on their surface that prevents direct adsorption of anionic particles. Although one would assume that this means that cationic particles should be distributed more avidly to macrophages of the RES, this is not necessarily true in the majority of cases. Particles carrying a cationic charge bind with little tissue specificity to all organs and to the blood vasculature on account of ionic interactions with anionic charges present on the glycoproteins present in vascular membranes. Somewhat preferential targeting of a more metabolically stable D-lysine capped poly-L-lysine dendrimer towards the liver, spleen and kidneys was observed in rats, suggesting some capacity of very small nanometer particles to target the RES (Boyd et al. 2006). This was not seen for dendrimers composed entirely of L-lysine as the dendrimer was rapidly degraded to free L-lysine that was subsequently incorporated into protein resynthetic pathways (Boyd et al. 2006). Thus in general, cationic particles have a somewhat limited chance for specific uptake via macrophages in either the RES or elsewhere.

Interestingly, however, increasing the surface cationic charge on liposomes from 0% to 25% surface cationic lipid has no influence on the uptake of the dose in either the liver, spleen or the lungs. At a 1:1 ratio of cationic lipid: uncharged lipid, however, uptake of liposomes via the lungs is increased by almost an order of magnitude, but again, accumulation in the liver and spleen does not change. The reason for this lung specific accumulation is not clear although a similar pattern of biodistribution has been observed with several cationic nanoparticles, where most of the injected dose is recovered in the lungs, with accumulation in the liver being the second most common fate of particles 10 min after initial IV administration (Al Jamal et al. 2009).

Anionic particles (both lipid and non-lipidic), as discussed previously, are substrates for opsonins and therefore are targeted more avidly towards liver and spleen macrophages. Furthermore, increasing the anionic strength of the particle increases both the degree of opsonisation and the extent of liver targeting. This was demonstrated using arylsulphonate and succinic acid capped polylysine dendrimers (Kaminskas et al. 2007). A generation 4 dendrimer containing 32 surface succinate groups (weakly acidic) did not show any evidence of opsonisation and was excreted via the urine (Fig. 7). A small generation 3 dendrimer containing 16 surface benzene sulphonate groups (stronger anion) showed some evidence of plasma opsonisation and was targeted more avidly towards the liver. Increasing anionic surface charge further by increasing the number of anionic groups to 32 benzene sulphonates on a generation 4 dendrimer resulted in progressively increasing degree of opsonisation plus increased uptake via the liver. The dendrimer with the strongest surface anionic charge (the generation 4 benzene disulphonate dendrimer) also showed a massive increase in uptake via the spleen such that the proportion of the injected dose recovered in the spleen was higher than the proportion of the dose in liver tissue (on a recovered dose per gram of tissue basis). A similar observation is seen with lipid colloids where increasing anionic surface charge increases uptake by the spleen more so than the liver.

3.4 *Effect of Particle Hydrophobicity*

The hydrophobicity of administered particles has a significant influence on the RES targeting of the system, in particular to the liver. For example, by modifying the structural hydrophobicity of dendrimers, the extent of liver targeting of systems with similar size and surface charge can be altered significantly. An example is a comparison between PAMAM-core dendrimers and more hydrophobic diaminobutane (DAB) – core dendrimers labeled with Gd contrast agent. Intravenous administration of the DAB-core dendrimer resulted in preferential accumulation of the MRI contrast agent in the liver 15 min after dosing, whereas PAMAM dendrimers of the same size were located in the blood vasculature and bladder/kidneys (Kobayashi et al. 2001b) (Fig. 8). Similarly, subcutaneous administration of Gd-DAB dendrimers resulted in better retention and resolution of the contrast agent in draining lymph nodes (Kobayashi et al. 2003, 2006). The reason for this effect may be due to increased capacity for hydrophobic interactions of the particles with plasma proteins, resulting in the recognition of particles as being opsonised by liver Kupffer cells. Another mechanism may be the receptor mediated recognition of more hydrophobic particles by macrophage receptors that recognize low density lipoproteins.

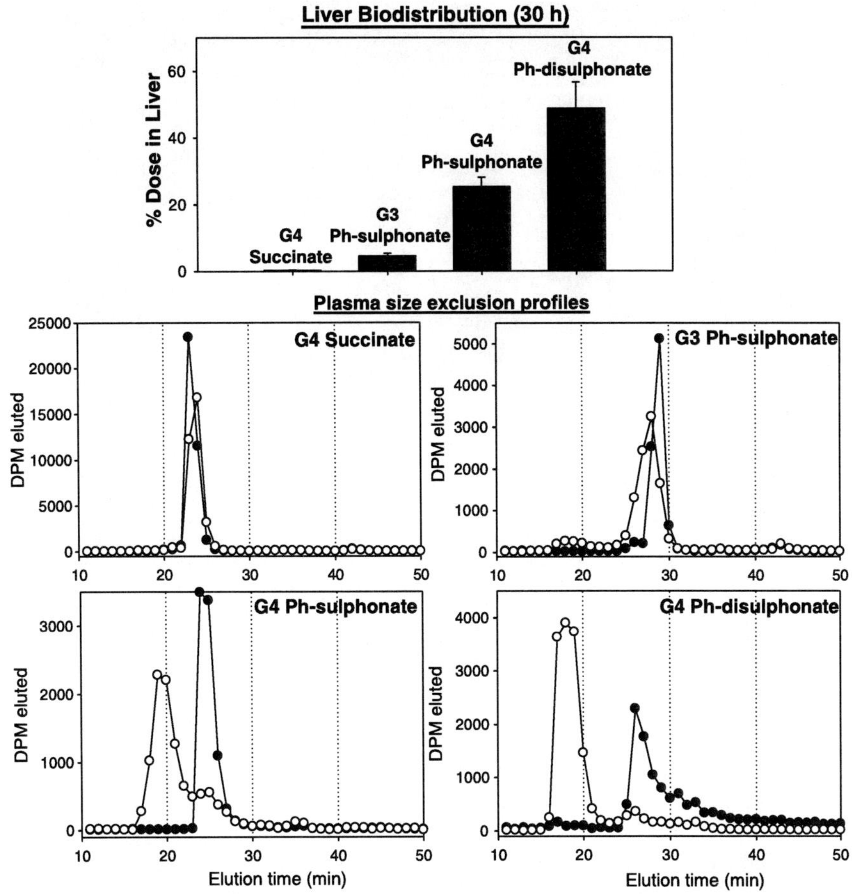

Fig. 7 Proportion of an IV dose of ^{3}H-labelled generation 3 (*G3*) and 4 (*G4*) polylysine dendrimers conjugated with succinic acid (*succinate*), benzene sulphonate (*Ph-sulphonate*) or benzene disulphonate (*Ph-disulphonate*) identified in the liver of rats 30 h after dosing (5 mg/kg, *top panel*) and size exclusion chromatography of each dendrimer diluted in phosphate buffered saline (*closed symbols*) or preincubated in plasma at 37°C for 1 h (*open symbols*). The shift in the ^{3}H-peak to an earlier elution time represents the formation of protein-bound dendrimer in plasma (Data taken from Kaminskas et al. (2007))

3.5 *Presentation of Ligands to Organs of the Reticuloendothelial System*

The previous discussion has focused mainly on particle properties that result in intrinsic targeting towards macrophages of the RES. However, in some instances macrophage targeting represents a therapeutic aim, rather than a negative consequence of various particle properties. Examples include drug delivery to macrophages involved in microbial or viral propagation (such as HIV infection), cancer

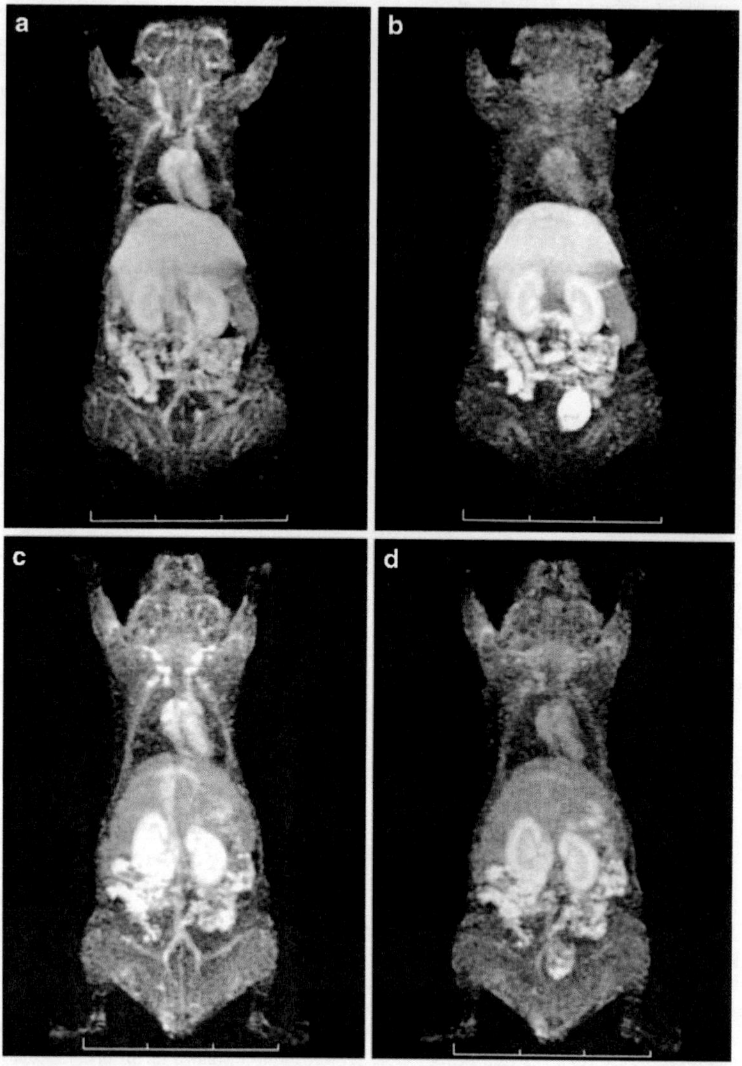

Fig. 8 Whole body 3D-micro MR imaging of mice injected with 0.033 mmol Gd/kg of generation 4 polypropyleneimine diaminobutane dendrimer conjugated with 64 IB4M-Gd (panel **a** and **b**) or generation 4 polyamidoamine dendrimer conjugated with 64 IB4M-Gd (panel **c** and **d**). Panels (**a**) and (**c**) represent imaged taken within 2 min of IV dosing of dendrimer and panels (**b**) and (**d**) represent images taken 10 min after IV injection (Reproduced from Kobayashi et al. (2001b). With permission)

or improved drug delivery to lymph nodes, specific delivery of toxic agents to prevent macrophage activity and delivery of antigens to promote immunization. For instance, Amphotericin B liposomes are used to target macrophage resident fungal infections. Each of these goals requires specific internalization within target macrophages and avoidance of other cell types. Thus, although targeting to RES macrophages can be achieved by the administration of nanomedicines containing the agent of interest that are readily opsonised and endocytosed, the use of targeting ligands can achieve this effect for non-RES targeted particles that otherwise have been optimized for the proposed delivery.

Macrophage targeting ligands are specifically ligands to receptors that are ether specifically or over-expressed on the surface of the target macrophages. In addition to targeting towards the specific cells that make up the RES, other ligands may be used to enhance the delivery and cellular internalization of materials to other cells within organs of the RES, such as hepatocytes. Galactose is an example of a hepatocyte targeting ligand. This section, however, is focused specifically on the targeted delivery of particles to reticuloendothelial cells (specifically the macrophages). Table 1 summarizes the different ligands and their target receptors that have been used to improve nanomedicine targeting towards macrophages.

Phosphatidylserine is commonly incorporated into liposomes to decrease the surface charge of the colloid or to effect targeting towards macrophages. In mice, targeting of phosphatidylserine liposomes to liver Kupffer cells occurs without the need for prior opsonisation by plasma components (Liu and Liu 1996). However,

Table 1 Ligands used to improve targeting of nanoparticles towards reticuloendothelial organs and macrophages and the target receptors

Ligand	Receptor	Conjugated nanomedicine	References
Phosphatidylserine	TIM-4, stablin-1 & 2, BAI-1 plus others	Liposomes	Liu et al. (1995a)
Mannose/mannan	Mannose receptors (multiple sub-types)	Liposomes, dendrimers, nanoparticles, niosomes, antigens. micelles, metal colloids, emulsions	Nag and Ghosh (1999); Kaur et al. (2008); Vyas et al. (2000); Irache et al. (2008)
RVG peptide	Nicotinic acetylcholine receptor	RVG peptide-siRNA	Kim et al. (2010)
Stromal cell derived factor 1	CXCR4	Peptide-DNA complex	Egorova et al. (2009)
Anti-HLA-DR	HLA-DR	Liposomes	Dufresne et al. (1999)
N-formyl-methionyl leucyl phenylalanine	Formyl peptide receptor 1	Nanoparticle, liposomes	Wan et al. (2008); Morikawa et al. (1988)
Fc region of antibody	Fc receptor	Opsonised nanomedicines	
Decadeoxyguanine	Scavenger receptor class 1	Liposomes	Rensen et al. (2006)
Tuftsin	Tuftsin receptor (uncloned)	Dendrimer, liposomes	Dutta et al. (2008); Agarwal et al. (1994)

in other species, prior opsonisation is often required to optimize uptake of liposomes by Kupffer cells, highlighting a species difference in the mechanism of macrophage targeting. Similarly, after subcutaneous injection of phosphatidylserine containing liposomes into rats the retention of the lymphatically available dose in regional lymph nodes is over 300% of the injected dose per gram, approximately three-fold higher than subcutaneous administration of phosphatidylcholine or phosphatidylglycerol based liposomes (Oussoren and Storm 1997).

By far the most common method of targeting drugs, DNA and cytotoxic agents to the RES is by conjugation of sugars, particularly mannose (mannan, mannose polymer) to the surface. These target the many mannose receptors present on the surface of macrophages and monocytes. Although in general good targeting can be achieved towards many organs of the RES via the use of sugars as targeting ligands, the cellular targets vary widely. This was demonstrated by Kawakami et al. (Kawakami et al. 2000) who prepared liposomes of approximately 90 nm in diameter with 5% galactose-cholesterol, mannose-cholesterol or fructose-cholesterol. Each of these liposomes rapidly distributed to the liver and spleen after IV dosing to mice when compared to control liposomes without sugar, but distribution of these liposomes into parenchymal (hepatocytes) versus non-parenchymal (Kupffer, sinusoidal and stellate) cells was 15.1:1, 0.5:1 and 0.2:1 when compared to control liposomes that were 1:1 (Fig. 9). Hence, the choice of conjugated sugar can dictate both the tissue biodistribution of the nanomedicine and the cellular disposition. Others have reported over tenfold greater accumulation of galactose-conjugated liposomes in hepatocytes than non-parenchymal cells, and seven to ten times greater accumulation of mannose-conjugated liposomes in hepatocytes compared

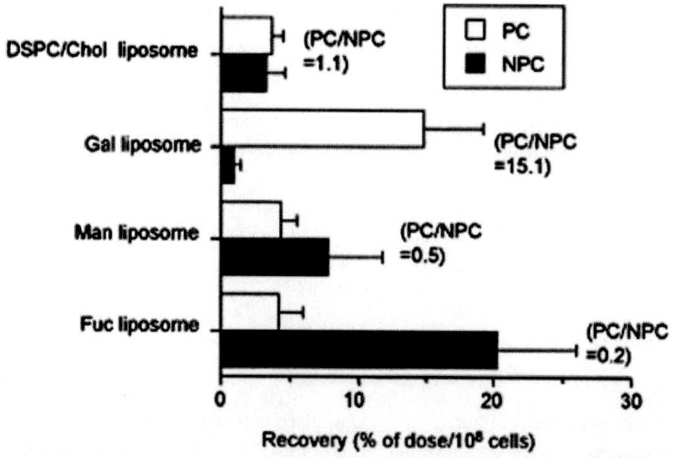

Fig. 9 Hepatocellular localization of ^{3}H CHE from *DSPC/cholesterol liposomes* (85–95 nm diameter) that were glycosylated, galactosylated, mannosylated or fucosylated 30 min after IV administration in mice. Data represent mean ± s.d. (n = 3) in parenchymal cells (*PC*) and non parenchymal cells (*NPC*) in the liver (Reproduced from Kawakami et al. (2000). With permission)

to non-parenchymal cells after IV administration in mice, depending on whether the liposomes were conventional or PEGylated (Nag and Ghosh 1999).

Mannose-conjugated generation 5 polypropyleneimine dendrimers also distribute preferentially and rapidly to the liver and kidneys (due to renal filtration) after IV administration in mice, and to a lesser extent in the spleen and lungs. By contrast, conjugation of lactose to dendrimers increased uptake via the liver and spleen (Agashe et al. 2007). It is therefore clear that galactose promotes cellular targeting towards hepatocytes, whereas mannose specifically targets liver and spleen resident macrophages after intravenous dosing.

Using a mannose targeting approach to improve delivery towards the lungs has been limited in success, however improved delivery of the anti-HIV drug didanosine into alveolar macrophages has been observed for mannosylated gelatin nanoparticles after IV administration in rats (Jain et al. 2008). Similarly, mannose conjugation improved the delivery of the gelatin nanoparticles into the spleen, liver and lymph nodes. Mannose conjugation has also been demonstrated to improve the delivery of the anti-HIV drug zidovudine entrapped within liposomes, to draining lymph nodes when compared to administration of the drug in non-targeted liposomes by subcutaneous administration (Kaur et al. 2008). In this respect, subcutaneous administration of mannose-targeted nanomedicines is expected to improve targeting and cellular internalization in both lymph node resident macrophages and in the lymphatic endothelium that similarly express mannose receptors that mediate lymphocyte and cancer cell trafficking.

4 Avoiding Reticuloendothelial Targeting

4.1 Reducing Surface Anionic Charge

As demonstrated previously, reduction in the extent and strength of surface anionic charge results in reduced particle opsonisation and subsequent uptake via the liver and spleen. This can be achieved either by changing the surface materials (either by incorporating less anionic lipids into colloid membranes or changing the surface functionality on nanoparticles) or by masking surface anionic charges. Lipids such as phosphatidylserine have a permanent anionic charge and incorporation of increasing amounts of these lipids into colloidal membranes increases the magnitude of the negative zeta potential and increases RES targeting. Alternatively, anionic groups may be masked by the inclusion of long, uncharged polymers into the particle surface. A classical example of this is the use of polyethylene glycol that coils around the particle surface, masking surface charge and neutralizing zeta potential. Since anionic charges are prone to recognition by plasma opsonins, a greater load of PEG (i.e., higher MW PEG) is needed to increase plasma residence and decrease RES targeting than for cationic particles. In theory, however, any uncharged polymer that can coat the particle surface has the potential to mask surface charges.

4.2 PEGylation

Polyethylene glycol (Fig. 10) was identified in the 1970s to be a highly biocompatible polymer. Since then, its use in improving the pharmacokinetic and biodistribution properties of nanomedicines has become widespread. The structure is composed of multiple repeating units of ethylene glycol that complex many water molecules resulting in the biological ‘masking’ properties of the polymer. The polymer is also relatively non-toxic; the administration of several grams per kg of body weight is required before toxicity becomes evident. Although it is slowly metabolized in the body, metabolism of PEG is not considered a major mechanism in the clearance of conjugated nanomedicines. However, for very large, long circulating systems, gradual metabolism of PEG chains with a resulting reduction in particle size and exposure of surface groups may eventually alter the biodistribution of the particles.

In the pharmaceutical industry, PEGylation has been used to increase the circulation times of proteins and liposomes, reduce RES uptake by liposomes and reduce the susceptibility of proteins to enzymatic degradation. Although PEGylation of proteins masks key receptor binding sites and therefore decreases *in vitro* activity compared to the native protein, *in vivo* therapeutic efficacy is generally increased as a result of the increased exposure of target receptors to the protein due to improved residence time in circulation. PEGylation works in a similar manner to mask biologically incompatible surface properties on nanomedicines, thus reducing both opsonisation and uptake via macrophages of the RES.

Although increasing particle size generally acts to increase RES targeting, conjugation of PEG, resulting in increased particle molecular weight and hydrodynamic size, does not have a similar effect. Increasing the size of conjugated nanomedicines via PEGylation instead acts to increase circulation times and reduce excretion via the urine. Increased eventual uptake by the RES (in particular the liver and spleen) for large PEGylated materials is more a consequence of increased exposure to macrophages that slowly phagocytose the material rather than increased uptake of larger materials. This is demonstrated by data showing that increasing the extent of PEGylation on polylysine dendrimers (resulting in increasing dendrimer sizes due to increased PEG molecular weight) results in slow uptake of the particles by the liver and spleen, such that approximately 12% of the injected dose of 68 kDa dendrimers with terminal plasma half lives of 3 days are recovered in the liver and spleen after 1 week (Kaminskas et al. 2008).

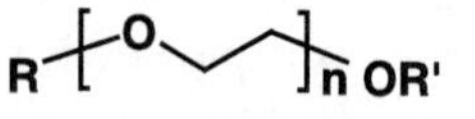

R - nanoparticle
R' - H (hydroxy PEG)
CH_3 (methoxy PEG)
Functional group attachment point

Fig. 10 Polyethylene glycol

4.3 Alternatives to PEGylation

In addition to PEG, several other materials have been used to improve plasma circulation times and avoid reticuloendothelial uptake of liposomes, including monosialoganglioside GM1, hydrogenated phosphatidylinositol and glucuronide conjugates. Studies in mice and *in vitro* in isolated murine bone marrow macrophages have demonstrated that the inclusion of only 5% GM1 into the liposome structure has the capacity to significantly reduce uptake of the liposome into RES organs (Allen et al. 1991a, b). This is reportedly due to the capacity of GM1 to prevent opsonisation of the liposome. *In vitro*, as the molar ratio of GM1 increases, so too does uptake of the liposomes by macrophages (Allen et al. 1991a). *In vivo*, uptake into the liver and spleen over 24 h can be halved when compared to liposomes of similar composition (Allen et al. 1991b). At molar ratios of GM1 or PEG in the range 5–10, however, PEG appears to be superior to GM1 *in vitro* at reducing the macrophage uptake of liposomes, although interestingly, this is not reflected by *in vivo* observations that show equivalent RES avoiding capacity (Allen et al. 1991a, b).

Although hydrogenated phosphatidylinositol (HPI) has not been widely incorporated into the design of long circulating liposomes, some research has been conducted by Alberto Gabizon who has shown in general that increasing the proportion of HPI into the liposome structure increases blood retention and decreases uptake by the liver and spleen up to a molar ratio of 41% HPI which results in increased uptake by the liver with a concomitant decrease in uptake by the spleen on account of the massive increase in anionic charge (Gabizon and Papahadjopoulos 1992). Thus, on account of the long circulating behavior of HPI-liposomes, Gabizon has demonstrated improved tumour biodistribution and anti-tumour efficacy of chemotherapeutic drug loaded liposomes composed of approximately 5 mol% of HPI in tumour bearing mice (Gabizon 1992; Gabizon et al. 1989; Gabizon and Papahadjopoulos 1988).

The glucuronide conjugate palmityl-D-glucuronide has demonstrated some capacity to improve the blood circulation time of liposomes in mice by reducing liver uptake, however it also increases the proportion of a dose taken up by the spleen by approximately 15%, 12 h after IV dosing (Oku et al. 1992). On account of the increased plasma blood exposure, tumour accumulation of modified liposomes is increased when compared to unmodified liposomes. However, this effect of the glucuronide is only seen in mice, since administration of palmityl-D-glucuronide modified liposomes in rats results in an increase in blood clearance and an increase in uptake by the liver and spleen, presumably via the action of complement factors (Liu et al. 1995b).

4.4 Choice of Lipids for Construction of Colloids

As described in the previous sections, certain structural features of a nanoparticle can render it a target for uptake by reticuloendothelial organs. This applies both to surface and structural features. Although a number of surface modifications have

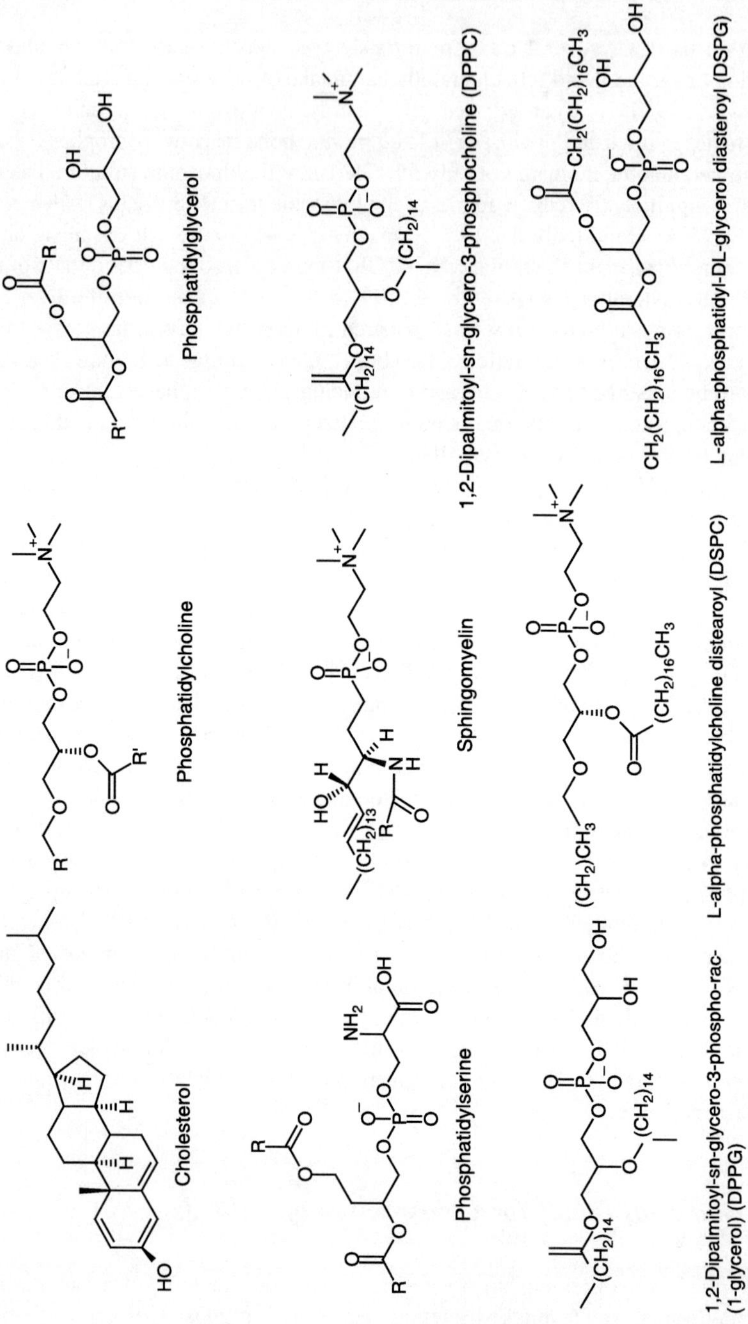

Fig. 11 Structure of lipids commonly used to formulate colloidal nanoparticles

proven useful in reducing the uptake of a variety of nanoparticles by the RES organs, the appropriate choice of lipid in the construction of the particle can similarly minimize or indeed increase the extent of RES targeting. This is particularly relevant for intravenous delivery systems that require extended plasma circulation times and subcutaneous delivery systems that may be used to improve drug uptake into the lymphatic system or specifically into regional lymph nodes. Much of the literature surrounding the influence of lipid composition on the affinity of colloids for the RES is based on liposomes. Hence, Fig. 11 depicts some of the lipids commonly used to generate liposomes.

As mentioned previously, the biodistribution of liposomes to the liver and spleen can be increased via the incorporation of phosphatidylserine into the lipid bilayer that increases the uptake of liposomes via fixed macrophages (Oussoren and Storm 1997). However, liposomes based entirely on DPPC also display avid uptake into both lymph nodes and the spleen after subcutaneous administration, even though most of the dose remains at the injection site after more than 2 days (Oussoren and Storm 1997). This effect is abrogated via the concurrent use of DPPC and cholesterol (Oussoren and Storm 1997). In general, however, increased plasma circulation times and minimized RES uptake appear to be achieved when using lipid compositions that include phosphatidylcholine, phosphatidylglycerol and sphingomyelin, although in each situation, RES uptake is generally best minimized by the incorporation of a hydrophilic or alternate biocompatible polymer into the outer surface (Oussoren and Storm 1997; Spanjer et al. 1986).

5 Summary

In the majority of applications of nanomedicines, extended blood circulation times are required. Consequently RES uptake is a major drawback in the use of nanoparticles as drug delivery systems, and increased uptake by the RES organs has the potential to be associated with increased delivery of drug to organs that may be sensitive to the toxic effects of the drug. As a general rule, the more 'foreign' the particle looks the greater the potential for RES uptake. Examples include increased particle surface charge (that may mediate particle opsonisation or increased adhesion to the surface of macrophages and other tissues), increased size and the use of structural materials that promote receptor mediated recognition of the particle by one or more cells within the RES. The flip side to this rule is in instances where improved nanoparticle targeting to particular cells within the RES, particularly macrophages, are desired. Examples of these are in the improved delivery of antigens to immune cells in order to enhance the immunization process or where delivery of antiretrovirals to macrophages may improve the resistance of the cells to viral attack or replication. Examples of such modifications have been given. However, where RES uptake is not desired, steps can be taken to minimize the susceptibility of nanoparticle systems for uptake via these organs. These include reducing surface charge and size or incorporating polymers and other functions into the surface of nanoparticles to improve their biocompatibility. To date, the most widely utilized method of avoiding RES uptake of nanoparticles is via the attachment of PEG chains onto the surface. However, several other systems are showing promise as alternatives to the use of PEG.

References

A. Agarwal, H. Kandpal, H.P. Gupta, N.B. Singh, C.M. Gupta, Tuftsin-bearing liposomes as rifampin vehicles in treatment of tuberculosis in mice, Antimicrobial Agents and Chemotherapy, 38 (1994) 588–593.

H.B. Agashe, A.K. Babbar, S. Jain, R.K. Sharma, A.K. Mishra, A. Asthana, M. Garg, T. Dutta, N.K. Jain, Investigations on biodiostribution of technetium-99 m-labelled carbohydrate-coated poly (propylene imine) dendrimers, Nanomedicine: Nanotechnology, Biology and Medicine, 3 (2007) 120–127.

W.T. Al Jamal, K.T. Al Jamal, A. Cakebread, J.M. Halket, K. Kostarelos, Blood circulation and tissue biodistribution of lipid-quantum dot (L-QD) hybrid vesicles intravenously administered in mice, Bioconjugate Chemistry, 20 (2009) 1696–1702.

T.M. Allen, G.A. Austin, A. Chonn, L. Lin, K.C. Lee, Uptake of liposomes by cultured mouse bone marrow macrophages: influence of liposome composition and size, Biochimica et Biophysica Acta, 1061 (1991a) 56–64.

T.M. Allen, C. Hansen, F. Martin, C. Redemann, A. Yau-Young, Liposomes containing synthetic lipid derivatives of poly(ethylene glycol) show prolonged circulation half lives in vivo, Biochimica et Biophysica Acta, 1066 (1991b) 29–36.

B.J. Boyd, L.M. Kaminskas, P. Karellas, G. Krippner, R. Lessene, C.J.H. Porter, Cationic poly-L-lysine dendrimers: Pharmacokinetics, biodistribution and evidence for metabolism and bioresorption after intravenous administration in rats., Molecular Pharmaceutics, 3 (2006) 614–627.

I. Dufresne, A. Desormeaux, J. Bestman-Smith, P. Gourde, M.J. Tremblay, M.G. Bergeron, Targeting lymph nodes with liposome bearing anti-HLA-DR-Fab' fragments, Biochimica et Biophysica Acta, 1421 (1999) 284–294.

T. Dutta, M. Garg, N.K. Jain, targeting of efavirenz loaded tuftsin conjugated poly(propyleneimine) dendrimers to HIV infected macrophages in vitro, European Journal of Pharmaceutical Sciences, 34 (2008) 181–189.

A. Egorova, A. Kiselev, M. Hakli, M. Ruponen, V. Baranov, A. Urtti, Chemokine-derived peptides as carriers for gene delivery to CXCR4 expressing cells, Journal of Gene Medicine, 11 (2009) 772–781.

A.A. Gabizon, Selective tumor localization and improved therapeutic index of anthracyclines encapsulated in long-circulating liposomes, Cancer Research, 52 (1992) 891–896.

A. Gabizon, D. Papahadjopoulos, Liposome frmulations with prolonged circulation time in blood and enhanced uptake by tumors, Proceedings of the National Academy of Science USA, 85 (1988) 6949–6953.

A. Gabizon, D. Papahadjopoulos, The role of surface charge and hydrophilic groups on liposome clearance in vivo, Biochimica et Biophysica Acta, 1103 (1992) 94–100.

A. Gabizon, R. Shiota, D. Papahadjopoulos, Pharmacokinetics and tissue distribution of doxorubicin encapsulated in stable liposomes with long circulation times, Journal of the National Cancer Institute, 81 (1989) 1484–1488.

A. Gabizon, D.C. Price, J. Huberty, R.S. Bresalier, D. Papahadjopoulos, Effect of liposome composition and other factors on the targeting of liposomes to experimental tumors: biodistribution and imaging studies, Cancer Research, 50 (1990) 6371–6378.

J.M. Irache, H.H. Salman, C. Gamazo, S. Espuelas, Mannose-targeted systems for the delivery of therapeutics, Expert Opinion in Drug Delivery, 5 (2008) 703–724.
S.K. Jain, Y. Gupta, A. Jain, A.R. Saxena, P. Khare, A. Jain, Mannosylated gelatin nanoparticles bearing an anti-HIV drug didanosine for site-specific delivery, Nanomedicine, 4 (2008) 41–48.
L.M. Kaminskas, B.J. Boyd, P. Karellas, S.A. Henderson, M.P. Giannis, G. Krippner, C.J. Porter, Impact of surface derivatisation of poly-L-lysine dendrimers with anionic arylsulphonate or succinate groups on intravenous pharmacokinetics and disposition, Molecular Pharmaceutics, 4 (2007) 949–961.
L.M. Kaminskas, B.J. Boyd, P. Karellas, G.Y. Krippner, R. Lessene, B. Kelly, C.J.H. Porter, The impact of molecular weight and PEG chain length on the systemic pharmacokinetics of PEGylated poly-L-lysine dendrimers, Molecular Pharmaceutics, 5 (2008) 449–463.
L.M. Kaminskas, B. Kelly, V. McLeod, B.J. Boyd, G.Y. Krippner, E.D. Williams, C.J.H. Porter, Pharmacokinetics and tumour disposition of PEGylated methotrexate conjugated poly-L-lysine dendrimers, Molecular Pharmaceutics, 6 (2009a) 1190–1204.
L.M. Kaminskas, J. Kota, V.M. McLeod, B.D. Kelly, P. Karellas, C.J.H. Porter, PEGylation of polylysine dendrimers improves absorption and lymphatic targeting following SC administration in rats, Journal of Controlled Release, 140 (2009b) 108–116.
C.D. Kaur, M.H. Nahar, N.K. Jain, Lymphatic targeting of zidovudine using surface-engineered liposomes, Journal of Drug Targeting, 16 (2008) 798–805.
S. Kawakami, J. Wong, A. Sato, Y. Hattori, F. Yamashita, M. Hashida, Biodistribution characteristics of mannosylated, fucosylated and galactosylated liposomes in mice, Biochimica et Biophysica Acta, 1524 (2000) 258–265.
S.S. Kim, C. Ye, P. Kumar, I. Chiu, S. Subramanya, P. Shankar, N. Manjunath, Targeted delivery of siRNA to macrophages for anti-inflammatory treatment, Molecular Therapeutics, 18 (2010) 993–1001.
H. Kobayashi, S. Kawamoto, T. Saga, N. Sato, A. Hiraga, J. Konishi, K. Togashi, M.W. Brechbiel, Micro-MR angiography of normal and intratumoral vessels in mice using dedicated intravascular MR contrast agents with high generation of polyamidoamine dendrimer core: reference to pharmacokinetic properties of dendrimer-based MR contrast agents, Journal of Magnetic Resonance Imaging, 14 (2001a) 705–713.
H. Kobayashi, S. Kawamoto, T. Saga, N. Sato, A. Hiraga, T. Ishimori, Y. Akita, M.H. Mamede, J. Konishi, K. Togashi, Novel liver macromolecular MR contrast agent with a polypropylenimine diaminobutyl dendrimer core: Comparison to the vascular MR contrast agent with the polyamidoamine dendrimer core, Magnetic Resonance in Medicine, 46 (2001b) 795–802.
H. Kobayashi, S. Kawamoto, P.L. Choyke, N. Sato, M.V. Knopp, R.A. Star, T.A. Waldmann, Y. Tagaya, M.W. Brechbiel, Comparison of dendrimer-based macromolecular contrast agents for dynamic micro-magnetic resonance lymphangiography, Magnetic Resonance in Medicine, 50 (2003) 758–766.
H. Kobayashi, S. Kawamoto, M. Bernardo, M.W. Brechbiel, M.V. Knopp, P.L. Choyke, Delivery of gadolinium-labeled nanoparticles to the sentinel lymph node: comparison of the sentinel node visualization and estimations of intra-nodal gadolinium concentration by the magnetic resonance imaging, Journal of Controlled Release, 111 (2006) 343–351.
D.C. Litzinger, A.M. Buiting, N. Van Rooijen, L. Huang, Effect of liposome size on the circulation time and intraorgan distribution of amphipathic poly(ethylene glycol)-containing liposomes, Biochimica et Biophysica Acta, 1190 (1994) 99–107.
F. Liu, D. Liu, Serum independent liposome uptake by mouse liver, Biochimica et Biophysica Acta, 1278 (1996) 5–11.
D. Liu, F. Liu, Y.K. Song, Recognition and clearance of liposomes containing phosphatidylserine are mediated by serum opsonin, Biochimica et Biophysica Acta, 1235 (1995a) 140–146.
D. Liu, F. Liu, Y.K. Song, Monosialoganglioside GM1 shortens the blood circulation of liposomes in rats, Pharmaceutical Research, 12 (1995b) 508–512.
S.M. Moghimi, Chemical camouflage of nanospheres with a poorly reactive surface: towards development of stealth and target-specific nanocarriers, Biochimica et Biophysica Acta, 1590 (2002) 131–139.

K. Morikawa, R. Nayar, I.J. Fidler, In vitro activation of tumoricidal properties in mouse macrophages using the chemotactic peptide N-formyl-methionyl-phenylalanine (FMLP) incorporated in liposomes, Cancer Immunology and Immunotherapy, 27 (1988) 1–6.

R. Mounzer, P. Shakarin, X. Papademetris, T. Constable, N.H. Ruddle, T.M. Fahmy, Dynamic imaging of lymphatic vessles and lymph nodes using a bimodal nanoparticulate contrast agent, Lymphatic Research and Biology, 5 (2007) 151–158.

A. Nag, P.C. Ghosh, Assessment of targeting potential of galactosylated and mannosylated sterically stabilized liposomes to different cell types of mouse liver, Journal of Drug Targeting, 6 (1999) 427–438.

N. Nimje, A. Agarwal, G.K. Saraogi, N. Lariya, G. Rai, H. Agarwal, G.P. Agarwal, Mannosylated nanoparticulate carriers of rifabutin for alveolar targeting, Journal of Drug Targeting, 17 (2009) 777–787.

N. Oku, Y. Namba, S. Okada, Tumor accumulation of novel RES-avoiding liposomes, Biochimica et Biophysica Acta, 1126 (1992) 255–260.

C. Oussoren, G. Storm, Lymphatic uptake and biodistributions of liposomes after subcutaneous injection: III. influence of surface modification with poly(ethyleneglycol), Pharmaceutical Research, 14 (1997) 1479–1484.

C. Oussoren, G. Storm, Liposomes to target the lymphatics by subcutaneous administration, Advanced Drug Delivery Reviews, 50 (2001) 143–156.

P.C. Rensen, J.C. Gras, E.K. Lindfors, K.W. van Dijk, J.W. Jukema, T.J. van Berkel, E.A. Biessen, Selective targeting of liposomes to macrophages using a ligand with high affinity for the macrophage scavenger receptor class A, Current Drug Discovery Technologies, 3 (2006) 135–144.

H.H. Spanjer, M. van Galen, F.H. Roerdink, J. Regts, G.L. Scherphof, Intrahepatic distribution of small unilamellar liposomes as a function of liposome lipid composition, Biochimica et Biophysica Acta, 863 (1986) 224–230.

L. Thiele, J.E. Deiederichs, R. Reszka, H.P. Merkle, E. Walter, Competitive adsorption of serum proteins at microparticles affects phagocytosis by dendritic cells, Biomaterials, 24 (2003) 1409–1418.

S.P. Vyas, Y.K. Katare, V. Mishra, V. Sihorkar, Ligand directed macrophage targeting of amphotericin B loaded liposomes, International Journal of Pharmaceutics, 210 (2000) 1–14.

L. Wan, X. Zhang, S. Pooyan, M.S. Palombo, M.J. Leibowitz, S. Stein, P.J. Sinko, Optimizing size and copy number for PEG-fMLF (N-formyl-methionyl-leucyl-phenylalanine) nanocarrier uptake by macrophages, Bioconjugate Chemistry, 19 (2008) 28–38.

Membrane Crossover by Cell-Penetrating Peptides: Kinetics and Mechanisms – From Model to Cell Membrane Perturbation by Permeant Peptides

Isabel D. Alves, Nicolas Rodriguez, Sophie Cribier, and Sandrine Sagan

Abstract Membrane-active peptides are a large family endowed with a wide pattern of biological activities (antimicrobial, viral fusion and infection, cell-penetrating or protein-transduction domain), which share the property of interacting with membranes and being internalized in eukaryotic cells. Apart from pinocytosis internalization pathways, these peptides have the capacity to re-organize lipid membranes and to lead to membrane fusion, disruption or pore formation. In this chapter, we focus on these membrane perturbation processes evoked by cell-penetrating peptides that have been widely studied with membrane models and in cultured cells.

Keywords Cell-penetrating peptide • Membrane • Permeation • Pinocytosis • Translocation

Abbreviations

AMP	Antimicrobial Peptide
Antp	antennapedia, homeoprotein
CHO	chinese hamster ovary cells
CPP	Cell-Penetrating Peptide
CS	Chondroitin Sulphate
DOPC	dioleoylphosphatidylcholine
DOPG	dioleoylphosphatidylglycerol
DPPC	dipalmitoyl phosphatidylcholine
DSC	Differential Scanning Calorimetry
ESR	Electron Spin Resonance spectroscopy
GUV	Giant Unilamellar Vesicle

I.D. Alves, N. Rodriguez, S. Cribier, and S. Sagan (✉)
Laboratory of Biomolecule, Universite Pierre et Marie Curie, Paris,
75252 Cedex 05, France
e-mail: sandrine.sagan@upmc.fr

A. Prokop (ed.), *Intracellular Delivery: Fundamentals and Applications*, Fundamental Biomedical Technologies 5, DOI 10.1007/978-94-007-1248-5_7,

HS	Heparan Sulphate
HSPG	Heparan Sulphate ProteoGlycans
ITC	Isothermal Titration Calorimetry
LUV	Large Unilamellar Vesicle
MAP	Membrane Active Peptide
NBD	Nitrobenzo-2-oxa-1,3-diazole
NMR	Nuclear Magnetic Resonance Spectroscopy
PEP-1	hepatite C virus related peptide, SGSWLRDVWDWICTVLTDFK-TWLQSKLDYKD-NH_2
P/L ratio	peptide over lipid ratio
Transportan	galanin/mastoparan chimeric peptide, GWTLNSAGYLLGKINLK-ALAALAKKIL-NH_2
Tat	Trans Activator of Transcription protein
Tat(46–58)	Tat derived peptide, GRKKRRQRRRPQ-NH_2

1 Introduction

Cellular signaling mechanisms in plants and animals include homeoprotein transduction, which is particularly important in developmental and physiological processes (Tassetto et al. 2005; Brunet et al. 2007). Homeoproteins have an important paracrine function, being secreted by and internalized into neighbored cells (Prochiantz and Joliot 2003; Joliot and Prochiantz 2008). Specialized peptide domains that are endowed with the property of membrane translocation have been identified in numerous different proteins (Lindgren et al. 2000; Prochiantz 2008). These peptides are grouped under the generic term of cell-penetrating peptides (CPPs) or, when they are derived from proteins, such as the Antennapedia homeoprotein or Tat transcription factors, protein transduction domains (PTD) (Hansen et al. 2008). These peptides have the ability to convey into cells conjugated cargo molecules that can give a positive biological or imaging read out of the intracellular localization of the peptide (Dietz and Bähr 2004; El-Andaloussi et al. 2004; Morris et al. 2008).

2 Amino Acid Composition of Cell-Penetrating Peptides

Dozens of different cell-penetrating peptides have now been reported, which derive from natural protein sequences or have been rationally designed (Hansen et al. 2008). Those peptides are indeed basic and/or amphipathic with a length of 10–20 amino acids. Some of these permeant peptides are pure basic sequences such as oligoarginine (R8, R9) (Futaki et al. 2001; Wender et al. 2001; Nakase et al. 2004) or Tat(48–59) (GRKKRRQRRRPQ) peptides (Weeks et al. 1995; Vives et al. 1997) when others are more hydrophobic such as transportan (WTLNSAGYLLGKINL-KAKAAKAKKIL).

2.1 *Necessity for a Peptide Secondary Structure for Internalization?*

It is clear that peptide – membrane interactions must be of fundamental importance for the internalization process. Therefore, whether the secondary structure of these peptides when interacting with membrane, plays a key role in the internalization properties has been widely studied. Moreover different peptide secondary structures have been reported for the same peptide. This is certainly a result of the different experimental conditions used in the studies regarding: peptide/lipid (P/L) ratios and concentrations used, buffer composition (e.g., ionic strength), temperature at which the experiments were performed as well as the method used to determine its structure.

Penetratin, the cell-penetrating sequence derived from the third helix of the Antennapedia homeodomain protein, has been shown to have a strong propensity for α-helix formation in lipid environments, which suggested first that the helical structure was necessary for internalization of the peptide (Magzoub et al. 2002; Letoha et al. 2003; Lindberg et al. 2003; Christiaens et al. 2004; Caesar et al. 2006; Clayton et al. 2006). But, a recent computational study on the molecular structure of penetratin, in interaction with lipid bilayers, and experiments with various phospholipids mixtures, have indicated a high structural polymorphism of penetratin (Polyansky et al. 2009). Penetratin could indeed adopt a α-helical, a β-strand or a β-turn conformation depending on the model membrane composition (Magzoub et al. 2002; Clayton et al. 2006; Su et al. 2008). In addition, it has been underlined that the α-helical conformation was not mandatory and could even be detrimental to the membrane translocation properties of penetratin (Derossi et al. 1996; Christiaens et al. 2004). Finally a recent study in living cells with high concentrations (25–50 μM) of penetratin has shown that the secondary structure of the peptide was found to be mainly random coil and beta-strand in the cytoplasm, and possibly assembling as beta-sheets in the nucleus (Ye et al. 2010). Ye and collaborators report no evidence of α-helical structure formation by penetratin, although it is possible (because of limitations with the signal intensity and the lateral spatial resolution (~0.5 μm) of the Raman microscopy methods) that the peptide could form α-helical or other transient conformations as it crosses the cell membrane (Ye et al. 2010).

Thus, whether a correlation exists between the capacity of a CPP to adopt a specific structure and its membrane translocation ability, is still a matter of debate. A recent study with ten different CPPs attempts to classify the peptides in three subgroups depending on their physicochemical properties (the secondary structure being one of them) and correlates those with different internalization pathways (Eiríksdóttir et al. 2010a, b). It has been suggested that the structural polymorphism and malleability of CPPs could be important for the membrane interaction and internalization route (Deshayes et al. 2008). An aspect that has been briefly evocated in the literature is the relevance of CPP self-assembly in the uptake mechanism. It follows that certain CPPs (penetratin, transportan, Pep-1, MAPs) can self-assemble, suggesting that they can be internalized as monomers or aggregates (Pujals et al. 2006).

Therefore, there is no clear relationship between the structure these peptides might adopt in solution or in contact with biological membranes and their ability to enter cells.

3 Binding of Cell-Penetrating Peptides to Membrane Components

3.1 Role of Proteoglycans

In the majority of cases, and independently of the internalization pathway of the CPP, the initial contact involves interactions between the CPP and cell-surface proteoglycans (PGs). Using model systems, the role of heparan sulphate proteoglycans (HSPGs) in CPP uptake has been investigated using isothermal titration calorimetry (ITC; Ziegler and Seelig 2004; Goncalves et al. 2005), plasmon resonance methods (Duchardt et al. 2009; Ram et al. 2008), ESR spectroscopy (Ghibaudi et al. 2005), and affinity chromatography (Fuchs and Raines 2004). Such studies point to considerably tight binding of CPPs to HSPGs such as heparan sulphates (HS), heparin and chondroitin sulfate B (CS) with dissociation constants in the low micromolar range. A higher affinity was observed for these HSPGs when comparing to anionic lipids. Although, the primary interaction between CPPs and HSPGs was considered to be electrostatic, it is also likely that hydrogen bonding occurs, taking into account the ability of the guanidium group (arginines are often present in CPPs) to form hydrogen bonds with sulfate and carboxylate groups.

3.2 Lipids

Taking into account the nature of CPPs, both electrostatic interactions between the positively charged amino acids and the lipid headgroups, and hydrophobic interactions between residues such as tryptophan and the lipid fatty acid region are possible. A strong electrostatic interaction can be established between the peptides and the lipids due to the large entropy gain that results from the release of counterions both at the level of the membrane (the Gouy-Chapman layer) (Zimm and Le Bret 1983) and the peptide (the Manning layer) (Manning 1969). Negative charges in the lipid can arise from the lipid headgroup itself in case of anionic lipids (phosphatidylglycerol, phosphatidylserine and phosphatidic acid) or from the phosphatidic groups of the fatty acids, which can establish strong ionic interactions with guanidium groups, often present in CPPs (Nakase et al. 2008). Even if the majority of the lipids present in the outer layer of eukaryotic cells are zwitterionic, anionic lipids are also present and their role may become relevant when they cluster in small domains, a process that is induced by CPPs and antimicrobial peptides

(Joanne et al. 2009; Epand and Epand 2009). Several studies have demonstrated the importance of an electrostatic recognition using different approaches. Therefore, to measure the affinity between the CPPs and lipid model systems and provide a thermodynamic characterization of such an interaction, ITC and plasmon resonance methods have been used (Goncalves et al. 2005; Binder and Lindblom 2003; Salamon et al. 2003; Alves et al. 2009; Henriques et al. 2010). ITC studies point to binding affinities in the low micromolar range and positive heat capacities indicating that electrostatics plays a major role in the interaction. Titration experiments with increasing amounts of anionic lipids show that at certain anionic lipid concentration and P/L ratios, penetratin is able to bind to both the outer and the inner leaflet presuming transbilayer distribution of penetratin. Since penetratin has not been associated with perturbations in membrane integrity, this indicates that, penetratin translocates through the vesicle membrane by an electroporation mechanism (this will be further discussed below) (Binder and Lindblom 2003). Plasmon resonance studies performed with penetratin indicate that binding to planar lipid bilayers is a fast and multistep process, primarily governed by electrostatic interactions followed by peptide insertion into the hydrophobic membrane core. The peptide also affected the amount of bound water, lipid-packing density, and bilayer thickness (the latter only at high peptide concentrations) accompanied by a decrease in membrane capacitance. A considerable enhancement of the binding was observed in the presence of anionic lipids (Salamon et al. 2003; Alves et al. 2009; Henriques et al. 2010). Improved binding affinities of penetratin to anionic lipids as compared with zwitterionic lipids (10–100 fold increase) have also been observed by following the intrinsic tryptophan fluorescence intensity of penetratin as a function of the lipid concentration (Christiaens et al. 2002).

4 Kinetics of Internalization in Cells

The kinetics of cell-penetrating peptide internalization has been studied by several groups. The reported data differ from one peptide to another depending on the cell type and the method used for peptide tracking.

One of the first studies has shown that ^{125}I-Biotinyl-transportan internalization was quick in Bowes melanoma cells and reached the steady-state after 20 min. Interestingly, the time course was found similar whatever the concentration (5–500 nM) of the peptide (Pooga et al. 1998). The same group reported the kinetics of internalization of penetratin (RQIKIWFQNRRMKWKK), transportan, Tat(48–60) and MAP (KLALKLALKALKAALKLA) (Hällbrink et al. 2001). The CPPs were labelled with a fluorescence quencher (3-nitrotyrosine) and were coupled to a pentapeptide cargo labeled with a fluorophore (2-amino benzoic acid) via a disulfide bond. The kinetics was recorded by following the increase in fluorescence intensity as the disulfide bridge is reduced into the intracellular milieu (Hällbrink et al. 2001). In those experimental conditions, the more hydrophobic the peptides (transportan and MAP) the faster they internalized in cells, the fluorescence plateau being reached after 15 min for transportan and after 1 h for penetratin. Similar results were obtained by Drin (Drin et al. 2003), using a NBD-labeled fluorescent penetratin analogue. The kinetics plateau was reached after 1 h incubation at 37°C of non-adherent human K562 leukemia cells with 1 μM NBD-penetratin. We have reported also kinetics of internalization at 37°C and 4°C of a biotin-labeled and photoactivatable penetratin analogue (Jiao et al. 2009). The kinetics was determined in CHO-K1 cells and CHO-pgA745 (GAG-deficient) cells. It was shown that at 37°C the plateau was reached after 1 h incubation of 5 μM penetratin with CHO-K1 and after 30 min with CHO-pgA745 (Jiao et al. 2009). Another study with a Tat-conjugated cargo also reported similar kinetics using a fluorescence assay (Cheung et al. 2009). In addition, Pep-1 (Ac-KETWWETWWTEWSQPKKKRKV-cysteamine), a rationally designed cell-penetrating peptide that can establish hydrophobic interactions with cargo molecules (thus that does not require a covalent link with these latter) is able to convey β-galactosidase into cells with a similar time course (Henriques et al. 2005). However, faster (in the range of seconds) and slower (tens of minutes) internalization kinetics could be measured for different cell-penetrating peptides (Eiríksdóttir et al. 2010a) using a releasable luciferin assay (Jones et al. 2006). These results led the authors to classify cell-penetrating peptides according to the internalization kinetics, which reflect their uptake pathways, as translocation for the fast kinetics and endocytosis for the slower one (Eiríksdóttir et al. 2010a).

However all these studies were done with a population of cells, thus the kinetics observed are not the kinetics of single cells but an average of the internalization kinetics in a population.

5 Pathways of Internalization in Cells

Regarding the internalization pathways of cell-penetrating peptides, there is a huge discrepancy between reported studies. The important point that is now spreading in literature is that any single chemical modification of the peptide sequence severely impacts the internalization pathways of the resulting compounds (Maiolo et al. 2005; El Andaloussi et al. 2007; Aussedat et al. 2008; Walter et al. 2009), as well as the cell-type (Mueller et al. 2008). These observations are quite understandable as any modification in the peptide or in membrane components must also affect peptide/membrane interactions.

Although still controversial for some cell-penetrating peptides, it is nonetheless quite clear that there are multiple internalization pathways for cell-penetrating peptides (Nakase et al. 2008; Jiao et al. 2009; Alves et al. 2010). It came out that the discrepancies between reported studies should have arisen from the experimental conditions used, principally the concentration of peptide and the chemicals used to inhibit internalization pathways, that could also have side-effects (Ivanov 2008).

All pinocytosis pathways have indeed been suggested for cell-penetrating peptide internalization, especially macropinocytosis (Jones 2007) that is a reported mechanism for penetratin (Amand et al. 2008), oligoarginine (Nakase et al. 2004, 2007),

Tat (Wadia et al. 2004), M918 (El-Andaloussi et al. 2007) and other permeant peptides (Sawant and Torchilin 2010).

Clathrin-mediated internalization was initially reported for Tat (Richard et al. 2003) but different results were obtained in further studies. It was first reported that knock down of clathrin-mediated endocytosis or knockout of caveolin-mediated endocytosis did not affect the ability of Tat to enter cells (Ter-Avetisyan et al. 2009). In addition, it was suggested that Tat could internalize at 4°C through a direct translocation mechanism (Jiao et al. 2009; Ter-Avetisyan et al. 2009).

For oligoarginine peptides, it was first reported that the peptide internalized through macropinocytosis (Nakase et al. 2004) or that direct translocation driven by the membrane potential occurred (Rothbard et al. 2004; 2005). It was further proposed by a single-molecule motion study that the mode by which octaarginine penetrates the cell membrane could be either a multi-mechanism uptake process or a mechanism different from passive diffusion and endocytosis (Lee et al. 2008). Other studies suggested that this peptide induces the formation of transient pores in cell membranes in the presence of an electrostatic potential gradient (Herce et al. 2009; Cahill 2010). A recent work reports that oligoarginine can change the lipid composition of cell membrane through the translocation in the outer membrane leaflet of sphingomyelinase and ceramide formation (Verdurmen et al. 2010).

Finally, internalization of the pAntp homeobox and of the derived penetratin peptide was originally described as a temperature and energy-independent process (Joliot et al. 1991; Derossi et al. 1994). Further studies suggested that penetratin enters via an endocytosis pathway rather than a translocation mechanism (Drin et al. 2003; Jones et al. 2005). A recent study highlighted the possibility that penetratin could internalize through transient pores and activate a resealing mechanism, known as a membrane repair response (Palm-Apergi et al. 2009)

6 Model Membrane Perturbation by CPP

To evaluate the energy-independent contribution to CPP uptake, the so-called direct translocation, several biophysical studies using different lipid models systems and a panoply of techniques have been employed in an attempt to elucidate the role of proteoglycans and lipids in the uptake mechanism as well as the peptide and lipid restructuration taking place upon their contact.

The direct translocation of CPPs through liposomes, has been investigated by several laboratories and conflicting results have been obtained. Uptake studies on giant unilamellar vesicles (GUVs) reported a transbilayer movement of penetratin (Thoren et al. 2000) or Tat(48–59) (Curnow et al. 2005), which contradicted studies on smaller lipid vesicles or planar membranes where these CPPs were found not to cross the membrane. Studies on LUVs have established that translocation is dependent on membrane potential and is modulated by the lipid composition (Terrone et al. 2003). One of the reasons for the divergent results may come from the different membrane curvature of the different lipid model systems used.

Despite controversial studies regarding the capacity of CPPs to penetrate through lipid bilayers or liposomal membranes (direct translocation), the initial contact with the cell membrane constitutes an important stage in the internalization process and has been investigated in depth. This initial interaction might be important just to increase the local peptide concentration in the surface before its uptake by either direct translocation (eventually leading to a deeper peptide penetration and lipid reorganization) or endocytosis.

Following the initial contact of CPPs with the cell membrane surface mainly driven by electrostatic interactions and the increase in the local peptide concentration, both peptide and lipid reorganization take place to allow peptide uptake by either endocytosis or direct translocation. The mode of action of the peptide in terms of lipid reorganization is dictated both by the CPP structure and the lipid composition of cellular or model systems. Taking into account the different peptide sequences, CPPs have been classified in three major classes:

1. Primary amphipathic such as transportan (Pooga et al. 1998), Pep-1 (Morris et al. 2001), they comprise sequentially hydrophobic and cationic domains and contain more than 20 amino acids, long enough to span the bilayer. They bind with strong affinities to both zwitterionic and anionic lipids suggesting that membrane interaction is dominated by hydrophobic interaction (Magzoub et al. 2001). They penetrate deeper than other CPPs in the membrane but without spanning the bilayer (Deshayes et al. 2004), the insertion is often accompanied by a secondary structure change. They have a tendency to self associate in the headgroup region. They often have antimicrobial activity and are rather difficult to distinguish from antimicrobial peptides as they can greatly perturb bilayer integrity, although less than AMP because they are less deeply inserted.
2. Secondary amphipathic such as penetratin (Derossi et al. 1996), KLAL (Dathe et al. 1996) and RL16 (Lamaziere et al. 2007), are shorter and display amphipathic property (evident when their amino acid sequence is depicted on a helical wheel) only through a change in their secondary structure upon lipid or HSPG contact. They possess poor affinity to neutral membranes, and their affinity is highly enhanced when anionic lipids are present due not only to electrostatic interaction but to a change in the peptide secondary structure (Binder and Lindblom 2003; Wieprecht et al. 2002). Despite the formation of an amphipathic structure by these peptides, the insertion in the bilayer is not marked and no membrane perturbation is usually observed at low anionic lipid content. Their binding leads to a change in the polar lipid headgroup orientation (Kichler et al. 2006; Roux et al. 1989). For tryptophan-containing cell-penetrating peptides, as examplified by pAntp analogues, in-cell studies support the hypothesis that hydrophobic interactions anchor those peptides in the membrane and might help their translocation into the cytosol (Le Roux et al. 1993; Fischer et al. 2002; Christiaens et al. 2004)
3. Non-amphipathic are generally shorter and comprise almost exclusively cationic amino acids such as R9. They do not bind lipid membranes unless they contain a high fraction of anionic phospholipids. Contrarily to amphipathic CPPs, direct translocation is not observed at low micromolar concentrations and at low

anionic lipid contents (Lamaziere et al. 2007; Thoren et al. 2005; Tiriveedhi and Butko 2007; Hitz et al. 2006). They do not induce liposome leakage or other types of membrane perturbation at low P/L ratios and concentrations around those required for biological uptake (1–10 μM) (Fuchs et al. 2004; Afonin et al. 2006). Additionally, no structure change is associated with their membrane binding and they are only superficially adsorbed on the membrane (Goncalves et al. 2005; Roux et al. 1988).

7 Mechanisms of CPP Direct Translocation

There is much evidence that direct translocation through the lipid membrane plays a significant role in CPP entry into cells. The relative importance of direct translocation and endocytosis seems dependent on conditions such as type of CPP, CPP concentration, temperature, cargo or cell type but the existence of the direct translocation pathways seems now ascertained. This leads to the need to explain by which mechanism(s) these highly soluble CPPs, all bearing several positive charges and few or no hydrophobic residues, can cross the hydrophobic core of the membrane bilayer. The order of magnitude of the activation energy for a naked CPP that would enter this hydrophobic core is given by the Born energy of an ion leaving an aqueous solution (relative permittivity 80) for the layer formed by the aliphatic lipid chains of the lipids (relative permittivity 2). This energy for a guadinium ion (radius r_G = 0.25 nm) is $\Delta E_{ion} = \frac{e^2}{8\pi\varepsilon_0 r_G}\left(\frac{1}{\varepsilon_c} - \frac{1}{\varepsilon_w}\right) \approx 60kT$ per ion (at T = 300K). The spontaneous entry of a naked CPP into the bilayer is thus highly unfavorable from a thermodynamical point of view. Several more refined mechanisms, presented below, have been proposed and experimentally backed up, and it is worth mentioning from the beginning that the question of the translocation mechanisms is currently still debated and may have no single answer, the direct translocation mechanisms being CPP or experimental condition dependent.

A first class of proposed mechanism is the neutralization of the positively charged CPP residues by some hydrophobic counterions that would simultaneously reduce the Born energy stated above and favor the solubilization of the CPP in the hydrophobic core of the membrane (Sakai and Matile 2003, Nishihara et al. 2005; Takeuchi et al. 2006; Wender et al. 2008) (Fig. 1). Several potential candidates for the role of amphipatic counterion lie in membranes such as anionic phospholipids or sulfated proteoglycans. The proof of concept has been given by Sakai and Matile (Sakai and Matile 2003) who showed that polyarginines (~80 residues) initially dissolved in an aqueous buffer can partition into chloroform when phosphatidylglycerol lipids were added. Rothbard and collaborators conducted a similar experiment with an arginine octamer labeled with fluorescein (Rothbard et al. 2004). When they added sodium laurate, the CPP migrated completely from water to an octanol phase. They also demonstrated the role of the two hydrogen bonds that a guadinium ion can form with a phosphate, sulphate or carboxylate group on its counterion. Mono- (resp. di-) methylation of the guadinium group of a CPP prevents formation of one (resp two) hydrogen bond between the guadinium and its counterion. For the fluorescinated arginine octamer it entailed a 80% (resp. 95%) decrease of its uptake into Jurkat cells (treatment : 5 min, 50 μM). In the framework of the solubilizing counterion mechanism, the strength of the non covalent bonds between a CPP and its counterions would therefore prove important. Interestingly this may give a rationale to the reported lower internalization efficiency of CPPs for which arginine residues were replaced by lysines (Mitchell et al. 2000).

Other proposed mechanisms (Fig. 1) are those that maintain, at least partly, the CPP in a polar environment (aqueous solution or polar layer of the membrane). Crossing the membrane then requires a transient reorganization of the bilayer such as the formation of a pore or the encapsulation of CPPs in an inverted micelle. Both options entail high local curvature of the lipids (on the rim of the pore or in/around the inverted micelle embedded in the bilayer) and a possible mismatch between the lipids ("void" on the rim of the inverted micelle). From an energetic point of view, the saving of the Born energy is balanced to some extent by the cost of the deformation of the bilayer that can amount to tens of kT (Siegel 1993; Glaser et al. 1988). Within these models, the transition is driven by the interaction of several peptides with the surface of the membrane and its consequences on the curvature and stability of the bilayer.

The inverted micelle model (Fig. 1) has been mainly proposed for penetratin (Derossi et al. 1996). In this model, cationic residues of penetratins interact with negatively charged phospholipids in the plasma membrane and subsequent interaction of Trp in the peptide with the hydrophobic membrane is thought to induce an invagination in the plasma membrane. The concomitant reorganization of the neighboring lipids results in formation of an inverted micelle, followed by release of peptide and cargo upon micelle disruption. This explanation for the penetratin translocation is supported by ^{31}P-NMR and differential scanning calorimetry experiments showing that penetratin favors the lamellar to hexagonal inverse transition for certain lipid compositions (Berlose et al. 1996; Alves et al. 2008).

A pore formation could occur through different paths (Fig. 1). One is referred to as the "carpet model". It consists of the binding of numerous peptides in the polar region of the membrane. These peptides destabilize the lipid assembly and lead to disruption of the bilayer and formation of pores. This model was first proposed for some AMPs (Shai 1999). But molecular dynamic simulations have suggested that it applies to TAT and arginine nonamer (Herce and Garcia 2007; Herce et al. 2009). The proposed mechanism was that these CPPs bound strongly to the phosphate and carbonyl groups of the phospholipids deep in the membrane just above the hydrophobic core. This destabilized the membrane and lead to the crossing of few CPPs immediately followed by the opening of aqueous pores. This later event has been experimentally confirmed by Herce and collaborators in the case of arginine nonamer by electrophysiological measurements on planar bilayers and cells. However the simulations have been criticized by Yesylevskyy and collaborators who found no pore formation evidenced by similar simulations (Yesylevskyy et al. 2009).

Nevertheless, they illustrate the possible complexity of the translocation mechanism: the simulations showed an accumulation of peptides at the boundary of the polar region that lead to bilayer destabilization and pore nucleation which was consistent with the carpet model. But they also suggested that the entry of the peptides was lead by the phosphate-basic residue interactions and that the first step of the pore nucleation is the entry of the CPP in the bilayer hydrophobic core to create an interaction with distant phosphates: this is more in tune with the concepts developed in the hydrophobic counterion model mentioned above. This emphasizes the fact that the classification presented here distinguishes models for clarity purposes but their borders may be permeable. Another illustration of this complexity is the lipid segregation model. It starts similarly to the carpet model with accumulation of peptides on the membrane surface. But if the peptide binds preferentially to certain lipids (such as anionic lipids), this binding is likely to entail domain formation or more generally a modification of the lateral organization of the lipids in the membrane. These domains may show packing defects at their boundaries and these sites are likely to be more favorable to peptide entry in the membrane or even act as nucleation sites for pore formation. This model again, suggested for AMPs (Epand et al. 2006), may be relevant for CPPs. Recent DSC measurements have shown that penetratin, by segregating cardiolipin in a DDPC/cardiolipin mixtures, was able to induce the formation of domains in an otherwise homogenous membrane (Joanne et al. 2009). Finally, another model related to the carpet model is the electroporation mechanism proposed by Lindblom and collaborators (Binder and Lindblom 2003): they propose that the destabilization of the membrane by the CPP carpet is due (in the case of penetratin) to the asymmetrical distribution of the charged CPPs between the outer and inner surfaces of the bilayer causing a transmembrane electrical field, which alters the lateral and the curvature stresses acting within the membrane.

Certain mechanisms include the formation of an aqueous pore following CPP addition to a membrane. This step is somewhat easier to check experimentally because it can manifest itself trough leakage of hydrophilic markers or ionic current flow. Lactate dehydrogenase release assays conducted on HeLa cells incubated 3 h with 100 μM with Tat, HIV1-rev-(34–50) or arginine octamers showed no significant leak. Absence of leakage of CHO cells treated with penetratin or RL16 (RRLRRLLRRLLRRLRR) was confirmed with a similar assay (1 h, 10 μM) (Joanne et al. 2009). However electrophysiological experiments conducted on DOPC: DOPG 3:1 planar bilayers and HUA smooth muscle cells in presence of arginine nonamers showed significant ionic current revealing membrane permeation (Herce et al. 2009). These results seem controversial but must be compared with the sensitivity of the technique in mind (electrophysiological measurements being more sensitive). It seemed to date that major leaking induced by CPPs can be ruled out while the possibility of transient, rapid, small aqueous pores cannot.

Finally, another question often mentioned in the literature regarding the mechanism of CPP entry is that of the driving force for the uptake of the peptides. This role is generally attributed to the transmembrane potential across cell membrane, a natural candidate for these polycationic peptides. Experiments on vesicules showing membrane potential dependant translocation of penetratin are consistent with this role of

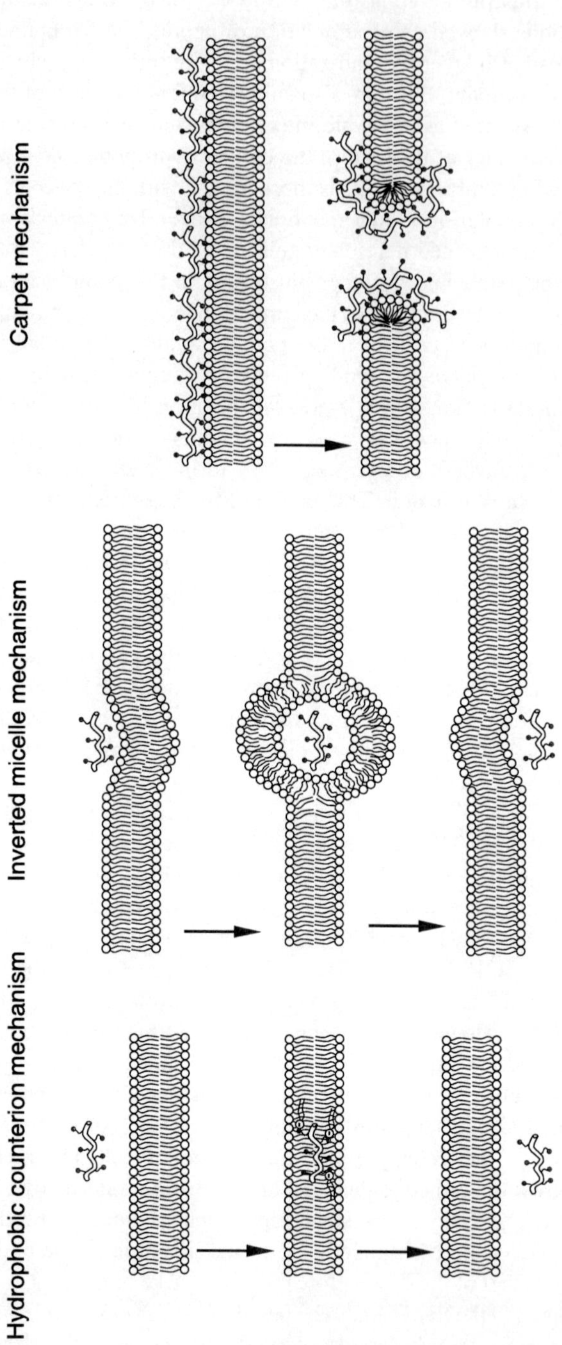

Fig. 1 Schematic representation of the translocation mechanisms reported in literature. Read the Sect. 7 for more details regarding these mechanisms

the potential (Terrone et al. 2003). Modulation of membrane potential trough variation of potassium concentration outside Jurkat cells also showed a strong impact on uptake of fluorescently labelled Tat49–57 or arginine octamer (Rothbard et al. 2004), and of biotin-labelled RW9 (R6W3) (Delaroche et al. 2007). Thus, the role of the potential in cell-penetrating peptide internalization needs to be included in translocation mechanisms. For example, in the frame of the hydrophobic counterion model, the sensitivity of a CPP to the potential requires the absence of neutralization of certain positive charges as suggested by Rothbard and collaborators (Rothbard et al. 2004). Intriguingly, crossing the membrane for a positive elementary charge amounts to a gain of ~5 kT for a –100 mV potential that appears at first sight weak compared to the Born energy of the charge. For models that appeal to aqueous pore formation, an obvious impact of the potential would be a proportional electromotive force on charges in the pore that may be supplemented by a promoting effect on pore formation (the later is not seen on the linear I-V relationship of the ionic current generated in a planar bilayer by arginine octamer in Herce et al. (2009)).

This review of possible pathways for CPP translocation illustrates that direct investigations of the mechanism are difficult because all mechanisms involve nanometric, rare, transient structures. However very significant progress has been recently made in the understanding of the complex interactions between membranes and CPPs and a clarification of the mechanisms of translocation is likely to occur in upcoming years.

References

Afonin S, Frey A, Bayerl S, Fischer D, Wadhwani P, Weinkauf S, Ulrich AS (2006) The cell-penetrating peptide TAT(48–60) induces a non-lamellar phase in DMPC membranes. Chemphyschem 7:2134–2142.

Alves ID, Goasdoué N, Correia I, Aubry S, Galanth C, Sagan, S, Lavielle S, Chassaing G (2008) Membrane interaction and perturbation mechanisms induced by two cationic cell penetrating peptides with distinct charge distribution. Biochim Biophys Acta. 1780:948–59.

Alves ID, Correia I, Jiao CY, Sachon E, Sagan S, Lavielle S, Tollin G, Chassaing G. (2009) The interaction of cell-penetrating peptides with lipid model systems and subsequent lipid reorganization: thermodynamic and structural characterization. J Pept Sci 15:200–209.

Alves ID, Jiao CY, Aubry S, Aussedat B, Burlina F, Chassaing G, Sagan S (2010) Cell biology meets biophysics to unveil the different mechanisms of penetratin internalization in cells. Biochim Biophys Acta. 2010 Feb 10. [Epub ahead of print]

Amand HL, Fant K, Nordén B, Esbjörner EK (2008) Stimulated endocytosis in penetratin uptake: effect of arginine and lysine. Biochem Biophys Res Commun. 2008 Jul 11;371(4):621–5. Epub 2008 May 13

Aussedat B, Dupont E, Sagan S, Joliot A, Lavielle S, Chassaing G, Burlina F (2008) Modifications in the chemical structure of Trojan carriers: impact on cargo delivery, Chem. Commun. 12:1398–1400.

Berlose JP, Convert O, Derossi D, Brunissen A, Chassaing G (1996) Conformational and associative behaviours of the third helix of antennapedia homeodomain in membrane-mimetic environments. Eur J Biochem 242:372–386.

Binder H, Lindblom G (2003) Charge-dependent translocation of the Trojan peptide penetratin across lipid membranes. Biophys J 85:982–995.

Brunet I, Di Nardo AA, Sonnier L, Beurdeley M, Prochiantz A (2007) The topological role of homeoproteins in the developing central nervous system. Trends Neurosci 6:260–267.

Caesar CE, Esbjörner EK, Lincoln P, Nordén B (2006) Membrane interactions of cell-penetrating peptides probed by tryptophan fluorescence and dichroism techniques: correlations of structure to cellular uptake. Biochemistry 45:7682–7692.

Cahill K (2010) Molecular electroporation and the transduction of oligoarginines. Phys Biol 7:1–14

Cheung JC, Kim Chiaw P, Deber CM, Bear CE (2009) A novel method for monitoring the cytosolic delivery of peptide cargo. J Control Release. 2009 Jul 1;137(1):2–7.

Christiaens B, Symoens S, Verheyden S, Engelborghs Y, Joliot A, Prochiantz A, Vandekerckhove J, Rosseneu M, Vanloo B (2002) Tryptophan fluorescence study of the interaction of penetratin peptides with model membranes. Eur J Biochem 269:2918–2926.

Christiaens B, Grooten J, Reusens M, Joliot A, Goethals M, Vandekerckhove J, Prochiantz A, Rosseneu M (2004) Membrane interaction and cellular internalization of penetratin peptides. Eur. J. Biochem 271, 1187–1197.

Clayton AH, Atcliffe BW, Howlett GJ, Sawyer WH (2006) Conformation and orientation of penetratin in phospholipid membranes. J Pept Sci 12:233–238.

Curnow P, Mellor H, Stephens DJ, Lorch M, Booth PJ (2005) Translocation of the cell-penetrating Tat peptide across artificial bilayers and into living cells. Biochem Soc Symp 72:199–209.

Dathe M, Schumann M, Wieprecht T, Winkler A, Beyermann M, Krause E, Matsuzaki K, Murase O, Bienert M (1996) Peptide helicity and membrane surface charge modulate the balance of electrostatic and hydrophobic interactions with lipid bilayers and biological membranes. Biochemistry 35:12612–12622.

Delaroche D, Aussedat B, Aubry S, Chassaing G, Burlina F, Clodic G, Bolbach G, Lavielle S, Sagan S (2007) Tracking a new cell-penetrating (W/R) nonapeptide, through an enzyme-stable mass spectrometry reporter tag. Anal Chem 79:1932–1938.

Derossi D, Joliot AH, Chassaing G, Prochiantz A (1994) The third helix of the Antennapedia homeodomain translocates through biological membranes. J Biol Chem 269:10444–50.

Derossi D, Calvet S, Trembleau A, Brunissen A, Chassaing G, Prochiantz A (1996) Cell internalization of the third helix of the Antennapedia homeodomain is receptor-independent. J Biol Chem 271 (1996) 18188–18193.

Deshayes S, Plenat T, Aldrian-Herrada G, Divita G, Le Grimellec C, Heitz F (2004) Primary amphipathic cell-penetrating peptides: structural requirements and interactions with model membranes. Biochemistry 43:7698–7706.

Deshayes S, Decaffmeyer M, Brasseur R, Thomas A (2008) Structural polymorphism of two CPP: an important parameter of activity. Biochim Biophys Acta 1778:1197–1205.

Dietz GP, Bähr M (2004) Delivery of bioactive molecules into the cell: the Trojan horse approach. Mol Cell Neurosci. 2:85–131.

Drin G, Cottin S, Blanc E, Rees AR, Temsamani J. (2003) Studies on the internalization mechanism of cationic cell-penetrating peptides. J Biol Chem 278:31192–31201.

Duchardt F, Ruttekolk IR, Verdurmen WP, Lortat-Jacob H, Bürck J, Hufnagel H, Fischer R, van den Heuvel M, Löwik DW, Vuister GW, Ulrich A, de Waard M, Brock R (2009) A cell-penetrating peptide derived from human lactoferrin with conformation-dependent uptake efficiency. J Biol Chem 284:36099–36108.

El-Andaloussi S, Jarver P, Johansson HJ, Langel U (2007) Cargo-dependent cytotoxicity and delivery efficacy of cell-penetrating peptides: a comparative study, Biochem J 407: 285–292.

Eiríksdóttir E, Mäger I, Lehto T, El Andaloussi S, Langel U. (2010a) Cellular internalization kinetics of (luciferin-)cell-penetrating peptide conjugates. Bioconjug Chem. 21:1662–1672.

Eiríksdóttir E, Konate K, Langel U, Divita G, Deshayes S (2010b) Secondary structure of cell-penetrating peptides controls membrane interaction and insertion. Biochim Biophys Acta. 1798:1119–1128.

El-Andaloussi S, Johansson HJ, Holm T, Langel U (2004) A novel cell-penetrating peptide, M918, for efficient delivery of proteins and peptide nucleic acids. Mol Ther 15:1820–1826.

Epand RF, Schmitt MA, Gellman SH, Epand RM (2006) Role of membrane lipids in the mechanism of bacterial species selective toxicity by two alpha/beta-antimicrobial peptides. Biochim Biophys Acta. 1758:1343–50.

Epand RM, Epand RF (2009) Lipid domains in bacterial membranes and the action of antimicrobial agents, Biochim Biophys Acta 1788:289–294.

Fischer R, Waizenegger T, Kohler K, Brock R (2002) A quantitative validation of fluorophore-labelled cell-permeable peptide conjugates: fluorophore and cargo dependence of import. Biochim Biophys Acta 1564, 365–374.

Fuchs SM, Raines RT (2004) Pathway for polyarginine entry into mammalian cells. Biochemistry 43:2438–2444.

Futaki S, Suzuki T, Ohashi W, Yagami T, Tanaka S, Ueda K, Sugiura Y (2001) Arginine-rich peptides. An abundant source of membrane-permeable peptides having potential as carriers for intracellular protein delivery, J. Biol. Chem 276:5836–5840.

Ghibaudi E, Boscolo B, Inserra G, Laurenti E, Traversa S, Barbero L, Ferrari RP (2005) The interaction of the cell-penetrating peptide penetratin with heparin, heparansulfates and phospholipid vesicles investigated by ESR spectroscopy. J Pept Sci. 11:401–409.

Goncalves E, Kitas E, Seelig J (2005) Binding of oligoarginine to membrane lipids and heparan sulfate: structural and thermodynamic characterization of a cell-penetrating peptide. Biochemistry 44:2692–2702.

Glaser RW, Leikin SL, Chernomordik LV, Pastushenko VF, Sokirko AI (1988) Reversible electrical breakdown of lipid bilayers: formation and evolution of pores. Biochim Biophys Acta 940:275–87.

Hallbrink M, Floren A, Elmquist A, Pooga M, Bartfai T, Langel U (2001) Cargo delivery kinetics of cell-penetrating peptides, Biochim. Biophys. Acta 1515:101–109.

Hansen M, Kilk K, Langel U (2008) Predicting cell-penetrating peptides. Adv Drug Deliv Rev 60(4–5):572–579.

Henriques ST, Costa J, Castanho MA (2005) Translocation of beta-galactosidase mediated by the cell-penetrating peptide pep-1 into lipid vesicles and human HeLa cells is driven by membrane electrostatic potential. Biochemistry 44:10189–10198.

Henriques ST, Castanho MA, Pattenden LK, Aguilar MI (2010) Fast membrane association is a crucial factor in the peptide pep-1 translocation mechanism: a kinetic study followed by surface plasmon resonance. Biopolymers 94:314–322.

Herce HD, Garcia AE (2007) Molecular dynamics simulations suggest a mechanism for translocation of the HIV-1 TAT peptide across lipid membranes. Proc Natl Acad Sci USA. 104:20805–10.

Herce HD, Garcia AE, Litt J, Kane RS, Martin P, Enrique N, Rebolledo A, Milesi V (2009) Arginine-rich peptides destabilize the plasma membrane, consistent with a pore formation translocation mechanism of cell penetrating peptides. Biophys J 97:1917–1925.

Hitz T, Iten R, Gardiner J, Namoto K, Walde P, Seebach D (2006) Interaction of alpha-and beta-oligoarginine-acids and amides with anionic lipid vesicles: a mechanistic and thermodynamic study. Biochemistry 45:5817–5829.

Ivanov AI (2008) Pharmacological inhibition of endocytic pathways: is it specific enough to be useful? Methods Mol Biol 440:15–33.

Jiao CY, Delaroche D, Burlina F, Alves ID, Chassaing G, Sagan S (2009) Translocation and endocytosis for cell-penetrating peptide internalization. J Biol Chem 284:33957–33965.

Joanne P, Galanth C, Goasdoué N, Nicolas P, Sagan S, Lavielle S, Chassaing G, El Amri C, Alves ID (2009) Lipid reorganization induced by membrane-active peptides probed using differential scanning calorimetry. Biochim Biophys Acta 1788:1772–81.

Joliot A, Triller A, Volovitch M, Pernelle C, Prochiantz A (1991) alpha-2,8-Polysialic acid is the neuronal surface receptor of antennapedia homeobox peptide. New Biol 3:1121–1134.

Joliot A, Prochiantz A (2008) Homeoproteins as natural Penetratin cargoes with signaling properties. Adv Drug Deliv Rev 60: 608–613.

Jones SW, Christison R, Bundell K, Voyce CJ, Brockbank SM, Newham P, Lindsay MA (2005) Characterisation of cell-penetrating peptide-mediated peptide delivery. Br J Pharmacol, 145, 1093–1102.

Jones LR, Goun EA, Shinde R, Rothbard JB, Contag CH, Wender PA (2006) Releasable luciferin-transporter conjugates: tools for the real-time analysis of cellular uptake and release. J Am Chem Soc 128:6526–6527.

Jones AT (2007) J Cell Mol Med 11:670–684. Macropinocytosis: searching for an endocytic identity and role in the uptake of cell penetrating peptides.

Kichler A, Mason AJ, Bechinger B (2006) Cationic amphipathic histidine rich peptides for gene delivery. Biochim Biophys Acta 1758:301–307.

Lamaziere A, Burlina F, Wolf C, Chassaing G, Trugnan G and Ayala-Sanmartin J (2007) Non-metabolic membrane tubulation and permeability induced by bioactive peptides. PLoS ONE 2:e201.

Lee HL, Dubikovskaya EA, Hwang H, Semyonov AN, Wang H, Jones LR, Twieg RJ, Moerner WE, Wender PA. (2008) Single-molecule motions of oligoarginine transporter conjugates on the plasma membrane of Chinese hamster ovary cells. J Am Chem Soc. 130:9364–9370.

Le Roux I, Joliot AH, Bloch-Gallego E, Prochiantz A, Volovitch M (1993) Neurotrophic activity of the Antennapedia homeodomain depends on its specific DNA-binding properties. Proc Natl Acad Sci U S A. 90:9120–9124.

Letoha T, Gaál S, Somlai C, Czajlik A, Perczel A, Penke B (2003) Membrane translocation of penetratin and its derivatives in different cell lines. J Mol Recognit 16:272–279.

Lindberg M, Biverståhl H, Gräslund A, Mäler L (2003) Structure and positioning comparison of two variants of penetratin in two different membrane mimicking systems by NMR. Eur J Biochem 270:3055–3063.

Lindgren M, Hallbrink M, Prochiantz A, Langel U (2000) Cell-penetrating peptides. Trends Pharmcol Sci 21:99–103.

Magzoub M, Kilk K, Eriksson LE, Langel U, Graslund A (2001) Interaction and structure induction of cell-penetrating peptides in the presence of phospholipid vesicles. Biochim Biophys Acta 1512:77–89.

Magzoub M, Eriksson LE, Graslund A (2002) Conformational states of the cell-penetrating peptide penetratin when interacting with phospholipid vesicles: effects of surface charge and peptide concentration. Biochim. Biophys. Acta1563, 53–63.

Maiolo JR, Ferrer M, Ottinger EA (2005) Effects of cargo molecules on the cellular uptake of arginine-rich cell-penetrating peptides, Biochim Biophys Acta 1712: 161–172.

Manning GS, Bresler EH, Wendt RP (1969) Irreversible Thermodynamics and Flow across Membranes. Science 166:1438.

Mitchell DJ, Kim DT, Steinman L, Fathman CG, Rothbard JB (2000) Polyarginine enters cells more efficiently than other polycationic homopolymers, J. Pept. Res. 56:318–325.

Morris MC, Depollier J, Mery J, Heitz F, Divita G (2001) A peptide carrier for the delivery of biologically active proteins into mammalian cells. Nat. Biotechnol. 19:1173–1176.

Morris MC, Deshayes S, Heitz F, Divita G (2008) Cell-penetrating peptides: from molecular mechanisms to therapeutics. Biol Cell 100:201–217.

Mueller J, Kretzschmar I, Volkmer R, Boisguerin P (2008) Bioconjug Chem 19:2363–2374. Comparison of cellular uptake using 22 CPPs in 4 different cell lines.

Nakase I, Niwa M, Takeuchi T, Sonomura K, Kawabata N, Koike Y, Takehashi M, Tanaka S, Ueda K, Simpson JC, Jones AT, Sugiura Y, Futaki S (2004) Cellular uptake of arginine-rich peptides: Roles for macropinocytosis and actin rearrangement. Mol Ther 10:1011–1022.

Nakase I, Tadokoro A, Kawabata N, Takeuchi T, Katoh H, Hiramoto K, Negishi M, Nomizu M, Sugiura Y, Futaki S (2007) Interaction of arginine-rich peptides with membrane-associated proteoglycans is crucial for induction of actin organization and macropinocytosis Biochemistry 46:492–501.

Nakase I, Takeuchi T, Tanaka G, Futaki S (2008) Methodological and cellular aspects that govern the internalization mechanisms of arginine-rich cell-penetrating peptides. Adv Drug Deliv Rev 60:598–607.

Nishihara M, Perret F, Takeuchi T, Futaki S, Lazar AN, Coleman AW, Sakai N, Matile S (2005) Arginine magic with new counterions up the sleeve. Org Biomol Chem. 3:1659–1669.

Palm-Apergi C, Lorents A, Padari K, Pooga M, Hällbrink M (2009) The membrane repair response masks membrane disturbances caused by cell-penetrating peptide uptake. FASEB J. 23:214–223.

Polyansky A A, Volynsky PE, Arseniev AS, Efremov RG (2009) Adaptation of a membrane-active peptide to heterogeneous environment. I. Structural plasticity of the peptide. J Phys Chem B 113:1107–1119.

Pooga M, Hallbrink M, Zorko M, Langel U (1998) Cell penetration by transportan. FASEB J 12:67–77.

Prochiantz A (2008) Protein and peptide transduction, twenty years later a happy birthday. Adv Drug Deliv Rev. 2008 Mar 1;60(4–5):448–51.

Prochiantz A, Joliot A. (2003) Can transcription factors function as cell-cell signalling molecules? Nat Rev Mol Cell Biol 10:814–819.

Pujals S, Fernández-Carneado J, López-Iglesias C, Kogan MJ, Giralt E (2006) Mechanistic aspects of CPP-mediated intracellular drug delivery: relevance of CPP self-assembly. Biochim Biophys Acta. 1758:264–279.

Ram N, Aroui S, Jaumain E, Bichraoui H, Mabrouk K, Ronjat M, Lortat-Jacob H, De Waard M (2008) Direct peptide interaction with surface glycosaminoglycans contributes to the cell penetration of maurocalcine. J Biol Chem 283:24274–24284.

Richard JP, Melikov K, Vives E, Ramos C, Verbeure B, Gait MJ, Chernomordik LV, Lebleu B (2003) Cell-penetrating Peptides – A reevaluation of the mechanism of uptake. J Biol Chem 278:585–590.

Rothbard JB, Jessop TC, Lewis RS, Murray BA, Wender PA (2004) Role of membrane potential and hydrogen bonding in the mechanism of translocation of guanidinium-rich peptides into cells. J Am Chem Soc 126:9506–9507.

Rothbard JB, Jessop TC, Wender PA (2005) Adaptive translocation: the role of hydrogen bonding and membrane potential in the uptake of guanidinium-rich transporters into cells. Adv Drug Deliv Rev 57:495– 504.

Roux M, Neumann JM, Bloom M, Devaux PF (1988) 2 H and 31P NMR study of pentalysine interaction with headgroup deuterated phosphatidylcholine and phosphatidylserine. Eur Biophys J 16:267–273.

Roux M, Neumann JM, Hodges RS, Devaux PF, Bloom M (1989) Conformational changes of phospholipid headgroups induced by a cationic integral membrane peptide as seen by deuterium magnetic resonance. Biochemistry 28:2313–2321.

Sakai N, Matile S (2003) Anion-mediated transfer of polyarginine across liquid and bilayer membranes. J Am Chem Soc 125:14348–14356.

Salamon Z, Lindblom G, Tollin G (2003) Plasmon-waveguide resonance and impedance spectroscopy studies of the interaction between penetratin and supported lipid bilayer membranes. Biophys J 84:1796–1807.

Sawant R, Torchilin V (2010) Intracellular transduction using cell-penetrating peptides. Mol Biosyst 6:628–640.

Shai Y (1999) Mechanism of the binding, insertion and destabilization of phospholipid bilayer membranes by alpha-helical antimicrobial and cell non-selective membrane-lytic peptides. Biochim Biophys Acta. 1462:55–70.

Siegel DP (1993) Energetics of intermediates in membrane fusion: comparison of stalk and inverted micellar intermediate mechanisms. Biophysical Journal 65:2124–2140.

Su Y, Mani R, Doherty T, Waring AJ, Hong M (2008) Reversible sheet-turn conformational change of a cell-penetrating peptide in lipid bilayers studied by solid-state NMR J Mol Biol 381:1133–1144.

Takeuchi T, Kosuge M, Tadokoro A, Sugiura Y, Nishi M, Kawata M, Sakai N, Matile S, Futaki S (2006) Direct and rapid cytosolic delivery using cell-penetrating peptides mediated by pyrenebutyrate ACS Chem Biol. 1:299–303.

Tassetto M, Maizel A, Osorio J, Joliot A (2005) Plant and animal homeodomains use convergent mechanisms for intercellular transfer, EMBO Rep 6:885–890.

Ter-Avetisyan G, Tünnemann G, Nowak D, Nitschke M, Herrmann A, Drab M, Cardoso MC (2009) Cell entry of arginine-rich peptides is independent of endocytosis. J Biol Chem 284:3370–3378.

Terrone D, Sang SL, Roudaia L, Silvius JR (2003) Penetratin and related cell-penetrating cationic peptides can translocate across lipid bilayers in the presence of a transbilayer potential. Biochemistry 42:13787–13799.

Thoren PE, Persson D, Karlsson M, Norden B (2000) The antennapedia peptide penetratin translocates across lipid bilayers — the first direct observation. FEBS Lett 482:265–268.

Thoren PE, Persson D, Lincoln P, Norden B (2005) Membrane destabilizing properties of cell-penetrating peptides. Biophys. Chem. 114:169–179.

Tiriveedhi V, Butko P (2007) A fluorescence spectroscopy study on the interactions of the TAT-PTD peptide with model lipid membranes. Biochemistry 46: 3888–3895.

Verdurmen WPR, Thanos M, Ruttekolk IR, Gulbins E, Brock R (2010) Cationic cell-penetrating peptides induce ceramide formation via acid sphingomyelinase: Implications for uptake. J Control Release 147:171–179.

Vives E, Brodin P, Lebleu BA (1997) truncated HIV-1 Tat protein basic domain rapidly translocates through the plasma membrane and accumulates in the cell nucleus. J Biol Chem 272:16010–16017.

Wadia JS, Stan RV, Dowdy SF (2004) Transducible TAT-HA fusogenic peptide enhances escape of TAT-fusion proteins after lipid raft macropinocytosis. Nat Med 10:310–315.

Walter C, Ott I, Gust R, Neundorf I (2009) Specific Labeling With Potent Radiolabels Alters the Uptake of Cell-Penetrating Peptides. Biopolymers 92:445–451.

Wender PA, Mitchell DJ, Pattabiraman K, Pelkey ET, Steinman L, Rothbard JB (2001) The design, synthesis, and evaluation of molecules that enable or enhance cellular uptake: Peptoid molecular transporters. Proc Nat Acad Sci 97:13003–13008.

Wender PA, Galliher WC, Goun EA, Jones LR, Pillow TH (2008) The design of guanidinium-rich transporters and their internalization mechanisms. Adv Drug Deliv Rev 60:452–72.

Weeks BS, Lieberman DM, Johnson B, Roque E, Green M, Loewenstein P, Oldfield EH, Kleinman HK (1995) Neurotoxicity of the human-immunodeficiency-virus type-1 TAT transactivator to PC12 cells requires the Tat amino-acid-49-58 basic domain. J Neurosci Res 42:34–40.

Wieprecht T, Beyermann M, Seelig J (2002) Thermodynamics of the coil alpha-helix transition of amphipathic peptides in a membrane environment: the role of vesicle curvature. Biophys Chem 96:191–201.

Ye J, Fox SA, Cudic M, Rezler EM, Lauer JL, Fields GB, Terentis AC (2010) Determination of penetratin secondary structure in live cells with Raman microscopy. J Am Chem Soc 132:980–988.

Yesylevskyy S, Marrink SJ, Mark AE (2009) Alternative mechanisms for the interaction of the cell-penetrating peptides penetratin and the TAT peptide with lipid bilayers. Biophys J. 97:40–9.

Ziegler A, Seelig J (2004) Interaction of the protein transduction domain of HIV-1 TAT with heparan sulfate: binding mechanism and thermodynamic parameters. Biophys J. 86:254–63.

Zimm BH, Le Bret M (1983) Counter-ion condensation and system dimensionality. J Biomol Struct Dyn 1:461–471.

Part II
Nanocarrier Formulation

Intracellular Delivery: A Multifunctional and Modular Approach

Rupa R. Sawant and Vladimir P. Torchilin

Abstract Intracellular delivery of drugs and nucleic acids has become one of the most widely explored areas of research. However, it has become increasingly evident that it is also necessary to control the nanocarrier's disposition within a cell. Much attention has been paid nowadays to control the distribution of the nanocarrier within the cell by using organelle targeted nanocarriers. In this review we have described various approaches developed in our laboratory for intracellular delivery of drugs and nucleic acids with lipid-based nanocarriers.

Keywords Cell penetrating peptides • Nanocarrier • Intracellular delivery • pH-sensitive • Organelle targeting

Abbreviations

EPR	enhanced permeability and retention
PEG	polyethylene glycol
PEG-PE	polyethylene glycol-phosphatidyl ethanolamine
CPPs	cell penetrating peptides
LL	Lipofectin® lipids
PTDs	protein Transduction Domains
HIV-1	human immunodeficiency virus type 1
pNP-PEG-PE	(*p*-nitrophenyl) carbonyl-PEG-PE
QDs	quantum dots
APC	antigen presenting cells
PC	egg phosphatidylcholine
Chol	cholesterol

R.R. Sawant and V.P. Torchilin (✉)
Department of Pharmaceutical Sciences and Center for Pharmaceutical Biotechnology and Nanomedicine, Northeastern University, Mugar Building, Room 312, 360 Huntington Avenue, Boston, MA 02115, USA
e-mail: v.torchilin@neu.edu

A. Prokop (ed.), *Intracellular Delivery: Fundamentals and Applications*, Fundamental Biomedical Technologies 5, DOI 10.1007/978-94-007-1248-5_8,

Rh	rhodamine
DDS	drug delivery system
STPP	stearyl triphenyl phosphonium
Rh-123	rhodamine-123
PCL	paclitaxel
LSD	lysosomal storage diseases
ERT	enzyme replacement therapy
RhB	octadecyl derivative of rhodamine B
C_{12}FDG	5-dodecanoylamino fluorescein di-β-D-galactopyranoside

1 Introduction

It is now well understood that nanocarrier-mediated drug delivery can control the disposition of a drug within the body. The term 'targeting' in nanocarrier-mediated drug delivery often involves binding of the nanocarrier to a cell-surface receptor (receptors that are preferentially expressed/over-expressed on the target cell) followed by internalization of the nanocarrier via the endocytic pathway. The problem, however, is that any nanocarrier entering the cell via the endocytic pathway becomes entrapped in the endosome and eventually ends up in the lysosome, where active degradation under the action of the lysosomal enzymes takes place. This problem is particularly critical for nucleic acids, peptidic drugs that are sensitive to degradation. As a result, only a small fraction of such an unaffected substance appears in the cell cytoplasm. So far, multiple but only partially successful attempts have been made to bring various macromolecular drugs and drug-loaded pharmaceutical carriers directly into the cell cytoplasm, bypassing the endocytic pathway. Methods such as microinjection or electroporation used for the delivery of membrane-impermeable molecules in cell experiments are invasive in nature and can damage the cellular membrane (Chakrabarti et al. 1989; Arnheiter and Haller 1988). Non-invasive methods such as the use of pH-sensitive carriers, including pH-sensitive liposomes (which at low pH inside endosomes destabilize endosomal membrane and liberate the entrapped drug into the cytoplasm) (Straubinger et al. 1985; Torchilin 2005b) and cell-penetrating molecules (Torchilin 2005b, 2007b; Sawant and Torchilin 2010) are much more efficient. These approaches assume that just the cell cytosol delivery is adequate for the final action of a drug or nucleic acid.

However, it has become increasingly evident that it is also necessary to control the nanocarrier's disposition within the cell. Many drugs must be delivered to specific cell organelles such as nuclei (the target for gene and antisense therapy), lysosomes (the target for the delivery of deficient lysosomal enzymes in therapy of lysosomal storage diseases), and mitochondria (the target for pro-apoptotic anticancer drugs) to exert their therapeutic action. Thus, the focus has now moved towards targeting the nanocarrier or its cargo to an individual cell organelle.

Liposomes (mainly, for water soluble drugs) and micelles (for poorly soluble drugs) can be considered as prototype nanocarriers. Liposomes are artificial phospholipid vesicles with sizes varying from 50 to 1,000 nm and greater, which can be loaded with a variety of drugs (Lasic 1993; Torchilin 2005b). Further, the addition of a polyethylene glycol (PEG) coating renders these liposomes long-circulating (Lasic and Martin 1995), which allows them to accumulate in various pathological areas with compromised (leaky) vasculature, such as tumors or infarcts. It has been shown that (Torchilin et al. 1996) the long-circulating liposomes can be made 'targeted', if antibodies or other specific binding molecules (ligands) have been attached to the water-exposed tips of the PEG chains (Torchilin et al. 2001a).

Micelles, including polymeric micelles, are also a popular and well-investigated pharmaceutical carrier due to their small size (10–100 nm), *in vivo* stability, ability to solubilize water insoluble anticancer drugs, and prolonged blood circulation times (Torchilin 2001, 2007a). The typical core-shell structure of polymeric micelles is formed by the self-assembly of amphiphilic block-copolymers consisting of hydrophilic and hydrophobic monomer units in aqueous media (Torchilin 2001). The use of special amphiphilic molecules as micelle-building blocks can also introduce the property of micelle extended blood half-life. Block-copolymer micelles can also be used to target their payload to specific tissues through either passive or active means. The passive targeting is due to the small micellar size which allows in spontaneous penetration into the interstitium of body compartments with a leaky vasculature (tumors and infarcts) by the enhanced permeability and retention (EPR) effect (Maeda et al. 2000; Torchilin 2001, 2007a; Maeda et al. 2009; Lukyanov et al. 2004). Active targeting of micelles can also be achieved by attachment of target-specific ligands to their surface (Torchilin 2001, 2007a).

We have been specifically interested in micelles made of PEG-phosphatidyl ethanolamine (PEG-PE), where, the use of lipid moieties as hydrophobic blocks capping hydrophilic polymer (such as PEG) chains provides the additional advantage of particle stability when compared with conventional amphiphilic polymer micelles due to the existence of two fatty acid acyls, which contribute considerably to an increase in the hydrophobic interactions between the polymeric chains of the micelle core (Lukyanov and Torchilin 2004). Such PEG-PE micelles demonstrate good stability, longevity in the blood and the ability to accumulate in the areas with a damaged or leaky vasculature (Lukyanov et al. 2004; Lukyanov and Torchilin 2004).

The liposomes and micelles can be assembled in 'modular fashion' by addition of various components such as cationic lipids, intracellular peptide-conjugated lipids, ligand-modified lipids and organelle-targeted lipid conjugates for tumor targeted or intracellular delivery. Various components of these 'modular system' can be further incorporated in one carrier (either liposomes or micelles) to build a 'multifunctional system' (e.g. combination of cell penetrating function, cancer cell targeting antibody and stimuli-sensitivity in one system) to perform various functions simultaneously or in orchestrated fashion.

In this review, we will discuss the approaches successfully developed in our laboratory for intracellular delivery of liposomes and lipid-core micelles, particularly, the use of cationic lipids, cell penetrating peptides (CPPs), and organelle-targeting ligands.

2 Intracellular Delivery of Lipid-Core Micelles with Cationic Lipids

Polymeric micelles cannot diffuse through the cell membrane but rather are internalized by endocytosis. Detailed reviews of the endocytotic pathways and endocytosis of nanocarriers can be found in (Conner and Schmid 2003; Mukherjee et al. 1997; Bareford and Swaan 2007). Following cell uptake, micelles are contained within acidic endosomes and are further directed to various transport pathways including fusion with lysosomes or exocytosis. Therefore, it is necessary to further improve the efficiency of drug-loaded micelles by enhancement of their intracellular delivery to compensate for excessive drug degradation in lysosomes as a result of the endocytosis-mediated capture of micelles by cells.

PEG-PE micelles carry a net negative charge (Lukyanov and Torchilin 2004), which can hinder their internalization by cells. Modification of PEG-PE micelles with positively charged lipids may improve the uptake of drug-loaded micelles by cells. Such positively charged micelles could also more readily escape from endosomes and enter the cytoplasm. To test these ideas, we have prepared paclitaxel-loaded micelles from mixture of PEG-PE and positively charged Lipofectin® lipids (LL) (Wang et al. 2005). The intracellular fate of paclitaxel-loaded PEG-PE/LL micelles and micelles prepared without the addition of the LL was investigated by fluorescence microscopy with BT-20 breast adenocarcinoma cells. Both fluorescently-labeled PEG-PE and PEG-PE/LL micelles were endocytosed by BT-20 cells (Fig. 1). However, with PEG-PE/LL micelles, endosomes appeared to be partially disrupted and released drug-loaded micelles into the cell cytoplasm, a result of the

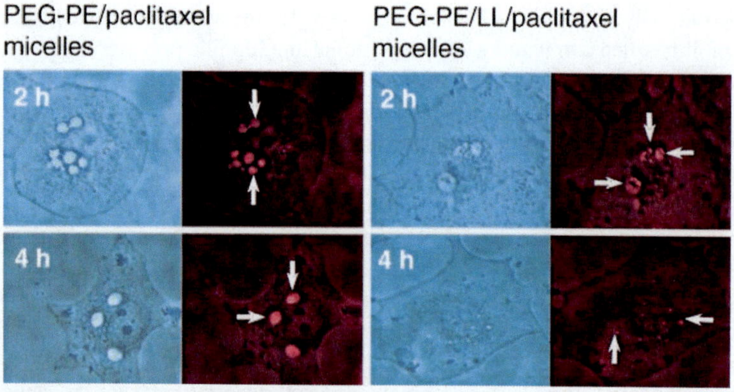

Fig. 1 Microscopy of BT-20 cells incubated with PEG-PE/ paclitaxel micelles and PEG-PE/LL/paclitaxel micelles for 2 and 4 h. Bright-field (*left images* in each pair) and fluorescence (*right images* in each pair). *Arrows* indicate fluorescent endosomes in cells incubated with PEG-PE/paclitaxel micelles for 2 h; partially degraded endosomes in cells incubated with PEG-PE/LL/paclitaxel micelles for 2 h; punctuate fluorescent structures in cells incubated with PEG-PE/LL/paclitaxel micelles for 4 h; larger (fused) endosomes in cells incubated with PEG-PE/paclitaxel micelles for 4 h (Modified from Torchilin 2005a)

de-stabilizing effect of the LL component on the endosomal membranes. After 4 h incubation, larger, fused fluorescent endosomal structures became apparent in the case of LL-free micelles, whereas cells incubated with PEG-PE/LL micelles had smaller punctuate fluorescent structures in the cytoplasm. Increased cytoplasmic delivery of paclitaxel was confirmed by the results of *in vitro* cytotoxicity studies against BT-20 cells (human breast carcinoma) and A2780 cells (human ovarian carcinoma). The paclitaxel-loaded PEG-PE/LL micelles were significantly more cytotoxic compared to that of free paclitaxel or paclitaxel delivered using noncationic LL-free PEG-PE micelles: in A2780 cancer cells, the IC50 values for free paclitaxel, paclitaxel in PEG-PE micelles, and paclitaxel in PEG-PE/LL micelles were 22.5, 5.8 and 1.2 μM, respectively. In BT-20 cancer cells, the IC50 values of the same preparations were 24.3, 9.5 and 6.4 μM, respectively.

However, use of cationic lipids has sometimes been associated with cytotoxicity, especially when used in the high amounts usually used for gene delivery (Torchilin et al. 2003a). Therefore, it is necessary to find a novel ligand that enhances both the cellular uptake and the escape from lysosomal degradation without cytotoxicity or immunogenicity.

3 Intracellular Delivery of Nanocarriers Using Cell Penetrating Peptides (CPPs)

A promising approach for the intracellular delivery that has emerged over the last decade is the use of CPPs (Schwarze et al. 1999). Many different short peptide sequences have been identified that promote transport of a variety of cargoes across the plasma membrane and deliver their payload intracellularly. This process is termed "protein transduction". Such proteins or peptides contain domains of less than 20 amino acids and are referred to as Protein Transduction Domains (PTDs) or CPPs, which are highly enriched with basic residues (Schwarze and Dowdy 2000).

TATp, the most frequently used CPP, is derived from the transcriptional activator protein encoded by human immunodeficiency virus type 1 (HIV-1) (Jeang et al. 1999). Authors Green (Green and Loewenstein 1988) and Frankel (Frankel and Pabo 1988) demonstrated that the 86-mer trans-activating transcriptional activator, Tat, protein encoded by HIV-1, was efficiently internalized by cells *in vitro* when introduced in the surrounding media. Later it was shown that residues 49–57 were responsible for membrane translocation, with the positive charge contributing largely to the transduction ability of TAT (Park et al. 2002).

TATp-mediated cytoplasmic uptake of plasmid DNA (Astriab-Fisher et al. 2002; Nguyen et al. 2008), nanoparticles (Zhao et al. 2002; Lewin et al. 2000; Rao et al. 2008), liposomes (Torchilin et al. 2001b; Fretz et al. 2004; Levchenko et al. 2003; Torchilin 2001; Zhao et al.) and micelles (Sethuraman and Bae 2007; Sawant et al. 2008) has been reported. A variety of uptake mechanisms appear to be involved in different systems, and in some cases, the mechanism is cell-type or cargo-specific (Zorko and Langel 2005). Smaller molecules attached to TATp seem to transduce directly into cells by energy independent electrostatic interactions and hydrogen bonding (Vives et al. 2003), but the larger cargos get into cells by an energy dependent macropinocytosis pathway (Wadia et al. 2004). Some examples relating to the intracellular delivery of pharmaceutical nanocarriers by CPPs are presented in Table 1.

Table 1 Examples of nanoparticles delivered using CPPs

Particle and CPP used	Result	References
Liposomes, TAT, antennapedia, octaarginine	Increased intracellular targeting of airway cells.	Cryan et al. (2006)
Liposomes, lipid-modified TATp	Increased cellular uptake in various cells.	Yagi et al. (2007)
Sterically stabilized liposomes, 200 nm, TATp coupled to the linker	Increased transfection in vitro and in vivo.	Torchilin et al. (2003a)
Liposomes, polyarginine	siRNA in R8-liposomes effectively inhibited the targeted gene and significantly reduced the proliferation of cancer cells.	Zhang et al. (2006)
Folic acid targeted, paclitaxel loaded, TATp-modified polymeric liposomes (FA-TATp-PLs)	Increased intracellular uptake of FA-TATp-PLs in both KB and A549 cells and superior toxicity in vivo in SCID mice bearing KB nasopharyngeal cancer.	Zhao et al. (2010)
CLIO (MION) particles, 41 nm, TATp	Mouse lymphocytes, human natural killer, HeLa, human hematopoietic CD34+, mouse neural progenitor C17.2, human lymphocytes CD4+, T- cells, B-cells, macrophages, immune cells, stem cells. For intracellular labeling, MRI, magnetic separation of homed cells, cell imaging.	Josephson et al. (1999), Lewin et al. (2000), and Dodd et al. (2001)
PEG-Polylactic acid micelles, 20–45 nm, TATp	Increased cytotoxicity of doxorubicin at acidic pH.	Sethuraman et al. (2008)
pH-sensitive polymeric micelles, 20–45 nm, TATp	Increased uptake at pH 6.6 compared to pH 7.4.	Sethuraman and Bae (2007)
Gold particles, 20 nm, TATp	Increased intracellular localization.	Tkachenko et al. (2004) and de la Fuente and Berry (2005)
Quantum dot-loaded polymeric micelles, 20 nm, TATp coupled to a linker	Concurrent imaging and distinguishing tumor vessels from perivascular cells and matrix made possible.	Stroh et al. (2005)
Nanocomplexes of PEI and DNA, TATp	Increased transfection in SH-SY5Y cells.	Suk et al. (2006)

(continued)

Table 1 (continued)

Particle and CPP used	Result	References
PEG-PEI conjugates, TATp	Increased transfection in mice.	Kleemann et al. (2005)
Boron carbide nanoparticles, TATp	Increased translocation in murine EL4 lymphoma cells, B16F10 melanoma cells. For boron neutron capture therapy.	Mortensen et al. (2006)
TATp liposomes DNA complexes	Increased transfection in vitro and in vivo.	Gupta et al. (2007)
TATp liposomes DNA complexes.	Increased transfection in vitro of the antigen presenting cells.	Pappalardo et al. (2009)
Low cationic liposomes–plasmid DNA complexes (lipoplexes) modified with TATp and/or with anti-myosin monoclonal antibody 2G4 (mAb 2G4) specific toward cardiac myosin	Increased transfection of hypoxic cardiomyocytes in vitro by the mAb 2G4-modified TATp lipoplexes. Increased accumulation of mAb 2G4-modified TATp lipoplexes in the ischemic rat myocardium and significantly enhanced transfection of cardiomyocytes in the ischemic zone in vivo.	Ko et al. (2009)

3.1 TATp-Modified Liposomes and Lipid-Core Micelles for Drug Delivery and Imaging

We have successfully modified liposomes and micelles with TATp using (*p*-nitrophenyl) carbonyl-PEG-PE (pNP-PEG-PE) (Torchilin et al. 2001a, 2003b). pNP-PEG-PE can be readily incorporated into liposomes and micelles *via* its phospholipid moiety, and it reacts easily with any amino group-containing substrate compound *via* its water-exposed pNP group to form a stable and non-toxic carbamate bond (Fig. 2). The reaction between the pNP group and the amino group of the ligand proceeds easily and quantitatively at pH 8.0, while excessive wfree pNP groups are readily eliminated by spontaneous hydrolysis. Some possible ways to attach CPPs to the micellar and liposomal surface are shown in Fig. 3.

An early study of liposomal delivery with a TATp showed that modification with TATp (47–57), enhanced delivery liposomes intracellularly to different cells, such as murine Lewis lung carcinoma (LLC) cells, human breast tumor (BT20) cells and rat cardiac myocytes (H9C2) (Torchilin et al. 2001b). The liposomes were tagged with TATp via the spacer, pNP-PEG-PE, at a density of a few hundreds of TATp per 200 nm liposome vesicle. These preparations of TATp-liposomes, which allowed for the direct contact of TATp residues with cells, displayed an enhanced liposome uptake by the cells. This suggested that the translocation of TATp-liposomes into cells requires direct, free interaction of TATp with the cell surface. Further studies on the intracellular trafficking of rhodamine-labeled TATp-liposomes loaded with FITC-dextran revealed that

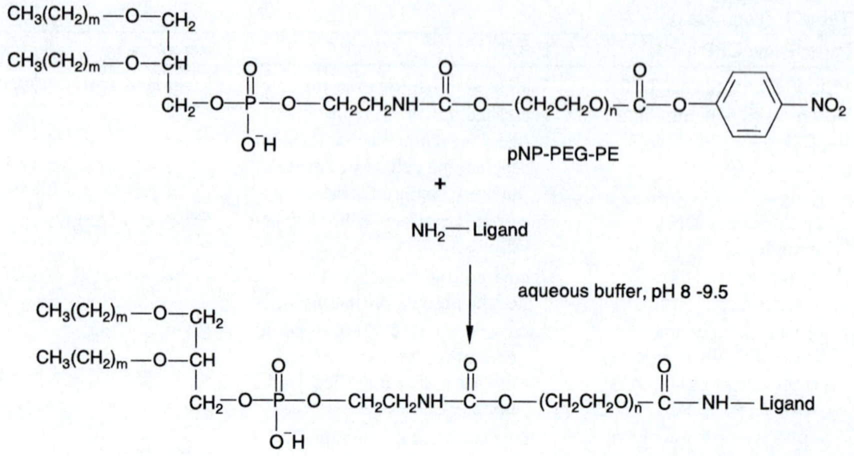

Fig. 2 Schematic of amino-group-containing ligand, such as peptide attachment using pNP-PEG-PE

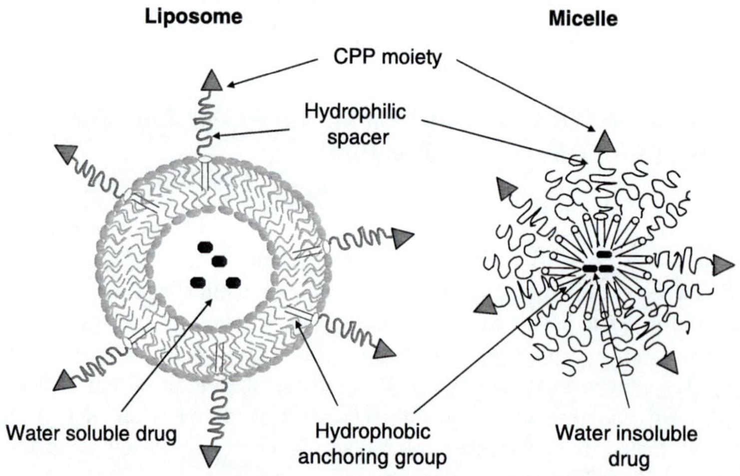

Fig. 3 CPP attached to liposomes and micelles by the insertion into the hydrophobic phase of liposome membrane or micelle inner core via a spacer arm (linker) with a hydrophobic "anchor"

TATp-liposomes remained intact in the cell cytoplasm at 1 h of translocation since the fluorescence of the intraliposomal (FITC-dextran) and membrane (rhodamine-PE) labels coincided (Fig. 4). After 2 h, they had migrated into the perinuclear zone and after several hours, the liposomes had completely disintegrated (Torchilin et al. 2003a).

One of the major obstacles to the use of TATp-mediated intracellular delivery of pharmaceutical nanocarriers is the lack of selectivity of TATp. This non-selectivity has generated concern about drug-induced toxic effects towards normal tissues. This suggested that intratumoral administration of TATp containing

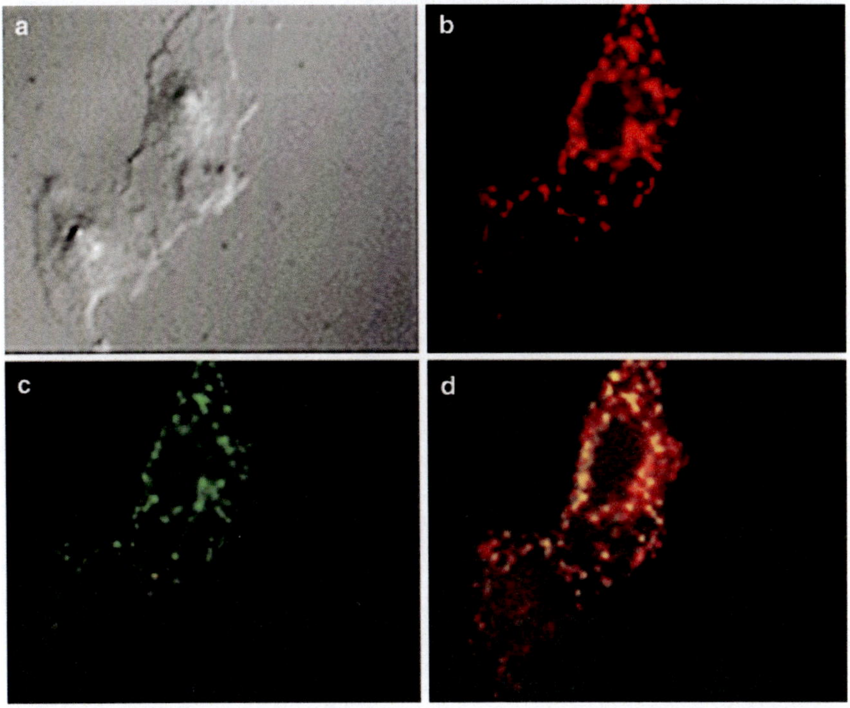

Fig. 4 Intracellular trafficking of rhodamine-PE- labeled and FITC-dextran-loaded TATp-liposomes within BT20 cells. Typical patterns of intracellular localization and integrity of TATp-liposome after 1 h. (**a**), DIC light; (**b**), DIC with an rhodamine filter; (**c**), DIC with an FITC filter; (**d**), DIC composite of (**b**)–(**c**). (Magnification, ×400) (Modified from Torchilin 2005a)

nanocarriers may serve as good solution to this problem for delivery of anticancer drugs at least in certain cases. We prepared and studied paclitaxel-loaded TATp containing PEG-PE micelles (Sawant and Torchilin 2009). Such paclitaxel-loaded micelles were prepared using PEG_{750}-PE as the main micelle-forming component with the addition of 2.5 mol% TATp-PEG_{1000}-PE to promote direct unhindered contact with cells.

The *in vitro* cell interaction of the TATp-bearing PEG-PE micelles was confirmed by fluorescence microscopy with 4T1 cells (Fig. 5). Plain micelles composed of PEG_{750}-PE demonstrated limited interaction with the cells (Fig. 5a). However, the use of the TATp-bearing PEG_{750}-PE micelles resulted in a strong interaction with the cells (Fig. 5b). This enhanced interaction also resulted in increased *in vitro* cytotoxicity against MCF-7 and 4T1 cells with paclitaxel-loaded TATp-bearing micelles compared to paclitaxel-loaded micelles without TATp at both 5 and 50 nM paclitaxel concentrations.

For *in vivo* studies, to avoid any unwanted distribution of paclitaxel-loaded TATp-micelles, micelles were injected intratumorally in mice and tumors were harvested after 48 h. Nuclear DNA fragmentation in tumor sections undergoing

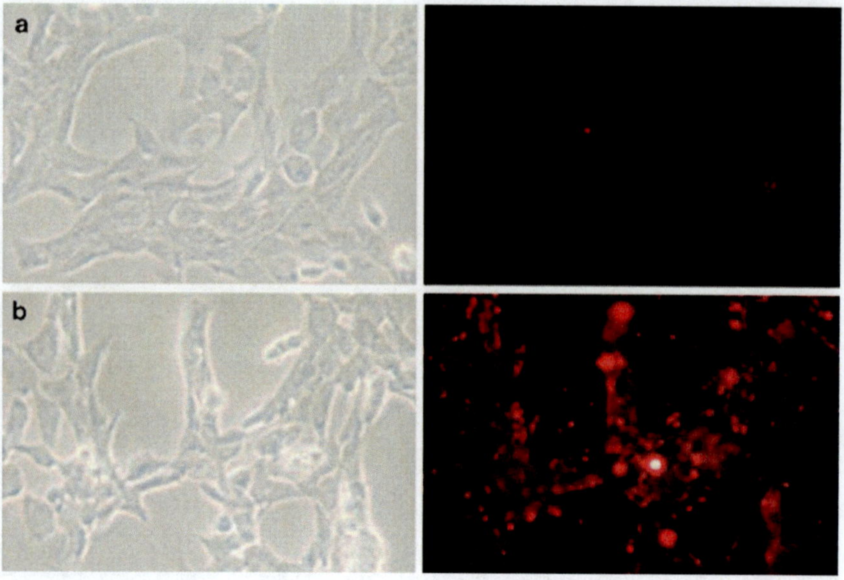

Fig. 5 *In vitro* interaction of rhodamine-PE labeled PEG-PE micelles with 4 T1 cells. Left panel shows the bright field and right panel shows the fluorescent microscopy of 4 T1 cells treated with rhodamine-PE labeled micelles. (**a**), rhodamine-PE :PEG_{750}-PE micelles; (**b**), rhodamine-PE: PEG_{750}-PE: TATp-PEG_{1000}-PE micelles. Magnification ×40 objective (Modified from Sawant and Torchilin 2009)

apoptosis was observed using the TUNEL assay with a DNA fragmentation kit. Very few TUNEL-positive cells were observed in tumors injected with free paclitaxel and paclitaxel-loaded micelles (Fig. 6c). However, significant apoptotic cell death was observed in tumors treated with paclitaxel-loaded TATp-bearing micelles (Fig. 6d).

Another application of CPPs involves the labeling of cells with semiconductor nanocrystals or quantum dots (QDs). QDs are now more popular than standard fluorophores for the study of tumor pathophysiology since they are photostable, more robust stable light emitters, and relatively insensitive to the wavelength of the excitation. They are also capable of distinguishing tumor vessels from both the perivascular cells and the matrix, with concurrent imaging. QDs trapped within PEG-PE micelles bearing a TATp-PEG-PE linker were used to label mouse endothelial cells *in vitro*. For *in vivo* tracking, bone marrow-derived progenitor cells labeled with TATp-bearing QD-containing micelles *ex vivo*, were injected in mice bearing a tumor in a cranial window model. It was possible to track the movement of labeled progenitor cells to the tumor endothelium, that may provide a path towards the understanding of the fine details of tumor neovascularization (Stroh et al. 2005).

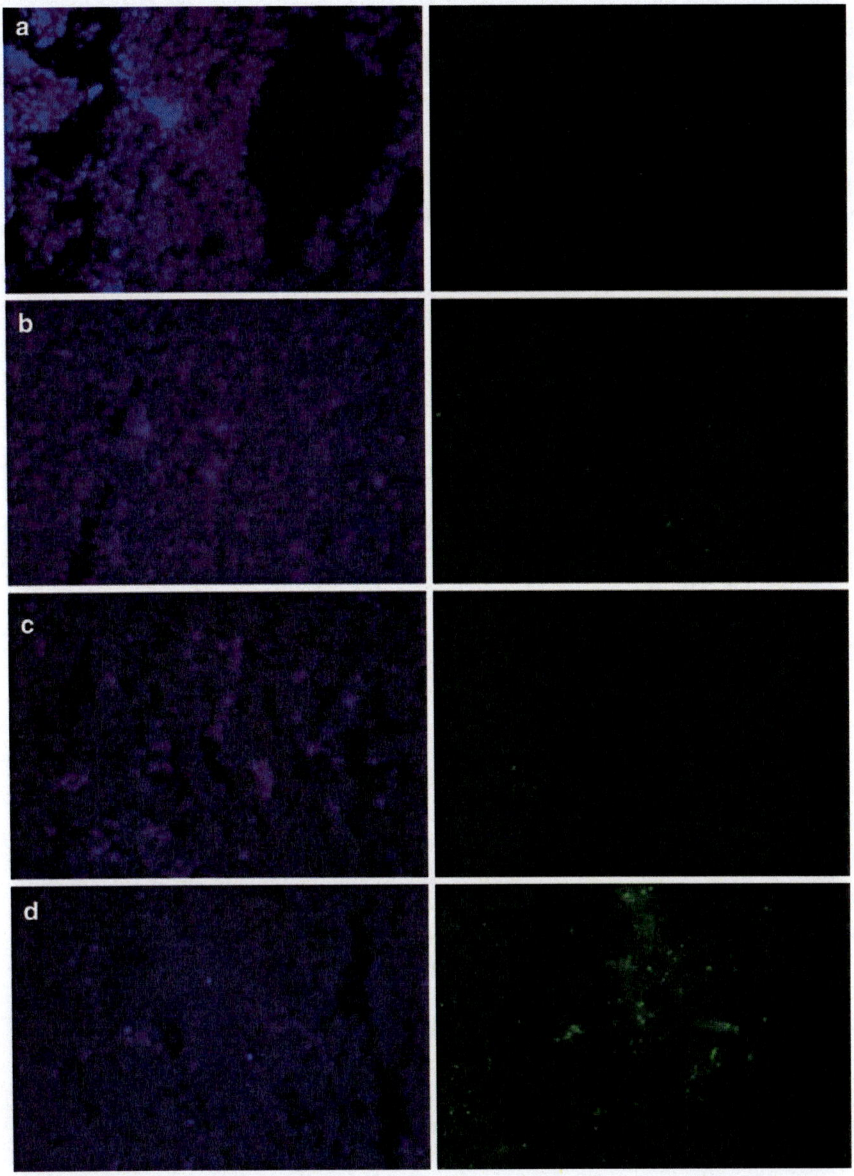

Fig. 6 Detection of apoptotic cells by fluorescence microscopy of frozen tumor sections. Apoptosis was determined by TUNEL. *The left panel* shows the sections stained with DAPI and the *right panel* shows TUNEL. (**a**), Negative control; (**b**), free paclitaxel; (**c**), paclitaxel-loaded micelles without TATp; (**d**), paclitaxel-loaded micelles with TATp. Magnification ×20 objective (Modified from Sawant and Torchilin 2009)

3.2 TATp- Modified Liposomes for Delivery of Nucleic Acids

Another exceptionally important yet challenging task is cellular delivery of nucleic acids. Various methods used to deliver these highly negatively charged biomolecules are associated with cellular toxicity or poor efficiency in certain types of cells (e.g. lipofectamnine or microinjection). Currently, liposomes and cationic polymers are used for transfection but they are also less efficient and often accompanied by high levels of toxicity.

Due to the size of plasmids and the high number of negative charges, non-covalent approach have been mostly used. It has been reported that TATp binds to DNA to form complexes which can be internalized through endocytosis (Sandgren et al. 2002). In our laboratory we prepared TATp–liposomes with the addition of a small quantity of a cationic lipid (DOTAP) and incubated with DNA to form stable non-covalent complexes with a model gene encoding for the enhanced-green fluorescent protein (pEGFP-N1) (Torchilin and Levchenko 2003). Such TATp–liposome–DNA complexes when incubated with mouse fibroblast NIH 3T3 and cardiac myocytes H9C2 showed substantially higher transfection *in vitro*, with lower cytotoxicity than the commonly used Lipofectin®.

We have also investigated the potential of TATp-modified liposomes to enhance the delivery of the model gene, pEGFP, to human brain tumor U-87 MG cells *in vitro* and in an intracranial tumor model in nude mice (Gupta et al. 2007). The size distribution of DNA-loaded TATp–liposomes was narrow (around 250 nm) and the DNA complexation was firm at lipid/DNA (+/–) charge ratios of 5 and higher. TATp–lipoplexes demonstrated an enhanced delivery of pEGFP to U-87 MG tumor cells *in vitro* at lipid/DNA (+/–) charge ratios of 5 and 10. *In vivo* transfection of intracranial brain tumors by intratumoral injections of TATp–lipoplexes showed an enhanced delivery of pEGFP selectively to tumor cells and subsequent effective transfection compared to plain plasmid-loaded lipoplexes. No transfection was observed in the normal brain adjacent to the tumor. Thus, TATp–lipoplexes can be used to augment the delivery of genes to tumor cells when injected intratumorally, without affecting the normal adjacent brain.

Another example is gene delivery into immunocompetent cells to modulate immune response. Antigen presenting cells (APC) are among the most important cells of the immune system since they link the innate and the adaptive immune responses, directing the type of immune response to be elicited. However, APC are very resistant to transfection. To increase the efficiency of APC transfection, we used liposome-based lipoplexes additionally modified with TATp for better intracellular delivery of a model pEGFP. pEGFP-bearing lipoplexes made of a mixture of egg phosphatidylcholine (PC): cholesterol (Chol): DOTAP (60:30:10 M ratio) with the addition of 2% mol of PEG-PE conjugate (plain-L) or TATp-PEG-PE (TATp-L) effectively protected the incorporated DNA from degradation. Uptake assays of rhodamine (rh)-labeled lipoplexes and transfections with the EGFP reporter gene were performed with APC derived from the mouse spleen.

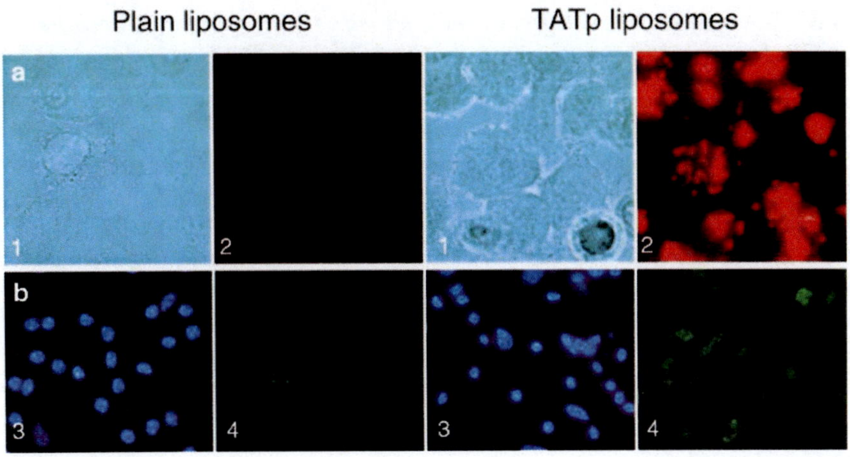

Fig. 7 (**a**) Liposome uptake and internalization of APC cultures incubated for 60 min with Rh-labeled plain-L and TATp-L (100×); (**b**), APC culture fluorescence microscopy 48 h post transfection with EGFP (40×). *1*: bright field, *2*: rh filter, *3*: Hoechst, *4*: EGFP (Modified from Pappalardo et al. 2009)

TATp-L-based lipoplexes significantly enhanced both the uptake and transfection of APC (Fig. 7) (Pappalardo et al. 2009).

We recently reported a double–targeted delivery system simultaneously capable of extracellular accumulation and intracellular penetration for gene therapy in the treatment of myocardial ischemia. We used low cationic liposome–plasmid DNA complexes (lipoplexes) modified with TATp and/or with monoclonal anti-myosin monoclonal antibody 2G4 (mAb 2G4) specific toward cardiac myosin, for targeted gene delivery to ischemic myocardium. *In vitro* transfection of both normoxic and hypoxic cardiomyocytes was enhanced by the presence of TATp determined by fluorescence microscopy and ELISA. The enhanced transfection with TATp-lipoplexes indicated that intracellular delivery mediated by TATp played an important role in the transfection of hypoxic as well as normoxic cells. The *in vitro* transfection was further enhanced by the additional modification with mAb 2G4 antibody in the case of hypoxic, but not normoxic cardiomyocytes. This can be explained by the additional mAb 2G4-mediated targeted delivery of the lipoplexes to the hypoxic cells because of the better binding of the lipoplexes with the hypoxically damaged cells due to exposure of intracellular cardiac myosin. However, we did not observe a synergism between TATp and mAb 2G4 ligands under our experimental conditions. In *in vivo* experiments, we clearly demonstrated an increased accumulation of mAb 2G4-modified TATp lipoplexes in the ischemic rat myocardium and significantly enhanced transfection of cardiomyocytes in the ischemic zone. Thus, the genetic transformation of normoxic and hypoxic cardiomyocytes can be enhanced by using lipoplexes modified with TATp and/or mAb 2G4 (Ko et al. 2009).

3.3 *TATp-Modified Liposomes and Micelles: A Multifunctional Approach*

An ideal nanoparticular drug delivery system (DDS) should be able to (1) specifically accumulate in the required organ or tissue, and then (2) penetrate target cells to deliver its load (drug or DNA) intracellularly. Organ or tissue (tumor, infarct) accumulation could be achieved by the passive targeting *via* the EPR effect (Maeda et al. 2000) assisted by prolonged circulation of such a nanocarrier (for example, as a result of its coating with protecting polymer such as PEG); or by antibody-mediated active targeting (Torchilin 2004) and (Jaracz et al. 2005), while the intracellular delivery could be mediated by certain internalizable ligands (folate, transferrin) (Gabizon et al. 2004) and (Widera et al. 2003) or by CPPs (Gupta et al. 2005). Ideally, such a DDS should simultaneously carry on its surface various active moieties, i.e. be multifunctional and possess the ability to "switch on" certain functions (such as intracellular penetration) only when necessary, for example under the action of local stimuli characteristic of the target pathological zone (first of all, increased temperature or lowered pH values characteristic of inflamed, ischemic, and neoplastic tissues). These "smart" DDS should be built in such a way that during the first phase of delivery, a non-specific cell-penetrating function is shielded by the function providing organ/tissue-specific delivery (sterically protecting polymer or antibody). Upon accumulating in the target, protecting polymer or antibody attached to the surface of the DDS *via* the stimuli-sensitive bond should detach under the action of local pathological conditions (abnormal pH or temperature) and expose the previously hidden second function to allow for the subsequent delivery of the carrier and its cargo inside cells (Fig. 8).

With this in mind, we prepared targeted long-circulating PEGylated liposomes and PEG–PE-based micelles possessing several functionalities (Sawant et al. 2006; Kale and Torchilin 2007a). First, such systems targeted a specific cell or organ by attaching the monoclonal antibody (infarct-specific antimyosin antibody 2G4 or cancer-specific antinucleosome antibody 2C5) to their surface *via* reactive pNP–PEG–PE moieties. Second, these liposomes and micelles were additionally modified with TATp moieties attached to the surface of the nanocarrier by using TATp-short PEG–PE derivatives. PEG–PE used for liposome surface modification or for micelle preparation was made degradable by inserting a pH-sensitive hydrazone bond between PEG and PE (PEG–Hz–PE). Under normal pH values, TATp functions on the surface of nanocarriers were "shielded" by the long PEG chains (pH-degradable PEG_{2000}–PE or PEG_{5000}–PE) or by long pNP–PEG–PE moieties used to attach antibodies to the nanocarrier (non-pH-degradable PEG_{3400}–PE or PEG_{5000}–PE). At pH 7.5–8.0, both liposomes and micelles demonstrated high specific binding with antibody substrates, but very limited internalization by NIH/3T3 or U-87 cells. However, upon brief incubation (15–30 min) at lower pH values (pH 5.0–6.0) nanocarriers lost their protective PEG shell because of acidic hydrolysis of the PEG–Hz–PE and were effectively internalized by cells *via* TATp moieties (Fig. 9a).

In vivo, TATp-modified pGFP-loaded liposomal preparations have been administered intratumorally in tumor-bearing mice, and the efficacy of tumor cell transfection

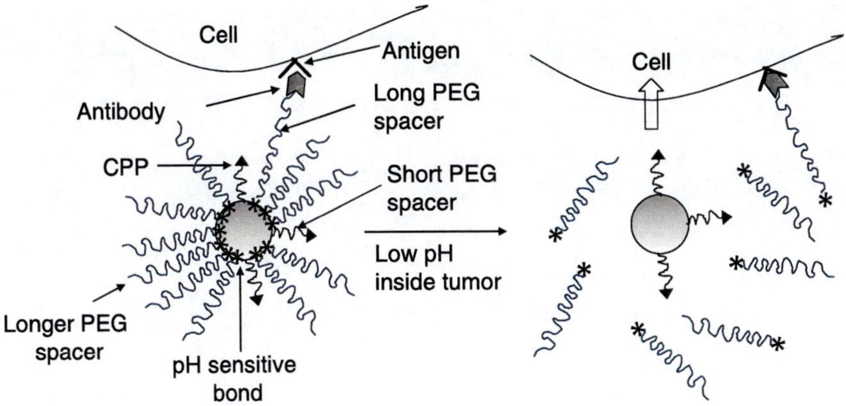

Fig. 8 The principle scheme of the action of stimuli-sensitive double-targeted nanocarriers. The surface of the nanocarrier is modified with a CPP via a relatively short spacer; with longer PEG chains; and with even longer PEG chains decorated at distal termini with a target-specific ligand (antibody). CPP is shielded with longer chains, while these PEG chains and PEG–antibody chains are attached to the surface via pH-sensitive bonds. The whole system is stable in the blood and accumulates in the tumor via the PEG-mediated EPR effect and via antibody-mediated targeting. Inside the tumor, protective PEG chains and PEG–antibody conjugates are detached from the surface because of fast hydrolysis of pH-sensitive bonds at the lowered intratumors pH, CPP becomes exposed and allows for the intracellular delivery

was assessed after 72 h. The administration of pGFP–TATp–liposomes with a non-pH-sensitive PEG coating resulted in minimal transfection of tumor cells because of steric hindrances for liposome-to-cell interaction created by the PEG coat, which shielded the surface-attached TATp. The administration of pGFP–TATp–liposomes with the low pH-detachable PEG resulted in the highly efficient transfection. The removal of PEG under the action of the decreased intratumoral pH led to the exposure of the liposome-attached TATp residues, enhanced penetration of the liposomes inside tumor cells and more effectively delivered the pGFP intracellularly (Fig. 9b) (Kale and Torchilin 2007b).

TATp-modified stimuli-sensitive polymeric micelles with an enhanced ability to interact with cells under acidified conditions have also been described in (Sethuraman and Bae 2007). These results can be considered as an important step in the development of tumor-specific stimuli-sensitive drug and gene delivery systems.

4 Subcellular Targeted Nanocarriers

The next step in the development of targeted nanocarriers would be designing of subcellular or organelle targeted nanocarriers to target at molecular receptor level (Rajendran and Knolker 2010; D'Souza and Weissig 2009; Torchilin 2006). Our focus has been on the development of nanocarrier systems targeted to mitochondria and lysosomes.

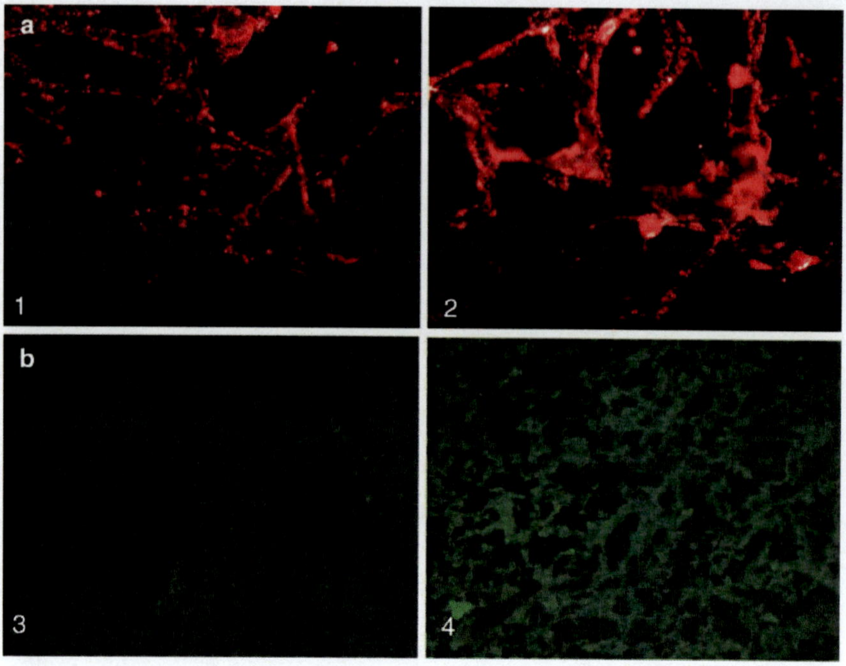

Fig. 9 (**a**), Fluorescence microscopy showing internalization of Rh-PE-labeled-TATp containing liposomes internalization by U87 MG astrocytoma. *1*: 9 mol % pH-non-sensitive PEG-PE at pH 7.4, *2*: 9 mol % pH-sensitive PEG-Hz-PE after incubation at pH 5.0 for 20 min. The cleavable (pH sensitive) PEG-PE-based TATp-containing liposomes kept at pH 7.4–8 show only a marginal association with cells while those preincubated for 20–30 min at pH 5.0 demonstrated a dramatically enhanced association with the cells (higher fluorescence). (**b**), Fluorescence microscopy images of LLC tumor sections from tumors injected with pGFP-loaded TATp-bearing liposomes. *3*: using a pH-non-cleavable PEG coat, *4*: with a low pH-cleavable PEG coat

Mitochondria represent an important target for intracellularly delivered drugs and DNA. Mitochondrial dysfunction contributes to a variety of human disorders, ranging from neurodegenerative diseases, obesity, diabetes, ischemia-reperfusion injury and cancer (Wallace 1999). The number of diseases is also found to be associated with defects of the mitochondria genome has grown significantly over the past decade. Mitochondria also play a key role in the complex apoptosis mechanism. The mechanism of paclitaxel induced apoptosis is believed to be by stabilization of microtubules of cells (Fan 1999; Wang et al. 2000). It has been also observed that there is a 24 h delay between paclitaxel-induced release of cytochrome *c* in intact cells versus cell-free system (Andre et al. 2002). This could be due to only few drug molecules reaching the mitochondria. Hence, we hypothesized that we could improve the apoptosis due to paclitaxel if we could delivery paclitaxel to mitochondria.

The mitochondrion has a major role in the metabolism of eukaryotic cells in the synthesis of ATP by oxidative phosphorylation via the respiratory chain. This process creates a transmembrane electrochemical gradient, which includes contributions from both a membrane potential (negative inside) and a pH difference (acidic outside). The membrane potential of mitochondria *in vitro* is between 180 and 200 mV, which is the maximum a lipid bilayer can sustain while maintaining its integrity (Murphy 1989). Positively charged molecules are attracted by mitochondria in response to the highly negative membrane potential, but most charged molecules cannot enter the mitochondrial matrix because the inner mitochondrial membrane is impermeable to polar molecules. However, certain amphiphile compounds are able to cross both mitochondrial membranes and accumulate in the mitochondrial matrix in response to the negative membrane potential. It has long been known that amphiphile compounds with delocalized cationic charge can accumulate in mitochondria (Weissig and Torchilin 2001). Rhodamine 123 (Rh-123), a stain for mitochondria in living cells, is the best known representative of this group (Chen et al. 1982). Mitochondrial accumulation of tetraphenylphosphonium chloride and other cationic aryl phosphonium salts was also demonstrated (Rideout et al. 1994). The mitochondrial accumulation and retention of dequalinium (DQA), a single-chain bola amphiphile with two delocalized positive charge centers was also demonstrated (Weissig and Torchilin 2001).

We have modified liposomes with using stearyl triphenyl phosphonium (STPP) to render them mitochondriotropic (Boddapati et al. 2008). *In vitro* STPP liposomes selectively accumulated in mitochondria of living cells. Also when loaded with ceramide as model drug, it elicited strong apoptotic response *in vivo* in 4T1 mammary carcinoma tumor-bearing mice at ceramide doses as low as 6 mg/kg in comparison with the 36 mg/kg or higher reported with non-targeted liposomes.

Recently, we prepared a novel mitochondria-targeted liposomal drug delivery system by the modification of the liposomal surface with Rh-123 (Biswas et al. 2010). A novel polymer was synthesized by conjugating the mitochondriotropic dye Rh-123, with the amphiphilic PEG-PE conjugate. The co-localization study with stained mitochondria (Fig. 10) as well as with the isolation of mitochondria of the cultured cells after their treatment with Rh123-liposomes showed a high degree of accumulation of the modified liposomes in the mitochondria.

To demonstrate that specific delivery of the drug to the desired subcellular compartment can significantly enhance drug action, the mitotic inhibitor, paclitaxel was used. Paclitaxel-loaded Rh123-liposomes (PCL-Rh123-L) produced significantly higher cytotoxicity than free paclitaxel or paclitaxel-loaded plain liposomes. An approximately 35–40% reduction of cell survival was observed with PCL-Rh123-L compared to non-targeted PCL formulations. Thus, Rh-123-modified liposomes target mitochondria efficiently and can facilitate the delivery of a therapeutic payload to mitochondria.

Lysosomes, acidic organelles responsible for recycling of cellular constituents, represent another important intracellular target for diseases such as lysosomal storage diseases (LSD). LSD is associated with the deficiency of certain lysosomal enzymes, which lead to accumulation of corresponding substrates in lysosomes (Futerman and van Meer 2004). These diseases pose a serious medical problem

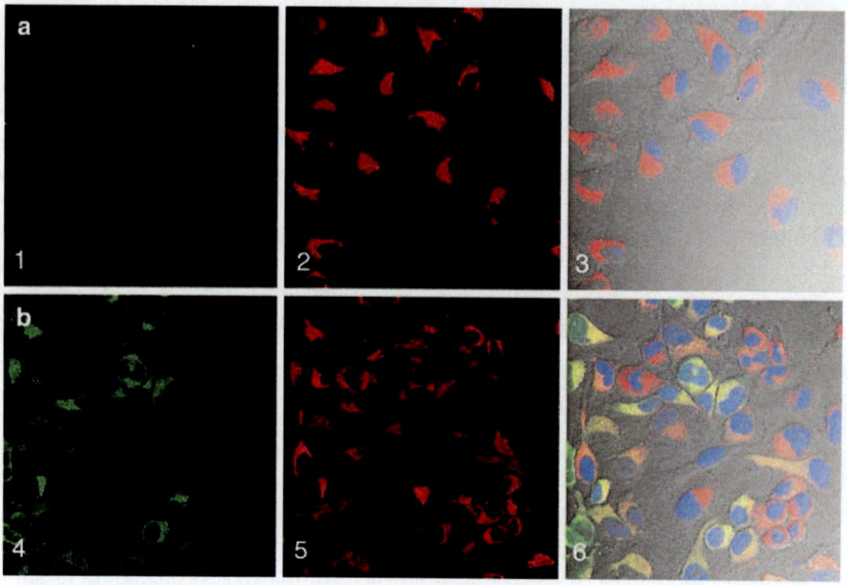

Fig. 10 Intracellular colocalization by fluorescence confocal microscopy. HeLa cells treated with (**a**), NBD-PE labeled plain liposomes; (**b**), Rh123-liposomes. *1,4*: Cell treatment with PL or Rh123-L in the green channel (Ex. 505 nm, Em. 530 nm); *2,5*: Cell staining treatment for visualization of mitochondria with Mitotracker deep red in the deep red channel (Ex. 644 nm, Em. 665 nm); *3,6*: Merged *left* and *middle* panels and Hoechst stained nuclei (Blue channel. Ex. 385 nm, Em,. 470 nm). *Yellow color* indicates co-localization of mitochondria and Rh123-L. Analysis of co-localization (Image J software) confirmed significant accumulation of targeted liposomes in the mitochondria (Pearson's coefficient 0.55, Mander's coefficient 0.75 for Rh123-L compared to Pearson's coefficient –0.083, Mander's coefficient 0.007 for plain liposomes)

(Grabowski 2008; Zarate and Hopkin 2008; van der Ploeg and Reuser 2008). The main approach for the treatment of LSD is enzyme replacement therapy (ERT) based on the administration of exogenous enzymes (Grabowski and Hopkin 2003). This procedure remains limited in use and expensive because of poor delivery and low stability of therapeutic enzymes. The use of liposome-immobilized enzymes for ERT, was understood long ago (Gregoriadis 1978). Lysosomes are also involved in the cellular apoptosis due to the lysosome-dependent cell death pathway (Kirkegaard and Jaattela 2009). Moderate permeabilization of lysosomal membranes can result in apoptosis of cancer cells (Boya et al. 2003). Thus, delivery of lysosome-destabilizing agents that cause cancer cell apoptosis may also benefit from lysosome-targeted carriers.

We recently attempted to develop a lysosome-targeted drug delivery system based on liposomes modified with a lysosome-specific ligand octadecyl derivative of rhodamine B (RhB). RhB is used to monitor membrane fusion (Hoekstra et al. 1984) and study lysosomal metabolism (Kuwana et al. 1995). It has been shown to specifically accumulate in the lysosomes of denervated skeletal muscle

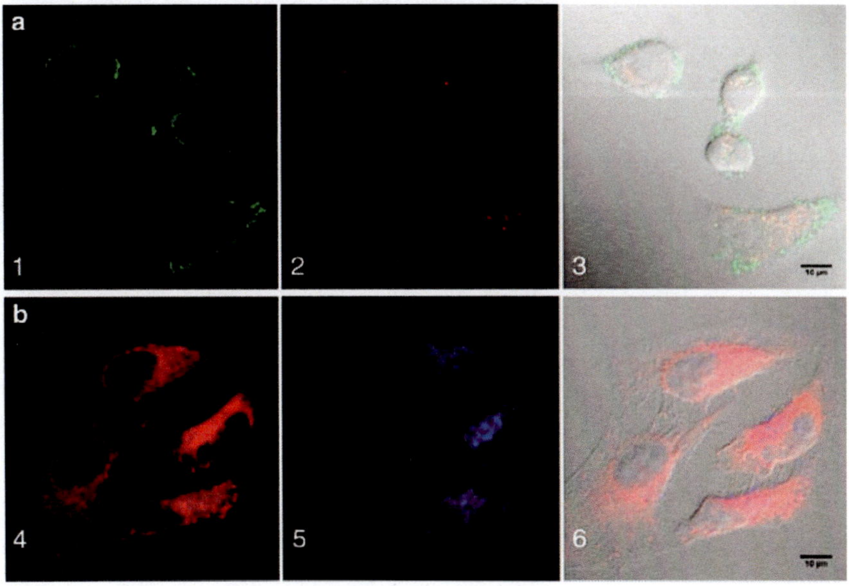

Fig. 11 Confocal microscopy of HeLa cells treated with (**a**), FITC-dextran-loaded liposomes; (**b**), RhB-modified FITC-dextran-loaded liposomes for 4 h. The treated cells were stained with lysosomal markers and analyzed by confocal microscopy. *1*: FITC-dextran-loaded plain liposomes (*green*), *2*: LysoTracker Red-stained lysosomes (*red*), *3*: Overlay of 1 and 2 images with their respective DIC image. *4*: RhB (*red*), *5*: anti-Lamp2 mAb-stained lysosomes (*blue*), *6*: Overlay of 4 and 5 images with their respective DIC image. Bar = 10 μm. Cell incubation with RhB1-modified FITC-dextran-loaded liposomes for 4 h led to the localization of RhB fluorescence mostly in the lysosomes with a high rate of the co-localization with the lysosomal marker (Pearson's correlation coefficient, PCC 0.7; Mander's overlap coefficient, MOP 0.8). The cells treated with the same concentration of FITC-dextran-loaded liposomes demonstrated much lower localization of FITC-dextran in the lysosomes (PCC – 0.1; MOP 0.2)

(Vult von Steyern et al. 1996). Novel acidic fluorescent probes based on rhodamine-B have also been described and used for the optical imaging of the intracellular H^+ (Zhang et al. 2009).

We prepared liposomes loaded with the model compound, FITC-dextran, and modified with RhB (Koshkaryev et al. 2010). Confocal microscopy demonstrated that RhB-liposomes co-localize well with the specific lysosomal markers, unlike plain liposomes (Fig. 11). The comparison of the FITC fluorescence of the lysosomes isolated by subcellular fractionation also showed that the efficiency of FITC-dextran delivery into lysosomes by RhB-modified liposomes was significantly higher compared to plain liposomes.

This was confirmed using 5-dodecanoylamino fluorescein di-β-D-galactopyranoside (C_{12}FDG)-loaded liposomes. C_{12}FDG, is a lipophilic substrate for the lysosomal β-galactosidase (Rotman et al. 1963). It was assumed that upon undergoing the endocytosis, C_{12}FDG-loaded liposomes would eventually deliver their contents into lysosomes, and the intra-lysosomal β-galactosidase will hydrolyze the non-fluorescent

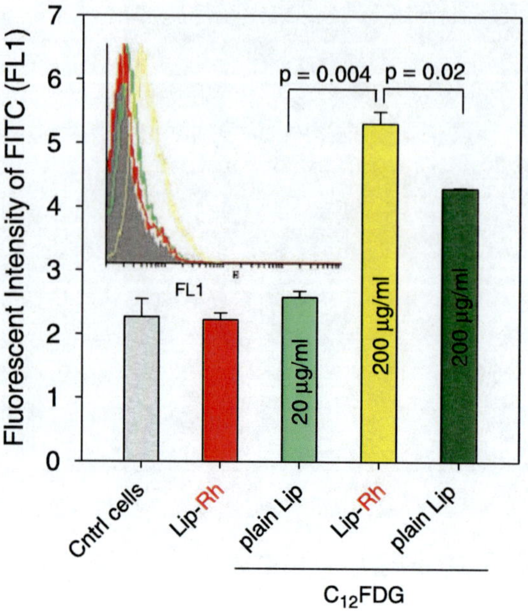

Fig. 12 Flow cytometry of lysosomal targeting by liposomes loaded with C_{12}FDG. HeLa cells were incubated with plain liposomes (20 and 200 μg/ml), Lip-RhB (200 μg/ml). The liposomes were loaded with C_{12}FDG (1.5% mol/mol), a fluorescent substrate for the intralysosomal β-galactosidase. After 4 h incubation with liposomes, the cells were washed and additionally incubated for 20 h with liposome-free DMEM. The fluorescence intensity of FITC (channel FL1) was determined by flow cytometry. Each value is the mean ± SD of 2 experiments

C_{12}FDG into the fluorescent C_{12}FITC, which will be retained inside lysosomes because of its lipophilic moiety. Thus, with a standard flow cytometry procedure, the lysosomal targeting can be quantified by following the fluorescence intensity of live, intact cells. We prepared plain and RhB-modified liposomes loaded with C_{12}FDG. The loading of C_{12}FDG into RhB-modified liposomes was approximately ten times less than into the plain liposomes. This decrease in the C_{12}FDG loading can be attributed to a stoichiometric competition between C_{12}FDG and RhB in the liposomal membrane leading to the partial loss of C_{12}FDG due to its shorter lipophilic moiety. Two different concentrations of plain liposomes (20 μg/ml and 200 μg/ml) were used for cell treatment to achieve the same amount of C_{12}FDG as with RhB-modified liposomes (200 μg/ml). The treatment of cells with different concentrations of C_{12}FDG-loaded plain liposomes (20 and 200 μg/ml) led to a dose-dependent increase in their FITC fluorescence relative to the control (untreated) cells (Fig. 12). These data suggest that endocytosed liposomes actually deliver C_{12}FDG into lysosomes. The cells treated with 200 μg/ml of RhB-modified liposomes demonstrated significantly increased C_{12}FITC fluorescence compared to the cells treated with both 20 and 200 μg/ml of the plain liposomes.

In any case, the development of organelle-specific drug delivery is only in its early stages; however, it might have important clinical future.

References

Andre N, Carre M, Brasseur G, Pourroy B, Kovacic H, Briand C, Braguer D (2002) Paclitaxel targets mitochondria upstream of caspase activation in intact human neuroblastoma cells. FEBS Lett 532 (1–2):256–260

Arnheiter H, Haller O (1988) Antiviral state against influenza virus neutralized by microinjection of antibodies to interferon-induced Mx proteins. Embo J 7 (5):1315–1320

Astriab-Fisher A, Sergueev D, Fisher M, Shaw BR, Juliano RL (2002) Conjugates of antisense oligonucleotides with the Tat and antennapedia cell-penetrating peptides: effects on cellular uptake, binding to target sequences, and biologic actions. Pharm Res 19 (6):744–754

Bareford LM, Swaan PW (2007) Endocytic mechanisms for targeted drug delivery. Adv Drug Deliv Rev 59 (8):748–758

Biswas S, Sawant RR, Koshkaryev A, Torchilin VP Novel Rhodamine 123-Conjugated Pharmaceutical Nanocarrier Targets Mitochondria. In: 37th Annual Meeting and Exposition of the Controlled Release Society, Portland, OR, USA, 2010

Boddapati SV, D'Souza GG, Erdogan S, Torchilin VP, Weissig V (2008) Organelle-targeted nanocarriers: specific delivery of liposomal ceramide to mitochondria enhances its cytotoxicity in vitro and in vivo. Nano Lett 8 (8):2559–2563

Boya P, Andreau K, Poncet D, Zamzami N, Perfettini JL, Metivier D, Ojcius DM, Jaattela M, Kroemer G (2003) Lysosomal membrane permeabilization induces cell death in a mitochondrion-dependent fashion. J Exp Med 197 (10):1323–1334

Chakrabarti R, Wylie DE, Schuster SM (1989) Transfer of monoclonal antibodies into mammalian cells by electroporation. J Biol Chem 264 (26):15494–15500

Chen LB, Summerhayes IC, Johnson LV, Walsh ML, Bernal SD, Lampidis TJ (1982) Probing mitochondria in living cells with rhodamine 123. Cold Spring Harb Symp Quant Biol 46 Pt 1:141–155

Conner SD, Schmid SL (2003) Regulated portals of entry into the cell. Nature 422 (6927):37–44

Cryan SA, Devocelle M, Moran PJ, Hickey AJ, Kelly JG (2006) Increased intracellular targeting to airway cells using octaarginine-coated liposomes: in vitro assessment of their suitability for inhalation. Mol Pharm 3 (2):104–112

D'Souza GG, Weissig V (2009) Subcellular targeting: a new frontier for drug-loaded pharmaceutical nanocarriers and the concept of the magic bullet. Expert Opin Drug Deliv 6 (11):1135–1148

de la Fuente JM, Berry CC (2005) Tat peptide as an efficient molecule to translocate gold nanoparticles into the cell nucleus. Bioconjug Chem 16 (5):1176–1180

Dodd CH, Hsu HC, Chu WJ, Yang P, Zhang HG, Mountz JD, Jr., Zinn K, Forder J, Josephson L, Weissleder R, Mountz JM, Mountz JD (2001) Normal T-cell response and in vivo magnetic resonance imaging of T cells loaded with HIV transactivator-peptide-derived superparamagnetic nanoparticles. J Immunol Methods 256 (1–2):89–105

Fan W (1999) Possible mechanisms of paclitaxel-induced apoptosis. Biochem Pharmacol 57 (11):1215–1221

Frankel AD, Pabo CO (1988) Cellular uptake of the tat protein from human immunodeficiency virus. Cell 55 (6):1189–1193

Fretz MM, Koning GA, Mastrobattista E, Jiskoot W, Storm G (2004) OVCAR-3 cells internalize TAT-peptide modified liposomes by endocytosis. Biochim Biophys Acta 1665 (1–2):48–56

Futerman AH, van Meer G (2004) The cell biology of lysosomal storage disorders. Nat Rev Mol Cell Biol 5 (7):554–565.

Gabizon A, Shmeeda H, Horowitz AT, Zalipsky S (2004) Tumor cell targeting of liposome-entrapped drugs with phospholipid-anchored folic acid-PEG conjugates. Adv Drug Deliv Rev 56 (8):1177–1192

Grabowski GA (2008) Phenotype, diagnosis, and treatment of Gaucher's disease. Lancet 372 (9645):1263–1271

Grabowski GA, Hopkin RJ (2003) Enzyme therapy for lysosomal storage disease: principles, practice, and prospects. Annu Rev Genomics Hum Genet 4:403–436

Green M, Loewenstein PM (1988) Autonomous functional domains of chemically synthesized human immunodeficiency virus tat trans-activator protein. Cell 55 (6):1179–1188

Gregoriadis G (1978) Liposomes in the therapy of lysosomal storage diseases. Nature 275 (5682):695–696

Gupta B, Levchenko TS, Torchilin VP (2005) Intracellular delivery of large molecules and small particles by cell-penetrating proteins and peptides. Adv Drug Deliv Rev 57 (4):637–651

Gupta B, Levchenko TS, Torchilin VP (2007) TAT peptide-modified liposomes provide enhanced gene delivery to intracranial human brain tumor xenografts in nude mice. Oncol Res 16 (8):351–359

Hoekstra D, de Boer T, Klappe K, Wilschut J (1984) Fluorescence method for measuring the kinetics of fusion between biological membranes. Biochemistry 23 (24):5675–5681

Jaracz S, Chen J, Kuznetsova LV, Ojima I (2005) Recent advances in tumor-targeting anticancer drug conjugates. Bioorg Med Chem 13 (17):5043–5054

Jeang KT, Xiao H, Rich EA (1999) Multifaceted activities of the HIV-1 transactivator of transcription, Tat. J Biol Chem 274 (41):28837–28840

Josephson L, Tung CH, Moore A, Weissleder R (1999) High-efficiency intracellular magnetic labeling with novel superparamagnetic-Tat peptide conjugates. Bioconjug Chem 10 (2):186–191

Kale AA, Torchilin VP (2007a) Design, synthesis, and characterization of pH-sensitive PEG-PE conjugates for stimuli-sensitive pharmaceutical nanocarriers: the effect of substitutes at the hydrazone linkage on the ph stability of PEG-PE conjugates. Bioconjug Chem 18 (2):363–370

Kale AA, Torchilin VP (2007b) Enhanced transfection of tumor cells in vivo using "Smart" pH-sensitive TAT-modified pegylated liposomes. J Drug Target 15 (7–8):538–545

Kirkegaard T, Jaattela M (2009) Lysosomal involvement in cell death and cancer. Biochim Biophys Acta 1793 (4):746–754

Kleemann E, Neu M, Jekel N, Fink L, Schmehl T, Gessler T, Seeger W, Kissel T (2005) Nano-carriers for DNA delivery to the lung based upon a TAT-derived peptide covalently coupled to PEG-PEI. J Control Release 109 (1–3):299–316

Ko YT, Hartner WC, Kale A, Torchilin VP (2009) Gene delivery into ischemic myocardium by double-targeted lipoplexes with anti-myosin antibody and TAT peptide. Gene Ther 16 (1):52–59

Koshkaryev A, Thekkedath R, Pagano C, Meerovich I, VP T Specific lysosomal targeting by liposomes modified with octadecyl-rhodamine B. In: 37th Annual Meeting and Exposition of the Controlled Release Society, Portland, OR, USA, 2010

Kuwana T, Mullock BM, Luzio JP (1995) Identification of a lysosomal protein causing lipid transfer, using a fluorescence assay designed to monitor membrane fusion between rat liver endosomes and lysosomes. Biochem J 308 (Pt 3):937–946

Lasic DD (1993) Liposomes: From Physics to Applications. Elsevier, Amsterdam

Lasic DD, Martin FJ (1995) Stealth Liposomes. CRC Press, Boca Raton

Levchenko TS, Rammohan R, Volodina N, Torchilin VP (2003) Tat peptide-mediated intracellular delivery of liposomes. Methods Enzymol 372:339–349

Lewin M, Carlesso N, Tung CH, Tang XW, Cory D, Scadden DT, Weissleder R (2000) Tat peptide-derivatized magnetic nanoparticles allow in vivo tracking and recovery of progenitor cells. Nat Biotechnol 18 (4):410–414

Lukyanov AN, Hartner WC, Torchilin VP (2004) Increased accumulation of PEG-PE micelles in the area of experimental myocardial infarction in rabbits. J Control Release 94 (1):187–193

Lukyanov AN, Torchilin VP (2004) Micelles from lipid derivatives of water-soluble polymers as delivery systems for poorly soluble drugs. Adv Drug Deliv Rev 56 (9):1273–1289

Maeda H, Bharate GY, Daruwalla J (2009) Polymeric drugs for efficient tumor-targeted drug delivery based on EPR-effect. Eur J Pharm Biopharm 71 (3):409–419

Maeda H, Wu J, Sawa T, Matsumura Y, Hori K (2000) Tumor vascular permeability and the EPR effect in macromolecular therapeutics: a review. J Control Release 65 (1–2):271–284

Mortensen MW, Bjorkdahl O, Sorensen PG, Hansen T, Jensen MR, Gundersen HJ, Bjornholm T (2006) Functionalization and cellular uptake of boron carbide nanoparticles. The first step toward T cell-guided boron neutron capture therapy. Bioconjug Chem 17 (2):284–290

Mukherjee S, Ghosh RN, Maxfield FR (1997) Endocytosis. Physiol Rev 77 (3):759–803

Murphy MP (1989) Slip and leak in mitochondrial oxidative phosphorylation. Biochim Biophys Acta 977 (2):123–141

Nguyen J, Xie X, Neu M, Dumitrascu R, Reul R, Sitterberg J, Bakowsky U, Schermuly R, Fink L, Schmehl T, Gessler T, Seeger W, Kissel T (2008) Effects of cell-penetrating peptides and pegylation on transfection efficiency of polyethylenimine in mouse lungs. J Gene Med 10 (11):1236–1246

Pappalardo JS, Quattrocchi V, Langellotti C, Di Giacomo S, Gnazzo V, Olivera V, Calamante G, Zamorano PI, Levchenko TS, Torchilin VP (2009) Improved transfection of spleen-derived antigen-presenting cells in culture using TATp-liposomes. J Control Release 134 (1):41–46

Park J, Ryu J, Kim KA, Lee HJ, Bahn JH, Han K, Choi EY, Lee KS, Kwon HY, Choi SY (2002) Mutational analysis of a human immunodeficiency virus type 1 Tat protein transduction domain which is required for delivery of an exogenous protein into mammalian cells. J Gen Virol 83 (Pt 5):1173–1181

Rajendran L, Knolker HJ (2010) Simons K Subcellular targeting strategies for drug design and delivery. Nat Rev Drug Discov 9 (1):29–42

Rao KS, Reddy MK, Horning JL, Labhasetwar V (2008) TAT-conjugated nanoparticles for the CNS delivery of anti-HIV drugs. Biomaterials 29 (33):4429–4438

Rideout D, Bustamante A, Patel J (1994) Mechanism of inhibition of FaDu hypopharyngeal carcinoma cell growth by tetraphenylphosphonium chloride. Int J Cancer 57 (2):247–253

Rotman B, Zderic JA, Edelstein M (1963) Fluorogenic substrates for beta-D-galactosidases and phosphatases derived from flurescein (3,6-dihydroxyfluoran) and its monomethylether. Proc Natl Acad Sci USA 50:1–6

Sandgren S, Cheng F, Belting M (2002) Nuclear targeting of macromolecular polyanions by an HIV-Tat derived peptide. Role for cell-surface proteoglycans. J Biol Chem 277 (41):38877–38883

Sawant R, Torchilin V (2010) Intracellular transduction using cell-penetrating peptides. Mol Biosyst 6 (4):628–640

Sawant RM, Hurley JP, Salmaso S, Kale A, Tolcheva E, Levchenko TS, Torchilin VP (2006) "SMART" drug delivery systems: double-targeted pH-responsive pharmaceutical nanocarriers. Bioconjug Chem 17 (4):943–949

Sawant RR, Sawant RM, Kale AA, Torchilin VP (2008) The architecture of ligand attachment to nanocarriers controls their specific interaction with target cells. J Drug Target 16 (7):596–600

Sawant RR, Torchilin VP (2009) Enhanced cytotoxicity of TATp-bearing paclitaxel-loaded micelles in vitro and in vivo. Int J Pharm 374 (1–2):114–118

Schwarze SR, Dowdy SF (2000) In vivo protein transduction: intracellular delivery of biologically active proteins, compounds and DNA. Trends Pharmacol Sci 21 (2):45–48

Schwarze SR, Ho A, Vocero-Akbani A, Dowdy SF (1999) In vivo protein transduction: delivery of a biologically active protein into the mouse. Science 285 (5433):1569–1572

Sethuraman VA, Bae YH (2007) TAT peptide-based micelle system for potential active targeting of anti-cancer agents to acidic solid tumors. J Control Release 118 (2):216–224

Sethuraman VA, Lee MC, Bae YH (2008) A biodegradable pH-sensitive micelle system for targeting acidic solid tumors. Pharm Res 25 (3):657–666

Straubinger RM, Duzgunes N, Papahadjopoulos D (1985) pH-sensitive liposomes mediate cytoplasmic delivery of encapsulated macromolecules. FEBS Lett 179 (1):148–154

Stroh M, Zimmer JP, Duda DG, Levchenko TS, Cohen KS, Brown EB, Scadden DT, Torchilin VP, Bawendi MG, Fukumura D, Jain RK (2005) Quantum dots spectrally distinguish multiple species within the tumor milieu in vivo. Nat Med 11 (6):678–682

Suk JS, Suh J, Choy K, Lai SK, Fu J, Hanes J (2006) Gene delivery to differentiated neurotypic cells with RGD and HIV Tat peptide functionalized polymeric nanoparticles. Biomaterials 27 (29):5143–5150

Tkachenko AG, Xie H, Liu Y, Coleman D, Ryan J, Glomm WR, Shipton MK, Franzen S, Feldheim DL (2004) Cellular trajectories of peptide-modified gold particle complexes: comparison of nuclear localization signals and peptide transduction domains. Bioconjug Chem 15 (3):482–490

Torchilin VP (2001) Structure and design of polymeric surfactant-based drug delivery systems. J Control Release 73 (2–3):137–172

Torchilin VP (2004) Targeted polymeric micelles for delivery of poorly soluble drugs. Cell Mol Life Sci 61 (19–20):2549–2559

Torchilin VP (2005a) Fluorescence microscopy to follow the targeting of liposomes and micelles to cells and their intracellular fate. Adv Drug Deliv Rev 57 (1):95–109

Torchilin VP (2005b) Recent advances with liposomes as pharmaceutical carriers. Nat Rev Drug Discov 4 (2):145–160

Torchilin VP (2006) Recent approaches to intracellular delivery of drugs and DNA and organelle targeting. Annu Rev Biomed Eng 8:343–375

Torchilin VP (2007a) Micellar nanocarriers: pharmaceutical perspectives. Pharm Res 24 (1):1–16

Torchilin VP (2007b) Tatp-mediated intracellular delivery of pharmaceutical nanocarriers. Biochem Soc Trans 35 (Pt 4):816–820

Torchilin VP, Levchenko TS (2003) TAT-liposomes: a novel intracellular drug carrier. Curr Protein Pept Sci 4 (2):133–140

Torchilin VP, Levchenko TS, Lukyanov AN, Khaw BA, Klibanov AL, Rammohan R, Samokhin GP, Whiteman KR (2001a) p-Nitrophenylcarbonyl-PEG-PE-liposomes: fast and simple attachment of specific ligands, including monoclonal antibodies, to distal ends of PEG chains via p-nitrophenylcarbonyl groups. Biochim Biophys Acta 1511 (2):397–411

Torchilin VP, Levchenko TS, Rammohan R, Volodina N, Papahadjopoulos-Sternberg B, D'Souza GG (2003a) Cell transfection in vitro and in vivo with nontoxic TAT peptide-liposome-DNA complexes. Proc Natl Acad Sci USA 100 (4):1972–1977

Torchilin VP, Lukyanov AN, Gao Z, Papahadjopoulos-Sternberg B (2003b) Immunomicelles: targeted pharmaceutical carriers for poorly soluble drugs. Proc Natl Acad Sci USA 100 (10):6039–6044

Torchilin VP, Narula J, Halpern E, Khaw BA (1996) Poly(ethylene glycol)-coated anti-cardiac myosin immunoliposomes: factors influencing targeted accumulation in the infarcted myocardium. Biochim Biophys Acta 1279 (1):75–83

Torchilin VP, Rammohan R, Weissig V, Levchenko TS (2001b) TAT peptide on the surface of liposomes affords their efficient intracellular delivery even at low temperature and in the presence of metabolic inhibitors. Proc Natl Acad Sci USA 98 (15):8786–8791

van der Ploeg AT, Reuser AJ (2008) Pompe's disease. Lancet 372 (9646):1342–1353

Vives E, Richard JP, Rispal C, Lebleu B (2003) TAT peptide internalization: seeking the mechanism of entry. Curr Protein Pept Sci 4 (2):125–132

Vult von Steyern F, Josefsson JO, Tagerud S (1996) Rhodamine B, a fluorescent probe for acidic organelles in denervated skeletal muscle. J Histochem Cytochem 44 (3):267–274

Wadia JS, Stan RV, Dowdy SF (2004) Transducible TAT-HA fusogenic peptide enhances escape of TAT-fusion proteins after lipid raft macropinocytosis. Nat Med 10 (3):310–315

Wallace DC (1999) Mitochondrial diseases in man and mouse. Science 283 (5407):1482–1488

Wang J, Mongayt D, Torchilin VP (2005) Polymeric micelles for delivery of poorly soluble drugs: preparation and anticancer activity in vitro of paclitaxel incorporated into mixed micelles based on poly(ethylene glycol)-lipid conjugate and positively charged lipids. J Drug Target 13 (1):73–80

Wang TH, Wang HS, Soong YK (2000) Paclitaxel-induced cell death: where the cell cycle and apoptosis come together. Cancer 88 (11):2619–2628

Weissig V, Torchilin VP (2001) Cationic bolasomes with delocalized charge centers as mitochondria-specific DNA delivery systems. Adv Drug Deliv Rev 49 (1–2):127–149

Widera A, Norouziyan F, Shen WC (2003) Mechanisms of TfR-mediated transcytosis and sorting in epithelial cells and applications toward drug delivery. Adv Drug Deliv Rev 55 (11):1439–1466

Yagi N, Yano Y, Hatanaka K, Yokoyama Y, Okuno H (2007) Synthesis and evaluation of a novel lipid-peptide conjugate for functionalized liposome. Bioorg Med Chem Lett 17 (9):2590–2593

Zarate YA, Hopkin RJ (2008) Fabry's disease. Lancet 372 (9647):1427–1435.

Zhang C, Tang N, Liu X, Liang W, Xu W, Torchilin VP (2006) siRNA-containing liposomes modified with polyarginine effectively silence the targeted gene. J Control Release 112 (2):229–239

Zhang W, Tang B, Liu X, Liu Y, Xu K, Ma J, Tong L, Yang G (2009) A highly sensitive acidic pH fluorescent probe and its application to HepG2 cells. Analyst 134 (2):367–371.

Zhao M, Kircher MF, Josephson L, Weissleder R (2002) Differential conjugation of tat peptide to superparamagnetic nanoparticles and its effect on cellular uptake. Bioconjug Chem 13 (4):840–844

Zhao P, Wang H, Yu M, Cao S, Zhang F, Chang J, Niu R (2010) Paclitaxel-loaded, folic-acid-targeted and TAT-peptide-conjugated polymeric liposomes: in vitro and in vivo evaluation. Pharm Res 27 (9):1914–1926

Zorko M, Langel U (2005) Cell-penetrating peptides: mechanism and kinetics of cargo delivery. Adv Drug Deliv Rev 57 (4):529–545

Poly(Alkyl Cyanoacrylate) Nanosystems

Julien Nicolas and Christine Vauthier

Abstract Poly(alkyl cyanoacrylate) nanosystems include various types of nanoparticles suitable to achieve in vivo drug delivery in a well controlled manner. Starting from general features on the synthesis of alkyl cyanoacrylate monomers and of their polymerization, this chapter aims to discuss how these monomers and corresponding polymers can be used to design drug delivery nanosystems with different properties.

Keywords Alkyl cyanoacrylate • Anionic polymerization • Copolymers • Free radical polymerization • Nanocapsules • Nanospheres • Zwitterionic polymerization

Abbreviations

$(PIB\text{-}CA)_3$	three-arm star cyanoacrylate-telechelic polyisobutylene
ACA	alkyl cyanoacrylate
AIBN	azobisisobutyronitrile
Apo B-100	apolipoprotein B-100
Apo E	apolipoprotein E
Apo E3	apolipoptotein E3
BBB	blood-brain barrier
BFs	breath figures method
DCC	*N,N'*-dicyclohexylcarbodiimide
ddI	didanosine
DXM	dexamethasone
ECA	ethyl cyanoacrylate

J. Nicolas and C. Vauthier (✉)
Physico-chimie, Pharmacotechnie, Biopharmacie, Univ Paris-Sud,
UMR 8612, Chatenay-Malabry F-92296, France
and
CNRS, Chatenay-Malabry F-92296, France
e-mail: christine.vauthier@u-psud.fr

A. Prokop (ed.), *Intracellular Delivery: Fundamentals and Applications*, Fundamental Biomedical Technologies 5, DOI 10.1007/978-94-007-1248-5_9,

HA	hyaluronic acid
HDCA	hexadecyl cyanoacrylate
IBCA	isobutyl cyanoacrylate
LDLR	low-density lipoprotein receptors
MCA	methyl cyanoacrylate
MePEG	methoxypoly(ethylene glycol)
MePEG-*b*-P*n*BCA	methoxypoly(ethylene glycol)-*b*-poly(*n*-butyl cyanoacrylate) diblock copolymer
MMA	methyl methacrylate
MPS	mononuclear phagocyte system
*n*BA	*n*-butyl acrylate
*n*BCA	*n*-butyl cyanoacrylate
NHS	*N*-hydroxysuccinimide
NIPAAm	*N*-isopropylacrylamide
NS	nanosphere
Nu^-	charged nucleophile
Nu	protic nucleophile
OCA	octyl cyanoacrylate
o-NC	oil-containing nanocapsule
P	radical
P(ECA-*co*-MMA)	poly(ethyl cyanoacrylate-*co*-methyl methacrylate) copolymer
P(HDCA-*co*-MePEGCA)	poly[(hexadecyl cyanoacrylate)-*co*-methoxypoly(ethylene glycol) cyanoacrylate] copolymer
P(HDCA-*co*-N_3PEGCA)	poly[(hexadecyl cyanoacrylate)-*co*-azidopoly(ethylene glycol) cyanoacrylate] copolymer
P(HDCA-*co*-NH_2PEGCA)	poly[(hexadecyl cyanoacrylate)-*co*-aminopoly (ethylene glycol) cyanoacrylate] copolymer
P(HDCA-*co*-RCA-*co*-MePEGCA)	poly[hexadecyl cyanoacrylate-*co*-rhodamine B cyanoacrylate-*co*-methoxypoly(ethylene glycol) cyanoacrylate] copolymer
P(HDCA-*co*-SelPEGCA)	poly[(hexadecyl cyanoacrylate)-*co*-selegiline-poly(ethylene glycol) cyanoacrylate] copolymer
PACA	poly(alkyl cyanoacrylate)
PECA	poly(ethyl cyanoacrylate)
PECA-*b*-PEG-*b*-PECA	poly(ethyl cyanoacrylate)-*b*-poly(ethylene glycol)-*b*-poly(ethyl cyanoacrylate) triblock copolymer
PEG	poly(ethylene glycol)
PHCA	poly(hexyl cyanoacrylate)

PIBCA	poly(isobutyl cyanoacrylate)
PIBCA-*b*-PEG	poly(isobutyl cyanoacrylate)-*b*-poly(ethylene glycol) block copolymer
PIBCA-*b*-PEG-*b*-PIBCA	poly(isobutyl cyanoacrylate)-*b*-poly(ethylene glycol)-*b*-poly(isobutyl cyanoacrylate) block copolymer
P*n*BCA	poly(*n*-butyl cyanoacrylate)
PNIPAAm	poly(*N*-isopropylacrylamide)
POCA	poly(octyl cyanoacrylate)
SelPEGCA	selegiline-poly(ethylene glycol) cyanoacetate
THF	tetrahydrofuran
w/o/w	water-in-oil-in-water
w/o	water-in-oil
w-NC	water-containing nanocapsule

1 Introduction

Poly(alkyl cyanoacrylate) (PACA) nanosystems include various types of nanoparticles suitable to achieve in vivo drug delivery in a well controlled manner (Vauthier et al. 2003). PACA nanosystems are one of the few polymer nanoparticles used as drug carriers which have reached clinical evaluation (Kattan et al. 1992; BioAlliance Pharma 2009; Zhou et al. 2009). Beside this, PACA nanosystems have been considered as potential carriers for many types of drugs; the literature is particularly abundant and includes reports on a wide range of methods for their synthesis (Vauthier et al. 2007; Nicolas and Couvreur 2009). Consequently, various types of nanoparticles were developed to answer the different drug delivery challenges. In turn, this required the development of suitable methods for the synthesis of the desired nanosystems. Large scale production was considered for the methods producing the nanoparticles used in clinical investigations. The aim of this chapter is to document synthetic pathways to achieve PACA nanosystems starting from general features on the synthesis of alkyl cyanoacrylate (ACA) monomers. Then the polymerization features of these monomers and their different polymerization methods will be discussed including those applied to synthesize copolymers with defined molecular architecture. The synthesis of PACA nanosystems will be considered in three subsections following the development of drug carriers with different degrees of specificity regarding their biodistribution after intravenous administration to animals. This also corresponded to the synthesis of nanoparticles with increasing degree of complexity. Synthesis methods from polymerization of ACA, from pre-formed polymers and methods of surface engineering are detailed in these subsections.

2 General Features and Synthesis of Alkyl Cyanoacrylate Monomers

Alkyl cyanoacrylates represent a class of very reactive monomers widely employed for biomedical purposes as surgical glues for the closure of skin wounds (Skeist and Miron 1977; Coover et al. 1990; King and Kinney 1999; Oowaki et al. 2000; Hollock 2001; Marcovich et al. 2001; Reece et al. 2001), embolitic material for endovascular surgery (Oowaki et al. 2000; Pollak and White 2001; Reece et al. 2001) and for nerves regenerating purposes (Merolli et al. 2010). Indeed, alkyl cyanoacrylate monomers exhibit a remarkable reactivity toward nucleophiles and have excellent adhesive properties resulting from the bonds of high strength they are able to form with most polar substrates, including living tissues and skin.

The synthesis of alkyl cyanoacrylates has been first described in the patent literature (Ardis 1949; Jeremias 1956; Joyner and Shearer 1956; Joyner and Hawkins 1955). The main strategy to achieve α-cyanoacrylates is based on a two-step procedure (Fig. 1). The corresponding alkyl cyanoacetate is reacted with formaldehyde in the presence of a basic catalyst, to form PACA oligomers by Knoevenagel condensation reaction, followed by their thermal depolymerization reaction which lead to the alkyl cyanoacrylate monomer. To prevent repolymerization events, protonic or Lewis acids with small amounts of a free-radical inhibitors are usually employed.

Slight improvements of this synthetic procedure were achieved, essentially by playing with the nature of the solvent mixture (Tseng et al. 1990; Sahadev et al. 2003), by applying a transesterification approach for making cyanoacrylates bearing longer alkyl ester chains (Malofsky and Badejo 2001), or by using pyrrolidine as a more efficient catalyst for the condensation step (Yadav et al. 2004).

Recently, a novel approach has been reported for the synthesis of cyanoacrylate derivatives (Kryger et al. 2010). Sonication of polymers containing a dicyanocyclobutane mechanophore moiety led to the selective cleavage of the mechanophore, thus releasing the corresponding cyanoacrylate monomers.

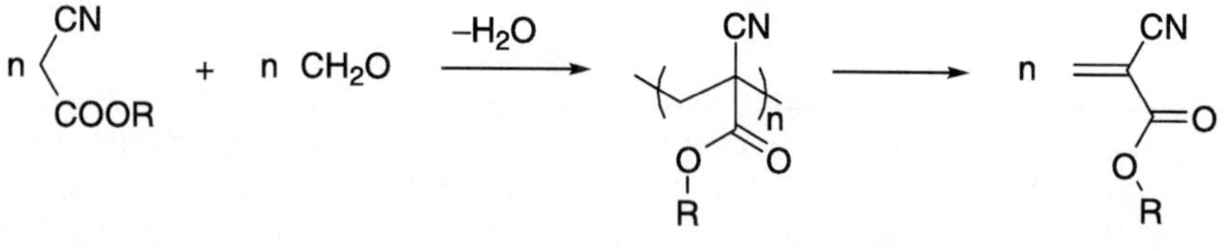

Fig. 1 Synthesis of alkyl cyanoacrylate monomer *via* Knoevenagel condensation reaction and subsequent thermal depolymerization

3 Polymerization in Homogeneous Media

Because of the presence of two powerful electro-withdrawing groups in the α-carbon of the double bond (i.e., ester and nitrile), alkyl cyanoacrylate monomers exhibit a very high reactivity toward nucleophiles such as anions (hydroxide, iodide, alcoholate, etc.) or weak bases (alcohol, amine, etc.), thus resulting in a very high polymerization rate. Traces of one of the above-mentioned compounds in the reaction medium can initiate the polymerization, which explains why batches of alkyl cyanoacrylates are usually stored with a small amount of acidic stabilizers (e.g., SO_2, sulfonic acid, etc.).

Depending on the experimental conditions and reactants, three different mechanisms are usually well-admitted for the polymerization of alkyl cyanoacrylates (Fig. 2): (i) anionic; (ii) zwitterionic and (iii) radical (Nicolas and Couvreur 2009). Those polymerizations are respectively initiated by charged nucleophiles (Nu^-), protic nucleophiles (Nu) and radicals ($P^\bullet$). However, due to the very high reactivity of alkyl cyanoacrylate monomers, anionic and zwitterionic polymerizations are highly predominant under conventional experimental conditions with respect to a radical process.

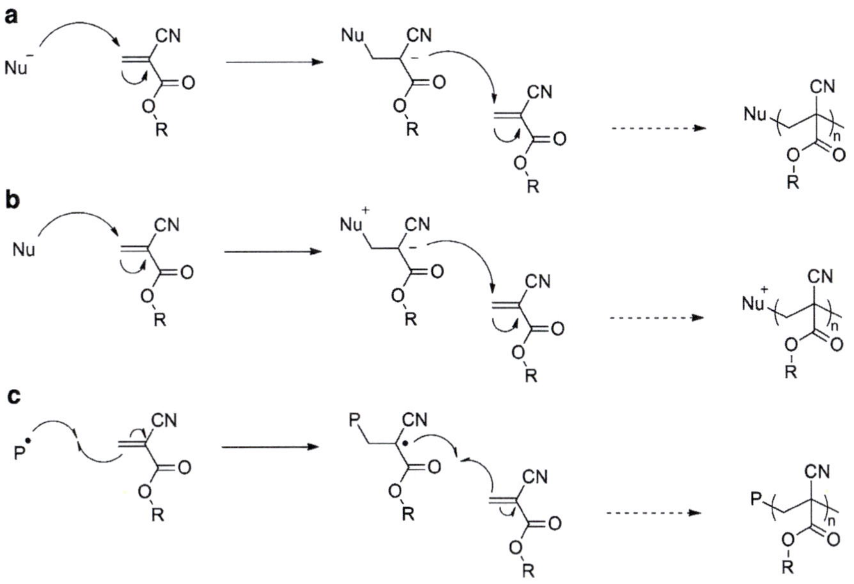

Fig. 2 Initiation and propagation steps involved during anionic (**a**), zwitterionic (**b**), and radical (**c**) polymerizations of alkyl cyanoacrylates initiated by a charged nucleophiles (Nu-), a protic nucleophile (Nu) and a radical (P•), respectively

3.1 *Homopolymerization and Random Copolymerization*

Homopolymers from alkyl cyanoacrylates (Fig. 3) has been well-studied since the late 1970s (Donnelly et al. 1977; Johnston and Pepper 1981a, b, c). The polymerization of ethyl cyanoacrylate (ECA) and *n*-butyl cyanoacrylate (*n*BA) was undertaken in tetrahydrofuran (THF) and initiated either by simple anions or by organic bases, leading to anionic or zwitterionic mechanism, respectively (Donnelly et al. 1977). Multiple parameters were varied in order to investigate their influence on the macromolecular properties of the resulting polymers (Pepper 1980; Johnston and Pepper 1981a, b, c; Pepper and Ryan 1983; Cronin and Pepper 1988; Eromosele et al. 1989; Johnson et al. 1995).

In particular for zwitterionic polymerization of *n*BCA, the influence of the nature of the initiator, the inhibiting species and the presence of water on the macromolecular characteristics of the polymer and on the polymerization kinetics was thoroughly investigated using several covalent organic bases (Pepper 1980; Johnston and Pepper 1981a, b, c; Pepper and Ryan 1983; Cronin and Pepper 1988; Eromosele et al. 1989). Regarding the anionic mechanism, several tetrabutyl ammonium salts were used as initiators for the polymerization of *n*BCA in THF and the best results were obtained from the hydroxide-based one (Eromosele and Pepper 1986, 1989a, b).

Copolymerization between cyanoacrylate derivatives, namely ECA and *n*BCA, was also performed in order to tune the glass transition temperature of the resulting materials. Copolymerizations were performed either by a piperidine-catalyzed bulk polymerization, leading to transparent brittle films, or by polymerization in aqueous medium in the presence of sodium bicarbonate, in order to obtain white powders (Denchev et al. 2008).

When a suitable inhibitor is introduced in the reaction medium (such as boron trifluoride-acetic acid complex, propane-1,3-sultone or acetic acid), anionic polymerization is made negligible during the timescale of the polymerization and allow a free-radical mechanism to be the main chain-extension process (Canale et al. 1960; Kinsinger et al. 1965; Otsu and Yamada 1967; Bevington et al. 1976;

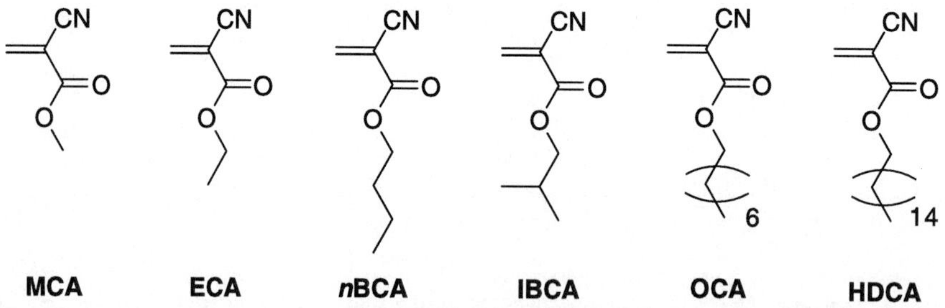

Fig. 3 Structure of methyl cyanoacrylate (MCA), ethyl cyanoacrylate (ECA), *n*-butyl cyanoacrylate (*n*BCA), isobutyl cyanoacrylate (IBCA), octyl cyanoacrylate (OCA) and hexadecyl cyanoacrylate (HDCA)

Kamachi et al. 1981; Yamada et al. 1983). This was successfully applied to bulk or solution homopolymerization and copolymerization of methyl cyanoacrylate (MCA) and ECA in the 30–60°C temperature range (Canale et al. 1960; Bevington et al. 1976; Yamada et al. 1983).

Alkyl cyanoacrylates were also copolymerized with methyl methacrylate (MMA) or styrene *via* a free-radical process in bulk or solution to yield random or alternating copolymers, respectively (Kinsinger et al. 1965). MCA or isobutyl cyanoacrylate (IBCA) were also copolymerized with difunctional alkyl cyanoacrylates and yielded crosslinked macromolecular adhesive compositions with superior mechanical properties than the noncrosslinked counterparts (Buck 1978). Free-radical bulk copolymerization of ECA and MMA was also undertaken to yield poly(ECA-*co*-MMA) (P(ECA-*co*-MMA)) random copolymers with various monomer compositions (Han and Kim 2009). Incorporation of MMA monomer units within the polymer structure conducted to both a better stability upon degradation and a lower glass transition temperature than PECA homopolymer. This can be seen as an easy method to tune the properties of the polymer depending on the application envisioned.

Recently, honeycomb-patterned PACA films were prepared from the chloroform solutions of ECA, *n*BCA or OCA by breath figures (BFs) method (Li et al. 2010). Condensed water droplets on the solution surface acted not only as templates to endow the ordered structure but also as initiators to trigger the polymerization of the monomer. After the polymerization started, the *in situ* formed polymer chains self-assembled around the water droplets, structuring PACA film with a hexagonal arrangement of holes. The cell proliferation assay revealed that: (i) porous morphology was more beneficial to Hela cell proliferation than the flat film and (ii) the longer the side-chain of the monomer, the better the biocompatibility.

3.2 *Macromolecular Architectures*

The high reactivity of alkyl cyanoacrylates tends to make the synthesis complex macromolecular architectures extremely difficult. However, several attempts succeeded in the preparation of sophisticated copolymers. For instance, diblock and triblock copolymers involving alkyl cyanoacrylates were prepared from triphenylphosphine end-capped monohydroxyl and dihydroxyl poly(ethylene glycol) (PEG) acting here as mono- or difunctional macroinitiators for the zwitterionic polymerization of IBCA. PIBCA-*b*-PEG diblock and PIBCA-*b*-PEG-*b*-PIBCA triblock copolymers were then prepared with variable compositions and different molar masses by simply playing with the initial stoichiometry (Choi et al. 1995). Similarly, a PECA-*b*-PEG-*b*-PECA triblock copolymer was synthesized by oxyanion-initiated polymerization from sodium difunctional alcoholate-terminated PEG macroinitiator (Lin et al. 2008).

Another approach consisted in the synthesis of amphiphilic comblike poly[(hexadecyl cyanoacrylate)-*co*-methoxypoly(ethylene glycol) cyanoacrylate]

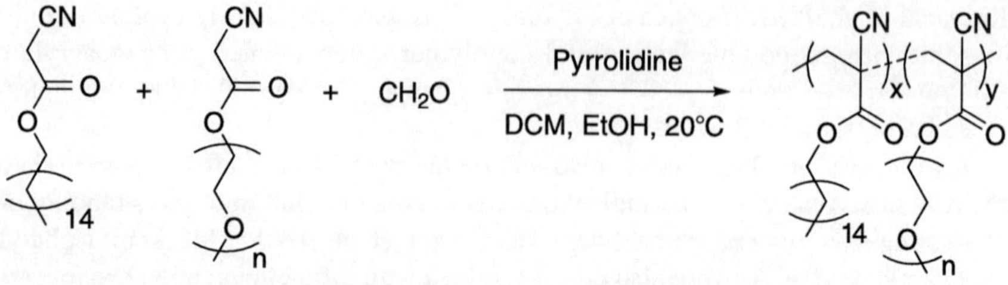

Fig. 4 Synthesis of poly[(hexadecyl cyanoacrylate)-*co*-methoxypoly(ethylene glycol) cyanoacrylate] [P(HDCA-*co*-MePEGCA)] copolymer *via* tandem Knoevenagel condensation-Michael addition reaction

Fig. 5 Scheme of poly [α-maleic anhydride-ω-methoxypoly(ethylene glycol)-*co*-ethyl cyanoacrylate] copolymer

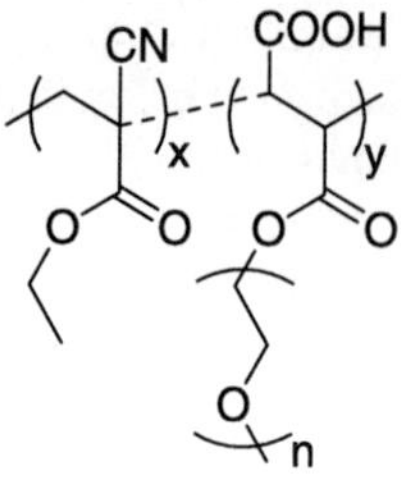

(P(HDCA-*co*-MePEGCA)) copolymer by tandem Knoevenagel condensation-Michael addition reaction between the corresponding cyanoacetate derivatives with formaldehyde in the presence of dimethylamine as the catalyst to build the polymeric backbone (Fig. 4). In this case, cyanoacrylate monomers are obtained *in situ* during the reaction and polymerization occurs spontaneously (Peracchia et al. 1997a).

Interestingly, poly[α-maleic anhydride-ω-methoxypoly(ethylene glycol)-*co*-ethyl cyanoacrylate] copolymers (Fig. 5) were prepared by radical solution copolymerization of poly(ethylene glycol) macromonomer and ECA at 60°C under azobisisobutyronitrile (AIBN) initiation (Deng et al. 2005; Xing et al. 2009). Due to its amphiphilic properties, this copolymer led to PEGylated nanoparticles upon self-assembly in aqueous solution.

By using protecting chemistry, a three-arm star cyanoacrylate-telechelic polyisobutylene $(PIB\text{-}CA)_3$ was obtained from the polymerization of isobutylene initiated by a tri-functional initiator (Fig. 6) (Kwon and Kennedy 2007a). This was followed by its termination with allyltrimethylsilane and an anti-Markovnikov addition of HBr. A masked cyanoacrylate moiety was then linked by esterification prior deprotection to release the cyanoacrylate group. The injection of $(PIB\text{-}CA)_3$ into living tissue yielded a bolus of crosslinked PIB rubber. This approach was also extended to the copolymerization of $(PIB\text{-}CA)_3$ with ECA initiated by nucleophiles or living tissues (Kwon and Kennedy 2007b).

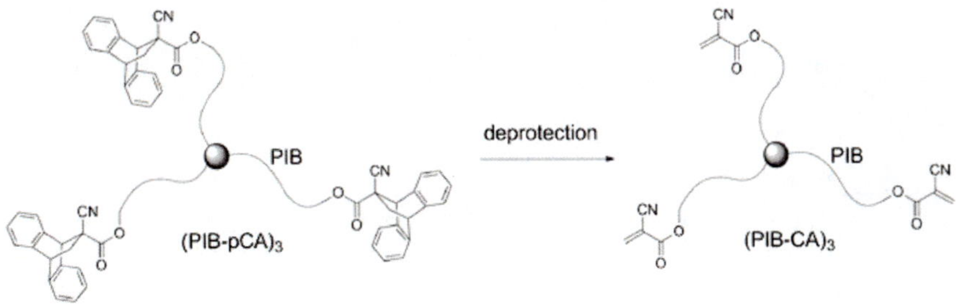

Fig. 6 Synthesis of three-arm star cyanoacrylate-telechelic polyisobutylene $(PIB\text{-}CA)_3$

4 Poly(Alkyl Cyanoacrylate) Nanoparticles

Another important application of PACA is related to the field of drug delivery due to their ability to form submicronic, biodegradable nanoparticles (Couvreur and Vauthier 1991), although PACA-based microparticles have also been reported (Lee et al. 2009; Li et al. 2009). Nanoparticle is a collective term which gathers different types of colloidal objects: (i) nanospheres, a matrix-type system constituted by the polymer in which the drug is homogeneously dispersed and (ii) nanocapsules, which are vesicular systems in which the drug is solubilized in a liquid core (either water or oil) surrounded by a thin polymer layer (Fig. 7) (Vauthier et al. 2003).

4.1 First Generation of Nanoparticles

4.1.1 Nanospheres

The first example in this field was reported in 1979 and consisted in a very simple process to generate *in situ* MCA or ECA nanospheres *via* the polymerization of the monomer into an acidic aqueous solution ($2<pH<3$) containing a nonionic or a macromolecular surfactant (Couvreur et al. 1979). This synthetic route was applied to various cyanoacrylate monomers and appeared to be a method of choice to prepare PACA nanoparticles for drug delivery purposes (Sullivan and Birkinshaw 2004; Arias et al. 2007, 2008b; Kisich et al. 2007; Kusonwiriyawong et al. 2008; Maksimenko et al. 2008; Wilson et al. 2008; Hekmatara et al. 2009; Kurakhmaeva et al. 2009). Numerous studies have also shown that macromolecular and colloidal characteristics of the resulting nanospheres can be easily modified, simply by playing on experimental parameters such as the nature and the concentration of the monomer and the surfactant, the pH, the polymerization temperature, or the concentration of the inhibitor (El-Egakey et al. 1983; Douglas et al. 1984; Vansnick et al. 1985; Seijo et al. 1990; Lescure et al. 1992; Yang et al. 2000; Dossi et al.

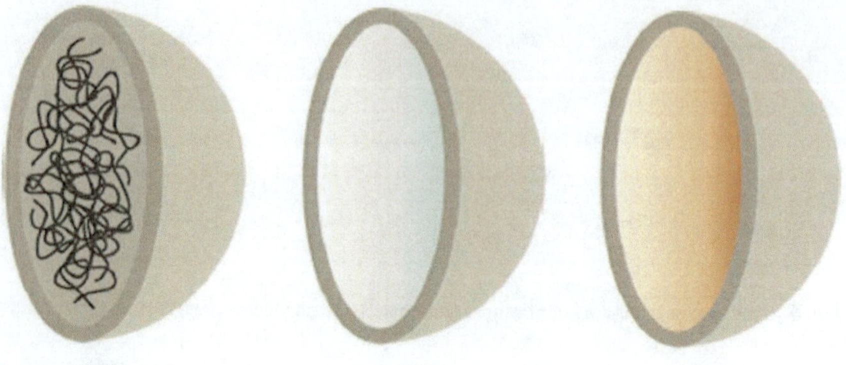

Fig. 7 Nanospheres (NS), water-containing nanocapsules (w-NC) and oil-containing nanocapsules (o-NC)

2010). Average diameters obtained by this process usually range from 50 to 300 nm, which is a well-adapted window for drug delivery purposes (Douglas et al. 1985; Alonso et al. 1990; Seijo et al. 1990; Shipulo et al. 2008). It has been recently shown that fluorophores can be loaded with high yields into poly(ECA) (PECA) nanoparticles prepared by anionic emulsion polymerization (Yordanov et al. 2009). This aqueous emulsion polymerization procedure was adapted to the synthesis of PACA nanospheres of PECA, poly(*n*butyl cyanoacrylate) (P*n*BCA), poly(hexyl cyanoacrylate) (PHCA) and poly(octyl cyanoacrylate) (POCA) with a magnetic core and loaded with Tegafur as the anticancer drug (Arias et al. 2008a). This method was scaled up to achieve production of clinical batches of PACA nanoparticles loaded with doxorubicine and those of PACA nanoparticles loaded with mitoxantrone evaluated in clinics for the treatment of hepatocellular carcinoma in human (BioAlliance Pharma 2009; Zhou et al. 2009).

P*n*BCA nanospheres were also prepared by miniemulsion and emulsion polymerization using molecular or macromolecular surfactants (Maitre et al. 2000; Cauvin et al. 2002; Limouzin et al. 2003; Weiss et al. 2007) or by the nanoprecipitation technique from the preformed polymer using pluronic F68 as a surfactant (He et al. 2008).

In order to increase the encapsulation yield of peptides into PACA nanoparticles, a copolymerization approach was developed in combination with a microemulsion process (Fig. 8). The peptide with the following sequence KAVYNFATM was derived with an acryloyl group and copolymerized with ECA in water-in-oil (w/o) microemulsion to yield stable peptide-loaded PECA nanoparticles of 250 nm in average diameter (Liang et al. 2008). In this approach, the peptide loading was twice-fold higher than that of with the unmodified peptide. In contrast, the same copolymerization performed using a micellar template only conducted to unstable aggregates.

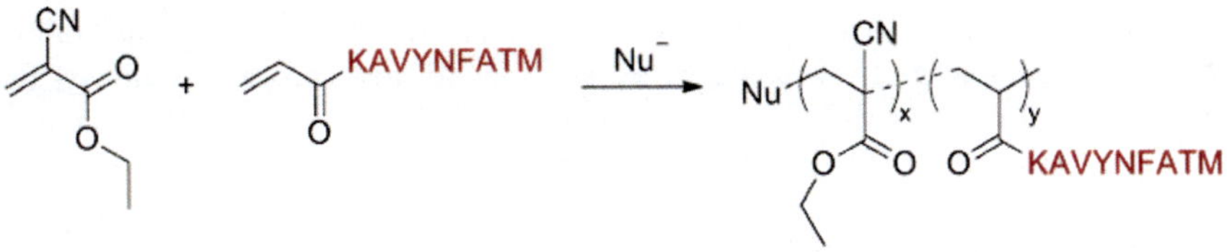

Fig. 8 Copolymerization mechanism of alkyl cyanoacrylate and acryloyl peptide

4.1.2 Nanocapsules

As opposed to nanospheres, PACA nanocapsules allowed a broader variety of drugs to be encapsulated. Indeed, oil-containing nanocapsules will be able to encapsulate hydrophobic drugs, whereas hydrophilic ones will be efficiently encapsulated into water-containing nanocapsules. Basically, the shell of the nanocapsule is formed by spontaneous anionic polymerization of alkyl cyanoacrylate occurring at the interface between the dispersed and the continuous phase during emulsion or microemulsion polymerization. The nature of the dispersed phase involved in a heterogeneous polymerization process governs the nature of the nanocapsules. Several studies reported the preparation of oil-containing PACA nanocapsules (Fallouh et al. 1986; Gallardo et al. 1993; Wohlgemuth and Mayer 2003; Bogdan et al. 2008) and highlighted the importance of the diffusion behavior of the organic solvent within the aqueous phase and the simultaneous precipitation of the polymer at the oil/water interface (Fallouh et al. 1986; Gallardo et al. 1993). Average diameters usually observed for such nanocapsules range from 200 to 350 nm (even though a miniemulsion process (Altinbas et al. 2006) allowed average diameters below 100 nm to be obtained) and are mainly governed by the nature and the concentration of the monomer, the amount of surfactant and oil as well as the speed of diffusion of the organic phase within the aqueous one. A drawback usually encountered during the preparation of nanocapsules is a polluting population of nanospheres resulting from a partial polymerization in the organic phase (Gallardo et al. 1993). This can be improved by an optimized ethanol/oil ratio (Gallardo et al. 1993; Aboubakar et al. 1999), the acidification of the organic phase (Wohlgemuth and Mayer 2003) or the inhibition of the polymerization in the organic phase by aprotic solvents (Puglisi et al. 1995).

Water-containing PACA nanocapsules were obtained by inverse (i.e., water-in-oil) emulsion or microemulsion polymerization. Depending on the nature of the surfactant and the starting system (emulsion or microemulsion), this process led to nanocapsules between 50 and 350 nm (Vranckx et al. 1996; Lambert et al. 2000; Watnasirichaikul et al. 2002; Krauel et al. 2005; Hillaireau et al. 2007). If required, a second step can be applied to transfer nanocapsules from an oil-dispersing medium to a water-dispersing medium and consisted in a centrifugation step of the nanocapsules onto an aqueous layer (Lambert et al. 2000; Hillaireau et al. 2006,

2007). Recently, poly(*N*-isopropylacrylamide) (PNIPAAm)/PECA composite hollow particles were successfully prepared by an emulsification process, followed by anionic polymerization of ECA and photopolymerization of *N*-isopropylacrylamide (NIPAAm) (Lee et al. 2010). The shell layer of the composite hollow particle was changed from semi-interpenetrating network to double shell layers by varying the mass ratio of NIPAAm to ECA monomers.

The synthesis of nanocapsules from preformed PACA polymers (called interfacial deposition) is also possible and consists of the addition of the polymer solution and a small amount of oil (which will constitute the oily core of the nanocapsules), into an aqueous solution of surfactant. The oil-containing nanocapsules were formed instantaneously by deposition of the polymer at the oil/water interface, which precipitates as a macromolecular shell (Barratt et al. 1994; Brigger et al. 2003; Mayer 2005).

Miniemulsion polymerization was also employed to prepare monodisperse DNA-containing, P*n*BCA nanocapsules by inverse miniemulsion polymerization (Musyanovych and Landfester 2008). It was shown that the nature of the continuous phase played the major role over the average diameter and the particle size distribution whereas the monomer concentration governed the shell thickness and the nanocapsule morphology.

4.2 *Second Generation: Surface Engineered and Stealth Nanoparticles*

In order to avoid quick accumulation of PACA nanoparticles into the liver and the spleen due to adsorption of opsonins onto their surface, which triggers the recognition of the mononuclear phagocyte system (MPS) by the macrophages, surface modified nanoparticles were developed. In this view, the major breakthrough was the grafting of PEG chains (the so-called PEGylation) (Stolnik et al. 1995; Storm et al. 1995) onto their surface which resulted in long-circulating drug delivery devices, also termed "stealth" nanoparticles.

Although simple adsorption of surfactant has been reported to prepared surface-modified PACA nanoparticles (Wilson et al. 2008; Kurakhmaeva et al. 2009), the following will only focus on covalent linkage between the core of the NPs and its coating.

4.2.1 Anionic and Zwitterionic Polymerization

The preparation of PACA nanospheres covered by PEG chains was achieved by using methoxypoly(ethylene glycol) (MePEG) or PEG as both stabilizer and initiator through zwitterionic emulsion polymerization of IBCA (Peracchia et al. 1997c, d). This led to PEG-coated PIBCA nanospheres where PEG chains exhibited different conformations; either stretched chains or loops depending on the number of terminal hydroxyl groups present in the PEG (Peracchia et al. 1997b).

Whereas early work reported the use of β-cyclodextrine as a surface-modifier/surfactant for the anionic emulsion polymerization of *n*BCA (Douglas et al. 1985), different kinds of polysaccharide were also used to replace traditional PEG coatings. It yielded stable 100–500 nm PACA nanospheres were obtained with surface properties governed by the nature of the polysaccharide (Yang et al. 2000; Bertholon et al. 2006b, 2005; Labarre et al. 2005; Bravo-Osuna et al. 2007b; Shirotake et al. 2008; Duan et al. 2009, 2010; Kulkarni et al. 2010). Interestingly, when covered by chitosan, the resulting cationic nanoparticles served as a colloidal scaffold for the preparation of nanoparticle/DNA complexes by the complex coacervation of nanoparticles with the DNA (Duan et al. 2009).

Anionic miniemulsion polymerization was recently employed to prepare PEGylated P*n*BCA nanocapsules with an oily core and loaded by paclitaxel (Zhang et al. 2008). The oily core was formed by a medium-chain triglyceride and the anionic polymerization of *n*BCA was initiated at the oil/water interface by MePEG via its alcoholate chain-end. MePEG-*b*-P*n*BCA block copolymers surrounding the oily core were thus formed *in situ*.

4.2.2 Free-Radical Polymerization

Free-radical emulsion polymerization of alkyl cyanoacrylates initiated by the polysaccharide/cerium IV (Ce^{4+}) ions redox couple was also undertaken leading to stable nanospheres (Chauvierre et al. 2003b). Whereas zwitterionic/anionic emulsion polymerizations gave polysaccharides compact loops at the nanoparticle surface due to the *in situ* synthesis of grafted copolymers, the free-radical process led to polysaccharides-based block copolymers and thus to hairy nanospheres due to a different initiation mechanism (Fig. 9) (Chauvierre et al. 2003a, 2004, 2007; Bertholon et al. 2006a, b; Bravo-Osuna et al. 2007a, c). One interest of making PACA nanoparticles by free radical polymerization is to obtain nanoparticles with a different conformational arrangement of the polysaccharide chains at the nanoparticle surface compared to that obtained with zwitterionic/anionic emulsion polymerization. Indeed, this was found critical to control interactions of the nanoparticles with serum proteins giving different complement activation pattern hence influencing the pharmacokinetics and the biodistribution of an associated drug. For instance, nanoparticles obtained by free radical polymerization behaved like stealth nanoparticles (Bertholon et al. 2006b; Alhareth 2010; De Martimprey et al. 2010; Vauthier et al. 2011).

It is noteworthy that free radical polymerization of alkyl cyanoacrylates can only be promoted at pH 1 by taking advantage of an extremely fast free-radical polymerization initiation (Chauvierre et al. 2003b, c). The low pH is mandatory to delay the zwitterionic/anionic polymerization initiation to let a narrow time window of 5 to 10 min to initiate the free radical polymerization. In the emulsion polymerization conditions this requirement is fulfilled at pH 1. The polysaccharide/cerium IV (Ce^{4+}) ions redox couple was found appropriate to initiate the free radical polymerization in

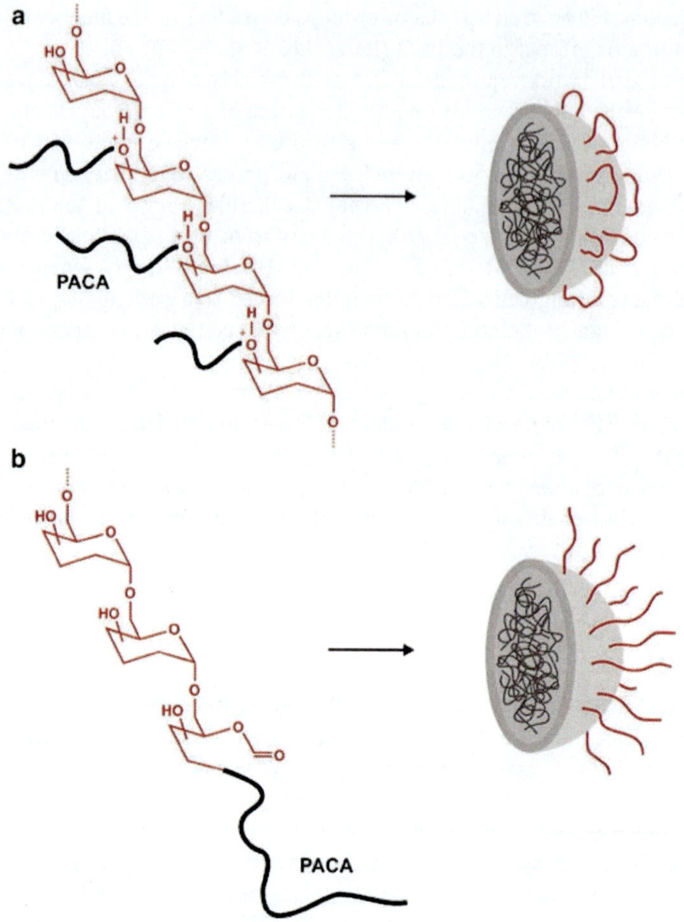

Fig. 9 Formation of grafted copolymers leading to compact loops by anionic emulsion polymerization of alkyl cyanoacrylates initiated by hydroxyl groups of dextran (**a**) and polysaccharides-based block copolymers leading to hairy nanospheres synthesized by redox free-radical emulsion polymerization of alkyl cyanoacrylates initiated by dextran/cerium IV (Ce4+) ions redox couple (**b**)

the short time window as the polymerization starts immediately after the addition of cerium IV ions in the polysaccharide and alkyl cyanoacrylate containing emulsion.

The method is versatile and made possible the preparation of hairy nanoparticles with various types of polysaccharides on the surface. The diameter can be tuned by varying the molecular weight of the polysaccharide and the polysaccharide/alkyl cyanoacrylate molar ratio (Bertholon et al. 2006b; He et al. 2009). For instance, stable nanoparticles with a diameter of 80 nm were prepared with heparin (Chauvierre et al. 2004). With dextran and chitosan, it was found that a minimal molar mass was required to insure the stability of the nanoparticles. Thus the minimal sizes of nanoparticles obtained with these polysaccharides were 137 and 170 nm using 10,000 Da dextran and 20,000 Da chitosan (Bertholon et al. 2006b). The size of nanoparticles synthesized with chitosan 20,000 g/mol can be reduced to 62 nm by addition of 3% pluronic F68 in the emulsion polymerization medium (De Martimprey et al. 2010). Using hyaluronic acid (HA), average diameters of the nanoparticles can be modulated from 290 to 325 nm by playing with the HA/*n*BCA molar ratio (He et al. 2009).

Finally, regarding the drug loading, chitosan coated nanoparticles can easily be loaded with siRNA by adsorption on preformed nanoparticles. These nanoparticles were able to deliver active siRNA to a subcutaneously implanted tumor after intravenous administration (De Martimprey et al. 2010). Paclitaxel was encapsulated with a maximal encapsulation efficiency of about 90%. In vitro release studies demonstrated that HA modification efficiently reduced the initial burst release in the first 10 h and provided a sustained release in the subsequent 188 h (He et al. 2009). Doxorubicin, another first line anticancer agent, could not be loaded in PIBCA nanoparticles by encapsulation because the drug is susceptible to oxidation by the cerium ions used to initiate the free radical polymerization. However, it could be associated by adsorption with the same loading performance and releasing features than those of the corresponding nanoparticles obtained by zwitterionic/anionic emulsion polymerization (Alhareth et al. 2011).

4.2.3 Self-Assembly of Preformed (co)Polymers

An alternative to the preparation of PEGylated PACA nanoparticles by direct polymerization in aqueous dispersed media is the use of preformed amphiphilic copolymers able to self-assemble in water (Choi et al. 1995; Peracchia et al. 1997a, 1998, 1999; Deng et al. 2005). Stable nanospheres in the 100–700 nm range were obtained from PIBCA-*b*-PEG diblock copolymers, either by nanoprecipitation or by emulsification/solvent evaporation (Choi et al. 1995). As explained in the **Chapter from Vauthier and Bouchemal** of this book, these methods can be adapted to produce large quantities of nanoparticles making possible their industrial development. It was shown that the copolymer compositions played an important role on the colloidal characteristics of the nanospheres. P(HDCA-*co*-MePEGCA) copolymers also demonstrated suitable amphiphilicity to form stable nanospheres of 100–200 nm in diameter upon nanoprecipitation which exhibited a biodegradable PHDCA core and a shell of excretable PEG chains (Peracchia et al. 1997a, 1998, 1999). Interestingly, it was shown that P(HDCA-*co*-MePEGCA) nanoparticles exhibited a significant ability to cross the blood-brain barrier (BBB), compared to non-PEGylated counterparts and nanoparticles with preadsorbed surfactants such as polysorbate 80 or poloxamine 908 (Brigger et al. 2002; Calvo et al. 2001, 2002; Garcia-Garcia et al. 2005a, b). This was explained by a specific adsorption of apolipoprotein E and B-100 (Apo E and B-100) onto these PEGylated nanospheres, leading to their translocation mediated by low-density lipoprotein receptors (LDLR) (Kim et al. 2007a, b, c). Following this discovery, the encapsulation into these nanoparticles of didanosine (ddI) prodrugs, which is a drug used to treat HIV infection, was then investigated in order to increase ddI concentration into the brain and to treat HIV-1 associated dementia (Bourgeois et al. 2009; Pierson et al. 2009).

PECA-*b*-PEG-*b*-PECA amphiphilic triblock copolymer was used to prepare dexamethasone (DXM)-loaded, PEGylated nanoparticles by the nanoprecipitation technique (Lin et al. 2008). Nanoparticles exhibited and average diameter below 100 nm and a high drug loading.

In order to circumvent the drawbacks usually encountered with the use of encapsulated fluorescent dyes into nanoparticles (i.e., leakage, burst effect, etc.), fluorescently-tagged, PEGylated nanoparticles were designed. The strategy was to incorporate the fluorophore during the synthesis of an amphiphilic PACA copolymer, by tandem Knoevenagel condensation-Michael addition reaction with HDCA, MePEGCA and a small amount of a cyanoacetate derivative based on either dansyl or rhodamine B as the fluorescent dye (Fig. 10) (Brambilla et al. 2010a). The resulting fluorescent nanoparticles showed suitable characteristics for *in vitro* imaging on human brain endothelial cells and their fluorescence signal was found extremely accurate, as opposed to a diffuse signal when a lipophilic dye is encapsulated. In addition, it was recently discovered that these nanoparticles were able to bind the amyloid β-peptide, a biomarker for Alzheimer's Disease, and to influence its aggregation kinetics (Brambilla et al. 2010b).

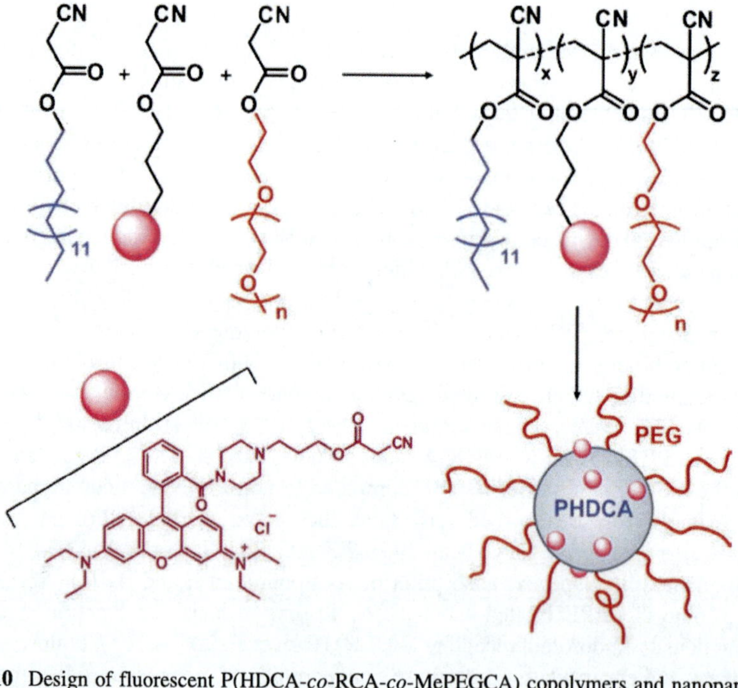

Fig. 10 Design of fluorescent P(HDCA-*co*-RCA-*co*-MePEGCA) copolymers and nanoparticles for cell imaging

The P(HDCA-*co*-MePEGCA) copolymer was also employed for the synthesis of PEGylated nanocapsules, either by interfacial deposition (Brigger et al. 2003) or by water-in-oil-in-water (w/o/w) double emulsion (Li et al. 2001a, b) In the latter case, water-containing nanocapsules of 140–150 nm were obtained.

4.3 Third Generation: Addressed Nanoparticles

The synthesis of efficient ligands-decorated colloidal devices for achieving specific cells targeting, based molecular recognition processes, has become one of the most exciting fields regarding drug delivery. Indeed, previous generations of PACA nanoparticles were unable to be efficiently addressed to the desired cells and the therapeutic activity of the encapsulated drug may be partly decreased.

The preparation of folate-decorated poly[(hexadecyl cyanoacrylate)-*co*-aminopoly(ethylene glycol) cyanoacrylate] [P(HDCA-*co*-NH_2PEGCA)] nanospheres (Fig. 11) to target the folate receptor, which is overexpressed at the surface of many tumor cells, was reported (Stella et al. 2000, 2007). For this purpose, a P(HDCA-*co*-NH_2PEGCA) copolymer was synthesized and self-assembled in water into stable 80 nm nanospheres, at the surface of which *N*-hydroxysuccinimide-folate (NHS-folate) was reacted *via* an amidation pathway. It was shown that the apparent affinity of the folate bound to the nanospheres towards its receptor appeared tenfold higher than the free folate in solution, because of the multivalency of the folate-decorated nanoparticles.

Curcumin-loaded P*n*BCA nanoparticles coated with Tween 80 (Mulik et al. 2009) were further decorated with apolipoptotein E3 (Apo E3) (Mulik et al. 2010). This strategy provided efficient photostability and cell uptake of curcumin by active targeting *via* receptor-mediated endocytosis. *In vitro* cell culture study showed enhanced therapeutic efficacy of the curcumin-loaded Apo E3-P*n*BCA nanoparticles against beta amyloid induced cytotoxicity in SH-SY5Y neuroblastoma cells compared to free curcumin.

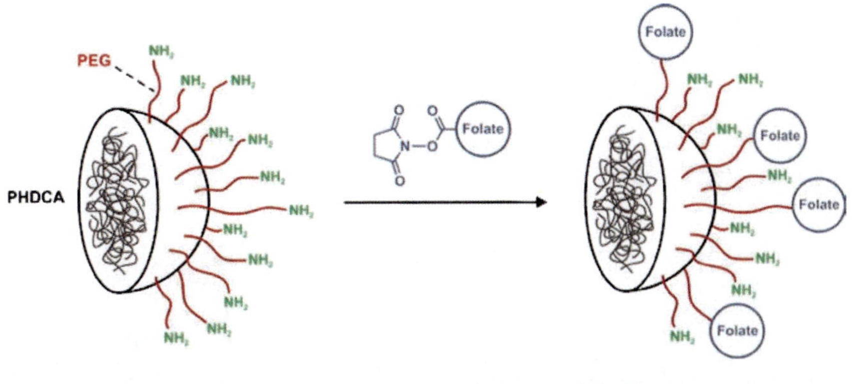

Fig. 11 Synthesis of poly[(hexadecyl cyanoacrylate)-*co*-aminopoly(ethylene glycol) cyanoacrylate] [P(HDCA-*co*-NH_2PEGCA)] copolymer *via* tandem Knoevenagel condensation-Michael addition reaction

Another promising route towards ligands-decorated PACA nanoparticles has been recently developed and was based on the Huisgen 1,3-dipolar cyclo-addition, the so-called "click chemistry" (Kolb et al. 2001). To achieve this goal, a poly[(hexadecyl cyanoacrylate)-*co*-azidopoly(ethylene glycol) cyanoacrylate] (P(HDCA-*co*-N_3PEGCA)) copolymer was synthesized. Model alkynes were then quantitatively coupled either to the P(HDCA-*co*-N_3PEGCA) copolymers in homogeneous medium followed by self-assembly in aqueous solution or directly at the surface of the preformed P(HDCA-*co*-N_3PEGCA) nanoparticles in aqueous dispersed medium, both yielding highly functionalized nanospheres (Nicolas et al. 2008).

The same research group extended this synthetic approach for the coupling of selegiline as a biologically active molecule against Parkinson and Alzheimer's disease. By a combination of click chemistry and *N,N'*-dicyclohexylcarbodiimide (DCC) -assisted coupling reaction, selegiline-poly(ethylene glycol) cyanoacetate (SelPEGCA) derivative was readily prepared and involved in the synthesis of poly[(hexadecyl cyanoacrylate)-*co*-selegiline-poly(ethylene glycol) cyanoacrylate] [P(HDCA-*co*-SelPEGCA)] copolymer. A co-nanoprecipitation process with poly[hexadecyl cyanoacrylate-*co*-rhodamine B cyanoacrylate-*co*-methoxypoly (ethylene glycol) cyanoacrylate] [P(HDCA-*co*-RCA-*co*-MePEGCA)] copolymer yielded PEGylated and fluorescent nanoparticles displaying selegiline moieties (Le Droumaguet et al. 2010).

4.4 Conclusion and Perspectives

As discussed in this chapter, a wide range of PACA nanosystems can be produced through different ways. Progresses in the comprehension of the polymerization mechanisms of the corresponding monomers, introduction of new methods of polymerization and synthesis of copolymers with well-defined macromolecular architectures made possible the engineering of sophisticated PACA nanoparticles. These important achievements led to the design of drug carriers suitable for optimizing the in vivo delivery of various types of drugs. The development of addressed PACA nanoparticles is also expanded and is paving the way to active targeting. It can be expected that the identification of new targets will continue to open opportunities for challenging construct developments based on PACA nanosystems.

References

Aboubakar, M., Puisieux, F., Couvreur, P., Deyme, M. & Vauthier, C. 1999. Study of the mechanism of insulin encapsulation in poly(isobutylcyanoacrylate) nanocapsules obtained by interfacial polymerization. Journal of Biomedical Materials Research, 47, 568–576.

Alhareth K. Association de la doxorubicine à des nanoparticles de poly(cyanoacrylate d'alkyle) prepares par la polymerization radicalaire en emulsion: pharmacocinétique et biodistribution. PhD Thesis of the University of Paris South, 29 September 2010.

Alhareth, K., Vauthier, C., Gueutin, C., Ponchel, G., Moussa, F. 2011. Doxorubicin Loading and In Vitro Release from Poly(alkylcyanoacrylate) Nanoparticles Produced by Redox Radical Emulsion Polymerization. Journal of Applied Polymer Science 119, 816–822.

Alonso, M. J., Sanchez, A., Torres, D., Seijo, B. & Vila-Jato, J. L. 1990. Joint effects of monomer and stabilizer concentrations on physicochemical characteristics of poly(butyl 2-cyanoacrylate) nanoparticles. Journal of Microencapsulation, 7, 517–526.

Altinbas, N., Fehmer, C., Terheiden, A., Shukla, A., Rehage, H. & Mayer, C. 2006. Alkyl cyanoacrylate nanocapsules prepared from mini-emulsions: a comparison with the conventional approach. Journal of Microencapsulation, 23, 567–581.

Ardis, A. E. 1949. Monomeric alkyl a-cyanoacrylates. Us patent application.

Arias, J. L., Gallardo, V., Ruiz, M. A. & Delgado, A. V. 2007. Ftorafur loading and controlled release from poly(ethyl-2-cyanoacrylate) and poly(butylcyanoacrylate) nanospheres. International Journal of Pharmaceutics, 337, 282–290.

Arias, J. L., Ruiz, M. A., Gallardo, V. & Delgado, A. V. 2008a. Tegafur loading and release properties of magnetite/poly(alkylcyanoacrylate) (core/shell) nanoparticles. Journal of Controlled Release, 125, 50–58.

Arias, J. L., Ruiz, M. A., Lopez-Viota, M. & Delgado, A. V. 2008b. Poly(alkylcyanoacrylate) colloidal particles as vehicles for antitumour drug delivery: A comparative study. Colloids and Surfaces, B: Biointerfaces, 62, 64–70.

Barratt, G., Puisieux, F., Yu, W. P., Foucher, C., Fessi, H. & Devissaguet, J. P. 1994. Anti-metastatic activity of MDP-L-alanyl-cholesterol incorporated into various types of nanocapsules. International Journal of Immunopharmacology, 16, 457–61.

Bertholon-Rajot, I., Labarre, D. & Vauthier, C. 2005. Influence of the initiator system, cerium-polysaccharide, on the surface properties of poly(isobutylcyanoacrylate) nanoparticles. Polymer, 46, 1407–1415.

Bertholon, I., Lesieur, S., Labarre, D., Besnard, M. & Vauthier, C. 2006a. Characterization of dextran-poly(isobutylcyanoacrylate) copolymers obtained by redox radical and anionic emulsion polymerization. Macromolecules, 39, 3559–3567.

Bertholon, I., Vauthier, C. & Labarre, D. 2006b. Complement Activation by Core-Shell Poly(isobutylcyanoacrylate)-Polysaccharide Nanoparticles: Influences of Surface Morphology, Length, and Type of Polysaccharide. Pharmaceutical Research, 23, 1313–1323.

Bevington, J. C., Jemmett, J. A. L. & Onyon, P. F. 1976. Polymerization of methyl a-cyanoacrylate. II. Conditions for radical polymerization. European Polymer Journal, 12, 255–7.

BioAlliance Pharma: www.bioalliancepharma.com: 08/12/2009 Doxorubicine Transdrug®: Amélioration significative de la durée de survie des patients atteints d'un carcinome hépatocellulaire avancé dans un essai clinique de phase II, 11 January 2011.

Bogdan, M., Nan, A., Pop, C. V. L., Barbu-Tudoran, L. & Ardelean, I. 2008. Preparation and NMR characterization of polyethyl-2-cyanoacrylate nanocapsules. Applied Magnetic Resonance, 34, 111–119.

Bourgeois, J., Pierson, L.-A., Nicolas, J., Lalanne, M., Couvreur, P. & Andrieux, K. 2009. Application of thermal analysis to the study of lipidic prodrug incorporation into nanocarriers. Journal of Thermal Analysis and Calorimetry, 98, 65–71.

Brambilla, D., Nicolas, J., Le Droumaguet, B., Andrieux, K., Marsaud, V., Couraud, P.-O. & Couvreur, P. 2010a. Design of fluorescently tagged poly(alkyl cyanoacrylate) nanoparticles for human brain endothelial cell imaging. Chemical Communications, 46, 2602–2604.

Brambilla, D., Verpillot, R., Taverna, M., De Kimpe, L., Le Droumaguet, B., Nicolas, J., Mantegazza, F., Canovi, M., Gobbi, M., Salmona, M., Nicolas, V., Scheper, W., Couvreur, P. & Andrieux, K. 2010b. Capillary electrophoresis with laser-induced fluorescence detection (CE-LIF) as a new protocol to monitor interaction between nanoparticles and the amyloid-β peptide. Analytical Chemistry, 82, 10083–10089.

Bravo-Osuna, I., Millotti, G., Vauthier, C. & Ponchel, G. 2007a. In vitro evaluation of calcium binding capacity of chitosan and thiolated chitosan poly(isobutyl cyanoacrylate) core-shell nanoparticles. International Journal of Pharmaceutics, 338, 284–290.

Bravo-Osuna, I., Ponchel, G. & Vauthier, C. 2007b. Tuning of shell and core characteristics of chitosan-decorated acrylic nanoparticles. European Journal of Pharmaceutical Sciences, 30, 143–154.

Bravo-Osuna, I., Vauthier, C., Farabollini, A., Palmieri, G. F. & Ponchel, G. 2007c. Mucoadhesion mechanism of chitosan and thiolated chitosan-poly(isobutyl cyanoacrylate) core-shell nanoparticles. Biomaterials, 28, 2233–2243.

Brigger, I., Armand-Lefevre, L., Chaminade, P., Besnard, M., Rigaldie, Y., Largeteau, A., Andremont, A., Grislain, L., Demazeau, G. & Couvreur, P. 2003. The stenlying effect of high hydrostatic pressure on thermally and hydrolytically labile nanosized carriers. Pharmaceutical Research, 20, 674–683.

Brigger, I., Morizet, J., Aubert, G., Chacun, H., Terrier-Lacombe, M. J., Couvreur, P. & Vassal, G. 2002. Poly(ethylene glycol)-coated hexadecylcyanoacrylate nanospheres display a combined effect for brain tumor targeting. Journal of Pharmacology and Experimental Therapeutics, 303, 928–936.

Buck, C. J. 1978. Unequivocal synthesis of bis(2-cyanoacrylate) monomers. I. Via anthracene adducts. Journal of Polymer Science, Polymer Chemistry Edition, 16, 2475–507.

Calvo, P., Gouritin, B., Chacun, H., Desmaele, D., D'angelo, J., Noel, J. P., Georgin, D., Fattal, E., Andreux, J. P. & Couvreur, P. 2001. Long-circulating PEGylated polycyanoacrylate nanoparticles as new drug carrier for brain delivery. Pharmaceutical Research, 18, 1157–1166.

Calvo, P., Gouritin, B., Villarroya, H., Eclancher, F., Giannavola, C., Klein, C., Andreux, J. P. & Couvreur, P. 2002. Quantification and localization of PEGylated polycyanoacrylate nanoparticles in brain and spinal cord during experimental allergic encephalomyelitis in the rat. European Journal of Neuroscience, 15, 1317–1326.

Canale, A. J., Goode, W. E., Kinsinger, J. B., Panchak, J. R., Kelso, R. L. & Graham, R. K. 1960. Methyl a-cyanoacrylate. I. Free-radical homopolymerization. Journal of Applied Polymer Science, 4, 231–6.

Cauvin, S., Sadoun, A., Dos Santos, R., Belleney, J., Ganachaud, F. & Hemery, P. 2002. Cationic polymerization of p-methoxystyrene in miniemulsion. Macromolecules, 35, 7919–7927.

Chauvierre, C., Labarre, D., Couvreur, P. & Vauthier, C. 2003a. Novel polysaccharide-decorated poly(isobutyl cyanoacrylate) nanoparticles. Pharmaceutical Research, 20, 1786–1793.

Chauvierre, C., Labarre, D., Couvreur, P. & Vauthier, C. 2003b. Radical Emulsion Polymerization of Alkylcyanoacrylates Initiated by the Redox System Dextran-Cerium(IV) under Acidic Aqueous Conditions. Macromolecules, 36, 6018–6027.

Chauvierre, C., Labarre, D., Couvreur, P. & Vauthier, C. 2003c. Plug-in spectrometry with optical fibers as a novel analytical tool for nanoparticles technology: Application to the investigation of the emulsion polymerization of the alkylcyanoacrylate. Journal of Nanoparticle Research 5: 365–371, 2003c.

Chauvierre, C., Leclerc, L., Labarre, D., Appel, M., Marden, M. C., Couvreur, P. & Vauthier, C. 2007. Enhancing the tolerance of poly (isobutylcyanoacrylate) nanoparticles with a modular surface design. International Journal of Pharmaceutics, 338, 327–332.

Chauvierre, C., Marden, M. C., Vauthier, C., Labarre, D., Couvreur, P. & Leclerc, L. 2004. Heparin coated poly(alkylcyanoacrylate) nanoparticles coupled to hemoglobin: a new oxygen carrier. Biomaterials, 25, 3081–3086.

Choi, Y. K., Bae, Y. H. & Kim, S. W. 1995. Block Copolymer Nanoparticles of Ethylene Oxide and Isobutyl Cyanoacrylate. Macromolecules, 28, 8419–21.

Coover, H. W., Dreifus, D. W. & O'connor, J. T. 1990. Cyanoacrylate adhesives. In: SKEIST, I. (ed.) Handbook of Adhesives. 3rd ed. New-York: Van Nostrand Reinhold.

Couvreur, P., Kante, B., Roland, M., Guiot, P., Bauduin, P. & Speiser, P. 1979. Polycyanoacrylate Nanocapsules as Potential Lysosomotropic Carriers – Preparation, Morphological and Sorptive Properties. Journal of Pharmacy and Pharmacology, 31, 331–332.

Couvreur, P. & Vauthier, C. 1991. Polyalkylcyanoacrylate Nanoparticles as Drug Carrier – Present State and Perspectives. Journal of Controlled Release, 17, 187–198.

Cronin, J. P. & Pepper, D. C. 1988. Zwitterionic polymerization of butyl cyanoacrylate by triphenylphosphine and pyridine. Makromolekulare Chemie, 189, 85–102.

De Martimprey, H.; Bertrand, J.R.; Malvy, C.; Couvreur, P. & Vauthier, C. 2010. New core-shell nanoparticules for the intravenous delivery of siRNA to experimental thyroid papillary carcinoma. Pharmaceutical Research, 27, 498–509.

Denchev, Z., Tomanova, M. & Lederer, A. 2008. On the anionic homo- and copolymerization of ethyl- and butyl-2-cyanoacrylates. Journal of Polymer Science, Part A: Polymer Chemistry, 46, 5142–5156.

Deng, L., Yao, C., Li, A. & Dong, A. 2005. Preparation and characterization of poly{[α-maleic anhydride-ω-methoxy-poly(ethylene glycol)]-co-(ethyl cyanoacrylate)} copolymer nanoparticles. Polymer International, 54, 1007–1013.

Donnelly, E. F., Johnston, D. S., Pepper, D. C. & Dunn, D. J. 1977. Ionic and zwitterionic polymerization of n-alkyl 2-cyanoacrylates. Journal of Polymer Science, Polymer Letters Edition, 15, 399–405.

Dossi, M., Storti, G. & Moscatelli, D. 2010. Synthesis of Poly(Alkyl Cyanoacrylates) as Biodegradable Polymers for Drug Delivery Applications. Macromolecular Symposia, 289, 124–128.

Douglas, S. J., Illum, L. & Davis, S. S. 1985. Particle-Size and Size Distribution of Poly(Butyl 2-Cyanoacrylate) Nanoparticles .2. Influence of Stabilizers. Journal of Colloid and Interface Science, 103, 154–163.

Douglas, S. J., Illum, L., Davis, S. S. & Kreuter, J. 1984. Particle-Size and Size Distribution of Poly(Butyl-2-Cyanoacrylate) Nanoparticles .1. Influence of Physicochemical Factors. Journal of Colloid and Interface Science, 101, 149–158.

Duan, J., Zhang, Y., Chen, W., Shen, C., Liao, M., Pan, Y., Wang, J., Deng, X. & Zhao, J. 2009. Cationic polybutyl cyanoacrylate nanoparticles for DNA delivery. Journal of Biomedicine and Biotechnology, Article ID 149254.

Duan, J., Zhang, Y., Han, S., Chen, Y., Li, B., Liao, M., Chen, W., Deng, X., Zhao, J. & Huang, B. 2010. Synthesis and in vitro/in vivo anti-cancer evaluation of curcumin-loaded chitosan/poly(butyl cyanoacrylate) nanoparticles. International Journal of Pharmaceutics, 400, 211–220.

El-Egakey, M. A., Bentele, V. & Kreuter, J. 1983. Molecular-Weights of Polycyanoacrylate Nanoparticles. International Journal of Pharmaceutics, 13, 349–352.

Eromosele, I. C. & Pepper, D. C. 1986. Free- and paired-ion propagation in the polymerization of butyl cyanoacrylate by tetrabutylammonium salts. Makromolekulare Chemie, Rapid Communications, 7, 639–43.

Eromosele, I. C. & Pepper, D. C. 1989a. Anionic polymerization of butyl cyanoacrylate by tetrabutylammonium salts. 1. Initiation processes. Makromolekulare Chemie, 190, 3085–94.

Eromosele, I. C. & Pepper, D. C. 1989b. Anionic polymerization of butyl cyanoacrylate by tetrabutylammonium salts. 2. Propagation rate constants. Makromolekulare Chemie, 190, 3095–103.

Eromosele, I. C., Pepper, D. C. & Ryan, B. 1989. Water effects on the zwitterionic polymerization of cyanoacrylates. Makromolekulare Chemie, 190, 1613–22.

Fallouh, N. A., Roblottreupel, L., Fessi, H., Devissaguet, J. P. & Puisieux, F. 1986. Development of a New Process for the Manufacture of Polyisobutylcyanoacrylate Nanocapsules. International Journal of Pharmaceutics, 28, 125–132.

Gallardo, M., Couarraze, G., Denizot, B., Treupel, L., Couvreur, P. & Puisieux, F. 1993. Study of the Mechanisms of Formation of Nanoparticles and Nanocapsules of Polyisobutyl-2-Cyanoacrylate. International Journal of Pharmaceutics, 100, 55–64.

Garcia-Garcia, E., Andrieux, K., Gil, S. & Couvreur, P. 2005a. Colloidal carriers and blood-brain barrier (BBB) translocation: A way to deliver drugs to the brain? International Journal of Pharmaceutics, 298, 274–292.

Garcia-Garcia, E., Gil, S., Andrieux, K., Desmaele, D., Nicolas, V., Taran, F., Georgin, D., Andreux, J. P., Roux, F. & Couvreur, P. 2005b. A relevant in vitro rat model for the evaluation of blood-brain barrier translocation of nanoparticles. Cellular and Molecular Life Sciences, 62, 1400–1408.

Han, M. G. & Kim, S. 2009. Controlled degradation of poly(ethyl cyanoacrylate-co-methyl methacrylate) (PECA-co-PMMA) copolymers. Polymer, 50, 1270–1280.

He, M., Zhao, Z., Yin, L., Tang, C. & Yin, C. 2009. Hyaluronic acid coated poly(butyl cyanoacrylate) nanoparticles as anticancer drug carriers. International Journal of Pharmaceutics, 373, 165–173.

He, W., Jiang, X. & Zhang, Z.-R. 2008. Preparation and evaluation of poly(butyl cyanoacrylate) nanoparticles for oral delivery of thymopentin. Journal of Pharmaceutical Sciences, 97, 2250–2259.

Hekmatara, T., Gelperina, S., Vogel, V., Yang, S.-R. & Kreuter, J. 2009. Encapsulation of water-insoluble drugs in poly(butyl cyanoacrylate) nanoparticles. Journal of Nanoscience and Nanotechnology, 9, 5091–5098.

Hillaireau, H., Le Doan, T., Besnard, M., Chacun, H., Janin, J. & Couvreur, P. 2006. Encapsulation of antiviral nucleotide analogues azidothymidine-triphosphate and cidofovir in poly(iso-butylcyanoacrylate) nanocapsules. International Journal of Pharmaceutics, 324, 37–42.

Hillaireau, H., Le Doan, T., Chacun, H., Janin, J. & Couvreur, P. 2007. Encapsulation of mono- and oligo-nucleotides into aqueous-core nanocapsules in presence of various water-soluble polymers. International Journal of Pharmaceutics, 331, 148–152.

Hollock, G. G. 2001. Expanded applications for octyl-2-cyanoacrylate as a tissue adhesive. Annals of Plastic Surgery, 46, 185–189.

Jeremias, C. G. 1956. Monomeric a-cyanoacrylates. Us patent application.

Johnson, L. K., Killian, C. M. & Brookhart, M. 1995. New Pd(II)- and Ni(II)-Based Catalysts for Polymerization of Ethylene and a-Olefins. Journal of the American Chemical Society, 117, 6414–6415.

Johnston, D. S. & Pepper, D. C. 1981a. Polymerization via macrozwitterions. 1. Ethyl and butyl cyanoacrylates polymerized by triethyl and triphenylphosphines. Macromolecular Chemistry and Physics, 182, 393–406.

Johnston, D. S. & Pepper, D. C. 1981b. Polymerization via macrozwitterions. 2. Ethyl and butyl cyanoacrylates polymerized by pyridine and poly(vinylpyridine). Macromolecular Chemistry and Physics, 182, 407–20.

Johnston, D. S. & Pepper, D. C. 1981c. Polymerization via macrozwitterions. 3. Ethyl and butyl cyanoacrylates polymerized by benzyldimethyl, triethyl, and tribenzylamines. Macromolecular Chemistry and Physics, 182, 421–35.

Joyner, F. B. & Hawkins, G. F. 1955. a-Cyanoacrylates. Us patent application.

Joyner, F. B. & Shearer, N. H., Jr. 1956. Monomeric a-cyanoacrylates. Us patent application.

Kamachi, M., Liaw, D. J. & Nozakura, S. 1981. Solvent effect on radical polymerization of methyl methacrylate. Polymer Journal, 13, 41–50.

Kattan, J., Droz, J.P., Couvreur, P ., Marino, J.P., Boutan-Laroze, A., Rougier, P., Brault, P., ranckx, H., Grognet, J.M., Morge, X. & Sancho- Garnier, H. 1992. Phase I clinical trial and pharmacokinetic evaluation of doxorubicin carried by polyisohexylcyanoacrylate nanoparticles. Investigation of New drugs 10, 191–199.

Kim, H. R., Andrieux, K., Delomenie, C., Chacun, H., Appel, M., Desmaele, D., Taran, F., Georgin, D., Couvreur, P. & Taverna, M. 2007a. Analysis of plasma protein adsorption onto PEGylated nanoparticles by complementary methods: 2-DE, CE and protein Lab-on-chip((R)) system. Electrophoresis, 28, 2252–2261.

Kim, H. R., Andrieux, K., Gil, S., Taverna, M., Chacun, H., Desmaele, D., Taran, F., Georgin, D. & Couvreur, P. 2007b. Translocation of poly(ethylene glycol-co-hexadecyl)cyanoacrylate nanoparticles into rat brain endothelial cells: Role of apolipoproteins in receptor-mediated endocytosis. Biomacromolecules, 8, 793–799.

Kim, H. R., Gil, S., Andrieux, K., Nicolas, V., Appel, M., Chacun, H., Desmaele, D., Taran, F., Georgin, D. & Couvreur, P. 2007c. Low-density lipoprotein receptor-mediated endocytosis of PEGylated nanoparticles in rat brain endothelial cells. Cellular and Molecular Life Sciences, 64, 356–364.

King, M. E. & Kinney, A. Y. 1999. Tissue adhesives: a new method of wound repair. The Nurse Practitioner, 24, 66–73.

Kinsinger, J. B., Panchak, J. R., Kelso, R. L., Bartlett, J. S. & Graham, R. K. 1965. Methyl a-cyanoacrylate. II. Copolymerization studies. Journal of Applied Polymer Science, 9, 429–37.

Kisich, K. O., Gelperina, S., Higgins, M. P., Wilson, S., Shipulo, E., Oganesyan, E. & Heifets, L. 2007. Encapsulation of moxifloxacin within poly(butyl cyanoacrylate) nanoparticles enhances efficacy against intracellular Mycobacterium tuberculosis. International Journal of Pharmaceutics, 345, 154–162.

Kolb, H. C., Finn, M. G. & Sharpless, K. B. 2001. Click chemistry: diverse chemical function from a few good reactions. Angewandte Chemie, International Edition, 40, 2004–2021.

Krauel, K., Davies, N. M., Hook, S. & Rades, T. 2005. Using different structure types of microemulsions for the preparation of poly(alkylcyanoacrylate) nanoparticles by interfacial polymerization. Journal of Controlled Release, 106, 76–87.

Kryger, M. J., Ong, M. T., Odom, S. A., Sottos, N. R., White, S. R., Martinez, T. J. & Moore, J. S. 2010. Masked Cyanoacrylates Unveiled by Mechanical Force. Journal of the American Chemical Society, 132, 4558–4559.

Kulkarni, P. V., Roney, C. A., Antich, P. P., Bonte, F. J., Raghu, A. V. & Aminabhavi, T. M. 2010. Quinoline-n-butylcyanoacrylate-based nanoparticles for brain targeting for the diagnosis of Alzheimer's disease. Wiley Interdisciplinart Reviews: Nanomedicine and Nanobiotechnology, 2, 35–47.

Kurakhmaeva, K. B., Djindjikhashvili, I. A., Petrov, V. E., Balabanyan, V. U., Voronina, T. A., Trofimov, S. S., Kreuter, J., Gelperina, S., Begley, D. & Alyautdin, R. N. 2009. Brain targeting of nerve growth factor using poly(butyl cyanoacrylate) nanoparticles. Journal of Drug Targeting, 17, 564–574.

Kusonwiriyawong, C., Lipipun, V., Zhang, Q. & Ritthidej, G. C. 2008. Poly([alpha]-butyl cyanoacrylate) nanoparticles for intracellular delivery of protein: Physicochemical properties, cytotoxicity study and cellular uptake in dendritic cells. Journal of Controlled Release, 132, e6-e8.

Kwon, Y. & Kennedy, J. P. 2007a. Polymerizability, copolymerizability, and properties of cyanoacrylate-telechelic polyisobutylenes I: three-arm star cyanoacrylate-telechelic polyisobutylene. Polymers for Advanced Technologies, 18, 800–807.

Kwon, Y. & Kennedy, J. P. 2007b. Polymerizability, copolymerizability, and properties of cyanoacrylate-telechelic polyisobutylenes II: copolymerization of three-arm star cyanoacrylate-telechelic polyisobutylene with ethyl cyanoacrylate. Polymers for Advanced Technologies, 18, 808–813.

Labarre, D., Vauthier, C., Chauvierre, U., Petri, B., Muller, R. & Chehimi, M. M. 2005. Interactions of blood proteins with poly(isobutylcyanoacrylate) nanoparticles decorated with a polysaccharidic brush. Biomaterials, 26, 5075–5084.

Lambert, G., Fattal, E., Pinto-Alphandary, H., Gulik, A. & Couvreur, P. 2000. Polyisobutylcyanoacrylate nanocapsules containing an aqueous core as a novel colloidal carrier for the delivery of oligonucleotides. Pharmaceutical Research, 17, 707–714.

Le Droumaguet, B., Souguir, H., Brambilla, D., Verpillot, R., Nicolas, J., Taverna, M., Couvreur, P. & Andrieux, K. 2010. Selegiline-Functionalized, PEGylated Poly(alkyl cyanoacrylate) Nanoparticles to Target the Amyloid-β Peptide. International Journal of Pharmaceutics, submitted.

Lee, S.-H., Baek, H.-H., Kim, J. H. & Choi, S.-W. 2009. Core-shell poly(D,L-lactide-co-glycolide)/poly(ethyl 2-cyanoacrylate) microparticles with doxorubicin to reduce initial burst release. Macromolecular Research, 17, 1010–1014.

Lee, Y. J., Cheong, I. W., Yeum, J. H. & Jeong, Y. U. 2010. Synthesis of poly(N-isopropylacrylamide)/poly(ethyl-2-cyanoacrylate) composite hollow particles. Macromolecular Research, 18, 208–211.

Lescure, F., Zimmer, C., Roy, D. & Couvreur, P. 1992. Optimization of Polyalkylcyanoacrylate Nanoparticle Preparation – Influence of Sulfur-Dioxide and Ph on Nanoparticle Characteristics. Journal of Colloid and Interface Science, 154, 77–86.

Li, X., Dai, H., Tan, S., Zhang, X., Liu, H., Wang, Y., Zhao, N. & Xu, J. 2009. Facile preparation of poly(ethyl alpha -cyanoacrylate) superhydrophobic and gradient wetting surfaces. Journal of Colloid and Interface Science, 340, 93–97.

Li, X., Wang, Y., Zhang, L., Tan, S., Yu, X., Zhao, N., Chen, G. & Xu, J. 2010. Fabrication of honeycomb-patterned polyalkylcyanoacrylate films from monomer solution by breath figures method. Journal of Colloid and Interface Science, 350, 253–259.

Li, Y.-P., Pei, Y.-Y., Zhou, Z.-H., Zhang, X.-Y., Gu, Z.-H., Ding, J., Zhou, J.-J., Gao, X.-J. & Zhu, J.-H. 2001a. Stealth polycyanoacrylate nanoparticles as tumor necrosis factor-a carriers: pharmacokinetics and anti-tumor effects. Biological & Pharmaceutical Bulletin, 24, 662–665.

Li, Y. P., Pei, Y. Y., Zhou, Z. H., Zhang, X. Y., Gu, Z. H., Ding, J., Zhou, J. J. & Gao, X. J. 2001b. PEGylated polycyanoacrylate nanoparticles as tumor necrosis factor-a carriers. Journal of Controlled Release, 71, 287–296.

Liang, M., Davies, N. M. & Toth, I. 2008. Increasing entrapment of peptides within poly(alkyl cyanoacrylate) nanoparticles prepared from water-in-oil microemulsions by copolymerization. International Journal of Pharmaceutics, 362, 141–146.

Limouzin, C., Caviggia, A., Ganachaud, F. & Hemery, P. 2003. Anionic polymerization of n-butyl cyanoacrylate in emulsion and miniemulsion. Macromolecules, 36, 667–674.

Lin, X., Zhou, R., Qiao, Y., Jin, F., Zhai, Y., Xing, J., Deng, L. & Dong, A. 2008. Poly(ethylene glycol)/poly(ethyl cyanoacrylate) amphiphilic triblock copolymer nanoparticles as delivery vehicles for dexamethasone. Journal of Polymer Science, Part A: Polymer Chemistry, 46, 7809–7815.

Maitre, C., Ganachaud, F., Ferreira, O., Lutz, J. F., Paintoux, Y. & Hemery, P. 2000. Anionic polymerization of phenyl glycidyl ether in miniemulsion. Macromolecules, 33, 7730–7736.

Maksimenko, O., Pavlov, E., Toushov, E., Molin, A., Stukalov, Y., Prudskova, T., Feldman, V., Kreuter, J. & Gelperina, S. 2008. Radiation sterilisation of doxorubicin bound to poly(butyl cyanoacrylate) nanoparticles. International Journal of Pharmaceutics, 356, 325–332.

Malofsky, B. & Badejo, I. T. 2001. Transesterification method and catalysts for making cyanoacrylates from cyanoacetates. Wo patent application.

Marcovich, R., Williams, A. L., Rubin, M. A. & Wolf, J. S. 2001. Comparison of 2-octyl cyanoacrylate adhesive, fibrin glue, and suturing for wound closure in the porcine urinary tract. Urology, 57, 806–810.

Mayer, C. 2005. Nanocapsules as drug delivery systems. The International Journal of Artificial Organs, 28, 1163–1171.

Merolli, A., Marceddu, S., Rocchi, L. & Catalano, F. 2010. In vivo study of ethyl-2-cyanoacrylate applied in direct contact with nerves regenerating in a novel nerve-guide. Journal of Materials Science: Materials in Medicine, 21, 1979–1987.

Mulik, R., Mahadik, K. & Paradkar, A. 2009. Development of curcuminoids loaded poly(butyl) cyanoacrylate nanoparticles: Physicochemical characterization and stability study. European Journal of Pharmaceutical Sciences, 37, 395–404.

Mulik, R. S., Monkkonen, J., Juvonen, R. O., Mahadik, K. R. & Paradkar, A. R. 2010. ApoE3 Mediated Poly(butyl) Cyanoacrylate Nanoparticles Containing Curcumin: Study of Enhanced Activity of Curcumin against Beta Amyloid Induced Cytotoxicity Using In Vitro Cell Culture Model. Molecular Pharmaceutics, 7, 815–825.

Musyanovych, A. & Landfester, K. 2008. Synthesis of poly(butylcyanoacrylate) nanocapsules by interfacial polymerization in miniemulsions for the delivery of DNA molecules. Progress in Colloid & Polymer Science, 134, 120–127.

Nicolas, J., Bensaid, F., Desmaele, D., Grogna, M., Detrembleur, C., Andrieux, K. & Couvreur, P. 2008. Synthesis of Highly Functionalized Poly(alkyl cyanoacrylate) Nanoparticles by Means of Click Chemistry. Macromolecules, 41, 8418–8428.

Nicolas, J. & Couvreur, P. 2009. Synthesis of poly(alkyl cyanoacrylate)-based colloidal nanomedicines. Wiley Interdisciplinart Reviews: Nanomedicine and Nanobiotechnology, 1, 111–127.

Oowaki, H., Matsuda, S., Sakai, N., Ohta, T., Iwata, H., Sadato, A., Taki, W., Hashimoto, N. & Ikada, Y. 2000. Non-adhesive cyanoacrylate as an embolic material for endovascular neurosurgery. Biomaterials, 21, 1039–1046.

Otsu, T. & Yamada, B. 1967. Determination of Q,e parameters of methyl a-cyanoacrylate. Makromolekulare Chemie, 110, 297–9.

Pepper, D. C. 1980. Kinetics and mechanism of zwitterionic polymerizations of alkyl cyanoacrylates. Polymer Journal, 12, 629–37.

Pepper, D. C. & Ryan, B. 1983. Initiation processes in polymerizations of alkyl cyanoacrylates by tertiary amines: inhibition by strong acids. Makromolekulare Chemie, 184, 383–94.

Peracchia, M. T., Desmaële, D., Couvreur, P. & D'angelo, J. 1997a. Synthesis of a Novel Poly(MePEG cyanoacrylate-co-alkyl cyanoacrylate) Amphiphilic Copolymer for Nanoparticle Technology. Macromolecules, 30, 846–851.

Peracchia, M. T., Fattal, E., Desmaele, D., Besnard, M., Noel, J. P., Gomis, J. M., Appel, M., D'angelo, J. & Couvreur, P. 1999. Stealth (R) PEGylated polycyanoacrylate nanoparticles for intravenous administration and splenic targeting. Journal of Controlled Release, 60, 121–128.

Peracchia, M. T., Vauthier, C., Desmaele, D., Gulik, A., Dedieu, J.-C., Demoy, M., D'angelo, J. & Couvreur, P. 1998. PEGylated nanoparticles from a novel methoxypolyethylene glycol cyanoacrylate-hexadecyl cyanoacrylate amphiphilic copolymer. Pharmaceutical Research, 15, 550–556.

Peracchia, M. T., Vauthier, C., Passirani, C., Couvreur, P. & Labarre, D. 1997b. Complement consumption by polyethylene glycol in different conformations chemically coupled to poly(isobutyl 2-cyanoacrylate) nanoparticles. Life Sciences, 61, 749–761.

Peracchia, M. T., Vauthier, C., Popa, M., Puisieux, F. & Couvreur, P. 1997c. Investigation of the formation of sterically stabilized poly(ethylene glycol/isobutyl cyanoacrylate) nanoparticles by chemical grafting of polyethylene glycol during the polymerization of isobutyl cyanoacrylate. S.T.P. Pharma Sciences, 7, 513–520.

Peracchia, M. T., Vauthier, C., Puisieux, F. & Couvreur, P. 1997d. Development of sterically stabilized poly(isobutyl 2-cyanoacrylate) nanoparticles by chemical coupling of polyethylene glycol. Journal of Biomedical Materials Research, 34, 317–326.

Pierson, L.-A., Nicolas, J., Lalanne, M., Brambilla, D., Marsaud, V., Nicolas, V., Ball, R., Stanimirovic, D., Couvreur, P. & Andrieux, K. 2009. Formulation of didanosine prodrugs into PEGylated poly(alkyl cyanoacrylate) nanoparticles and uptake by brain endothelial cells. Journal of Nanoneuroscience, 1, 174–183.

Pollak, J. S. & White, R. I., Jr. 2001. The use of cyanoacrylate adhesives in peripheral embolization. Journal of Vascular and Interventional Radiology, 12, 907–913.

Puglisi, G., Fresta, M., Giammona, G. & Ventura, C. A. 1995. Influence of the Preparation Conditions on Poly(Ethylcyanoacrylate) Nanocapsule Formation. International Journal of Pharmaceutics, 125, 283–287.

Reece, T. B., Maxey, T. S. & Kron, I. L. 2001. A prospectus on tissue adhesives. American Journal of Surgery, 182, 40S–44S.

Sahadev, C. C., Krishna, C. C. G., Ravindranath, K., Reddy, N. J. & Sahadev, N. 2003. A process for preparation of high purity octyl and isoamyl cyanoacrylates. In patent application.

Seijo, B., Fattal, E., Roblottreupel, L. & Couvreur, P. 1990. Design of Nanoparticles of Less Than 50 Nm Diameter – Preparation, Characterization and Drug Loading. International Journal of Pharmaceutics, 62, 1–7.

Shipulo, E. V., Lyubimov, I. I., Maksimenko, O. O., Vanchugova, L. V., Oganesyan, E. A., Sveshnikov, P. G., Biketov, S. F., Severin, E. S., Heifets, L. B. & Gel'perina, S. E. 2008. Development of a nanosomal formulation of moxifloxacin based on poly(butyl-2-cyanoacrylate). Pharmaceutical Chemistry Journal, 42, 145–149.

Shirotake, S., Tomita, Y. U., Yoshizaki, S. & Okuda, K. 2008. Preparation of cyanoacrylate nanoparticles using monosaccharides or disaccharides. Chemical & Pharmaceutical Bulletin, 56, 137–138.

Skeist, I. & Miron, J. 1977. Adhesive composition. In: Bikales, N. (ed.) Encyclopedia of Polymer Science and Engineering. New-York: Wiley-Interscience.

Stella, B., Arpicco, S., Peracchia, M. T., Desmaële, D., Hoebeke, J., Renoir, M., D'angelo, J., Cattel, L. & Couvreur, P. 2000. Design of Folic Acid-Conjugated Nanoparticles for Drug Targeting. Journal of Pharmaceutical Sciences, 89, 1452–1464.

Stella, B., Marsaud, V., Arpicco, S., Geraud, G., Cattel, L., Couvreur, P. & Renoir, J.-M. 2007. Biological characterization of folic acid-conjugated poly(H2NPEGCA-co-HDCA) nanoparticles in cellular models. Journal of Drug Targeting, 15, 146–153.

Stolnik, S., Illum, L. & Davis, S. S. 1995. Long Circulating Microparticulate Drug Carriers. Advanced Drug Delivery Reviews, 16, 195–214.

Storm, G., Belliot, S. O., Daemen, T. & Lasic, D. D. 1995. Surface Modification of Nanoparticles to Oppose Uptake by the Mononuclear Phagocyte System. Advanced Drug Delivery Reviews, 17, 31–48.

Sullivan, C. O. & Birkinshaw, C. 2004. In vitro degradation of insulin-loaded poly(n-butylcyanoacrylate) nanoparticles. Biomaterials, 25, 4375–4382.

Tseng, Y. C., Hyon, S. H. & Ikada, Y. 1990. Modification of synthesis and investigation of properties for 2-cyanoacrylates. Biomaterials, 11, 73–9.

Vansnick, L., Couvreur, P., Christiaensleyh, D. & Roland, M. 1985. Molecular-Weights of Free and Drug-Loaded Nanoparticles. Pharmaceutical Research, 36–41.

Vauthier, C., Dubernet, C., Fattal, E., Pinto-Alphandary, H. & Couvreur, P. 2003. Poly(alkylcyanoacrylates) as biodegradable materials for biomedical applications. Advanced Drug Delivery Reviews, 55, 519–548.

Vauthier, C., Labarre, D. & Ponchel, G. 2007. Design aspects of poly(alkylcyanoacrylate) nanoparticles for drug delivery. Journal of Drug Targeting, 15, 641–663.

Vauthier, C., Persson, B., Lindner, P. & Cabane, B. 2011. Protein adsorption and complement activation for di-block copolymer nanoparticles. Biomaterials, 32, 1646–1656.

Vranckx, H., Demoustier, M. & Deleers, M. 1996. A new nanocapsule formulation with hydrophilic core: Application to the oral administration of salmon calcitonin in rats. European Journal of Pharmaceutics and Biopharmaceutics, 42, 345–347.

Watnasirichaikul, S., Rades, T., Tucker, I. G. & Davies, N. M. 2002. Effects of formulation variables on characteristics of poly (ethylcyanoacrylate) nanocapsules prepared from w/o microemulsions. International Journal of Pharmaceutics, 235, 237–246.

Weiss, C. K., Ziener, U. & Landfester, K. 2007. A route to nonfunctionalized and functionalized poly(n-butylcyanoacrylate) nanoparticles: preparation in miniemulsion. Macromolecules, 40, 928–938.

Wilson, B., Samanta, M. K., Santhi, K., Kumar, K. P. S., Paramakrishnan, N. & Suresh, B. 2008. Poly(n-butyl cyanoacrylate) nanoparticles coated with Polysorbate 80 for the targeted delivery of rivastigmine into the brain to treat Alzheimer's disease. Brain Research, 1200, 159–168.

Wohlgemuth, M. & Mayer, C. 2003. Pulsed field gradient NMR on polybutylcyanoacrylate nanocapsules. Journal of Colloid and Interface Science, 260, 324–331.

Xing, J., Deng, L., Li, J. & Dong, A. 2009. Amphiphilic poly{[alpha -maleic anhydride-w-methoxy-poly(ethylene glycol)]-co-(ethyl cyanoacrylate)} graft copolymer nanoparticles as carriers for transdermal drug delivery. International Journal of Nanomedicine, 4, 227–232.

Yadav, J. S., Reddy, B. V. S., Basak, A. K., Visali, B., Narsaiah, A. V. & Nagaiah, K. 2004. Phosphane-catalyzed knoevenagel condensation: A facile synthesis of alpha-cyanoacrylates and alpha-cyanoacrylonitriles. European Journal of Organic Chemistry, 2004, 546–551.

Yamada, B., Yoshioka, M. & Otsu, T. 1983. Determination of absolute rate constants for radical polymerization and copolymerization of ethyl a-cyanoacrylate in the presence of effective inhibitors against anionic polymerization. Makromolekulare Chemie, 184, 1025–33.

Yang, S. C., Ge, H. X., Hu, Y., Jiang, X. Q. & Yang, C. Z. 2000. Formation of positively charged poly(butyl cyanoacrylate) nanoparticles stabilized with chitosan. Colloid and Polymer Science, 278, 285–292.

Yordanov, G., Bedzhova, Z. & Dushkin, C. 2009. Poly(ethyl cyanoacrylate) nanoparticles loaded with two organic fluorophores: preparation and physicochemical characterization. Nanoscience & Nanotechnology, 9, 65–67.

Zhang, Y., Zhu, S., Yin, L., Qian, F., Tang, C. & Yin, C. 2008. Preparation, characterization and biocompatibility of poly(ethylene glycol)-poly(n-butyl cyanoacrylate) nanocapsules with oil core via miniemulsion polymerization. European Polymer Journal, 44, 1654–1661.

Zhou, Q., Sun, X., Zeng, L., Liu, J. & Zhang, Z. 2009. A randomized multicenter phase II clinical trial of mitoxantrone-loaded nanoparticles in the treatment of 108 patients with unresected hepatocellular carcinoma. Nanomedicine, 5, 419–423.

Amphiphilic Block Copolymer Based Nanocarriers for Drug and Gene Delivery

Xiao-Bing Xiong and Afsaneh Lavasanifar

Abstract The use of amphiphilic block copolymers (ABC)s in experimental medicine and pharmaceutical sciences has a long history and is expecting rapid development. Poly(ethylene oxide)–*block*–poly(amino acid) and poly(ethylene oxide)–*block*–polyesters represented the most important two types of ABCs for development of nanocarriers for drug/gene delivery. In this chapter, we provided an update on several chemical strategies used to enhance the properties of nanoscopic core/shell structures formed from self assembly of ABCs, namely polymeric micelles. Versatility of polymer chemistry in ABCs provides unique opportunities for tailoring polymeric micelles for optimal properties in gene and drug delivery. Chemical modification of the polymer structure in the micellar core through introduction of hydrophobic or charged moieties, conjugation of drug compatible groups, core cross-linking has led to enhanced stability for the micellar structure and sustained or pH-sensitive drug release. The modification of polymeric micellar surface with specific ligands (carbohydrates, peptides, antibodies) has shown benefits in enhancing the recognition of carrier by selective cells leading to improved drug and gene delivery to the desired targets. Research in drug delivery by polymeric vesicles is still in its infancy, but a similar principle on the importance and benefit of chemical flexibility of block copolymers in improving the delivery properties of polymeric vesicles can also be envisioned. The demanding challenge of the future research in this field is to find the right carrier architecture and optimum polymer chemistry that can improve the delivery of sophisticated and complex therapeutic agents (e.g., poorly soluble drugs, proteins and genes) to their cellular and intracellular targets.

Keywords Polymeric micelles • Drug delivery • Nanocarriers • Amphiphilic block copolymers • Gene delivery • Drug targeting

X.-B. Xiong and A. Lavasanifar (✉)
Faculty of Pharmacy and Pharmaceutical Sciences, University of Alberta,
Edmonton, AB T6G 2N8, Canada
e-mail: alavasanifar@pharmacy.ualberta.ca

A. Prokop (ed.), *Intracellular Delivery: Fundamentals and Applications*, Fundamental Biomedical Technologies 5, DOI 10.1007/978-94-007-1248-5_10,

Abbreviations

AFM	atomic force microscopy
APCs	antigen presenting cells
AEDP	3-[(2-aminoethyl) dithio] propionic acid
bPEI-Chol	cholesteryl modified bPEI
bPEI-PA	palmitic acid modified bPEI
bPEIs	branched PEIs
BSA	bovine serum albumin
C2C12	murine myoblast
CaP	calcium phosphate
CDs	cyclodextrins
α-CD	α-cyclodextrin
β-CD	β-cyclodextrin
βCDPs	β-cyclodextrin-containing polymers
CDPRs	cyclodextrins-based polypseudorotaxanes
CFTR	cystic fibrosis transmembrane conductance regulator
CHO	chinese hamster ovarian
CL	cationic liposomes
CLIP-1	rac-[(2,3-dioctadecyloxypropyl)(2-hydroxyethyl)]-dimethylammonium chloride
CLIP-6	rac-[2(2,3-dihexadecyloxypropyl-oxymethyloxy)ethyl] trimethylammonium bromide
CLIP-9	rac-[2(2,3-dihexadecyloxypropyl-oxysuccinyloxy) ethyl]-trimethylammonium
CLSM	confocal laser scanning microscopy
CNT	carbon nanotube
CPNP-DNA	DNA-loaded calcium phosphate nanoparticles
CS	chitosan
DA-CS	deoxycholic acid modified chitosan
DC-Chol	3β-[N-(dimethylaminoethane)carbamoyl]cholesterol
DEAE-dextran	diethylaminoethyl-dextran
DEX-SP	spermine-conjugated dextran
DEX	dexamethasone
DMAEMA	2-(dimethylamino)ethyl methacrylate
DMRIE	1,2-dimyristyloxypropyl-3-dimethyl-hydroxyethyl ammonium bromide
DOGS	dioctadecylamidoglicylspermine
DOPE	dioleoylphosphatidylethanolamine
DOPC	dioleoylphosphatidyl choline
DORI	1,2-dioleoyl-3-dimethyl-hydroxyethyl ammonium bromide
DORIE	1,2-dioleyloxypropyl-3-dimethyl-hydroxyethyalm-3-dimethyl-hydroxyethyl ammonium bromide
DORIE-HB	1,2-dioleyloxypropyl-3-dimethyl-hydroxybutyl ammonium bromide

DORIE-HP	1,2-dioleyloxypropyl-3-dimethyl-hydroxypropyl ammonium bromide
DORIE-Hpe	1,2-dioleyloxypropyl-3-dimethyl-hydroxypentyl ammonium bromide
DOSPA	2,3-dioleoyloxy-N-[2(sperminecarboxamido)ethyl]-N,N-dimethyl-1-propanammonium trifluoroacetate
DOTAP	N-[1-(2,3-dioleoyloxy)propyl]-N,N,N-trimethylammonium chloride
DOTMA	1,2-dioleyloxypropyl-3-trimethyl ammonium bromide
DOX	doxorubicin
DPRIE	1,2-dipalmityloxypropyl-3-dimethyl-hydroxyethyl ammonium bromide
DSP	dithiobis(succinimidylpropionate)
DSRIE	1,2-disteryloxypropyl-3-dimethyl-hydroxyethyl ammonium bromide
DT	diphtheria toxin
DTBP	dimethyl-3,3′-dithiobispropionimidate·2HCl
EDAC	1-Ethyl-3-(3-dimethylaminopropyl)carbodiimide
F-PE	partially fluorinated glycerophosphoethanolamine
FRET	fluorescence resonance energy transfer (or Förster resonance energy transfer)
GA-CS	lactobionic acid bearing galactose conjugated chitosan
gal-PEI	galactosylated PEI
GC	glucocorticoid
GMP	2-deoxyguanosine-5-monophosphate
GFP	green fluorescent protein
HEK293	human embryonic kidney
His-PLL	histidylated polylysine
HLA-B7	histocompatibility complex protein
HMW-PEI	high molecular weight PEI
HP-β-CD	(2-hydroxypropyl)-β-cyclodextrin
HP-γ-CD	(2-hydroxypropyl)-γ-cyclodextrin
Huh7 cells	human hepatoma cell lines
KanaChol	3b–[6′-kanamycin-carbamoyl]cholesterol
LDH	layered double hydroxides
LMW-PEI	low molecular weight PEI
lPEI	linear PEI
MCD	methyl-β-cyclodextrin
MDAMB-231	human breast carcinoma cells
MEND	multifunctional envelope-type nano device
MnMEIO	manganese-doped magnetism-engineered iron oxide
MNPs	magnetic nanoparticles
mRNA	microRNA
MRI	magnetic resonance imaging
MSNs	mesoporous silica nanoparticles
MTT	3-(4,5-dimethylthiazol-2-yl)-2,5-diphenyltetrazolium bromide
MVL5	N1-[2-1S-1-[3-aminopropylamino]-4-[di-3-aminopropylamino] butylcarboxa midoethyl]-3, 4-dioleyloxybenzamide

MWCNTs	multiwalled CNTs
NIRF	near-infrared in vivo optical imaging
NHS	N-hydroxysuccinimide
NLS	nuclear localization signal
NOD	nonobese diabetic
NPCs	nuclear pore complexes
n-PAE	network-type poly(amino ester)
NSAIDs	nonsteroidal anti-inflammatory drugs
ODNs	oligonucleotides
PA	palmitic acid
PAE	poly(β-amino esters)
PAGA	poly[α-(4-aminobutyl)-L-glycolic acid]
PAMAM	polyamidoamine
PAMAM-α-CD	α-CD conjugated PAMAM
PAsp(DET)	polyaspartamide derivative bearing 1,2-diaminoethane side chains
PBLA	poly(β-benzyl L-aspartate)
PCI	photochemical internalization
PCs	triester phosphatidylcholines
PDMAEMA	poly[2-(dimethylamino)ethyl methacrylate]
pDNA	plasmid DNA
PEG	polyethylene glycol
PEG-PAA	PEG-*b*-poly(aspartic acid)
PEG-*b*-PLL	poly(ethylene glycol)-*b*-poly(L-lysine)
PEG-Ada	1-adamantanemethylamine (Ada) conjugated PEG
PEI	polyethylenimine
PEI25	branched PEI of 25 kDa
PEI-CDs	cyclodextrin modified PEIs
PEO-PPO-PEO	poly(ethylene oxide)-*b*-poly(propylene oxide)-*b*-poly(ethylene oxide)
PHP	poly(4-hydroxy-L-proline)
PHPMA	poly[N-(2-hydroxypropyl)methacrylamide]
PKA	protein kinase
PLA	polylactide
PLL	poly(L-lysine)
PLL-PA	palmitic acid conjugated PLL
PLGA	poly(lactide-co-glycolide)
PMA	phorbol myristate acetate
PMDS	poly(N-methyldietheneamine sebacate)
PNIPAm	poly(N-isopropylacrylamide)
POD	PEG-diorthoester-lipid conjugate
POEs	poly(ortho esters)
PPA	polyphosphoramidate
PPA-EA	polyphosphoramidate with ethylenediamine
PPA-SP	polyphosphoramidate bearing a spermidine side chain
PPE-EA	poly(2-aminoethyl propylene phosphate)

PPG	polypropylene glycol
PPPs	polyphosphazenes
PPP-DMAE	polyphosphazenes with 2-dimethylaminoethanol side groups
PPP-DMAEA	polyphosphazenes with 2-dimethylaminoethylamine side groups
PSI	poly(L-succinimide)
QDs	quantum dots
RES	reticuloendothelial system
RGD	Arg-Gly-Asp
SANS	small angle neutron scattering
SERS	surface enhanced Raman spectroscopy
SPIO	superparamagnetic iron oxide
SWCNTs	single-walled CNTs
TM-Bz-CS	quaternized N-(4-N,N-dimethylaminobenzyl) chitosan
TM-CS	trimethylated CS oligomers

1 Introduction

The use of amphiphilic block copolymers (ABC)s in experimental medicine and pharmaceutical sciences has a long history and is experiencing rapid development. ABCs have been used as safer replacements for surfactants in the solubilization of poorly soluble drugs, as stabilizing agents in the formulation of coarse and colloidal dispersions, as gels providing depot or bioavailable formulations and core/shell association colloids for nano-scale drug and gene delivery (Lavasanifar et al. 2002a; Lemieux et al. 2000; Osada and Kataoka 2006; Bijsterbosch et al. 1999). The rapid development in the application of ABCs is primarily due to several degrees of freedom in their chemistry that may be used to design specific carriers for specific delivery applications. For instance, the size of both the hydrophilic and the hydrophobic section can be varied at will; the molecular weight of the polymer can be changed within a wide range while maintaining constant hydrophilic-lipophilic balances; the properties and function of ABCs at interface, e.g. oil-water, can be controlled through changes in polymer structure, and more importantly, both hydrophilic and hydrophobic section can be chemically engineered towards specific application. In addition to their application in the design of nano-delivery systems, ABCs are shown to have biological effects, themselves (Kabanov et al. 2003; Zastre et al. 2004; Todd et al. 1998).

The aim of this chapter is to provide an overview on the pharmaceutical application of ABC based nanostructures. The emphasis will be placed on poly(ethylene oxide)–*block*–poly(amino acids) and poly(ethylene oxide)–*block*–polyester based polymeric micelles. Recent advancements in polymeric micellar drug/gene targeting will be highlighted and related nanostructures formed from ABCs will be briefly discussed.

2 Polymeric Micelles Drug Delivery

Polymeric micelles are core/shell nano-delivery systems formed through self assembly of ABCs in an aqueous environment (Fig. 1). Polymeric micelles have been widely used for solubilization of poorly water soluble drugs, depot drug release and targeted drug delivery (Aliabadi and Lavasanifar 2006; Mahmud et al. 2007; Sutton et al. 2007). They have been the subject of several extensive reviews in recent years (Croy and Kwon 2006; Lavasanifar et al. 2002b; Xiong et al. 2006; Osada and Kataoka 2006).

Among different structures, micelles consisting of poly(ethylene oxide)-poly(propylene oxide) PEO–PPO, PEO–poly(ester)s and PEO–poly(L-amino acid)s (PEO–PLAA) are used more extensively (Aliabadi and Lavasanifar 2006; Sutton et al. 2007; Kwon and Forrest 2006). The widespread application of polymeric micelles in drug delivery is linked to their unique core-shell architecture in which the hydrophobic core creates a space for the encapsulation of hydrophobic drugs, proteins or DNA; and the hydrophilic shell masks the hydrophobic core from the biological milieu. Since hydrophilic shell minimizes protein adsorption on micelles, polymeric micelles have the ability to evade non-specific capture by the reticuloendothelial system (RES). The size of polymeric micelles is also above the threshold of kidney filtration. Thus, polymeric micelles can lodge several types of molecules, demonstrate prolonged circulation in blood (Nishiyama and Kataoka 2006; Kwon and Forrest 2006; Aliabadi and Lavasanifar 2006; Kataoka et al. 1993) (Fig. 2), and even get a chance for accumulation in sites with leaky vasculature (e.g., solid tumors and inflammation sites) because of the enhanced permeability and retention (EPR) effect (Kwon and Forrest 2006; Aliabadi and Lavasanifar 2006) (Fig. 2). New blood vessels formed in tumor have large gaps in their endothelium. Some tumor vessels even have a defective cellular lining composed of disorganized, loosely connected, branched, overlapping, or spouting endothelial cells (Maeda 2001; Maeda and Matsumura 1989; Muggia 1999). The presence of leaky vasculature at

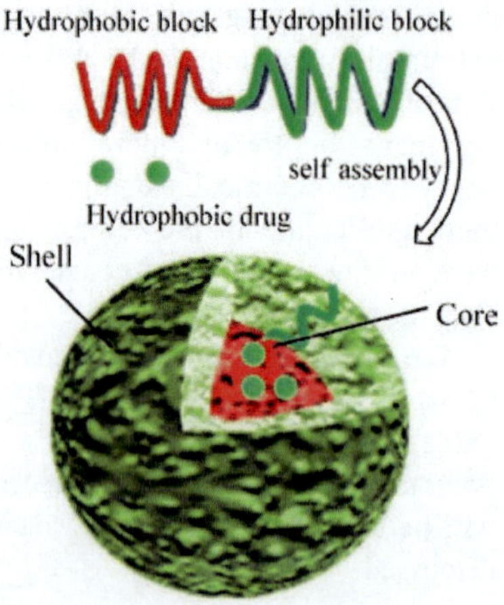

Fig. 1 Schematic representation of the self assembly of amphiphilic block copolymers into polymeric micelles

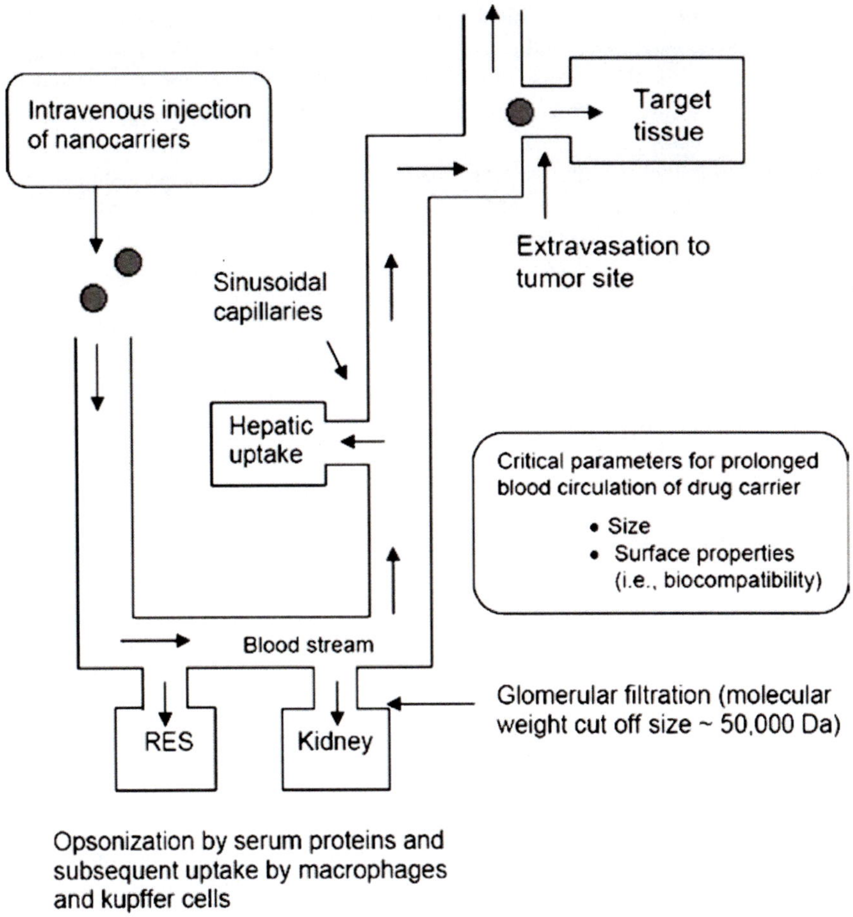

Fig. 2 Proposed fate of polymeric micelles after intravenous administration (Reproduced from Nishiyama and Kataoka 2006 with permission)

tumor facilitates the extravasation of nanocarriers. The permeated nanocarrier usually gets trapped in tumor because the lymphatic system that drains fluids out of other organs is dysfunctional in tumors. This phenomenon, known as the EPR effect, is believed to be the reason for the passive accumulation of polymeric micelles and stealth liposomes in solid tumors (Matsumura et al. 2004).

2.1 *First Generation Polymeric Micelles for Drug Delivery*

Polymeric micellar systems developed so far can be categorized to three classes: micelle-forming polymer-drug conjugates, micellar nano-containers, and polyion complex micelles (PIC)s. In micelle-forming polymer-drug conjugates, drug is incorporated and stabilized within the carrier through formation of chemical bonds between the functional group(s) of the polymeric backbone and the drug (Aliabadi and Lavasanifar 2006). When solubilization of drugs in the polymeric micelles is achieved via hydrophobic interactions or hydrogen bonds between the core-forming block and drug, the resulting system is called polymeric micellar nano-containers. Polymeric micellar nano-containers may be prepared by the direct addition and incubation of drug with block copolymers in an aqueous environment, only if the block copolymer and the drug are water soluble. The method, however, is not very efficient in terms of drug-loading levels and not feasible for most block copolymer/drug structures. Instead, physical incorporation of drugs into polymeric micellar nano-containers is usually accomplished through dialysis, oil/water emulsion, solvent evaporation, co-solvent evaporation, or freeze drying method (Aliabadi and Lavasanifar 2006). Finally, in (PIC)s, drug incorporation is promoted through electrostatic interactions between oppositely charged polymer/drug combinations. Neutralization of the charge on the core-forming block will trigger self assembly of the PIC and further stabilization of the complex within the hydrophobic environment of the micellar core (Aliabadi and Lavasanifar 2006). PICs have been investigated for the delivery of different therapeutic moieties that carry charge (drugs (Nishiyama et al. 1999, 2003), peptides (Yuan et al. 2005) and DNA (Yuan et al. 2005; Wakebayashi et al. 2004a, b; Itaka et al. 2003)). Drug release from polymeric micellar drug conjugates, nano-containers and PIC micelles is governed by different mechanisms (Fig. 3).

To date, seven polymeric micellar formulations have reached preclinical and clinical trials (Table 1) (Aliabadi and Lavasanifar 2006; Mahmud et al. 2007; Sutton et al. 2007; Matsumura 2008).

A micellar nano-container of doxorubicin (DOX) composed of PEO–poly (L-aspartate) with conjugated DOX (PEO–P(Asp–DOX)) containing physically encapsulated DOX, namely NK911 (Fig. 4), is one of the few polymeric micellar formulations with a favorable pharmacokinetics for passive drug targeting (Danson et al. 2004). A PEO–poly(propylene oxide)–PEO (PEO–PPO–PEO), i.e., Pluronic® micellar DOX, namely SP1049C, has shown good encapsulation, but similar pharmacokinetic profile to that free DOX in human (Hamaguchi et al. 2005b; Kim et al. 2004).

Development of polymeric micellar nano-containers for the delivery of paclitaxel (PXT) has also been pursued. Both PEO–poly(D-lactide) (PEO–PDLA) and PEO–p(L-Asp) micellar formulations of PXT were successful in increasing the water solubility of PXT (Hamaguchi et al. 2005b). However, except for the NK105 (the PEO–poly(4-phenyl-1-butanoate) (L-aspartamide) formulation) (Fig. 5), other polymeric micelles failed to show any benefit over PXT commercial formulation, Taxol®, in passive PTX targeting (Zhang et al. 2005b).

Conjugation of drugs to PEO–poly(ester)s has been accomplished by functionlization of the poly(ester) end, followed by reaction with drugs (Fig. 6) (Yoo and Park 2001; Hirano et al. 1979). In this approach, only one drug molecule is introduced for each copolymer molecule. In our group, DOX was conjugated to the PEO–poly(ε-caprolactone) (PEO–PCL) backbone by the reaction between

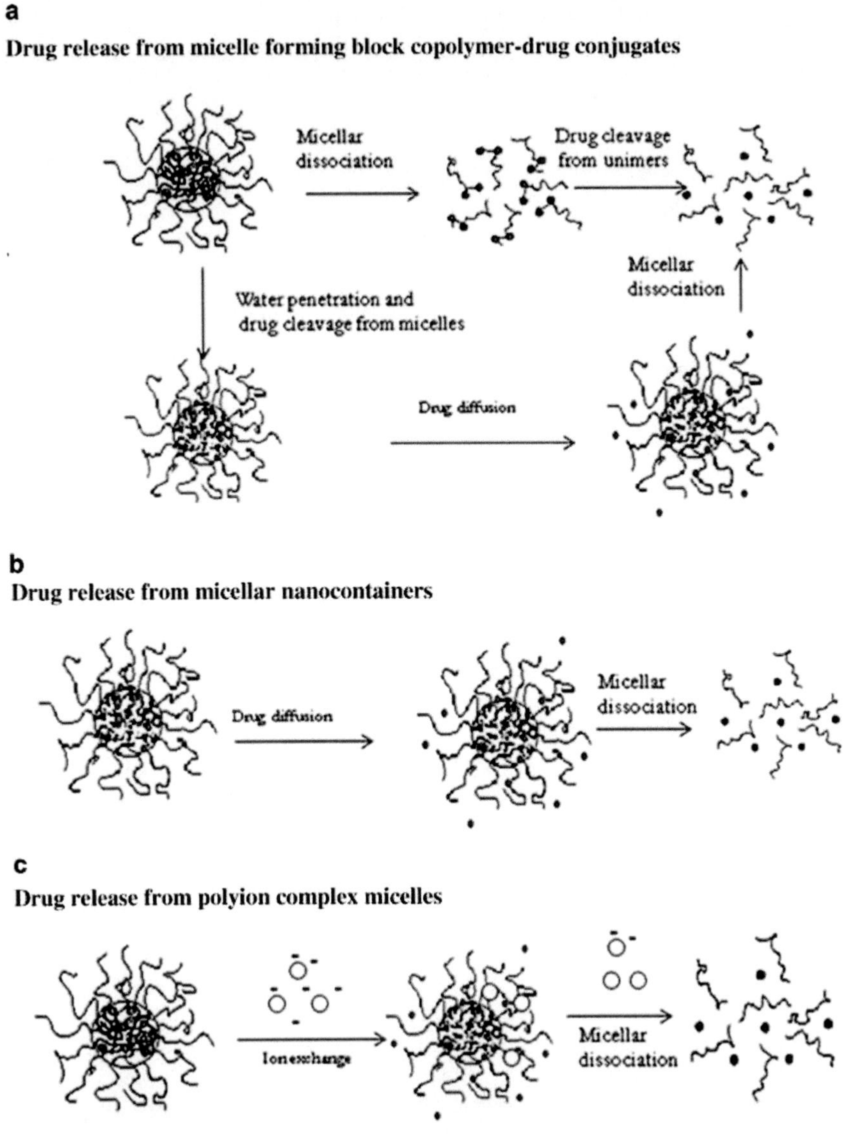

Fig. 3 General scheme for the release of incorporated molecule from polymeric micellar drug conjugates; micellar nano-containers and polyion complex (PIC) micelles (Adopted from Aliabadi and Lavasanifar 2006 with permission)

the –COOH side groups in PEO–poly(α-carboxyl-ε-caprolactone) (PEO–PCCL) and amine group of DOX. At early studies the degree of DOX conjugation reached 20% (molar ratios).

PEO–PLAA block copolymers have also been extensively used for chemical conjugation of drugs to the core-forming block via ester, amide, hydrazone, or disulfide

Table 1 Polymeric micellar delivery systems in clinical trials

Trade name	Polymer category	Incorporated drug	Progress phase	References
NK911	PEO–PLAA	DOX/chemically conjugated and physically loaded	II	(Matsumura 2008; Matsumura et al. 2004)
NK105	PEO–PLAA	Paclitaxel/physically loaded	II	(Hamaguchi et al. 2005a)
NC-6004	PEO–PLAA	Cisplatine/to form complex	I/II[a]	(Uchino et al. 2005b)
SP1049C	Pluronic	DOX/physically loaded	III[a]	(Sharma et al. 2008)
PAXCEED®	PEO–poly(ester)	Paclitaxel/physically loaded	I/II[a]	(Zhang et al. 2005a)
Genexol®-PM	PEO–poly(ester)	Paclitaxel/physically loaded	II/III/IV[a]	(Kim et al. 2007; Lee et al. 2008)
NK012	PEO–p(Glu)	7-Ethyl-10-hydroxy camptothecin/ chemically conjugated	I/II[a]	(Koizumi et al. 2006)

[a] Updated from http://clinicaltrials.gov as of August 2010

bonds (Fig. 7). Ringsdorf et al. reported on the preparation of micelle-forming conjugates of cyclophosphamide (CP) sulfide and PEO–poly(L-lysine) (PEO–poly(L-Lys)) (Yokoyama et al. 1991; Kataoka et al. 1993). Simultaneous conjugation of fatty acids to the polymeric back bone was used to increase the thermodynamic stability of this system. This formulation was found efficient in the stabilization of active CP metabolite, *in vivo*, and caused a fivefold increase in the life span of L1210 tumor bearing mice even at a reduced CP-equivalent dose. Block copolymer drug conjugates of PEO–P(L-Asp–DOX) have been developed by Kataoka's group (Bae et al. 2003a). DOX was covalently conjugated to the side chain of the P(L-Asp) segment by an amide bond between the carboxylic group in P(L-Asp) and the primary amine group of DOX. However, the amide bound was too stable and did not provide efficient drug release. In further studies, DOX was conjugated to the P(L-Asp) through a hydrazone linker that is stable under physiological pH, but cleavable under acidic pH of endosomes/lysosomes (Bae et al. 2005b). The PEO–p(Asp-Hyd–DOX) showed increased maximum tolerable dose, and increased blood and tumor DOX levels in C_{26} tumor bearing mice (Nishiyama et al. 1999, 2003). In addition, the pH sensitive micellar DOX conjugate showed better therapeutic efficacy than free drug at corresponding maximum tolerable doses.

Incorporation of cisplatin (CDDP) into the PEO–P(L-Asp) and PEO–P(L-Glutamine) (PEO–P(L-Glu)) micelles to form polymer-drug complex has been achieved (Fig. 8) (Uchino et al. 2005a). Release of CDDP from the complexes is triggered in the presence of high concentrations of salt in medium, which will replace the complexed drug. The PEO–P(L-Glu) formulation of CDDP is currently in clinical trials in Japan under the name of NC-6004 (Table 1) (Ward et al. 2002; Lee et al. 2005b; Katayose and Kataoka 1997, 1998).

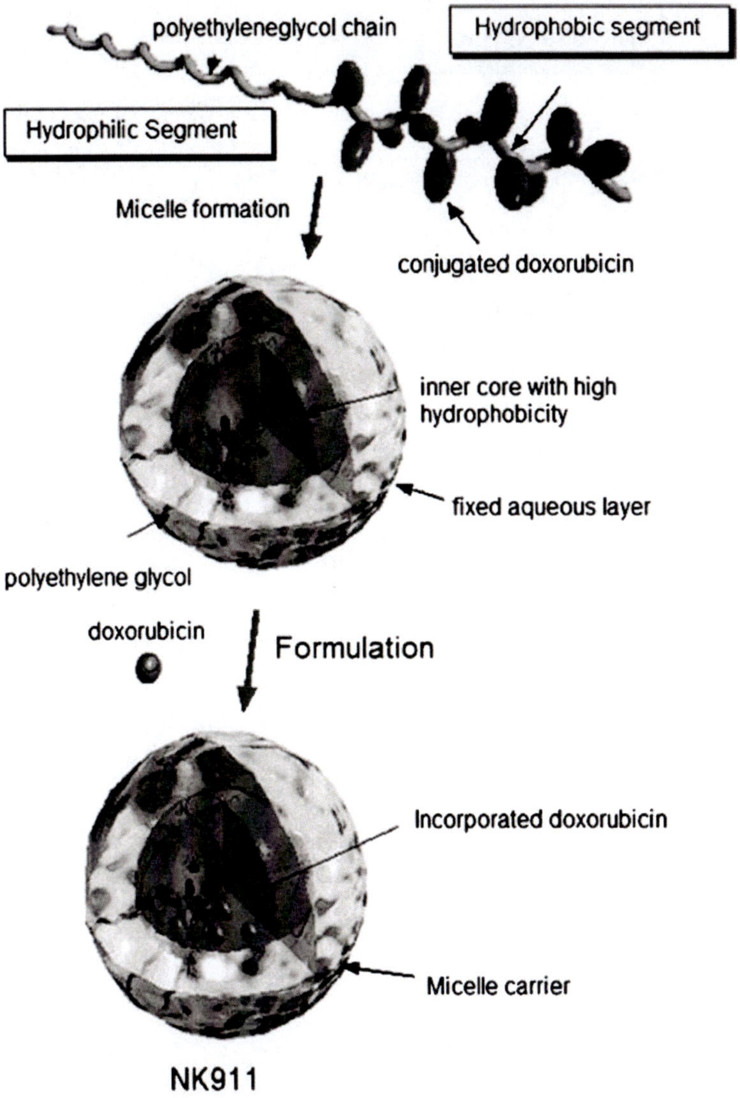

Fig. 4 Schematic model of NK911 (Reproduced from Nakanishi et al. 2001 with permission)

2.2 *Stimuli-Responsive Drug Delivery by Polymeric Micelles*

Response to ultrasound – Application of ultrasound as external stimuli that can trigger drug release from polymeric micelles and/or enhance cellular internalization of both free and encapsulated drug at the diseased site has been pursued (Gao et al. 2004, 2005b; Husseini et al. 2005; Rapoport 2004). Exposure to ultrasound leads to faster drug release from micelles, possibly because of increased drug diffusion

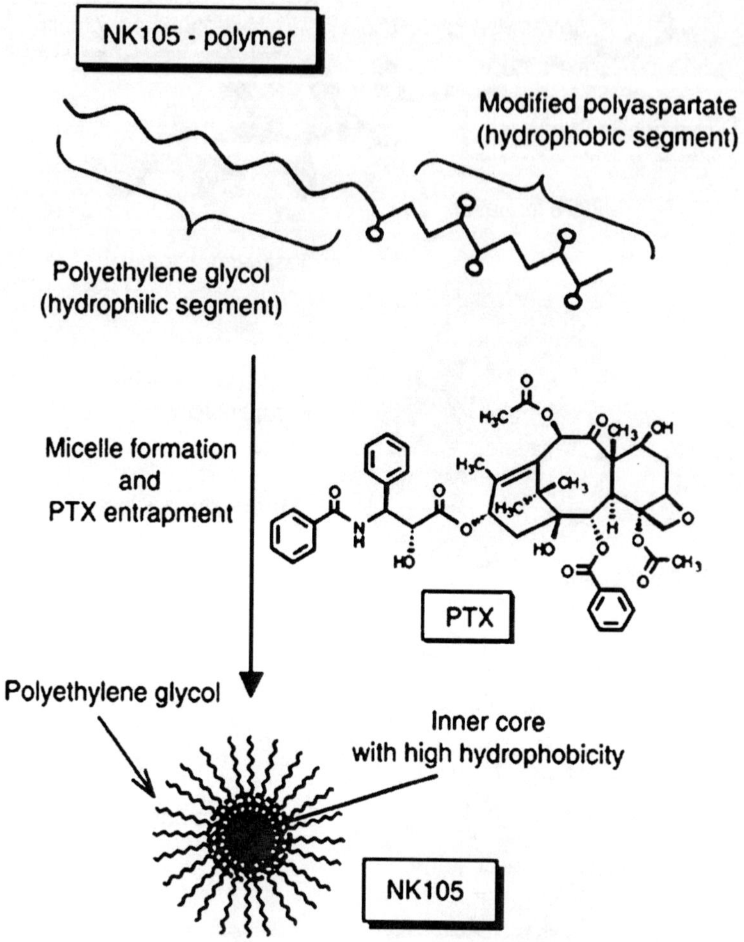

Fig. 5 Schematic model and chemical structure of NK105 (Reproduced from Hamaguchi et al. 2005a with permission)

rate from the micellar core, ultrasound-induced rapid micellar dissociation and re-association, or ultrasound-induced formation of transient collapsed cavitation (Fig. 9) (Pitt et al. 2004; Husseini et al. 2002; Munshi et al. 1997). Application of ultrasound was shown to increase the cellular internalization and cytotoxicity of both free and Pluronic® P105 micellar formulation of DOX against Human Leukemia HL-60 cells (Husseini et al. 2000; Rapoport et al. 2002). Rapoport et al. demonstrated that the cellular uptake and nuclear association of encapsulated DOX to be substantially enhanced after exposure to ultrasound leading to an increased cytotoxicity of DOX loaded Pluronic® micelles in resistant A2780/ADR cells (Marin et al. 2001; Rapoport 2004; Rapoport et al. 2004; Nelson et al. 2002). Pitt et al. reported on the *in vivo* antitumor efficacy of Plurogel formulations of DOX

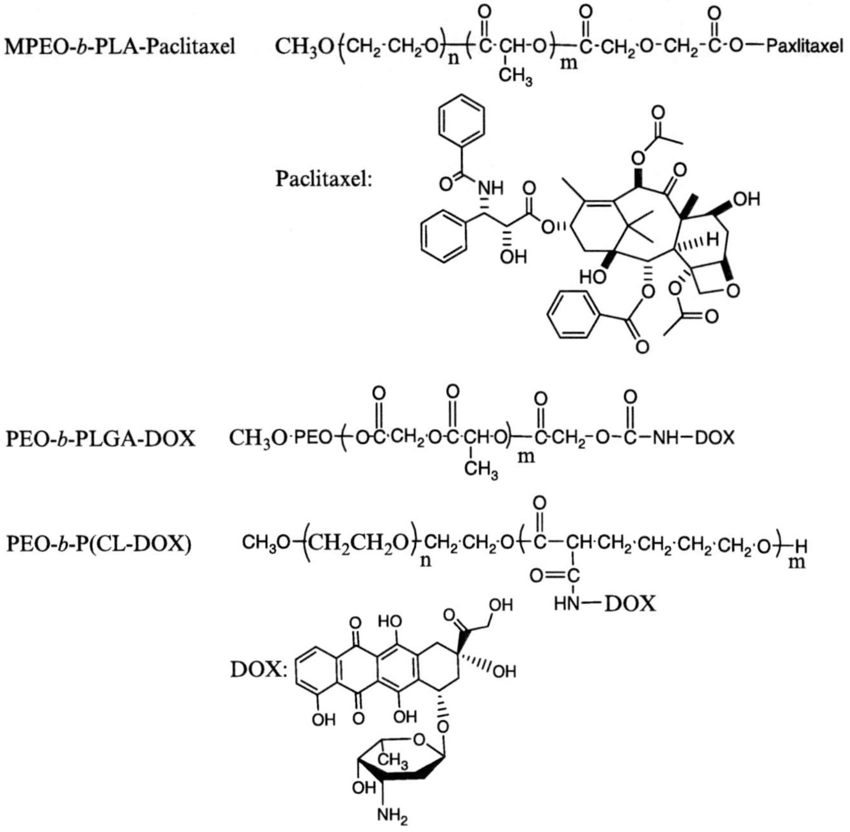

Fig. 6 Example structures of PEO–polyester-anticancer drug conjugates

injected intravenously in rats bearing solid DHD/K12/TRb colorectal tumors on both legs where the tumor on one side was exposed to ultrasound waves. Application of low-frequency ultrasound significantly reduced tumor size (Pitt et al. 2004).

pH sensitive polymeric micelles – Various polymeric micelles have been designed that can respond to the acidic pH and show rapid drug release, drug cleavage or micellar dissociation (Torchilin 2006; Bae et al. 2003b). The pH of the extracellular space of most solid tumors is below 7.2, while the pH of normal blood pH is 7.4 (Yatvin et al. 1984). The pH of tumor interstitial fluid can further be depressed to values as low as 6.2 by the administration of either glucose or sodium bicarbonate (Collins et al. 1989). After cellular uptake by endocytosis, the micellar carrier may end up in endosomes/lysosomes that exhibit acidic pHs around 5.0–5.5 (Tang et al. 2003). The decreased pH can be used as an internal stimulus triggering preferential drug release at tumor site for delivery systems that bear pH sensitive groups or linkages.

Two strategies are generally used to provide pH sensitivity in polymeric micelles. The first strategy involves the selective protonation of acid sensitive

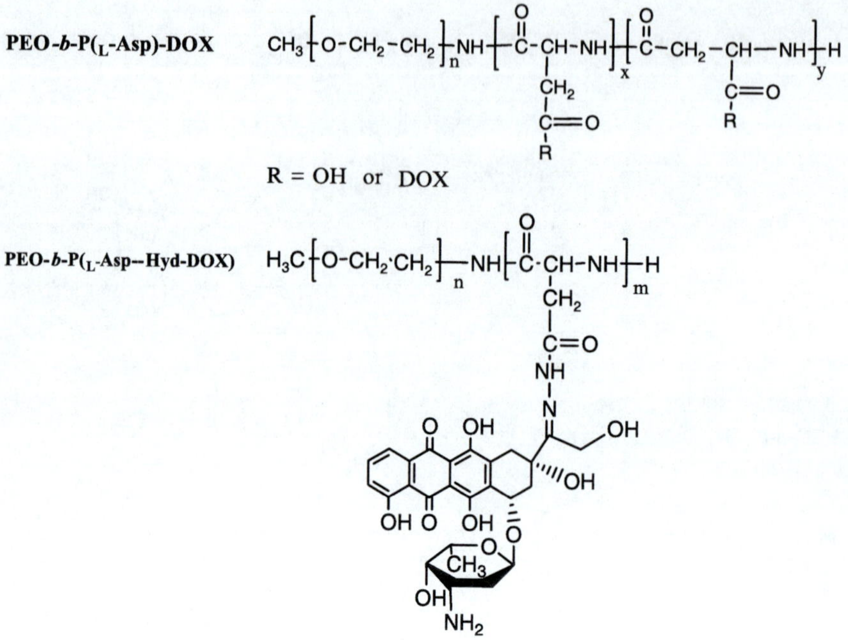

Fig. 7 Example structures of PEO–PLAA-anticancer drug conjugates

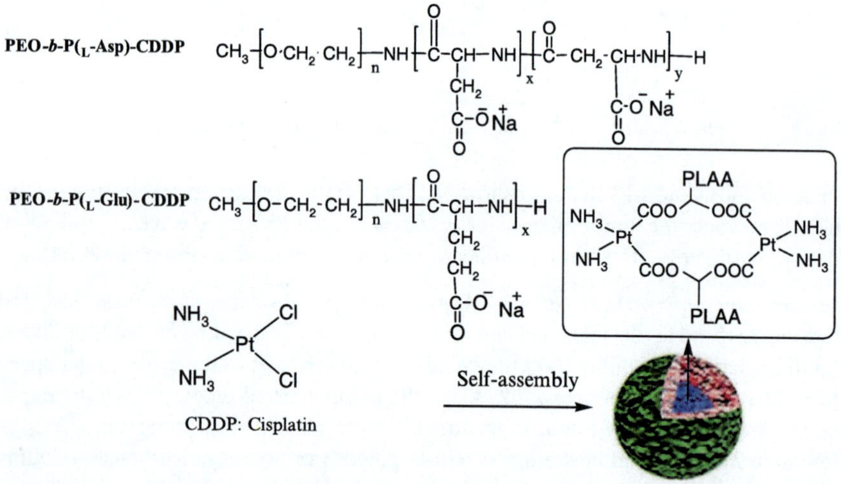

Fig. 8 Formation of complex between PEO–PLAA with CDDP and their self assembly

component of the block copolymers at acidic pHs (e.g., for intratumoral drug delivery), or deprotonation of base sensitive groups at basic pHs (e.g., for intestinal drug delivery). In this category, the polymers mostly bear ionizable groups of a weak acid (carboxylic acid) or weak base (amino group) (Benns et al. 2000;

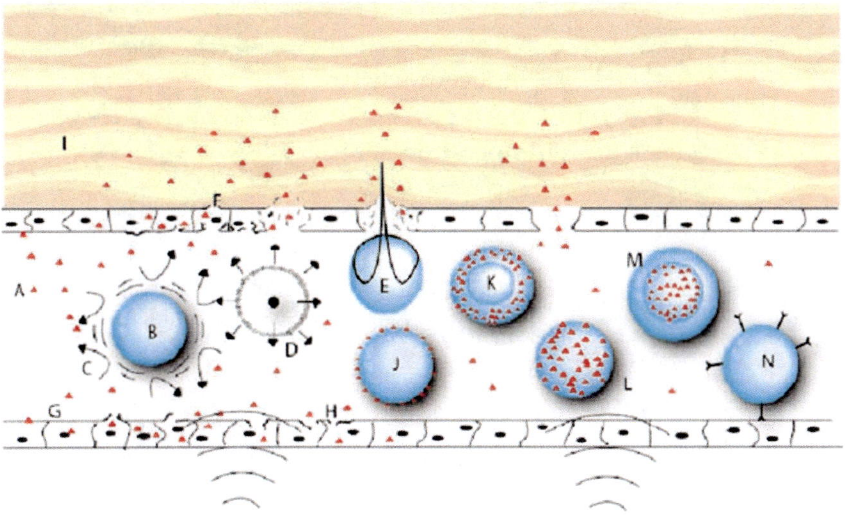

Fig. 9 Possible mechanisms for the role of ultrasound in enhanced drug delivery (reproduced from Pitt et al. 2004 with permission). (**a**) therapeutic agent (triangles); (**b**) gas bubble undergoing stable cavitation; (**c**) microstreaming around cavitating bubble; (**d**) collapse cavitation emitting a shock wave; (**e**) asymmetrical bubble collapse producing a liquid jet that pierces the endothelial lining; (**f**) completely pierced and ruptured cell; (**g**) non-ruptured cells with increased membrane permeability due to insonation; (**h**) cell with damaged membrane from microstreaming or shock wave; (**i**) extravascular tissue; (**j**) thin-walled microbubble decorated with agent on surface; (**k**) thick-walled microbubble with agent in lipophilic phase; (**l**) micelle with agent in lipophilic phase; (**m**) Nanocarriers with therapeutic agent; (**n**) vesicle decorated with targeting moieties

Wang and Huang 1984; Lee et al. 2003b; Gao et al. 2005c) (Fig. 10). The second strategy involves formation of an acid labile linkage between the therapeutic agent and the micelle forming copolymer, which leads to the pH dependant cleavage of drug from micellar nano-conjugates at the pH of tumor extracellular or endosomal environment (Yoo et al. 2002; Bae et al. 2003b; Gillies and Frechet 2003; Gillies et al. 2004; Gao et al. 2005a; Han et al. 2003).

Thermoresponsive polymeric micelles – Local hyperthermia is receiving increasing attention as a tool to promote selective delivery of drugs to solid tumors (Crile 1962). Many normal mammalian cells start to damage at about 42°C. Therefore the aim of thermoresponsive drug delivery is to achieve therapeutic effect just a few degrees above physiological temperature (van der Zee 2002). The architecture of vasculature in solid tumors is chaotic resulting in regions with hypoxic and low pH which are not found in normal tissue. These environmental factors make the tumor cells more sensitive to hyperthermia (Huang et al. 1978; Yatvin et al. 1978; Takei et al. 1995; Okano et al. 1993, 1995). Local hyperthermia might cause preferential drug uptake by the tumor cells through: (i) promoting local drug release at higher temperatures, (ii) increasing the local blood flow; (iii) increasing endothelial permeability promoting enhanced accumulation of colloidal carrier, (iv) increasing the permeability or susceptibility of target cells to the released drug, and (v) increasing the cellular interaction and uptake of the colloidal carrier (Hales et al. 2004; Takei et al. 1993).

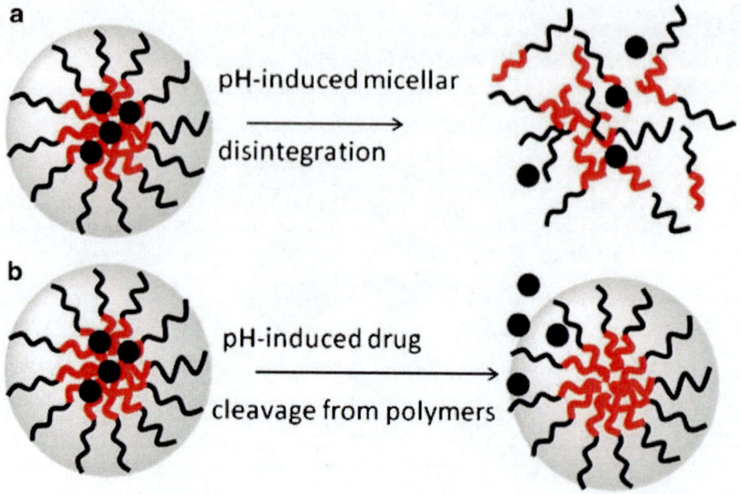

Fig. 10 Schematic illustration of two types of pH-sensitive micelles. (**a**) pH-induced disintegration of polymeric micelles; (**b**) pH-induced drug cleavage from polymers

The feasibility of thermo-sensitive micelles for targeting of anticancer drugs to tumors by dual mechanisms, i.e., passive accumulation/aggregation and the enhanced release of loaded drugs by an hyperthermic condition induced externally has been shown (Kim and Park 2002; Choi et al. 2006). One of the most important parameters for the design of appropriate thermoresponsive delivery systems is optimization of their lower critical solution concentration (LCST) to temperatures that are only a few degrees above body temperature. Poly(N-isopropylacrylamide) (PNiPAAm) is well known to exhibit thermo-reversible phase transition at 32°C (Chung et al. 1998, 1999). Incorporation of hydrophilic groups in PNiPAAm raised its LCST. The enhancement is more pronounced when the hydrophilic groups (e.g. amino or hydroxyl group) are at the end of the chain due to the stronger hydrogen bonding with water (Chung et al. 1998; Takei et al. 1993; Yoshida et al. 1994). Accordingly incorporation of hydrophobic groups (e.g. alkyl group) shifts the LCST to lower temperatures and the degree of hydrophobic contribution also depends on the location of the hydrophobic group (Cammas et al. 1997; Kohori et al. 1998).

Okano et al. reported on the micellization of PNiPAAm–polystyrene (PNiPAAm–Pst) and PNiPAAm–poly(d, l-lactide) (PNiPAAm-PDLLA) block copolymers (Fig. 11) (Kohori et al. 1998). The synthesized block copolymers exhibited LCSTs lower than that of pure PNiPAAm at polymer concentration below critical micelle concentration (CMC), but had an LCST of 32°C when their concentration was above the CMC as the outer hydrated PNiPAAm shell of micelles behaved like free PNiPAAm (Nakayama and Okano 2005). Above the LCST, the PNiPAAm outer shell collapses and these micelles aggregate by hydrophobic surface association.

Thermo-responsive PNiPAAm–co–dimethylacrylamide–PDLLA (P(NiPAAm-co-DMAAm)–PDLLA) micelles showing phase transition at 40°C in phosphate

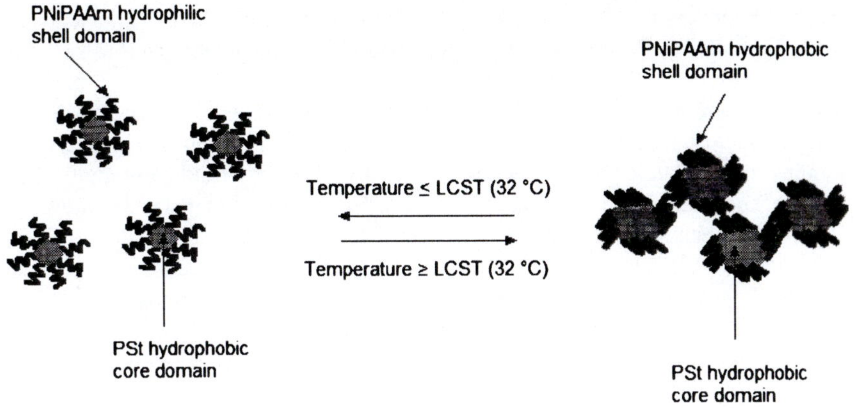

Fig. 11 Aggregation of PNiPAAm–PSt micelle due to hydrophobic surface association above the LCST

buffered saline were used for targeted DOX delivery (Chung et al. 2000; Kohori et al. 1999). Micelles released five times more DOX at 42.5°C than 37°C. DOX-loaded micelles did not show much cytotoxicity against bovine aorta endothelial cells at 37°C, but showed high cytotoxicity at 42.5°C.

2.3 *Ligand Decorated Polymeric Micelles for Active Targeting*

Monoclonal antibodies (mAb)s – Chemical conjugation of mAbs or their Fab fragments to the shell of polymeric micelles can produce immunomicelles. Kabanov et al. developed Pluronic P-85 micelles conjugated to murine polyclonal antibody against α_2 glycoprotein (α_2-GP) to deliver neuroleptic agent haloperidol to brain (Kabanov et al. 1989). Antibody C225 against epidermal growth factor receptors was coupled to the terminus of a DOX-bound PEO–P(L-Glu) and showed enhanced cytotoxicity on A431 cells compared to free DOX (Vega et al. 2003). Torchilin et al. developed tumor-targeted phosphalipid-based immunomicelles by chemical conjugation of mAb 2C5 and 2G4 to the micellar surface. PXT-loaded 2C5-immunomicelles exhibited a superior efficiency compared to PXT-loaded plain micelles or free drug against human breast cancer MCF-7 cells *in vitro*. ^{111}In-labeled 2C5 immunomicelles revealed significantly higher accumulation in LLC tumor-bearing mice than plain micelles. Accordingly, PXT loaded 2C5 immunomicelles exhibited higher drug accumulation in LLC tumors in mice which resulted in higher therapeutic efficacy compared to free PXT and PXT-loaded in plain micelles (Torchilin et al. 2003).

Carbohydrates – Polymeric micelles decorated with glucose, galactose, mannose and lactose have been reported (Choi et al. 1998; Jule et al. 2003; Nagasaki et al. 2001). Carbohydrate receptors like asialoglycoprotein (ASGP) on hepatocytes and mannose receptors on kupffer and liver endothelial cells are known to play a crucial role in biorecognition (Yasugi et al. 1999; Nagasaki et al. 1998). The interaction of the sugar installed PEO–PDLLA micelles with lectin was investigated to simulate binding to glyco-receptors. Binding to Con A was investigated for mannose-micelles and interaction with *Ricinus communis* agglutinin (RCA-1) was assessed for galactose and lactose (Nagasaki et al. 2001). In general, lactose-micelles exhibited rapid binding to the protein bed. An increase in the ligand density resulted in multivalent binding and diminished dissociation kinetics. Development of galactose-modified PEO–poly(γ-benzyl-L-glutamate) (galactose-PEO–(PEO–PBLG) micelles and their application for PXT in ASGP expressing HepG2 cells have been reported (Cho et al. 2001). Kataoka et al. also reported on the development of lactose conjugated PEO–poly(2-(dimethylamino)ethyl methacrylate) (PEO–PAMA) that can form PIC micelles with plasmid DNA (pDNA) and efficiently transfect HepG2 cells (Wakebayashi et al. 2004b).

Folic acid – Folic acid (folate) has become an attractive ligand for targeting cancer cells because its receptor is overexpressed in many human cancers. It is an essential vitamin for the biosynthesis of nucleotide bases and is consumed in elevated quantities by proliferating cells. Folate has been covalently attached to liposomes, polymer conjugates and nanoparticles (Hilgenbrink and Low 2005; Reddy et al. 2005). The poor water-soluble drug PXT was physically loaded into the folate-modified PEO–PCL micelles (Park et al. 2005). Folate–PEO–PCL micelles exhibited significantly higher cytotoxicity compared to unmodified micelles against human breast adenocarcinoma (MCF-7), human uterine cervix adenocarcinoma (HeLa 229) cells, whereas conjugation of folate on the micellar carrier had no significant effect on the cytotoxicity of PEO–PCL micellar PXT in normal human fibroblasts (HF). Park et al. developed a mixed polymeric micellar system composed of folate–PEO–poly(D, L-lactic-co-glycolic acid) (folate–PEO–PLGA) and PEO–PLGA–DOX conjugates (Yoo and Park 2004). The mixed micelles exhibited higher cytotoxicity than DOX, suggesting an increased intracellular transportation of DOX to cells through folate receptor medicated endocytosis. Systemic administration of this formulation significantly retarded tumor growth compared to unmodified micelles. Biodistribution studies also indicated an increased accumulation of DOX in tumor tissue. Conjugation of folic acid to PEO–P(L-Lys) has been pursued to target therapeutic proteins or other negatively charged bioactive macromolecules (Hwa Kim et al. 2005).

RGD peptides – Peptides containing the RGD sequence can recognize integrins overexpressed on the endothelial cells of the tumor neovasculature or metastatic tumor cells. Gao et al. developed cyclic (Arg-Gly-Asp-d-Phe-Lys) (cRGDfK) modified PEO–PCL micelles that can selectively deliver drugs to angiogenic tumor endothelial cells (Fig. 12a) (Nasongkla et al. 2004). Confocal laser scanning microscopy exhibited 30 times higher accumulation of DOX-loaded in cRGDfK-modified micelles compared to unmodified micelles in human Kaposis sarcoma tumor endothelial SLK cells.

Our group has developed RGD modified PEO–PCL micelles containing physically or chemically incorporated DOX (Fig. 12b) (Xiong et al. 2007a). Higher uptake

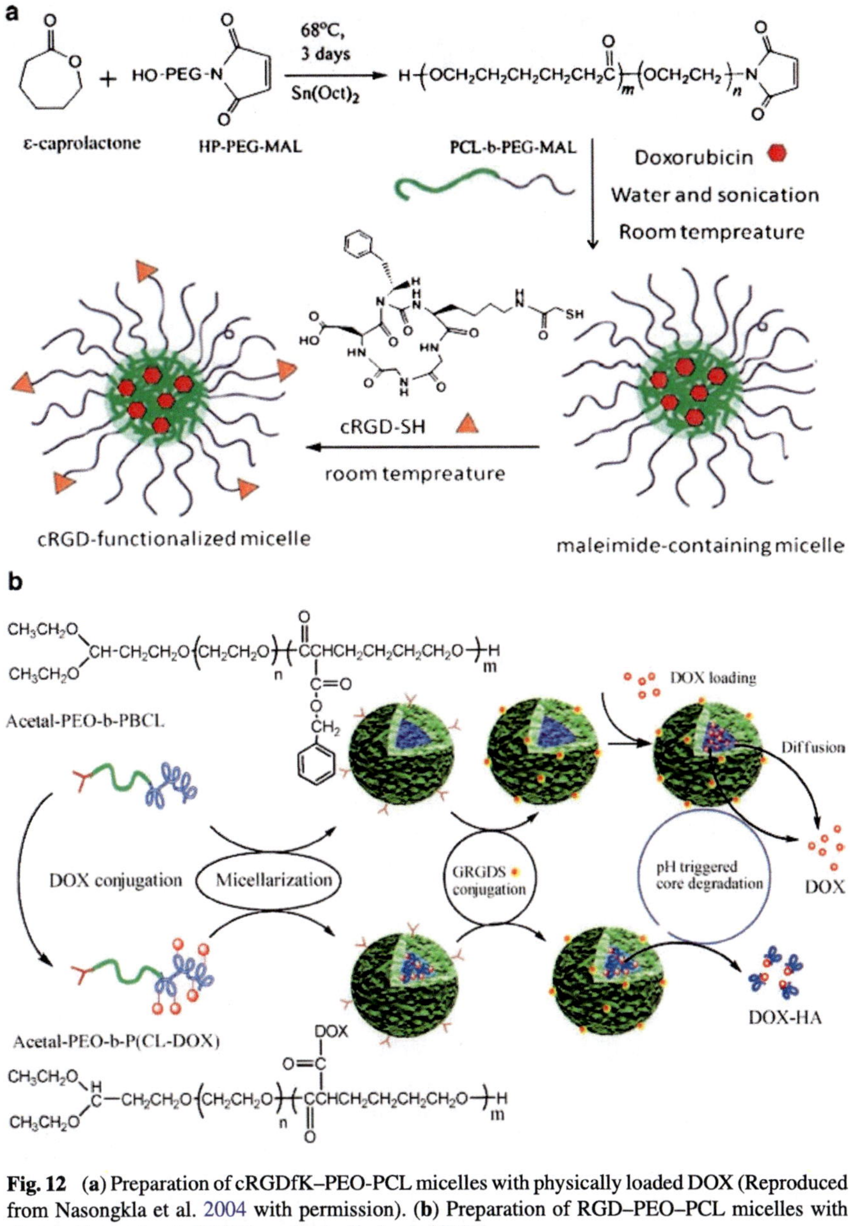

Fig. 12 (**a**) Preparation of cRGDfK–PEO-PCL micelles with physically loaded DOX (Reproduced from Nasongkla et al. 2004 with permission). (**b**) Preparation of RGD–PEO–PCL micelles with chemically conjugated DOX or physically loaded DOX

after 3 h incubation with mouse melanoma B16–F10 cells was observed for GRGDS modified micelles compared to unmodified micelles. RGD modified micelles bearing conjugated DOX demonstrated higher cytotoxicity on B16–F10 cells than the control conjugate without the RGD modification (Xiong et al. 2007b, 2008a).

Transferrin – Human transferrin is a relatively large serum glycoprotein (80 KDa). Transferrin receptors are known to be elevated in various types of cancer cells. The level of elevation in transferrin receptor is also known to correlate with the proliferative ability of the tumor cells. Kabanov et al. proposed delivery of phosphorothioate oligonucleotides (SODNs) by PIC micelles formed with transferrin-conjugated PEO–polyethylenimine (PEO–PEI). Compared to unmodified micelles, fluorescent-labeled transferrin-micelles showed a significantly higher accumulation in resistant human oral epidermoid carcinoma (KBv) cells. Delivery of antisense SODNs against expression of P-glycoprotein (P-gp) human *mdr1*-mRNA by transferrin–PEO–PEI nano-carriers exhibited a significantly higher inhibition of P-gp efflux system in resistant MCF-7/ADR cells that over-expresses P-gp compared to unmodified micelles (Vinogradov et al. 1999). Complexes were generated by mixing of plasmid DNA, linear PEI, PEO–PEI, and transferrin–PEO–PEI to provide a ligand for receptor-mediated cell uptake (Kursa et al. 2003). For complexes containing the luciferase reporter gene the highest expression was found in tumor tissue of mice. An optimum formulation for *in vivo* application containing plasmid DNA encoding for the tumor necrosis factor (TNF-α), inhibited tumor growth in three murine tumor models.

2.4 Multifunctional Polymeric Micelles

Multifunctional polymeric micelles are designed to bear a combination of structural components required for various targeting strategies on an individual carrier in order to enhance the selectivity of the incorporated drug for the target site (Torchilin 2006). Development of multifunctional polymeric micelles having cancer specific targeting ligands on the surface and anticancer drug conjugated through acid-cleavable bond in the core, i.e., folate–PEO–p(Asp–Hyd–DOX), was reported by Kataoka et al. (Bae et al. 2005a). The folate-bound micelles can be guided to cancer cells and get internalized through receptor mediated endocytosis. After micellar entry to the cells, hydrazone bonds are cleaved by endosomal acidic environment (pH 5–6).

The development of multidrug resistance (MDR) and severe side effects of anticancer drugs require custom-designed targeted carriers to improve the cancer chemotherapeutic efficacy. The high affinity of anthracyclines (e.g. DOX) towards chromosomal DNA has been the main mechanism for their antitumor activity against sensitive tumors, while highly lipophilic and unaminated anthracyclines led to selective accumulation in cytoplasm, and demonstrated the ability to circumvent MDR which are most likely associated with their ability to reduce P-glycoprotein (P-gp) recognition, to localize in cytoplasm, to affect subcellular targets other than nuclear DNA, or to act as self modulators of MDR (Lampidis et al. 1997; Guillemard and Uri Saragovi 2004). Bearing these in mind, our group used PEO–PCL based polymer to design multifunctional polymeric micelles which have peptide-decorated shell for tumor targeting and pH-sensitive core for acid-triggered drug release to effectively deliver intact DOX to sensitive cancer cells and derivatized DOX to resistant cancer cells (Xiong et al. 2010a). Briefly, DOX was chemically conjugated to the micellar core by amide or hydrazone linkages, while the RGD4C moieties specifically homing to integrin αvβ3 receptors expressed by cancer cells (e.g. MDA435/LCC6 sensitive and resistant cells) were used to functionalize the micellar shell (Fig. 13a, b). These targeted micelles showed markedly increased cellular uptake mediated by RGD4C modification. In particular, pH-triggered drug release of intact DOX and derivatized DOX from hydrazone and amide bound DOX-conjugates, respectively led to changed cellular distribution and cytotoxicity of the carried DOX against both cell lines. As expected, subcellular distribution revealed that targeted micelles containing hydrazone-linked DOX showed preferential accumulation in nucleus in sensitive cells, while micelles containing amide-linked DOX showed preferential DOX accumulation in mitochondria in resistant cells after 24 h of incubation (Fig. 13c). SCID mice bearing with human sensitive and resistant tumors have been optimally treated by the RGD4C-functionalized micelles with hydrazone-linked and amide-linked DOX, respectively. The results point to the potential of these multifunctional polymeric micelles as the custom-designed carrier for personalized cancer chemotherapy.

Including a combination of an ultra pH-sensitive polymer and cell penetrating TAT protein in the structure of polymeric micelles by Bae et al. led to the preparation of another multifunctional polymeric micellar system (Sethuraman and Bae 2007). The delivery system consisted of TAT-attached PEO–poly(L–lactic acid) (PEO–PLLA) micelles and the TAT shield, i.e. an ultra pH-sensitive PEO–poly(methacryloyl sulfadimethoxine) (PEO–PSD). At pH 7.4, the anionic PSD is complexed with cationic TAT of the micelles via electrostatic interaction so that the non-specific cell penetrating TAT peptides on micelles are shielded by PEO–PSD. The PSD will become neutral below pH 7.0. Therefore, deshielding of TAT micelle is triggered by the lower pH of the tumor milieu, exposing TAT to the cells.

Multifunctional mixed micelles of folate–PEO–poly(L-Histidine) (folate–PEO–P(L-His)) and PEO—PLLA were developed also by the latter group (Lee et al. 2003a). The mixed micelles were able to recognize the minute differences in pH and provide enhanced release below pH 7.4. The superior cytotoxicity of this system at tumor pH (6.8–7.2) compared to physiological pH (7.4) was demonstrated in MCF-7 cells. The combined mechanisms of pH-triggered release and active internalization by folate receptor improved the *in vitro* cytotoxicity of encapsulated DOX surpassing that of free drug in sensitive MCF-7 cells (Lee et al. 2003a). This system was effective in sensitizing MCF-7 resistant cells to DOX achieving similar IC_{50} to that of free DOX in sensitive cell line. The multifunctional polymeric micellar carrier has shown superior *in vivo* efficacy to that of free DOX or DOX encapsulated in pH sensitive micelles without folate (Lee et al. 2005a).

The same research group has reported on the formation of multifunctional mixed micelles of PEO–P(L-His) and PLLA–PEO–P(L-His)–biotin (Lee et al. 2005c). In this system, biotin on the micellar surface was used to enhance tumor specificity. P(L-His) section was used as a pH sensitive segment that can hypothetically cover

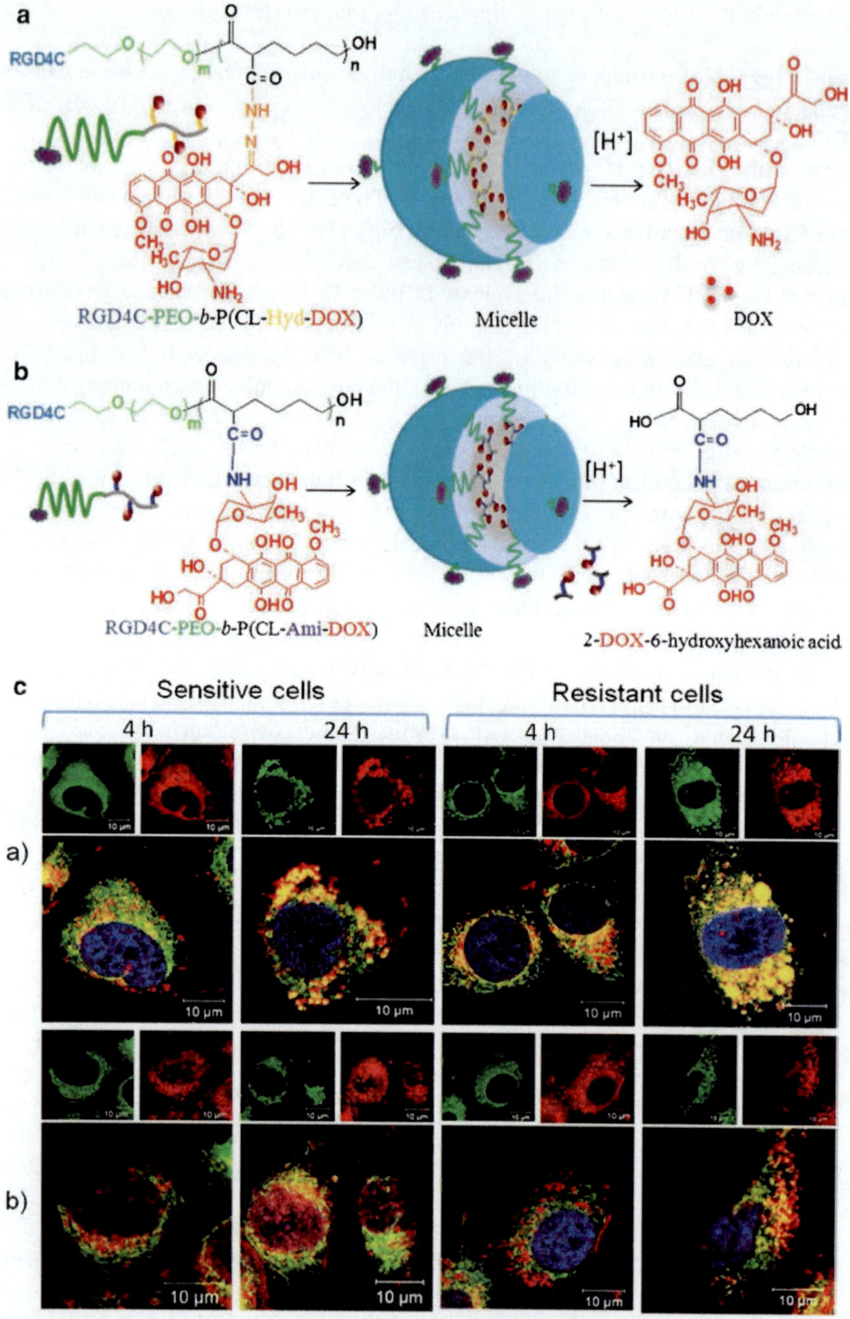

Fig. 13 Self assembly of (**a**) RGD4C–PEO–P(CL–Hyd–DOX) and (**b**) RGD4C–PEO–P(CL–Ami–DOX) copolymers into targeted micelles. (**c**) Endosome, mitochondrial versus nuclear distribution of DOX as part of (*a*) RGD4C–PEO–P(CL–Ami–DOX) and (*b*) RGD4C–PEO–P(CL–Hyd–DOX) micelles in MDA-435/LCC6^{WT} and MDA-435/LCC6^{MDR} cells. Cells were incubated with (*a*) RGD4C–PEO–P(CL–Ami–DOX) micelles and (*b*) RGD4C–PEO–P(CL–Hyd–DOX) micelles

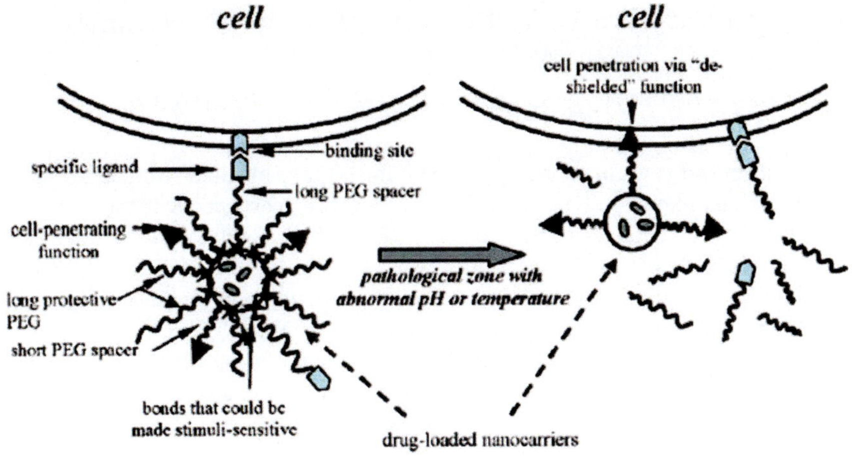

Fig. 14 Interaction of the multifunctional pH-responsive pharmaceutical nanocarrier with the target cell. Local stimuli-dependent removal of protecting PEG chains or mAb-PEG moieties allows for the direct interaction of the cell penetrating peptide moiety with the cell membrane (Reproduced from Sawant et al. 2006 with permission)

biotin residues at pH 7.4 when it is unionized, but take protons at pH 6.8–7.4, stretch and expose biotin for interaction with cancer cells. The micelles also showed pH-dependent dissociation, causing the enhanced release of DOX from the carrier in early endosomal pH.

Double-targeted immunomicelles were developed by Torchilin et al. (Fig. 14) (Sawant et al. 2006). The mixed micelles consisted of monoclonal antimyosin antibody 2G4 conjugated to PEO–phosphatidylethanolamine (PEO–PE) via p-Nitrophenylcarbonyl–PEO–PE (pNP–PEO–PE); nonspecific cell penetrating peptide (e.g. TATp) or biotin attached to PE (TAT–PE or biotin–PE) or short PEO–PE; and PEO–Hyd–PE. Under physiological pH, biotin and TATp on the surface of nanocarriers were shielded by long protecting PEO chains or by even longer pNP–PEO–PE used to attach antibodies to the nanocarrier. At pH 7.4–8.0, micelles demonstrated high specific binding with 2G4 antibody substrate (myosin), but very limited binding on biotin substrate (avidin) or internalization by NIH/3T3 or U-87 cells mediated by TATp on micelles. However, upon 15–30 min incubation at pH 5.0–6.0, nanocarriers lost their protective PEO shell because of acidic hydrolysis of PEO–Hyd–PE and acquired the ability for strong interaction with an avidin column or effective internalization by cells via TATp.

Fig. 13 (continued) for 4 and 24 h. Confocal images of cells stained with MitoTracker® green (*green*) and DAPI (*blue*) were then taken and the pictures were superimposed. Pink color shows localization of DOX (*red*) in nucleus (*blue*), while yellow color is an indication for localization of DOX (*red*) in mitochondria (*green*) or endo/lysosomes (Reproduced from Xiong et al. 2010a with permission)

3 Polymeric Micelles for Delivery of Genetic Therapeutics

3.1 Requirements for Effective Gene/siRNA Transfection

The past several years have witnessed evolution of gene medicine from experimental technology into new clinic strategies for developing new genetic therapeutics for a wide range of human disorders such as cancer, AIDS, neurological disorders and cardiovascular disorders. These include using nucleic acid-based compounds such as plasmids containing transgenes for gene therapy, oligonucleotides, ribozymes, DNAzymes, aptamers, and small interfering RNAs (siRNAs). The successful clinical application of this nucleic acid-based therapeutics in particular for gene and siRNA greatly relies on an effective delivery system for systemic administration. So far, various nanomedicine platforms have been developed as viral or non-viral vectors to meet these challenges for an effective gene or siRNA delivery.

There are several requirements for an effective gene/siRNA delivery system. Generally, the delivery system needs to able to efficiently encapsulate gene/siRNA, actively target sites of interest, escape from endosome/lysosomes, and finally release siRNA intracellularly. In addition to accumulation within desired tissues, it also requires gene to be delivered to the nucleus and siRNA to be delivered into the cytoplasm. In particular, it requires the ability of a delivery system to escape endosome after cellular endocytosis and effective delivery of nucleic acid based drugs to their specific targets. Endosomal escape is believed to be achieved through 'proton sponge effect' (Fig. 15a) (Behr 1997; Pack et al. 2005). It is hypothesized that polymers with buffering capacities between 7.2 and 5.0, such as PEI and imidazole-containing polymers, could buffer the endosome and potentially induces its rupture. Polycations are believed to increase osmotic pressure in the endosome by the so-called 'proton sponge effect' resulting in endosomal disruption and gene release (Behr 1997; Boussif et al. 1995).

Polymeric micelles composed of a hydrophilic and polycationic segments have attracted attention as non-viral gene/siRNA vectors due to their unique core/shell structure and highly tunable characteristics. PEO–PLAA based copolymers have been extensively investigated to develop gene/siRNA delivery. Recently, by introduction cationic groups on the polyester block, PEO–polyesters have been emerging as important polymeric micellar materials. Generally, neutralization of the positive charge on the polycation by the negatively charged DNA will lead to micellization (Fig. 15b).

3.2 Development of PIC Micelles from PEO–PLAA for Gene Delivery

Existence of several functional groups on a PLAA block offers the advantages for PEO—PLAA to be used for gene/siRNA delivery. PEO–PLAA based gene vectors can be smartly engineered to meet the challenges for DNA/siRNA delivery due to

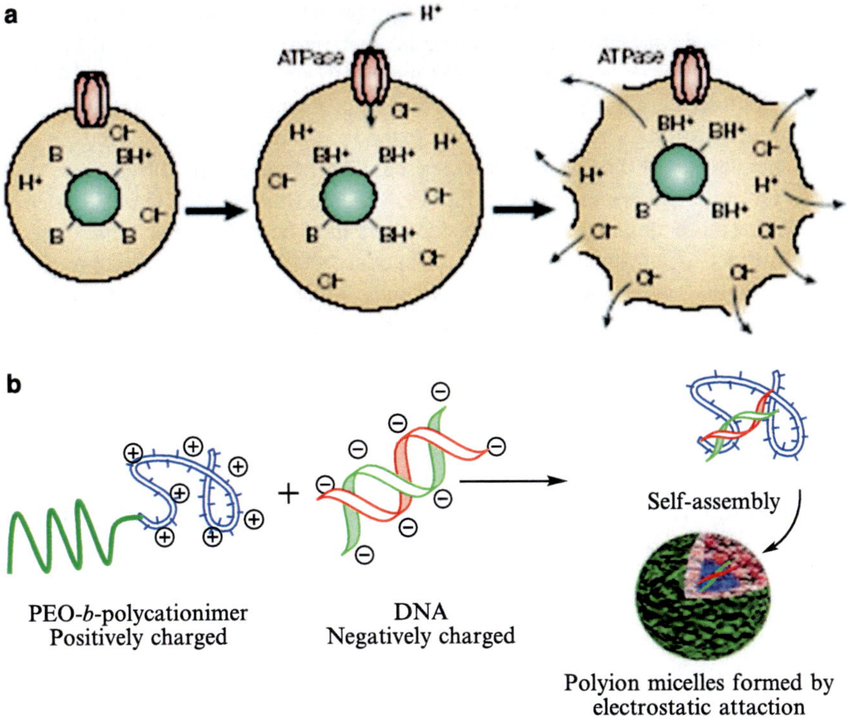

Fig. 15 (**a**) Schematic of the proton-sponge mechanism. Protonation of the proton-sponge polymer (*green*) causes increased influx of protons (and counter-ions) into endocytic vesicles. Increasing osmotic pressure causes the vesicle to swell and rupture (Reproduced from Pack et al. 2005 with permission). (**b**) Schematic illustration of formation of polyion complex (PIC) micelles from amphiphilic copolymer and DNA

their easy synthesis and flexible chemistry in their structures. For example, PEO–poly(L-lysine) (PEO–PLL) micelles have been extensively used for application in gene delivery because the amine-contained core-forming block can form a stable polyion complex with DNA through an electrostatic interaction (Ward et al. 2002; Lee et al. 2005b; Katayose and Kataoka 1997, 1998; Kakizawa et al. 1999, 2001; Harada et al. 2001; Miyata et al. 2004; Meister and Anderson 1983; Behr 1997; Boussif et al. 1995). PEO—PLL micelles were injected via superior mesenteric vein and showed sustained gene expression in the liver for 3 days. However, PEO—PLL has high CMC and may dissociate upon dilution.

To increase the stability of PEO–PLL micelles against dissociation, PEO–PLL micelles with crosslinked core were prepared by substituting certain fractions of the lysine residues in the PLL core with thiol groups which can readily form disulfide cross-links with other sulfide substituted PEO–PLL polymers and develop a network structure in the micelle core after DNA complexation (Kakizawa et al. 1999, 2001; Harada et al. 2001). The cross-linked core of the micelles can be cleaved inside the cell in the presence of reducing glutathione abundant in the cytoplasm but not in blood compartment (Meister and Anderson 1983). However, introduction of thiol groups was found to decreases the electrostatic association sites for the interaction of PLL and DNA. To deal with this problem, Traut's reagent was used to introduce the crosslinking thiol groups to the PLL and, meanwhile, avoid the loss of charge density of the block copolymer segment (Miyata et al. 2004).

To promote endosome escape, PEO–PLAA-based block copolymers that contain two or more amino groups in the side chain have been designed to use 'proton sponge effect' (Kanayama et al. 2006; Itaka et al. 2004). The primary amine group at the distal end of the side chain showed a high pKa and was used for complexation with negatively charged siRNA or DNA. The secondary amine of the side chain located closer to the polymeric backbone showed a lower pKa and was expected to provide buffering capacity for proton sponge effect. These block catiomers, were prepared by the aminolysis of PEO–poly(β-benzyl-L-aspartate) (PEO-PBLA) with either diethylenetriamine (DET), 4-methyldiethylenetriamine (MDET) or *N,N*-diethyldiethylenetriamine (DEDET). Both the PEO–P[Asp(MDET)] and PEO–P[Asp(DEDET)] polyplex micelles showed an appreciably lower transfection than the PEO–P[Asp(DET)] polyplex micelles. The optimum transfection efficiency was obtained by DET substituted PEO–PBLA micelles. In a separate study, dipropylenetriamine (DPT) was reacted with PEO–PBLA. The polyion complex of this system with siRNA has shown superior transfection efficiency over lipid-based commercial vector for siRNA delivery i.e., RNAiFect. Free polycations in such polyplexes substantially contribute to efficient transfection due to proton sponge effect but also introduce toxic effects.

To reduce the free-polycation mediated toxic effects for in vivo gene delivery, PEO–PLL based triblock polymers were designed to have multiple functions in the core. Briefly, the triblock copolymer consisted of PEO, poly((3-morpholinopropyl) aspartamide) (PMPA) as the low pKa segment with buffering capacity, and PLL as the high pKa segment to condense DNA (Itaka et al. 2004). When plasmid DNA was encapsulated in PEO–PMPA–PLL it revealed one order of magnitude higher transfection efficiency than PEO–PLL, which was comparable to the transfection efficiency of plasmid DNA encapsulated PEI at the corresponding negative to positive (N/P) ratio, without showing appreciable cytotoxicity. Synthesis of tri-block copolymers of PEO–PEI–PBLG has also been reported where hydrophobicity of PBLG was used to induce micellization where PEI was used to incorporate plasmid DNA effectively (Tian et al. 2005).

To achieve active targeting, ligand-modified PEO–PLAA micelles were developed for gene delivery. For cancer-targeted gene therapy, cyclic RGD peptide (c(RGDfK)), which specifically recognize integrin αvβ3 receptors expressed in cancer and its vascular endothelial cells as well, was conjugated at the end of PEG–PLL block copolymer (Oba et al. 2008). The c(RGDfK)–PEG–PLys/pDNA polyplex micelle showed remarkably increased transfection efficiency compared with non-targeted polyplex micelles against HeLa cells which express αvβ3 integrins. The results from cellular uptake and intracellular distribution study suggested that c(RGDfK)-modified micelles not only increased cellular uptake but enhanced intracellular trafficking of pDNA towards the perinuclear region via αvβ3 integrin receptor mediated endocytosis, leading to increased transfection efficiency for Hela cells. In another study, lactose-modified polymeric micelle encapsulating pDNA targeted to hepatocytes, which possess abundant ASGP receptors that recognize lactose moieties was developed for liver targeting (Wakebayashi et al. 2004b). The lactose micelle was significantly more efficient to transfect HepG2 cells (hepatoma) than the micelle-lacking lactose. A competitive assay using asialofetuin (ASF), a natural ligand against ASGP receptors, inhibited uptake of the lactose-installed micelle, suggesting that ASGP receptor-mediated endocytosis is a major pathway for the cellular uptake of the lactosylated micelle.

3.3 *Development of PIC Micelles from PEO–Polyesters for Gene/siRNA Delivery*

Polyesters are biodegradable and biocompatible polymers and have a history of safe application in absorbable biomedical devices such as sutures. Compared to PLAA, polyester tends to be more hydrophobic and micelles formed from PEO-b-polyester are more stable upon dilution after intravenous injection. PEO–polyesters such as PEO-b-PLGA and PEO–PCL have been extensively explored in drug delivery. However, the lack of cationic moieties and untailorable polyester block in their structures limits their use for gene or siRNA delivery. Although a few of papers reported using PLGA-based particulate carriers to encapsulate antisense oligonucleotides or plasmid DNA, the encapsulation efficiency of gene is often low and the energy or complicated procedures are required to increase DNA encapsulation which results in some denaturation of the DNA. Therefore, there is a need to modify PEO-b-polyesters to enable them suitable for gene/siRNA delivery.

Modifications of PEO–polyesters have been made either by attaching cationic blocks or moieties at the end of polyester block or functionalization of the polyester block with cationic moieties. These cationic moieties would provide anchoring site for genetic cargoes. Amine-containing group was attached to the end of PEO–PCL to prepare amine-terminated PEO–PCL (PEO–PMCL) which can effectively encapsulate DNA by electrostatic interaction to self assemble into micelles (Jang et al. 2006). Compared to physically encapsulated DAN in PEO–PCL micelles, DAN encapsulated PEO–PMCL micelles showed remarkably increased encapsulation efficiency and transfection efficiency of DNA toward normal human fibroblast cells without introducing obvious cytotoxicity. Triblock copolymer of PEO, PCL and PEI was synthesized by grafting the activated PEO–PCL onto the hyperbranched PEI (Shuai et al. 2003) (Fig. 16a). The resulted polymers are completely water soluble or form micelles depending on the polymer composition. Complexation of plasmid DNA with various copolymers can form micelle-like particles of 200 nm in diameters with neutral surface (Fig. 16b). With well-controlled composition over these copolymers, pDNA complexed in PEO–PCL–PEI with high CMC or good solubility showed enhanced DNA binding and

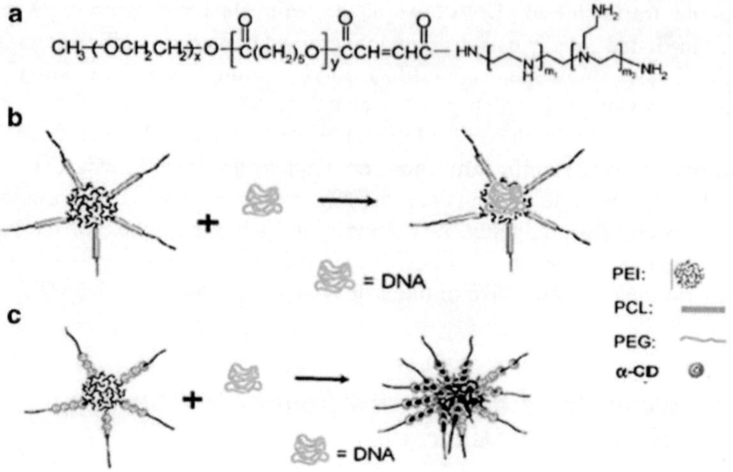

Fig. 16 PEO-b-PCL-PEI micelles for plasmid DAN delivery. (**a**) Structure of PEO–PCL–PEI micelles. (**b**) PEO–PCL–PEI/pDNA micelles (**c**) pDNA complexed PEO–PCL–PEI micelles with α-CD to decrease the hydrogen bonding led to increased gene transfection efficiency

condensation and excellent transgene expression efficiency comparable to PEI/DNA polyplexes but better biocompatibility and potential biodegradability but lower cytotoxicity than PEI. However, PEO–PCL–PEI with high PEO graft density and long PCL blocks showed very poor gene expression due to hydrophobic PCL and hydrogen bonding between PCL and PEI. Therefore, α-cyclodextrin (α-CD) was used to break the hydrogen bonds and dissolute PCL block as well by formation of inclusion complexes between α-CD and PCL blocks(Shuai et al. 2005). Addition of α-CD led to increased gene transfection efficiency and excellent biocompatibility for the delivery system for in vivo use (Fig. 16c).

Instead of introducing new cationic block or group to the end of PCL, our group has grafted various polyamine moieties to PCL block for siRNA delivery (Xiong et al. 2008b). After introducing carboxylic groups to caprolactone units of PCL block, we synthesized a novel family of PEO–PCL based polycationic copolymers, i.e., PEO–PCL with grafted spermine (PEO–P(CL–SP)), tetraethylenepentamine (PEO–P(CL–TP)), or N,N-dimethyldipropylenetriamine (PEO–P(CL–DP)) (Fig. 17a). These polyamine-grafted PEO–PCL copolymers demonstrated good compatibility and were able to effectively bind siRNA, self-assemble into micelles, protect siRNA from degradation by nuclease and release complexed siRNA efficiently in the presence of low concentrations of polyanionic heparin. siRNA formulated in PEO–P(CL–SP) and PEO–P(CL–TP) micelles showed efficient cellular uptake through endocytosis by MDA435/LCC6 cells transfected with *MDR-1* gene, which encodes for the expression of P-glycoprotein (P-gp). Importantly, the siRNA formulated in PEO–P(CL–SP) and PEO–P(CL–TP) micelles demonstrated effective endosomal escape after cellular uptake (Fig. 17b). *MDR1* targeted siRNA formulated in PEO–P(CL–SP) and PEO–P(CL–TP) micelles exhibited efficient gene

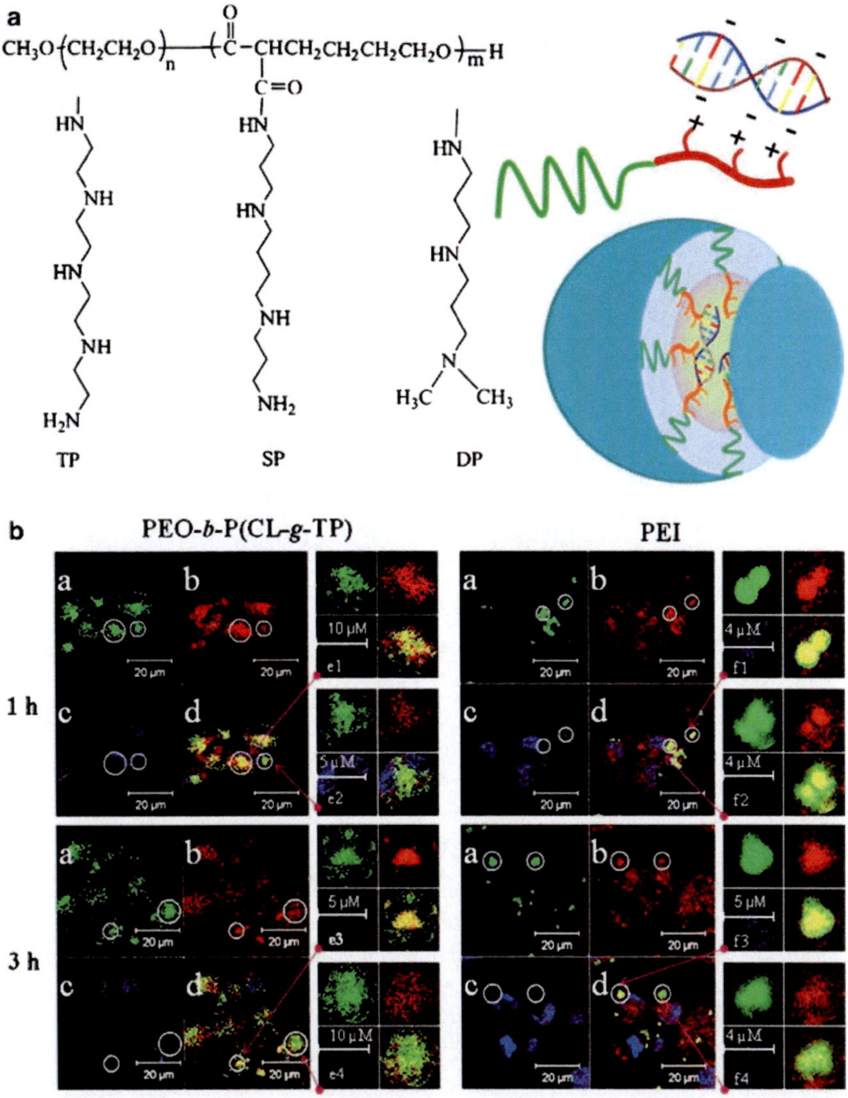

Fig. 17 (**a**) Structures of PEO–P(CL–polyamine) copolymers and siRNA complexed PEO–P(CL–polyamine) micelles. (**b**) Assessment of endosomal escape for PEO-*b*-P(CL-*g*-TP) and PEI after intracellular uptake by endocytosis upon 1 and 3 h incubation by confocal microscopy. The cells were treated with FAM-labeled siRNA formulated in PEO-*b*-P(CL-*g*-TP) and PEI (*green, panel a*); Lysosomes and nuclei were stained by LysoTracker (*red, panel b*) and DAPI (*blue, panel c*), respectively, and the images were merged in *panel d*. The endosomes/lysosomes in cells treated with PCI micelles and particles in different phases were magnified in e and f, respectively (Reproduced from Xiong et al. 2008b with permission)

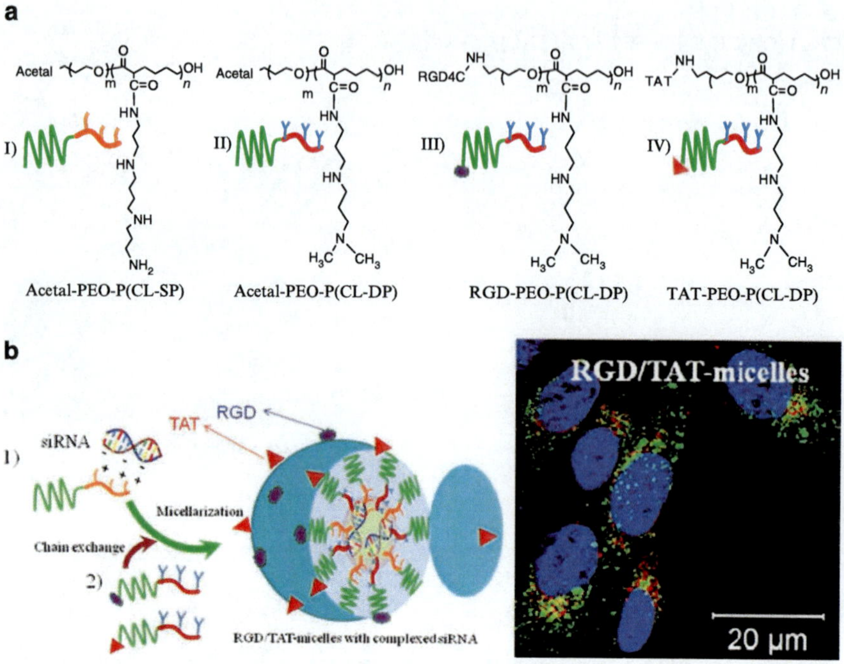

Fig. 18 Virus-mimetic polymeric micelles for siRNA delivery. (**a**) Structures of ligand-attached PEO–P(CL–polyamine) copolymers. (**b**) RGD/TAT-micelles with complexed siRNA. (**c**) Endosome escape of RGD/TAT–PEO–P(CL–SP)/siRNA micelles after incubation with MDA435/LCC6 resistant cells for 4 h (*green*: FAM-siRNA, *red*: Lysotracker red, and *blue*: DAPI) (Reproduced from Xiong et al. 2010b with permission)

silencing for P-gp expression. However, high concentration of siRNA needs to be used to achieve P-gp silencing which might bring off-target gene silencing.

To increase siRNA delivery efficiency and achieve cancer targeting as well, PEO–PCL micelles with structural features mimicking that of viral vectors were developed and their potential for siRNA delivery was assessed (Fig. 18) (Xiong et al. 2010b). The micellar shell composed of PEO was decorated with virus-related peptides such as integrin αvβ3 targeting peptide (RGD4C) and/or cell penetrating peptide (TAT), while the micellar core composed of PCL was modified with a polycation (spermine) for siRNA binding and protection. The peptide-functionalized micelles especially those with dual functionality (RGD/TAT-micelles) demonstrated increased cellular uptake and effective endosomal escape of siRNA compared to unmodified micelles (NON-micelles) in MDA435/LCC6 resistant cells. Transfection of mdr1 siRNA formulated in peptide-modified micelles led to P-gp down regulation both at the mRNA and protein level, and therefore increased DOX accumulation in the cytoplasm and nucleus. Compared to RGD- or TAT-micelles, RGD/TAT-micellar siRNA complexes produced improved cellular uptake, P-gp silencing, DOX cellular accumulation, DOX nuclear localization and DOX induced cytotoxicity in MDA435/LCC6 cells, pointing to the potential of RGD/TAT-functionalized virus-like micelles as promising carriers for efficient delivery to reverse the P-gp mediated multidrug resistance.

4 Polymeric Micelles for Nonparental Drug Administration

Research on the potential benefit of polymeric micelles in improving the absorption and bioavailability of drugs through non-intravenous routes of administration is in its infancy (Bromberg 2008; Malik et al. 2007). Polymeric micelles can enhance the oral absorption of encapsulated drugs in different ways (Mathot et al. 2006; Sant et al. 2005) including an increase in the apparent solubility of hydrophobic drugs; protecting the encapsulated drug against degradation in gastrointestinal tract; prolonging drug residence time through insertion of mucoadhesive shells, facilitating the transport of the encapsulated drug through the epithelial barrier by fluid phase endocytosis; bypassing or inhibiting the efflux pumps; and targeting specific carriers for receptor mediated transport. In addition to oral route, attention has been paid to the application of polymeric micelles through transdermal and ocular administration (Tu et al. 2007; Liaw and Lin 2000; Gupta et al. 2007). Pluronic F-127, has been widely used to provide a transdermal delivery through enhancement of penetration in the stratum corneum and the viable epidermis layer (Ruiz et al. 2007). The utility of Pluronic F127 for ocular delivery of pilocarpin has also been assessed (Desai and Blanchard 1998).

5 Concluding Remarks and Perspectives

In this chapter, we provided an update on several chemical strategies used to enhance the properties of nanoscopic core/shell structures formed from self assembly of ABCs, namely polymeric micelles. Clearly, versatility of polymer chemistry in ABCs provides unique opportunities for tailoring polymeric micelles for optimal properties in gene and drug delivery. Chemical modification of the polymer structure in the micellar core through introduction of hydrophobic or charged moieties, conjugation of drug compatible groups, core cross-linking has led to enhanced stability for the micellar structure and sustained or pH-sensitive drug release. The modification of polymeric micellar surface with specific ligands (carbohydrates, peptides, antibodies) have shown benefit in enhancing the recognition of carrier by selective cells leading to improved drug and gene delivery to the desired targets. Importantly, their self-assembly nature in aqueous environment offers the opportunity to scale-up the production of the micelle formulation to meet potential clinical application.

Despite that the importance and benefit of chemical flexibility of block copolymers in improving the delivery properties of polymeric vesicles can also be envisioned, research in drug delivery by polymeric nanocarriers is still in its infancy.

The demanding challenge of the future research in this field is to find the right carrier architecture and optimum polymer chemistry that can improve the delivery of sophisticated and complex therapeutic agents (e.g., poorly soluble drugs, proteins and genes) to their cellular and intracellular targets. For a potential clinical application, barriers like high drug loading, stable drug encapsulation, controlled drug release, and micelle-cell interaction should be addressed, a large-scale good manufacturing practice (GMP) production procedure should be developed, and the safety profile needs to be established.

References

Aliabadi, H. M. & Lavasanifar, A. (2006) Polymeric micelles for drug delivery. *Expert Opin Drug Deliv*, 3, 139–62.

Bae, Y., Fukushima, S., Harada, A. & Kataoka, K. (2003a) Design of environment-sensitive supramolecular assemblies for intracellular drug delivery: Polymeric micelles that are responsive to intracellular pH change. *Angewandte Chemie-International Edition*, 42, 4640–4643.

Bae, Y., Fukushima, S., Harada, A. & Kataoka, K. (2003b) Design of environment-sensitive supramolecular assemblies for intracellular drug delivery: polymeric micelles that are responsive to intracellular pH change. *Angew Chem Int Ed Engl*, 42, 4640–3.

Bae, Y., Jang, W. D., Nishiyama, N., Fukushima, S. & Kataoka, K. (2005a) Multifunctional polymeric micelles with folate-mediated cancer cell targeting and pH-triggered drug releasing properties for active intracellular drug delivery. *Molecular Biosystems*, 1, 242–250.

Bae, Y., Nishiyama, N., Fukushima, S., Koyama, H., Yasuhiro, M. & Kataoka, K. (2005b) Preparation and biological characterization of polymeric micelle drug carriers with intracellular pH-triggered drug release property: tumor permeability, controlled subcellular drug distribution, and enhanced in vivo antitumor efficacy. *Bioconjug Chem*, 16, 122–30.

Behr, J. P. (1997) The proton sponge: A trick to enter cells the viruses did not exploit. *Chimia*, 51, 34–36.

Benns, J. M., Choi, J. S., Mahato, R. I., Park, J. S. & Kim, S. W. (2000) pH-sensitive cationic polymer gene delivery vehicle: N-Ac-poly(L-histidine)-graft-poly(L-lysine) comb shaped polymer. *Bioconjug Chem*, 11, 637–45.

Bijsterbosch, H. D., Cohen Stuart, M. A. & Fleer, G. J. (1999) Effect of Block and Graft Copolymers on the Stability of Colloidal Silica. *J Colloid Interface Sci*, 210, 37–42.

Boussif, O., Lezoualch, F., Zanta, M. A., Mergny, M. D., Scherman, D., Demeneix, B. & Behr, J. P. (1995) A Versatile Vector for Gene and Oligonucleotide Transfer into Cells in Culture and in-Vivo – Polyethylenimine. *Proceedings of the National Academy of Sciences of the United States of America*, 92, 7297–7301.

Bromberg, L. (2008) Polymeric micelles in oral chemotherapy. *J Control Release*, 128, 99–112.

Cammas, S., Suzuki, K., Sone, C., Sakurai, Y., Kataoka, K. & Okano, T. (1997) Thermo-responsive polymer nanoparticles with a core-shell micelle structure as site-specific drug carriers. *Journal of Controlled Release*, 48, 157–164.

Cho, C. S., Kobayashi, A., Takei, R., Ishihara, T., Maruyama, A. & Akaike, T. (2001) Receptor-mediated cell modulator delivery to hepatocyte using nanoparticles coated with carbohydrate-carrying polymers. *Biomaterials*, 22, 45–51.

Choi, S. H., Lee, S. H. & Park, T. G. (2006) Temperature-sensitive pluronic/poly(ethylenimine) nanocapsules for thermally triggered disruption of intracellular endosomal compartment. *Biomacromolecules*, 7, 1864–70.

Choi, Y. H., Liu, F., Park, J. S. & Kim, S. W. (1998) Lactose-poly(ethylene glycol)-grafted poly-L-lysine as hepatoma cell-targeted gene carrier. *Bioconjugate Chemistry*, 9, 708–718.

Chung, J. E., Yokoyama, M., Aoyagi, T., Sakurai, Y. & Okano, T. (1998) Effect of molecular architecture of hydrophobically modified poly(N-isopropylacrylamide) on the formation of thermoresponsive core-shell micellar drug carriers. *J Control Release*, 53, 119–30.

Chung, J. E., Yokoyama, M. & Okano, T. (2000) Inner core segment design for drug delivery control of thermo-responsive polymeric micelles. *Journal of Controlled Release*, 65, 93–103.

Chung, J. E., Yokoyama, M., Yamato, M., Aoyagi, T., Sakurai, Y. & Okano, T. (1999) Thermoresponsive drug delivery from polymeric micelles constructed using block copolymers of poly(N-isopropylacrylamide) and poly(butylmethacrylate). *J Control Release*, 62, 115–27.

Collins, D., Maxfield, F. & Huang, L. (1989) Immunoliposomes with different acid sensitivities as probes for the cellular endocytic pathway. *Biochim Biophys Acta*, 987, 47–55.

Crile, G. (1962) Selective Destruction of Cancers after Exposure to Heat. *Annals of Surgery*, 156, 404–407.

Croy, S. R. & Kwon, G. S. (2006) Polymeric micelles for drug delivery. *Curr Pharm Des*, 12, 4669–84.

Danson, S., Ferry, D., Alakhov, V., Margison, J., Kerr, D., Jowle, D., Brampton, M., Halbert, G. & Ranson, M. (2004) Phase I dose escalation and pharmacokinetic study of pluronic polymer-bound doxorubicin (SP1049C) in patients with advanced cancer. *Br J Cancer*, 90, 2085–91.

Desai, S. D. & Blanchard, J. (1998) In vitro evaluation of pluronic F127-based controlled-release ocular delivery systems for pilocarpine. *J Pharm Sci*, 87, 226–30.

Gao, Z.-G., Fain, H. D. & Rapoport, N. (2005a) Controlled and targeted tumor chemotherapy by micellar-encapsulated drug and ultrasound. *Journal of Controlled Release*, 102, 203–222.

Gao, Z., Fain, H. D. & Rapoport, N. (2004) Ultrasound-enhanced tumor targeting of polymeric micellar drug carriers. *Mol Pharm*, 1, 317–30.

Gao, Z. G., Fain, H. D. & Rapoport, N. (2005b) Controlled and targeted tumor chemotherapy by micellar-encapsulated drug and ultrasound. *J Control Release*, 102, 203–22.

Gao, Z. G., Lee, D. H., Kim, D. I. & Bae, Y. H. (2005c) Doxorubicin loaded pH-sensitive micelle targeting acidic extracellular pH of human ovarian A2780 tumor in mice. *Journal of Drug Targeting*, 13, 391–397.

Gillies, E. R. & Frechet, J. M. J. (2003) A new approach towards acid sensitive copolymer micelles for drug delivery. *Chemical Communications*, 1640–1641.

Gillies, E. R., Jonsson, T. B. & Frechet, J. M. (2004) Stimuli-responsive supramolecular assemblies of linear-dendritic copolymers. *J Am Chem Soc*, 126, 11936–43.

Guillemard, V. & Uri Saragovi, H. (2004) Prodrug chemotherapeutics bypass p-glycoprotein resistance and kill tumors in vivo with high efficacy and target-dependent selectivity. *Oncogene*, 23, 3613–21.

Gupta, H., Jain, S., Mathur, R., Mishra, P., Mishra, A. K. & Velpandian, T. (2007) Sustained ocular drug delivery from a temperature and pH triggered novel in situ gel system. *Drug Deliv*, 14, 507–15.

Hales, M., Barner-Kowollik, C., Davis, T. P. & Stenzel, M. H. (2004) Shell-cross-linked vesicles synthesized from block copolymers of Poly(D,L-lactide) and Poly(N-isopropyl acrylamide) as thermoresponsive nanocontainers. *Langmuir*, 20, 10809–17.

Hamaguchi, T., Matsumura, Y., Suzuki, M., Shimizu, K., Goda, R., Nakamura, I., Nakatomi, I., Yokoyama, M., Kataoka, K. & Kakizoe, T. (2005a) NK105, a paclitaxel-incorporating micellar nanoparticle formulation, can extend in vivo antitumour activity and reduce the neurotoxicity of paclitaxel. *Br J Cancer*, 92, 1240–6.

Hamaguchi, T., Matsumura, Y., Suzuki, M., Shimizu, K., Goda, R., Nakamura, I., Nakatomi, I., Yokoyama, M., Kataoka, K. & Kakizoe, T. (2005b) NK105, a paclitaxel-incorporating micellar nanoparticle formulation, can extend in vivo antitumour activity and reduce the neurotoxicity of paclitaxel. *British Journal of Cancer*, 92, 1240–1246.

Han, S. K., Na, K. & Bae, Y. H. (2003) Sulfonamide based pH-sensitive polymeric micelles: physicochemical characteristics and pH-dependent aggregation. *Colloids and Surfaces A: Physicochemical and Engineering Aspects*, 214, 49–59.

Harada, A., Togawa, H. & Kataoka, K. (2001) Physicochemical properties and nuclease resistance of antisense-oligodeoxynucleotides entrapped in the core of polyion complex micelles composed of poly(ethylene glycol)-poly(L-lysine) block copolymers. *European Journal of Pharmaceutical Sciences*, 13, 35–42.

Hilgenbrink, A. R. & Low, P. S. (2005) Folate receptor-mediated drug targeting: from therapeutics to diagnostics. *J Pharm Sci*, 94, 2135–46.

Hirano, T., Klesse, W. & Ringsdorf, H. (1979) Polymeric Derivatives of Activated Cyclophosphamide as Drug Delivery Systems in Anti-Tumor Chemotherapy – Pharmacologically Active Polymers. 20. *Makromolekulare Chemie-Macromolecular Chemistry and Physics*, 180, 1125–1131.

Huang, L., Ozato, K. & Pagano, R. E. (1978) Interactions of Phospholipid Vesicles with Murine Lymphocytes .1. Vesicle-Cell Adsorption and Fusion as Alternate Pathways of Uptake. *Membrane Biochemistry*, 1, 1–25.

Husseini, G. A., Diaz De La Rosa, M. A., Richardson, E. S., Christensen, D. A. & Pitt, W. G. (2005) The role of cavitation in acoustically activated drug delivery. *J Control Release*, 107, 253–61.

Husseini, G. A., El-Fayoumi, R. I., O'neill, K. L., Rapoport, N. Y. & Pitt, W. G. (2000) DNA damage induced by micellar-delivered doxorubicin and ultrasound: comet assay study. *Cancer Lett*, 154, 211–6.

Husseini, G. A., Rapoport, N. Y., Christensen, D. A., Pruitt, J. D. & Pitt, W. G. (2002) Kinetics of ultrasonic release of doxorubicin from pluronic P105 micelles. *Colloids and Surfaces B: Biointerfaces*, 24, 253–264.

Hwa Kim, S., Hoon Jeong, J., Joe, C. O. & Gwan Park, T. (2005) Folate receptor mediated intracellular protein delivery using PLL-PEG-FOL conjugate. *J Control Release*, 103, 625–34.

Itaka, K., Kanayama, N., Nishiyama, N., Jang, W. D., Yamasaki, Y., Nakamura, K., Kawaguchi, H. & Kataoka, K. (2004) Supramolecular nanocarrier of siRNA from PEG-based block catiomer carrying diamine side chain with distinctive pK(a) directed to enhance intracellular gene silencing. *Journal of the American Chemical Society*, 126, 13612–13613.

Itaka, K., Yamauchi, K., Harada, A., Nakamura, K., Kawaguchi, H. & Kataoka, K. (2003) Polyion complex micelles from plasmid DNA and poly(ethylene glycol)-poly(L-lysine) block copolymer as serum-tolerable polyplex system: physicochemical properties of micelles relevant to gene transfection efficiency. *Biomaterials*, 24, 4495–506.

Jang, J. S., Kim, S. Y., Lee, S. B., Kim, K. O., Han, J. S. & Lee, Y. M. (2006) Poly(ethylene glycol)/poly(epsilon-caprolactone) diblock copolymeric nanoparticles for non-viral gene delivery: the role of charge group and molecular weight in particle formation, cytotoxicity and transfection. *J Control Release*, 113, 173–82.

Jule, E., Nagasaki, Y. & Kataoka, K. (2003) Lactose-installed poly(ethylene glycol)-poly(d,l-lactide) block copolymer micelles exhibit fast-rate binding and high affinity toward a protein bed simulating a cell surface. A surface plasmon resonance study. *Bioconjug Chem*, 14, 177–86.

Kabanov, A. V., Batrakova, E. V. & Miller, D. W. (2003) Pluronic block copolymers as modulators of drug efflux transporter activity in the blood-brain barrier. *Adv Drug Deliv Rev*, 55, 151–64.

Kabanov, A. V., Chekhonin, V. P., Alakhov, V., Batrakova, E. V., Lebedev, A. S., Melik-Nubarov, N. S., Arzhakov, S. A., Levashov, A. V., Morozov, G. V., Severin, E. S. & et al. (1989) The neuroleptic activity of haloperidol increases after its solubilization in surfactant micelles. Micelles as microcontainers for drug targeting. *FEBS Lett*, 258, 343–5.

Kakizawa, Y., Harada, A. & Kataoka, K. (1999) Environment-sensitive stabilization of core-shell structured polyion complex micelle by reversible cross-linking of the core through disulfide bond. *Journal of the American Chemical Society*, 121, 11247–11248.

Kakizawa, Y., Harada, A. & Kataoka, K. (2001) Glutathione-sensitive stabilization of block copolymer micelles composed of antisense DNA and thiolated poly(ethylene glycol)-block-poly(L-lysine): A potential carrier for systemic delivery of antisense DNA. *Biomacromolecules*, 2, 491–497.

Kanayama, N., Fukushima, S., Nishiyama, N., Itaka, K., Jang, W. D., Miyata, K., Yamasaki, Y., Chung, U. I. & Kataoka, K. (2006) A PEG-based biocompatible block catiomer with high buffering capacity for the construction of polyplex micelles showing efficient gene transfer toward primary cells. *ChemMedChem*, 1, 439–44.

Kataoka, K., Kwon, G. S., Yokoyama, M., Okano, T. & Sakurai, Y. (1993) Block-Copolymer Micelles as Vehicles for Drug Delivery. *Journal of Controlled Release*, 24, 119–132.

Katayose, S. & Kataoka, K. (1997) Water-soluble polyion complex associates of DNA and poly(ethylene glycol)-poly(L-lysine) block copolymer. *Bioconjug Chem*, 8, 702–7.

Katayose, S. & Kataoka, K. (1998) Remarkable increase in nuclease resistance of plasmid DNA through supramolecular assembly with poly(ethylene glycol)-poly(L-lysine) block copolymer. *J Pharm Sci*, 87, 160–3.

Kim, D. W., Kim, S. Y., Kim, H. K., Kim, S. W., Shin, S. W., Kim, J. S., Park, K., Lee, M. Y. & Heo, D. S. (2007) Multicenter phase II trial of Genexol-PM, a novel Cremophor-free, polymeric micelle formulation of paclitaxel, with cisplatin in patients with advanced non-small-cell lung cancer. *Ann Oncol*, 18, 2009–14.

Kim, M. R. & Park, T. G. (2002) Temperature-responsive and degradable hyaluronic acid/Pluronic composite hydrogels for controlled release of human growth hormone. *J Control Release*, 80, 69–77.

Kim, T. Y., Kim, D. W., Chung, J. Y., Shin, S. G., Kim, S. C., Heo, D. S., Kim, N. K. & Bang, Y. J. (2004) Phase I and pharmacokinetic study of Genexol-PM, a cremophor-free, polymeric micelle-formulated paclitaxel, in patients with advanced malignancies. *Clin Cancer Res*, 10, 3708–16.

Kohori, F., Sakai, K., Aoyagi, T., Yokoyama, M., Sakurai, Y. & Okano, T. (1998) Preparation and characterization of thermally responsive block copolymer micelles comprising poly(N-isopropylacrylamide-b-DL-lactide). *J Control Release*, 55, 87–98.

Kohori, F., Sakai, K., Aoyagi, T., Yokoyama, M., Yamato, M., Sakurai, Y. & Okano, T. (1999) Control of adriamycin cytotoxic activity using thermally responsive polymeric micelles composed of poly(N-isopropylacrylamide-co-N,N-dimethylacrylamide)-b-poly(–lactide). *Colloids and Surfaces B: Biointerfaces*, 16, 195–205.

Koizumi, F., Kitagawa, M., Negishi, T., Onda, T., Matsumoto, S., Hamaguchi, T. & Matsumura, Y. (2006) Novel SN-38-incorporating polymeric micelles, NK012, eradicate vascular endothelial growth factor-secreting bulky tumors. *Cancer Res*, 66, 10048–56.

Kursa, M., Walker, G. F., Roessler, V., Ogris, M., Roedl, W., Kircheis, R. & Wagner, E. (2003) Novel shielded transferrin-polyethylene glycol-polyethylenimine/DNA complexes for systemic tumor-targeted gene transfer. *Bioconjug Chem*, 14, 222–31.

Kwon, G. K. & Forrest, M. L. (2006) Amphiphilic block copolymer micelles for nanoscale drug delivery. *Drug Development Research*, 67, 15–22.

Lampidis, T. J., Kolonias, D., Podona, T., Israel, M., Safa, A. R., Lothstein, L., Savaraj, N., Tapiero, H. & Priebe, W. (1997) Circumvention of P-GP MDR as a function of anthracycline lipophilicity and charge. *Biochemistry*, 36, 2679–85.

Lavasanifar, A., Samuel, J. & Kwon, G. S. (2002a) The effect of fatty acid substitution on the in vitro release of amphotericin B from micelles composed of poly(ethylene oxide)-block-poly(N-hexyl stearate-L-aspartamide). *J Control Release*, 79, 165–72.

Lavasanifar, A., Samuel, J. & Kwon, G. S. (2002b) Poly(ethylene oxide)-block-poly(L-amino acid) micelles for drug delivery. *Adv Drug Deliv Rev*, 54, 169–90.

Lee, E. S., Na, K. & Bae, Y. H. (2003a) Polymeric micelle for tumor pH and folate-mediated targeting. *Journal of Controlled Release*, 91, 103–113.

Lee, E. S., Na, K. & Bae, Y. H. (2005a) Doxorubicin loaded pH-sensitive polymeric micelles for reversal of resistant MCF-7 tumor. *J Control Release*, 103, 405–18.

Lee, E. S., Na, K. & Bae, Y. H. (2005b) Doxorubicin loaded pH-sensitive polymeric micelles for reversal of resistant MCF-7 tumor. *Journal of Controlled Release*, 103, 405–418.

Lee, E. S., Na, K. & Bae, Y. H. (2005c) Super pH-sensitive multifunctional polymeric micelle. *Nano Lett*, 5, 325–9.

Lee, E. S., Shin, H. J., Na, K. & Bae, Y. H. (2003b) Poly(L-histidine)-PEG block copolymer micelles and pH-induced destabilization. *J Control Release*, 90, 363–74.

Lee, K. S., Chung, H. C., Im, S. A., Park, Y. H., Kim, C. S., Kim, S. B., Rha, S. Y., Lee, M. Y. & Ro, J. (2008) Multicenter phase II trial of Genexol-PM, a Cremophor-free, polymeric micelle formulation of paclitaxel, in patients with metastatic breast cancer. *Breast Cancer Res Treat*, 108, 241–50.

Lemieux, P., Vinogradov, S. V., Gebhart, C. L., Guerin, N., Paradis, G., Nguyen, H. K., Ochietti, B., Suzdaltseva, Y. G., Bartakova, E. V., Bronich, T. K., St-Pierre, Y., Alakhov, V. Y. & Kabanov, A. V. (2000) Block and graft copolymers and NanoGel copolymer networks for DNA delivery into cell. *J Drug Target*, 8, 91–105.

Liaw, J. & Lin, Y. (2000) Evaluation of poly(ethylene oxide)-poly(propylene oxide)-poly(ethylene oxide) (PEO-PPO-PEO) gels as a release vehicle for percutaneous fentanyl. *J Control Release*, 68, 273–82.

Maeda, H. (2001) SMANCS and polymer-conjugated macromolecular drugs: advantages in cancer chemotherapy. *Advanced Drug Delivery Reviews*, 46, 169–85.

Maeda, H. & Matsumura, Y. (1989) Tumoritropic and lymphotropic principles of macromolecular drugs. *Critical Reviews in Therapeutic Drug Carrier Systems*, 6, 193–210.

Mahmud, A., Xiong, X. B., Aliabadi, H. M. & Lavasanifar, A. (2007) Polymeric micelles for drug targeting. *J Drug Target*, 15, 553–84.

Malik, D. K., Baboota, S., Ahuja, A., Hasan, S. & Ali, J. (2007) Recent advances in protein and peptide drug delivery systems. *Curr Drug Deliv*, 4, 141–51.

Marin, A., Muniruzzaman, M. & Rapoport, N. (2001) Mechanism of the ultrasonic activation of micellar drug delivery. *J Control Release*, 75, 69–81.

Mathot, F., Van Beijsterveldt, L., Preat, V., Brewster, M. & Arien, A. (2006) Intestinal uptake and biodistribution of novel polymeric micelles after oral administration. *J Control Release*, 111, 47–55.

Matsumura, Y. (2008) Poly (amino acid) micelle nanocarriers in preclinical and clinical studies. *Adv Drug Deliv Rev*, 60, 899–914.

Matsumura, Y., Hamaguchi, T., Ura, T., Muro, K., Yamada, Y., Shimada, Y., Shirao, K., Okusaka, T., Ueno, H., Ikeda, M. & Watanabe, N. (2004) Phase I clinical trial and pharmacokinetic evaluation of NK911, a micelle-encapsulated doxorubicin. *Br J Cancer*, 91, 1775–81.

Meister, A. & Anderson, M. E. (1983) Glutathione. *Annual Review of Biochemistry*, 52, 711–760.

Miyata, K., Kakizawa, Y., Nishiyama, N., Harada, A., Yamasaki, Y., Koyama, H. & Kataoka, K. (2004) Block catiomer polyplexes with regulated densities of charge and disulfide cross-linking directed to enhance gene expression. *Journal of the American Chemical Society*, 126, 2355–2361.

Muggia, F. M. (1999) Doxorubicin-polymer conjugates: further demonstration of the concept of enhanced permeability and retention.[comment]. *Clinical Cancer Research*, 5, 7–8.

Munshi, N., Rapoport, N. & Pitt, W. G. (1997) Ultrasonic activated drug delivery from Pluronic P-105 micelles. *Cancer Lett*, 118, 13–9.

Nagasaki, Y., Okada, T., Scholz, C., Iijima, M., Kato, M. & Kataoka, K. (1998) The reactive polymeric micelle based on an aldehyde-ended poly(ethylene glycol)/poly(lactide) block copolymer. *Macromolecules*, 31, 1473–1479.

Nagasaki, Y., Yasugi, K., Yamamoto, Y., Harada, A. & Kataoka, K. (2001) Sugar-installed block copolymer micelles: their preparation and specific interaction with lectin molecules. *Biomacromolecules*, 2, 1067–70.

Nakanishi, T., Fukushima, S., Okamoto, K., Suzuki, M., Matsumura, Y., Yokoyama, M., Okano, T., Sakurai, Y. & Kataoka, K. (2001) Development of the polymer micelle carrier system for doxorubicin. *J Control Release*, 74, 295–302.

Nakayama, M. & Okano, T. (2005) Polymer terminal group effects on properties of thermoresponsive polymeric micelles with controlled outer-shell chain lengths. *Biomacromolecules*, 6, 2320–7.

Nasongkla, N., Shuai, X., Ai, H., Weinberg, B. D., Pink, J., Boothman, D. A. & Gao, J. M. (2004) cRGD-functionalized polymer micelles for targeted doxorubicin delivery. *Angewandte Chemie-International Edition*, 43, 6323–6327.

Nelson, J. L., Roeder, B. L., Carmen, J. C., Roloff, F. & Pitt, W. G. (2002) Ultrasonically activated chemotherapeutic drug delivery in a rat model. *Cancer Research*, 62, 7280–7283.

Nishiyama, N. & Kataoka, K. (2006) Nanostructured Devices Based on Block Copolymer Assemblies for Drug Delivery: Designing Structures for Enhanced Drug Function *Advances in Polymer Science*, 193, 67–101.

Nishiyama, N., Okazaki, S., Cabral, H., Miyamoto, M., Kato, Y., Sugiyama, Y., Nishio, K., Matsumura, Y. & Kataoka, K. (2003) Novel cisplatin-incorporated polymeric micelles can eradicate solid tumors in mice. *Cancer Research.*, 63, 8977–83.

Nishiyama, N., Yokoyama, M., Aoyaga, T., Okano, T., Sakurai, Y. & Kataoka, K. (1999) Preparation and characterization of self-assembled polymer-metal complex micelle from cis-dichlorodiammineplatinium (II) and poly(ethylene glycole)-poly(aspartic acid) block copolymer in an aqueous medium. *Langmuir*, 15, 377–383.

Oba, M., Aoyagi, K., Miyata, K., Matsumoto, Y., Itaka, K., Nishiyama, N., Yamasaki, Y., Koyama, H. & Kataoka, K. (2008) Polyplex micelles with cyclic RGD peptide ligands and disulfide cross-links directing to the enhanced transfection via controlled intracellular trafficking. *Mol Pharm*, 5, 1080–92.

Okano, T., Yamada, N., Okuhara, M., Sakai, H. & Sakurai, Y. (1995) Mechanism of cell detachment from temperature-modulated, hydrophilic-hydrophobic polymer surfaces. *Biomaterials*, 16, 297–303.

Okano, T., Yamada, N., Sakai, H. & Sakurai, Y. (1993) A novel recovery system for cultured cells using plasma-treated polystyrene dishes grafted with poly(N-isopropylacrylamide). *J Biomed Mater Res*, 27, 1243–51.

Osada, K. & Kataoka, K. (2006) Drug and gene delivery based on supramolecular assembly of PEG-polypeptide hybrid block copolymers. *Peptide Hybrid Polymers.*

Pack, D. W., Hoffman, A. S., Pun, S. & Stayton, P. S. (2005) Design and development of polymers for gene delivery. *Nature Reviews Drug Discovery*, 4, 581–593.

Park, E. K., Kim, S. Y., Lee, S. B. & Lee, Y. M. (2005) Folate-conjugated methoxy poly(ethylene glycol)/poly(epsilon-caprolactone) amphiphilic block copolymeric micelles for tumor-targeted drug delivery. *J Control Release*, 109, 158–68.

Pitt, W. G., Husseini, G. A. & Staples, B. J. (2004) Ultrasonic drug delivery–a general review. *Expert Opin Drug Deliv*, 1, 37–56.

Rapoport, N. (2004) Combined cancer therapy by micellar-encapsulated drug and ultrasound. *Int J Pharm*, 277, 155–62.

Rapoport, N., Marin, A., Luo, Y., Prestwich, G. D. & Muniruzzaman, M. D. (2002) Intracellular uptake and trafficking of Pluronic micelles in drug-sensitive and MDR cells: effect on the intracellular drug localization. *J Pharm Sci*, 91, 157–70.

Rapoport, N. Y., Christensen, D. A., Fain, H. D., Barrows, L. & Gao, Z. (2004) Ultrasound-triggered drug targeting of tumors in vitro and in vivo. *Ultrasonics*, 42, 943–50.

Reddy, J. A., Allagadda, V. M. & Leamon, C. P. (2005) Targeting therapeutic and imaging agents to folate receptor positive tumors. *Curr Pharm Biotechnol*, 6, 131–50.

Ruiz, M. A., Clares, B., Morales, M. E. & Gallardo, V. (2007) Preparation, rheological study, and characterization of an organogel as a system for transdermal release of active principles. *Pharm Dev Technol*, 12, 637–44.

Sant, V. P., Smith, D. & Leroux, J. C. (2005) Enhancement of oral bioavailability of poorly water-soluble drugs by poly(ethylene glycol)-block-poly(alkyl acrylate-co-methacrylic acid) self-assemblies. *J Control Release*, 104, 289–300.

Sawant, R. M., Hurley, J. P., Salmaso, S., Kale, A., Tolcheva, E., Levchenko, T. S. & Torchilin, V. P. (2006) "SMART" drug delivery systems: double-targeted pH-responsive pharmaceutical nanocarriers. *Bioconjug Chem*, 17, 943–9.

Sethuraman, V. A. & Bae, Y. H. (2007) TAT peptide-based micelle system for potential active targeting of anti-cancer agents to acidic solid tumors. *J Control Release*, 118, 216–24.

Sharma, A. K., Zhang, L., Li, S., Kelly, D. L., Alakhov, V. Y., Batrakova, E. V. & Kabanov, A. V. (2008) Prevention of MDR development in leukemia cells by micelle-forming polymeric surfactant. *J Control Release*, 131, 220–7.

Shuai, X. T., Merdan, T., Unger, F. & Kissel, T. (2005) Supramolecular gene delivery vectors showing enhanced transgene expression and good biocompatibility. *Bioconjugate Chemistry*, 16, 322–329.

Shuai, X. T., Merdan, T., Unger, F., Wittmar, M. & Kissel, T. (2003) Novel biodegradable ternary copolymers hy-PEI-g-PCL-b-PEG: Synthesis, characterization, and potential as efficient non-viral gene delivery vectors. *Macromolecules*, 36, 5751–5759.

Sutton, D., Nasongkla, N., Blanco, E. & Gao, J. (2007) Functionalized micellar systems for cancer targeted drug delivery. *Pharm Res*, 24, 1029–46.

Takei, Y. G., Aoki, T., Sanui, K., Ogata, N., Okano, T. & Sakurai, Y. (1993) Temperature-responsive bioconjugates. 2. Molecular design for temperature-modulated bioseparations. *Bioconjug Chem*, 4, 341–6.

Takei, Y. G., Aoki, T., Sanui, K., Ogata, N., Sakurai, Y. & Okano, T. (1995) Temperature-modulated platelet and lymphocyte interactions with poly(N-isopropylacrylamide)-grafted surfaces. *Biomaterials*, 16, 667–73.

Tang, Y., Liu, S. Y., Armes, S. P. & Billingham, N. C. (2003) Solubilization and controlled release of a hydrophobic drug using novel micelle-forming ABC triblock copolymers. *Biomacromolecules*, 4, 1636–45.

Tian, H. Y., Deng, C., Lin, H., Sun, J. R., Deng, M. X., Chen, X. S. & Jing, X. B. (2005) Biodegradable cationic PEG-PEI-PBLG hyperbranched block copolymer: synthesis and micelle characterization. *Biomaterials*, 26, 4209–4217.

Todd, C. W., Balusubramanian, M. & Newman, M. J. (1998) Development of adjuvant-active nonionic block copolymers. *Adv Drug Deliv Rev*, 32, 199–223.

Torchilin, V. P. (2006) Multifunctional nanocarriers. *Adv Drug Deliv Rev*, 58, 1532–55.

Torchilin, V. P., Lukyanov, A. N., Gao, Z. & Papahadjopoulos-Sternberg, B. (2003) Immunomicelles: targeted pharmaceutical carriers for poorly soluble drugs. *Proc Natl Acad Sci U S A*, 100, 6039–44.

Tu, J., Pang, H., Yan, Z. & Li, P. (2007) Ocular permeability of pirenzepine hydrochloride enhanced by methoxy poly(ethylene glycol)-poly(D, L-lactide) block copolymer. *Drug Dev Ind Pharm*, 33, 1142–50.

Uchino, H., Matsumura, Y., Negishi, T., Koizumi, F., Hayashi, T., Honda, T., Nishiyama, N., Kataoka, K., Naito, S. & Kakizoe, T. (2005a) Cisplatin-incorporating polymeric micelles (NC-6004) can reduce nephrotoxicity and neurotoxicity of cisplatin in rats. *British Journal of Cancer*, 93, 678–87.

Uchino, H., Matsumura, Y., Negishi, T., Koizumi, F., Hayashi, T., Honda, T., Nishiyama, N., Kataoka, K., Naito, S. & Kakizoe, T. (2005b) Cisplatin-incorporating polymeric micelles (NC-6004) can reduce nephrotoxicity and neurotoxicity of cisplatin in rats. *Br J Cancer*, 93, 678–87.

Van Der Zee, J. (2002) Heating the patient: a promising approach? *Annals of Oncology*, 13, 1173–1184.

Vega, J., Ke, S., Fan, Z., Wallace, S., Charsangavej, C. & Li, C. (2003) Targeting doxorubicin to epidermal growth factor receptors by site-specific conjugation of C225 to poly(L-glutamic acid) through a polyethylene glycol Spacer. *Pharmaceutical Research*, 20, 826–832.

Vinogradov, S., Batrakova, E., Li, S. & Kabanov, A. (1999) Polyion complex micelles with protein-modified corona for receptor-mediated delivery of oligonucleotides into cells. *Bioconjug Chem*, 10, 851–60.

Wakebayashi, D., Nishiyama, N., Itaka, K., Miyata, K., Yamasaki, Y., Harada, A., Koyama, H., Nagasaki, Y. & Kataoka, K. (2004a) Polyion complex micelles of pDNA with acetal-poly(ethylene glycol)-poly(2-(dimethylamino)ethyl methacrylate) block copolymer as the gene carrier system: physicochemical properties of micelles relevant to gene transfection efficacy. *Biomacromolecules*, 5, 2128–36.

Wakebayashi, D., Nishiyama, N., Yamasaki, Y., Itaka, K., Kanayama, N., Harada, A., Nagasaki, Y. & Kataoka, K. (2004b) Lactose-conjugated polyion complex micelles incorporating plasmid DNA as a targetable gene vector system: their preparation and gene transfecting efficiency against cultured HepG2 cells. *J Control Release*, 95, 653–64.

Wang, C. Y. & Huang, L. (1984) Polyhistidine mediates an acid-dependent fusion of negatively charged liposomes. *Biochemistry*, 23, 4409–16.

Ward, C. M., Pechar, M., Oupicky, D., Ulbrich, K. & Seymour, L. W. (2002) Modification of pLL/DNA complexes with a multivalent hydrophilic polymer permits folate-mediated targeting in vitro and prolonged plasma circulation in vivo. *Journal of Gene Medicine*, 4, 536–547.

Xiong, X. B., Aliabadi, H. M. & Lavasanifar, A. (2006) PEO-modified Poly(L-amino acid) micelles for drug delivery, CRC Press.

Xiong, X. B., Ma, Z. S., Lai, R. & Lavasanifar, A. (2010a) The therapeutic response to multifunctional polymeric nano-conjugates in the targeted cellular and subcellular delivery of doxorubicin. *Biomaterials*, 31, 757–768.

Xiong, X. B., Mahmud, A., Uludag, H. & Lavasanifar, A. (2007a) Conjugation of Arginine-Glycine-Aspartic Acid Peptides to Poly(ethylene oxide)-b-poly(epsilon-caprolactone) Micelles for Enhanced Intracellular Drug Delivery to Metastatic Tumor Cells. *Biomacromolecules.*

Xiong, X. B., Mahmud, A., Uludag, H. & Lavasanifar, A. (2007b) Novel RGD-Functionlized Polymer-Drug Conjugate for Targeted Drug Delivery. *Proceedings of the annual meeting of the Controlled Release Society, July 2007, US.*

Xiong, X. B., Mahmud, A., Uludag, H. & Lavasanifar, A. (2008a) Multifunctional polymeric micelles for enhanced intracellular delivery of doxorubicin to metastatic cancer cells. *Pharmaceutical Research*, in press.

Xiong, X. B., Uludag, H. & Lavasanifar, A. (2010b) Virus-mimetic polymeric micelles for targeted siRNA delivery. *Biomaterials*, 31, 5886–5893.

Xiong, X. B., Uludağ, H. & Lavasanifar, A. (2008b) Biodegradable amphiphilic poly(ethylene oxide)-block-polyesters with grafted polyamines as supramolecular nanocarriers for efficient siRNA delivery. *Biomaterials.*

Yasugi, K., Nakamura, T., Nagasaki, Y., Kato, M. & Kataoka, K. (1999) Sugar-installed polymer micelles: Synthesis and micellization of poly(ethylene glycol)-poly(D,L-lactide) block copolymers having sugar groups at the PEG chain end. *Macromolecules*, 32, 8024–8032.

Yatvin, M. B., Cree, T. C., Tegmo-Larsson, I. M. & Gipp, J. J. (1984) Liposomes as drug carriers in cancer therapy: hyperthermia and pH sensitivity as modalities for targeting. *Strahlentherapie*, 160, 732–40.

Yatvin, M. B., Weinstein, J. N., Dennis, W. H. & Blumenthal, R. (1978) Design of liposomes for enhanced local release of drugs by hyperthermia. *Science*, 202, 1290–3.

Yokoyama, M., Okano, T., Sakurai, Y., Ekimoto, H., Shibazaki, C. & Kataoka, K. (1991) Toxicity and antitumor activity against solid tumors of micelle-forming polymeric anticancer drug and its extremely long circulation in blood. *Cancer Res*, 51, 3229–36.

Yoo, H. S., Lee, E. A. & Park, T. G. (2002) Doxorubicin-conjugated biodegradable polymeric micelles having acid-cleavable linkages. *J Control Release*, 82, 17–27.

Yoo, H. S. & Park, T. G. (2001) Biodegradable polymeric micelles composed of doxorubicin conjugated PLGA-PEG block copolymer. *J Control Release*, 70, 63–70.

Yoo, H. S. & Park, T. G. (2004) Folate-receptor-targeted delivery of doxorubicin nano-aggregates stabilized by doxorubicin-PEG-folate conjugate. *J Control Release*, 100, 247–56.

Yoshida, R., Sakai, K., Okano, T. & Sakurai, Y. (1994) Modulating the phase transition temperature and thermosensitivity in N-isopropylacrylamide copolymer gels. *J Biomater Sci Polym Ed*, 6, 585–98.

Yuan, X., Harada, A., Yamasaki, Y. & Kataoka, K. (2005) Stabilization of lysozyme-incorporated polyion complex micelles by the omega-end derivatization of poly(ethylene glycol)-poly(alpha,beta-aspartic acid) block copolymers with hydrophobic groups. *Langmuir*, 21, 2668–74.

Zastre, J., Jackson, J. & Burt, H. (2004) Evidence for modulation of P-glycoprotein-mediated efflux by methoxypolyethylene glycol-block-Polycaprolactone amphiphilic diblock copolymers. *Pharm Res*, 21, 1489–97.

Zhang, B., Maiti, A., Shively, S., Lakhani, F., Mcdonald-Jones, G., Bruce, J., Lee, E. B., Xie, S. X., Joyce, S., Li, C., Toleikis, P. M., Lee, V. M. & Trojanowski, J. Q. (2005a) Microtubule-binding drugs offset tau sequestration by stabilizing microtubules and reversing fast axonal transport deficits in a tauopathy model. *Proc Natl Acad Sci U S A*, 102, 227–31.

Zhang, X., Li, Y., Chen, X., Wang, X., Xu, X., Liang, Q., Hu, J. & Jing, X. (2005b) Synthesis and characterization of the paclitaxel/MPEG-PLA block copolymer conjugate. *Biomaterials*, 26, 2121–8.

Stimuli-Responsive Polymersomes

Min-Hui Li

Abstract Polymer vesicles, commonly called polymersomes, are spherical shell structures in which an aqueous compartment is enclosed by a bilayer membrane made from amphiphilic block copolymers. Compared to liposomes, their low molecular weight analogues, polymersomes have many superior properties (higher toughness, better stability, tailorable membrane properties), which make them attractive candidates for applications including drug delivery, diagnosis, nanoreactors and templates for micro- or nano-structured materials. Many potential applications require the ability to control the release of substances encapsulated in the interior compartment and /or in the hydrophobic core of membrane. To address this goal, polymersomes have been developed in which a specific stimulus destabilises the vesicular structure. The responsiveness is mainly achieved via proper hydrophobic block design. In this chapter we review the most promising approaches to make stimuli-responsive polymersomes that permit the controlled release of encapsulated contents. Chemical stimuli including hydrolysis, oxidation, reduction and pH change, and physical stimuli including temperature, light, magnetic field, electric field, osmotic shock and ultrasonic wave are discussed, on emphasizing their effects on the chemical and physical structure of the amphiphilic copolymers.

Keywords Polymersomes • Responsive • Amphiphilic block copolymers • Chemical stimuli • Physical stimuli

1 Introduction

Polymersomes, or polymer vesicles (Discher and Eisenberg 2002a; Hammer and Discher 2001), are macromolecular homologues of liposomes, or lipid vesicles. They have spherical shell structures in which an aqueous compartment is enclosed

M.-H. Li[1,2,3] (✉)
[1]Institut Curie, Centre de recherche; [2]CNRS, UMR 168; [3]Université Pierre et Marie Curie, 26 rue d'Ulm, F-75248 Paris, France
e-mail: min-hui.li@curie.fr

A. Prokop (ed.), *Intracellular Delivery: Fundamentals and Applications*, Fundamental Biomedical Technologies 5, DOI 10.1007/978-94-007-1248-5_11,

by a bilayer membrane made of amphiphilic block copolymers. Because of the high molecular weight of the block copolymers, the polymer bilayer has a higher thickness (10–30 nm) than that of the lipid blilayer (3–5 nm). Moreover, the critical micelle concentrations C_{CMC} as well as the amphiphile exchange rates between aggregates (generally proportional to C_{CMC}) are lower for polymersomes than those for liposomes. These confer polymersomes greater toughness and superior stability than liposomes. Polymersomes are therefore excellent candidates for drug carriers. Their interior cavity can be used for the encapsulation of hydrophilic substances such as hydrophilic therapeutics, while the hydrophobic bilayer membrane can trap hydrophobic moieties such as hydrophobic therapeutics or dye molecules for biomedical imaging (Discher and Ahmed 2006). More interestingly, polymer chemistry enables almost unlimited molecular design of polymersomes (Taton and Gnanou 2006). Their membrane properties can be extensively tailored by variation of chemical structures of each polymer component (Napoli et al. 2006; Mabrouk et al. 2009b; Battaglia and Ryan 2005). Targeted transport can be achieved by taking advantage of the many possibilities to end-functionalize the copolymers (Lin et al. 2004; Meng et al. 2005; Broz et al. 2005; Ben-Haim et al. 2008; Christian et al. 2007; Hammer et al. 2008; Zupancich et al. 2009; Robbins et al. 2010; Nehring et al. 2009; Geng et al. 2006; Demirgoz et al. 2009; van Dongen et al. 2008; Opsteen et al. 2007; Sun et al. 2009). The controlled release of therapeutic substances can be integrated through the use of copolymers with blocks that respond to external or internal stimuli in the treated disease sites (Meng et al. 2009; Li and Keller 2009; Du and O'Reilly 2009; Onaca et al. 2009).

In order to achieve effective intracellular drug delivery, the polymersomes should be engineered to (1) survive in biological fluids and extracellular space, (2) bind to cell surface, (3) escape or survive in the endocytic pathway, (4) release drug when the cytosol or another desired intracellular compartment is reached. Development of "smart" polymersomes, i.e., stimuli-responsive polymersomes bearing a protective coat, site-specific targeting ligands and a cell-penetrating function, is the current research trend in this field. In this review, we will focus on the stimuli-responsive properties of polymersomes made from amphiphilic block copolymers. The responsiveness is mainly achieved via proper hydrophobic block design so that the hydrophobic core of the membrane can be altered or destroyed under the action of chemical and physical stimuli. Before the detail discussion of this issue, we first briefly discuss the choice of hydrophilic block, the structural requirement of block copolymer for polymersome formation, and the techniques employed for polymersome preparation, and strategies to get controlled release.

1.1 Hydrophilic Block Choice

Biocompatible and flexible hydrophilic polymers are usually utilized as the hydrophilic block of the amphiphilic block copolymers in order to design non-fouling and non-antigenic (stealth) polymersomes for a long circulation time in the bloodstream (Knop et al. 2010). The most used hydrophilic polymer is polyethylene glycol (PEG, also known as polyethylene oxide, PEO), which is a neutral polymer and confer the steric stabilization to polymersomes (Lasic and Papahadjopoulos 1998; Semple et al. 1998). Other neutral polymers have also been developed, such as poly(2-methyloxazoline) (PMOXA) (Nardin et al. 2000), which also shows low non-specific protein binding ability (Woodle et al. 1994) and poly[2-(methacryloyloxy)ethyl phosphorylcholine] (PMPC), a promising hydrophilic block reported recently (Du et al. 2005; Du and Armes 2009). Several hydrophilic polymers attempted first for stealth liposomes, including poly[N-(2-hydroxypropyl)methacrylamide] (Whiteman et al. 2001), poly-N-vinylpyrrolidones (Torchilin et al. 2001), polyvinyl alcohol (Takeuchi et al. 2001) and amino-acid-based biodegradable polymer (Romberg et al. 2007), can also be used to design the hydrophilic block of polymersomes. Hydrophilic peptide-based polymers have attracted considerable interests for their high biocompatibility (Kukula et al. 2002; Bertin et al. 2010).

The polyacrylic acid (PAA) is a typical non-neutral hydrophilic block for polymersome preparation (Zhang and Eisenberg 1995b), which is anionic in physiological pH (=7.4). Kataoka's group has shown that anionic polymer spherical micelles had significant promise over neutral spherical micelles in evading the mononuclear phagocytic system of liver and spleen that is primarily responsible for the clearance of foreign particles (Yamamoto et al. 2001). A recent research in Discher's group (Christian et al. 2010) has taken advantage of this anionic profile to make very long-circulating polymersomes with a red cell-like surface charge. As a matter of fact, the life span of healthy human red blood cells is of 100 days, while the polymersomes with neutral PEG hydrophilic shell circulate for 1–2 days. The red blood cells' external "glycocalyx" contains sialic acid which imparts a negative surface charge of approximately −15 mV. Discher's group has achieved this surface charge of polymersomes by blending an anionic diblock copolymer, poly(acrylic acid)-*b*-polybutadiene (PAA-*b*-PBD), with a neutral diblock copolymer (PEG-*b*-PBD) under physiological conditions.

1.2 Structural Consideration of Block Copolymers for Polymersome Formation

The unfavourable contact between water and the hydrophobic part of amphiphilic molecules leads to their self-assembly into small aggregates in water with a hydrophobic core and a hydrophilic shell. The aggregates display rich polymorphism, the simplest structure being spherical micelles. If only isotropic fluidlike arrangements are considered in the hydrophobic core and in the hydrophilic shell, the self-organizing structures can be partially predicted on the basis of concepts developed for small amphiphilic molecules, such as the 'spontaneous curvature' or the 'critical packing parameter', defined as $p = v/al$, where v is the hydrophobic volume, a is the optimal interfacial area and l is the length of the hydrophobic block normal to the interface.(Israelachvili 2005) For $p \leq 1/3$, surfactants are predicted to assemble into spherical micelles; for $1/3 \leq p \leq 1/2$, cylindrical micelles are expected; and for $1/2 \leq p \leq 1$, lamellar aggregates or vesicles should form spontaneously. The packing parameter is not strictly a geometrical parameter, since it is dependent on a number of other factors such as thermodynamics and interactions. In the case of coil-coil polymer amphiphiles, for example, the consideration of the packing parameter is more complicated and should be connected with the macromolecular character of each block, including the entropy contribution and the interaction between water and the hydrophilic polymer block. It is often noted that the morphology of self-assembled block copolymer aggregates is governed primarily by three components of the free energy of aggregation: (*i*) core–chain stretching, (*ii*) interfacial energy, and (*iii*) intercoronal chain interactions (Gennes 1978; Zhang and Eisenberg 1998; Halperin et al. 1992). These components depend on the chemical structure of the copolymer, the hydrophilic/hydrophobic ratio, the copolymer concentration in solution, the solvent properties such as the type of organic solvent and the ratio of organic solvent/water, salt concentration, solution pH, temperature or shear rate. Extensive studies have been made on prototypical amphiphilic block copolymers, such as poly(acrylic acid)-*b*-polystyrene (PAA-*b*-PS) (Zhang and Eisenberg 1995a, 1996), PEG-*b*-PS (Yu and Eisenberg 1998; Bhargava et al. 2006), PEO-*b*-PBD (Won et al. 1999; Jain and Bates 2003; Discher and Eisenberg 2002b), polyethyleneglycol-*b*-polycaprolactone (PEG-*b*-PCL) (Meng et al. 2003; Ghoroghchian et al. 2006; Ahmed and Discher 2004), poly(methacrylic acid)-*b*-polybutadiene (PMAA-*b*-PBD) (Fernyhough et al. 2009). Spherical micelles, cylindrical micelles and bilayer vesicles have all been observed in these coil-coil polymer systems. An appropriate phospholipids-like hydrophilic/hydrophobic ratio is often used as a practical guide to prepare block copolymers which form vesicles. But this is not universal to all systems and should be taken with care. For example, vesicles were formed directly in water by PEO-*b*-PBD as its hydrophilic to total mass ratio f was around 35% (a phospholipids-like ratio), while vesicles were also obtained by PAA-*b*-PS with a very short hydrophilic block (e.g. $f < 20\%$, even at 4%) in the mixture of water and dioxane (Eisenberg 1999).

If the organization in the hydrophobic core and/or in the hydrophilic shell is not isotropic fluidlike, but crystalline or strongly bound by hydrogen bonding or $\pi-\pi$ stacking etc, considerable changes will take place in the micellar aggregates (Antonietti and Forster 2003). Rigid rods and tubules instead of fluid cylindrical micelles and planar bilayers are, for example, obtained in these kinds of small amphiphilic molecules. In the case of coil-rod block copolymers, the shape anisotropy and additional order (introduced by liquid crystalline, crystalline structures and secondary peptide structures such as α helices or β sheets) in the rod-like block also influences considerably the self-assembly (Pinol et al. 2007; Jia et al. 2009; Jenekhe and Chen 1998; Massey et al. 2000; Kukula et al. 2002; Bellomo et al. 2004; Burkoth et al. 2000; Jiang et al. 2007a). In conclusion, the polymersome formation depends on the chemical and physical structure of the block copolymers.

1.3 Polymersome Preparation

Polymersome preparation requires the mutual diffusion of water into the bulk block copolymer and vice versa (Battaglia and Ryan 2006). In general, all reported methods for liposome preparation can also be used for polymersome preparation (Szoka and Papahadjopoulos 1978; Gregoriadis 1992; Angelova and Dimitrov 1986; Pautot et al. 2003a). In preparation protocols, the contact between water and polymer can be achieved directly or with the aid of organic solvent if the hydrophobic block is glassy at the preparation temperature. For block copolymers with hydrophobic blocks with a low Tg, such as PEO-*b*-PBD ($T_{g,PBD}$ ~ −90°C to −8°C according to their relative 1,4- and 1,2-content), vesicles can be formed by direct hydration techniques assisted by sonication or electrical field (Dimova et al. 2002). In contrast, block copolymers with a glassy hydrophobic block, such as PAA-*b*-PS ($T_{g,PS}$ ~ 100°C), often require an organic co-solvent to fluidize the polymer layers. Typically, a polymer solution is first prepared in an organic solvent and the solvent is then gradually exchanged with water (Zhang and Eisenberg 1995b). It belongs to a more general nanoprecipitation method based on the interfacial deposition due to the displacement of a solvent with the non-solvent. Recently, microfluidic and micro-patterning technology have opened some fascinating ways to prepare polymersomes with controlled size and efficient encapsulation (Hauschild et al. 2005; Shum et al. 2008; Stachowiak et al. 2009; Howse et al. 2009).

Polymersome preparation methods discussed above naturally lead to a symmetric copolymer distribution between both leaflets of bilayer membrane. One special method, called inverted emulsion which is first developed by Weitz' group for liposomes (Pautot et al. 2003c), permits to obtain asymmetric vesicles by independent assembly of the inner and outer leaflets of the vesicle (Pautot et al. 2003b). Briefly, the inner monolayer is first formed via the emulsification of water droplets in oil containing the first amphiphile of interest. The outer monolayer is then formed by the centrifugation of the water droplets stabilized by the first amphiphile through the monolayer of the second amphiphile at the interface between a second oil solution (containing the second amphiphile of interest) and a water solution.

1.4 Strategies for Controlled release

The major advantage of polymersomes is that they can be used as carriers of hydrophilic substances (in the interior compartment) as well as hydrophobic substances (in the membrane) offering a cocktail-treatment or a diagnostic-therapy combination in biomedical applications. A crucial question here is how to release the active substances when and where they are needed. In general, the continuous loss of encapsulated substance via diffusion is very slow as a result of the considerable thickness of the polymer membrane. In some cases, more selective permeability has been achieved through the use of special chemical structures for the block copolymer (Battaglia et al. 2006), (Vriezema et al. 2007), or by incorporating channel proteins into the polymersome membranes (Choi and Montemagno 2005). In most applications it is highly desirable to be able to control the release of encapsulated substance by triggering a change in membrane properties of polymersomes via the action of a stimulus. The present chapter focuses on this issue of controlled release and reviews the most promising approaches to create stimuli-responsive polymersomes. So far, two strategies have been followed to achieve the controlled disassembly of polymersomes. The first one exploits the vast possibilities of chemical synthesis to develop polymer membranes sensitive to the chemical stimuli, including hydrolysis, oxidation reaction, reduction reaction and pH change. The second strategy, which also takes advantage of the chemical diversity of polymers, uses physical stimuli, such as temperature variation, light, magnetic fields, electric field, osmotic shock or ultrasonic wave, to remotely destroy the polymersomes. Stimuli-responsive polymersomes are then reviewed in detail according to these two strategies by dividing the stimuli into chemical and physical stimuli.

2 Polymersomes Responsive to Chemical Stimuli

2.1 *Response to Hydrolytic Degradation*

Biodegradable polyester-based polymersomes have been made from polyethyleneglycol-*b*-polylactic acid (PEG-*b*-PLA) (Ahmed and Discher 2004), (Meng et al. 2003), polyethyleneglycol-*b*-polycaprolactone (PEG-*b*-PCL) (Ghoroghchian et al. 2006), and polyethyleneglycol-*b*-poly(γ-methyl-ε-caprolactone) (PEG-*b*-PMCL) (Zupancich et al. 2006). Under physiological conditions (pH = 7.4), these polyesters degrade by hydrolysis. However, polyester hydrolysis is accelerated by low pH, which may be useful given the acidic environment in tumors and endolysosomes. To provide controllable degradation and adjustable release times ranging from hours to weeks, polymersomes were formed by blending PEG-*b*-PLA and PEG-*b*-PCL with inert PEO-*b*-PBD (Ahmed and Discher 2004). These polymersomes were loaded with two anticancer drugs. Doxorubicin (water soluble) was loaded in the aqueous interior of the polymersomes while paclitaxel (water insoluble) was included in the hydrophobic layer of the membrane. The loaded polymersomes were degraded in vivo and drug release occurred with a time scale of a day. This degradation and release was shown to be coupled with the phase transition behaviour of the block copolymer amphiphiles (see Fig. 1). Polyester hydrolysis occurs preferentially at the chain end, thereby increasing the hydrophilic/hydrophobic ratio of PEG-polyester chains and preferred curvature of the self-assembly. If we discuss with the packing parameter *p*, the hydrophobic chain shortening induces a decrease of *p* value from 1 to 1/2 to 1/3. The comparatively short hydrophobic blocks of the degraded chains are unstable in a bilayer. Instead, they tend to segregate, congregate, and ultimately induce hydrophilic (i.e., PEG-lined) pores and eventually the vesicular carriers disintegrate into mixed micellar assemblies. These polymersomes are a promising method for multi-drug delivery.

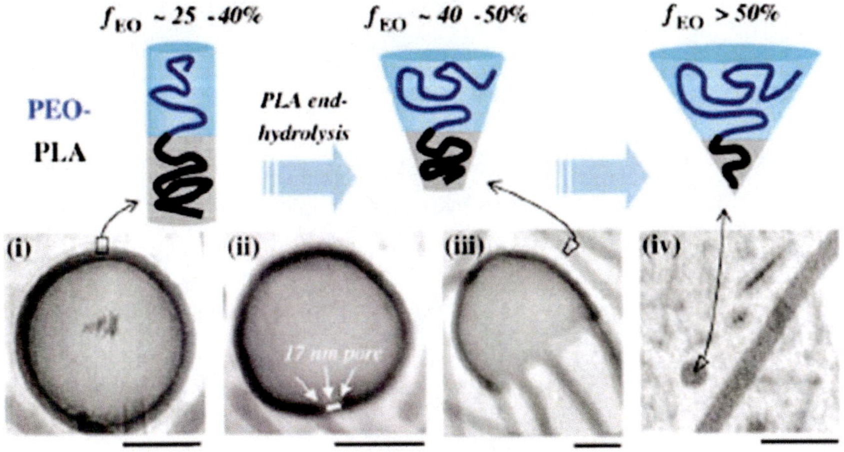

Fig. 1 Representative time series showing the breakdown structures formed as PEG-*b*-PLA polymersomes transform from vesicles to worm-like micelles. The scale bars in the cryo-TEM images are 100 nm. (Reproduced from ref. Ahmed et al. 2006)

2.2 *Response to Oxidation Reaction*

Instead of hydrolyzing the hydrophobic polymer block, Hubbell and co-workers changed the value of hydrophilic/hydrophobic ratio via oxidation of the hydrophobic block. This strategy could potentially exploit the oxidative environment present in sites of inflammation as well as within endolysosomes. They developed oxidation-responsive polymersomes from PEO-*b*-PPS-*b*-PEO (PPS: polypropylene sulphide) triblock copolymers (Napoli et al. 2004a, b, 2005), and showed that exposure to either aqueous H_2O_2 or H_2O_2 from glucose-oxidase/glucose/oxygen system oxidized the hydrophobic polypropylene sulphide layer, thereby transforming it into hydrophilic poly(propylene sulphoxide) and poly(propylene sulphone). Oxidation thus increased the hydrophilic percentage in the amphiphilic system, thereby causing a transition from polymersomes to wormlike micelles, spherical micelles, and eventually soluble oxidized copolymers. This transition took about 10 h in 10-vol%H_2O_2 aqueous solution and around 300 h in 0.03-vol% H_2O_2. These oxidation-responsive polymersomes may find applications as nanocontainers in drug delivery, biosensing and biodetection.

2.3 *Response to Reduction Reaction*

Reduction-sensitive materials containing disulphide bonds have been used to produce reduction-responsive polymersomes. Hubbell and co-workers recently reported polymersomes based on diblock copolymers PEG-SS-PPS, where a

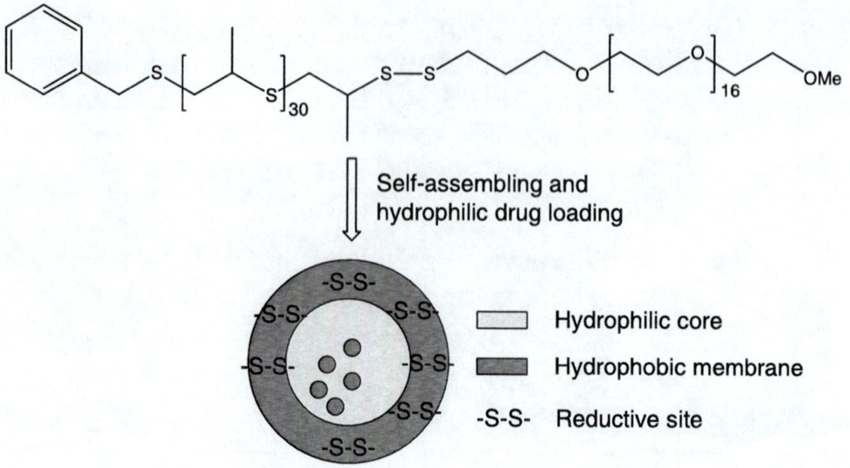

Fig. 2 Polymersomes formed from the block copolymer, PEG_{17}-SS-PPS_{30}, where the hydrophilic poly(ethylene glycol) (PEG) and the hydrophobic poly(propylene sulfide) (PPS) are connected with a reductive disulfide group. (Reproduced from ref. Cerritelli et al. 2007)

reduction-sensitive disulphide link (-S-S-) was placed between the two blocks (see Fig. 2) (Cerritelli et al. 2007). Intracellular glutathione or cysteine can reductively cleave these links and thereby destabilise the system. Other reductive compounds such as dithiothreitol and tris(2-carboxyethyl)phosphine can also cleave the disulfide bonds (Li et al. 2006b).

Polymersomes formed from the block copolymer PEG-SS-PPS were demonstrated to break down in the presence of intracellular concentrations of cysteine. In cellular experiments, uptake, disruption, and release were observed within 10 min of exposure to cells, well within the time frame of early endosome and endolysosomal processing. This system presents obvious advantages over the hydolysis and the oxidation responsive polymersomes reviewed above. The hydrolysis process is not nearly as fast (>hours) and the low pH needed to accelerate the hydrolysis is encountered only within the lysosomal compartment of the endosomal-lysosomal processing pathway. Oxidative conditions are also not encountered until the vesicles are processed within the lysosome. In both cases, the release can be too late for sensitive biological macromolecules, i.e. within the less desirable lysosome, where biomolecular drugs are exposed to very harsh conditions, rather than the more desirable endosome. In contrast, reduction-responsive polymersomes can rapidly burst within the early endosome, releasing their contents prior to exposure to the harsh conditions encountered after lysosomal fusion. This system may be useful in cytoplasmic delivery of biomolecular drugs such as peptides, proteins, oligonucleotides, and DNA.

2.4 pH Response

2.4.1 pH Responsive Polymersomes Formed from Synthetic Block Copolymers

Vesicles containing polymer blocks with solubility responding to changes in pH present additional opportunities for controlled release. A pH-response can be obtained either by using polyacid blocks (e.g. PAA) whose ionization status can be changed by a pH variation, or by using polybase blocks (e.g. PVP: polyvinyl pyridine) that can be rendered water soluble by protonation at low pH. One distinct advantage of pH-triggered release is the fast response of the system. For example, the acid-induced change for the polybase blocks solubility occur almost instantaneously, whereas the hydrolysis of PLA or PCL, occurs over time scales ranging from minutes to days, even when it is acid-catalyzed.

We will first discuss systems based on polyacid. Since the early work of Eisenberg's group (Zhang and Eisenberg 1995b) on mapping the phase diagram of "crew-cut" PAA-*b*-PS in dilute organic/aqueous solution, Liu & Eisenberg (Liu and Eisenberg 2003) have shown rapid pH-triggered inversion of amphiphilic triblock copolymer vesicles of PAA-*b*-PS-*b*-P4VP in organic/aqueous solution mixtures. For vesicles formed from a PAA-*b*-PBD diblock copolymer in water, Discher's group (Geng et al. 2005) observed that a sudden increase in pH induced the rapid (~minutes) transition of vesicles into worms and spheres. Chiu's group has recently reported an interesting work on polymersomes with pH-responsive transmembrane channels (Chiu et al. 2008). The copolymers used to form vesicles were not block copolymers but rather random copolymers of acrylic acid (AAc) and distearin acrylate (DSA), that were obtained from partial transesterification of poly(N-acryloxysuccinimide) (poly(NAS)) with distearin (a lipid graft) followed by thorough hydrolysis of the un-reacted NAS to AAc units. Using a double emulsion technique in a water/oil/water system and a copolymer with an average molecular weight of 2.97×10^5 g mol^{-1} and a composition of 9.1 mol% DSA, they prepared large polymersomes that contained small polymersomes and had a pH of 4.0–5.5 within the interior aqueous compartment (see Fig. 3). When the pH was increased to 6.5, the vesicles became permeable to hydrophilic solutes. The authors suggested that un-ionized AAc-rich regions in the hydrophobic bilayer regions of distearate grafts (parallel to the aligned lipid chains) could act as pH-responsive channels. When the pH was increased to 6.5, AAc ionization would occur and the resulting abrupt disruption of hydrogen bonds and hydrophobic association of un-ionized AAc would create permeable channels. This system is an elegant example of vesicles that were equipped with transmembrane channels without requiring the incorporation of channel-forming proteins. Similarly, Eisenberg's group (Yu et al. 2009) has also reported polymersomes with pH-induced reversible change of permeability to water and to proton. They call them "breathing" polymerosmes, which are prepared from triblock copolymer poly(ethylene oxide)$_{45}$-*b*-polystyrene

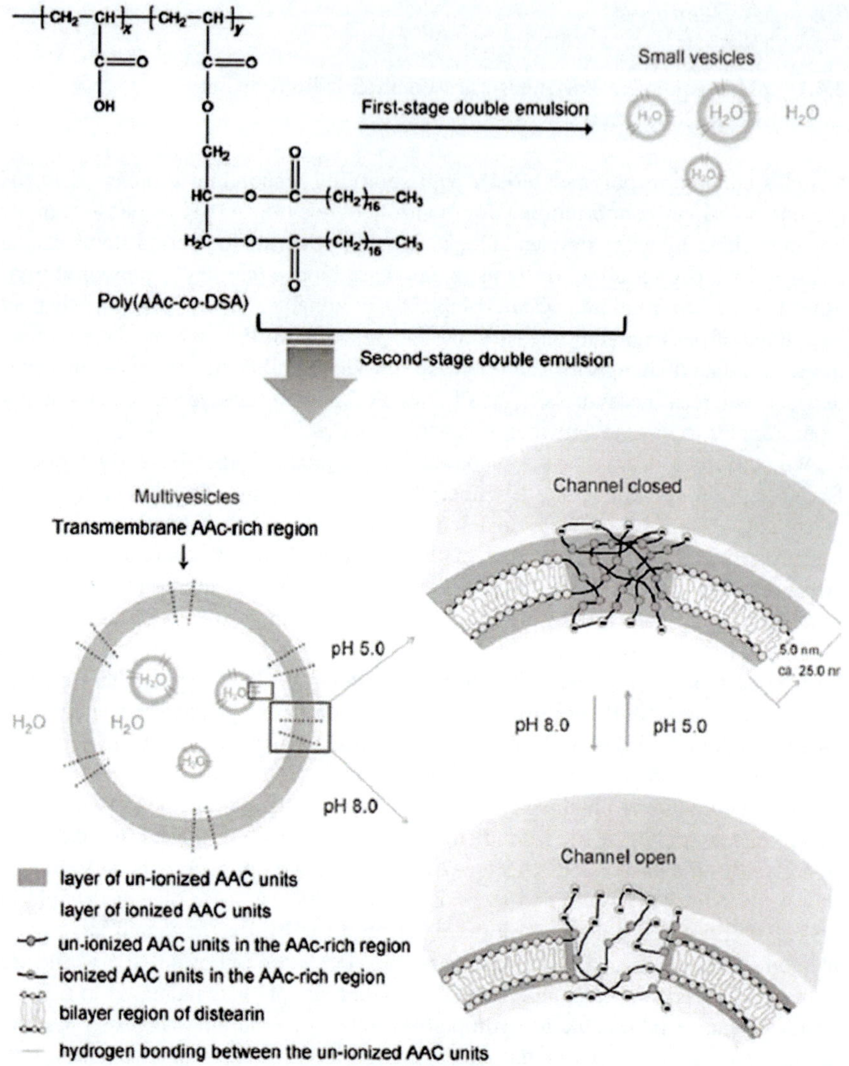

Fig. 3 Illustration of multi-vesicular assemblies produced via a two-stage double emulsion of poly(AAc-*co*-DSA). The resulting vesicles are equipped with pH-responsive transmembrane channels. (Reproduced from ref. Chiu et al. 2008)

$_{130}$-*b*-poly(2-diethylaminoethyl methacrylate)$_{120}$ (PEO$_{45}$-*b*-PS$_{130}$-*b*-PDEA$_{120}$). Self-assembly into vesicles was carried out at a pH of ca. 10.4. The vesicle wall was shown by cryo-TEM to consist of a sandwich of two external ca. 4 nm thick continuous PS layers and one ca. 17 nm thick PDEA layer in the middle (Fig. 4). As the pH decreases, both the vesicle size and the thickness of all three layers increase. The increase of the thickness of the intermediate PDEA layer arises from the protonation and hydration, but the swelling is constrained by the PS layers.

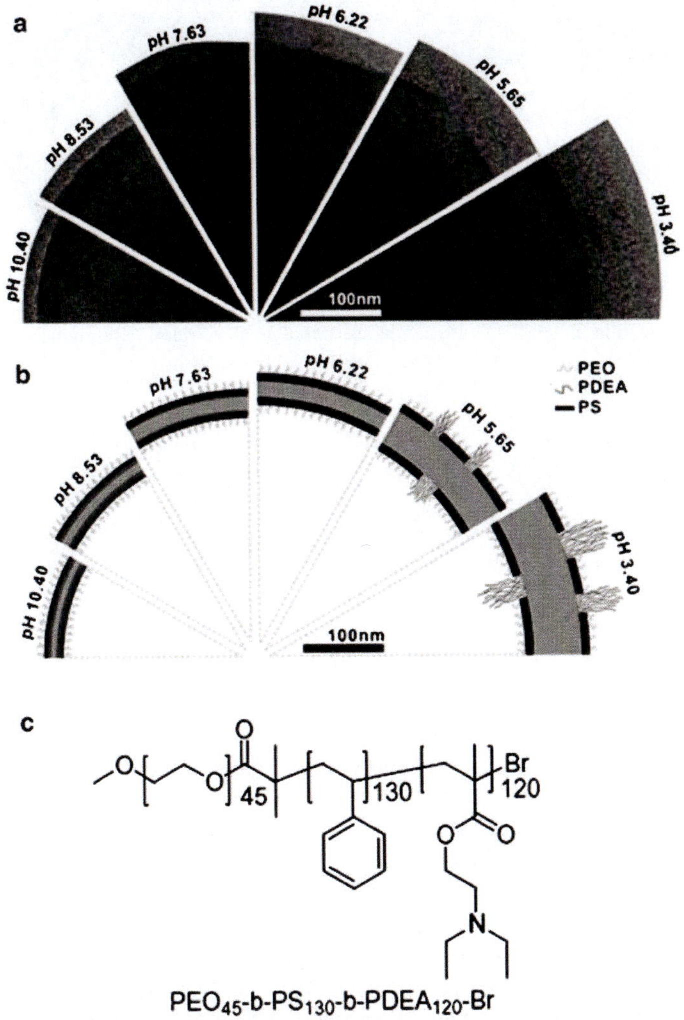

Fig. 4 (**a**) cryo-TEM images of vesicle wall structures at pH 10.4, 8.53, 7.63, 6.22, 5.65, and 3.40, respectively; (**b**) schematic illustrations of the three-layered wall structures at corresponding pH values; (**c**) the chemical structure of the triblock copolymer used to form the breathing polymersomes. (Reproduced from ref Yu et al. 2009)

The increase of the thickness of the two PS layers is a result of an increasing incompatibility and an accompanying sharpening of the interface between the PS layers and the PDEA layer. Starting at a pH slightly below 6, progressive swelling of the PDEA layer with decreasing pH induces a cracking of the two PS layers and also a sharp increase of the vesicle size and the wall thickness.

For applications in chemical and gene therapy, pH-induced release based on polybase blocks may permit the delivery of drugs and genes into the cytosol via endolysosomal acidification and escape. Battaglia et al. recently prepared biomimetic pH

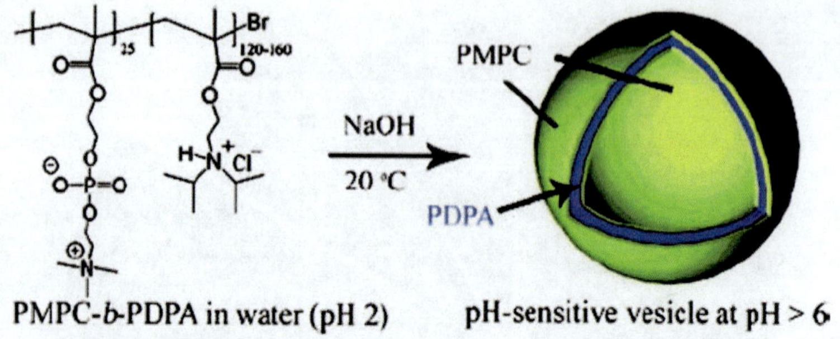

Fig. 5 Formation of PMPC-*b*-PDPA block copolymer vesicles. (Reproduced from ref. Du et al. 2005)

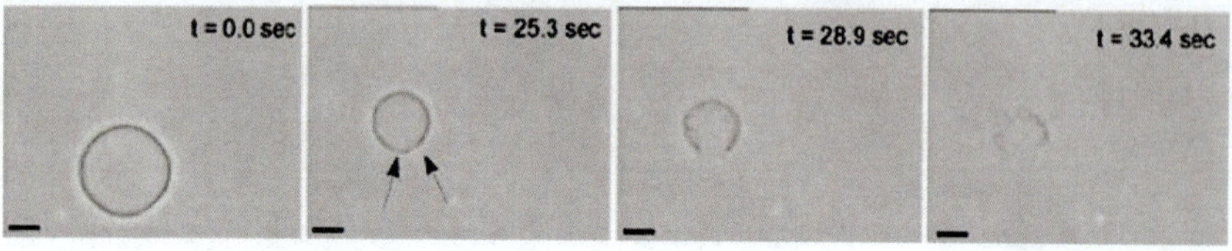

Fig. 6 Sequence of microscopy images showing the dissolution of a giant polymersome of P2VP-*b*-PEO upon addition of dilute acetic acid. (Reproduced from ref. Borchert et al. 2006)

sensitive polymersomes for efficient DNA encapsulation (Lomas et al. 2007) using pH-sensitive diblock copolymers, poly(2-(methacryloyloxy)ethylphosphorylcholine)-*b*-poly(2-diisopropylamino)ethyl methacrylate) (PMPC-*b*-PDPA, see Fig. 5), developed in Armes' group (Du et al. 2005).

PMPC is a phospholipid-like, biocompatible, and stealthy hydrophilic polymer, while PDPA is a polybase that has a pH-dependent solubility in water (pK_a 5.8–6.6). Such diblock copolymers form stable vesicles at physiological pH but rapidly dissociate at around pH 5–6. DNA was encapsulated within these PMPC-*b*-PDPA polymersomes at neutral pH, and lowering the solution pH then caused the disruption of the polymersomes and the formation of DNA-copolymer complexes.

Förster's group has used poly(2-vinylpyridine)-*b*-polyethylene oxide (P2VP-*b*-PEO) copolymers to produce another example of fast, complete release of polymersome contents via pH-induced dissolution (Borchert et al. 2006). P2VP is also a polybase that is insoluble in water under neutral and alkaline conditions. However, when the pH is below 5, P2VP is protonated and becomes water-soluble. Fig. 6 shows the rapid dissolution of a P2VP-*b*-PEO giant polymersome within around 30 sec resulting from the addition of dilute acetic acid.

Du and Armes have reported that variation of pH could be used to tune the membrane permeability of cross-linked polymer vesicles in THF/water mixtures (Du and Armes 2005). These vesicles are formed by a pH-responsive, hydrolytically self-crosslinkable copolymer, poly(ethylene oxide)-*b*-poly(2-(diethylamino) ethyl methacrylate-*stat*-3-(trimethoxysilyl)propyl methacrylate) (PEO-*b*-P(DEA-*stat*-TMSPMA)). DEA homopolymer is water soluble at low pH and becomes insoluble above pH 7. In the cross-linked hydrophobic membranes of polymersomes, a decrease in solution pH protonates the DEA residues which causes in turn the membrane swelling and the permeability increasing.

2.4.2 pH Responsive Polymersomes Formed from Peptide-Based Polymersomes

Polypeptides are a special class of building blocks for vesicle-forming amphiphilic block copolymers because of their stimuli-responsiveness (to pH or temperature), secondary structures, functionalities, and biocompatibility (Schlaad 2006), (Carlsen and Lecommandoux 2009). Nevertheless, there are only a limited number of examples of polymersomes made from polypeptide-containing block copolymers and these examples can be divided into two main families. The first family is composed of hybrid block copolymers, where the polypeptide is the hydrophilic block and a classical synthetic polymer, such as polybutadiene (PBD) or polyisoprene (PI), is the hydrophobic block. The second family includes co-polypeptides in which both the hydrophilic and hydrophobic blocks are polypeptides. Polymersomes formed by hybrid block copolymers with a polypeptide as the hydrophobic block and a synthetic polymer as the hydrophilic one have not yet been reported.

Polymersomes have been formed in aqueous solutions using polybutadiene-*b*-poly(L-glutamic acid) (PBD-*b*-PGlu, 17–54 mol% glutamate) (Kukula et al. 2002; Checot et al. 2002, 2005), polybutadiene-*b*-poly(L-lysine) (PBD_{165}-*b*-$PLys_{88}$ and PBD_{107}-*b*-$PLys_{27}$) (Sigel et al. 2007; Gebhardt et al. 2008) and polyisoprene-*b*-poly(L-lysine). In these systems, a change in pH or temperature can induce a change in the secondary structure of the hydrophilic polypeptide corona (charged coil, α-helix or β-sheet). The secondary structure of poly(L-glutamic acid) (a polypeptide with –COOH side groups) changes from a charged coil at high pH ($pH>6$) to an α-helix at low pH ($pH<5$), while that of poly(L-lysine) (a polypeptide with $-NH_2$ side groups) changes from an α-helix at high pH ($pH=11$) to a charged coil at low pH ($pH=6$). It has been reported that the transition from a charged coil to a α-helix conformation in the hydrophilic corona did not change the morphology of the polymersomes, but did cause a large decrease in their size (hydrodynamic radius, by 20–50%). Savin's group has also reported that in basic conditions ($pH=10.9$), a temperature increase induced a transition from an α-helix to a β-sheet conformation in the poly(L-lysine) corona of polymersomes made from PBD_{107}-*b*-$PLys_{27}$ (Gebhardt et al. 2008). Consequently, the hydrodynamic radius of the polymersomes increased by a factor of two (from 70 nm to 140 nm) when the temperature was raised from 40°C to 63°C. This block copolymer is a good example of a dual-responsive system sensitive to both pH and temperature changes.

Similar pH and temperature dual-responsive polymersomes were also formed from a series of triblock ABA copolypeptides (poly(L-lysine)-*b*-poly(γ-benzyl-L-glutamate)-*b*-poly(L-lysine) (PLys-*b*-PBGlu-*b*-PLys)) in which the block ratios ranged widely (Iatrou et al. 2007). For example, the hydrodynamic radius of polymersomes made by $PLys_{134}$-*b*-$PBGlu_{64}$-*b*-$PLys_{134}$ was of 129 nm at $pH=7.4$ and $T=25°C$ (PLys in charged coil); this value decreased to 92.5 nm at $pH=11.7$ and

T=25°C (PLys in α-helix); it increased further to 148 nm at the same pH=11.7 but at higher temperature T=37°C (PLys in β-sheet). In all these conditions, the neutral hydrophobic central block PBGlu retained an α-helix conformation.

While the secondary structure change in the polypeptide hydrophilic corona induces a change in polymersome size, the mechanism has not been resolved and is the subject of vigorous discussion (Checot et al. 2002; Sigel et al. 2007). The release process seems not to be directly connected with these size changes. Conversely, the effects of secondary structure on polymersome assemblies should be pronounced when the stimuli-responsive polypeptide constitutes the hydrophobic layer of the membrane. These effects are well illustrated by the following two examples.

Deming's group has introduced non-ionic block copolypeptides of L-leucine and ethylene glycol-modified L-lysine residues, PELys-*b*-PLeu [poly(N_ε-2-(2-(2-methoxyethoxy)ethoxy)acetyl –L-lysine)-*b*-poly(L-leucine)] (Bellomo et al. 2004). These copolypeptides have been shown to self-assemble into bilayer vesicles whose size and structures are dictated primarily by the ordered conformations (rod-like or α-helical) of the peptides segments. pH sensitivity can be achieved by replacing 70% of the L-leucine in the hydrophobic domain with L-lysine. At pH=9, the conformation of the hydrophobic block remains α-helical and vesicles form. However, at pH=3, protonation of the lysine residues enhances their hydrophilicity and simultaneously destabilizes the α-helical structure of the leucine-rich domain because of electrostatic repulsion. This helix-to-coil transition destabilises the vesicular assembly, leading to porous membranes or complete dissociation of the structures (see Fig. 7).

Rodriguez-Hernandez and Lecommandoux have used a zwitterionic block copolypeptide, poly(L-glutamic acid)-*b*-poly(L-lysine) ($PGlu_{15}$-*b*-$PLys_{15}$), to form

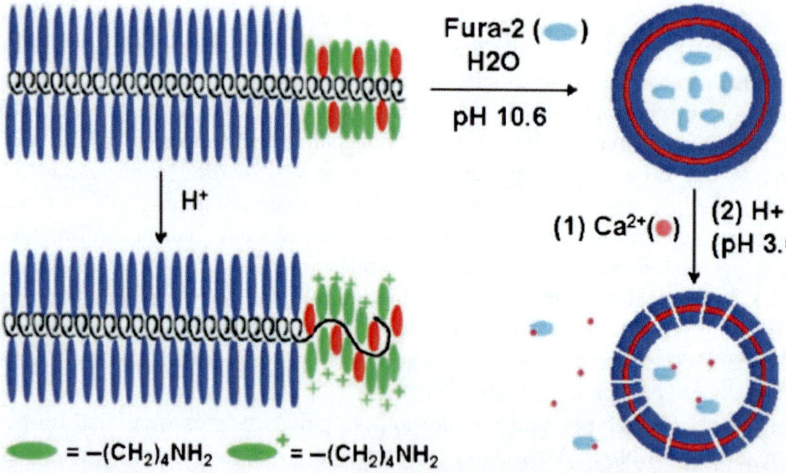

Fig. 7 pH-responsive polymer vesicles formed from PELys-*b*-P($Leu_{0.3}$-*co*-$Lys_{0.7}$)$_{40}$. When the pH was lowered by addition of HCl, the release of entrapped Fura-2 dye took place within seconds. (Reproduced from ref. Bellomo et al. 2004)

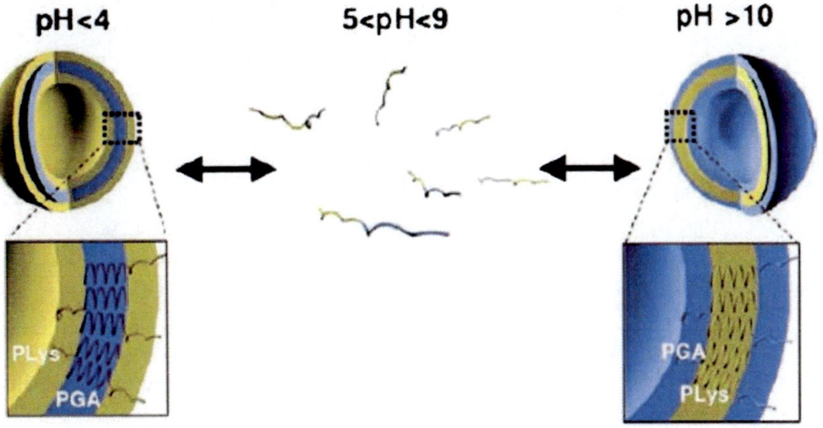

Fig. 8 Schematic representation of the self-assembly of vesicles from the diblock copolymer, $PGlu_{15}$-*b*-$PLys_{15}$ (PGlu is noted as PGA in the sketch). (Reproduced from ref. Rodriguez-Hernandez and Lecommandoux 2005)

schizophrenic vesicles (see Fig. 8) (Rodriguez-Hernandez and Lecommandoux 2005). Using this polyacid-b-polybase, the vesicles can be reversibly produced as a function of pH. At low pH, the poly(L-glutamic acid) with helical structure constitutes the hydrophobic part of the membrane. At high pH, this hydrophobic part is destabilized and becomes hydrophilic because of its transition to the charged coil conformation; instead, the deprotonated poly(L-lysine) takes its place to form the hydrophobic part with its rodlike α-helical structure.

These last two copolypeptide systems are very promising candidates for macromolecular nanobiotechnologies. For drug delivery *in vivo*, the pH of transition (around 3) for lysine is not optimal but other aminoacids (for example, histidine, $pK_a = 6.0$) could be substituted to the lysine to adjust the pH range.

The pH-dependent permeability and reversible structural transition of polyion complex vesicles (PICsomes) in aqueous media was also recently reported by Kataoka and coworkers (Kishimura et al. 2009). At first, the aqueous solution properties of PEG_{45}-$P(Asp\text{-}AP)_{75}$ and PEG_{45}-$PAsp_{75}$, where, P(Asp-AP) stands for poly-[(5-aminopentyl)-α,β-aspartamide] and PAsp for poly(α,β-aspartic acid), were analyzed by potentiometric titration. The pKa values of PEG-P(Asp-AP) and PEG-PAsp were calculated to be 10.47 and 4.88, respectively. These titration results revealed that both block copolymers are equally charged at around physiological pH (pH 7.8, ionization degree = 96%). After mixing the two copolymers, PICsomes form with a mean diameter of 2 μm, these vesicle structures maintain their structure at pH 7.4 for more than 48 h and only dissociate into small particles upon lowering the pH to 5.7. Interestingly, guest molecules can be trapped through this process which suggests that PICsomes can deliver, release and also trap their cargoes by sensing acidic conditions of the intracellular endosomal compartments.

3 Polymersomes Responsive to Physical Stimuli

3.1 *Temperature Responsive Polymersomes*

In the previous section, we mentioned that polymersomes made from PBD-*b*-PLys (Gebhardt et al. 2008) and from PLys-*b*-PBGlu-*b*-PLys (Iatrou et al. 2007) are temperature responsive in basic solutions. With a temperature increase, polymersome size increased and the poly(L-lysine) block underwent a conformational transition from an α-helix to a β-sheet structure. In the following examples (Li et al. 2006a, 2007; Qin et al. 2006), a well-known thermo-responsive polymer, poly(*N*-isopropylacrylamide) (PNIPAM), was used to produce temperature responsive polymersomes. In aqueous solution, PNIPAM undergoes a reversible phase transition from hydrophilic to hydrophobic at a lower critical solution temperature (LCST, typically around 32°C). Hydrophilic-hydrophilic block copolymers can then be designed using this kind of thermo-responsive block. When the solution temperature is higher than the LCST, the thermo-responsive block becomes hydrophobic and the block copolymer amphiphilic, which can self-assemble to form polymersomes. Below the LCST, the thermo-responsive block is turned hydrophilic again and the polymersomes dissociate to rapidly release of encapsulated substances.

McCormick and co-workers (Li et al. 2006a, 2007) have recently prepared hydrophilic-hydrophilic block copolymers, poly(*N*-(3-aminopropyl)methacrylamide hydrochloride)-*b*-poly(*N*-isopropylacrylamide) (PAMPA-*b*-PNIPAM) and poly(2-(dimethylamino)ethyl methacrylate-*b*-poly(*N*-isopropylacrylamide) (PDMAEMA-*b*-PNIPAM) (see Fig. 9), by reversible addition-fragmentation chain-transfer (RAFT) polymerization. These researchers reported that these water-soluble block copolymers could directly from vesicular aggregates in water and that aggregate

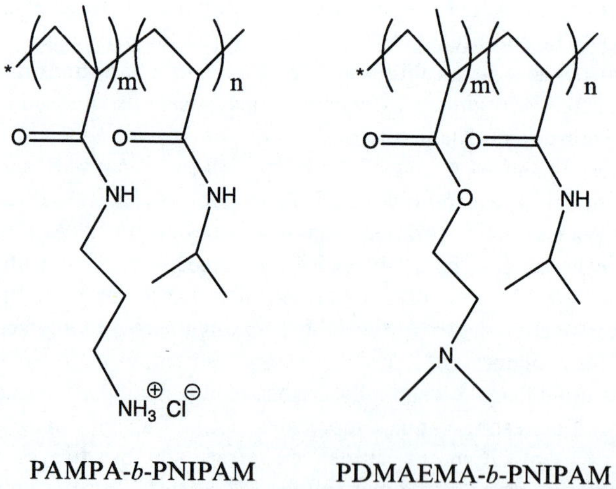

Fig. 9 Hydrophilic-hydrophilic block copolymers with a temperature responsive block PNIPAM

formation could be reversed by changing the solution temperature with respect to the LCST of PNIPAM (32°C or higher, depending on the molecular weight of both blocks). The aggregates could also be trapped by ionic cross-linking of the hydrophilic PAMPA block, either through the addition of an oppositely charged polyelectrolyte that causes inter-polyelectrolyte complexation or by forming gold nanoparticles from the reduction of $NaAuCl_4$ complexed to amine groups in the PDMAEMA.

Qin et al. described polymersome formation using a well-defined and monodisperse ($M_w/M_n < 1.2$) block copolymer PEO-*b*-PNIPAM (Qin et al. 2006). The block copolymer was synthesized by RAFT polymerization and polymersomes were grown in an aqueous solution above the LCST of PNIPAM. This family of biocompatible block copolymers is a promising system for drug delivery because the LCST can be adjusted to a value near or slightly above physiological temperatures.

Our group developed a new class of polymersomes in which the hydrophobic part is a liquid crystal (LC) polymer (Yang et al. 2005, 2006). It is well known that liquid crystal systems excel as responsive systems and can respond to multiple stimuli including temperature, light, electric and magnetic fields. If this responsiveness could be retained in the liquid crystal membrane, we speculated that liquid crystal polymersomes would have potential as multi-responsive, smart polymersomes. Recently, we studied the structural changes in liquid crystal polymersomes triggered by changes in temperature using small angle neutron scattering (SANS), cryo-TEM, SEM and high sensitivity DSC (Hocine et al. 2011). PEG-*b*-PA444 and PEG-*b*-PMAazo444 (Fig. 10), two block copolymers with side-on nematic hydrophobic blocks, were used to form vesicles with a bilayer membrane thickness of 10–15 nm at room temperature (Fig. 11a) . Upon heating the membrane thickness, d, started to increase dramatically from a temperature (~55°C) above T_g but below T_{NI} of the LC polymer block, and reached up to 120 nm at $T > T_{NI}$ ($T_{NI} \sim 80$–85°C) (Fig. 11b). The vesicles transformed into thick-walled capsules. The thickness of the membrane was inconsistent with a bilayer structure and surprisingly the capsules were stable even for temperatures above T_{NI}. As the PEG chains should partially dehydrate with temperature, we propose that the membrane reorganized into a structure consisting of microphase separated LC and PEG domains. Analysis of changes in structural parameters such as the internal aqueous volume and the polymer membrane volume suggest that capsule scission and fusion also occurred during this transition. Substance release would be accompanied by capsule scission.

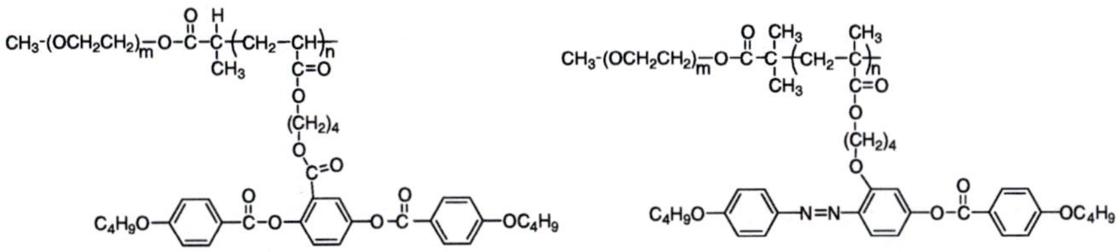

PEG-*b*-PA444 PEG-*b*-PMAazo444

Fig. 10 Molecular structures of liquid crystal copolymers PEG-*b*-PA444 and PEG-*b*-PMAazo444

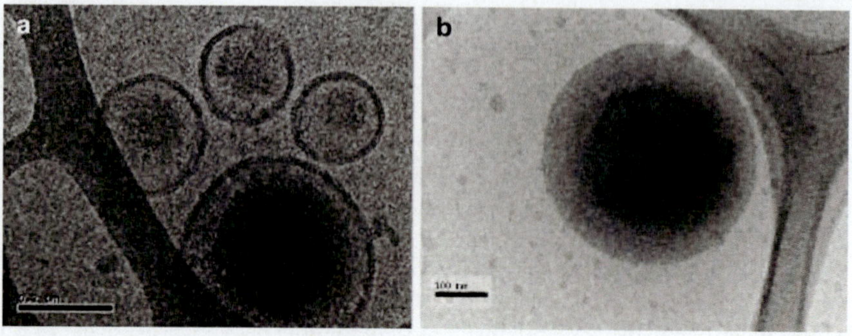

Fig. 11 (**a**) Cryo-TEM image of PEG-*b*-PMAazo444 vesicles at room temperature. Scale: 200 nm. (**b**) Cryo-TEM image of PEG-*b*-PMAazo444 vesicles heated for 1 h at 90°C. Scale: 100 nm

3.2 *Light Responsive Polymersomes*

Stable, light-responsive polymersomes are attractive because the release of entrapped species can be rapidly induced at specific time and location via exposure to light. While pH, oxidation and reduction-responsive systems can also respond quickly, they require that the chemical environment be modified by the addition of acids or bases or other reagents. These environment changes may not necessarily compatible with the drug targeting sites (in the case of applications in drug delivery) or with other applications such as controlled chemical reactions in microfluidics. In contrast, light is a remote stimulus that does not require any chemical environmental change and can be applied locally.

The first examples of light responsive polymersomes are built by incorporating photo-responsive protein channels (such as Bacteriorhodopsin) in the membrane of polymersomes (Choi and Montemagno 2005). However, this case will not be further discussed in this review because the responsiveness does not result from a change of the membrane properties.

3.2.1 Systems with Photoactive Groups in Their Polymer Structures

The exposure of some photoactive groups to light can generate reversible structural changes, thereby directly changing the hydrophilic-hydrophobic balance without addition of other reagents. Typical groups that display photochemically-induced transitions include azobenzenes (change of dipole moment, size and shape), triphenylmethane leucohydroxide or triphenylmethane nitrile (generation of charges), spyrobenzopyran (formation of zwitterionic species) and cinnamoyl (photodimerization) (see Fig. 12). These transitions can further induce changes in the optical, mechanical and chemical properties of the system containing the chromophore. Currently, these groups, especially azo derivatives, are intensively investigated to implement light sensitivity in polymers (Pieroni et al. 1998; Ikeda et al. 2007).

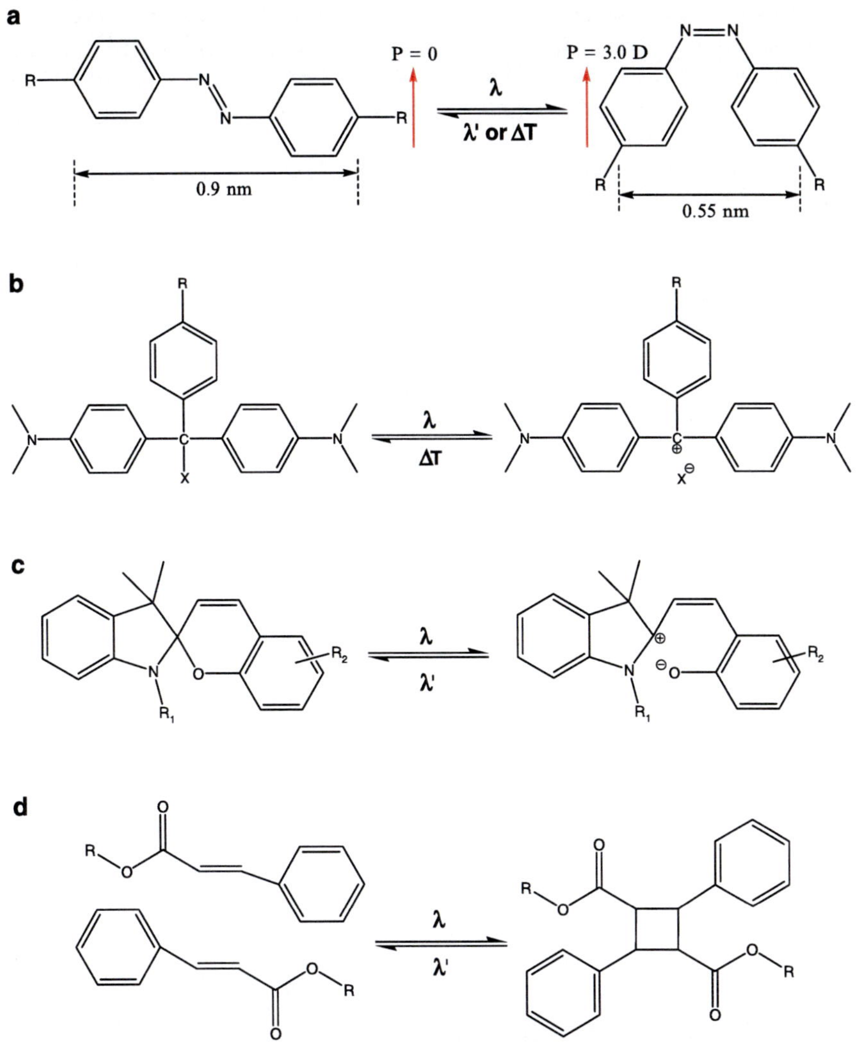

Fig. 12 Examples of chromophores that display photochemically- induced transitions. (**a**) Reversible *trans* (*left*) and *cis* (*right*) photo-isomerization of azobenzene. (**b**) Dissociation of triphenylmethane leucoderivatives into an ion pair under ultraviolet irradiation. (**c**) Reversible photo-isomerization of spirobenzopyran derivatives. (**d**) Reversible photodimerization of cinnamoyl group

These light sensitive triggers could, in principle, be incorporated into amphiphilic copolymers suitable for polymer vesicle formation.

Azobenzene derivatives are the most studied photoactive groups. The azo group can undergo reversible isomerization between the *trans* and *cis* configurations by light and heat (see Fig. 12a). The *trans* isomer is thermodynamically more stable in common conditions. The *cis* isomer obtained by light (usually UV) irradiation is thermodynamically less stable and therefore it isomerises slowly back to *trans* form (typically over several hours). However, irradiation (usually by visible light) reduces the back reaction time to minutes. The photoisomerization of azobenzene chromophores has no side reactions and the wavelength to induce the transformation can be tuned by incorporating substituents in the chromophores.

The polarity, shape and size changes that azobenzene undergoes after isomerization modify significantly the structures and properties of azobenzene-containing polymer blocks.

Zhao's group (Wang et al. 2004), (Tong et al. 2005) prepared amphiphilic diblock copolymers composed of hydrophilic poly(acrylic acid) and hydrophobic polymethacylate with azobenzene side groups. They showed that these copolymers self-assembled in dioxane/water mixtures to form photoresponsive polymer vesicles of 100–200 nm. Upon UV irradiation, the majority of the vesicular aggregates disappeared. The disruption of these morphological structures appeared to be fully reversible as the aggregates reformed following subsequent illumination with visible light. Solubilization of the hydrophobic block in dioxane/water mixtures caused by the polarity change of azobenzene after *trans*-to-*cis* isomerization is claimed to be responsible for vesicle disruption.

We have recently reported photo-responsive polymersomes (del Barrio et al. 2010) base on amphiphilic PEG-*b*-azodendron block copolymer with a hydrophobic block based on the fourth generation of aliphatic polyester dendrons functionalized at the periphery with cyanoazobenzene mesogens (**PEG45-AZO16**). The polymersomes in water were prepared by nanoprecipitation with dioxane/water co-solvents followed by extensive dialyse against water. Wrinkles, and even rupture, in the membrane of the polymersome were detected upon UV irradiation of polymersome dispersion in water (Fig. 13).

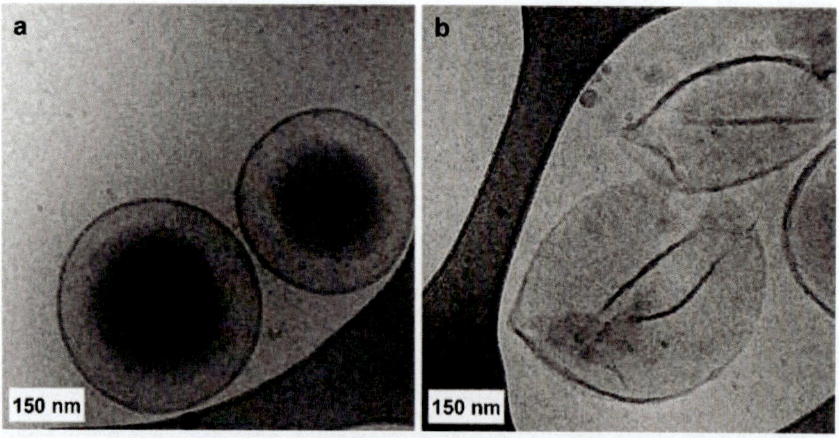

Fig. 13 Cryo-transmission electron micrograph of the polymersomes formed in water by the PEG-*b*-azodendron block copolymer **PEG45-AZO16** (**a**) before and (**b**) after illumination with 360 nm unpolarized light (150 mW/cm^2) for 35 min

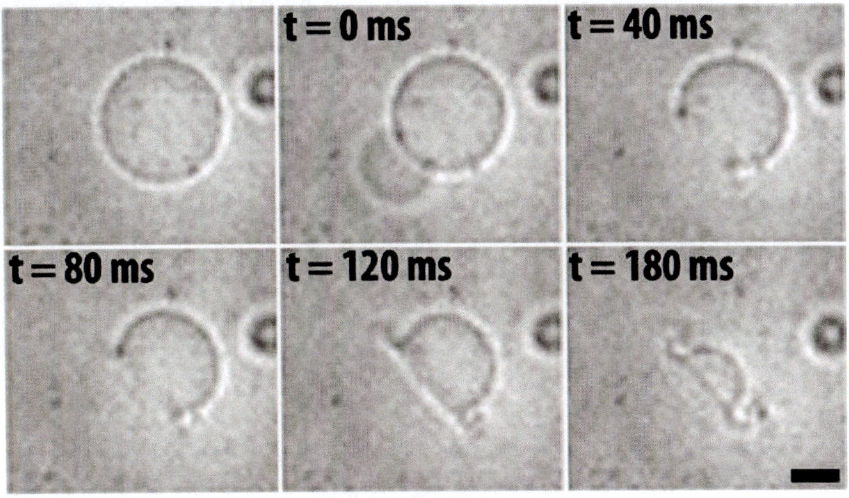

Fig. 14 Snapshots of an ePBD-iPAzo polymersome bursting under UV illumination. Bright-field images were taken using a high-speed digital camera. The first image shows the vesicle prior to illumination. Time t=0 corresponds to pore nucleation. The expulsion of sucrose solution is visible as the pore nucleated. The other images correspond to pore growth and clearly show outward spirals (scale bar=5 μm)

Our group has further achieved the polymersome bursting induced by UV light (Mabrouk et al. 2009a). Exposure to UV illumination around 360 nm caused vesicle rupture which was completed in less than a few hundreds of milliseconds (Fig. 14). The basic principle to rapidly burst the polymersome is to induce frustration in the membrane via a remote stimulus. To implement this approach, we prepared asymmetric polymersomes in which each leaflet consisted of a different type of diblock copolymer: one copolymer was insensitive to any remote stimulus (PEG-*b*-PBD, or PBD for simplicity) while the hydrophobic moiety of the second copolymer was a light sensitive liquid crystalline polymer PEG-*b*-PMAazo444 (PAzo). The mesogenic unit in this side-on LC polymer block contains an azobenzene group, which undergoes a *trans*-to-*cis* configurational transition upon UV exposure. The azobenzene shape change after isomerization induces a nematic (N) to isotropic (I) transition in the LC polymer (Li et al. 2000) causing a conformational change of the chain from a rod to a coil (Cotton and Hardouin 1997; Li et al. 2003). Figure 15a, b shows the chemical structures of the two selected copolymers and a cartoon of the LC copolymer conformation in the membrane both in the absence and in the presence of UV light for polymersomes ePBD-iPAzo (external leaflet=PBD, inner leaflet=PAzo). Starting from a thin, cigar-like shape corresponding to N state (Yang et al. 2005), UV irradiation transforms the LC hydrophobic block to a coil characterized by an increased molecular area. The induced area difference between the two polymer monolayers, i.e., the creation of spontaneous curvature, triggers membrane rupture and

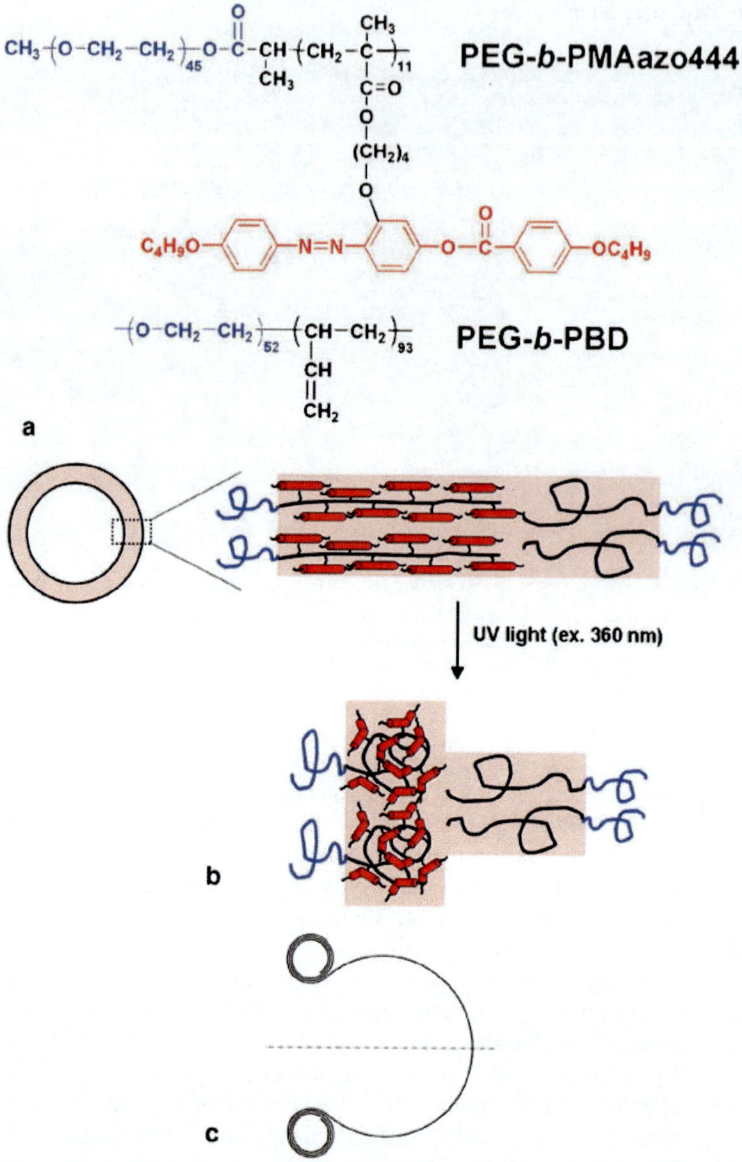

Fig. 15 Copolymers and bilayer conformation. (**a**) Chemical structures of the two selected copolymers, PEG-*b*-PBD and PEG-*b*-PMAzo444. (**b**) Cartoon of a polymersome and cartoon depicting the conformation of both copolymers within the bilayer for an ePBD-iPAzo vesicle. The PEG-*b*-PBD copolymer is always in a coil-coil state. In the absence of UV light, the hydrophobic LC block of the PEG-*b*-PMAzo444 copolymer has a rod-like conformation (corresponding to a nematic state). Under UV illumination, isomerization of the mesogenic groups induces a conformational change of the polymer backbone to a disordered, isotropic state. The net effect of UV exposure is two-fold: at the molecular scale, the projected area of the LC block is increased; at the mesoscopic scale, the spontaneous curvature of the bilayer is increased. (**c**) Schematic representation of pore opening driven by outward curling (for ePBD-iPAzo)

polymersomes bursting. Figure 15c shows the outward curling rim expected to be generated by the change of spontaneous curvature in the membrane where the inner leaflet is light-responsive. This is actually observed in the giant asymmetric polymersomes as shown in Fig. 14. The giant asymmetric polymersomes (>few microns in diameter) were prepared by the method of inverted emulsion. The polymersome bursting takes place also if the inner leaflet is inert and the external leaflet is light-responsive, but with inward curling rim during the vesicle opening. These results highlight a new general strategy to create stimuli-responsive polymersomes based on the fabrication of asymmetric membranes, and driven by a change in membrane spontaneous curvature. While UV light was the stimulus used for this study, temperature, electric or magnetic fields could also act as remote stimuli provided that one of the two leaflets of the membrane is composed of suitably designed copolymers sensitive to these physical stimuli. This flexibility, combined with the low permeability of polymer bilayers, ensures a wide range of potential applications in the fields of drug delivery, cosmetics and material chemistry.

Trans-cis isomerization of azobenzene was also used to modify (normally increase) the LCST of thermosensitive polymers because the *cis* conformers is more polar and hydrophilic (Sugiyama and Sono 2001; Desponds and Freitag 2003; Ravi et al. 2005; Sin et al. 2005). These polymers could potentially be used to produce temperature and light dual-responsive polymersomes.

An interesting system containing triphenylmethane nitrile (Fig. 12b) was described by the group of X. Zhang (Jiang et al. 2007b), as illustrated Fig. 16, which consists of a PEG chain linked to a group of hydrophobic Malachite green. This PEG-*b*-bis(4-dimethylaminophenyl)phenyl methyl nitrile is able to self assemble into vesicles in water. Under UV irradiation, the photochromic group ionizes in its cationic form and the molecules become soluble in water. Consequently vesicles are disassembled. The molecules can recover their neutral form by heat treatment and free chains are reassembled again into vesicles.

Photoactive polymers bearing triphenylmethane leucohydroxide (Kono et al. 1995) and nitrocinnamate (Yuan et al. 2005) have been used to prepare light responsive microcapsules for controlled encapsulation and release of substance. Spirobenzopyran derivatives (Fig. 12c) have been used to create water-soluble polymers that associate into aggregates under UV irradiation (Konak et al. 1997). We foresee that these photoactive groups are also potentially useful for light-responsive polymersome formation providing that proper chemical design is made for block copolymers.

The last example, described as follows, with photoactive groups incorporated in their polymer structures is rather different from the systems discussed above. It is a polymersome susceptible to UV-induced degradation (Katz et al. 2010). Photolabile 2-nitrophenylalanine (2NPA) was used to join the hydrophilic PEG block and hydrophobic PCL block (Fig. 17a). Exposure of the polymersomes formed by this 2NPA linked PEG-*b*-PCL to 365 nm light enable the cleavage between PEG and

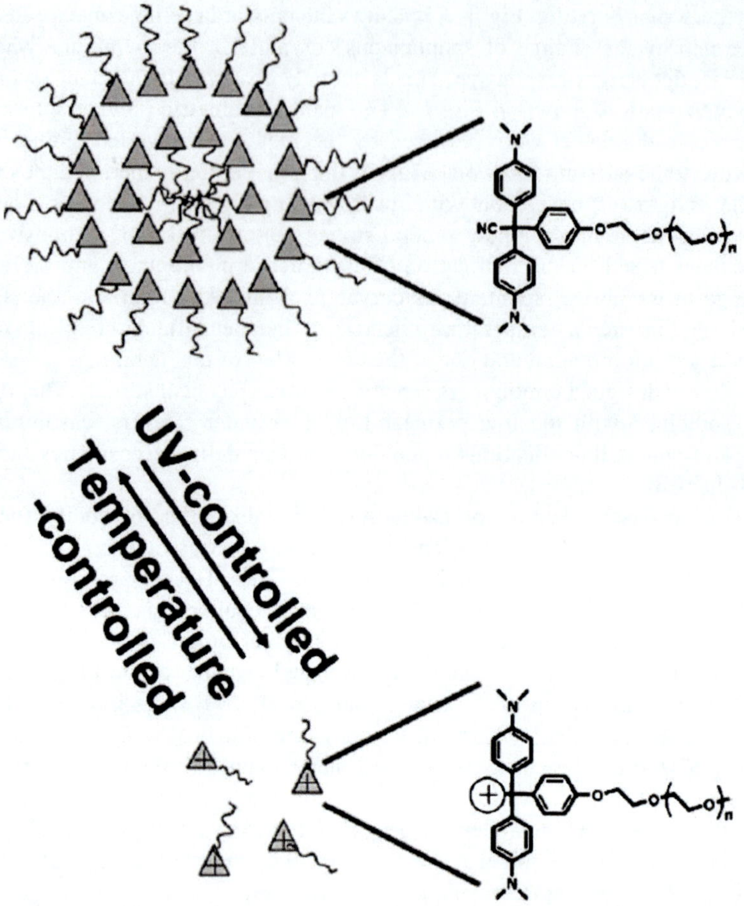

Fig. 16 Schematic illustration of vesicles made from PEG and malachite green and their disassembly under UV illumination. (Reproduced from ref. Jiang et al. 2007b)

PCL blocks. As PEG is liberated from the membrane by UV light, there is excess hydrophobic material that must be stabilized by the remaining PEG shell. This results in membrane thickening, decreasing overall size of the polymersome, and flux of aqueous contents out of the polymersome (Fig. 17b, c). Ultimately, there is insufficient PEG surface-coverage to stabilize the PCL shell and the insoluble PCL crashes out of solution and aggregates.

The above system is an example that light (a physical stimuli) induce irreversible chemical structure changes in light-responsive polymers. The chemical structure changes can also be produced by light in photo-inert polymers with the aid of photosensitizers as discussed below.

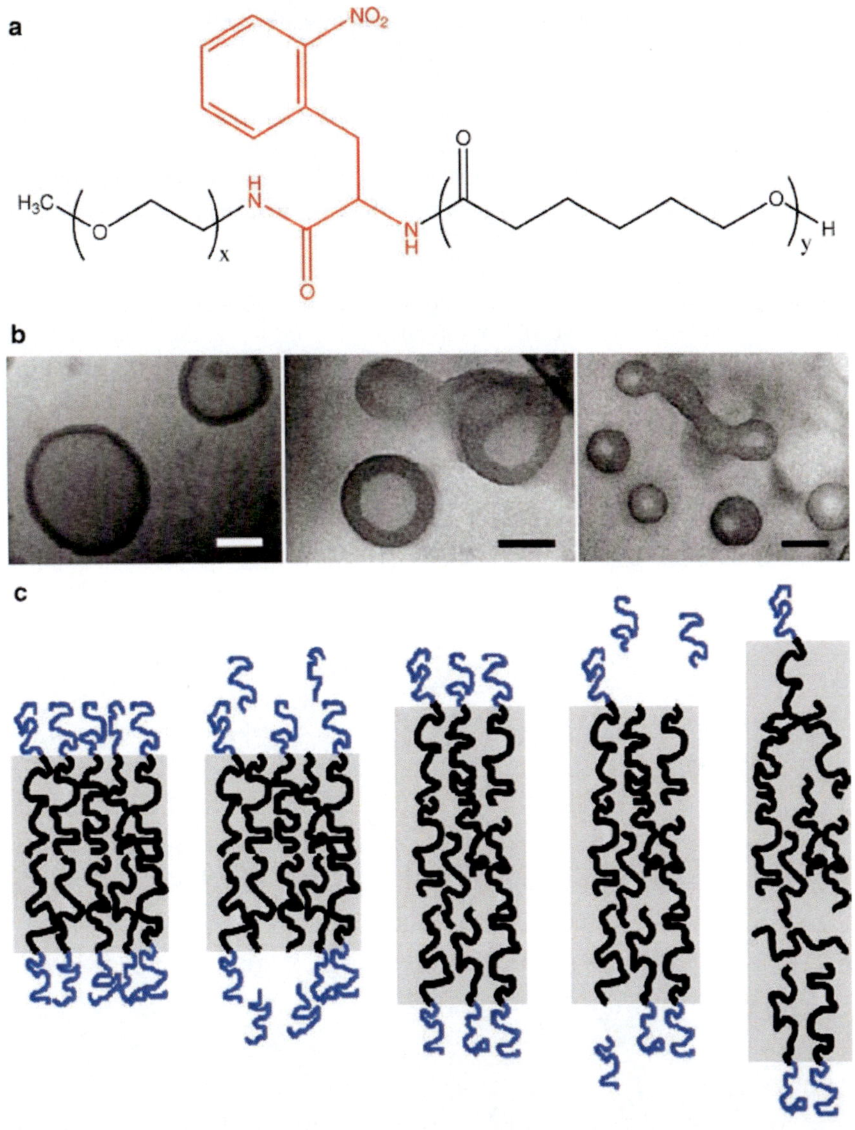

Fig. 17 (**a**) Chemical structure of 2NPA linked PEG-*b*-PCL. (**b**) Cryo-TEM images of 2NPA polymersomes (*left to right*) before, during and after 6 h of UV exposure. Scale bars = 100 nm. (**c**) Schematic representation of membrane structures to explain the Cryo-TEM images. (Reproduced from Katz et al. 2010)

3.2.2 Composite Systems with Classical Polymersomes Charged with Photosensitizers

The group of Dmochowski has first reported photoinitiated destruction of composite porphyrin-protein polymersomes (Robbins et al. 2009). The polymersomes were formed by incorporating a protein in the aqueous interior and a meso-to-meso ethyn-bridged bis[(porphinato)zinc] (PZn_2) chromophore in the membrane polymersomes made from classical photo-inert PEO-*b*-PBD. Confocal laser scanning microscopy (CLSM) imaging of polymersomes loaded with both ferritin and PZn2 at excitation wavelengths (488, 543, or 633 nm, where PZn2 absorbs strongly) caused many of the vesicles to undergo irreversible morphological changes ranging from formation of new bends or "arms" and budding of smaller vesicles to total polymersome destruction (Fig. 18). Similar results were seen during imaging by widefield fluorescence microscopy using a mercury arc lamp. Even though the mechanism of photodestruction is not elucidated, it may be possible to harness light-activated vesicle destruction for in vivo targeted drug delivery, given the established exceptional NIR absorptivity of PZn2 and closely related chromophores.

Our group has prepared a binary system of polymersome and chlorine e6 (Ce6) (Mabrouk et al. 2010). Ce6 (see Fig. 19) is a classical chlorine photosensitizer,

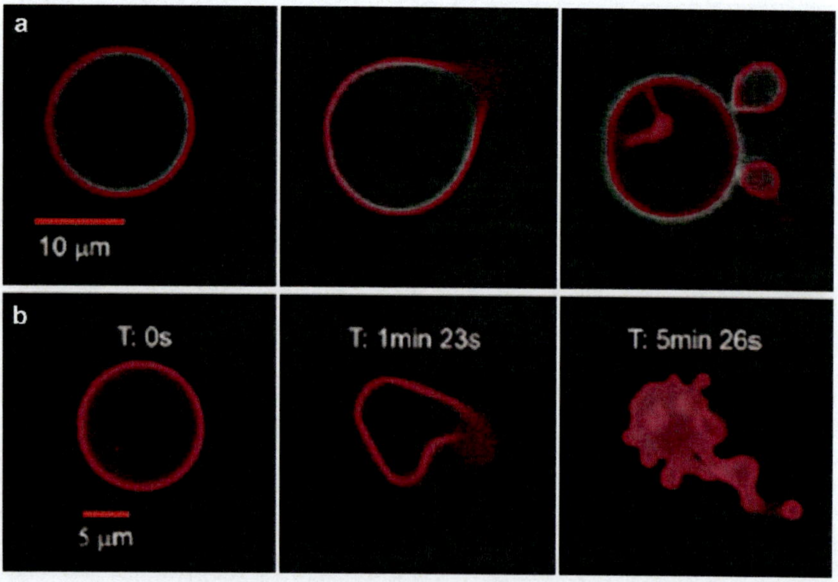

Fig. 18 Confocal micrographs of polymersomes that membrane-disperse PZn2 (purple) and encapsulate HSAF obtained in continuous scanning mode. (**a**) BODIPY-FL-labeled HSAF (green, 3 mg/mL)+PZn2 vesicle, imaged using two lasers simultaneously (488, 543 nm). HSAF is the horse spleen iron-free apoferritin, and BODIPY-FL a neutral dye. Images proceed in time, left to right, over a period of ~5 min. (**b**) Unlabeled HSAF (1.5 mg/mL)+PZn2 vesicle. Vesicle imaged using three lasers simultaneously (488,543, 633 nm). (Reproduced from Robbins et al. 2009)

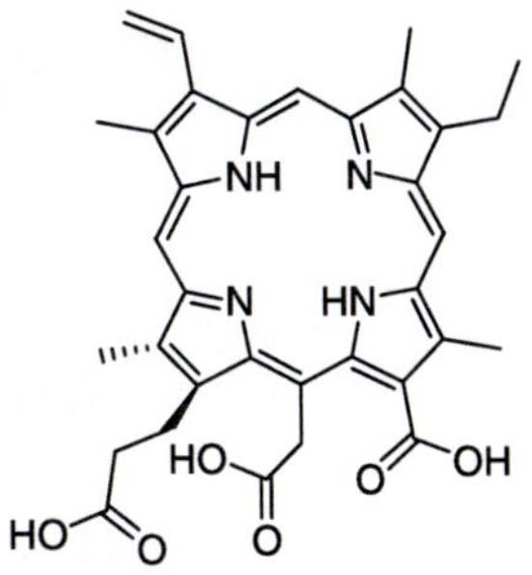

Fig. 19 Chemical structure of the photosensitizer Chlorine e6

which has been used in photodynamic therapy (PDT) because upon light exposure Ce6 generates highly reactive oxygen species (ROS) including free radicals and/or singlet oxygen $^1O_2$. In the core of a lipid bilayer, carbon-carbon double bonds of unsaturated lipids have been shown to be privileged sites for $^1O_2$ and radical reactions that lead to cellular membrane alteration. The selected photo-inert copolymer is again the prototypical PEO-*b*-PBD, which contains a large number of carbon–carbon double bonds that are preferred sites of photo-oxidation. Ce6 is an amphiphilic molecule and can be loaded in the membrane as well as in the aqueous internal compartment. In these polymersomes loaded with Ce6, we have observed a complex sequence of light-induced morphological changes. Using micromechanical assays based on micropipette manipulation, we have quantitatively monitored the different phases of the photo-response, which include membrane area variation, osmotic swelling, membrane cross-linking and vesicle deflation. In addition, these morphological changes can be adjusted by the concentration of Ce6 (see Fig. 20). We have thus gained insight into the complex cascade of chemical reactions involved in photosensitization. Our findings suggest that composite chlorine-copolymer vesicles may be used as a new class of light-sensitive drug carriers.

3.3 *Polymersomes Responsive to Magnetic Field*

The magnetic field is a very promising stimulus in drug delivery, because it is easy to be applied remotely and permits also to combine together the medical diagnosis and therapy due to the development of MRI (magnetic resonance imaging) technique. The common strategy is to incorporate magnetic nanoparticles (e.g., γ-Fe_2O_3) into the vesicle structures. Iron oxide nanoparticles, namely magnemite (γ-Fe_2O_3) or magnetite (Fe_3O_4), with particle sizes of 4–10 nm have drawn special interest for biomedical applications. They are referred to as superparamagnetic iron oxide nanoparticles (SPIONs) due to their superparamagnetic properties and small size. The resulting superparamagnetic vesicles have potential applications because of the specific properties of magnetic iron oxide in the fields of biomedicine and biotechnology: e.g., manipulation by an external magnetic field gradient, radio-frequency heating for cancer therapy, and labelling of organs in magnetic resonant imaging.

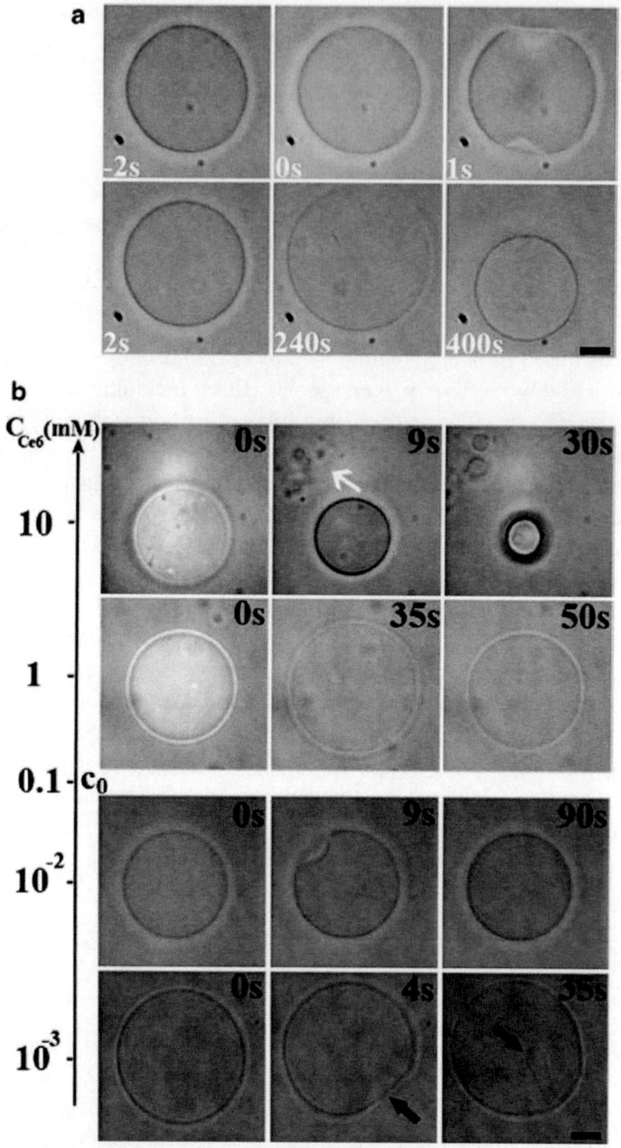

Fig. 20 (**a**) Video sequence showing the morphological changes of Ce6-loaded PEO-*b*-PBD polymersomes under UV exposure. The bulk concentration of Ce6 is $c_0 = 0.1$ mM. Irradiation starts at time $t = 0$. Scale bar = 5 μm. (**b**) Representative UV-induced morphological changes of Ce6-loaded PEO-*b*-PBD polymersomes for different bulk concentrations of Ce6. The black arrows show the formation of membrane invaginations during UV illumination. Scale bar = 5 μm

Here we will emphasize the vesicle structure changes induced by the magnetic field for the purpose of controlled release.

Magnetic nanoparticles were firstly incorporated in liposomes (small molecular vesicles) either in its aqueous compartment (Sabaté et al. 2008; Zhang et al. 2005) or in its lipid bilayer via magnetized polymers (Kamaly et al. 2007; Leclercq et al. 2003). Such magnetic liposomes can be used to target therapeutic molecules to a specific site when exposed to a magnetic field. Once the drug-loaded magnetic liposomes reach the target, the drug can also be released by radio-frequency heating (hyperthermia) of SPIONs with a thermosensitive lipid bilayer. Temperature-sensitive lipid bilayer frequently include dipalmitoylphosphatidylcholine (DPPC) as the key component, since liposomes usually become leaky at a gel-to-liquid crystalline phase transition that takes place at 41°C (Yatvin et al. 1978; Jeong et al. 2009). Pradhan et al. (2010) developed folate receptor targeted thermosensitive magnetic liposomes (MagFolDox), which are designed to combine features of biological and physical (magnetic) drug targeting for use in magnetic hyperthermia-triggered drug release (Fig. 21). The optimized liposome formulation had

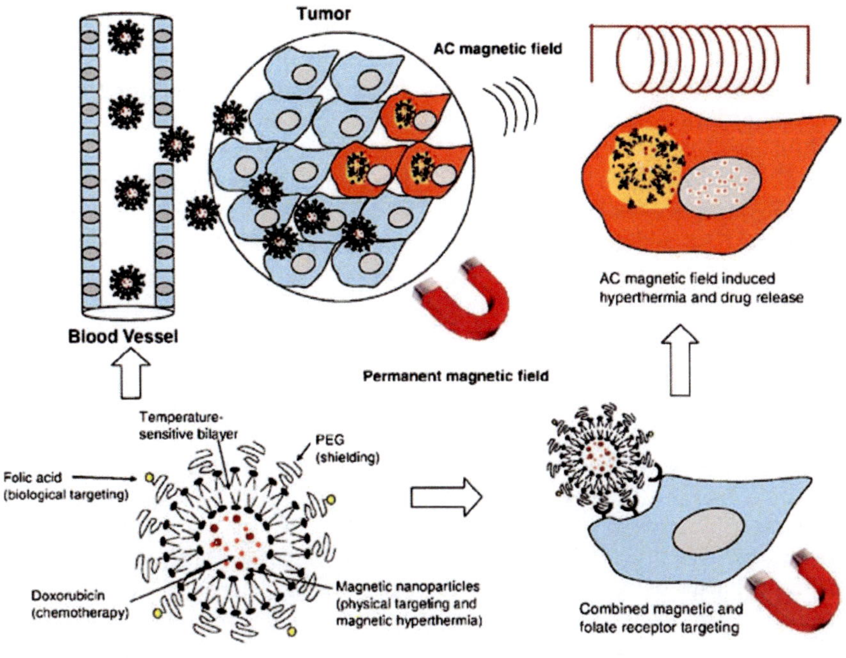

Fig. 21 The figure shows the concept of a multifunctional drug carrier responsive to magnetic field. It is a folate-receptor-targeted and temperature-sensitive magnetic liposome containing doxorubicin, which can be targeted physically by magnetic field and biologically by folic acid to tumor cells. Drug release will be triggered by hyperthermia upon local application of an AC magnetic field on the tumor tissue. (Reproduced from Pradhan et al. 2010)

the composition of DPPC: cholesterol: DSPE-PEG2000: DSPE-PEG2000-folate at 80: 20: 4.5: 0.5M ratio. Magnetic hyperthermia at 42.5°C and 43.5°C synergistically increased the cytotoxicity of MagFolDox. The results suggest that an integrated concept of biological and physical drug targeting, triggered drug release and hyperthermia based on magnetic field influence can be used advantageously for thermo-chemotherapy of cancers.

It is nearly impossible to incorporate inorganic nanoparticles within the membrane of liposomes because of their very thin membrane thickness (3–5 nm). However, this becomes possible, even still challenging, with polymersomes due to the thick membrane (5—30 nm) and toughness resulting from polymer characters. The soft magnetic shells are especially promising for drug delivery, because their internal compartment is available for encapsulation of water-soluble species and the possible toxicity arising from cells directly contacting the iron oxide would not be an issue because it is embedded in the copolymer. Application of a magnetic field could trigger the transient opening of the bilayer and the release of an encapsulated content. The pioneer work on magnetic polymersomes has been made by Lecommandoux & Sandre et al. (Lecommandoux et al. 2005). They have prepared polymersomes of PGA_{56}-*b*-PBD_{48} loaded with surfactant-coated γ-Fe_2O_3 nanoparticles in the layer of PBD blocks, and studied their structural transformation under magnetic field by small angle neutron scattering (SANS). Anisotropic SANS data detected with a 2-dimensional detector provide experimental evidence of the capability to modify the shape of these hybrid membranes in response to a magnetic field of an intensity as low as 290 G. Analyses of the anisotropic SANS patterns (at intermediate wave vector q range) have shown that the portions of membranes mostly affected by the magnetic field are those with their normal vector parallel to the field. The membrane becomes stretched in these portions (decrease of the apparent membrane thickness) or almost equivalently the nanoparticles move away from the magnetic poles.

Our group worked on the LC polymersomes PEG-*b*-PA444 and PEG-*b*-PAazo444 that contain diamagnetic mesogens (Hocine et al. 2011). We hoped to achieve the structural changes triggered by magnetic field using the intrinsic positive diamagnetic response of the LC polymers. Interestingly, for temperatures between 65 and 80°C, the application of a magnetic field of 1.4 T can increase the membrane thickness by up to 50% without significantly changing the inner volume of vesicles. This structural change is consistent with the alignment of LC domains under magnetic field. While thickness changes could happen in polymersomes loaded with magnetic nanoparticles as discussed above, to the best of our knowledge, this is the first time that an increase in membrane thickness has been observed in a pure polymer vesicle system.

We envision that magnetic field triggered drug release from SIONs-loaded magnetic polymersomes would be achieved in the near future provided that the hydrophobic block is composed of suitably designed liquid crystalline polymer or crystalline polymer with proper transition temperature by analogy with the gel-to-liquid crystalline phase transition in lipid DPPC.

3.4 *Responses to Electric Field, Osmotic Shock and Ultrasonic Wave*

The effects of electric field on giant polymersomes of PEO-*b*-PBD (Aranda-Espinoza et al. 2001; Bermudez et al. 2003) and on giant liposomes (Vlahovska et al. 2009; Riske and Dimova 2005, 2006; Dimova et al. 2007) have been studied by the group of Discher and by the group of Dimova, respectively. The stimulation by an electric field is not specific to a given amphiphile. The response of the vesicles to the electric field and the destruction of the membrane are due to an increase tension of the vesicle directly induced by the electric field. The hydrophobic core of the membrane thickness d and dielectric constant ε (low compared to that of the aqueous environment) behaves as a capacitance C (typically in the order of $\mu F.cm^{-2}$). The charging time for the membrane depends on the environmental conductivity inside and outside the vesicle, and typically takes the order of a fraction of μs in saline solution. Consequently, if the pulse duration supply is larger than the charging time of the capacitance, the transmembrane potential V thus generated can be considered constant and only dependent on the applied electric field E, the radius R of the vesicle and the angle θ between the field direction and the normal to the membrane: $V = RE\cos(\theta)$. This potential creates an electro-compressive stress perpendicular to the plane of the membrane $\sigma = 1/2CV^2$. The increase in electric field applied resulting in a net increase in tension of the membrane. There is a threshold electric potential for which the vesicle tension reaches its lysis value and consequently the vesicle disintegrates. This experiment is illustrated in Fig. 22. In practice, the critical transmembrane potential is of order of $V_c = 1–10$ V. The critical constraints are then in the order of $\sigma_c = 10$ (liposomes) to 30 (polymersomes) $pN.nm^{-1}$. Nevertheless, it should be noticed that the magnitude of the electric field necessary to achieve polymersome lysis is kV/cm. Apply these fields through the human body for drug delivery would be likely accompanied by severe side effects.

The responsive polymersomes to osmotic shock has been reported by the group of Weitz (Shum et al. 2008). Polymersomes encapsulating 1-hydroxypyrene-3,6,8-trisulfonate sodium (HPTS) were produced from PEG-*b*-PLA block copolymer using microfluidic fabrication. Although the membrane is impermeable to the small HPTS salts, water molecules can diffuse in and out of the polymersomes. The osmotic pressure, π_{osm}, is related to the concentration of solutes,

$$\pi_{osm} = cRT$$

where *c* is the molar concentration of the solutes, *R* is the gas constant and *T* is the temperature. Due to osmotic pressure difference, water diffuses from regions with a low salt concentration to regions with a higher concentration. Osmotic pressure can therefore be used to tune the sizes of the polymersomes. If the osmotic pressure change is sudden and large, the resulting shock can break the polymersomes. The kinetics of the polymersomes' response following a large osmotic shock is too fast to visualize; the process is therefore slowed down for visualization by gradually

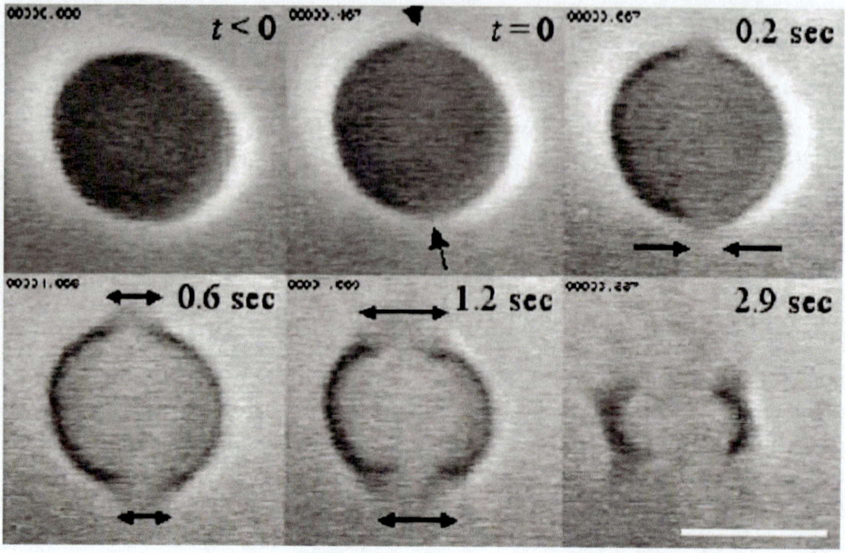

Fig. 22 Phase contrast imaging of pore growth with time during a PEG-*b*-PBD polymersome bursting under electric field. The first image shows the polymersome prior to application of electric field ($t<0$). Time $t=0$ corresponds to the formation of two pores at both vertical poles, indicated by the vertical arrows in the second image (from left to right). The applied electric field is in the vertical direction. Time sequence is $t=0.2$ sec, $t=0.6$ sec, $t=1.2$ sec and $t=2.9$ sec in other images, where the horizontal double arrows show the growth of the pores. Scale bar = 10 μm. (Reproduced from ref. Bermudez et al. 2003)

increasing polyvinyl alcohol (PVA) concentration (c_0 = 10 wt%) in the continuous phase through water evaporation. As water evaporates, the PVA concentration in becomes higher and higher outside of the polymersomes and so water is squeezed out from the inside of the polymersomes. As a result, the polymersome becomes smaller, and its wall buckles, as shown in Fig. 23. This provides a simple trigger for the release of the encapsulated fluorescent HPTS. By tuning the properties of the polymersome wall, it might also be possible to adjust the level of osmotic shock required to break the polymersomes. Alternatively, release can also be triggered by diluting the continuous phase and thus reducing its osmotic pressure. This simple triggered release mechanism makes polymersomes a promising candidate for encapsulation and release of actives.

Ultrasound is gaining attention as a therapeutic tool in addition to its use in diagnostics. Recently, Hammer's group (Pangu et al. 2010) reported ultrasonically induced release from nanosized polymersomes made from PEO-*b*-PBD copolymers under ultrasound at 20 kHz. Leakage of a fluorescent dye from vesicle core was measured to study the permeation. Ultrasound causes significant leakage from the core above threshold intensity, suggesting that leakage is governed by acoustic cavitation. Size measurements and direct visualization of vesicles show that ultrasound does not completely rupture them into fragments but causes transient poration (Fig. 24). The extent of leakage inversely depends on membrane thickness and

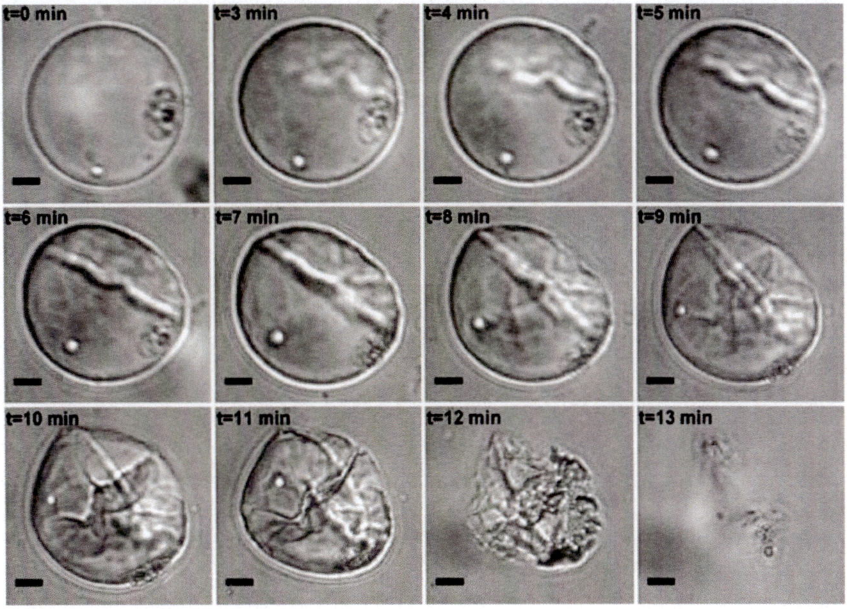

Fig. 23 Bright-field microscope images showing the shrinkage and breakage of a PEG(5,000)-*b*-PLA(5,000) polymersome after an osmotic shock. As a result of water expulsion from its inside, the polymersome shrinks and buckles. Scale bar is 10 μm. (Reproduced from Shum et al. 2008)

directly depends on sonication time and intensity. The strong dependence of membrane permeation on ultrasonic and chemical parameters suggests the possibility of tailoring polymersome features to tune the desired extent of membrane destabilization. Even though the exact mechanism of pore formation and release under an ultrasonic field need to be elucidated, the results indicate that ultrasound has potential as a therapeutic tool for drug delivery from polymer vesicles.

4 Concluding Remarks

In Nature, supra-molecular nano or micro-objects formed by the self-assembly of amphiphilic molecules or macromolecules are omnipresent (e.g. cells, liposomes, viral capsids etc.) and come in a wide range of shapes, sizes and functions. Synthetic amphiphilic copolymers are an attractive biomimetic approach that has been widely used to produce nano or micro assemblies such as polymer vesicles, polymer micelles, nanofibers and nanotubes. Applications in drug delivery have driven much of this research. Other opportunities also exist for these polymer assemblies including use in diagnosis, as nanoreactors and as templates for nanomaterial preparation. In this paper, we have focused on polymer vesicle structures, i.e. polymersomes. Many applications require the ability to selectively and

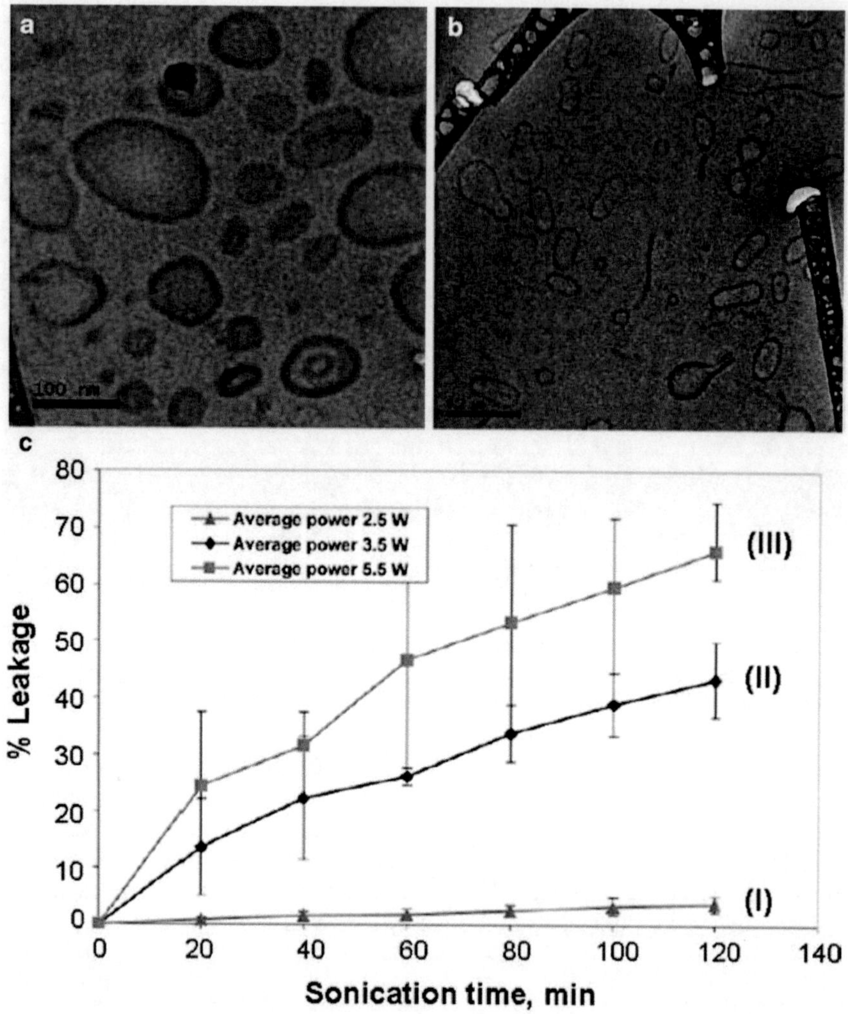

Fig. 24 Cryo-TEM images of polymersomes (**a**) before and (**b**) after sonication. Scale bar = 100 nm in (**a**) and 200 nm in (**b**). Percentage of ANTS (8-Aminonaphthalene-1, 3, 6-trisulfonic acid, disodium salt) release as a function of time where the average sonication power during sonication cycles as recorded on sonic dismembrator dial was (**I**) 2.5 W, (**II**) 3.5 W, and (**III**) 5.5 W. (Reproduced from Pangu et al. 2010)

controllably destabilise the polymer vesicle structures so as to liberate the substances encapsulated in the interior compartment and/or in the membrane. To achieve this goal, stimuli-responsive polymersomes have been developed. In addition, the utilization of biocompatible polymers and biocompatible stimuli is essential for *in vivo* applications.

We have discussed in this review all the most promising approaches to make stimuli-responsive polymersomes, including non-biocompatible ones, for the proof of concept and for the larger applications of polymersomes. Classical chemical stimuli such as pH changes, hydrolysis, oxidation or reduction reaction, are used to trigger a change in the hydrophilic-hydrophobic balance of the amphiphilic copolymers, which in turn destabilises the vesicular structure either by forming leaking pores or causing the vesicle to burst. Physical stimuli such as temperature, light, magnetic field or ultrasonic wave are also of great potential interest, because they don't require any chemical environmental change and can be applied remotely and/ or locally. Different mechanisms are involved in the destruction or deformation of polymersomes under physical stimuli. We believe the use of the physical stimuli will provide the impetus for the development of new stimuli-responsive polymer vesicles.

References

Ahmed, F. & Discher, D. (2004) Self-porating polymersomes of PEG-PLA and PEG-PCL: hydrolysis-triggered controlled release vesicles. *Journal of Controlled Release*, 96, 37–53.

Ahmed, F., Pakunlu, R., Brannan, A., Bates, F., Minko, T. & Discher, D. (2006) Biodegradable polymersomes loaded with both paclitaxel and doxorubicin permeate and shrink tumors, inducing apoptosis in proportion to accumulated drug. *Journal of Controlled Release*, 116, 150–158.

Angelova, M. I. & Dimitrov, D. S. (1986) Liposome electroformation. *Faraday Discussions*, 81, 303-+.

Antonietti, M. & Forster, S. (2003) Vesicles and liposomes: A self-assembly principle beyond lipids. *Advanced Materials*, 15, 1323–1333.

Aranda-Espinoza, H., Bermudez, H., Bates, F. & Discher, D. (2001) Electromechanical limits of polymersomes. *Physical Review Letters*, 87, 208301.

Battaglia, G. & Ryan, A. (2005) Bilayers and interdigitation in block copolymer vesicles. *Journal of the American Chemical Society*, 127, 8757–8764.

Battaglia, G., Ryan, A. & Tomas, S. (2006) Polymeric vesicle permeability: A facile chemical assay. *Langmuir*, 22, 4910–4913.

Battaglia, G. & Ryan, A. J. (2006) Pathways of polymeric vesicle formation. *Journal of Physical Chemistry B*, 110, 10272–10279.

Bellomo, E., Wyrsta, M., Pakstis, L., Pochan, D. & Deming, T. (2004) Stimuli-responsive polypeptide vesicles by conformation-specific assembly. *Nature Materials*, 3, 244–248.

Ben-Haim, N., Broz, P., Marsch, S., Meier, W. & Hunziker, P. (2008) Cell-specific integration of artificial organelles based on functionalized polymer vesicles. *Nano Letters*, 8, 1368–1373.

Bermudez, H., Aranda-Espinoza, H., Hammer, D. & Discher, D. (2003) Pore stability and dynamics in polymer membranes. *Europhysics Letters*, 64, 550.

Bertin, A., Hermes, F. & Schlaad, H. (2010) Biohybrid and Peptide-Based Polymer Vesicles. *Polymer Membranes/Biomembranes*. Berlin, Springer-Verlag Berlin.

Bhargava, P., Zheng, J. X., Li, P., Quirk, R. P., Harris, F. W. & Cheng, S. Z. D. (2006) Self-assembled polystyrene-block-poly(ethylene oxide) micelle morphologies in solution. *Macromolecules*, 39, 4880–4888.

Borchert, U., Lipprandt, U., Bilang, M., Kimpfler, A., Rank, A., Peschka-Suss, R., Schubert, R., Lindner, P. & Forster, S. (2006) pH-Induced Release from P2VP PEO Block Copolymer Vesicles. *Langmuir*, 22, 5843–5847.

Broz, P., Benito, S. M., Saw, C., Burger, P., Heider, H., Pfisterer, M., Marsch, S., Meier, W. & Hunziker, P. (2005) Cell targeting by a generic receptor-targeted polymer nanocontainer platform. *Journal of Controlled Release*, 102, 475–488.

Burkoth, T., Benzinger, T., Urban, V., Morgan, D., Gregory, D., Thiyagarajan, P., Botto, R., Meredith, S. & Lynn, D. (2000) Structure of the -amyloid (10–35) fibril. *J. Am. Chem. Soc*, 122, 7883–7889.

Cameron, N. S., Corbierre, M. K. & Eisenberg, A. (1999) Asymmetric amphiphilic block copolymers in solution: a morphological wonderland. *Can. J. Chem.*, 77, 1311–1326.

Carlsen, A. & Lecommandoux, S. (2009) Self-assembly of polypeptide-based block copolymer amphiphiles. *Current Opinion in Colloid & Interface Science*, 14, 329–339.

Cerritelli, S., Velluto, D. & Hubbell, J. (2007) PEG-SS-PPS: reduction-sensitive disulfide block copolymer vesicles for intracellular drug delivery. *Biomacromolecules*, 8, 1966–1972.

Chécot, F., Brûlet, A., Oberdisse, J., Gnanou, Y., Mondain-Monval, O. & Lecommandoux, S. (2005) Structure of polypeptide-based diblock copolymers in solution: Stimuli-responsive vesicles and micelles. *Langmuir*, 21, 4308–4315.

Checot, F., Lecommandoux, S., Gnanou, Y. & Klok, H. A. (2002) Water-soluble stimuli-responsive vesicles from peptide-based diblock copolymers. *Angewandte Chemie-International Edition*, 41, 1339–1343.

Chiu, H., Lin, Y., Huang, Y., Chuang, C. & Chern, C. (2008) Cover Picture: Polymer Vesicles Containing Small Vesicles within Interior Aqueous Compartments and pH-Responsive Transmembrane Channels (Angew. Chem. Int. Ed. 10/2008). *Angewandte Chemie International Edition*, 47.

Choi, H. & Montemagno, C. (2005) Artificial organelle: ATP synthesis from cellular mimetic polymersomes. *Nano Letters*, 5, 2538–2542.

Christian, D. A., Garbuzenko, O. B., Minko, T. & Discher, D. E. (2010) Polymer Vesicles with a Red Cell-like Surface Charge: Microvascular Imaging and in vivo Tracking with Near-Infrared Fluorescence. *Macromolecular Rapid Communications*, 31, 135–141.

Christian, N. A., Milone, M. C., Ranka, S. S., Li, G. Z., Frail, P. R., Davis, K. P., Bates, F. S., Therien, M. J., Ghoroghchian, P. P., June, C. H. & Hammer, D. A. (2007) Tat-functionalized near-infrared emissive polymersomes for dendritic cell labeling. *Bioconjugate Chemistry*, 18, 31–40.

Cotton, J. P. & Hardouin, F. (1997) Chain conformation of liquid-crystalline polymers studied by small-angle neutron scattering. *Progress in Polymer Science*, 22, 795–828.

Del Barrio, J., Oriol, L., Sanchez, C., Serrano, J.-L., Di Cicco, A., Keller, P. & Li, M.-H. (2010) Self-Assembly of Linear- Dendritic Diblock Copolymers: From Nanofibers to Polymersomes. *Journal of the American Chemical Society*, 132, 3762–3769.

Demirgoz, D., Pangburn, T. O., Davis, K. P., Lee, S., Bates, F. S. & Kokkoli, E. (2009) PR_b-targeted delivery of tumor necrosis factor-alpha by polymersomes for the treatment of prostate cancer. *Soft Matter*, 5, 2011–2019.

Desponds, A. & Freitag, R. (2003) Synthesis and characterization of photoresponsive N-isopropylacrylamide cotelomers. *Langmuir*, 19, 6261–6270.

Dimova, R., Riske, K., Aranda, S., Bezlyepkina, N., Knorr, R. & Lipowsky, R. (2007) Giant vesicles in electric fields. *Soft Matter*, 3, 817–827.

Dimova, R., Seifert, U., Pouligny, B., Förster, S. & Döbereiner, H. (2002) Hyperviscous diblock copolymer vesicles. *The European Physical Journal E: Soft Matter and Biological Physics*, 7, 241–250.

Discher, D. & Ahmed, F. (2006) Polymersomes. *The Annual Review of Biomedical Engineering*, 8, 323–341.

Discher, D. & Eisenberg, A. (2002a) Polymer vesicles. *Science*, 297, 967–973.

Discher, D. & Eisenberg, A. (2002b) Polymer vesicles. *Science*, 297, 967.

Du, J. & Armes, S. (2005) pH-responsive vesicles based on a hydrolytically self-cross-linkable copolymer. *J. Am. Chem. Soc*, 127, 12800–12801.

Du, J. & O'reilly, R. (2009) Advances and challenges in smart and functional polymer vesicles. *Soft Matter*, 5, 3544–3561.

Du, J. Z. & Armes, S. P. (2009) Preparation of Biocompatible Zwitterionic Block Copolymer Vesicles by Direct Dissolution in Water and Subsequent Silicification within Their Membranes. *Langmuir*, 25, 9564–9570.

Du, J. Z., Tang, Y. P., Lewis, A. L. & Armes, S. P. (2005) pH-sensitive vesicles based on a biocompatible zwitterionic diblock copolymer. *Journal of the American Chemical Society*, 127, 17982–17983.

Fernyhough, C., Ryan, A. & Battaglia, G. (2009) pH controlled assembly of a polybutadiene–poly (methacrylic acid) copolymer in water: packing considerations and kinetic limitations. *Soft Matter*, 5, 1674–1682.

Gebhardt, K., Ahn, S., Venkatachalam, G. & Savin, D. (2008) Role of secondary structure changes on the morphology of polypeptide-based block copolymer vesicles. *Journal of Colloid And Interface Science*, 317, 70–76.

Geng, Y., Ahmed, F., Bhasin, N. & Discher, D. (2005) Visualizing worm micelle dynamics and phase transitions of a charged diblock copolymer in water. *Journal of Physical Chemistry B-Condensed Phase*, 109, 3772–3779.

Geng, Y., Discher, D. E., Justynska, J. & Schlaad, H. (2006) Grafting short peptides onto polybutadiene-block-poly(ethylene oxide): A platform for self-assembling hybrid amphiphiles. *Angewandte Chemie-International Edition*, 45, 7578–7581.

Gennes, P.-G. D. (1978) Macromolecules and liquid crystals: reflexions on certain lines of research. *Solid State Physics*, Suppl.14, 1–18.

Ghoroghchian, P., Li, G., Levine, D., Davis, K., Bates, F., Hammer, D. & Therien, M. (2006) Bioresorbable vesicles formed through spontaneous self-assembly of amphiphilic poly (ethylene oxide)-block-polycaprolactone. *Macromolecules*, 39, 1673–1675.

Gregoriadis, G. (1992) *Liposome technology*, CRC.

Halperin, A., Tirrell, M. & Lodge, T. P. (1992) Tethered chains in polymer microstructures. *Advances in Polymer Science*, 100, 31–71.

Hammer, D. & Discher, D. (2001) Synthetic cells-self-assembling polymer membranes and bioadhesive colloids. *Annual Review of Materials Research*, 31, 387–404.

Hammer, D. A., Robbins, G. P., Haun, J. B., Lin, J. J., Qi, W., Smith, L. A., Ghoroghchian, P. P., Therien, M. J. & Bates, F. S. (2008) Leuko-polymersomes. *Faraday Discussions*, 139, 129–141.

Hauschild, S., Lipprandt, U., Rumplecker, A., Borchert, U., Rank, A., Schubert, R. & Förster, S. (2005) Direct preparation and loading of lipid and polymer vesicles using inkjets. *Small*, 1, 1177–1180.

Hocine, S., Brûlet, A., Jia, L., Yang, J., Di Cicco, A., Bouteiller, L. & Li, M.-H. (2011) Structural changes in liquid crystal polymer vesicles induced by temperature variation and magnetic fields. *Soft Matter*, in press.

Howse, J., Jones, R., Battaglia, G., Ducker, R., Leggett, G. & Ryan, A. (2009) Templated formation of giant polymer vesicles with controlled size distributions. *Nature Materials*, 8, 507–511.

Iatrou, H., Frielinghaus, H., Hanski, S., Ferderigos, N., Ruokolainen, J., Ikkala, O., Richter, D., Mays, J. & Hadjichristidis, N. (2007) Architecturally induced multiresponsive vesicles from well-defined polypeptides. Formation of gene vehicles. *Biomacromolecules*, 8, 2173–2181.

Ikeda, T., Mamiya, J. & Yu, Y. L. (2007) Photomechanics of liquid-crystalline elastomers and other polymers. *Angewandte Chemie-International Edition*, 46, 506–528.

Israelachvili, J. (2005) Intermolecular and Surface Forces: With Applications to Colloidal and Biological Systems (Colloid Science). *London: Academic Press*.

Jain, S. & Bates, F. S. (2003) On the origins of morphological complexity in block copolymer surfactants. *Science*, 300, 460–464.

Jenekhe, S. & Chen, X. (1998) Self-assembled aggregates of rod-coil block copolymers and their solubilization and encapsulation of fullerenes. *Science*, 279, 1903.

Jeong, S., Yi, S., Lim, S., Park, S., Jung, J., Woo, H., Song, S., Kim, J. & Lee, J. (2009) Enhancement of radiotherapeutic effectiveness by temperature-sensitive liposomal 1-methylxanthine. *International Journal of Pharmaceutics*, 372, 132–139.

Jia, L., Cao, A., Levy, D., Xu, B., Albouy, P. A., Xing, X. J., Bowick, M. J. & Li, M. H. (2009) Smectic polymer vesicles. *Soft Matter*, 5, 3446–3451.

Jiang, H. Z., Guler, M. O. & Stupp, S. I. (2007a) The internal structure of self-assembled peptide amphiphiles nanofibers. *Soft Matter*, 3, 454–462.

Jiang, Y., Wang, Y., Ma, N., Wang, Z., Smet, M. & Zhang, X. (2007b) Reversible self-organization of a UV-responsive PEG-terminated malachite green derivative: Vesicle formation and photo-induced disassembly. *Langmuir*, 23, 4029–4034.

Kamaly, N., Kalber, T., Ahmad, A., Oliver, M., So, P., Herlihy, A., Bell, J., Jorgensen, M. & Miller, A. (2007) Bimodal paramagnetic and fluorescent liposomes for cellular and tumor magnetic resonance imaging. *Bioconjugate Chemistry*, 19, 118–129.

Katz, J. S., Zhong, S., Ricart, B. G., Pochan, D. J., Hammer, D. A. & Burdick, J. A. (2010) Modular Synthesis of Biodegradable Diblock Copolymers for Designing Functional Polymersomes. *Journal of the American Chemical Society*, 132.

Kishimura, A., Liamsuwan, S., Matsuda, H., Dong, W. F., Osada, K., Yamasaki, Y. & Kataoka, K. (2009) pH-dependent permeability change and reversible structural transition of PEGylated polyion complex vesicles (PICsomes) in aqueous media. *Soft Matter*, 5, 529–532.

Knop, K., Hoogenboom, R., Fischer, D. & Schubert, U. (2010) Poly (ethylene glycol) in Drug Delivery: Pros and Cons As Well as Potential Alternatives. *Angewandte Chemie International Edition*, 49, 6288–6308.

Konak, C., Rathi, R., Kopec Kova, P. & Kopecek, J. (1997) Photoregulated association of water-soluble copolymers with spirobenzopyran-containing side chains. *Macromolecules*, 30, 5553–5556.

Kono, K., Nishihara, Y. & Takagishi, T. (1995) Photoresponsive permeability of polyelectrolyte complex capsule membrane containing triphenylmethane leucohydroxide residues. *Journal of Applied Polymer Science*, 56, 707–713.

Kukula, H., Schlaad, H., Antonietti, M. & Forster, S. (2002) The formation of polymer vesicles or "peptosomes" by polybutadiene-block-poly(L-glutamate)s in dilute aqueous solution. *Journal of the American Chemical Society*, 124, 1658–1663.

Lasic, D. & Papahadjopoulos, D. (1998) *Medical applications of liposomes*, Elsevier Science Ltd.

Leclercq, F., Cohen-Ohana, M., Mignet, N., Sbarbati, A., Herscovici, J., Scherman, D. & Byk, G. (2003) Design, synthesis, and evaluation of gadolinium cationic lipids as tools for biodistribution studies of gene delivery complexes. *Bioconjugate Chemistry*, 14, 112–119.

Lecommandoux, S., Sandre, O., Chécot, F., Rodriguez-Hernandez, J. & Perzynski, R. (2005) Magnetic nanocomposite micelles and vesicles. *Advanced Materials*, 17, 712–718.

Li, M.-H., Auroy, P. & Keller, P. (2000) An azobenzene-containing side-on liquid crystal polymer. *Liquid crystals*, 27, 1497–1502.

Li, M.-H. & Keller, P. (2009) Stimuli-responsive polymer vesicles. *Soft Matter*, 5, 927–937.

Li, M.-H., Keller, P., Li, B., Wang, X. & Brunet, M. (2003) Light-Driven Side-On Nematic Elastomer Actuators. *Advanced Materials*, 15, 569--572.

Li, Y., Lokitz, B. & Mccormick, C. (2006a) Thermally responsive vesicles and their structural locking through polyelectrolyte complex formation. *Angewandte Chemie International Edition*, 45, 5792–5795.

Li, Y., Smith, A., Lokitz, B. & Mccormick, C. (2007) In Situ Formation of Gold-"Decorated" Vesicles from a RAFT-Synthesized, Thermally Responsive Block Copolymers. *Macromolecules*, 40, 8524–8526.

Li, Y. T., Lokitz, B. S., Armes, S. P. & Mccormick, C. L. (2006b) Synthesis of reversible shell cross-linked micelles for controlled release of bioactive agents. *Macromolecules*, 39, 2726–2728.

Lin, J. J., Silas, J. A., Bermudez, H., Milam, V. T., Bates, F. S. & Hammer, D. A. (2004) The effect of polymer chain length and surface density on the adhesiveness of functionalized polymersomes. *Langmuir*, 20, 5493–5500.

Liu, F. & Eisenberg, A. (2003) Preparation and pH triggered inversion of vesicles from poly (acrylic acid)-block-polystyrene-block-poly (4-vinyl pyridine). *J. Am. Chem. Soc*, 125, 15059–15064.

Lomas, H., Canton, I., Macneil, S., Du, J., Armes, S., Ryan, A., Lewis, A. & Battaglia, G. (2007) Biomimetic pH sensitive polymersomes for efficient DNA encapsulation and delivery. *Advanced Materials*, 19, 4238–43.

Mabrouk, E., Bonneau, S., Jia, L., Cuvelier, D., Li, M.-H. & Nassoy, P. (2010) *Soft Matter*, 6, 4863–4875.

Mabrouk, E., Cuvelier, D., Brochard-Wyart, F., Nassoy, P. & Li, M.-H. (2009a) Bursting of sensitive polymersomes induced by curling. *Proceedings of the National Academy of Sciences*, 106, 7294–7298.

Mabrouk, E., Cuvelier, D., Pontani, L., Xu, B., Lévy, D., Keller, P., Brochard-Wyart, F., Nassoy, P. & Li, M. (2009b) Formation and material properties of giant liquid crystal polymersomes. *Soft Matter*, 5, 1870–1878.

Massey, J., Temple, K., Cao, L., Rharbi, Y., Raez, J., Winnik, M. & Manners, I. (2000) Self-assembly of organometallic block copolymers: The role of crystallinity of the core-forming polyferrocene block in the micellar morphologies formed by poly (ferrocenylsilane-b-dimethylsiloxane) in n-alkane solvents. *J. Am. Chem. Soc*, 122, 11577–11584.

Meng, F., Hiemstra, C., Engbers, G. & Feijen, J. (2003) Biodegradable polymersomes. *Macromolecules*, 36, 3004–3006.

Meng, F., Zhong, Z. & Feijen, J. (2009) Stimuli-Responsive Polymersomes for Programmed Drug Delivery. *Biomacromolecules*, 10, 197–209.

Meng, F. H., Engbers, G. H. M. & Feijen, J. (2005) Biodegradable polymersomes as a basis for artificial cells: encapsulation, release and targeting. *Journal of Controlled Release*, 101, 187–198.

Napoli, A., Bermudez, H. & Hubbell, J. A. (2005) Interfacial reactivity of block copolymers: Understanding the amphiphile-to-hydrophile transition. *Langmuir*, 21, 9149–9153.

Napoli, A., Boerakker, M., Tirelli, N., Nolte, R., Sommerdijk, N. & Hubbell, J. (2004a) Glucose-oxidase based self-destructing polymeric vesicles. *Langmuir*, 20, 3487–3491.

Napoli, A., Sebok, D., Senti, A. & Meier, W. (2006) Block Copolymer Vesicles. *Block Copolymers in Nanoscience*.

Napoli, A., Valentini, M., Tirelli, N., Müller, M. & Hubbell, J. (2004b) Oxidation-responsive polymeric vesicles. *Nature Materials*, 3, 183–189.

Nardin, C., Hirt, T., Leukel, J. & Meier, W. (2000) Polymerized ABA triblock copolymer vesicles. *Langmuir*, 16, 1035–1041.

Nehring, R., Palivan, C. G., Casse, O., Tanner, P., Tuxen, J. & Meier, W. (2009) Amphiphilic Diblock Copolymers for Molecular Recognition: Metal-Nitrilotriacetic Acid Functionalized Vesicles. *Langmuir*, 25, 1122–1130.

Onaca, O., Enea, R., Hughes, D. W. & Meier, W. (2009) Stimuli-Responsive Polymersomes as Nanocarriers for Drug and Gene Delivery. *Macromolecular Bioscience*, 9, 129–139.

Opsteen, J. A., Brinkhuis, R. P., Teeuwen, R. L. M., Lowik, D. & Van Hest, J. C. M. (2007) "Clickable" polymersomes. *Chemical Communications*, 3136–3138.

Pangu, G., Davis, K., Bates, F. & Hammer, D. (2010) Ultrasonically Induced Release from Nanosized Polymer Vesicles. *Macromolecular Bioscience*, 10, 546–554.

Pautot, S., Frisken, B., Cheng, J., Xie, X. & Weitz, D. (2003a) Spontaneous formation of lipid structures at oil/water/lipid interfaces. *Langmuir*, 19, 10281–10287.

Pautot, S., Frisken, B. & Weitz, D. (2003b) Engineering asymmetric vesicles. *Proceedings of the National Academy of Sciences*, 100, 10718–10721.

Pautot, S., Frisken, B. & Weitz, D. (2003c) Production of unilamellar vesicles using an inverted emulsion. *Langmuir*, 19, 2870–2879.

Pieroni, O., Fissi, A. & Popova, G. (1998) Photochromic polypeptides. *Progress in Polymer Science*, 23, 81–123.

Pinol, R., Jia, L., Gubellini, F., Levy, D., Albouy, P. A., Keller, P., Cao, A. & Li, M. H. (2007) Self-assembly of PEG-b-Liquid crystal polymer: The role of smectic order in the formation of nanofibers. *Macromolecules*, 40, 5625–5627.

Pradhan, P., Giri, J., Rieken, F., Koch, C., Mykhaylyk, O., Döblinger, M., Banerjee, R., Bahadur, D. & Plank, C. (2010) Targeted temperature sensitive magnetic liposomes for thermo-chemotherapy. *Journal of Controlled Release*, 142, 108–121.

Qin, S., Geng, Y., Discher, D. E. & Yang, S. (2006) Temperature-controlled assembly and release from polymer vesicles of poly (ethylene oxide)-block-gpoly (N-isopropylacrylamide). *Advanced materials(Weinheim)*, 18, 2905–2909.

Ravi, P., Sin, S., Gan, L., Gan, Y., Tam, K., Xia, X. & Hu, X. (2005) New water soluble azobenzene-containing diblock copolymers: synthesis and aggregation behavior. *Polymer*, 46, 137–146.

Riske, K. & Dimova, R. (2005) Electro-deformation and poration of giant vesicles viewed with high temporal resolution. *Biophysical Journal*, 88, 1143–1155.

Riske, K. & Dimova, R. (2006) Electric pulses induce cylindrical deformations on giant vesicles in salt solutions. *Biophysical Journal*, 91, 1778–1786.

Robbins, G., Jimbo, M., Swift, J., Therien, M., Hammer, D. & Dmochowski, I. (2009) Photoinitiated Destruction of Composite Porphyrin Protein Polymersomes. *Journal of the American Chemical Society*, 131, 3872–3874.

Robbins, G. P., Saunders, R. L., Haun, J. B., Rawson, J., Therien, M. J. & Hammer, D. A. (2010) Tunable Leuko-polymersomes That Adhere Specifically to Inflammatory Markers. *Langmuir*, 26, 14089–14096.

Rodriguez-Hernandez, J. & Lecommandoux, S. (2005) Reversible Inside Out Micellization of pH-responsive and Water-Soluble Vesicles Based on Polypeptide Diblock Copolymers. *J. Am. Chem. Soc*, 127, 2026–2027.

Romberg, B., Metselaar, J., Baranyi, L., Snel, C., Bünger, R., Hennink, W., Szebeni, J. & Storm, G. (2007) Poly (amino acid) s: promising enzymatically degradable stealth coatings for liposomes. *International Journal of Pharmaceutics*, 331, 186–189.

Sabaté, R., Barnadas-Rodríguez, R., Callejas-Fernández, J., Hidalgo-Álvarez, R. & Estelrich, J. (2008) Preparation and characterization of extruded magnetoliposomes. *International Journal of Pharmaceutics*, 347, 156–162.

Schlaad, H. (2006) Solution properties of polypeptide-based copolymers. *Peptide Hybrid Polymers.* Berlin, Springer-Verlag Berlin.

Semple, S., Chonn, A. & Cullis, P. (1998) Interactions of liposomes and lipid-based carrier systems with blood proteins: Relation to clearance behaviour in vivo. *Advanced Drug Delivery Reviews*, 32, 3–17.

Shum, H., Kim, J. & Weitz, D. (2008) Microfluidic Fabrication of Monodisperse Biocompatible and Biodegradable Polymersomes with Controlled Permeability. *Journal of the American Chemical Society*, 130, 9543–9549.

Sigel, R., Losik, M. & Schlaad, H. (2007) pH responsiveness of block copolymer vesicles with a polypeptide corona. *Langmuir*, 23, 7196–7199.

Sin, S., Gan, L., Hu, X., Tam, K. & Gan, Y. (2005) Photochemical and thermal isomerizations of azobenzene-containing amphiphilic diblock copolymers in aqueous micellar aggregates and in film. *Macromolecules*, 38, 3943–3948.

Stachowiak, J., Richmond, D., Li, T., Brochard-Wyart, F. & Fletcher, D. (2009) Inkjet formation of unilamellar lipid vesicles for cell-like encapsulation. *Lab on a Chip*, 9, 2003–2009.

Sugiyama, K. & Sono, K. (2001) Characterization of photo and thermoresponsible amphiphilic copolymers having azobenzene moieties as side groups. *Journal of Applied Polymer Science*, 81, 3056–3063.

Sun, G., Fang, H., Cheng, C., Lu, P., Zhang, K., Walker, A., Taylor, J. & Wooley, K. (2009) Benzaldehyde-Functionalized Polymer Vesicles.

Szoka, F. & Papahadjopoulos, D. (1978) Procedure for preparation of liposomes with large internal aqueous space and high capture by reverse-phase evaporation. *Proceedings of the National Academy of Sciences of the United States of America*, 75, 4194.

Takeuchi, H., Kojima, H., Yamamoto, H. & Kawashima, Y. (2001) Evaluation of circulation profiles of liposomes coated with hydrophilic polymers having different molecular weights in rats. *Journal of Controlled Release*, 75, 83–91.

Taton, D. & Gnanou, Y. (2006) Guidelines for Synthesizing Block Copolymers. *Block Copolymers in Nanoscience*, 9–35.

Tong, X., Wang, G., Soldera, A. & Zhao, Y. (2005) How can azobenzene block copolymer vesicles be dissociated and reformed by light? *Journal of Physical Chemistry B*, 109, 20281–20287.

Torchilin, V., Levchenko, T., Whiteman, K., Yaroslavov, A., Tsatsakis, A., Rizos, A., Michailova, E. & Shtilman, M. (2001) Amphiphilic poly-N-vinylpyrrolidones synthesis, properties and liposome surface modification. *Biomaterials*, 22, 3035–3044.

Van Dongen, S. F. M., Nallani, M., Schoffelen, S., Cornelissen, J., Nolte, R. J. M. & Van Hest, J. C. M. (2008) A block copolymer for functionalisation of polymersome surfaces. *Macromolecular Rapid Communications*, 29, 321–325.

Vlahovska, P., Graci, R., Aranda-Espinoza, S. & Dimova, R. (2009) Electrohydrodynamic model of vesicle deformation in alternating electric fields. *Biophysical Journal*, 96, 4789–4803.

Vriezema, D., Garcia, P., Sancho, EMSP14, Oltra, N., Hatzakis, N., Kuiper, S., Nolte, R., Rowan, A., Van & Hest, J. (2007) Positional Assembly of Enzymes in Polymersome Nanoreactors for Cascade Reactions We thank Dr. E. Pierson for guidance with the fluorescence microscope and NWO-ACTS and Synthon BV for financial support. *Angewandte Chemie*, 119.

Wang, G., Tong, X. & Zhao, Y. (2004) Preparation of azobenzene-containing amphiphilic diblock copolymers for light-responsive micellar aggregates. *Macromolecules*, 37, 8911–8917.

Whiteman, K., Subr, V., Ulbrich, K. & Torchilin, V. (2001) Poly (HPMA)-coated liposomes demonstrate prolonged circulation in mice. *Journal of Liposome Research*, 11, 153–164.

Won, Y. Y., Davis, H. T. & Bates, F. S. (1999) Giant wormlike rubber micelles. *Science*, 283, 960–963.

Woodle, M. C., Engbers, C. M. & Zalipsky, S. (1994) New amphipatic polymer lipid conjugates forming long-circulating reticuloendothelial system-evading liposomes. *Bioconjugate Chemistry*, 5, 493–496.

Yamamoto, Y., Nagasaki, Y., Kato, Y., Sugiyama, Y. & Kataoka, K. (2001) Long-circulating poly(ethylene glycol)-poly(D,L-lactide) block copolymer micelles with modulated surface charge. *Journal of Controlled Release*, 77, 27–38.

Yang, J., Lévy, D., Deng, W., Keller, P. & Li, M.-H. (2005) Polymer vesicles formed by amphiphilic diblock copolymers containing a thermotropic liquid crystalline polymer block. *Chemical Communications*, 2005, 4345–4347.

Yang, J., Pinol, R., Gubellini, F., Levy, D., Albouy, P., Keller, P. & Li, M.-H. (2006) Formation of polymer vesicles by liquid crystal amphiphilic block copolymers. *Langmuir*, 22, 7907–7911.

Yatvin, M., Weinstein, J., Dennis, W. & Blumenthal, R. (1978) Design of liposomes for enhanced local release of drugs by hyperthermia. *Science*, 202, 1290.

Yu, K. & Eisenberg, A. (1998) Bilayer morphologies of self-assembled crew-cut aggregates of amphiphilic PS-b-PEO diblock copolymers in solution. *Macromolecules*, 31, 3509–3518.

Yu, S., Azzam, T., Rouiller, I., & Eisenberg, A. (2009) "Breathing" Vesicles, *J. AM. CHEM. SOC*, 131, 10557–10566.

Yuan, X., Fischer, K. & Schärtl, W. (2005) Photocleavable microcapsules built from photoreactive nanospheres. *Langmuir*, 21, 9374–9380.

Zhang, J., Zhang, Z., Yang, H., Tan, Q., Qin, S. & Qiu, X. (2005) Lyophilized paclitaxel magnetoliposomes as a potential drug delivery system for breast carcinoma via parenteral administration: in vitro and in vivo studies. *Pharmaceutical Research*, 22, 573–583.

Zhang, L. & Eisenberg, A. (1995a) Aggregates of Polystyrene-b-poly (acrylic acid) Block Copolymers. *Science*, 268, 23.

Zhang, L. & Eisenberg, A. (1995b) Multiple morphologies of "crew-cut" aggregates of polystyrene-b-poly (acrylic acid) block copolymers. *Science*, 268, 1728–1731.

Zhang, L. & Eisenberg, A. (1996) Multiple morphologies and characteristics of "crew-cut" micelle-like aggregates of polystyrene-b-poly (acrylic acid) diblock copolymers in aqueous solutions. *J. Am. Chem. Soc*, 118, 3168–3181.

Zhang, L. F. & Eisenberg, A. (1998) Formation of crew-cut aggregates of various morphologies from amphiphilic block copolymers in solution. *Polymers for Advanced Technologies*, 9, 677–699.

Zupancich, J., Bates, F. & Hillmyer, M. (2006) Aqueous dispersions of poly (ethylene oxide)-b-poly (–methyl- -caprolactone) block copolymers. *Macromolecules*, 39, 4286–4288.

Zupancich, J. A., Bates, F. S. & Hillmyer, M. A. (2009) Synthesis and Self-Assembly of RGD-Functionalized PEO-PB Amphiphiles. *Biomacromolecules*, 10, 1554–1563.

Silica-Based Nanoparticles for Intracellular Drug Delivery

Sandrine Quignard, Sylvie Masse, and Thibaud Coradin

Abstract Silica-based nanoparticles have recently raised a great deal of attention as possible drug carriers. Such an interest is driven by the possibility to control their size, the chemical composition and the porous structure as well as to easily modify their surface with a wide range of biologically-relevant functionalities, favoring colloidal stability, long-time blood circulation and even specific targeting. Drug loading can be performed during particle formation but, at this time, the most popular method relies on the impregnation of pre-formed mesoporous colloids. Strategies to control drug delivery via bio-responsive pore capping are also developed. However, despite an increasing number of in vitro and in vivo studies related to the interaction of silica particles with cells and animals, their biocompatibility is still an issue, especially if applications in intracellular drug delivery are foreseen.

Keywords Silica • Organosilanes • Mesoporous particles • Hybrid materials

Abbreviations

FITC	fluorescein isothiocyanate
HSN	hollow silica nanoparticles
MSN	mesoporous silica nanoparticles
PEG	polyethylene glycol
PEI	polyethylene imine
PLGA	poly(lactic-co-glycolic acid)
PLLA	poly-L-lactic acid
PVP	polyvinylpyrrolidone

S. Quignard, S. Masse, and T. Coradin (✉)
CNRS, Laboratoire de Chimie de la Matière Condensée de Paris,
Collège de France, UPMC Univ Paris 06, 11 Place Marcelin Berthelot,
F-75005 Paris, France
e-mail: thibaud.coradin@upmc.fr

A. Prokop (ed.), *Intracellular Delivery: Fundamentals and Applications*, Fundamental Biomedical Technologies 5, DOI 10.1007/978-94-007-1248-5_12,

RES	reticuloendothelial system
siRNA	short-interfering RNA
TEOS	tetraethoxysilane

1 Introduction

Chemistry at the nano-bio interface is one of the most promising challenges of the 2000s, as a more and more sophisticated set of tools can be used to take advantage of the unique properties of nanomaterials. Indeed, their physical and chemical properties deviate significantly from the bulk properties of such materials, and a perfect control of chemistry allow to design a wide range of nanostructures of controlled size, shape, dispersity and surface chemistry. The complexity of these structures can go from simple spheres to nanotubes or nanowires, going through core-shell particles, hollow capsules, prisms, Janus particles, etc. Also their surface could be differentially functionalized on their inner and outer surfaces, opening a wide field for applications in drug delivery and medical therapy.

One of the first significant attempts to conjugate inorganic nanoparticles to biological molecules came out in the 1990s, when semiconductor quantum dots were coupled to antibodies in order to target them to specific biomolecules into the cells (Bruchez et al. 1998). These experiments demonstrate the efficiency of brightly luminescent quantum dots for intracellular imaging, that make these nanocrystals probes to be considered as novel candidates that could be, in some cases, more competitive than existing fluorophores. New strategies for delivery of short-interfering RNA (siRNA) to cells using nanoparticles were also developed by coating semiconductor quantum dots with proton-absorbing polymeric layers that allow the nanoparticle and its siRNA cargo to perform the functions of cellular penetration, endosomal release, carrier unpacking and intracellular transport (Yezhelyev et al. 2008). Other authors showed that a high packing density of siRNA on the surface of gold particles inhibit its degradation by nucleases and promotes its uptake by HeLa cells (Giljohann et al. 2009).

Among inorganic particles, silica colloids represent a very unique situation. First their synthesis and further functionalization is rather simple and highly versatile, as discussed in several reviews (Wang et al. 2008; Jin et al. 2009; Taylor-Pashow et al. 2010). Second, silica is usually considered as a non-toxic material, a property that it shares with only a few other inorganic materials such as calcium phosphate, calcium carbonate and iron oxides. It is therefore not surprising that silica nanoparticles have been widely studied for biomedical applications: imaging, drug delivery, ultra-trace detection, DNA/RNA delivery or therapy.

In particular, the evaluation of silica-based nanoparticles as drug carriers has become very popular among the last 5 years or so (Slowing et al. 2008; Di Pasqua et al. 2008; Rosenholm et al. 2010). Over this period, silica particles of increasing complexity, in terms of structure, chemical composition and biological responses have been designed. Such a complexity reflects the many requirements of drug vectors: colloidal stability, biocompatibility, long-term circulation, specific targeting, drug loading and release,...In parallel, an impressive set of data related to the interactions of particles with cells in vitro and, to a lesser extent, related to their in vivo fate has been accumulated, that are necessary before any clinical studies, not to mention practical applications, can be foreseen.

This is the scope of this chapter to provide the readers with an up-to-date overview of these progresses from a chemical perspective. A first section gathers some key information about the synthesis and physico-chemical properties of silica-based nanoparticles. The current knowledge about in vitro and in vivo behavior of these particles is also presented. A second section describes the different chemical strategies that have been developed to turn silica nanoparticles into drug carriers. In the context of the present book, a third section is specifically dedicated to intracellular delivery. As a conclusion, some current fundamental and practical challenges in this area are presented and critically discussed.

2 Silica-Based Nanoparticles: Synthesis and Reactivity in Biological Conditions

2.1 *The Chemistry of Silica*

The process of silica formation from solution occurs following an inorganic polymerization reaction (Iler 1979). Two monomers, silicic acid $Si(OH)_4$, condense with each other to form a Si-O-Si siloxane bond, giving rise to the $(OH)_3$-Si-O-Si-$(OH)_3$ dimers with departure of a water molecule. As polymerization proceeds, higher oligomers are formed, with a strong tendency to form cyclic species. The reaction continues until particles, 2–3 nm in diameter are formed. These particles consist of hydrated silica with low condensation degree (*i.e.* with internal porosity). The pK_a of polysilicic acid is *ca.* 6 so that the surface of the particle is negatively charged above pH 3, consisting of silanol Si-OH and silanolate Si-Si-O- groups. If silica formation occurs near pH 3, these particles bear a low surface charge and tend to aggregate, forming in many cases a silica gel. Otherwise, these particles continue to grow either by further monomer addition or by flocculation, but, at any time of this process, percolation can occur, forming a gel again. Only above pH 8, where surface charge is high enough, do stable silica particles can be obtained.

However, the condensation reaction is always balanced by the reverse reaction, i.e. hydrolysis (Icopini et al. 2005). This reaction is also pH dependent as it is favored in the presence of silanolate (Vogelsberger et al. 1992). Thus at pH 9 or above, dissolution is largely favored over condensation, preventing silica formation but the latter process becomes predominant below this value. Indeed, this equilibrium is reflected in silica solubility and is therefore temperature-dependent. For instance, in pure water near neutral pH, amorphous silica solubility is *ca.* 50 ppm at 10°C and *ca.* 130 ppm at 25°C. In parallel, the reaction kinetics are also influenced by temperature and by specific surface area (either surface/volume ratio for particles or porosity for bulk materials) (Rimer et al. 2007).

The presence of additives in the reaction media may have several impacts on these processes. Inorganic salts tend to reduce the stability of silica particle sols due to surface charge screening, decrease solubility of silica and increase dissolution rate (Wallace et al. 2010). Polymers may favor or limit silica condensation as function of their charge and concentration. They can also coat the surface of silica particles, decreasing the availability of silanol groups and therefore decreasing their dissolution rate.

2.2 *Strategies for Silica Nanoparticle Formation*

2.2.1 Pure Silica Nanoparticle Formation

Nowadays, the so-called Stöber process is undoubtedly the most popular route to obtain colloidal silica (Stöber et al. 1968). It is based on the growth of silica nanoparticles from silicon alkoxides (mainly tetraethoxysilane, TEOS) in basic media (ammonia solution) in the presence of an alcohol as a co-solvent. The basic conditions are required to obtain fast condensation and particle stabilization. The alcohol favors the initial solubilization of the silicon alkoxide precursor and probably limits particle growth by increasing the interfacial energy. Despite its apparent simplicity, the Stöber process requires a careful control of process parameters (volume, stirring, mixing rate) to get access to monodisperse populations, especially in the small size domain (i.e. below ca. 50 nm). Therefore many modifications of the initial procedure are available to improve the size distribution (Bogush et al. 1988; Rao et al. 2005). For instance, a seed-growth approach was developed that starts from well-defined small colloids that are further grown in a TEOS/ammonia solution (Chang et al. 2005) (Fig. 1). A recent report also shows that soluble silicates can be used to form silica nanoparticles by simple ethanol addition (Jung et al. 2010). This emphasizes that the main parameter controlling the particle size is the interfacial energy. Thus, it is not surprising that silica nanoparticles have also been grown within emulsion systems in the presence of surfactants (Arriagada and Osseo-Asare 1995; Gan et al. 1996). Finally, silica particles can also be obtained using spray-drying techniques (Iskandar et al. 2003) but this approach has been mainly applied to hybrid and porous nanomaterials, as described below (Baccile et al. 2003)

2.2.2 Hybrid Silica Nanoparticle Formation

A first approach to convert pure silica nanoparticles into hybrid nanomaterials is through the grafting of organic functions using organosilanes. A wide variety of these silanes are commercially available, from hydrocarbon to amine- or thiol-terminated chains. Although this approach has been extensively used (see Sect. 3), the precise structure of the resulting surface is far from being understood (de Monredon-Senani

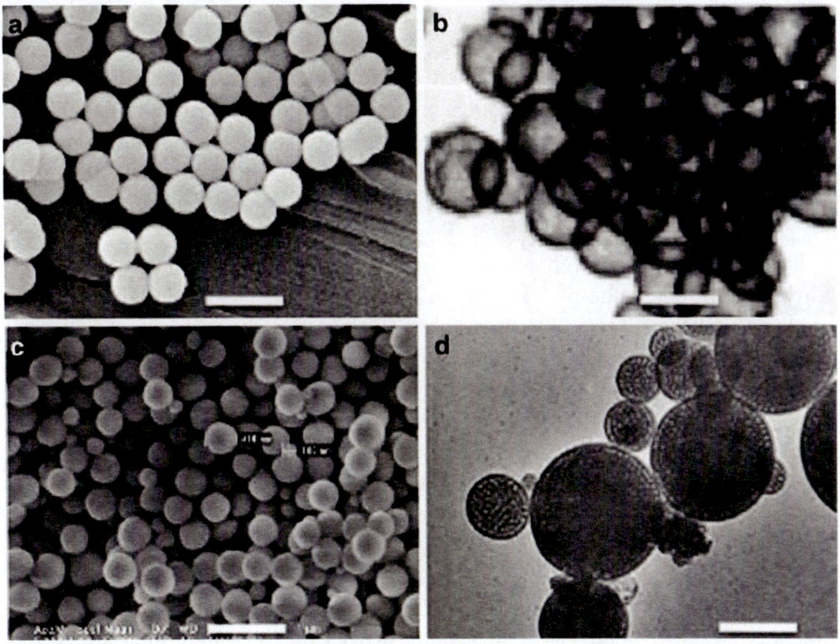

Fig. 1 Silica nanoparticles : (**a**) SEM of plain silica nanoparticles (scale bar = 2 μm) (Reprinted from (Chang et al. 2005), Copyright (2005), with permission from Elsevier), (**b**) TEM of liposils (scale bar = 200 nm) (Reprinted from (Bégu et al. 2007), Copyright (2007), with permission from Elsevier), (**c**) SEM of PLLA@silica homogeneous particles (scale bar = 1 μm) (From Zalzberg and Avnir 2008. Copyright Springer Science + Business Media, LLC 2008), (**d**) TEM of mesoporous particles (scale bar = 300 nm)

et al. 2009). In particular, it is important that some water is present in the reaction media to allow the hydrolysis of the silane before it can condense on the silica surface. On the other hand, there is a reasonable probability for self-condensation (i.e. organosilanes react with each other rather than with the silanol groups on the surface) resulting in the grafting of a silane chain rather than a one-to-one grafting. When the desired function is not available, it is possible to perform a coupling reaction on the initial silane or on the pre-grafted function. Many considerations should be taken into account when selecting the most convenient approach. In particular, the first approach must be performed in water-free conditions that can be a very drastic limit. In principle, silica nanoparticles can also be coated by polymers by simple adsorption. To improve the stability of the coating, the particle surface may be pre-grafted. However, the most efficient method relies on the grafting of a monomer followed by controlled polymerization (Radhakrishnan et al. 2006). To complete this overview, it is worth noting that it was possible to coat silica particles with lipid bilayers exhibiting high stability (Van Shooneveld et al. 2008).

The above-described approaches lead to core-shell structures where the organic functions are localized on the particle surface. The mirror situation, where a silica shell is surrounding an organic or bio-organic core, was also described (Zou et al. 2008).

In this situation, it is important to work at low silica concentration to avoid gel formation in the whole reaction volume. Thus, there should be a strong affinity of the organic seed for silica, as showed for gelatin nanoparticles whose positive charge attract negatively-charged silica precursors (Allouche et al. 2006). However, the silica condensation may lead to structural constraints on the templating particle so that it is also important to control the kinetics of the silica shell formation. This was clearly demonstrated for the design of so-called liposils consisting of silica-coated liposomes (Bégu et al. 2007) (Fig. 1).

It is also possible to prepare homogenous nanocomposite particles. One possibility is to introduce the polymer in the initial silica precursor solution within an emulsion, as shown for poly-L-lactic acid (PLLA) or polyethylene glycol (PEG) (Sertchook et al. 2007; Zalzberg and Avnir 2008) (Fig. 1). Alternatively, the use of spray-drying techniques was found very efficient to prepare bionanocomposite associating silica and alginic acid, a polysaccharide, with a large variety of composition and size (Yang and Coradin 2008). Compared to the previous hybrid structure, it is very important to have a limited interaction between the polymer and silica to avoid precipitation during gelation.

A specific case of hybrid colloids is the family of mesostructured silica nanoparticles. These particles consist of an organic templating mesophase, mainly self-assembled surfactants or amphiphilic polymers, trapped in a silica network (Soler Illia et al. 2002). Noticeably, such systems are not much used as such but as colloidal precursors to ordered mesoporous nanoparticles, obtained after calcination or template extraction (Fig. 1). In fact, the formation of mesostructured materials is a complex process which is mainly based on the simultaneous self-assembly of the organic templating phase and the condensation of the inorganic phase. The chemical routes to mesostructured silica nanoparticles are not significantly different from that used for pure silica particle growth. In particular, many studies have been focusing on the use of the spray-drying technique (Andersson et al. 2004). This is due to the fact that this process induces a fast evaporation step from precursors in solution to solid particles. This step has a strong impact on the formation of the organic-inorganic mesostructure. An improved understanding and control of this process, called Evaporation-Induced Self-Assembly (Brinker et al. 1999), has allowed to get access to a wide variety of silica particles with different sizes, pore size and structure and even to prepare mixed oxide phases in a single step (Boissiere et al. 2010).

2.3 Silica Behavior in Biologically-Relevant Environment

2.3.1 Acellular Media

The typical preparation of silica nanoparticles results in an aqueous suspension at relatively high silica concentration (0.1 M–1 M), low ionic strength (below 0.01 M, to avoid aggregation) stored at room temperature or below. When put in simulated (in vitro) or effective (in vivo) human body conditions, a strong change in physico-chemical conditions occurs, especially higher temperature (37°C) and ionic strength (ca. 0.1 M). As mentioned above, this leads to a higher solubility that is reached by a fast dissolution reaction. But the most important parameter is the silica concentration, which is typically below 1 mg/mL (0.015 M). As a comparison, the solubility of silica in these conditions is ca. 0.1 mg/mL so that the relative fraction of soluble silica becomes significant. Indeed, higher ionic strength can also induce nanoparticle aggregation. However, biological media also contain a considerable amount of proteins that will adsorb either unspecifically or specifically (during opsonization) (Lord et al 2006). The surface of silica nanoparticles being negatively charged, the first process will take place for positively-charged proteins and the second is very likely to occur as it is know to be specifically effective for anionic surface. As mentioned earlier, this adsorption will tend to decrease the rate of solubilization (Finnie et al. 2009).

An increase in porosity, especially in the case of mesoporous particles, can have two effects (Bass et al 2007; Lu et al. 2010; Izquierdo-Barba et al. 2010) First, because of higher specific surface area the dissolution rate will be increased when compared to plain particles. Secondly, because pores are present on the surface, less surface silanol groups are present. As a consequence, the surface charge is decreased, that should favor particle aggregation in purely inorganic media. In parallel, it may also decrease the available sites for protein binding in biological fluids. Indeed, every procedure that leads to a modification of particle surface will modify its behavior in solution. In many cases, long chain hydrophilic groups have been grafted to limit opsonization phenomena (Cauda et al. 2010). At the same time, because this procedure also decreases the density of surface silanol groups, it should also decrease particle solubility, as indicated earlier. Accordingly, homogeneous nanocomposite particles exhibit different properties from plain silica colloids because the (bio)-organic component is also present on the particle surface. In particular, their solubility depends strongly on the solubility of the additive. For instance, the dissolution of alginate/silica particles has been studied both in solution and within cells, showing that the biopolymer solubilization occurs before that of silica, leading to a rearrangement of the particle internal structure (Yang and Coradin 2008; Boissière et al. 2006).

2.3.2 In Vitro Interactions with Cells

The toxicity of silica-based materials has been studied for a very long time, but most of the studies were related to asbestos fiber-induced silicosis in lungs tissues (Bagchi 1992). As early as the 60's, it was found that silica particles could interfere with the integrity of biological membranes, probably by interaction between the negatively-charged surface and the positively-charged groups of proteins or phospholipids (Depasse 1977). Further studies demonstrate that silica particles could impact on alveolar macrophages that produce specific proteins triggering fibroblast proliferation and therefore formation of fibrous tissues (Arcangeli et al. 1990). It was also suggested that not only the particle surface but also its solubilization products could be responsible for toxicity (Erdogu and Hasirci 1998). In parallel, silica particles have long been used in the formulation of oral drugs, although recent concerns have been raised about their capacity to induce sarcoidosis, that was previously observed for crystalline silica particle only (Sola et al. 2009).

Nevertheless, most recent studies related to silica particle cytotoxicity were triggered by the possibility to use silica-based silica particles for systemic drug delivery. A representative list of cell/particle systems that have been used for *in vitro* studies have been gathered on Table 1.

Table 1 Examples of in vitro evaluation of silica nanoparticles

Cell	Particle structure	Particle size (nm)	Refs
Embryonic kidney cells	Plain	20, 50	Wang et al. (2009)
Hepatic cells	Plain	20, 50, 85	Ye et al. (2010)
Keratinocytes	Plain	7, 10–20	Gong et al. (2010)
Endothelial cells	Plain	20	Liu and Sun (2010)
Lung epithelial	Plain	10, 80	Akhtar et al. (2010)
Primary microglia	Plain	150–200	Choi et al. (2010)
Jurkat cells	Plain and mesoporous-amine-coated	300	Tao et al. (2009)
Human neuroblastoma	Mesoporous-amine- and thiol-coated	250	Di Pasqua et al. (2008)
Monocyte-derived dendritic cells	Mesoporous	270	Vallhov et al. (2007)
HeLa	Mesoporous	30–280	Lu et al. (2009)
	Mesoporous-ODN intercalator-coated	220, 300	Vivero-Escoto et al. (2010a)
Human red blood cells (HRB)	Plain and mesoporous-PEG-coated	25, 40, 95, 155, 225	Lin and Haynes (2010)
HRB/leukemia-derived macrophages	Mesoporous-PEG-coated	150	He et al. (2010)
Human mesothelial/ Mouse peritoneal macrophage/Mouse myoblast (muscle) cells	Mesoporous	100–150, 700–800	Hudson et al. (2008)
Human fibroblasts (HF)	Plain	50, 80	Zhang et al. (2010)
HF/tumor cells	Hybrid chitosan/silica	150–200	Chang et al. (2007)
3 T3-L1 fibroblasts	Plain	20, 30, 240	Barnes et al. (2008)
	Plain	10, 30, 80, 400	Park et al. (2010)
	Hybrid gelatin/silica	400-500	Allouche et al. (2006)
HF/Chinese Hamster Ovarian	Mesoporous	100, 500	Trewyn et al. (2008)
Lung cancer cells	Crystalline	15, 46	Lin et al. (2006)
Mouse embryo fibroblasts	Crystalline	20	Yang et al. (2009)

Taken together, these data suggest that silica nanoparticles impact on cell viability is three-fold: (1) surface interaction, (2) internalization and (3) solubility. Considering the first point, detrimental interactions exist between surface silanol groups and cell membrane, that can lead to cell damage via oxidative stress and even cell lysis. As a result, for plain silica nanoparticles, toxicity increases with decreasing particle size. For mesoporous materials, these interactions decrease with increasing porosity. Interestingly, a recent study showed that the mesoporosity of some silica particles tends to collapse with time, increasing the plain surface of the colloid and therefore its toxicity. Along the same lines, grafting the particle surface with cationic groups (such as amino) or neutral hydrophilic groups (such as PEG) or preparing hybrid particles incorporating cationic polymers (such as chitosan) decreases the observed toxicity. Indeed, all these effects were shown to be dose- and cell-dependent (Fig. 2).

Considering the second point, internalization processes will be discussed to a large extent in the Sect. 4. At this stage, it is just important to emphasize that no

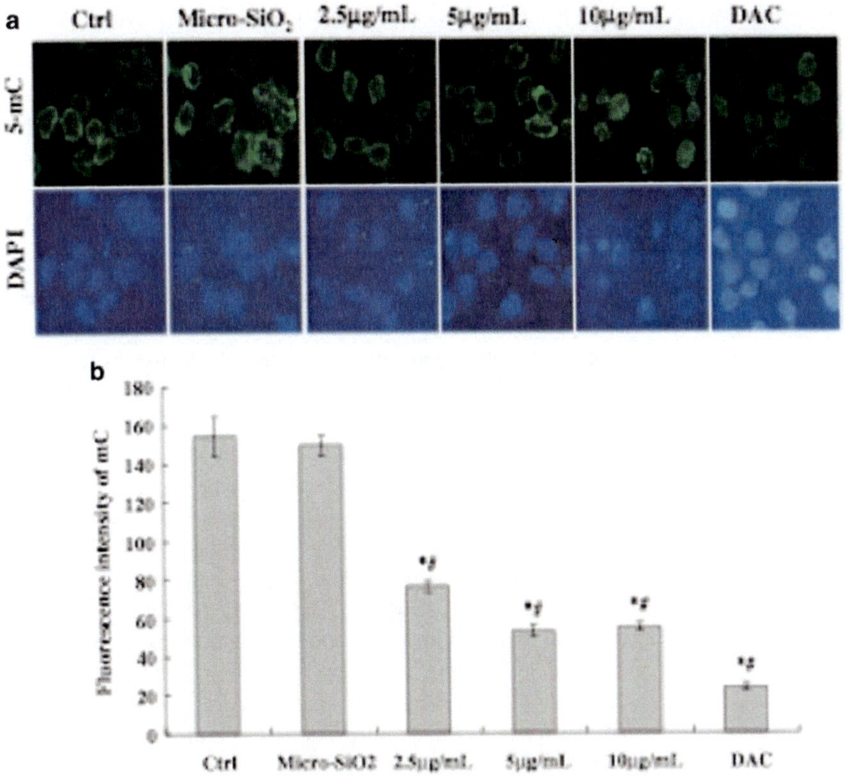

Fig. 2 Analysis of silica nanoparticles (15 nm) toxicity via global DNA methylation in HaCaT cells: (**a**) Anti-5-methylcytosine (5-mC) immunofluorescence (*green*). Nuclei were counterstained with DAPI (*blue*), (**b**) Quantitative representation of fluorescence of 5-mC (Reprinted from (Gong et al. 2010), Copyright (2010), with permission from Elsevier)

genotoxicity has been reported to date on pure silica particles, suggesting that the cell nucleus is not reached by these particles.

Coming to the third point, related to the influence of silica solubility, it must be admitted that it suffers from a lack of available data, although the impact of soluble silica released during particle dissolution on cell viability is often mentioned. One reason is probably that silicic acid, $Si(OH)_4$, the product of silica dissolution, is a very reactive species that is very likely to combine with other components of the biological environment, such as proteins.

2.3.3 In Vivo Toxicity and Biodistribution of Silica Particles

Most recent in vivo studies of silica nanoparticles concern intravenous injections in mice. In the first time following injection, particles are found in the blood circulation system. When a single infusion of plain particles with size ranging from 1,000 nm to 30 nm was performed, an increase in the level of several proteins related to acute inflammatory response (haptaglobin, C-reactive protein, serum amyloid A) was observed. The intensity of this rise increases with decreasing particle size, so that 30 nm particles showed high toxicity, and was dose-dependent (Higashisaka et al. 2011). Interaction of particles with blood circulation, as monitored by an increase in mean arterial pressure and pulse arterial pressure, was also observed after repeated infusions (Galagudza et al. 2010). Noticeably, rapid (<15 min) animal death was also reported when mesoporous particles were infused at high doses, a fact that may be attributed to remaining toxic surfactants (Hudson et al. 2008).

In a second step, particles circulate within the blood flow before being transported to organs. This can happen either due to the reticuloendothelial system (RES), i.e. monocytes and macrophages that are responsible for the clearance of bacteria and colloidal particles, or due to direct deposition. The first mechanism is related to opsonization mechanisms in which specific serum proteins are adsorbed on the particle surface, triggering the macrophage response. The extent of this phenomenon can be evaluated by two facts: (i) the blood circulation time and (ii) the favored biodistribution of particles in specific organs (liver, spleen, lungs) where the RES is particularly active (Burns et al. 2009; Xie et al. 2010). In the case of silica particles, it was shown that they are found located within a few hours mainly in the liver, that may suffer from noticeable injuries (Nishimori et al. 2009; Galagudza et al. 2010; Kumar et al. 2010; Souris et al. 2010). However, depending on the nature of the particle surface, they can also be found in spleen and lungs, and sometimes in skin. Higher amounts of silica particles in these organs are found after 24 h but after 120 h the particles are detected in stomach and intestine (Fig. 3). After 360 h, nearly total clearance was observed. Here again, the particle charge has a strong influence, as positively-charged particle are found in urine and feces after 30 min whereas negatively-charged colloids are still present in the gastrointestinal tract after 3 days.

It is worth noting that other internalization routes, such as intranasal, subcutaneous and intraperitoneal, were also studied (Hudson et al. 2008). Intranasal injections showed very similar results compared to intravenous. Subcutaneous injections showed very limited effects of silica particles whereas significant mortality was observed after

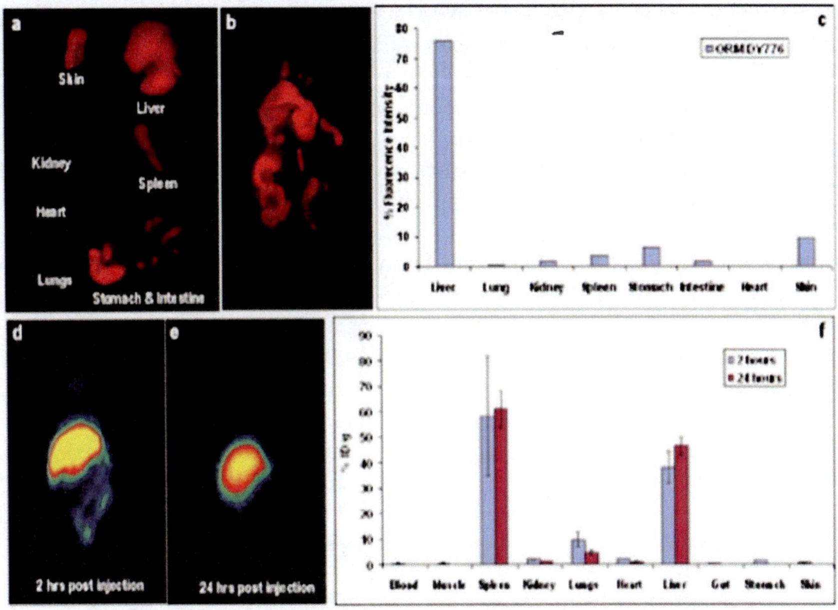

Fig. 3 Biodistribution of silica nanoparticles by the measurement of the fluorescence (**a–c**) and radioactivity (**d–f**) from different organs of the mice injected with DY776 conjugated and ^{124}I labeled silica nanoparticles respectively: (**a**) fluorescence acquired from the individual organs, (**b**) whole body fluorescence imaging of the mice 24 h postinjection, (**c**) quantitative estimation of fluorescence acquired from different organs of the mice, (**d** and **e**) PET images of the mice injected with ^{124}I-silica nanoparticles 2 and 24 h postinjection, respectively, and (**f**) quantitative radioactivity measurements from the individual organs of the mice. (Reproduced with permission from Kumar et al. 2010. Copyright 2010 American Chemical Society)

intraperitoneal infusion. On this basis, it was suggested that silica particles are 100 times more toxic than poly(lactic-co-glycolic acid) (PLGA) nanoparticles.

To close this section, it is worth noting that studies performed on zebrafish embryo showed no uptake (silica particles being located on the chorion, a porous membrane surrounding the fertilized egg) and therefore no mortality nor physiological deformation (Fent et al. 2010). This enlightens an important issue related to possible silica particle internalization through the skin, that is still to be elucidated (Park et al. 2010).

3 Silica-Based Nanomaterials for Drug Delivery

3.1 Drug Encapsulation in Plain Particles

The simplest method is to entrap hydrophilic functional molecules within the silica matrix *via* non-covalent interactions. This procedure is commonly used to incorporate luminescent dyes into the silica matrix. The main drawback is the possible leakage of the fluorophore out of the nanoparticle. Nevertheless, it is worth mentioning an emulsion process described by Barbé et al. for encapsulation and release of drugs (Barbé et al. 2004), that was recently applied to silica microcapsule-based protein delivery (Finnie et al. 2006). However, neither the Stöber method nor the reverse microemulsion process previously described is able to efficiently incorporate hydrophobic molecules into the silica matrix. To solve this problem, organically-modified silica nanoparticles can be used (Bharali et al. 2005). Hydrophobic drugs are first dissolved in the organic phase of the reverse microemulsion, that allows to partition more effectively into the aqueous water droplets and are more efficiently trapped inside the silica nanoparticles by hydrolysis of the silane precursor. Some recent experimental results have demonstrated that photosensitizers retain their functions of generating singlet oxygen after being embedded inside silica matrix (Kim et al. 2007) (Fig. 4). Moreover, on the one hand, the amount of singlet oxygen generated by the nanoparticle-doped photosensitizer is comparable to that produced by the same amount of free photosentisizer, and on the other hand, the negative effects of photosensitizers – such as toxicity or hydrophobicity – are reduced when embedded in silica.

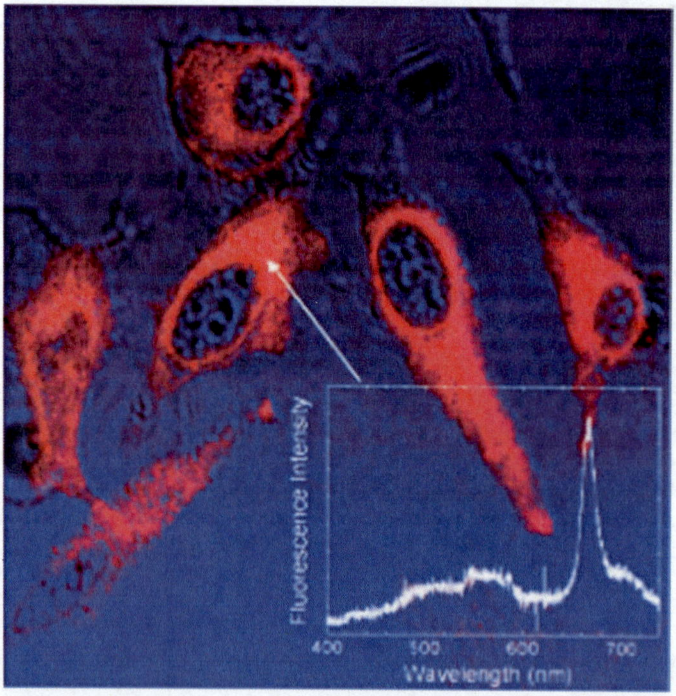

Fig. 4 Merged transmission (*blue*) and two-photon excited fluorescence (*red*) images of HeLa cells, stained with nanoparticles co-encapsulating 1.1 wt.% HPPH/20 wt.% BDSA. Inset: Localized two-photon fluorescence spectrum from the cytoplasm of the stained cell. The excitation wavelength is 850 nm. (Reproduced with permission from Kim et al. 2007. Copyright 2007 American Chemical Society)

3.2 *Surface Modification of Silica Particles*

As mentioned earlier, proper surface modification of silica nanoparticles can improve their drug delivery capability and efficiency. This can be performed by chemical binding or physical adsorption. A typical modifier is an antibody molecule decorating a carboxyl-functionalized silica nanoparticle. In that case, the carboxyl groups on the silica surface form amide bonds with the primary amine groups of antibodies. The immobilized antibody can then more effectively direct the nanoparticle toward a specific tissue or cell type. Another example of the chemical binding approach to immobilize recognition elements is the formation of disulfide bonds. However, the electrostatic forces that govern the physical adsorption of biocompatible macromolecules are often the most simple and efficient method for immobilization. Many macromolecules, such as PEG, proteins or lipids can be physically adsorbed onto nanoparticles surface to decorate them with a variety of functional groups (Van Shooneveld et al. 2008; Thierry et al. 2008).

Despite dye-doped silica nanoparticles can be linked to biorecognition elements such as antibodies or DNA molecules via physical adsorption onto the surface particle, covalent attachment of such molecules is preferred, not only to avoid desorption from the particle surface, but also to get a better control of the biorecognition function localization. So, an alternative way relies on the stable incorporation of imaging or therapeutic agents *via* covalent bonding using trialkoxysilane-derived molecules that contain suitable organic moieties. These molecules can be incorporated into the silica matrix through silanol linkages during particle synthesis, leading to stable hybrid material with uniform agents throughout the nanoparticle that are protected from the environment. The silica nanoparticles can also be post-synthetically modified by reacting with the trialkoxysilane precursor. This post-synthesis grafting is particularly useful for modifying the particle surface with selected agents that are not stable during the silica particle synthesis. During the post-coating step, the particle surface first needs to be modified with suitable functional groups such as thiol, amine or carboxyl groups.

For the Stöber nanoparticle synthesis, surface modification is usually done after nanoparticle synthesis to avoid potential secondary nucleation, but co-condensation is also possible: through the use of respective organosilane, organic functionalities such as alkyl, thiol, amino, cyano/isocyano, vinyl/alkyl, organophosphine or aromatic groups can be incorporated into the pore walls of the silica network. Surface modification of microemulsion nanoparticles can be achieved in the same manner or *via* direct hydrolysis and co-condensation of TEOS and other organosilanes present in the microemulsion solution (Deng et al. 2000). This method allows a multi-step sequence procedure, very useful for imaging or therapeutic cargoes incorporation. After attachment of the desired functionality to the silica nanoparticle, further modifications can be undertaken, through the additional functional group, by the way of conjugation chemistry. For instance, an amine-modified particle can be reacted with various carboxylate-containing molecules to form a stable amide bond. During conjugation with biomolecules, the colloidal stability of the nanoparticles in solution

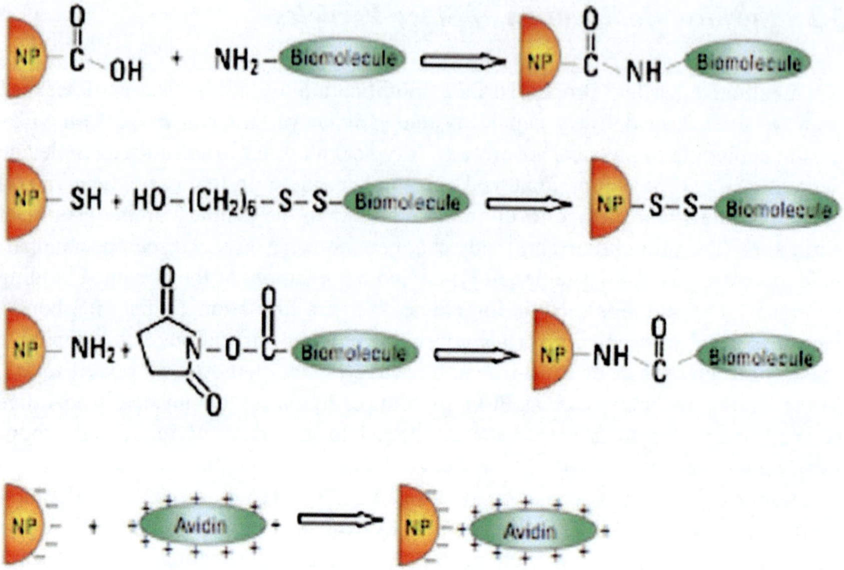

Fig. 5 Schematic illustration of bioconjugation strategies applied to silica nanoparticles (Reproduced with permission from Wang et al. 2006. Copyright 2006 American Chemical Society)

could be altered due to electrostatic interactions with charged functional groups. For instance, a post-coating with amine-containing organosilane compounds could be neutralized by the negative surface charge of nanoparticles at neutral pH, so that the overall charge of the material will drop down, leading to colloidal instability and then to severe particle aggregation in aqueous medium. To circumvent this drawback, inert negatively-charged organosilanes could be introduced, such as phosphonates, that will play the role of dispersing agents acting during the post-grafting. As a consequence, the particles, that will have a net negative charge, will be again well dispersed in aqueous solution (Santra 2004). After surface modification with different functional groups, nanoparticles can then act as scaffolds for the grafting of biological moieties such as DNA oligonucleotides, aptamers, antibodies or peptides, etc. As examples, carboxyl-modified nanoparticles, having pendant carboxylic groups, are good candidates for covalent coupling of proteins or other amine-containing biomolecules; thiol-functionalized nanoparticles can immobilize disulfide-modified oligonucleotides *via* disulfide-coupling chemistry; amine-modified nanoparticles can be coupled to a wide variety of haptens and drugs via succinimidyl esters and iso(thio)cyanates (Wang et al. 2006) (Fig. 5).

3.3 *Mesoporous Silica Nanoparticles (MSN)*

Different kinds of drugs were evaluated for release by mesoporous silica materials. Vallet-Regi and co-workers have demonstrated the release of ibuprofen (Vallet-Regi 2006), erythromycin (Doadrio et al. 2006) and alendronate (Balas et al. 2006), using different strategies to modify the control of drug release. Tamanoi et al demonstrated the release of camptothecin, a hydrophobic anti-cancer drug, from MSN (Lu et al. 2007). The delivery into malignant cells, growth inhibition and then cells death was confirmed. The enhanced bioavailability of itraconazole, a model drug, was confirmed when included within the silica porous matrix (Mellaerts et al. 2008). Tested in dogs and rabbits, the MSN material can be considered as a promising carrier to perform oral availability for drugs with extremely low water solubility.

MSN-based stimuli-responsive systems using a concept of gate-keeping were first developed by V. S.-Y. Lin and his group (Slowing et al. 2008). This redox-controlled drug delivery system is based on MSN capped with cadmium sulphide (CdS) (Fig. 6). In this system, CdS was chemically attached to MSN through a disulfide linker, which is chemically labile and that could be cleaved with various disulfide reducing agents, such as dithiothreitol (DTT) and mercaptoethanol (ME). The working principle of the stimuli-controlled release drug/gene delivery system was tested on imaging agents such as fluorescein, Texas Red or rhodamine B in order to perform the conditions of guest molecules to be released under adapted gate-opening trigger concentration and to demonstrate the performance of these systems. The loading is usually in the order of hundreds milligrams of guest molecule per gram of MSN. This type of drug delivery system with "zero premature release" performance is particularly useful when the cargo to be delivered is toxic, like anti-cancer drugs. These materials are potentially candidates for releasing drug/gene "at will", i.e. to control precisely the location and timing of the release of drug. Through this method, the controlled release to live cells of vancomycin and adenosine triphosphate by such CdS-MSN system was successfully demonstrated (Lai et al. 2003).

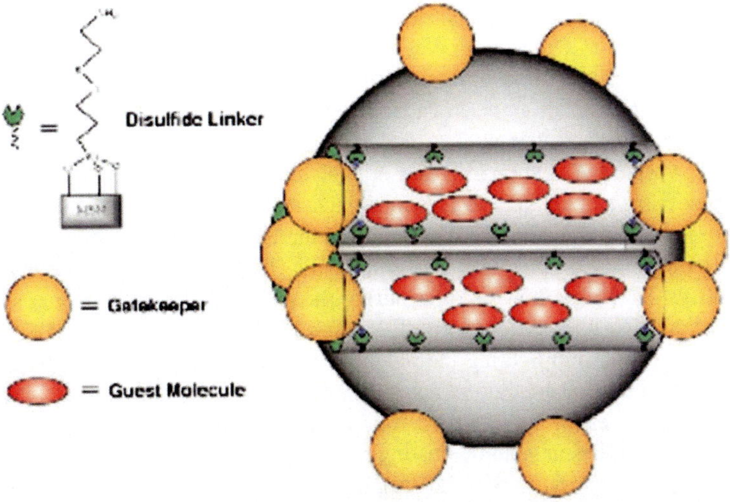

Fig. 6 Schematic illustration of the CdS-based gate-keeper strategies in mesoporous particles (Reprinted from (Slowing et al. 2008), Copyright (2008), with permission from Elsevier)

3.4 Hollow Silica Nanoparticles (HSN)

Hollow silica nanoparticles (Caruso et al. 1998) have also been used as drug carriers. They possess much higher loading capacity for drug molecules than the plain ones, but weaker interactions between the drug molecules and the silica matrix, making the control of release more delicate. The two main methodologies for synthesizing hollow silica nanoparticles are first, the template-based synthesis and second, the dissolution-growth process of solid silica nanoparticles. The first method generally utilises a solid nanoparticle composed of metal, metal oxide or polymer as a core template. The silica shell is then grown on the template and the core is removed by an etching process, dissolution or calcination. The second method is relatively simple since the hollow cores are produced by etching the silica nanoparticles with basis such as sodium hydroxide (NaOH) or sodium borohydride ($NaBH_4$), using polyvinylpyrrolidone (PVP) molecule as a surface protective agent. Thus, the inner part of the silica nanoparticle is selectively etched, leaving room for the drug molecule to be encapsulated. The key factors that govern the drug-release kinetic for hollow silica nanoparticles include cavity size, shell thickness and type of surface functional groups (Yang et al. 2008). For instance, PEG-coated hollow silica particles were found to release drug molecules at a much slower rate than the hydroxyl- or amino- functionalized counterparts (Liu et al. 2007a), probably due to partial blockage of pores by PEG molecules at the surface of the nanoparticle. Thicker shells also allow drug to be released more slowly and over a longer delay, but the drug-loading capacity is concomitantly decreased due to reduced cavity size. More generally, physical-chemical parameters such as temperature and pH can also influence the drug release rate.

Porosity is also a factor of interest, since drugs could be incorporated in the inner core and in the internal porosity of the shell. Li et al. developed a new method for preparing such particles with a porous silica shell structure via the sol-gel route and using inorganic $CaCO_3$ nanoparticles as templates (Li et al. 2004). When removed, the inner core can also offer place to a drug model such as Brilliant Blue F (Fig. 7). The first attempts made on such carriers exhibit typical sustained release without any burst effect, since a slow release was observed for 1,140 min compared to only 10 min for usual SiO_2 nanoparticles. Liu et al. explored the *in-vitro* release of fluorescein isothiocyanate (FITC), a drug model that can be easily detected by fluorescent spectroscopy (Liu et al. 2007b). They showed that the release time courses of FITC-doped silica nanocapsules with very thin shells (in the range 3–10 nm) and large cavities (ca. 70% of hollow core), ensuring the encapsulation of a great amount of drug, are significantly different from the free FITC control, with a FITC release peak at ca. 1.5 hr, highlighting the potential applications of such particles in drug release. By the same method, Chen et al. carried out another $CaCO_3$-templated silica nanocapsules for the *in-vitro* cefradine delivery (an antibacterial agent) (Chen et al. 2004). In this study, an interesting release profile, proceeding in three stages, was explained by the authors as the drug release from the surface, the pore channels in the wall, and the inside hollow part of the particles, respectively, markedly encouraging the drug delivery application for such material.

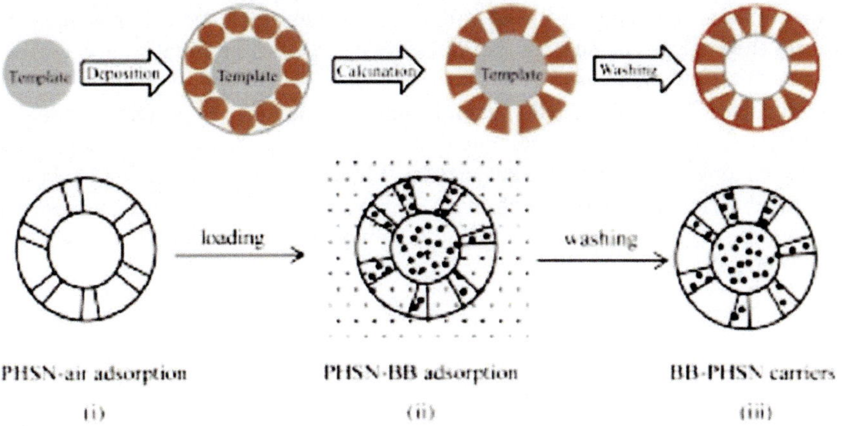

Fig. 7 Schematic illustration of the preparation of porous hollow silica spheres (HSN) and further immobilisation of *Brilliant Blue* (BB). (Reprinted from (Li et al. 2004), Copyright (2004), with permission from Elsevier)

4 Intracellular Uptake and Fate of Silica-Based Nanoparticles

One of the interests of silica-based drug delivery system is the possibility to control and manipulate intracellular uptake of the particles and their accumulation for an extended period of time. To achieve such a goal the versatile chemistry of silica is exploited to graft several functional groups such as stimuli-responsive gatekeepers to control the release of the loaded drug, cell-specific moieties to achieve a targeted delivery, cell-penetrating peptides to enhance the uptake efficiency, protecting polymer layers, or ligands for endosomal escape strategies…(Vivero-Escoto et al. 2010b). All of these modifications are aimed at obtaining an efficient drug delivery system, which means that the device must be able to: (i) reach its target without being prematurely eliminated (role of surface coating), (ii) be internalized (functionalization for enhanced endocytosis or direct uptake), (iii) reach the right location (cellular and intracellular targeting) and (iv) present no toxicity for the healthy cells (control of the long-term fate of nanoparticles).

4.1 Avoiding Premature Elimination

As mentioned earlier, coating silica nanoparticles has proved to be useful to enhance systemic circulating half-time. Some polymers such as PEG (polyethylene glycol) or carbohydrates can be used for this purpose because it creates a protective layer (Mailaender and Landfester. 2009). The polymer shell created on the surface of the particles makes them more hydrophilic and decreases unspecific adsorption of proteins. If the particles have no opsonising proteins adsorbed on the surface they are not recognized and eliminated by macrophages.

Thus, the particles remain in the circulatory system and have better chances to reach their target. Uncoated nanoparticles are less efficiently internalized because of their tendency to agglomerate into large aggregates that are less endocytosed. This could be the result of interactions with the proteins or the salt of the physiological media that would reduce repulsive interactions between particles. Among the polymer-coated particles, PEI-coated nanoparticles (PEI = polyethyleneimine) are easily solubilised in physiological conditions and do not form large aggregates, probably due to the high charge density obtained with the coating (Fuller et al. 2008).

4.2 Improving Nanoparticles Uptake

Cellular uptake of silica-based nanoparticles is governed by parameters such as cell line, morphology of the particles, electrostatic interactions between cellular membrane and nanoparticles, surface functionalization of the particles. The mechanism of internalization depends on these parameters that can be tuned to favour one pathway upon the others.

4.2.1 Influence of Morphology (Size and Shape)

In order to develop intracellular drug delivery system, the morphology of the nanoparticles should be controlled to optimize the extent of internalization depending on the aim of the system. Different studies show that, as a general trend, spherical particles are more easily internalized than rod-shape or tubular particles, but this depends on the cell-line considered (Trewyn et al. 2008) (Fig. 8). One hypothesis to explain such a difference is the so-called "wrapping time" parameter. Indeed, if the particles are internalized through endocytosis, a membrane must be formed around the particle before the uptake process can take place and wrapping a tubular shape is slower than for a spherical shape. But this has to be moderated taking into account the aspect ratio of the particles. Indeed, short rods seem to be as easily internalized as spherical nanoparticles. Another point not to miss is that apart from the influence on the uptake efficiency, the morphology of the nanoparticles and more particularly their aspect ratio can strongly influence the cell cytoskeleton and the viability (Huang et al. 2010). Indeed some studies have proven that rod-shape silica particles, once internalized, have an effect on the F-actin network as well as a critical impact on cytotoxicity (by enhancing apoptosis). Size is also a critical parameter because it will strongly influence the type of pathway followed for the internalization. Silica nanoparticles can be obtained with a size varying from a few nanometers to hundreds of nanometers, depending on the synthesis chosen. Generally, particles smaller than 200 nm will be preferentially internalized through endocytosis, particles bigger than 200 nm will either be endocytosed or not

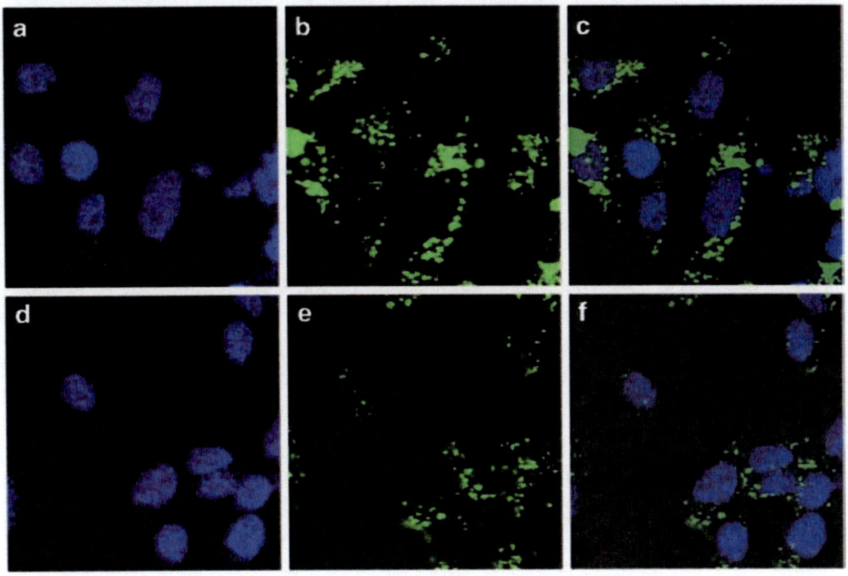

Fig. 8 Fluorescence confocal micrographs of CHO cells incubated with 50 gmL − 1 Tubular-FITC-MSN (**a–c**) and Spherical-FITC-MSN (**d–f**). (**a** and **d**) Fluorescence image excited at 340 nm to visualize the cell nuclei stained with DAPI. (**b** and **e**) Fluorescence image excited at 488 nm to visualize the FITC doped MSN that have been internalized by cells. (**c** and **f**) Overlaid micrographs of (**a** and **b**) and (**d** and **e**), respectively. (Reprinted from Trewyn et al. 2008, Copyright (2008), with permission from Elsevier)

internalized at all, but this size limit is cell line dependent. Overall, it is better to have small round particles, sufficiently stabilized to avoid the formation of large aggregates.

4.2.2 Influence of Surface Charge

Recent studies (Slowing et al. 2006) have shown that the uptake of positively charged mesoporous silica nanoparticles was more efficient than the uptake of negative particles, due to the net negative charge of cell membranes. These studies have also enlighten that negatively charged particles were more able to escape endosomes within 6 h, probably due to the "proton sponge effect" that supposes that the more negatively charge particles have a better buffering capacity which is a key factor for endosomal escape. Thus both factors must be taken into account while designing silica-based drug delivery systems. Another point is that positive nanoparticles are more subject to non-specific adhesion of proteins or non-specific interaction with the cells. But surface charge is not the only tunable parameter and different groups can be grafted on the surface of the silica-based device to facilitate receptor-mediated endocytosis, e.g. folate groups have proved to facilitate folate-receptor mediated endocytosis in human cancer cells (Slowing et al. 2006). This will be developed later in this section.

4.2.3 Influence of Surface Functionality

To deliver a drug loaded in silica-based nanoparticles, there are two possible ways of entry in the cell. The first one is the internalization of the nanoparticles via endocytosis that will require to take into account the endosomal acidic environment and a strategy for endosomal escape. The influence of surface charge on the ability of the particles to escape the endosome has already been mentioned but endosomal escape can also be tuned by grafting different groups on the surface of the nanoparticles, for instance amine containing polymers, that have the ability to buffer endosomes thus affecting the osmotic gradient. This will eventually lead to the release of the endosomal cargo into the cytoplasm (Rosenholm et al. 2009). Coating silica particles with polyethyleneimine is easily and rapidly obtained by adding the particles to a stirring solution of the polymer in sodium acetate buffer. This polymer has another interest: amine groups are positively charged at physiological pH, thus leading to positively charged nanoparticles. As the cell membrane is negatively charged, a natural affinity between the positive particles and the membrane can lead to some adherence of the nanoparticles to the membrane, thus enhancing their internalization.

In vitro studies have also used photochemically-induced endosomal release of silica nanoparticles (Sauer et al. 2010). Here the principle is a little different because the photosensitizer group (disulfonated porphyrin derivative meso-tetraphenyl porphine) is not grafted to the particles but incubated with the cells. This molecule will be endocytosed and bound to the membrane of the endosomes. Under light irradiation, the molecule is activated and generates a singlet oxygen that will lead to endosome disruption through oxidation of the different compounds of the membrane. Such a technique could be used for intracellular delivery if enough control is gained on the targeting of the photosensitizer. Another possibility is a direct delivery to the cytoplasm using cell-penetrating peptides (e.g. Tat) grafted to the particles (Santra 2004; Mao et al. 2010) (Fig. 9). The later pathway has the advantage to avoid any endosomal escape strategy and to possibly preserve the cargo from the endosome environment. Furthermore, some drug will have to be delivered specifically to definite organelles to maximize their efficiency and direct delivery would minimize the loss of bioactivity. Tat peptide can be covalently grafted via water-soluble carbodiimide reagents onto the surface of carboxyl-coated silica nanoparticles. One key parameter is to adjust the number of Tat grafted to the size of the particles. Indeed the bigger the particle, the more peptides will be needed to permit membrane permeation. Of course, Tat-peptide is not the only cell-penetrating peptide available and, depending on the peptide, the internalization mechanism may vary.

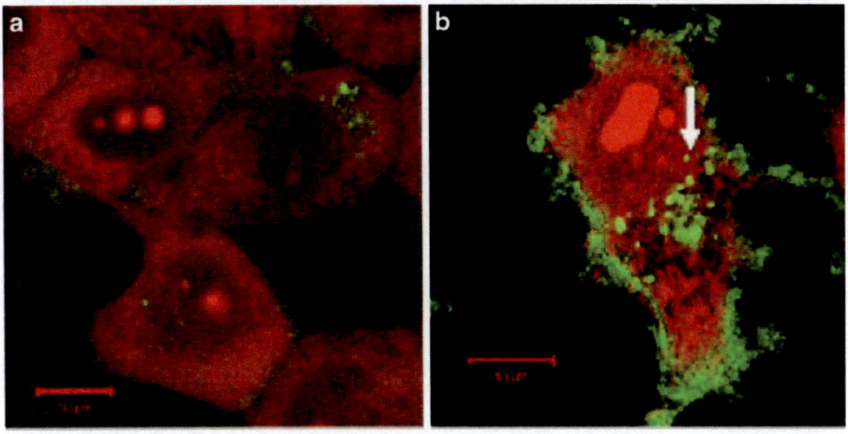

Fig. 9 HepG2 cells (labeled with EB to show *red color*) incubated for 6 h with (**a**) FITC-SiO_2-NH_2 and (**b**) FITC–SiO_2–Tat3.5 particles. Bar is 10 μm. (Reprinted from Mao et al. 2010, Copyright (2010), with permission from Elsevier)

4.3 Controlled and Targeted Delivery

4.3.1 Tuning Silica-Based System to Control the Intracellular Delivery

To answer the problem of drugs side effects, controlled drug delivery devices are being developed. These systems will prevent drug leakage or premature release, thus diminishing the necessary dose administered. Such systems can be obtained by grafting stimuli-responsive tethers on silica-based nanoparticles. These tethers will act as gatekeepers that could be triggered either by an intracellular signal (such as pH, reducing environment, competitive binding, enzymatic activity) or an external signal (such as light, ultrasound, electromagnetic field, temperature). For instance, silica nanoparticles with pH-sensitive gates have been developed (Cauda et al. 2010). This device is based on the sealing of the mesopores in acidic conditions via an interaction between two different grafted groups: sulfonate and amine. When pH reaches a physiological value, the previous interaction ceases to exist leading to the gradual opening of the pores and the release of the encapsulated drug. But practical problems can be encountered while designing such devices and for instance the pore size has to be adjusted in order to obtain an effective gate because if the pore is too large, the release takes place with no incidence of the pH. This implies that the gating system must be matched with the pore size opening. Another example would be to use cyclodextrin as a glutathione sensitive gatekeeper (Kim et al. 2010). The principle is the same: cyclodextrin groups are covalently grafted on the external surface of mesoporous silica nanoparticles via a disulfide bond and mask the opening of the pores. When the particles are in a glutathione containing environment (e.g. in the targeted cell) the disulfide bond is cut, releasing the cyclodextrin and opening the pore. Thus such systems enable a control of the release and no premature release. Coupled to a targeting moiety, such device could be used for specific and controlled drug delivery.

4.3.2 Targeted Delivery (Cellular and Intracellular)

One of main issues in biomedical sciences nowadays is to develop drug delivery systems that can be specifically targeted to the desired site. These systems are mainly based on recognition with binding receptor of the targeted cells. Different chemical structures can be grafted to silica nanoparticles to obtain these selective interactions, as for example antibodies, peptides or specific ligands. The use of folate groups has been mentioned previously. Such ligands are useful to target cancer cells that possess folate-receptors in larger numbers than healthy cells (Rosenholm et al. 2009). Experiments on porous hybrid silica nanoparticles, functionalized with PEI and folic acid have shown that the number of particles internalized per cell is significantly larger in cells expressing high levels of folate receptors, as well as the proportion of cells that have internalized particles (Fig. 10). Furthermore, this kind of functionalized silica particles have proven to have a specific targeting capacity with no cytotoxicity detected. Folic acid is grafted via carbodiimide/ succinimide coupling of the carboxylic acid group of folic acid with the amine groups on the surface of the PEI-coated particles, in a buffer at acidic pH. However, one limitation appears in using folate-targeted particles that is the low rate of escape of the vector after receptor-mediated endocytose (Breunig et al. 2008). Using these biomolecule conjugations with functionalized silica particles, different antibodies (e.g. monoclonal antibodies) or peptides can be grafted to target specific cells. Cellular targeting is extensively studied *in vitro* and different ways have been developed but yet there are very few reports of *in vivo* testing (Lu et al. 2010).

Apart from cellular targeting, intracellular targeting can also be considered with delivery of the drug to a specific organelle. Once the particles have escaped the endosome, they can either remain into the cytosol or diffuse specifically towards mitochondria or nucleus depending on the action site aimed for the drug. The nucleus is a major subcellular structure and has to be targeted for gene delivery. To answer this issue, mesoporous silica nanoparticles have been successfully combine with a dendrimer assembly that act as gatekeeper for the gene transfection system trapped in the pores (Radu et al. 2004). The dendrimer (generation two of poly(amidoamine)) was grafted on an isocyanatopropyl functionalised particle and efficiently bind plasmid DNA vectors. Furthermore, this hybrid material is able to protect DNA against enzymatic cleavage and induce expression of the DNA vector in the targeted cells.

4.4 Intracellular Fate of the Nanoparticles

While conceiving silica-based drug delivery system, the fate of the internalized particles must be evaluated. There are three possibilities: the particles can be degraded

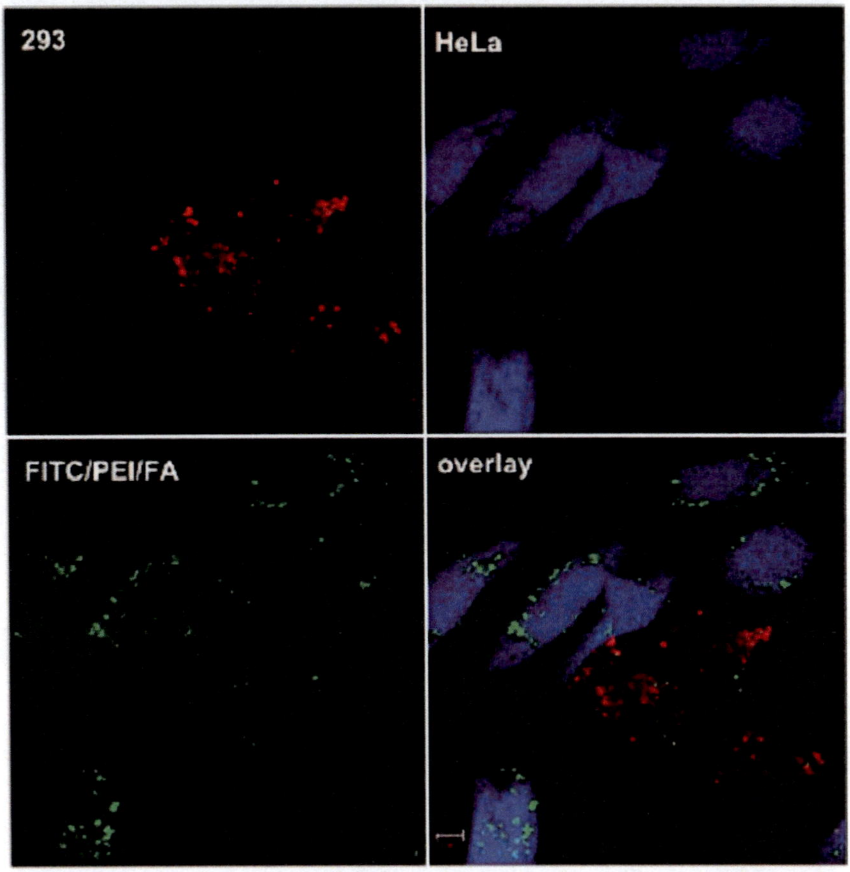

Fig. 10 Specific endocytosis of FITC/PEI/FolicAcid-functionalized silica nanoparticles in coculture of HeLa and 293 cells. The cells were labeled with *blue* CMAC (HeLa) or CellTracker *Red* (293) and plated together overnight prior to incubation with the particles for 4 h. After incubation the extracellular fluorescence was quenched by trypan blue and the endocytosed particles with FITC-label (*green*) inside blue- or red-labeled cells were detected by confocal microscopy. Scale bar 10 μm. (Reproduced with permission from (Rosenholm et al. 2009). Copyright 2009 American Chemical Society)

by enzymes in the lysosomes, the system can be exocytosed either directly or after partial degradation, or the particles can accumulate inside the cells, potentially leading to side effects such as formation of reactive oxygen species, affecting the cell cycle, influencing cytokine synthesis… (Hu et al. 2009). At this time, very few data has been gathered on this topic. Fluorescent mesoporous nanoparticles have been shown to escape endolytic vesicles and resist lysosomal degradation (Huang et al. 2005). In the case of biopolymer-silica hybrid nanoparticles, it was observed that gelatin or alginic acid could be dissolved intracellularly within 3 T3-fibroblasts, leaving silica nanograins in the vesicles (Boissière et al. 2006; Allouche et al. 2006). However, nothing is known about further degradation/exocytose of these particles.

5 Conclusions

Taken together, this large amount of data and achievements suggest that silica-based nanoparticles are promising candidates for intracellular drug delivery. Their main advantages are the control of particle size and porosity as well as the very versatile functionalization chemistry. Their main drawback is clearly related to their biocompatibility but this term covers several aspects that deserve to be discussed. First, it worth underlining that, like all other materials, silica may coexist in several forms: colloidal, polymer and monomer of decreasing size but of increasing reactivity. Second, several in vivo environments should be distinguished: body fluids, cell membrane and intracellular space. Most available data concern silica in the colloidal state in body fluids and at the vicinity of cell membrane. In all these situations, it was suggested that a suitable surface functionalization allowed the decrease of the detrimental effect of the silica particle. Considering colloidal silica within cells, there is a strong lack of relevant information. In particular, to our knowledge, no long-term studies of the fate of internalized particles are available. Turning our attention to soluble forms of silica, only one paper has addressed the influence of silicic acids on cells, and only in vitro (Linthicum 2001). This study indicated that these species could be internalized, leading to intracellular damage. It can therefore be suggested that these two fundamental aspects deserve to be studied in more details.

From a practical point of view, it is also necessary to consider several factors. First, the problem of colloidal stability is an important issue both during storage but also when sterilization is to be undertaken. For instance, it is worth noting that commercial solutions of silica nanoparticles contain stabilizers that may not be fully compatible with in vivo applications. From an economical perspective, the large-scale production of plain silica nanoparticles, as such or with simple organic coatings is already existing. However, to our knowledge, this is not true for mesoporous particles. Considering further functionalization of these carriers, it is worth mentioning that organosilanes are quite expensive and that most promising multifunctional particles involve several synthetic steps that may not be easy to scale-up. Overall, it implies an extensive adaptation of the silica industry to biomedical constraints. Such an adaptation will be triggered only if (i) silica-based particles demonstrate strong benefits compared to more traditional (bio)-organic nanoparticles and (ii) a better understanding of their in vivo behavior is gained.

Interestingly, recent bibliographic data suggest that most reports in this area concern the elaboration of silica-based novel carriers (> 500 papers in the 2009–2010 period from Web of Science®) whereas biological evaluations of silica particles are still scarce (ca. 60 papers over the same period). A better balance between these two approaches must now be reached, especially through the development of a true “biochemistry” of silica.

References

Akhtar MJ, Ahmed M, Kumar S, Siddiqui H, Patil G, Ashquin M, Ahmad I (2010) Nanotoxicity of pure silica mediated through oxidant generation rather than glutathionne depletion in human lung epithelial cells. Toxicol 276:95–102

Allouche J, Boissière M, Hélary C, Livage J, Coradin T (2006) Biomimetic core-shell gelatine/silica nanoparticles: a new example of biopolymer-based nanocomposites. J Mater Chem 16: 3121–3126

Andersson N, Alberius PCA, Pedersen JS, Bergstrom L (2004)Structural features and adsorption behaviour of mesoporous silica particles formed from droplets generated in a spraying chamber. Microporous Mesoporous Mater 72:175–183

Arcangeli G, Rotella CM, Giuliano G, Arcangeli P (1990) Induction of cell from patients with silicosis by silica crystals. Effects of cell-derived factors and silica particles on human lung fibroblasts proliferation and biosynthesis. Int J Immunopathol Pharmacol 3:51–61

Arriagada FJ, Osseo-Asare K (1995) Synthesis of nanosize silica in a nonionic water-in-oil micro-emulsion: effects of the water/surfactant molar ratio and ammonia concentration. J Colloid Interf Sci 211:210–220

Baccile N, Grosso D, Sanchez C (2003) Aerosol generated mesoporous silica particles. J Mater Chem 13:3011–3016

Bagchi N (1992) What makes silica toxic ? Br J Ind Med 49:163–166

Balas F, Manzano M, Horcajada P, Vallet-Regi M (2006) Confinement and controlled release of bisphosphonates on ordered mesoporous silica-based materials, J Am Chem Soc 128:8116–8117

Barbé C, Bartlett J, Kong L, Finni K, Lin HQ, Larkin M, Calleja S, Bush A, Calleja G (2004) Silica particles: a novel drug-delivery system. Adv Mater 16:1959–1966

Barnes CA, Elsaesser A, Arkusz J et al (2008) Reproducible comet assay of amorphous silica nanoparticles detects no genotoxicity. Nano Lett 8:3069–3074

Bass JD, Grosso D, Boissière C, Coradin T, Sanchez C (2007) The stability of mesoporous oxide and mixed-metal oxide materials under biologically-relevant conditions. Chem Mater 19: 4349–4354

Bharali DJ, Klejbor I, Stachowiak EK, Dutta P, Roy I, Kaur N, Bergey EJ, Prasad PN, Stachowiak MK (2005) Organically modified silica nanoparticles: A nonviral vector for in-vivo gene delivery and expression in the brain, Proc Natl Acad Sci USA 102:11539–11544

Bégu S, Aubert Pouëssel Lerner DA, Charnay C, Tourné-Péteilh C, Devoisselle JM (2007) Liposil, a promising composite material for drug storage and release. J Controlled Release 118:1–6

Bogush GH, Tracy MA, Zukoski IV CF (1988) Preparation of monodisperse silica particles: control of size and mass fraction. J Non-Cryst Solids 104:95–106

Boissiere C, Grosso D, Chumonnot A, Nicole L, Sanchez C (2010) Aerosol route to functional nanostructured inorganic and hybrid porous materials. Adv Mater DOI: 10.1002/adma.201001410

Boissière M, Meadows PJ, Brayner R, Hélary C, Livage J, Coradin T (2006) Turning biopolymer particles into hybrid capsules: the example of silica/alginate nanocomposites. J Mater Chem 16:1178–1182

Breunig M, Bauer S, Goepferich A (2008) Polymers and nanoparticles: intelligent tools for intra-cellular targeting? Eur J Pharm Biopharm 68:112–128

Brinker CJ, Lu Y, Sellinger A, Fan H (1999) Evaporation-Induced Self-Assembly: nanostructures made easy. Adv Mater 11:579–585

Bruchez M, Moronne M, Gin P, Weiss S, Alivisatos AP (1998) Semiconductor nanocrystals as fluorescent biological labels, Science 281:2013–2016

Burns AA, Vider J, Ow H et al (2009) Fluorescent silica nanoparticles with efficient urinary excretion for nanomedicine. Nano Lett 9:442–448

Caruso F, Caruso RA, Möhwald H (1998) Nanoengineering of inorganic and hybrid hollow spheres by colloidal templating, Science 282:1111–1114

Cauda V, Argyo C, Bein T (2010) Impact of different PEGylation patterns on the long-term bio-stability of colloidal mesoporous silica nanoparticles. J Mater Chem 20:8693–8699

Chang JS, Chang KLB, Hwang DF, Kong ZL (2007) In vitro cytotoxicity of silica nanoparticles at high concentrations strongly depends on the metabolic activity type of the cell line. Environ Sci Technol 41:2064–2068

Chang, SM, Lee M, Kim WS (2005) Preparation of large monodispersed spherical silica particles using seed particle growth. J Colloid Interf Sci 286:536–542

Chen JF, Ding HM, Wang JX, Shao L (2004) Preparation and characterization of porous hollow silica nanoparticles for drug delivery application, Biomaterials 25:723–727

Choi J, Zheng Q, Katz HE, Guilarte TR (2010) Silica-based nanoparticle uptake and cellular response by primary microglia. Env Health Persp 118:589–595

De Monredon-Senani S, Bonhomme C, Ribot F, Babonneau F (2009) Covalent grafting of organo-alkoxysilanes on silica surfaces in water-rich medium as evidenced by ^{29}Si NMR. J Sol-Gel Sci Technol 50:152–157

Deng G, Markowitz MA, Kust PR, Gaber BP (2000) Control of surface expression of functional groups on silica particles, Mater Sci Eng C 11:165–172

Depasse J (1977) Comparison between two hypotheses about the physicochemical basis of the toxicity of silica. J Colloid Interf Sci 60:414–415

Di Pasqua AJ, Sharma KK, Shi YL, Toms BB, Ouellette W, Dabrowiak JC, Asefa T (2008) Cytotoxicity of mesoporous silica nanomaterials. J Inorg Biochem 102:1416–1423

Doadrio JC, Sousa EMB, Izquierdo-Barba I, Doadrio AL, Perez-Pariente J, Vallet-Regi M (2006) Functionalization of mesoporous materials with long alkyl chains as a strategy for controlling drug delivery pattern, J Mater Chem 16:462–466

Erdogu G, Hasirci V (1998) An overview of the role of mineral solubility in silicosis and abestosis. Environ Res Sect A 78:38–42

Fent K, Weisbrod CJ, Wirth-Heller A, Pieles U (2010) Assessment of uptake and toxicity of fluorescent silica nanoparticles in zebrafish (Danio rerio) early life stages. Aquatic Toxicol 100:218–228

Finnie KS, Jacques DA, McGann MJ, Blackford MG, Barbé CJ (2006) Encapsulation and controlled release of biomolecules from silica microparticles. J Mater Chem 16:4494–4498.

Finnie KS, Waller DJ, Perret FL, Krause-Heuer AM, Lin HQ, Hanna JV, Barbé CJ (2009) Biodegradability of sol-gel silica microparticles for drug delivery. J Sol-Gel Sci Technol 49:12–18

Fuller J, Zugates G, Ferreira L, Ow H, Nguyen N, Wiesner U, Langer R (2008) Intracellular delivery of core-shell fluorescent silica nanoparticles. Biomaterials 29:1526–1532

Galagudza MM, Korolev DV, Sonin DL, Postnov VN, Papayan GV, Uskov IS, Belozertseva AV, Shlyakhto EV (2010) Targeted drug delivery into reversibly injured myocardium with silica nanoparticles: surface, functionalization, natural biodistribution and acute toxicity. Int J Nanomed 5:231–237

Gan LM, Zhang K, Chew CH (1996) Preparation of silica nanoparticles from sodium orthosilicate in inverse microemulsions. Colloid Surf A 110:199–206

Giljohann DA, Seferos DS, Prigodich AE, Patel PC, Mirkin CA (2009) Gene regulation with polyvalent siRNA-nanoparticle conjugates, J Am Chem Soc 131:2072–2073

Gong C, Tao G, Yang L, Liu J, Liu Q, Zhuang Z (2010) SiO_2 nanoparticles induce global genomic hypomethylation in HaCaT cells. Biochem Biophy Res Commun 397:397–400

He Q, Zhan J, Shi J, Zhu Z, Zhang L, Bu W, Guo L, Chen Y (2010) The effect of PEGylation of mesoporous silica nanoparticles on the binding of serum proteins and cellular responses. Biomaterials 31:1085–1092

Higashisaka K, Yoshioka Y, Yamashita K et al. (2011) Acute phase proteins as biomarkers for predicting the exposure and toxicity of nanomaterials. Biomaterials 32:3–9

Hu L, Mao Z, Gao C (2009) Colloidal particles for cellular uptake and delivery. J. Mater. Chem. 19:3108–3115

Huang DM, Hung Y, Ko BS et al. (2005) Highly efficient cellular labeling of mesoporous nanoparticles in human mesenchymal stem cells: implication for stem cell tracking. FASEB J 19:2014–2016

Huang X, Teng X, Chen D, Tang F, He J (2010) The effect of the shape of mesoporous silica nanoparticles on cellular uptake and cell function. Biomaterials 31:438–448

Hudson SP, Padera RF, Langer R, Kohane DS (2008) The biocompatibility of mesoporous silicates. Biomaterials 29:4045–4055

Icopini GA, Brantley SL, Heaney PJ (2005) Kinetics of silica oligomerization and nanocolloid formation as a function of pH and ionic strength at 25°C. Geochim Cosmochim Acta 69:293–303

Iler RK (1979), The Chemistry of Silica, Wiley, New York

Iskandar F, Gradon L, Okuyama K (2003) Control of the morphology of nanostructured particles prepared by the spray drying of a nanoparticle sol. J Colloid Interf Sci 265:296–303

Izquierdo-Barba I, Colilla M, Manzano M, Vallet-Regi M (2010) In vitro stability of SBA-15 under physiological conditions. Microporous Mesoporous Mater 132:442–452

Jin YH, Li AZ, Hazelton SG, Liang S, John CL, Selid PD, Pierce DT, Zhao JX (2009) Amorphous silica nanohybrids: synthesis, properties and applications, Coord Chem Rev 253:2998–3014

Jung CY, Kim JS, Chang TS, Kim ST, Lim HJ, Koo SM (2010) One step synthesis of structurally-controlled silicate particles from sodium silicate particles from sodium silicates using a simple precipitation process. Langmuir 26:5456–5461

Kumar R, Roy I, Ohulchanskky TY, Vathy LA, Bergey EJ, Sajjad M, Prasad PN (2010) In vivo biodistribution and clearance studies using multimodal organically-modified silica nanoparticles. ACS Nano 4:699–708

Kim H., Kim S., Park C., Lee H., Park HJ, Kim C (2010) Glutathione-induced intracellular release of guests from mesoporous silica nanocontainers with cyclodextrin gatekeepers. Adv Mater 22:4280–4283

Kim S, Ohulchanskyy TY, Pudavar HE, Pandey RK, Prasad PN (2007) Organically modified silica nanoparticles co-encapsulating photosensitizing drug and aggregation-enhanced two-photon absorbing fluorescent dye aggregates for two-photon photodymamic therapy, J Am Chem Soc 129:2669–2675

Lai CY, Trewyn BG, Jeftinija DM, Jeftinija K, Xu S, Jeftinija S, Lin VSY (2003) A mesoporous silica nanosphere-based carrier system with chemically removable CdS nanoparticle caps for stimumi-responsive controlled released of neurotransmitters and drug molecules, J Am Chem Soc 125:4451–4459

Li Z, Wen L, Shao L, Chen J (2004) Fabrication of porous hollow silica nanoparticles and their applications in drug release control, J. Control Release 98:245–254

Lin W, Huang Y, Zhou XD, Ma Y (2006) In vitro toxicity of silica nanoparticles in human lung cancer cells. Toxicol Appl Pharm 21:252–259

Lin YS, Haynes CL (2010) Impacts of mesoporous silica nanoparticle size, pore ordering and pore integrity on hemolytic activity. J Am Chem Soc 132:4834–4842

Linthicum DS (2001) Untrastructural effects of silicic acid on primary lung fibroblasts in tissue culture. Tissue Cell 33:514–523

Liu X, Sun J (2010) Endothelial cells dysfunction induced by silica nanoparticles through oxidative stress via JNK/P53 and NF-kB pathways. Biomaterials 31:8198–8209

Liu Y, Miyoshi H, Nakamura M (2007a) Nanomedicine for drug delivery and imaging : A promising avenue for cancer therapy and diagnosis using targeted functional nanoparticle, Int J Cancer 120:2527–2537

Liu Y, Miyoshi H, Nakamura M (2007b) Novel drug delivery system of hollow mesoporous silica capsules with thin shells: Preparation and fluorescein isothiocyanate (FITC) release kinetics, Colloids Surf B 58:180–187

Lord MS, Cousins BG, Doherty PJ, Whitelock JM, Simmons A, Williams RL, Milthorpe BK (2006) The effect of silica nanoparticulate coatings on serum protein adsorption and cellular response. Biomaterials 27:4856–4862

Lu F, Wu SH, Hung Y, Mou CY (2009) Size effect on cell uptake in well-suspended, uniform mesoporous silica nanoparticles. Small 5:1408–1413

Lu J, Liong M, Fink ZI, F. Tamanoi F (2007) Mesoporous silica nanoparticles as a delivery system for hydrophobic anticancer drug, Small 3:1341–1346

Lu J, Liong M, Li Z, Zink J, Tamanoi F (2010) Biocompatibility, biodistribution, and drug-delivery efficiency of mesoporous silica nanoparticles for cancer therapy in animals. Small. 6:1794–1805

Mailaender V, Landfester K (2009) Interaction of Nanoparticles with Cells. Biomacromolecules 10:2379–2400

Mao Z, Wan L, Hu L, Ma L, Gao C (2010) Tat peptide mediated cellular uptake of SiO_2 submicron particles. Colloids Surf B 75:432–440

Mellaerts R, Mols R, Jammaer JAG, Aerts CA, Annaert P, Humbeeck JV, Van den Mooter G, Augustijns P, Martens JA (2008) Increasing the oral bioavailability of the poorly water soluble drug itraconazole with ordered mesoporous silica, Eur J Pharm Biopharm 69:223–230

Nishimori H, Kondoh M, Isoda K, Tsunoda S, Tsutsumi Y, Yagi K (2009) Histological analysis of 70 nm silica particles-induced chronic toxicity in mice. Eur J Pharm Biopharm 72:626–629

Park YH, Kim JN, Jepong SH et al (2010) Assessment of dermal toxicity of nanosilica using culture keratinocytes, a human skin equivalent model and an in vivo model. Toxicol 267:178–181

Radhakrishnan B, Ranjan R, Brittain WJ (2006) Surface initiated polymerizations from silica nanoparticles. Soft Matter 2:386–396

Radu DR, Lai CY, Jeftinija K, Rowe EW, Jeftinija S, Lin VSY (2004) A polyamidoamine dendrimer-capped mesoporous silica nanosphere-based gene transfection reagent, J Am Chem Soc 126:13216–13217

Rao KS, El-Hami K, Kodaki T, Matsushige K (2005) A novel method for the synthesis of silica nanoparticles. J Colloid Interf Sci 289:125–131

Rimer JD, Trofymluk O, Navrotsky A, Lobo RF, Vlachos DG (2007) Kinetic and thermodynamic studies of silica nanoparticle dissolution. Chem Mater 19:4189–4197

Rosenholm J, Meinander A, Peuhu E, Niemi R, Eriksson J, Sahlgren C, Lindén M (2009) Targeting of porous hybrid silica nanoparticles to cancer cells. ACS Nano 3:197–206

Rosenholm JM, Sahlgren C, Linden M (2010) Towards multifunctional, targeted drug delivery systems using mesoporous silica nanoparticles- opportunities & challenges. Nanoscale 2:1870–1883

Santra S (2004) TAT conjugated, FITC doped silica nanoparticles for bioimaging applications, Chem Commun 24:2810–2811

Sauer A, Schlossbauer A, Ruthardt N, Cauda V, Bein T, Bräuchle C (2010) Role of Endosomal Escape for Disulfide-Based Drug Delivery from Colloidal Mesoporous Silica Evaluated by Live-Cell Imaging. Nano Lett 10:3684–3691

Sertchook H, Elimelech H, Makarov C, Khalfin R, Cohen Y, Shuster M, Babonneau F, Avnir D (2007) Composite particles of polyethylene@silica. J Am Chem Soc 129:98–108.

Sola R, Boj M, Hernandez-Felix S, Camprubi M (2009) Silica in oral drugs as a possible sarcoidosis-inducing antigen. Lancet 373:1943–1944

Slowing II, Trewyn BG, Lin VSY (2006) Effect of surface functionalisation of MCM-41-type mesoporous silica nanoparticles on the endocytosis by human cancer cells. J Am Chem Soc 128:14792–14793

Slowing II, Vivero-Escoto JL, Wu CW, Lin VSY (2008) Mesoporous silica nanoparticles as controlled release drug delivery and gene transfection carriers. Adv Drug Deliv Rev 60: 1278–1288

Soler Illia GJ, Sanchez C, Lebeau B, Patarin J (2002) Chemical strategies to design textured materials: from microporous and mesoporous oxides to nanonetworks and hierarchical structures. Chem Rev 102:4093–4138

Souris JS, Lee CH, Cheng SH, Chen CT, Yang CS, Ho JA, Mou CY, Lo LW (2010) Surface charge-mediated rapid hepatobiliary excretion of mesoporous silica nanoparticles. Biomaterials 31:5564–5574

Stöber W, Fink A, Bohn E (1968) Controlled growth of monodisperse silica spheres in the micron size range. J Colloid Interf Sci 26:62–69
Tao Z, Toms BB, Goddisman J, Asefa T (2009) Mesoporosity and functional group dependent endocytosis and cytotoxicity of silica nanomaterials. Chem Res Toxicol 22:1869–1880
Taylor-Pashow KML, Rocca JD, Huxford R, Lin W (2010) Hybrid nanomaterials for biomedical applications, Chem Commun 46:5832–5849
Thierry B, Zimmer L, McNiven S, Finnie K, Barbé C, Griesser HJ (2008) Electrostatic self-assembly of PEG copolymers onto porous silica nanoparticles, Langmuir 24:8143–8150
Trewyn BG, Nieweg JA, Zhao Y, Lin VSY (2008) Biocompatible mesoporous silica nanoparticles with different morphologies for animal cell membrane penetration. Chem Engin J 137:23–29
Vallhov H, Gabrielsson S, Stromme M, Scheynius A, Garcia-Bennett AE (2007) Mesoporous silica particles induce size dependent effects on human dendritic cells. Nano Lett 7: 3576–3582
Vallet-Regi M (2006) Ordered mesoporous materials in the context of drug delivery systems and bone tissue engineering, Chem Eur J 12:5934–5943
Van Shooneveld MM, Vucic E, Koole R, Zhou Y, Stocks J, Cormode DP, Tang CY, Gordon RE, Nicolay K, Meijerink A, Fayad ZA, Mulder WJM (2008) Improved biocompatibility and pharmacokinetics of silica nanoparticles by means of a lipid coating: a multimodality investigation. Nano Lett 8:2517–2525
Vivero-Escoto JL, Slowing II, Lin VSY (2010a) Tuning the cellular uptake and cytotoxicity properties of oligonucleotide intercalator-functionalized mesoporous silica nanoparticles with human cervical cancer cells HeLa. Biomaterials 31:1325–1333
Vivero-Escoto JL, Slowing II, Trewyn BG, Lin VSY (2010b) Mesoporous Silica Nanoparticles for Intracellular Controlled Drug Delivery. Small 6:1952–1967
Vogelsberger W, Seidel A, Rudakoff G (1992) The solubility of silica in water. J Chem Soc Faraday Trans 88:473–476
Wallace AF, Gibbs, GV, Dove PM (2010) Influence of ion-associated water on the hydrolysis of Si-O bonded interactions. J Phys Chem A 114:2534–2542
Wang F, Gao F, Lan M, Yuan H, Huang Y, Liu J (2009) Oxidative stress contributes to silica nanoparticle-induces cytotoxicity in human embryonic kidney cells. Toxicol in Vitro 23:808–815
Wang L, Wang K, Santra S, Zhao X, Hilliard LR, Smith JE, Wu Y, Tan W (2006) Watching silica nanoparticles glow in the biological world, Anal Chem 78:646–654
Wang L, Zhao W, Tan W (2008) Bioconjugated silica nanoparticles, Nano Res 1:99–115
Xie G, Sun J, Zhong G, Shi L, Zhang D (2010) Biodistribution and toxicity of intravenously administered silica nanoparticles in mice. Arch Toxicol 84:183–190
Yang H, Liu C, Yang D, Zhang H, Xi Z (2009) Comparative study of cytotoxicity, oxidative stress and genotoxicity induced by four typical nanomaterials: the role of particle size, shape and composition. J Appl Toxicol 29:69–78
Yang J, Lee J, Kang J, Lee K, Suh J, Yoon Y, Haam S (2008) Hollow silica nanocontainers as drug delivery vehicles, Langmuir 24:3417–3421
Yang Y, Coradin T (2008) A green route to silica nanoparticles with controlled size and structure. Green Chem 10:183–190
Ye Y, Liu J, Sun L, Chen M, Lan M (2010) Nano-SiO_2 induces apoptosis via activation of p53 and Bax mediated by oxidative stress in human hepatic cell line. Toxicol in Vitro 24:751–758
Yezhelyev MV, Qi L, O'Regan RM, Nie S, Gao X (2008) Proton-sponge coated quantum dots for siRNA delivery and intracellular imaging, J Am Chem Soc 130:9006–9012
Zalzberg L, Avnir D (2008) Biocompatible hybrid particles of poly(L-lactic acid)@silica. J Sol-Gel Sci Technol 48:47–50
Zhang Y, Hu L, Yu D, Gao C (2010) Influence of silica particle internalization on adhesion and migration of human dermal fibroblasts. Biomaterials 31:8465–8474
Zou H, Wu S, Shen J (2008) Preparation of silica-coated poly(styrene-co-4-vinylpyridine) particles and hollow particles. Langmuir 24:10453–10461

Magnetic Nanoparticles for Biomedicine

Ivo Šafařík, Kateřina Horská, and Mirka Šafaříková

Abstract Biocompatible materials exhibiting different types of response to external magnetic field have already found many important applications in various areas of biosciences, biotechnology, medicine, environmental technology etc. In most cases they can be described as composite materials, where the magnetic properties are caused by the presence of iron oxides nano- or microparticles. Such materials can be efficiently separated from difficult-to-handle samples and targeted to the desired place, applied as contrast agents for magnetic resonance imaging or used to generate heat during exposure to alternating magnetic field.

Keywords Maghemite • Magnetic iron oxides • Magnetic particles • Magnetic separation • Magnetite

1 Introduction

Biocompatible materials exhibiting response to external magnetic field have found many interesting applications in various areas of biosciences and biotechnology, including different medical disciplines. The broad family of magnetic field-controllable materials includes both nano- and microparticles, high aspect ratio structures (nanotubes, nanowires), thin films, etc. Ferrofluids (magnetic

I. Šafařík (✉)
Department of Nanobiotechnology, Institute of Nanobiology and Structural Biology of GCRC, Academy of Sciences of the Czech Republic, Na Sádkách 7, 370 05 České Budějovice, Czech Republic
and
Regional Centre of Advanced Technologies and Materials, Palacký University, Šlechtitelů 11, 783 71, Olomouc, Czech Republic
e-mail: ivosaf@yahoo.com

K. Horská and M. Šafaříková
Department of Nanobiotechnology, Institute of Nanobiology and Structural Biology of GCRC, Academy of Sciences of the Czech Republic, Na Sádkách 7, 370 05 České Budějovice, Czech Republic

A. Prokop (ed.), *Intracellular Delivery: Fundamentals and Applications*, Fundamental Biomedical Technologies 5, DOI 10.1007/978-94-007-1248-5_13,

fluids), magnetorheological fluids, magnetic polymers, magnetic inorganic materials, magnetically modified biological structures, magnetic particles with bound biomolecules etc. can serve as typical examples. In many cases magnetically responsive composite materials consist of small magnetic particles (most often formed by magnetite, maghemite or various ferrites), usually in the nanometer to micrometer range, dispersed in a polymer, biopolymer or inorganic matrix; alternatively magnetic particles can be adsorbed on the outer surface of diamagnetic particles (Safarik and Safarikova 2009b).

Magnetic properties of such materials enable their applications in numerous areas (Arruebo et al. 2007; Safarik and Safarikova 2009a), namely:

- Magnetically responsive nano- and microparticles and other relevant materials can be selectively separated (removed) from the complex samples using an external magnetic field (e.g. using an appropriate magnetic separator, permanent magnet, or electromagnet). This process is very important for bioapplications due to the fact that absolute majority of biological materials have diamagnetic properties which enable efficient selective separation of magnetic materials.
- Magnetic particles can be targeted to the desired place and kept there using an external magnetic field. These properties can be used e.g. for sealing the rotating objects or in the course of magnetic drug targeting.
- Magnetic particles can generate heat when subjected to high frequency alternating magnetic field; this phenomenon is employed especially during magnetic fluid hyperthermia (e.g., for cancer treatment).
- Magnetic iron oxides nanoparticles generate a negative T2 contrast during magnetic resonance imaging thus serving as efficient contrast agents.
- Magnetorheological fluids exhibit great increase of apparent viscosity when subjected to a magnetic field.
- Magnetic nano- and microparticles can be used for magnetic modification of diamagnetic biological materials (e.g. cells or plant-derived materials), organic polymers and inorganic materials, and for magnetic labeling of biologically active compounds (e.g. antibodies, enzymes, aptamers etc.).

In most cases synthetic (laboratory-produced) magnetically responsive nano- and microparticles and related structures have been developed, however, biologically produced magnetic particles (e.g., magnetosomes produced by magnetotactic bacteria) have been successfully used for selected bioapplications (Arakaki et al. 2008). This short review chapter shows typical examples of biocompatible magnetic materials synthesis and typical examples of their biomedical applications.

2 Synthesis of Magnetic Nanoparticles

Many chemical procedures have been used to synthesize magnetic nano- and microparticles applicable for bioapplications, such as classical coprecipitation, reactions in constrained environments (e.g., microemulsions), sol-gel syntheses, sonochemical and microwave reactions, hydrothermal reactions, hydrolysis and thermolysis of precursors, flow injection syntheses, electrospray syntheses and mechanochemical processes (Laurent et al. 2008; Lin et al. 2006; Zheng et al. 2010).

The simplest and most efficient chemical pathway to obtain magnetic particles is probably the coprecipitation technique. Iron oxides, either in the form of magnetite (Fe_3O_4) or maghemite (γ-Fe_2O_3), are usually prepared by aging stoichiometric mixture of ferrous and ferric salts in aqueous alkaline medium. The chemical reaction of Fe_3O_4 formation is usually written as follows:

$$Fe^{2+} + 2Fe^{3+} + 8OH^- \rightarrow Fe_3O_4 + 4H_2O$$

However, magnetite (Fe_3O_4) is not very stable and is sensitive to oxidation which results in the formation of maghemite (γ-Fe_2O_3).

The main advantage of the coprecipitation process is that a large amount of nanoparticles can be synthesized. However, the control of particle size distribution is limited. The addition of chelating organic anions (carboxylate or α-hydroxy carboxylate ions, such as citric, gluconic, or oleic acids) or polymer surface complexing agents (dextran, carboxydextran, starch, or polyvinyl alcohol) during the formation of magnetite can help to control the size of the nanoparticles. According to the molar ratio between the organic ion and the iron salts, the chelation of these organic ions on the iron oxide surface can either prevent nucleation and then lead to larger particles or inhibit the growth of the crystal nuclei, leading to small nanoparticles (Berger et al. 1999; Laurent et al. 2008).

Classical coprecipitation method generates particles with a broad size distribution. Synthesis of iron oxide nanoparticles with more uniform dimensions can be performed in synthetic and biological nanoreactors, such as water-swollen reversed micellar structures in non-polar solvents, apoferritin protein cages, dendrimers, cyclodextrins, and liposomes (Laurent et al. 2008).

Hydrothermal syntheses of magnetite nanoparticles are performed in aqueous media in reactors or autoclaves where the pressure can be higher than 2,000 psi (ca 13.8 MPa) and the temperature can be above 200°C. In this process, the reaction conditions, such as solvent, temperature, and time, usually have important effects on the products. The particle size of magnetite powders increased with a prolonged reaction time and the higher water content resulted in the precipitation of larger magnetite particles (Laurent et al. 2008).

The sol-gel process is a suitable wet route to the synthesis of nanostructured metal oxides. This process is based on the hydroxylation and condensation of molecular precursors in solution, originating a "sol" of nanometric particles. Further condensation and inorganic polymerization leads to a three-dimensional metal oxide network denominated wet gel. Because these reactions are performed at room temperature, further heat treatments are needed to acquire the final crystalline state. The main parameters that influence the kinetics, growth reactions, hydrolysis, condensation reactions, and consequently, the structure and properties of the gel are solvent, temperature, nature and concentration of the salt precursors employed, pH, and agitation (Laurent et al. 2008).

The polyol process is a versatile chemical approach for the synthesis of nano- and microparticles with well-defined shapes and controlled sizes. Selected polyols (for example, polyethylene glycol) used as solvents exhibit high dielectric constants, and can dissolve inorganic compounds. In addition, due to their relatively high boiling points, they offer a wide operating-temperature range for preparing inorganic compounds. Polyols also serve as reducing agents as well as stabilizers to control particle growth and prevent interparticle aggregation (Laurent et al. 2008). Non-aggregated magnetite nanoparticles (7 nm in diameter) were synthesized during the reaction of triethylene glycol with $Fe(acac)_3$ at an elevated temperature (Cai and Wan 2007).

Recently a novel synthesis of magnetite nanoparticles based on a flow injection synthesis (FIS) technique has been developed. The technique consists of continuous or segmented mixing of reagents under laminar flow regime in a capillary reactor. The FIS technique has shown some advantages, such as a high reproducibility because of the plug-flow and laminar conditions, a high mixing homogeneity, and an opportunity for a precise external control of the process. The obtained magnetite nanoparticles had a narrow size distribution in the range of 2–7 nm (Laurent et al. 2008; Salazar-Alvarez et al. 2006).

Spray and laser pyrolysis, typical representatives of aerosol technologies, are continuous chemical processes allowing for high rate production of nanoparticles. In spray pyrolysis, a solution of ferric salts and a reducing agent in organic solvent is sprayed into a series of reactors, where the aerosol solute condenses and the solvent evaporates. The resulting dried residue consists of particles whose size depends upon the initial size of the original droplets. Maghemite particles with size ranging from 5 to 60 nm with different shapes have been obtained using different iron precursor salts in alcoholic solution (Laurent et al. 2008).

A wide variety of chemical reactions accelerated by microwave irradiation of reactants have been observed. Recently a simple, quick and cost effective microwave method to prepare relatively uniform magnetite nanoparticles (80 ± 5 nm) directly from Fe^{2+} salts has been developed; the formation of magnetic nanoparticles using microwave method requires only a few seconds or minutes. Also magnetite nanoparticles doped with silver nanoparticles can be prepared using this procedure (Zheng et al. 2010).

Nanosized magnetite powders can also be synthesized via a mechanochemical reaction. Ball milling of ferrous and ferric chlorides with sodium hydroxide led to a mixture of magnetite and sodium chloride. To avoid agglomeration, the excess of NaCl is usually added to the precursor before ball milling. To prepare different size of particles, the as-milled powders were annealed at temperatures ranging from 100°C to 800°C for 1 h in appropriate atmosphere (Lin et al. 2006).

3 Stabilization of Magnetic Particles

To obtain biocompatible magnetically responsive materials it is usually necessary to stabilize the prepared iron oxide nanoparticles by appropriate modification of their surface or by their incorporation into appropriate biocompatible matrix.

The modified magnetic nanoparticles should be stable against aggregation in both a biological medium and a magnetic field.

Several compounds with carboxylic, phosphate and sulfate functional groups are known to bind to the surface of magnetic particles and stabilize them. Citric acid can be successfully used to stabilize water-based magnetic fluids (ferrofluids) by coordinating via one or two of the carboxyl residues; this leaves at least one carboxylic acid group exposed to the solvent, which should be responsible for making the surface negatively charged and hydrophilic. Other ferrofluids can be stabilized by ionic interactions, using e.g., perchloric acid or tetramethylammonium hydroxide (Berger et al. 1999; Laurent et al. 2008).

In most cases biocompatible (bio)polymers are used for magnetic particles stabilization and modification. Ideal natural or synthetic polymeric materials used for particles stabilization should have several important properties; they should be biocompatible and for many applications also biodegradable, non-toxic, non-thrombogenic, non-immunogenic and inexpensive. The "ideal" magnetically responsive (bio)polymer biocompatible composite nanoparticles should have the following typical properties: particles diameter below 100 nm, stability in blood, no activation of neutrophils, no platelet aggregation, avoidance of the reticuloendothelial system, noninflamatory behavior, prolonged circulation time, possible immobilization of appropriate biologically active compounds (e.g., antibodies) and scalable and cost-effective production (Lockman et al. 2002). Dextran, a polysaccharide polymer composed exclusively of α-D-glucopyranosyl units with varying degrees of chain length and branching has often been used as a polymer coating mostly because of its excellent biocompatibility. The formation of magnetite in the presence of dextran 40,000 was reported for the first time in 1980s (Molday and Mackenzie 1982). The same procedure has been used for the preparation of approved magnetic resonance contrast agents Ferumoxtran-10 (Combidex, Sinerem); this material has a small hydrodynamic diameter (15–30 nm), and shows a prolonged blood residence time, which allows this USPIO ("Ultrasmall superparamagnetic iron oxides") to access macrophages located in deep and pathologic tissues (such as lymph nodes, kidney, brain, osteoarticular tissues, etc.). Other common biopolymer coatings are formed e.g. by carboxymethylated dextran, carboxydextran, starch, chitosan, alginate, arabinogalactan or glycosaminoglycan, while polyethylene glycol (PEG) and polyvinyl alcohol (PVA) represent biocompatible synthetic polymers (Laurent et al. 2008).

Magnetic nanoparticles often form a magnetic part of magnetically responsive composite microparticles formed from various synthetic polymers, biopolymers, inorganic materials, microbial cells or plant materials (Safarik and Safarikova 2009b). Superparamagnetic monodisperse microparticles composed of polystyrene matrix with entrapped maghemite nanoparticles (approx. 8 nm in diameter; Fonnum et al. 2005), known as Dynabeads (Invitrogen), have been used in enormous amount of bioapplications, especially in molecular biology, cell biology, microbiology and protein separation.

4 Examples of Biomedical Applications of Magnetic Particles

Magnetically responsive nano- and microparticles have many already established and potential applications in various areas of biosciences, biotechnology and environmental technology. Biomedicine related applications are mainly based on the utilization of selected properties, namely magnetic separation, magnetic targeting, heat production and MRI contrast increase.

4.1 In Vitro Application of Magnetic Particles

Magnetic separation is a very simple, fast, efficient and gentle tool for the rapid isolation of target molecules, cell organelles and cells from complex biological mixtures and crude samples such as blood, bone marrow, tissue homogenates, urine, stools and other biological materials. The molecules and cells isolated by magnetic separation are usually pure, viable and unaltered. The gentle test tube magnetic separation is the technology of choice when there is a need for high yields of pure and biologically active compounds and biological structures in small scale (Šafařík and Šafaříková 1999, 2004).

Magnetic separations have been mainly employed for the immunomagnetic separation (IMS) of eukaryotic target cells (e.g., cancer cells, stem cells or T-lymphocytes). The progress of this procedure has been enabled by the development of monoclonal antibodies and the improved characterization of cells specific antigens. It is now possible to separate many specific cell types from mixed populations in blood, bone marrow and other sources without cell loss or damage. By positive isolation the cells of interest are magnetically labeled by specific particles and subsequently isolated for analysis. For some downstream applications (e.g., mRNA or DNA isolation) the beads can remain attached to the isolated cells. For other applications (e.g., FACS analysis) magnetic microbeads have to be removed from the cell surface. By negative isolation (= depletion), unwanted cells are removed prior to analysis of the remaining population. Primary coated Dynabeads and some other magnetic particles are ready-to-use products for a wide variety of cell surface markers. In addition, secondary coated magnetic particles offer an excellent possibility to make beads with the reactivity of choice using mouse, rat, sheep or rabbit antibodies (Šafařík and Šafaříková 1999).

The magnetic separation procedure can be scaled up if large quantities of living cells are required. Several commercially available devices enable to separate target cells from larger volume of blood or cells suspensions. The CliniMACS System (Miltenyi Biotec, Germany) is based on MACS Technology which employs MACS MicroBeads (nano-sized superparamagnetic particles approximately 50 nm in diameter, composed of an iron dextran matrix coupled to specific antibodies) and a magnetic separation unit. CliniMACS permits the automated separation of cells on a clinical-scale level in a closed and sterile system. The key components of the CliniMACS Plus Instrument are the integrated microcomputer, the magnetic separation unit, the peristaltic pump and various pinch valves. The Isolex® 300 System (Baxter, USA) is a semi-automated magnetic cell separation system designed to select and isolate CD34 positive cells, ex *vivo,* from mobilized peripheral blood using anti-CD34 monoclonal antibody and superparamagnetic microspheres (Dynabeads). The device consists of an instrument for separating Dynabeads from mobilized peripheral blood mononuclear cell (MNC) suspensions and an associated disposable set for providing the fluid path.

IMS of important pathogenic bacteria (*Salmonella, Legionella, Listeria* and verotoxigenic strains of *Escherichia coli* (e.g., O157, EPEC/VTEC O103, O111 and O145)) and protozoan parasites (*Cryptosporidium, Giardia*) is efficiently used in medical, food and water microbiology. Both micrometer and nanometer scale magnetic particles with immobilized specific antibodies are used as labels. IMS enables to shorten the detection time which is a very important feature from the medical point of view. As an alternative to IMS, magnetic particles with immobilized lectins have been used for similar purpose (Šafařík and Šafaříková 1999).

Magnetic particles with immobilized annexin V have been employed for the simple and efficient separation of apoptotic cells from normal culture. This procedure is based on the fact that annexin V is a Ca^{2+}-dependent phospholipid binding protein with high affinity for phosphatidylserine (PS), which is redistributed from the inner to the outer plasma membrane leaflet in apoptotic or dead cells. Once on the cell surface, PS becomes available for binding to annexin V and any of its conjugates.

Magnetic targeting of nucleic acids bound to magnetic particles into the recipient cells is the basis of magnetofection which is a simple and highly efficient transfection method. The magnetic iron oxide nanoparticles are usually coated with specific cationic molecules which can associate with the gene vectors (DNA, siRNA, ODN, virus, etc.). The magnetic particles are then concentrated on the target cells by the influence of an external magnetic field generated by magnets. The cellular uptake of the genetic material is accomplished by endocytosis and pinocytosis, two natural biological processes. Consequently, membrane architecture and structure stay intact, in contrast to other physical transfection methods that damage the cell membrane. The nucleic acids are then released into the cytoplasm by different mechanisms depending upon the formulation used. Coupling magnetic nanoparticles to gene vectors of any kind results in a dramatic increase of the uptake of these vectors and consequently high transfection efficiency. Transfected cells can be separated from the non-transfected ones using an appropriate magnetic separation technique (Plank et al. 2003; Scherer et al. 2002; Safarik and Safarikova 2009a).

Magnetic separation of nucleic acids and proteins enables direct isolation and purification of target biomolecules from difficult-to-handle biological samples, such as cell homogenates or body fluids. Magnetic separation techniques have several advantages in comparison to traditional separation procedures; the laboratory-scale process is very simple, and all steps of the purification can take place in one test tube without expensive liquid chromatography systems (Berensmeier 2006; Šafařík and Šafaříková 2004).

Magnetic particles can be efficiently used for the detection and determination of target analytes in clinical biochemistry. Many devices have been developed such as electrochemical immunoassay systems for the simultaneous measurements of several proteins, flow injection analysis (FIA) employing enzymes immobilized on magnetic particles, magnetic separation immunoassay for digoxin with flow injection fluorescence detection, sequential injection analysis with a chemiluminescence detector for the determination of vitellogenin or immunoassay for sequential injection analysis (SIA) (Aguilar-Arteaga et al. 2010). Efficient preconcentration of target analytes from large volumes of water or biological samples can be achieved using Magnetic solid phase extraction (Šafaříková and Šafařík 1999).

4.2 *In Vivo Application of Magnetic Particles*

Magnetic drug targeting employs magnetic biocompatible nanoparticles containing a drug which could be injected intravenously, transported to a site of action (e.g., cancerous tumor or arterial blockage) and be retained at the site by application of a magnetic field gradient. This form of drug delivery is advantageous in that a specific site in the body can be targeted by the magnetic field gradient, the doses required for systemic drug delivery are reduced, localized drug levels can be increased significantly with reduced potential toxic side effects at non-targeted tissues, and a prolonged release of high localized drug concentrations at a required site can be obtained (Vatta et al. 2006).

Magnetic fluid hyperthermia, based on the fact that subdomain biocompatible magnetic particles produce heat through various kinds of energy losses during application of external AC magnetic field, is a promising approach to cancer therapy due to the heating of the target tissue to the temperatures between 42°C and 46°C that generally reduces the viability of cancer cells and increases their sensitivity to chemotherapy and radiation. Unlike chemotherapy and radiotherapy, hyperthermia itself has fewer side effects. Different types of magnetic biocompatible nanocomposites such as dextran-stabilized magnetic fluid, other types of biocompatible magnetic fluids, aminosilane-modified nanoparticles, cationic magnetoliposomes or affinity magnetoliposomes have been used for hyperthermia treatment (Goya et al. 2008; Safarik and Safarikova 2009b).

Superparamagnetic particles are used as magnetic resonance imaging (MRI) contrast agent in diagnostics applications. Particulate magnetic contrast agents include ultrasmall particles (USPIO – "Ultrasmall superparamagnetic iron oxides"; diameter between 10 and 40 nm), small particles (SPIO – "Small superparamagnetic iron oxides"; diameter between 60 and 150 nm), and oral (large) particles (diameter between 300 nm and 3.5 μm). Commercially available iron oxides based contrast agents are usually stabilized by dextran, carboxymethyl dextran, carboxydextran or styrene divinylbenzene copolymer (Laurent et al. 2008).

A crucial aspect of successful cell transplantation is tracking and monitoring the grafted cells in the transplant recipient. To screen cells *in vivo*, superparamagnetic iron-oxide nanoparticles have been used to label the cells, enabling subsequent MRI visualization *in vivo,* due to the selective shortening of T_2-relaxation time, leading to a hypointense (dark) signal. MRI, as a noninvasive method, may then be used not only to evaluate whether the cells have been successfully engrafted, but also monitor the time course of cell migration and their survival in the targeted tissue. This information may further help to optimize the transplantation procedure in terms of the number of required cells, the method or site of cell administration and the therapeutic time window after injury during which transplantation will be most effective (Syková and Jendelová 2005).

Detailed information about *in vivo* applications of magnetic particles has been described in other chapters of this book.

5 Conclusions

Magnetic nano- and microparticles have been intensively studied for many years. Magnetically responsive biocompatible materials represent an extremely important group of stimuli responsive materials with high potential both in research and application area. Various areas of biomedicine have already found a substantial benefit due to the application of magnetic nano- and microparticles, both for *in vitro* and *in vivo* procedures. In fact, many different types of magnetic materials are available, however, only a small part of them can be used for possible biomedical applications. Further progress in this area could be expected if cost effective biocompatible magnetic particles would become available. Safety and biocompatibility studies of magnetically responsive materials, in particular long-term toxicity studies, have to be carried out. The potential of magnetic nanomaterials will probably expand when complex magnetic nanoparticles and drugs containing (nano)systems will be constructed, enabling simultaneously their magnetic navigation, MRI detection and heat production.

Acknowledgements This research was supported by the Ministry of Education of the Czech Republic (projects OC 157 – Action COST 868 and OC 09052 – Action COST MP0701), by the Ministry of Industry and Trade of the Czech Republic (Project No. 2A-1TP1/094), and by the Research Aim of the Institute of Systems Biology and Ecology (AV0Z60870520).

References

Aguilar-Arteaga, K., Rodriguez, J.A., and Barrado, E. (2010). Magnetic solids in analytical chemistry: A review. Anal. Chim. Acta *674*, 157–165.

Arakaki, A., Nakazawa, H., Nemoto, M., Mori, T., and Matsunaga, T. (2008). Formation of magnetite by bacteria and its application. J. Royal Soc. Interface *5*, 977–999.

Arruebo, M., Fernandez-Pacheco, R., Ibarra, M.R., and Santamaria, J. (2007). Magnetic nanoparticles for drug delivery. Nano Today *2*, 22–32.

Berensmeier, S. (2006). Magnetic particles for the separation and purification of nucleic acids. Appl. Microbiol. Biotechnol. *73*, 495–504.

Berger, P., Adelman, N.B., Beckman, K.J., Campbell, D.J., Ellis, A.B., and Lisensky, G.C. (1999). Preparation and properties of an aqueous ferrofluid. J. Chem. Educ. *76*, 943–948.

Cai, W., and Wan, J.Q. (2007). Facile synthesis of superparamagnetic magnetite nanoparticles in liquid polyols. J. Colloid Interface Sci. *305*, 366–370.

Fonnum, G., Johansson, C., Molteberg, A., Morup, S., and Aksnes, E. (2005). Characterisation of Dynabeads (R) by magnetization measurements and Mossbauer spectroscopy. J. Magn. Magn. Mater. *293*, 41–47.

Goya, G.F., Grazu, V., and Ibarra, M.R. (2008). Magnetic nanoparticles for cancer therapy. Curr. Nanosci. *4*, 1–16.

Laurent, S., Forge, D., Port, M., Roch, A., Robic, C., Elst, L.V., and Muller, R.N. (2008). Magnetic iron oxide nanoparticles: Synthesis, stabilization, vectorization, physicochemical characterizations, and biological applications. Chem. Rev. *108*, 2064–2110.

Lin, C.R., Chu, Y.M., and Wang, S.C. (2006). Magnetic properties of magnetite nanoparticles prepared by mechanochemical reaction. Mater. Lett. *60*, 447–450.

Lockman, P.R., Mumper, R.J., Khan, M.A., and Allen, D.D. (2002). Nanoparticle technology for drug delivery across the blood-brain barrier. Drug Devel. Ind. Pharm. *28*, 1–13.

Molday, R.S., and Mackenzie, D. (1982). Immunospecific ferromagnetic iron-dextran reagents for the labeling and magnetic separation of cells. J. Immunol. Methods *52*, 353–367.

Plank, C., Schillinger, U., Scherer, F., Bergemann, C., Remy, J.S., Krotz, F., Anton, M., Lausier, J., and Rosenecker, J. (2003). The magnetofection method: Using magnetic force to enhance gene delivery. Biol. Chem. *384*, 737–747.

Šafařík, I., and Šafaříková, M. (1999). Use of magnetic techniques for the isolation of cells. J. Chromatogr. B *722*, 33–53.

Šafařík, I., and Šafaříková, M. (2004). Magnetic techniques for the isolation and purification of proteins and peptides. BioMagn Res Technol *2*, Article No. 7.

Safarik, I., and Safarikova, M. (2009a). Magnetic nano- and microparticles in biotechnology. Chem. Papers *63*, 497–505.

Safarik, I., and Safarikova, M. (2009b). Magnetically responsive nonocomposite materials for bioapplications. Solid State Phenomena *151*, 88–94.

Šafaříková, M., and Šafařík, I. (1999). Magnetic solid-phase extraction. J. Magn. Magn. Mater. *194*, 108–112.

Salazar-Alvarez, G., Muhammed, M., and Zagorodni, A.A. (2006). Novel flow injection synthesis of iron oxide nanoparticles with narrow size distribution. Chem. Eng. Sci. *61*, 4625–4633.

Scherer, F., Anton, M., Schillinger, U., Henkel, J., Bergemann, C., Kruger, A., Gansbacher, B., and Plank, C. (2002). Magnetofection: enhancing and targeting gene delivery by magnetic force in vitro and in vivo. Gene Ther. *9*, 102–109.

Syková, E., and Jendelová, P. (2005). Magnetic resonance tracking of implanted adult and embryonic stem cells in injured brain and spinal cord. Ann. N. Y. Acad. Sci. *1049*, 146–160.

Vatta, L.L., Sanderson, R.D., and Koch, K.R. (2006). Magnetic nanoparticles: Properties and potential applications. Pure Appl. Chem. *78*, 1793–1801.

Zheng, B.Z., Zhang, M.H., Xiao, D., Jin, Y., and Choi, M.M.F. (2010). Fast microwave synthesis of Fe_3O_4 and Fe_3O_4/Ag magnetic nanoparticles using Fe^{2+} as precursor. Inorg. Mater. *46*, 1106–1111.

Biosynthesis of Metallic Nanoparticles and Their Applications

Adam Schröfel and Gabriela Kratošová

Abstract Biosynthesis and biofabrication of the metallic NPs have became an important approach to NP preparation. They are not only equal with the chemical or physical methods, but also offer quite a few assets compared to classical tacks. In this review, we present comprehensive overview of existing published records, which include clear and realistic application of biosynthesized metallic NPs. Our survey covers NP utilization from biosorption and catalysis to medicinal and sensing applications. Moreover, we add current review references and comparison (or synergy) with chemical and physical methods.

Keywords Biosynthesis • Nanoparticles • Biosorption • Bioremediation • Catalysis • Precious metal

Abbreviations

NP	nanoparticle
AuNP	gold nanoparticle
AgNP	silver nanoparticle
PdNP	palladium nanoparticle
PEI	polyethylenimine
CP	chlorophenol
PCB	polychlorinated bifenyl
PBDE	polybrominated diphenyl ether
TCPP	tris(chloroisopropyl)phosphate
G+	Gram positive
G–	Gram negative

A. Schröfel (✉) and G. Kratošová
Nanotechnology Centre, VŠB-Technical University in Ostrava,
17. listopadu 15, 708 33, Ostrava, Czech Republic
e-mail: adam.schrofel@vsb.cz

A. Prokop (ed.), *Intracellular Delivery: Fundamentals and Applications*, Fundamental Biomedical Technologies 5, DOI 10.1007/978-94-007-1248-5_14,

γ-HCH	γ-hexachlorocyclohexane (lindane)
ICM	iodinated contrast media
TCE	trichloroethylene
bioPd	biofabricated palladium
PEM	polymer electrolyte membrane
4-NP	*p*-nitrophenol; 4-nitrophenol
SERS	surface enhanced Raman scattering
IR	infra red
BM	bacterial magnetosome
PTP	tyrosine phosphatase
PVDF	polyvinylidene fluoride
GCE	glassy carbon electrode
QD	quantum dot

1 Introduction

"Green" approach toward nanotechnology research became widely used in last few years. Inspiration by nature and processes inside the living organisms can produce new opportunities and perspectives for whole wide nanotechnology branch of research and industry. Particularly, synthesis of NPs and nanostructures can easily profit from usage of nature equipment located in cells. Right attitude towards world of biomolecules, function groups, enzymes and other important factors leads to large number of advantages compared to common chemical or mechanical methods.

Scientific interest in the NPs originates from their unique and variable properties. They create connecting link between bulk material and individual atoms or molecules. When the bulk materials have constant physical properties, the same materials have been uncovered to exhibit different interesting properties when studied in the nanoscale. For instance, NPs have a much larger surface compared to bulk materials. Generally, it follows that they have higher reactivity and different chemical properties. Also, the wavelength of NPs is similar to the wavelength of light. This results in unique optical properties (NPs are transparent).

Possible applications of metallic NPs are taking place in varied areas such as electronics, coating technology, packaging, cosmetics, biosensing, medicine and they will be discussed in corresponding sections.

2 Preparation of Nanoparticles

There are three main approaches for the metallic NP synthesis: physical, chemical and biological. From the structural point of view, we can describe methods as "bottom-up" and "top-down". Bottom-up approach is way to assemble the final structure atom by atom, molecule by molecule. The building units are formed at first and, subsequently, formed into NPs or nanostructures – the final product. Some of the major assets of the bottom-up approach are resulting NPs without structural defects and with homogenous chemical compositions. Using "top-down" techniques, starting bulk material is reduced in size by mean of mechanical or chemical methods. The major drawback of this approach is often imperfection of the obtained structure. The surface defects have significant impact on physical properties and surface chemistry behavior.

2.1 Physical Methods

Commonly utilized are physical (or mechanical) methods such as attrition, evaporation/condensation, laser ablation and pyrolysis. In attrition, particles are ground by a size-reducing mechanism (e.g. ball mill). Resulted NPs are critically affected by the used starting material, time of drilling and medium which is used as atmosphere. This method is typical representative of "top-down" approach. Inert gas condensation can be used in atom creation (e.g. for the preparation small clusters of gold atoms). As a result of interatomic collisions with the gas atoms in the chamber, the evaporated metal atoms lose their kinetic energy and condense in the form of small crystals (Lee et al. 2007). Laser ablation method is used to produce NPs by using the pulsed laser irradiation on the metal target in liquid or gas environment (Yang 2007). In pyrolysis, the precursor solution is atomized into series of droplet "reactors". These very small droplets are introduced by a carrier air gas into a hot-wall region under atmospheric pressure conditions. The solvent in the droplets evaporates inside the furnace, and the remaining solutes cause precipitation, thermal decomposition and intraparticle reactions to form product particles (Widiyastuti et al. 2010).

The advantage of these methods is narrow particle size distribution of the produced NPs, while its limitation is the need for expensive equipment (lasers etc.). Also the production rate is lower compare to chemical methods. Higher energy consumption for maintaining the pressure and temperature conditions used in the aforementioned procedures are an additional handicap.

2.2 Chemical Methods

The second route is a chemical approach (usually "bottom-up" methods), particularly the wet chemical procedure – the metal ions of dissolved precursor (e.g. $AgNO_3$) is reduced in defined conditions which allows the subsequent formation of small clusters or aggregates of metal atoms (Khomutov and Gubin 2002; Oliveira et al. 2005). Similar procedures can be used also for creating of metallic molecules such as sulfides or oxides (Seoudi et al. 2010). Specifically, methods such as the sonochemical method (Salkar et al. 1999; Shchukin et al. 2010), the polyol process, the solvent-reduction method (Bonet et al. 2000), the template method

(Fukuoka et al. 2002), and the reverse micelles method (Yadav et al. 2003) have been developed for the preparation of NPs.

There are also several modified chemical methods including seed-mediated growth where small particles (produced by other techniques like irradiation) are utilized as seeds and fresh metallic ions are reduced by reducing agent and grow along the surface of the seed particle (Samanta et al. 2010). Reduction agent usage differs depending on the purpose, required properties, types, sizes. For instance, the first reduction of gold salt by the so-called Turkevich method was introduced in 1951, when a sodium citrate reduction of $HAuCl_4$ was used for synthesis of stable gold nanoparticles (AuNPs) (Nguyen et al. 2010). There are also other reducing agents such as $NaBH_4$ (Wagner et al. 2008), methoxypolyethylene glycol (Mallick et al. 2004), stannous chloride (Vaskelis et al. 2007) and ascorbic acid (Wagner and Köhler 2005). Amine or hydroxyl-containing molecules such as branched poly(ethyleneimine) (Note et al. 2006), azacryptand, amino acid (Selvakannan et al. 2004) or chitosan (Shih et al. 2009) were also reported as a suitable reducing agents for metallic NP preparation.

The chemical methods are relatively inexpensive for high volume, usually easy to perform, and very variable. However, their disadvantages include toxic chemicals usage, likely contamination from precursor chemicals or reducing agents, and also formation of dangerous and hazardous reaction byproducts.

2.3 *Biological Methods*

Because of certain aforementioned limitations of physical and chemical method, there is an increasing need to develop methods, which will be nontoxic, fast, high-yield, energy saving (occur under normal conditions – normal air pressure and temperature), and environmentally benign. Consequently, researchers have turned to biological systems for inspiration. Bioprocesses mediated by living organisms (employing their cells, enzymes, transport chains etc.) therefore became important for metallic NP synthesis. For this purpose, we have a vast variety of organisms in nature such as viruses, bacteria, yeast, fungi, algae, plants and plant products at our disposal.

The ability to form inorganic materials by many organisms either intra- or extracellularly has been well known for almost 30 years (Wilbur and Simkiss 1979; De Stefano et al. 2008). Many biotechnological applications, such as the remediation of toxic metals, employ microorganisms such as bacteria (Pérez-de-Mora et al. 2006). Therefore, many microorganisms (e.g. fungi, bacteria) were found as possible nanofacilities for NP fabrication. These nature-derived processes contributed and led to development of relatively new biosynthesis methods for fabrication of nano- and microscale inorganic materials by microbes and other living organisms (Ahmad et al. 2003). Until now, a large number of both unicellular and multicellular organisms have been known to produce intracellular or extracellular metallic NPs (Thakkar et al. 2010).

Biosynthesis of NPs is becoming an emerging consequence of overlap between nano- and biotechnology. In the last few years it has received attention due to its potential to develop environmentally benign technologies in material science. But in fact, this type of NP synthesis method is also a "chemical" approach. Living cells are extremely complex system with thousands of molecules. These molecules have varied functional groups (such as hydroxyl, amine etc.) which each can possibly facilitate metal reduction. Therefore, it is very difficult to describe a specific place or process responsible directly for NP growth. This can result in certain drawbacks for biosynthesis methods. The resulting matter is usually mixture of cells (cell debris) and NPs, accompanying with thousands of metabolic products and other molecules. Frequently it is very complicated to separate the tiny product particles from the cell debris. Moreover, surrounding matrix and capping proteins contribute to NP stability (Lynch and Dawson 2008) and can influence their properties. Among the other disadvantages of precursors (such as $AgNO_3$) is their toxicity to the target organisms. Therefore, this does not allow the usage of higher concentrations of the salts.

In this article, we provide a brief overview of the current research worldwide on the use of organisms such as bacteria, cyanobacteria and actinomycetes (both prokaryotes), as well as algae, yeast, fungi and plants (eukaryotes) in the biosynthesis of metal NPs with emphasize on their applications.

3 Applications of Metallic Nanoparticles

Although current research results show a wide field for biosynthesized NPs, we can segment these applications into several groups. The following division is based primarily on the purpose of biofabricated NPs (even though the chemical composition, shape and source organisms will be mentioned too).

3.1 Biosorption

Different organisms have ability to change metal oxidation state and concomitantly deposit resulting metal compounds and zero-valent metals on the cell surface or inside their cells. A variety of biomaterials have been known for a long time to bind the precious metals (including algae, fungi, bacteria, actinomycetes, yeast etc.). along with some biopolymers and biowaste materials (Table 1).

In particular, recovery of precious metals like gold, silver, palladium, and platinum is interesting due to their increasing market prices and various industrial applications. Conventional technologies (e.g. ion exchange, chemical binding, surface precipitation) which been have been developed for the recovery of such metals are neither efficient nor economically attractive. Biosorption represents a biotechnological innovation as well as a cost effective tool for recovery of precious metals from

Table 1 Biosorption

NP	Organism used	Application	Reference
Au	*Nitzschia obtusa, Navicula minima*	Bioaccumulation	Chakraborty et al. (2006)
Au	*Lyngbya majuscula, Spirulina subsalsa, Rhizoclonium hieroglyphicum*	Bioaccumulation, biorecovery	Chakraborty et al. (2009)
Au	*Fucus vesiculosus*	Bioaccumulation, biorecovery	Mata et al. (2009)
Ag	*Pleurotus platypus*	Biosorption	Das et al. (2010)
Pt	*E. coli*	Biosorption, biorecovery by incineration	Won et al. (2010)
Au	*Sargassum sp.*	Biosorption, biorecovery by incineration	Sathishkumar et al. (2010a)

aqueous solutions. In particular the microbial mechanisms involved in the biosorption and bioaccumulation processes have been extensively studied in natural environments, and researchers have recently gained interest in the applications of microbe–metal interactions in biotechnology, nanotechnology or material engineering. The connection between the recently discovered ability of NP biosynthesis and long-term investigated biosorption is apparent. Since the field of biosorption is wide, there is abundance of suitable and quality literature and reviews (Arief et al. 2008; Das 2010; Gadd 2009, 2010; Hennebel et al. 2009a; Chojnacka 2010; Wang and Chen 2009; Lesmana et al. 2009; Barakat 2010; Kavamura and Esposito 2009; Volesky 2007; Vijayaraghavan and Yun 2008). Furthermore, and without aspiration for more detailed probe, we will discuss some examples and current trends of metallic NPs biosorption application, specifically with regard to their application for metal bioaccumulation and recovery, waste remediation, soil and water treatment. Additionally we will discuss the NPs formation or ion bioreduction process. For more exhaustive analysis of biosorbed and biofabricated palladium and platinum catalysts see also Sect. 3.2.

As instance of noble metals biorecovery, Chakraborty et al. (2006) described experiments of Au bioaccumulation with two diatom strains. These unicellular algae organisms are one of the most abundant amongst the species both in marine and fresh water ecosystems on Earth. Due to low detection limit and also with regard to biosorption of other radioactive heavy metals in previous studies, gold radionuclide ^{198}Au was used. In subsequent study (Chakraborty et al. 2009), AuNP formation process was described and comparison in biorecovery abilities between prokaryotic and eukaryotic algal genera was performed. Gold biosorption and bioreduction with another brown alga *Fucus vesiculosus* was also reported (Mata et al. 2009), describing pH dependence and stages of the bioreduction process. Results of these studies indicate that live algal biomass may be a viable and cost effective for biorecovering of gold.

Bioaccumulation of silver ions Ag(I) from the solution or wastewater is reported e.g. by Das et al. (2010), accompanied also with kinetics studies and thermodynamic calculations on sorption of silver ions on gilled macrofungus *Pleurotus platypus*. This paper represents a modern and innovative approach for the study of interactions between biomass and metal ions.

Concrete example of platinum recovery is presented by Won et al. (2010) by means of biosorption and subsequent incineration of PEI modified biomass (prepared by attaching PEI onto the surface of inactive *E. coli* biomass). Wastewater containing platinum used for the recovery study was obtained from an industrial laboratory for inductively coupled plasma (ICP). Recovery efficiency of platinum in ash after incineration was over 98.7%. Similar study with gold solution and easily accessible biomass of *Sargassum sp.* (Sathishkumar et al. 2010a) confirms recovery efficiency more than 90%.

3.2 Catalysis Applications

Based on the basic knowledge of inorganic catalysis, noble metal NP catalysts are very attractive when compared to bulk catalyst – they have a high surface to volume ratios and their surface atoms are very active. For more knowledge and information about the noble metal nanocatalysists in colloidal solutions or adsorbed on different supports, we recommend numerous review articles, which have been published for many different types of organic and inorganic reactions (Narayanan 2010; Narayanan and El-Sayed 2008; Roucoux et al. 2008; Thibault-Starzyk et al. 2008; Kumar et al. 2004; Shiju and Guliants 2009).

3.2.1 Biosorption and Biosynthesis of Palladium Nanoparticles in Organic and Inorganic Catalysis

The first large group of biosynthesized nanomaterials with catalytic activity is represented by palladium NPs. Baxter-Plant et al. (2003; 2004) reported usage of cell surface of three different species of *Desulfovibrio* (G– sulfate-reducing bacteria) for manufacturing the novel bioinorganic catalyst via reduction process of Pd(II) to Pd(0). Although the presence of reducing agent is necessary (e.g. in form of H_2) for the Pd(0) genesis, reduction process is critically influenced by the bacteria presence and we can indicate this biosorption process as a biosynthesis. On the other hand, reduction in the absence of cells does not lead to the formation of Pd(0) NPs (Bunge et al. 2010). This catalyst on "palladised cells" was used for reductive dehalogenation of chlorophenol (CP) and selected polychlorinated biphenyl (PCB) types. The same organism was used for dehalogenation of the other environmentally prevalent PCBs and polybrominated diphenyl ether (Harrad et al. 2007). The versatility of "bioPd" catalyst is also demonstrated in various reactions including dehalogenation of flame retardants such polybrominated diphenyl ether (PBDE) or tris(chloroisopropyl)phosphate (TCPP). Authors also compare effectiveness between biocatalyst, chemically reduced Pd(II) and commercial Pd(0) catalysts. Although chemically reduced Pd(II) and commercial Pd(0) were more effective debromination agents, "bioPd" dechlorinated TCPP was five times more effective than using commercial Pd(0) catalyst (Deplanche et al. 2009) (Table 2).

Table 2 Catalysis applications

NP	Organism used	Application	Reference
Pd	*Desulfovibrio vulgaris, D. desulfuricans*	Dehalogenation of CP and PCBs	Baxter-Plant et al. (2003)
Pd	*D. desulfuricans*	Dehalogenation of CP and PCBs	Baxter-Plant et al. (2004)
Pd	*D. desulfuricans*	Dehalogenation PCBs and polybrominated diphenyl ether	Harrad et al. (2007)
Pd	*D. desulfuricans*	Dehalogenation of flame retardant materials	Deplanche et al. (2009)
Pd	*D. desulfuricans, Rhodobacter sphaeroides*	Dehalogenation of (PCBs) and penta-CP	Redwood et al. (2008)
Pd	*D. desulfuricans, Bacillus sphaericus*	Hydrogenation of itaconic acid	Creamer et al. (2007)
Pd	*D. desulfuricans, B. sphaericus*	Hydrogenation, reduction and selective dehalogenation in non-aqueous solvents	Creamer et al. (2008)
Pd	*R. capsulatas, Arthrobacter oxidans*	Hydrogenation of 2-Butyne-1,4-diol	Wood et al. (2010)
Pd	*Cupriavidus necator, Pseudomonas putida*	Suzuki–Miyaura and Mizoroki–Heck reactions	Sobjerg et al. (2009)
Pd	*C. necator*	Catalysis of C—C bond formation	Gauthier et al. (2010)
Pd	*C. necator, P. putida, Paracoccus denitrificans*	Hydrogen production from hypophosphite	Bunge et al. (2010)
Pd	*Gardenia jasminoides* Ellis	Hydrogenation of p-nitrotoluene	Jia et al. (2009)
Pd	*Shewanella oneidensis*	Dehalogenation of chlorophenol and PCBs	De Windt et al. (2005)
Pd	*S. oneidensis*	Dehalogenation of perchlorate and PCBs	De Windt et al. (2006)
Pd	*S. oneidensis*	Dechlorination of lindane	Mertens et al. (2007)
Pd	*S. oneidensis*	Dechlorination of TCE, membrane reactor	Hennebel et al. (2009b)
Pd	*S. oneidensis*	Dechlorination of TCE, fixed bed reactor	Hennebel et al. (2009c)
Pd	*S. oneidensis*	Degradation process for diatrizoate, ICM	Hennebel et al. (2010)
Pd, Pt	*D. desulfuricans*	Continuous reduction Cr(VI) to Cr(III)	Mabbett et al. (2005)
Pd	*D. vulgaris, D. desulfuricans*	Reduction of Cr(VI) to Cr(III)	Humphries et al. (2006)

(continued)

Table 2 (continued)

NP	Organism used	Application	Reference
Pd	*Serratia sp.* (NCIMB)	Reduction of Cr(VI) to Cr(III)	Beauregard et al. (2010)
Pd	*E. coli* mutant strains	Reduction of Cr(VI) to Cr(III)	Deplanche et al. (2010)
Pd	*Clostridium pasteurianum*	Reduction of Cr(VI) to Cr(III)	Chidambaram et al. (2010)
Pt	waste yeast biomass	Fuel cell; energy production	Dimitriadis et al. (2007)
Pd, Pt	*D. desulfuricans*	Fuel cell; energy production	Yong et al. (2007)
Pd	*D. desulfuricans, E. coli, C. metallidurans*	Waste biorefining, fuel cells	Yong et al. (2010)
Pd	*E. coli* MC4100 (parent), mutant (IC007)	Fuel cell; energy production	Orozco et al. (2010)
Pd	*S. oneidensis*	Fuel cell; energy production	Ogi et al. (2011)
Au	*Sesbania drummondii*	Reduction of 4-nitrophenol	Sharma et al. (2007)
Au	*Cacumen platycladi*	Reduction of 4-nitrophenol	Huang et al. (2009)
Ag	*Sepia esculenta* cuttle-bone organic matrix	Reduction of 4-nitrophenol	Jia et al. (2008)

Using of *Desulfovibrio desulfuricans* in comparison with other bacterial strains has been also demonstrated: Redwood et al. (2008) reported comparison of catalytic efficiency of and *Rhodobacter sphaeroides* in dehalogenation of PCBs and penta-CP. Gram negative (G–) and Gram positive (G+) bacterial strains *D. desulfuricans* and *Bacillus sphaericus* took place as Pd(II) reducing agent for catalysis of itaconic (methylene succinic) acid (Creamer et al. 2007). Remarkably, the same research group published experiments in non-aqueous solvents (methanol). Specifically, experiments leading to hydrogenations of 4-azidoaniline hydrochloride and 3-nitrostyrene, and hydrogenolysis (reductive debromination) of 1-bromo-2-nitrobenzene were conducted (Creamer et al. 2008).

Another type of G– bacteria, *Shewanella oneidensis*, was also used for biofabrication of Pd(0) catalyst (with H_2, formate, lactate, pyruvate or ethanol as electron donors) for dehalogenation purpose (De Windt et al. 2005). The obtained bioPd(0) NPs had the ability to reductively dehalogenate (PCB) congeners in aqueous and sediment matrices from anonymous industrial plant. Moreover, the aforementioned paper offers a comparison with commercially available palladium powders. Further studies of *S. oneidensis* show differences between catalytic reactivity of Pd(0) crystals formed on viable or non-viable biomass. The relatively large and densely covering Pd(0) crystals (non-viable biomass) exhibited high catalytic reactivity towards hydrophobic molecules such as polychlorinated biphenyls. In contrast, the smaller and more dispersed nanocrystals on a viable bacterial carrier were catalytically active towards anionic pollutant perchlorate (De Windt et al. 2006).

S. oneidensis bacterial strain was also used for removal of the pesticide lindane (γ-hexachlorocyclohexane or γ-HCH) by catalytic reduction of γ-HCH to benzene (as more efficient than with commercial powdered Pd(0) – Mertens et al. 2007). The same study introduces a membrane reactor technology suitable for dechlorination of γ-HCH polluted wastewater at a low-flux synthetic dialysis membrane. Similar implementation of membrane reactor was introduced for degradation process of diatrizoate – iodinated contrast media (ICM). Although currently applied techniques such as advanced oxidation processes exhibit only limited removal efficiencies of ICM, work by Hennebel et al. (2010) showed that membrane contactors with encapsulated biogenic NPs can be effective for contaminated water treatment. Topic of reactor technology for "bioPd" catalysts is further pursued in works dealing with dechlorination of trichloroethylene (TCE) in a pilot-scale membrane reactor (Hennebel et al. 2009b) and dechlorination of TCE by encapsulated palladium NPs in a fixed bed reactor (Hennebel et al. 2009c). Polyurethane cubes empowered with "bio-Pd" were implemented in a fixed bed reactor for the treatment of water containing TCE. This study shows that the influent recycle configuration resulted in a cumulative removal of 98% TCE after 22 h (with ethane as main reaction product). The same reactor in a flow through configuration achieved removal rates up to 1,059 mg TCE g $(Pd)^{-1} \cdot day^{-1}$.

Feasibility of another organisms for reduction of Pd(II) to Pd(0) for organic catalysis was also demonstrated by other studies. Bacterial strains *Rhodobacter capsulatas* and *Arthrobacter oxidans* were employed in "bioPd" formation for partial hydrogenation of 2-butyne-1,4-diol to 2-butene-1,4-diol (Wood et al. 2010). This "bioPd" was proven to be a highly selective catalyst for partial hydrogenation reactions. Bunge et al. (2010) tested possibilities of three bacterial strains (*Cupriavidus necator, Pseudomonas putida, Paracoccus denitrificans*) on bioPd(0) catalysis of hydrogen production from hypophosphite and further discuss the hypothetical mechanism of bacterial reduction of Pd(II) to Pd(0). Remarkably, Pd(0) catalysts fabricated by the organisms mentioned above were used also for catalysis of Suzuki–Miyaura and Mizoroki–Heck reactions (briefly C—C bond formation) by Sobjerg et al. (2009) and Gauthier et al. (2010). The enormous importance of these reactions for organic synthesis may be confirmed by the long-anticipated Nobel prize in Chemistry 2010 for their discoverers – Richard F. Heck, Ei-ichi Negishi and Akira Suzuki (more about these reactions and NPs in review article by Narayanan (2010)). Moreover, aforementioned studies also contribute to the hot issue of metal waste management and waste recovery.

Interestingly, Jia et al. (2009) published bioreduction method – reduction of palladium chloride by water crude extract – with plant *Gardenia jasminoides* Ellis'. Abilities of this "bioPd(0)" nanocatalyst were tested and documented on hydrogenation reaction of p-nitrotoluene. The catalysts showed a conversion of 100% under conditions of 5 MPa, 150°C for 2 h. The selectivity of the product – p-methylcyclohexylamine – achieved 26.3%. The "bioPd(0)" catalyst was recycled five times without any agglomeration and with highly maintained activity.

It is also well known that aforementioned bacterial species such as *Shewanella alga, Pseudomonas putida* or *Desulfovibrio vulgaris* (Mabbett et al. 2002) may be used to biologically treat contaminated wastewaters by reduction of Cr(VI) – known as carcinogen and mutagen – to Cr(III) – relatively non-toxic and non-carcinogenic form. Nevertheless, some studies showed that "bio-Pd(0)" is more efficient at Cr(VI) reduction than live cells of *D. desulfuricans* or chemically reduced Pd(II), using hydrogen as the electron donor (Mabbett and Macaskie 2002). Pd(0) mediates hemolytic bond cleavage of H_2, with the production of radical H^*, which can then donate its electron to Cr(VI). Continuous-flow studies using *D. vulgaris* Bio-Pd(0) with agar as the immobilization matrix were also investigated (Humphries et al. 2006), showing the effect of Bio-Pd(0) loading, inlet Cr(VI) concentration, and flow rate on the efficiency of Cr(VI) reduction. Mabbett et al. (2005) presents possibility of mixed-metal-bioPd(O) catalysts employing *D. desulfuricans*, Pd(II) and Pt(IV) or industrial waste leachates (contains e.g. Rh, Cu, Fe, Al, Pt). Two flow-through reactor systems were also compared by aforementioned work. Similar experiments were performed by Beauregard et al. (2010) using *Serratia sp.* and formate as the electron donor. Remarkably, Cr species concentrations within the reactor were controlled by spatial mapping using magnetic resonance imaging technique ($Cr(VI)_{(aq)}$ is non-paramagnetic while $Cr(III)_{(aq)}$ is paramagnetic).

Moreover, Chidambaram et al. (2010) published experiments where the electron donor is substituted by fermentation process (fermentatively produce hydrogen in presence of glucose) in bacteria *Clostridium pasteurianum*, which also serves for reduction Pd(II) ions to form PdNPs "bio-Pd(0)" that primarily precipitated on the cell wall and in the cytoplasm. Finally, the most scientifically explored organism in the world, *Escherichia coli* (or its mutants), contribute to the "bioPd(0)" catalyst knowledge. Experiments with three types of hydrogenases encoded by bacterial DNA were performed by Deplanche et al. (2010), based on optimal catalytic activity in Cr(VI)/Cr(III) system.

3.2.2 "BioPd(0)" and "BioPt(0)" as a Fuel Cells Electro-Catalysts

Similar approach to utilization of microorganisms with ability to enzymatically reduce and absorb palladium, platinum and other precious metals was also used for manufacturing of a bio-fuel cell for power production. Since fuel cells have been identified as a future technology to power motor vehicles, generators and portable electronic device, authors recommend overall review papers (Andújar and Segura 2009; Winter and Brodd 2005) for more information and historical context in fuel cells topic.

In work by Yong et al. (2007) Pt(0) and Pd(0) bio-accumulated by *D. desulfuricans* was applied onto carbon paper and tested as anodes in a polymer electrolyte membrane (PEM) fuel cell for power production and compared to commercial fuel cell grade Pt catalyst. A similar strategy is also suggested using yeast-based biomass, immobilized in polyvinyl alcohol cryogels, for the manufacture of fuel cell Pt(0). This is then used to generate electrical energy from renewable sources such as glucose and ethanol (Dimitriadis et al. 2007). Finally, the dried biomass-supported palladium (*Shewanella oneidensis*) was tested as an anode catalyst in a PEM fuel cell for power production. It was shown to have a maximum power generation comparable to the commercial catalyst (Ogi et al. 2011).

Fusion of waste biorefining and cheap nanocatalyst for fuel cells and power generation employing *D. desulfuricans, E. coli* and *C. metallidurans*, carbon paper and proton exchange membrane fuel cell was recently published (Yong et al. 2010). Using an *E. coli* MC4100 strain, a mixed metallic catalyst was manufactured from an industrial processing waste. This mixed-metal biocatalyst gave approximately 50% of the power output compared to commercial or "bioPd" *D. desulfuricans* catalyst. Electrical energy production efficiency of biocatalyst fabricated by aforementioned *E. coli* (parent) strain and its derived mutant strain IC007 (as well as a comparison with *D. desulfuricans*) is further discussed by Orozco et al. (2010).

Another electrocatalysis application with modified electrodes is also mentioned in Sect. 3.5.1.

3.2.3 Catalysis of 4-Nitrophenol Reduction Reactions

Presence of toxic pollutants such as nitro-aromatic compounds in soil and water is a result of incomplete combustion of fossil fuels and their usage as chemical feedstock for synthesis of explosives, pesticides, herbicides, dyes, pharmaceuticals, etc. The headlong and reckless utilization of these pollutants in the past has resulted in wide-ranging environmental pollution. The usage of biosynthesized NPs capable to catalyze degradation of, among other chemicals, nitro-aromatics (and then together with microbial remediation), would be a great contribution to this particularly topical issue (take a look at the following reviews for more information – Kulkarni and Chaudhari 2007; Liotta et al. 2009; Lewis et al. 2004; Guimarães et al. 2010).

As the first report of a NP-bearing biomatrix directly reducing a toxic pollutant 4-nitrophenol (*p*-nitrophenol; 4-NP), Sharma et al. (2007) published experiments of growth of *Sesbania* seedlings in chloroaurate Au(III) solution. This procedure resulted in the accumulation of gold with the formation of stable AuNPs in plant tissues. The catalytic effectiveness of the biomass with Au(0)NPs was documented by the reduction of aqueous 4-nitrophenol (4-NP).

Remarkably, extensive research was performed using 21 traditional Chinese medicinal plant and herb species (Huang et al. 2009). After classification into four categories including leaves, flowers, fruits, and grasses, effectiveness of the protocol in producing AuNPs was demonstrated usually after 30 min of incubation with aqueous $HAuCl_4$. Potential application of the these biogenic AuNPs as catalysts was exhibited in *Cacumen Platycladi* (which exhibited biosynthesis of very small and monodisperse NPs). Catalytic reduction of 4-NP showed excellent catalytic performance compared to the aforementioned study (Sharma et al. 2007).

For instance of AgNPs, Jia et al. (2008) reported the use of a cuttlebone-derived organic matrix (from *Sepia esculenta*) as scaffold and reducer for the formation of AgNPs. The resulting composite was applied to catalyze the reduction of 4-NP. Possibilities of separation from the liquid-phase reaction and reusability in more cycles have been also reported.

3.3 Medical Applications

The application of metallic NPs in the medical and biopharmaceutical fields are both numerous and promising. AgNPs can be utilized as infection protection (wound coatings, bone cements and implants) or prophylactic environment (paints, disinfectants) due to their antibacterial effects (see specialized section [Sect. 3.4] for detail information). Other qualities of silver include regenerative properties (skin regeneration) and wound-healing ability (dressing for burns and ulcers) (Chaloupka et al. 2010). Another review article by Nair and Laurencin (2007) is dedicated to therapeutic applications of AgNPs, particularly to their antibacterial, antiviral, antifungal properties (remarkably, with regard to properties of bulk silver). Additionally, wound healing ability and the potential usage for surface enhanced Raman scattering (SERS) and metal enhanced fluorescence (both powerful tools for detection and identification) is discussed (Table 3).

The authors also recommend two recent review articles dealing with nanogold pharmaceutical applications. The first article by Alanazi et al. (2010) describes properties such as surface plasmon absorption, surface plasmon light scattering, and the biosensing, diagnostic and therapeutic applications of AuNPs. The second review by Patra et al. (2010) concerns the specific application and fabrication of AuNPs for targeted therapy in pancreatic cancer.

Also magnetic NPs appear as very promising for targeted drug delivery or hyperthermia applications. For further reading, see the prominent review by Pankhurst et al. (2003) and its continuance (Pankhurst et al. 2009) or other useful studies by the following: (Cherukuri et al. 2010; Adarsh et al. 2010; Veiseh et al. 2010; Chen and Schluesener 2008). On the other hand, we strongly disagree with the statement and references dedicated to medical applications of biosynthesized metallic NPs in paper by Rodríguez-Carmona and Villaverde (2010). Cited literature is misleading and, excluding one study, has nothing to do with proposed applications.

Despite wide possibilities and theoretical application, only few records can be found in recent literature. As previously mentioned, the obvious drawback of

Table 3 Medical applications

NP	Organism used	Application	Reference
Au	*Cymbopogon citratus* – lemon grass	NIR tunable absorption; coating technology and hyperthermia of cancer cells	Shankar et al. (2005)
Au	*Psidium guajava* leaf	Antidiabetic study, inhibition of PTP1B	Basha et al. (2010)
Ag	Tea leaf extract	Biocompatible	Moulton et al. (2010)
Fe_3O_4	*Magnetospirillum gryphiswaldense*	Drug carrier	Sun et al. (2008)
Ag	*Aspergillus niger*	Wound healing activity	Sundaramoorthi et al. (2009)
Au	Egg shell	Blood serum glucose sensor	Zheng et al. (2011)

biological methods is the need to purify the sample and extract the NPs. This is due to potential pathogens or poisons that might contaminate material and is particularly important for medical applications. Despite these problems, we can still find some useful studies and potential applications.

Biocompatibility is a very important property for all material possibly dealing with medical usage in living organisms. Moulton et al. (2010) reported biosynthesis experiments with tea leaf extract as a reducing and capping agent for AgNP fabrication. Evaluation of mitochondrial function to assess cell viability and membrane integrity in human keratinocytes showed that the AgNPs were nontoxic. This may be attributed to the tea antioxidants on the NP surface. Although this method of synthesis appears to be promising based on the initial *in vitro* studies, they need to be followed by future *in vivo* tests to accurately evaluate the biocompatibility.

Wound healing activity of AgNPs, synthesized extracellularly using *Aspergillus niger*, was evaluated on rat model for case of excision wound and thermal wound (Sundaramoorthi et al. 2009). Researchers illustrated the efficient antimicrobial property of AgNPs and also confirmed the ability of nanosilver to modulate the cytokines involved in wound healing.

Shape of NPs can critically influence their properties, especially the optical. Gold nanotriangles with tunable size were biofabricated by a simple method involving the reduction of aqueous gold ions with an extract of the lemongrass plant (Shankar et al. 2005). Interestingly, absorption in the near-infrared region of the electromagnetic spectrum is "expected to be of application" in hyperthermia of cancer cells and in infra red (IR) – absorbing optical coatings.

As an elegant application of bacterial magnetosomes (BM), Sun et al. (2008) presented experiments leading to employment of BM from *Magnetospirillum gryphiswaldense* as a chemotherapy drug carrier (doxorubicin was loaded isolated and cleaned BMs using a bifunctional crosslinker). As drug efficiency and toxicity may be significantly altered by structural modification, evaluation of the drug effect (drug coupled with BM) was also performed. The antitumor effects of doxorubicin loaded BMs were evaluated by HL60 and EMT-6 carcinoma cells. They were cytotoxic to the cancer cells, inhibited cancer cell proliferation and suppressed the mRNA levels of the significant oncogene *c-myc*. The drug releasing process from BMs was also monitored. The assets of this approach (compare to artificial magnetic particles) may be the ability to carry larger amounts of drug, ease for preparation and dispersion, high stability and more uniformity. Being surrounded with membrane that consists of lipids and proteins, purified and sterilized magnetosomes were not toxic to mouse fibroblasts in vitro (Li et al. 2007). This indicates the advantage of biocompatibility.

Another medical application deals with diabetes, particularly with inhibion of enzyme protein tyrosine phosphatase (PTP), type PTP1B. Disturbance of the normal balance of PTP function has been implicated as the source of several human diseases, including diabetes, cancer, and inflammation (Tonks 2003). The rapid formation of AuNPs with guavanoic acid from the leaf extract of *Psidium guajava* was reported by Basha et al. (2010). These were used in an antidiabetic PTP 1B inhibitory assay and showed significant inhibitory effect with an IC_{50} of 1.14 μg/mL.

Zheng et al. (2011) reported the biofabrication of AuNPs (by means of egg shell membrane) and their application in the glucose biosensor for blood glucose determination (further information in Sect. 3.5.1 and Zheng et al. (2010a)). The biosensor has been also applied to measure the glucose content in human blood serum samples (showing agreement with a standard routine medical spectrophotometric test method).

3.4 Antimicrobial Applications

Although this chapter may have been included in discussion of medical usages (Sect. 3.3), authors have decided to create a new section to address the increasingly large amount of studies being published in this area (Table 4).

Table 4 Antimicrobial applications

NP	Organism used	Application	Reference
Ag	*P. aeruginosa*	Activity vs. G+ and G– bacteria	Suresh et al. (2010)
Ag	*Penicilium sp.*	Activity vs. G+ and G– bacteria	Maliszewska and Puzio (2009)
Ag	*Pleurotus sajor caju*	Activity vs. G+ and G– bacteria	Nithya and Ragunathan (2009)
Ag	*Bipolaris nodulosa*	Activity vs. G+ and G– bacteria	Saha et al. (2010)
Ag	*Streptomyces sp*	Activity vs. G+ and G– bacteria	Shirley et al. (2010)
Ag	*Aspergillus oryzae* var. *viridis*	Activity vs. *S. aureus*	Binupriya et al. (2009)
Au	*Trichoderma viride*	Activity vs. VRSA	Fayaz et al. (2011)
Ag	*T. viride*	Vegetable and fruit preservation	Fayaz et al. (2009)
Ag	*Aspergillus clavatus*	Activity vs. MRSA, MRSE	Saravanan and Nanda (2010)
Ag	*Garcinia mangostana* (Mangosteen) leaf	Activity against *E. coli, S. aureus*	Veerasamy et al. (2011)
Ag	*Candida albicans, E. coli, B. cereus, P. fluorescens*	Activity against *E. coli*	Ul'berg et al. (2010)
Ag	*Phoma glomerata*	Synergy with antibiotics	Birla et al. (2009)
Ag	*Opuntia ficus-indica*	Effect in combination with antibiotics	Gade et al. (2010)
Ag	*T. viridae*	Effect in combination with antibiotics	Fayaz et al. (2010a)
Ag	*Fusarium acuminatum*	Activity vs. G+ and G– bacteria	Ingle et al. (2008)

(continued)

Table 4 (continued)

NP	Organism used	Application	Reference
Ag	Varied microalgae species	Activity vs. G+ and G– bacteria	Merin et al. (2010)
Ag	*Acalypha indica* leaf	Activity vs *E. coli, Vibrio cholerae*	Krishnaraj et al. (2010)
Ag	*Fusarium oxysporum*	Cotton fabrics incorporated with AgNPs	Durán et al. (2007)
Ag	*Fusarium solani*	Cotton fabrics incorporated with AgNPs	El-Rafie et al. (2010)
Ag	*Azadirachta indica* (Neem) leaf	Bactericidal effect in cotton cloth against *E. coli*	Tripathi et al. (2009)
Ag	*Cinnamon zeylanicum* bark	Activity against *E. coli* BL-21	Sathishkumar et al. (2010a)
Ag	*Curcuma longa* tuber	Immobilization on cotton cloth	Sathishkumar et al. (2010a, b)
Ag	*Eucalyptus citriodora, Ficus bengalensis*	Antibacterial activity against *E. coli,* loaded on the cotton fibres	Ravindra et al. (2010)
Ce	*Leptothrix discophora, Pseudomonas putida*	Activity against bacteriophage UZ1	De Gusseme et al. (2010a)
Ag	*Lactobacillus fermentum*	Activity against bacteriophage UZ1	De Gusseme et al. (2010b)
Ag	*Amylomyces rouxii*	Antifungal and antibacterial activity	Musarrat et al. (2010)
Ag	*S. hygroscopicus*	Antifungal and antibacterial activity	Sadhasivam et al. (2010)
Ag	*A. niger*	Antifungal and antibacterial activity	Jaidev and Narasimha (2010)
Ag	*Sesuvium portulacastrum* callus and leaf	Antifungal and antibacterial activity	Nabikhan et al. (2010)
Ag	*Alternaria alternate*	Activity in combination with fluconazol	Gajbhiye et al. (2009)
Au	Genus *Musa* – banana peel	Antifungal and antibacterial activity	Bankar et al. (2010)
Au	*Rhizopus oryzae*	Antifungal and antibacterial activity	Das et al. (2009)
Ag	*Aspergillus clavatus*	Antifungal and antibacterial activity	Verma et al. (2010)
Ag	*Solanum torvum*	Antifungal and antibacterial activity	Govindaraju et al. (2010)

Due to the outbreak of infectious diseases caused by different pathogenic bacteria and fungi and the development of antibiotic or metal resistant strains, there is increasing need to find new antibacterial products. Although different types of nanomaterials like titanium, copper, magnesium or alginate have promising antibacterial properties, Au and AgNP have showed best efficiency against bacteria, viruses and fungi (Rai et al. 2009). The broad-spectrum antimicrobial properties of metallic NPs (mostly silver and gold) encourage their use as disinfectants in purification processes (medicine, water and air), food production, cosmetics, clothing, and numerous household products (Marambio-Jones and Hoek 2010).

In this section, we will illustrate the antibacterial, antiviral and antifungal effects of biosynthesized NPs. This application promises to be very beneficial for both the industrial and medical fields. For further reading on AgNPs utilization and silver mechanism of action, see aforementioned review articles – Marambio-Jones and Hoek (2010), Rai et al. (2009), Nair and Laurencin (2007) or review papers by Sharma et al. (2009), Cho et al. (2005).

3.4.1 Antibacterial Activity

Bacteria have different membrane and cell wall structures. Therefore, we can classify them generally as Gram-negative (G–) or Gram-positive (G+). The key component of the membrane, peptidoglycan, is a decisive factor in the membrane organization. G– bacteria has only a thin peptidoglycan layer (~2–3 nm) between their two membranes, while G+ bacteria lack the outer membrane (is substituted by thick peptidoglycan layer).

Morones et al. (2005) published study regarding possible interactions between AgNPs and G– bacteria. Small NPs disturb the function of the membrane (such as permeability or respiration) by attaching to it's surface and, subsequently, penetrate the cell and cause further damage by interacting with the DNA.

Spectrum of organisms used for biosynthesis of NPs with antibacterial effect varies from bacteria, fungi and alga to leaf, root, bark and tuber extracts of higher plants and trees. As one of the first records, Ingle et al. (2008) reported a mycosynthesis of silver antibacterial NPs with biological activity against different human pathogens including multidrug resistant and highly pathogenic bacteria such as *Staphylococcus aureus, Salmonella typhi, Staphylococcus epidermidis*, and *Escherichia coli*. Similarly, fungal strain *Aspergillus clavatus* was used for extracellular biosynthesis of stable AgNPs with antibacterial activity against methicillin (antibiotics) resistant *Staphylococcus aureus* and *Staphylococcus epidermidis* (Saravanan and Nanda 2010). Antibacterial activity against *Staphylococcus aureus* KCCM 12256 was also observed in case of AgNPs biosynthesized by filamentous mold *Aspergillus oryzae* (Binupriya et al. 2009). Bioreductive synthesis of nanosized Ag particles was performed using live and dead cell filtrates with NP size varying from 5 to 50 nm. Another phytopathogenic fungal specie *Bipolaris nodulosa* can serve as reducing agent for silver nitrate reduction with resulting Ag NPs active against *Bacillus subtilis, Bacillus cereus, Pseudomonas aeruginosa, Proteus vulgaris, Escherichia coli* and *Micrococcus luteus* pathogens (Saha et al. 2010). Ag NPs biosynthesized by gilled mushroom specie Pleurotus sajor-caju (Nithya and Ragunathan 2009) can serve against *Pseudomonas aeruginosa, Escherichia coli* (G–) and *Staphylococcus aureus* (G+). Identical bacterial species were used in a similar study (Maliszewska and Puzio 2009) employing the famous genus of ascomycetous fungi, *Penicillium sp*, with major importance in the natural environment as well as food and drug production.

Fungal plant pathogen *Phoma glomerata* was employed in synthesis of Ag NP and, together with antibiotics, proved effective against *Escherichia coli, Staphylococcus aureus, Pseudomonas aeruginosa* (Birla et al. 2009). Synthesized NPs showed comprehensive bactericidal activity against the aforementioned G– and G+ bacterial species and enhanced the antimicrobial activity of used antibiotics (ampicillin, gentamycin, streptomycin and vancomycin). Interestingly by using gold, a mold species *Trichoderma viride* (widely used as bio-fungicide) was used to biosynthesize vancomycin bound NPs and exhibited activity against vancomycin resistant *Staphylococcus aureus*, vancomycin sensitive *S. aureus* and *E. coli* (Fayaz et al. 2011). Additionally, all experiments were performed as comparison between vancomycin bound AuNPs and vancomycin as such.

Streptomyces sp. bacterially derived AgNPs were reported (Shirley et al. 2010) as biologically active against seven species of both G+ and G– bacteria (*Staphylococcus aureus, S. epidermidis, E. coli, S. typhi, Pseudomonas aeruginosa, Klebsiella pneumonia, Proteus vulgaris)*. Also, already known metal reducing G– bacteria *Shewanella oneidensis* was used for silver nanocrystallites biofabrication (Suresh et al. 2010). Bacterial toxicity assessments showed that prepared biogenic Ag NPs have a greater bactericidal activity on *E. coli*, *S. oneidensis*, and *B. subtilis* strains than chemically synthesized colloidal-AgNPs. Ul'berg et al. (2010) reported usage of four bacterial species leading to bio-AgNPs active against *E. coli*.

Photosynthetic organisms are in antimicrobial biofabrication represented by algae, plants and trees. Four species of marine microalgae (normal and microwave irritated) were used in comparison and assessment of antimicrobial properties of resulting AgNPs against human pathogens *Escherichia coli, Klebsiella sp, Proteus vulgaricus, Pseudomonas aeruginosa* (Merin et al. 2010). Also higher plants can take a place in NP synthesis. Leafs extract of Garcinia mangostana (Mangosteen) were employed in AgNPs biofabrication and the antibacterial assays were done on human pathogenic *E. coli* and *Staphylococcus aureus* by standard disc diffusion method with considerable results (Veerasamy et al. 2011). Krishnaraj et al. (2010) investigated biosynthesis of AgNPs and its activity on water borne bacterial pathogens (*E. coli* and *Vibrio cholerae*). During the antibacterial experiments, alteration in membrane permeability and respiration of the AgNP treated bacterial cells were recorded.

Gade et al. (2010) reported *Opuntia ficus-indica* mediated synthesis of colloidal AgNPs and their antimicrobial assessment in combination with commercially available antibiotics (the maximum activity was demonstrated by ampicillin followed by streptomycin and vancomycin). Similarly, the extracellular biosynthesis of AgNPs from silver nitrate solution by fungus *Trichoderma viride* is reported (Fayaz et al. 2010a). Increasing of their antimicrobial activities with various antibiotics against gram-positive and gram-negative bacteria was described. Although antibacterial activities of ampicillin, kanamycin, erythromycin, and chloramphenicol were increased in the presence of AgNPs against test strains, ampicillin showed the highest enhancing effect.

Also food-storage possibilities of biofabricated NPs were examined using the same organism (Fayaz et al. 2009). Antibacterial activities of AgNP-incorporated sodium alginate films were tested against *E. coli* ATCC 8739 and *S. aureus* ATCC 6538 strains – disk method exhibited antibacterial activity against both G+ and G– bacteria. Antimicrobial coating was applied to carrot and pear surface and conservation impact was examined compare to untreated samples.

3.4.2 Antifungal and Combined Activity

Although there are reports of biosynthesized NPs with antifungal activity, they usually exhibit it only in combination with antibacterial activity. It is in this context, then, that they will be mentioned in the following text.

Mycelia-free water extracts from *Amylomyces rouxii* facilitated the production of stable, monodispersed and spherical AgNPs (size range of 5–27 nm). Biosynthesized AgNPs exhibited antimicrobial activity against bacterial (*Shigella dysenteriae type I, Staphylococcus aureus, Citrobacter sp., E. coli, Pseudomonas aeruginosa, Bacillus subtilis*) as well as fungal (*Candida albicans, Fusarium oxysporum*) species. Biological reduction of aqueous silver ions by extracellular components of *Streptomyces hygroscopicus* (Sadhasivam et al. 2010) resulted in AgNPs which significantly inhibited the growth of medically important pathogenic gram-positive bacteria (*Bacillus subtilis, Enterococcus faecalis*), gram-negative bacteria (*Escherichia coli* and *Salmonella typhimurium*) and yeast (*Candida albicans*). The colloidal AgNPs biosynthesized with filtrate from *Aspergillus niger* inhibited the growth of the fungus itself (seeded in the nutrient agar plate). Potential antifungal activity was due to inactivation of sulfhydryl groups in the fungal cell wall and disruption of membrane bound enzymes and lipids which causes the cell lysis (Jaidev and Narasimha 2010). Antibacterial activity against both G+ (*Staphylococcus sp., Bacillus sp.*) and G– (*E. coli*) bacterial species was observed. Similar results were obtained employing *Aspergillus clavatus* (against *Candida albicans, Pseudomonas fluorescens* and *Escherichia coli*) by Verma et al. (2010). Also, Govindaraju et al. (2010) published study utilizing *Solanum torvum* as mediator for biosynthesis of AgNPs eliciting antibacterial activity against pathogenic bacteria *Pseudomonas aeruginosa*, *Staphylococcus aureus*, pathogenic fungi *Aspergillus flavus* and *Aspergillus niger*.

Interestingly, extracts from tissue culture-derived callus and leaf of the saltmarsh plant *Sesuvium portulacastrum* were used for AgNPs growth (Nabikhan et al. 2010). The antibacterial activity (*Pseudomonas aeruginosa, Staphylococcus aureus, Listeria monocytogenes, Micrococcus luteus, Klebsiella pneumoniae*) was more distinct compared to antifungal (*Alternaria alternata, Penicillium italicum, Fusarium equisetii, Candida albicans*) activity. Moreover, antimicrobial activity was enhanced when polyvinyl alcohol was added as a stabilizing agent (in comparison with samples prepared with distilled water).

Study of antifungal properties against large group of fungal species (*Phoma glomerata, Phoma herbarum, Fusarium semitectum, Trichoderma sp., Candida albicans*) in combination with triazole antifungal drug fluconazol was published by Gajbhiye et al. (2009). Biosynthesized AgNPs (by phyto-pathogenic fungus *Alternaria alternata*) enhanced antifungal activity of fluconazole against the test fungi (showing maximum inhibition against *C. albicans*, followed by *P. glomerata* and *Trichoderma sp.*). However, no significant enhancement was found against *P. herbarum* and *F. semitectum*.

Last but not least, formation of AuNPs with antifungal activity was also described. As a contribution to the water hygiene and treatment management, Das et al. (2009) describes simple on step procedure to obtain potable water free of pathogens and pesticides. AuNPs (10 nm average) were produced on the surface of fungus *Rhizopus oryzae* and showed antimicrobial activity against several G– and G+ pathogenic bacteria as well as the yeasts *Saccharomyces cerevisiae* and *Candida albicans*. Simulated contaminated water containing organophosphate pesticides (malathion, parathion, chloropyrifos, and dimethoate) along with E. coli was treated with gold bionanoconjugate and successful removal of contaminants was monitored by means standard disk method (*E. coli*) and gas chromatography analysis (pesticides). AuNPs with antibacterial (*Citrobacter kosari, Escherichia coli, Proteus valgaris, Pseudomonas aeruginosa, Enterobacter aerogenes and Klebsiella sp.*) and antifungal (*Candida albicans*) activity were also obtained employing banana peel extract (Bankar et al. 2010).

3.4.3 Antiviral Activity

Although humankind is waging an ongoing war against viruses in such wide ranging fields as medicine and agriculture, there has been only one recorded study about antiviral activity of biosynthesized NPs. Although not of the biosynthesized nature, the post-infected anti-HIV-1(BaL) activities of AgNPs (prepared chemically, 10 nm) toward Hut/CCR5 cells (cells derived from a human T cell line, which express the chemokine receptor CCR5) were evaluated by Sun et al. (2005). When compared to the control sample, AgNPs showed dose-dependent anti-retrovirus activities and showed high activity (at 50 mM – 98%) in inhibiting HIV-1 replication (for comparison, the AuNPs exhibited relatively low anti-HIV-1 activities 6–20%). This is an interesting example of the use of strictly chemically fabricated NPs. Remarkably, De Gusseme et al. (2010a) published study of virus removal by biogenic rare earth element cerium (produced by addition of aqueous Ce(III) to actively growing cultures of either freshwater manganese-oxidizing bacteria *Leptothrix discophora* or *Pseudomonas putida*). A model organisms for the antiviral assay was bacteriophage UZ1 (bacteriophage specific for common pathogenic bacterium *Enterobacter aerogenes*).

In study by the same research group (De Gusseme et al. 2010b), *Lactobacillus fermentum* served as a reducing agent and carrier matrix for AgNPs. The antiviral qualities of biogenic AgNPs was confirmed in water containing aforementioned bacteriophage UZ1 (bacteriophage specific for common pathogenic bacterium *Enterobacter aerogenes*) and murine norovirus 1 (a model organism for human noroviruses). For continuous disinfection capability in water environment, the biogenic material was applied to electropositive filter (NanoCeram) and exhibited higher antiviral activity in comparison with the results obtained with the original filter.

3.4.4 Antibacterial Fabrics and Cloth

Another interesting utilization for biosynthesized NPs is immobilization on cotton cloth or cotton fibers. This approach demonstrates the possible use of such cloth in disinfection or sterilization.

As the first record, Durán et al. (2007) reported the extracellular production of AgNPs by fungus *Fusarium oxysporum* and antimicrobial effect of the NPs incorporated in cotton fabrics against *Staphylococcus aureus*. Moreover, effluent from impregnated fabrics (after several washing cycles) was treated with the suspension of *Chromobacterium violaceum*, metal binding bacteria, to reabsorb released NPs. Using a similar procedure and organism, fungi *Fusarium solani*, El-Rafie et al. (2010) prepared AgNPs and applied them to cotton fabrics with and without binder. The bleached cotton fabrics were padded through silver colloidal bath and squeezed with laboratory padder. Following the incorporation of a binder, after 20 washing cycles the material still exhibited effectiveness in antibacterial activity against *Staphylococcus aureus* and *E. coli*.

Azadirachta indica (Neem) is genus from mahogany family *Meliaceae*. Tripathi et al. (2009), studied the biosynthetic production of AgNPs by aqueous extract of Neem leaves and their immobilization on cotton cloth. Subsequently, utilizing standard disk method (including effect of consecutive washing in distilled water), their bactericidal effect against *E. coli* was observed. NP incorporation into cotton disks was performed by three approaches: (a) centrifuging the disks with liquid extract containing biosynthesized NPs; (b) *in-situ* coating process during synthesis, and (c) coating with dried and purified NPs. Antibacterial effect against E. coli BL-21 strain was also tested with AgNPs prepared by means of phyto-reductive extract and powder of *Cinnamon zeylanicum* (Sathishkumar et al. 2009) and *Curcuma longa* tuber (Sathishkumar et al. 2010b). The second work presents immobilization of AgNPs on cotton cloth. NPs were resuspended in water or polyvinylidene fluoride (PVDF) and sprayed over the pre-sterilized white cotton cloth in aseptic condition. PVDF immobilized cloth exhibited less antibacterial activity. However, consecutive washing drastically reduced antibacterial effectiveness of AgNPs immobilized in sterile water.

Topic of antibacterial NP containing cloth can be very promising for biofabrication approach and whole of nanotechnology. For further reading see following reviews: Dastjerdi and Montazer (2010), Perelshtein et al. (2008), Simoncic and Tomsic (2010).

3.5 *Electrochemical and Sensing Applications*

Metallic NPs are at the centre of intense research because an understanding of their surface chemistry might play a key role in effective utilization of technologies such as nanosensor, biosensor, electrocatalysis, nanodevice and nanoelectrochemistry (Guo and Wang 2007). From the view of electroanalytical chemistry, more attention has been paid to AuNPs because of their good biological compatibility, excellent conducting capability and high surface-to-volume ratio (Daniel and Astruc 2004). Usage of AuNPs in electrochemical interfaces has contributed to new vigor in electrochemistry. Development of new techniques and different electrode modified strategies may potentially enhance analytical selectivity and sensitivity of commonly used facilities (Xiong et al. 2007; Nguyen et al. 2011).

While not yet containing any direct applications, there are several studies which are potentially remarkable for their discussion of electrochemical properties. Biosynthesis of ferroelectric BaTiO3 NPs in assistance of *Lactobacillus sp.* was reported by Jha and Prasad (2010). After modification with PVDF, resulting nanocomposite exhibited enhancement in dielectric properties. AuNPs biosynthesized by alkalothermophilic actinomycetes *Thermomonospora curvata*, *Thermomonospora fusca*, and *Thermomonospora chromogena* and stabilized by cross-linker glutaraldehyde have potential usage as biosensor enhancer (Torres-Chavolla et al. 2010). Shilov et al. (2010) investigated electro-physical characteristics (cell ζ-potential, surface conductivity, electrophoretic mobility, dispersion of cell conductivity) of yeast cell with silver precipitate.

3.5.1 Sensors

Zheng et al. (2010b) reported biosynthesis of Au–Ag alloy NPs by yeast cells and their application to electrochemical vanillin sensing. Sensitive vanillin sensor based on glassy carbon electrode (GCE), modified by Au–Ag alloy NPs, was able to enhance the electrochemical response of vanillin. Electrochemical investigations confirmed a linear increase of the vanillin oxidation peak current at the sensor with its concentration in the range of 0.2–50 μM (detection limit 40 nm). Constructed sensor was successfully applied to the determination of vanillin in samples of vanilla bean and vanilla tea. This approach suggests possible replacement of commonly used methods in vanillin monitoring system (chromatography, capillary electrophoresis etc.).

Interestingly, Zheng et al. (2010a) also published green biosynthesis method for AuNPs based on usage of natural biomaterial, eggshell membrane ("fresh eggs were bought from a local supermarket in Hong Kong"). AuNPs on the eggshell surface were used to immobilize glucose oxidase (by cross-linking method with glutaraldehyde – Pingarrón et al. 2008) on the GCE for detection of glucose in solution. Enzyme activity of glucose oxidase was enhanced by the presence of highly conductive AuNPs. Constructed biosensor showed linear response to glucose concentration ranged from 20 μM to 0.80 mM (with a detection limit of 17 μM) and has been successfully applied to detect the glucose level in glucose injections. Glucose sensor used for blood serum based on eggshell membrane and AuNPs (Zheng et al. 2011) is mentioned in medicinal applications in medical section (Sect. 3.3).

Even though it pertains to semi-metals, we will also mention a study dealing with synthesis of semiconductor selenium NPs employing cells of bacteria *Bacillus subtilis*. Two kinds (spherical, 1D-trigonal) of Se nanomaterial crystals with good adhesive ability and biocompatibility were employed as enhancing materials for hydrogen peroxide (H_2O_2) horseradish peroxidase biosensor, with the detection limit 8×10^{-8} M (for H_2O_2 concentration). Different types of Se crystals had no significant difference in sensor usage. Due to the obtained results, selenium nanomaterials GCE can be a promising instrument for applications dealing with the detection of H_2O_2 in food, pharmaceutical, clinical, industrial and environmental analyses (Wang et al. 2010).

3.5.2 Electrochemical Applications and Properties

Du et al. (2007) demonstrated in communication the bioreduction of aqueous Au(III) ions by the *E. coli* DH5α bacterial strain. AuNPs bound to the surface of the bacteria were used for application in direct electron transfer of protein hemoglobin (glass carbon electrode (GCE) coated by protein layer as well as the AuNPs biocomposite). Cyclic voltametry experiments were reported for different electrodes at scan rates of 0.1 V/s in pH 7.0 phosphate buffer solution. Although there are no obvious redox peaks at the blank electrodes (GCE, *E. coli*-GCE, hemoglobin-*E. coli*-GCE and AuNPs-*E. coli*-GCE), a pair of redox peaks (with formal potential of −0.325 V vs. Ag/AgCl reference electrode) was observed at AuNPs-hemoglobin-*E. coli*-GCE. These results proved the electron transfer between hemoglobin and GCE provided by means of the AuNPs modified electrode.

Similarly, extracellular synthesis of AuNPs using plant *Scutellaria barbata* as the reducing agent was observed (Wang et al. 2009). The obtained AuNPs were modified on the GCE and enhanced the electronic transmission rate between the electrode and the 4-NP.

Remarkably, biosynthesized CdSNPs were also used for construction of an ideal diode (Kowshik et al. 2002). Semiconducting NPs were biofabricated by *Schizosaccharomyces pombe* and were confirmed to have a Wurtzite ($Cd_{16}S_{20}$)-type structure. Diode was fabricated by means of tin-doped indium oxide coated glass substrate. This structure was spin coated by thin film of poly-phenylene vinylene (p-type material) and with washed *S. pombe* CdSNPs (n-type material) respectively. Silver contacts were also added. Obtained diode operated at low voltage and had forward current value, which makes the structure suitable for use as an ideal diode.

3.6 Optical, Bio-Imaging, Bio-Labeling Applications

The metallic nanocrystals are held at the center due to their photo-induced nonlinear optical properties. In particular, the unique optical properties associated with NPs and their composite materials include a high- or low-refractive index, high transparency, novel photoluminescence properties, photonic crystal, and plasmon resonance (Iskandar 2009). In nanoregime (hundreds to thousands of atoms) optical and electro-optical properties of materials can be tuned by varying the physical size of the crystal, leading to new phenomena, such as surface plasmon resonance in Au and AgNPs and the size dependent band gap of semiconductor (Talapin et al. 2009). Scientists are therefore able to tailor the electronic structure and properties without introducing any changes in the sample chemical composition. Methods for NP synthesis that allow control of NP characteristics, including size distribution, morphology, crystallinity, purity, and composition are of particular note (Iskandar 2009). Several methods for the synthesis of NPs and their composite materials have been reported previously. However, to be feasible for utilization in industrial scale, the process needs to be simple, low-cost, and able to operate continuously with a high production rate. Therefore, utilization of the biosynthesis approach may be beneficial (Table 5).

Non-pathogenic, fast-growing fungus *Trichoderma viride* (habited in dead organic materials) was used to biosynthesis small (2–4 nm), highly dispersed AgNPs (Fayaz et al. 2010b). Interestingly, photoluminescence measurements

Table 5 Electrochemical and sensing applications

NP	Organism used	Application	Reference
Ag	*Trichoderma viride*	Blue orange emission – photoluminiscence	Fayaz et al. (2010b)
Ag	*Parthenium hysterophorus*	Photoluminescence	Sarkar et al. (2010)
Ag	*Coriandrum sativum* leaf	Reverse sat. absorption, optical limiting	Sathyavathi et al. (2010)
CdTe	*Saccharomyces cerevisiae*	CdTe QDs for biolabeling and biosensing	Bao et al. (2010a)
CdTe	*Escherichia coli*	CdTe QDs for biolabeling and biosensing	Bao et al. (2010b)
Au-Ag	*Saccharomyces cerevisiae*	Vanillin sensor	Zheng et al. (2010a)
Au	Eggshell	Glucose sensor	Zheng et al. (2010b)
Se	*Bacillus subtilis*	H_2O_2 sensor	Wang et al. (2010)
CdS	*Schizosaccharomyces pombe*	Construction of ideal diode	Kowshik et al. (2002)
Au	*Escherichi coli*	Direct electrochemistry of hemoglobin	Du et al. (2007)
Au	*Scutellaria barbata*	Direct electrochemistry of 4-NP	Wang et al. (2009)

showed an emission in the range of 320–520 nm (fall in blue-orange region). This fact indicates such a method of AgNPs preparation as suitable for future bio-imaging and labeling application. Sarkar et al. (2010) published a similar study utilizing the flowering plant species *Parthenium hysterophorus* and AgNPs.

A simple, fast, and economical biological procedure using *Coriandrum sativum* leaf extract to synthesize AgNPs was reported by Sathyavathi et al. (2010). These NPs have an important application in nonlinear optics. Nonlinear refraction and absorption coefficients were measured using Z-scan technique with laser pulses. AgNPs were found to exhibit strong reverse saturable absorption, which has been identified as the main mechanism responsible for optical limiting. For more detailed information about optical limiting, see study by the same research group (Porel et al. 2007).

Extracellular synthesis cadmium telluride CdTe quantum dots (QDs) with tunable fluorescence emission employing *Saccharomyces cerevisiae* cells was published by Bao et al. (2010a). Fabricated CdTe QDs with uniform size (2–3.6 nm) were protein-capped, which makes them highly soluble in water. A similar approach (Bao et al. 2010b) was used for CdTe QDs by means of *E. coli*. Size-tunable optical properties were confirmed in both cases by ultraviolet–visible, photoluminescence, X-ray diffraction and transmission electron microscopy, with fluorescence emission from 488 to 551 nm. Moreover, QDs functionalized with folic acid and were used to image cultured cervical cancer cells *in vitro* (Bao et al. 2010b). The biosynthesized QDs may therefore have potential in broad bio-imaging and bio-labeling applications.

Shankar et al. (2005) published formation of gold nanotriangles with interesting absorption in the near-infrared region (see Sect. 3.3)

3.7 *Further Applications and Properties*

Throughout our study, we encountered various works that, while relevant, did not fit neatly in to any of aforementioned sections. Therefore, we chose several studies and publications with interesting applications or properties of biofabricated NPs for additional consideration (Table 6).

Table 6 Further applications and properties

NP	Organism used	Application	Reference
CuAlO2	*Humicola sp.*	Difficult-to-synthesize nanoparticles	Ahmad et al. (2007)
Au	*Emblica Officinalis*	Phase transfer, transmetallation	Ankamwar et al. (2005a)
Au	*Tamarindus indica* leaf	Nanotriangles, vapor sensing	Ankamwar et al. (2005b)
Au-Ag-Cu	*Brassica juncea* seed	Alloys	Haverkamp et al. (2007)
PbS, ZnS	Different *cocci* and *bacillus* cells	Hollow nanostructures; light harvesting and photocatalytic properties	Zhou et al. (2009)

Fabrication of otherwise difficult-to-synthesize NPs is undoubtedly an asset of bio methods. Ahmad et al. (2007) reported such a synthesis – multifunctional $CuAlO_2$ NPs (fungus based).

Phase transfer of biosynthesizes Au and AgNPs was published in a paper by Ankamwar et al. (2005a). NPs were fabricated using *Emblica officinalis* (amla, Indian Gooseberry) fruit extract. The experiments contained also their subsequent phase transfer to an organic solution (methanol) and the transmetallation reaction of hydrophobized AgNPs with hydrophobized chloroaurate ions (resulted in AuNPs).

Different shapes of resulting NPs also play a role in determining their properties. Tamarind leaf extract served as the reducing agent for the synthesis of gold nanotriangles ranged 20–40 nm in thickness (Ankamwar et al. 2005b). The effect of different organic solvent vapors like methanol, benzene and acetone on the conductivity of AuNPs triangles was investigated. Current-voltage characteristics were determined in presence of aforementioned organic solvent vapors and observation suggests possible application as chemical sensors. Biofabrication of anisotropic gold nanotriangles was also reported by Verma et al. (2010). On the other hand, Xie et al. (2007) performed experiments resulting in silver nanoplates employing green algal species *Chlorella vulgaris.*

Fabrication of alloys without sophisticated equipment or appropriately high temperature (Haverkamp et al. 2007) demonstrated the ability to synthesis the Au–Ag–Cu class of alloy by means of the *Brassica juncea* seed. Similarly, studies dealing with bimetallic Au-Ag alloy biosynthesis processes employing fungi were also published (Senapati et al. 2005; Sawle et al. 2008).

Morph-biotemplates, hollowed, former-bacterial cells, coated by chalcogenide NPs also represent an innovative approach to material science (Zhou et al. 2009). Moreover, these PbS and ZnS structures, prepared by the sonochemical method in presence of different *cocci* and *bacillus* (rod) templates, exhibited light harvesting and photocatalytic properties. Hollow structures possess superior photocatalytic activity to their solid counterparts during photocatalytic degradation of acid fuchsine and can be used as electromagnetic wave absorbers, ultraviolet shielding materials, photocatalysts or solar cells.

4 Summary, Conclusions, and Outlook

In this review, recent advances in the application of biosynthesized metallic NPs have been addressed. Applications have been categorized into a wide spectrum of sections such as catalysis, medicine, disinfection, sensors etc. When possible, comparison with the common chemical and physical synthesis approaches was investigated and discussed. In addition, the introduction of the relatively novel biosynthesized metallic NPs phenomenon into medicine, electrochemistry, optics and material science can reinforce their unique functions and properties, resulting in new methods and strategies for applied research and industrial utilization.

Although research in the field has been dramatically increasing since the beginning of the decade, biofabrication methods and approaches are still on the horizon of genuine applied research. This rapidly spreading, interdisciplinary field will require great research efforts from biochemists, physicists, biologists and materials scientists. A greater understanding of particular application mechanisms (how such technology will be utilized in the real world) will lead to mark crucial assets and drawbacks of bio-methods. In order to fully exploit biosynthesized metallic NPs, better stabilization techniques, more efficient material definition, better isolation processes and more effective immobilization possibilities should be designed and developed. To address these issues, the novel methods (organisms, substrates) should be incorporated and stabilization and capping mechanisms must be further investigated. Last but not least, possibilities of varied biosynthesized shapes (e.g. disk, plate, sponge, star, flake, urchin, prisms, wires and rods) are also quite promising.

Based on this study, we anticipate further development of biosynthesis methods and their utilization in many pertinent and advantageous real-world applications. In combination with other NP preparation methods, further investigation of processes and an expansion of potential applications will result in the deeper development of bionanotechnology.

Acknowledgments This work was supported by Ministry of Education, Youth and Sports of Czech Republic (research grant MSM 6198910016).

References

Adarsh S, et al. (2010) Synthesis and applications of magnetic nanoparticles for biorecognition and point of care medical diagnostics. Nanotechnology 21 (44):442001

Ahmad A, Jagadale T, Dhas V, Khan S, Patil S, Pasricha R, Ravi V, Ogale S (2007) Fungus-Based Synthesis of Chemically Difficult-To-Synthesize Multifunctional Nanoparticles of CuAlO2. Advanced Materials 19 (20):3295–3299. doi:10.1002/adma.200602605

Ahmad A, Mukherjee P, Senapati S, Mandal D, Khan MI, Kumar R, Sastry M (2003) Extracellular biosynthesis of silver nanoparticles using the fungus Fusarium oxysporum. Colloids and Surfaces B: Biointerfaces 28 (4):313–318. doi:10.1016/s0927-7765(02)00174-1

Alanazi FK, Radwan AA, Alsarra IA (2010) Biopharmaceutical applications of nanogold. Saudi Pharmaceutical Journal 18 (4):179–193. doi:10.1016/j.jsps.2010.07.002

Andújar JM, Segura F (2009) Fuel cells: History and updating. A walk along two centuries. Renewable and Sustainable Energy Reviews 13 (9):2309–2322. doi:10.1016/j.rser.2009.03.015

Ankamwar B, Damle C, Ahmad A, Sastry M (2005a) Biosynthesis of Gold and Silver Nanoparticles Using Emblica Officinalis Fruit Extract, Their Phase Transfer and Transmetallation in an Organic Solution. Journal of Nanoscience and Nanotechnology 5:1665–1671. doi:10.1166/jnn.2005.184

Ankamwar B, Chaudhary M, Sastry M (2005b) Gold Nanotriangles Biologically Synthesized using Tamarind Leaf Extract and Potential Application in Vapor Sensing. Synthesis and Reactivity in Inorganic, Metal-Organic, and Nano-Metal Chemistry 35 (1):19–26

Arief VO, Trilestari K, Sunarso J, Indraswati N, Ismadji S (2008) Recent Progress on Biosorption of Heavy Metals from Liquids Using Low Cost Biosorbents: Characterization, Biosorption Parameters and Mechanism Studies. Clean-Soil Air Water 36 (12):937–962. doi:10.1002/clen.200800167

Bankar A, Joshi B, Ravi Kumar A, Zinjarde S (2010) Banana peel extract mediated synthesis of gold nanoparticles. Colloids and Surfaces B: Biointerfaces 80 (1):45–50. doi:10.1016/j.colsurfb.2010.05.029

Bao H, Hao N, Yang Y, Zhao D (2010a) Biosynthesis of biocompatible cadmium telluride quantum dots using yeast cells. Nano Research 3 (7):481–489. doi:10.1007/s12274-010-0008-6

Bao H, Lu Z, Cui X, Qiao Y, Guo J, Anderson JM, Li CM (2010b) Extracellular microbial synthesis of biocompatible CdTe quantum dots. Acta Biomaterialia 6 (9):3534–3541. doi:10.1016/j.actbio.2010.03.030

Barakat MA (2010) New trends in removing heavy metals from industrial wastewater. Arabian Journal of Chemistry In Press, Corrected Proof. doi:10.1016/j.arabjc.2010.07.019

Basha KS, Govindaraju K, Manikandan R, Ahn JS, Bae EY, Singaravelu G (2010) Phytochemical mediated gold nanoparticles and their PTP 1B inhibitory activity. Colloids and Surfaces B: Biointerfaces 75 (2):405–409. doi:10.1016/j.colsurfb.2009.09.008

Baxter-Plant VS, Mikheenko IP, Macaskie LE (2003) Sulphate-reducing bacteria, palladium and the reductive dehalogenation of chlorinated aromatic compounds. Biodegradation 14 (2):83–90. doi:10.1023/a:1024084611555

Baxter-Plant VS, Mikheenko IP, Robson M, Harrad SJ, Macaskie LE (2004) Dehalogenation of chlorinated aromatic compounds using a hybrid bioinorganic catalyst on cells of <i>Desulfovibrio desulfuricans</i>. Biotechnology Letters 26 (24):1885–1890. doi:10.1007/s10529-004-6039-x

Beauregard DA, Yong P, Macaskie LE, Johns ML (2010) Using non-invasive magnetic resonance imaging (MRI) to assess the reduction of Cr(VI) using a biofilm–palladium catalyst. Biotechnology and Bioengineering 107 (1):11–20. doi:10.1002/bit.22791

Binupriya AR, Sathishkumar M, Yun S-I (2009) Myco-crystallization of Silver Ions to Nanosized Particles by Live and Dead Cell Filtrates of Aspergillus oryzae var. viridis and Its Bactericidal Activity toward Staphylococcus aureus KCCM 12256. Industrial & Engineering Chemistry Research 49 (2):852–858. doi:10.1021/ie9014183

Birla SS, Tiwari VV, Gade AK, Ingle AP, Yadav AP, Rai MK (2009) Fabrication of silver nanoparticles by Phoma glomerata and its combined effect against Escherichia coli, Pseudomonas aeruginosa and Staphylococcus aureus. Letters in Applied Microbiology 48 (2):173–179. doi:10.1111/j.1472-765X.2008.02510.x

Bonet F, Tekaia-Elhsissen K, Sarathy KV (2000) Study of interaction of ethylene glycol/PVP phase on noble metal powders prepared by polyol process. Bulletin of Materials Science 23 (3):165–168

Bunge M, Søbjerg LS, Rotaru A-E, Gauthier D, Lindhardt AT, Hause G, Finster K, Kingshott P, Skrydstrup T, Meyer RL (2010) Formation of palladium(0) nanoparticles at microbial surfaces. Biotechnology and Bioengineering 107 (2):206–215. doi:10.1002/bit.22801

Creamer NJ, Deplanche K, Snape TJ, Mikheenko IP, Yong P, Samyahumbi D, Wood J, Pollmann K, Selenska-Pobell S, Macaskie LE (2008) A biogenic catalyst for hydrogenation, reduction and selective dehalogenation in non-aqueous solvents. Hydrometallurgy 94 (1–4):138–143. doi:10.1016/j.hydromet.2008.05.029

Creamer NJ, Mikheenko IP, Yong P, Deplanche K, Sanyahumbi D, Wood J, Pollmann K, Merroun M, Selenska-Pobell S, Macaskie LE (2007) Novel supported Pd hydrogenation bionanocatalyst for hybrid homogeneous/heterogeneous catalysis. Catalysis Today 128 (1–2):80–87. doi:10.1016/j.cattod.2007.04.014

Daniel M-C, Astruc D (2004) Gold Nanoparticles: Assembly, Supramolecular Chemistry, Quantum-Size-Related Properties, and Applications toward Biology, Catalysis, and Nanotechnology. Chemical Reviews 104 (1):293–346. doi:10.1021/cr030698+

Das D, Das N, Mathew L (2010) Kinetics, equilibrium and thermodynamic studies on biosorption of Ag(I) from aqueous solution by macrofungus Pleurotus platypus. J Hazard Mater 184 (1–3):765–774. doi:10.1016/j.jhazmat.2010.08.105

Das N (2010) Recovery of precious metals through biosorption – A review. Hydrometallurgy 103 (1–4):180–189. doi:10.1016/j.hydromet.2010.03.016

Das SK, Das AR, Guha AK (2009) Gold Nanoparticles: Microbial Synthesis and Application in Water Hygiene Management. Langmuir 25 (14):8192–8199. doi:10.1021/la900585p

Dastjerdi R, Montazer M (2010) A review on the application of inorganic nano-structured materials in the modification of textiles: Focus on anti-microbial properties. Colloids and Surfaces B: Biointerfaces 79 (1):5–18. doi:10.1016/j.colsurfb.2010.03.029
De Gusseme B, Du Laing G, Hennebel T, Renard P, Chidambaram D, Fitts JP, Bruneel E, Van Driessche I, Verbeken K, Boon N, Verstraete W (2010a) Virus Removal by Biogenic Cerium. Environmental Science & Technology 44 (16):6350–6356. doi:10.1021/es100100p
De Gusseme B, Sintubin L, Baert L, Thibo E, Hennebel T, Vermeulen G, Uyttendaele M, Verstraete W, Boon N (2010b) Biogenic Silver for Disinfection of Water Contaminated with Viruses. Applied and Environmental Microbiology 76 (4):1082–1087. doi:10.1128/aem.02433-09
De Stefano M, De Stefano L, Congestri R (2008) Functional morphology of micro- and nanostructures in two distinct diatom frustules. Superlattices and Microstructures 46 (1–2):64–68. doi:10.1016/j.spmi.2008.12.007
De Windt W, Aelterman P, Verstraete W (2005) Bioreductive deposition of palladium (0) nanoparticles on Shewanella oneidensis with catalytic activity towards reductive dechlorination of polychlorinated biphenyls. Environmental Microbiology 7 (3):314–325. doi:10.1111/j.1462-2920.2005.00696.x
De Windt W, Boon N, Van den Bulcke J, Rubberecht L, Prata F, Mast J, Hennebel T, Verstraete W (2006) Biological control of the size and reactivity of catalytic Pd(0) produced by Shewanella oneidensis. Antonie van Leeuwenhoek 90 (4):377–389. doi:10.1007/s10482-006-9088-4
Deplanche K, Caldelari I, Mikheenko IP, Sargent F, Macaskie LE (2010) Involvement of hydrogenases in the formation of highly catalytic Pd(0) nanoparticles by bioreduction of Pd(II) using Escherichia coli mutant strains. Microbiology 156 (9):2630–2640. doi:10.1099/mic.0.036681-0
Deplanche K, Snape TJ, Hazrati S, Harrad S, Macaskie LE (2009) Versatility of a new bioinorganic catalyst: Palladized cells of Desulfovibrio desulfuricans and application to dehalogenation of flame retardant materials. Environmental Technology 30 (7):681–692
Dimitriadis S, Nomikou N, McHale AP (2007) Pt-based electro-catalytic materials derived from biosorption processes and their exploitation in fuel cell technology. Biotechnology Letters 29 (4):545–551. doi:10.1007/s10529-006-9289-y
Du L, Jiang H, Liu X, Wang E (2007) Biosynthesis of gold nanoparticles assisted by Escherichia coli DH5[alpha] and its application on direct electrochemistry of hemoglobin. Electrochemistry Communications 9 (5):1165–1170. doi:10.1016/j.elecom.2007.01.007
Durán N, Marcato PD, De Souza GIH, Alves OL, Esposito E (2007) Antibacterial effect of silver nanoparticles produced by fungal process on textile fabrics and their effluent treatment. Journal of Biomedical Nanotechnology 3 (2):203–208. doi:10.1166/jbn.2007.022
El-Rafie MH, Mohamed AA, Shaheen TI, Hebeish A (2010) Antimicrobial effect of silver nanoparticles produced by fungal process on cotton fabrics. Carbohydrate Polymers 80 (3):779–782. doi:10.1016/j.carbpol.2009.12.028
Fayaz AM, Balaji K, Girilal M, Kalaichelvan PT, Venkatesan R (2009) Mycobased Synthesis of Silver Nanoparticles and Their Incorporation into Sodium Alginate Films for Vegetable and Fruit Preservation. Journal of Agricultural and Food Chemistry 57 (14):6246–6252. doi:10.1021/jf900337h
Fayaz AM, Balaji K, Girilal M, Yadav R, Kalaichelvan PT, Venkatesan R (2010a) Biogenic synthesis of silver nanoparticles and their synergistic effect with antibiotics: a study against gram-positive and gram-negative bacteria. Nanomedicine: Nanotechnology, Biology and Medicine 6 (1):103–109
Fayaz AM, Girilal M, Mahdy SA, Somsundar SS, Venkatesan R, Kalaichelvan PT (2011) Vancomycin bound bigenic gold nanoparticles: A different perspective for development of anti VRSA agents. Process Biochemistry 46 (3):636–641. doi:10.1016/j.procbio.2010.11.001
Fayaz AM, Tiwary CS, Kalaichelvan PT, Venkatesan R (2010b) Blue orange light emission from biogenic synthesized silver nanoparticles using Trichoderma viride. Colloid Surf B-Biointerfaces 75 (1):175–178. doi:10.1016/j.colsurfb.2009.08.028
Fukuoka A, Araki H, Sakamoto Y, Sugimoto N, Tsukada H, Kumai Y, Akimoto Y, Ichikawa M (2002) Template Synthesis of Nanoparticle Arrays of Gold and Platinum in Mesoporous Silica Films. Nano Letters 2 (7):793–795. doi:10.1021/nl0256107

Gadd GM (2009) Biosorption: critical review of scientific rationale, environmental importance and significance for pollution treatment. Journal of Chemical Technology and Biotechnology 84 (1):13–28. doi:10.1002/jctb.1999

Gadd GM (2010) Metals, minerals and microbes: geomicrobiology and bioremediation. Microbiology-Sgm 156:609–643. doi:10.1099/mic.0.037143-0

Gade A, Gaikwad S, Tiwari V, Yadav A, Ingle A, Rai M (2010) Biofabrication of Silver Nanoparticles by Opuntia ficus-indica: In vitro Antibacterial Activity and Study of the Mechanism Involved in the Synthesis. Current Nanoscience 6 (4):370–375. doi:10.2174/157341310791659026

Gajbhiye M, Kesharwani J, Ingle A, Gade A, Rai M (2009) Fungus-mediated synthesis of silver nanoparticles and their activity against pathogenic fungi in combination with fluconazole. Nanomedicine: Nanotechnology, Biology and Medicine 5 (4):382–386. doi:10.1016/j.nano.2009.06.005

Gauthier D, Søbjerg LS, Jensen KM, Lindhardt AT, Bunge M, Finster K, Meyer RL, Skrydstrup T (2010) Environmentally Benign Recovery and Reactivation of Palladium from Industrial Waste by Using Gram-Negative Bacteria. ChemSusChem 3 (9):1036–1039. doi:10.1002/cssc.201000091

Govindaraju K, Tamilselvan S, Kiruthiga V, Singaravelu G (2010) Biogenic silver nanoparticles by Solanum torvum and their promising antimicrobial activity. Journal of Biopesticides 3 (1 Special Issue):394–399

Guimarães BCM, Arends JBA, van der Ha D, Van de Wiele T, Boon N, Verstraete W (2010) Microbial services and their management: Recent progresses in soil bioremediation technology. Applied Soil Ecology 46 (2):157–167. doi:10.1016/j.apsoil.2010.06.018

Guo S, Wang E (2007) Synthesis and electrochemical applications of gold nanoparticles. Analytica Chimica Acta 598 (2):181–192. doi:10.1016/j.aca.2007.07.054

Harrad S, Robson M, Hazrati S, Baxter-Plant VS, Deplanche K, Redwood MD, Macaskie LE (2007) Dehalogenation of polychlorinated biphenyls and polybrominated diphenyl ethers using a hybrid bioinorganic catalyst. Journal of Environmental Monitoring 9 (4):314–318

Haverkamp R, Marshall A, Agterveld D (2007) Pick your carats: nanoparticles of goldsilvercopper alloy produced in vivo. Journal of Nanoparticle Research 9 (4):697–700. doi:10.1007/s11051-006-9198-y

Hennebel T, De Corte S, Vanhaecke L, Vanherck K, Forrez I, De Gusseme B, Verhagen P, Verbeken K, Van der Bruggen B, Vankelecom I, Boon N, Verstraete W (2010) Removal of diatrizoate with catalytically active membranes incorporating microbially produced palladium nanoparticles. Water Research 44 (5):1498–1506. doi:10.1016/j.watres.2009.10.041

Hennebel T, De Gusseme B, Boon N, Verstraete W (2009a) Biogenic metals in advanced water treatment. Trends in Biotechnology 27 (2):90–98. doi:10.1016/j.tibtech.2008.11.002

Hennebel T, Simoen H, De Windt W, Verloo M, Boon N, Verstraete W (2009b) Biocatalytic dechlorination of trichloroethylene with bio-palladium in a pilot-scale membrane reactor. Biotechnology and Bioengineering 102 (4):995–1002. doi:10.1002/bit.22138

Hennebel T, Verhagen P, Simoen H, Gusseme BD, Vlaeminck SE, Boon N, Verstraete W (2009c) Remediation of trichloroethylene by bio-precipitated and encapsulated palladium nanoparticles in a fixed bed reactor. Chemosphere 76 (9):1221–1225. doi:10.1016/j.chemosphere.2009.05.046

Huang JL, Wang WT, Lin LQ, Li QB, Lin WS, Li M, Mann S (2009) A General Strategy for the Biosynthesis of Gold Nanoparticles by Traditional Chinese Medicines and Their Potential Application as Catalysts. Chemistry-an Asian Journal 4 (7):1050–1054. doi:10.1002/asia.200900064

Humphries AC, Mikheenko IP, Macaskie LE (2006) Chromate reduction by immobilized palladized sulfate-reducing bacteria. Biotechnology and Bioengineering 94 (1):81–90. doi:10.1002/bit.20814

Chakraborty N, Banerjee A, Lahiri S, Panda A, Ghosh AN, Pal R (2009) Biorecovery of gold using cyanobacteria and an eukaryotic alga with special reference to nanogold formation - a novel phenomenon. Journal of Applied Phycology 21 (1):145–152. doi:10.1007/s10811-008-9343-3

Chakraborty N, Pal R, Ramaswami A, Nayak D, Lahiri S (2006) Diatom: A potential bio-accumulator of gold. Journal of Radioanalytical and Nuclear Chemistry 270 (3):645–649. doi:10.1007/s10967-006-0475-0

Chaloupka K, Malam Y, Seifalian AM (2010) Nanosilver as a new generation of nanoproduct in biomedical applications. Trends in Biotechnology 28 (11):580–588. doi:10.1016/j.tibtech.2010.07.006

Chen X, Schluesener HJ (2008) Nanosilver: A nanoproduct in medical application. Toxicology Letters 176 (1):1–12. doi:10.1016/j.toxlet.2007.10.004

Cherukuri P, Glazer ES, Curley SA (2010) Targeted hyperthermia using metal nanoparticles. Advanced Drug Delivery Reviews 62 (3):339–345. doi:10.1016/j.addr.2009.11.006

Chidambaram D, Hennebel T, Taghavi S, Mast J, Boon N, Verstraete W, van der Lelie D, Fitts JP (2010) Concomitant Microbial Generation of Palladium Nanoparticles and Hydrogen To Immobilize Chromate. Environmental Science & Technology 44 (19):7635–7640. doi:10.1021/es101559r

Cho K-H, Park J-E, Osaka T, Park S-G (2005) The study of antimicrobial activity and preservative effects of nanosilver ingredient. Electrochimica Acta 51 (5):956–960. doi:10.1016/j.electacta.2005.04.071

Chojnacka K (2010) Biosorption and bioaccumulation - the prospects for practical applications. Environment International 36 (3):299–307. doi:10.1016/j.envint.2009.12.001

Ingle A, Gade A, Pierrat S, Sönnichsen C, Rai M (2008) Mycosynthesis of Silver Nanoparticles Using the Fungus Fusarium acuminatum and its Activity Against Some Human Pathogenic Bacteria. Current Nanoscience 4:141–144. doi:10.2174/157341308784340804

Iskandar F (2009) Nanoparticle processing for optical applications - A review. Advanced Powder Technology 20 (4):283–292. doi:10.1016/j.apt.2009.07.001

Jaidev LR, Narasimha G (2010) Fungal mediated biosynthesis of silver nanoparticles, characterization and antimicrobial activity. Colloids and Surfaces B: Biointerfaces 81 (2):430–433. doi:10.1016/j.colsurfb.2010.07.033

Jha AK, Prasad K (2010) Ferroelectric BaTiO3 nanoparticles: Biosynthesis and characterization. Colloids and Surfaces B: Biointerfaces 75 (1):330–334. doi:10.1016/j.colsurfb.2009.09.005

Jia L, Zhang Q, Li Q, Song H (2009) The biosynthesis of palladium nanoparticles by antioxidants in Gardenia jasminoides Ellis : long lifetime nanocatalysts for p -nitrotoluene hydrogenation. Nanotechnology 20 (38):385601

Jia XP, Ma XY, Wei DW, Dong J, Qian WP (2008) Direct formation of silver nanoparticles in cuttlebone-derived organic matrix for catalytic applications. Colloids and Surfaces a-Physicochemical and Engineering Aspects 330 (2–3):234–240. doi:10.1016/j.colsurfa.2008.08.016

Kavamura VN, Esposito E (2009) Biotechnological strategies applied to the decontamination of soils polluted with heavy metals. Biotechnology Advances 28 (1):61–69. doi:10.1016/j.biotechadv.2009.09.002

Khomutov GB, Gubin SP (2002) Interfacial synthesis of noble metal nanoparticles. Materials Science and Engineering: C 22 (2):141–146. doi:10.1016/s0928-4931(02)00162-5

Kowshik M, Deshmukh N, Vogel W, Urban J, Kulkarni SK, Paknikar KM (2002) Microbial synthesis of semiconductor CdS nanoparticles, their characterization, and their use in the fabrication of an ideal diode. Biotechnology and Bioengineering 78 (5):583–588. doi:10.1002/bit.10233

Krishnaraj C, Jagan EG, Rajasekar S, Selvakumar P, Kalaichelvan PT, Mohan N (2010) Synthesis of silver nanoparticles using Acalypha indica leaf extracts and its antibacterial activity against water borne pathogens. Colloids and Surfaces B: Biointerfaces 76 (1):50–56. doi:10.1016/j.colsurfb.2009.10.008

Kulkarni M, Chaudhari A (2007) Microbial remediation of nitro-aromatic compounds: An overview. J Environ Manage 85 (2):496–512. doi:10.1016/j.jenvman.2007.06.009

Kumar R, Ghosh A, Patra CR, Mukherjee P, Sastry M (2004) Gold Nanoparticles Formed within Ordered Mesoporous Silica and on Amorphous Silica. In: Zhou B, Hermans S, Somorjai G (eds) Nanoparticle in Catalysis, vol 1. Kluwer Academic/Plenum Publishers, New York, pp 111–136

Lee KY, Hwang J, Lee YW, Kim J, Han SW (2007) One-step synthesis of gold nanoparticles using azacryptand and their applications in SERS and catalysis. Journal of Colloid and Interface Science 316 (2):476–481. doi:10.1016/j.jcis.2007.07.076

Lesmana SO, Febriana N, Soetaredjo FE, Sunarso J, Ismadji S (2009) Studies on potential applications of biomass for the separation of heavy metals from water and wastewater. Biochemical Engineering Journal 44 (1):19–41. doi:10.1016/j.bej.2008.12.009

Lewis TA, Newcombe DA, Crawford RL (2004) Bioremediation of soils contaminated with explosives. J Environ Manage 70 (4):291–307. doi:10.1016/j.jenvman.2003.12.005

Li X, Jiang W, Sun JB, Wang GL, Guan F, Li Y (2007) Purified and sterilized magnetosomes from Magnetospirillum gryphiswaldense MSR-1 were not toxic to mouse fibroblasts in vitro. Letters in Applied Microbiology 45 (1):75–81. doi:10.1111/j.1472-765X.2007.02143.x

Liotta LF, Gruttadauria M, Di Carlo G, Perrini G, Librando V (2009) Heterogeneous catalytic degradation of phenolic substrates: Catalysts activity. J Hazard Mater 162 (2–3):588–606. doi:10.1016/j.jhazmat.2008.05.115

Lynch I, Dawson KA (2008) Protein-nanoparticle interactions. Nano Today 3 (1–2):40–47. doi:10.1016/s1748-0132(08)70014-8

Mabbett AN, Lloyd JR, Macaskie LE (2002) Effect of complexing agents on reduction of Cr(VI) by Desulfovibrio vulgaris ATCC 29579. Biotechnology and Bioengineering 79 (4):389–397. doi:10.1002/bit.10361

Mabbett AN, Macaskie LE (2002) A new bioinorganic process for the remediation of Cr(VI). Journal of Chemical Technology & Biotechnology 77 (10):1169–1175. doi:10.1002/jctb.693

Mabbett AN, Sanyahumbi D, Yong P, Macaskie LE (2005) Biorecovered Precious Metals from Industrial Wastes: Single-Step Conversion of a Mixed Metal Liquid Waste to a Bioinorganic Catalyst with Environmental Application. Environmental Science & Technology 40 (3):1015–1021. doi:10.1021/es0509836

Maliszewska I, Puzio M (2009) Extracellular biosynthesis and antimicrobial activity of silver nanoparticles. Acta Physica Polonica A 116:S160–S162

Mallick K, Witcomb MJ, Scurrell MS (2004) Polymer stabilized silver nanoparticles: A photochemical synthesis route. Journal of Materials Science 39 (14):4459–4463. doi:10.1023/b:jmsc.0000034138.80116.50

Marambio-Jones C, Hoek EMV (2010) A review of the antibacterial effects of silver nanomaterials and potential implications for human health and the environment. Journal of Nanoparticle Research 12 (5):1531–1551. doi:10.1007/s11051-010-9900-y

Mata YN, Torres E, Blazquez ML, Ballester A, Gonzalez F, Munoz JA (2009) Gold(III) biosorption and bioreduction with the brown alga Fucus vesiculosus. J Hazard Mater 166 (2–3):612–618. doi:10.1016/j.jhazmat.2008.11.064

Merin DD, Prakash S, Bhimba BV (2010) Antibacterial screening of silver nanoparticles synthesized by marine micro algae. Asian Pacific Journal of Tropical Medicine 3 (10):797–799. doi:10.1016/s1995-7645(10)60191-5

Mertens B, Blothe C, Windey K, De Windt W, Verstraete W (2007) Biocatalytic dechlorination of lindane by nano-scale particles of Pd(0) deposited on Shewanella oneidensis. Chemosphere 66 (1):99–105. doi:10.1016/j.chemosphere.2006.05.018

Morones JR, Elechiguerra JL, Camacho A, Holt K, Kouri JB, Ramírez JT, Yacaman MJ (2005) The bactericidal effect of silver nanoparticles. Nanotechnology 16 (10):2346

Moulton MC, Braydich-Stolle LK, Nadagouda MN, Kunzelman S, Hussain SM, Varma RS (2010) Synthesis, characterization and biocompatibility of "green" synthesized silver nanoparticles using tea polyphenols. Nanoscale 2 (5):763–770

Musarrat J, Dwivedi S, Singh BR, Al-Khedhairy AA, Azam A, Naqvi A (2010) Production of antimicrobial silver nanoparticles in water extracts of the fungus *Amylomyces rouxii* strain KSU-09. Bioresource Technology 101 (22):8772–8776. doi: 10.1016/j.biortech.2010.06.065

Nabikhan A, Kandasamy K, Raj A, Alikunhi NM (2010) Synthesis of antimicrobial silver nanoparticles by callus and leaf extracts from saltmarsh plant, Sesuvium–portulacastrum L. Colloids and Surfaces B: Biointerfaces 79 (2):488–493. doi:10.1016/j.colsurfb.2010.05.018

Nair LS, Laurencin CT (2007) Silver nanoparticles: Synthesis and therapeutic applications. Journal of Biomedical Nanotechnology 3 (4):301–316. doi:10.1166/jbn.2007.041

Narayanan R (2010) Recent Advances in Noble Metal Nanocatalysts for Suzuki and Heck Cross-Coupling Reactions. Molecules 15 (4):2124–2138. doi:10.3390/molecules15042124

Narayanan R, El-Sayed M (2008) Some Aspects of Colloidal Nanoparticle Stability, Catalytic Activity, and Recycling Potential. Topics in Catalysis 47 (1):15–21. doi:10.1007/s11244-007-9029-0

Nguyen DT, Kim D-J, Kim K-S (2011) Controlled synthesis and biomolecular probe application of gold nanoparticles. Micron 42 (3):207–227. doi:10.1016/j.micron.2010.09.008

Nguyen DT, Kim D-J, So MG, Kim K-S (2010) Experimental measurements of gold nanoparticle nucleation and growth by citrate reduction of HAuCl4. Advanced Powder Technology 21 (2):111–118. doi:10.1016/j.apt.2009.11.005

Nithya R, Ragunathan R (2009) Synthesis of silver nanoparticles using Pleurotus sajor caju and its microbial study. Digest Journal of Nanomaterials and Biostructures 4 (4):623–629

Note C, Kosmella S, Koetz J (2006) Poly(ethyleneimine) as reducing and stabilizing agent for the formation of gold nanoparticles in w/o microemulsions. Colloids and Surfaces A: Physicochemical and Engineering Aspects 290 (1–3):150–156. doi:10.1016/j.colsurfa.2006.05.018

Ogi T, Honda R, Tamaoki K, Saitoh N, Konishi Y (2011) Direct room-temperature synthesis of a highly dispersed Pd nanoparticle catalyst and its electrical properties in a fuel cell. Powder Technology 205 (1–3):143–148. doi:10.1016/j.powtec.2010.09.004

Oliveira MM, Ugarte D, Zanchet D, Zarbin AJG (2005) Influence of synthetic parameters on the size, structure, and stability of dodecanethiol-stabilized silver nanoparticles. Journal of Colloid and Interface Science 292 (2):429–435. doi:10.1016/j.jcis.2005.05.068

Orozco R, Redwood M, Yong P, Caldelari I, Sargent F, Macaskie L (2010) Towards an integrated system for bio-energy: hydrogen production by <i>Escherichia coli</i> and use of palladium-coated waste cells for electricity generation in a fuel cell. Biotechnology Letters 32 (12):1837–1845. doi:10.1007/s10529-010-0383-9

Pankhurst QA, et al. (2003) Applications of magnetic nanoparticles in biomedicine. Journal of Physics D: Applied Physics 36 (13):R167

Pankhurst QA, et al. (2009) Progress in applications of magnetic nanoparticles in biomedicine. Journal of Physics D: Applied Physics 42 (22):224001

Patra CR, Bhattacharya R, Mukhopadhyay D, Mukherjee P (2010) Fabrication of gold nanoparticles for targeted therapy in pancreatic cancer. Advanced Drug Delivery Reviews 62 (3):346–361. doi:10.1016/j.addr.2009.11.007

Perelshtein I, Applerot G, Perkas N, Guibert G, Mikhailov S, Gedanken A (2008) Sonochemical coating of silver nanoparticles on textile fabrics (nylon, polyester and cotton) and their antibacterial activity. Nanotechnology 19 (24):245705

Pérez-de-Mora A, Burgos P, Madejón E, Cabrera F, Jaeckel P, Schloter M (2006) Microbial community structure and function in a soil contaminated by heavy metals: effects of plant growth and different amendments. Soil Biology and Biochemistry 38 (2):327–341. doi:10.1016/j.soilbio.2005.05.010

Pingarrón JM, Yáñez-Sedeño P, González-Cortés A (2008) Gold nanoparticle-based electrochemical biosensors. Electrochimica Acta 53 (19):5848–5866. doi:10.1016/j.electacta.2008.03.005

Porel S, Venkatram N, Rao DN, Radhakrishnan TP (2007) In situ synthesis of metal nanoparticles in polymer matrix and their optical limiting applications. Journal of Nanoscience and Nanotechnology 7 (6):1887–1892. doi:10.1166/jnn.2007.736

Rai M, Yadav A, Gade A (2009) Silver nanoparticles as a new generation of antimicrobials. Biotechnology Advances 27 (1):76–83. doi:10.1016/j.biotechadv.2008.09.002

Ravindra S, Murali Mohan Y, Narayana Reddy N, Mohana Raju K (2010) Fabrication of antibacterial cotton fibres loaded with silver nanoparticles via "Green Approach". Colloids and Surfaces A: Physicochemical and Engineering Aspects 367 (1–3):31–40. doi:10.1016/j.colsurfa.2010.06.013

Redwood MD, Deplanche K, Baxter-Plant VS, Macaskie LE (2008) Biomass-supported palladium catalysts on Desulfovibrio desulfuricans and Rhodobacter sphaeroides. Biotechnology and Bioengineering 99 (5):1045–1054. doi:10.1002/bit.21689

Rodríguez-Carmona E, Villaverde A (2010) Nanostructured bacterial materials for innovative medicines. Trends in Microbiology 18 (9):423–430. doi:10.1016/j.tim.2010.06.007

Roucoux A, Nowicki A, Philippot K (2008) Rhodium and Ruthenium Nanoparticles in Catalysis. Nanoparticles and Catalysis. Wiley-VCH Verlag GmbH & Co. KGaA. doi:10.1002/9783527621323.ch11

Sadhasivam S, Shanmugam P, Yun K (2010) Biosynthesis of silver nanoparticles by Streptomyces hygroscopicus and antimicrobial activity against medically important pathogenic microorganisms. Colloids and Surfaces B: Biointerfaces 81 (1):358–362. doi:10.1016/j.colsurfb.2010.07.036

Saha S, Sarkar J, Chattopadhyay D, Patra S, Chakraborty A, Acharya K (2010) Production of silver nanoparticles by a phytopathogenic fungus Bipolaris nodulosa and its antimicrobial activity. Digest Journal of Nanomaterials and Biostructures 5 (4):887–895

Salkar RA, Jeevanandam P, Aruna ST, Koltypin Y, Gedanken A (1999) The sonochemical preparation of amorphous silver nanoparticles. Journal of Materials Chemistry 9 (6):1333–1335

Samanta S, Pyne S, Sarkar P, Sahoo GP, Bar H, Bhui DK, Misra A (2010) Synthesis of silver nanostructures of varying morphologies through seed mediated growth approach. Journal of Molecular Liquids 153 (2–3):170–173. doi:10.1016/j.molliq.2010.02.008

Saravanan M, Nanda A (2010) Extracellular synthesis of silver bionanoparticles from Aspergillus clavatus and its antimicrobial activity against MRSA and MRSE. Colloids and Surfaces B: Biointerfaces 77 (2):214–218. doi:10.1016/j.colsurfb.2010.01.026

Sarkar R, Kumbhakar P, MIitra AK (2010) Green synthesis of silver nanoparticles and its optical properties. Digest Journal of Nanomaterials and Biostructures 5 (2):491–496

Sathishkumar M, Mahadevan A, Vijayaraghavan K, Pavagadhi S, Balasubramanian R (2010a) Green Recovery of Gold through Biosorption, Biocrystallization, and Pyro-Crystallization. Industrial & Engineering Chemistry Research 49 (16):7129–7135. doi:10.1021/ie100104j

Sathishkumar M, Sneha K, Won SW, Cho CW, Kim S, Yun YS (2009) Cinnamon zeylanicum bark extract and powder mediated green synthesis of nano-crystalline silver particles and its bactericidal activity. Colloids and Surfaces B: Biointerfaces 73 (2):332–338. doi:10.1016/j.colsurfb.2009.06.005

Sathishkumar M, Sneha K, Yun Y-S (2010b) Immobilization of silver nanoparticles synthesized using Curcuma longa tuber powder and extract on cotton cloth for bactericidal activity. Bioresource Technology 101 (20):7958-7965. doi:10.1016/j.biortech.2010.05.051

Sathyavathi R, Krishna MB, Rao SV, Saritha R, Rao DN (2010) Biosynthesis of Silver Nanoparticles Using Coriandrum Sativum Leaf Extract and Their Application in Nonlinear Optics. Advanced Science Letters 3 (2):138–143. doi:10.1166/asl.2010.1099

Sawle BD, Salimath B, Deshpande R, Bedre MD, Prabhakar BK, Venkataraman A (2008) Biosynthesis and stabilization of Au and Au–Ag alloy nanoparticles by fungus, Fusarium semitectum. Science and Technology of Advanced Materials 9 (3):035012

Selvakannan PR, Mandal S, Phadtare S, Gole A, Pasricha R, Adyanthaya SD, Sastry M (2004) Water-dispersible tryptophan-protected gold nanoparticles prepared by the spontaneous reduction of aqueous chloroaurate ions by the amino acid. Journal of Colloid and Interface Science 269 (1):97–102. doi:10.1016/s0021-9797(03)00616-7

Senapati S, Ahmad A, Khan MI, Sastry M, Kumar R (2005) Extracellular Biosynthesis of Bimetallic Au–Ag Alloy Nanoparticles. Small 1 (5):517–520. doi:10.1002/smll.200400053

Seoudi R, Shabaka A, Eisa WH, Anies B, Farage NM (2010) Effect of the prepared temperature on the size of CdS and ZnS nanoparticle. Physica B: Condensed Matter 405 (3):919–924. doi:10.1016/j.physb.2009.10.015

Shankar SS, Rai A, Ahmad A, Sastry M (2005) Controlling the Optical Properties of Lemongrass Extract Synthesized Gold Nanotriangles and Potential Application in Infrared-Absorbing Optical Coatings. Chemistry of Materials 17 (3):566–572. doi:10.1021/cm048292g

Sharma NC, Sahi SV, Nath S, Parsons JG, Gardea-Torresdey JL, Pal T (2007) Synthesis of plant-mediated gold nanoparticles and catalytic role of biomatrix-embedded nanomaterials. Environmental Science & Technology 41 (14):5137–5142. doi:10.1021/es062929a

Sharma VK, Yngard RA, Lin Y (2009) Silver nanoparticles: Green synthesis and their antimicrobial activities. Advances in Colloid and Interface Science 145 (1–2):83–96. doi:10.1016/j.cis.2008.09.002

Shchukin DG, Radziuk D, Mohwald H (2010) Ultrasonic Fabrication of Metallic Nanomaterial and Nanoalloys. In: Annual Review of Materials Research, Vol 40, vol 40. Annual Review of Materials Research. pp 345–362. doi:10.1146/annurev-matsci-070909-104540

Shih C-M, Shieh Y-T, Twu Y-K (2009) Preparation of gold nanopowders and nanoparticles using chitosan suspensions. Carbohydrate Polymers 78 (2):309–315. doi:10.1016/j.carbpol.2009.04.008

Shiju NR, Guliants VV (2009) Recent developments in catalysis using nanostructured materials. Applied Catalysis A: General 356 (1):1–17. doi:10.1016/j.apcata.2008.11.034

Shilov V, Voitenko E, Marochko L, Podol'skaya V (2010) Electric characteristics of cellular structures containing colloidal silver. Colloid Journal 72 (1):125–132. doi:10.1134/s1061933x10010138

Shirley AD, Dayanand A, Sreedhar B, Dastager SG (2010) Antimicrobial activity of silver nanoparticles synthesized from novel Streptomyces species. Digest Journal of Nanomaterials and Biostructures 5 (2):447–451

Simoncic B, Tomsic B (2010) Structures of Novel Antimicrobial Agents for Textiles - A Review. Textile Research Journal 80 (16):1721–1737. doi:10.1177/0040517510363193

Sobjerg LS, Gauthier D, Lindhardt AT, Bunge M, Finster K, Meyer RL, Skrydstrup T (2009) Bio-supported palladium nanoparticles as a catalyst for Suzuki-Miyaura and Mizoroki-Heck reactions. Green Chemistry 11 (12):2041–2046

Sun J-B, Duan J-H, Dai S-L, Ren J, Guo L, Jiang W, Li Y (2008) Preparation and anti-tumor efficiency evaluation of doxorubicin-loaded bacterial magnetosomes: Magnetic nanoparticles as drug carriers isolated from Magnetospirillum gryphiswaldense. Biotechnology and Bioengineering 101 (6):1313–1320. doi:10.1002/bit.22011

Sun RWY, Chen R, Chung NPY, Ho CM, Lin CLS, Che CM (2005) Silver nanoparticles fabricated in Hepes buffer exhibit cytoprotective activities toward HIV-1 infected cells. Chem Commun (40):5059–5061. doi:10.1039/b510984a

Sundaramoorthi C, Kalaivani M, Mathews DM, Palanisamy S, Kalaiselvan V, Rajasekaran A (2009) Biosynthesis of silver nanoparticles from Aspergillus niger and evaluation of its wound healing activity in experimental rat model. International Journal of PharmTech Research 1 (4):1523–1529

Suresh AK, Pelletier DA, Wang W, Moon JW, Gu BH, Mortensen NP, Allison DP, Joy DC, Phelps TJ, Doktycz MJ (2010) Silver Nanocrystallites: Biofabrication using Shewanella oneidensis, and an Evaluation of Their Comparative Toxicity on Gram-negative and Gram-positive Bacteria. Environmental Science & Technology 44 (13):5210–5215. doi:10.1021/es903684r

Talapin DV, Lee J-S, Kovalenko MV, Shevchenko EV (2009) Prospects of Colloidal Nanocrystals for Electronic and Optoelectronic Applications. Chemical Reviews 110 (1):389–458. doi:10.1021/cr900137k

Thakkar KN, Mhatre SS, Parikh RY (2010) Biological synthesis of metallic nanoparticles. Nanomedicine-Nanotechnology Biology and Medicine 6 (2):257–262. doi:10.1016/j.nano.2009.07.002

Thibault-Starzyk Fdr, Daturi M, Bazin P, Marie O (2008) NO Heterogeneous Catalysis Viewed from the Angle of Nanoparticles. Nanoparticles and Catalysis. Wiley-VCH Verlag GmbH & Co. KGaA. doi:10.1002/9783527621323.ch16

Tonks NK (2003) PTP1B: From the sidelines to the front lines! FEBS Letters 546 (1):140–148. doi:10.1016/s0014-5793(03)00603-3

Torres-Chavolla E, Ranasinghe RJ, Alocilja EC (2010) Characterization and Functionalization of Biogenic Gold Nanoparticles for Biosensing Enhancement. Nanotechnology, IEEE Transactions on 9 (5):533–538

Tripathi A, Chandrasekaran N, Raichur AM, Mukherjee A (2009) Antibacterial Applications of Silver Nanoparticles Synthesized by Aqueous Extract of Azadirachta Indica (Neem) Leaves. Journal of Biomedical Nanotechnology 5 (1):93–98. doi:10.1166/jbn.2009.038

Ul'berg Z, Podol'skaya V, Voitenko E, Grishchenko N, Yakubenko L (2010) Formation and biological activity of preparations based on microorganisms and colloidal silver. Colloid Journal 72 (1):66–73. doi:10.1134/s1061933x10010096

Vaskelis A, Tarozaite R, Jagminiene A, Tamasiunaite LT, Juskenas R, Kurtinaitiene M (2007) Gold nanoparticles obtained by Au(III) reduction with Sn(II): Preparation and electrocatalytic properties in oxidation of reducing agents. Electrochimica Acta 53 (2):407–416. doi:10.1016/j.electacta.2007.04.008

Veerasamy R, Xin TZ, Gunasagaran S, Xiang TFW, Yang EFC, Jeyakumar N, Dhanaraj SA (2011) Biosynthesis of silver nanoparticles using mangosteen leaf extract and evaluation of their antimicrobial activities. Journal of Saudi Chemical Society 15 (2):113–120. doi:10.1016/j.jscs.2010.06.004

Veiseh O, Gunn JW, Zhang M (2010) Design and fabrication of magnetic nanoparticles for targeted drug delivery and imaging. Advanced Drug Delivery Reviews 62 (3):284–304. doi:10.1016/j.addr.2009.11.002

Verma VC, Kharwar RN, Gange AC (2010) Biosynthesis of antimicrobial silver nanoparticles by the endophytic fungus Aspergillus clavatus. Nanomedicine: Nanotechnology, Biology and Medicine 5 (1):33–40

Vijayaraghavan K, Yun Y-S (2008) Bacterial biosorbents and biosorption. Biotechnology Advances 26 (3):266–291. doi:10.1016/j.biotechadv.2008.02.002

Volesky B (2007) Biosorption and me. Water Research 41 (18):4017–4029. doi:10.1016/j.watres.2007.05.062

Wagner J, Köhler JM (2005) Continuous Synthesis of Gold Nanoparticles in a Microreactor. Nano Letters 5 (4):685–691. doi:10.1021/nl050097t

Wagner J, Tshikhudo TR, Köhler JM (2008) Microfluidic generation of metal nanoparticles by borohydride reduction. Chemical Engineering Journal 135 (1):S104–S109. doi:10.1016/j.cej.2007.07.046

Wang J, Chen C (2009) Biosorbents for heavy metals removal and their future. Biotechnology Advances 27 (2):195–226. doi:10.1016/j.biotechadv.2008.11.002

Wang T, Yang L, Zhang B, Liu J (2010) Extracellular biosynthesis and transformation of selenium nanoparticles and application in H2O2 biosensor. Colloids and Surfaces B: Biointerfaces 80 (1):94–102. doi:10.1016/j.colsurfb.2010.05.041

Wang Y, He X, Wang K, Zhang X, Tan W (2009) Barbated Skullcup herb extract-mediated biosynthesis of gold nanoparticles and its primary application in electrochemistry. Colloids and Surfaces B: Biointerfaces 73 (1):75–79. doi:10.1016/j.colsurfb.2009.04.027

Widiyastuti W, Balgis R, Iskandar F, Okuyama K (2010) Nanoparticle formation in spray pyrolysis under low-pressure conditions. Chemical Engineering Science 65 (5):1846–1854. doi:10.1016/j.ces.2009.11.026

Wilbur KM, Simkiss K (1979) Chapter 2.3 Carbonate Turnover and Deposition by Metazoa. In: Trudinger PA, Swaine DJ (eds) Studies in Environmental Science, vol Volume 3. Elsevier, pp 69–106

Winter M, Brodd RJ (2005) What Are Batteries, Fuel Cells, and Supercapacitors? (Chem. Rev. 2003, 104, 4245−4269. Published on the Web 09/28/2004.). Chemical Reviews 105 (3):1021–1021. doi:10.1021/cr040110e

Won SW, Mao J, Kwak I-S, Sathishkumar M, Yun Y-S (2010) Platinum recovery from ICP wastewater by a combined method of biosorption and incineration. Bioresource Technology 101 (4):1135–1140. doi:10.1016/j.biortech.2009.09.056

Wood J, Bodenes L, Bennett J, Deplanche K, Macaskie LE (2010) Hydrogenation of 2-Butyne-1,4-diol Using Novel Bio-Palladium Catalysts. Industrial & Engineering Chemistry Research 49 (3):980–988. doi:10.1021/ie900663k

Xie J, Lee JY, Wang DIC, Ting YP (2007) Silver Nanoplates: From Biological to Biomimetic Synthesis. ACS Nano 1 (5):429–439. doi:10.1021/nn7000883

Xiong X, Lidstrom ME, Parviz BA (2007) Microorganisms for MEMS. Microelectromechanical Systems, Journal of 16 (2):429–444

Yadav OP, Palmqvist A, Cruise N, Holmberg K (2003) Synthesis of platinum nanoparticles in microemulsions and their catalytic activity for the oxidation of carbon monoxide. Colloids and Surfaces A: Physicochemical and Engineering Aspects 221 (1–3):131–134. doi:10.1016/s0927-7757(03)00141-9

Yang GW (2007) Laser ablation in liquids: Applications in the synthesis of nanocrystals. Progress in Materials Science 52 (4):648–698. doi:10.1016/j.pmatsci.2006.10.016

Yong P, Mikheenko I, Deplanche K, Redwood M, Macaskie L (2010) Biorefining of precious metals from wastes: an answer to manufacturing of cheap nanocatalysts for fuel cells and power generation via an integrated biorefinery? Biotechnology Letters 32 (12):1821–1828. doi:10.1007/s10529-010-0378-6

Yong P, Paterson-Beedle M, Mikheenko IP, Macaskie LE (2007) From bio-mineralisation to fuel cells: biomanufacture of Pt and Pd nanocrystals for fuel cell electrode catalyst. Biotechnology Letters 29 (4):539–544. doi:10.1007/s10529-006-9283-4

Zheng B, Qian L, Yuan H, Xiao D, Yang X, Paau MC, Choi MMF (2010a) Preparation of gold nanoparticles on eggshell membrane and their biosensing application. Talanta 82 (1):177–183. doi:10.1016/j.talanta.2010.04.014

Zheng B, Xie S, Qian L, Yuan H, Xiao D, Choi MMF (2011) Gold nanoparticles-coated eggshell membrane with immobilized glucose oxidase for fabrication of glucose biosensor. Sensors and Actuators B: Chemical 152 (1):49–55. doi:10.1016/j.snb.2010.09.051

Zheng D, Hu C, Gan T, Dang X, Hu S (2010b) Preparation and application of a novel vanillin sensor based on biosynthesis of Au-Ag alloy nanoparticles. Sensors and Actuators B: Chemical 148 (1):247–252. doi:10.1016/j.snb.2010.04.031

Zhou H, Fan F, Han T, Li X, Ding J, Zhang D, Guo Q, Ogawa H (2009) Bacteria-based controlled assembly of metal chalcogenide hollow nanostructures with enhanced light-harvesting and photocatalytic properties. Nanotechnology 20 (8):085603

Nanocrystals: Production, Cellular Drug Delivery, Current and Future Products

Rainer H. Müller, Ranjita Shegokar, Sven Gohla, and Cornelia M. Keck

Abstract Drug nanocrystals are a formulation principle for systemic and also intracellular delivery of poorly soluble drugs. Their production by bottom up techiques (precipitation – hydrosols, Nanomorph) and by top down techniques (bead milling – NanoCrystal®, high pressure homogenization – DissoCubes®, NANOEDGE®) is briefly described, representing the first generation of nanocrystals. The second generation, the smartCrystal®, is produced by combination processes. They are featured by e.g. increased physical stability and/or smaller sizes (<100 nm), favourable when exposed to the destabilizing electrolytes in biological fluids and for uptake by cells by pinocytosis. The lab scale processes were successfully transferred to industrial scale by using discontinuous bead mills and high capacity homogenizers (top down), precipitation can be performed by static blenders. According to the nanotoxicological classification system (NCS), the nanocrystals belong to class I, being highly tolerable. They can be produced using only regulatorily accepted excipients. Both ease the way to the patient and market. Nanotoxicity studies confirm the good tolerability. The nanocrystal products on the market are no direct intracellular delivery systems. They transport drug to the biological barrier and then promote penetration and permeation of drugs in molecular form through barriers and cellular membranes (cellular delivery mechanism I). Formulations based on the cellular uptake of nanocrystals are still in development (cellular delivery mechanism II). Examples are i.v. targeting to endothelial cells of the blood-brain barrier and the loading of blood cells (monocytes, erythrocytes) to use these cells as transport vehicles for the nanocrystals. By now, very little work has been done to study and actively modulate the intracellular fate of nanocrystals.

Keywords Nanocrystals • Saturation solubility • Dissolution velocity • Dermal delivery • Oral delivery • Intravenous delivery • Nanotoxicity

R.H. Müller, R. Shegokar, S. Gohla, and C.M. Keck (✉)
Department of Pharmaceutics, Biopharmaceutics and NutriCosmetics,
Institute of Pharmacy/Pharmaceutical Technology, Freie Universität Berlin,
Kelchstr. 31, 12169 Berlin, Germany
e-mail: cornelia.keck@fh-kl.de

A. Prokop (ed.), *Intracellular Delivery: Fundamentals and Applications*, Fundamental Biomedical Technologies 5, DOI 10.1007/978-94-007-1248-5_15,

1 Introduction

Soluble drugs can reach their target in the body (organ, cell, cellular compartment) by a simple diffusion process of the drug molecules in the body fluids. In contrast to this, poorly soluble drugs – due to their insolubility in body fluids – need to have a carrier which carries them to the target, e.g. particulate nanocarriers. Diffusion in the body fluids is a non-specific, i.e. non-directed process. Soluble drug molecules distribute regarding distribution velocity and organ pattern according to their molecular properties (e.g. molecular weight, diffusion coefficient, log P value, permeability of membranes for the specific molecule, clearance). In contrast to this, incorporation of a molecule in a nanocarrier allows to target it to specific sites in the body, certain organs, specific cells within this organ or ideally to a specific cell compartment.

Such nanocarriers are e.g. micelles, polymeric nanoparticles or lipidic systems such as nanoemulsions, liposomes, transfersomes, niosomes or lipid nanoparticles made from solid lipids (e.g. solid lipid nanoparticles (SLN) and nanostructured lipid carriers (NLC) (Müller, Shegokar, and Keck, in press). A basic problem of these nanocarriers is that very often the loading capacity for drug is relatively low, especially when drugs are just bound to the surface of a nanocarrier (e.g. dalargin to the surface of polymeric nanoparticles (Kreuter et al. 1997)). This is a very pronounced problem when the aim is to target. Only a limited fraction of the administered carriers reach the target (assumed 10%). If the loading capacity is low (e.g. 10%), only a small fraction of the totally administered drug reaches the organ which might not be enough to reach a therapeutic level for cure of the disease. Therefore a nanocarrier would be ideal having a loading capacity of close to 100%.

This is realized by the nanocrystals. They are particles consisting of drug only without any matrix material as e.g. in polymeric nanoparticles (polymers) or liposomes (phospholipids). They are only stabilized by an adsorbed surfactant layer or sterically stabilizing polymer layer. Such adsorption layers on nanoparticles are typically 2–5 nm, maximum about 10 nm in thickness. Considering cubic nanocrystals of 500 nm size and the adsorption layer as part of the nanocrystal, this corresponds to maximum just 6% of the volume. Practically the nanocrystals can simplified be considered as nanocarriers with 100% loading capacity.

Major prerequisites for use of a nanocarrier in therapy are the possibility of controlled production, ability to produce on large industrial scale, accordance with regulatory requirements (e.g. status of excipients), ability for efficient delivery of drugs, ideally target to the interior of cells, and in case of nanocarriers the increasing need of data to prove the absence of nanotoxicity. Ideally the nanocarriers used should belong to class I of the nanotoxicological classification system (Müller, Gohla, and Keck, in press). These issues will be presented and discussed within this book chapter.

2 Definitions and Special Properties

Drug nanocrystals are particles consisting of pure drug only, and being by definition in the nano size range, i.e. below 1,000 nm to a few nm. They are in the crystalline state, due to this they are of cuboid shape. Trade names are NanoCrystal® (élan, prev. Nanosystems), DissoCubes® (SkyePharma, prev. PharmaSol) and smartCrystal® (Soliqs/Abbott, prev. PharmaSol). Drug nanoparticles can also be amorphous. In this case, in a strict sense they should not be called nanocrystals, trade name is Nanomorph (Soliqs/Abbott). Because of the amorphous state and lack of ordered, periodically repeating structure, they are spherical (Fig. 1).

There are different views how big a nanoparticle is. The above definition is based on dimensional considerations, i.e. the complete nanometer range, and typically used by most pharmacists. The US Food and Drug Administration (FDA), the National Nanotechnology Initiative (NNI) and the new European Cosmetic regulations consider nanoparticles as particles below 100 nm (Regulation (EC) No 1223/2009; National Nanotechnology Initiative (NNI) 2001). This can be rationalized by the increased toxicity risk when going below this size threshold (cf. 5, NCS classes). Some colloid scientists consider a "real" nanoparticle below about 20 nm.

In case the nanocrystals or amorphous drug nanoparticles are dispersed in a liquid, this is called "nanosuspension". Nanocrystals can be dispersed in either water or in non-aqueous dispersion media. Examples for non-aqueous media are oils, paraffins, liquid polyethylene glycols (PEGs) but also solid PEGs. In the latter case it is a solid nanosuspension or solid nanodispersion, analogous to classical solid dispersions of particles or solid dispersions of solutes.

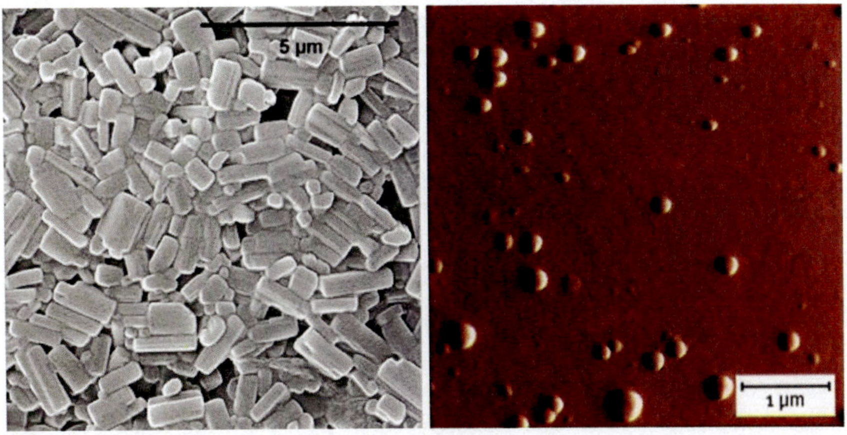

Fig. 1 Crystalline nanocrystals with cuboid shape (*left*, Courtesy by Böhm (1999), Modified) and nanoMorph amorphous drug nanoparticles of spherical shape due to lack of ordered crystal structure (*right*, Courtesy by Soliqs, Ludwigshafen, Germany)

Special properties of nanocrystals are an increased saturation solubility c_s and an increased surface area A, both leading to an increased dissolution velocity dc/dt according to the Noyes-Whitney equation (Buckton and Beezer 1992). The increase in saturation solubility leads to increased concentration gradients at biological barriers and membranes (e.g. gut wall, skin, barriers such as blood-brain barrier), and subsequently to increased penetration into or permeation across (Keck and Müller 2010). The increased dissolution velocity has advantages, but implies the problem that the nanocrystals might be dissolved before reaching the cellular target. How to deal with this problem is discussed below. Another important feature is the adhesiveness to surfaces. Identical to any other nanomaterial, nanocrystals stick to surfaces due to increased interaction of their large surface with substrates. This adhesion process is very reproducible, and the reason for better pharmacological performance, e.g. minimized variation in bioavailability when using nanocrystals as delivery system (Liversidge and Conzentino 1995; Liversidge and Cundy 1995). The physical background of the nanocrystal properties is described in detail in (Müller et al. 2003).

3 Process Technology

3.1 *Bottom Up Processes*

In a bottom up process, one starts from a small unit and in the process the size is increased. In case of nanocrystals one starts from a molecular solution of the drug, the molecules are aggregated to form particles in the nanometer size range. Practically in most of these processes we have a classical precipitation. A solvent containing the drug is added to a non-solvent. The solvent can be water, or organic solvents, the non solvent can be fluids miscible with the solvent (e.g. ethanol, acetone) or even supercritical fluids, typically carbon dioxide.

The so called hydrosols by Sucker are generated by classical precipitation, they are crystalline (List and Sucker 1988). Applying special precipitation conditions leads to amorphous nanoparticles, a process developed by Auweter and co-workers at BASF Germany. Food products on the market based on this technology are carotenoid powders (Auweter et al. 2002). In the pharmaceutical area, amorphous precipitation is performed by Soliqs for its product NanoMorph. There is quite a variety of other precipitation methods described in the literature, e.g. high-gravity controlled precipitation technology (Chen et al. 2009) and flash nanoprecipitation (Bénet et al. 2002; Johnson et al. 2006). Also a number of supercritical fluid processes is applicable, which would actually require a review on their own. Therefore it is referred to (Byrappa et al. 2008).

Controlled precipitation is a little bit tricky to run, costly because solvents might need to be removed, solvent residues need to be controlled, with regard to many aspects more complex than some top down processes. This is the reason why there are no currently marketed pharmaceutical products for therapy made with this technology.

However, this should not lead to the conclusion that bottom up processes are not industrially viable! On the contrary, they might be the basis for the third generation of drug nanocrystals/amorphous drug nanoparticles. By precipitation very small sizes (<<100 nm) are accessible, which are not or only very difficult to access by top down technologies. They are of high interest especially for intracellular delivery because they can be internalized by endocytosis/pinocytosis, that means by many cells in the body and not only by macrophages.

3.2 Top Down Processes – First Generation Nanocrystals

In the top down processes one starts from a larger size unit and reduces the size to the nanometer range. Typically wet milling processes are applied. They can be differentiated in low energy milling and high energy milling processes, i.e. bead (pearl, ball) milling and high pressure homogenization. They are used for production of nanocrystals which can be referred to as the first generation: NanoCrystal®, DissoCubes® and NANOEDGE®.

The NanoCrystals® are produced by bead milling (Liversidge et al. 1992). A macrosuspension of the drug powder is fed to a bead mill, containing small hard milling beads (e.g. 0.2–0.4 mm diameter, made from zirconium oxide or hard polystyrene). The beads are moved by an agitator or by rotating the milling chamber itself, and the drug crystals are ground between the moving beads (Merisko-Liversidge 2002). The system can be discontinuous (mill filled with complete batch), or continuous by pumping macrosuspension through the mill. The lab scale milling process typically lasts a few hours, relatively low energy is applied for a longer time.

In the high energy process of homogenization (DissoCubes® (Müller et al. 1999)) the macrosuspension is passed through the homogenizer (piston-gap, jet stream) at pressure of 1,500 bar and higher. Typically 10–20 passes through the homogenizer need to be applied. In the first generation, the dispersion medium of the macrosuspension was water, cavitation was considered as major cause for size reduction. In the NANOEDGE® process, a pre-step of precipitation is performed, then the precipitated suspension is exposed to a high energy step, typically high pressure homogenization (Kipp et al. 2003). In a further development, precipitation can be performed in a counter flow process (Kipp et al. 2005).

3.3 Top Down Processes – Second Generation Nanocrystals

In further developments, the nanocrystal production was improved regarding:

- running of process (e.g. less homogenization cycles, production of final dosage forms)
- smaller particle sizes (<100 nm)
- physical properties of nanosuspension (improved physical stability, long-term and against electrolytes).

This was achieved by using combination processes, but also varying the dispersion medium. Water was replaced by non-aqueous dispersion media, or water mixtures (Nanopure®, Müller et al. (2000)). Oily nanosuspensions can directly be filled into capsules for oral delivery. Combination processes optimize in a first step the drug material to be more fragile in the second step of high pressure homogenization, or to yield more physically stable nanocrystals. Combinations are:

H42: spray-drying of drug solution and subsequent homogenization (Möschwitzer 2005)

H69: precipitation of suspension just before entering or in the cavitation zone of the homogenizer (Müller and Möschwitzer 2007)

H96: lyophilisation of drug solution and subsequent homogenization (Möschwitzer and Lemke 2007)

CT: bead milling followed by homogenization (Petersen 2006)

These different technologies are unified under the trade name smartCrystal® by Soliqs/Germany, offered as a toolbox of different processes for tailor-making of nanocrystals.

4 Process Scale Up

Lack of scale up ability is one of the reasons for the failure of interesting nanocarriers – feasible on lab scale – to enter the market. First of all, the process itself needs to be scaleable, secondly the process needs to be able to be qualified and validated, the production lines have to be regulatorily acceptable, and of course cost-effective (sufficient capacity per hour, limited manpower required for process, yield, costs of excipients). The costs of the production line are less critical, because it is a unique investment. Below the processes are discussed possessing from our point of view presently highest commercial potential, being already used in products or being closest to products. Therefore the supercritical processes are not considered.

4.1 Precipitation

Precipitation itself is basically a simple process when the precipitation conditions are established. Critical can be the scale up, because precipitation velocity is quite different on lab scale (e.g. 1 L volume) or on large scale in a 1,000 L container. Mixing of solvent and non-solvent is much faster in a small volume. The Ostwald-Mier region is passed very fast, which yields small crystals. A major development step, easing the scale up tremendously, is the static blenders. Precipitation takes place within the small volume of the static blender, being of similar size than the lab scale beaker. Lab scale conditions are imitated on large scale production.

Solvent and non-solvent are pumped via separate tubes into the static blender. The blender contains buffles leading to fast blending and precipitation. The system can be run continuously from two supply containers (Fig. 2, upper).

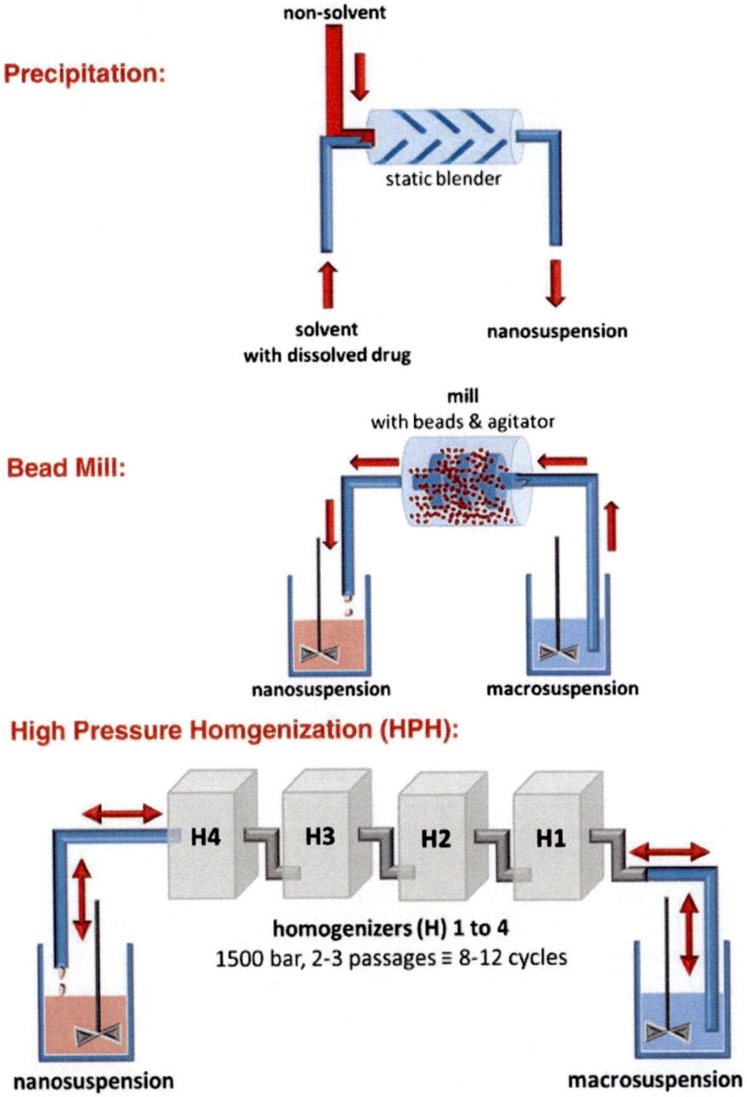

Fig. 2 Large scale production of nanocrystals (explanations cf. text): *upper* – solvent with drug and non solvent are mixed in a static blender; *middle* – macrosuspension is passed multiply from two containers through a bead mill; *lower* – four homogenizers in line for nanocrystal production, one pass is equivalent to four homogenization cycles

4.2 Bead Milling

Scale up of the bead mill in a discontinuous production is only feasible to a limited extend, because the weight of the beads (density zirconium oxide: approx. 6 g/mL (Zirconium dioxide – Wikipedia)) would lead to a too high weight of the mill. It needs to be also considered that about 70% of the mill volume is filled by beads, leaving only about 30% volume for adding the macrosuspension. A relatively high mill weight results compared to a low volume of suspension for processing. Therefore the continuous process is applied in pharmaceutical industry. The suspension is passed multiply through the mill (Fig. 2, middle). Very often, with just one passage a mean diameter in the nanometer range is obtained. The number of passages depends on the product requirements. About three to four passages are typically required to obtain a mean diameter below 500 nm. To minimize the content of crystals >1 μm (e.g. for i.v. injectables) about six to ten passages are needed. Typically after 10 passages, latest 15 passages very often the maximum dispersivity is reached. The perfectness of the crystals increases with decreasing size because the crystals break preferentially at their imperfections. At maximum dispersivity the energy is not sufficient to break these more perfect crystals. Therefore additional passages do not lead to further size reduction. On the contrary additional energy input can promote aggregation because the energy put in cannot be used any more for size reduction. This energy is now available to overcome the repulsive forces between crystals and to aggregate them.

4.3 High Pressure Homogenization

Homogenizers with a homogenization capacity of 1,000 L/h and more are commercially available. This is very favourable for large scale production and can compensate the need of 10 or even 20 passages through the homogenizer (= homogenization cycles) for the product. Compared to other equipment, the homogenizers are relatively low cost. To accelerate the process, four homogenizers can be placed in line, one passage through this line is equivalent to four homogenization cycles. The pumping velocity of the four homogenizers is electronically coordinated (Fig. 2, lower). Assuming a batch size of about half a ton, roughly 500 L, and ten production cycles on a homogenizer line with 1,000 L/h per homogenizer, one passage through this line is equivalent to four cycles and takes 30 min homogenization time. For the equivalence of ten cycles 1 h and 15 min are required, plus preparation and cleaning time.

Important for the output is also the solid concentration of the suspension to be processed. For both bead milling and homogenization concentrations in the range 10–20% are in most cases processable without sincere problems. With the bead mill sometimes concentrations up to about 40% are processable (depending on particle size of the starting material, affecting the viscosity). However, more viscous suspensions are more difficult to process. Therefore at the end of the day it might be faster to have longer milling time but using a less concentrated, less viscous suspension.

5 Nanotoxicity and Regulatory Aspects

A key issue in nanocarrier development since decades is the regulatory status of the excipients used. In addition, in recent years potential nanotoxicty is coming increasingly into the focus of the consumer, triggered by often uncritical, simplifying reports in the popular press. Nanotoxocity is sincerely an issue, because by transferring material to the nanodimension its physico-chemical properties and subsequently its interaction with the biological environment, mainly the cells, change.

Many nanocarriers originating from the lab are made from materials which are not accepted for use by the authorities, e.g. newly synthesized polymers to give the nanoparticles a special performance. Industry is very reluctant to invest in expensive toxicity studies, especially when the result is questionable with a new compound, thus hindering these nanocarriers to enter clinical trials and the market. The situation is not that much different, when excipients are used which are accepted by authorities, but not accepted for the respective purpose. The classical example is the poly lactic glycolic acid polymer (PLG). They are on the pharmaceutical market as microparticles for injection (e.g. Enantone Depot, Decapeptyl Depot). However, when making small particles from this polymer, pronounced cytotoxicty was found in macrophage cultures (Smith and Hunneyball 1986). In contrast to 50–100 μm microparticles, particles with a size of a few μm and nanoparticles can be taken up by macrophages. Intracellular degradation leads to release of lactic and glycolic acids, causing at too high concentration the observed cytotoxicity – despite being both physiological compounds in the body.

In contrast to many other nanocarriers, the drug nanocrystals can be made from regulatorily accepted excipients only. They consist just of the drug and stabilizers. One can choose from a broad range of stabilizers regulatorily accepted in formulations for the respective administration route (dermal, oral, intravenous (i.v.), intramuscular, etc.). Examples range from various polymers for oral administration (e.g. polyvinyl alcohol (PVA), polyvinyl pyrrolidon (PVP)) including the basically membrane-toxic sodium dodecyl sulphate (SDS) to i.v. accepted lecithins, Tween 80, Poloxamer 188, low molecular weight PVP. This gives a big advantage to this delivery system regarding use in the clinic and entering the market. Of course, the question of potential nanotoxicity still needs to be discussed.

According to the nanotoxicological classification system (NCS, Fig. 3) different levels of tolerability and toxicity are distinguished, based on the particles size and the degree of biodegradability. There are two size classes of the NCS, nanoparticles in the range of about 100–1,000 nm can only be taken up by macrophages, possess therefore only access to a limited number of body cells, having therefore a lower toxicological risk. Nanoparticles below 100 nm can enter any cell of the body by

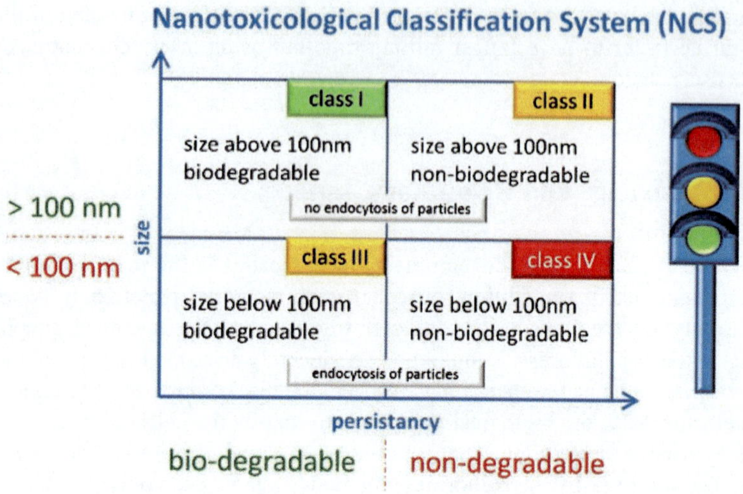

Fig. 3 The nanoparticles are differentiated in class I–IV with increasing toxicological risk, based on size (<100 nm, 100–1,000 nm) and biodegradability/non-biodegradability (i.e. persistency in the body) (Modified after Müller, Gohla, and Keck, submitted)

endocytosis/pinocytosis, having therefore a higher potential risk. Nanoparticles which are biodegradable in the body will disappear after some time, thus potential undesired effects are often limited to the time of existence of the particles. Biodegradable nanoparticles are therefore also a lower risk class. Non-biodegradable particles such as e.g. fullerens and carbon nanotubes (CNT) – often discussed as potential drug delivery systems – will stay forever. Non-biodegradable particles can cause continuing irritations, thus being excluded for the use as drug delivery system. Based on these considerations, the nanocrystals are in the lowest risk class I, or in class III depending on the size above or below 100 nm.

It should be kept in mind that also a biodegradable nanoparticle can cause toxic effects. For example during their life time biodegradable nanoparticles can be taken up by cells of the immune system and can cause irritation/activation of the immune system. Therefore also in this case assessment of potential nanotoxicity risks is meaningful. Assumingly because of the a priori assumed lack of nanotoxicty of the nanocrystals, there are very few reports about cytotoxicity investigations. Oral nanocrystal products are on the pharmaceutical market since 10 years, but no systematic investigations are published. Good tolerability can be assumed because mankind lives for centuries with drug nanocrystals present in the gastrointestinal tract (GIT). Each drug crystal orally administered will reduce in size during its dissolution, to a few µm and finally to the nanometer range prior to its complete dissolution. Nanosuspensions were mainly investigated regarding treatment efficiency, in vitro and in vivo, less looking at toxic effects. For example, the efficacy of atovaquone nanosuspensions was investigated against toxoplasma gondii in vitro

(Schöler et al. 2000) and in vivo (Schöler et al. 2001). Recently toxicity investigations were published on the new phospholipase A_2 inhibitors PX-13 and PX-18 nanocrystals (Pardeike and Müller 2010). The nanocrystals were invesitgated using the EPISKIN Test and the HET-CAM test to study the eye irritation potential. The nanosuspensions were found to be dermally safe, they were not or only slightly irritant to the eye.

It was also found that nanocrystals can even reduce irritancy to cell layers, e.g. the gastric wall. Nanocrystals of naproxen were shown to decrease the gastric irritancy (Liversidge and Conzentino 1995) compared to the drug powder. Intraperitoneally injected azodicarbonamide (ADA) was not irritant as nanosuspension, much better tolerated than irritant micrometer crystals (unpublished data). The underlying mechanism could be that the drug is more evenly, finely distributed about the walls of the gut or the peritoneum, similar to the improved tolerability of pellets compared to tablets loaded with irritating drugs. In addition, size and form of the drug crystals might also affect the tolerability (e.g. long needles compared to small cubes, e.g. similar to the toxicity of needle-like carbon nanotubes).

As a general aspect, nanocrystals normally change the pharmacokinetic profile of actives (i.e. higher c_{max}, shorter t_{max}). This can lead locally to a higher drug exposure of cells (e.g. kidney at higher plasma concentrations). This effect is well described for the nephrotoxicity of Amphotericin B. The nephrotoxicity is higher in formulations leading to higher concentrations of free Amphotericin B in the blood (Amphotericin B injectable solution versus Ambisome® liposomes). Definitely, there is a need for further closer examination of cellular effects of nanocrystals.

6 Nanocrystal Products on the Market – "Cellular Delivery Mechanism I"

6.1 *Cosmetic/Dermal Market*

Normally the cosmetic market watches carefully developments in pharmaceutical labs and industry to identify technologies and carrier systems with potential use in cosmetics. The classical example are the liposomes which appeared first on the cosmetic market (1986, product Capture by Dior) before entering the pharmaceutical market around 1990. In case of the nanocrystals this was different. The nanocrystals appeared first on the pharmaceutical market in 2000 (product RAPAMUNE® company Wyeth). The potential of nanocrystals for dermal delivery was realized a few years later (Petersen 2006). The first products were placed on the market in 2007 (line Juvedical, company Juvena).

In these products the delivery of the cosmetic active is not achieved by intracellular uptake of the nanocrystals. Based on the physical properties of the nanocrystals, they deliver molecules to the cell which dissolve from their surface, which we call delivery mechanism I. In delivery mechanism II, delivery of active to the cell takes place by uptake of the nanocrystals themselves (cf. 7). Delivery mechanism I consists of the steps (Fig. 4, upper):

1. Formation of a supersaturated solution around the crystal, thus
2. creating a high concentration gradient between nanocrystal and target cell, and
3. fast replacement of diffused molecules by very fast continuing dissolution from the large nanocrystal surface (= depot).

Supersaturated solution, concentration gradient and dissolution are much faster compared to micrometer crystals, thus explaining the superior delivery properties

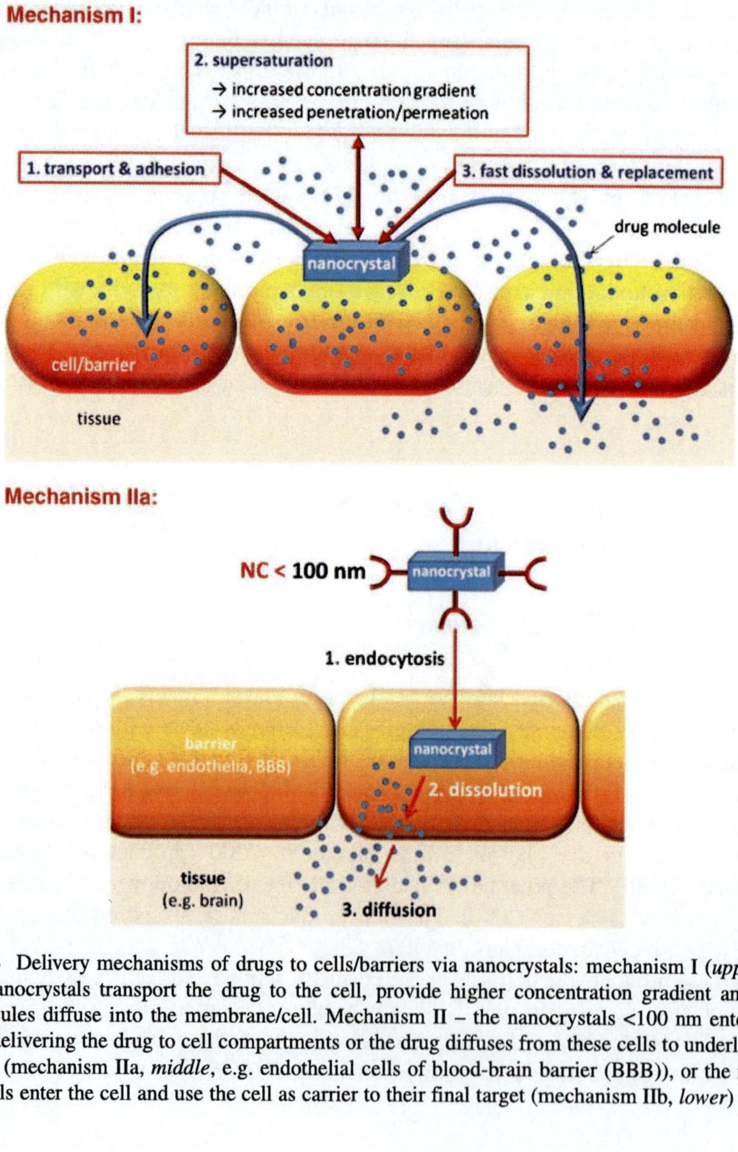

Fig. 4 Delivery mechanisms of drugs to cells/barriers via nanocrystals: mechanism I (*upper*) – the nanocrystals transport the drug to the cell, provide higher concentration gradient and the molecules diffuse into the membrane/cell. Mechanism II – the nanocrystals <100 nm enter the cell, delivering the drug to cell compartments or the drug diffuses from these cells to underlaying tissue (mechanism IIa, *middle*, e.g. endothelial cells of blood-brain barrier (BBB)), or the nano-crystals enter the cell and use the cell as carrier to their final target (mechanism IIb, *lower*)

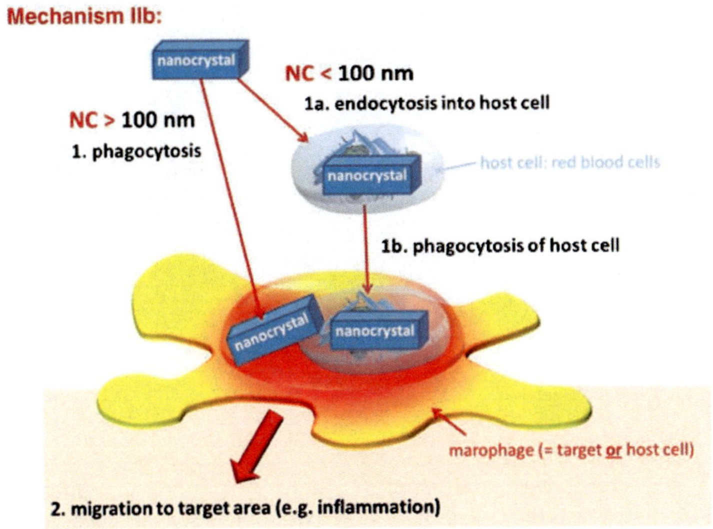

Fig. 4 (continued)

of nanocrystals. The mechanism will be explained in detail using the example of rutin and hesperidin (Petersen 2006).

Delivery of actives to the skin or any other target site can be studied by measuring concentrations at the target site or ideally by measuring the resulting effect. Measuring the effect assesses the overall performance of a delivery formulation, especially being essential when comparing two molecule derivatives (e.g. original molecule and derivative, e.g. glucoside). Rutin and hesperidin are poorly soluble antioxidants, the solubility is so low that incorporation as normal micrometer powder into a cream will lead to no effect. Therefore in the study by Petersen soluble rutin was made by glucosidation. The rutin-glucoside was compared in its antioxidative capacity to nanocrystals of the original poorly soluble rutin and in addition to hesperidin (Petersen 2006). In addition a formulation with alpha tocopherol acetate was run, which should make the skin more sensitive. To quantify the activity in the skin (= effect), the sun protection factor (SPF) was determined in humans under UV exposure. Antioxidant activity increases the SPF. Measurement of penetrated concentrations (e.g. tesa stripping test, biopsy) would have been less conclusive, because it does not consider potential different activities of the molecules in the cell (original versus glucoside). The overall performance of a formulation is a function of delivery efficiency and activity of the delivered type of molecule.

The rutin glucoside was dissolved at a concentration of 5.0% in the dermal test formulation. The nanocrystal formulation contained a depot of nanocrystals and a dissolved concentration of rutin of about 0.01%, i.e. 500 times less dissolved active. In vivo the rutin glucoside formulation increased the SPF by 27%, the rutin nanocrystal dermal formulation by 59%, i.e. two times increase at 500 times less dissolved molecules. Simplified it could be stated that the nanocrystal formulation

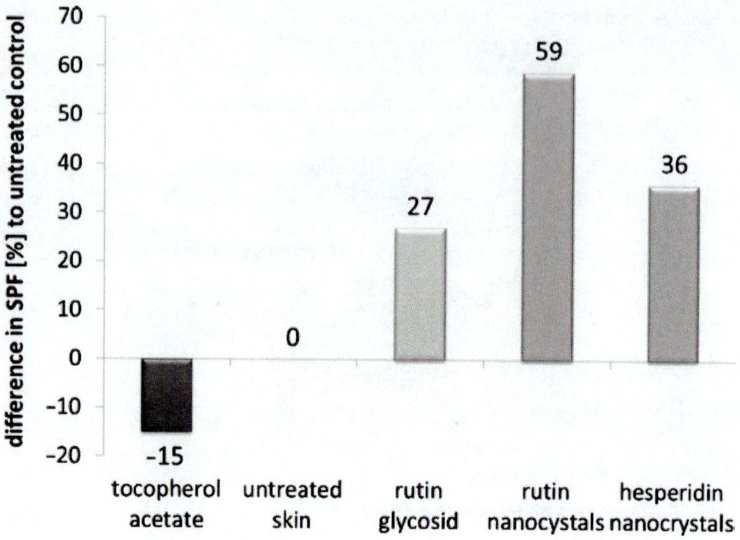

Fig. 5 Change in % of SPF of human skin after treatment with formulations containing alpha tocopherol acetate, water soluble rutin glucoside, rutin nanocrystals and hesperidin nanocrystals (After Petersen 2006, SPF of untreated skin was set = 100%)

has a 2×500 = 1,000 fold higher activity. A similar performance was observed for hesperidin (increase of SPF by 36%), meanwhile launched in the product platinum rare (la prairie). The alpha tocopherol acetate reduced the SPF (Fig. 5).

Explanation of the mechanism of action (Fig. 6):

1. Glucosidation made the molecule more hydrophilic, thus more water soluble. 5% could be dissolved which provided a very high concentration gradient between dermal formulation and skin. However, the hydrophilic derivative likes the hydrophilic environment in the dermal formulation, and stays there. In addition, the hydrophilic glucoside has less penetration ability than the more hydrophobic rutin.
2. The more hydrophobic rutin has a priori better penetration ability, the nanocrystals provided a supersaturated solution and consequently a concentration gradient obviously high enough for a sufficient penetration causing the antioxidant effect in the skin. Rutin penetrated into the skin was immediately replaced from rutin dissolving from the nanocrystal depot.
3. It can be assumed that the original lipophilic molecule has a higher affinity to the respective binding sites in the cell than the hydrophilic glucoside, thus being superior in antioxidant activity.
4. The observed increase in SPF is a superposition of delivery ability of the formulation and the intracellular effect of the molecule. The sum of penetration and efficiency is higher for the nanocrystal formulation, despite the 500 times less dissolved rutin in the formulation.

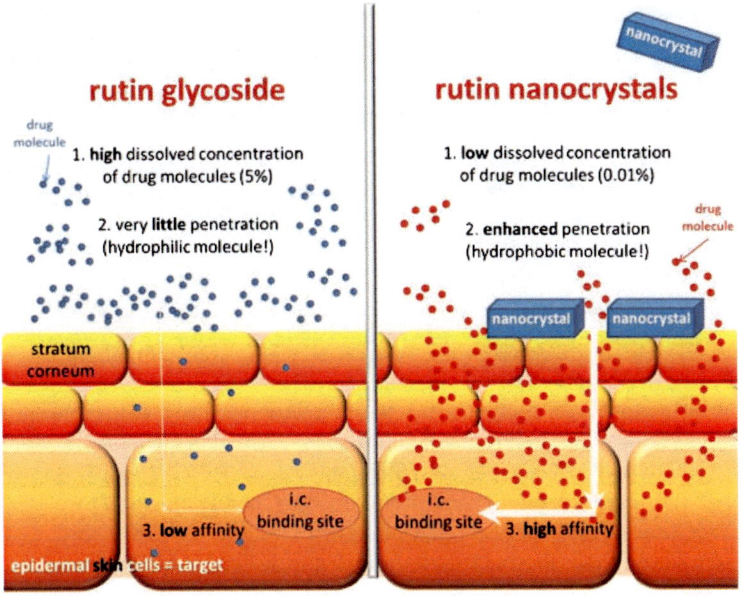

Fig. 6 Mechanism of improved dermal action of rutin nanocrystals (*right*) versus water soluble rutin glucoside (*left*, explanation cf. text)

A similar but less pronounced increase in SPF was observed for the hesperidin nanocrystals. The important conclusion is, that the nanocrystal technology opens the perspective to use new classes of molecules, e.g. the plant molecules such as flavonoids. They could not be used before because of low solubility and limited penetration of water soluble derivatives. The rutin glucoside shows some effect, but it needs to be kept in mind that 5% are needed in the formulation, and the effect being less (hesperidin), or only half of the effect of the nanocrystals (rutin). The same formulation principle can of course also be applied to pharmaceutical dermal formulations.

6.2 *Pharmaceutical Market*

The products on the pharmaceutical market (Table 1) exploit the same cell delivery mechanism I as the dermal products, but using the oral administration route, only one product is an injectable (Table 1). However, the delivery advantage is often combined with an increased patient convenience. Patients using more patient-friendly dosage forms show a higher compliance compared to less user friendly formulations. The principles for oral delivery are the same as for dermal. Nanocrystals provide a higher concentration gradient at the barrier, the barrier skin is replaced by

Table 1 Nanocrystal products on the pharmaceutical market: trade name, drug, producer, nanoparticle technology

Product	Drug	Company	Nanoparticle technology
Rapamune®	Sirolimus	Wyeth	Elan nanocrystals®
Emend®	Aprepitant	Merck	Elan nanocrystals®
Tricor®	Fenofibrate	Abbott	Elan nanocrystals®
Megace® ES	Megestrol acetate	PAR pharmaceutical	Elan nanocrystals®
Invega® Sustenna™	Paliperidone palmitate	Janssen	Elan nanocrystals®
Triglide™	Fenofibrate	Shionogi pharma	SkyePharma IDD®-P technology

the barrier gut wall. In addition, the nanocrystals provide a fast replacement of molecules permeated. Therefore nanocrystals are a suitable formulation for class II drugs of the biopharmaceutical classification system (BCS). Class II drugs penetrate well (similar to lipophilic rutin), but dissolve slowly, the dissolution velocity is the rate limiting step for oral bioavailability.

The first product on the market was Rapamune (sirolimus) (Müller and Junghanns 2006). The nanocrystals in a tablet have a higher bioavailability than a solution of sirolimus. In addition the solution is less patient friendly, also required special storage conditions (fridge) and complicated reconstitution. The tablet can be stored at room temperature. The second product was Emend (aprepitant, pellets in a capsule). The drug aprepitant is absorbed within an absorption window in the upper intestinal tract. The nanocrystals provide a sufficiently high dissolution velocity to exploit this absorption window for efficient oral delivery of this drug, not achievable with a classical formulation. The most successful nanocrystal product is Tricor (fenofibrate, tablet) launched by Abbott. Fenofibrate shows a 35% higher absorption in the fed state. The nanocrystal formulation removed the big difference in bioavailability between fed and non-fed state. In addition the dose could be reduced from 54/160 mg to 48/145 mg. Sales are meanwhile more than one billion USD per year, a nano block buster. Triglide is the competitive product of fenofibrate, but produced with high pressure homogenization.

Remarkable is the product Megace *ES* (*E*nhanced *S*tability), a formulation of megestrol acetate. It is a suspension with the shelf life of a pharmaceutical product. It demonstrates nicely that physically long-term stable nanosuspension can be produced – despite their state of high dispersitivity, high interfacial energy and risk of Ostwald ripening. The previous oral suspension showed a food effect on the bioavailability. The nanosuspension reduced the difference between fed and non-fed, in addition the nanosuspension is less viscous and requires only ¼ of the administration volume – three patient convenient factors increasing compliance.

Invega® Sustenna® is the first injectable nanocrystal product, approved in 2009. As once-a-month release formulation of the drug paliperidone palmitate, it is injected intramuscularly, thus avoiding the problems of i.v. injected nanosuspensions. The aqueous nanosuspension delivers the drug in a small volume, conveniently to inject.

7 Targeted Nanocrystals in Development – "Cellular Delivery Mechanism II"

In these developments the nanocrystals deliver the drug to the cells by internalization. The nanocrystals are taken up, and dissolve inside the cell (= cell delivery mechanism II). The administration route used by now is intravenous injection in form of aqueous nanosuspensions. These nanosuspensions need to be made isotonic by addition of glycerol. Addition of NaCl has to be avoided, because this reduces the zeta potential of the nanocrystals and causes subsequently aggregation. Furthermore the nanosuspensions need to be sterile, either made sterile by terminal sterilization or be produced aseptically.

At the beginning i.v. nanosuspensions were developed with the aim to replace toxicologically problematic excipients in existing i.v. formulations on the market. Examples are Taxol and Sporanox. Taxol contains paclitaxel solubilised with Cremophor EL causing sometimes anaphylactic shocks during administration (Strachan 1981; Dye and Watkins 1980). Sporanox contains itraconazol made soluble by inclusion into hydroxypropyl cyclodextrin (HP-CD). The HP-CD can cause nephrotoxicity (Szejtli 1988). The technological aim was to produce nanosuspensions of both drugs, whereas the nanocrystals are stabilized by well tolerated stabilizers, e.g. lecithins or Poloxamer 188.

Paclitaxel nanosuspensions could be successfully produced, nanocrystal size about 300 nm. Stabilizers used were well tolerable phospholipon 90 and various Poloxamers (Böhm 1999; Böhm et al. 1997). However after i.v. administration of the nanosuspension the pharmacokinetic was completely different to the solution Taxol. The drug nanocrystals were recognized as being foreign to the body and taken up the macrophages of liver and spleen. The same was observed for an injected itraconazole nanosuspension (Rabinow et al. 2007). With regard to the original development aim of a generic product, this was a failure but the data demonstrate nicely the possibility to target drugs via nanocrystals to the cells of the mononuclear phagocytic system (MPS). Of high interest are for example anti-HIV drugs to target to viruses residing in the macrophages, e.g. as shown for the drug nevirapine (Müller, Shegokar, and Keck in press).

There are two ways to imitate the pharmacokinetics of injected solutions. Firstly the nanocrystals can be made small enough that they are dissolved before "meeting" the macrophages. It was shown that i.v. injected 897 nm oridonin nanosuspensions accumulated in the liver, whereas 103 nm nanocrystals showed a pharmacokinetics similar to a solution (Gao et al. 2008). Secondly, the nanocrystal surface can be modified analogue to the stealth liposomes generating stealth nanocrystals. A stealth surface avoids the adsorption of e.g. opsonins which leads to the recognition by the macrophages. Pre-requisite of this concept is that the stealth properties on the nanocrystal surface remain during the dissolution process of the nanocrystals in the blood. The stealth properties can be checked in vitro by analysing the protein adsorption patterns in plasma and in serum (Lück et al. 1998; Lind et al. 2001; Göppert and Müller 2003). The analytical tool is two-dimensional polyacrylamide gel electrophoresis (2-DE, 2D-PAGE) (Blunk et al. 1993) or two-dimensional differential in gel electrophoresis (2D-DIGE). At least a negative selection is possible identifying surfaces which adsorb opsonins, thus minimizing animal experiments.

Cells of the MPS are a relatively easy target, but the complexity starts when a certain MPS population should be targeted, e.g. lung macrophages. After i.v. injection recognized particles are cleared mainly by the liver macrophages, up to 90% of the injected does within 5 s, 2–5% by the spleen and only a few % by the lung macrophages (Müller 1991). Avoiding e.g. the uptake by the liver macrophages and directing the nanocrystals to macrophage subpopulations is a first challenge. One approach could be to use opsonins specific to macrophage subpopulations (Roubin and Zolla-Pazner 1979). For other cellular target sides, recognition by the MPS cells needs to be avoided completely and simultaneously a homing device attached to the surface to localize the particles at the target cells. After i.v. injection only target cells are accessible which can be reached via the blood stream.

Kreuter et al. found that Tween 80 stabilized i.v. injected polymeric nanoparticles could deliver the drug dalargin to the brain (Kreuter et al. 1997, 2003). As mechanism was identified that after injection apolipoprotein E in the blood bound to the particle surface and mediated the adherence to the endothelial cells of the blood-brain barrier (BBB) (Müller et al. 2001). For paclitaxel-loaded polymeric nanoparticles could be shown, that the drug was released in the endothelial cells and diffused from here into the brain (Gelperina et al. 2002; Kreuter 2001). However, the basic problem was that only a small part of the injected particle mass reached the brain (e.g. loss to the liver) and that the drug loading of the particles was relatively low. This resulted in low drug concentrations in the brain. It would be desirable to use a nanocarrier with a very high loading capacity, i.e. using drug nanocrystals.

This was realized with buparvaqone nanocrystal suspensions. In vitro it could be shown that the nanocrystals adsorbed apolipoprotein E. They were tested using a toxoplasmosis animal model. After i.v. injection into mice, the parasites could be completely eradicated in the brain (Schöler et al. 2001). However, cure of the animals was not achieved (Schöler 2001) under the study design applied, which can be a function of the design and/or some parasites still residing somewhere in the body. The drug loading with the nanocrystals was much higher than with polymeric nanoparticles, but after injection there was loss of drug by dissolution during their travel time to the blood-brain barrier. To minimize this drug loss, the nanocrystals should be coated with a thin polymer layer. The surface properties of the polymer layer can be designed this way that apolipoprotein E is adsorbed preferentially, may be in higher amounts than on Tween 80-stabilzed nanocrystals. For example, the nanocrystals could be coated with the polymer of the polymeric nanoparticles by Kreuter which proved efficient in targeting the endothelial cells (i.e. poly(butyl) cyanoacrylate) (Schroeder et al. 1998; Alyautdin et al. 1995).

Another concept is to use a "taxi" for the nanocrystals to deliver them into the brain. The taxi can be macrophages in the blood, which extravasate and travel e.g. to sites of inflammation, including crossing the blood-brain barrier. This concept was exploited by various research groups, e.g. Barrett Rabinow et al. (Dou et al. 2009), and in a modified version by Bäumler et al. (Staedtke et al. 2010). They loaded Amphotericin B nanosuspension (AmB-NS) into human red blood cells (RBCs). The AmB-NS-RBCs were then taken up by phagocytosis by leukocytes, which are the main effector arms of antifungal defense. The leukocytes carry drugs also in areas of inflammation. The loading of the RBCs with nanoparticles should be preferentially performed ex vivo in purified cell populations, also the phagocytosis by the leukocytes. This avoids competitive uptake (e.g. in vivo after injection of nanosuspension by MPS organs, ex vivo by other cells present in the suspension). In case the nanocrystals should be loaded into cells which have no phagocytic capacity (e.g. RBC), the nanocrystals need to be small enough to be internalized by pinocytosis, i.e. below 100 nm. This was performed for example with Amphotericin B nanocrystals which were 65 nm in size.

Basically three targeting levels can be differentiated:

1. a specific organ
2. a certain cell population within the organ
3. and ideally a specific compartment within the cells,

whereas level three is the most challenging. To our knowledge very little work has been done to study the intracellular fate of nanocrystals.

The reason for this is surely the problem of simultaneous distribution and dissolution of the nanocrystals inside the cell. One approach to tackle this problem is the use of fluorescent nanocrystals, and studying simultaneously dissolution and resulting drug distribution inside the cell. With different nanocrystals different drug distributions should be achievable. This is definitely one field of investigation in the future.

8 Conclusions and Perspectives

Delivery to target cells and subsequent intracellular delivery by internalization of particles is under investigation for more than half a century, dating back to the 1950s. This involves identification of mechanisms to localize nanocarriers in target cells or cell compartments, and in parallel the development of suitable nanocarriers, preferentially usable in patients at the end of the day.

At the beginning targeting mechanisms investigated were very simple, e.g. in the 1960s effect of charge on i.v. injected particles (Wilkins and Myres 1966). Meanwhile the targeting mechanisms got rather complex, e.g. via antibodies or via modulation of the protein adsorption patterns in the blood to enrich the particle in a target cell. In future it will get even more complex when considering the conformation of these proteins and the role of the conformation in cellular uptake.

Parallel went the development of nanocarriers. Using the number of nanocarriers for drug delivery on the market for treatment of patients as a performance measure, the success was limited. The most important nanocarriers in the second half of the last century were i.v. nanoemulsions and liposomes. In 2000 the nanocrystals appeared on the market. From our point of view, the nanocrystals have clear advantages compared to many of the "academic" nanocarriers presently under development. These are besides others accepted status of excipients, well tolerable (NCS class I/III), easy and cost-effective large scale production, proven make-ability of products. Tricor® is the first nano block buster, making the nanocrystals to the most successful nanotechnology by now.

The next steps will be to combine sophisticated targeting approaches with the nanocrystals as carrier and to exploit much more the opportunities in controlled intracellular delivery.

Acknowledgement The authors thank Abbott GmbH und Ko. KG and its drug delivery company Soliqs/Ludwigshafen in Germany for the kind permission to reproduce the NanoMorph picture in Fig. 1, right.

References

Alyautdin R, Gothier, D., Petrov, V., Kharkevich, D., Kreuter, J. (1995) Analgesic activity of the hexapeptide dalargin adsorbed on the surface of polysorbate 80-coated poly(butyl cyanoacrylate) nanoparticles. European Journal of Pharmaceutics and Biopharmaceutics 41:44–48

Auweter H, Bohn H, Heger R, Horn D, Siegel B, Siemensmeyer K (2002) Precipitated water-insoluble colorants in colloid disperse form. United States Patent 6,494,924

Bénet N, Muhr H, Plasari E, Rousseaux JM (2002) New technologies for the precipitation of solid particles with controlled properties. Powder Technology 128 (2–3):93–98

Blunk T, Hochstrasser DF, Sanchez JC, Müller BW, Müller RH (1993) Colloidal carriers for intravenous drug targeting: plasma protein adsorption patterns on surface-modified latex particles evaluated by two-dimensional polyacrylamide gel electrophoresis. Electrophoresis 14 (12):1382–1387

Böhm BHL (1999) Herstellung und Charakterisierung von Nanosuspensionen als neue Arzneiform für Arzneistoffe mit geringer Bioverfügbarkeit. PhD-thesis, Freie Universität Berlin,

Böhm BHL, Behnke D, Müller RH Production of paclitaxel nanosuspensions by high-pressure homogenisation. In: Int. Symp. Control. Rel. Bioact. Mater. 24, 1997. pp 927–928

Buckton G, Beezer AE (1992) The relationship between particle size and solubility. International Journal of Pharmaceutics 82 (3):R7-R10

Byrappa K, Ohara S, Adschiri T (2008) Nanoparticles synthesis using supercritical fluid technology – towards biomedical applications. Advanced Drug Delivery Reviews 60 (3):299–327

Chen J-F, Wang Y-H, Guo F, Wang X-M, Zheng C (2009) Synthesis of Nanoparticles with Novel Technology: High-Gravity Reactive Precipitation. Industrial & Engineering Chemistry Research 39 (4):948–954

Dou H, Grotepas CB, McMillan JM, Destache CJ, Chaubal M, Werling J, Kipp J, Rabinow B, Gendelman HE (2009) Macrophage delivery of nanoformulated antiretroviral drug to the brain in a murine model of neuroAIDS. J Immunol 183 (1):661–669

Dye D, Watkins J (1980) Suspected anaphylactic reaction to Cremophor EL. Br Med J 280 (6228):1353

Gao L, Zhang D, Chen M, Duan C, Dai W, Jia L, Zhao W (2008) Studies on pharmacokinetics and tissue distribution of oridonin nanosuspensions. Int J Pharm 355 (1–2):321–327

Gelperina SE, Khalansky AS, Skidan IN, Smirnova ZS, Bobruskin AI, Severin SE, Turowski B, Zanella FE, Kreuter J (2002) Toxicological studies of doxorubicin bound to polysorbate 80-coated poly(butyl cyanoacrylate) nanoparticles in healthy rats and rats with intracranial glioblastoma. Toxicology Letters 126 (2):131–141

Göppert TM, Müller RH (2003) Plasma protein adsorption of Tween 80- and poloxamer 188-stabilized solid lipid nanoparticles. Journal of Drug Targeting 11 (4):225–231
Johnson BK, Saad W, Prud'homme RK (2006) Nanoprecipitation of Pharmaceuticals Using Mixing and Block Copolymer Stabilization. In: Polymeric Drug Delivery II - ACS Symposium Series, vol 924. American Chemical Society, pp 278–291
Keck CM, Müller RH (2010) smartCrystals – Review of the Second Generation of Drug Nanocrystals. In: Torchilin VP, Amiji MM (eds) Handbook of Materials for Nanomedicine. Imperial College Press, London, p Pan Stanford Publishing
Kipp JE, Wong JCT, Doty MJ, Rebbeck CL (2003) Microprecipitation Method For Preparing Submicron Suspensions. United States Patent 6,607,784
Kipp JE, Wong JCT, Doty MJ, Werling J, Rebbeck CL, Brynjelsen S (2005) Method for preparing submicron particle suspensions. 6,884,436
Kreuter J (2001) Nanoparticulate systems for brain delivery of drugs. Adv Drug Deliv Rev 47 (1):65–81
Kreuter J, Petrov VE, Kharkevich DA, Alyautdin RN (1997) Influence of the type of surfactant on the analgesic effects induced by the peptide dalargin after its delivery across the blood-brain barrier using surfactant-coated nanoparticles. Journal of Controlled Release 49 (1):81–87
Kreuter J, Ramge P, Petrov V, Hamm S, Gelperina SE, Engelhardt B, Alyautdin R, von Briesen H, Begley DJ (2003) Direct evidence that polysorbate-80-coated poly(butylcyanoacrylate) nanoparticles deliver drugs to the CNS via specific mechanisms requiring prior binding of drug to the nanoparticles. Pharmaceutical Research 20 (3):409–416
Lind K, Kresse M, Müller RH (2001) Comparison of protein adsorption patterns onto differently charged hydrophilic superparamagnetic iron oxide particles obtained in vitro and ex vivo. Electrophoresis 22 (16):3514–3521
List M, Sucker H. (1988) Pharmaceutical colloidal hydrosols for injection. GB Patent 2,200,048
Liversidge GG, Conzentino P (1995) Drug particle size reduction for decreasing gastric irritancy and enhancing absorption of naproxen in rats. Int J Pharm 125 (2):309–313
Liversidge GG, Cundy KC (1995) Particle size reduction for improvement of oral bioavailability of hydrophobic drugs: I. Absolute oral bioavailability of nanocrystalline danazol in beagle dogs. Int J Pharm 125 (1):91–97
Liversidge GG, Cundy KC, Bishop JF, Czekai DA (1992) Surface modified drug nanoparticles. United States Patent 5,145,684
Lück M, Paulke BR, Schröder W, Blunk T, Müller RH (1998) Analysis of plasma protein adsorption on polymeric nanoparticles with different surface characteristics. Journal of Biomedical Materials Research 39 (3):478–485
Merisko-Liversidge E Nanocrystals: Resolving Pharmaceutical Formulation Issues associated with poorly water-soluble Compounds. In: Marty JJ (ed) Particles, Orlando, 2002. Marcel Dekker,
Möschwitzer JP (2005) Method for the production of ultra-fine submicron suspensions. DE, 10 2005 011 786.4
Möschwitzer J, Lemke A (2007) Method for carefully producing ultrafine particle suspensions and ultrafine particles and use thereof. PCT/EP2006/003377
Müller RH (1991) Colloidal Carriers for Controlled Drug Delivery and Targeting. Wissenschaftliche Verlagsgesellschaft mbH, CRC Press, Stuttgart, Boston
Müller RH, Shegokar R, Keck CM (in press) 20 Years of Lipid Nanoparticles (SLN, NLC): present state of development & industrial applications. Curr Drug Discov Technol PMID: 21291409
Müller RH, Gohla S, Keck CM (submitted) State of the Art of Nanocrystals – special features, production, nanotoxicology aspects & intracellular delivery. European Journal of Pharmaceutics and Biopharmaceutics
Müller RH, Junghanns J-U (2006) Drug nanocrystals/nanosuspensions for the delivery of poorly soluble drugs. In: Torchilin VP (ed) Nanoparticulates as Drug Carriers. 1 edn. Imperial College Press, London, pp 307–328
Müller RH, Möschwitzer J (2007) Method and device for producing very fine particles and coating such particles. PCT/EP2006/009930

Müller RH, Becker R, Kruss B, Peters K (1999) Pharmaceutical nanosuspensions for medicament administration as systems with increased saturation solubility and rate of solution. United States Patent 5,858,410

Müller RH, Mäder K, Krause K (2000) Method for controlled production of ultrafine microparticles and nanoparticles. PCT/EP2000/006535

Müller RH, Lück M, Kreuter J (2001) Medicament excipient particles for tissue-specific application of a medicament. United States Patent 6,288,040

Müller RH, Jacobs, C., Kayser, O. (2003) DissoCubes – a novel formulation for poorly soluble and poorly bioavailable drugs. In: Rathbone MJ, Hadgraft, J., Roberts, M. S. (ed) Modified-Release Drug Delivery Systems. Marcel Dekker, pp 135–149

National Nanotechnology Initiative (NNI) (2001) What is Nanotechnology? http://www.nano.gov.

Regulation (EC) No 1223/2009 of the European Parliament and of the Council of 30 November 2009 on cosmetic products (2009). Official Journal of the European Union L342:59–209

Pardeike J, Müller RH (2010) Nanosuspensions: a promising formulation for the new phospholipase A2 inhibitor PX-18. Int J Pharm 391 (1–2):322–329

Petersen RD (2006) Nanocrystals for use in topical fomulations and method of production thereof. PCT/EP2007/009943

Rabinow B, Kipp J, Papadopoulos P, Wong J, Glosson J, Gass J, Sun CS, Wielgos T, White R, Cook C, Barker K, Wood K (2007) Itraconazole IV nanosuspension enhances efficacy through altered pharmacokinetics in the rat. Int J Pharm 339 (1–2):251–260

Roubin R, Zolla-Pazner S (1979) Markers of macrophage heterogeneity. I. Studies of macrophages from various organs of normal mice. European Journal of Immunology 9 (12):972–978

Schöler N (2001) Solid lipid nanoparticles (SLN) and nanosuspensions: In vitro cytotoxicity and therapeutic use in the model of reactivating toxoplasmosis in mice. Dissertation, PhD thesis, Freie Universität Berlin,

Schöler N, Krause C, Kayser O, Müller RH, Hahn H, Liesenfeld O In vitro efficacy of atovaquone nanosuspensions against toxoplasma gondii. In: 3rd World Meeting on Pharmaceutics, Biopharmaceutics and Pharmaceutical Technology APGI/APV, Berlin, 2000. pp 437–438

Schöler N, Krause, K., Kayser, O., Müller, R. H., Hartwig, H., Hahn, H., Liesenfeld, O. (2001) Atovaquone nanosuspensions show excellent therapeutic effect in a new murine model of reactivated toxoplasmosis. Antimicrobial Agents & Chemotherapy:1771–1779

Schroeder U, Sommerfeld P, Sabel BA (1998) Efficacy of oral dalargin-loaded nanoparticle delivery across the blood-brain barrier. Peptides 19 (4):777–780

Smith A, Hunneyball lM (1986) Evaluation of poly(lactic acid) as a biodegradable drug delivery system for parenteral administration. International Journal of Pharmaceutics 30 (2–3):215–220

Staedtke V, Brahler M, Müller A, Georgieva R, Bauer S, Sternberg N, Voigt A, Lemke A, Keck C, Möschwitzer J, Bäumler H (2010) In vitro inhibition of fungal activity by macrophage-mediated sequestration and release of encapsulated amphotericin B nanosupension in red blood cells. Small 6 (1):96–103

Strachan EB (1981) Case report--suspected anaphylactic reaction to Cremophor El. SAAD Dig 4 (9):209

Szejtli J (1988) Cyclodextrin technology. Kluwer Academic Publishers,

Wilkins DJ, Myres PA (1966) Studies on the relationship betweeen the electrophoretic properties of colloides and their blood clearance and organ distribution in the rat. British Journal of Experimental Pathology 47:568–576

Zirconium dioxide. Wikipedia

Processing and Scale-up of Polymeric Nanoparticles

Christine Vauthier and Kawthar Bouchemal

Abstract This chapter presents methods of nanoparticle processing based on the use of preformed polymers. It will discuss the basic principle of the different methods, the scale up and the methods for preparing the polymer nanoparticles for storage including purification, drying sterilization and eventually concentration. The last part of the chapter will discuss the performance, the application and the present limitation of the processing methods considered here.

Keywords Emulsification-solvent diffusion • Emulsification-solvent evaporation • Gelation • Nanocapsules • Nanoprecipitation • Nanospheres • Polymer micelles • Scale-up

Abbreviations

ACN	acetonitrile
CMC	critical micellar concentration
DMAc	N-N-dimethylacetamide
DMF	dimethylformamide
O/O	Oil in oil emulsion
O/W	oil in water emulsion
PEC	polyelectrolyte complexes
PEG	poly(ethylene glycol)
PEI	poly(ethylenimine)

C. Vauthier (✉)
Physico-chimie, Pharmacotechnie, Biopharmacie, Univ Paris-Sud, UMR 8612, F-92296 Chatenay-Malabry, France
and
CNRS, Chatenay-Malabry F-92296, France
e-mail: christine.vauthier@u-psud.fr

K. Bouchemal
Physico-chimie, Pharmacotechnie, Biopharmacie, UMR CNRS 8612, Univ Paris-Sud, F-92296 Chatenay-Malabry, France

A. Prokop (ed.), *Intracellular Delivery: Fundamentals and Applications*, Fundamental Biomedical Technologies 5, DOI 10.1007/978-94-007-1248-5_16,

PEO	poly(oxyethylene)
PLL	poly(lysine)
THF	tetrahydrofuran
W/O/O	water in oil in oil multiple emulsion
W/O/W	water in oil in water multiple emulsion
W/O	water in oil emulsion

1 Introduction

Nanoparticles have become common tools in research to improve drug efficiency in vivo by a better control of the biodistribution. Proof of concept is now well established and the use of nanomaterials to deliver drugs is about to revolution treatments of severe diseases. It is noteworthy that a few systems have reached the clinics for the treatment of cancer including TransDrug® and one already marketed formulation, Abraxane®. Due to the variety of drug candidates, many polymer nanoparticles were developed requiring different methods of processing. The aim of this chapter is to summarize the different methods of processing polymer nanoparticles and to discuss their performance regarding their potential and limitations for a scale up and for drug encapsulation. The principles of the preparation methods will be described first. Considerations about the scaling up will be presented in the next part followed by general methods of treatments applied after preparation to purify, sterilize, dry and condition the nanoparticles. The different functionalities which may be interested to integrate in the nanoparticles will be discussed in the last part of the chapter.

2 Principles of Methods of Preparation of Polymer Nanoparticles

Two classes of methods can be identified depending whether the obtaining of nanoparticles is achieved from a polymer solution or from an emulsion. The following presentation of the general principles for nanoparticle preparation from polymers will include this subdivision. Only methods of nanoparticle preparations from polymers will be presented in this chapter. The methods based on polymerizations are discussed in another chapter (See **chapter by Nicolas and Vauthier in this book**). Schemes of the different types of nanoparticles are illustrated in Fig. 1.

2.1 Methods Based on the Conversion of a Polymer Solution into Nanoparticle Dispersions

Dilute solutions of polymers can be converted into nanoparticle dispersions taking advantages of solubility properties of the dissolved polymer to precipitate as

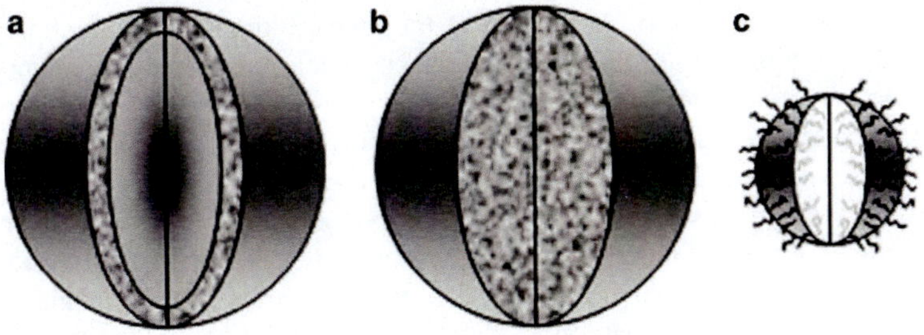

Fig. 1 Scheme of the different types of nanoparticles including nanocapsules (**a**), nanospheres (**b**) and polymer micelles (**c**). Nanocapsules are reservoir type particles including a cavity surrounded by a polymer envelope. Nanospheres are plain nanoparticles. Polymer micelles are formed by self association of amphiphilic polymers

nanoparticles (nanoprecipitation) or of its capacity to form nanogels, nanosized polyelectrolyte complexes (PEC) and micelles. In all cases, formation of particles in the nanometer size range occurred in well controlled conditions which can be studied through systematic approaches including the establishment of a phase diagram.

2.1.1 Obtaining Nanoparticles by Nanoprecipitation or Solvent Displacement

Nanoprecipitation or solvent displacement methods are now well established methods producing nanoparticles or polymer micelles from a polymer solution. They are based on the induction of the precipitation of a polymer thanks to the displacement of the polymer solvent by a non solvent which is miscible with the polymer solvent (Fessi et al. 1989; Thioune et al. 1997; Murakami et al. 1999; Legrand et al. 2007). In practice, a solution of polymer is prepared by dissolving the polymer in one of its solvent. Then, the polymer solution is added in a non solvent which is miscible to the polymer solvent allowing polymer colloids to form. Polymers with no or poor amphiphilic properties precipitate to form nanoparticles in a narrow window of the phase diagram considering the composition in polymer-solvent- non solvent of the nanoprecipitation system (Stainemesse et al. 1995). In general, the produced nanoparticles are well defined in size with a narrow size distribution. As requirements for the success of the methods, (i) the solution of polymer should be rather diluted, (ii) the polymer solvent and the non solvent should be miscible, (iii) the polymer solvent should easily be removed at the end of the preparation. Thus, the preferred polymer solvents are acetone, ethanol and THF while the preferred non solvent is water. Suitable polymers include many of the polymers suggested as material for the development of nanoparticulate drug delivery systems including main types of polyesters (poly(lactic acid), poly(lactide-co-glycolide), poly(epsilon caprolactone)), ethylcellulose, new polymer candidates like poly(benzylglutamic acids) and all the corresponding copolymers with poly(ethyleneglycol) moiety (Avgoustakis

2004; Barbosa et al. 2008). At the end of the preparation, the polymer solvent is removed by evaporation or ultrafiltration. Nanoparticles produced by nanoprecipitation are generally characterized by a diameter ranging from 200 to 300 nm. This size range is rather narrow and was found to result from the precipitation of polymer chains with well define molecular weight (Legrand et al. 2007).

Several theoretical works have been developed to explain the formation of the nanoparticles from the nucleation of supersaturated zones in the solution which form during mixing the polymer solution with the non solvent (Johnson and Prud'homme 2003; Ganachaud and Katz 2005; Lince et al. 2008; Aubry et al. 2009). Optimized conditions for the production of nanoparticles with the highest yield of nanoparticle production are fulfilled when the polymer is dissolved in a teta solvent at a concentration comprised in the dilute regimen which means that the polymer chains are surrounded by enough solvent so that they remain separated in the solution (Legrand et al. 2007). In other words, in the optimal conditions polymer molecules do not overlap with each other and remain independent in the solution. By increasing polymer concentration in the solution above the critical interpenetrating concentration C*, polymer molecules overlap each others promoting the formation of aggregates instead of individual nanoparticles. To adjust solvency properties some authors have suggested to use binary blends of solvents such as acetone with small amount of added water (Thioune et al. 1997), or blends of ethanol and acetone (Murakami et al. 1999). Another optimization parameter includes the molecular weight of the polymer. Indeed, nanoparticles were found to be formed by polymer chains with define molecular weight. It was shown that all polymer chains with a mean molecular weight outside the optimal molecular weight precipitate as aggregates. Thus, by choosing polymers with optimal molecular weight, i.e. molecular weight of polymer chains found in nanoparticles, yield of nanoparticle production can be optimized while formation of aggregates can be avoided. Nanoprecipitation methods suit well to prepare drug carriers incorporating lipophilic drugs. Indeed, in general, the drug to be encapsulated in the nanospheres produced by this technique is simply added in the polymer solution (Niwa et al. 1993; Murakami et al. 2000; Chorny et al. 2002; Peltonen et al. 2004).

Although the main type of nanoparticles prepared by nanoprecipitation is nanospheres, the method can easily be adapted to produce nanocapsules by adding a small amount of oil in the polymer solution (Fessi et al. 1989). During mixing of the polymer solution with the non solvent of the polymer, the oil splits as tiny droplets around which the polymer precipitates to form the nanocapsule shell. The dispersion phenomenon of the oil was explained by the "Ouzo or Pastis effect" elucidated by Vitale and Katz (2003).

It is noteworthy that polymer micelles can be prepared by nanoprecipitation of amphiphilic copolymers (Trivedi and Kompella 2010). In general, polymer micelles are formed spontaneously in a solution just because the concentration of the amphiphilic molecule is above the Critical Micellar Concentration (CMC). Although polymer micelles form spontaneously, artefacts can be used to promote their formation from amphiphilic polymers with low CMC in water and to favour drug entrapment with a high payload (Fournier et al. 2004; Gaucher et al. 2010a, b;

Joralemon et al. 2010). The two major methods of preparation of drug loaded polymer micelles are based either on nanoprecipitation or on dialysis. The polymer micelles are to be produced in a continuous aqueous phase. In general, the solubility of the amphiphilic polymer is higher in organic solvents compared to water. The use of an organic solvent miscible with water promotes the dissolution of unimers of the polymer helping the formation of well structured polymer micelles when solubility conditions change by addition of large amount of water such as in the nanoprecipitation method (Johnson and Prud'homme 2003; Forrest et al. 2008) or by the progressive replacement of the organic solvent by water through a dialysis against water (Kim et al. 1999; Lee et al. 2003; Huh et al. 2005; Park et al 2005). Eventually, the polymer solution is slightly diluted by the addition of a limited amount of water prior to dialyse against water (Jie et al. 2005; He et al. 2007; Kang and Leroux 2004). Typical organic solvents are chosen among acetonitrile (ACN), tetrahydrofuran (THF), dimethylformamide (DMF), or N-N-dimethylacetamide (DMAc). The drug to be encapsulated in the polymer micelles is added in the polymer solution before induction of the micelle formation achieved by modifications of polymer solubility conditions.

2.1.2 Obtaining Nanoparticles Through Controlled Gelation

Gels are a three-dimensional polymer network swollen by solvent. Several polymers form gels either by cooling down a solution prepared at hot temperature or by adding small molecules which crosslink polymer chains through chemical linkage or by physical interactions. Obtaining nanoparticles from a polymer solution through a gelation process was described inducing gel formation by physical interactions. Typical polymers are charged polysaccharides dissolved at a low concentration in an aqueous solvent. The addition of a small molecule bearing a low number of opposite charges induces formation of gel nanoparticles through ionic gelation. This can easily be followed by measuring the viscosity of the polysaccharide solution which dropped down when the nanoparticles formed, by electron microscopy and by light scattering allowing to measure the size of the nanoparticles formed (Vauthier et al. 1994; Vauthier and Couvreur 2000). In general, the nanosized gels form at low concentrations of both polymer and gelling agent.

This method was applied with alginate which gelify with calcium ions. Consolidation of nanogels formed with calcium can be achieved by addition of a positively charged polyelectrolyte. Poly-lysine was used in the earlier development but most of the recent works are considering chitosan as another suitable polyelectrolyte candidate (Rajaonarivony et al. 1993; Vauthier and Couvreur 2000; Douglas and Tabrizian 2005). Alginate nanoparticles produced by controlled gelation can easily be loaded with oligonucleotides. They protect oligonucleotides from degradation by nucleases (Aynié et al. 1999). This system was also highly investigated for the delivery of oral formulation of peptides including insulin (Li et al. 2007; Sarmento et al. 2007). Chitosan based nanoparticles can be obtained by gelation with tripolyphosphate in presence of PEG to achieve stability of the nanoparticles (Calvo et al. 1997). Chitosan nanoparticles were designed as delivery systems for macromolecules (Janes et al. 2001; Brunel et al. 2010).

2.1.3 Formation of Nanoparticles from Polyelectrolyte Complexes

Another method for the production of nanoparticles is based on formation of polyelectrolyte complexes (PEC). In this method, two polymers of opposite charges are brought together to interact and form aggregates in the nanosize range (Berger et al. 2004; Schatz et al. 2004; Goycoolea et al. 2009; Oyarzun-Ampuero et al. 2009; Woitiski et al. 2009a, b). Originally, this approach was suggested to develop drug carriers for nucleic acid delivery. Complexes were prepared by mixing polycations like poly(lysine) (PLL) or poly(ethylenimine) (PEI) and nucleic acids which are negatively charged macromolecules (Boussif et al. 1995; Coll et al. 1999; Jeong et al. 2007; Sun and Zhang 2010). The main difficulty with this method was to produce nanoparticles which remained stable over time. Stability of nanoparticles can be improved by using block copolymers including a poly(ethylene glycol) (PEG) moiety combined with either the polycation or the nucleic acid. The nanoparticles obtained with these copolymers are sterically stabilized. The PEG chains form a corona at the surface of the core of the nanoparticles formed by the poly(electrolyte) complex between the nucleic acid and the polycation (Kabanov et al. 2005; Jeong et al. 2007; Joralemon et al. 2010).

Recent studies provide relevant methodologies to prepare stable and well define nanoparticles by mixing polyelectrolytes of opposite charges. A pioneer work was given by the group of Delair who drawn a concept in which one polyelectrolyte is a guest for the second polyelectrolyte of opposite charge which is the host. They clearly demonstrated that the host and guest balance control the formation of well define nanoparticles with either positive or negative charges (Schatz et al. 2004; Drogoz et al. 2007). This approach is now successfully applied by many authors using various polymers (Berger et al. 2004; Schatz et al. 2004; Oyarzun-Ampuero et al. 2009; Goycoolea et al. 2009; Woitiski et al. 2009a, b; Mao et al. 2006; Sun et al. 2008). Briefly, the polyelectrolyte with the larger molecular weight is considered as the host while the polyelectrolyte of opposite charge with a much lower molecular weight is consider as the guest. To obtain nanoparticles, the guest polyelectrolyte is added into a solution containing the host polyelectrolyte. Parameters such as the respective concentration of the two polyelectrolytes, and the ratio of their molecular weight are affecting the characteristics of the produced nanoparticles (Drogoz et al. 2007).

Nanoparticles made of polyelectrolyte complexes are largely developed for hydrophilic macromolecular drugs. These nanoparticles are suitable for all types of nucleic acids (Brunel et al. 2010; Sun and Zhang 2010). Peptides and proteins which are ampholyte molecules are a second category of drug molecules which are associated with such nanoparticles (Mao et al. 2006; Woitiski et al. 2009b). Several works are considering their use for the development of an oral formulation of insulin while others are considering these nanoparticles for the delivery of antigens entering the formulation of vaccines (Mao et al. 2006; Weber et al. 2010; Woitiski et al. 2009b).

2.2 *Methods Based on the Formation of an Emulsion*

The second class of methods requires the preparation of an emulsion prior to the formation of nanoparticles. This part of the chapter will present methods for the preparation of the required emulsion and then the different methods used to convert the emulsion into dispersion of polymer nanoparticles.

2.2.1 Preparation of Emulsions Suitable for Nanoparticle Preparation

The preparation of the emulsion is critical in the sense that it will influence the size of the polymer particles at the end of the preparation. It consists in mixing two immiscible phases together, the polymer solution being the dispersed phase of the emulsion. The sense of the emulsion, i.e. water in oil (W/O) or oil in water (O/W), depends on the nature of the polymer and solubility properties. Oil in oil (O/O) emulsions were introduced to improve drug incorporation performance in nanoparticles (Nah et al. 2008). In general, simple emulsions are used to prepare nanospheres. Preparation of nanocapsules can be achieved from multiple emulsions water in oil in water (W/O/W) (Lu et al. 1999; Bilati et al. 2003). In this latter case, the polymer is dissolved in the intermediate oily phase while the drug is dissolved in the aqueous inner phase. More complex emulsions were recently proposed to produce double walled particles (Zheng 2009). Although the technique is presently applied to produce microparticles which are larger particles than that of the nanometer range but it can be expected that such methods of emulsification will soon be applied to produce nanoparticles as well.

Several methods are suitable to prepare the required simple emulsions straight from the two separate phases. This includes ultrasounds, extrusion through microporous and nanoporous materials, microfluidic systems and co-solvent assisted emulsification (Ouzo effect) (Table 1).

For all other methods, the two phases are premixed before being submitted to specific treatments providing the emulsion of the desired characteristics. For instance, very small size emulsions, i.e. nano-emulsion can be generated from a bi-continuous system prepared with poly(ethylene glycol) (PEG) containing surfactants through a phase inversion method (Anton et al. 2008). The conversion of the bi-continuous system into an oil-in-water nano-emulsion is induced by simultaneous dilution and temperature drop. This method requires a very low energy to achieve the formation of the desired emulsion. Depending on the characteristics of the PEG surfactant, it can be applied at temperature ranging from 40°C to 90°C.

Table 1 Methods used to prepare thin emulsions straight from the two phases

Ultrasounds (Vila et al. 2002; Perez et al. 2001; Freitas et al. 2005)	An ultrasound probe is used to mix together the oil and water phases of the emulsion. 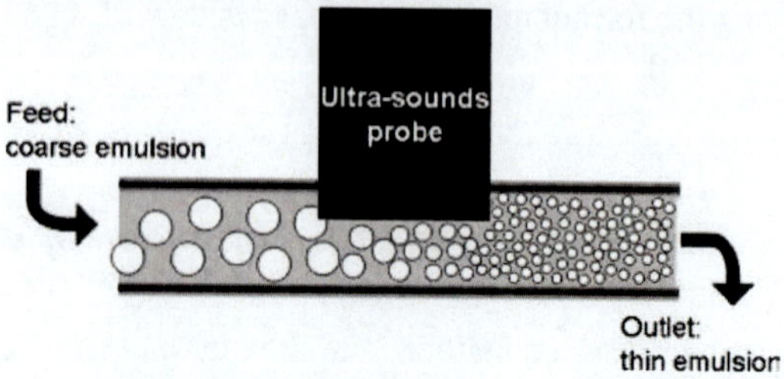
Extrusion through porous materials (Charcosset and Fessi 2005)	The polymer and drug containing phase is extruded through a porous membrane into the continuous phase placed on the other side of the membrane. Although this method is suitable to prepare simple emulsions, double emulsions can also be prepared from simple emulsions extruded into a new continuous phase. 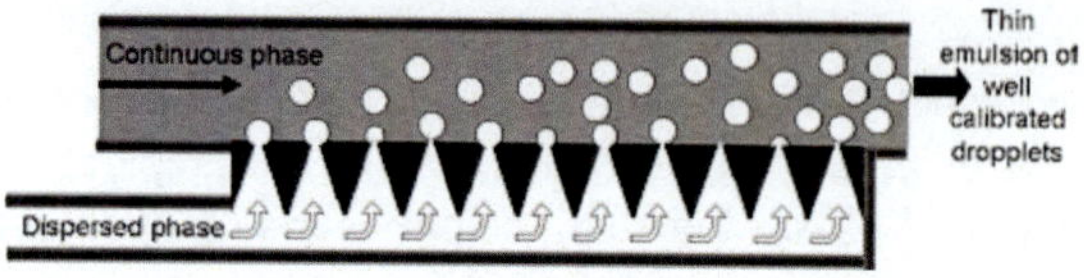
Microfluidic systems (Abate and Weitz 2009)	The two phases are mixed together through a well organized circulation in a microfluidic device composed of nanochannels. Emulsion droplets of well define diameter are formed. This method is suitable to produce double emulsion by adding more channel in the microfluidic device. 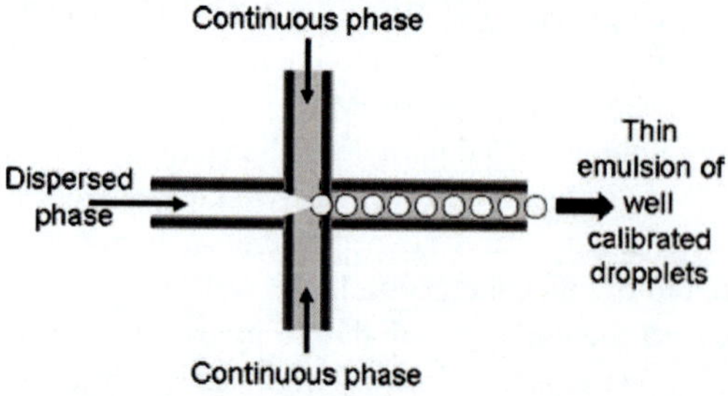
Co-solvent assisted emulsification: Ouzo effect (Vitale and Katz 2003; Ganachaud and Katz 2005)	The polymer, the drug and a small amount of oil are dissolved in an organic solvent miscible with water. By mixing this organic phase with water under gentle conditions, oil droplets form while the solvent diffuses in the aqueous phase. 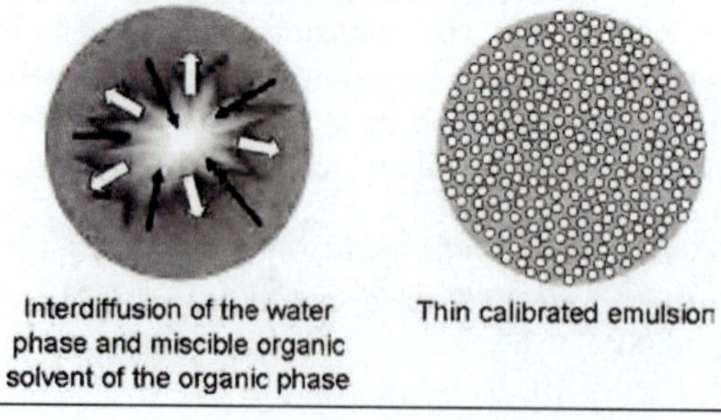

Other methods consist in reducing size of the droplets of a rough emulsion using special machines including colloidal mills, microfluidizer and high pressure homogeniser (Table 2) (Urban et al. 2006; Anton et al. 2008). By using a microfluidizer, the diameter of the droplets, D, formed in the final emulsion depends on the viscosity of the organic solution of polymer as indicated by the Eq. 1 (Koshy et al. 1988; Walstra 1983).

$$D = \eta^{\alpha} \tag{1}$$

In this equation, η corresponds to the viscosity of the polymer solution, α is a coefficient which depends on many parameters of the system including the nature of the polymer. The alpha coefficients were determined for ethylcellulose and poly(lactic acid) using ethylacetate as the organic solvent. The values were 0.05 and 0.28 respectively (Desgouilles et al. 2003).

Table 2 Apparatus used to reduce size of emulsion droplets from a rough emulsion (Adapted from Urban et al. 2006)

<table>
<tr><td>Microfluidizer (Bodmeier and Huagang 1990; Desgouilles et al. 2003)</td><td>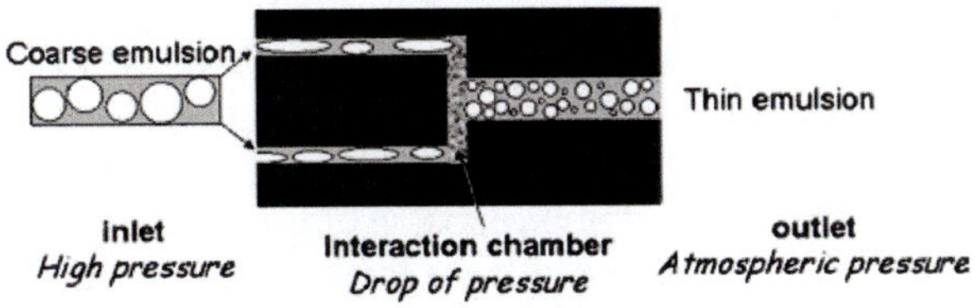
</td></tr>
<tr><td>Colloidal mill (Stork et al. 2003; Mabile et al. 2003)</td><td>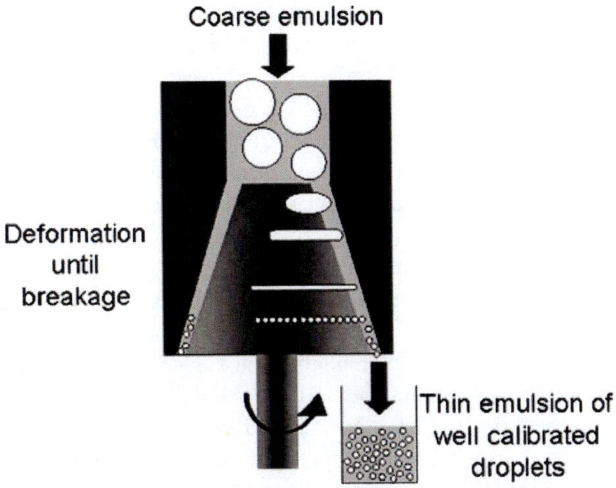
</td></tr>
<tr><td>High pressure homogenizer (Shegokar et al. 2010)</td><td>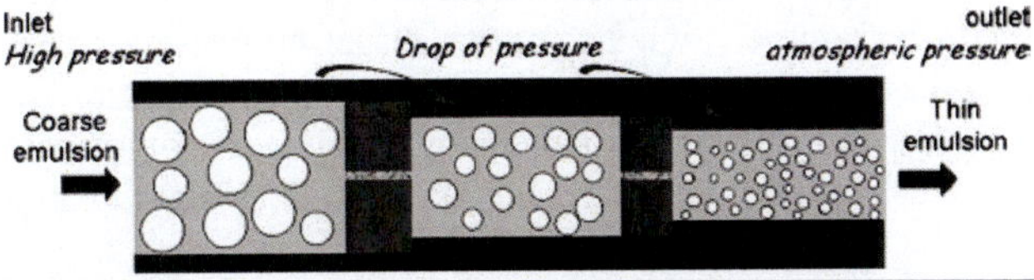
</td></tr>
</table>

Multiple emulsions can be prepared with microfluidizer (Sani et al. 2009), high pressure homogeniser (Lamprecht et al. 1999, 2000) and ultrasounds (Perez et al. 2001) by re-emulsifying the simple emulsion prepared above into a new continuous phase.

Surfactants such as pluronic and span are often used to prepare stable emulsions. However new polymer surfactants were synthesized to formulate suitable emulsions for nanoparticle preparation. In general these polymer surfactants include polymer chains of the same nature than that composing the nanoparticle. The hydrophilic moiety required to achieve the stability of the nanoparticles which protruded in the aqueous phase is composed either of polyethylene glycol (Gref et al. 1994; Bazile et al. 1995; Avgoustakis 2004; Qui and Bae 2006) or of various polysaccharides (Lemarchand et al. 2003; Chauvierre et al. 2004).

2.2.2 Conversion of an Emulsion into Nanoparticle Dispersion

Several artefacts can be used to convert the emulsion into nanoparticle dispersion. For instance, it can be achieved by inducing the precipitation of the polymer dissolved in the emulsion droplets. The simplest methods inducing polymer precipitation consists in removing the solvent contained in the emulsion droplets by evaporation of the solvent composing the dispersed phase of O/W and W/O/W emulsions (Avgoustakis 2004; Vauthier and Bouchemal 2009). This corresponds to the original method of preparation of pseudolatexes by Gurny et al. (1981). To achieve this, it is required that the emulsion was prepared with an organic volatile solvent such as dichloromethane and ethyl-acetate. Because solvent evaporation is a slow process, formation of nanoparticles takes more then a few minutes. Polymer concentration increases gradually over time in the emulsion droplets thanks to solvent removal until polymer starts precipitation (Desgouilles et al. 2003). Diameter and size distribution of the nanoparticles depends on the diameter and size distribution of the emulsion droplets. However, no general relationship can be drawn between droplet and nanoparticle size because this depends on the extend of emulsion droplet coalescence occurrence during the solvent removal step which in turn depends on surface active properties of the polymer and may be affected by the presence of surfactant.

A second method consists in a rapid extraction of the solvent contained in the emulsion droplets by diffusion toward the continuous phase of the emulsion. This method was named the emulsification-solvent extraction method or the emulsification-solvent diffusion method. Typically, suitable conditions are obtained by dilution of the emulsion by adding more continuous phase if the polymer solvent contained in emulsion droplets is partly miscible with the continuous phase (i.e. ethylacetate for instance). Instead of diluting the continuous phase, solvents in which both the continuous and the dispersed phase are miscible can be added to displace the polymer solvent from the dispersed phase towards the continuous phase causing polymer precipitation (Leroux et al. 1995; Quintanard-Guerrero et al. 1999; Perez et al. 2001; Guinebretiere et al. 2002; Moinard-Chécot et al. 2006, 2008).

In contrast with the previous methods, nanoparticles form within milliseconds (Moinard-Chécot et al. 2008). The so called emulsification-inverse salting out method is a specific case of preparation of nanoparticles based on solvent extraction. The emulsion is prepared with a solution of polymer dissolved in acetone which is dispersed in a concentrated aqueous solution of electrolytes or mono/disaccharides. After simple dilution of the emulsion with an excess of water, the electrolyte or saccharide concentration in solute dropped down inducing instantaneous diffusion of acetone out of the emulsion droplets towards the continuous phase of the emulsion. As a consequence, the polymer precipitates as nanoparticles (Allémann et al. 1992; Ibrahim et al. 1992).

Alternative methods to polymer precipitation are based on gelation of polymers contained in the emulsion droplets of water-in-oil emulsions. Suitable polymers used to prepare nanoparticles by this method are polysaccharides. For instance, agarose nanoparticles are obtained by decreasing the temperature of the emulsion prepared at high temperature allowing the polysaccharide to solubilise (Wang and Wu 1997). Pectine nanoparticles are prepared from two emulsions, one containing the polysaccharide and the second a basic aqueous dispersed phase. By mixing the two emulsions together, gelation of pectine contained in the emulsion droplets of the first emulsion is promoted thanks to an increase of pH of the emulsion droplets. This gelation produces the nanospheres (Tokumitsu et al. 1998).

3 Pilot Scale Production of Nanoparticles

Only few articles report preparation of large amount of nanoparticles for drug delivery applications. Indeed, nothing is known about transposition from laboratory preparation to production of clinical batches for doxorubicin loaded poly(alkylcyanoacrylate) nanoparticles, Transdrug®, used in phase II/III clinical trials (BioAlliance Pharma 2010). No more is known about the production of a marketed formulation of paclitaxel-loaded albumin nanoparticles, nab-paclitaxel or Abraxane®, used in clinics in the USA for the treatment of metastatic breast cancer (Hawkin et al. 2008; Petrelli et al. 2010). The most detailed data about the scale up of production of pharmaceutical grade nanoparticle dispersions are only available on the pilot scale production of nanoparticles prepared by the emulsification-solvent diffusion method and by the nanoprecipitation method. By extension, the scale up approach developed for the method of emulsification-solvent diffusion was also applied to produce large batches of nanoparticles by the emulsification-reverse salting out method. As a reminder, a pilot-scale production is intermediate between the laboratory and the industrial scale production. Pilot size production is aimed to simulate the closest as possible industrial production and hence needs to integrate all parameters that need to be optimized before reaching the industrial production. This topic was recently extensively reviewed by the authors (Vauthier and Bouchemal 2009).

3.1 *Pilot-Scale Production of Nanoparticles by Emulsification-Solvent Diffusion Method*

Large scale production of nanocapsules by emulsification-solvent diffusion method was developed up to a production of batches of 15 L each (Colombo et al. 2001; Galindo-Rodriguez et al. 2005). The pilot set up comprises several vessels for the preparation of the different solutions which are connected through a system of flexible tubing to a main reactor. Transfer of solutions from the vessels to the reactor occurred by simple gravity. The rational behind this design was to reproduce as much as possible fluid motions produced in the laboratory scale set up (60 mL) (Galindo-Rodriguez et al. 2005). By introducing a few modifications, this pilot set up can be used for the production of nanoparticles by the emulsification-reverse salting out method (Galindo-Rodriguez et al. 2005).

In both cases, agitation is an important process parameter to consider during the preparation of the emulsion (Colombo et al. 2001). It greatly influences the size of the nanoparticles produced at the end of the procedure. Optimal conditions of agitation were found by using stirring rates above 1,000 rpm for the emulsification–solvent diffusion method and 790 rpm for the emulsification-reverse salting-out method. By applying the optimal conditions, the methods are reproducible and provide with nanoparticle dispersions with a narrow size distribution (Galindo-Rodriguez et al. 2005).

3.2 *Pilot-Scale Production of Nanoparticles by a One Step Procedure Based on the Nanoprecipitation of a Polymer*

Success of the nanoprecipitation method resides on the way the organic and the aqueous phases mix together resulting in the precipitation of the polymer as nanoparticles. Progresses in the comprehension of nanoprecipitation phenomena highlighted that it is controlled by a few key parameters. It was established that the time of mixing must be faster than the time required to induce nanoparticle formation (Johnson and Prud'homme 2003). Ideal conditions are obtained at the beginning of the mixing of the two phases. In conditions found in the lab scale production, the amounts of organic solution of polymer and that of the non-solvent are small. Ideal conditions of nanoprecipitation are fulfilled at the beginning of the mixing and are almost maintained during mixing of the total amounts of solutions used to produce small batch of nanoparticles in small reactors (lab scale production of a few mL). However, production of large batches of nanoparticles by nanoprecipitation in a reactor is not suitable as the conditions for nanoprecipitation change from the beginning to the end of the two phases mixing. To maintain a homogenous condition for nanoprecipitation along mixing time, the two phases need to be brought together in the same conditions all along the process. Briançon et al. (1999) have suggested to inject the two phases in a mixing device which is continuously feed at specific flow rates with each phase including the polymer solution and the non solvent of the polymer. Simultaneously, the nanoparticle dispersion flows out of the mixing device through the outlet. The mixing conditions provided by the mixing device are constant whatever the volume of nanoparticle dispersion produced. This can be achieved thanks to the small volume of each phase processed instantaneously and continuously in the mixing device. Therefore, the mixing device is the central piece of the production set-up and allows to maintain homogenous starting condition all along the addition of the two phases whatever the total volume is. This first mixing device was designed on the basis of a T-shape tube (Briançon et al. 1999; Tewa-Tagne et al. 2007; Colombo et al. 2001; Galindo-Rodriguez et al. 2005). Several mixing devices have been designed since then by different groups with capacities varying from a few microlitres to a few millilitres (Fig. 2) (Liu et al. 2008; Akbulut et al. 2009; Lince et al. 2009; Tuereli et al. 2009). Simulations of the mixing dynamic flow occurring in the mixing device help optimizing operating conditions of nanoprecipitation (Gavy et al. 2009; Lince et al. 2009). The whole

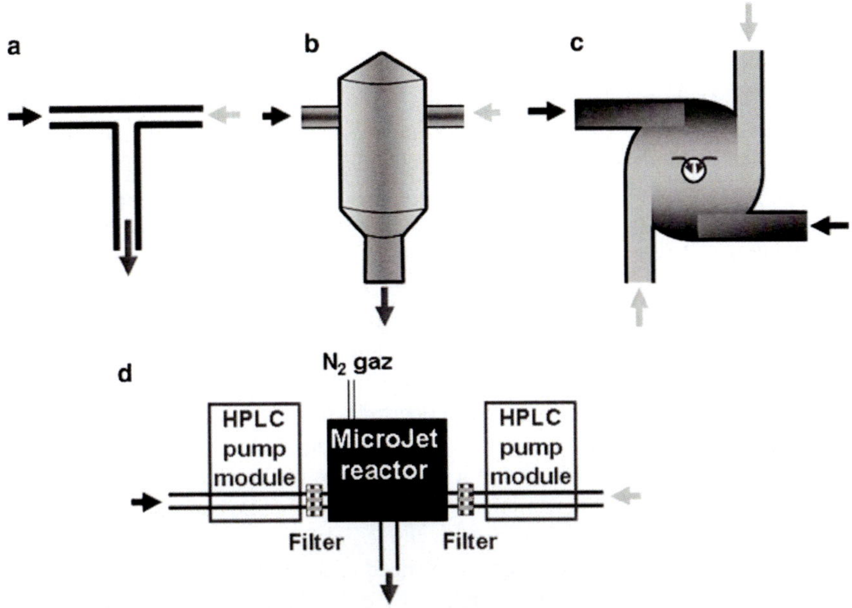

Fig. 2 Mixing devices used to produce large batches of nanoparticles by nanoprecipitation or solvent displacement methods. (**a**) T shape mixing device according to Briançon et al. (1999), diameters of the inlets can vary; (**b**) Smart mixer and reactor according to Lince et al. (2009), diameters of the inlets can vary; (**c**) Vortex jet mixer according to Liu et al. (2008) and Akbulut et al. (2009), the nanoparticle dispersion is evacuated from the center of the device; (**d**) Microjet reactor set-up according to Tuereli et al. (2009). *Black arrows* indicate the inlet of the mixing device feed with polymer solution, *light gray arrows* indicate the inlet of the mixing device feed with the polymer non solvent and medium *gray arrows* indicate the outlet of the mixing device where the nanoparticle dispersions are collected

pilot plant for the production of nanoparticles by nanoprecipitation includes one reservoir (can be several litres) for the polymer solution, one reservoir (can be several litres as well) with the polymer non-solvent, a receiver of large capacity (can be several litres), the mixing device and pumps used to feed the mixing device with the polymer solution on the one hands and the non-solvent on the other hands with perfectly controlled flow rates. A pilot plan built with reservoirs of 3 L each and a T-shape mixing device can convert 5 g of polymer into nanoparticles in 2 h with a very good reproducibility (Galindo-Rodriguez et al. 2005; Tewa-Tagne et al. 2007)

4 Treatment of Nanoparticles and Preparation for Storage

After synthesis, several types of treatments may be necessary to apply on nanoparticle dispersions including purification, sterilization and preparation for storage.

Purification is often required to remove traces of impurities such as residual organic solvent, excess of surfactants, salts and large polymer aggregates (Vauthier and Bouchemal 2009). Volatile organic solvents can be removed by evaporation under reduced pressure. Although this method can easily be applied on small amounts of nanoparticle dispersions, ultrafiltration (Allemann et al. 1993), diafiltration (Tishchenko et al. 2003) and cross-flow microfiltration (Allemann et al. 1993; Limayem et al. 2004; Quintanar-Guerrero et al. 1998) are suitable methods to treat large volumes of nanoparticle dispersions. Filtration can be used to remove aggregates (Govender et al. 1999; Murakami et al. 1999). Centrifugations and ultracentrifugations can be used to separate nanoparticles from the dispersing medium which retains all excess of reagent not included in the nanoparticles during preparation (Bouchemal et al. 2004, 2006; Govender et al. 1999; Calvo et al. 1997; Sahoo et al. 2002; Nguyen et al. 2003; Lambert et al. 2000). Dialysis (Chauvierre et al. 2003) and gel filtration (Beck et al. 1990) are alternative methods when difficulties of dispersion of nanoparticles arise after centrifugation.

In general, the solid content of nanoparticle dispersions as prepared by the previously described methods is rather low (Desgouilles et al. 2003; Legrand et al. 2007). In some case, it is so low that it compromises the application as drug delivery system as the volume to administer to reach a therapeutic concentration in drug in vivo is much above the maximal volume tolerated for the administration. Therefore, nanoparticle dispersions may need to be concentrated prior to their in vivo administration. Straightforward methods to increase nanoparticle concentration are freeze drying, spray drying, ultracentrifugation and solvent evaporation. Although these methods are suitable for several types of nanoparticles, the main problem resides in the formation of aggregates which form when nanoparticles come in contact with each other (Abdelwahed et al. 2006; Avgoustakis 2004; Bozdag et al. 2005; Vauthier et al. 2008). Methods given the more satisfactory results without any risk of causing nanoparticle aggregation are based on dialysis and ultrafiltration. Difficulties which might be encountered with simple ultrafiltration (i.e. clogged membranes with nanoparticles) can be resolved using diafiltration or tangential filtration methods (Limayem et al. 2004). Concentration of nanoparticle dispersion based on dialysis is a gentle method in which the water from the nanoparticle dispersion is removed thanks to the application of an osmotic pressure on the dialysing bag using a solution of high molecular weight polymer as counter dialysing medium. The dispersions can be concentrated up to factors of 50 without producing aggregation of the nanoparticles and in a couple of hours (Vauthier et al. 2008).

Nanoparticle dispersions for parenteral administration need to be sterilized. Autoclaving is not always suitable because nanoparticles may be modified and drug may loose their activity (Rollot et al. 1986; Masson et al. 1997; Boess et al. 1996). Autoclaving can be used to sterilize nanoparticles made of chitosan-carboxymethyl dextran polyelectrolyte complexes (Lin et al. 2009). Gamma irradiations also need to be applied with caution because it can significantly modify the characteristics of the polymer composing the nanoparticles and the drug hence the initial performance of the drug delivery system (Sintzel et al. 1997; Masson et al. 1997; Athanasiou et al. 1996). However, this technique was found suitable to sterilize doxorubicin-loaded poly(butyl cyanoacrylate) (PBCA) nanoparticles (Maksimenko et al. 2008) and vaccine nanoparticles made of poly(anhydride) and containing Brucella ovis antigen (Da Costa Martinez et al. 2009). Both types of nanoparticles showed an excellent stability to irradiation without radiolysis of the polymer and with good performance preserving the integrity of the drug/antigen activity. Sterilization by filtration can be applied only on nanoparticle dispersions with a diameter below 0.22 μm which is a major limitation (Memisoglu-Bilensoy and Hincal 2006). High Hydrostatic Pressure treatment (HHP) proved its efficacy to destroy vegetative forms of microorganisms found in nanoparticle dispersions (Brigger et al. 2003). However, this technique still needs improvement to make possible elimination of bacteria spores before if can be validated as a suitable method of sterilization for nanoparticles. It could be interesting to also explore other sterilization methods such as those based on the use of gaz plasma and ethylene oxide which were recently applied on gold nanoparticles developed for biomedical applications (França et al. 2010). In the absence of suitable satisfactory method of sterilization, preparations must be done in sterile environment which complicates the development for clinical applications.

As valuable for many pharmaceutical products, storage of nanoparticle drug delivery systems under a dried form would be the preferred method. In general, nanoparticles are obtained in a liquid medium after preparation. Transformation of the liquid dispersion into a dried powder can be achieved either by lyophilisation (Nemati et al. 1992; De Chasteigner et al. 1996; De Jaeghere et al. 1999; Abdelwahed et al. 2006) or spray-drying (Tewa-Tagne et al. 2007) which are both suitable at industrial scale. Considering freeze drying, it may be required to add cryoprotectants to the nanoparticle dispersion to prevent aggregation of nanoparticles during the reconstitution of the dispersion from the dried powder prior to use. Optimization of the freeze drying cycle improves the quality of the freeze dried product (Patapoff and Overcashier 2002; Abdelwahed et al. 2006). In contrast with freeze drying, spray drying is a continuous process. Other advantages are the low prize and the speed compared to the freeze drying process. Spray drying required addition of drying auxiliary compounds in the nanoparticle dispersion prior to be processed in the spray drier (Müller et al 2000; Tewa-Tagne et al. 2007).

5 Integration of Required Functionalities

Challenging functions need to be associated to nanoparticles to be applied as drug carriers. (1) The polymers should be non toxic, biocompatible and biodegradable if the nanoparticles are designed for parenteral administration. This considerably limits the number of suitable polymers (Vauthier and Bouchemal 2009). (2) Drug association must be efficient. Many types of drugs are interesting to associate with nanoparticle drug carriers. This includes small and large molecules as well as hydrophilic and hydrophobic compounds. By combining the choice of polymer composing nanoparticles and method of nanoparticle preparation almost all types of drugs can be associated with nanoparticles based on their physico-chemical properties. Use of cyclodextrin may improve drug loading if necessary (Duchene et al. 1999). However, remaining challenges are to increase the amount of drug actually associated with the nanoparticles (drug payload) and to reduce drug leakage from the nanoparticles before they reach the target site (Li and Huang 2008). (3) Nanoparticles should transport active drug from the site of administration to target site. In general, drugs associated with nanoparticles are well protected against degradation. However, the most challenging and motivating part of the development of nanoparticle drug carriers is to achieve transport to a specific target. The method is to add specific equipments on the drug carrier surface. This equipment has a dual role. First, it insures nanoparticles to remain well dispersed including in biological media. Second, it confers specific capacity of the carrier to recognize the pharmacological target. As obvious, one of these equipments is the targeting moiety which allows the nanoparticles to recognize target tissue and target cells with a high specificity. For instance, nanoparticles can be decorated with folic acid to target cancer cells over-expressing the folic acid receptor at the cell surface (Xia and Low 2010) but antibodies can also be used to this aim (Nobs et al. 2004a). Nanoparticle platforms with cyclodextrin and biotin residues at the surface were created to facilitate further attachment of targeting moieties (Gref et al. 2003; Nobs et al. 2004b). In addition to the targeting moiety and for all nanoparticles designed to target tissues outside the mononuclear phagocyte system, nanoparticle surface must be masked with a protective coating conferring the nanoparticles the required stability in the blood and a low capacity to activate the complement system (Vonabourg et al. 2006; Vauthier et al. 2011). Typical material used are poly(ethylene glycol) and polysaccharides (Li and Huang 2010; Romberg et al. 2008; Labarre et al. 2005). The chain density should be high enough to hamper accessibility of the nanoparticle surface to large proteins such as those implicated in the activation of the complement system (Gref et al. 2000; Vauthier et al. 2011). With polysaccharide coatings, an additional effect of the chain conformation was highlighted

(Labarre et al. 2005; Bertholon et al. 2006). (4) Once the drug carrier has reached the target site, the cargo should be released. The main challenge in this task is that the drug will be retained by the formulation all along the way from the site of administration down to the target site. Indeed, burst release from nanoparticles is often pointed out as a problem which compromises the delivery of enough amount of drug to the target because of the premature leak of the carrier. New formulation approaches integrate stimuli responsive materials allowing triggered drug release. Thus, the drug remains associated with the carrier until it is triggered by local in vivo variation of either temperature, pH or electrolyte concentrations due to the physiopathology of the targeted area. The drug release can also be triggered from the outside of the body using the action of a magnetic field or of the illumination by light (Bawa et al. 2009). Design of nanoparticles with such properties are feasible with the recent development of stimuli-responding polymers (Stuart et al. 2010) but lots of efforts are still needed to make these materials acceptable for a safe use in drug delivery formulations. (5) By integrating both drug delivery properties and contrast agent properties in a single nanoparticle, this more complex nanoparticulate drug delivery system is ready to enter the era of theragnostic. A couple of prototypes have been designed and it can be expected that multifunctional nanoparticles will be the next step in the development of drug delivery systems (Schärtl 2010).

6 Conclusion and Perspectives

Numerous methods can be used to produce nanoparticles from preformed polymers to improve drug delivery. Several methods are ready for large scale production. Purification, freeze drying and sterilization still need improvements to make possible treatment of large batches and to prevent aggregation. Future developments of nanoparticles for drug delivery are going towards multifunctional nanoparticulate systems eventually integrating stimuli responsive functionalities. The methods described in this chapter should easily be applicable with minor modifications with the new materials to produce the next generation of nanocapsules and nanospheres.

References

Abate A.R., Weitz D.A. High order multiple emulsions formed in poly(dimethylsiloxane) microfluidics. *Small* **5**:2030–2032 (2009)

Abdelwahed W., Degobert G., Stainmesse S., Fessi H. Freeze-drying of nanoparticles: Formulation, process and storage considerations. *Adv Drug Deliv Rev.* **58**:1688–1713 (2006).

Akbulut M., Ginart P., Gindy M.E., Theriault C., Chin KH, Soboyejo W., Prud'homme R.K. Generic method of preparing multifunctional fluorescent nanoparticles using flash nanoprecipitation. *Adv. Funct. Mater.* **19**:718–725 (2009).

Allémann E., Gurny R., Doelker E. Preparation of aqueous polymeric nanodispersions by a reversible salting-out process: influence of process parameters on particle size. *Int. J. Pharm.* **87**:247–253 (1992).

Allémann E., Doelker E., Gurny R. Drug loaded poly(lactic acid) nanoparticles produced by a reversible salting-out process: Purification of an injectable dosage form. *Eur. J. Pharm. Biopharm.* **39**:13–18 (1993).

Anton N., Benoit J.P., Saulnier P. Design and production of nanoparticles formulated from nano-emulsion templates-a review. *J Control Release.* **128**:185–199 (2008).

Athanasiou K.A., Niederauer G.G., Agrawal C.M. Sterilization, toxicity, biocompatibility and clinical applications of polylactic acid/polyglycolic acid copolymers. *Biomaterials.* **17**:93–102 (1996).

Aubry J., Ganachaud F., Cohen, Addad J.P., Cabane B. Nanoprecipitation of Polymethylmethacrylate by Solvent Shifting: 1.Boundaries. *Langmuir* **25**:1970–1979 (2009).

Avgoustakis K. Pegylated poly(lactide) and poly(lactide-co-glycolide) nanoparticles: preparation, properties and possible applications in drug delivery. *Curr Drug Deliv.* **1**:321–333 (2004).

Aynié I., Vauthier C. Chacun H., Fattal E., Couvreur P. Sponge-like alginate nanoparticles as a new system for the delivery of antisense oligonucleotides. *Antisens and Nucleic Acid Drug Development*, **9**:301–312 (1999).

Barbosa M.E., Bouteiller L., Cammas-Marion S., Montembault V., Fontaine L., Ponchel G. Synthesis and ITC characterization of novel nanoparticles constituted by poly(gamma-benzyl L-glutamate)-beta-cyclodextrin. *J Mol Recognit.* **21**:169–78 (2008).

Bawa P., Pillay V., Choonara Y.E., du Toit L.C. Stimuli-responsive polymers and their applications in drug delivery. *Biomed Mater.* **4**:022001 (2009).

Bazile D., Prud'homme C., Bassoullet M.T., Marlard M., Spenlehauer G., Veillard M. Stealth Me.PEG-PLA nanoparticles avoid uptake by the mononuclear phagocytes system. *J Pharm Sci.* **84**:493–498 (1995).

Beck P., Scherer D., Kreuter J. Separation of drug-loaded nanoparticles from free drug by gel filtration. *J. Microencapsul.* **7**:491–496 (1990).

Berger J., Reist M., Mayer J.M., Felt O., Gurny R. Structure and interactions in chitosan hydrogels formed by complexation or aggregation for biomedical applications. *Eur J Pharm Biopharm.* **57**:35–52 (2004).

Bertholon I., Vauthier C., Labarre D. Complement Activation by Core-Shell Poly(isobutylcyanoacrylate)-Polysaccharide Nanoparticles: Influences of Surface Morphology, Length, and Type of Polysaccharide. *Pharm Res.* **23**:1313–1323 (2006).

Bilati U., Allémann E., Doelker E. Sonication parameters for the preparation of biodegradable nanocapsules of controlled size by the double emulsion method. *Pharm Dev Technol.* **8**:1–9 (2003).

BioAlliance Pharma web site at "http://www.bioalliancepharma.com/fre/R-D/Projets" consulted 15 December 2010

Bodmeier R., Huagang C. Indomethacin polymeric nanosuspensions prepared by microfluidization. *J Control Release.* **12**:223–233 (1990).

Boess C., Bögl K.W. Influence of Radiation Treatment on Pharmaceuticals-A Review: Alkaloids, Morphine Derivatives, and Antibiotics. *Drug Dev Ind Pharm.* http://www.informaworld.com/smpp/title~content=t713597245~db=all~tab=issueslist~branches=22 - v22**22**:495–529 (1996).

Bouchemal K., Briançon S., Perrier E., Fessi H., Bonnet I., Zydowicz N. Synthesis and characterization of polyurethane and poly (ether urethane) nanocapsules using a new technique of interfacial polycondensation combined to spontaneous emulsification. *Int J Pharm.* **269**:89–100 (2004).

Bouchemal K., Couenne F., Briançon S., Fessi H., Tayakout M. Stability studies on colloidal suspensions of Polyurethane nanocapsules. *J. Nanosci. Nanotechnol.* **6**:3187–3192 (2006).

Boussif O., Lezoualc'h F., Zanta M.A., Mergny M.D., Scherman D., Demeneix B., Behr J.P. A versatile vector for gene and oligonucleotide transfer into cells in culture and in vivo: polyethylenimine. *Proc Natl Acad Sci U S A.* **92**:7297–7301 (1995).

Bozdag S., Dillen K., Vandervoort J., Ludwig A. The effect of freeze drying with cryoprotectants and gamma-irradiation sterilization on the characteristics of ciprofloxacin HCl-loaded poly(D,L-lactideglycolide) nanoparticles. *J Pharm Pharmacol.* **57**:699–707 (2005).

Briançon S., Fessi H., Lecompte F., Lieto J. Study of an original production process of nanoparticles by precipitation. *Récents Progrès en Génie des Procédés.* **13**:157–164 (1999).

Brigger I., Armand-Lefevre L., Chaminade P., Besnard M., Rigaldie Y., Largeteau A., Andremont A., Grislain L., Demazeau G., Couvreur P. The stenlying effect of high hydrostatic pressure on thermally and hydrolytically labile nanosized carriers. *Pharm Res.* **20**:674–683 (2003).

Brunel F., Véron L., David L., Domard A., Verrier B., Delair T. Self-assemblies on chitosan nanohydrogels. *Macromol Biosci.* **10**:424–432 (2010).

Calvo P., Remuñan-López C., Vila-Jato J.L., Alonso M.J. Chitosan and chitosan/ethylene oxide-propylene oxide block copolymer nanoparticles as novel carriers for proteins and vaccines. *Pharm Res.* **14**:1431–1436 (1997).

Charcosset C., Fessi H. A new process for drug loaded nanocapsules preparation using a membrane contactor. *Drug Dev Ind Pharm.* **31**:987–992 (2005).

Chauvierre C., Labarre D., Couvreur P., Vauthier C. Novel polysaccharide-decorated poly(isobutyl cyanoacrylate) nanoparticles. *Pharm Res.* **20**:1786–1793 (2003).

Chauvierre C, Labarre D, Couvreur P, Vauthier C. A new approach for the characterization of insoluble amphiphilic copolymers based on their emulsifying properties. *Coll Polym Sci.* **282**: 1097–1104 (2004).

Chorny M., Fishbein I., Danenberg H.D., Golomb G. Study of the drug release mechanism from tyrphostin AG-1295-loaded nanospheres by in situ and external sink methods. *J. Control. Release.* **83**:389–400 (2002).

Coll J.L., Chollet P., Brambilla E., Desplanques D., Behr J.P., Favrot M. In vivo delivery to tumors of DNA complexed with linear polyethylenimine. *Hum Gene Ther.* **10**:1659–1666 (1999).

Colombo A.P., Briancon S., Lieto J., Fessi H. Project design and use of a pilot plant for nanocapsule production. *Drug Dev Ind Pharm.* **27**:1063–1072 (2001).

Da Costa Martinez R., Gamazo C., Irache J.M. Design and influence of gamma-irradiation on the biopharmaceutical properties of nanoparticles containing an antigenic complex from Brucella ovis. *Eur J Pharm Sci.* **37**:563–572 (2009).

De Chasteigner S., Cavé G., Fessi H., Devissaguet J.P., Puisieux F. Freeze-drying of Itraconazole-loaded nanosphere suspensions : a feasibility study. *Drug Dev Res.* **38**:116–124 (1996).

De Jaeghere F., Allémann E., Leroux J.-C., Stevels W., Feijen J., Doelker E., Gurny R. Formulation and lyoprotection of poly (Lactic acid-co-ethylene oxide) nanoparticles: influence on physical stability and in vitro cell uptake. *Pharm Res.* **16**:859–866 (1999).

Desgouilles D., Vauthier C., Bazile D., Vacus J., Grossiord J.L., Veillard M. Couvreur P. The design of nanoparticles obtained by solvent evaporation : A comprehensive study. *Langmuir.* **19**:9504–9510 (2003)

Douglas K.L., Tabrizian M. Effect of experimental parameters on the formation of alginate-chitosan nanoparticles and evaluation of their potential application as DNA carrier. *J Biomater Sci Polym Ed.* **16**:43–56 (2005).

Drogoz A., David L., Rochas C., Domard A., Delair T. Polyelectrolyte complexes from polysaccharides: formation and stoichiometry monitoring. *Langmuir.* **23**:10950–8 (2007).

Duchêne D., Ponchel G., Wouessidjewe D. Cyclodextrins in targeting. Application to nanoparticles. *Adv Drug Deliv Rev.* **36**:29–40 (1999).

Fessi H., Puisieux F., Devissaguet J-P., Ammoury N., Benita S. Nanocapsule formation by interfacial deposition following solvent displacement. *Int J Pharm.* **55**:R1- R4 (1989).

Forrest M.L., Yanez J.A., Remsberg C.M., Ohgami Y., Kwon G.S., Davies N.M. Paclitaxel prodrugs with sustained release and high solubility in poly(ethylene glycol)-b-poly(ε-caprolactone) micelle nanocarriers: pharmacokinetic disposition, tolerability, and cytotoxicity. *Pharm Res.* **25**:194–206 (2008).

Fournier E., Dufresne M.H., Smith D.C., Ranger M., Leroux J.C. A novel one-step drug-loading procedure for water-soluble amphiphilic nanocarriers. *Pharm Res.* **21**:962–968 (2004).

França A., Pelaz B., Moros M., Sánchez-Espinel C., Hernández A., Fernández-López C., Grazú V., de la Fuente J.M., Pastoriza-Santos I., Liz-Marzán L.M., González-Fernández A. Sterilization matters: consequences of different sterilization techniques on gold nanoparticles. *Small*. **6**:89–95 (2010).

Freitas S., Rudolf B., Merkle H.P., Gander B. Flow-through ultrasonic emulsification combined with static micromixing for aseptic production of microspheres by solvent extraction. *Eur J Pharm Biopharm*. **61**:181–187 (2005).

Galindo-Rodriguez S.A., Puel F., Briançon S., Allémann E., Doelker E., Fessi H. Comparative scale-up of three methods for producing ibuprofen-loaded nanoparticles. *Eur J Pharm Sci*. **25**:357–367 (2005).

Ganachaud F., Katz J.L. Nanoparticles and Nanocapsules Created Using the Ouzo Effect: Spontaneous Emulsification as an Alternative to Ultrasonic and High-Shear Devices *ChemPhysChem*. **6**:209–216 (2005).

Gaucher G., Marchessault R.H., Leroux J.C. Polyester-based micelles and nanoparticles for the parenteral delivery of taxanes. *J Control Release*. **143**:2–12 (2010a).

Gaucher G., Satturwar P., Jones M.C., Furtos A., Leroux J.C. Polymeric micelles for oral drug delivery. *Eur J Pharm Biopharm*. **76**:147–158 (2010b).

Gavi E., Marchisio D.L., Barresi A.A. CDF modelling of polycaprolactone nanoparticles precipitation via solvent-displacement for pharmaceutical applications. Proceedings of the 8th World Congress of Chemical Engineering, 23–29 August 2009, Montreal, Canada. http://www.wcce8.org/index.html and http://archivos.labcontrol.cl/wcce8/offline/techsched/manuscripts%5Cvilu6w.pdf. Accessed 15 November 2010

Govender T., Stolnik S., Garnett M. C., Illum L., Davis S.S. PLGA nanoparticles prepared by nanoprecipitation: drug loading and release studies of a water soluble drug. *J Control Release*. **57**:171–185 (1999).

Goycoolea F.M., Lollo G., Remuñán-López C., Quaglia F., Alonso M.J. Chitosan-Alginate Blended Nanoparticles as Carriers for the Transmucosal Delivery of Macromolecules. *Biomacromolecules*. 2009 Jun 22. [Epub ahead of print]

Gref R., Minamitake Y., Peracchia M.T., Trubetskoy V., Torchilin V., Langer R. Biodegradable long-circulating polymeric nanospheres. *Science*. **263**:1600–1603 (1994).

Gref R., Lück M., Quellec P., Marchand M., Dellacherie E., Harnisch S., Blunk T., Müller R.H. 'Stealth' corona-core nanoparticles surface modified by polyethylene glycol (PEG): influences of the corona (PEG chain length and surface density) and of the core composition on phagocytic uptake and plasma protein adsorption. *Colloids Surf B Biointerfaces*. **18**:301–313 (2000).

Gref R., Couvreur P., Barratt G., Mysiakine E. Surface-engineered nanoparticles for multiple ligand coupling. *Biomaterials* **24**:4529–4537 (2003).

Guinebretière S., Briancon S., Fessi H., Teodorescu V.S., Blanchin M.G. Nanocapsules of biodegradable polymers: preparation and characterization by direct high resolution electron microscopy. *Mater. Sci. Eng. C*. **21**:137–142 (2002).

Gurny R., Peppas N.A., Harrington D.D., Banker G.S. Development of biodegradable and injectable lattices for controlled release potent drugs. *Drug Dev Ind Pharm*, **7**:1–25 (1981)

Hawkins M.J., Soon-Shiong P., Desai N. Protein nanoparticles as drug carriers in clinical medicine. *Adv Drug Deliv Rev*. **60**:876–885 (2008).

He G., Ma L.L., Pan J., Venkatraman S. ABA and BAB type triblock copolymers of PEG and PLA: a comparative study of drug release properties and "stealth" particle characteristics. *Int J Pharm*. **334**:48–55 (2007).

Huh K.M., Lee S.C., Cho Y.W., Lee J., Jeong J.H., Park K. Hydrotropic polymer micelle system for delivery of paclitaxel. *J Control Release*. **101**:59–68 (2005).

Ibrahim H., Bindschaedler C., Doelker E., Buri P., Gurny R. Aqueous nanodispersions prepared by a salting-out process. *Int. J. Pharm*. **87**:239–246 (1992).

Janes K.A., Calvo P., Alonso M.J. Polysaccharide colloidal particles as delivery systems for macromolecules. *Adv Drug Deliv Rev*. **47**:83–97 (2001).

Jeong J.H., Kim S.W., Park T.G. Molecular design of functional polymers for gene therapy. *Prog. Polym. Sci*. **32**:1239–1274 (2007).

Jie P., Venkatraman S.S., Min F., Freddy B.Y., Huat G.L. Micelle-like nanoparticles of star-branched PEO-PLA copolymers as chemotherapeutic carrier. *J Control Release*. **110**:20–33 (2005).

Johnson B.K., Prud'homme R.K. Mechanism for rapid self-assembly of block copolymer nanoparticles. *Phys Rev Lett.* **91**:118302 (2003).

Joralemon M.J., McRae S., Emrick T. PEGylated polymers for medicine: from conjugation to self-assembled systems. *Chem Commun (Camb)*. **46**:1377–1393 (2010).

Kabanov A., Zhu J., Alakhov V. Pluronic block copolymers for gene delivery. *Adv Genet.* **53**:231–261 (2005).

Kang N., Leroux J.-C. Triblock and star-block copolymer of N-(2-hydroxypropyl) methacrylamide or N-vinyl-2-pyrrolidone and D L-lactide: synthesis and self assembling properties in water. *Polymer* **45**:8967–8980 (2004).

Kim J.H., Emoto K., Iijima M, Nagasaki Y., Aoyagi T., Okano T., Sakurai Y., Kataoka K. Core-stabilized polymeric micelle as potential drug carrier: increased solubilization of Taxol. *Polym Adv. Technol.* **10**:647–654 (1999).

Koshy A., Das T. R., Kumar R. Effect of surfactants on drop breakage in turbulent liquid dispersions. *Chem Eng Sci.* **43**:649–654 (1988).

Labarre D., Vauthier C., Chauvierre C., Petri B., Müller R., Chehimi M.M. Interactions of blood proteins with poly(isobutylcyanoacrylate) nanoparticles decorated with a polysaccharidic brush. *Biomaterials*. **26**:5075–84 (2005).

Lambert G., Fattal E., Pinto-Alphandary H., Gulik A., Couvreur P. Polyisobutylcyanoacrylate nanocapsules containing an aqueous core as a novel colloidal carrier for the delivery of oligonucleotides. *Pharm Res.* **17**(6):707–714 (2000).

Lamprecht A., Ubrich N., Hombreiro Pérez M., Lehr C., Hoffman M., Maincent P. Biodegradable monodispersed nanoparticles prepared by pressure homogenization-emulsification. *Int J Pharm.* **184**:97–105 (1999).

Lamprecht A., Ubrich N., Hombreiro Pérez M., Lehr C., Hoffman M., Maincent P. Influences of process parameters on nanoparticle preparation performed by a double emulsion pressure homogenization technique. *Int J Pharm.* **196**:177–182 (2000).

Lee S.C., Kim C., Kwon I.C., Chung H., Jeong S.Y. Polymeric micelles of poly(2-ethyl-2-oxazoline)-block-poly(ε-caprolactone) copolymer as a carrier for paclitaxel. *J Control Release* **89**:437–446 (2003).

Legrand P., Lesieur S., Bochot A., Gref R., Raatjes W., Barratt G., Vauthier C. Influence of polymer behaviour in organic solution on the production of polylactide nanoparticles by nanoprecipitation. *Int. J. Pharm.* **344**:33–43 (2007).

Lemarchand C., Couvreur P., Besnard M., Costantini D., Gref R. Novel polyester-polysaccharide nanoparticles. *Pharm Res.* **20**:1284–1292 (2003).

Leroux J.C., Allemann E., Doelker E., Gurny R. New approach for the preparation of nanoparticles by an emulsification-diffusion method. *Eur. J. Pharm. Biopharm.* **41**(1):14–18 (1995).

Li S.D., Huang L. Pharmacokinetics and biodistribution of nanoparticles. *Molecular pharmaceutics* **5**:496–504 (2008).

Li S.D., Huang L. Stealth nanoparticles: high density but sheddable PEG is a key for tumor targeting. *J Control Release.* **145**:178–181 (2010).

Li T., Shi X.W., Du Y.M., Tang Y.F. Quaternized chitosan/alginate nanoparticles for protein delivery. *J Biomed Mater Res A.* **83**:383–390 (2007).

Limayem I., Charcosset C., Fessi H. Purification of nanoparticle suspensions by a concentration/diafiltration process. *Sep Purif Technol.* **38**:1–9 (2004).

Lin Y.S., Renbutsu E., Morimoto M., Okamura Y., Tsuka T., Saimoto H., Okamoto Y., Minami S. Preparation of stable chitosan-carboxymethyl dextran nanoparticles. *J Nanosci Nanotechnol.* **9**:2558–2565 (2009).

Lince F., Marchisio D.L., Barresi A.A. Strategies to control the particle size distribution of poly-epsilon-caprolactone nanoparticles for pharmaceutical applications. *J Colloid Interface Sci.* **322**:505–515 (2008).

Lince F., Marchisio D.L., Barresi A.A. Smart mixers and reactors for the production of pharmaceutical nanoparticles: proof of concept. 13th European Conference on Mixing, London, 14–17 April 2009.

Liu Y., Cheng C., Liu Y., Prud'homme R.K., Fox R.O. Mixing in a multi-inlet vortex mixer (MIVM) for flash nano-precipitation. *Chem Eng Sci.* **63**:2829–2842 (2008).

Lu Z,. Bei J., Wang S. A method for the preparation of polymeric nanocapsules without stabilizer. *J Control Release.* **61**:107–112 (1999).

Mabille C., Leal-Calderon F., Bibette J., Schmitt V. Monodisperse fragmentation in emulsions: Mechanisms and kinetics. *Europhys Lett.* **61**:708–714 (2003).

Maksimenko O., Pavlov E., Toushov E., Molin A., Stukalov Y., Prudskova T., Feldman V., Kreuter J., Gelperina S. Radiation sterilisation of doxorubicin bound to poly(butyl cyanoacrylate) nanoparticles. *Int J Pharm.* **356**:325–332 (2008).

Mao S., Bakowsky U., Jintapattanakit A., Kissel T. Self-assembled polyelectrolyte nanocomplexes between chitosan derivatives and insulin. *J Pharm Sci.* **95**:1035–1048 (2006).

Masson V., Maurin F., Fessi H., Devissaguet J.P. Influence of sterilization processes on poly (ε-caprolactone) nanospheres. *Biomaterials.* **18**:327–335 (1997).

Memisoglu-Bilensoy E., Hincal A.A. Sterile, injectable cyclodextrin nanoparticles: Effects of gamma irradiation and autoclaving. *Int J Pharm.* **311**:203–208 (2006).

Moinard-Chécot D., Chevalier Y., Briançon S., Fessi H., Guinebretière S. Nanoparticles for drug delivery: review of the formulation and process difficulties illustrated by the emulsion-diffusion process. *J. Nanosci. Nanotechnol.* **6(9–10)**:2664–2681 (2006).

Moinard-Chécot D., Chevalier Y., Briançon S., Beney L., Fessi H. Mechanism of nanocapsules formation by the emulsion-diffusion process. *J Colloid Interface Sci.* **317**:458–468 (2008).

Müller C.R., Bassani V.L., Pohlmann A.R., Michalowski C.B., Petrovick P.R., Guterres S.S. Preparation and characterization of spray-dried nanocapsules. *Drug Dev. Ind. Pharm.* **26**:343–347 (2000).

Murakami H., Kobayashi M., Takeuchi H., Kawashima Y. Preparation of poly(DL-lactide-co-glycolide) nanoparticles by modified spontaneous emulsification solvent diffusion method. *Int J Pharm.* **187**(2):143–152 (1999)

Murakami H., Kobayashi M., Takeuchi H., Kawashima Y. Further application of a modified spontaneous emulsification solvent diffusion method to various types of PLGA and PLA polymers for preparation of nanoparticles. *Powder Technol.* **107**:137–143 (2000).

Nah J.W., Jung T.R., Jeong Y.L., Jang M.K. Biodegradable nanoparticles of poly(DL-lactide-co-glycolide) encapsulating ciprofloxacin HCl having an extended-released property and manufacturing method thereof. World Patent 054042 (2008)

Nemati F., Cavé G.N., Couvreur P. Lyophilization of substances with low water permeability by a modification of crystallized structures during Freezing. Proceedings of the 6th International Congress of Pharmaceutical Technology Assoc. Pharm. Galenique Ind., Chatenay Malabry, APGI, Paris Vol. 3, 1992, pp. 487–493.

Nguyen C.A., Allemann E., Schwach G., Doelker E., Gurny R. Synthesis of a novel fluorescent poly (DL,-lactide) endcapped with 1-pyrenebutanol used for the preparation of nanoparticles. *Eur J Pharm Sci.* **20**:217–222 (2003).

Niwa T., Takeuchi T., Hino T., Kunou N., Kawashima Y. Preparations of biodegradable nanospheres of water-soluble and insoluble drugs with d,llactide/glycolide copolymer by a novel spontaneous emulsification solvent diffusion method, and the drug release behavior. *J. Control. Release.* **25**:89–98 (1993).

Nobs L., Buchegger F., Gurny R., Allemann E. Current methods for attaching targeting ligands to liposomes and nanoparticles. *J Pharm Sci.* **93**:1980–92 (2004a).

Nobs L., Buchegger F., Gurny R., Allemann E. Poly(lactic acid) nanoparticles labeled with biologically active Neutravidin for active targeting. *Eur J Pharm Biopharm.* **58**:483–490 (2004b).

Oyarzun-Ampuero F.A., Brea J., Loza M.I., Torres D., Alonso M.J. Chitosan-hyaluronic acid nanoparticles loaded with heparin for the treatment of asthma. *Int J Pharm.* **381**:122–129 (2009).

Park E.K., Kim S.Y., Lee S.B., Lee Y.M. Folate-conjugated methoxy poly(ethylene glycol)/poly(ε-caprolactone) amphiphilic block copolymeric micelles for tumor targeted drug delivery. *J Control Release* **109**:158–168 (2005).

Patapoff T.W., Overcashier D.E. The importance of freezing on lyophilization cycle development. *Biopharm.* **3**:16–21 (2002).

Peltonen L., Anitta J., Hyvönen S., Kajalainen M., Hirvonen J. Improved entrapment efficiency of hydrophilic drug substance during nanoprecipitation of poly(l)lactide nanoparticles. *AAPS Pharm Sci Tech.* **5**:1–6 (2004).

Perez C., Sanchez A., Putnam D., Ting D., Langer R., Alonso M.J. Poly(lactic acid)-poly(ethylene glycol) nanoparticles as new carriers for the delivery of plasmid DNA. *J Control Release.* **75**:211–224 (2001).

Petrelli F., Borgonovo K., Barni S. Targeted delivery for breast cancer therapy: the history of nanoparticle-albumin-bound paclitaxel. *Expert Opin Pharmacother.* **11**:1413–1432 (2010).

Qiu L.Y., Bae Y.H. Polymer architecture and drug delivery. *Pharm Res.* **23**(1):1–30 (2006).

Quintanar-Guerrero D., Ganem-Quintanar A., Allemann E., Fessi H., Doelker E. Influence of the stabilizer coating layer on the purification and freeze-drying of poly(D,L-lactic acid) nanoparticles prepared by an emulsion-diffusion technique. *J Microencapsul.* **15**:107–119 (1998).

Quintanar-Guerrero D., Allémann E., Fessi H., Doelker E. Pseudolatex preparation using a novel emulsion-diffusion process involving direct displacement of partially water-miscible solvents by distillation. *Int. J. Pharm.* **188**:155–164 (1999).

Rajaonarivony M., Vauthier C., Couarraze G., Puisieux F., Couvreur P. Development of a new drug carrier made from alginate. *J. Pharm. Sci.* **82**:912–918 (1993).

Rollot J., Couvreur P., Roblot-Treupel L., Puisieux F. Physicochemical and morphological characterization of polyisobutylcyanoacrylate nanocapsules. *J Pharm Sci.* **75**:361–364 (1986).

Romberg B., Hennink W.E., Storm G. Sheddable coatings for long-circulating nanoparticles. *Pharm Res.* **25**:55–71 (2008).

Sahoo S.K., Panyam J., Prabha S., Labhasetwar V. Residual polyvinyl alcohol associated with poly (DL,-lactidecoglycolide) nanoparticles affects their physical properties and cellular uptake. *J Control Release.* **82**:105–114 (2002).

Sani S.N., Das N.G., Das S.K. Effect of microfluidization parameters on the physical properties of PEG-PLGA nanoparticles prepared using high pressure microfluidization. *J Microencapsul.* **26**:556–561 (2009).

Sarmento B., Ferreira D.C., Jorgensen L., van de Weert M. Probing insulin's secondary structure after entrapment into alginate/chitosan nanoparticles. *Eur J Pharm Biopharm.* **65**:10–17 (2007).

Schärtl W. Current directions in core-shell nanoparticle design. *Nanoscale.* **2**:829–843 (2010).

Schatz C., Domard A., Viton C., Pichot C., Delair T. Versatile and efficient formation of colloids of biopolymer-based polyelectrolyte complexes. *Biomacromolecules.* **5**:1882–1892 (2004).

Shegokar R., Singh K.K., Müller R.H. Production & stability of stavudine solid lipid nanoparticles-From lab to industrial scale. *Int J Pharm.* In Press 2010 Aug 18. [Epub ahead of print]

Sintzel M.B., Merklia A., Tabatabay C., Gurny R. Influence of irradiation sterilization on polymers used as drug carriers : A review. *Drug Dev Ind Pharm.* **23**:857–878 (1997).

Stainemesse S., Orecchioni A.M., Nakache E., Puisieux F., Fessi H. Formation and stabilization of a biodegradable polymeric colloidal suspension of nanoparticles. *Colloid and Polym. Sci.* **273**:505–511 (1995).

Stork M., Tousain R.L., Wieringa J.A., Bosgra O.H. A MILP approach to the optimization of the operation procedure of a fed-batch emulsification process in a stirred vessel. *Comp Chem Eng.* **27**:1681–1691 (2003).

Stuart M.A., Huck W.T., Genzer J., Müller M., Ober C., Stamm M., Sukhorukov G.B., Szleifer I., Tsukruk V.V., Urban M., Winnik F., Zauscher S., Luzinov I., Minko S. Emerging applications of stimuli-responsive polymer materials. *Nat Mater.* **9**:101–113 (2010).

Sun W., Mao S., Mei D., Kissel T. Self-assembled polyelectrolyte nanocomplexes between chitosan derivatives and enoxaparin. *Eur J Pharm Biopharm.* **69**:417–425 (2008).

Sun X., Zhang N. Cationic polymer optimization for efficient gene delivery. *Mini Rev Med Chem.* **10**:108–125 (2010).
Tewa-Tagne P., Briançon S., Fessi H. Preparation of redispersible dry nanocapsules by means of spray-drying: development and characterisation. *Eur J Pharm Sci.* **30**:124–135 (2007).
Thioune O., Fessi H., Devissaguet J.P., Puisieux F. Preparation of pseudolatex by nanoprecipitation: Influence of the solvent nature on intrinsic viscosity and interaction constant. *Int. J. Pharm.* **146**:233–238 (1997).
Tishchenko G., Hilke R., Albrecht W., Schauer J., Luetzow K., Pientka Z., Bleha M. Ultrafiltration and microfiltration membranes in latex purification by diafiltration with suction. *Sep Purif Technol.* **30**:57–68 (2003).
Tokumitsu H., Ichikawa H., Fukumori Y., Hiratsuka J., Sakurai Y., Kobayashi T. Preparation of gadopentenate-loaded nanoparticles for gadolinium neutron capture therapy of cancer using a novel emulsion droplet coalescence technique. *Proc. 2nd world meeting APGI/APV*, Paris, 25–28 Mai 1998. 641–642 (1998).
Trivedi R., Kompella U.B. Nanomicellar formulations for sustained drug delivery: strategies and underlying principles. *Nanomedicine* (Lond) **5**:485–505 (2010).
Tuereli A.E., Penth B., Langguth P., Boumstuemmler B., Kraemer J. Preparation of drug nanoparticles using microjet reactor technology. Proceedings of the 2nd PharmaSciFair, 8–12 July 2009, Nice, France
Urban K., Wagner G., Schaffner D., Röglin D., Ulrich J. Rotor-stator and disc systems for emulsufication processes. *Chem Eng Technol* **29**:24–31 (2006).
Vauthier C., Bouchemal K. Methods for the preparation and manufacture of polymeric nanoparticles. *Pharm Res.* **26**:1025–1058 (2009).
Vauthier C., Couvreur P. Development of nanoparticles made of polysaccharides as novel drug carrier systems. In "Handbook of Pharmaceutical Controlled Release Technology", D.L. Wise Ed., Marcel Dekker Inc. New-York, USA (2000) chap.21, pp. 413–429
Vauthier C., Rajaonarivony M., Couarraze G., Couvreur P., Puisieux F. Characterization of alginate pregel by rheological investigation. *Eur. J. Pharm. Biopharm.* **40**:218–222 (1994).
Vauthier C., Cabane B., Labarre D. How to concentrate nanoparticles and avoid aggregation ? *Eur J Pharm Biopharm.* **69**:466–475 (2008).
Vauthier C., Persson B., Lindner P, Cabane B. Protein adsorption and complement activation for di-block copolymer nanoparticles. *Biomaterials.* **32**:1646–1656 (2011).
Vila A., Sánchez A., Tobío M., Calvo P., Alonso M.J. Design of biodegradable particles for protein delivery. *J Control Release.* **78**:15–24 (2002).
Vitale S.A., Katz J.L. Liquid Droplet Dispersions Formed by Homogeneous Liquid-Liquid Nucleation: "The Ouzo Effect" *Langmuir* **19**:4105–4110 (2003).
Vonabourg A., Passirani C., Saulnier P., Benoit J.P. Parameters influencing the stealthiness of colloidal drug delivery systems. *Biomaterials.* **27**:4356–4373 (2006).
Walstra P. In *Encyclopedia of Emulsion Technology, 1 Basic Theory*; Becher, P., Ed.; Marcel Dekker Inc.: New York, 1983; Vol. 1, Chapter 2, pp 57–127.
Wang N., Wu X.S. Preparation and characterization of agarose hydrogel nanoparticles for protein and peptide drug delivery. *Pharm Dev Technol,* **2**:135–142 (1997).
Weber C., Drogoz A., David L., Domard A., Charles M.H., Verrier B., Delair T. Polysaccharide-based vaccine delivery systems: Macromolecular assembly, interactions with antigen presenting cells, and in vivo immunomonitoring. *J Biomed Mater Res A.* **93**:1322–1334 (2010).
Woitiski C.B., Neufeld R.J., Ribeiro A.J., Veiga F. Colloidal carrier integrating biomaterials for oral insulin delivery: Influence of component formulation on physicochemical and biological parameters. *Acta Biomater.* **5**:2475–2484 (2009a).
Woitiski C.B., Veiga F., Ribeiro A., Neufeld R. Design for optimization of nanoparticles integrating biomaterials for orally dosed insulin. *Eur J Pharm Biopharm.* **73**:25–33 (2009b).
Xia W., Low P.S. Folate-targeted therapies for cancer. *J Med Chem.* **53**:6811–6824 (2010).
Zheng W. A water-in-oil-in-oil-in-water (W/O/O/W) method for producing drug-releasing, double-walled microspheres. *Int J Pharm.* **374**:90–95 (2009).

Part III
Medical Applications

Magnetic Resonance Tracking of Stem Cells with Iron Oxide Particles

Eddy S.M. Lee, Brian K. Rutt, Nicholas M. Fisk, Shih-Chang Wang, and Jerry Chan

Abstract Stem cells transplantation is a promising therapy for numerous diseases where transplanted cells repair or replace damaged host tissue. While their efficacy and optimal delivery is under intense investigation, there lies a pivotal question seeking the whereabouts of the cells after transplantation. Imaging techniques have emerged in recent years, both to enable monitoring of stem cell location in patients and to improve the reliability of animal experimentation. Magnetic resonance imaging (MRI) allows tracking of stem cells tagged with magnetic nanoparticle labels

E.S.M. Lee (✉)
Richard M. Lucas Center for Imaging, Radiology Department, Stanford University,
1201 Welch Road, Stanford, CA 94305–5488, USA
and
UQ Centre for Clinical Research, University of Queensland, Brisbane, QLD 4029, Australia
e-mail: eddysmlee@gmail.com

B.K. Rutt
Richard M. Lucas Center for Imaging, Radiology Department, Stanford University,
1201 Welch Road, Room PS064, Stanford, CA 94305–5488, USA

N.M. Fisk
UQ Centre for Clinical Research, University of Queensland, Brisbane, QLD 4029, Australia

S.-C. Wang
Discipline of Medical Imaging, University of Sydney, Sydney, NSW 2006, Australia
and
Department of Radiology, Westmead Hospital, Wentworthville, NSW 2145, Australia

J. Chan
UQ Centre for Clinical Research, University of Queensland, Brisbane, QLD 4029, Australia
and
Experimental Fetal Medicine Group, Department of Obstetrics and Gynecology,
Yong Loo Lin School of Medicine, National University of Singapore,
5 Lower Kent Ridge Road, Singapore 119074, Singapore
and
Department of Reproductive Medicine, KK Women's and Children's Hospital,
100 Bukit Timah Road, Singapore 229899, Singapore

A. Prokop (ed.), *Intracellular Delivery: Fundamentals and Applications*, Fundamental Biomedical Technologies 5, DOI 10.1007/978-94-007-1248-5_17,

prior to transplantation, but is restricted by the inability of stem cells to incorporate sufficient label. This review addresses the optimisation of stem cell tagging with iron oxide particles to improve MR tracking, alternative cell labelling techniques using gene transfer, and the translational applications of cellular imaging.

Keywords Magnetic nanoparticle • Stem cell tracking • MRI • SPIO

Abbreviations

MRI	magnetic resonance imaging
ESC	embryonic stem cells
MSC	mesenchymal stem cells
Gd-DTPA	gadolinium diethylenetriaminopentaacetic acid
SPIO	superparamagnetic iron oxide particles
USPIO	ultrasmall SPIO
MION	monocrystalline iron oxide nanocompound
VSOP	very small superparamagnetic iron oxide particles
AMNP	anionic magnetic nanoparticle
IV	intravenously
CLIO	cross-linked iron oxide particles
Ab	antibody
MGIO	microgel iron oxide particles
EPC	endothelial progenitor cells
PLL	poly-L-lysine
Tf	transferrin
TfR	transferrin receptor
NSC	neural stem cells
DC	dendritic cells
CGMP	clinical grade manufacturing practice
BrdU	5-bromo-2-deoxyuridine
BBZ	bis benzamide
SDR	static dephasing regime

1 Stem Cell Therapy

Stem cell therapy is a rapidly emerging field of regenerative medicine where transplanted cells directly replace and/or ameliorate the repair of damaged host tissue. The adult body possesses natural, albeit rare, stem cells, either remaining from fetal development or acquired through microchimeric events such as pregnancies (Lee et al. 2010a). Although endogenous stem cells participate in tissue repair during injuries or diseases, they are often insufficient in number to obtain effective regeneration and repair, thus requiring augmentation by stem cell transplantation to improve clinical outcomes.

Stem cells are immature cells that possess the ability to self-renew and to differentiate into various cell types. They can be broadly classified into three categories based on their capacity for differentiation. Totipotent stem cells, such as the zygote or cells from early embryos (1–3 days post fertilisation), have the ability for each cell to develop into a complete individual. Pluripotent stem cells can form all three germ layers of the body (endoderm, mesoderm and ectoderm), an example of which are embryonic stem cells (ESC) isolated from the inner cell mass of blastocyst (5–14 days). Multipotent stem cells are committed cells that can still form a number of other tissues, but not all three germ layers. An example of a multipotent stem cell is the haemopoietic stem cell which can derive both lymphoid and myeloid lineage blood cell types.

Recent developments in the understanding of multipotent stem cells from non-embryonic sources have sparked renewed excitement in this field. Multipotent cells, such as the mesenchymal stem cells (MSC), appear to possess greater plasticity than dictated by established paradigms of embryonic development (Phinney and Prockop 2007). As MSC can differentiate from primitive cells into mature cell types, they can be used for cell replacement therapy, tissue engineering, regenerative medicine and as vehicles for gene therapy (Gafni et al. 2004). Unlike ESC which are generated with difficulty and only with the sacrifice of human embryos, multipotent cells from adult or discarded fetal-placental tissue are subject to fewer practical and ethical issues. These attributes of multipotent cells render them promising candidates for future clinical use.

Several clinical trials are underway for the treatment of various diseases such as ischemic stroke (Bang et al. 2005), skeletal dysplasia (Horwitz et al. 2001), spinal cord injury (Callera and de Melo 2007) and myocardial infarction (Meyer et al. 2006). The key determinants for the translation of such therapies are safety and efficacy. While the therapeutic capability of cell therapy is under investigation, detecting the location of transplanted cells is critical to the evaluation of transplantation route, timing, dosage and cell type. Temporal and spatial information on the cells informs their engraftment efficiency and functional capacity, facilitating the monitoring and optimising of the therapeutic process in patients. When applied to animal experimentation, serial *in vivo* tracking of cells, as opposed to sacrificing animals at several cross sectional time points for histology, improves the reliability of results while reducing animal numbers.

2 *In Vivo* Imaging Modalities

A number of cellular imaging modalities are under investigation, but only a few are clinically relevant (see Table 1) (Arbab and Frank 2008). Clinical translation requires a modality that provides sufficient imaging resolution and depth of

Table 1 Modalities of cellular imaging adapted from Arbab et al. (Arbab and Frank 2008)

	MRI						Optical			
Modality	1H	19F	CEST[a]	CT[b]	PET[c]	SPECT[d]	IM[e]	FRI[f]	BLI[g]	US[h]
Spatial resolution	10–100 μm	>100 μm	>100 μm	<10 μm	1–2 mm (animal scanner) 3–5 mm (clinical)		1 μm	2–3 mm	2–3 mm	50 μm
Depth of penetration	Unlimited	Unlimited	Unlimited	Unlimited	Unlimited	Unlimited	0.4 mm	<1 cm	1 cm	Unlimited
Sensitivity[i]	1–1,000	7,500	25,000	Poor	Hundreds	No data	No data	No data	10	1
Reporter gene[j]	In progress	No	Yes	No	Yes	No	Yes	No	Yes	No
Longevity	Good	Good	Excellent	Good	Excellent	Poor	Excellent	Good	Excellent	Good
Label approach	Exogenous label	Exogenous label	Reporter gene	Exogenous label	Reporter gene	Decaying isotope	Reporter gene	Exogenous label	Reporter gene	Exogenous label
Cellular quantification	Possible	Yes	Possible	Possible	Possible	Yes	No	No	Possible	Possible
Ionising radiation	No	No	No	Yes	Yes	Yes	No	No	No	No
Clinical translation	Yes	Yes	Yes	Possible	Yes	Yes	Limited	No	No	Yes
Characteristics	Gd, Mn, iron oxide	19F	Ln or peptides	Iodine	18F, 11C, 15O, 64Cu	111In, 123I, 99mTc	GFP	GFP, NIR fluoro-chrome	Luciferins	Perfluoro-carbon
	Soft tissue contrast	Good specificity	Good specificity		Cyclotron needed	Limited duration		Rapid molecular diagnostics	Gene expression	μbubble interventional

[a] CEST Chemical exchange-dependent saturation transfer
[b] CT computer tomography
[c] PET positron emission tomography
[d] SPECT single photon emission computed tomography
[e] IM intravital microscopy
[f] FRI fluorescence reflectance imaging
[g] BLI bioluminescence imaging
[h] US ultrasound
[i] Sensitivity: the minimum number of cells detectable
[j] Reporter gene: whether transgenic cells can carry a reporter gene that generates contrast

penetration through the human body, while minimising ionising radiation exposure to be at least within safety limits.

Compared to other modalities, ultrasound is inexpensive and widely available but the imaging physics underpinning these features limit its application in cellular imaging. The inherent trade-off between resolution and depth with ultrasound typically restricts ultrasound to mammography and the imaging of subcutaneous regions, limbs and neck. However, advances in endocavitary ultrasound offer the potential for effective ultrasound interrogation of the prostate and female pelvic organs (Aigner et al. 2009).

Optical techniques offer superior resolution and multi-spectral acquisition, albeit at limited depth of penetration; a limitation that is acceptable in applications such as intravital microscopy. Donor cells can be identified with fluorescent transgenic materials, such as reporter genes or xenogeneic antibodies. Transgeneic cells that express luciferase can be located with bioluminescence imaging in rodents (Nguyen Huu et al. 2007). However, the history of drug approval suggests transgeneic and xenogeneic materials are unlikely to receive clinical approval in the foreseeable future. Optical methods are thus likely to remain an imaging modality for small animals or for superficial visualisation in patients.

With modalities that require ionising radiation such as computerized tomography (CT), positron emission tomography (PET) and single photon emission computed tomography (SPECT), patient exposure to ionising radiation needs to be minimised. The period of monitoring with SPECT is restricted when using practical isotopes such as ^{111}In. A promising modality is PET, as it is not limited by depth, however, it has poor spatial resolution and generally suffers from very short tracer isotope half-lives. Notwithstanding this, gene transfection methods have been devised to overcome the latter limitation. Prior to transplantation, donor cells are transduced with herpes simplex virus thymidine kinase genes such that they can then be tracked by positron emitters (Cao et al. 2006). Recent advancements have focused on improving spatial resolution and efficiency with miniature PET detectors (Park et al. 2007), pinhole inserts for gamma cameras (Wu et al. 2008) and better signal processing (Fontaine et al. 2007). CT scans, even with newer advances in dual energy or spectral imaging using iterative reconstruction protocols, are too insensitive to detect cellular labels in the concentrations achievable *in vivo*.

There has been much interest in cellular tracking by MRI – its balance of spatial resolution and sensitivity, as well as its lack of ionising radiation, makes it an attractive and practical modality for cell detection. By comparison, MRI edges out PET in terms of spatial resolution, but demonstrates significantly lower label detection sensitivity than PET and also suffers from a lack of specificity due to the challenges of identifying the labelled cells against the high and complex background signal in the typical ^{1}H MR image. Researchers have been working to improve specificity by using spectrally-distinct agents such as ^{19}F (Kadayakkara et al. 2010), and chemical exchange-dependent saturation transfer agents (Gilad et al. 2009), which in theory can acquire signal only from the cells and without interference from background. Other attempts have applied specialised sequences for optimal cellular detection (Heyn et al. 2005) or simply increased the amount of contrast agent per cell.

3 MRI with Iron Oxide Particles

The most sensitive method of detecting cells by MRI currently is through the use of iron oxide particles. Their development as MRI contrast agents took place more than two decades ago. Then, gadolinium diethylenetriaminopentaacetic acid (Gd-DTPA) was a commonly used contrast agent (Saini et al. 1987), but did not improve the detection of liver carcinoma in clinical studies (Carr et al. 1984). In fact, metastases were often obscured due to the extracellular distribution of Gd-DTPA into both neoplastic and normal liver (Saini et al. 1986). This limitation was largely due to the poor temporal resolution of the spin echo imaging sequences clinically available at that time; multiphasic imaging sequences having not yet been developed. This limitation has since been overcome by the development and implementation of fast multi-echo breath-held spoiled gradient echo and motion-compensated fast spin echo sequences. Today, gadolinium-based dynamic multiphasic contrast enhancement remains the key method for the detection and characterisation of liver masses with MRI in most centres. Because of the limitations of Gd-enhanced T1-weighted MRI in the late 1980s and early 1990s, superparamagnetic iron oxide particles (SPIO) attracted attention as an alternative for this application because of their large moment in a magnetic field, absence of remnant magnetization when the field is removed (superparamagnetism) and unique biodistribution (Mendonca and Lauterbur 1986; Saini et al. 1987).

When injected intravenously, these particles home to the reticuloendothelial system (Saini et al. 1987) and are taken up by macrophages in the liver, known as Kupffer cells. The presence of SPIO causes darkening of normal hepatic tissue on T2-weighted MR images, while liver carcinomas and metastases remain bright. This application of SPIO led to the 1996 FDA approval of ferumoxide (US trade name: Feridex®, Europe trade name: Endorem®, research name: AMI-25, generic name: ferumoxide), which is administered by dilution in 5% dextrose and slow infusion. Ferumoxide is a dextran-coated composite particle with a hydrated diameter of 120–180 nm, comprising multiple 5 nm iron oxide nanoparticle cores (Raynal et al. 2004). In 2001, European market approval was granted to a smaller SPIO, ferucarbotran (trade name: Resovist®, research name: SHU 555A, generic name: ferucarbotran), which is carboxydextran-coated, measures 62 nm in hydrated diameter and comprises 5 nm iron oxide nanoparticle cores (Reimer et al. 1995; Reimer and Balzer 2003). Compared to ferumoxide, ferucarbotran offers the advantage of delivery by undiluted, bolus injection. It is also less likely to cause nausea and other mild symptoms that are commonplace with ferumoxide.

Smaller than SPIO is a class of particles known as ultrasmall SPIO (USPIO), with diameters less than 30 nm (Weissleder et al. 1990b). Ultrasmall SPIO provides a mixture of paramagnetic blood pool positive enhancement and reticuloendothelial cell negative enhancement, and has been injected intravenously to differentiate between darkened normal lymph nodes and metastatic ones that remain bright (Weissleder et al. 1990a). A USPIO undergoing clinical trials is ferumoxtran (trade name: Sinerem/Combidex®, research name: AMI-227, generic name: ferumoxtran) which has a 5 nm iron oxide core and 17–20 or 21–30 nm hydrated diameter as a result of its dextran coat (Sharma et al. 1999; Jung and Jacobs 1995). A preclinical particle of similar size to USPIO is the monocrystalline iron oxide nanocompound (MION) developed by Massachusetts General Hospital (Shen et al. 1993). Each particle has a single 4–5 nm core coated with dextran, resulting in a hydrated diameter of 20 nm. Two other ultra-small preclinical particles that are coated with monomers are the citrate-coated very small superparamagnetic iron oxide particles (VSOP, 5 nm core, 8 nm hydrated diameter) (Taupitz et al. 2000) and the dimercaptosuccinic acid-coated anionic magnetic nanoparticle (AMNP, 8 nm core, 30 nm hydrated diameter) (Wilhelm and Gazeau 2008).

MRI of inflammatory processes is feasible using the ability of macrophages to take up iron oxide particles *in vivo*. Macrophages involved in atherosclerotic plaques have been imaged following an intravenous injection of SPIO (Ruehm et al. 2001; von Zur Muhlen et al. 2007). After USPIO injection, particle uptake was predominantly in ruptured and rupture-prone atherosclerotic lesions, differentiating such lesions from intact ones (Kooi et al. 2003). Iron oxide particles can also be used for assessing neuroinflammation following ischemic stroke, by differentiating infiltrating macrophages from resident ones. By intravenously (IV) injecting rodents with SPIO 24 h prior to imaging, MR images showed darkened stroke periphery at day 6 or darkened stroke core at day 8 post-stroke. The specific darkening was due to iron-laden macrophages infiltrating the stroke site (Kleinschnitz et al. 2003). Interestingly, when SPIO/USPIO were delivered IV 2–5 h post-stroke, the stroke periphery darkened as well but this was due to focal accumulation of SPIO/USPIO in the occluded vessels (Kleinschnitz et al. 2005; Wiart et al. 2007). Therefore, IV administration of SPIO/USPIO can monitor both ischemic development and the infiltration of macrophages into the stroke site in animal models (Bendszus et al. 2007), leading to clinical phase II trials of the particles in neuroinflammatory imaging (Saleh et al. 2004).

4 Cellular Labelling with Iron Oxide Particles

Another method of MR cell tracking, first reported two decades ago, is through the *ex vivo* labelling of cells with iron oxide particles prior to their transplantation (Ghosh et al. 1990). Particles were introduced into the culture medium of cells and incubated for hours to encourage particle uptake by endocytosis into intracellular compartments. Labelled cells appeared as hypo-intense regions on MR, similar to those that have taken up intravenously-injected SPIO. In the early 1990s, neural tissues (Norman et al. 1992) and T cells (Yeh et al. 1993) were labelled with iron oxide particles and transplanted into healthy animals, demonstrating that labelled cells can be visualized *in vivo*. The signal void created by iron oxide labelling extends well beyond the cell membrane, making it easier to identify the cells in the image (Pintaskeet et al. 2006).

Since its early successes, more cell types have been labelled with iron oxide particles, including monocytes, glioma cells, oligodendrocytes progenitors (Bulte and Kraitchman 2004), macrophages, mesenchymal stem cells (Wang et al. 2010) and endothelial progenitors (Lee et al. 2010b). Despite the apparent ability to label virtually any cell type, the detection of target cells has been restricted by the number of particles taken up. The quantity of iron oxide attached to a cell is a direct determinant of cell detectability. With a greater amount of iron oxide loading per cell, there is less emphasis on hardware requirements and imaging duration to achieve the required SNR and resolution for cellular detection (Heyn et al. 2005).

Labelling appears to be relatively harmless to the cells – cell survival, proliferation, function and even differentiation are usually minimally affected, provided the intracellular iron load is not excessive. Although high concentrations of particles in the culture medium increases cellular iron quantity, excessive concentrations have also been associated with free radical generation, decrease in cell proliferation, apoptosis (van den Bos et al. 2003) and decrease in motility (Nohroudi et al. 2010). Cell labelling can be achieved by binding iron oxide particles onto the cellular surface or by incorporating them into the intracellular space. The latter method is favoured as particles at the cell surface can hinder cell-cell interaction by blocking cell surface receptors. Even when particles are internalised, it is important that they do not alter the properties or functions of the cell, particularly with high amounts of iron loading. Hence, labelled cells are often compared with intact cells in terms of viability, proliferation, gene expression, and in addition, the differentiation capacity for stem cells (Lee et al. 2009; Balakumaran et al. 2010).

The quantity of iron oxide particles internalised by cells, typically expressed as picograms of iron per cell, determines their MR detectability. High iron loading can thus relax the hardware requirements to detect low cell numbers. When labelling phagocytic cells like macrophages, large amounts of iron loading (61 pg/cell) can be achieved with clinically-available SPIO (Heyn et al. 2006). However, labelling of non-phagocytic cells such as MSC with SPIO has been less efficient, with approximately only 9 pg/MSC (Mailander et al. 2008), and may require alterations of the labelling method to achieve higher iron loading.

Iron oxide loading of cells can be increased by using particles with high content of iron oxide, a characteristic that depends on the synthesis method. The internalisation of particles is also affected by the interaction at the particle-cellular interface, which is determined by the physicochemical properties of the particles (chemical composition, size/geometry, surface charge, coating/ligands, aggregation status), the cell type (professional phagocytes versus other cell types), as well as the labelling medium.

4.1 Particle Surface Composition and Charge

The iron oxide particles in biomedical applications are often coated with dextran or other formulations. The coating material, at times carrying electrostatic charges, can mediate both inter-particle and particle-cellular interactions. While in a dispersing medium, particles are in Brownian motion due to random collision from water molecules, where they experience both interparticle attractive and repulsive forces. When attractive forces, such as van der Waal's forces or magnetic dipole-dipole interactions from residual magnetic moment dominate, the particles form irregular aggregates.

The polymer coating around iron oxide particles provides repulsive forces between particles in the form of electrostatic repulsion or steric hindrance to enable stable suspension in water. Clinically-available SPIO such as ferumoxide or ferucarbotran are well suspended in water due to the steric hindrance offered by their dextran or dextran-derivative coating (Hunter 1995).

Various particle coatings have been engineered to enhance or avoid specific particle-cellular interactions (Nath et al. 2004). Both macromolecules and monomers have been used as coating material. An example of macromolecular coating is the 'PEGylation' of iron oxide to allow for a long circulation time by avoiding opsonisation and uptake by the macrophages of the reticuloendothelial system (Lee et al. 2006). The PEGylated surface is hydrophilic, electrostatically neutral and without hydrogen donor or acceptors, which explains the lack of attractive forces between PEG and opsonising proteins.

Particles that are stabilised by monomer coating remain suspended due to the electrostatic repulsion between particles. Charge polarity can also affect particle uptake by cells. Positively and negatively-charged particles with diameters similar to SPIO (100 nm) were compared in terms of uptake by Hela cells (Harush-Frenkel et al. 2007). The anionic polylactide-based particles provide poorer uptake than their aminated, cationic cousins. It was shown that the cationic particles used the clathrin- and caveolae-mediated endocytic pathways, but not the anionic ones. Interestingly, when the clathrin/caveolae pathway was inhibited, Hela cells took up the cationic particle through a possibly compensatory, macropinocytic pathway. Such data suggests that at least some cell types have more than one mechanism by which they can be labelled with various iron particles, and that the mechanism used may vary depending on the labelling conditions and the particle composition.

In addition to charge polarity, the zeta potential, an indication of particle surface with respect to bulk medium, plays an important role in the efficiency of particle uptake by cells. Citrate-coated VSOP (8 nm) showed improved uptake by macrophages compared to uncharged particles of similar size (Fleige et al. 2002). Supporting this observation, anionic dimercaptosuccinic acid-coated AMNP (−30 mV in zeta potential) showed greater uptake by macrophages and Hela cells compared to dextran-coated ferumoxtran (of similar size) which carries no charge (Wilhelm et al. 2003; Jung 1995). Comparison of zeta potentials between iron oxide particles is difficult due to the indirect determination of zeta potential from electrokinetic techniques, a process highly dependent on the approximation of particle rigidity (as a result of different coatings) (Di Marco et al. 2007). Further comparison of particles with similar sizes and coatings but differing zeta potentials is required to understand the effect of particle charge on labelling efficiency.

4.2 *Surface Ligand*

Another method to improve cellular uptake is to encourage receptor-mediated endocytosis (RME) using biological ligands. Attachment of HIV tat peptide to aminated, dextran-cross-linked iron oxide particles (CLIO-tat, ~45 nm) improved particle uptake by human CD34+, human CD4+ and mouse splenocytes (4–8 pg/cell) compared to CLIO alone (<2 pg/cell) when labelled at 0.1 mg Fe/ml (Lewin et al. 2000). However, the particles were located in the cell nucleus instead of the usual endosomal location of other particles. This raises the question as to whether or not CLIO-tat is more likely than other particles to alter cellular function through interfering with nuclear activity. Instead of using peptides, it is possible to direct receptor-mediated endocytosis with antibodies (Ab) as well. The transferrin receptor is abundant on numerous cell types. Tagging SPIO with anti-transferrin Ab improved the iron loading of human haemopoietic progenitor cells (9.8 pg/cell) over the use of SPIO alone (2.4 pg/cell) (Daldrup-Link et al. 2003). These particles were internalised in contrast to another report where transferrin-coupled SPIO are only bound to the surface of human dermal fibroblast (Berry et al. 2004). Another antibody under exploration is CD11c, which increased SPIO uptake in dendritic cells (Ahrens et al. 2003).

An inherent problem with Ab-directed RME is the species-specific attachment to the target receptor. This requires each Ab-SPIO to be tailored to the target species and cell type. Moreover, from a clinical perspective, since Abs are typically developed in animals such as mice, this method will have to contend with regulations on xenogeneic contamination of transplants. Finally, the production of antibodies remains very expensive, and pharmacological doses required to obtain clinically useful *in vivo* labelling of cells would be prohibitive. For these reasons, it is expected that such methods could only be used *in vitro* in the foreseeable future.

4.3 Particle Size

As demonstrated in studies on a range of cell types, the rate and quantity of cellular uptake of particles depends not only on particle composition but also on the particle diameter. Early studies in macrophages showed that the volume of particles taken up increased in tandem with increasing particle diameter, with maximal uptake reached at around 2 μm (Pratten and Lloyd 1986; Tabata and Ikada 1988). As diameters increased from 2 to 4.6 μm, the volume of particles taken up actually decreased. This may not be surprising if we consider that the larger of such particles approach the size of whole cells. . It was also shown that leukocytes took up more styrene-based particles of diameter 500 nm–1 μm than particles of other sizes (Kawaguchi et al. 1986). A similar result was shown where both macrophages (Zhang et al. 2001) and monocytes (Metz et al. 2004) took up greater masses of 60 nm carboxydextran-coated SPIO than similarly coated but smaller USPIOs.

There have been only a few studies of non-phagocytic cells on how particle size affects uptake. When labelled with carboxydextran-coated particles between 17 and 65 nm, lung carcinoma cells take up greater masses of the larger particles (Matuszewski et al. 2005). Non-phagocytic cells tend to show maximum uptake at a lower size range compared to macrophages. Rejman et al demonstrated that uptake into murine melanoma cells decreased as polystyrene particles diameter increased from 50 to 500 nm, and was absent when 1 μm particles were used (Rejman et al. 2004). In a study labelling human T-cells with particles of 33 nm or larger diameters, uptake increased with diameter peaking at 107 nm before decreasing with diameter from 207 nm to 1.4 μm (Thorek and Tsourkas 2008). However, it is difficult to infer the pattern of T-cell uptake with particle diameters as Thorek *et al*'s study used a variety of particle coatings at different sizes: dextran (33–107 nm), styrene (207–289 nm) and silica (1.4 μm) coated particles.

It is possible to use transfection agents during labelling to generate particle aggregates of varying sizes. However, results relating aggregate size to? uptake have been contradictory as aggregate surface composition and charge among particles greatly differed in these experiments (Matuszewski et al. 2005; Song et al. 2007). Eliciting the relationship between particle size and cellular uptake requires a set of particles of similar surface composition. Size-dependent uptake of microgel iron oxide particles (MGIO) was demonstrated between the diameters of 87 and 766 nm with 600 nm MGIO providing the maximum loading of 33 pg/MSC, which is three- to sixfold greater than with ferucarbotran (Lee et al. 2009). Interestingly, endothelial progenitor cells (EPC) labelled with either 600 nm MGIO or ferucarbotran resulted in similar quantity of cellular iron (26–28 pg/EPC) (Lee et al. 2010b). Taken together, these results suggest that uptake is size-dependent in some cell types, including monocytes, lung carcinoma cells and melanoma cells. In addition, macrophages and MSC have shown maximum uptake efficiency at specific diameters. However, it remains to be determined if certain cells such as EPC are insensitive particle diameters while controlling for particle coating or surface charge.

4.4 Labelling Duration and Concentration

Current evidence shows that the number of particles internalised increases and saturates with increasing labelling duration and concentration, independent of particle type, size or cell type. The uptake of sub-100 nm dextran-, 200–300 nm styrene- and 1.4 μm silica-coated particles increased with incubation duration but was saturated by 4 h with non-phagocytic T-cells (Thorek and Tsourkas 2008). The same study also demonstrated that the saturating concentration varied for different particle sizes. Labelling concentration is commonly expressed as mass of iron per unit volume of labelling medium, for example mg Fe / ml or simply mg/ml. The uptake for sub-100 nm dextran particle saturates at 0.025 mg/ml while 200–300 nm styrene particles and 1.4 μm silica particles are saturated at 0.1 mg/ml. When 30 nm AMNP was used to label Hela cells and macrophages, the saturating durations were 5 and 10 h respectively while the saturating concentration was 0.5 mg/ml for both cell types, further demonstrating the variability of labelling efficiency between cell types (Wilhelm et al. 2003). The longest reported labelling duration was 72 h where ferumoxide was incorporated into bone marrow cells (Jendelova et al. 2003). Most studies are kept to a maximum labelling duration of 24 h and a maximum concentration 0.2 mg/ml, as it has been shown that long duration and high concentration, in particular the combination of both, can result in cytotoxicity (van den Bos et al. 2003; Neri et al. 2008).

4.5 Electroporation and Transfection Agents

Inspired by DNA transfection technologies, electroporation and transfection agents (TA) have been used to improve uptake of particles into cells. When ferumoxide labelling was assisted by electroporation, iron loading of 11.5 pg/rat MSC was achieved compared to 3 pg/rat MSC with simple incubation alone (Walczak et al. 2005). However, this technique requires careful tuning of the electrical parameters to avoid affecting cell viability (Daldrup-Link et al. 2005).

Poly-L-lysine (PLL), a common cationic TA, binds with DNA to form complexes of <100 nm in diameters. By complexing PLL with ferumoxide prior to their incubation with hMSC, uptake between 13 pg/cell (Arbab et al. 2003b) and 16 pg/cell (Frank et al. 2003) have been reported. As an alternative to PLL, protamine sulphate, an FDA-approved TA, has been complexed with ferumoxide to achieve loading of 11 pg/hMSC (Arbab et al. 2004a). However, no comparison against the uptake by simple incubation was done in these studies.

Despite the promising cellular uptake achieved with TA, the method has a few potential disadvantages. The use of ferumoxide-PLL labelling inhibited chondrogenesis (Kostura et al. 2004), although differentiation overall was unaffected when hMSC were labelled with ferumoxide-protamine (Arbab et al. 2004b). It was suggested that the detrimental effect was a result of using PLL as the TA, rather than the presence of ferumoxide (Bulte et al. 2004), notwithstanding a lack of detailed study to examine the effect of PLL and protamine sulphate alone on the chondrogeneicity of MSC.

Despite an apparent increase in iron loading, it has been reported that TA-mediated labelling may result in only surface attachment to the cell rather than internalisation of SPIO (Montet-Abou et al. 2007). In that study, the use of TA lipofactamine resulted in SPIO attaching to the surface, in contrast to protamine sulphate which allowed the complete internalisation of ferumoxide in mouse neural progenitor cells but not in other cell types. These results suggest that the internalisation of SPIO may be dependent on the type of TA, labelling condition and cell type. Another distinctive feature of TA-mediated labelling is the cellular uptake with respect to the iron concentration of the labelling medium. Labelling with naïve ferumoxide or other SPIO results in a phenomenon where uptake increases with iron concentration of the labelling medium and saturates at high labelling concentrations. PLL-mediated particle uptake on the other hand, varies non-linearly with the labelling concentration (Kim 2008), and may present difficulty in optimising cell labelling conditions for different cell types.

This observation may be due to the dynamic nature of the complex formation between TA and SPIO, which is less understood compared to the complex formation between TA and DNA. The formation of PLL and DNA complexes, depending on the relative amounts of the constituents, results in structures that range from spheres to rods and toroids (Liu et al. 2001) and change dynamically in size over several hours (Lai and van Zanten 2001). The complex formed by TA and SPIO has a hydrated diameter that depends on the relative amount of its constituents and may precipitate at certain ratios (Montet-Abou et al. 2007). Therefore, the use of TA-mediated labelling requires careful optimisation of labelling conditions for each cell type to be labelled, and has potential effects on cell physiology that are difficult to predict.

4.6 *Labelling Medium*

During labelling of the cells, physicochemical properties of particles are influenced by the microenvironment. Studies on the effects of the labelling medium have shown that the use of 10% serum increases the uptake of ferumoxide compared to serum-free medium (Rogers and Basu 2005). Similar observations were reported when monocrystalline iron oxide particles (MION), opsonised by plasma, purified and incubated with either macrophages or cancer cells, resulted in a six- and two-fold increase in uptake respectively as compared to naïve MION (Moore et al. 1997). It was suggested that opsonised MION were taken up by scavenger and complement receptors of macrophages while naïve MION entered cells through fluid-phase endocytosis. This suggestion was supported by reports that the macrophage uptake of albumin-coated particles is C3b-complement-dependent (a component of opsonisation) (Roser et al. 1998). In contrast, opsonised AMNP showed reduced uptake by Hela cells (1 pg/cell) compared to naïve AMNP (16 pg/cell) (Wilhelm et al. 2003). It was also reported that macrophage uptake of polystyrene and phenylated polyacrolein between 0.5 and 4.5 μm was higher in serum-free media (Tabata and Ikada 1988). The presence of serum in medium can thus alter uptake of particles depending on the cell type. Indeed, it was observed that cell types with large cytoplasm-nucleus size ratio are insensitive to the presence of serum during uptake of poly-L-lysine complexed SPIO, as opposed to enhanced uptake by hematopoietic-derived cells with small cytoplasm-nucleus ratio in serum-free media (Arbab et al. 2003a).

5 Transgenic Methods of Cellular Labelling

Encouraging cells to produce iron oxide particles intra-cellularly is an alternative to labelling with exogenous particles. Living organisms possess specialised iron storage mechanisms involving ferritin, transferrin and transferrin receptors. Expression of these proteins in certain cells may be accompanied by an increase in intracellular iron storage, which will allow the cells to be detected through MRI. One application for this technology is the use of these proteins as markers of expression during gene therapy. These proteins can also be applied to the imaging of endogenous gene expression during development and pathogenesis in transgenic animals.

5.1 *Transferrin*

Free, systemic iron is captured by the plasma protein transferrin (Tf), while iron-loaded Tf is transferred into cells after binding to the transferrin receptor (TfR). This receptor is naturally abundant on some cell types and has been utilised to improve the uptake of Tf-conjugated SPIO (Daldrup-Link et al. 2003). When human Tf-conjugated MION were administered IV, the particles bound specifically to rat gliosarcoma cells transfected to express human TfR (Weissleder et al. 2000; Moore et al. 2001). Therefore, TfR can be a model marker for MR visualisation of *in vivo* gene expression.

5.2 *Ferritin*

Ferritin is a ubiquitous and highly-conserved iron-binding protein (Harrison et al. 1996). In vertebrates, cytosolic ferritin is a heteropolymer composed of variable proportions of H and L subunits. The H subunit has ferroxidase activity that promotes iron oxidation and incorporation, while the L subunit facilitates the activity of the H-chain by offering sites for iron nucleation and mineralization. Twenty-four ferritin subunits assemble to form the apoferritin shell. Each apoferritin molecule of 450 kDa can sequester up to 4,500 iron atoms, depending on the tissue type and physiologic status of the cell. Abnormality of iron metabolism has been detected by MRI due to the elevation in ferritin and iron storage (Grabill et al. 2003). The T2 relaxation properties of iron-loaded ferritin have also been studied *in vitro* (Gossuin et al. 2000).

An early study on the transfection of cells to overexpress H-chain ferritin showed upregulation of TfR accompanied with increased intracellular iron (Cozzi et al. 2000). In a separate study, the change in cellular iron was detected as a 7% change in MR relaxation rate when the ferritin gene was switched on (Cohen et al. 2005). Inducible ferritin expression was also detected in the liver and heart of transgenic fetal mice, with 25% change in relaxation rates (Cohen et al. 2007).

Poor sensitivity is a limitation of these attempts to image gene expression, as turning on the ferritin gene increases cellular iron content by only 0.2 pg/cell (Cohen et al. 2005). Sensitivity has not been improved by the co-transfection of both ferritin and TrR genes which yielded an iron content increase of only 14 fg/cell (Deans et al. 2006). However, studies are ongoing to improve the sensitivity of imaging ferritin expression by increase relaxation rate for the same iron quantity per cell through controlled aggregation of ferritin (Bennett et al. 2008).

5.3 *MagA*

Magnetotactic bacteria are motile prokaryotes that synthesize intracellular magnetic structures and move in relation to the earth's magnetic field (Bazylinski and Frankel 2004). The synthesized structures, magnetosomes, consist of multiple iron oxide crystals surrounded by a lipid membrane. One of the genes identified to be involved in magnetosome production is *magA*. When 293FT cells, a common human packaging cell line, were transfected with *magA*, the cells produced sub-micron sized magnetosomes that contained multiple iron oxide crystals each measuring 3–5 nm in diameter (Zurkiya et al. 2008). Induction of the gene *magA* increased intracellular iron by 0.55 pg/cell and produced up to a fourfold change in measured relaxation rate. The use of *magA* is in its infancy, but current results suggest that it is a promising MR reporter.

6 Applications of Cellular MRI

The development of *in vivo* cellular tracking by MRI has generated much interest. The balance between spatial resolution and sensitivity, as well as its lack of ionising radiation, makes MRI a practical and translatable modality for cellular detection. Not only has MRI been used to track the movement and incorporation of administered cells in animal models of diseases, but also clinical trials of MR cellular tracking have begun.

6.1 *Animal Models of Diseases*

Tissue engineering has progressed from early use of acellular artificial scaffolds to the incorporation of cells or tissue onto biocompatible materials prior to their implantation, improving tissue regeneration. With the integration of imaging techniques, the participation of cells in tissue repair can be monitored. The approach by various research groups is similar: label cells with a clinical iron oxide contrast agent, seed the labelled cells on an engineered implant (the sequence of the first two steps can be reversed) and introduce the cell-loaded implant into an animal model. Ferumoxide-labelled MSC on collagen scaffolds have been imaged *in vitro* (Terrovitis et al. 2006), as have labelled MSC within gelatine sponges *in vivo* (Ko et al. 2007). By labelling aortic smooth muscle cells with USPIO prior to seeding them onto a polymeric vascular graft, researchers were able to evaluate the graft performance *in vivo* (Nelson et al. 2008). The cells were detected to have successfully remained in the scaffold vicinity for up to several weeks. The success of MR monitoring, however, may be limited by the MR properties of implants used. Ferrous metallic implants and voluminous, water-impenetrable implants may cause susceptibility artefacts, and may not provide sufficient contrast to identify hypo-intense cells (Nelson et al. 2008).

Labelled stem cells, directly injected into myocardial infarcts can also be monitored. Ferumoxide-labelled (Kraitchman et al. 2003) and micron-size iron oxide particles (MPIO)-labelled (Stuckey et al. 2006; Hill et al. 2003) allogeneic MSC have been injected intramyocardially into infarcted hearts and their location studied up to 16 weeks in various animal models. In a different study, intravenously-injected iron oxide-labelled, allogeneic MSC homed to the myocardial infarct in a canine model and persisted for at least 1 week (Kraitchman et al. 2005). However, the successful identification of hypo-intense cells may be complicated by haemorrhage from microvascular obstruction during infarct reperfusion (van den Bos et al. 2006). Haemorrhage causes susceptibility-induced signal voids that appear similar to iron oxide induced ones.

The homing of intravenously-injected MSC has been demonstrated in other disease models as well. The arrival of ferumoxide-labelled MSC at injured arteries (Gao et al. 2007) and atherosclerotic plaques (Qiu et al. 2007) was monitored by MRI. The use of MRI tracking to optimise cellular delivery was also demonstrated. Ferumoxide-labelling provided evidence of better MSC engraftment when cells were given through the internal carotid artery, compared to intravenously in a rodent stroke model (Walczak et al. 2008).

A number of cell types have been shown to migrate towards the injured central nervous system. MSC labelled with MPIO were shown to migrate towards white matter (Chen et al. 2010) or spinal cord injury (Gonzalez-Lara et al. 2010). Ferumoxide-labelled MSC (Jendelova et al. 2003), ferumoxide-labelled neural stem cells (NSC) (Zhu et al. 2006) or USPIO-labelled ESC (Hoehn et al. 2002) were transplanted and shown to migrate from the contralateral hemisphere and across the corpus callosum to the infarct periphery. By optimising MSC labelling with 600 nm MGIO, the cells can be tracked in a rat stroke model using modest, standard clinical hardware over 19 days (Fig. 1) (Lee et al. 2009). The same

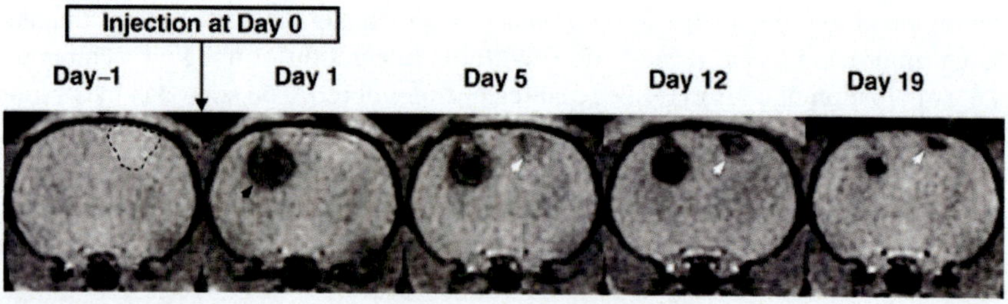

Fig. 1 *In vivo* MR tracking of MSC labelled with 600 nm MGIO to stroke lesion. A focal cortical stroke (*dashed boundary*) was induced at Day 2 and 2 × 10^4 MSC were transplanted contralaterally on Day 0 (*black arrowhead*). An area of hypo-intensity appeared in the stroke lesion (*white arrowheads*) noticeable at Day 5, and increased over time to Day 19. Images were acquired with gradient echo sequence – field of view: 5 cm; matrix: 160 × 160 and zero-filled to 512 × 512; voxel dimensions: 313 μm × 313 μm × 1.5 mm; repetition time/echo time/flip angle/number of excitations : 280 ms/20 ms/20°/15, as 10 two-dimensional slices (unpublished data)

phenomenon was demonstrated using NSC labelled with a gadolium-dextran T2 agent (Modo et al. 2002). When transplanted to the ipsilateral side of the stroke, ferumoxide-labelled NSC were reported to migrate along the infarct periphery (Guzman et al. 2007). In a Huntington's disease model where lesions were induced by quinolinic acid, ferumoxide-labelled MSC showed a distinct migration route toward the lesions (Sadan et al. 2008).

6.2 *Clinical Studies of Cellular MRI*

Although there have only been four completed clinical trials of MRI cellular tracking to date, another four active studies are ongoing or recruiting, according to the NIH database *clinicaltrial.gov*. In one of the completed studies, autologous dendritic cells (DC) were labelled with ferumoxide or ^{111}In-oxine prior to injection into the lymph nodes of 11 stage-III melanoma patients (de Vries et al. 2005). Post-injection verification by MRI was found to be superior to ultrasound-guided injection. In three out of eight patients, MRI showed unsuccessful intranodal delivery, an observation corresponding to the lack of DC migration to the draining lymph nodes. Verified by histology of resected nodes, the detection of migrated DC to draining lymph nodes was more reliable by MRI than scintigraphy. An inherent advantage of MRI is the anatomical localisation of DC at injection sites or after migration. Scintigraphy is a superior quantification modality and ^{111}In-oxine labelling may remain a useful co-label with MRI contrast agents until MRI quantification matures. The other clinical trial involved patients with traumatic, open head injuries where loose neural tissue is cultured in a media known to proliferate NSC (Zhu et al. 2006). Following labelling with ferumoxide, the NSC were injected intra-cerebrally at four sites around the lesion. Weekly MRI assessments subsequently showed that the lesion periphery began to darken progressively while hypo-intensity at the injection site faded from 1 to 3 weeks post-transplantation. This was not observed in the control patient that received unlabelled NSC. As the patients survived the trauma and transplantation, histology has not been possible. The third completed study stands out with the use of nonclinical 4.5 μm particles (Callera and de Melo 2007). Human CD34+ cells with surface-bound particles were tracked over time to the traumatic brain injury in ten patients, an observation absent in the six control patients who were given only particles without cells. Although no adverse effect was observed over 35 days post-transplantation, the attachment of cell graft with particles of comparable sizes, in particular nonbiodegradable, nonclinical particles, raises concern about the long term safety in patients. In the most recently completed study, four diabetic patients were given ferucarbotran-labelled islet cells intraportally (Toso et al. 2008). Labelled islet cells were functionally competent as observed by the insulin independence of all four patients. However, fewer hypointense spots were observed compared to the number of islet cells injected, prompting the postulation that the labelling efficiency was sufficient only for the MR-visibility of islet cell clusters, but not individual cells.

These four clinical studies show that iron oxide labels appear to be safe and that clinical detection of the labelled cells is possible. To achieve the goal of clinical cellular MRI, more off-label use of clinically approved contrast agents, such as ferumoxide or ferucarbotran, in cellular therapy trials is needed. Currently, the FDA requires preclinical studies to be performed in experimental models of diseases using labelled and unlabelled cells along with sham controls to assess the toxicity and any alteration in morbidity and mortality (Arbab and Frank 2008). Serum chemistry analysis, histology correlation and experimentation in a clinical grade manufacturing practice (CGMP) facility are foreseeable requirements in the future.

7 Challenges of Cellular MRI

Cellular MRI faces several challenges in the pursuit of a cell marking method that can be translated to clinical use. Being non-phagocytic, stem cells do not internalise sufficient quantities of clinically-available SPIO by simple incubation. We have reviewed the optimization of uptake through particle design, labelling conditions and the use of transfection agents or electroporation while ensuring that cellular functions are retained. The other challenges of this technique are the dilution of label, false positivity, cellular quantification and positive contrast generation.

7.1 Label Dilution

A drawback of using exogenous particles is the dilution of labels as cells undergo division *in vivo*. The cellular labels are split among the two daughter cells, resulting in approximately halving of the iron quantity per cell after each division (Fig. 2). Although the dilution of label will affect the visibility of cell grafts that undergo rapid division, it is an acceptable limitation to cell types, such as MSC, that generally do not proliferate extensively post-transplantation.

7.2 Label Transfer

In spite of sufficient cellular iron loading, the route and timing of cellular delivery may have significant effects on cell survival and engraftment. Cellular graft destruction by hostile host immune responses may result in transfer of labels to the host immune cells. When cells are tagged with exogenous fluorescent labels such as 5-bromo-2-deoxyuridine (BrdU) and bis benzamide (BBZ) prior to transplantation, label transfer to host macrophages can potentially cause erroneous diagnoses of cellular engraftment or differentiation (Pawelczyk et al. 2006; Burns et al. 2006; Coyne et al. 2006). Iron oxide MRI labels face the same problem. Hence, secondary

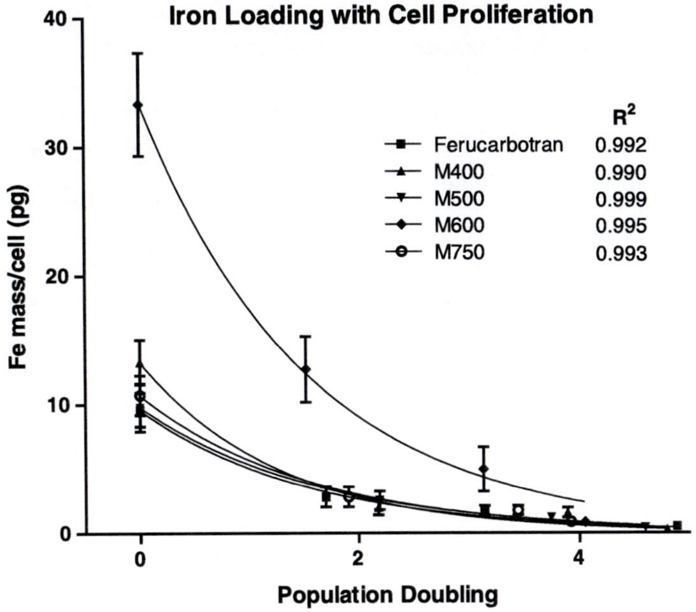

Fig. 2 Halving of cellular iron when MSC divide. The measurements of cellular iron with respect to the number of times the cell count has doubled are shown as points. The R^2 values are the exponential decay fittings to the points with a half-life of 1 population doubling. These results show that when MSC are labelled with ferucarbotran or 400–750 nm MGIO, the iron mass per cell approximately halves when the cells undergo one population doubling (unpublished data)

methods of identifying donor cells by means of FISH or sex mismatch may prove to be an essential procedure to verify MRI observations in preliminary experiments.

7.3 *Cellular Quantification*

Quantification of cells through the use of radioactive methods such as scintigraphy is well-established. Quantification of cells by iron oxide MRI on the other hand is in its infancy, but reports to date have been promising. As particle diameter increases, its relaxation rate increases and saturates at a constant value, as defined by the static dephasing regime (SDR) (Lee et al. 2010b). When cells compartmentalise small particles such as SPIO or USPIO, their MR relaxation is similar to that of large magnetic spheres operating in the SDR (Bowen et al. 2002; Ziener et al. 2005). Since the relaxation rate of cells is constant when in the SDR, the R2* relaxation rate is directly proportional to the number of cells per voxel when iron mass per cell is known (Bowen et al. 2002).

The accurate measurement of R2* relaxation rate is beset with artefacts from macroscopic magnetic susceptibility of tissue-air interfaces, leading to overestimates of R2*. A number of methods have been proposed to compensate for susceptibility artefacts, including increasing spatial resolution, alternating slice selection, tailoring rf pulses or utilizing 3D z-shimming. A method that combined 3D high-resolution ΔB_0 maps and image post-processing was reported to provide improved R2* measurements that would have been a twofold overestimate if left uncorrected (Dahnke and Schaeffter 2005).

Besides improving cellular iron loading, improvements in hardware and image processing can increase detection sensitivity. Nevertheless, it remains a considerable challenge to detect small numbers of cells against native low signals from, for example, haemosiderin or haemorrhagic artefacts, or the presence of metals such as calcium. Furthermore, the detection of labelled cells is limited by partial volume effects, which can be minimized by decreasing voxel sizes at the sacrifice of signal-to-noise ratios (Heyn et al. 2005). A post-processing algorithm based on phase map cross correlation was proposed for robust identification of labelled cells even in images with low SNR (Mills et al. 2008). Deoxyhemoglobin in small, slow flowing vessels is another source of native hypo-intensity that can have similar appearances to labelled cells, especially in 2D images. The problem can be circumvented by the administration of paramagnetic contrast agent prior to cellular imaging (Anderson et al. 2005) or the use of carbogen inhalation (95% O_2 and 5% CO_2) to allow vasodilation by CO_2 (Himmelreich et al. 2005).

Acknowledgements EL was supported by project grants from Australian NHMRC, and JC received salary support from Singapore NMRC (CSA/012/2009). Appreciation is expressed to Debbie Ng for proofreading of the manuscript.

References

Ahrens, E. T., Feili-Hariri, M., Xu, H., Genove, G. & Morel, P. A. 2003. Receptor-mediated endocytosis of iron-oxide particles provides efficient labeling of dendritic cells for in vivo MR imaging. *Magnetic Resonance in Medicine,* 49, 1006–1013.

Aigner, F., Pallwein, L., Mitterberger, M., Pinggera, G. M., Mikuz, G., Horninger, W. & Frauscher, F. 2009. Contrast-enhanced ultrasonography using cadence-contrast pulse sequencing technology for targeted biopsy of the prostate. *BJU International,* 103, 458–463.

Anderson, S. A., Glod, J., Arbab, A. S., Noel, M., Ashari, P., Fine, H. A. & Frank, J. A. 2005. Noninvasive MR imaging of magnetically labeled stem cells to directly identify neovasculature in a glioma model. *Blood,* 105, 420–5.

Arbab, A. S., Bashaw, L. A., Miller, B. R., Jordan, E. K., Bulte, J. W. & Frank, J. A. 2003a. Intracytoplasmic tagging of cells with ferumoxides and transfection agent for cellular magnetic resonance imaging after cell transplantation: methods and techniques. *Transplantation,* 76, 1123–1130.

Arbab, A. S., Bashaw, L. A., Miller, B. R., Jordan, E. K., Lewis, B. K., Kalish, H. & Frank, J. A. 2003b. Characterization of biophysical and metabolic properties of cells labeled with superparamagnetic iron oxide nanoparticles and transfection agent for cellular MR imaging. *Radiology,* 229, 838–846.

Arbab, A. S. & Frank, J. A. 2008. Cellular MRI and its role in stem cell therapy. *Regen Med,* 3, 199–215.

Arbab, A. S., Yocum, G. T., Kalish, H., Jordan, E. K., Anderson, S. A., Khakoo, A. Y., Read, E. J. & Frank, J. A. 2004a. Efficient magnetic cell labeling with protamine sulfate complexed to ferumoxides for cellular MRI. *Blood,* 104, 1217–1223.

Arbab, A. S., Yocum, G. T., Kalish, H., Jordan, E. K., Anderson, S. A., Khakoo, A. Y., Read, E. J. & Frank, J. A. 2004b. Feridex-protamine sulfate labeling does not alter differentiation of mesenchymal stem cells. *Blood,* 104, 3412–3413.

Balakumaran, A., Pawelczyk, E., Ren, J., Sworder, B., Chaudhry, A., Sabatino, M., Stroncek, D., Frank, J. A. & Robey, P. G. 2010. Superparamagnetic Iron Oxide Nanoparticles Labeling of Bone Marrow Stromal (Mesenchymal) Cells Does Not Affect Their "Stemness". *PLoS One,* 5, e11462.

Bang, O. Y., Lee, J. S., Lee, P. H. & Lee, G. 2005. Autologous mesenchymal stem cell transplantation in stroke patients. *Ann Neurol,* 57, 874–82.

Bazylinski, D. A. & Frankel, R. B. 2004. Magnetosome formation in prokaryotes. *Nat Rev Microbiol,* 2, 217–30.

Bendszus, M., Kleinschnitz, C. & Stoll, G. 2007. Iron-Enhanced MRI in Ischemic Stroke: Intravascular Trapping Versus Cellular Inflammation. *Stroke,* 38, e12.

Bennett, K. M., Shapiro, E. M., Sotak, C. H. & Koretsky, A. P. 2008. Controlled Aggregation of Ferritin to Modulate MRI Relaxivity. *Biophys. J.,* 95, 342–351.

Berry, C. C., Charles, S., Wells, S., Dalby, M. J. & Curtis, A. S. 2004. The influence of transferrin stabilised magnetic nanoparticles on human dermal fibroblasts in culture. *International Journal of Pharmaceutics,* 269, 211–225.

Bowen, C. V., Zhang, X., Saab, G., Gareau, P. J. & Rutt, B. K. 2002. Application of the static dephasing regime theory to superparamagnetic iron-oxide loaded cells. *Magn Reson Med,* 48, 52–61.

Bulte, J. W. & Kraitchman, D. L. 2004. Monitoring cell therapy using iron oxide MR contrast agents. *Curr Pharm Biotechnol,* 5, 567–584.

Bulte, J. W., Kraitchman, D. L., Mackay, A. M., Pittenger, M. F., Arbab, A. S., Yocum, G. T., Kalish, H., Jordan, E. K., Anderson, S. A., Khakoo, A. Y., Read, E. J. & Frank, J. A. 2004. Chondrogenic differentiation of mesenchymal stem cells is inhibited after magnetic labeling with ferumoxides. *Blood,* 104, 3410–3413.

Burns, T. C., Ortiz-Gonzalez, X. R., Gutierrez-Perez, M., Keene, C. D., Sharda, R., Demorest, Z. L., Jiang, Y., Nelson-Holte, M., Soriano, M. & Nakagawa, Y. 2006. Thymidine Analogs Are Transferred from Prelabeled Donor to Host Cells in the Central Nervous System After Transplantation: A Word of Caution. *Stem Cells,* 24, 1121–1127.

Callera, F. & De Melo, C. M. 2007. Magnetic Resonance Tracking of Magnetically Labeled Autologous Bone Marrow CD34+ Cells Transplanted into the Spinal Cord via Lumbar Puncture Technique in Patients with Chronic Spinal Cord Injury: CD34+ Cells' Migration into the Injured Site. *Stem Cells and Development,* 16, 461–466.

Cao, F., Lin, S., Xie, X., Ray, P., Patel, M., Zhang, X., Drukker, M., Dylla, S. J., Connolly, A. J. & Chen, X. 2006. In vivo visualization of embryonic stem cell survival, proliferation, and migration after cardiac delivery. *Circulation,* 113, 1005–14.

Carr, D. H., Brown, J., Bydder, G. M., Steiner, R. E., Weinmann, H. J., Speck, U., Hall, A. S. & Young, I. R. 1984. Gadolinium-DTPA as a contrast agent in MRI: initial clinical experience in 20 patients. *American Journal of Roentgenology,* 143, 215–224.

Chen, A., Siow, B., Blamire, A. M., Lako, M. & Clowry, G. J. 2010. Transplantation of magnetically labeled mesenchymal stem cells in a model of perinatal brain injury. *Stem Cell Research,* 5, 255–266.

Cohen, B., Dafni, H., Meir, G., Harmelin, A. & Neeman, M. 2005. Ferritin as an endogenous MRI reporter for noninvasive imaging of gene expression in C6 glioma tumors. *Neoplasia,* 7, 109–17.

Cohen, B., Ziv, K., Plaks, V., Israely, T., Kalchenko, V., Harmelin, A., Benjamin, L. E. & Neeman, M. 2007. MRI detection of transcriptional regulation of gene expression in transgenic mice. *Nat Med,* 13, 498–503.

Coyne, T. M., Marcus, A. J., Woodbury, D. & Black, I. B. 2006. Marrow stromal cells transplanted to the adult brain are rejected by an inflammatory response and transfer donor labels to host neurons and glia. *Stem Cells,* 24, 2483–92.

Cozzi, A., Corsi, B., Levi, S., Santambrogio, P., Albertini, A. & Arosio, P. 2000. Overexpression of Wild Type and Mutated Human Ferritin H-chain in HeLa Cells IN VIVO ROLE OF FERRITIN FERROXIDASE ACTIVITY. *Journal of Biological Chemistry,* 275, 25122–25129.

Dahnke, H. & Schaeffter, T. 2005. Limits of detection of SPIO at 3.0 T using T2 relaxometry. *Magn Reson Med,* 53, 1202–6.

Daldrup-Link, H. E., Rudelius, M., Oostendorp, R. A., Jacobs, V. R., Simon, G. H., Gooding, C. & Rummeny, E. J. 2005. Comparison of iron oxide labeling properties of hematopoietic progenitor cells from umbilical cord blood and from peripheral blood for subsequent in vivo tracking in a xenotransplant mouse model XXX1. *Academic Radiology,* 12, 502–510.

Daldrup-Link, H. E., Rudelius, M., Oostendorp, R. A., Settles, M., Piontek, G., Metz, S., Rosenbrock, H., Keller, U., Heinzmann, U. & Rummeny, E. J. 2003. Targeting of hematopoietic progenitor cells with MR contrast agents. *Radiology,* 228, 760–7.

De Vries, I., Lesterhuis, W. J., Barentsz, J. O., Verdijk, P., Van, K. J., Boerman, O. C., Oyen, W. J., Bonenkamp, J. J., Boezeman, J. B. & Adema, G. J. 2005. Magnetic resonance tracking of dendritic cells in melanoma patients for monitoring of cellular therapy. *Nat Biotechnol,* 23, 1407–1413.

Deans, A. E., Wadghiri, Y. Z., Bernas, L. M., Yu, X., Rutt, B. K. & Turnbull, D. H. 2006. Cellular MRI contrast via coexpression of transferrin receptor and ferritin. *Magn Reson Med,* 56, 51–9.

Di Marco, M., Sadun, C., Port, M., Guilbert, I., Couvreur, P. & Dubernet, C. 2007. Physicochemical characterization of ultrasmall superparamagnetic iron oxide particles (USPIO) for biomedical application as MRI contrast agents. *Int J Nanomedicine,* 2, 609–622.

Fleige, G., Seeberger, F., Laux, D., Kresse, M., Taupitz, M., Pilgrimm, H. & Zimmer, C. 2002. In vitro characterization of two different ultrasmall iron oxide particles for magnetic resonance cell tracking. *Invest Radiol,* 37, 482–8.

Fontaine, R., Viscogliosi, N., Semmaoui, H., B, F., Lemieux, F., T, M. A., Michaud, J. B., B, P., Cadorette, J. & Pepin, C. M. 2007. Digital signal processing applied to crystal identification in Positron Emission Tomography dedicated to small animals. *Nuclear Inst. and Methods in Physics Research, A,* 571, 385–388.

Frank, J. A., Miller, B. R., Arbab, A. S., Zywicke, H. A., Jordan, E. K., Lewis, B. K., Bryant, L. H. & Bulte, J. W. 2003. Clinically Applicable Labeling of Mammalian and Stem Cells by Combining Superparamagnetic Iron Oxides and Transfection Agents. *Radiology,* 228, 480–487.

Gafni, Y., Turgeman, G., Liebergal, M., Pelled, G., Gazit, Z. & Gazit, D. 2004. Stem cells as vehicles for orthopedic gene therapy. *Gene Ther,* 11, 417–26.

Gao, F., Kar, S., Zhang, J., Qiu, B., Walczak, P., Larabi, M., Xue, R., Frost, E., Qian, Z. & Bulte, J. W. 2007. MRI of intravenously injected bone marrow cells homing to the site of injured arteries. *NMR Biomed,* 20, 673–81.

Ghosh, P., Hawrylak, N., Broadus, J., Greenough, W. T. & Lauterbur, P. C. Year. NMR imaging of transplanted iron oxide-labelled cells in the rat brain. *In,* 1990. 1193.

Gilad, A. A., Van Laarhoven, H. W. M., Mcmahon, M. T., Walczak, P., Heerschap, A., Neeman, M., Van Zijl, P. C. M. & Bulte, J. W. M. 2009. Feasibility of concurrent dual contrast enhancement using CEST contrast agents and superparamagnetic iron oxide particles. *Magnetic Resonance in Medicine,* 61, 970–974.

Gonzalez-Lara, L., Xu, X., Hofstetrova, K., Pniak, A., Chen, Y., Mcfadden, C., Martinez-Santiesteban, F., Rutt, B., Brown, A. & Foster, P. 2010. The Use of Cellular Magnetic Resonance Imaging to Track the Fate of Iron-Labeled Multipotent Stromal Cells after Direct Transplantation in a Mouse Model of Spinal Cord Injury. *Molecular Imaging and Biology,* 1–10.

Gossuin, Y., Roch, A., Muller, R. N. & Gillis, P. 2000. Relaxation induced by ferritin and ferritin-like magnetic particles: The role of proton exchange. *Magnetic Resonance in Medicine,* 43, 237–243.

Grabill, C., Silva, A. C., Smith, S. S., Koretsky, A. P. & Rouault, T. A. 2003. MRI detection of ferritin iron overload and associated neuronal pathology in iron regulatory protein-2 knockout mice. *Brain Research,* 971, 95–106.

Guzman, R., Uchida, N., Bliss, T. M., He, D., Christopherson, K. K., Stellwagen, D., Capela, A., Greve, J., Malenka, R. C., Moseley, M. E., Palmer, T. D. & Steinberg, G. K. 2007. Long-term monitoring of transplanted human neural stem cells in developmental and pathological contexts with MRI. *Proc Natl Acad Sci USA,* 104, 10211–6.

Harrison, P. M. & Arosio, P. 1996. The ferritins: molecular properties, iron storage function and cellular regulation. *BBA-Bioenergetics*, 1275, 161–203.

Harush-Frenkel, O., Debotton, N., Benita, S. & Altschuler, Y. 2007. Targeting of nanoparticles to the clathrin-mediated endocytic pathway. *Biochemical and Biophysical Research Communications,* 353, 26–32.

Heyn, C., Bowen, C. V., Rutt, B. K. & Foster, P. J. 2005. Detection threshold of single SPIO-labeled cells with FIESTA. *Magn Reson Med,* 53, 312–20.

Heyn, C., Ronald, J. A., Mackenzie, L. T., Macdonald, I. C., Chambers, A. F., Rutt, B. K. & Foster, P. J. 2006. In Vivo Magnetic Resonance Imaging of Single Cells in Mouse Brain with Optical Validation. *Magn Reson Med,* 55, 23.

Hill, J. M., Dick, A. J., Raman, V. K., Thompson, R. B., Yu, Z. X., Hinds, K. A., Pessanha, B. S., Guttman, M. A., Varney, T. R. & Martin, B. J. 2003. Serial cardiac magnetic resonance imaging of injected mesenchymal stem cells. *Circulation,* 108, 1009.

Himmelreich, U., Weber, R., Ramos-Cabrer, P., Wegener, S., Kandal, K., Shapiro, E. M., Koretsky, A. P. & Hoehn, M. 2005. Improved stem cell MR detectability in animal models by modification of the inhalation gas. *Mol Imaging,* 4, 104–9.

Hoehn, M., Kustermann, E., Blunk, J., Wiedermann, D., Trapp, T., Wecker, S., Focking, M., Arnold, H., Hescheler, J. & Fleischmann, B. K. 2002. Monitoring of implanted stem cell migration in vivo: A highly resolved in vivo magnetic resonance imaging investigation of experimental stroke in rat. *Proceedings of the National Academy of Sciences,* 99, 16267–16272.

Horwitz, E. M., Prockop, D. J., Gordon, P. L., Koo, W. W., Fitzpatrick, L. A., Neel, M. D., Mccarville, M. E., Orchard, P. J., Pyeritz, R. E. & Brenner, M. K. 2001. Clinical responses to bone marrow transplantation in children with severe osteogenesis imperfecta. *Blood,* 97, 1227–31.

Hunter, R. J. 1995. *Foundations of colloid science,* Oxford, Clarendon Press.

Jendelova, P., Herynek, V., Decroos, J., Glogarova, K., Andersson, B., Hajek, M. & Sykova, E. 2003. Imaging the fate of implanted bone marrow stromal cells labeled with superparamagnetic nanoparticles. *Magn Reson Med,* 50, 767–76.

Jung, C. W. 1995. Surface properties of superparamagnetic iron oxide MR contrast agents: ferumoxides, ferumoxtran, ferumoxsil. *Magnetic Resonance Imaging,* 13, 675–691.

Jung, C. W. & Jacobs, P. 1995. Physical and chemical properties of superparamagnetic iron oxide MR contrast agents: Ferumoxides, ferumoxtran, ferumoxsil. *Magnetic Resonance Imaging,* 13, 661–674.

Kadayakkara, D. K. K., Janjic, J. M., Pusateri, L. K., Young, W.-B. & Ahrens, E. T. 2010. In vivo observation of intracellular oximetry in perfluorocarbon-labeled glioma cells and chemotherapeutic response in the CNS using fluorine-19 MRI. *Magnetic Resonance in Medicine,* 64, 1252–1259.

Kawaguchi, H., Koiwai, N., Ohtsuka, Y., Miyamoto, M. & Sasakawa, S. 1986. Phagocytosis of latex particles by leucocytes. I. Dependence of phagocytosis on the size and surface potential of particles. *Biomaterials,* 7, 61–66.

Kim, H.-S. Year. Comparison of the Biological Properties of Human Mesenchymal Stem Cells Labeled with Different Iron Oxide Nanoparticles. *In,* 2008 Nice, France. Poster No. 1071

Kleinschnitz, C., Bendszus, M., Frank, M., Solymosi, L., Toyka, K. V. & Stoll, G. 2003. In Vivo Monitoring of Macrophage Infiltration in Experimental Ischemic Brain Lesions by Magnetic Resonance Imaging. *Journal of Cerebral Blood Flow & Metabolism,* 23, 1356–1361.

Kleinschnitz, C., Sch, A., N, I., Horn, T., Frank, M., Solymosi, L., Stoll, G. & Bendszus, M. 2005. In vivo detection of developing vessel occlusion in photothrombotic ischemic brain lesions in the rat by iron particle enhanced MRI. *Journal of Cerebral Blood Flow & Metabolism,* 25, 1548–1555

Ko, I. K., Song, H. T., Cho, E. J., Lee, E. S., Huh, Y. M. & Suh, J. S. 2007. In vivo MR imaging of tissue-engineered human mesenchymal stem cells transplanted to mouse: a preliminary study. *Ann Biomed Eng,* 35, 101–8.

Kooi, M. E., Cappendijk, V. C., Cleutjens, K., Kessels, A. G., Kitslaar, P., Borgers, M., Frederik, P. M., Daemen, M. & Van, E. J. 2003. Accumulation of Ultrasmall Superparamagnetic Particles of Iron Oxide in Human Atherosclerotic Plaques Can Be Detected by In Vivo Magnetic Resonance Imaging. Am Heart Assoc

Kostura, L., Kraitchman, D. L., Mackay, A. M., Pittenger, M. F. & Bulte, J. W. 2004. Feridex labeling of mesenchymal stem cells inhibits chondrogenesis but not adipogenesis or osteogenesis. *NMR in Biomedicine,* 17, 513–517.

Kraitchman, D. L., Heldman, A. W., Atalar, E., Amado, L. C., Martin, B. J., Pittenger, M. F., Hare, J. M. & Bulte, J. W. 2003. In vivo magnetic resonance imaging of mesenchymal stem cells in myocardial infarction. *Circulation,* 107, 2290–3.

Kraitchman, D. L., Tatsumi, M., Gilson, W. D., Ishimori, T., Kedziorek, D., Walczak, P., Segars, W. P., Chen, H. H., Fritzges, D. & Izbudak, I. 2005. Dynamic imaging of allogeneic mesenchymal stem cells trafficking to myocardial infarction. *Circulation,* 112, 1451–61.

Lai, E. & Van Zanten, J. H. 2001. Monitoring DNA/Poly-l-Lysine Polyplex Formation with Time-Resolved Multiangle Laser Light Scattering. *Biophysical Journal,* 80, 864–873.

Lee, E. S., Bou-Gharios, G., Seppanen, E., Khosrotehrani, K. & Fisk, N. M. 2010a. Fetal Stem Cell Microchimerism: Natural Born Healers or Killers? *Molecular Human Reproduction,* [Epub ahead of print].

Lee, E. S., Chan, J., Shuter, B., Tan, L. G., Chong, M. S. K., Ramachandra, D. L., Dawe, G. S., Ding, J., Teoh, S. H., Beuf, O., Briguet, A., Tam, K. C., Choolani, M. & Wang, S.-C. 2009. Microgel Iron Oxide Nanoparticles For Tracking Human Fetal Mesenchymal Stem Cells Through Magnetic Resonance Imaging. *Stem Cells,* 27, 1921–1931.

Lee, E. S., Shuter, B., Chan, J., Chong, M. S. K., Ding, J., Teoh, S.-H., Beuf, O., Briguet, A., Tam, K. C., Choolani, M. & Wang, S.-C. 2010b. The use of microgel iron oxide nanoparticles in studies of magnetic resonance relaxation and endothelial progenitor cell labelling. *Biomaterials,* 31, 3296–3306.

Lee, H., Lee, E., Kim, K., Jang, N. K., Jeong, Y. Y. & Jon, S. 2006. Antibiofouling polymer-coated superparamagnetic iron oxide nanoparticles as potential magnetic resonance contrast agents for in vivo cancer imaging. *J Am Chem Soc,* 128, 7383–9.

Lewin, M., Carlesso, N., Tung, C. H., Tang, X. W., Cory, D., Scadden, D. T. & Weissleder, R. 2000. Tat peptide-derivatized magnetic nanoparticles allow in vivo tracking and recovery of progenitor cells. *Nature biotechnology,* 18, 410–414.

Liu, G., Molas, M., Grossmann, G. A., Pasumarthy, M., Perales, J. C., Cooper, M. J. & Hanson, R. W. 2001. Biological Properties of Poly-l-lysine-DNA Complexes Generated by Cooperative Binding of the Polycation. *Journal of Biological Chemistry,* 276, 34379–34387.

Mailander, V., Lorenz, M. R., Holzapfel, V., Musyanovych, A., Fuchs, K., Wiesneth, M., Walther, P., Landfester, K. & Schrezenmeier, H. 2008. Carboxylated Superparamagnetic Iron Oxide Particles Label Cells Intracellularly Without Transfection Agents. *Molecular Imaging and Biology,* 10, 138–146.

Matuszewski, L., Persigehl, T., Wall, A., Schwindt, W., Tombach, B., Fobker, M., Poremba, C., Ebert, W., Heindel, W. & Bremer, C. 2005. Cell Tagging with Clinically Approved Iron Oxides: Feasibility and Effect of Lipofection, Particle Size, and Surface Coating on Labeling Efficiency. *Radiology,* 235, 155–161.

Mendonca, D. M. & Lauterbur, P. C. 1986. Ferromagnetic particles as contrast agents for magnetic resonance imaging of liver and spleen. *Magnetic resonance in medicine: official journal of the Society of Magnetic Resonance in Medicine/Society of Magnetic Resonance in Medicine,* 3, 328–330.

Metz, S., Bonaterra, G., Rudelius, M., Settles, M., Rummeny, E. J. & Daldrup-Link, H. E. 2004. Capacity of human monocytes to phagocytose approved iron oxide MR contrast agents in vitro. *Eur Radiol,* 14, 1851–8.

Meyer, G. P., Wollert, K. C., Lotz, J., Steffens, J., Lippolt, P., Fichtner, S., Hecker, H., Schaefer, A., Arseniev, L., Hertenstein, B., Ganser, A. & Drexler, H. 2006. Intracoronary Bone Marrow

Cell Transfer After Myocardial Infarction: Eighteen Months' Follow-Up Data From the Randomized, Controlled BOOST (BOne marrOw transfer to enhance ST-elevation infarct regeneration) Trial. *Circulation,* 113, 1287–1294.

Mills, P. H., Wu, Y. J., Ho, C. & Ahrens, E. T. 2008. Sensitive and automated detection of iron-oxide-labeled cells using phase image cross-correlation analysis. *Magnetic Resonance Imaging*

Modo, M., Cash, D., Mellodew, K., Williams, S. C., Fraser, S. E., Meade, T. J., Price, J. & Hodges, H. 2002. Tracking transplanted stem cell migration using bifunctional, contrast agent-enhanced, magnetic resonance imaging. *Neuroimage,* 17, 803–11.

Montet-Abou, K., Montet, X., Weissleder, R. & Josephson, L. 2007. Cell internalization of magnetic nanoparticles using transfection agents. *Mol Imaging,* 6, 1–9.

Moore, A., Josephson, L., Bhorade, R. M., Basilion, J. P. & Weissleder, R. 2001. Human transferrin receptor gene as a marker gene for MR imaging. *Radiology,* 221, 244–50.

Moore, A., Weissleder, R. & Bogdanov, A. 1997. Uptake of Dextran-Coated Monocrystalline Iron Oxides in Tumor Cells and Macrophages. *Journal of Magnetic Resonance Imaging*, 7, 1140–1145.

Nath, N., Hyun, J., Ma, H. & Chilkoti, A. 2004. Surface engineering strategies for control of protein and cell interactions. *Surface Science,* 570, 98–110.

Nelson, G. N., Roh, J. D., Mirensky, T. L., Wang, Y., Yi, T., Tellides, G., Pober, J. S., Shkarin, P., Shapiro, E. M., Saltzman, W. M., Papademetris, X., Fahmy, T. M. & Breuer, C. K. 2008. Initial evaluation of the use of USPIO cell labeling and noninvasive MR monitoring of human tissue-engineered vascular grafts in vivo. *FASEB J.,* 22, 3888–3895.

Neri, M., Maderna, C., Cavazzin, C., Deidda-Vigoriti, V., Politi, L. S., Scotti, G., Marzola, P., Sbarbati, A., Vescovi, A. L. & Gritti, A. 2008. Efficient In Vitro Labeling of Human Neural Precursor Cells with Superparamagnetic Iron Oxide Particles: Relevance for In Vivo Cell Tracking. *Stem Cells,* 26, 505–516.

Nguyen Huu, S., Oster, M., Uzan, S., Chareyre, F., Aractingi, S. & Khosrotehrani, K. 2007. Maternal neoangiogenesis during pregnancy partly derives from fetal endothelial progenitor cells. *Proceedings of the National Academy of Sciences,* 104, 1871–1876.

Nohroudi, K. A., S.; Berhorn, T.; Addicks, K.; Hoehn, M.; Himmelreich, U. 2010. In Vivo MRI Stem Cell Tracking Requires Balancing of Detection Limit and Cell Viability. *Cell Transplantation,* 19, 431–441

Norman, A. B., Thomas, S. R., Pratt, R. G., Lu, S. Y. & Norgren, R. B. 1992. Magnetic resonance imaging of neural transplants in rat brain using a superparamagnetic contrast agent. *Brain Res,* 594, 279–83.

Park, S. J., Leslie, R. W., Huh, S., Kagan, H., Honscheid, K., Burdette, D., Chesi, E., Lacasta, C., Llosa, G. & Mikuz, M. 2007. A prototype of very high-resolution small animal PET scanner using silicon pad detectors. *Nuclear Inst. and Methods in Physics Research, A,* 570, 543–555.

Pawelczyk, E., Arbab, A. S., Pandit, S., Hu, E. & Frank, J. A. 2006. Expression of transferrin receptor and ferritin following ferumoxides–protamine sulfate labeling of cells: implications for cellular magnetic resonance imaging. *NMR in Biomedicine,* 19, 581–592.

Phinney, D. G. & Prockop, D. J. 2007. Concise review: mesenchymal stem/multipotent stromal cells: the state of transdifferentiation and modes of tissue repair – current views. *Stem Cells,* 25, 2896–902.

Pintaske, J., Muller-Bierl, B. & Schick, F. 2006. Geometry and extension of signal voids in MR images induced by aggregations of magnetically labelled cells. *Phys Med Biol,* 51, 4707–4718.

Pratten, M. K. & Lloyd, J. B. 1986. Pinocytosis and phagocytosis: the effect of size of a particulate substrate on its mode of capture by rat peritoneal macrophages cultured in vitro. *Biochim Biophys Acta,* 881, 307–13.

Qiu, B., Gao, F., Walczak, P., Zhang, J., Kar, S., Bulte, J. W. & Yang, X. 2007. In vivo MR imaging of bone marrow cells trafficking to atherosclerotic plaques. *J Magn Reson Imaging,* 26, 339–43.

Raynal, I., Prigent, P., Peyramaure, S., Najid, A., Rebuzzi, C. & Corot, C. 2004. Macrophage endocytosis of superparamagnetic iron oxide nanoparticles: mechanisms and comparison of ferumoxides and ferumoxtran-10. *Invest Radiol,* 39, 56–63.

Reimer, P. & Balzer, T. 2003. Ferucarbotran (Resovist): a new clinically approved RES-specific contrast agent for contrast-enhanced MRI of the liver: properties, clinical development, and applications. *European Radiology,* 13, 1266–1276.

Reimer, P., Rummeny, E. J., Daldrup, H. E., Balzer, T., Tombach, B., Berns, T. & Peters, P. E. 1995. Clinical results with Resovist: a phase 2 clinical trial. *Radiology,* 195, 489–96.

Rejman, J., Oberle, V., Zuhorn, I. S. & Hoekstra, D. 2004. Size-dependent internalization of particles via the pathways of clathrin- and caveolae-mediated endocytosis. *Biochem J,* 377, 159–69.

Rogers, W. J. & Basu, P. 2005. Factors regulating macrophage endocytosis of nanoparticles: implications for targeted magnetic resonance plaque imaging. *Atherosclerosis,* 178, 67–73.

Roser, M., Fischer, D. & Kissel, T. 1998. Surface-modified biodegradable albumin nano-and microspheres. II: effect of surface charges on in vitro phagocytosis and biodistribution in rats. *European Journal of Pharmaceutics and Biopharmaceutics,* 46, 255–263.

Ruehm, S. G., Corot, C., Vogt, P., Kolb, S. & Debatin, J. F. 2001. Magnetic Resonance Imaging of Atherosclerotic Plaque With Ultrasmall Superparamagnetic Particles of Iron Oxide in Hyperlipidemic Rabbits. Am Heart Assoc

Sadan, O., Shemesh, N., Barzilay, R., Bahat-Stromza, M., Melamed, E., Cohen, Y. & Offen, D. 2008. Migration of neurotrophic factors-secreting mesenchymal stem cells towards a quinolinic acid lesion as viewed by MRI. *Stem Cells*

Saini, S., Stark, D. D., Brady, T. J., Wittenberg, J. & Ferrucci, J. J. 1986. Dynamic spin-echo MRI of liver cancer using Gadolinium-DTPA: animal investigation. *AJR Am J Roentgenol,* 147, 357–62.

Saini, S., Stark, D. D., Hahn, P. F., Wittenberg, J., Brady, T. J. & Ferrucci, J. J. 1987. Ferrite particles: a superparamagnetic MR contrast agent for the reticuloendothelial system. *Radiology,* 162, 211–216.

Saleh, A., Schroeter, M., Jonkmanns, C., Hartung, H. P., Modder, U. & Jander, S. 2004. In vivo MRI of brain inflammation in human ischaemic stroke. *Brain,* 127, 1670.

Sharma, R., Saini, S., Ros, P. R., Hahn, P. F., Small, W. C., De, L. E., Stillman, A. E., Edelman, R. R., Runge, V. M. & Outwater, E. K. 1999. Safety profile of ultrasmall superparamagnetic iron oxide ferumoxtran-10: phase II clinical trial data. *Journal of magnetic resonance imaging: JMRI,* 9, 291.

Shen, T., Weissleder, R., Papisov, M., Bogdanov, J. A. & Brady, T. J. 1993. Monocrystalline iron oxide nanocompounds (MION): physicochemical properties. *Magn Reson Med,* 29, 599–604.

Song, M., Moon, W. K., Kim, Y., Lim, D., Song, I. C. & Yoon, B. W. 2007. Labeling efficacy of superparamagnetic iron oxide nanoparticles to human neural stem cells: comparison of ferumoxides, monocrystalline iron oxide, cross-linked iron oxide (CLIO)-NH2 and tat-CLIO. *Korean J Radiol,* 8, 365–71.

Stuckey, D. J., Carr, C. A., Martin-Rendon, E., Tyler, D. J., Willmott, C., Cassidy, P. J., Hale, S. J., Schneider, J. E., Tatton, L. & Harding, S. E. 2006. Iron Particles for Noninvasive Monitoring of Bone Marrow Stromal Cell Engraftment into, and Isolation of Viable Engrafted Donor Cells from, the Heart. *Stem Cells,* 24, 1968–1975.

Tabata, Y. & Ikada, Y. 1988. Effect of the size and surface charge of polymer microspheres on their phagocytosis by macrophage. *Biomaterials,* 9, 356–362.

Taupitz, M., Schnorr, J., Abramjuk, C., Wagner, S., Pilgrimm, H., Hunigen, H. & Hamm, B. 2000. New generation of monomer-stabilized very small superparamagnetic iron oxide particles (VSOP) as contrast medium for MR angiography: preclinical results in rats and rabbits. *J Magn Reson Imaging,* 12, 905–11.

Terrovitis, J. V., Bulte, J. W., Sarvananthan, S., Crowe, L. A., Sarathchandra, P., Batten, P., Sachlos, E., Chester, A. H., Czernuszka, J. T. & Firmin, D. N. 2006. Magnetic resonance imaging of ferumoxide-labeled mesenchymal stem cells seeded on collagen scaffolds-relevance to tissue engineering. *Tissue Eng,* 12, 2765–75.

Thorek, D. L. & Tsourkas, A. 2008. Size, charge and concentration dependent uptake of iron oxide particles by non-phagocytic cells. *Biomaterials,* 29, 3583–3590.

Toso, C., Vallee, J. P., Morel, P., Ris, F., Demuylder-Mischler, S., Lepetit-Coiffe, M., Marangon, N., Saudek, F., James Shapiro, A. M., Bosco, D. & Berney, T. 2008. Clinical Magnetic Resonance Imaging of Pancreatic Islet Grafts After Iron Nanoparticle Labeling. *American Journal of Transplantation,* 8, 701–706.

Van Den Bos, E. J., Baks, T., Moelker, A. D., Kerver, W., Van, G. R., Van, G. W., Duncker, D. J. & Wielopolski, P. A. 2006. Magnetic resonance imaging of haemorrhage within reperfused myocardial infarcts: possible interference with iron oxide-labelled cell tracking? *Eur Heart J,* 27, 1620–6.

Van Den Bos, E. J., Wagner, A., Mahrholdt, H., Thompson, R. B., Morimoto, Y., Sutton, B. S., Judd, R. M. & Taylor, D. A. 2003. Improved efficacy of stem cell labeling for magnetic resonance imaging studies by the use of cationic liposomes. *Cell Transplant,* 12, 743–56.

Von Zur Muhlen, C., Von, E. D., Bassler, N., Neudorfer, I., Steitz, B., Petri-Fink, A., Hofmann, H., Bode, C. & Peter, K. 2007. Superparamagnetic iron oxide binding and uptake as imaged by magnetic resonance is mediated by the integrin receptor Mac-1 (CD11b/CD18): implications on imaging of atherosclerotic plaques. *Atherosclerosis,* 193, 102–11.

Walczak, P., Kedziorek, D. A., Gilad, A. A., Lin, S. & Bulte, J. W. 2005. Instant MR labeling of stem cells using magnetoelectroporation. *Magn Reson Med,* 54, 769–774.

Walczak, P., Zhang, J., Gilad, A. A., Kedziorek, D. A., Ruiz-Cabello, J., Young, R. G., Pittenger, M. F., Van, Z. P., Huang, J. & Bulte, J. W. 2008. Dual-Modality Monitoring of Targeted Intraarterial Delivery of Mesenchymal Stem Cells After Transient Ischemia. *Stroke,* 39, 1569.

Wang, L., Neoh, K.-G., Kang, E.-T., Shuter, B. & Wang, S.-C. 2010. Biodegradable magnetic-fluorescent magnetite/poly(dl-lactic acid-co-[alpha],[beta]-malic acid) composite nanoparticles for stem cell labeling. *Biomaterials,* 31, 3502–3511.

Weissleder, R., Elizondo, G., Wittenberg, J., Lee, A. S., Josephson, L. & Brady, T. J. 1990a. Ultrasmall superparamagnetic iron oxide: an intravenous contrast agent for assessing lymph nodes with MR imaging. *Radiology,* 175, 494–8.

Weissleder, R., Moore, A., Mahmood, U., Bhorade, R., Benveniste, H., Chiocca, E. A. & Basilion, J. P. 2000. In vivo magnetic resonance imaging of transgene expression. *Nature medicine,* 6, 351.

Weissleder, R., Wittenberg, J., Rabito, C. A. & Bengele, H. H. 1990b. Ultrasmall Superparamagnetic Iron Oxide: Characterization of a New Class of Contrast Agents for MR Imaging'. *Radiology,* 175, 489–493.

Wiart, M., Davoust, N., Pialat, J. B., Desestret, V., Moucharaffie, S., Cho, T. H., Mutin, M., Langlois, J. B., Beuf, O. & Honnorat, J. 2007. MRI Monitoring of Neuroinflammation in Mouse Focal Ischemia. *Stroke,* 38, 131.

Wilhelm, C., Billotey, C., Roger, J., Pons, J. N., Bacri, J. C. & Gazeau, F. 2003. Intracellular uptake of anionic superparamagnetic nanoparticles as a function of their surface coating. *Biomaterials,* 24, 1001–1011.

Wilhelm, C. & Gazeau, F. 2008. Universal cell labelling with anionic magnetic nanoparticles. *Biomaterials*

Wu, H., Pal, D., O, S. J. & Tai, Y.-C. 2008. A Feasibility Study of a Prototype PET Insert Device to Convert a General-Purpose Animal PET Scanner to Higher Resolution. *J Nucl Med,* 49, 79–87.

Yeh, T. C., Zhang, W., Ildstad, S. T. & Ho, C. 1993. Intracellular labeling of T-cells with superparamagnetic contrast agents. *Magn Reson Med,* 30, 617–25.

Zhang, X., Bowen, C. V., Gareau, P. & Rutt, B. K. Year. Quantitative Analysis of SPIO and USPIO Uptake Rate by Macrophages: Effects of Particle Size, Concentration, and Labeling Time. *In:* Proc. Intl. Soc. Mag. Reson. Med, 2001. 880.

Zhu, J., Zhou, L. & Xingwu, F. G. 2006. Tracking Neural Stem Cells in Patients with Brain Trauma. *New England Journal of Medicine,* 355, 2376.

Ziener, C. H., Bauer, W. R. & Jakob, P. M. 2005. Transverse relaxation of cells labeled with magnetic nanoparticles. *Magn Reson Med,* 54, 702–6.

Zurkiya, O., Chan, A. W. S. & Hu, X. 2008. MagA is sufficient for producing magnetic nanoparticles in mammalian cells, making it an MRI reporter. *Magnetic Resonance in Medicine,* 59, 1225–1231.

Nanoparticles for the Oral Administration of Cancer Therapies

Socorro Espuelas, Maite Agüeros, Irene Esparza, and Juan M. Irache

Abstract Oral delivery remains the preferred route for drug administration thanks to its patient convenience and compliance. However, many anticancer drugs are unsuitable for oral formulations. This chapter describes the major obstacles to cancer drugs gastrointestinal absorption (namely low solubility, poor membrane permeability or extensive pre-systemic metabolism) and strategies to circumvent them, with special attention to nanocarriers. Until now we have proved an enhancement in the oral bioavailability of entrapped anticancer drugs. The remaining challenge is designing nanoparticles able to specifically target the drugs to tumor cells anywhere in the body after oral administration. To this aim, it will be necessary more studies clarifying the correlation between nanoparticles properties and their interaction with the gut mucosa, their transport through and their final outcome. The possibilities of oral administration of other modalities of cancer treatment, as immunotherapy, are also discussed.

Keywords Cancer therapy • Oral administration • Bioavailability • Nanocarriers • Cancer immunotherapy • P-gp

Abbreviations

5-FU	5-fluorouracil
ABC	ATP-binding cassette
Ag	Antigen
COX-2	Cyclooxygenase 2
CT	Cholera toxin
CYP	Cytochrome P450

S. Espuelas, M. Agüeros, I. Esparza, and J.M. Irache (✉)
Department of Pharmaceutics and Pharmaceutical Technology,
School of Pharmacy, University of Navarra, Pamplona 31080, Spain
e-mail: jmirache@unav.es

A. Prokop (ed.), *Intracellular Delivery: Fundamentals and Applications*, Fundamental Biomedical Technologies 5, DOI 10.1007/978-94-007-1248-5_18,

DCs	Dendritic cells
FURD	5-fluorouridine
i.v.	intravenous
LT	Heat labile enterotoxin
MDR	multidrug resistance
MHC	Major histocompatibility complex
MLNs	Mesenteric lymph nodes
MMT	montmorillonite
o/w	oil in water
PEG	poly(ethylenglycol)
P-gp	P-glycoprotein
PLGA	Poly(d,l-lactide-co-glycolide)
SA-NP	albumin coated-poly(anhydride) NP
s.c.	subcutaneous
SEDDS	Self-emulsifying DDS
SLC	solute carrier
S-SEDDS	Supersaturable SEDDS
TAA	Tumor-associated antigen
VEGF	Vascular Endothelial Growth Factor

1 Introduction

Oncology is one of the few areas of medicine where most patients are treated intravenously (i.v.) rather than receiving oral drugs. In the past, the new anti-cancer therapies focused mainly on parenteral drug delivery, in part because this route bypassed the solubility and permeability problems associated with the gastrointestinal route. In addition, parenteral administration was considered relatively straightforward and compatible with the cytotoxic action of most chemotherapies. With this type of administration, the aim is to deliver the maximal tolerated dose of drug to optimize cell kill in a single episode, followed by a several week period to allow bone marrow recovery (Weingart et al. 2008).

However, the recent past years has seen an accelerating expansion of the development of oral anticancer drugs, which will continue to growth in the coming years. In 2008, experts estimated that more than one quarter of the 400 antineoplastic agents in the pipeline were planned as oral drugs (Kuppens et al. 2005).

Oral chemotherapy is attractive because of its convenience and ease of administration, particularly in the palliative setting (Payne 1992; Liu et al. 1997). It is also especially appropriate where prolonged drug exposure is desirable. At present, many current anti-cancer therapies are cytostatic in nature and thus are optimally effective when given chronically for a continuous tumour exposure (Pasquier et al. 2010). This mechanism of action virtually requires oral daily and prolonged therapies (Weingart et al. 2008). On the other hand, from the economic standpoint, oral administration is more attractive because it reduces the cost for hospitalization and infusion equipment supplies (Terwogt et al. 1999), what makes it a "dominant strategy" in pharmacoeconomic terms (Irshad 2010).

However, despite the numerous advantages described above, the oral bioavailability of current cancer drugs is limited by multiple factors such as rapid degradation, low solubility, poor membrane permeability or extensive pre-systemic metabolism (Schellens et al. 2000). Their efflux by P-glycoprotein and/or intestinal and hepatic metabolism by cytochrome P450 metabolizing enzyme system have recognized as major determinants of oral bioavailability of common anticancer drugs (Zhang and Benet 2001).

Strategies to solve these handicaps are the synthesis of more soluble structural analogs, prodrugs, the concomitant administration of cancer drugs with P-glycoprotein or CYP inhibitors and the use of particulate carriers systems. In the literature, several nanoparticle-based formulations developed for oral delivery applications have succeeded in protecting the encapsulated drugs, promoting their absorption through the gastrointestinal mucosa and enhancing their oral bioavailability (Cai et al. 2010).

This paper describes the major obstacles to cancer drugs gastrointestinal absorption and strategies to circumvent them, with special attention to nanocarriers. We also briefly analyse the factors affecting the interaction of these carriers with the gut mucosa, their transport through and their final outcome. Finally, the possibilities of oral administration of other modalities of cancer treatment, as immunotherapy, are also discussed.

There are several excellent recent reviews that describe the goodness of nanocarriers for enhancement of oral bioavailability of drugs (Roger et al. 2010a) or as adjuvants in mucosal vaccination with diverse therapeutic applications (Chadwick et al. 2010). This chapter is only focused in nanocarriers intended for oral administration of cancer therapies.

2 Main Obstacles to Oral Delivery of Cancer Chemotherapy

Orally administered, most anticancer drugs are highly hampered to reach the target cells. In fact, the bioavailability of these drugs may be limited by their solubility and/or permeability through the gut epithelia. In the former a low aqueous solubility and slow dissolution rate may decrease the amount of solubilised drug able to reach the absorptive membrane (Pasquier et al. 2010). The drug has to be sufficiently and rapidly dissolved in the aqueous fluids of the gastrointestinal tract before absorption through the epithelial cells. Poor dissolution properties, as observed for lipophilic molecules may importantly limit the fraction of drug absorbed (Ganta et al. 2010). In the latter, the permeability of a molecule may be negatively affected when the drug is substrate of the efflux transport systems or by the presence of hydrophilic or ionisable groups in its chemical structure. Another important factor that can negatively affect the bioavailability of a molecule is its sensibility to the harsh conditions of the gastrointestinal tract (Roger et al. 2010a) or its degradation by enzymatic complex located in the enterocytes (pre-systemic metabolism) (Barrett et al. 1996)

(Fig. 1a-1) (Bernkop-Schnurch and Schmitz 2007) or in the liver (first-pass metabolism). In any case, overcoming these hurdles is the major challenge in developing oral cancer therapy.

In general, drugs are orally absorbed by a transcellular route (through the cell); although, in some cases, a paracellular pathway and a carrier-mediated mechanism can also occur. The transcellular transport of a molecule can take place by a passive mechanism (Fig. 1a-5) or be mediated by a specific carrier (Fig. 1a-6). In the first case, the molecule have to show the adequate physicochemical properties (such as

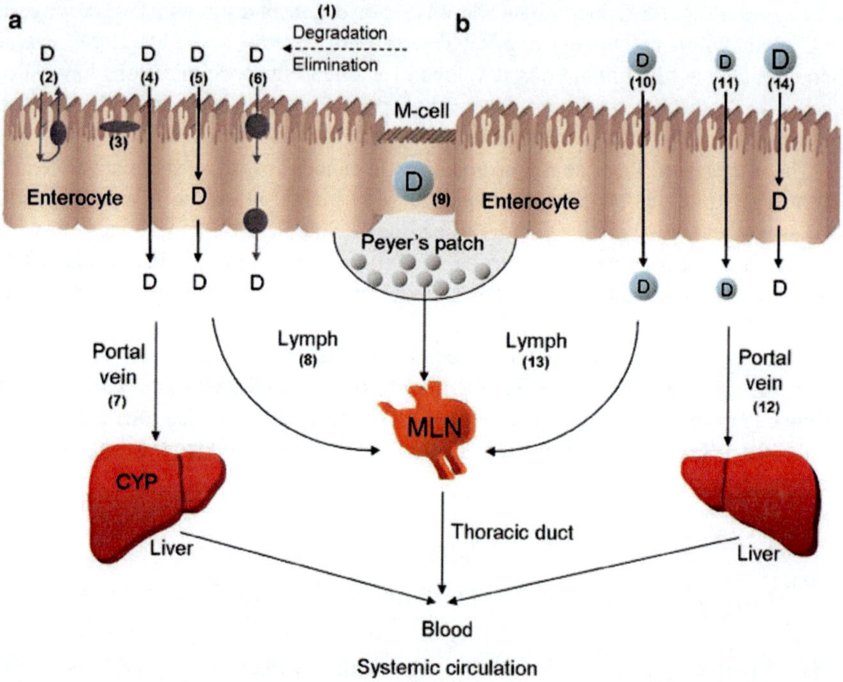

Fig. 1 Intestinal absorption of drugs (**a**) or drug-associated nanoparticles (**b**). In the intestinal gut, drug can suffer enzymatic or chemical degradation (*1*). The drug once is dissolved in the aqueous fluids of the gastrointestinal tract, must diffuse through the mucosa and reach the intestinal epithelium in order to be absorbed. Some drugs will not arrive to the systemic circulation due to efflux mechanism by P-gp (*2*) or intestinal degradation by CYP (*3*).The physico-chemical properties of the drug will determine the mechanism of absorption: paracellular diffusion (*4*); transcellular passive diffusion (*5*); carrier-mediated transcellular transport (*6*). After absorption, some drugs go to blood and then to the liver through the portal vein (*7*). Other drugs go to the mesenteric lymph nodes and reach the blood systemic circulation by the thoracic duct (*8*). The association of drug to nanoparticles (**b**) can protect it from degradation and avoid the efflux mechanisms. These nanocarriers can cross the epithelium either via M-cells (*9*) or via enterocytes by endocytic pathways (*10* and *11*) and they will reach the systemic circulation via portal vein and liver (*12*) or via lymphatic system (*13*). However, in general most of nanocarriers designed for oral administration do not cross the epithelium themselves; instead prolong the contact of the drug with the mucosa (i.e. bioadhesive nanoparticles), increasing the bioavailability (*14*)

lipophilicity, polarity and molecular size) to cross both the apical and basolateral membrane as well as the lipophilic barriers of cells (Hunter and Hirst 1997). The transcellular transport can be performed across enterocytes or, in a lesser extent, M-cells. M-cells are specialized phagocytic enterocytes found in the follicle-associated epithelium of the Peyer's patches (Clark et al. 2001). They are specialised in the capture and transport of bacteria, virus, macromolecules and particles from the gut lumen to immune cells across the epithelial barrier, and thus are important in stimulating mucosal immunity. Unlike their neighbouring cells (enterocytes), they have the unique ability to take up compounds from the lumen of the small intestine via endocytosis or phagocytosis and, then, deliver them via transcytosis to dendritic cells (DCs) and lymphocytes located in an unique pocket-like structure on their basolateral side. M-cells, which possess broader microfolds, differ from normal enterocytes in that they lack microvilli on their apical surface and that they are far less abundant (Clark et al. 2001; Corr et al. 2008).

The paracellular transport or the transference between adjacent cells (Fig. 1a-4) appear to be restricted to small hydrophilic molecules (approximately <350 Da) (Gaucher et al. 2010). This extracellular route across the epithelium is based on the establishment of an electrochemical gradient derived from differences in concentration, electrical potential and/or hydrostatic pressure between the two sides of the epithelium. Nevertheless, the total contribution of the paracellular pathway to general drug transport is, in general, very discrete (Artursson et al. 2001; Palm et al. 1996).

Drugs that are structural analogues of natural compounds can gain entry into the cell by carrier-mediated mechanisms (Fig. 1a-6). These influx transporters can increase the drug absorption by specific binding of the compound with the specific carrier, which is located on the surface of the cellular lipid bilayer (Blanchette et al. 2004). Compounds that are substrates of these transporters exhibit higher intestinal absorption than expected for a mechanism of transport by diffusion. There are more than 400 membrane transporters, belonging to two major superfamilies: the ATP binding cassette (ABC) and the solute carrier (SLC) families (Sugano et al. 2010). However, their substrate specificity is not absolute, and carrier-mediated transport is only available to a certain number of drugs.

On the other hand, drug efflux pumps like P-glycoprotein (P-gp) (Fig. 1a-2) play a major role in the absorption of many drugs. P-glycoprotein is best known for its ability to transport drug substrates out of cells in a variety of tissues, including the intestine (Sparreboom et al. 1997). The intestinal P-glycoprotein is the 170-kDa product of the human MDR1 gene and member of the ABC family of transport proteins (Sparreboom et al. 1997). Both the expression and the function of P-glycoprotein have been related to considerable variability in oral drug absorption (Hunter and Hirst 1997; Hunter et al. 1993). It is present in the columnar epithelial cells (enterocytes) of the lower gastrointestinal tract (Varma et al. 2006). Given its apical distribution on the enterocyte, P-gp is exquisitely positioned to limit the absorption of substances that the cell perceives as harmful. Unlike some conventional transporter, P-gp does not transport a specific substrate but a wide variety of chemically most diverse compounds (Neudeck et al. 2004). Among others, many important anticancer drugs currently in use are substrates of the P-gp such as doxorubicin, daunorubicin, idarubicin, paclitaxel, docetaxel, vincristine, vinblastine, topotecan, etoposide and teniposide (Breedveld et al. 2006).

Furthermore, it has been proposed that intestinal P-gp would be functionally linked to the cytochrome P-450 (more particularly the CYP3A4) and both compounds would display complementary and synergistic roles in the pharmacokinetics of drug absorption and elimination (Bardelmeijer et al. 2000a) (Fig. 1a-3). The P-gp activity would increase the residence of the drug in the gut and, that, would enable more exposure to and more extensive metabolism by CYP3A4 (Kivisto et al. 2004). This fact is supported for at least, the following evidences: (i) the two proteins are co-localized within the digestive tract and within enterocytes, (iii) they share many of the same substrates and (iii) they are co-inducible in response to at least some xenobiotics (Watkins 1997).

Apart from a poor membrane uptake, an extensive pre-systemic metabolism may also be responsible of an important decrease of the drug bioavailability. This pre-systemic metabolism in the gastrointestinal tract is related to the presence of the following enzymes: (i) luminally secreted enzymes (i.e. pepsins, trypsin, chymotrypsin, elastase and carboxypeptidase A/B), (ii) brush border membrane bound enzymes such as various carboxypeptidases and aminopeptidases, and (iii) cytosolic enzymes. In this context, cytochrome P450 (CYP) is the main oxidative drug metabolizing enzyme system (Schellens et al. 2000). CYP consist of approximately 40 P450s, among which the isoenzyme CYP3A4 is of special interest because it has a very broad substrate specificity (van Waterschoot et al. 2009). This enzyme is the principal CYP isoenzyme found in the enterocytes, where it serves as an important barrier for xenobiotics to enter into the circulation (Bardelmeijer et al. 2000a; Thummel 2007; Wacher et al. 2001). In fact, this enzyme is involved in the metabolism of at least 50% of drugs (Lin and Lu 1998; Guengerich 1999). As indicated previously, many anticancer agents are extensively metabolyzed by CYPs, such as the taxanes paclitaxel and docetaxel, epipodophyllotoxins (etoposide, teniposide), *vinca* alkaloids (vincristine, vinblastine, vindesine) and anthracyclines (doxorubicin, daunorubicin, idarubicin) (Schellens et al. 2000). In addition, food constituents can profoundly inhibit CYP3A, making significative variations in the bioavailability between patients (van Herwaarden et al. 2007).

Regardless the mechanism of absorption and efflux mechanisms, the enterocyte uptake leads to transport via blood capillaries and Peyer patches uptake through intestinal lymphatics, from where drugs go to the blood via the mesenteric nodes and the thoracic lymph duct (Fig. 1a-8). Lipophilic drugs can also gain access to the intestinal lymphatic system via association with developing lipoproteins into the enterocytes (Porter et al. 2007; Stremmel 1988).

The first-pass metabolism by the liver is the last obstacle for drug bioavailability (Fig. 1a-7). This process refers to the loss of the drug due its biotransformation in the liver before reaching the systemic circulation. In this organ the cytochrome P-450 family of enzymes is the main responsible for the oxidation of drugs and xenobiotics (Boobis and Davies 1984). The rate of drug elimination in the liver depends on hepatic flow rate, intrinsic capacity of the enzymes to metabolize the drug, and drug binding to plasma proteins (Agoram et al. 2001).

Only drugs that are absorbed via the intestinal lymphatic system are essentially protected from hepatic first pass metabolism since the mesenteric lymph, in contrast to portal blood, enters the systemic circulation directly without first passing through the liver (Trevaskis et al. 2008; Hussain et al. 2001). Another advantage of this route of delivery is the lymphatic exposure to drug concentrations orders of magnitude higher than that attained in systemic blood (Trevaskis et al. 2008; Hussain et al. 2001).

3 Strategies to Improve Oral Bioavailability of Anticancer Drugs

In spite of the different developments in progress, to date, the majority of anticancer drugs have to be administered via parenteral routes. A common strategy to overcome solubility and permeability limitations of anticancer drugs consist on chemical modification in order to make the drug more hydrophilic and/or stable face to enzymatic attack. On the other hand, the encapsulation of the drug in nanocarriers can be applied in order to protect the loaded drug and try to promote its permeability across the mucosal membrane. They include lipid nanospheres, nanoemulsions or polymer nanoparticles.

3.1 Prodrugs

Prodrugs are compounds that need to be transformed before exhibiting their pharmacological activity. The term prodrug was introduced in the late 1950s by Albert who defined it as a pharmacologically inactive derivative of a parent drug molecule that requires spontaneous or enzymatic transformation within the body to convert to the active drug molecule (Albert 1958). Prodrug strategies are typically employed to reduce toxicity, enhance the solubility or the stability, modify lipophilicity, permeability, and tissue specificity of the parent drug (Singh et al. 2008).

In general, for those prodrugs designed to increase the bioavailability, it is preferred a relatively low activation of the prodrug into the antitumor agent in the blood or liver, thereby preventing acute toxic effects (Connors 1986). Ideally, after absorption, the drug should be slowly released and activated only in the tumor site where the parent drug can execute its mode of action. However, due to the reactivity of most antitumor drugs, a limitation of this slow-releasing prodrug concept is that frequently "healthy" tissues are also affected. Another drawback of prodrugs activated by enzymes with low catalytic efficiencies, may be the metabolism of these prodrugs by competing enzymes into inactive metabolites (Boddy and Yule 2000).

5-fluorouracil (5-FU) is a good example of anticancer drug for which a relatively wide variety of prodrug systems have been evaluated. Among others, 5-FU prodrugs that have been studied include 5-Fluoro-Pyrimidinone (5-FP) (Guo et al. 1995),

Ftorafur® (Kawata et al. 1984), 5-FU glucuronide (Guerquin-Kern et al. 2000), and 5′-DFUR and capacitabine (Armstrong and Diasio 1980; Thorne et al. 2009). All of these prodrugs were designed to overcome the rapid breakdown of 5-FU in the gastrointestinal tract and its gut toxicity.

5-FP is activated to 5-fluorouracile by aldehyde oxidase; although, no improved antitumor activity compared with 5-FU has been observed (Guo et al. 1995). In addition, aldehyde oxidase is widely distributed within the body and, thus, it would not be an ideal candidate for the development of prodrugs (Murray et al. 1995). Ftorafur® or Tegafur® is a slow-releasing prodrug which is activated to 5-FU by cytosolic thymidine phosphorylase and microsomal P450 (Ohashi et al. 2010; Takiuchi and Ajani 1998), whereas 5-FU glucuronide is activated by β-glucuronidase (Guerquin-Kern et al. 2000). Again, P450 and β-glucuronidase are also widely distributed and, thus, the activation of the prodrug would also occur in nontumor tissues (Murray et al. 1995). Probably, the most interesting 5-FU prodrug is capecitabine (Xeloda®) which is a derivative of 5′-deoxy-5-fluorouridine (5′-DFUR) with a significantly lower gastrointestinal toxicity. This prodrug, which is currently clinically used, is sequentially activated into 5-FU by three enzymes, i.e., carboxylesterase, cytidine deaminase, and thymidine phosphorylase (Miwa et al. 1998). This last enzyme is up-regulated in breast, ovarian, colorectal, and gastric cancers (Cole et al. 1999).

3.2 P-Glycoprotein Inhibitors

Many anticancer drugs are substrates by P-gp. The inhibition of this efflux pump has been a common strategy to overcome the low oral bioavailability of many anticancer drugs. Inhibition of P-gp may be tackled by (i) co-administration of agents known as P-gp substrates in order to act as inhibitors of the transporter, (ii) development of novel agents that are non P-gp substrates, and (iii) design of formulations that allow the drug to bypass efflux pump transport.

The first attempt to inhibit the activity of intestinal P-gp was the co-administration of other substrates of this transporter. In this context, drugs such as verapamil (Kruijtzer et al. 2002) or cyclosporine A (Demeule et al. 1999) were proposed. Both compounds can compete with the drug of interest by P-gp, increasing the possibility of drug absorption. However, due to the usually low binding affinities of these drugs for the intestinal P-gp, large doses are required to be effective. As a consequence, unacceptable toxicity events can be observed (Bansal et al. 2009a; Benet et al. 2004). In addition, these P-gp inhibitors also show a concomitant inhibition of cytochrome P-450 and, thus, pharmacokinetic interactions can occur (Bansal et al. 2009a). In summary, co-administration of an anticancer drug and a non-selective P-glycoprotein inhibitor resulted in decreased clearance and increased toxicity of the treatment (Benet et al. 2004).

The second generation of P-gp inhibitors was initially developed to reverse multidrug resistance. Many of these compounds (i.e. the cyclosporin A analog valspodar) were derivatives of the first generation, however, as the original compounds, these

P-glycoprotein inhibitors also affected the activity of CYP3A4, limiting their clinical applicability (Fischer et al. 1998). More recently, a third-generation of compounds was synthetised in order to specifically interact with the P-glycoprotein transporter (Breedveld et al. 2006; Hyafil et al. 1993; Dantzig et al. 1996). In this context, some agents such as elacridar (GF120918), zosuquidar (LY335979) and tariquidar (XR9576) have been developed (Breedveld et al. 2006; Hyafil et al. 1993; Dantzig et al. 1996). Nevertheless, their usefulness in vivo is limited due to toxicological issues such as cardiotoxicity or immunosuppressive effects (Bardelmeijer et al. 2000b; Woo et al. 2003).

Therefore, several authors are focused in the identification of more effective and safe P-gp inhibitors without pharmacological activity (Bansal et al. 2009b). In this context, recently, it has been reported that some pharmaceutical excipients could inhibit the function of intestinal P-gp. These excipients offer a number of advantages including their well-known use in the formulation of parenteral and enteral medicines, safety and regulatory acceptance (Buggins et al. 2007). Among others, surfactants (i.e. Pluronic derivatives), co-solvents (i.e. polyethylenglycols) and solubilising agents (i.e. cyclodextrins) have been proposed. Under specific circumstances, these compounds may interact with the polar head of the lipid bilayers modifying the membrane structure by inserting themselves between the lipophilic tails of the bilayers. These membrane perturbations would induce a certain degree of fluidization of the lipid membrane affecting, as a consequence, the activity of the P-gp (Lo 2003). In the specific case of Pluronics, Batrakova and Kabanov showed that these surfactants would modulate the P-gp by inhibiting the ATPase activity resulting in ATP depletion (Batrakova and Kabanov 2008). In any case, it appears that moderate lipophilicity of surfactants (with HLB <20) is a key factor which determines the inhibitory potency (Lo 2003; Batrakova et al. 2003).

Ishiwaka and Arima, also reported the inhibitory effect of dimethyl-β-cyclodextrin, hydroxypropyl-β-cyclodextrin and other β-cyclodextrin derivatives in the activity of the CYP3A complex (Ishikawa et al. 2005) and the P-gp (Arima et al. 2001). The mechanism by which these oligosaccharides inhibit P-gp activity would be due to a modulation of the membrane microenvironment, mediated by a removal process of cholesterol and phospholipids in the P-gp's immediate environment (Arima et al. 2001), leading to a transient impairment of the transmembrane protein activity (Lo 2003).

Other pharmaceutical excipients, such as poly(ethylene glycol)s (PEG) and some thiolated polymers (i.e. thiolated chitosan and thiolated carbomer) also display a specific ability to inhibit P-gp activity (Foger et al. 2006).

3.3 *Nanocarriers*

Nanocarriers offer several advantages for oral drug delivery. They can protect the drug against degradation and promote its absorption through the gut mucosa. According to their physico-chemical properties, nanoparticles can diffuse within the mucus and increase the permeability of the drug across the intestinal barrier (Roger et al. 2010a). In some cases, nanodevices themselves can cross the epithelial membrane and translocate the drug cargo from the lumen to the blood or lymph (Fig. 1b). However, in general, nanoparticles are not absorbed, but at least can prolong the time of contact with the mucosa or epithelia and, therefore, increase the drug concentration gradient from the lumen to the epithelia (Fig. 1b-14). The fate of nanoparticles in the mucosa and in subcellular domains is dependent on their size and their surface properties. Very small particles (up to 200 nm) seem to be endocytosable by enterocytes (Fig. 1b-11). Particles between 200 nm and 3 μm are preferentially uptaked by phagocytic M-cells in the Peyer patches, although they have not been detected further than mesenteric lymph nodes (Fig. 1b-9). Particles higher than 3 μm have been found in follicle-associated epithelia but they did not show passage to associated lymphoid tissues (Des Rieux et al. 2006). The uptake of nanoparticles by M-cells produces higher accumulation of the drug in the local lymph nodes from it can be directly transported to the systemic circulation without first passing through the liver. This route of delivery can enhance the activity of immunomodulatory drugs or vaccines or drugs aimed to target the lymphoid tissue (Hussain et al. 2001). However, for drugs that need to reach the blood circulation, their transport across absorptive epithelial cells produce a higher improvement in their oral bioavailability.

The uptake of the particles by enterocytes could be by clathrine or caveolae-mediated or independent endocytosis (Fig. 1b-10 and 11). Particles captured by clathrine-mediated endocytosis are localized in the endolysosomes where they are degraded by enzymes although some of them can escape from endolysosome vesicles and to be released by exocytosis outside enterocytes via the basolateral cell side. Also caveolae-mediated endocytosis could lead to direct exocytosis of the carrier on the basolateral side of enterocytes. If nanoparticles are exocyted by the enterocytes, the impermeability of the vascular endothelium combined with the large inter-endothelial gaps present in the lymphatic endothelium would facilitate the entry of these carriers in the lymphatic system rather than the blood capillaries (Cai et al. 2010).

Thus, either through enterocytes or M-cells, nanoparticles that cross the epithelium and remained intact would reach the bloodstream through the lymphatic system (Fig. 1b-13). To the best of our knowledge, it has not been detected intact nanoparticles in the bloodstream after oral delivery although an increase in the drug bioavailability has been frequently observed.

3.3.1 Lipophilic Devices

The presence of lipids in the diet, as well as the use of lipophilic formulations, can influence and modulate drug absorption, especially water-insoluble compounds (Porter et al. 2007). Lipid-based formulations can increment the dissolution rate of the drug by a modification of the interfacial tension between the drug and the dissolution medium. The particles of hydrophobic drugs offer high surface resistance which prevents its solubilization. In contrast, formulations containing lipids as drug carriers are capable of reducing the surface tension of the drug and, then, improve its dissolution. However, these drugs (especially weakly basic drugs) may come out of the formulation due to solubilization in the gastric fluid and may precipitate in the intestinal fluid on gastric emptying. Thus, the maximum advantage of a lipid formulation could only be drawn if the drug remains in lipid solution throughout its residence in the gastrointestinal tract. However, the performance of lipid formulations and the fate of the drug in the gastrointestinal tract depend on three different phenomenons that occur simultaneously: dispersion, dilution and digestion of the formulation. Dispersion of the formulation due to gastric agitation results in increased exposure of the inner surfaces of the formulation, dilution in the gastric fluid results in partitioning of the drug in the external fluid phase and enzymatic breakdown of the lipid in the gastrointestinal tract can result in change in their composition, structure and potential loss of their solvent capacity. The digestion of the lipid vehicle in the presence of gastric lipase results in reduction in their solubilization capacity (Mohsin et al. 2009; Kaukonen et al. 2004). However, the precipitated drug may be resolubilized by both the exogenous and the endogenous lipids in the form of colloidal species in the intestinal region which would depend on several factors, like the lipophilicity of the drug, the nature of the colloidal phases produced on digestion of the different lipids and the kinetics of drug transfer between the digesting formulation and the colloidal phases produced (Chakraborty et al. 2009).

Nanoemulsions

Nanoemulsions and self-emulsified drug delivery systems (SEDDS) have been extensively used for oral administration of lipophilic compounds. Nanoemulsions are heterogeneous systems consisting of two immiscible liquids in which one liquid is dispersed as droplets within the other liquid (Singh and Vingkar 2008). Self-nanoemulsified drug delivery systems are isotropic mixtures of oil, surfactant, cosurfactant and drug that form fine oil-in-water (o/w) nanoemulsions upon mild agitation in aqueous media (Nazzal et al. 2002). They are stabilized by an interfacial film of surfactant molecules with a typical droplet size lower than 100 nm, which guarantees efficient absorption of the droplets (Cai et al. 2010). These systems improve the oral bioavailability of lipophilic drugs by different mechanisms including drug solubilisation and protection against enzymatic degradation. In addition, the small size of these droplets increases the interphase between the lipophilic droplet and the aqueous medium of the gut facilitating a homogeneous and wide distribution of the drug along the gastrointestinal tract.

However these systems usually need the presence of large amount of surfactant agents, which can be of interest to modify the permeability of the membrane (Gao 2003, 2009); although, at the same time, may induce gastrointestinal side-effects. In an attempt to reduce the surfactant side-effects and achieve rapid absorption of poorly soluble drugs, it has been proposed a new class of supersaturable formulations based on SEDDS formulations (S-SEDDS) (Gao 2003, 2009). The S-SEDDS contain a reduced amount of surfactant and a thickening agent to temporarily stabilize a supersaturated amount of the drug. In this context, Gao and collaborators used a typical SEDDS (formulated with Cremophor® EL and a mixture of ethanol and glyceryl dioleate) mixed with hydroxypropylmethylcellulose for the oral delivery of paclitaxel. This formulation displayed a fourfold higher bioavailability of the anticancer drug that when formulated as Taxol®. Co-administration of this S-SEDDS-paclitaxel formulation with cyclosporine A (as inhibitor of the P-gp) resulted in significantly higher absolute bioavailability of paclitaxel (threefold higher than without the inhibitor of the P-gp) (Gao et al. 2003).

Lipid Nanoparticles

Solid lipid nanoparticles are devices made from lipids that are solid at room temperature. In the body, they can be erosioned and degraded by bile salts and pancreatic lipase (Roger et al. 2010a). In principle, these systems are specifically designed for the controlled release of lipophilic compounds such as paclitaxel (Cai et al. 2010). Thus, paclitaxel-loaded lipid nanocapsules demonstrated a threefold increase in the oral bioavailability of this anticancer drug in comparison with Taxol® orally administered. This effect on the oral paclitaxel bioavailability would be directly related to the presence of the surfactant Solutol® HS15, which would inhibit the P-glycoprotein efflux system. Besides, lymphatic transport of paclitaxel and the nanocapsule structure might also be beneficial for the oral permeability of paclitaxel (Roger et al. 2010b; Peltier et al. 2006). However, in some cases, solid lipid nanoparticles can present an insufficient drug loading capacity due to drug expulsion after polymorphic transition during storage (Almeida and Souto 2007).

3.3.2 Polymer Nanoparticles

Nanoparticles that combine bioadhesive properties with a certain inhibitory capacity of the P-gp extrusion system and cytochrome P-450 complex can be an adequate strategy for oral chemotherapy. Then, a high therapeutic efficacy and less side effects of the drug can thus be expected. Another advantage from these polymer nanoparticles can be their inherent properties to control the release of the loaded drug, which may be of interest to obtain sustained and prolonged release of the drug.

In this way, it was reported a novel bioadhesive drug delivery system based on the combination between poly(d,l-lactide-co-glycolide) nanoparticles and montmorillonite (PLGA/MMT) for the oral delivery of paclitaxel. Paclitaxel-loaded PLGA/MMT nanoparticles were prepared by the emulsion/solvent evaporation method and montmorillonite was incorporated in the formulation as a matrix material component, which also plays the role of a co-emulsifier in the nanoparticle preparative process. Besides, these nanoparticles were coated with trastuzumab (human epidermal growth factor receptor-2 antibody or HER2). Trastuzumab "decoration" of nanoparticles significantly increased their cellular uptake by a human breast cancer cell line (SK-BR-3) and Caco-2 (Sun et al. 2008; Dong and Feng 2005).

This novel formulation is expected to possess extended residence time in the gastrointestinal tract; however, no in vivo pharmacokinetic investigation was conducted to show the feasibility and efficiency.

Another interesting strategy could be the use of polymer nanoparticles based on the copolymer of methylvinylether and maleic anhydride (Gantrez AN from Isp. Corp.). This copolymer permits the preparation of nanoparticles under "soft" conditions and the resulting carriers have demonstrated a high ability to develop bioadhesive interactions within the gastrointestinal tract (Arbos et al. 2002). More important, the surface of these poly(anhydride) nanoparticles can be easily modified by simple incubation with different excipients or ligands in order to modify their fate within the gastrointestinal tract (Arbos et al. 2003; Salman et al. 2005; Yoncheva et al. 2005). In this context, some compounds such as albumin (Arbos et al. 2003) or chitosan [data not published] permit to modify the distribution of these nanoparticles within the gut, whereas some excipients such as cyclodextrins or polyethylenglycols can be of interest to inhibit the activity of P-gp and the activity of CYP3A4 (Agueros et al. 2009).

In this context, using 5-fluorouridine (FURD) as model, the potential of serum albumin-coated poly(methylvinylether -co- maleic anhydride) nanoparticles (SA-NP) were evaluated. These nanoparticles were capable to specifically target the stomach mucosa and develop strong bioadhesive interactions in this part of the gut after their oral administration in laboratory animals (Arbos et al. 2003). From urine data, the FURD bioavailability when loaded in either SA-NP or in uncoated nanoparticles was about 79% and 21%, respectively. For the control oral solution this parameter was calculated to be only 11%. These values correlated well with the presence of FURD metabolite in the lumen of animals (Arbos et al. 2004).

More recently, and in a different approach, the combination between poly(anhydride) nanoparticles and cyclodextrins have been demonstrated effective to significantly improve the oral bioavailability of paclitaxel (see Fig. 2) (Agueros et al. 2010). Cyclodextrins have been used for a double purpose: as a solubilising agent of paclitaxel, which it is an extremely lipophilic compound, and for the capability of these oligosaccharides to disturb and inhibit the activity of the intestinal P-glycoprotein (Ishikawa et al. 2005; Arima et al. 2001).

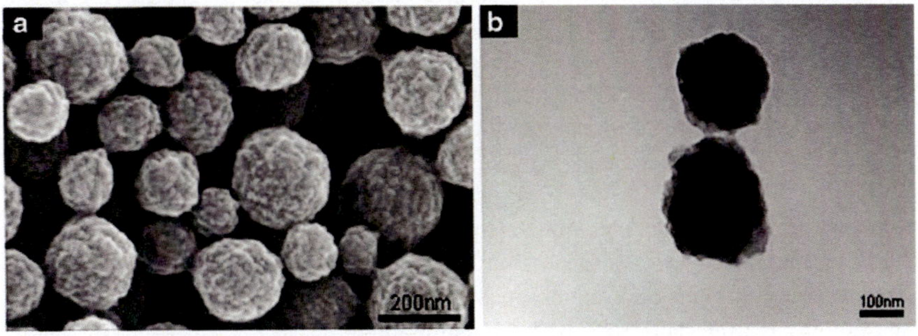

Fig. 2 Microphotographs of paclixatel loaded cyclodextrin-poly(anhydride) nanoparticles obtained by scanning electron microscopy (**a**) and transmission electron microscopy (**b**)

After the oral administration of these nanoparticles to laboratory animals, the paclitaxel plasma levels were characterized by a plateau close to the C_{max} spanning from T_{max} till 24 h post-administration. These sustained levels of the anticancer drug corresponded with a relative oral bioavailability of about 80% (Agueros et al. 2010). This fact would be due to the combination of both bioadhesive and inhibitory properties of these cyclodextrin-poly(anhydride) nanoparticles. In fact, the nanoparticles would be capable to approach the paclitaxel-cyclodextrin complex till the surface of the mucosa where these carriers would remain immobilised in intimate contact with the absorptive membrane. Then, nanoparticles would progressively release their content to the gut environment. At this moment, the paclitaxel-cyclodextrin complex would dissociate yielding the free drug and the oligosaccharide molecules. Under these conditions, the anticancer drug would be rapidly absorbed whereas the cyclodextrins would interact with lipidic components of the membrane disturbing the activity of the efflux pump and cytochrome P450.

4 Cancer Immunotherapy: Implications of Gut and Tumor Immunology in the Design of Oral Cancer Vaccines

Cancer immunotherapy is an attractive alternative, or addition, to conventional cancer chemotherapy treatments and their study has increased significantly in the past two decades. The idea behind them focused primarily on manipulating the patient's own immune system to destroy cancer cells. Significant advantages to this modality of cancer treatment are its ability to (i) induce specific killing of tumor cells (with minimal detriment to healthy, non-tumor cells), (ii) systemically stimulate antitumor immune responses that can target primary and secondary metastasis and (iii) result in immunological memory that would provide long-term protection against possible future tumor recurrences (Blattman and Greenberg 2004; Higgins et al. 2009).

There are several modalities of cancer immunotherapy as the administration of tumor-cell targeting monoclonal antibodies, adjuvant cytokine treatment and vaccines. These strategies of antitumor therapy are unlikely to be effective when orally administered. First, monoclonal antibodies, cytokines or Tumor-Associated Antigen (TAA) as cancer vaccines are labile proteins that are easily degraded in the gastrointestinal tract by the low pH or digestive enzymes. Furthermore, what immune response concerns, intestinal mucosa is characterized by the constitutive and abundant presence of prostaglandine E2, TGF-β and IL-10 that favour the differentiation of DCs and T-cells towards a regulatory phenotype (producing IL-4, IL-10 and TGF-β). This fact leads to systemic tolerance (Malik et al. 2010). Furthermore, cancer vaccines must also be effective in overcoming tumor- associated immune dysfunction/suppression and may need to adopt the role of proinflammatory cytokines and chemokines in inducing tumoricidal effector functions and homing (Lasaro and Ertl 2010; Mocellin and Nitti 2008).

The tendency of oral vaccination to induce tolerance with immunosuppressive environment that frequently surrounded tumors could explain that few cancer vaccines have been orally administered. The exceptions are the recombinant bacteria and viral vectors expressing TAA that naturally infect oral mucosa like Salmonella. These microorganisms have been described capable to overcome the default status of the intestinal immune system and promote the production of IL-12 by DCs in Peyer's patches and MLNs (Mowat 2003).

In general, a successful vaccination requires two components: an antigen and an adjuvant. When designing oral anticancer vaccines, the adjuvant, an amplifier of the response induced by the antigen itself, should be extremely potent in order to surpass both the oral tolerance and the tumor suppressive environment.

Synthetic particles have been widely investigated, on the one hand, as adjuvants for mucosal vaccination (Chadwick et al. 2010; Des Rieux et al. 2006) with applications other than cancer therapy and, on the other hand, as adjuvants for parenteral cancer vaccination (Xiang et al. 2008; Krishnamachari et al. 2010). Orally administered, the carriers offer a number of advantages that well justify their application as adjuvant for oral vaccination. They protect the antigen (Ag) cargo from degradation. Moreover, as previously described these carriers preferentially target the M-cells in the follicle-associated epithelium, where are transferred to local DCs. These DCs present the Ag directly to T-cells in the Peyer's patch or migrate to mesenteric lymph nodes and could augment or not the mucosal and systemic immune response (Primard et al. 2010; Slutter et al. 2009).

In spite of preferential targeting of particles by M-cells the studies show that, in general, a very low fraction of administered particles bind to the gastrointestinal tract. Particles in the range of 500-nm or/and hydrophobic were better uptaken than the larger or more hydrophilic ones (Brayden 2001). In order to improve the adhesion of the particles to M-cells, a common strategy has been the coating of the particulate Ag with lectins that selectively bind M-cell specific surface carbohydrates (Jepson et al. 2004). Other interesting approach has been the covering of the particles with compounds involved in the process of micro-organisms colonization like flagellin, trying to mimic the bacteria Salmonella (Irache et al. 2010). However, glycocalix of mucus barrier act as a size selective barrier that limits the accessibility to the targeted particles towards cell surface receptor.

Beside strategies to improve the deposit of particles in the immune inductive (Peyer's patch) sites of the gastrointestinal tract, nanoparticles based vaccines should be interpreted by the immune system as an invasive pathogen in order to generate immune response. Although it has been described that particles activate NALP3 inflammosome, only the TLR triggering can produce local inflammation that induce complete maturation of DCs in the Peyer's patch. As a result, these activated DCs after taking up antigen (signal 1), produce IL-12 (signal 3) and present antigenic fragment to T-cells with appropriate co-stimulation (signal 2). These DCs prime Th-cells results into their activation and differentiation. Keeping this idea in mind, CpG motifs and mutant of two bacterial toxins, heat labile enterotoxin (LT) and cholera toxin (CT), have been combined with particles and evaluated as oral vaccines against some microorganisms (Chen et al. 2010).

Once effector specific T-cells are generated they must migrate, recognize and destroy tumoral cells. Particles targeting DCs from Peyer's patches tend to induce gut-homing properties in T-cells. Moreover, tumors have developed immunosuppressive mechanisms as down-regulation of MHC molecules or production of immunosuppressive soluble factors as TGF-β, IL-10, VEGF or COX-2 that can make the vaccine ineffective (Kalinski 2009).

In agreement with the extremely potency required for an oral cancer vaccine to be effective, the previous literature usually proposed the use of recombinant and attenuated bacteria and viral vectors expressing TAA that naturally infect oral mucosa like Salmonella (Zhu et al. 2010) or Papilomavirus and/or are also able to invade others organs following oral inoculation like plant virus nanoparticles (Rae et al. 2005). The attributes adhesion, invasion and danger signalization are naturally presented in these natural carriers and should be molecularly defined for the entire synthetic construction in artificial cancer nanovaccines. It must be feasible and in fact, either administered by oral or subcutaneous route, a nanoemulsion encapsulating MAGE1-HSP70 and SEA complex protein was able to delay tumor growth and defer tumor occurrence of mice challenged with B16-MAGE-1 tumor cells. The encapsulation of the Ag complex in nanoemulsion was significantly more important for the oral than s.c. administration, indicating the importance of nanocarrier in mucosal vaccination (Ge et al. 2009). However, most mice-tumor models usually did not recreate the complexity and immunosupressive environment found in human tumors.

5 Conclusions and Future Perspectives

The gut performs a hurt locker to cancer therapy. In this moment, none strategy is able to target an anticancer drug specifically to distant cancer cells after oral administration. However, we have made some progress and pharmaceutical nanodevices have shown certain ability to improve the oral bioavailability of anticancer drugs by overcoming solubility problems, exerting protection from degradation, prolonging their interaction with the mucosa, bypassing P-glycoprotein extrusion and the first-pass metabolism and facilitating translocation of the drug through the epithelia from the gut lumen to the blood or lymph.

The physicochemical characteristics of the particles, mainly size, surface and specific targeting, will critically determine the principal route of adsorption (if it takes place) and their final outcome although the most documented and common route of uptake is via the M-cell rich layer of Peyer's patches. The possibility of transport across the intestinal epithelial cells would be restricted to very small particles (50–100 nm). Either uptake by enterocytes or M-cells, particles eventually reach the mesenteric lymph nodes, but there are no evidences of further spread. Thus, the anticancer drug will be released from the carrier and will arrive into the bloodstream since the lymphatic system avoiding the passage through the liver. This route of delivery could enhance the bioavailability of anticancer drugs with high first-pass metabolism or to be useful for improving the local accumulation of therapeutics in intestinal lymph nodes.

Nanoparticles that attach to the surface of the enterocytes (without uptake) and, if necessary, avoid P-gp extrusion and CYP metabolism could increase the absorption of the free drug directly to the blood or lymph. This strategy should be preferable in order to improve the oral bioavailability of the drug because the proportion of M-cells in the intestinal epithelia is very low. However, it does not avoid toxicity cancer drug concerns because free distribution of the drug to organs other than tumour after blood systemic circulation arrival.

Other modality of cancer treatment, the cancer immunotherapy, could also profit the targeting of M-cells with particles. However, it must face up the natural tendency of oral mucosa to induce tolerance besides the immunosupressive mechanisms of tumour itself. This fact adds an unjustifiable factor of complexity in the design of effective cancer vaccines that need to be rethought.

It will be necessary more and deeper studies with in vivo imaging techniques to clarify the correlation between nanoparticle properties and their interaction with the GI mucosa, their transport through and their final outcome. In this context, the strategies used by microorganisms capable to cross the intestinal epithelia and invade other organs should be extrapolated to nanocarriers design.

References

Agoram B, Woltosz WS, Bolger MB (2001) Predicting the impact of physiological and biochemical processes on oral drug bioavailability. Adv Drug Deliv Rev 50 Suppl 1:S41–67. doi:S0169409X0100179X [pii]

Agueros M, Ruiz-Gaton L, Vauthier C, Bouchemal K, Espuelas S, Ponchel G, Irache JM (2009) Combined hydroxypropyl-beta-cyclodextrin and poly(anhydride) nanoparticles improve the oral permeability of paclitaxel. Eur J Pharm Sci 38 (4):405– 413. doi:S0928-0987(09)00269-3 [pii] 10.1016/j.ejps.2009.09.010

Agueros M, Zabaleta V, Espuelas S, Campanero MA, Irache JM (2010) Increased oral bioavailability of paclitaxel by its encapsulation through complex formation with cyclodextrins in poly(anhydride) nanoparticles. J Control Release 145 (1):2–8. doi:S0168-3659(10)00217-8 [pii] 10.1016/j.jconrel.2010.03.012

Albert A (1958) Chemical aspects of selective toxicity. Nature 182 (4633):421–422

Almeida AJ, Souto E (2007) Solid lipid nanoparticles as a drug delivery system for peptides and proteins. Adv Drug Deliv Rev 59 (6):478–490. doi:S0169-409X(07)00043-9 [pii] 10.1016/j.addr.2007.04.007

Arbos P, Arangoa MA, Campanero MA, Irache JM (2002) Quantification of the bioadhesive properties of protein-coated PVM/MA nanoparticles. Int J Pharm 242 (1–2):129–136. doi:S0378517302001825 [pii]

Arbos P, Campanero MA, Arangoa MA, Renedo MJ, Irache JM (2003) Influence of the surface characteristics of PVM/MA nanoparticles on their bioadhesive properties. J Control Release 89 (1):19–30. doi:S016836590300066X [pii]

Arbos P, Campanero MA, Arangoa MA, Irache JM (2004) Nanoparticles with specific bioadhesive properties to circumvent the pre-systemic degradation of fluorinated pyrimidines. J Control Release 96 (1):55–65. doi:10.1016/j.jconrel.2004.01.006 S0168365904000288 [pii]

Arima H, Yunomae K, Hirayama F, Uekama K (2001) Contribution of P-glycoprotein to the enhancing effects of dimethyl-beta-cyclodextrin on oral bioavailability of tacrolimus. J Pharmacol Exp Ther 297 (2):547–555

Armstrong RD, Diasio RB (1980) Metabolism and biological activity of 5′-deoxy-5-fluorouridine, a novel fluoropyrimidine. Cancer Res 40 (9):3333–3338

Artursson P, Palm K, Luthman K (2001) Caco-2 monolayers in experimental and theoretical predictions of drug transport. Adv Drug Deliv Rev 46 (1–3):27–43. doi:S0169-409X(00)00128-9 [pii]

Bansal T, Akhtar N, Jaggi M, Khar RK, Talegaonkar S (2009) Novel formulation approaches for optimising delivery of anticancer drugs based on P-glycoprotein modulation. Drug Discov Today 14 (21–22):1067-1074. doi:S1359-6446(09)00254-2 [pii] 10.1016/j.drudis.2009.07.010

Bansal T, Jaggi M, Khar RK, Talegaonkar S (2009) Emerging significance of flavonoids as P-glycoprotein inhibitors in cancer chemotherapy. J Pharm Pharm Sci 12 (1):46–78

Bardelmeijer HA, van Tellingen O, Schellens JH, Beijnen JH (2000) The oral route for the administration of cytotoxic drugs: strategies to increase the efficiency and consistency of drug delivery. Invest New Drugs 18 (3):231–241

Bardelmeijer HA, Beijnen JH, Brouwer KR, Rosing H, Nooijen WJ, Schellens JH, van Tellingen O (2000) Increased oral bioavailability of paclitaxel by GF120918 in mice through selective modulation of P-glycoprotein. Clin Cancer Res 6 (11):4416–4421

Barrett JS, Szego P, Rohatagi S, Morales RJ, De Witt KE, Rajewski G, Ireland J (1996) Absorption and presystemic metabolism of selegiline hydrochloride at different regions in the gastrointestinal tract in healthy males. Pharm Res 13 (10):1535–1540

Batrakova EV, Kabanov AV (2008) Pluronic block copolymers: evolution of drug delivery concept from inert nanocarriers to biological response modifiers. J Control Release 130 (2):98–106. doi:S0168-3659(08)00221-6 [pii] 10.1016/j.jconrel.2008.04.013

Batrakova EV, Li S, Alakhov VY, Miller DW, Kabanov AV (2003) Optimal structure requirements for pluronic block copolymers in modifying P-glycoprotein drug efflux transporter activity in bovine brain microvessel endothelial cells. J Pharmacol Exp Ther 304 (2):845–854. doi:10.1124/jpet.102.043307

Benet LZ, Cummins CL, Wu CY (2004) Unmasking the dynamic interplay between efflux transporters and metabolic enzymes. Int J Pharm 277 (1–2):3–9. doi:10.1016/j.ijpharm.2002.12.002 S0378517304001255 [pii]

Bernkop-Schnurch A, Schmitz T (2007) Presystemic metabolism of orally administered peptide drugs and strategies to overcome it. Curr Drug Metab 8 (5):509–517

Blanchette J, Kavimandan N, Peppas NA (2004) Principles of transmucosal delivery of therapeutic agents. Biomed Pharmacother 58 (3):142–151. doi:10.1016/j.biopha.2004.01.006 S0753332204000228 [pii]

Blattman JN, Greenberg PD (2004) Cancer immunotherapy: a treatment for the masses. Science 305 (5681):200–205. doi:10.1126/science.1100369 305/5681/200 [pii]

Boddy AV, Yule SM (2000) Metabolism and pharmacokinetics of oxazaphosphorines. Clin Pharmacokinet 38 (4):291–304

Boobis AR, Davies DS (1984) Human cytochromes P-450. Xenobiotica 14 (1–2):151–185. doi:10.3109/00498258409151404

Brayden DJ (2001) Oral vaccination in man using antigens in particles: current status. Eur J Pharm Sci 14 (3):183–189. doi:S0928098701001750 [pii]

Breedveld P, Beijnen JH, Schellens JH (2006) Use of P-glycoprotein and BCRP inhibitors to improve oral bioavailability and CNS penetration of anticancer drugs. Trends Pharmacol Sci 27 (1):17–24. doi:S0165-6147(05)00310-X [pii] 10.1016/j.tips.2005.11.009

Buggins TR, Dickinson PA, Taylor G (2007) The effects of pharmaceutical excipients on drug disposition. Adv Drug Deliv Rev 59 (15):1482–1503

Cai Z, Wang Y, Zhu LJ, Liu ZQ (2010) Nanocarriers: a general strategy for enhancement of oral bioavailability of poorly absorbed or pre-systemically metabolized drugs. Curr Drug Metab 11 (2):197–207

Chadwick S, Kriegel C, Amiji M (2010) Nanotechnology solutions for mucosal immunization. Adv Drug Deliv Rev 62 (4-5):394-407. doi:S0169-409X(09)00353-6 [pii] 10.1016/j.addr.2009.11.012

Chakraborty S, Shukla D, Mishra B, Singh S (2009) Lipid--an emerging platform for oral delivery of drugs with poor bioavailability. Eur J Pharm Biopharm 73 (1):1–15. doi:S0939-6411(09)00166-0 [pii] 10.1016/j.ejpb.2009.06.001

Chen W, Patel GB, Yan H, Zhang J (2010) Recent advances in the development of novel mucosal adjuvants and antigen delivery systems. Hum Vaccin 6 (9). doi:11561 [pii]

Clark MA, Jepson MA, Hirst BH (2001) Exploiting M cells for drug and vaccine delivery. Adv Drug Deliv Rev 50 (1–2):81-106. doi:S0169-409X(01)00149-1 [pii]

Cole C, Foster AJ, Freeman S, Jaffar M, Murray PE, Strafford IJ (1999) The role of thymidine phosphorylase/PD-ECGF in cancer chemotherapy: a chemical perspective. Anticancer Drug Des 14 (5):383–392

Connors TA (1986) Prodrugs in cancer chemotherapy. Xenobiotica 16 (10–11):975–988. doi:10.3109/00498258609038977

Corr SC, Gahan CC, Hill C (2008) M-cells: origin, morphology and role in mucosal immunity and microbial pathogenesis. FEMS Immunol Med Microbiol 52 (1):2–12. doi:FIM359 [pii] 10.1111/j.1574-695X.2007.00359.x

Dantzig AH, Shepard RL, Cao J, Law KL, Ehlhardt WJ, Baughman TM, Bumol TF, Starling JJ (1996) Reversal of P-glycoprotein-mediated multidrug resistance by a potent cyclopropyldibenzosuberane modulator, LY335979. Cancer Res 56 (18):4171–4179

Demeule M, Laplante A, Sepehr-Arae A, Beaulieu E, Averill-Bates D, Wenger RM, Beliveau R (1999) Inhibition of P-glycoprotein by cyclosporin A analogues and metabolites. Biochem Cell Biol 77 (1):47–58

des Rieux A, Fievez V, Garinot M, Schneider YJ, Preat V (2006) Nanoparticles as potential oral delivery systems of proteins and vaccines: a mechanistic approach. J Control Release 116 (1):1–27. doi:S0168-3659(06)00402-0 [pii] 10.1016/j.jconrel.2006.08.013

Dong Y, Feng SS (2005) Poly(d,l-lactide-co-glycolide)/montmorillonite nanoparticles for oral delivery of anticancer drugs. Biomaterials 26 (30):6068–6076. doi:S0142-9612(05)00256-5 [pii] 10.1016/j.biomaterials.2005.03.021

Fischer V, Rodriguez-Gascon A, Heitz F, Tynes R, Hauck C, Cohen D, Vickers AE (1998) The multidrug resistance modulator valspodar (PSC 833) is metabolized by human cytochrome P450 3A. Implications for drug-drug interactions and pharmacological activity of the main metabolite. Drug Metab Dispos 26 (8):802–811

Foger F, Schmitz T, Bernkop-Schnurch A (2006) In vivo evaluation of an oral delivery system for P-gp substrates based on thiolated chitosan. Biomaterials 27 (23):4250–4255. doi:S0142-9612(06)00268-7 [pii] 10.1016/j.biomaterials.2006.03.033

Ganta S, Devalapally H, Amiji M (2010) Curcumin enhances oral bioavailability and anti-tumor therapeutic efficacy of paclitaxel upon administration in nanoemulsion formulation. J Pharm Sci 99 (11):4630–4641. doi:10.1002/jps.22157

Gao P, Rush BD, Pfund WP, Huang T, Bauer JM, Morozowich W, Kuo MS, Hageman MJ (2003) Development of a supersaturable SEDDS (S-SEDDS) formulation of paclitaxel with improved oral bioavailability. J Pharm Sci 92 (12):2386–2398. doi:10.1002/jps.10511

Gao P, Akrami A, Alvarez F, Hu J, Li L, Ma C, Surapaneni S (2009) Characterization and optimization of AMG 517 supersaturatable self-emulsifying drug delivery system (S-SEDDS) for improved oral absorption. J Pharm Sci 98 (2):516–528. doi:10.1002/jps.21451

Gaucher G, Satturwar P, Jones MC, Furtos A, Leroux JC (2010) Polymeric micelles for oral drug delivery. Eur J Pharm Biopharm 76 (2):147–158. doi:S0939-6411(10)00156-6 [pii] 10.1016/j.ejpb.2010.06.007

Ge W, Li Y, Li ZS, Zhang SH, Sun YJ, Hu PZ, Wang XM, Huang Y, Si SY, Zhang XM, Sui YF (2009) The antitumor immune responses induced by nanoemulsion-encapsulated MAGE1-HSP70/SEA complex protein vaccine following peroral administration route. Cancer Immunol Immunother 58 (2):201–208. doi:10.1007/s00262-008-0539-9.

Guengerich FP (1999) Cytochrome P-450 3A4: regulation and role in drug metabolism. Annu Rev Pharmacol Toxicol 39:1–17. doi:10.1146/annurev.pharmtox.39.1.1

Guerquin-Kern JL, Volk A, Chenu E, Lougerstay-Madec R, Monneret C, Florent JC, Carrez D, Croisy A (2000) Direct in vivo observation of 5-fluorouracil release from a prodrug in human tumors heterotransplanted in nude mice: a magnetic resonance study. NMR Biomed 13 (5):306-310. doi:10.1002/1099-1492(200008)13:5<306::AID-NBM639>3.0.CO;2-P [pii]

Guo X, Lerner-Tung M, Chen HX, Chang CN, Zhu JL, Chang CP, Pizzorno G, Lin TS, Cheng YC (1995) 5-Fluoro-2-pyrimidinone, a liver aldehyde oxidase-activated prodrug of 5-fluorouracil. Biochem Pharmacol 49 (8):1111–1116. doi:0006-2952(95)98508-7 [pii]

Higgins JP, Bernstein MB, Hodge JW (2009) Enhancing immune responses to tumor-associated antigens. Cancer Biol Ther 8 (15):1440–1449. doi:9133 [pii]

Hunter J, Hirst BH (1997) Intestinal secretion of drugs. The role of P-glycoprotein and related drug efflux systems in limiting oral drug absorption. . Adv Drug Deliv Rev 25 (2-3):129–157. doi:10.1016/S0169-409X(97)00497-3

Hunter J, Jepson MA, Tsuruo T, Simmons NL, Hirst BH (1993) Functional expression of P-glycoprotein in apical membranes of human intestinal Caco-2 cells. Kinetics of vinblastine secretion and interaction with modulators. J Biol Chem 268 (20):14991–14997

Hussain N, Jaitley V, Florence AT (2001) Recent advances in the understanding of uptake of microparticulates across the gastrointestinal lymphatics. Adv Drug Deliv Rev 50 (1–2):107–142. doi:S0169-409X(01)00152-1 [pii]

Hyafil F, Vergely C, Du Vignaud P, Grand-Perret T (1993) In vitro and in vivo reversal of multidrug resistance by GF120918, an acridonecarboxamide derivative. Cancer Res 53 (19):4595–4602

Irache JM, Salman HH, Gomez S, Espuelas S, Gamazo C (2010) Poly(anhydride) nanoparticles as adjuvants for mucosal vaccination. Front Biosci (Schol Ed) 2:876–890. doi:108 [pii]

IRSHAD S (2010) Considerations when choosing oral chemotherapy: identifying and responding to patient need. European Journal of Cancer Care 19:5–11. doi:10.1111/j.1365-2354.2010.01199.x

Ishikawa M, Yoshii H, Furuta T (2005) Interaction of modified cyclodextrins with cytochrome P-450. Biosci Biotechnol Biochem 69 (1):246–248. doi:JST.JSTAGE/bbb/69.246 [pii]

Jepson MA, Clark MA, Hirst BH (2004) M cell targeting by lectins: a strategy for mucosal vaccination and drug delivery. Adv Drug Deliv Rev 56 (4):511–525. doi:10.1016/j.addr.2003.10.018 S0169409X03002333 [pii]

Kalinski P (2009) Dendritic cells in immunotherapy of established cancer: Roles of signals 1, 2, 3 and 4. Curr Opin Investig Drugs 10 (6):526–535

Kaukonen AM, Boyd BJ, Porter CJ, Charman WN (2004) Drug solubilization behavior during in vitro digestion of simple triglyceride lipid solution formulations. Pharm Res 21 (2):245–253

Kawata S, Minami Y, Tarui S, Marunaka T, Okamoto M, Yamano T (1984) Cytochrome P-450-dependent oxidative cleavage of 1-(tetrahydro-2-furanyl)-5-fluorouracil to 5-fluorouracil. Jpn J Pharmacol 36 (1):43–49

Kivisto KT, Niemi M, Fromm MF (2004) Functional interaction of intestinal CYP3A4 and P-glycoprotein. Fundam Clin Pharmacol 18 (6):621–626. doi:FCP291 [pii] 10.1111/j.1472-8206.2004.00291.x

Krishnamachari Y, Geary SM, Lemke CD, Salem AK (2010) Nanoparticle Delivery Systems in Cancer Vaccines. Pharm Res. doi:10.1007/s11095-010-0241-4

Kruijtzer CM, Beijnen JH, Schellens JH (2002) Improvement of oral drug treatment by temporary inhibition of drug transporters and/or cytochrome P450 in the gastrointestinal tract and liver: an overview. Oncologist 7 (6):516–530

Kuppens IE, Breedveld P, Beijnen JH, Schellens JH (2005) Modulation of oral drug bioavailability: from preclinical mechanism to therapeutic application. Cancer Invest 23 (5):443–464

Lasaro MO, Ertl HC (2010) Targeting inhibitory pathways in cancer immunotherapy. Curr Opin Immunol 22 (3):385–390. doi:S0952-7915(10)00069-5 [pii] 10.1016/j.coi.2010.04.005

Lin JH, Lu AY (1998) Inhibition and induction of cytochrome P450 and the clinical implications. Clin Pharmacokinet 35 (5):361–390

Liu G, Franssen E, Fitch M, Warner E (1997) Patient preference for oral versus intravenous palliative chemotherapy. J Clin Oncol 24 (15):110–115

Lo YL (2003) Relationships between the hydrophilic-lipophilic balance values of pharmaceutical excipients and their multidrug resistance modulating effect in Caco-2 cells and rat intestines. J Control Release 90 (1):37–48. doi:S0168365903001639 [pii]

Malik B, Goyal AK, Mangal S, Zakir F, Vyas SP (2010) Implication of gut immunology in the design of oral vaccines. Curr Mol Med 10 (1):47–70. doi:CMM # 16 [pii]

Miwa M, Ura M, Nishida M, Sawada N, Ishikawa T, Mori K, Shimma N, Umeda I, Ishitsuka H (1998) Design of a novel oral fluoropyrimidine carbamate, capecitabine, which generates 5-fluorouracil selectively in tumours by enzymes concentrated in human liver and cancer tissue. Eur J Cancer 34 (8):1274–1281. doi:S0959804998000586 [pii]

Mocellin S, Nitti D (2008) Therapeutics targeting tumor immune escape: towards the development of new generation anticancer vaccines. Med Res Rev 28 (3):413–444. doi:10.1002/med.20110

Mohsin K, Long MA, Pouton CW (2009) Design of lipid-based formulations for oral administration of poorly water-soluble drugs: precipitation of drug after dispersion of formulations in aqueous solution. J Pharm Sci 98 (10):3582–3595. doi:10.1002/jps.21659

Mowat AM (2003) Anatomical basis of tolerance and immunity to intestinal antigens. Nat Rev Immunol 3 (4):331–341. doi:10.1038/nri1057 nri1057 [pii]

Murray GI, Melvin WT, Burke MD (1995) Cytochrome P450 expression in tumours. J Pathol 176 (3):323–324

Nazzal S, Smalyukh, II, Lavrentovich OD, Khan MA (2002) Preparation and in vitro characterization of a eutectic based semisolid self-nanoemulsified drug delivery system (SNEDDS) of ubiquinone: mechanism and progress of emulsion formation. Int J Pharm 235 (1–2):247–265. doi:S0378517302000030 [pii]

Neudeck BL, Loeb JM, Faith NG, Czuprynski CJ (2004) Intestinal P glycoprotein acts as a natural defense mechanism against Listeria monocytogenes. Infect Immun 72 (7):3849–3854. doi:10.1128/IAI.72.7.3849-3854.2004 72/7/3849 [pii]

Ohashi Y, Watanabe T, Sano M, Koyama H, Inaji H, Suzuki T (2010) Efficacy of oral tegafur-uracil (UFT) as adjuvant therapy as compared with classical cyclophosphamide, methotrexate, and 5-fluorouracil (CMF) in early breast cancer: a pooled analysis of two randomized controlled trials (N.SAS-BC 01 trial and CUBC trial). Breast Cancer Res Treat 119 (3):633–641. doi:10.1007/s10549-009-0635-3

Palm K, Luthman K, Ungell AL, Strandlund G, Artursson P (1996) Correlation of drug absorption with molecular surface properties. J Pharm Sci 85 (1):32–39. doi:10.1021/js950285r 10.1021/js950285r [pii]

Pasquier E, Kavallaris M, Andre N (2010) Metronomic chemotherapy: new rationale for new directions. Nat Rev Clin Oncol 7 (8):455–465. doi:nrclinonc.2010.82 [pii] 10.1038/nrclinonc.2010.82

Payne SA (1992) A study of quality of life in cancer patients receiving palliative chemotherapy. Soc Sci Med 35 (12):1505–1509. doi:10.1016/0277-9536(92)90053-S

Peltier S, Oger JM, Lagarce F, Couet W, Benoit JP (2006) Enhanced oral paclitaxel bioavailability after administration of paclitaxel-loaded lipid nanocapsules. Pharm Res 23 (6):1243–1250. doi:10.1007/s11095-006-0022-2

Porter CJ, Trevaskis NL, Charman WN (2007) Lipids and lipid-based formulations: optimizing the oral delivery of lipophilic drugs. Nat Rev Drug Discov 6 (3):231–248. doi:nrd2197 [pii] 10.1038/nrd2197

Primard C, Rochereau N, Luciani E, Genin C, Delair T, Paul S, Verrier B (2010) Traffic of poly(lactic acid) nanoparticulate vaccine vehicle from intestinal mucus to sub-epithelial immune competent cells. Biomaterials 31 (23):6060–6068. doi:S0142-9612(10)00504-1 [pii] 10.1016/j.biomaterials.2010.04.021

Rae CS, Khor IW, Wang Q, Destito G, Gonzalez MJ, Singh P, Thomas DM, Estrada MN, Powell E, Finn MG, Manchester M (2005) Systemic trafficking of plant virus nanoparticles in mice via the oral route. Virology 343 (2):224–235. doi:S0042-6822(05)00468-X [pii] 10.1016/j.virol.2005.08.017

Roger E, Lagarce F, Garcion E, Benoit JP (2010) Biopharmaceutical parameters to consider in order to alter the fate of nanocarriers after oral delivery. Nanomedicine (Lond) 5 (2):287–306. doi:10.2217/nnm.09.110

Roger E, Lagarce F, Garcion E, Benoit JP (2010) Reciprocal competition between lipid nanocapsules and P-gp for paclitaxel transport across Caco-2 cells. Eur J Pharm Sci 40 (5):422–429. doi:S0928-0987(10)00160-0 [pii] 10.1016/j.ejps.2010.04.015

Salman HH, Gamazo C, Campanero MA, Irache JM (2005) Salmonella-like bioadhesive nanoparticles. J Control Release 106 (1–2):1–13. doi:S0168-3659(05)00115-X [pii] 10.1016/j.jconrel.2005.03.033

Schellens JH, Malingre MM, Kruijtzer CM, Bardelmeijer HA, van Tellingen O, Schinkel AH, Beijnen JH (2000) Modulation of oral bioavailability of anticancer drugs: from mouse to man. Eur J Pharm Sci 12 (2):103–110. doi:S0928-0987(00)00153-6 [pii]

Singh KK, Vingkar SK (2008) Formulation, antimalarial activity and biodistribution of oral lipid nanoemulsion of primaquine. Int J Pharm 347 (1–2):136–143. doi:S0378-5173(07)00552-2 [pii] 10.1016/j.ijpharm.2007.06.035

Singh Y, Palombo M, Sinko PJ (2008) Recent trends in targeted anticancer prodrug and conjugate design. Curr Med Chem 15 (18):1802–1826

Slutter B, Plapied L, Fievez V, Sande MA, des Rieux A, Schneider YJ, Van Riet E, Jiskoot W, Preat V (2009) Mechanistic study of the adjuvant effect of biodegradable nanoparticles in mucosal vaccination. J Control Release 138 (2):113–121. doi:S0168-3659(09)00299-5 [pii] 10.1016/j.jconrel.2009.05.011

Sparreboom A, van Asperen J, Mayer U, Schinkel AH, Smit JW, Meijer DK, Borst P, Nooijen WJ, Beijnen JH, van Tellingen O (1997) Limited oral bioavailability and active epithelial excretion of paclitaxel (Taxol) caused by P-glycoprotein in the intestine. Proc Natl Acad Sci U S A 94 (5):2031–2035

Stremmel W (1988) Uptake of fatty acids by jejunal mucosal cells is mediated by a fatty acid binding membrane protein. J Clin Invest 82 (6):2001–2010. doi:10.1172/JCI113820

Sugano K, Kansy M, Artursson P, Avdeef A, Bendels S, Di L, Ecker GF, Faller B, Fischer H, Gerebtzoff G, Lennernaes H, Senner F (2010) Coexistence of passive and carrier-mediated processes in drug transport. Nat Rev Drug Discov 9 (8):597–614. doi:nrd3187 [pii] 10.1038/nrd3187

Sun B, Ranganathan B, Feng SS (2008) Multifunctional poly(D,L-lactide-co-glycolide)/montmorillonite (PLGA/MMT) nanoparticles decorated by Trastuzumab for targeted chemotherapy of breast cancer. Biomaterials 29 (4):475–486. doi:S0142-9612(07)00790-9 [pii] 10.1016/j.biomaterials.2007.09.038

Takiuchi H, Ajani JA (1998) Uracil-tegafur in gastric carcinoma: a comprehensive review. J Clin Oncol 16 (8):2877–2885

Terwogt JM, Schellens JH, Huinink WW, Beijnen JH (1999) Clinical pharmacology of anticancer agents in relation to formulations and administration routes. Cancer Treat Rev 25 (2):83–101. doi:10.1053/ctrv.1998.0107 S0305-7372(98)90107–4 [pii]

Thorne SH, Barak Y, Liang W, Bachmann MH, Rao J, Contag CH, Matin A (2009) CNOB/ChrR6, a new prodrug enzyme cancer chemotherapy. Mol Cancer Ther 8 (2):333–341. doi:1535-7163.MCT-08-0707 [pii] 10.1158/1535-7163.MCT-08-0707

Thummel KE (2007) Gut instincts: CYP3A4 and intestinal drug metabolism. J Clin Invest 117 (11):3173–3176. doi:10.1172/JCI34007

Trevaskis NL, Charman WN, Porter CJ (2008) Lipid-based delivery systems and intestinal lymphatic drug transport: a mechanistic update. Adv Drug Deliv Rev 60 (6):702–716. doi:S0169-409X(07)00315-8 [pii] 10.1016/j.addr.2007.09.007

van Herwaarden AE, Wagenaar E, van der Kruijssen CM, van Waterschoot RA, Smit JW, Song JY, van der Valk MA, van Tellingen O, van der Hoorn JW, Rosing H, Beijnen JH, Schinkel AH (2007) Knockout of cytochrome P450 3A yields new mouse models for understanding xenobiotic metabolism. J Clin Invest 117 (11):3583–3592. doi:10.1172/JCI33435

van Waterschoot RA, Rooswinkel RW, Wagenaar E, van der Kruijssen CM, van Herwaarden AE, Schinkel AH (2009) Intestinal cytochrome P450 3A plays an important role in the regulation of detoxifying systems in the liver. FASEB J 23 (1):224–231. doi:fj.08-114876 [pii] 10.1096/fj.08-114876

Varma MV, Perumal OP, Panchagnula R (2006) Functional role of P-glycoprotein in limiting peroral drug absorption: optimizing drug delivery. Curr Opin Chem Biol 10 (4):367–373. doi:S1367-5931(06)00084-6 [pii] 10.1016/j.cbpa.2006.06.015

Wacher VJ, Salphati L, Benet LZ (2001) Active secretion and enterocytic drug metabolism barriers to drug absorption. Adv Drug Deliv Rev 46 (1–3):89–102. doi:S0169-409X(00)00126-5 [pii]

Watkins PB (1997) The barrier function of CYP3A4 and P-glycoprotein in the small bowel. Adv Drug Deliv Rev 27 (2–3):161–170. doi:S0169-409X(97)00041-0 [pii]

Weingart SN, Brown E, Bach PB, Eng K, Johnson SA, Kuzel TM, Langbaum TS, Leedy RD, Muller RJ, Newcomer LN, O'Brien S, Reinke D, Rubino M, Saltz L, Walters RS (2008) NCCN Task Force Report: Oral chemotherapy. J Natl Compr Canc Netw 6 Suppl 3:S1–14

Woo JS, Lee CH, Shim CK, Hwang SJ (2003) Enhanced oral bioavailability of paclitaxel by coadministration of the P-glycoprotein inhibitor KR30031. Pharm Res 20 (1):24–30

Xiang SD, Scalzo-Inguanti K, Minigo G, Park A, Hardy CL, Plebanski M (2008) Promising particle-based vaccines in cancer therapy. Expert Rev Vaccines 7 (7):1103–1119. doi:10.1586/14760584.7.7.1103

Yoncheva K, Lizarraga E, Irache JM (2005) Pegylated nanoparticles based on poly(methyl vinyl ether-co-maleic anhydride): preparation and evaluation of their bioadhesive properties. Eur J Pharm Sci 24 (5):411–419. doi:S0928-0987(04)00318-5 [pii] 10.1016/j.ejps.2004.12.002

Zhang Y, Benet LZ (2001) The gut as a barrier to drug absorption: combined role of cytochrome P450 3A and P-glycoprotein. Clin Pharmacokinet 40 (3):159–168

Zhu X, Cai J, Huang J, Jiang X, Ren D (2010) The treatment and prevention of mouse melanoma with an oral DNA vaccine carried by attenuated Salmonella typhimurium. J Immunother 33 (5):453–460. doi:10.1097/CJI.0b013e3181cf23a6

Nanoparticles for Photodynamic Therapy Applications

Régis Vanderesse, Céline Frochot, Muriel Barberi-Heyob, Sébastien Richeter, Laurence Raehm, and Jean-Olivier Durand

Abstract Photodynamic therapy has emerged as an alternative to chemo- and radiotherapy for the treatment of certain types of cancer. Nanoparticles have been used to improve the delivery and the efficiency of the photosensitizer as they allow its encapsulation without loss of activity. Using targeting strategies, they can also allow the selective accumulation of the photosensitizer in the cancer tissues. In this review, based on the chemical nature of the nanoparticles, the different methods for their syntheses are described from the pioneering works to the latest achievements. Indeed the nanosystems can be conjugated to a biomolecule or an antibody to target receptors over-expressed in cancer cells and/or angiogenic vascular endothelial cells. Numerous *in vivo* and *in vitro* applications have been described. Multifunctional nanoplatforms combining several applications within the same nano-object emerge as potential important theranostic tools.

Keywords Nanoparticles • Photosensitizers • Photodynamic therapy • Reactive oxygen species • Nanomedicine

---

R. Vanderesse
Laboratoire de Chimie-Physique Macromoléculaire, UMR CNRS-INPL 7568, Nancy Université, 1 rue Grandville, BP20451, 54001 Nancy Cedex, France

C. Frochot
Laboratoire Réactions et Génie des Procédés, UPR 3349, Nancy-Université, 1 rue Grandville, BP20451, 54001 Nancy Cedex, France

M. Barberi-Heyob
Centre de Recherche en Automatique de Nancy (CRAN), UMR 7039, Nancy-Université, CNRS, Centre Alexis Vautrin Brabois, Avenue de Bourgogne, 54511 Vandoeuvre-les-Nancy Cedex, France

S. Richeter, L. Raehm (✉), and J.-O. Durand
Equipe Chimie Moléculaire et Organisation du Solide, Institut Charles Gerhardt Montpellier, UMR 5253 CNRS-UM2-ENSCM-UM1, CC1701 Place Eugène Bataillon, 34095, Montpellier Cedex 05, France
e-mail: lraehm@univ-montp2.fr

A. Prokop (ed.), *Intracellular Delivery: Fundamentals and Applications*, Fundamental Biomedical Technologies 5, DOI 10.1007/978-94-007-1248-5_19,

List of Abbreviations

AA	Ascorbic acid
ABDA	9,10-Anthracenediyl-bis(methylene)dimalonic acid
ABMD	Disodium 9,10-anthracenediyl-bis-(methylene)dimalonic acid
ADPA	Anthracene-9,10-dipropionic acid
APTS	3-(Aminopropyl)triethoxysilane
ALA	5-Aminolevulinic acid
AOT	Sodium bis-(2-ethylhexyl) sulfosuccinate
AuNPs	Gold nanoparticles
BDSA	9,10-bis (4′-(4″-Aminostyryl)styryl)anthracene
CLSM	Confocal laser scanning microscopy
CMC	Carboxymethylcellulose
DID	Dioctadecyl tetramethyl indodicarbocyanine chloro benzene
DMSO	Dimethylsulfoxide
DNA	Deoxyribonucleic acid
DPBF	1,3 Diphenylisobenzofuran
DXP	(N,N′)-bis(2-6-dimethylphenyl)perylene-3,4,9,10-tetracarbodiimide
EPR	Enhanced permeability and retention
FDA	Food and Drug Administration
FITC	Fluoresceine isothiocyanate
FRET	Förster Resonance Energy Transfer
FTIR	Fourier transform infrared spectroscopy
GM	Göppert-Mayer
HB	Hypocrellin B
HP	Hematoporphyrin
HPPH	2-Devinyl-2-(1-hexyloxyethyl)pyropheophorbide
IP	Iodobenzylpyropheophorbide
IPS	Iodobenzyl-pyro-silane
IR	Infrared
MB	Methylene blue
MPA	Mercaptopropionic acid
MRSA	Methicillin-resistant *Staphylococcus aureus*
MSN	Mesoporous silica nanoparticles
MRI	Magnetic resonance imaging
MTAP	*meso*-Tetra (o-amino phenyl) porphyrin
m-THPC	*meso* Meta-tetra(hydroxyphenyl)chlorin
mTHPP	*meso*-Tetrahydroxyphenyl porphyrin
MTT	(3-(4,5-Dimethylthiazol-2-yl)-2,5-diphenyltetrazolium bromide
MWNTs	Multiwalled carbon nanotubes
NIR	Near Infrared
NPs	Nanoparticles
ORMOSIL	Organically modified silica
PBS	Phosphate buffered saline
PC	Phthalocyaninedihydroxyle

Pc	Phthalocyanine
Pc 4	Silicon phthalocyanine 4
Pc 19	Silicon phthalocyanine 219
PDT	Photodynamic therapy
PEG	Polyethylene glycol
PFHA	Perfluoroheptanoic acid
PpIX	Protoporphyrin IX
PS	Photosensitizer
PdTPP	Pd-*meso*-Tetra(4-carboxyphenyl) porphyrin
PHPP	2,7,12,18-Tetramethyl-3,8-di(1-propoxyethyl)-13,17-bis-(3-hydroxypropyl) porphyrin
PSS	Polystyrene sulfonic acid
RB	Rose Bengal
RNA	Ribonucleic acid
ROS	Reactive Oxygen Species
SWNTs	Single walled carbon nanotubes
TPP	Meso-tetraphenyl porphyrin
RNO	*N, N*-dimethyl-4-nitrosoaniline
SEM	Scanning Electron Microscopy
TEM	Transmission electron microscopy
TPA	Two-photon absorption
TEOS	Tetraethoxysilane
UV-Vis	Ultraviolet-visible
XRD	X-ray diffraction
ZnPc	Zinc(II) phthalocyanine

1 Introduction

Over the last few years, photodynamic therapy (PDT) has emerged as an alternative to chemo- and radiotherapy for the treatment of various diseases including cancer (Maisch 2009; Robertson et al. 2009). It involves the use of light and photosensitizers (PS) that accumulate in the tumor tissue. Photosensitizer are drugs that can transfer their energy from their triplet excited state to neighboring oxygen molecules (Ortel et al. 2009) when activated by light of a specific wavelength. Singlet oxygen ($^1O_2$) and other cytotoxic Reactive Oxygen Species (ROS) are formed and lead to the destruction of cancer cells by both apoptosis and necrosis. The efficiency of PDT can be related to the formation of ROS and it is commonly accepted that $^1O_2$ is the main cytotoxic species that destroys tumour cells. PS described so far in the literature present several disadvantages. Mainly, they are hydrophobic or have a limited solubility in water and therefore aggregate in aqueous media such as the blood compartment which leads to the modification of their photophysical properties and particularly to the decrease of $^1O_2$ quantum yield. Moreover, in order to avoid destruction of healthy cells, PS are required to accumulate selectively in tumor cells. Even if many efforts are focused on developing third generation PS with a covalently linked vector to target receptors over-expressed in cancer cells, very few have been evaluated for clinical applications mainly because their *in vivo* selectivity was not high enough (Taquet et al. 2007). In order to address these issues, PS have been encapsulated in nanoparticles (NPs). Indeed, encapsulation should lead to the administration of the PS in monomeric form and without loss of the photophysical properties. More importantly, nanoparticles can lead to selective accumulation of the PS in cancer tissue due to the enhanced permeability and retention (EPR) effect of tumor tissues (Hoffman 2008). Therefore, nanoparticles have been suggested to improve the efficiency of PDT. Liposomes, micelles, polymer-based nanoparticles and dendrimers which can swell or change with the conditions (e.g. temperature, pH) will not be described here: they have been reviewed recently (Allison et al. 2008; Bechet et al. 2008; Chatterjee et al. 2008). We will focus on nanoparticles possessing a three dimensional rigid matrix such as metallic, metal-oxide, semi-conductor-based nanoparticles or carbon nanotubes. Metallic nanoparticles (gold) have been used as PDT drug carriers, and photothermal therapy, leading to a synergetic effect. Metal oxide nanoparticles (Fe_3O_4) are promising vectors for theranostics (Magnetic Resonance Imaging, MRI, and PDT). ZnO and TiO_2 nanoparticles are able to generate ROS when irradiated with energetic UV-blue light. Alternatively, the use of a photosensitizer with those nanoparticles allows irradiation with red light which penetrates deeper inside tissues and creates less photo-damages. A variety of precursors and methods are available for the syntheses of silica-based nanoparticles, allowing flexibility and thus numerous PDT drugs to be encapsulated. Moreover, particles size, shape, porosity and mono-dispersibility can be easily controlled during their preparation (Wang et al. 2004). Semi-conductors-based nanoparticles (quantum dots QD, silicon nanoparticles) can generate ROS by direct photo-irradiation. The formation of ROS was also demonstrated by conjugation of QD with photosensitizers through Förster Resonance Energy Transfer (FRET) mechanism. Carbon nanotubes have been shown to be good delivery agents of PDT drugs by electrostatic or covalent attachment of the photosensitizer with the nanotube. All these nanosystems could be conjugated to a biomolecule or an antibody to target receptors over-expressed in cancer cells, which enhance the selectivity of the treatment. The very recent developments of nanosystems concern up-conversion or two-photon PDT which opens new promising ways to treat deeper tissue regions (Kim et al. 2007; Velusamy et al. 2009). Most of the studies with nanosystems have been carried out *in vitro*, but *in vivo* PDT is currently being investigated (He et al. 2009). Another field which grows rapidly consists in combining several applications within one nano-object, such as MRI, fluorescence imaging and PDT (Tada et al. 2007; Lai et al. 2008; Liu et al. 2008a). Such multifunctional nanoparticles emerge as important theranostic nano-objects. For clinical applications of the present nanosystems further important studies are required: toxicology, biocompatibility, biodegradabililty, pharmacokinetics, biodistribution, clearance from the body, dosimetry. These studies which are under way will open new perspectives for the future of nanomedicine.

2 Gold Nanoparticles

In the majority of experimental approaches, photosensitizers have a limited solubility in water. Thus, these molecules require a vehicle to deliver the drug to the tumor tissue, and gold nanoparticles (AuNPs) can play this role. One major advantage of AuNPs is their chemical inertness and their biocompatibility. The functionalization of the surface of the AuNPs with the PS requires the existence of interactions between the two components. These interactions can be covalent bonds or weak interactions such as van der Walls interactions and electrostatic interactions.

2.1 Phthalocyanines Gold Nanoparticles Conjugates

Phthalocyanine (Pc) derivatives are tetrapyrrolic macrocycles which can be used for PDT. These molecules upon irradiation can transfer their excited state energy to the ground state of molecular oxygen to produce singlet oxygen species which are highly cytotoxic. Hone et al. synthesized small AuNPs (2–4 nm diameter) stabilized with a zinc(II) phthalocyanine derivative bearing a mercaptoundecyl group attached to the surface of the nanoparticle through the formation of stable S–Au bonds (Hone et al. 2002) (Fig. 1). AuNP-phthalocyanine conjugates are able to generate singlet oxygen upon irradiation with enhanced quantum yields compared to the free phthalocyanine. Camerin et al. also compared the *in vivo* efficacy of the C11Pc molecule and the AuNP-C11Pc conjugates for the photodynamic therapy of amelanotic melanoma cells (Camerin et al. 2010) (Fig. 2). C11Pc molecules and the AuNP-C11Pc conjugates are not water-soluble and dispersed in a Cremophor EL emulsion before injection. Biodistribution studies performed on C57 mice bearing a subcutaneously transplanted amelanotic melanoma showed that cancer tissues are more selectively targeted by the AuNP-C11Pc conjugates compared to the free

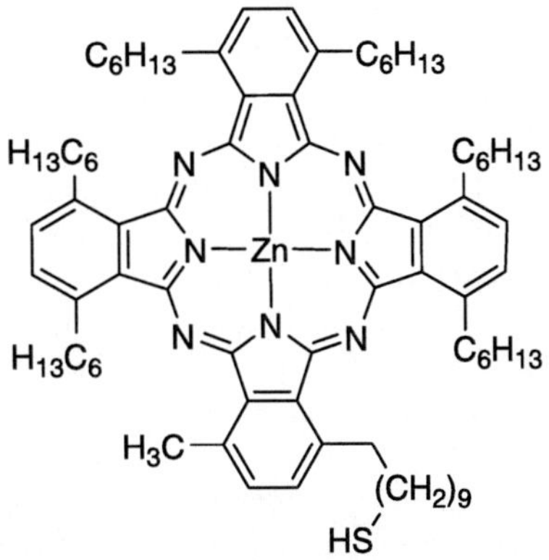

Fig. 1 1,4,8,11,15,18-Hexaethyl-22-methyl-25-(11-mercaptoundecyl) phthalocyaninato zinc

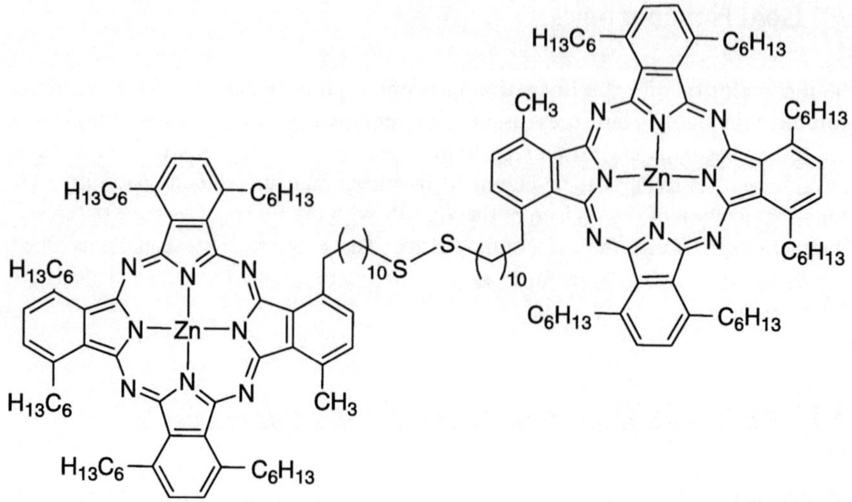

Fig. 2 C11Pc photosensitizer

C11Pc, and that the administration of C11Pc *via* AuNPs is beneficial for a longer blood circulation and a larger accumulation in most organs. As vascular damages were identified by electron microscopy, an anti-angiogenic action promoted by the AuNP-C11Pc conjugates was suggested to explain the efficacy of PDT.

Water-soluble AuNPs have been obtained by stabilization with polyethylene glycol (PEG). PEG molecules are also able to shield the AuNP surface against biomolecules. PEGylated AuNPs can preferentially accumulate in tumor tissues by passive accumulation through the leaky vascular system of tumors. Cheng et al. have developed the synthesis of ~5 nm diameter PEGylated AuNPs (Cheng et al. 2008) and studied their use as nanocarrier for the silicon phthalocyanine 4 (*Pc* 4), which is a nontoxic, hydrophobic photosensitizer approved by the FDA for clinical trials (Fig. 3). They showed that 5 nm diameter PEGylated AuNPs can be loaded with ~30 molecules of *Pc* 4 through N–Au bonds. Although N–Au bonds are not very stable (6 kcal/mol), the AuNP-*Pc* 4 conjugates appeared to be stable for several months in water, thanks to the PEG molecules at the surface which stabilized the *Pc* 4 molecules through van der Walls interactions and protected the drug from the aqueous media. Drug release experiments in a water-toluene two-phases system showed that *Pc* 4 molecules are released within 2 h, and thus can potentially accumulate in apolar cellular sites such as cell membranes. Pegylated AuNP did not transfer into the organic phase. Cell viability experiments (HeLa cells) was greater than 95% after incubation of the AuNP-*Pc* 4 conjugates (1 μM) in the dark for 24 h, and irradiation at λ_{exc} >500 nm induced photokilling of the cells with >90% efficiency. Uptake kinetics studies into HeLa cells showed that the pegylated AuNP-*Pc* 4 conjugates were taken up within the first 2 h. Only a few AuNPs were taken up into the cells. *Pc* 4 molecules were mainly localized in mitochondria and

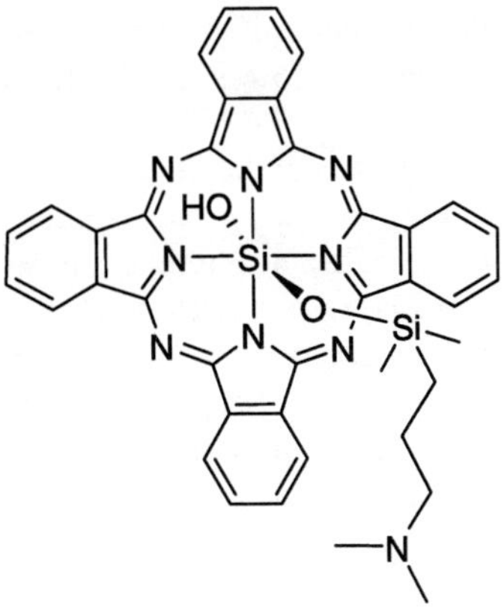

Fig. 3 Encapsulated silicon phthalocyanine 4 (*Pc* 4)

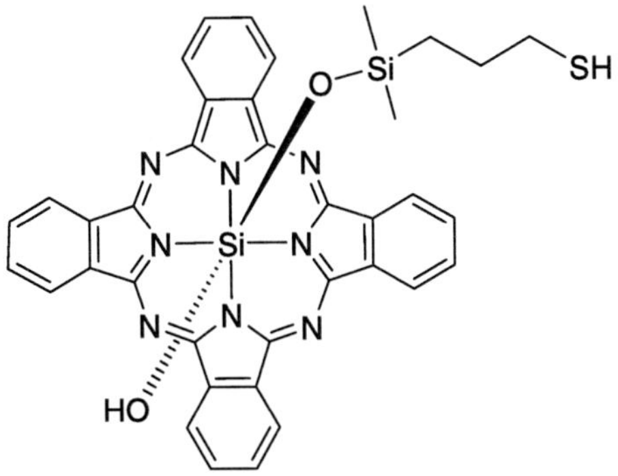

Fig. 4 Silicon phthalocyanine 219 (*Pc* 219)

cytoplasm. Knowing that few AuNPs were taken up into the cells, this implied that most *Pc* 4 molecules were released from the NPs before cellular uptake and were delivered in the cells without the NPs at the cell membrane surface. *In vivo* experiments on nude mice showed that the AuNP-*Pc* 4 conjugates were accumulated within 2 h at the tumor site. *Pc* 4 molecules were also observed in the lung and the kidneys (observation usually made with phthalocyanine photosensitizers). The PEGylated AuNP nanocarriers accelerated the *Pc* 4 administration by two orders of magnitude without apparent side effects and modification of the distribution pattern of *Pc* 4 (similar to that of *Pc* 4 free), and the effectiveness of the PDT treatment can be deduced by observing a decrease of the size of the tumor which became necrotic. Cheng et al. also showed that the delivery of the drug, and thus the PDT efficiency are strongly affected by the nature of the bond between the photosensitizer and the AuNPs (Cheng et al. 2010). Here, AuNPs act as delivery vehicles for loosely bound *Pc* 4 molecules which are then released in cancer cells. They carried out the same kind of experiment with *Pc* 219 which is the thiolated analogue of *Pc* 4. (Fig. 4) The S–Au bonds (47 kcal/mol) are much more stable compared to the N–Au bonds, and thus the released *Pc* 219 molecules should be very low compared to the release of

Pc 4 molecules. This was confirmed by drug release experiments in a water-toluene two-phases system which showed that the release of *Pc* 219 was >> 10 times slower and very few *Pc* 219 molecules were detected in HeLa cells. This is due to the low dissociation of the *Pc* 219 molecules from the AuNPs. Moreover, the AuNP-*Pc* 219 conjugates did not show appreciable phototoxicity. Compared to the system with *Pc* 4 molecules, a different cellular uptake pathway and a different localization were observed with AuNP-*Pc* 219 conjugates. Thiolated *Pc* 219 molecules are located in cell vesicles formed by the uptake of the AuNP-*Pc* 219 conjugates. Moreover, the AuNP-*Pc* 219 conjugates did not show appreciable phototoxicity. The absence of PDT effect observed with these AuNP-*Pc* 219 conjugated was explained by the formation of these vesicles containing the AuNP-*Pc* 219 conjugates. These results highlighted the importance of the nature of interactions between the nanocarriers and the drug to the delivery and efficacy of a cancer drug.

2.2 *Purpurin-18 and 5-Aminolevulinic Acid Gold Nanoparticles Conjugates*

Purpurin-18 is a chlorin molecule, i.e. a porphyrin derivative in which one pyrrole is reduced, extracted from the green algae *Spirulina maxima* and possessing a carboxylic acid function. By treating this molecule with choline hydroxide, Demberelnyamba et al. obtained the cholinium-purpurin-18-carboxylate salt (Demberelnyamba et al. 2008) (Fig. 5). This water-soluble photosensitizer was coupled to synthesize AuNPs without adding any particular reducing agents and CTAB. Here, the purpurin acted both as a reducing and a protective agent. These systems are also promising candidates for further PDT applications. Thus, weak interactions such as electrostatic interactions can be used to obtain PS-AuNPs conjugates. An interesting example has been reported by Oo et al. (2008). In PDT, an alternative to the administration of exogenous photosensitizers is to stimulate the cellular synthesis of endogenous ones. 5-Aminolaevulinic acid (5-ALA, Fig. 6) is a metabolic precursor in the biosynthesis of haem and protoporphyrin IX PpIX

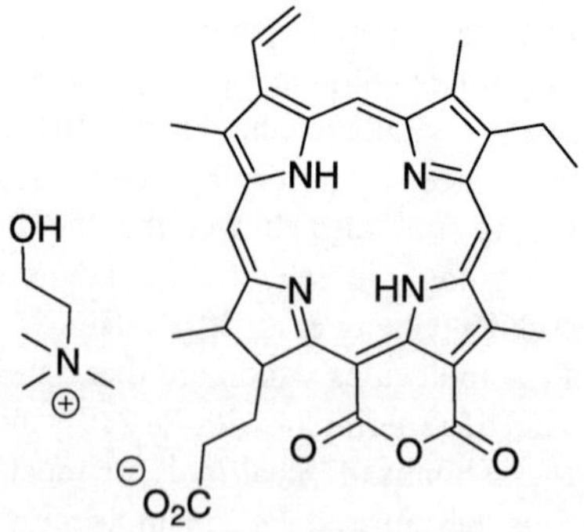

Fig. 5 Cholinium-purpurin 18 (Chol-Pu-18)

Fig. 6 5-Aminolevulinic acid (5-ALA) and its corresponding zwitterionic form

Fig. 7 Protoporphyrin IX (Pp IX)

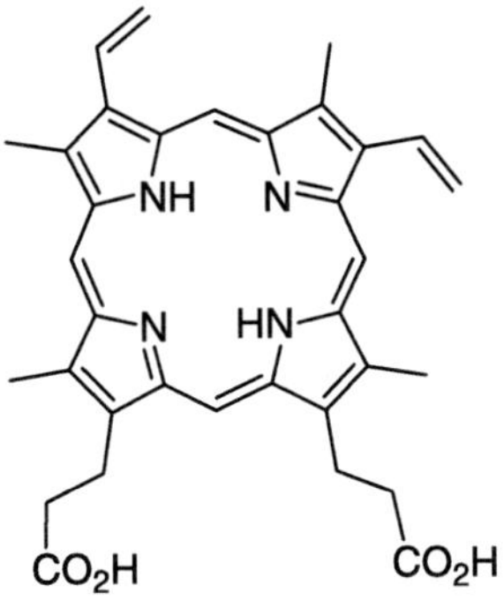

(Fig. 7) is the immediate precursor to haem, and thus PpIX can be considered as a natural photosensitizer. 5-Aminolevulinic and Photofrin® are the most widely used PS in clinical practice. However, its zwitterionic nature at physiological pH and its hydrophilicity limit its penetration through cell membranes. Oo et al. used AuNps as nano-carriers for target delivery of 5-ALA to cancer cells. 5-ALA molecules were immobilized at the surface of AuNPs through electrostatic interactions between the carboxylate group of 5-ALA and positively charged surface of AuNPs. Incubation experiments of 5-ALA-AuNPs conjugates with normal cells (fibroblasts) or cancer cells (fibrosarcoma cells) showed their biocompatibility with negligible cytotoxic effects. A greater uptake was observed for the cancer cells (more than threefold) and maximum accumulation of PpIX was observed 24 h after the addition of 5-ALA-AuNPs conjugates (fivefold higher in cancer cells). ROS production was investigated after 4 h of incubation and 1 min of light irradiation. ROS production was twofold higher in cancer cells treated with of 5-ALA-AuNPs conjugates and the survival percentage of cancer cells decreased of about 30% *versus* 60% with 5-ALA alone. Moreover, co-culture experiments using both cell lines showed that interestingly cancer cells were selectively destroyed. The higher ROS production in the presence of 5-ALA-AuNPs conjugates may be explained by the surface plasmonic effect of AuNPs and the enhancement of the photocurrent between the AuNPs and the PpIX after light irradiation leading to the energy transfer to PpIX molecules.

Photothermia therapy can be used to increase the effectiveness of other cancer treatments. In particular, it can increase the cellular uptake of oxygen molecules which is essential for PDT. Synthesis, characterization, cellular uptake and photodynamic activity of gold nanorods coupled to a photosensitizer have been described by Kuo et al. (2009, 2010). They coated gold nanorods with a cetyltrimethylammonium

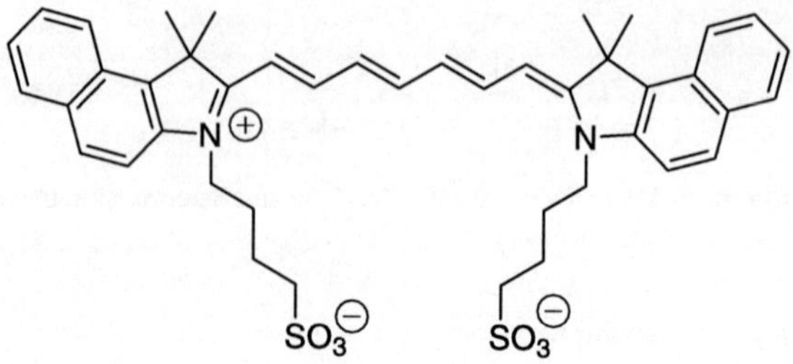

Fig. 8 Indocyanine green

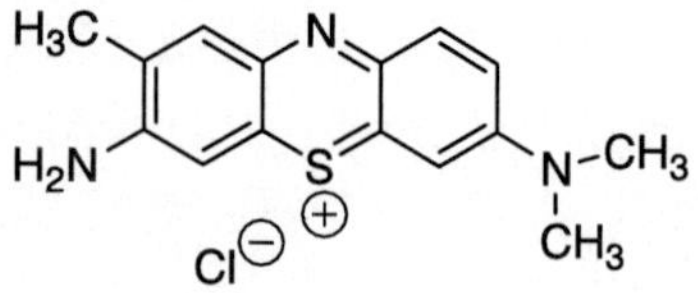

Fig. 9 Toluidine blue O

bromide surfactant coating and conjugated indocyanine green (Fig. 8) or toluidine blue O (Fig. 9) on the surface of the Au *via* respectively PSMA (poly(styrene-*alt*-maleic acid)) or PAA (polyacrylic acid). In both cases, the authors showed that the nanorods have been successfully prepared and can be used as photodynamic and hyperthermia agents with improved photodestruction efficacy compared to PDT or hyperthermia alone.

3 Silica Nanoparticles

3.1 Non Covalent Encapsulation of PS in Silica Nanoparticles

3.1.1 Silica Nanoparticles for PDT

Following the pioneering work of Prasad and Kopelman groups (Roy et al. 2003; Yan and Kopelman 2003), several researchers have tried to encapsulate PS inside silica nanoparticles in order to vectorize them in cancer cells. In 2003, the precursor work of Kopelman lead to the encapsulation of *meso* meta-tetra(hydroxyphenyl) chlorin (*m*-THPC, Fig. 10), or Foscan®, a medicinal product used clinically, into silica nanoparticles (Yan and Kopelman 2003).

To avoid the leaching of the PS from the silica matrix, the synthesis was performed by first using aminopropyltriethoxysilane as the reagent, then by initiating hydrolysis polycondensation of tetramethylorthosilicate. Increased amount of hydrogen bonds between PS and amino groups were probably involved to explain the higher efficiency of the encapsulation process than the Stöber procedure. An average nanoparticle diameter of 180 nm was calculated by SEM. A good spectral (UV-vis) correspondance between free and embedded *m*-THPC was observed. To

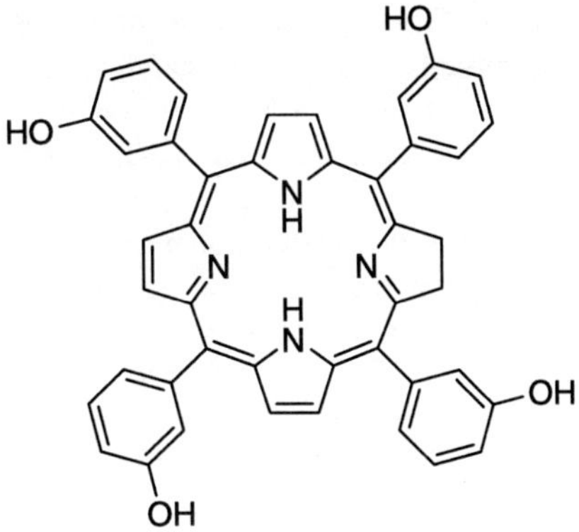

Fig. 10 *meso* Meta-tetra(hydroxyphenyl)chlorin (*m*-THPC) (Reproduced by permission of The Royal Society of Chemistry)

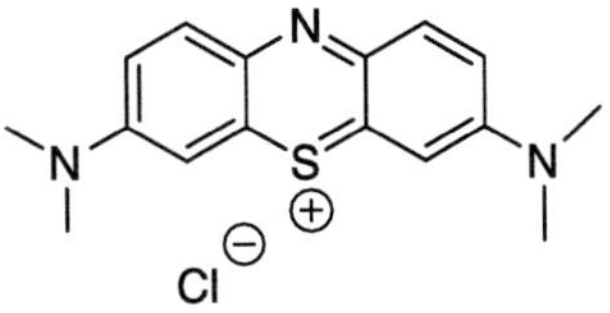

Fig. 11 Methylene blue (MB) (Reproduced by permission of The Royal Society of Chemistry)

characterize $^1O_2$ formation, anthracene-9,10-dipropionic acid (ADPA) was used as it is easily converted to an endoperoxide and the fluorescence of ADPA was followed *vs* the time of irradiation. The rate constant of the reaction of $^1O_2$ with ADPA was determined for PS in solution and for PS embedded inside the nanoparticles. This rate was higher for the nanoparticles, suggesting that $^1O_2$ production from nanoparticles exceeded that of free PS.

The same group published the encapsulation of methylene blue (MB, Fig. 11), a promising drug for PDT applications (Tang et al. 2005b). Intravenous administration of MB has been approved by the Food and Drug Administration (FDA) for methemoglobinemia. However, clinical use of methylene blue (MB) is limited partly because of the poor penetration of this drug in the cellular compartment of the tumor. Furthermore, methylene blue is usually inactivated *via* the reduction of the cation to the neutral leukomethylene blue which has negligible photodynamic activity. Encapsulation of methylene blue inside silica nanoparticles is a way to vectorize it and protect it towards degradations.

Two methods were performed for its encapsulation. The Stöber procedure was successful for the immobilization of the PS inside pure silica, giving nanoparticles with a diameter of 190 nm. Alternatively organically modified silica (ORMOSIL) nanoparticles were prepared by combining methyltrimethoxysilane and phenyltrimethoxysilane as precursors. This procedure resulted in nanoparticles with a diameter of 160 nm, revealing a hydrophobic outer shell and a hydrophilic inner core. The loading of PS was higher with the Stöber procedure than with ORMOSIL nanoparticles. However, reaction of $^1O_2$ with ADPA after irradiation of the nanoparticles at 650 nm showed a higher kinetic rate with ORMOSIL than with Stöber nanoparticles, although the production of $^1O_2$ with the Stöber nanoparticles was higher. The microenvironment in the double shell matrix of ORMOSIL could explain this result. Those nanoparticles were however not tested *in vitro*.

Other types of silica particles have been used for the encapsulation of PS for PDT. Wei and collaborators (Zhou et al. 2008) have elaborated porous hollow silica nanospheres of 130 nm diameter. The nanoparticles were prepared by careful hydrolysis of N-(β-aminoethyl)-α-aminopropyl triethoxysilane catalyzed by ammonia. The embedding rather than surface adsorption of hypocrellin A (Fig. 12) was demonstrated by fluorescence quenching experiments.

The particles were found to have good light and thermal stabilities. Such silica nanoparticles were efficiently taken up by HeLa tumor cells and presented more PDT efficiency than hypocrellin A. Higher $^1O_2$ generation quantum yield was found for the encapsulated PS than for free hypocrellin A.

The same collaborators later reported (Zhou et al. 2010) the preparation of hypocrellin A nanoparticles by reprecipitation followed by encapsulation of these particles using N-(β-aminoethyl)-α-aminopropyl triethoxysilane to form 110 nm silica nanovehicles. The resulting embedded hypocrellin A also showed superior light-stability and singlet oxygen generation ability than free hypocrellin A. Fluorescence images of Hela cells demonstrated the active cellular uptake of the drug-doped nanovehicles into the tumor cells. As the authors evidenced a mitochondria localization, light irradiation led to its membrane photodestruction.

In 2003, the Prasad group also prepared ORMOSIL nanoparticles for the entrapment of 2-devinyl-2-(1-hexyloxyethyl)pyropheophorbide (HPPH, Fig. 13), a photosensitizer in phase I/II clinical trials for oesophagal cancer (Roy et al. 2003).

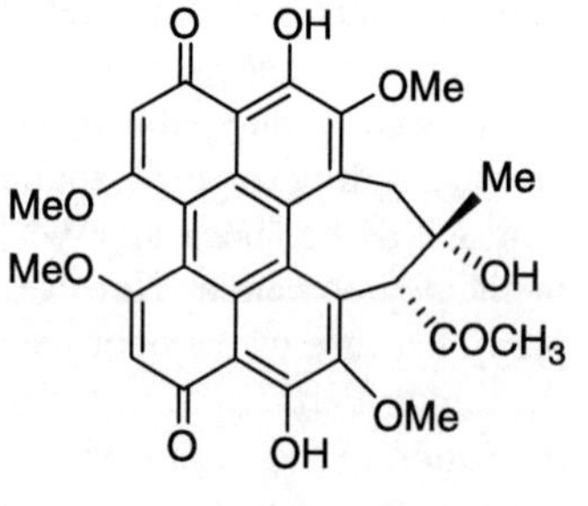

Fig. 12 Hypocrellin A (Reproduced by permission of The Royal Society of Chemistry)

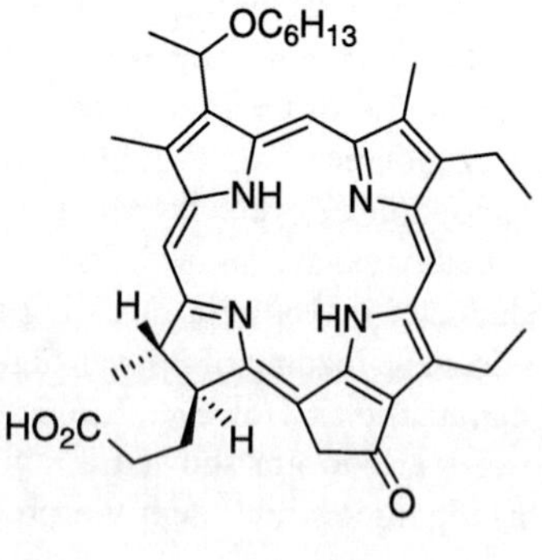

Fig. 13 2-Devinyl-2-(1-hexyloxyethyl)pyropheophorbide (HPPH) (Reproduced by permission of The Royal Society of Chemistry)

Microemulsion technique was performed with AOT and butanol as surfactants and vinyltriethoxysilane and aminopropyltriethoxysilane as the reagents for the gelation procedure. The nanoparticles exhibited a diameter of 30 nm which is smaller than in previously described procedures. $^1O_2$ production measured by its emission at 1,270 nm and by reaction with ADPA was very efficient. The nanoparticles were taken up by HeLa and UCI-107 cancer cells as confirmed by fluorescence of HPPH at 665 nm. The nanoparticles were very efficient in killing cancer cells after irradiation at 650 nm as analyzed by MTT assay.

Qian et al. encapsulated Protoporphyrin IX (PpIX, Fig. 7) in silica nanoparticles (Qian et al. 2009b), following Prasad's method. Monodispersed and spherical nanoparticles with a 25 nm diameter were obtained. Absorption and fluorescence spectra of free PpIX in DMSO and encapsulated PpIX were similar, even if PpIX encapsulating nanoparticles had a non negligible scattering efficiency (due to the 25 nm diameter) and the fluorescence was 2 nm red-shifted due to the different surroundings. $^1O_2$ was detected indirectly with 1,3 diphenylisobenzofuran (DPBF). *In vitro* PDT was performed on HeLa cells by a 532 nm light source irradiation (2 mW/cm^2, 2 min). After 8 min of irradiation, changes could be observed in the HeLa cells and the cell structures were destroyed.

The team of Reddi investigated the encapsulation of *m*-THPC into organic-modified silica nanoparticles of 33±9 nm diameter following Prasad's method (Compagnin et al. 2009). They showed that *m*-THPC was entrapped in a monomeric form and produced $^1O_2$ with a high efficiency. In aqueous media with high salt concentrations, the nanoparticles underwent aggregation and precipitation but their stability could be preserved in the presence of foetal bovine serum (3%). The cellular uptake of *m*-THPC (Fig. 10) entrapped in organically modified silica nanoparticles was shown to be mediated by serum proteins. The *in vitro* studies were realized on human oesophageal cancer cells KYSE 510 and the entrapped *m*-THPC was compared to *m*-THPC diluted into the standard solvent ethanol/ poly(ethyleneglycol) 400/water (20:30:50, by vol). In the absence of light, the two formulations did not affect the viability of the cells up to a *m*-THPC concentration of 1.75 mM. Phototoxicity was determined after 24 h dark incubation and irradiation with a light (600–700 nm, PTL Penta quartz halogen lamp, 2 mW cm^{-2}) of 0.12 J.cm^{-2}. All cells were killed after incubation with 1.25 μM *m*-THPC. The dose-response curves obtained by incubating the cells with increasing concentrations of *m*-THPC delivered by the two formulations were identical but the nanoparticles formulation reduced the cellular uptake of *m*-THPC by about 50% in comparison to standard solvent. With an elegant Förster Resonance Energy Transfer (FRET) approach, using a cyanine as an acceptor covalently linked to the nanoparticles, the authors demonstrated that *m*-THPC is transferred from the nanoparticles to serum proteins of the medium. To prevent this transfer to proteins it was necessary to coat the nanoparticles surface with poly(ethylene glycol).

Another type of PS based on hydrophobic silicon phthalocyanine *Pc* 4 (Fig. 3) was encapsulated in ORMOSIL by the method originally developed by Prasad's group with some modifications (replacement of AOT by Tween 80 as surfactant) which led to nanoparticles with a 25–30 nm diameter (Zhao et al. 2009). Compared to the free drug, encapsulation not only improved the aqueous solubility, stability and delivery of the drug but the PDT efficiency as well. Indeed cell viability measurements on A-375 and B16-F10 melanoma cells demonstrated that the nanoparticles were more photocytotoxic than the free drug. Apoptosis was the major pathway of cell death. Evaluation of the level of intracellular protein-derived peroxides in A-375 melanoma cells after internalization of the nanoparticles and PDT suggested a type II mechanism (phototoxicity through formation of $^1O_2$). Tracking with specific markers showed that more nanoparticles were localized in mitochondria and lysosomes than free PS.

Simon et al. recently published (Simon et al. 2010) the synthesis and properties of PpIX, (Fig. 7) ORMOSIL nanoparticles prepared by Prasad's method, and performed *in vitro* and *in vivo* studies. *In vitro* experiments showed that for the tested NPs (10, 25 and 60 nm), no significant difference of internalization kinetic and total amount of NPs uptake (around 200 fmol per 1,000 cells after 1 h of incubation) could be observed. The exact mechanism of internalization remains unknown, but both active and a high proportion of passive processes were involved, and it was cell-dependant. PpIX silica NPs effects on cell survival were tested on the HCT 116 colon cancer cell line. The results obtained showed that 3 h incubation with PpIX silica NPs and 20 min of illumination (50 $\mu W.cm^{-2}$) gave the best results for cancer cells destructions, i.e. EC_{50} at 0.44 μM for all sizes of NPs. Moreover, in all tumor types, PpIX silica NPs were more efficient (more than one log of difference) than free PpIX. For both systems, a loss of efficiency was found 5 h after incubation (no residual activity after 24 h). The cell clearance of PpIX was proved to be responsible of the loss of efficiency and the clearance process seems to begin early, before 2 h.

There is a strong relationship between the sites of subcellular localization of the PpIX and photodamage to nearby organelles involved in cell death. The intracellular accumulation of PpIX silica NPs takes place in the cytoplasm of cells. Quantification of cell internalization could be realized by measuring the ROS generation by fluorescence spectroscopy, since the ROS generation is correlated to the presence of PpIX silica NPs in both HCT 116 and HT-29 colon cancer cells. Higher ROS amounts cause larger cell damages leading to cell destruction.

In vivo studies on nude tumor-bearing mice were performed to observe the biodistribution of PpIX silica NPs loaded with DID (dioctadecyl tetramethyl indodicarbocyanine chloro benzene) tracer through whole body optical imaging. Images of the biodistribution were recorded after NPs tail vein injection (from 0 to 24 h). For the three cancer models studied (HCT 116, A 549 and glioblastoma), a high tumor uptake of PpIX silica NPs was observed, but the maximal accumulations were reached at different times: 2 h for glioblastoma, 16 h for A549 and 20 h for HCT 116 models. A549 and HCT 116 tumor models behaved similarly, i.e. a slow accumulation to reach a maximum until 24 h. For glioblastoma, the NPs uptake took place quickly after the injection to reach the maximum after 2 h, while a marked decrease was observed immediately later. NPs also accumulate in reticulo-endothelial system such as liver, for which a very high accumulation was observed (higher than that within the tumor mass), but also in spleen, lymph nodes, ovaries, lungs, and adrenal glands. The fluorescence intensities decreased over time in these organs, except for the liver. No urine elimination was observed throughout 24 h.

Recently, the Prasad group used ORMOSIL nanoparticles, embedded with HPPH (Fig. 13) but covalently iodine-concentrated (Kim et al. 2009b). These particles were prepared using (3-iodopropyl)-trimethoxysilane and vinyltriethoxysilane. The properties of these nanoparticles were compared with noniodinated nanoparticles (prepared from vinyltriethoxysilane) and it was shown that the intraparticle external heavy-atoms significantly enhanced the efficiency of $^1O_2$ generation and the *in vitro* PDT efficiency. Indeed, the quantity of $^1O_2$ generated by the iodinated nanoparticles was found to be increased by roughly 1.7 times over the non iodinated nanoparticles. Cell viability assay was carried out for RIF-1 tumor cells and nanoparticles with iodine were the most efficient for PDT applications.

Wang and collaborators entrapped methylene blue (Fig. 11) in a phosphonate-terminated silica matrix by controlled synchronous hydrolysis of tetraethoxysilane and trihydroxyl silyl propyl methyl phosphonate in a water-in-oil microemulsion (He et al. 2009). Spherical nanoparticles were obtained with a mean diameter of 105 nm. $^1O_2$ generation was evaluated using DPBF and estimated to be 0.049. *In vitro* PDT efficiency was investigated on HeLa cells. Cell photocytotoxicity was evaluated to 90% using 1 mg/mL methylene blue doped nanoparticles upon 30 min light exposure (635 nm laser, 27.5 mW/cm^2). Real-time *in vivo* imaging experiments on BALB/c nude mice bearing MB-encapsulating nanoparticles were performed. The methylene blue doped nanoparticles were injected intravenously in nude mice for near-infrared *in vivo* imaging. Moreover, 12 h post-injection, PDT experiments (635 nm laser, 500 mW/cm^2) have been realized with success, inducing a necrosis process.

Yang and collaborators (2010) prepared MSN encapsulating hypocrellin B (HB, Fig. 14) as a PDT agent and coated with a lipid layer. The particles were prepared using a base-catalyzed sol-gel method. The particles were calcinated and the HB were impregnated in the particles. A coat of lipids (1-palmitoyl-2-[6-[7-nitro-2-1,3-benzoxadiazol-4-yl)amino]hexanoyl]-*sn*-glycero-3-phosphocholine, 1,2-dimyristoyl-*sn*-glycero-3-phophocholine and 1,2-dimyristoyl-*sn*-glycero-3-phosphate monosodium salt) was prepared by drying the mixed lipid solution and reconstituting it as vesicles following a conventional method. The vesicles were then adsorbed

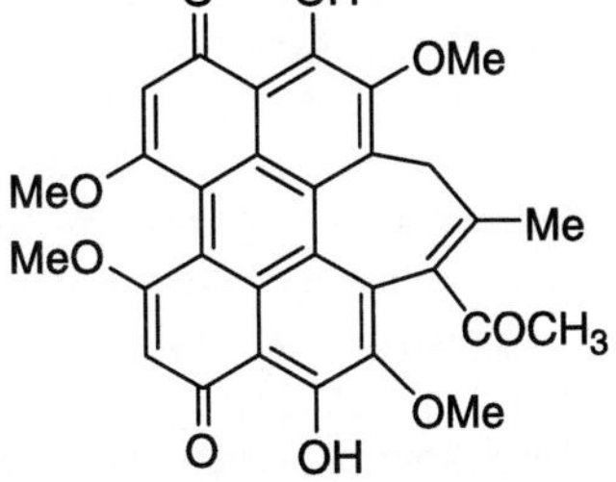

Fig. 14 Hypocrellin B (HB)

on the surface of the particles. The HB incorporating particles were incubated *in vitro* with human breast carcinoma MCF-7 cells. CLSM observation showed that the particles were indeed internalized into the cells. The phototoxicity of HB in the lipid-covered MSN were investigated for PDT *in vitro*. The cell activity decreased to 44.6% after light exposure (15 min, 480 nm), showing the potential of these particles for PDT applications.

3.1.2 Multifunctional Silica Nanoparticles for PDT and Diagnosis

A marked advantage of using nanoparticles is that they can be used as multi-tasks platforms. Indeed, their chemical composition can be modulated to, (i) couple or encapsulate photoactivable units for PDT treatment, (ii) target the nanoparticles to tumor cells or neovascularization by coupling vector units as well as, (iii) use them as a diagnostic agent by incorporating a contrast enhancer, (iv) use them for other therapies such as hyperthermia. Moreover, as with biodegradable nanoparticles it is possible to play with the hydrophilic/hydrophobic balance to increase their plasmatic lifetime. For biomedical applications, multifunctional nano-objects combine two or more functions, such as fluorescent markers or photothermal therapy agents with MRI contrast agents or hyperthermia therapy agents.

An example of a single particle platform that combines two functions has been described by Rossi and collaborators (Tada et al. 2007) in 2007. Silica-coated magnetic nanoparticles containing methylene blue (Fig. 11) as PS have been prepared and therefore combine therapy (PDT) and diagnostic (MRI contrast agent) possibilities. The particles were composed of silica spheres of about 30 nm diameter containing 11 nm-diameter magnetic particles. The PS, methylene blue, was added to the silica precursor tetraorthosilicate during the growth of the silica layer and was therefore entrapped in the silica matrix. The magnetic cores were prepared by coprecipitation of Fe^{2+}/Fe^{3+} ions under alkaline conditions followed by stabilization with tetraethylammonium oxide. The immobilized drug could generate $^1O_2$, detected by its characteristic phosphorescence decay curve in the near infra-red and by a chemical method using DPBF to trap $^1O_2$. The encapsulation of the PS in the silica led to a $^1O_2$ quantum yield of $3 \pm 2\%$, while the quantum yield of methylene blue free in acetonitrile was 50%. This difference in $^1O_2$ quantum yield values was also reported by Tang et al. (2005a). This can be due to scattering of nanoparticles, local sequestration of $^1O_2$ and/or intrinsic lower encapsulated methylene blue $^1O_2$ quantum yield. The magnetization curve confirmed the superparamagnetic behavior of the particle.

Another example of potentially interesting magnetic nanocarriers for PDT was reported by Zhou and collaborators (Liu et al. 2008a). Fe_3O_4 nanoparticles were coated with a silica layer using tetraethoxysilane in a cottonseed oil using reverse microemulsion method in presence of purpurin-18 (Fig. 15) as PS.

Nanoparticles with a 20–30 nm diameter were obtained. The characterization of the PS in particle was achieved by UV-visible spectroscopy. The generation of $^1O_2$ was followed by *N, N*-dimethyl-4-nitrosoaniline (RNO) bleaching assay and was found to be less effective for the encapsulated purpurin than for the free PS.

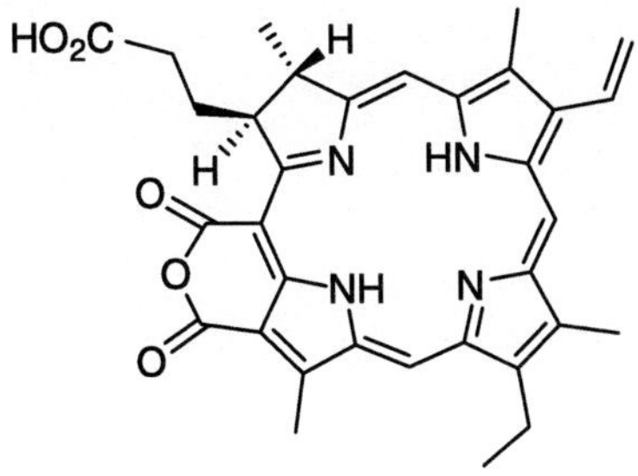

Fig. 15 Purpurin-18 (Reproduced by permission of The Royal Society of Chemistry)

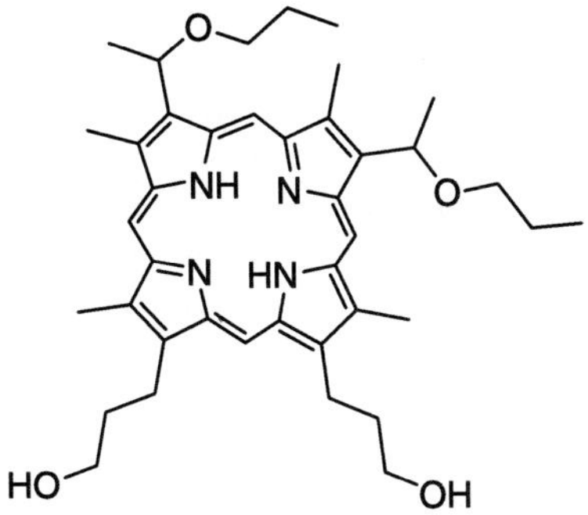

Fig. 16 2,7,12,18-Tetramethyl-3, 8-di(1-propoxyethyl)-13, 17-bis-(3-hydroxypropyl) porphyrin (PHPP) (Reproduced by permission of The Royal Society of Chemistry)

Silica covered Fe_3O_4 magnetic nanoparticles including 2,7,12,18-tetramethyl-3,8-di(1-propoxyethyl)-13,17-bis-(3-hydroxypropyl)porphyrin (PHPP, Fig. 16) as PS were described by Chen, Zhou and co-workers (Chen et al. 2009). Fe_3O_4 nanoparticles were prepared by co-precipitation and were about 10 ± 2 nm in diameter. AOT/1-butanol/water micelles method was combined with sol-gel method in presence of PHPP to coat the magnetic particles with silica. The resulting particles were about 20–30 nm in diameter but some agglomeration could be observed, as shown by TEM images. The encapsulation of the PS was checked by FT-IR measurements and photoluminescence of PHPP. The efficiency of PHPP encapsulation was estimated by UV-visible spectroscopy after destruction of Fe_3O_4 cores of the nanoparticles and was found to be around 21%. The RNO-bleaching method allowed to evaluate a significant $^1O_2$ generation. *In vivo* photodynamic efficacy was evaluated on SW480 colon carcinoma cells. No significant dark toxicity was detected for all tested concentrations until 80 μmol/L. PDT experiments were performed on the same cell lines and significant PDT efficacy (40% cell viability lost) was found (24 h incubation time, 10 min, 488 nm laser irradiation, 4.35 J/cm^2). The potential of the magnetic core for imaging and therapy is yet to be tested.

Zinc(II) phthalocyanines (ZnPc, Fig. 17) are second generation photosensitizers useful for PDT applications. However, the low solubility of these compounds result in the formation of aggregates in aqueous media, and thus in a loss of efficiency. Kim et al. (2009a) included ZnPc into Fe_3O_4-loaded mesoporous silica nanoparticles (MSN). MSN have been used for various biomedical applications such as cell

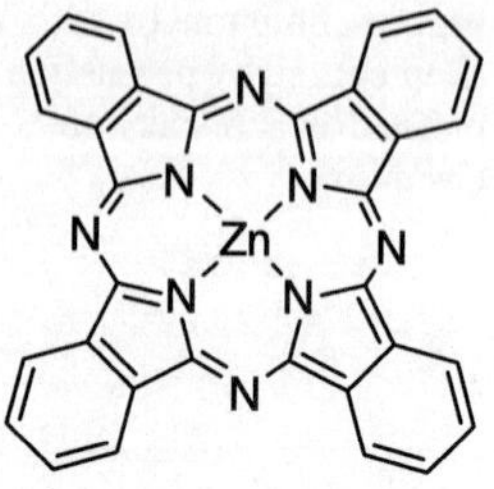

Fig. 17 Zinc (II) phthalocyanine (ZnPc) (Reproduced by permission of The Royal Society of Chemistry)

markers or drug and gene delivery platform. MSN possess advantages such as large surface area and pore volume, as well as uniform pore size. They are both a suitable carrier for hydrophobic molecules and a way to protect the PS from degradation. ZnPc – and Fe_3O_4-loaded MSN (ZnPc or Fe_3O_4/MSN) were prepared through a *one-pot* procedure under basic conditions in the presence of cetyltrimethylammonium bromide CTAB as structurating agent and Fe_3O_4 nanoparticles. Near spherical or rounded cubic nanoparticles were obtained with a size in the range of 60–120 nm. According to narrow-angle X-ray diffraction pattern of MSN and ZnPc or Fe_3O_4/MSN, a slightly disordered mesoporous structure was observed. A widening of the pore size from 3.6 to ~4 nm was also calculated. Wide-angle X-ray diffraction pattern of ZnPc or Fe_3O_4/MSN displayed the characteristic peaks of the amorphous silica matrix and the cubic inverse spinel Fe_3O_4. Magnetic measurements showed a superparamagnetic behavior at 1.0 T, which is enough for MRI and magnetic hyperthermia applications. UV-visible spectroscopy showed the presence of non-aggregated ZnPc molecules entrapped in the silica matrix. However, experiments concerning singlet oxygen generation or PDT still have to be carried out. Drug delivery experiments were performed with ibuprofen as guest molecules. The loading of ibuprofen was realized by dispersing and stirring MSN or ZnPc/Fe_3O_4/MSN in a solution of ibuprofen in hexane and the release of ibuprofen in distilled water could be monitored by UV-visible spectroscopy. It appeared that ZnPc/Fe_3O_4/MSN showed a more delayed and steady release of ibuprofen compared to MSN. This effect could be attributed to the interaction between ibuprofen and the ZnPc entrapped in the silica matrix.

Liu et al. (2009) described the preparation of photodynamic drug nanocarriers consisting of magnetite and photosensitizer (PHPP, Fig. 14), surrounded by a silica shell through a sol-gel process in a micellar media. TEM analyses showed nanocarriers with a nearly spherical shape in the range of 30–50 nm which tend to aggregate. XRD analyses showed the typical patterns of magnetite in addition to the broad diffraction band at ca. $2\theta = 20°$ corresponding to the amorphous silica matrix. The photosensitizer entrapped in the silica matrix was detected by FT-IR and UV-visible spectroscopy. Upon irradiation using 637 nm laser, $^1O_2$ was produced and released from the nanocarriers. The RNO-bleaching method, allowed the determination the $^1O_2$ generating efficiency. Compared to the free PHPP, the results obtained showed that: (1) the photosensitivity of the nanocarriers was less but still effective for potential PDT applications and (2) the presence of Fe_3O_4 and silica shell had no critical effect on the production and release of $^1O_2$.

NIR-to-visible up-conversion nanomaterials emit higher energy photons after absorbing lower energy photons. Up-conversion, in which excitation light at a longer wavelength produces emission at a shorter wavelength is very promising since it becomes possible to excite the photosensitizers in the near IR. Qian et al. (2009a) described the synthesis of mesoporous-silica-coated $NaYF_4$@silica nanoparticles with a core-shell structure containing ZnPc (Fig. 17) molecules as photosensitizer in the mesoporous silica matrix. The absorption peak of ZnPc located at 670 nm overlaps with the red emission peak of the up-conversion of the $NaYF_4$ nanocrystals, what is a suitable situation for the activation of ZnPc and the further production of $^1O_2$. $NaYF_4$:Yb/Er nanocrystals (35 nm width × 60 nm length) were coated by an amorphous silica shell (10 ± 1.5 nm) *via* a microemulsion method, and then by a mesoporous silica layer (11 ± 1.5 nm) onto the $NaYF_4$:Yb/Er@silica nanoparticles. N_2-absorption/desorption isotherm confirmed the mesoporous nature of the outer silica shell and the surface area was found to be 770 $m^2 g^{-1}$. Pore size distribution has a mean size of 2 nm. After the incubation of the nanoparticles with MB49 bladder cancer cells for 24 h, the subsequent intracellular localization of the nanoparticles could be realized using a confocal microscope equipped with a 980 nm near IR laser. Fluorescence of the nanoparticles was mainly observed in the cytoplasm. The introduction of a PS such as ZnPc into the mesoporous silica shell was realized by soaking the nanoparticles in a concentrated solution of ZnPc in pyridine. Then, the release of ZnPc in different solvent was studied. It appeared that ZnPc was released within 1 h in ethanol, but no release was observed in deionized water, Phosphate Buffered Saline (PBS) or cell culture media. The production of $^1O_2$ for further PDT applications was shown by exposing the $NaYF_4$/ZnPc nanoparticles to near IR laser irradiation in the presence of 9,10-anthracenediyl-bis(methylene)dimalonic acid (ABDA). $^1O_2$ formation could be observed with the $NaYF_4$/ZnPc nanoparticles whereas no production was detected the $NaYF_4$ nanoparticles without ZnPc. *In vitro* experiments showed that the viability of cells incubated with the $NaYF_4$ZnPc nanoparticles and exposed to NIR irradiation for 5 min was significantly lower than those incubated with ZnPc-free $NaYF_4$ nanoparticles. Further work (Guo et al. 2010a) demonstrated the efficacy of these particles for inducing cytotoxicity and its mechanism by $^1O_2$-induced apoptosis. The intracellular uptake and photodynamic effect of these upconversion nanoparticles were studied on MB49 murine bladder cancer cells, revealing a time- and concentration-dependent accumulation. A maximum of particles were found in the cells after a 6 h incubation time. There is an increased accumulation of nanoparticles in the cells when they are incubated with increasing concentrations (between 200 and 600 μg/mL). Their photodynamic efficiency in activating the loaded ZnPc upon irradiation with 980 nm NIR light was confirmed in living cells. The cytotoxic effect of the released oxygen was demonstrated. Indeed, the cell viability for MB49-PSA cells treated with ZnPc loaded $NaYF_4$ nanoparticles was found to be around 60% without irradiation and around 20% with a 5 min irradiation (500 mW).

Another approach has been published by Zhang et al. (2007). $NaYF_4/Yb^{3+}$, Er^{3+} nanoparticles have been used as photon up-converting nanoparticles and coated

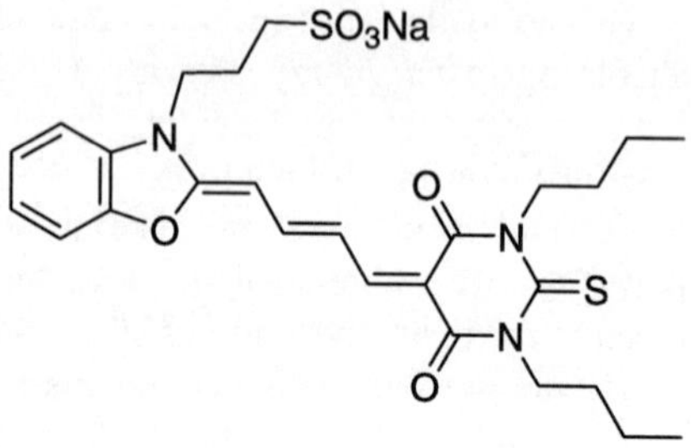

Fig. 18 Merocyanine 540 (Reproduced by permission of The Royal Society of Chemistry)

with a PS (Merocyanine 540, Fig. 18)-encapsulating porous thin layer of silica. A mouse monoclonal antibody, anti-MUC1/episialin which is highly specific towards MCF-7/AZ cancer cells was covalently attached to the silica surface. The generation of $^1O_2$ was detected chemically using ADPA. *In vitro* PDT tests were performed to confirm the PDT cytotoxicity of the particles towards MCF-7/AZ breast cancer cells, thus demonstrating the potential of these IR-region excitable nanoparticles.

Two-photon absorption (TPA)-induced excitation of photosensitizers is another promising approach to increase light penetration. Indeed, the photosensitizers can absorb simultaneously two less-energetic photons, the excitation in the NIR region avoids tissue absorbing or scattering and induce a deeper light penetration in the tissue.

Velusamy et al. (2009) synthesized new quadrupolar chromophores for two-photon absorption (Fig. 19) with a very high two-photon absorption cross-section of 7080 GM at 800 nm in toluene and a fluorescence quantum yield of 0.25. The compound was insoluble in water and was thus encapsulated in SiO_2 nanoparticles following the reverse microemulsion method. The nanoparticles were of a 60 nm diameter. The fluorescence quantum yield dropped to 0.08 with a red shift of λ_{em}. Quantum yield of $^1O_2$ generation of the nanoparticles was determined to be 0.51 in D_2O. Those nanoparticles were incubated with macrophages and shown to be tightly bound to the membrane with no toxicity as determined by MTT assay. Macrophage cells were then irradiated with a two-photon laser at 2.6 W/cm^{-2} (800 nm) for 3 min. The results indicated a significant and NP-dose responsive cell death, thus demonstrating the suitability of the nanoparticles for two-photon PDT.

The Prasad group described the synthesis (Kim et al. 2007) of organically modified silica nanoparticles co-encapsulating HPPH (Fig. 13) and an excess amount of BDSA (9,10-bis (4′-(4″-aminostyryl)styryl)anthracene, Fig. 20), a highly two-photon active molecule (TPA cross section in the aggregated state at 750 nm = 217 GM) as a donor. The size of nanoparticles entrapping 1.1 wt.% HPPH and 1.1 wt.% HPPH / 20 wt.% BDSA was 25 ± 7 nm and 30 ± 6 nm respectively. HPPH absorption in nanoparticles had significant overlap with the fluorescence of BDSA aggregates which enabled an efficient energy transfer through FRET mechanisms. After indirect two-photon excitation (850 nm) the authors proved that the energy of the NIR light is efficiently up-converted by BDSA aggregates to excite HPPH. After

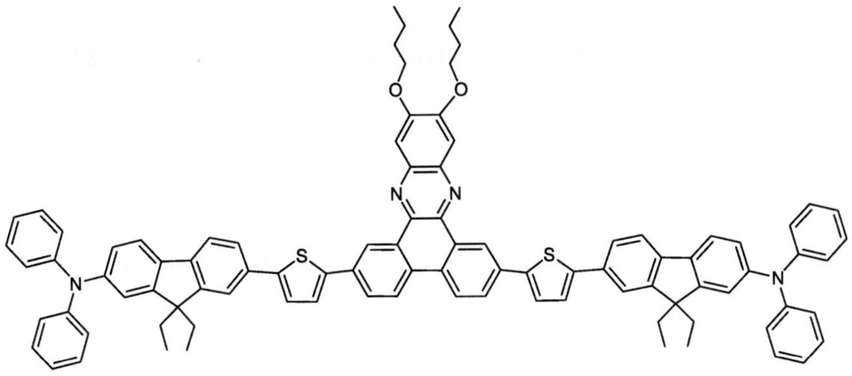

Fig. 19 2-Photons PS (Reproduced by permission of The Royal Society of Chemistry)

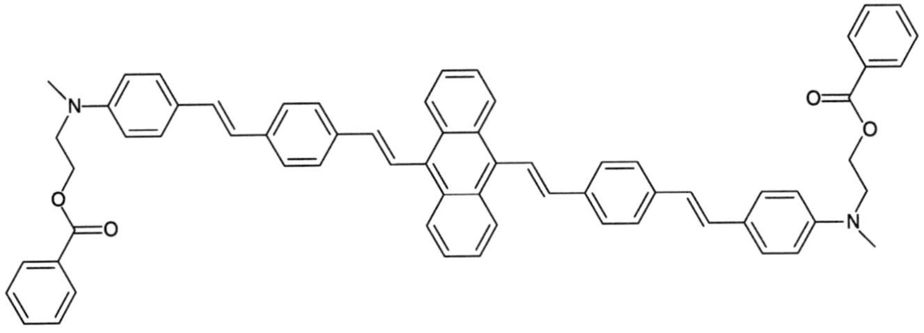

Fig. 20 9,10-Bis (4′-(4″-aminostyryl)styryl)anthracene (BDSA) (Reproduced by permission of The Royal Society of Chemistry)

excitation at 532 nm in D_2O, the characteristic $^1O_2$ emission at 1,270 nm was observed under photoexcitation of the dispersion of nanoparticles co-encapsulating HPPH and BDSA. For the two-photon sensitization of $^1O_2$ in water, ADPA was used and the authors could also observe the formation of $^1O_2$.

Hela cells were treated with the nanoparticles and two-photon laser scanning microscopic images revealed an intense fluorescence signal from the cytoplasm. The intracellular FRET efficiency was estimated up to 36%. Drastic changes of the morphology of the treated cells were observed in relation to the formation of $^1O_2$ after two-photon excitation. As a conclusion, this new concept of two-photon induced intra-particle FRET offers a simple and proper methodology for developing formulations with better photophysical properties.

Febvay et al. (2010) described a drug delivery system where light-activated disruption of intracellular vesicles occurs after internalization of biofunctionalized mesoporous silica nanoparticles loaded with fluorescent dye Alexia 456 as cargo. Size-tunable monodispersed MSN have been functionalized with streptavidin *via* PEG spacer arms in order to target cancer cells expressing P-glycoprotein, a transporter protein responsible for multidrug resistance in many tumours. These MSN can be loaded with different types of cargo and can mediate cytosolic release of cell-impermeable molecules in P-gp-expressing cells *via* light mediated endosomal breakage. Indeed, the exposure to light of the MSN at the dye's excitation wavelength leads to ROS-mediated membrane damage and therefore to the specific delivery of the cargo.

3.2 Covalent Encapsulation of PS in Silica Nanoparticles

When physically entrapped inside the silica network, the PS can be prematurely released from the carrier which can lead to a reduced efficiency of treatment and to side-effects. Covalent coupling of the PS inside or at the surface of the nanoparticles is expected to overcome these drawbacks.

Davydenko et al. (2006) studied the sensibilization of C_{60} fullerene anchored through covalently bonded spacers to hydrophilic non porous pyrogenic silica nanoparticles of 10 nm diameter modified by γ-aminopropyl groups. The ability of fullerene to generate active forms of oxygen was demonstrated using photosensitive organic molecules. No *in vitro* or *in vivo* studies have been done, but these first results suggested that these systems might be useful as fluorescent probes and active species for PDT.

Rose Bengal (RB, Fig. 21) decorated nanoparticles were used for inactivation of gram-positive bacteria by Zhang and collaborators (Guo et al. 2010b). Rose Bengal (4,5,6,7-tetrachloro-2′,4′,5′,7′-tetraiodofluorosceine in disodium salt), a well-known anionic photosensitizing molecule was covalently conjugated to the amine functionalized surface of silica nanoparticles prepared by hydrolysis of TEOS in reverse microemulsion. The grafted particles were characterized using TEM, showing particles in the range of 50–80 nm and FTIR, showing the presence of RB on the particles. The characterization of RB, free or decorated on SiO_2 particles, as a source of singlet oxygen under illumination, was performed through the photobleaching of ADPA in solution. The quantum yield of generating singlet oxygen of the particles was found to be 0.60 in D_2O (0.75 for the free RB). The *in vitro* photodynamic inactivation of the RM-modified nanoparticles were investigated in gram-positive bacteria, using methicillin-resistant *Staphylococcus aureus* (MRSA)

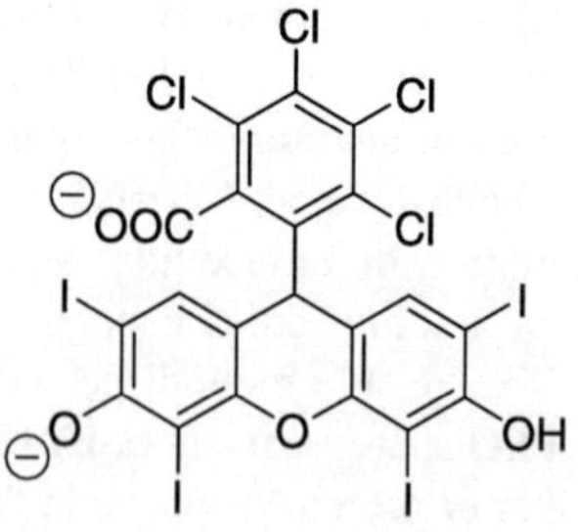

Fig. 21 Rose Bengal (RB)

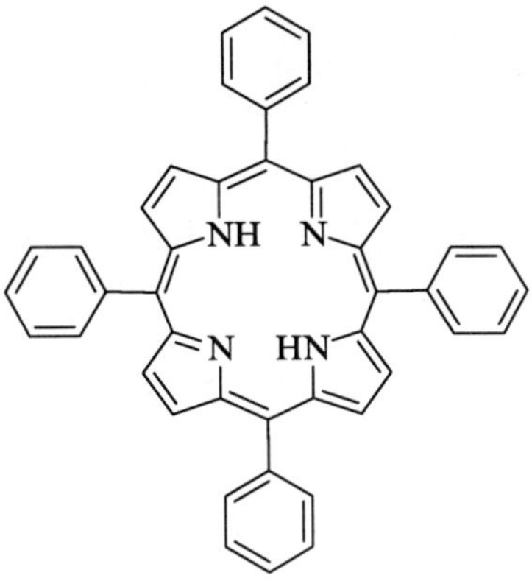

Fig. 22 *Meso*-tetraphenyl porphyrin, TPP

and *Staphylococcus epidermis*. The grafted particles were shown to be more efficient than free RB in inactivating gram-positive bacteria, with a two orders of magnitude reduction in the viability count.

pH-responsive silica nanoparticles for controllable $^1O_2$ generation have been developed by Zhang and collaborators (Li et al. 2010b). A hydrophobic photosensitizer (meso-tetraphenyl porphyrin, TPP, Fig. 22) and a pH indicator (Bromocresol Purple or Bromothymol Blue) were simultaneously encapsulated in 50 nm diameter organically modified silica nanoparticles *via* an oil-in-water microemulsion method. Both covalently encapsulated pH indicators have pKa values near the physiological pH (6.4 and 7.1 respectively). It was demonstrated that, in basic conditions (pH=10.0), the encapsulated pH indicator absorbed light competitively and thus restricted sensitizer excitation. In acidic conditions (pH=5.0), the blue shifted absorption of the pH indicator allowed the efficient excitation of the sensitizer. The photodynamic activity of the encapsulated TPP was proved using the chemical trapping of AMDA. The decrease in absorbance values of AMDA at pH=5 is 6.3 times larger than at pH=10 after 90 min irradiation, which underlines the difference of photodynamic activity between acidic and basic conditions. Interstitial fluid of different kinds of tumors is known to have a lower pH (6.15–6.80) than normal tissue (7.0–7.4), which makes it a good target for a potential clinical application of this system as a more specific anticancer drug.

Rossi and collaborators have covalently encapsulated PpIX by coupling the acid groups with aminopropyltriethoxysilane (Rossi et al. 2008) (Fig. 23). The Stöber procedure was used and led to particles of 77±12 nm. $^1O_2$ generation was characterized by its phosphorescence spectra at 1,270 nm and by reaction with DPBF. The efficiency of $^1O_2$ generation in acetonitrile was determined to be 0.90±0.06 for the nanoparticles. This efficiency was higher than the free porphyrin in CCl_4 solution ($^1O_2$ quantum yield of 0.77). These nanoparticles were not tested *in vitro*.

Prasad and collaborators first reported the covalent incorporation of PS molecules into ORMOSIL nanoparticles (Ohulchanskyy et al. 2007). Iodobenzyl-pyrosilane (IPS, Fig. 24), a precursor for ORMOSIL was synthesized with the linked PS iodobenzylpyropheophorbide (IP).

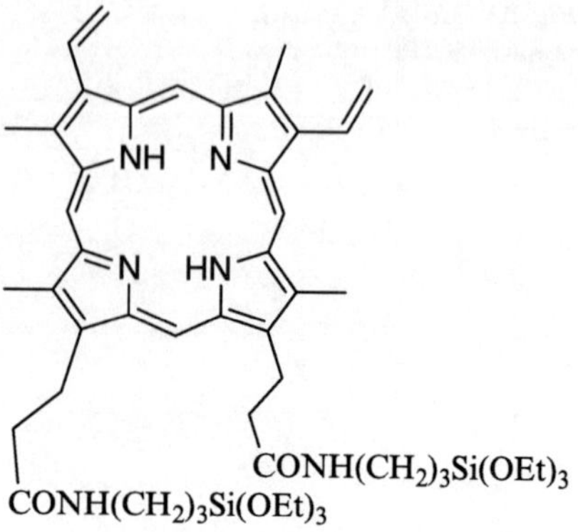

Fig. 23 Dimethyl-8,13-divinyl-3,7,12,17-tetramethyl-21 *H*,23 *H*-porphyrine-2,18-dipropylamidopropyltriethoxysilane (silyl-PpIX) (Reproduced by permission of The Royal Society of Chemistry)

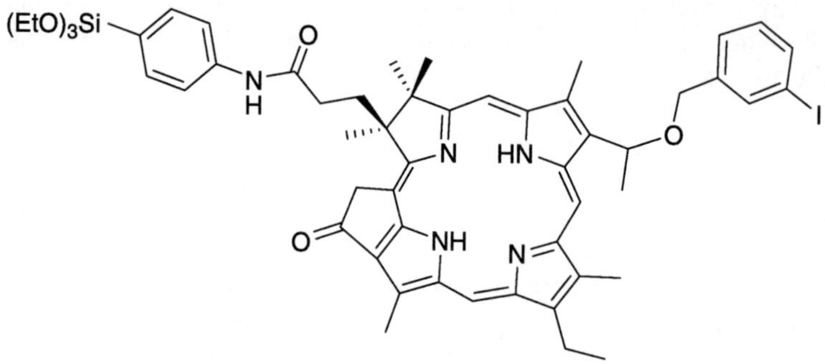

Fig. 24 Iodobenzyl-pyro-silane (IPS) (Reproduced by permission of The Royal Society of Chemistry)

The coprecipitation of IPS with the commonly used ORMOSIL precursor vinyltriethoxysilane in the nonpolar core of Tween-80/water microemulsion led to the formation of a monodispersed aqueous dispersion of ORMOSIL with covalently linked IP. The particles were of about 20 nm diameter. Since the covalently incorporated PS molecules were found to retain their spectroscopic and functional properties such as absorption or fluorescence independantly of the IPS/vinyltriethoxysilane ratio used, the authors concluded that no aggregation occurred and that it could be possible to increase the relative content of IPS within the nanoparticles to obtain more $^1O_2$. To be sure that the generated $^1O_2$ was not deactivated within the nanoparticles, they used the $^1O_2$ mediated bleaching of ADPA. They proved that $^1O_2$ was mainly deactivated outside the nanoparticles. Upon photoirradiation, cytotoxic $^1O_2$ molecules were generated. Therefore, the nanoparticles were tested for cellular uptake *in vitro*, on Colon-26 cells and proved to exhibit a manifest phototoxic effect on the cultured cells proportional to the cellular uptake. In addition, the highest photodynamic effect was associated to the highest content of the PS within the nanoparticles, thus demonstrating the potential of the nanoparticles for PDT.

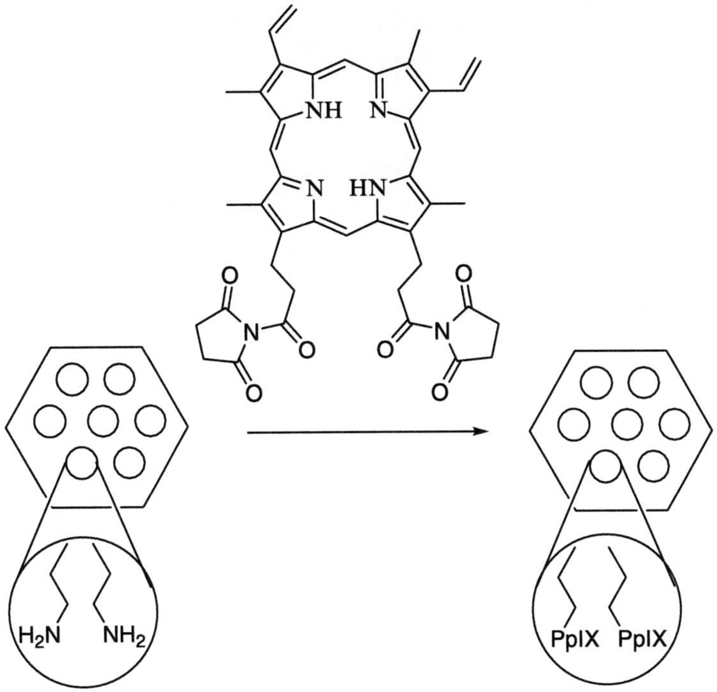

Fig. 25 PpIX-modified pores of MSN (Reproduced by permission of The Royal Society of Chemistry)

Mou and collaborators have reported conjugation of PpIX (Fig. 7) with MSN through covalent bonding to yield PpIX-modified MSN (Fig. 25) for PDT studies (Tu et al. 2009).

These particles were obtained by using classic MCM-41 type direct microemulsion synthesis, leading to 110 nm diameter objects. The MSN were then functionalized with 3-aminopropyltrimethoxysilane, allowing the grafting of activated PpIX (Fig. 7). The porous nature of the particles lead to site-isolation of the PS which avoided self-quenching of aggregated PpIX. The absorption spectrum of the PS-grafted nanoparticles was similar to that of the free PpIX, showing no change of the chromophore upon entrapment in the MSN. Moreover, samples of PpIX grafted to MSN were shown to have a better integrity under laser irradiation compared to free PpIX. $^1O_2$ generation was checked both by monitoring the $^1O_2$ luminescence and by using DPBF as a detector. Dark cytotoxicity was estimated for HeLa cells and the cell viability was not affected at dosage up to 80 μg/mL. Uptake of the PpIX-modified particles by HeLa cells was quite efficient (higher than 70% for 20–80 μg/mL concentrations after a total of 6 h incubation). Phototoxicity was evaluated after a 6 min tungsten lamp irradiation (115 mW/cm^2, 80 μg/mL). Both concentration- and irradiation time-dependent cell viability were observed. Depending on the tested concentration, both necrosis and apoptosis process could be observed.

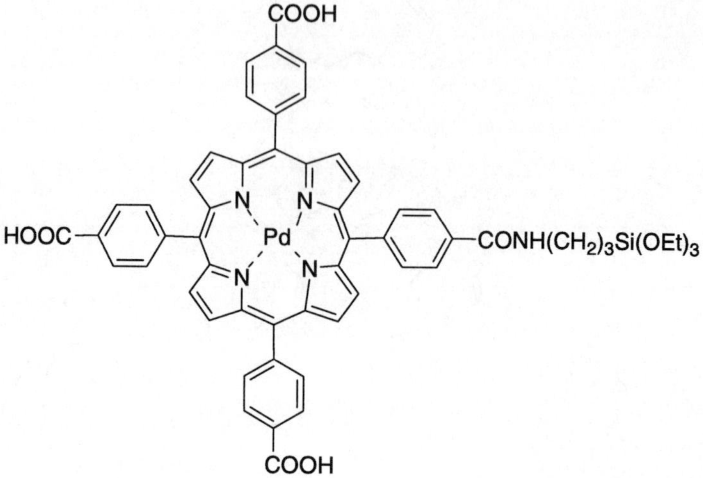

Fig. 26 Pd-*meso*-tetra(4-carboxyphenyl) porphyrin (PdTPP) (Reproduced by permission of The Royal Society of Chemistry)

The same group prepared MSN then grafted a phosphorescent Pd-*meso*-tetra(4-carboxyphenyl) porphyrin (PdTPP, Fig. 26) as the PS (Cheng et al. 2009). This metalloporphyrin is frequently used to measure oxygen distribution in tissues *via* oxygen-dependent quenching of phosphorescence.

$^1O_2$ generation formed by photoirradiation of the nanoparticles was measured both by direct luminescence at 1,270 nm in deuterated water and ethanol and by its reaction with DPBF. A significant photo-induced cytotoxicity at 532 nm was observed for MDA-MB-231 breast cancer cells. However, the reason of this photo-induced toxicity was not very clear but could be explained by the high concentration of PdTPP in the MSN as well as the fact that MSN facilitated the endocytotic cell uptake that dramatically increased the intracellular density of $^1O_2$ upon photoirradiation.

Further progress in the field of silica nanoparticles for PDT was achieved when multifunctional particles were functionalized by a biomolecule able to specifically target cancer cells. Active targeting represents an obvious improvement for PDT. Until now, most of the efforts in the development of tumor targeting-photosensitizers have focused on the targeting of markers over-expressed by tumor cells themselves even if anti-vascular strategies are being more and more important (Bechet et al. 2008; Taquet et al. 2007).

In order to specifically target cancer cells, MSN were functionalized by a biomolecule able to target cancer cells. As specific bioreceptors are over-expressed at the surface of cancer cells in many tumors, functionalizing the particle in order to target these receptors would enhance the uptake of the nanoparticles by these cells. Novel MSN (Brevet et al. 2009) were synthesized combining covalent anchoring of the water soluble photosensitizer (Fig. 27) to the mesoporous silica

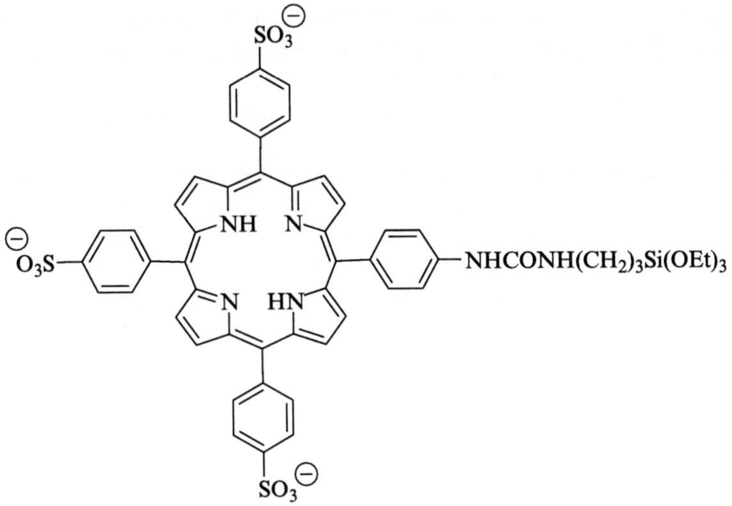

Fig. 27 Water-soluble porphyrin (Reproduced by permission of The Royal Society of Chemistry)

Fig. 28 APTS-HP (Reproduced by permission of The Royal Society of Chemistry)

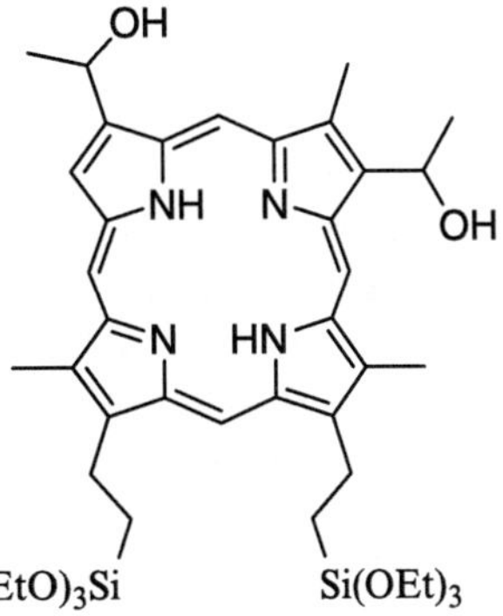

matrix and targeting of cancer cells with mannose attached on the surface of MSN. Targeting breast cancer cells (MDA-MB-231) with mannose was essential to get a high PDT efficiency. Indeed the involvement of mannose receptors in the active endocytosis of mannose-functionalized MSN was demonstrated.

In the promising field of using silica nanoparticles as multifunctional platform for imaging and therapy, Zhang et al. (2009) described the synthesis of a highly efficient multifunctional nanocomposite material that contains both a nonporous FITC-doped silica core for fluorescence imaging and a mesoporous silica shell containing hematoporphyrin (HP) as a photosensitizer. Non porous FITC@SiO_2 core were synthesized *via* a water-in-oil reverse microemulsion. The hematoporphyrin conjugated mesoporous silica nanoparticles were synthesized with TEOS and APTS-HP (Fig. 28) *via* a sol-gel O/W co-condensation reaction. The non porous silica core isolated the fluorescent dye from the external environment to avoid photobleaching. On the contrary, the mesoporous silica shell allowed the diffusion of $^1O_2$ outside of the nanoparticle. TEM images revealed a monomodal particle size distribution around 57 nm for mesoporous core-shell nanoparticles and 37 nm for silica cores. Disodium 9,10-anthracenediyl-bis-(methylene)dimalonic acid (ABMD) was used to chemically check the capacity of such material to produce $^1O_2$. A significant decrease of ABMD absorption for SiO_2@HP and HP alone was observed, but for the same amount of HP, the decrease was more important for SiO_2@HP or FITC@SiO_2@HP than for HP alone, due to a higher stability of PS covalently linked in a silica shell. Moreover, the FITC inside the non-porous silica shell was well protected from the damage by $^1O_2$ (no photobleaching). Finally, the imaging capability was demonstrated by cell imaging on HO-8910 PM cells. Confocal microscopy on such cells showed that most of the nanoparticles were localized in the membrane of the cell and a few inside the cell.

Hsiao and collaborators reported the synthesis of a multifunctional nanomaterial conjugating a functionalized iridium complex with a Fe_3O_4/SiO_2 core/shell nanocomposite (Lai et al. 2008). The magnetic core provides capability for MRI. The silica shell allows the embedment of the PS, with sufficient space for oxygen diffusion. A third row transition metal (Ir(III)) complex was chosen as the PS to serve both as a photosensitizer and as a luminescent moiety. A hydrophilic iridium (III) complex (Fig. 29) was developed to generate a good affinity with the silica shell.

The size of the nanoparticles was determined by TEM, 55±5 nm for Fe_3O_4/SiO_2(Ir) nanoparticles compared to 12±1 nm for Fe_3O_4. Upon aeration, both emission intensity and observed lifetime decreased drastically suggesting an O_2 quenching process presumably resulting in $^1O_2$ production. PDT experiments were performed on HeLa cells. The cellular uptake and imaging ability were confirmed by MRI revealing that the cellular-uptake was dose-responsive and that the uptake could be imaged by MRI at low concentration (6.25 μg mL^{-1}). The cytotoxicity of the particles towards HeLa cells was demonstrated upon irradiation. Evidence of $^1O_2$-induced apoptosis was found. This is the first example of a multifunctional

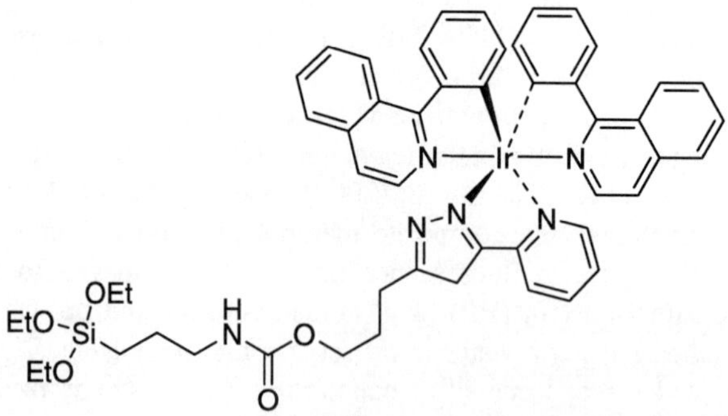

Fig. 29 Ir(III) complex (Reproduced by permission of The Royal Society of Chemistry)

particle, combining MRI, phosphorescence and PDT properties. The main drawback of this approach is the 366 nm irradiation, which is not really compatible with a good penetration of light in the tissue.

Lo and collaborators have achieved the tri-functionalization of mesoporous silica nanoparticles. They have incorporated a near-infra-red fluorescent contrast agent into the silica framework of MSN to obtain traceable imaging of nanoparticles. A palladium-porphyrin photosensitizer (PdTPP, Fig. 26) was incorporated in the pores of the nanoparticles to enable PDT. Targeting the over-expressed $\alpha_v\beta_3$ integrins of cancer cells was achieved by grafting the surface of the particle with cRGDyK peptides. These particles were shown to exhibit strong affinity for, and uptake by U87-MG glioblastoma cells (which over-express $\alpha_v\beta_3$ integrins) compared to MCF-7 human breast cancer cells (which are $\alpha_v\beta_3$ integrins deficient). *In vitro* evaluation of the theranostic platform demonstrated targeting specificity and minimal collateral damages as well as an important cell death after irradiation.

4 Semi-conductors Nanoparticles (Quantum Dots and Silica Nanoparticles)

4.1 Quantum Dots (QDs)

QDs have already been explored as potential agent for PDT and numerous papers have been published. In this part, we will just give few examples of recent studies. A review of Biju et al. (2010) described the different quantum dots that have been used so far for PDT applications. Indeed, exceptional photostability of QDs in addition with their broad absorption, their surface coating that can be functionalized to make them water soluble and biocompatible and their large two-photon absorption cross-section make them ideal candidates for PDT under the condition that they have good efficiency to produce ROS by direct irradiation or energy transfer to acceptor molecules. The combination of QDs and photosensitizers enabled the use of an excitation wavelength where the photosensitizer alone does not absorb. Thanks to the broad absorption spectrum of QDs, it is easy to find an excitation wavelength to activate the photosensitizer.

The first demonstration of a combination of QD with PDT photosensitizer with FRET (Förster Resonance Energy Tranfer) to facilitate the excitation of a phthalocyanine was described in 2003 by Samia et al. (2003) Indeed, QDs can be considered as energy donors and the possibility for energy transfer between QD and cell molecules has a potential to induce generation of ROS although results in the literature have been contradictory (Juzenas et al. 2008). Some studies find significant ROS production from QDs and others none (Ipe et al. 2005). It appears that QDs with CdSe (Chan et al. 2006) and CdTe (Choi et al. 2007) are very efficient at ROS generation while core/shell CdSe/ZnS does not produce significant ROS by itself (Ipe et al. 2005).

Many authors used FRET mechanism to produce singlet oxygen between a QD and a photosensitizer covalently or not covalently coupled. The systems are composed of CdSe, CdSe/CdS/ZnS, CdSe/ZnS and CdTe QDs as energy donors and porphyrins, chlorins, phthalocyanines, inorganic complexes or organic dyes as energy acceptors.

For example, Shi et al. (2006) proved that a water-soluble CdTe QD with 2-aminoethanethiol as surface stabilizer was not able to produce singlet oxygen by itself but did after excitation of a *meso*-tetra(4-sulfonatophenyl)porphyrin dihydrochloride (Fig. 30) bound to CdTe QD *via* electrostatic interaction. Jhonsi and Renganathan (2010) investigated the photoinduced interaction of water soluble thioglycolic acid capped CdTe Qds with porphyrins (Fig. 30). The porphyrins were found to be adsorbed in the surface of the QDs. Negatively charged porphyrins were involved in the energy transfer mechanism whereas positively charges ones involved electron transfer from QDs to porphyrin. The neutral porphyrin did not have any interaction with QD.

Tsay et al. (2007) coated chlorin e6 (Fig. 31) on the surface of CdSe/CdS/ZnS QDs which provided $^1O_2$ at 31% efficiency.

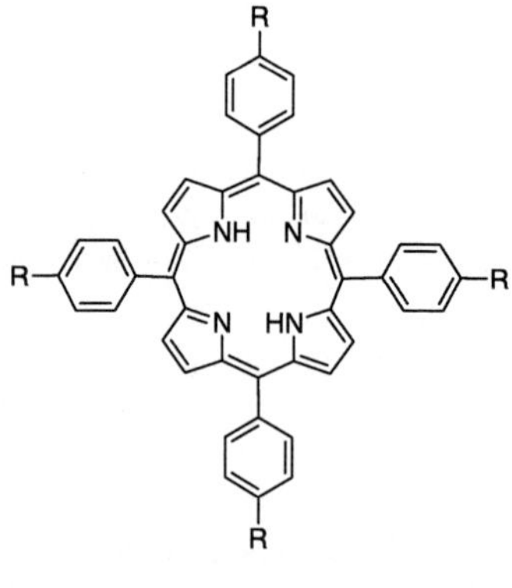

R = H meso-tetraphenylporphyrin (TPP)

R = $SO_3^{\ominus}$ meso-tetrakis(4-sulphonatophenyl)porphyrin (TSPP)

R = $COO^{\ominus}$ meso-tetrakis(4-carboxyphenyl)porphyrin (TCPP)

R = CH_3 meso-tetra(4-N-methylpyridyl)porphyrin (TMPyP)

Fig. 30 Porphyrins used by Jhonsi and Renganathan

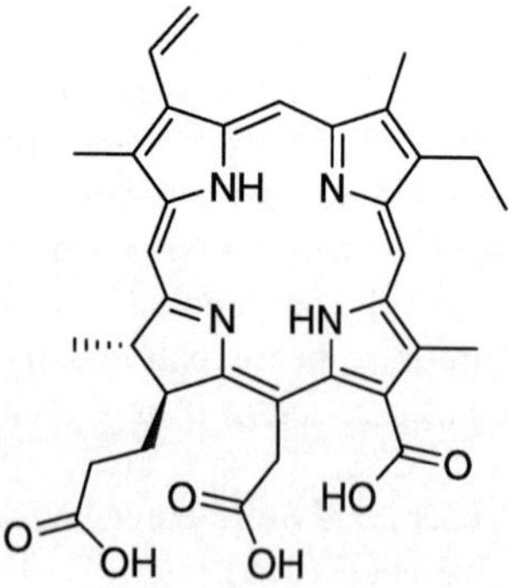

Fig. 31 Chlorin e6

Phthalocyanines have been also widely investigated due to their high triplet quantum yield and long-living triplet state. For example, Burda and co-workers (Dayal et al. 2006) investigated energy transfer and $^1O_2$ production by varying donor-acceptor distance, number of silicon *Pc* coupled to the QDs, terminal functional group in *Pc*. Some other groups were interested in tetrasulfonated aluminum *Pc* (Fig. 32) (Idowu et al. 2008) or octacarboxymetallophthalocyanines (Moeno et al. 2010).

The linking of QDs to low symmetry *Pc*s has been reported for the first time very recently by Chidawanyika et al. (2010). CdTe mercaptopropionic acid (MPA) QDs have been linked to a tris [9(10),16(17),23(24)-tert-butyl]imidophtalocyaninato Zinc (II) *Pc* (Fig. 33) and the authors showed that the linkage increased the FRET efficiency. The team of Nyokong (Erdogmus et al. 2010; Moeno et al. 2010) developed a system with zinc phthalocyanines (Fig. 34) and MPA capped CdTe QDs. They could observe a slight increase in the triplet state quantum yields and lifetime

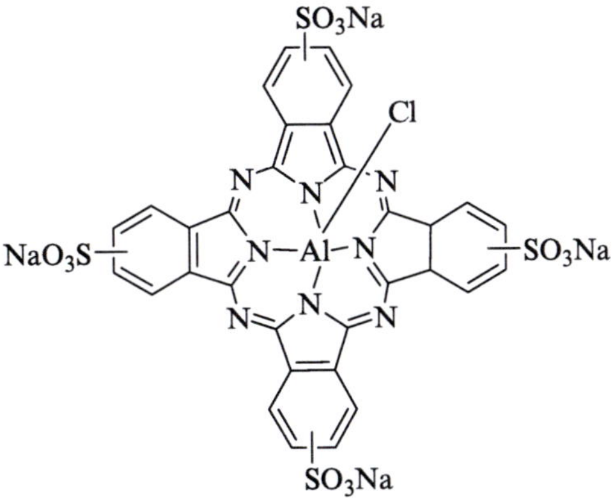

Fig. 32 Tetrasulfonated aluminum *Pc*

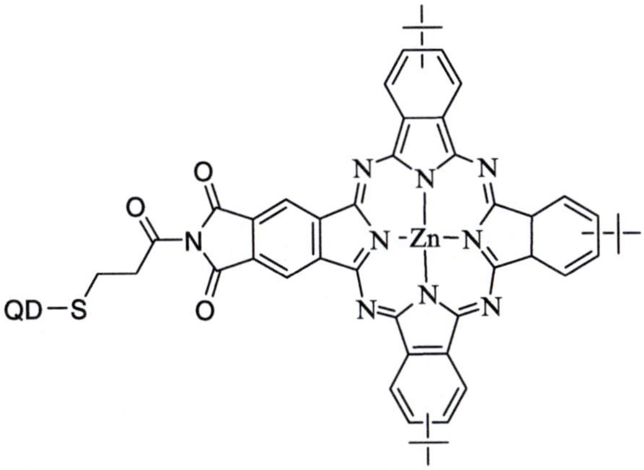

Fig. 33 Tris [9(10),16(17),23(24)-tert-butyl]imidophtalocyaninato Zinc (II) *Pc*

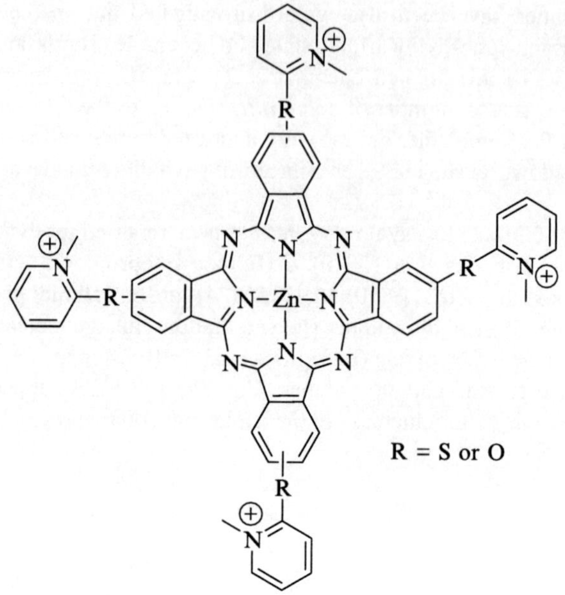

Fig. 34 Zinc phthalocyanine used by Nyokong

Fig. 35 Phthalocyanines used by Britton et al.

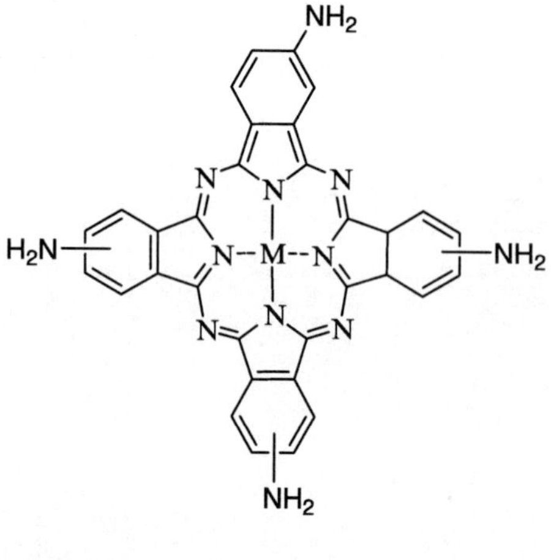

in the presence of QDs and energy transfer. The same team reported the chemically linked QDs with zinc phthalocyanines (Britton et al. 2009) or with indium tetraamino phthalocyanines (Britton et al. 2010) (Fig. 35). In all these studies, no *in vitro* results are given even if these systems seem to be good candidates for PDT.

Examples of organic and inorganic dyes for the preparation of QD-PS systems are respectively Merocyanine 540 (Narayanan et al. 2008) (Fig. 18) and iridium complexes (Hsieh et al. 2006) (Fig. 36). More recently, Rakovich et al. (2010)

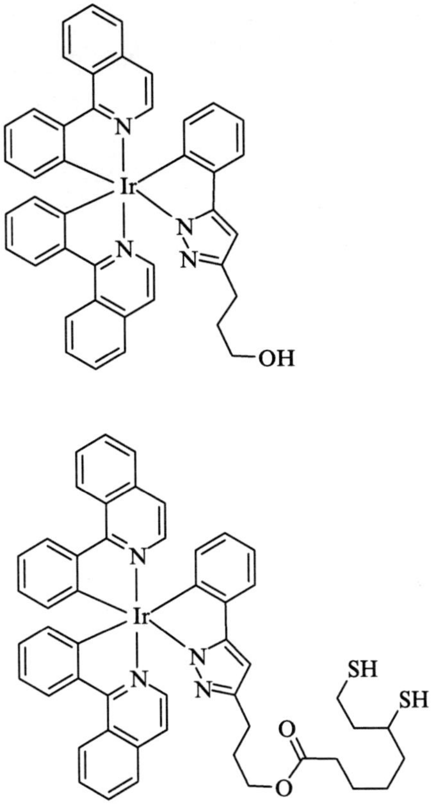

Fig. 36 Iridium complexes used by Hsieh et al.

proposed a novel CdTe Qds-methylene blue hybrid photosensitizer. Methylene blue (Fig. 11) adsorbs at the nanocrystal surface and the dye quenches the QD luminescence both by charge transfer and FRET mechanisms, leading to an increase of efficiency of MB to produce $^1O_2$. Some preliminary *in vitro* studies have been realized on Hela cell but the results demonstrate only a small increase in cell kill efficiency for the methylene blue-QDs mixture.

Without using FRET mechanism, Samia *et al.* have also demonstrated for the first time the emission of $^1O_2$ at 1,270 nm generated after photoactivation of hydrophobic capped CdSe QDs under excitation at 488 nm with $^1O_2$ quantum yield of around 0.05 in toluene. Cooper et al. studied the potential of CdSe/ZnS QDs for ROS production (Cooper et al. 2010) ROS assays in solution were performed by using sensor green while ROS generation inside cells were done by using 5-(and 6-)-chloromethyl-2′,7′-dichlorodihydrofluorescein diacetate, acetyl ester (Fig. 37). They showed that CdSe/ZnS make ROS although negligible singlet oxygen. When conjugated to dopamine, CdSe/ZnS QDs produces singlet oxygen. The authors suggested that this migth be due to the generated superoxide formed most probably both in reduction and oxidation processes namely in the reaction of photogenerated

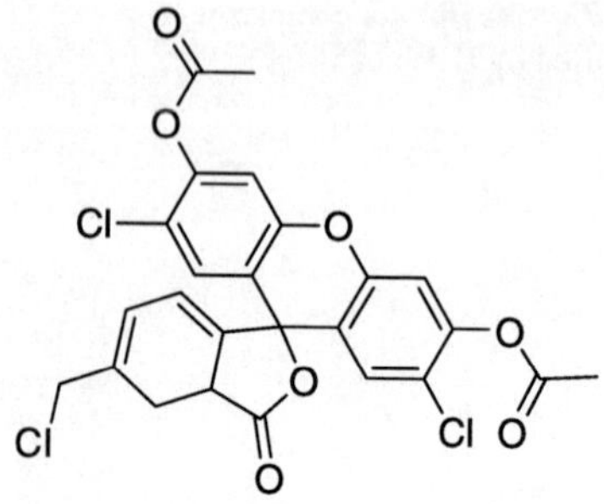

Fig. 37 5-Chloromethyl-2′, 7′-dichlorodihydrofluorescein diacetate

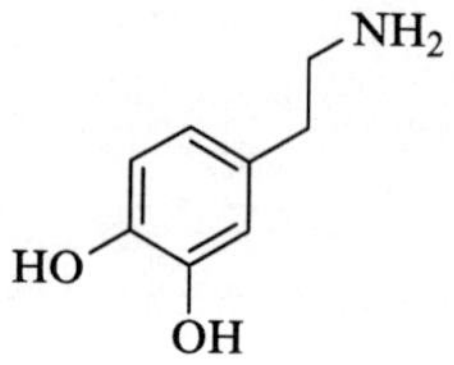

Fig. 38 Dopamine

electrons or in the reaction of positively-charged dopamine radicals with oxygen, respectively. Anas et al. (2008) also found formation of $^1O_2$ and that prolonged photoactivation of a CdSe ZnS QD-plasmid DNA conjugate at 512 nm resulted in the breakage and damage of DNA. Clarke et al. (2006) reported that photoactivation of QD-dopamine complex internalized in A9 cells results also in DNA damage due to the production of $^1O_2$ (Fig. 38).

4.2 Silicon Nanoparticles

Silicon nanosystems have been recently shown to be efficient photoluminescence emitter under light excitation as well as $^1O_2$ generator. Two major pathways have been described for their synthesis. Kovalev's method (Timoshenko et al. 2006) consists of obtaining silicon nanocrystals using electrochemical etching of silicon wafers in a hydrofluoric acid solution. This treatment led to porous silicon layers which were mechanically milled (ultrasonic bath). The grain size of the powder was several micrometers but the size of the nanocrystallites in the powder formed from them was equal to 3–4 nm. The suspension of the powder was prepared in water. Photoluminescence was observed by excitation at 337 nm. Photoluminescence was quenched when the suspension was saturated with O_2 indicating transfer to O_2 molecules. The silicon nanocrystals were then used to treat living mouse fibroblasts. 80% of cancer cells death was observed when a concentration of particles of 2.5 g/L with an irradiation of 1 h using a mercury-vapor light with a glass filter possessing a pass band of 350–600 nm, demonstrating the efficiency of silicon particles for PDT applications.

The second method used for the synthesis of silicon nanoparticles consists of a laser ablation of a silicon wafer in water by using a titane-sapphire laser (Rioux et al. 2009). The exposure of a silicon wafer to a femtosecond radiation for 30 min led to the formation of nanoparticles of mean size around 2.4 nm. The suspension was stable for months in water. Irradiation of the suspension at 266 or 523 nm led to the observation of $^1O_2$ luminescence. The $^1O_2$ production rate was estimated as 10% of the rate of photofrin. Si nanoparticles were also subjected to a 532 nm pulsed laser for 1 h at 200 mW/cm^2. No signs of photobleaching were observed indicating a higher stability than conventional organic photosensitizers.

5 Carbon Nanotubes

Single (SWNTs) or multiwalled carbon nanotubes (MWNTs) can be used as delivery agents for PDT photosensitizers. Carbon nanotubes present several advantages: it has been showed that they are internalized by mammalian cells (Pantarotto et al. 2004) through endocytosis (Kam et al. 2004) even if other mechanisms can occur (Kostarelos et al. 2007), they are excreted from the body rapidly and present no significant cytotoxicity (Liu et al. 2008c), they can be modified chemically (Hirsch 2002; Qu et al. 2002) but non covalent associations are also possible (Chen et al. 2001). The combination of carbon nanotubes to organic molecules can lead to the formation of unique composites with enhanced chemical and/or physical properties (Niyogi et al. 2002). SWNT have been also used as efficient fluorescence quenching probes (Chen et al. 2003; Kam et al. 2004).

Zhu et al. (2008) have designed a new kind of $^1O_2$ production probe by linking non-covalently a single strand DNA aptamer – chlorin e6 PS moiety (Fig. 31) with a single-walled carbon nanotubes (SWNT) as it has already been described in a recent review (Verhille et al. In Press). Zhu et al. supposed that they could be good $^1O_2$ quenchers. Meanwhile, interactions of SWNTs with biomolecules, such as proteins and DNA, have been intensively studied and applied to biosensing and as intracellular transporters. This indicates SWNTs may also protect DNA probes from digestion by nuclease. Aptamers are synthetic DNA/RNA probes that can recognize and bind their targets with high affinity and specificity (Ellington and Szostak 1990; Tuerk and Gold 1990). Thrombin aptamers have been widely investigated for the past decade. The aptamer coupled to a chlorin e6 (Fig. 31) was able to perform $^1O_2$ production upon PS irradiation. Due to the π-stacking between the aptamer and SWNT, the energy transfer occurs between the PS and the SWNT, leading to 98% $^1O_2$ quenching. DNA aptamers are good linkers to biological targets; they can bind biomolecules so strongly that the π-stacking, occurring between the SWNT and the photodynamic probe breaks off, releasing the chlorin from the $^1O_2$ quencher. This was illustrated by the chlorin e6 fluorescence which increased up to 20-fold after the addition of 2.0 μM thrombin. $^1O_2$ production was restored when the construct probe was put in presence of thrombin and then irradiated at chlorin e6 maximum absorption wavelength (404 nm). The high specificity of

Fig. 39 PS used by Erbas et al.

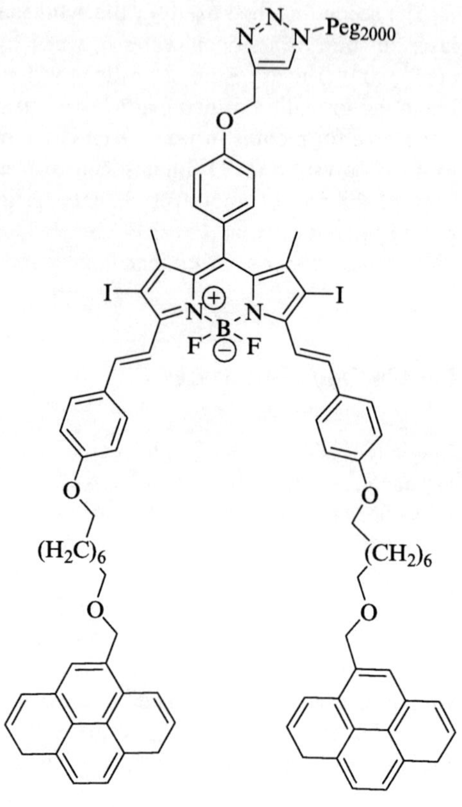

aptamers toward one biological target, and the quenching efficiency of the probe are many qualities to make out a relevant PDT agent.

In a recent paper, Erbas et al. (2009) developed non-covalent functionalized SWNTs as delivery agents for novel bodipy-based potential PDT photosensitizers. Their molecule was composed of PEG2000, distyryl-bodipy, intersystem crossing promoter heavy atoms (iodines at 2 and 6 positions of the bodipy core) and 2 pyrene subsituents for-π stacking interactions with the SWNTs (Fig. 39). A low decrease of $^1O_2$ production of the compound non covalently attached to the SWNT was observed, compared to the free one.

Even more recently, do Nascimento et al. (2010) modified SWNTs with organic dyes such as phenosafranine and Nile Blue (Figs. 40 and 41) that can act as photosensitizers in energy and electron transfer reactions to be applied in PDT. The authors realized a covalent assembly between SWNTs that have been oxidized to introduce COOH groups and the photosensitizers. Cytotoxicity and photocytoxicity effects were tested on BHK-21 cell line from mouse fibroblasts. The data for dye-modified SWNT composites in the presence or absence of light show that the SWNT-Nile Blue (4 μg/mL) composite described photodynamic efficiency and a reduced dye photobleaching by auto-oxidation.

Fig. 40 Phenosafranine

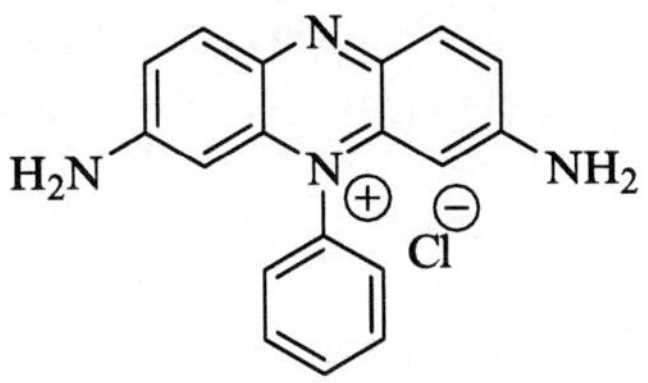

Fig. 41 Nile Blue

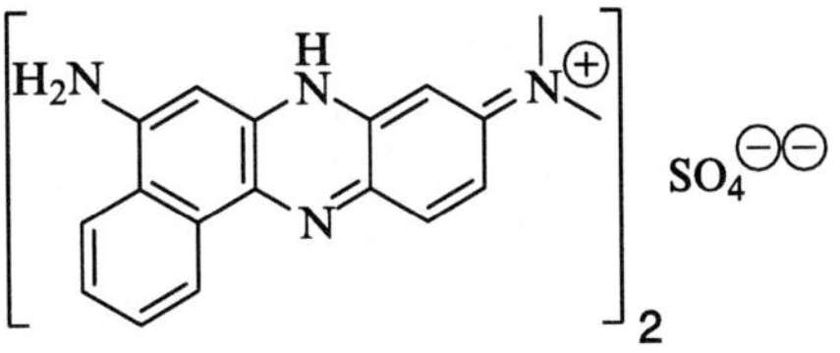

6 ZnO/TiO_2 Nanoparticles

6.1 TiO_2 Nanoparticles

Inorganic nanoparticles are rapidly emerging in biomedical applications as drug delivery carriers, biosensors, imaging diagnosis and disease therapeutics (Klostranec and Chan 2006). The inorganic nanoparticles that have been extensively explored in the field of biomedicine cover a variety of materials and diverse shapes, such as carbon nanotubes, silica nanospheres, metal nanoshells, and semiconductor quantum dots.

The application of TiO_2 nanoparticles in life science is attracting more and more attention since the first report of photocatalytic disinfection by Matsunaga et al. (1985). In recent years, TiO_2 nanoparticles were applied in the field of phototherapy of malignant cells, and have been regarded as the potential photosensitizing agents for PDT due to their unique phototoxic effect upon the irradiation (Cai et al. 1991, 1992; Fujishima et al. 2000; Sakai et al. 1994; Wamer et al. 1997; Zhang and Sun 2004).

The TiO_2 nanoparticles are novel photo-effecting material with a band gap of 3.23 eV for anatase and 3.06 eV for rutile polymorph of TiO_2, respectively. Under the irradiation of UV light with the energy higher than that of the band gap of TiO_2, i.e. the light wavelength shorter than 385 and 400 nm for anatase and rutile polymorphs, respectively, the electrons in the valence band of TiO_2 can be excited to the conduction band, creating the pairs of photo-induced electron and hole (Ashikaga et al. 2000). These photo-induced electrons and holes present strong reduction and oxidation properties. In aqueous environment, the photo-induced holes can react with hydroxyl ions or water to form powerful oxidizing radicals, such as hydroxyl radicals OH·, perhydroxyl radicals HO_2 (Fujishima et al. 2000) and superoxide anion (O_2^-) which can destroy the structure of various organic molecules (Fukahori et al. 2003; Henderson 1997; Semenikhin et al. 1999; Wu et al. 2006).

Cai et al. reported that photo-excited TiO_2 particles induced a photodynamic activity on HeLa cells and T-24 human bladder cancer cells (Cai et al. 1992; Kubota et al. 1994). They checked that the ROS generated by the photoexcited TiO_2 acted as a cytotoxin on the cell membrane and in the cytoplasm. The photo-cytotoxicity of TiO_2 was obtained by UV light irradiation with a wavelength region 300–400 nm that directly excites TiO_2. The penetration depth of UV light into human tissues is however limited to a few millimeters. Being largely absorbed by the skin, UV light cannot reach at the depth of the tissues where cancer exists (Kurwa and Barlow 1999).

Lei et al. studied the photodynamic activity of ascorbic acid-modified TiO_2 nanoparticles upon visible illumination (Lei et al. 2009; Takahashi et al. 1998). Though both TiO_2 and ascorbic acid (AA) are non cytotoxic in the dark or under visible irradiation, ascorbic acid-modified TiO_2 nanoparticles were found to exhibit efficient DNA and cancer cell damage ability under visible light irradiation, opening a new pathway for developing highly efficient PDT sensitizers. To investigate the photodynamic properties of AA-modified TiO_2 *in vitro*, human breast cancer MCF-7 cells were used. An 1-h irradiation (>550 nm) led to viability reduction from 50% to 80% in the cases of AA-modified TiO_2. In contrast, AA or TiO_2 alone exhibited negligible phototoxicity under the same conditions. The low dark cytotoxicity and remarkable phototoxicity of AA-modified TiO_2 demonstrated its promising potential as a versatile photosensitizing system in PDT application.

Yet, the TiO_2 nanoparticles still present some drawbacks for a potential clinical use, such as insufficient selectivity and low efficiency resulted from lack of cell-specific accumulation of TiO_2 in cancer cells. In addition, a relatively high concentration of TiO_2 nanoparticles is needed (commonly higher than 50 μg/mL). Hence, it is necessary to modify TiO_2 nanoparticles with some biomolecules that can specifically bind with cancer cells for improving selectivity and efficiency. Recently, the high affinity and specificity of antibodies to antigens have been used to improve the performance of photosensitizing agents (Takahashi et al. 1998). For example, the monoclonal antibodies have been conjugated with some porphyrin-based photosensitizers to improve the specificity of PDT agents (Liang et al. 2006). In PDT, use of visible light-sensitive dyes (photosensitizers) such as hematoporphyrin derivatives has been extensively studied. Among sensitizers, chlorin e6 (Fig. 31), absorbing at 670 nm is an ideal dye. Ogura et al. showed the photocytotoxicity of conjugated chlorin and chlorin e6 in Hela cells and found high toxicity over non-conjugated dyes (Hirohara et al. 2005; Ogura et al. 2005).

Tokuoka's team assessed also photocytotoxicity of dye-sensitized TiO_2 particles as an anticancer agent for PDT (Tokuoka et al. 2006). They prepared nanoparticles by mixing neutral aqueous sol of nanocrystalline TiO_2 (anatase) with water and a dilute solution of chlorin e6 trisodium salt as a dye. These TiO_2 nanocrystalline particles (dye-TiO_2 particles) were administered into a cancer murine thymic lymphoma (EL-4). Irradiation using polychromatic light (550–750 nm) to excite chlorin e6 significantly damaged the EL-4 cells. A higher cell-killing effect was found for the dye-TiO_2 particle than for chlorin e6 alone. The survival rate in the presence of dye–TiO_2 gradually decreased with increasing dye concentration. In contrast, the survival rate in the TiO_2-free chlorin e6 system remained 100%, indicating that no cytotoxic effect occured with TiO_2 alone in the dark as well as under visible light irradiation. The measured photocytotoxicity would be caused by reactive oxygen species, for example, singlet oxygen, superoxide, and hydroxy radical, generated by the photoexcited dye–TiO_2 particles although the nature of reactive oxygen species involved has not been specified. It was expected, however, that the positively charged TiO_2 nanocrystals have higher affinity to tumor cells, which are more negatively charged than normal cells (Miyasaka and Kijitori 2004).

Electroporation (Labanauskiene et al. 2007; Lambreva et al. 2004; Nolkrantz et al. 2002; Underhill et al. 2003; Xu et al. 2007b) is an important bioengineering technique that has been used to deliver genes or anticancer drugs into the cytoplasm through the micropores on the cell membrane produced through electric stimulation and has been successfully utilized in photodynamic therapy. Xu et al. reported for the first time that the combination of the electroporation and the conjugation of the TiO_2 nanoparticles with the monoclonal antibody could improve the photokilling selectivity and efficiency of photoexcited TiO_2 on cancer cells in PDT because the conjugation of the TiO_2 nanoparticles with monoclonal antibodies could increase the photokilling selectivity of TiO_2 nanoparticles to cancer cells and the electroporation could accelerate the delivery speed of the TiO_2 nanoparticles to cancer cells.

They firstly conjugated TiO_2 nanoparticles with a specific antibody against the carcinoembryonic antigen (Hefta et al. 1998) of human LoVo cancer cells. Electroporation was used to improve the delivery of antibody–TiO_2 bioconjugates into the cancer cells.

To obtain their nanoparticles, anti-carcinoembryonic antigen antibody was mixed to fluorescein isothiocyanate-dimethylsulfoxide solution to obtain a FITC-antibody solution which was added to a pH adjusted TiO_2 suspension (ratio of 800 μL antibody solution with 1 mg TiO_2 nanoparticles). The selective binding of the FITC-antibody–TiO_2 conjugate was compared between LoVo cancer cells and TE353.sk normal cells. They observed that after the incubation in the culture medium containing the FITC-antibody–TiO_2 conjugate for 30 min, LoVo cancer cells clearly exhibit a green fluorescence light, in contrast TE353.sk normal cells did not show any fluorescence, indicating that the FITC-antibody–TiO_2 conjugate bind only to the LoVo cancer cells.

Using the combination of PDT and electroporation, 100% human LoVo cancer cells were photodestroyed within 90 min, *versus* 39% for normal cells under ultraviolet (UV) light (365 nm). Furthermore, the combination might be used to photokill other types of cancer cells, by using other antibodies.

In 2010, Lopez et al. developed novel nanostructured TiO_2 and SiO_2 based biocatalysts, with 3–4 wt.% of Pt (Lopez et al. 2010b). In fact, since the pioneer work of Rosenberg in 1965, Pt complexes have been widely used in cancer therapy (Rosenberg et al. 1965, 1969). From then on, numerous attempts to obtain new platinum complexes with better therapeutic performances than cis-platin have been made (Kovala-Demertzi et al. 2007; Marković et al. 2005).

H_2PtCl_6/TiO_2 and H_2PtC16/SiO_2 were prepared by a sol-gel process using TEOS and titanium butoxide as precursors. Hydrolysis was carried out at neutral pH and $H_2PtCl_6.6H_2O$ was used as platinum precursor (Lopez et al. 1993, 1999). Titania was further functionalized with a neurotransmitter in order to develop an analogy with cells (Lopez-Goerne 2007). The obtained gels were dried to form solid nanoparticles. The materials exhibit a high surface area together with a broad pore size distribution. The method of synthesis allowed obtaining high dispersed platinum metal nanoparticles. *In vitro* DNA reactivity test of the biocatalysts were carried out by electrophoresis and formation of DNA adducts was observed. The most active biocatalyst was H_2PtCl_6/SiO_2. These biocatalysts were also tested in an experimental model of C6 brain tumours in Wistar rats. Administration of the material was made by stereotactic brain surgery to place it directly in the malignant tissue. A significant decrease in tumour size and weight as well as morphologic changes in cancer cells were observed. The platinum compound coordinates to DNA, probably by forming bonds with DNA through hydrolysis of the phosphate group. The success of the *in vivo* treatment is related to the modification of the surface properties of the titania support enhancing the contact of the solid with the cell membrane, which permits to deliver the Pt compounds at the damaged site. These particular features, besides the high surface area of sol gel materials, lead to a major availability of platinum, making possible to use lower platinum doses than systemic administration.

The same group investigated the possible synergy of zinc phthalocyanines (ZnPc, Fig. 17) supported on TiO_2 nanoparticles and probed their *in vitro* photoactivity using visible light, on cancer cells and Leishmania parasites (Lopez et al. 2010a). They studied the photosensitizing effect of ZnPc, nano-TiO_2, and ZnPc-TiO_2 conjugate, against a panel of tumor and nontumor mammalian cells and on promastigote forms of Leishmania parasites. As was expected, nano-TiO_2 alone under visible light irradiation was not phototoxic for the cells; in contrast, ZnPc treatment at the same condition was photoactive for all the studied cells and parasites. The composite ZnPc-TiO_2 was not phototoxic for *L. Chagasi* or *L. Panamensis* promastigotes; however, it was active against tumor and nontumor mammalian cells but less than the pure ZnPc PS. ZnPc-TiO_2 was internalized by the cells at a lower level than ZnPc. The localization of ZnPc-TiO_2 and ZnPc were observed in mitochondrial cytoplasm. No fluorescence signal was observed in human-derived fibroblasts exposed to ZnPc-TiO_2.

6.2 *ZnO Nanoparticles*

Thanks to its potential of non-toxicity, zinc oxide (ZnO) is a candidate for the materials applicable for biomedicine (Dorfman et al. 2006b, 2006a; Liu et al. 2008b). Furthermore, among the PDT agents, ZnO is a material of particular interest due to its unique optical and electronic properties, which has been widely used for device applications including transducers, phosphors and varistors (Van De Walle 2000). Currently, as a well known photosensitizer, ZnO nanoparticles could absorb over a larger fraction of UV zone. ZnO is a direct band gap semiconductor with a band gap of 3.4 eV. ZnO nanostructures typically have a near-band-edge emission in the UV region and a defect-related visible emission in the blue to green region (Lin et al. 2005; Liu et al. 2008b; Cheng et al. 2005). With UV illumination, ZnO photocatalyst in aqueous solution can generate ROS such as hydroxyl radical, hydrogen peroxide and superoxide, which makes it possible and high efficient for the application of ZnO nanoparticles in the decomposition of some organic compounds and induced cytotoxicity against some mutated cells (Brunner et al. 2006; Wang et al. 2007; Xu et al. 2007a). Considering the possibility of combined application of nano-sized ZnO particles with anticancer drugs in PDT, Guo et al. (2008) studied the synergistic potency of the high reactivity of ZnO nanoparticles with daunorubicin, one of the most important anticancer drugs in the clinic treatment of acute leukemia and solid tumors (Beguin et al. 1997; Lowis et al. 2006), to inhibit multidrug resistance of drug-resistant leukemia K562/A02 cells and facilitate the relevant drug delivery efficiency. They explored the cytotoxic effect of anticancer drug daunorubicin on leukemia cancer cells in the absence and presence of ZnO nanoparticles *via* fluorescence microscopy, UV–Vis absorption spectroscopy and electrochemical analysis. Their observations demonstrate the apparently enhanced cellular uptake of daunorubicin for both leukemia cell lines in the presence of the different sized ZnO nanoparticles. They demonstrated (Li et al. 2010a) that different-sized ZnO nanoparticles exposed to SMMC-7721 cancer cells could exert dose-dependent cytotoxicity suppression *in vitro*, and this cytotoxicity of nanoparticles was time- and dose- dependant, while the size-depended effect was not clear in the scope from 20 to 100 nm. UV irradiation could readily enhance the proliferation suppression ability of ZnO nanoparticles on cancer cells. More importantly, these effects were size dependent, while the smaller the nanoparticle size, the higher the cytotoxicity of cancer cell proliferation caused by ZnO nanoparticle. Meanwhile, if ZnO nanoparticles combined with daunorubicin, cytotoxicity of daunorubicin for SMMC-7721 cancer cells was especially enhanced. These observations suggest that ZnO nanoparticles could play an important role in the PDT and have the great potential and promising applications in clinical and biomedical engineering.

Liu et al. reported synthesis of ZnO-porphyrin conjugates and their potential applications in PDT of cancer (Liu et al. 2008b). ZnO nanoparticles were synthesized and conjugated to the porphyrin derivative, *meso*-tetra (o-amino phenyl) porphyrin (MTAP, Fig. 42). The key step of the synthesis is the ligation of MTAP to Zn *via* a cysteine. The surface of ZnO nanoparticles was first modified to contain –COOH group. The isoelectric point of ZnO is 9 and Zn^{2+} is dominant on the surface of the ZnO nanoparticles. The –SH group of L-cysteine can react with Zn^{2+} to form Zn(II) complex ZnO–(L-cysteine), since Zn–S bond is stronger than Zn–O bond. ZnO–(L-cysteine) contains –COO– group which can conjugate with the primary amine group of MTAP, when appropriately activated (Wang et al. 2002).

They investigated the fluorescence resonance energy transfer following UV irradiation and the energy transfer rate was measured to be as high as 83%. Their observations indicate that ZnO–(L-cysteine)-MTAP nanoparticle conjugates are efficient photodynamic agents that can be induced by 365 nm UV light.

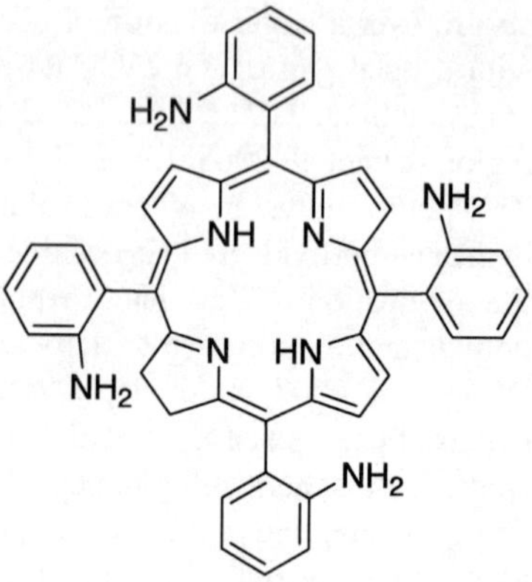

Fig. 42 *Meso*-tetra (o-amino phenyl) porphyrin (MTAP)

Human carcinoma NIH:OVCAR-3 cells were exposed to ZnO, MTAP, or ZnO–(L-cysteine)-MTAP under dark conditions or with exposure to UV (365 nm, 30 min) and viability subsequently measured. Metabolic activity was 98% of control with ZnO alone, 100% with MTAP alone, and 98% with ZnO–(L-cysteine)-MTAP conjugates under dark conditions, i.e., neither of the conditions affected cell viability. Using the same conditions but with UV light exposure, metabolic activity was 98% with ZnO alone, 75% with MTAP, and 10% with ZnO–(L-cysteine)-MTAP conjugates. These results suggest that (1) ZnO nanoparticles have relatively little cytotoxic potential in NIH:OVCAR-3 cells under either dark or UV-illuminated conditions, (2) UV irradiation alone (30 min, 365 nm) has relatively little effect on cell viability, (3) the porphyrin derivative MTAP alone has little cytotoxic potential under dark conditions but elicits moderate cytotoxicity with UV, and (4) ZnO–(L-cysteine)-MTAP conjugates can elicit substantial cytotoxicity, compared to either ZnO or MTAP alone, when exposed to UV.

In 2009, Hiromitsu et al. studied the photoluminescence properties and the energy transfer efficiency of ZnO nanoparticle–tetraphenylporphyrin (TPP, Fig. 22) systems. Four kinds of tetraphenylporphyrins, i.e. 5,10,15,20-tetraphenylporphyrin (H_2TPP), 5-(4-aminophenyl)-10,15,20 triphenylporphyrin ($H_2TPP(NH_2)$), 5,10-bis(4-aminophenyl)-15,20-diphenylporphyrin ($H_2TPP(NH_2)_2$), and [5,10-bis(4-aminophenyl)-15,20-diphenylporphyrinato] zinc(II) ($ZnTPP(NH_2)_2$) were used instead of MTAP studied by Liu et al. (2008b). Among these, $H_2TPP(NH_2)$, $H_2TPP(NH_2)_2$ and $ZnTPP(NH_2)_2$ have one or two p-aminophenyl groups instead of four o-aminophenyls in MTAP. The ZnO–(L-cysteine)-TPP conjugates were prepared using the standard cross linking method (Liu et al. 2008b; Staros 1982).

Photoluminescence properties of the ZnO–(L-cysteine)–porphyrin systems dispersed in diethylene glycol were studied. The central hydrogen atoms of the metal-free TPPs were replaced by a Zn atom by an aging of 2–4 weeks, which altered the Q-band features of the TPPs significantly. For ZnO–(L-cysteine)-$H_2TPP(NH_2)_2$ and ZnO–(L-cysteine)–$ZnTPP(NH_2)_2$, new emission bands were generated at 540 and 577 nm, the origin of which is unknown at present. The energy transfer efficiency from the photoexcited ZnO to the porphyrin was the same for ZnO–(L-cysteine)–H_2TPP,

ZnO–(L-cysteine)–$H_2TPP(NH_2)$ and ZnO-$H_2TPP(NH_2)$, and the efficiency became smaller by diluting the system by adding diethylene glycol. This clearly indicates that the energy transfer occurs by the collisions of the ZnO particles with porphyrins in the dispersion. ZnO and TPP do not bound with each other because L-cysteine is not bound to the ZnO particles.

When ZnO nanoparticles are combined with drugs or photosensitizers, the related cytotoxicity of anticancer agents on cancer cells seems enhanced, suggesting that ZnO nanoparticles could play an important role in drug delivery. This may offer the possibility of the great potential and promising applications of the ZnO nanoparticles in clinical and biomedical areas like photodynamic cancer therapy and others.

7 Miscellaneous

7.1 Mg-Al Hydroxides

Kantonis et al. (2007) proposed Mg-Al layered double hydroxides nanohybrids to produce biocompatible nanomaterials for application in PDT. Protoporphyrin IX (Fig. 7) was thus encapsulated and used as the PDT drug. To maintain high oxygen concentration in solution, the authors have also co-encapsulated perfluoroheptanoic acid (PFHA), a non-toxic perfluorocarbon rapidly eliminated from the body. Indeed, perfluorocarbons can dissolve 20 times more oxygen than water. PFHA formed a bilayer between the inorganic layers of the nanohybrid, in addition to the intercalation of PpIX. 10% (w/w) of PpIX were immobilized in the nanohybrid. The photocatalytic activity was shown by oxidation of imidazole, 2,3-dimethyl-2-butene and linoleic acid. The oxidation of hydrophobic linoleic acid was much faster with PFHA-PpIX-modified nanohybrid than PpIX alone whereas the oxidation of water soluble imidazole was much faster with PpIX. Thus, Mg-Al layered double hydroxides nanohybrids are promising for PDT.

7.2 Calcium Phosphate

Epple's group (Epple et al. 2010) have functionalized Ca-phosphate nanoparticles with different polymers and photosensitizers were incorporated into that layer. The calcium phosphate nanoparticles were prepared by rapid precipitation from water with polymers to stabilize the nanoparticles. The polymers used were polystyrene sulfonic acid (PSS), polyallylamine hydrochloride, carboxymethylcellulose (CMC), polyethylene imine. The layer by layer technique was used to coat the negatively charged nanoparticles (PSS and CMC) with polyallylamine hydrochloride in the case of PSS, polyethylene imine in the case of CMC. mTHPP (Fig. 43) and Methylene Blue (Fig. 11) were used as photosensitizers. Whatever

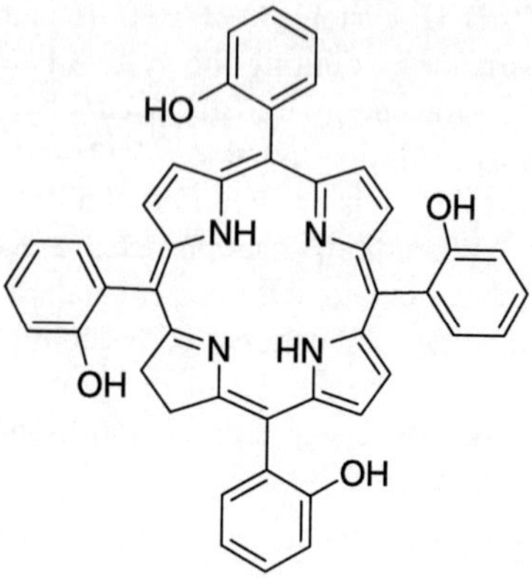

Fig. 43 *Meso*-tetrahydroxyphenyl porphyrin, mTHPP

the nanoparticles, the PDT results on HT29 cancer cells line showed a moderate activity whereas a good performance with HIG-82 Synoviocytes was observed. For J774A.1 macrophages, nanoparticles were also toxic in the dark, probably due to a lethal uptake of calcium. A good phototoxicity was also observed against the bacterial strain *S. Aureus* (Gram positive) both with positively and negatively charged nanoparticles loaded with mTHPP. Against *P. aeruginosa* (Gram negative) good phototoxicity was observed only with positively charged nanoparticles loaded with mTHPP. At higher concentrations, methylene-blue loaded nanoparticles where active against *S. Aureus*.

7.3 Zeolites

Zeolites are crystalline microporous alumino-silicates whose nanoparticles are very promising for PDT applications. Strassert and co-workers (2009) proposed the use of zeolite L nanocrystals 50 nm in both length and diameter for PDT against antibiotic-resistant bacteria. The nanocrystal were loaded with *(N,N′)*-bis(2-6-dimethylphenyl)perylene-3,4,9,10-tetracarbodiimide (DXP, Fig. 44), which was encapsulated by gas-phase insertion, for imaging and to label the cells. The DXP-loaded zeolite-L nanocrystals were further functionalized with tetra-tert-butyl-substituted Si^{IV}, phthalocyaninedihydroxyle (PC, Fig. 45) by axial attachment of the macrocycle's central silicon atom to OH groups of the zeolite L surface. No energy transfer occurred from DXP to PC. The surface of the multi-functional zeolites was coated with amino-groups to promote the adhesion to bacteria. Indeed the zeolites are then positively charged, favoring adhesion to the negatively charged membranes of the bacteria. The PDT activity of the zeolite nanocrystals was tested on *E. Coli* and *N. Gonorrhoeae* bacteria. Irradiation for 2.5 h between 570 and 900 nm (3 mW cm^{-2}, light dose 27 J cm^{-2}) led to the inactivation of the cells (95% efficiency).

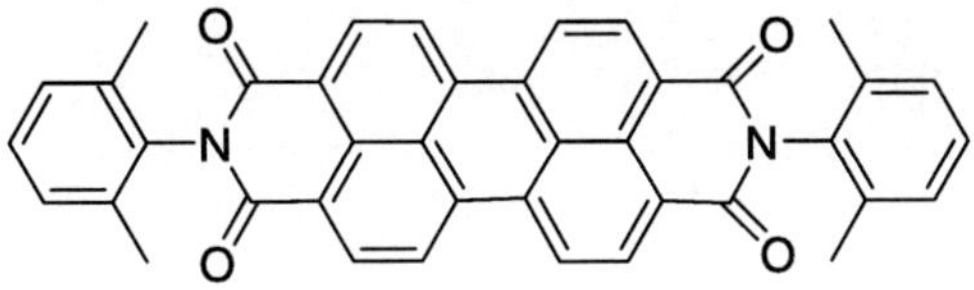

Fig. 44 (N,N′)-bis(2-6-dimethylphenyl)perylene-3,4,9,10-tetracarbodiimide (DXP)

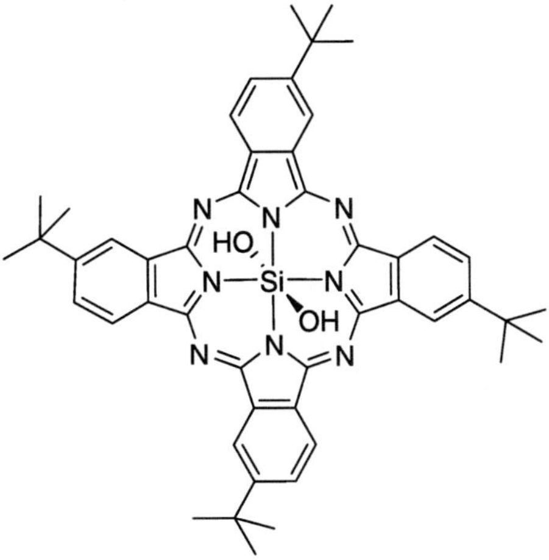

Fig. 45 Phthalocyaninedihy droxyle (PC)

Hocine et al. (doi:10.1016/j.ijpharm.2010.10.004) described the preparation of silicalite (crystalline silica zeolite) nanoparticles with diameters below 100 nm. Those nanoparticles covalently encapsulated water-soluble porphyrins, at low amount inside the silica matrix. Those nanoparticles were able to generate singlet oxygen. After incubation of the silicalite nanoparticles with MDA-MB-231 breast cancer cells for 24 h, and irradiation with red light (650 nm), significant cancer cell death (45–50%) was observed.

7.4 *Iron Oxide (Fe_3O_4)*

Macrophage has emerged as a key biological, imaging and therapeutic target for atherosclerosis and antimacrophage therapeutic strategies are becoming prevalent. Among them, the team of McCarthy et al. (2006, 2010) used systemic modulation of macrophage activity and localized PDT. They described in 2006 the synthesis and biological study of dextran-coated magnetofluorescent nanoparticles conjugated to 5-(4-carboxyphenyl)-10,15,20-triphenyl-2,3-dihydrochlorin (Fig. 46) as a photosensitizer and Alexa Fluor 750 as a fluorophore. Cellular uptake and light-induce phototoxicity were investigated in both the RAW 264.7 murine macrophage cell line and transformed human macrophages (U937 cells). These nanoobjects

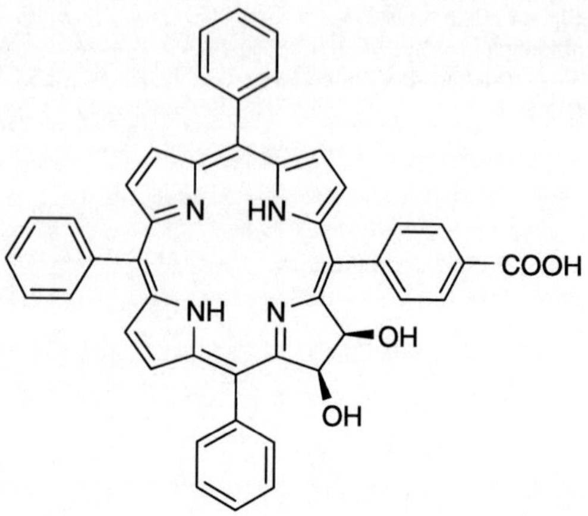

Fig. 46 5-(4-carboxyphenyl)-10,15,20-triphenyl-2,3-dihydrochlorin

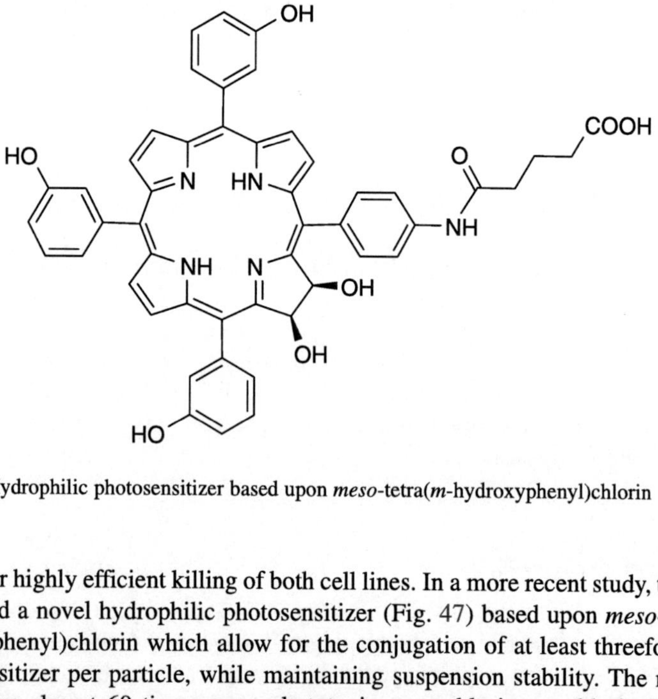

Fig. 47 Hydrophilic photosensitizer based upon *meso*-tetra(*m*-hydroxyphenyl)chlorin

allows for highly efficient killing of both cell lines. In a more recent study, the same team used a novel hydrophilic photosensitizer (Fig. 47) based upon *meso*-tetra(*m*-hydroxyphenyl)chlorin which allow for the conjugation of at least threefold more photosensitizer per particle, while maintaining suspension stability. The nanoparticles were almost 60 times more phototoxic on a chlorin-per-chlorin basis and 1,900-fold more toxic on a molecular level compared to classical chlorin e6. The *in vivo* localization to atherosclerotic lesions was examined in apolipoprotein E deficient (apoE-/-) mice. The agent clearly accumulated within the plaques, particularly in areas rich in macrophages and foam cells. The nanoparticles induced extensive apoptosis upon irradiation at 650 nm and the skin phototoxicity measured was also substantially reduced compared to conventional photosensitizers. While most therapeutic approaches have involved the systemic administration of cytotoxic drugs, focal therapy such as PDT allows for real control of the localization of the therapeutic effect.

8 Conclusion

Simple and proper methodologies to develop formulations of nanoparticles applicable for PDT applications are currently being developed. Depending of the synthetic method, nanoparticles of different controlled size, porosity and properties can be elaborated. Moreover, addition of cancer biomarkers can guide the particle to specifically targeted cancer cells resulting in less damage of healthy cells. Recent developments concern the use of silica nanoparticles that can be excited by IR or two-photon in order to treat tissues more deeply. New researches are also focused on multifunctional nanoparticles that could serve both as diagnosis tools and therapeutic agents. Thus, we believe that nanoparticles gathering (ii) covalent coupling of the photosensitizer, (ii) specific core that allow for example MRI for diagnosis or drug encapsulation and (ii) a vector may be an important advancement for the future of PDT.

References

Allison, R. R., Mota, H. C., Bagnato, V. S. & Sibata, C. H. (2008) Bio-nanotechnology and photodynamic therapy – State of the art review. *Photodiagnosis and Photodynamic Therapy*, 5, 19–28.

Anas, A., Akita, H., Harashima, H., Itoh, T., Ishikawa, M. & Biju, V. (2008) Photosensitized breakage and damage of DNA by CdSe-ZnS quantum dots. *J. Phys. Chem. B*, 112, 10005–10011.

Ashikaga, T., Wada, M., Kobayashi, H., Mori, M., Katsumura, Y., Fukui, H., Kato, S., Yamaguchi, M. & Takamatsu, T. (2000) Effect of the photocatalytic activity of TiO_2 on plasmid DNA. *Mutation Research – Genetic Toxicology and Environmental Mutagenesis*, 466, 1–7.

Bechet, D., Couleaud, P., Frochot, C., Viriot, M.-L., Guillemin, F. & Barberi-Heyob, M. (2008) Nanoparticles as vehicles for delivery of photodynamic therapy agents. *Trends in Biotechnology*, 26, 612–621.

Beguin, Y., Sautois, B., Forget, P., Bury, J. & Fillet, G. (1997) Long term follow-up of patients with acute myelogenous leukemia who received the daunorubicin, vincristine, and cytosine arabinoside regimen. *Cancer*, 79, 1351–1354.

Biju, V., Mundayoor, S., Omkumar, R. V., Anas, A. & Ishikawa, M. (2010) Bioconjugated quantum dots for cancer research: Present status, prospects and remaining issues. *Biotechnology Advances*, 28, 199–213.

Brevet, D., Gary-Bobo, M., Raehm, L., Richeter, S., Hocine, O., Amro, K., Loock, B., Couleaud, P., Frochot, C., Morere, A., Maillard, P., Garcia, M. & Durand, J. O. (2009) Mannose-targeted mesoporous silica nanoparticles for photodynamic therapy. *Chem. Commun.*, 1475–1477.

Britton, J., Antunes, E. & Nyokong, T. (2009) Fluorescence studies of quantum dots and zinc tetraamino phthalocyanine conjugates. *Inorganic Chemistry Communications*, 12, 828–831.

Britton, J., Antunes, E. & Nyokong, T. (2010) Fluorescence quenching and energy transfer in conjugates of quantum dots with zinc and indium tetraamino phthalocyanines. *Journal of Photochemistry and Photobiology a-Chemistry*, 210, 1–7.

Brunner, T. J., Wick, P., Manser, P., Spohn, P., Grass, R. N., Limbach, L. K., Bruinink, A. & Stark, W. J. (2006) In vitro cytotoxicity of oxide nanoparticles: Comparison to asbestos, silica, and the effect of particle solubility. *Environmental Science and Technology*, 40, 4374–4381.

Cai, R., Hashimoto, K., Itoh, K., Kubota, Y. & Fujishima, A. (1991) Photokilling of malignant cells with ultrafine TiO_2 powder. *Bulletin of the Chemical Society of Japan*, 64, 1268–1273.

Cai, R., Kubota, Y., Shuin, T., Sakai, H., Hashimoto, K. & Fujishima, A. (1992) Induction of cytotoxicity by photoexcited TiO_2 particles. *Cancer Research*, 52, 2346–2348.

Camerin, M., Magaraggia, M., Soncin, M., Jori, G., Moreno, M., Chambrier, I., Cook, M. J. & Russell, D. A. (2010) The in vivo efficacy of phthalocyanine-nanoparticle conjugates for the photodynamic therapy of amelanotic melanoma. *Eur J Cancer*, 46, 1910–8.

Chan, W. H., Shiao, N. H. & Lu, P. Z. (2006) CdSe quantum dots induce apoptosis in human neuroblastoma cells via mitochondrial-dependent pathways and inhibition of survival signals. *Toxicol. Lett.*, 167, 191–200.

Chatterjee, D. K., Fong, L. S. & Zhang, Y. (2008) Nanoparticles in photodynamic therapy: An emerging paradigm. *Adv. Drug. Deliv. Rev.*, 60, 1627–1637.

Chen, R. J., Bangsaruntip, S., Drouvalakis, K. A., Kam, N. W. S., Shim, M., Li, Y. M., Kim, W., Utz, P. J. & Dai, H. J. (2003) Noncovalent functionalization of carbon nanotubes for highly specific electronic biosensors. *Proc. Natl. Acad. Sci. U. S. A.*, 100, 4984–4989.

Chen, R. J., Zhang, Y. G., Wang, D. W. & Dai, H. J. (2001) Noncovalent sidewall functionalization of single-walled carbon nanotubes for protein immobilization. *J. Am. Chem. Soc.*, 123, 3838–3839.

Chen, Z. L., Sun, Y., Huang, P., Yang, X. X. & Zhou, X. P. (2009) Studies on Preparation of Photosensitizer Loaded Magnetic Silica Nanoparticles and Their Anti-Tumor Effects for Targeting Photodynamic Therapy. *Nanoscale Res. Lett.*, 4, 400–408.

Cheng, H. M., Hsu, H. C., Chen, S. L., Wu, W. T., Kao, C. C., Lin, L. J. & Hsieh, W. F. (2005) Efficient UV photoluminescence from monodispersed secondary ZnO colloidal spheres synthesized by sol-gel method. *Journal of Crystal Growth*, 277, 192–199.

Cheng, S.-H., Lee, C.-H., Yang, C.-S., Tseng, F.-G., Mou, C.-Y. & Lo, L.-W. (2009) Mesoporous silica nanoparticles functionalized with an oxygen-sensing probe for cell photodynamic therapy: potential cancer theranostics. *J. Mater. Chem.*, 19, 1252–1257.

Cheng, Y., Samia, A. C., Li, J., Kenney, M. E., Resnick, A. & Burda, C. (2010) Delivery and Efficacy of a Cancer Drug as a Function of the Bond to the Gold Nanoparticle Surface. *Langmuir*, 26, 2248–2255.

Cheng, Y., Samia, A. C., Meyers, J. D., Panagopoulos, I., Fei, B. W. & Burda, C. (2008) Highly efficient drug delivery with gold nanoparticle vectors for in vivo photodynamic therapy of cancer. *J. Am. Chem. Soc.*, 130, 10643–10647.

Chidawanyika, W., Litwinski, C., Antunes, E. & Nyokong, T. (2010) Photophysical study of a covalently linked quantum dot-low symmetry phthalocyanine conjugate. *Journal of Photochemistry and Photobiology a-Chemistry*, 212, 27–35.

Choi, A. O., Cho, S. J., Desbarats, J., Lovric, J. & Maysinger, D. (2007) Quantum dot-induced cell death involves Fas upregulation and lipid peroxidation in human neuroblastoma cells. *J Nanobiotechnology*, 5, 1.

Clarke, S. J., Hollmann, C. A., Zhang, Z. J., Suffern, D., Bradforth, S. E., Dimitrijevic, N. M., Minarik, W. G. & Nadeau, J. L. (2006) Photophysics of dopamine-modified quantumdots and effects on biological systems. *Nature Materials*, 5, 409–417.

Compagnin, C., Bau, L., Mognato, M., Celotti, L., Miotto, G., Arduini, M., Moret, F., Fede, C., Selvestrel, F., Echevarria, I. M. R., Mancin, F. & Reddi, E. (2009) The cellular uptake of meta-tetra (hydroxyphenyl)chlorin entrapped in organically modified silica nanoparticles is mediated by serum proteins. *Nanotechnology*, 20.

Cooper, D. R., Dimitrijevic, N. M. & Nadeau, J. L. (2010) Photosensitization of CdSe/ZnS QDs and reliability of assays for reactive oxygen species production. *Nanoscale*, 2, 114–121.

Davydenko, M. O., Radchenko, E. O., Yashchuk, V. M., Dmitruk, I. M., Prylutskyy, Y. I., Matishevska, O. P. & Golub, A. A. (2006) Sensitization of fullerene C60 immobilized at silica nanoparticles for cancer photodynamic therapy. *J. Mol. Liq.*, 127, 145–147.

Dayal, S., Krolicki, R., Lou, Y., Qiu, X., Berlin, J. C., Kenney, M. E. & Burda, C. (2006) Femtosecond time-resolved energy transfer from CdSe nanoparticles to phthalocyanines. *Applied Physics B-Lasers and Optics*, 84, 309–315.

Demberelnyamba, D., Ariunaa, M. & Shim, Y. K. (2008) Newly synthesized water soluble cholinium-purpurin photosensitizers and their stabilized gold nanoparticles as promising anticancer agents. *International Journal of Molecular Sciences*, 9, 864–871.

Do Nascimento, G. M., De Oliveira, R. C., Pradie, N. A., Lins, P. R. G., Worfel, P. R., Martinez, G. R., Di Mascio, P., Dresselhaus, M. S. & Corio, P. (2010) Single-wall carbon nanotubes modified with organic dyes: Synthesis, characterization and potential cytotoxic effects. *Journal of Photochemistry and Photobiology a-Chemistry*, 211, 99–107.

Dorfman, A., Kumar, N. & Hahm, J. I. (2006a) Highly sensitive biomolecular fluorescence detection using nanoscale ZnO platforms. *Langmuir*, 22, 4890–4895.

Dorfman, A., Kumar, N. & Hahm, J. I. (2006b) Nanoscale ZnO-enhanced fluorescence detection of protein interactions. *Advanced Materials*, 18, 2685–2690.

Ellington, A. D. & Szostak, J. W. (1990) INVITRO SELECTION OF RNA MOLECULES THAT BIND SPECIFIC LIGANDS. *Nature*, 346, 818–822.

Epple, M., Ganesan, K., Heumann, R., Klesing, J., Kovtun, A., Neumann, S. & Sokolova, V. (2010) Application of calcium phosphate nanoparticles in biomedicine. *J. Mater. Chem.*, 20, 18–23.

Erbas, S., Gorgulu, A., Kocakusakogullari, M. & Akkaya, E. U. (2009) Non-covalent functionalized SWNTs as delivery agents for novel Bodipy-based potential PDT sensitizers. *Chem. Commun.*, 4956–4958.

Erdogmus, A., Moeno, S., Litwinski, C. & Nyokong, T. (2010) Photophysical properties of newly synthesized fluorinated zinc phthalocyanines in the presence of CdTe quantum dots and the accompanying energy transfer processes. *Journal of Photochemistry and Photobiology a-Chemistry*, 210, 200–208.

Febvay, S., Marini, D. M., Belcher, A. M. & Clapham, D. E. (2010) Targeted Cytosolic Delivery of Cell-Impermeable Compounds by Nanoparticle-Mediated, Light-Triggered Endosome Disruption. *Nano Lett.*, 10, 2211–2219.

Fujishima, A., Rao, T. N. & Tryk, D. A. (2000) Titanium dioxide photocatalysis. *Journal of Photochemistry and Photobiology C: Photochemistry Reviews*, 1, 1–21.

Fukahori, S., Ichiura, H., Kitaoka, T. & Tanaka, H. (2003) Capturing of bisphenol A photodecomposition intermediates by composite TiO_2-zeolite sheets. *Applied Catalysis B: Environmental*, 46, 453–462.

Guo, D., Wu, C., Jiang, H., Li, Q., Wang, X. & Chen, B. (2008) Synergistic cytotoxic effect of different sized ZnO nanoparticles and daunorubicin against leukemia cancer cells under UV irradiation. *Journal of Photochemistry and Photobiology B: Biology*, 93, 119–126.

Guo, H. C., Qian, H. S., Idris, N. M. & Zhang, Y. (2010a) Singlet oxygen-induced apoptosis of cancer cells using upconversion fluorescent nanoparticles as a carrier of photosensitizer. *Nanomedicine-Nanotechnology Biology and Medicine*, 6, 486–495.

Guo, Y. Y., Rogelj, S. & Zhang, P. (2010b) Rose Bengal-decorated silica nanoparticles as photosensitizers for inactivation of gram-positive bacteria. *Nanotechnology*, 21.

He, X., Wu, X., Wang, K., Shi, B. & Hai, L. (2009) Methylene blue-encapsulated phosphonate-terminated silica nanoparticles for simultaneous in vivo imaging and photodynamic therapy. *Biomaterials*, 30, 5601–9.

Hefta, L. J. F., Neumaier, M. & Shively, J. E. (1998) Kinetic and affinity constants of epitope specific anti-carcinoembryonic antigen (CEA) monoclonal antibodies for CEA and engineered CEA domain constructs. *Immunotechnology*, 4, 49–57.

Henderson, M. A. (1997) Complexity in the decomposition of formic acid on the TiO_2(110) surface. *Journal of Physical Chemistry B*, 101, 221–229.

Hirohara, S., Obata, M., Ogata, S. I., Ohtsuki, C., Higashida, S., Ogura, S. I., Okura, I., Takenaka, M., Ono, H., Sugai, Y., Mikata, Y., Tanihara, M. & Yano, S. (2005) Cellular uptake and photocytotoxicity of glycoconjugated chlorins in HeLa cells. *Journal of Photochemistry and Photobiology B: Biology*, 78, 7–15.

Hirsch, A. (2002) Functionalization of single-walled carbon nanotubes. *Angew. Chem. Int. Ed.*, 41, 1853–1859.

Hocine, O., Gary-Bobo, M., Brevet, D., Maynadier, M., Fontanel, S., Raehm, L., Richeter, S., Loock, B., Couleaud, P., Frochot, C., Charnay, C., Derrien, G., Smaïhi, M., Sahmoune, A., Morère, A., Maillard, P., Garcia, M. & Durand, J.-O. (doi:10.1016/j.ijpharm.2010.10.004) Silicalites and Mesoporous Silica Nanoparticles for photodynamic therapy. *Int. J. Pharma.*

Hoffman, A. S. (2008) The origins and evolution of "controlled" drug delivery systems. *J. Control. Release*, 132, 153–163.

Hone, D. C., Walker, P. I., Evans-Gowing, R., Fitzgerald, S., Beeby, A., Chambrier, I., Cook, M. J. & Russell, D. A. (2002) Generation of cytotoxic singlet oxygen via phthalocyanine-stabilized gold nanoparticles: A potential delivery vehicle for photodynamic therapy. *Langmuir*, 18, 2985–2987.

Hsieh, J. M., Ho, M. L., Wu, P. W., Chou, P. T., Tsai, T. T. & Chi, Y. (2006) Iridium-complex modified CdSe/ZnS quantum dots; a conceptual design for bifunctionality toward imaging and photosensitization. *Chem. Commun.*, 615–617.

Idowu, M., Chen, J. Y. & Nyokong, T. (2008) Photoinduced energy transfer between water-soluble CdTe quantum dots and aluminium tetrasulfonated phthalocyanine. *New Journal of Chemistry*, 32, 290–296.

Ipe, B. I., Lehnig, M. & Niemeyer, C. M. (2005) On the generation of free radical species from quantum dots. *Small*, 1, 706–709.

Jhonsi, M. A. & Renganathan, R. (2010) Investigations on the photoinduced interaction of water soluble thioglycolic acid (TGA) capped CdTe quantum dots with certain porphyrins. *Journal of Colloid and Interface Science*, 344, 596–602.

Juzenas, P., Chen, W., Sun, Y. P., Coelho, M. A. N., Generalov, R., Generalova, N. & Christensen, I. L. (2008) Quantum dots and nanoparticles for photodynamic and radiation therapies of cancer. *Adv. Drug. Deliv. Rev.*, 60, 1600–1614.

Kam, N. W. S., Jessop, T. C., Wender, P. A. & Dai, H. J. (2004) Nanotube molecular transporters: Internalization of carbon nanotube-protein conjugates into mammalian cells. *J. Am. Chem. Soc.*, 126, 6850–6851.

Kantonis, G., Trikeriotis, M. & Ghanotakis, D. F. (2007) Biocompatible protoporphyrin IX-containing nanohybrids with potential applications in photodynamic therapy. *Journal of Photochemistry and Photobiology a-Chemistry*, 185, 62–66.

Kim, H. J., Shin, K. J., Han, M. K., An, K., Lee, J. K., Honma, I. & Kim, H. (2009a) One-pot synthesis of multifunctional mesoporous silica nanoparticle incorporated with zinc(II) phthalocyanine and iron oxide. *Scripta Materialia*, 61, 1137–1140.

Kim, S., Ohulchanskyy, T. Y., Bharali, D., Chen, Y. H., Pandey, R. K. & Prasad, P. N. (2009b) Organically Modified Silica Nanoparticles with Intraparticle Heavy-Atom Effect on the Encapsulated Photosensitizer for Enhanced Efficacy of Photodynamic Therapy. *Journal of Physical Chemistry C*, 113, 12641–12644.

Kim, S., Ohulchanskyy, T. Y., Pudavar, H. E., Pandey, R. K. & Prasad, P. N. (2007) Organically Modified Silica Nanoparticles Co-encapsulating Photosensitizing Drug and Aggregation-Enhanced Two-Photon Absorbing Fluorescent Dye Aggregates for Two-Photon Photodynamic Therapy. *J. Am. Chem. Soc.*, 129, 2669–2675.

Klostranec, J. M. & Chan, W. C. W. (2006) Quantum dots in biological and biomedical research: Recent progress and present challenges. *Advanced Materials*, 18, 1953–1964.

Kostarelos, K., Lacerda, L., Pastorin, G., Wu, W., Wieckowski, S., Luangsivilay, J., Godefroy, S., Pantarotto, D., Briand, J. P., Muller, S., Prato, M. & Bianco, A. (2007) Cellular uptake of functionalized carbon nanotubes is independent of functional group and cell type. *Nature Nanotechnology*, 2, 108–113.

Kovala-Demertzi, D., Galani, A., Kourkoumelis, N., Miller, J. R. & Demertzis, M. A. (2007) Synthesis, characterization, crystal structure and antiproliferative activity of platinum(II) complexes with 2-acetylpyridine-4-cyclohexyl-thiosemicarbazone. *Polyhedron*, 26, 2871–2879.

Kubota, Y., Shuin, T., Kawasaki, C., Hosaka, M., Kitamura, H., Cai, R., Sakai, H., Hashimoto, K. & Fujishima, A. (1994) Photokilling of T-24 human bladder cancer cells with titanium dioxide. *British Journal of Cancer*, 70, 1107–1111.

Kuo, W. S., Chang, C. N., Chang, Y. T., Yang, M. H., Chien, Y. H., Chen, S. J. & Yeh, C. S. (2010) Gold Nanorods in Photodynamic Therapy, as Hyperthermia Agents, and in Near-Infrared Optical Imaging. *Angew. Chem. Int. Ed.*, 49, 2711–2715.

Kuo, W. S., Chang, C. N., Chang, Y. T. & Yeh, C. S. (2009) Antimicrobial gold nanorods with dual-modality photodynamic inactivation and hyperthermia. *Chem. Commun.*, 4853–4855.

Kurwa, H. A. & Barlow, R. J. (1999) The role of photodynamic therapy in dermatology. *Clinical and Experimental Dermatology*, 24, 143–148.

Labanauskiene, J., Gehl, J. & Didziapetriene, J. (2007) Evaluation of cytotoxic effect of photodynamic therapy in combination with electroporation in vitro. *Bioelectrochemistry*, 70, 78–82.

Lai, C.-W., Wang, Y.-H., Lai, C.-H., Yang, M.-J., Chen, C.-Y., Chou, P.-T., Chan, C.-S., Chi, Y., Chen, Y.-C. & Hsiao, J.-K. (2008) Iridium-complex-functionalized Fe3O4/SiO2 core/shell nanoparticles: a facile three-in-one system in magnetic resonance imaging, luminescence imaging, and photodynamic therapy. *Small*, 4, 218–224.

Lambreva, M., Glück, B., Radeva, M. & Berg, H. (2004) Electroporation of cell membranes supporting penetration of photodynamic active macromolecular chromophore dextrans. *Bioelectrochemistry*, 62, 95–98.

Lei, W., Zhou, Q., Li, Z., Wang, X. & Zhang, B. (2009) Photodynamic activity of ascorbic acid-modified TiO_2 nanoparticles upon visible illumination (>550 nm) *Chemistry Letters*, 38, 1138–1139.

Li, J., Guo, D., Wang, X., Wang, H., Jiang, H. & Chen, B. (2010a) The photodynamic effect of different size ZnO nanoparticles on cancer cell proliferation in vitro. *Nanoscale Research Letters*, 5, 1063–1071.

Li, Z. B., Wang, J. G., Chen, J. R., Lei, W. H., Wang, X. S. & Zhang, B. W. (2010b) pH-responsive silica nanoparticles for controllable O-1(2) generation. *Nanotechnology*, 21.

Liang, G., Wang, L., Yang, Z., Koon, H., Mak, N., Chang, C. K. & Xu, B. (2006) Using enzymatic reactions to enhance the photodynamic therapy effect of porphyrin dityrosine phosphates. *Chemical Communications*, 5021–5023.

Lin, K. F., Cheng, H. M., Hsu, H. C., Lin, L. J. & Hsieh, W. F. (2005) Band gap variation of size-controlled ZnO quantum dots synthesized by sol-gel method. *Chemical Physics Letters*, 409, 208–211.

Liu, F., Zhou, X., Chen, Z., Huang, P., Wang, X. & Zhou, Y. (2008a) Preparation of purpurin-18 loaded magnetic nanocarriers in cottonseed oil for photodynamic therapy. *Mater. Lett.*, 62, 2844–2847.

Liu, Y., Zhang, Y., Wang, S., Pope, C. & Chen, W. (2008b) Optical behaviors of ZnO-porphyrin conjugates and their potential applications for cancer treatment. *Applied Physics Letters*, 92, Article number 143901.

Liu, Z., Davis, C., Cai, W. B., He, L., Chen, X. Y. & Dai, H. J. (2008c) Circulation and long-term fate of functionalized, biocompatible single-walled carbon nanotubes in mice probed by Raman spectroscopy. *Proc. Natl. Acad. Sci. U.S.A.*, 105, 1410–1415.

Liu, F., Zhou, X., Ni, S., Wang, X., Y., Z. & Chen, Z. (2009) Preparation and properties of Photosensitizer Loaded Magnetic Nanocarriers. *Current Nanoscience*, 5, 293–296.

Lopez-Goerne, T. M. (2007) Sol-gel nanostructured titania reservoirs for use in the controlled release of drugs in the central nervous system WO2007141590.

Lopez, T., Bosch, P., Moran, M. & Gomez, R. (1993) Pt/SiO_2 sol-gel catalysts: Effects of pH and platinum precursor. *Journal of Physical Chemistry*, 97, 1671–1677.

López, T., Gómez, R., Pecci, G., Reyes, P., Bokhimi, X. & Novaro, O. (1999) Effect of pH on the incorporation of platinum into the lattice of sol-gel titania phases. *Materials Letters*, 40, 59–65.

Lopez, T., Figueras, F., Manjarrez, J., Bustos, J., Alvarez, M., Silvestre-Albero, J., Rodríguez-Reinoso, F., Martínez-Ferre, A. & Martínez, E. (2010a) Catalytic nanomedicine: A new field in antitumor treatment using supported platinum nanoparticles. In vitro DNA degradation and in vivo tests with C6 animal model on Wistar rats. *European Journal of Medicinal Chemistry*, 45, 1982–1990.

Lopez, T., Ortiz, E., Alvarez, M., Navarrete, J., Odriozola, J. A., Martinez-Ortega, F., Páez-Mozo, E. A., Escobar, P., Espinoza, K. A. & Rivero, I. A. (2010b) Study of the stabilization of zinc phthalocyanine in sol-gel TiO_2 for photodynamic therapy applications *Nanomedicine: Nanotechnology, Biology, and Medicine DOI:* 10.1016/j.nano.2010.04.007.

Lowis, S., Lewis, I., Elsworth, A., Weston, C., Doz, F., Vassal, G., Bellott, R., Robert, J., Pein, F., Ablett, S., Pinkerton, R. & Frappaz, D. (2006) A phase I study of intravenous liposomal daunorubicin (DaunoXome) in paediatric patients with relapsed or resistant solid tumours. *British Journal of Cancer*, 95, 571–580.

Maisch, T. (2009) A new strategy to destroy antibiotic resistant microorganisms: antimicrobial photodynamic treatment. *Mini-Rev. Med. Chem.*, 9, 974–983.

Marković, M., Knežević, N., Momćilović, M., Grgurić-Šipka, S., Harhaji, L., Trajković, V., Stojković, M. M., Sabo, T. & Miljković, D. (2005) [Pt(HPxSC)Cl3], a novel platinum(IV) compound with anticancer properties. *European Journal of Pharmacology*, 517, 28–34.

Matsunaga, T., Tomoda, R., Nakajima, T. & Wake, H. (1985) Photoelectrochemical sterilization of microbial cells by semiconductor powders. *FEMS Microbiology Letters*, 29, 211–214.

Mccarthy, J. R., Jaffer, F. A. & Weissleder, R. (2006) A macrophage-targeted theranostic nanoparticle for biomedical applications. *Small*, 2, 983–987.

Mccarthy, J. R., Korngold, E., Weissleder, R. & Jaffer, F. A. (2010) A Light-Activated Theranostic Nanoagent for Targeted Macrophage Ablation in Inflammatory Atherosclerosis. *Small*, 6, 2041–2049.

Miyasaka, T. & Kijitori, Y. (2004) Low-temperature fabrication of dye-sensitized plastic electrodes by electrophoretic preparation of mesoporous TiO_2 layers. *Journal of the Electrochemical Society*, 151.

Moeno, S., Antunes, E., Khene, S., Litwinski, C. & Nyokong, T. (2010) The effect of substituents on the photoinduced energy transfer between CdTe quantum dots and mercapto substituted zinc phthalocyanine derivatives. *Dalton Transactions*, 39, 3460–3471.

Narayanan, S. S., Sinha, S. S. & Pal, S. K. (2008) Sensitized emission from a chemotherapeutic drug conjugated to CdSe/ZnS QDs. *J. Phys. Chem. C*, 112, 12716–12720.

Niyogi, S., Hamon, M. A., Hu, H., Zhao, B., Bhowmik, P., Sen, R., Itkis, M. E. & Haddon, R. C. (2002) Chemistry of single-walled carbon nanotubes. *Acc. Chem. Res.*, 35, 1105–1113.

Nolkrantz, K., Farre, C., Hurtig, K. J., Rylander, P. & Orwar, O. (2002) Functional screening of intracellular proteins in single cells and in patterned cell arrays using electroporation. *Analytical Chemistry*, 74, 4300–4305.

Ogura, S. I., Yazaki, K., Yamaguchi, K., Kamachi, T. & Okura, I. (2005) Localization of poly-L-lysine-photosensitizer conjugate in nucleus. *Journal of Controlled Release*, 103, 1–6.

Ohulchanskyy, T. Y., Roy, I., Goswami, L. N., Chen, Y., Bergey, E. J., Pandey, R. K., Oseroff, A. R. & Prasad, P. N. (2007) Organically Modified Silica Nanoparticles with Covalently Incorporated Photosensitizer for Photodynamic Therapy of Cancer. *Nano Lett.*, 7, 2835–2842.

Oo, M. K. K., Yang, X., Du, H. & Wang, H. (2008) 5-aminolevulinic acid-conjugated gold nanoparticles for photodynamic therapy of cancer. *Nanomedicine*, 3, 777–786.

Ortel, B., Shea, C. R. & Calzavara-Pinton, P. (2009) Molecular mechanisms of photodynamic therapy. *Front. Biosci.*, 14, 4157–4172.

Pantarotto, D., Briand, J. P., Prato, M. & Bianco, A. (2004) Translocation of bioactive peptides across cell membranes by carbon nanotubes. *Chem. Commun.*, 16–17.

Qian, H. S., Guo, H. C., Ho, P. C. L., Mahendran, R. & Zhang, Y. (2009a) Mesoporous-Silica-Coated Up-Conversion Fluorescent Nanoparticles for Photodynamic Therapy. *Small*, 5, 2285–2290.

Qian, J., Gharibi, A. & He, S. (2009b) Colloidal mesoporous silica nanoparticles with protoporphyrin IX encapsulated for photodynamic therapy. *J. Biomed. Opt.*, 14, 014012/1-014012/6.

Qu, L. W., Martin, R. B., Huang, W. J., Fu, K. F., Zweifel, D., Lin, Y., Sun, Y. P., Bunker, C. E., Harruff, B. A., Gord, J. R. & Allard, L. F. (2002) Interactions of functionalized carbon nanotubes with tethered pyrenes in solution. *Journal of Chemical Physics*, 117, 8089–8094.

Rakovich, A., Savateeva, D., Rakovich, T., Donegan, J. F., Rakovich, Y. P., Kelly, V., Lesnyak, V. & Eychmuller, A. (2010) CdTe Quantum Dot/Dye Hybrid System as Photosensitizer for Photodynamic Therapy. *Nanoscale Res. Lett.*, 5, 753–760.

Rioux, D., Laferriere, M., Douplik, A., Shah, D., Lilge, L., Kabashin, A. V. & Meunier, M. M. (2009) Silicon nanoparticles produced by femtosecond laser ablation in water as novel contamination-free photosensitizers. *J. Biomed. Opt.*, 14, 5.

Robertson, C. A., Evans, D. H. & Abrahamse, H. (2009) Photodynamic therapy (PDT): A short review on cellular mechanisms and cancer research applications for PDT. *J. Photochem. Photobiol., B*, 96, 1–8.

Rosenberg, B., Van Camp, L. & Krigas, T. (1965) Inhibition of cell division in Escherichia coli by electrolysis products from a Platinum electrode. *Nature*, 205, 698–699.

Rosenberg, B., Van Camp, L., Trosko, J. & Mansour, V. H. (1969) Platinum compounds: a new class of potent antitumor agents. *Nature*, 222, 385–386.

Rossi, L. M., Silva, P. R., Vono, L. L. R., Fernandes, A. U., Tada, D. B. & Baptista, M. S. (2008) Protoporphyrin IX Nanoparticle Carrier: Preparation, Optical Properties, and Singlet Oxygen Generation. *Langmuir*, 24, 12534–12538.

Roy, I., Ohulchanskyy, T. Y., Pudavar, H. E., Bergey, E. J., Oseroff, A. R., Morgan, J., Dougherty, T. J. & Prasad, P. N. (2003) Ceramic-based nanoparticles entrapping water-insoluble photosensitizing anticancer drugs: a novel drug-carrier system for photodynamic therapy. *J. Am. Chem. Soc.*, 125, 7860–7865.

Sakai, A. H., Cai, R. X., Yoshioka, T., Kubota, Y., Hashimoto, K. & Fujishima, A. (1994) Intracellular Ca^{2+} concentration change of T24 cell under irradiation in the presence of TiO_2 ultrafine particles. *Biochimica et Biophysica Acta – General Subjects*, 1201, 259–265.

Samia, A. C. S., Chen, X. B. & Burda, C. (2003) Semiconductor quantum dots for photodynamic therapy. *J. Am. Chem. Soc.*, 125, 15736–15737.

Semenikhin, O. A., Kazarinov, V. E., Jiang, L., Hashimoto, K. & Fujishima, A. (1999) Suppression of surface recombination on TiO_2 anatase photocatalysts in aqueous solutions containing alcohol. *Langmuir*, 15, 3731–3737.

Shi, L. X., Hernandez, B. & Selke, M. (2006) Singlet oxygen generation from water-soluble quantum dot-organic dye nanocomposites. *J. Am. Chem. Soc.*, 128, 6278–6279.

Simon, V., Devaux, C., Darmon, A., Donnet, T., Thiénot, E., Germain, M., Honnorat, J., Duval, A., Pottier, A., Borghi, E., Levy, L. & Marill, J. (2010) Pp IX Silica Nanoparticles Demonstrate Differential Interactions with<i>In Vitro</i>Tumor Cell Lines and<i>In Vivo</i>Mouse Models of Human Cancers. *Photochem. Photobiol.*, 86, 213–222.

Staros, J. V. (1982) N-hydroxysulfosuccinimide active esters: Bis(N-hydroxysulfosuccinimide) esters of two dicarboxylic acids are hydrophilic, membrane-impermeant, protein cross-linkers. *Biochemistry*, 21, 3950–3955.

Strassert, C. A., Otter, M., Albuquerque, R. Q., Hone, A., Vida, Y., Maier, B. & De Cola, L. (2009) Photoactive Hybrid Nanomaterial for Targeting, Labeling, and Killing Antibiotic-Resistant Bacteria. *Angew. Chem. Int. Ed.*, 48, 7928–7931.

Tada, D. B., Vono, L. L. R., Duarte, E. L., Itri, R., Kiyohara, P. K., Baptista, M. S. & Rossi, L. M. (2007) Methylene Blue-Containing Silica-Coated Magnetic Particles: A Potential Magnetic Carrier for Photodynamic Therapy. *Langmuir*, 23, 8194–8199.

Takahashi, M., Ueno, A., Uda, T. & Mihara, H. (1998) Design of novel porphyrin-binding peptides based on antibody CDR. *Bioorganic and Medicinal Chemistry Letters*, 8, 2023–2026.

Tang, W., Xu, H., Kopelman, R. & Philbert, M. A. (2005a) Photodynamic characterization and in vitro application of methylene blue-containing nanoparticle platforms. *Photochem Photobiol*, 81, 242–9.

Tang, W., Xu, H., Kopelman, R. & Philbert, M. A. (2005b) Photodynamic characterization and in vitro application of methylene blue-containing nanoparticle platforms. *Photochem. Photobiol.*, 81, 242–249.

Taquet, J. P., Frochot, C., Manneville, V. & Barberi-Heyob, M. (2007) Phthalocyanines covalently bound to biomolecules for a targeted photodynamic therapy. *Current Medicinal Chemistry*, 14, 1673–1687.

Timoshenko, V. Y., Kudryavtsev, A. A., Osminkina, L. A., Vorontsov, A. S., Ryabchikov, Y. V., Belogorokhov, I. A., Kovalev, D. & Kashkarov, P. K. (2006) Silicon nanocrystals as photosensitizers of active oxygen for biomedical applications. *Jetp Letters*, 83, 423–426.

Tokuoka, Y., Yamada, M., Kawashima, N. & Miyasaka, T. (2006) Anticancer effect of dye-sensitized TiO_2 nanocrystals by polychromatic visible light irradiation. *Chemistry Letters*, 35, 496–497.

Tsay, J. M., Trzoss, M., Shi, L. X., Kong, X. X., Selke, M., Jung, M. E. & Weiss, S. (2007) Singlet oxygen production by peptide-coated quantum dot-photosensitizer conjugates. *J. Am. Chem. Soc.*, 129, 6865–6871.

Tu, B. H.-L., Lin, Y.-S., Hung, Y., Lo, L.-W., Chen, Y.-F. & Mou, C.-Y. (2009) In vitro Studies of Functionalized Mesoporous Silica Nanoparticles for Photodynamic Therapy. *Adv. Mater.*, 21, 172–174.

Tuerk, C. & Gold, L. (1990) SYSTEMATIC EVOLUTION OF LIGANDS BY EXPONENTIAL ENRICHMENT – RNA LIGANDS TO BACTERIOPHAGE-T4 DNA-POLYMERASE. *Science*, 249, 505–510.

Underhill, M. F., Coley, C., Birch, J. R., Findlay, A., Kallmeier, R., Proud, C. G. & James, D. C. (2003) Engineering mRNA translation initiation to enhance transient gene expression in Chinese hamster ovary cells. *Biotechnology Progress*, 19, 121–129.

Van De Walle, C. G. (2000) Hydrogen as a cause of doping in zinc oxide. *Physical Review Letters*, 85, 1012–1015.

Velusamy, M., Shen, J. Y., Lin, J. T., Lin, Y. C., Hsieh, C. C., Lai, C. H., Lai, C. W., Ho, M. L., Chen, Y. C., Chou, P. T. & Hsiao, J. K. (2009) A New Series of Quadrupolar Type Two-Photon Absorption Chromophores Bearing 11,12-Dibutoxydibenzo[a,c]-phenazine Bridged Amines; Their Applications in Two-Photon Fluorescence Imaging and Two-Photon Photodynamic Therapy. *Adv. Func. Mater.*, 19, 2388–2397.

Verhille, M., Couleaud, P., Vanderesse, R., Brault, D., Barberi-Heyob, M. & Frochot, C. (In Press) Modulation of photosensitization processes for an improved targeted photodynamic therapy. *Current Medical Chemistry*.

Wamer, W. G., Yin, J. J. & Wei, R. R. (1997) Oxidative damage to nucleic acids photosensitized by titanium dioxide. *Free Radical Biology and Medicine*, 23, 851–858.

Wang, H., Xie, C., Zhang, W., Cai, S., Yang, Z. & Gui, Y. (2007) Comparison of dye degradation efficiency using ZnO powders with various size scales. *Journal of Hazardous Materials*, 141, 645–652.

Wang, S., Gao, R., Zhou, F. & Selke, M. (2004) Nanomaterials and Singlet Oxygen Photosensitizers: Potential Applications in Photodynamic Therapy. *J. Mater. Chem.*, 14, 487–493.

Wang, S., Mamedova, N., Kotov, N. A., Chen, W. & Studer, J. (2002) Antigen/Antibody Immunocomplex from CdTe Nanoparticle Bioconjugates. *Nano Letters*, 2, 817–822.

Wu, C. H., Huang, K. S. & Chern, J. M. (2006) Decomposition of acid dye by TiO_2 thin films prepared by the sol-gel method. *Industrial and Engineering Chemistry Research*, 45, 2040–2045.

Xu, F., Zhang, P., Navrotsky, A., Yuan, Z. Y., Ren, T. Z., Halasa, M. & Su, B. L. (2007a) Hierarchically assembled porous ZnO nanoparticles: Synthesis, surface energy, and photocatalytic activity. *Chemistry of Materials*, 19, 5680–5686.

Xu, J., Sun, Y., Huang, J., Chen, C., Liu, G., Jiang, Y., Zhao, Y. & Jiang, Z. (2007b) Photokilling cancer cells using highly cell-specific antibody-TiO_2 bioconjugates and electroporation. *Bioelectrochemistry*, 71, 217–222.

Yan, F. & Kopelman, R. (2003) The embedding of meta-tetra(hydroxyphenyl)-chlorin into silica nanoparticle platforms for photodynamic therapy and their singlet oxygen production and ph-dependent optical properties. *Photochem. Photobiol.*, 78, 587–591.

Yang, Y., Song, W. X., Wang, A. H., Zhu, P. L., Fei, J. B. & Li, J. B. (2010) Lipid coated mesoporous silica nanoparticles as photosensitive drug carriers. *Physical Chemistry Chemical Physics*, 12, 4418–4422.

Zhang, A. P. & Sun, Y. P. (2004) Photocatalytic killing effect of TiO_2 nanoparticles on Ls-174-t human colon carcinoma cells. *World Journal of Gastroenterology*, 10, 3191–3193.

Zhang, P., Steelant, W., Kumar, M. & Scholfield, M. (2007) Versatile Photosensitizers for Photodynamic Therapy at Infrared Excitation. *J. Am. Chem. Soc.*, 129, 4526–4527.

Zhang, R. R., Wu, C. L., Tong, L. L., Tang, B. & Xu, Q. H. (2009) Multifunctional Core-Shell Nanoparticles as Highly Efficient Imaging and Photosensitizing Agents. *Langmuir*, 25, 10153–10158.

Zhao, B. Z., Yin, J. J., Bilski, P. J., Chignell, C. F., Roberts, J. E. & He, Y. Y. (2009) Enhanced photodynamic efficacy towards melanoma cells by encapsulation of Pc4 in silica nanoparticles. *Toxicology and Applied Pharmacology*, 241, 163–172.

Zhou, J., Zhou, L., Dong, C., Feng, Y., Wei, S., Shen, J. & Wang, X. (2008) Preparation and photodynamic properties of water-soluble hypocrellin A-silica nanospheres. *Mater. Lett.*, 62, 2910–2913.

Zhou, L., Liu, J.-H., Zhang, J., Wei, S.-H., Feng, Y.-Y., Zhou, J.-H., Yu, B.-Y. & Shen, J. (2010) A new sol-gel silica nanovehicle preparation for photodynamic therapy in vitro. *International Journal of Pharmaceutics*, 386, 131–137.

Zhu, Z., Tang, Z. W., Phillips, J. A., Yang, R. H., Wang, H. & Tan, W. H. (2008) Regulation of singlet oxygen generation using single-walled carbon nanotubes. *J. Am. Chem. Soc.*, 130, 10856–10857.

Nanoparticles Enhanced Hyperthermia

Qian Wang and Jing Liu

Abstract Nano-hyperthermia is regarded as a promising alternative to conventional thermal ablation. Many nanoparticles with specific physical properties in electrical, magnetic, acoustic, optical or thermal features have been tried to induce various enhanced hyperthermia, aiming to significantly improve the treatment efficiency of conventional heating. This article is dedicated to present a comprehensive review on the thermal properties, targeted heating mechanisms, and heat generation behaviors of typical nanoparticles. In addition, potential combination of the nanoparticles with radiotherapy and/or chemotherapy will also be discussed.

Keywords Hyperthermia • Nanoparticle • Nanomedicine • Thermal ablation • Physical therapy

Abbreviations

CNT	Carbon nanotube
CVD	Chemical vapor deposition
GNP	Gold nanoparticle
HF	High frequency
HIFU	High intensity focused ultrasound
LF	Low frequency
LITT	Laser-induced thermotherapy
MF	Middle frequency
MGC	Magnetic glass-ceramic
MNP	Magnetic nanoparticle
MRI	Magnetic resonance imaging

Q. Wang and J. Liu (✉)
Department of Biomedical Engineering, School of Medicine,
Tsinghua University, Beijing 100084, P. R. China
e-mail: jliubme@tsinghua.edu.cn

A. Prokop (ed.), *Intracellular Delivery: Fundamentals and Applications*, Fundamental Biomedical Technologies 5, DOI 10.1007/978-94-007-1248-5_20,

MW	Microwave
MWNT	Multi-walled carbon nanotubes
NIR	Near-infrared
PTT	Photothermal therapy
PVA	Polyvinyl alcohol
RF	Radiofrequency
SAR	Specific absorption rate
SPM	Superparamagnetic
SWNT	Single-walled carbon nanotube
UCA	Ultrasound contrast agents
UHF	Ultra high frequency
VHF	Very high frequency

1 Background

Hyperthermia (Liu and Deng 2008; Alpard et al. 1996), one of newly emerging approaches currently being tested for cancer therapy, is known as "green therapy" for cancer treatment compared with conventional resection surgery, chemotherapy, and radiotherapy. It is a procedure of heating tissues to a higher temperature level than that of normal tissues in order to kill tumor (Alpard et al. 1996). There are generally two kinds of heating strategies. One is mild hyperthermia performed between 41°C and 46°C to stimulate the immune response for non-specific immunotherapy of cancers. Another is thermal ablation (more than 46°C, even much higher) leading to tumor destruction by direct cell necrosis, coagulation or carbonization. In a conventional hyperthermia, however, due to difficulty to precisely control heating, low accuracy in temperature measurement and possibility of damaging healthy tissues due to overheating, it often leads to survival, transfer, diffusion, or even re-generation of tumor cells. To tackle such tough technical barriers, nano-hyperthermia was recently identified as a promising way to improve conventional heating. Its basic idea is to load nanoparticles with specific functions into tissues to help achieve the desired heating on the target tissues. This is because the nanoparticles can help induce much larger heat and thus higher temperature. Particularly, selectively loading particles to carefully designed tissue site will enable a conformal heating. This would expand the scope of conventional hyperthermia heating equipment. In addition, it can administrate a more accurate treatment on cells and in sub-cellular scale. In general, controlling temperature range, heating time, and heating rate can be well improved after loading nanoparticles.

Since Gilchrist et al. (1957) proposed the concept of magnetically mediated hyperthermia and proved that magnetic particles can be selectively deposited in the body and heat tumor tissue, researchers throughout the world have made unremitting efforts. Particularly in recent years, Jordan et al. successfully lead nano-particle enhanced hyperthermia to clinical trials. In addition, many labs are also joining along this research direction. Meanwhile, other heating approaches such as near infrared and laser, etc., are also being tried in order to establish new nano hyperthermia approach.

In tumor hyperthermia, the major target is to deliver energy to tumor to improve its temperature. Up to now, various heating methods have been developed, such as RF, MW and laser heating, etc. (Fig. 1). As is well known, when the size of a kind of material reduces to nano-level, its properties in electronic, magnetic, acoustics, optics, thermal and chemical properties will change. Therefore, suitable nanoparticles can help improve the effect of specific means of traditional tumor hyperthermia.

In this chapter, we mainly review the nanoparticles enhanced magnetic, optical and RF hyperthermia, and discuss several possible methods in ultrasound, thermal chemical hyperthermia and direct heat delivery hyperthermia, then analyze several challenges in developing nanoparticle enhanced hyperthermia, such as particle delivery, monitoring, combined effects with radiotherapy and chemotherapy, etc.

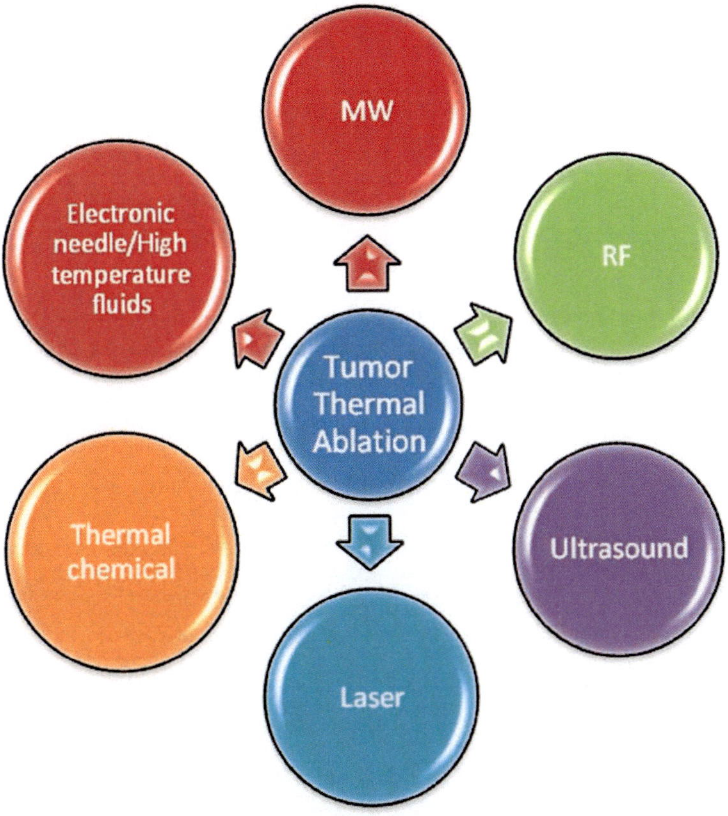

Fig. 1 Typical heating strategies for tumor thermal ablation

2 Nanoparticles Enhanced Magnetic Hyperthermia

2.1 *Basic Features*

MNPs are a major class of nanoscale materials with the potential to revolutionize current clinical diagnostic and therapeutic techniques. As early as 1957, Gilchrist et al. advanced the concept of magnetic targeting hyperthermia, which however received few attentions for a rather long time. Only when magnetic nanomaterials developed rapidly was some progress made in this area. Now, MNPs are being actively used in tumor hyperthermia and as the contrast agents in MRI (Duguet et al. 2006) and as carriers for targeted drug delivery. The goal of magnetic hyperthermia is to heat specifically and exclusively the local tumor region by means of the magnetic losses of nanoparticles in an external alternating magnetic field, without damaging the surrounding healthy tissues. In addition, MNPs will aggregate to form a larger diameter of the polymer, which effectively blocked the blood vessels around the tumor at its position to play a role in vascular thrombosis. This can cut off the blood supply of tumor. What's more, the magnetic field could inhibit the growth of cancer tissues through some mechanisms, such as impacting biological magnetic fields of tumor tissue, disrupting its blood and oxygen supply, changing the functions of cancer cell membrane so that affecting the material exchange, inhibition of cancer cell proliferation and enhancing the body immune function.

2.1.1 Enhanced Magnetism of MNPs

When MNPs are placed in an external magnetic field, the magnetic field is locally augmented through atomistic mechanisms in the magnetic material. The magnetization M is written in terms of the applied field H as (Wetzel and Fink 2001)

$$M(H) = \chi(H)H \tag{1}$$

where, χ is called magnetic susceptibility. The magnetization M is not linearly related to the applied field H.

Magnetic induction B is proportional to the sum of the applied H field and the material magnetization M, that is

$$B = \mu_0(H + M) \tag{2}$$

where, μ_0 is the magnetic permeability of free space, whose exact value is defined as $\mu_0 = 4\pi \times 10^{-7} H / M$.

The material's magnetic properties depend on its magnetic susceptibility, which is defined by the ratio of the induced magnetization to the applied magnetic field. According to the direction and strength of H, magnetic materials can be sorted into several kinds (Classes of Magnetic Materials; Rishi 2008): diamagnetic, paramagnetic, ferromagnetic, antiferromagnetic and ferrimagnetic. Diamagnetic is due to the non-cooperative behavior of orbiting electrons when exposed to an external magnetic field. Diamagnetic substances are composed of atoms which have no net magnetic moments, while in paramagnetic materials, some of the atoms or ions in the material, due to unpaired electrons in partially filled orbital, have a net magnetic moment even without external magnetic field. Besides, the magnetization is related to the temperature and their susceptibility is inversely proportional to the absolute temperature. Unlike paramagnetic materials, the atomic moments in ferromagnetic materials exhibit very strong interactions produced by electronic exchange forces. Spontaneous magnetization is the distinct characteristics of ferromagnetic materials. Ferromagnetic materials have specific maximum temperature called Curie temperature (T_c). For safe and effective tumor hyperthermia therapy, it is of great significance to maintain the temperature in the entire tumor at the optimal value during the treating procedure. Such particles can be effectively heated at low temperatures, but they stop absorbing energy once their temperature reaches T_c and the heating would stop. As the particles temperature decrease below T_c, magnetic ordering re-establishes and the heating resumes. Different magnetic materials have different T_c, therefore, by varying synthesis parameters of particles and post-treatment, one can utilize particles with suitable T_c in magnetic hyperthermia (Kuznetsov et al. 2007). The macro-features of antiferromagnetic materials are similar with paramagnet. Their susceptibilities are very small at the high temperature and increase as the temperature decreases. When reaching certain temperature, the susceptibilities reach the maximum. Antiferromagnets are composed of two kinds of exactly equal but opposite sub-lattices, which results in no net magnetic moment. Compared with antiferromagnets, the two kinds of opposite sub-lattices are not equal, so ferrimagnets have a net magnetic moment and their behaviors are similar to ferromagnets. The differences between several kinds of magnetic materials are shown in Table 1.

Table 1 Comparison between several different magnetic materials

	Diamagnet	Paramagnet	Ferromagnet	Antiferromagnet	Ferrimagnet
Permanent magnetic moment	No	Yes	Yes	No	Yes
Order of magnitude	Weak	Weak	Strong	Changing as the temperature	Strong
Susceptibility	Negative	Positive	Positive	Positive	Positive
Magnetic properties after removing external field	Disappear	Disappear	Retain	Disappear	Retain
Typical materials	Bi, Cu, Ag, Au	Transition elements, lanthano, aluminum, platinum	Fe, Co, Ni and many of their alloys	Non-metallic compounds such as MnO	Magnetite

2.1.2 Better Targeting of MNPs

The separation of tumor cells from healthy cells is a vital problem in tumor hyperthermia. One must protect healthy cells from hurt to the greatest extent during raising temperature of the tumor cells. MNPs show a feasibility to fulfill these needs through combination with tumor cells. Firstly, they have controllable sizes which are smaller than or comparable to a cell, so they can enter the human body and get close to the area of interest. Nanoparticles must pass capillary endothelium before reaching targeted location. The majority of nanoparticles with size bigger than 7 μm are obstructed by pulmonary capillary. The liver and spleen uptake remaining particles with diameters above 100 nm size, while those smaller than 100 nm are swallowed by osteoclast. These features allow nanoparticles to reach corresponding organ by adjusting their size. This is termed by passive targeting. Secondly and importantly, they can be coated with biological molecules, such as dextran, PVA and phospholipids, to make them to bind to or interact with the tumor cells so that providing a controllable means of tagging them. Besides, the MNPs can be controlled by the external magnetic field gradient to gather at the tumor which can result in interaction with target tissues as soon as possible.

2.1.3 Heat Generation Mechanism

In this part, we will discuss how the MNPs convert other forms of energy into thermal energy.

Electromagnetic fields E and H can affect heating of tissues. The electrocaloric effect is caused by electric field component E, based on the Maxwell equations and thermodynamic relations. The magnetocaloric effect is caused by variations in H. These effects are mainly determined by the tissue dielectric and magnetic properties of substances, respectively. Because magnetism of biological objects is negligibly small, biologically compatible nontoxic MNPs (based on magnetite and so on) are used to strengthen the influence of an external magnetic field.

Hysteresis, eddy current, Neel paramagnetic switching, and friction from Brownian rotation are the four possible mechanisms to generate heat for magnetic materials exposed to an alternating magnetic field (Rosensweig 2002; Hergt et al. 2006). It is quite possible that all four mechanisms may contribute to the total heat generated by a particular magnetic sample in an alternating magnetic field. But it is expected that only one or two of the mechanisms will dominate. This is determined by the properties of the magnetic material, its environment and the magnetic field.

Considering a spherical particle of diameter D_0 and resistivity ρ_Ω in the presence of a uniform periodic magnetic field of strength $B = B_p \sin \omega t$, where B_p is the field amplitude and ω is the frequency of the field (in radians per second), the power of eddy current heating is

$$P_E = \frac{(\pi f B_P D_0)^2}{20\rho_\Omega} \tag{3}$$

where, $f \equiv \omega / 2\pi$ (in cycles per second).

A number of important trends can be observed from this equation. The eddy current power increases with frequency and sphere diameter, while it decreases with increasing resistivity. In an alternating magnetic field the induction of eddy currents in the patient's body may lead to an unwanted heating at both cancerous and healthy tissue reducing in this way the selectivity of the therapy. Therefore one can minimize the eddy current power by using lower frequencies, smaller particles, and higher resistivity materials.

If frequency of the external magnetic field could not cause significant eddy current heat, the MNPs will produce heat by a large number of magnetic hysteresis losses. For multi-domain ferro- or ferri-magnetic materials, heating is induced due to hysteresis losses (Mornet et al. 2004). As limited by the applied magnetic field strength in hyperthermia, MNPs can not be fully magnetized, leading to the greatest heat production by magnetic hysteresis losses. Generally, only 25% of maximum is available.

Hysteresis loss (Hergt et al. 2008) is proportional to the frequency, three power of amplitude of the external magnetic field and hysteresis loop constant, which is related to the kinds of material. Ignoring the heat caused by the eddy current effect and the natural resonance, heat generated per unit volume due to ferromagnetic hysteresis loss in the alternating magnetic field is

$$P_{FM} = \mu_0 f \oint \mathrm{H} d\mathrm{M} \tag{4}$$

As the decrease of the size of nanoparticles within the single domain range, their coercivity and remanence become bigger and would produce more heat. However, when reduced to below critical size of nanoparticles, their coercivity and remanence drop to zero. At this time, the MNPs show SPM without hysteresis, and produce heat through Neel relaxation mechanism (Mornet et al. 2004; de Chatel et al. 2009). Compared with paramagnetism, the magnetic moment of SPM particles is not of a single atom but of the entire particle, resulting in a much higher susceptibility value than that for simple paramagnetism. The magnetic moments of SPM nanoparticles are randomly reoriented by the thermal energy of their environment and do not display magnetism in the absence of a magnetic field. Unlike ferrimagnetic materials, they do not aggregate after exposure to a magnetic field. Aggregation can hinder the body's efforts to remove the nanoparticles. Therefore, SPM nanoparticles are ideal candidates for hyperthermia cancer treatment.

The magnetic susceptibility of SPM can be expressed as

$$\chi = \chi' + i\chi'' \tag{5}$$

where, χ' and χ'' depend on the frequency of external magnetic field. In this case, SPM is determined by (Chan et al. 1993)

$$P_{SFM} = \mu_0 \pi f \chi'' \mathrm{H}^2 \tag{6}$$

Experiments show that SPM materials need smaller strength of electromagnetic field than that of ferromagnetic, which therefore can improve cancer treatment effect.

For a motionless sample of magnetic fluids in a low-frequency field (f < 1 MHz), a general expression of magnetic relaxation loss reads as

$$P = \mu_0 \pi f \chi_0 \mathrm{H}^2 \frac{\omega\tau}{1+(\omega\tau)^2} \tag{7}$$

Heating can also be caused by the friction from Brownian rotation within a carrier liquid. In this case, the energy barrier for reorientation of a particle is determined by rotational friction within the surrounding liquid (Mornet et al. 2004).

During magnetic hyperthermia, produced heat is measured by the SAR in order to compare the efficiency. A large number of parameters, such as size, distribution, shape and chemical composition of particles, frequency and amplitude of the magnetic field, etc. all contribute to SAR. Thus, it is very difficult to obtain a general theoretical expression of SAR for magnetic hyperthermia.

2.2 *Typical Materials*

According to their structures, MNPs used in magnetic induction hyperthermia can be broadly divided into the magnetic fluid, magnetic liposomes and glass-ceramic nano-magnetic particles.

2.2.1 Magnetic Fluids

Magnetic fluids, or called ferrofluids, a liquid containing MNPs, are a stable colloidal suspension of sub-domain magnetic particles in the liquid carrier. Therefore, magnetic fluid possesses magnetism as solid magnetic materials, but also flowability as a liquid, which can be controlled by the external magnetic field. Magnetic fluids can absorb electromagnetic energy into heat in alternating magnetic field. Magnetic induction hyperthermia with magnetic fluids attracts more and more attentions because of its several advantages. First of all, magnetic fluids are up-taken more easily by the tumor cells and dispersed evenly in the tumor domain to heat cells. They can enter the daughter cells with division of tumor cells, which is more lethal. Furthermore, MNPs distribute uniformly under the control of alternating magnetic field, expanding the treatment area gradually and heating surrounding cancer cells around the original particles, thus saving the use of nanomaterials. Besides, magnetic fluids possess better biocompatibility.

2.2.2 Magnetoliposomes

Magnetoliposomes are produced by mixing magnetic material in liposome. The liposome is an artificial class of lipid globules, which can carry a variety of hydrophilic, hydrophobic and amphiphilic substances. It has poor targeting property, and if inserting MNPs in lipid, Magnetoliposomes can carry the drugs to reach the tumor under the guidance of the magnetic field in vitro. Since the magnetic particles in Magnetoliposomes have special absorption rate on the electromagnetic wave, they can strongly absorb the energy to improve temperature in the alternating magnetic field, and this inhibit tumor growth, or even make the tumor disappeared.

2.2.3 Magnetic Glass-Ceramic Nanoparticles

MGC nanoparticles are produced by mixing iron oxide particles and glass ceramics. This kind of material can satisfy the need of tumor hyperthermia, especially for deep tumors. The coercivities of ferromagnetic glasses are much higher than that of magnetic fluids. One reason is because glasses have larger internal stresses which hinder the magnetic moment rotation. In addition, another reason is the magnetic grains are separated by glasses and wollastonites without magnetism.

According to elements species, the MNPS can be classified into the following categories.

2.2.4 Iron Oxide Particles and Other Ferrous Metals

In magnetic hyperthermia, the majority is focused on Fe_3O_4 and γFe_2O_3. Many researchers have studied their toxicity and biocompatibility. Feasibility in animal experiments has also been proven and stepped into clinical trials (Johannsen et al. 2007a).

SPM iron oxide nanoparticles, possess magnetic moment of 300–400 emu/cm^3 generally. Varanda et al have synthesized Fe_3O_4-coated $Fe_{55}Pt_{45}$ nanoparticles in a core/shell structure with high magnetization close to the $\alpha - Fe$ (1,000 emu/cm^3) and low coercivity (~60 Oe). In addition, the nanoparticles present a slightly oxide layer on surface favoring their functionalization as shown by the realized dextran molecules coating (Varanda et al. 2008).

Other ferrous metals and their alloy used in magnetic hyperthermia include $MeFe_2O_3$ (Me is Co, Mn, Ni and so on), NiCoFe, FeCo and NiFe metals.

Iron oxides are relatively safe and can regularly be excreted, while nickel and cobalt magnetic materials have certain degrees of physiological toxicity. Their saturation magnetization gradually increased by the order but stability decreased, which is determined by their Gibbs free energy.

MnZn ferrites have high permeability and low magnetic losses at high frequencies and the properties are largely dependent on their microstructure. La^{3+} can change the orientations of crystal growth of nano-MnZn ferrites, then decrease the saturation magnetization of La^{3+} doped sample and even show SPM property (Gu et al. 2004). Silica-coated lanthanum-strontium manganite particles with $La_{0.76}Sr_{0.24}MnO_{3+\delta}$ stoichiometric formula, due to its Curie temperature at ~40°C, has been established as a potential candidate for self-regulated power-absorbing and temperature-controlling materials in tumor hyperthermia treatments (Uskokovic et al. 2006).

Monodispersed ternary $Mn_{0.5}Zn_{0.5}Fe_2O_4$ ferrite nanoparticles synthesized using thermal decomposition of metal acetylacetonate in the presence of a high temperature boiling point solvent and fatty acids are SPM at room temperature, and therefore can be applied in hyperthermia (Parekh et al. 2006).

The SPM gadonanotubes also show promise for targeted magnetic hyperthermia (Sithararnan and Wilson 2006). Furthermore, non-spherical particles such as the gadonanotubes are yet to be tested as hyperthermia agents. And it is possible that their unique rod-shaped geometry may augment performance through non-axial Brownian relaxation.

2.2.5 Iron-Gold Nanoparticles

MNPs (Fe,Fe_3O_4,$\gamma-Fe_2O_3$) can be coated by another inert shell, such as SiO_2 (Pereira et al. 2010), gold (Ravel et al. 2002; Tamer et al. 2010), Ag (Choi et al. 2010; Ding et al. 2010), polymer (De La Cruz-Montoya and Rinaldi 2010) and their hybrids (Zhou et al. 2010). SPM iron-gold (Fe@Au) nanoparticles use gold as the shell on the surface of the core. Gold is diamagnetic, and the presence of a gold coating inhibits oxidation and allows the particles to be stable. Besides, the gold produces no toxicity for human body. Therefore this kind of core/shell structure can be used in tumor hyperthermia. Their SPM behavior has been confirmed by magnetic measurements at various temperatures (4.2–280K) and by Mössbauer spectra measurements. Fe and metallic Au in Fe@Au nanoparticles exist as composite nanoparticles, not as separate particles. The Au-coated particles exhibit a surface plasma resonance peak that red-shifts from 520 to 680 nm. The gold-coat protects the magnetic core against oxidation, without drastic reduction of the magnetic properties. The obtained hyperfine parameters also confirmed the absence of iron oxides (Fe_3O_4, Fe_2O_3) (Zelenakova et al. 2008; Ban et al. 2005; Jafari et al. 2010).

2.2.6 Carbon Encapsulated Iron

Metallic iron, due to higher saturation magnetization (Huber 2005) in comparison to iron oxides, could act better if oxidization in ambient or biological conditions is avoided (Klingeler et al. 2008). Recently, scientists have found a promising way to overcome this problem, which is to coat the iron with a carbon shell by insertion of the material inside of CNTs. Various studies have shown that, independent of the synthesis technique, the carbon encapsulated iron is efficiently protected by the surrounding shells and its magnetic properties are retained (Klingeler et al. 2008). In addition, carbon shells may act as multi-functional containers which can be filled with different materials. Additional material such as a nanothermometer in Fe-CNT would increase the potential of CNT as hyperthermia agents. Attaching functional elements to the outer shell of CNT increase their biocompatibility (Klingeler et al. 2008). Recent studies of cytotoxic effects of Fe-CNT on cells indicated no significant toxicity of Fe-CNT prepared by the same synthesis route (Klingeler et al. 2008).

The feasibility of Fe-filled CNT for hyperthermia has also been demonstrated. In addition to Fe, Fe_3O_4 conjugated CNT was found to experience hyperthermia heating under alternating electromagnetic field (Shi et al. 2009). Krupskaya et al. observed a different magnetic response of Fe-CNT powder compared to Fe-CNT dispersed in aqueous solution. Ferromagnetic Fe-CNT in powder does not show any hysteresis when being dispersed in liquid (Krupskaya et al. 2009). In addition, other MNPs (Ni, Co) coated by carbon have also been produced by high pressure chemical vapour deposition (Tang et al. 2009; El-Gendy et al. 2009; Liu et al. 2004).

Besides, the carbon structures are also measured to understand the magnetic properties (Dumé 2004). They are widely believed to have ferromagnetic properties, but the effects are so weak that physicists are not sure if the magnetism is due to tiny amounts of iron-rich impurities, or if it is an intrinsic property of the carbon. In 2002 Coey's group proposed that ferromagnetic nanocrystals in the sample induced a magnetic moment in the carbon via proximity effects (Dumé 2004).

2.3 *Latest Progress in Clinical Trials*

There are many preclinical researches in vitro and in vivo using animal models for the application of magnetic nano-hyperthermia. However, only several groups have carried out the clinical tests and the tumors refer to prostate cancer, melanoma, glioblastoma and so on. The first treatment system for patients has been developed at the Charité- Universitätsmedizin Berlin (Gneveckow et al. 2004). Table 2 summarizes some of these studies. The MNP used in their test is magnetic fluid MFLAS (MagForce Nanotechnologies AG, Berlin, Germany) which consists of aminosilane coated SPM iron oxide nanoparticles (core diameter: 15 nm) dispersed in water, with an iron concentration of 112 mg/mL. These clinical tests proved the potential of magnetic hyperthermia, and many patients prolonged their survival time although they are dead of disease.

3 Nanoparticles Enhanced Photothermal Therapy

3.1 *Basic Feature*

PTT is a class of tumor hyperthermia which converts optical energy into heat to abate the tumor. LITT has been proven as an effective means in this area (Zhou and Liu 2004). It mainly uses visible and NIR light to deliver heat. However, due to limitation of tissue penetration of the band laser, LITT is only efficient for small, localized tumor, or serves as adjuvant of chemotherapy, radiotherapy and photodynamic therapy. One solution to address such issue is to add some optical

Table 2 Several clinical studies on magnetic nano-hyperthermia

Tumor cells	IAMFL (mL)	NOT	NOP	T_{max} (°C)	TS	Result	Temperature measurement	First therapy	Reference
Prostate	11.4 (4–14)	6	10	42.7 (40.5–51.3)	15.8 (1.6–45.3)	–	Fiberoptic thermometry probes	Curative	Johannsen et al. (2007a, b)
	8.5 (6.0–12.5)	6	8	41.5 (40.9–45.4)	29.7 (15.9–48.7)	–	–	R	Wust et al. (2006)
Glioblastoma	(4.5, 4.6)	6	2	(49.5, 65.6)	(36.3, 50)	DOD	Fiberoptic thermometry probes	S+R+C	van Landeghem et al. (2009)
	0.2 (0.1–0.7)	6(4–10)	14	44.6 (42.4–49.5)	12.1 (4–41.2)	4DWD 9DOD 1Alive	Fiberoptic thermometry probes	S+C	Maier-Hauff et al. (2007)

DOD dead of disease, *DWD* dead without disease, *NOT* number of treatment, *NOP* number of patients, *IAMFL* injected amount magnetic fluid, *TS* tumor size, *S* surgery, *R* radiotherapy, *C* chemotherapy absorption dyes to selectively strengthen tumor thermal damage (Chen et al. 1995). Recent researches find that quite a few nanoparticles including GNPs have the ability to absorb mightily the NIR irradiation to generate sufficient heat to cause cell death. Compared to traditional dyes, their metal surfaces are less sensitive to chemical corrosion and photobleaching, and have much higher heat absorption intensity.

GNPs received most attention as photosensitizers used in PTT because of their good absorption property, which offer lots of opportunities to make them suitable for use in tumor hyperthermia. Above all, such material has a strong optical extinction peak known as surface plasma resonance (SPR) that can be varied by control of particle morphology and light in the resonance band can be efficiently absorbed by them. Therefore they can enhance the photothermal effect under irradiation of NIR light. Besides, GNPs are easily synthesized to precisely control the morphology, size, structure and chemical composition, which determines that GNPs would be connected to its target molecule such as antibodies or coated with PEG on the surface. GNPs are easily modified and its IR absorption peaks will not change after that. Bits immunogenicity and clearance rate will be reduced. GNPs connected targeting molecules can specifically kill the tumor cells without affecting normal cells. In addition, GNP is safe for treating human cancer because of its little toxicity, good biocompatibility and small side effects. GNPs have much morphology include nanospheres, nanoshells, nanocages, and nanorods. Laser with longer wavelength would obtain deeper penetration and cause less harm to human body. Different shapes of GNPs would produce different and adjustable optical extinction peaks, which can be seen by the color of different morphology and size of GNP suspensions. Controlling the morphology and size of nanoparticle tailors them to achieve maximum absorption under the irradiation of light with different frequencies.

The nanoparticles are conjugated to special antibodies as targeted labels which will combine with receptors over pressed by types of cancer cells. Table 3 shows various antibodies conjugated with gold nanostructures and CNTs used in PTT.

Various gold structures are believed to be nontoxic (Villiers et al. 2010) even at high concentration because gold is a kind of inert metal in the body (Sabadell et al. 1999). In fact, gold colloids have been used to treat arthritis (Tarsy 1941) for over 100 years.

In addition to gold nanostructures, CNTs, owing to its good absorption characteristics, have also been utilized as photothermal enhancers under irradiation of NIR light (Kam et al. 2005). The toxicity of CNTs has been a major concern for the researchers. In the studies of CNTs used in biomedical applications, they are considered to be safe for the body. Gold-plated CNTs can enhance NIR contrast (~102-fold) and preliminary in vitro viability tests show that golden carbon nanotubes have minimal toxicity (Kim et al. 2009). The CNTs can be produced (Karthikeyan et al. 2009) by techniques like arc discharge (Sun et al. 2007), laser ablation (Scott et al. 2002), CVD (Tao et al. 2007) and super-growth CVD (Hata et al. 2004), etc.

In order to prevent proteins from depositing on the nanoparticles, which would result in their being captured by the immune system and dragged out of the bloodstream into the liver or spleen, scientists have made efforts on coating them with a layer of nontoxic PEG. This impedes the adsorption of proteins, in effect disguising

Table 3 Various antibodies conjugated with nanostructures used in PTT

Nano structure	Antibody	Receptor	Tumor cells
SWNT	Folate moiety	Folate receptor	Murine mammary tumor line EMT6 cells (Zhou et al. 2009) breast, lung, endometrium, cervix, ovary, and renal carcinoma cells (Weitman et al. 1992; Bagnoli et al. 2003)
Nanocage	Anti-HER2 antibody	Epidermal growth factor receptors (EGFR)	Breast cancer cells (Chen et al. 2007; Au et al. 2008)
Gold nanosphere	Monoclonal F19 Antibodies		Human pancreatic cancer tissue (Eck et al. 2008)
Gold nanosphere	Herceptin® (monoclonal antibody that binds HER2/neu)		Human breast tumors (Copland et al. 2004)
SWNT	Anti-CD22		Human Burkitt's lymphoma cells in vitro (Chakravarty et al. 2008)

the nanoparticles so that the immune system cannot recognize them, which allows them to prolong retention time in the circulation to accumulate in tumors (Lipka et al. 2010).

3.2 *Typical Materials and Optical Properties*

3.2.1 Gold Nanospheres

Nanospheres are the simplest structures that can range in diameter from 2 nm to several 100 nm (Cherukuri et al. 2010a). The peaks are red-shifted with the increase of nanosphere diameters. Spherical GNPs with 100 nm in diameter can only reach the peak at 580 nm, while skin, tissues, and hemoglobin have a transmission window from 650 up to 900 nm with a peak transmission at approximately 800 nm. This barrier was solved with other shapes.

3.2.2 Gold Nanoshells

Gold nanoshells, one of core-shell materials, refer to a special structure of nano-materials composed of a dielectric core (eg, silica) and gold formed a uniform coating on its core. Compared with gold nanosphere, the core-shell structure with the same diameter results in a red-shift and this resonance frequency can be tuned by changing the shell thickness to core diameter ratio. By adjusting the proportion of core-shell, the range of optical resonance frequency can cover visible to infrared region including near infrared region (700–1,300 nm) with the best penetrability for biological tissue.

3.2.3 Gold Nanorods

Gold nanorods, because of their adjustable aspect ratio, have chemical and optical anisotropy. As a new class of nanoparticles, nanorods have unique optical properties compared with other structure. They have both transverse and longitudinal SPR double-peaks. The transverse SPR, which is due to an electronic oscillation across the width of the rod, is effectively of the same nature as the plasma resonance of simple gold nanospheres. However, the longitudinal one, which is due to oscillation of electrons in the long direction of the rod, provides a much larger extinction coefficient. The longitudinal SPR is red-shifted relative to the transverse one and can be tuned by controlling aspect ratio. Therefore, the gold nanorods biological probe can match NIR laser wavelength and allow the near infrared light to penetrate the deep tissue.

3.2.4 Gold Nanocages

Nanocages are hollow porous GNPs ranging in size from 10 to over 150 nm. They are created by reacting silver nanoparticles with chloroauric acid in an aqueous solution (Lu et al. 2007; Chen et al. 2006). Gold nanocages also absorb light and heat up, killing surrounding cancer cells. Importantly, compared with other gold nanostructures, the gold nanocages have a more conveniently and precisely fine-tunable SPR peak by varying the wall thickness relative to the overall dimension that extends into the NIR where the optical attenuation caused by blood and soft tissue is essentially negligible (Skrabalak et al. 2007). Skrabalak et al. have found that changing the amount of metal precursor added to the suspension of Ag nanocubes is a simple means of tuning both the composition and the localized surface plasma resonance of the metal nanocages (Skrabalak et al. 2008). Besides, they are also biocompatible and present a well-established surface for easy functionalization.

The key to PTT is the gold nanostructures' ability to efficiently absorb light and convert it to heat. Compared with nanoshell, both the nanorods and nanocages are stronger light absorbers due to their higher ratio of light absorption to scattering.

3.2.5 Carbon Nanotubes

CNTs are made from sheets of carbon, usually one-atom thick, and then folded into tubular structures. These impossibly tiny cylinders are extremely efficient in conducting heat. There are various types of nanotubes, such as SWNTs and MWNTs. SWNTs, a class of prototypical one-dimensional nanomaterials, possess exceptional electronic and mechanical properties (Jorio et al. 2008). Most SWNTs have a diameter close to 1 nm, with a tube length that can be many millions of times longer. Alkali-halide salts, particularly potassium bromide, can reduce the photothermal emission from SWNTs (Manning et al. 2003), so trapping salts inside SWNTs and coating SWNTs with the salt has a more pronounced impact on reducing photothermal emission and therefore enhancing optical energy absorption by SWNTs.

MWNT consists of multiple rolled layers of graphite, and Russian Doll model (Galanov et al. 2002) and Parchment model can be used to describe the structures of MWNT. Compared with the specific resonance absorptions of SWNTs, MWNTs behave as highly efficient dipole antennae with broad absorption spectra, which resulting in rendering them amenable to stimulation by a range of NIR energy sources (Burlaka et al. 2010). Additionally, MWNTs as the metallic tubes can be expected to absorb significantly more NIR irradiation because, per weight, SWNTs contain both metallic and semiconducting ones. CNTs have extraordinary strength and unique electrical (Minati et al. 2010), optical (Zhang et al. 2010; Lefrant et al. 2010) and thermal conduction properties (Savin et al. 2009).

3.2.6 Graphene

Grapheme is a one-atom-thick planar sheet. Since it was first created in 2004 by Geim and Novoselov at University of Manchester, researchers are amazed by its unique properties. Recently, Acik et al. reported reduced graphene oxide can strongly absorb infrared radiation (Acik et al. 2010). This implied that such kind of material may have the possibility to provide better performance in PTT in the future.

3.3 Latest Progress in Clinical Trials

3.3.1 DNA-Encasement Improves Carbon Nanotubes

The CNTs will stick to each other and self-associate, which limits the utility of CNTs for in vivo use. The solution is to bind DNA to the CNTs. The studies on DNA-encasement of SWNTs have demonstrated their feasibility and effect (Hughes et al. 2007). For MWCNTs, it is more difficult to interact with DNA because they have larger diameters than SWNTs and present a surface with a large radius of curvature. Supratim et al have demonstrated that DNA-encasement enhance heat production efficiency relative to non-DNA-encasement following NIR irradiation of MWNTs, and DNA-encased MWNTs can be used to safely eradicate a tumor mass in vivo. Upon irradiation of DNA-encased MWNTs, heat is generated with a linear dependence on irradiation time and laser power (Ghosh et al. 2009).

3.3.2 Theoretical Modeling on Temperature Fields During Gold Nanoshells Enhanced Optical Hyperthermia

There are many experiments in vivo or vitro on the nanoparticles used in the PTT currently. However, there is a strong lack of understanding the energy distribution and the transient temperature field in the tissues due to addition of nanoparticles. Recent efforts were made on theoretical modeling the optical and temperature fields during gold nanoshells enhanced hyperthermia through combining the Monte-Carlo simulation strategy and the Pennes bioheat transfer equation. Effects of size, concentration of the nanoshells to the heating behaviors were predicted and the method would serve well for future treatment planning for the nanoshells enhanced LITT on target tumor tissues (Wang et al. 2010).

4 Nanoparticles Enhanced Electrical Hyperthermia

4.1 Basic Features

RF ablation is another type of tumor hyperthermia therapy relying on the administration of RF irradiation. RF ablation inserts an electrode into the tumor and applies a RF current, which induces agitation of the ions within the tissue, leading to frictional heating. As is well known, for a given RF power output, the power deposition at each point in space is strongly dependent on the local electrical conductivity. RF frequency and amplitude, and the electrical conductivity of the tissue determined the SAR which is expressed as

$$SAR = \int_{sample} \sigma |E|^2 / \rho dr \tag{8}$$

Thus, improving the conductivity around the tumor tissue is the way to enhance the RF ablation effect. it has been demonstrated that altering electrical conductivity of tissue through saline injection prior to or during RF ablation can increase RF tissue coagulation volume (Lobo et al. 2004).

Recently, it was further demonstrated that nanoparticles with high electrical conductivity can be used as effective adjuvants to induce hyperthermia in vitro and in vivo upon irradiation with an external 13.56 MHz RF field (Gach et al. 2010). The presence of nanoparticles can be targeted to malignant cells and provide a possibility of non-invasive RF treatment, compared with traditional RF treatment which will insert an electrode into the tumor.

RF ablation is now used in clinical practice to treat some malignant tumors, yet nanoparticles enhanced radiofrequency is still a new method and need more studies.

4.2 Types of Nanoparticles Enhancing the Conductivity

Nanoparticles with good conductivity properties, such as GNPs and CNTs, are studied to provide the possibility of more efficient heating under RF field exposure recently.

4.2.1 Carbon Nanotubes

SWNTs can act efficiently to convert RF irradiation into heat, which is possible due to their resistive conductivity property (Gannon et al. 2007; Cherukuri et al. 2010b; Day et al. 2009). For SWNTs, the rate of heating depends on the concentration. Gannon et al. injected SWNTs or control solutions into Hepatic VX2 tumors of rabbits and the rabbits were treated in the RF field. The RF field induced lethal thermal injury to the malignant cells under efficient heating of aqueous suspensions of SWNTs (Gannon et al. 2007).

4.2.2 Gold Nanoparticles

GNPs within a tumor can significantly increases the tissue temperature upon exposure to RF energy other than used in the photothermal therapy as the optical sensors (Cardinal et al. 2008; Gannon et al. 2008; Curley et al. 2008; Glazer et al. 2010). Increasing GNPs concentration and increasing power of RF field both contributed to increased temperature and rate of heating (Gannon et al. 2008).

Cardinal et al. (2008) directly injected the GNPs into the tumor of rats in vivo and exposed at 35 W, demonstrating significant temperature increases and thermal injury at subcutaneous injection sites as compared to water injection controls. Evan et al. (Glazer et al. 2010) use 20-nm gold particle conjugated to cetuximab, which is an EGFR-1 antibody over expressed by pancreatic carcinoma cells in an RF field with a generator power of 200 W for 5 min. The results showed that targeted cells internalized a sufficient amount of antibody-conjugated GNPs to induce injury in a noninvasive RF field, while the non-targeted cells did not.

When comparing the heating rates under RF field exposure, GNPs seem to provide a more efficient system over SWNTs. This suggests that GNPs may be better conductors of RF energy than SWNTs and may provide a better application for RF tumor ablation.

5 Nanoparticles Enhanced Ultrasound Hyperthermia

Ultrasonic heating method, whose energy is easy to produce and focused, spreading non-destructively within the organization, has become an important heating means in tumor hyperthermia. Traditional ultrasound hyperthermia treatment generally applies in superficial tumors, while the HIFU (Lynn et al. 1942; Zhu et al. 2009) is for the treatment of deep tumors.

Since tumor hyperthermia with many different heating sources, such as IR and RF irradiation, can be enhanced by the presence of nanoparticles, people believe that ultrasonic heating method also has the possibility to connect with nanoparticles with fairish acoustic properties. Now there are many nanoparticles used as UCA (Calliada et al. 1998; Kornmann et al. 2010). Some of them are not good agents, but can bring favorable influence on ultrasonic nano-hyperthermia. Like MNPs, the ultrasonic nanoparticles also can be used as the enhancer of imaging and hyperthermia treatment simultaneously. The studies of Razansky et al. showed that the maximal temperature elevation obtained during 300 s experiments was 21°C, in a 10 mL volume target containing UCA, insonified by unfocused 3.2 MHz continuous wave at spatial average intensity of 1.1 W/cm^2. The results also suggest that higher frequencies are more efficiently absorbed by commonly used UCA (Razansky et al. 2006a). Based on theoretical predictions, they designed experiments to measure the dissipation and heating effects of encapsulated UCA in a well-controlled and calibrated environment (Razansky et al. 2006b).

The study of Jozefczak et al. presented acoustic properties of water-based biocompatible fluids in which magnetite particles (Fe_3O_4) were coated with sodium oleate and polysaccharide dextran and showed the attenuation coefficient of ultrasonic wave (Jozefczak and Skumiel 2007).

6 Thermochemical Heating Methods

All of the heating means, RF heating, microwave heating, ultrasonic or laser heating, have some deficiencies more or less. The common drawbacks are complicated equipments, high cost and complex surgical operation. In recent years, researchers have initiated a new method of tumor hyperthermia- the targeted thermochemical ablation therapy, which breaks inherent heating methods in traditional methods and do not require complex equipment, so as to achieve a cost-effective treatment of tumors.

Thermochemical reaction heating method is based on chemical reactions to release plenty of heat to the target tissues. By delivering appropriate exothermic materials into the tumor site, tremendous heat will be released from the chemical reactions among these substances with their tissue environment. Researchers have suggested two typical thermochemical ablation methods, acid-alkali reaction enabled thermal ablation (Deng and Liu 2007) and alkali metal enabled thermal ablation (Rao and Liu 2008; Bauer et al. 2008; Rao et al. 2008). It was proved through various experiments that the thermochemical reaction heating method can ensure releasing high-intensity heat only in the target site, while ensure the surrounding tissue away from injury by heat mechanical puncture. In the study of Rao et al. (2008), some fundamental mechanisms related to the antitumor effects of sodium and its controllability as an economic, safe and efficient thermochemical agent in targeted tumor treatment were disclosed. Followed their work, Cressman et al. evaluated exothermic permanganate redox chemistry for utility in tumour ablation and proved it a powerful system with potential to increase tumor temperature. The maximum temperature could reach 97.4°C in the experiments. They also used different reagent volumes, concentrations, injection order, and substrate choice to study the parametric effect over the reaction (Cressman et al. 2010).

Similarly, nanoparticles can also be combined with thermochemical reaction method to enhance the effect of cancer treatment. For one hand, nanoparticles could be used as the catalysts to accelerate the reactions. They have small size and large ratio of surface area to volume. And the bonding states and the valence states for surface atoms are different with that of inner atoms resulting in increased surface activity. Therefore, nanocatalytic activity and selectivity are much higher than that of traditional catalysts. For example, CuO at micro-level does not have catalytic activity for decomposition of H_2O_2, while when the particle size is reduced to nanometer grade, it shows high catalytic activity. Besides, metal nanoparticles can fracture catalytically H–H, C–C, C–H and C–O bond, so that speeding up response time. Wu et al. showed that for the ethane oxidation dehydrogenation reaction, nano-NiO catalyst play better role than conventional NiO catalyst at lower temperature (Cao et al. 2005). Bennett loaded nano-Pd on TiO2 and catalyzed 1-hexene hydrogenation reaction to produce more hexane than conventional palladium catalysts (Bennett 1989).

On the other hand, some nanoparticles could be used as mediums involved in chemical reactions to produce heat. Nano-materials have small size, large percentage of surface atoms, superior adsorption ability and high surface reactivity. Metal nanocrystals are easy to oxidize; even corrosion-resistant nitride ceramic materials are also unstable when their diameters reduce to nanoscale.

7 Other Heating Methods Through Delivering Heat Directly

The common purpose for all the heating methods is to convert other forms of energy into heat at tumor tissue. Therefore, if one could inject heat energy directly, the effect may be better.

The electrothermal needle mainly used in traditional Chinese medicine is such a heating method. It is originated from fire needle, and can insert directly into the central of tumor. When energy reaches 165 cal and lasts for 40 min, the central temperature can reach 56°C while the surrounding tissue stays at only 44°C ~ 47°C. Apart from this, the temperature of electrothermal needle can adjust by changing electric current, so it can extend the treatment time to enhance the effect. Electrothermal needle method is more easily operated than other hyperthermia approaches to reduce tumor proliferation and metastasis. Maybe there are still other ways similar to electrothermal needle to provide heat energy directly to the tumor, and this method will probably become a new means for cancer hyperthermia.

Besides, one can release the heat energy with the help of transportation of high temperature fluids. It consists of two forms: one is hyperthermia probe technology based on closed high-temperature steam heating; another is based on percutaneous fluid injection. The former belongs to closed heating, whose purpose is to transport high-temperature fluid such as steam into the target tumor tissue, while to protect surrounding healthy tissue with vacuum insulation. Yu et al. developed a new minimally invasive tumor hyperthermia probe using high temperature water vapor (Yu et al. 2004) and evaluated it theoretically and experimentally. The latter directly injects the high-temperature fluids into the target tissue to raise its temperature.

The nanoparticles have special properties. One can choose some nanoparticles with better thermal conductivity and load them into the tumor, and then these nanoparticles could improve the thermal conductivity of the tumor, so that increase the tumor temperature much faster. In recent years, the thermal conductivity of nanoparticle-fluid mixture has been reported by many authors (Wang et al. 1999; Shalkevich et al. 2010; Le Goff et al. 2008; Tian and Yang 2008). The nanoparticles can be suspended into heat transfer fluids such as water to increase the thermal conductivities relative to those of the base fluids. The nanoparticles have been studied include Al_2O_3, CuO (Wang et al. 1999), Fe_3O_4 (Tsai et al. 2009), Cu (Xuan and Li 2003), etc., and some of them are safe for human body.

8 Challenging Issues

8.1 *Comparison Among Different Heating Strategies*

From the above review, it is known that heat generation sources refer to a wide range of radiation spectrum in the forms of RF, MW, IR, and ultrasound, chemical reaction, fluids or steam with high temperature. Each method has its own heat transfer principle, advantages and disadvantages. Facing so many alternatives, how can the clinicians make the best choice to treat tumor? The answers lie in the comparisons among these methods.

Figure 2 shows the division of spectrum for the electronic magnetic wave. The lower the frequency is, the deeper it penetrates into the tissues. Therefore, RF, which is a subset of electromagnetic radiation with a frequency below 300 MHz, has largest penetration possibility among the electronic magnetic wave heating methods. RF is divided by LF, MF, HF, and VHF. For capacitive RF, HF and VHF can reach any deep tissue of human body; while HF is mainly used on superficial muscle for inductive heating. LF and MF are applied for interstitial heating. The usual bands are 40.68, 13.56 and 8 MHz.

MW refers to the wave with a frequency of 300 MHz to 300 GHz, where UHF is mainly used for heating. The MW and RF have an overlap, and low

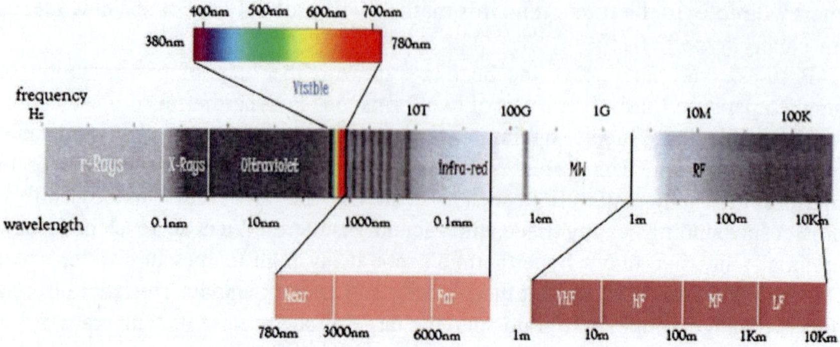

Fig. 2 Electromagnetic spectrum

frequency of the heating bands can reach 100 MHz of VHF, while the high frequency reaches 3,000 MHz. The accepted MW frequencies are generally 2.45 GHz, 915 and 434 MHz.

The wavelength band from 1 mm to 780 nm is infrared light, where IR is below 3,000 nm. The wavelength range of 780 to 380 nm is visible light. The laser heating method makes use of IR and visible light.

Heating is a process of absorbing energy during spread of electromagnetic wave. When human is irradiated under the RF field, the free electrons, ions and dipoles, will move resulting in ohm-heat. At the same time, vibrations among the molecules produce parts of heat. The principle of MW heating methods is similar to that of RF, but the vibration heat holds the uppermost.

Ultrasound frequency for heat treatment is 0.5–5 MHz; generally 1 MHz. Ultrasound wave is a king of mechanical wave. It produces thermal energy according to vibration friction. The higher ultrasound intensity is, the faster temperature rises; and the higher the frequency is, the more heat tissues will absorb. Therefore, treating superficial tumors uses high-frequency ultrasound to prevent the tissue below tumor from heating, while deep tumors treatment uses low-frequency to prevent surface overheating. Besides, the non-thermal effect brought about by ultrasound is also a research focus.

The domain of action, advantages and disadvantages are considered to determine the choice among various heating methods (Table 4) (Li and Hu 1995; Lin et al. 1997; Peng and Zhao 2002). In nanoparticle enhanced hyperthermia, the toxicity, biocompatibility, heat efficiency of the nanoparticles will also affect the results. Heating methods based on electrothermal needle, high temperature fluids or thermochemistry are in the initial stage, so there are still many problems associated with application need to be clarified.

Although each kind of energy has its advantages, one can still envisage with a bold hand that there would be a new device which can emit waves with different frequency energy according to need. That means, we can combine RF and MW devices into one multifunction device, and then change frequency according to practical need.

Table 4 Comparison among various heating methods for tumor hyperthermia

Heating source	Principle	Advantages	Disadvantages	Sphere of action
RF	Electrothermal effect	Wide action scope and depth; relatively simple equipment; require no shielded rooms; easy temperature measure	Uneven heat distribution; subcutaneous fat overheating	Large shallow tumor; deep tumors for abdomen, pelvis, chest and limbs; interstitial heating for large tumors, such as brain, neck, liver and breast
MW	Magnetocaloric effect	High temperature; no fat overheating; relatively safe	Larger muscle attenuation coefficient; limited penetration depth; difficult temperature measure	Superficial tumor for breast, head and neck, limbs; intraluminal heating for esophagus, cervix, rectum, nasopharynx, urethra, prostate; interstitial heating for brain
Laser (IR and visible light)	Absorb light energy	Less harm to the human body; enter into the body cavity with the help of endoscope	Small scope, depth is not enough	Small and superficial tumor ; possibility of carbonization at local tissue
Ultrasound	Vibration and friction	No fat overheating; easy temperature measure; better penetration and focus performance	Limited by air cavity, significant pain in the interface between soft tissue and bone	Superficial and deep tumor, such as soft tissue tumor, pelvic tumor, brain tumor
Electrothermal needle/ high temperature fluids	Heat conduction	Safe heating source; simple equipment and cheap; no dependence on dielectric and conductivity constant	Small action scope for single heating source	Superficial and deep tumor
Thermochemistry	Exothermic chemical reaction	Simple and cheap; high heat energy	Safety is under study	Embryonic stage

8.2 Transport Issue and Hyperthermia Dosage

Ability for tissue to achieve the desired temperature largely depends on the transport mode and the dose of loading nanoparticles.

The typical transport modes for nanoparticles mainly include arterial injection (Kuznetsov et al. 2001), intravenous injection (Yanase et al. 1997), direct injection (Yanase et al. 1998a) subcutaneous injection (Yanase et al. 1998b) and so on. The metabolism of tumor tissue is exuberant, so the vascular system around tumor is very different from that of normal tissues. The gaps of adjacent endothelial cells reach 600 ~ 800 nm and tumor impairs lymphatic system. Therefore, the nanoparticles will exudation from the gap and focus on the tumor cells after injection. In addition, for MNPs, the external magnetic field can be used to control the movement and determine the position of particles so that to send them to target cells. Importing from arterial can achieve both artery embolism and hyperthermia, and intravenous injection is relatively simple but has poor targeting. Direct injection and subcutaneous injection are only applicable to superficial tumors. Different injection methods have direct influence on nanoparticle doses for use. For example, the required doses through direct injections are much smaller than that through arterial and intravenous injections. In nano hyperthermia, the concentration of loading particles should reach certain value to achieve obvious temperature ascent.

On the other hand, too much nanoparticles can not be injected into human body, which will delay in the body can subsequently generate some injury. Generally, the concentration for MNPs is better to be within 5 ~ 10 mg/cm (Pankhurst et al. 2003).

When nanoparticles reach the targeted location, they will integrate with the tumor cells. Some of them may get into the cells. In other words, the cells swallow the nanoparticles. The remaining particles will gather on the surface of the cells. With different structures, some nanoparticles will adhere to the surface separately whereas some nanoparticles will assemble into agglomerate on the surface of cells. Different conjunction means with cells contribute differently in generating heat. For a long time, scientists thought that nanoparticles swallowed by tumor cells, called intracellular hyperthermia, are the most effective way to kill the tumor cells. However, some other researchers question this concept. Moroz et al. (2002) considered that the results of intracellular hyperthermia can not exactly confirm its very large contribution to the hyperthermia, because there are still some nanoparticles existing at healthy cells, thus extracellular heat may also contribute to the results. Rabin (2002) questioned that whether intracellular hyperthermia was superior to extracellular hyperthermia through heat transfer simulation. His results demonstrated that a single cell uptaking a large amount of nanoparticles can not reach the hyperthermia temperature. Only when the cells uptake a large number of nanoparticles together and form a cell cluster with diameter greater than 1 mm can it reach the effective temperature higher than 41°C. If the nanoparticles numbers of intracellular and extracellular are equal, there is no reason to believe intracellular hyperthermia can achieve better effect, unless intracellular nanoparticles can induce new chemical effects or cause mechanical damage. But there is still no empirical evidence.

8.3 Combination with Imaging Monitor

Another important usage of nanoparticle is to enhance medical imaging as contrast agents. There are already many studies in this aspect (Kohler et al. 2006; Sun et al. 2008; Sharma and Chen 2009). Some nanoparticles can be used as contrast agents because of their dependences on temperature. Therefore, in hyperthermia, maybe one can detect the temperature distribution information of interest domain with the help of these thermo sensitive particles which can help reveal the tissue destruction details. By all appearances, an idealistic case lies in that the chosen nanoparticles should possess both excellent heating and image enhancing functions. Thus, when nanoparticles finish raising temperature of the tissue, they can also serve as mediums to diagnose state of disease and therapeutic effect, thereby providing planning guidance for clinicians.

8.4 Combination with Chemotherapy and Radiotherapy

Hyperthermia can be combined with chemotherapy or radiotherapy to achieve better therapeutic effect, since heating makes tumor cells more sensitive to the chemotherapy or radiotherapy (Saniei 2009; Horsman and Overgaard 2007; Issels 2008; Ben-Yosef et al. 2004; Hurwitz et al. 2007). Combining them with nano hyperthermia can help satisfy the required selective destruction of malignant cells without causing damages to healthy cells.

Nanoparticles can serve as carriers to deliver drug to tumor tissue, which provides a new way to combine drug therapy with hyperthermia. As reported in existing researches, nanoparticles integrated with chemotherapeutics and thermal seeds can realize enhanced drug release in tumor cells and selective heat deposition simultaneously. This is a particularly attractive feature, which meets the demands for homogenous heating within targeted tissue and effective drug release at the same time. Now there are many experiments on thermochemotherapy using nanoparticles integrated with some drugs, such as carboplatin-Fe@C-loaded chitosan nanoparticles (Li et al. 2009), nanosized As_2O_3/Fe_3O_4 complex (Du et al. 2009), etc.

Hypoxia is a key intrinsic resistance to radiation therapy. Therefore, any strategy that mitigates tissue hypoxia could potentially overcome radio resistance and promote the effects of radiation therapy. For example, gold nanoshell-mediated hyperthermia can cause an increase in tumor perfusion that reduces the radio resistant hypoxic fraction. Then a subsequent radiation will induce vascular disruption with extensive tumor necrosis (Diagaradjane et al. 2008).

9 Conclusions

Now, nanoparticles enhanced hyperthermia is developing rather rapidly as a promising alternative to conventional hyperthermia. With incomparable merits, they can be a critical tool to enhance the thermal ablation effect. So far, many heating sources are studied to different extent combing nanoparticles with suitable properties, where magnetic particles enabled hyperthermia and PTT with GNPs developed more mature or even stepped into clinical trials, and RF ablation with nanoparticles has come into notice with emerging researches reported on CNT or gold nanomaterial enhanced RF thermal destruction of tumor. Meanwhile, there are also quite a few fundamental or practical challenges raised. But with all the efforts ever or will be made from academics and industries, nanoparticles enhanced hyperthermia is expected to have ever promising future until finally lend itself to the war with human killer – cancer.

Acknowledgment This work is partially supported by the NSFC under grant No.81071255 and Tsinghua-Yue-Yuen Medical Sciences Fund.

References

Liu J., Deng ZS. Physics of Tumor Hyperthermia (In Chinese). Beijing, Science Publisher, 2008.

Alpard SK, Vertrees RA, Tao W, Deyo DJ, Brunston RL, Jr., Zwischenberger JB. Therapeutic hyperthermia. Perfusion (London) 11(6): 425–435, 1996.

Gilchrist RK, Medal R, Shorey WD, Hanselman RC, Parrott JC, Taylor CB. Selective inductive heating of lymph nodes. Annals of Surgery 146(4): 596–606, 1957.

Duguet E, Mornet S, Vasseur S, Grasset F, Veverka P, Goglio G, Demourgues A, Portier J, Pollert E. Magnetic nanoparticle design for medical applications. Progress in Solid State Chemistry 34(2–4): 237–47, 2006.

Wetzel ED, Fink BK. Feasibility of Magnetic Particle Films for Curie Temperature-Controlled Processing of Composite Materials. Report supported by Storming Media. 2001.

Classes of Magnetic Materials. http://www.irm.umn.edu/hg2m/hg2m_b/hg2m_b.html

Rishi W. Fabrication and characterization of granular magnetic gold nanoparticles. The University of Texas at Arlington, 2008.

Kuznetsov OA, Kuznetsov AA, Leontiev VG, Brukvin VA, Vorozhtsov GN, Kogan BY, Shlyakhtin OA, Yunin AM, Tsybin OI. Local radiofrequency-induced hyperthermia using CuNi nanoparticles with therapeutically suitable Curie temperature. Journal of Magnetism and Magnetic Materials 311(1): 197–203, 2007.

Rosensweig RE. Heating magnetic fluid with alternating magnetic field. Journal of Magnetism and Magnetic Materials 252(1–3): 370–374, 2002.

Hergt R, Dutz S, Muller R, Zeisberger M. Magnetic particle hyperthermia: Nanoparticle magnetism and materials development for cancer therapy. Journal of Physics-Condensed Matter 18(38): S2919–S2934, 2006.

Mornet S, Vasseur S, Grasset F, Duguet E. Magnetic nanoparticle design for medical diagnosis and therapy. Journal of Materials Chemistry 14(14): 2161–2175, 2004.

Hergt R, Dutz S, Roder M. Effects of size distribution on hysteresis losses of magnetic nanoparticles for hyperthermia. Journal of Physics: Condensed Matter 20(38): 385214 (12 pp.), 2008.

de Chatel PF, Nandori I, Hakl J, Meszaros S, Vad K. Magnetic particle hyperthermia: Neel relaxation in magnetic nanoparticles under circularly polarized field. Journal of Physics: Condensed Matter 124202 (8 pp.), 2009.

Chan DCF, Kirpotin DB, Bunn PA. Synthesis and evaluation of colloidal magnetic iron-oxides for the site-specific raduifrequency-induced hyperthermia of cancer. Journal of Magnetism and Magnetic Materials 122(1–3): 374–378, 1993.

Johannsen M, Gneveckow U, Taymoorian K, Thiesen B, Waldofner N, Scholz R, Jung K, Jordan A, Wust P, Loening SA. Morbidity and quality of life during thermotherapy using magnetic nanoparticles in locally recurrent prostate cancer: Results of a prospective phase I trial. International Journal of Hyperthermia 23(3): 315–323, 2007.

Varanda LC, Imaizumi M, Santos FJ, Jafelicci M. Iron oxide versus Fe55Pt45/Fe3O4: Improved magnetic properties of core/shell nanoparticles for biomedical applications. IEEE Transactions on Magnetics 44(11): 4448–4451, 2008.

Gu YY, Tan XP, Liang SQ, Sang SB. Effects of La3+ doping on MnZn ferrite nanoscale particles synthesized by hydrothermal method. Journal of Central South University of Technology 11(2): 166–168, 2004.

Uskokovic V, Kosak A, Drofenik M. Preparation of silica-coated lanthanum-strontium manganite particles with designable curie point, for application in hyperthermia treatments. International Journal of Applied Ceramic Technology 3(2): 134–143, 2006.

Parekh K, Upadhyay RV, Belova L, Rao KV. Ternary monodispersed Mn0.5Zn0.5Fe2O4 ferrite nanoparticles: preparation and magnetic characterization. Nanotechnology 17(24): 5970–5975, 2006.

Sithararnan B, Wilson LJ. Gadonanotubes as new high-performance MRI contrast agents. International Journal of Nanomedicine 1(3): 291–295, 2006.

Pereira C, Pereira AM, Quaresma P, Tavares PB, Pereira E, Araujo JP, Freire C. Superparamagnetic gamma-Fe2O3@SiO2 nanoparticles: a novel support for the immobilization of VO(acac)2. Dalton Trans 39(11): 2842–54, 2010.

Ravel B, Carpenter EE, Harris VG. Oxidation of iron in iron/gold core/shell nanoparticles. Journal of Applied Physics 91(10): 8195–8197, 2002.

Tamer U, Gundogdu Y, Boyaci IH, Pekmez K. Synthesis of magnetic core-shell Fe3O4-Au nanoparticle for biomolecule immobilization and detection. Journal of Nanoparticle Research 12(4): 1187–1196, 2010.

Choi JY, Kim K, Shin KS. Surface-enhanced Raman scattering inducible by recyclable Ag-coated magnetic particles. Vibrational Spectroscopy 53(1): 117–120, 2010.

Ding H, Shen CM, Hui C, Xu ZC, Li C, Tian Y, Shi XZ, Gao HJ. Synthesis and properties of Au-Fe3O4 and Ag-Fe3O4 heterodimeric nanoparticles. Chinese Physics B 19(6): 066102 (6 pp.), 2010.

De La Cruz-Montoya E, Rinaldi C. Synthesis and characterization of polymer nanocomposites containing magnetic nanoparticles. Journal of Applied Physics 107(9): 2010.

Zhou L, Gao C, Xu WJ. Robust Fe3O4/SiO2-Pt/Au/Pd magnetic nanocatalysts with multifunctional hyperbranched polyglycerol amplifiers. Langmuir 26(13): 11217–11225, 2010.

Zelenakova A, Zelenak V, Degmova J, Kovac J, Sedlackova K, Kusy M, Sitek J. The iron-gold magnetic nanoparticles: preparation,characterization and magnetic properties. Reviews on Advanced Materials Science 18(6): 501–504, 2008.

Ban ZH, Barnakov YA, Golub VO, O'Connor CJ. The synthesis of core-shell iron@gold nanoparticles and their characterization. Journal of Materials Chemistry 15(43): 4660–4662, 2005.

Jafari T, Simchi A, Khakpash N. Synthesis and cytotoxicity assessment of superparamagnetic iron-gold core-shell nanoparticles coated with polyglycerol. Journal of Colloid and Interface Science 345(1): 64–71, 2010.

Huber DL. Synthesis, properties, and applications of iron nanoparticles. Small 1(5): 482–501, 2005.

Klingeler R, Hampel S, Buchner B. Carbon nanotube based biomedical agents for heating, temperature sensoring and drug delivery. International Journal of Hyperthermia 24(6): 496–505, 2008.

Shi DL, Cho HS, Huth C, Wang F, Dong ZY, Pauletti GM, Lian J, Wang W, Liu GK, Bud'ko SL, Wang LM, Ewing RC. Conjugation of quantum dots and Fe3O4 on carbon nanotubes for medical diagnosis and treatment. Applied Physics Letters 95(22): 2009.

Krupskaya Y, Mahn C, Parameswaran A, Taylor A, Kramer K, Hampel S, Leonhardt A, Ritschel M, Buchner B, Klingeler R. Magnetic study of iron-containing carbon nanotubes: Feasibility for magnetic hyperthermia. Journal of Magnetism and Magnetic Materials 321(24): 4067–4071, 2009.

Tang NJ, Lu LY, Zhong W. Synthesis and magnetic properties of carbon-coated Ni/SiO2 core/shell nanocomposites. Science in China Series G (Physics, Mechanics and Astronomy) 31–4, 2009.

El-Gendy AA, Ibrahim EMM, Khavrus VO, Krupskaya Y, Hampel S, Leonhardt A, Buchner B, Klingeler R. The synthesis of carbon coated Fe, Co and Ni nanoparticles and an examination of their magnetic properties. Carbon 47(12): 2821–2828, 2009.

Liu Y, Ling J, Wei U, Zhang XG. Effective synthesis of carbon-coated Co and Ni nanocrystallites with improved magnetic properties by AC arc discharge under an N-2 atmosphere. Nanotechnology 15(1): 43–47, 2004.

Dumé B. Carbon nanotubes go magnetic. Physicsworld.com, 2004.

Gneveckow U, Jordan A, Scholz R, Bruss V, Waldofner N, Ricke J, Feussner A, Hildebrandt B, Rau B, Wust P. Description and characterization of the novel hyperthermia- and thermoablation-system MFH (R) 300 F for clinical magnetic fluid hyperthermia. Medical Physics 31(6): 1444–1451, 2004.

Johannsen M, Gneueckow U, Thiesen B, Taymoorian K, Cho CH, Waldofner N, Scholz R, Jordan A, Loening SA, Wust P. Thermotherapy of prostate cancer using magnetic nanoparticles: Feasibility, imaging, and three-dimensional temperature distribution. European Urology 52(6): 1653–1662, 2007.

Wust P, Gneveckow U, Johannsen M, Bohmer D, Henkel T, Kahmann F, Sehouli J, Felix R, Ricke J, Jordan A. Magnetic nanoparticles for interstitial thermotherapy - feasibility, tolerance and achieved temperatures. International Journal of Hyperthermia 22(8): 673–685, 2006.

van Landeghem FKH, Maier-Hauff K, Jordan A, Hoffmann KT, Gneveckow U, Scholz R, Thiesen B, Bruck W, von Deimling A. Post-mortem studies in glioblastoma patients treated with thermotherapy using magnetic nanoparticles. Biomaterials 30(1): 52–57, 2009.

Maier-Hauff K, Rothe R, Scholz R, Gneveckow U, Wust P, Thiesen B, Feussner A, von Deimling A, Waldoefner N, Felix R, Jordan A. Intracranial thermotherapy using magnetic nanoparticles combined with external beam radiotherapy: Results of a feasibility study on patients with glioblastoma multiforme. Journal of Neuro-Oncology 81(1): 53–60, 2007.

Zhou JH, Liu J. Numerical study on 3-D light and heat transport in biological tissues embedded with large blood vessels during laser-induced thermotherapy. Numerical Heat Transfer Part a-Applications 45(5): 415–449, 2004.

W.R.Chen RLA, K.E.Bartels et al. Photothermal effects on murine mammary tumors using indocyanine green and an 808-nm diode laser: an in vivo efficacy study. Cancer Lett. 94(125–131, 1995.

Zhou FF, Xing D, Ou ZM, Wu BY, Resasco DE, Chen WR. Cancer photothermal therapy in the near-infrared region by using single-walled carbon nanotubes. Journal of Biomedical Optics 14(2): 2009.

Weitman SD, Weinberg AG, Coney LR, Zurawski VR, Jennings DS, Kamen BA. Cellular-localization of the folate receptor-potential role in drug toxicity and folate homeostasis. Cancer Research 52(23): 6708–6711, 1992.

Bagnoli M, Canevari S, Figini M, Mezzanzanica D, Raspagliesi F, Tomassetti A, Miotti S. A step further in understanding the biology of the folate receptor in ovarian carcinoma. Gynecologic Oncology 88(1): S140–S144, 2003.

Chen JY, Wang DL, Xi JF, Au L, Siekkinen A, Warsen A, Li ZY, Zhang H, Xia YN, Li XD. Immuno gold nanocages with tailored optical properties for targeted photothermal destruction of cancer cells. Nano Letters 7: 1318–1322, 2007.

Au L, Zheng DS, Zhou F, Li ZY, Li XD, Xia YN. A quantitative study on the photothermal effect of immuno gold nanocages targeted to breast cancer cells. Acs Nano 2(8): 1645–1652, 2008.

Eck W, Craig G, Sigdel A, Ritter G, Old LJ, Tang L, Brennan MF, Allen PJ, Mason MD. PEGylated gold nanoparticles conjugated to monoclonal F19 antibodies as targeted labeling agents for human pancreatic carcinoma tissue. ACS Nano 2(11): 2263–2272, 2008.

Copland JA, Eghtedari M, Popov VL, Kotov N, Mamedova N, Motamedi M, Oraevsky AA. Bioconjugated gold nanoparticles as a molecular based contrast agent: Implications for imaging of deep tumors using optoacoustic tomography. Molecular Imaging and Biology 6(5): 341–349, 2004.

Chakravarty P, Marches R, Zimmerman NS, Swafford ADE, Bajaj P, Musselman IH, Pantano P, Draper RK, Vitetta ES. Thermal ablation of tumor cells with anti body-functionalized single-walled carbon nanotubes. Proceedings of the National Academy of Sciences of the United States of America 105(25): 8697–8702, 2008.

Villiers CL, Freitas H, Couderc R, Villiers MB, Marche PN. Analysis of the toxicity of gold nano particles on the immune system: Effect on dendritic cell functions. Journal of Nanoparticle Research 12(1): 55–60, 2010.

Sabadell C, Casado E, Izquierdo J, Olive A. Pneumonitis by gold salts in psoriasic arthritis. Revista Espanola de Reumatologia 26(1): 27–28, 1999.

Tarsy JM. Gold colloid and colloids of other heavy metals in the treatment of rheumatoid arthritis. Journal of Laboratory and Clinical Medicine 26: 1918–1924, 1941.

Kam NWS, O'Connell M, Wisdom JA, Dai HJ. Carbon nanotubes as multifunctional biological transporters and near-infrared agents for selective cancer cell destruction. Proceedings of the National Academy of Sciences of the United States of America 102: 11600–11605, 2005.

Kim JW, Galanzha EI, Shashkov EV, Moon HM, Zharov VP. Golden carbon nanotubes as multimodal photoacoustic and photothermal high-contrast molecular agents. Nature Nanotechnology 4(10): 688–694, 2009.

Karthikeyan S, Mahalingam P, Karthik M. Large scale synthesis of carbon nanotubes. E-Journal of Chemistry 6(1): 1–12, 2009.

Sun X, Bao WR, Lv YK, Deng JQ, Wang XY. Synthesis of high quality single-walled carbon nanotubes by arc discharge method in large scale. Materials Letters 61(18): 3956–3958 2007.

Scott CD, Arepalli S, Nikolaev P, Smalley RE. Growth mechanisms for single-wall carbon nanotubes a laser ablation process (vol 72, pg 573, 2001). Applied Physics a-Materials Science & Processing 74(1): 11–11, 2002.

Tao XY, Zhang XB, Sun FY, Cheng JP, Liu F, Luo ZQ. Large-scale CVD synthesis of nitrogen-doped multi-walled carbon nanotubes with controllable nitrogen content on a CoxMg1-xMoO4 catalyst. Diamond and Related Materials 16(3): 425–430, 2007.

Hata K, Futaba DN, Mizuno K, Namai T, Yumura M, Iijima S. Water-assisted highly efficient synthesis of impurity-free single-waited carbon nanotubes. Science 306(5700): 1362–1364, 2004.

Lipka J, Semmler-Behnke M, Sperling RA, Wenk A, Takenaka S, Schleh C, Kissel T, Parak WJ, Kreyline WG. Biodistribution of PEG-modified gold nanoparticles following intratracheal instillation and intravenous injection. Biomaterials 31(25): 6574–6581, 2010.

Cherukuri P, Glazer ES, Curleya SA. Targeted hyperthermia using metal nanoparticles. Advanced Drug Delivery Reviews 62(3): 339–345, 2010.

Lu XM, Au L, McLellan J, Li ZY, Marquez M, Xia YN. Fabrication of cubic nanocages and nanoframes by dealloying Au/Ag alloy nanoboxes with an aqueous etchant based on Fe(NO3) (3) or NH4OH. Nano Letters 7(6): 1764–1769, 2007.

Chen JY, McLellan JM, Siekkinen A, Xiong YJ, Li ZY, Xia YN. Facile synthesis of gold-silver nanocages with controllable pores on the surface. Journal of the American Chemical Society 128(46): 14776–14777, 2006.

Skrabalak SE, Chen J, Au L, Lu X, Li X, Xia Y. Gold nanocages for biomedical applications. Advanced Materials 19(20): 3177–3184, 2007.

Skrabalak SE, Chen JY, Sun YG, Lu XM, Au L, Cobley CM, Xia YN. Gold Nanocages: Synthesis, Properties, and Applications. Accounts of Chemical Research 41(12): 1587–1595, 2008.

Jorio A., G. Dresselhaus, M. S., Dresselhaus G. Carbon Nanotubes: Advanced Topics in the Synthesis, Structure, Properties and Applications. Springer, 2008.

Manning TJ, Taylor L, Purcell J, Olsen K. Impact on the photothermal emission from single wall nanotubes by some alkali halide salts. Carbon 41(14): 2813–2818, 2003.

Galanov BA, Galanov SB, Gogotsi Y. Stress-strain state of multiwall carbon nanotube under internal pressure. Journal of Nanoparticle Research 4(3): 207–214, 2002.

Burlaka A, Lukin S, Prylutska S, Remeniak O, Prylutskyy Y, Shuba M, Maksimenko S, Ritter U, Scharff P. Hyperthermic effect of multi-walled carbon nanotubes stimulated with near infaraed irradiation for anticancer therapy: in vitro studies. Experimental Oncology 32(1): 48–50, 2010.

Minati L, Speranza G, Bernagozzi I, Torrengo S, Toniutti L, Rossi B, Ferrari M, Chiasera A. Investigation on the electronic and optical properties of short oxidized multiwalled carbon nanotubes. Journal of Physical Chemistry 114(25): 11068–11073, 2010.

Zhang XL, Liu ZB, Zhao X, Zhou WY, Tian JG. Nonlinear optical properties of hydroxyl groups modified multi-walled carbon nanotubes. Chemical Physics Letters 494(1–3): 75–79, 2010.

Lefrant S, Buisson JP, Mevellec JY, Baibarac M, Baltog I. Linear and Non-Linear Optical Properties of Carbon Nanotubes. Molecular Crystals and Liquid Crystals 522: 472–479, 2010.

Savin AV, Hu BB, Kivshar YS. Thermal conductivity of single-walled carbon nanotubes. Physical Review B 80(19): 2009.

M. Acik, G. Lee, C. Mattevi, M. Chhowalla, K. Cho, Y. J. Chabal. Unusual infrared-absorption mechanism in thermally reduced graphene oxide. Nature Materials 9: 840–845, 2010.

Hughes ME, Brandin E, Golovchenko JA. Optical absorption of DNA-carbon nanotube structures. Nano Letters 7(5): 1191–1194, 2007.

Ghosh S, Dutta S, Gomes E, Carroll D, D'Agostino R, Olson J, Guthold M, Gmeiner WH. Increased Heating Efficiency and Selective Thermal Ablation of Malignant Tissue with DNA-Encased Multiwalled Carbon Nanotubes. ACS Nano 3(9): 2667–2673, 2009.

Wang X, Gao X, Liu J. Monte-Carlo simulation on gold nanoshells enhanced laser interstitial thermal therapy on target tumor. Journal of Computational and Theoretical Nanoscience 7(6): 1025–1031, 2010.

Lobo SM, Afzal KS, Ahmed M, Kruskal JB, Lenkinski RE, Goldberg SN. Radiofrequency ablation: Modeling the enhanced temperature response to adjuvant NaCl pretreatment. Radiology 230(1): 175–182, 2004.

Gach HM, Balachandrasekaran A, Nair T. RF hyperthermia using conductive nanoparticles. Proceedings of the SPIE-The International Society for Optical Engineering 754841 (11 pp.), 2010.

Gannon CJ, Cherukuri P, Yakobson BI, Cognet L, Kanzius JS, Kittrell C, Weisman RB, Pasquali M, Schmidt HK, Smalley RE, Curley SA. Carbon nanotube-enhanced thermal destruction of cancer cells in a noninvasive radiofrequency field. Cancer 110(12): 2654–2665, 2007.

Cherukuri P, Glazer ES, Curleya SA. Targeted hyperthermia using metal nanoparticles. Advanced Drug Delivery Reviews 62(3): 2010.

Day ES, Morton JG, West JL. Nanoparticles for Thermal Cancer Therapy. ASME Journal of Biomechanical Engineering 131(7): 2009.

Cardinal J, Klune JR, Chory E, Jeyabalan G, Kanzius JS, Nalesnik M, Geller DA. Noninvasive radiofrequency ablation of cancer targeted by gold nanoparticles. Surgery 144(2): 125–132, 2008.

Gannon CJ, Patra CR, Bhattacharya R, Mukherjee P, Curley SA. Intracellular gold nanoparticles enhance non-invasive radiofrequency thermal destruction of human gastrointestinal cancer cells. Journal of Nanobiotechnology 6: 1–9, 2008.

Curley SA, Cherukuri P, Briggs K, Patra CR, Upton M, Dolson E, Mukherjee P. Noninvasive radiofrequency field-induced hyperthermic cytotoxicity in human cancer cells using cetuximab-targeted gold nanoparticles. Journal of Exp Ther Oncol 7(4): 313–326, 2008.

Glazer ES, Bs KLM, Zhu CH, Curley SA. Pancreatic carcinoma cells are susceptible to non-invasive radio frequency fields after treatment with targeted gold nanoparticles. Surgery 148(2): 319–324, 2010.

Lynn JG, Zwemer RL, Chick AJ, Miller AE. A new method for the generation and use of focused ultrasound in experimental biology. Jour Gen Physiol 26((2)): 179–193, 1942.

Zhu H, Zhou K, Zhang L, Jin CB, Peng S, Yang W, Li KQ, Su HB, Chen WZ, Bai J, Wu F, Wang ZBA. High intensity focused ultrasound (HIFU) therapy for local treatment of hepatocellular carcinoma: Role of partial rib resection. European Journal of Radiology 72(1): 160–166, 2009.

Calliada F, Campani R, Bottinelli O, Bozzini A, Sommaruga MG. Ultrasound contrast agents – Basic principles. European Journal of Radiology 27: S157–S160, 1998.

Kornmann LM, Reesink KD, Reneman RS, Hoeks APG. Critical appraisal of targeted ultrasound contrast agents for molecular imaging in lager arteries. Ultrasound in Medicine and Biology 36(2): 181–191, 2010.

Razansky D, Einziger PD, Adam DR. Enhanced heat deposition using ultrasound contrast agent - Modeling and experimental observations. Ieee Transactions on Ultrasonics Ferroelectrics and Frequency Control 53(1): 137–147, 2006.

Razansky D, Adam DR, Einziger PD. Experimental study of ultrasound contrast agent mediated heat transfer for therapeutic applications. AIP Conference Proceedings 829(1): 318–22, 2006.

Jozefczak A, Skumiel A. Study of heating effect and acoustic properties of dextran stabilized magnetic fluid. Journal of Magnetism and Magnetic Materials 311(1): 193-196, 2007.

Deng ZS, Liu J. Minimally invasive thermotherapy method for tumor treatment based on an exothermic chemical reaction. Minimally Invasive Therapy and Allied Technologies 16: 341–346, 2007.

Rao W, Liu J. Tumor thermal ablation therapy using alkali metals as powerful self-heating seeds. Minimally Invasive Therapy & Allied Technologies 17(1): 43–49, 2008.

Bauer JC, Chen X, Liu QS, Phan TH, Schaak RE. Converting nanocrystalline metals into alloys and intermetallic compounds for applications in catalysis. Journal of Materials Chemistry 18(3): 275–282, 2008.

Rao W, Liu J, Zhou YX, Yang Y, Zhang H. Anti-tumor effect of sodium-induced thermochemical ablation therapy. International Journal of Hyperthermia 24(8): 675–681, 2008.

Cressman ENK, Tseng HJ, Talaie R, Henderson BM. A new heat source for thermochemical ablation based on redox chemistry: Initial studies using permanganate. International Journal of Hyperthermia 26(4): 327–337, 2010.

Cao XD, Chen T, Wu Y, Zhang JF, Weng WZ, Wan HL. Preparation of Zr-promoted nano-NiO catalyst and its catalytic performance for oxidative dehydrogenation of ethane to ethylene. Chinese Journal of Catalysis 26(2): 148–152, 2005.

Bennett. Application of Titanium Dioxide catlaysis. AdyCatal 36: 5–9, 1989.

Yu TR, Zhou YX, Liu J. Development of a new mini-invasive tumour hyperthermia probe using high-temperature water vapour. Journal of Medical Engineering & Technology 28(4): 167–177, 2004.

Wang XW, Xu XF, Choi SUS. Thermal conductivity of nanoparticle-fluid mixture. Journal of Thermophysics and Heat Transfer 13(4): 474–480, 1999.

Shalkevich N, Escher W, Burgi T, Michel B, Si-Ahmed L, Poulikakos D. On the Thermal Conductivity of Gold Nanoparticle Colloids. Langmuir 26(2): 663–670, 2010.

Le Goff C, Ben-Abdallah P, Domingues G, El Moctar AO. Enhanced thermal conductivity in nanofluids under the action of oscillating force fields. Journal of Nanoparticle Research 10(7): 1115–1120, 2008.

Tian WX, Yang RG. Phonon transport and thermal conductivity percolation in random nanoparticle composites. Computer Modeling in Engineering & Sciences 123–41, 2008.

Tsai TH, Kuo LS, Chen PH, Yang CT. Thermal conductivity of nanofluid with magnetic nanoparticles. PIERS 2009 Beijing: Progress in Electromagnetics Research Symposium, Proceedings I and II 1344–1347, 2009

Xuan Y, Li Q. Investigation on convetive heat transfer and flow features of nanofluids. Journal of Heat Transfer 125: 151–155, 2003.

Li DJ, Hu ZS. Tumor Hyperthermia (In Chinese). Zhengzhou, Zhengzhou University Press, 1995.

Lin SY, LI RY. Principles, Methods and Clinical of Modern Tumor Hyperthermia (In Chinese). Xueyuan Press, 1997.

Peng N, Zhao BD. Clinical Tumor Hyperthermia (In Chinese). People's Medical Press, 2002.

Kuznetsov AA, Filippov VI, Alyautdin RN, Torshina NL, Kuznetsov OA. Application of magnetic liposomes for magnetically guided transport of muscle relaxants and anti-cancer photodynamic drugs. Journal of Magnetism and Magnetic Materials 225(1–2): 95–100, 2001.

Yanase M, Shinkai M, Honda H, Wakabayashi T, Yoshida J, Kobayashi T. Intracellular hyperthermia for cancer using magnetite cationic liposomes: Ex vivo study. Japanese Journal of Cancer Research 88(7): 630–632, 1997.

Yanase M, Shinkai M, Honda H, Wakabayashi T, Yoshida J, Kobayashi T. Intracellular hyperthermia for cancer using magnetite cationic liposomes: An in vivo study. Japanese Journal of Cancer Research 89(4): 463–469, 1998.

Yanase M, Shinkai M, Honda H, Wakabayashi T, Yoshida J, Kobayashi T. Antitumor immunity induction by intracellular hyperthermia using magnetite cationic liposomes. Japanese Journal of Cancer Research 89(7): 775–782, 1998.

Pankhurst QA, Connolly J, Jones SK. Application of magnetic nanoparticles in biomedicine. . Journal of Physics D:Applied Physics 36: 167–181, 2003.

Moroz P, Jones SK, Gray BN. Magnetically mediated hyperthermia: current status and future directions. International Journal of Hyperthermia 18(4): 267–284, 2002.

Rabin Y. Is intracellular hyperthermia superior to extracellular hyperthermia in the thermal sense? International Journal of Hyperthermia 18(3): 194–202, 2002.

Kohler N, Sun C, Fichtenholtz A, Gunn J, Fang C, Zhang MQ. Methotrexate-immobilized poly(ethylene glycol) magnetic nanoparticles for MR imaging and drug delivery. Small 2(6): 785–792, 2006.

Sun C, Lee JSH, Zhang MQ. Magnetic nanoparticles in MR imaging and drug delivery. Advanced Drug Delivery Reviews 60(11): 1252–1265, 2008.

Sharma R, Chen CJ. Newer nanoparticles in hyperthermia treatment and thermometry. Journal of Nanoparticle Research 11(3): 671–689, 2009.

Saniei N. Hyperthermia and cancer treatment. Heat Transfer Engineering 30(12): 915–917, 2009.

Horsman MR, Overgaard J. Hyperthermia: A potent enhancer of radiotherapy. Clinical Oncology 19(6): 418–426, 2007.

Issels RD. Hyperthermia adds to chemotherapy. European Journal of Cancer 44(17): 2546–2554, 2008.

Ben-Yosef R, Vigler N, Inbar M, Vexler A. Hyperthermia combined with radiation therapy in the treatment of local recurrent breast cancer. Israel Medical Association Journal 6(7): 392–395, 2004.

Hurwitz ND, Hansen JL, Prokopios-Davos S, Manola J, Hynynen K, Bornstein BA, Topulos GP, Kaplan ID. Hyperthermia combined with radiation in treatment of locally advanced prostate cancer: Long-term results of DFCI 94–153. International Journal of Radiation Oncology Biology Physics 69(3): S318–S318, 2007.

Li FR, Yan WH, Guo YH, Qi H, Zhou HX. Preparation of carboplatin-Fe@C-loaded chitosan nanoparticles and study on hyperthermia combined with pharmacotherapy for liver cancer. International Journal of Hyperthermia 25(5): 383–391, 2009.

Du YQ, Zhang DS, Liu H, Lai RS. Thermochemotherapy effect of nanosized As2O3/Fe3O4 complex on experimental mouse tumors and its influence on the expression of CD44v6, VEGF-C and MMP-9. BMC Biotechnology 9(84): 2009.

Diagaradjane P, Shetty A, Wang JC, Elliott AM, Schwartz J, Shentu S, Park HC, Deorukhkar A, Stafford RJ, Cho SH, Tunnell JW, Hazle JD, Krishnan S. Modulation of in vivo tumor radiation response via gold nanoshell-mediated vascular-focused hyperthermia: Characterizing an integrated antihypoxic and localized vascular disrupting targeting strategy. Nano Letters 8(5): 1492–1500, 2008.

Non-viral Gene Therapy

Jianxiang Zhang, Xiaohui Li, Liping Lou, Xiaodong Li, Yi Jia, Zhe Jin, and Yuxuan Zhu

Abstract Gene therapy has been considered to be a powerful approach for the prevention and/or treatment of a variety of diseases from genetic disorders, infections, to cancer. The success of gene therapy in the clinic is largely limited currently, mainly due to the lack of safe and efficient delivery vectors. Despite the high transfection efficiency, viral vectors encounter the vital toxicity issues and production problems. Increasing endeavors have been therefore directed towards the development of non-viral systems with the advantages of low immunogenicity and toxicity, ease in manufacturing and mass production, low cost, excellent stability, reduced vector size limitations, high flexibility regarding the size of transgenes to be delivered, and diverse chemical designs for constructing vectors with multiple functions. In this chapter, we summarized most of the synthetic non-viral systems currently employed for gene therapy, including lipid and polymer-based vectors, nanomaterials such as magnetic nanoparticles, quantum dots, gold/silica nano-structures, carbon nanotubes, calcium phosphate nanoparticles, and layered double hydroxides/clays, as well as multifunctional nanosystems among them. Particular attention has been paid on synthetic polymers and the related nanomedicines. Selected clinical trials of gene therapy using non-viral vectors as well as the future development in this rapidly growing field were briefly discussed.

J. Zhang (✉), X. Li (✉), Y. Jia, and Y. Zhu
Department of Pharmaceutics, College of Pharmacy, Third Military Medical University, 30 Gaotanyan Zhengjie, Shapingba District, Chongqing 400038, China
e-mail: jxzhang1980@gmail.com; lpsh008@yahoo.com.cn

Z. Jin
Company 13, Brigade of Cadets, Third Military Medical University, 30 Gaotanyan Zhengjie, Shapingba District, Chongqing 400038, China

L. Lou
Polytechnic Institute of NYU, 6 MetroTech Center, Brooklyn, NY 11201, USA

X. Li
Department of Oral and Maxillofacial Surgery, The Affiliated Stomatology Hospital, College of Medicine, Zhejiang University, Hangzhou 310006, Zhejiang, China

A. Prokop (ed.), *Intracellular Delivery: Fundamentals and Applications*, Fundamental Biomedical Technologies 5, DOI 10.1007/978-94-007-1248-5_21,

Keywords Biodegradable polymer • Calcium phosphate • Carbon nanotube • Cationic lipid • Cellular trafficking • Cellular uptake • Chitosan • Clinical trial • Collagen • Cyclodextrin • Dendrimer • Dextran • Endocytosis • Endosomal escape • Gene expression • Gene packaging • Gene therapy • Ligand • Lipoplex • Liposome • Magnetic nanoparticle • Multifunctional nanoparticle • Nanomaterial • Nanomedicine • Nanoparticle • Nanorod • Natural polymer • Nuclear localization • Nonviral vector • PEGylation • Photoresponsive • pH-sensitive • Poly(β-amino ester) • Polycation • Polyethylenimine • Poly(lactide-co-glycolide) • Poly(L-lysine) • Poly(ortho ester) • Polyphosphazene • Polyplex • Protein • Quantum dot • Redox-responsive • Self-assembly • Silica nanoparticle • Stimuli-responsive polymer • Synthetic polymer • Transfection efficiency • Targeting therapy • Targeting unit • Temperature-responsive

Abbreviations

AFM	Atomic force microscopy
APCs	Antigen presenting cells
AEDP	3-[(2-Aminoethyl) dithio] propionic acid
bPEI-Chol	Cholesteryl modified bPEI
bPEI-PA	Palmitic acid modified bPEI
bPEIs	Branched PEIs
BSA	Bovine serum albumin
C2C12	Murine myoblasts
CaP	Calcuim phosphates
CDs	Cyclodextrins
α-CD	α-Cyclodextrin
β-CD	β-Cyclodextrin
βCDPs	β-Cyclodextrin-containing polymers
CDPRs	Cyclodextrins-based polypseudorotaxanes
CFTR	Cystic fibrosis transmembrane conductance regulator
CHO	Chinese hamster ovarian
CL	Cationic liposomes
CLIP-1	Rac-[(2,3-dioctadecyloxypropyl)(2-hydroxyethyl)]- dimethylammonium chloride
CLIP-6	Rac-[2(2,3-dihexadecyloxypropyl-oxymethyloxy)ethyl] trimethylammonium bromide
CLIP-9	Rac-[2(2,3-dihexadecyloxypropyl-oxysuccinyloxy) ethyl]-trimethylammonium
CLSM	Confocal laser scanning microscopy
CNT	Carbon nanotube
CPNP-DNA	DNA-loaded calcium phosphate nanoparticles
CS	Chitosan
DA-CS	Deoxycholic acid modified chitosan
DC-Chol	3β-[N-(dimethylaminoethane)carbamoyl]cholesterol

DEAE-dextran	Diethylaminoethyl-dextran
DEX-SP	Spermine-conjugated dextran
DEX	Dexamethasone
DMAEMA	2-(Dimethylamino)ethyl methacrylate
DMRIE	1,2-Dimyristyloxypropyl-3-dimethyl-hydroxyethyl ammonium bromide
DOGS	Dioctadecylamidoglicylspermin
DOPE	Dioleoylphosphatidylethanolamine
DOPC	Dioleoyl phosphatidyl choline
DORI	1,2-Dioleoyl-3-dimethyl-hydroxyethyl ammonium bromide
DORIE	1,2-Dioleyloxypropyl-3-dimethyl-hydroxyethyalm-3-dimethyl-hydroxyethyl ammonium bromide
DORIE-HB	1,2-Dioleyloxypropyl-3-dimethyl-hydroxybutyl ammonium bromide
DORIE-HP	1,2-Dioleyloxypropyl-3-dimethyl-hydroxypropyl ammonium bromide
DORIE-Hpe	1,2-Dioleyloxypropyl-3-dimethyl-hydroxypentyl ammonium bromide
DOSPA	2,3-Dioleoyloxy-N-[2(sperminecarboxamido)ethyl]-N,N-dimethyl-1-propanammonium trifluoroacetate
DOTAP	N-[1-(2,3-dioleoyloxy)propyl]-N,N,N-trimethylammonium chloride
DOTMA	1,2-Dioleyloxypropyl-3-trimethyl ammonium bromide
DOX	Doxorubicin
DPRIE	1,2-Dipalmityloxypropyl-3-dimethyl-hydroxyethyl ammonium bromide
DSP	Dithiobis(succinimidylpropionate)
DSRIE	1,2-Disteryloxypropyl-3-dimethyl-hydroxyethyl ammonium bromide
DT	Diphtheria toxin
DTBP	Dimethyl-3,3'-dithiobispropionimidate·2HCl
EDAC	1-Ethyl-3-(3-dimethylaminopropyl)carbodiimide
F-PE	Partially fluorinated glycerophosphoethanolamine
FRET	Fluorescence resonance energy transfer (or Förster resonance energy transfer)
GA-CS	Lactobionic acid bearing galactose conjugated chitosan
gal-PEI	Galactosylated PEI
GC	Glucocorticoid
GMP	2-Deoxyguanosine-5-monophosphate
GFP	Green fluorescent protein
HEK293	Human embryonic kidney
His-PLL	Histidylated polylysine
HLA-B7	Histocompatibility complex protein
HMW-PEI	High molecular weight PEI
HP-β-CD	(2-Hydroxypropyl)-β-cyclodextrin

HP-γ-CD	(2-Hydroxypropyl)-γ-cyclodextrin
Huh7 cells	Human hepatoma cell lines
KanaChol	3b–[6'-Kanamycin-carbamoyl] cholesterol
LDH	Layered double hydroxides
LMW-PEI	Low molecular weight PEI
lPEI	Linear PEI
MCD	Methyl-β-cyclodextrin
MDAMB-231	Human breast carcinoma cells
MEND	Multifunctional envelope-type nano device
MnMEIO	Manganese-doped magnetism-engineered iron oxide
MNPs	Magnetic nanoparticles
mRNA	MicroRNA
MRI	Magnetic resonance imaging
MSNs	Mesoporous silica nanoparticles
MTT	3-(4,5-Dimethylthiazol-2-yl)-2,5-diphenyltetrazolium bromide
MVL5	N1-[2-1S-1-[3-aminopropylamino]-4-[di-3-aminopropylamino] butylcarboxa midoethyl]-3 4-dioleyloxybenzamide
MWCNTs	Multiwalled CNTs
NIRF	Near-infrared in vivo optical imaging
NHS	N-Hydroxysuccinimide
NLS	Nuclear localization signals
NOD	Nonobese diabetic
NPCs	Nuclear pore complexes
n-PAE	Network-type poly(amino esters)
NSAIDs	Nonsteroidal anti-inflammatory drugs
ODNs	Oligonucleotides
PA	Palmitic acid
PAE	Poly(β-amino esters)
PAGA	Poly[α-(4-aminobutyl)-L-glycolic acid]
PAMAM	Polyamidoamine
PAMAM-α-CD	α-CD conjugated dendrimer
PAsp(DET)	Polyaspartamide derivative bearing 1,2-diaminoethane side chains
PBLA	Poly(β-benzyl L-aspartate)
PCI	Photochemical internalization
PCs	Triester phosphatidylcholines
PDMAEMA	Poly[2-(dimethylamino)ethyl methacrylate]
pDNA	Plasmid DNA
PEG	Polyethylene glycol
PEG-PAA	PEG-*b*-poly(aspartic acid)
PEG-*b*-PLL	Poly(ethylene glycol)-*b*-poly(L-lysine)
PEG-Ada	1-Adamantanemethylamine (Ada) conjugated PEG
PEI	Polyethylenimine
PEI25	Branched PEI of 25 kDa
PEI-CDs	Cyclodextrin modified PEIs

PEO-PPO-PEO	Poly(ethylene oxide)-*b*-poly(propylene oxide)-*b*-poly(ethyleneoxide)
PHP	Poly(4-hydroxy-L-proline)
PHPMA	Poly[N-(2-hydroxypropyl)methacrylamide]
PKA	Protein kinase
PLA	Polylactide
PLL	Poly(L-lysine)
PLL-PA	Palmitic acid conjugated PLL
PLGA	Poly(lactide-co-glycolide)
PMA	Phorbol myristate acetate
PMDS	Poly(N-methyldietheneamine sebacate)
PNIPAm	Poly(N-isopropylacrylamide)
POD	PEG-diorthoester-lipid conjugate
POEs	Poly(ortho esters)
PPA	Polyphosphoramidate
PPA-EA	Polyphosphoramidate with ethylenediamine
PPA-SP	Polyphosphoramidate bearing a spermidine side chain
PPE-EA	Poly(2-aminoethylpropylene phosphate)
PPG	Polypropylene glycol
PPPs	Polyphosphazenes
PPP-DMAE	Polyphosphazenes with 2-dimethylaminoethanol side groups
PPP-DMAEA	Polyphosphazenes with 2-dimethylaminoethylamine side groups
PSI	Poly(L-succinimide)
QDs	Quantum dots
RES	Reticuloendothelial system
RGD	Arg-Gly-Asp
SANS	Small angle neutron scattering
SERS	Surface enhanced Raman spectroscopy
SPIO	Superparamagnetic iron oxide
SWCNTs	Single-walled CNTs
TM-Bz-CS	Quaternized N-(4-N,N-dimethylaminobenzyl) chitosan
TM-CS	Trimethylated CS oligomers

1 Introduction

Gene therapy has been considered to be a powerful approach for the treatment or prevention of a variety of diseases from genetic disorders, infections, to cancer, by producing bioactive agents or stopping abnormal functions of the cells such as genetic disorder or uncontrollable proliferation. The success of gene therapy in the clinic, however, is still limited due to the lack of safe and efficient delivery vectors. Although over 400 clinical trials were carried out during the last 15 years, the majority of them rely on viruses such as retroviruses, adenoviruses, adeno-associated viruses, lentiviruses, pox viruses, alphaviruses, and herpes viruses, to deliver the desired gene (Smith 1995; Breyer et al. 2001; Young et al. 2006). Despite their high transfection efficiency, the application of viral vectors to human body is often frustrated by potentially serious toxicity and production problems. The death of a patient in a gene therapy trial using adenoviral vectors has accelerated the research on non-viral vectors (Hollon 2000; Glover et al. 2005). Compared with viral vectors, non-viral systems offer a number of advantages such as low immunogenicity and toxicity, ease in manufacturing and mass production, low cost, stability, reduced vector size limitations, high flexibility regarding the size of the transgene delivered, and diverse chemical designs for constructing vectors with multiple functions (Li and Huang 2006; Wong et al. 2007).

A variety of effective nonviral gene delivery approaches including naked DNA injection (Wolff et al. 1990), physical techniques such as electroporation or gene gun (Yang et al. 1990; Mathiesen 1999), and synthetic vectors (Luo and Saltzman 2000) have been developed. Synthetic vectors are among the most extensively studied systems for gene delivery. In general, synthetic vectors are materials that can electrostatically bind DNA or RNA, condense the genetic materials into nanostructured particles, protect the genes and mediate cellular entry. The complexes of plasmid DNA (pDNA) or RNA with cationic lipids and polymers are known as lipoplexes and polyplexes, respectively. The use of cationic lipids for gene delivery was first introduced by Felgner in 1987 (Felgner et al. 1987). Due to their relatively high efficiency, vectors based on cationic lipids such as 'Lipofection' have been routinely used in both *in vitro* and *in vivo* gene delivery studies, as well as in some clinical trials for gene therapy. Synthetic polymers, like diethylaminoethyl-dextran (DEAE-dextran), have been used for *in vitro* gene transfection studies since the 1960s (Vaheri and Pagano 1965). Progress in polymer chemistry makes it possible to engineer diverse polymers with a plethora of different architectures such as linear, branched, and dendritic, for gene delivery applications. Through delicate molecular design, chemically defined polymer carriers can be powerfully armed with multiple functions required for efficient gene delivery while maintaining biocompatibility, facile manufacturing, and robust and stable formulation. As a result, polymer vectors have great potential for human gene therapy (Putnam 2006; Wagner and Kloeckner 2006). Most recently, nonviral vectors based on nanomaterials, including magnetic nanoparticles, quantum dots, gold or silica nanostructures, and other inorganic/organic nano-hybrids, have gained significant attention. These nano-carriers due to their special characters such as well-controlled size and morphology at nanoscale, multimodal imaging, and targetability, have the potential to be developed to new gene vectors, or elucidate the transport events involved in gene therapy from molecular, cellular, tissue to whole body level. Furthermore, integrating the functions owed by these nanostructures into one platform offers the opportunity to create new generations of multifunctional vectors (Sanvicens and Marco 2008).

Nonviral vectors are generally viewed as less efficacious than the viral methods, especially for *in vivo* applications. However, recent developments suggest that the gene delivery by some synthetic vectors has reached the efficiency and expression duration that are clinically meaningful. Indeed, some synthetic vehicles have been developed to clinical trials. In this chapter, we reviewed most of the synthetic non-viral systems currently employed for gene therapy, including lipid and polymer-based vectors, nanomaterials (e.g., magnetic nanoparticles, quantum dots, gold/silica nanostructures, carbon nanotubes, calcium phosphate nanoparticles, and layered double hydroxides/clays) vehicles, as well as multifunctional nano-systems among them. Particular emphasis has been laid on synthetic polymers and the related delivery systems. Selective clinical trials of gene therapy using nonviral vectors were briefly reviewed as well.

2 Gene Delivery Barriers

To achieve successful transgene expression, a series of extracellular and intracellular transport barriers such as DNA protection, internalization, intracellular trafficking and nuclear transport, need to be overcome by delivery vectors. Viral vectors have already showed their great success in addressing each challenge. Synthetic vectors, however, lack one or several of the necessary functions. Understanding the important barriers encountered by delivery vectors is a prerequisite to design and discover more efficient carriers for gene therapy.

2.1 Gene Packaging

Whereas the naked DNA can be introduced into cells via electrotransfer (Andre and Mir 2004), gene gun inoculation (Fynan et al. 1993a), direct injection into target tissue (Wolff et al. 1990), or hydrodynamic injection (Zhang et al. 1999), their inherent disadvantages largely limit the translation of these procedures to the clinical practice. Packaging therapeutic genes can protect them from degradation by nucleolytic enzymes, and condense their bulky structures to facilitate cellular internalization.

Complexation of DNA with most of the synthetic cationic vectors is based on electrostatic forces between negative phosphates along DNA and positive charges displayed on vector materials. This electrostatically mediated self-assembly condenses DNA into small, compact structures. The process of condensation is considered to be entropically driven (Bloomfield 1997). The size and morphology of resulting particles are dependent on the type and structure of cationic materials used, and preparation conditions such as DNA concentration, type and pH of buffer and N/P ratio. In the case of cationic lipids, most of the different techniques of preparing liopsomes have been applied, such as hydration of a lipid film, dehydration-rehydration, ethanolic injection, reverse-phase evaporation, or the detergent dialysis technique (de Ilarduya et al. 2010). The obtained liposomes are then mixed with DNA to form lipoplexes. In this step, the component concentration, temperature, and kinetics of mixing should be carefully selected to control the formation of lipoplexes and their physicochemical characters that may influence the final transfection activity. For polymer vectors, their molecular weight and molecular weight distribution also affect the physicochemical characteristics of polyplexes. The structure and morphology of polyplexes are thought to be kinetically controlled (Lai and van Zanten 2001), which is often related to the mixing order (for example, adding polycation to DNA solution or DNA to polycation solution). Although the presence of cationic moieties is necessary for complexation, the positive charges of lipoplexes or polyplexes are responsible for their cytotoxicity, at least to some degree (Lv et al. 2006). Furthermore, the electrostatic forces should be carefully modulated so that the payload genes can be disassociated at the site of action.

Alternatively, therapeutic genes can be incorporated into nano- or micro-particles based on biodegradable polymers like polyesters, polyanhydrides or poly(ortho esters). DNA-loaded microparticles can be produced by methods frequently adopted to fabricate microparticles for sustained drug delivery (Jilek et al. 2005). Their size can vary from nanometer to micrometer length scale, depending on preparation techniques and process parameters. Potential DNA degradation during fabrication and release, and low DNA bioavailability are the main issues for these sustained delivery vectors. These problems can be partly addressed by adsorbing DNA onto microparticles that have been peripherally decorated with cationic moieties (Singh et al. 2000; Kasturi et al. 2005). In addition, a recent study by Discher's group has demonstrated that siRNA and oligonucleotides (ODNs) can also be incorporated into polymersomes assembled from neutral amphiphiles (Kim et al. 2009c).

2.2 *Cellular Entry*

2.2.1 Non-specific Uptake

Sulfated proteoglycans are among the most negatively charged components of the cell, which are considered to be the targets of nonviral complexes carrying positive charges. For cellular uptake, the cationic moieties of lipoplexes or polyplexes first bind to anionic proteoglycans by electrostatic interactions. Study by Mounkes et al. demonstrated that lipoplexes could not transfer the payload DNA in Raji cells that do not express proteoglycans (Mounkes et al. 1998). Mislick and Baldeschwieler have showed optimal gene transfer and expression in HeLa cells by polylysine/polyplexes when the N/P ratio was above one, while a N/P ratio <1 led to the low level of gene expression (Mislick and Baldeschwieler 1996). Significant reduction in luciferase expression was observed after the treatment of HeLa cells with sodium, a potent inhibitor of proteoglycan sulfation. Additionally, transfection was also dramatically inhibited by heparin and heparan sulfate as well as chondroitin sulfate B. Other unidentified cell surface components may also mediate the binding and subsequent uptake of nonviral vectors.

The cellular uptake of nonviral vectors may be accomplished mainly through endocytic pathways such as clathrin-mediated, caveolae-mediated, or macropinocytic routes. Whereas drugs that can inhibit both clathrin- and caveolae-mediated endocytosis reduced the transfection efficiency of lipoplexes composed of the cationic amphiphile SAINT-2 and dioleoylphosphatidylethanolamine (DOPE), considering the fact that HepG2 cells lack caveolae, these results reveal a clathrin-mediated route in these cells (Zuhorn et al. 2002). Transfection by these lipoplexes could be strongly inhibited by cholesterol depletion with methyl-β-cyclodextrin (MCD). Additionally, the internalized lipoplexes colocalized with transferrin (used as a marker for clathrin-coated endosomes) in early endocytic compartments and that lipoplexes internalization was inhibited in potassium-depleted cells and in cells overexpressing dominant negative Eps15 mutants. All these results may point to the fact that the cellular internalization of lipoplexes mainly proceed via the clathrin-mediated pathway.

The endocytosis of polyplexes, however, is more complicated, which is thought to be dependent on the cell type and the polymers used. Pichon et al. studied the internalization and trafficking of polyplexes derived from histidylated polylysine (His-PLL) and pDNA encoding the luciferase in HepG2 cells (Gonçalves et al. 2004). The employed polyplexes exhibited an average size of 110 nm with a ζ-potential of 18 mV. Analysis on the fluorescence intensity of fluorescence labeled polyplexes (F-His-PLL or F-pDNA) using flow cytometry showed the uptake was dose-, temperature-, and time-dependent, indicating endocytosis mediated uptake. Treatment using inhibitors of clathrin-dependent pathway, such as hypertonic medium, cytosol acidification, and chlorpromazine, significantly inhibited the uptake of polyplexes by 40–50%, indicative of clathrin-dependent and –independent uptake mechanisms. Depletion of cholesterol using MCD, lovastatin, and filipin III, suppressed the polyplexes uptake by 50%. The presence of both chlorpromazine and filipin III could completely abolish the uptake. This suggests that the uptake occurred via both clathrin-dependent and clathrin-independent pathways. It was also found that actin filaments and microtubules were involved in the uptake of polyplexes, by treatment with cytochalasin D and nocodazole, which induce actin microfilament depolymeirzation and disruption of microtubules, respectively. Treatment with 1 μm of phorbol myristate acetate (PMA), a protein kinase C activator that can stimulate macropinocytosis by increasing membrane ruffles, was able to stimulate internalization of polyplexes by 130%. Furthermore, the uptake of polyplexes was inhibited by dimethylamiloride (which can inhibit Na^+/H^+ exchanger and macropinocytic processes) as well as by wortmannin (a inhibitor of phosphatidyl inositol-3-phosphate kinase required for completion of macropinocytosis) by about 40%. Taking together, these results suggest the involvement of macropinocytosis in the initial internalization of polyplexes and also indicate a role of cholesterol in macropinosome formation. Since HepG2 cells do not express caveolins, the major constituents of caveolae, caveolae-mediated uptake of polyplexes in HepG2 can be excluded. Studies on the cellular uptake of PEI/DNA polyplexes by Behr and coworkers (Kopatz et al. 2004), however, have proposed the following internalizing process: (1) electrostatic binding of cationic polyplexes to syndecan HSPG; (2) polyplex-induced syndecan clustering into cholesterol-rich rafts; and (3) actin-mediated engulfment of polyplexes ('phagocytosis').

2.2.2 Specific Uptake

Numerous lipoplexes and polyplexes have been designed to promote cellular entry via receptor-mediated endocytosis, in which a panoply of targeting ligands have been investigated, including endogenous ligands (such as transferrin and RGD peptide), carbohydrates (galactose, mannose, lactose and etc.), antibodies (like antiCD3 and antiEGF), and cell penetrating peptides (such as HIV Tat and polyarginine) (Roth and Sundaram 2004; Wong et al. 2007).

More often, the targeted lipoplexes enter the cell through the pathway that is involved for the targeting ligand. A ligand that induces clathrin-mediated endocytosis should likewise induce the same endocytotic pathway when incorporated into lipoplexes. This point has been demonstrated by Brahimi-Horn et al., based on Arg-Gly-Asp (RGD) decorated lipoplexes (Colin et al. 2000). The integrin targeting lipoplexes were considered to be internalized via clathrin-coated pits and vesicles, as observed by TEM. The uptake was inhibited by K^+ depletion (which inhibits receptor mediated endocytosis by removing clathrin from the plasma membrane) and hypertonic medium (which blocks clathrin-coated pit formation). However, studies by Düzgüneş et al. have showed transferrin targeted lipoplexes (Tf-lipoplexes) deliver transgenes by endocytosis primarily via a non-receptor-mediated mechanism (Simoes et al. 1999). They proposed that the Tf-mediated enhancement in transfection of Tf-lipoplexes is mainly by two different mechanisms: (1) by triggering internalization of the Tf-lipoplexes through both non-coated pit (presumably phagocytosis or pinocytosis) and coated pit mediated endocytosis (after binding of the lipoplexes to non-specific receptors); and (2) by promoting intracytoplasmic gene delivery in a pH-dependent manner. Taking together, these results may imply the exact mechanism involved in the cellular uptake of targeted lipoplexes is related to both ligand type and the physicochemical characters of lipoplexes.

Polyplexes derivatized with targeting ligands are often internalized by receptor-mediated endocytosis. The successful targeting of polyplexes depends on the conjugation chemistry, the spacer group between ligand and polyplex, the strength of ligand-receptor binding, and the number of targeting ligands per polyplex. The ligand-receptor interaction should not be disrupted by conjugation. Schaffer and Lauffenburgen have found that for EGF–biotin/avidin–polylysine conjugates, increasing cross-linker (between the EGF and biotin) length can improve binding affinity, while the short crosslinkers interfered with EGF/EGF receptor binding (Schaffer and Lauffenburger 1998). There also exists a balance between the specific ligand-receptor interactions and nonspecific electrostatic binding to the cell surface. The specific targeting was found only within a narrow window of polymer/DNA ratios near electroneutrality. In addition, there is typically an optimal ligand valency, due to the saturation of both receptor binding and the internalization machinery (Schaffer and Lauffenburger 1998). All the various parameters should be carefully optimized to achieve efficient cell-specific targeting.

2.3 Endosomal Escape

The cellular endosome is a major barrier to efficient gene transfer. Endosome escape mechanism is a necessary feature of nonviral complexes, especially targeted vectors. Some peptides such as diphtheria toxin (DT) translocation domain undergoes, melittin, and GALA peptide (mainly comprised of a Glu-Ala-Leu-Ala repeat) can induce bending of the membrane and contiguity of the bilayer leaflets, thus opening a pore in the membrane and facilitating release of endosomal contents (Medina-Kauwe et al. 2005). Complexes conjugated with these functional peptides can therefore escape endosome via the pore formation mechanism. For instance, incorporation of melittin into PEI polyplexes led to 700-fold increase in gene transfer of luciferase reporter gene (Ogris et al. 2001). The GALA mediated enhancement of transfection was demonstrated by dendrimer-based vectors (Haensler and Szoka 1993). Inspired by the endosomolytic capability of these peptides, an alternative class of polymers with specific alkyl groups has been developed. Among them, poly(propylacrylic acid) was found to be the most promising one for gene delivery (Murthy et al. 1999; Cheung et al. 2001).

For lipoplexes, it has been proposed a fusion and/or flip-flop mechanism for endosomolysis (Xu and Szoka 1996). After their endocytosis, the lipoplexes destabilize the membrane, inducing flipping of anionic lipids from the cytoplasm to face the endosomal lumen. Then, the anionic lipids form charge neutral ion pairs with cationic lipids of the complex, thus displacing the DNA and releasing it into the cytoplasm. This mechanism was experimentally demonstrated by fluorescence resonance energy transfer (FRET) and fluorescence microscopy (Zelphati and Szoka 1996).

A ‘proton-sponge’ mechanism has been attributed to the endosomal escape of polyplexes based on cationic polymers such as polyethylenimine (PEI) and polyamidoamine (PAMAM) (Haensler and Szoka 1993; Boussif et al. 1995). Most of the synthetic polymers used for gene delivery are polycations with amino groups. These amino groups can be protonated during endosome acidification, thus offering a high buffering capacity, commonly referred to as the ‘proton-sponge’ effect. Massive proton accumulation results in high chloride ion influx into the endosome, causing osmotic swelling and eventual endosomolysis and release of its contents into the cytosol.

2.4 Cytosolic Transport and Nuclear Import

The cytoskeleton is thought to have dramatic effect on the intracellular trafficking of endocytosed gene delivery complexes. Studies on lipoplexes based transfection showed that microtubule disruption can enhance gene transfer level. By incorporating microtubule depolymerizing drugs (such as vinblastine) into lipoplexes, Wang and MacDonald found approximately 30-fold increase in the transfection efficiency in vascular smooth muscle cells (Wang and MacDonald 2004). Other microtubule disrupting agents, including colchicine, vincristine, nocodazole and podophyllotoxin, could also enhance gene transfer. This may be attributed to the inhibition of lipoplexes transport to lysosomes, since intact microtubules are necessary for endosomes to translocate from the plasma membrane to lysosomes (Matteoni and Kreis 1987; Rosette and Karin 1995).

On the other hand, the diffusion of naked DNA larger than 250 bp can be severely impeded by the mesh-like structure of the cytoskeleton (Dauty and Verkman 2005). Accordingly, it may be extremely difficult for lipoplexes or polyplexes to transport to the nucleus by free diffusion. To minimize retention time within the cytosol and promote transport towards and into the nucleus, various ligands that can promote cytonuclear translocation were integrated into non-viral vectors, to facilitate the shuttling of these delivery systems. The frequently employed ligands include nuclear localization signals (NLS), NLS-containing proteins, and carbohydrates (Wong et al. 2007). The cationic NLSs can directly condense DNA to form compact complexes to achieve effective gene transfer in a variety of human and nonhuman cell lines (Kichler et al. 2000). Alternatively, NLSs may be introduced into polyplexes (or lipoplexes) by conjugating to a polymer (or lipid in the case of lipoplexes) vector (Moffatt et al. 2006). The effectiveness of NLSs, however, is affected by a variety of factors, such as the NLS sequence, the size and type of DNA, the method employed for NLS incorporation, the number of NLSs per polyplex (or lipoplex), and the type of vectors used (Bremner et al. 2004). In addition, nuclear proteins that contain NLS sequence are able to shuttle the attached DNA into nucleus (Mesika et al. 2005). Lectins, the carbohydrate receptors, provide new targets for directing the cytonulcear transport of nonviral vectors (Monsigny et al. 2004). Whereas they possess the advantages of ease of synthesis, biocompatibility, and lower immunogenicity, the use of carbohydrate ligands to target the nucleus still needs more systemic studies.

More recently, it has been found that glucocorticoids such as dexamethasone may facilitate the shuttling of synthetic vectors into the nucleus. Glucocorticoid receptor, a nuclear receptor, is mainly located in cytoplasm in the absence of its ligand. In the presence of ligands, however, the glucocorticoid receptor may bind to the ligand and the receptor-ligand complex is translocated to the nucleus (Adcock and Caramori 2001). Upon exposure to glucocorticoids, nuclear pore complexes can be induced to transiently but markedly dilate (Shahin 2006). Therefore, in the presence of glucocorticoids, the transport of DNA-containing complexes into nucleus may be facilitated. Dexamethasone facilitated gene transfer has been demonstrated based on DNA conjugates and polyplexes (Rebuffat et al. 2001, 2002; Choi and Lee 2005).

Besides the use of various ligands, cell mitosis is another issue can be considered for nuclear entry of synthetic vectors. During cell division, there may be a temporary breakdown of the nuclear envelope, which can considerably enhance the gene transfection of lipoplexes and polyplexes as well as recombinant adenovirus (Brunner et al. 2000). Nevertheless, this strategy is clinically limited due to the postmitotic character of most cells in the body.

2.5 Gene Expression

It is still unclear how and when DNA delivered by nonviral vectors is released. It is commonly thought that genomic DNA displaces the DNA condensing agents in the nucleus, thus releasing the therapeutic gene and enabling gene expression (Remy et al. 1995). Upon arrival into the nucleus, exogenous genes must be expressed to perform their therapeutic effects. Episomal transgene retention may result in short term expression, while transgenic insertion into the host genome is always necessary for a stable, long term expression. Several types of DNA elements have been employed to direct genome insertion, including transposons, site-specific elements, and transcription-regulating promoters among them (Yant et al. 2000; Held et al. 2005; Auricchio et al. 2002). Transposon-mediated genome integration has the drawback of random insertion into the host genome, which may elicit unwanted side effects. This can be overcome via site specific integration using site-specific elements such as ΦC31 integrase or hybrid systems containing DNA binding domain. However, the limitations of these strategies have spurred scientists to explore more efficient approaches, such as transcription-regulating promoters, to achieve desirable transgene expression.

3 Gene Vectors Based on Synthetic Polymers

3.1 Polycations with Peptide Backbone

3.1.1 Polylysine

Poly(L-lysine) (PLL, as shown in Table 1) is available in a large variety of molecular weights, which are generally synthesized by the polymerization of N-carboxyanhydride of ε-(benzyoxycarbonyl)-L-lysine combined with a subsequent de-protection(Deming 2006). PLL is one of the first cationic polymers used for gene delivery investigation (Zauner et al. 1998). The epsilon amino groups of PLL are positively charged at physiological pH, and therefore they can ionically interact with the negatively charged phosphate groups of DNA or RNA, condensing the related gene species into polyplexes. Depending on preparation methods and formulations, the structure of polyplexes could be toroid, rod-like, or spherical particle (Wagner et al. 1991; Perales et al. 1994). PLL itself, however, is a poor delivery vector for DNA. Even in the presence of chloroquine, only moderate transfection was achieved (Pack et al. 2005). Various targeting ligands, on the other hand, can greatly enhance *in vitro* and *in vivo* delivery efficiency of PLL-DNA polyplexes. For examples, asialoorosomucoid glycoprotein(Wu and Wu 1987), transferring (Wagner et al. 1990), sugars like lactose and mannose (Erbacher et al. 1995; Ferkol et al. 1996), folate (Mislick et al. 1995), RGD-containing peptides (Hart et al. 1995) and antibodies (Suh et al. 2001) have been covalently conjugated

Table 1 Polycations with peptide backbone

Polymer	Chemical structure
Poly(L-lysine) (PLL)	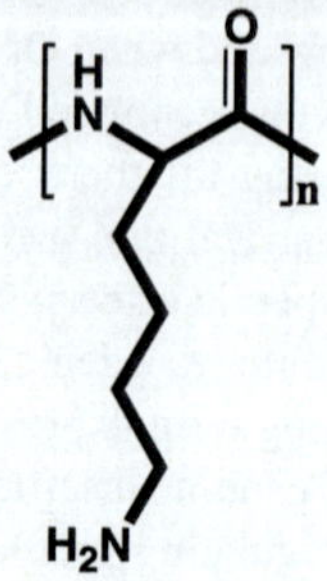
Polyaspartamide derivative bearing 1,2-diaminoethane side chains [PAsp(DET)]	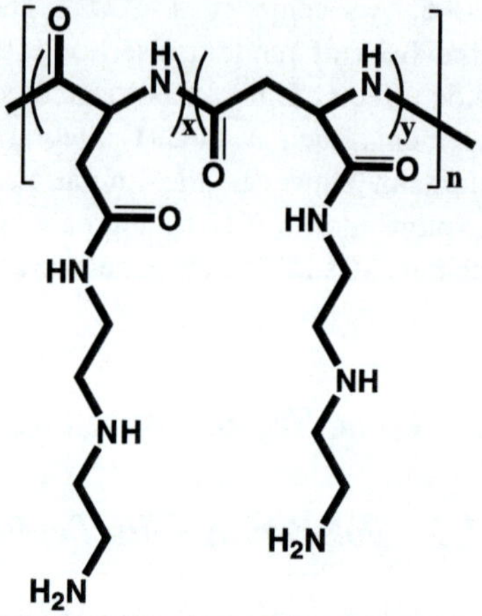

onto PLL for targeting specific cells. Although some encouraging results were obtained in early studies, the rather low transfection efficiency has limited the clinical applications of PLL-based polyplexes, which is ascribed to the poor escape from the endocytic pathway.

The safety issue is another factor that affects *in vivo* applications of PLL in gene therapy. PLL has been found to be not inert, and it may influence many cellular processes such as the activation of phospholipases and microtubule-associated protein two kinase (Shier et al. 1984; Kyriakis and Avruch 1990), membrane stability and permeability (Kornguth and Stahmann 1961; Elferink 1985) as well cell division (Kundahl et al. 1981). In addition, some studies have shown the immunogenicity of PLL, in a molecular weight dependent manner (Plank et al. 1996).

3.1.2 PLL Derivatives

To develop PLL-based gene carriers with enhanced transfection efficacy as well as lower toxicity, variety of modifications have been performed (Liu et al. 2010) (Table 2). For example, palmitic acid conjugated PLL (PLL-PA) was employed for

Table 2 Various chemical modifications on PLL

Molecules used for modifications	Reference
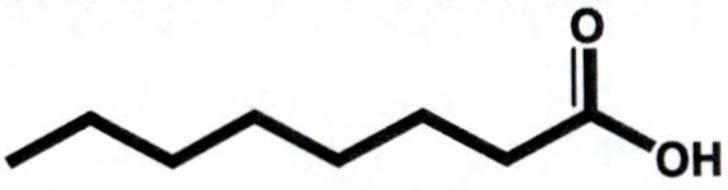 Caprylic acid	Incani et al. (2009)
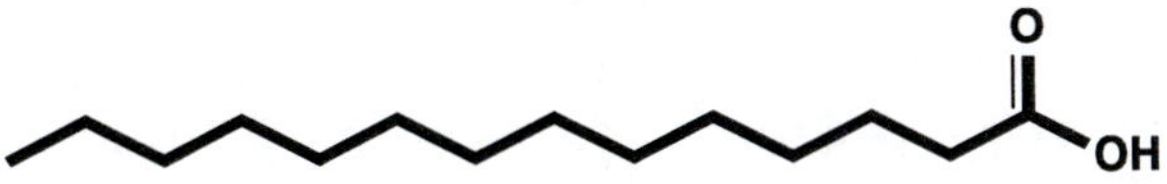 Myristic acid	Incani et al. (2009)
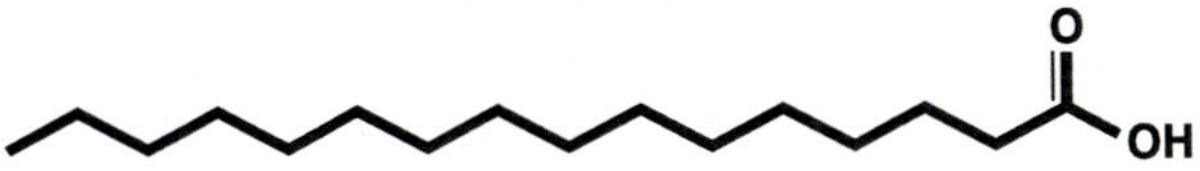 Palmitic acid	Incani et al. (2007)
 Stearic acid	Incani et al. (2009)
 Oleic acid	Incani et al. (2009)
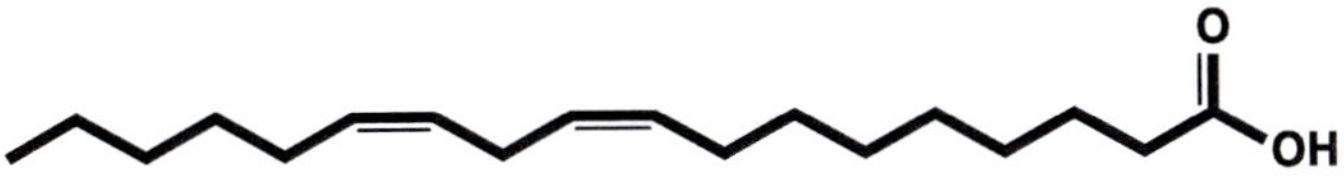 Linoleic acid	Incani et al. (2009)
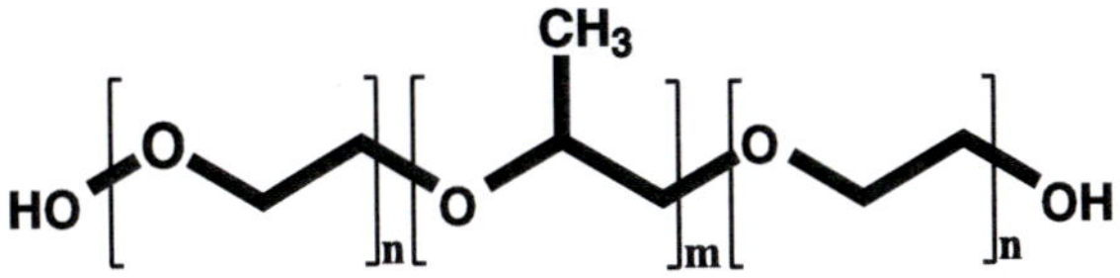 Pluronic	Jeon et al. (2003)
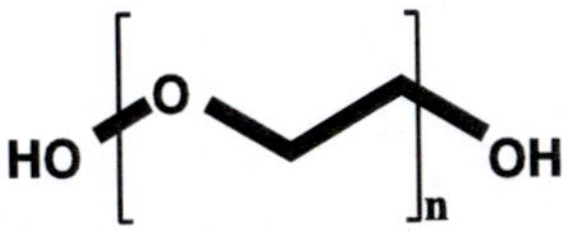 PEG	Katayose and Kataoka (1997)

delivering plasmid DNA (pDNA) into bone marrow stromal cells (Incani et al. 2007). Compared with the precursor PLL, the transfection efficiency of PLL-PA was greatly enhanced (~5-fold). PLL-PA was also found to be able to effectively deliver pDNA into human skin fibroblasts *in vitro*, which was more efficient than that of PEI, Lipofectamine 2000, and even an adenovirus employed by the same researchers (Abbasi et al. 2007). Lipid-substituted PLLs were also synthesized using NHS esters of fatty acids, in order to establish a structure-function relationship (Abbasi et al. 2008). Both lipid type and substitution degree was thought to influence the transfection efficiency. PLL substituted with lipids that are longer than caprylic acid were effective DNA carriers for skin fibroblasts. Higher substitution of lipids also offered more efficient PLL-based vectors. Myristic, palmitic, and stearic acid conjugated PLLs were suggested to be the most effective vehicles (Incani et al. 2009).

In addition to modifications using small molecules, polymers have been employed for tailoring the characteristics of PLL vectors. Grafting PLL using Pluronic was shown to give rise to polymers with enhanced efficiency and lower cytotoxicity (Jeon et al. 2003). On the other hand, conjugating PLL with polyethylene glycol (PEG) led to diblock copolymers (PEG-*b*-PLLs) that can form water-soluble polyion complex (PIC) assemblies with DNA. The introduction of PEG endowed the resulting PIC micelles with great colloidal stability, and can protect the payload DNA from DNase I attacking (Katayose and Kataoka 1997). Further studies revealed that pDNA-loaded PIC micelles exhibited practical gene transfection efficacy, particularly under serum-containing medium. In the presence of 100 μM chloroquine, PIC micellar formulations showed a gene expression comparable to the lipofection toward cultured 293 cells (Itaka et al. 2003).

3.1.3 Polyaspartamides and Their Derivatives

Through a facile and quantitative aminolysis of poly(β-benzyl L-aspartate) (PBLA) in the presence of oligoamines, Kataoka's group have developed a new type of polycations, i.e. polyaspartamides (Nakanishi et al. 2007). Among these polymers, polyaspartamide derivative bearing 1,2-diaminoethane side chains [PAsp(DET)] (Table 1) was demonstrated to be the most efficient one, with transfection activity comparable or even higher than that of PEI25 (Masago et al. 2007). Importantly, this polymer exhibited negligible cytotoxicity. Further mechanism exploration indicated that the pH-selective membrane destabilization profile of PAsp(DET) mainly contribute to the amplified transfection and minimized toxicity (Miyata et al. 2008b). PEGylation of PAsp(DET) could endow the resultant polyplexes with improved stability for systemic administration (Kanayama et al. 2006). Moreover, the efficiency of PEGylated PAsp(DET) could be further optimized by introducing lysine unit (Miyata et al. 2007, 2008a) or incorporating disulfide linkage (Takae et al. 2008). Through systemic administration, the lysine-optimized polycation could achieve clear *in vivo* transfection of the EGFP gene in fibrotic pancreatic adenocarcinoma of BALB/c nude mice (Miyata et al. 2008a). The polyplex micelles PEGylated PAsp(DET) with disulfide linkage showed both a 1~3 orders of magnitude higher gene transfection efficiency and a more rapid onset of gene expression than those based on PEGylated PAsp(DET) without disulfide linkages, due to much more effective endosomal escape based on the PEG detachment in endosome (Takae et al. 2008).

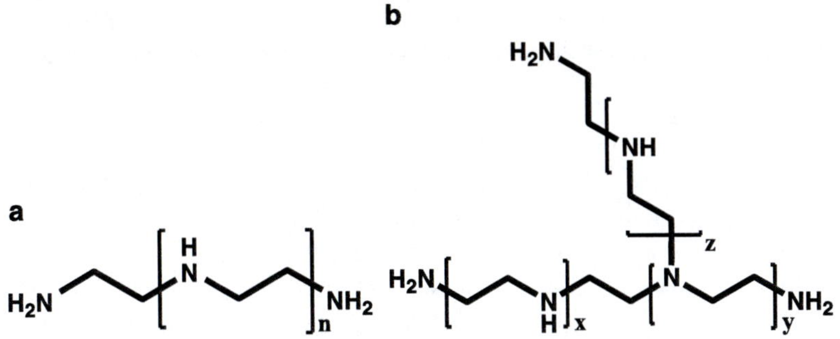

Fig. 1 Linear (**a**) and branched (**b**) polyethylenimine (PEI)

3.2 *Polyethylenimine (PEI) and Its Modifications*

3.2.1 PEI

PEI is a synthetic polymer that has been widely used in processes such as water purification, paper production, and shampoo manufacturing. Two main forms of this polymer are commercially available, i.e. linear and branched, and both of them can be conveniently synthesized by the polymerization of aziridine monomers (Fig. 1) (Godbey et al. 1999). There are PEIs of various molecular weights, from <1,000 Da to 1,600 kDa, provided by various manufacturers. Since the pioneer work by Behr and coworkers (Boussif et al. 1995), there have be increasing interests to develop non-viral vectors based on PEI and its derivatives (Godbey et al. 1999; Kircheis et al. 2001b; Neu et al. 2005; Demeneix and Behr 2005). Among the various synthetic vectors, PEIs have shown particularly promising efficacy in transfections in both *in vitro* and *in vivo* experiments. Broadly available studies have substantiated that the branched PEIs (bPEIs) possess significantly higher transfection efficiency for variety of cell lines. As a consequence, they have frequently served as the gold standard to measure the efficacy of newly developed synthetic vectors in a variety of studies. For example, bPEI of 25 kDa (PEI25) is the most frequently utilized positive control for gene transfection experiments of synthetic materials. The high cationic charge density is the most prominent feature of bPEI. There is one nitrogen atom in every three atoms of bPEI, and the amine groups covers the primary, secondary, and tertiary ones, each with the potential to be protonated. These structural characters make bPEI effective for buffering over a very wide pH range(Boussif et al. 1995).

The structure of PEI dominates the physicochemical and biological properties of the polyplexes with DNA or RNA to a large extent, and therefore it is closely related to the efficiency for gene transfection. Studies by Kissel' group have revealed that higher branched PEIs exhibited a stronger condensation capacity for DNA, formed smaller polyplexes, and induced the expression of pDNAs to a higher extent than less branched polymers(Fischer et al. 2002). However, increasing the branching degree is found to increase the *in vitro* cytotoxicity as well as the *in vitro* haemolysis effects on erythrocytes. In addition to the molecular architecture, the molecular weight of PEI severely influences the polyplex characteristics. Generally, increasing the molecular weight (MW) of PEI leads to polyplexes with smaller size, while a decrease in molecular weight may dramatically impair its capability for condensation. Compared with the high molecular weight PEI (HMW-PEI), a thorough DNA condensation of low molecular weight PEI (LMW-PEI) occurred at higher N/P ratios (the molar ratio of PEI nitrogen-to-DNA phosphate) (Kunath et al. 2003b). Whereas a considerable results have shown the higher transfection efficiency of PEI as its MW increased, HMW-PEI encounter a pronounced toxicity both in cell culture and *in vivo* and a variety of undesired unspecific interactions with the biological components. Great efforts have been therefore directed towards creating PEI derivatives integrated with higher transfection efficacy and excellent biocompatibility.

3.2.2 Cross-Linking of Low Molecular Weight PEI

Low molecular weight PEI (LMW-PEI) is generally regarded as poor gene vectors, despite its low toxicity. Cross-linking of LMW-PEI has been considered to be a simple but effective approach to discover pseudo-HMW-PEIs with low toxicity and higher transgene efficacy. Theoretically, any molecule with multiple functional groups (≥2) that are able to react with amine groups can function as a cross-linker to synthesize high molecular weight polymers. A larger number of cross-linkers, including dithiobis(succinimidylpropionate) (DSP) and dimethyl-3,3'-dithiobispropionimidate·2HCl (DTBP), methylthiirane, diacrylate, glutadialdehyde, cyclodextrins (CDs) and their derivatives among them, have been utilized to produce HMW-PEIs (Table 3).

Gosselin et al. cross-linked 800 Da PEI using DSP and DTBP, and the disulfide bonds in the resulting high molecular weight products could be reduced in intracellular reducing microenvironments to release their low molecular weight counterparts that can be cleared more easily from the body (Gosselin et al. 2001). *In vitro* evaluation using the luciferase reporter gene showed significantly higher transfection efficiency to Chinese hamster ovary (CHO) cells, compared with that of 800 Da PEI. The efficacy was dependent on the cross-linking reagent, the extent of conjugation, and the N/P ratio. At higher N/P ratios, the transgene efficiency of cross-linked PEIs was even comparable to that of PEI25. This may be the first report regarding the *in vitro* transfection study on cross-linked LMW-PEI. DSP has also been used by Kissel's group to cross-link LMW-PEI (Neu et al. 2006). It was found that polyplexes prepared by the cross-linked PEI displayed significantly reduce interactions with major blood components like albumin and erythrocytes. No transfection results, however, were reported by these researchers. Through the ring-opening reaction of LMW-PEI (800 Da) with methylthiirane, disulfide cross-linked PEI (7,100~8,400 Da based on viscosity measurements) can be synthesized (Peng et al. 2008). In comparison to HMW-PEI

Table 3 Cross-linking agents for low molecular weight PEI

Agents used for the cross-linking of LMW-PEI	Reference
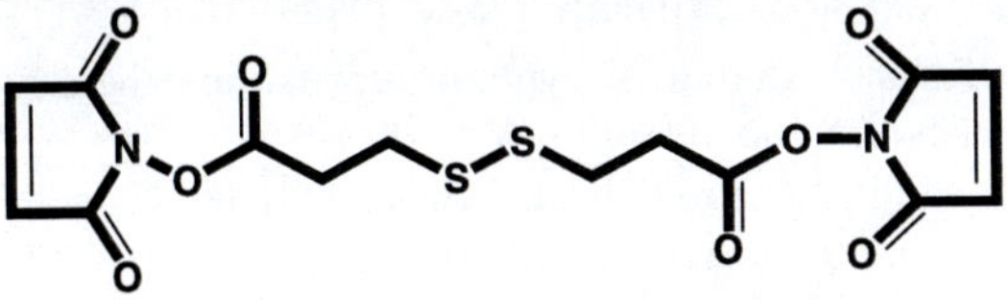 Dithiobis(succinimidylpropionate) (DSP)	Gosselin et al. (2001)
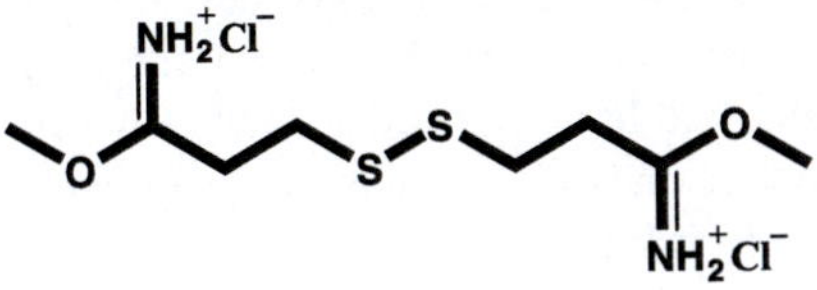 Dimethyl-3,3'-dithiobispropionimidate·2HCl (DTBP)	Gosselin et al. (2001)
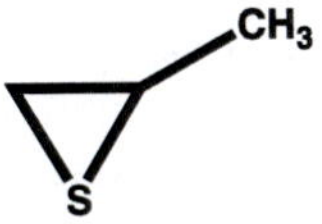 Methylthiirane	Peng et al. (2008), Peng et al. (2009)
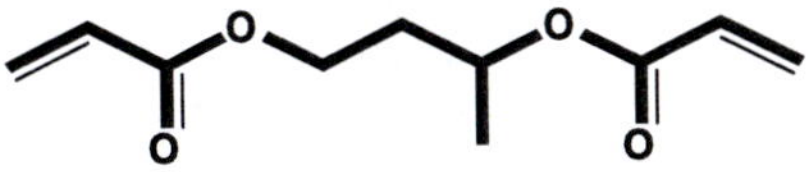 1,3-Butanediol diacrylate	Forrest et al. (2003)
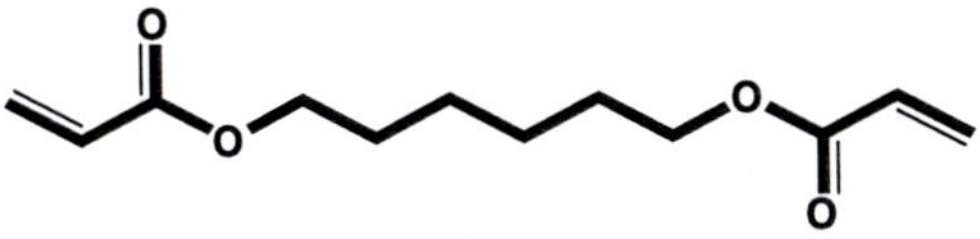 1,6-Hexanediol diacrylate	Forrest et al. (2003)
 Glutadialdehyde	Kim et al. (2005b)
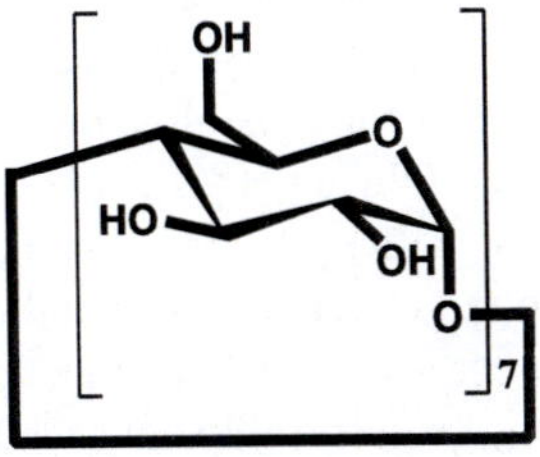 β-Cyclodextrin	Tang et al. (2006)
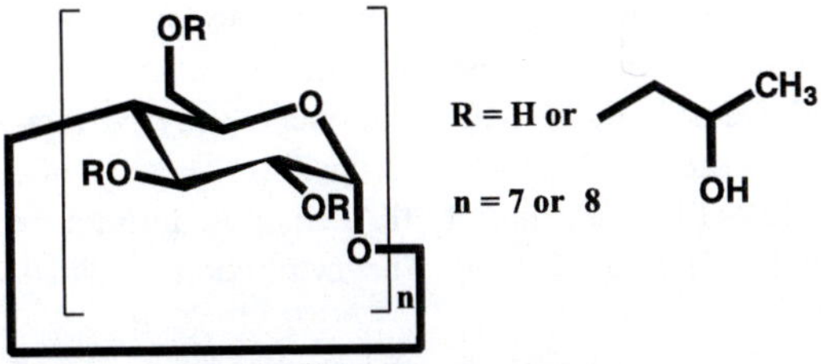 R = H or; n = 7 or 8 (2-Hydroxypropyl)-β-cyclodextrin	Huang et al. (2006)

(25 kDa), these disulfide-containing PEIs displayed a lower cytotoxicity and higher gene transfection efficiency. Fore these types of polycations, their cytotoxicity and transfection efficacy are dependent on both disulfide density and MW of raw PEIs (Peng et al. 2009). Among the polymers derived from PEIs of 800 Da, 1,800 Da, and 25 kDa, the cross-linked polymer based on 800 Da-PEI gave the best results.

The biodegradable derivatives of 14–30 kDa were synthesized by cross-linking 800 Da bPEI using 1,3-butanediol diacrylate or 1,6-hexanediol diacrylate (Forrest et al. 2003). These polymers exhibited low cytotoxicity, and they can mediate the luciferase gene expression more than one order of magnitude more efficiently than commercially available PEI25, according to the cell culture experiments using human breast carcinoma cells (MDAMB-231), and murine myoblasts (C2C12). Kim and colleagues synthesized HMW-PEI (13–23.7 kDa) using glutadialdehyde as a cross-linker (Kim et al. 2005b). Due to the presence of Schiff base units, these synthesized PEIs are acid-labile, which can be rapidly degraded into nontoxic LMW-PEI in acidic endosome. Compared with their precursor PEI (1,800 Da), the acid-labile PEIs exhibited dramatically higher transfection efficiency to 293 Tcells and A7R5 cells. In addition, their cytotoxicity was greatly lower than that of PEI25, despite of the slightly lower efficacy in comparison to the latter.

CDs are a family of naturally occurring oligosaccharides constructed from α(1→4)-linked glucose units, which have been widely used for drug formulations (Davis and Brewster 2004). β-CD has been used recently to prepare polycation via the cross-linking of LMW-PEI (600 Da). This new polymer with improved biocompatibility could mediate effective gene transfection in cultured neurons and in the central nervous system (Tang et al. 2006). Biocompatible derivatives of cyclodextrins, i.e. (2-hydroxypropyl)-β-cyclodextrin (HP-β-CD) and (2-hydroxypropyl)-γ-cyclodextrin (HP-γ-CD) were also employed to prepare cross-linked PEI, and a 600 Da PEI served as the raw material (Huang et al. 2006). Cell culture experiments using SKOV-3 cells indicated that the new polymers showed significantly lower cytotoxicity and 1.5–1.7 fold higher transfection efficiency than that of PEI25. Most recently, HP-γ-CD cross-linked 600 Da PEI coupled with MC-10 oligopeptide containing a sequence of Met-Ala-Arg-Ala-Lys-Glu that can target to HER2 (human epidermal growth factor receptor 2, which is often over expressed in many breast and ovary cancers) was synthesized (Huang et al. 2010). This vector showed very low cytotoxicity, strong targeting specificity to HER2 receptor, and high efficiency for delivering reporter genes to target cells *in vitro* and *in vivo*. Moreover, the new polycation could more efficiently deliver a therapeutic IFN-α gene to tumor-bearing nude mice, implementing a significantly enhanced anti-tumor effect, in comparison to PEI25.

On the other hand, LMW-PEI has also been grafted onto polymers such as poly(aspartic acid), poly(glutamic acid), dextran and chitosan, to achieve effective gene delivery (Table 4). These anionic or neutral polymers may be regarded as pseudo-macromolecular cross-linkers. By the ring opening reaction of poly(L-succinimide) (PSI) with LMW-PEI (600 and 1,200 Da), pseudo-branched PEIs (PSI-PEIs) were synthesized (Yu et al. 2009a). The cytotoxicity against 293 T,

Table 4 Polymers used for conjugating LMW-PEI

Polymers for LMW-PEI conjugation	Reference
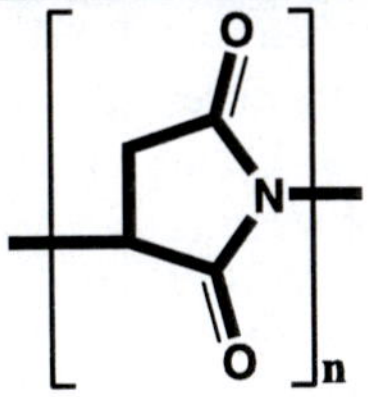 Poly(L-succinimide) (PSI)	Yu et al. (2009a, b)
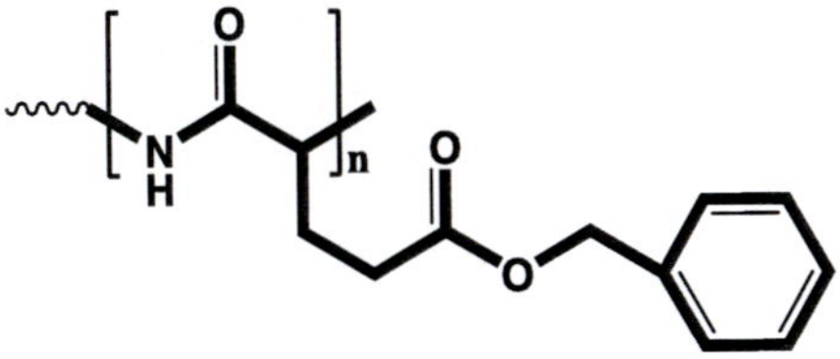 Poly(γ-benzyl L-glutamate) derivative	Wen et al. (2009)
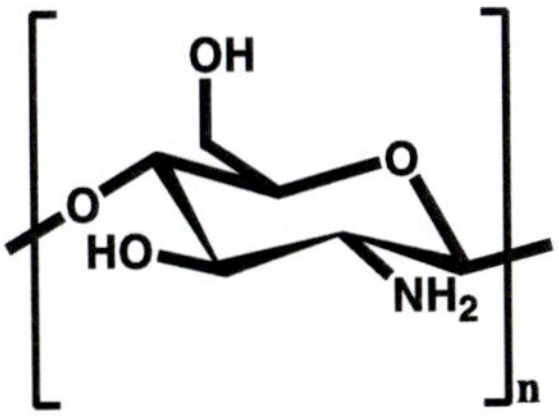 Chitosan	Wong et al. (2006), Jiang et al. (2007), Lu et al. (2008)
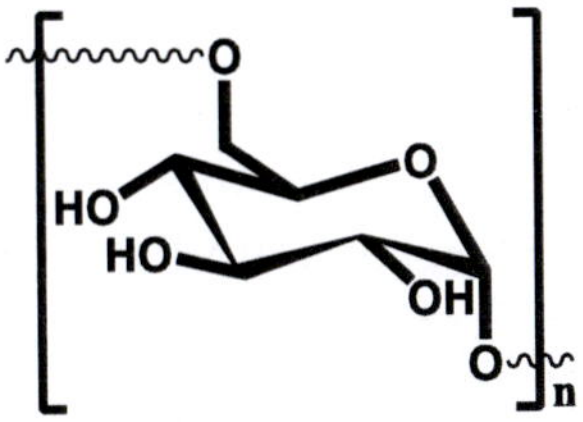 Dextran	Sun et al. (2008a, b)
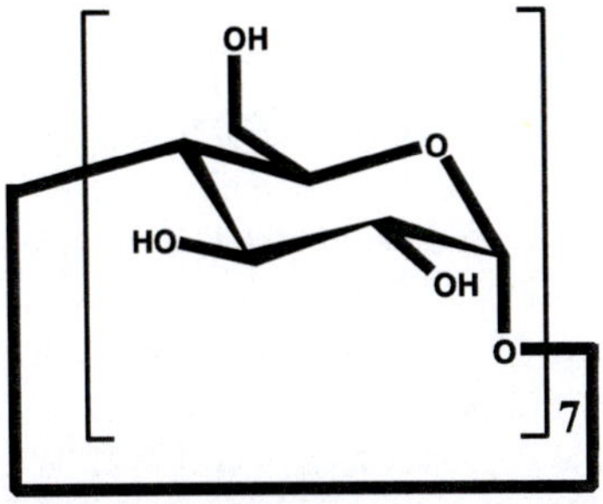 β-Cyclodextrin	Pun et al. (2004a), Forrest et al. (2005)

HeLa, and HepG2 cells of PSI-PEI/DNA complexes was rather lower than that based on PEI25, especially at high N/P ratios. The most efficient gene transfection of PSI-PEI/DNA complexes was even slightly higher than that of PEI25 in the same cell lines. Decreasing in the transfection efficiency was observed as the molecular weight of parent PEI increased from 600 to 1,200. PSI-PEI analogues synthesized using linear LMW-PEI of 423 Da has also been demonstrated to be effective for *in vivo* gene delivery and expression(Yu et al. 2009b). Wen et al. conjugated 423 Da PEI to a poly(glutamic acid) derivative by aminolysis (Wen et al. 2009). *In vitro* studies revealed that the newly synthesized polymers displayed markedly higher transfection efficiency than PEI25 in Hela, HepG2, Bel 7,402, and 293 cell lines. Importantly, polyplexes based on these polycations were more tolerable to serum compared with those originated from PEI25 and Lipofectamine 2000. LMW-PEIs were also covalently coupled to chitosan, and the synthesized polymers with low cytotoxicity showed significantly higher transfection efficiency in both *in vitro* and *in vivo* tests (Wong et al. 2006; Jiang et al. 2007; Lu et al. 2008). Through a hexamethylenediisocyanate spacer, LMW-PEI (800 Da) could be grafted onto dextran to give cationic polymers with effective vectors. *In vivo* studies, however, have not been reported so far to these carriers, although some promising *in vitro* results achieved (Sun et al. 2008a, b). For these LMW-PEI grafted cationic polymers, the main chain structure should have profound effects on the transfection behaviors of resulting polyplexes.

In general, cross-linking of LMW-PEI can produce pseudo-HMW-PEI with great structural diversity and multiple functionalities. For example, polymers with linear, graft or dendrimer-like architecture can be obtained, using linear or branched LMW-PEIs as initial materials. All the cross-linked products showed improved characteristics in terms of transfection efficiency and toxicity, at least to some extent, when compared to their precursor PEIs or PEI25. Additional functions, such as pH and chemical (such as intracellular glutathione) responsiveness can be integrated to the resulting polymers to offer better performances for gene delivery. However, it remains a challenge currently for delicate structure control over the cross-linked PEIs, and batch to batch variation is frequently observed. This makes it difficult to establish a useful structure-function relationship for designing and developing much more effective and safe PEI-based vectors for clinical applications.

3.2.3 Ligand-PEI Conjugates

To replace the non-specific electrostatic interactions between polyplexes and target cells with a cell-specific interaction, different types of ligands such as sugars, peptides, proteins and antibodies have been used for modifying PEI. Galactose residues were conjugated to PEI25, to target hepatocytes which express asialoglycoprotein receptor. Polyplexes of galactosylated PEI (gal-PEI) showed comparable transfection efficiencies at low substitution degrees (3.5%), an enhancement in efficiency with a higher conjugation of 5% (Kunath et al. 2003a; Zanta et al. 1997). When the amount of galactose was further increased to 31%, however, a decrease in transfection efficiency was observed, which is likely attributed to steric shielding effects. The cytotoxicity of both gal-PEI and polyplexes significantly decreased with increasing galactosylation. Other ligands that have been used to target PEI/DNA polyplexes include mannose(Diebold et al. 1999), RGD-containing peptide (Erbacher et al. 1999), transferring (Kircheis et al. 1997), epidermal growth factor (Blessing et al. 2001), anti-CD3 antibody (O'Neill et al. 2001), and anti-platelet endothelial cell adhesion molecule antibody (Li et al. 2000). Interestingly, the conjugation of negatively charged transferring protein to PEI is able to reduce the surface positive charge of PEI/DNA polyplexes, thereby produce shielded formulations without PEGylation (Kircheis et al. 2001a).

Furthermore, PEGylation and ligand conjugation can be combined to give polymers with even more excellent performance for gene transfections. Ligands like galactose (Sagara and Kim 2002), transferrin (Kursa et al. 2003), epidermal growth factor(Lee et al. 2002), and folate (Benns et al. 2002) have been used for targeting PEGylated PEI/DNA polyplexes. These ligands can be conjugated to the PEG chains of PEGylated PEI either before or after polyplex formation. Alternatively, the ligands can be covalently coupled to PEI, while the formed ligand-PEI complexes are then modified with PEG chains. The related studies have demonstrated that combination of PEGylation and cell targeting can favorably modify the behaviors (such as decreased toxicity, prolonged circulation time, and tissue-selective accumulation) of PEI/DNA polyplexes after system administration.

3.2.4 PEGylation of PEI

Despite its efficacy, *in vivo* applications of PEI for gene therapy have been largely baffled by its toxicity (Kichler 2004). The LD50 (median lethal dose) of a linear PEI (lPEI, 22 kDa) in Balb/C mice is 4 mg/kg. The systemic toxicity of PEI is considered to, at least in part, relate to the unspecific interactions with erythrocytes and other blood components such as albumin, fibrinogen and complement C3, mainly owing to the excess positive charges of the formed polyplexes (Ogris et al. 1999).

Conjugates of PEI with PEG (~5,000 Da) were shown reduced unspecific binding capacity of PEI itself. Polyplexes based on these 'shielding' conjugates displayed improved solubility and stability as well as strongly reduced *in vitro* and *in vivo* toxicity (Table 5) (Ogris et al. 2003). Shielding the positive surface charges provides the polyplexes with longer half-lives in circulation, due to the reduced opsonization and reticuloendothelial system (RES) uptake (Kircheis et al. 1999). As a result, these 'stealth' polyplexes allow the injection of highly concentrated formulations. As a successful case, the polymer of lPEI (22 kDa) conjugated with PEG (~5,000 Da) has shown significant levels of transgene expression in a mice model post-nasal instillation (Kichler et al. 2002). PEGylation, however, may sacrifice the DNA-binding capability of modified PEI, sterically hinder interactions of the polyplexes with the target cells, and therefore may impair the transfection efficiency of resulting formulations.

Table 5 Hydrophilic modifications of PEI

Hydrophilic moieties used for PEI modifications	Reference
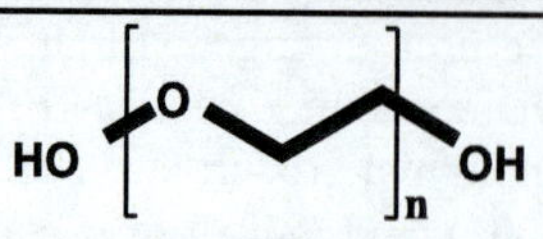 PEG	Kichler et al. (2002), Ogris et al. (2003)
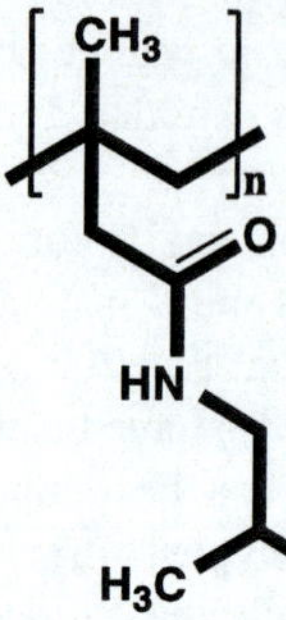 Poly(N-(2-hydroxypropyl)methacrylamide) (PHPMA)	Oupicky et al. (2002)
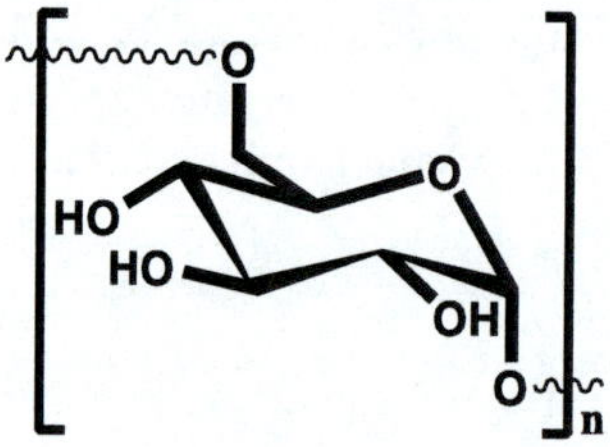 Dextran	Tseng and Jong (2003), Tseng et al. (2004, 2005)
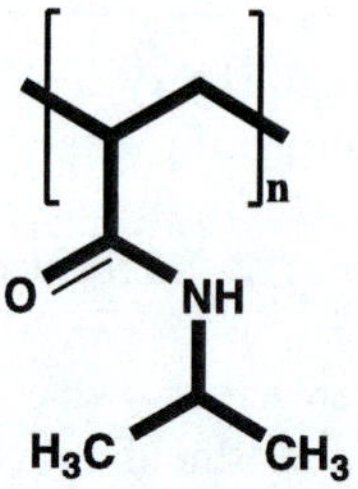 Poly(N-isopropylacrylamide) (PNIPAm)	Twaites et al. (2004, 2005), Zintchenko et al. (2006), Lavigne et al. (2007)
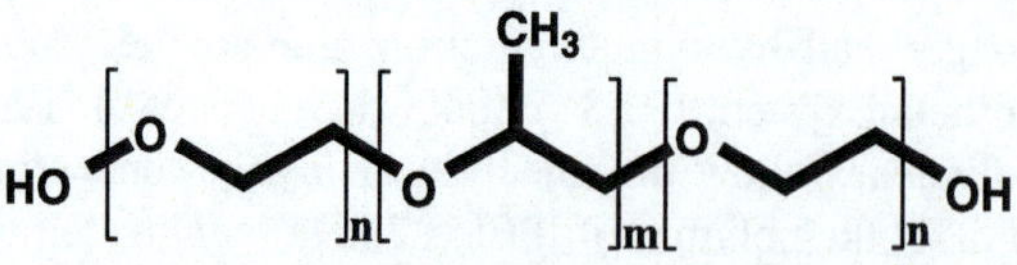 Pluronic	Ochietti et al. (2002), Belenkov et al. (2004)
Gly-Ile-Gly-Ala-Val-Leu-Lys-Val-Leu-Thr-Thr-Gly-Leu-Pro-Ala-Leu-Ile-Ser-Trp-Ile-Lys-Arg-Lys-Arg-Gln-Gln-NH_2 Melittin	Ogris et al. (2001)

3.2.5 Other Hydrophilic Modifications on PEI

PEI derivatives have been synthesized by conjugating with hydrophilic polymers other than PEG, in an attempt to develop PEI-based polyplex formulations with increased solubility, sufficient stabilization and prolonged *in vivo* circulation as well as enhanced endosomolytic activity and improved nuclear transport. Table 5 summarizes the selected examples about the polymers employed for PEI decoration.

A multivalent poly[N-(2-hydroxypropyl)methacrylamide] (PHPMA) bearing reactive ester groups has been used for the surface modification of preformed PEI/DNA polyplexes (Oupicky et al. 2002). The aqueous solubility and systemic circulation times of the laterally stabilized complexes were enhanced compared with those without PHPMA modification.

Tseng and coworkers systemically investigated the effect of dextran modification of bPEI on the transfection mechanisms (Tseng and Jong 2003; Tseng et al. 2004, 2005). Polyplexes prepared using the modified polycations showed significantly diminished cytotoxicity and improved stability against bovine serum albumin (BSA). Unfortunately, transfection using green fluorescent protein (GFP) indicated that the modified polymers were less efficient than the unmodified PEI. The impaired condensation capability, buffering capacity, cellular entry and the integrity of DNA-polymer polyplexes were proposed to be responsible for the decreased transgene expression.

Poly(N-isopropylacrylamide) (PNIPAm) is a water-soluble polymer that exhibits a lower critical solution temperature (LCST) at about 32°C. Generally, copolymers containing PNIPAm segment are also thermosensitive. To achieve a temperature-regulated gene transfection, studies have been conducted to modify PEI using PNIPAm (Twaites et al. 2004, 2005; Zintchenko et al. 2006; Lavigne et al. 2007). Temperature-dependent transfection activities were observed for polyplexes based on PNIPAm/PEI copolymers. These results indicate that changes in molecular structures triggered by carefully designed stimuli during intracellular trafficking may be used to enhance gene delivery.

Pluronic, also termed 'Poloxamer' is triblock copolymers of poly(ethylene oxide)-*b*-poly(propylene oxide)-*b*-poly(ethylene oxide) (PEO-PPO-PEO). Despite their amphiphilicity, they are essentially water soluble polymers. The hydrophobic PPO block in Pluronic is considered to be able to stabilize the polyplexes of polycation and DNA, and enhance the intracellular translocation of Pluronic molecules. PEI conjugates with Pluronic 85 could target the payload oligonucleotide (ODN) to the hepatocytes of the liver, different from the kidney accumulation mediated by PEGylated PEI (Ochietti et al. 2002). In addition, Pluronic (P85) modified PEI (PEI-P85) could improve the solubility of PEI/DNA complexes (Belenkov et al. 2004). Polyplexes formed by PEI-P85 and Ku86 antisense oligonucleotide were highly stable in aqueous dispersions. *In vivo* therapeutic effect was implemented using this new formation in athymic nude mice bearing subcutaneous human HT29 colon adenocarcinoma xenografts. The applications of Pluronic copolymers for gene delivery can refer to three reviews(Lemieux et al. 2000; Kabanov et al. 2002; Bromberg et al. 2006).

CDs, especially β-CD, have been used for modifying PEIs (PEI-CDs) to lower their cytotoxicity (Pun et al. 2004a; Forrest et al. 2005). PEI-CDs showed significantly enhanced *in vitro* transfection levels for several cell lines compared with PEI25. After they were further stabilized using a 1-adamantanemethylamine (Ada) conjugated PEG (PEG-Ada) via host-guest interactions between β-CD units and Ada groups, the pDNA-PEI-CD polyplexes clearly mediated transgene expression in tissues like liver, kidney and lung of immune-compromised female nude mice bearing subcutaneous PC3 tumors (Pun et al. 2004a). In addition to formulations for systemic administration, a recent study revealed that polyplexes formed by pDNA and PEI-CD could be used for constructing multilayer structures which were able to mediate in situ gene transfection (Hu et al. 2010). Besides the reduction in toxicity, modification using CDs can endow the resulting polyplexes with flexible modulability via inclusion complex formation. For instance, *in vitro* and *in vivo* stability, prolonged bloodstream circulation as well as targeting capability can be achieved using this strategy, which have been well demonstrated by Davis' group (Davis 2009).

For non-viral delivery vectors, the limited endosomolytic capacity is one of the important factors responsible for their relative low transfection efficiency compared with viral systems. Covalently linking melittin to PEI25 could significantly increase the intracytoplasmic release of DNA, due to the membrane lytic activity of melittin (Ogris et al. 2001). Compared with PEI25, the transfection efficiency of this conjugate was strongly increased within a broad range of cell lines (such as different tumor cells, primary hepatocytes, and human umbilical vein endothelial cells). It was found that melittin not only could increase the endosomal release of DNA-loaded polyplexes, but also could enhance their intranuclear transport. The non-immunogenic all-(D)-melittin, a stereoisomer of natural all-(L)-melittin, is also effective to mediate enhanced gene expression similar to the natural one (Boeckle et al. 2005). All-(D)-melittin conjugates of PEI displayed up to 160-fold higher luciferase activity than unmodified PEI. The site of melittin linkage strongly influenced the membrane-destabilizing activities of both conjugates and polyplexes. C-terminus conjugated PEI (C-mel-PEI) is highly lytic at neutral pH, and therefore increased doses of C-mel-PEI polyplexes elicited high toxicity. In contrast, N-terminus conjugated polycation (N-mel-PEI) was less lytic at physiological pH but retained higher lytic activity than C-mel-PEI at endosomal pH. This apparently implies that N-terminus conjugation of melittin is preferred for a better transfection. To overcome the inherent lytic activity of melittin at neutral pH that may provoke high cytotoxicity, acidic modification of melittin was implemented by replacing neutral glutamines with glutamic acid residues. Through this protocol, the lytic activity of C-mel-PEI at endosomal pH was greatly improved (Boeckle et al. 2006).

3.2.6 Modification Using Hydrophobic and Amphiphilic Segments

Hydrophobic modification of cationic polymers has been demonstrated to be a useful strategy to developing new gene vectors with improved delivery characteristics (Incani et al. 2010). Since lipids are the main component of cell membrane, incorporation of hydrophobic moieties may result in additional hydrophobic interaction between polyplexes and cell membranes, which in turn facilitates the delivery of genetic payload into cells. In addition, the amphiphilicity of hydrophobically modified polycations possess the self-assembly potential for constructing complex structures with advanced functions. Table 6 lists the frequently used hydrophobic moieties.

Table 6 Hydrophobic modification of PEI

Hydrophobic moieties for PEI modifications	Reference
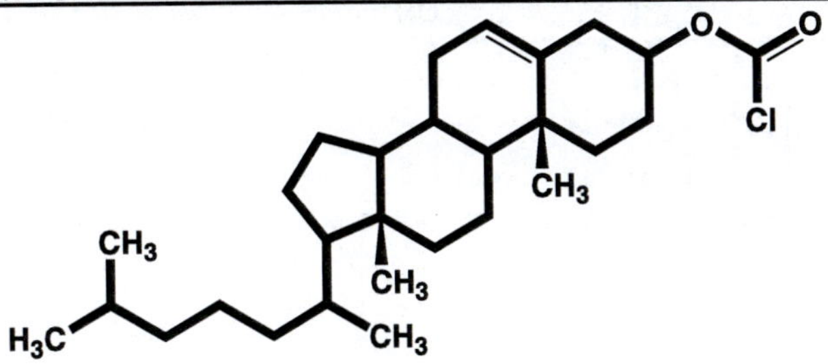 Cholesteryl chloroformate	Kim et al. (2001a), Han et al. (2001), Wang et al. (2002a)
 Myristate	Kim et al. (2001a)
 Dodecyl iodide	Thomas and Klibanov (2002)
 Hexadecyl iodide	Thomas and Klibanov (2002)
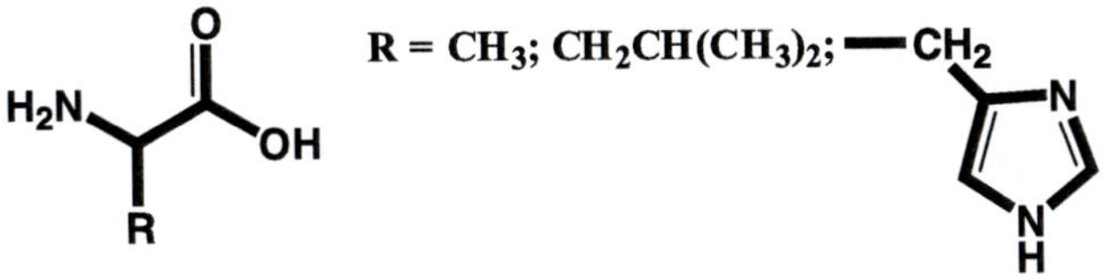 Alanine; Leucine; Histidine	Thomas and Klibanov (2002)
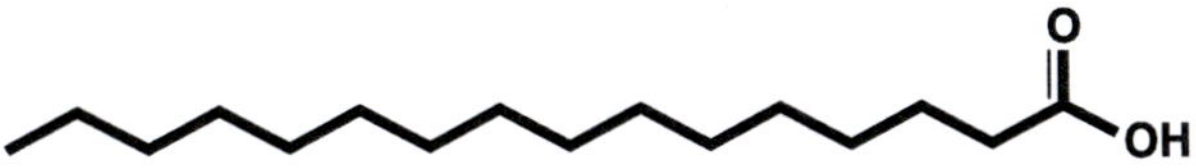 Palmitic acid	Brownlie et al. (2004), Incani et al. (2007)
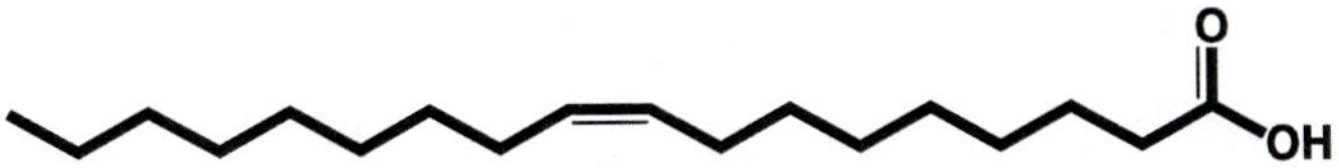 Oleic acid	Alshamsan et al. (2009)
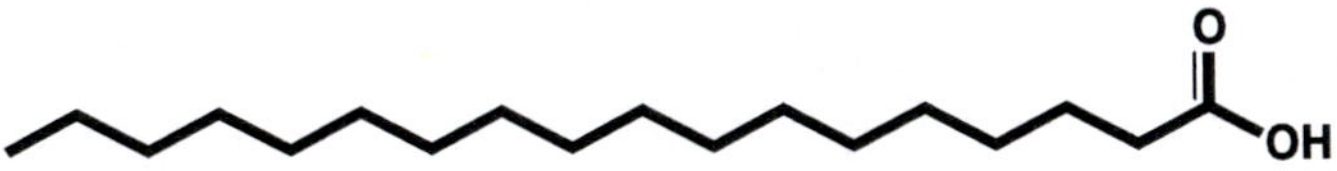 Stearic acid	Alshamsan et al. (2009)
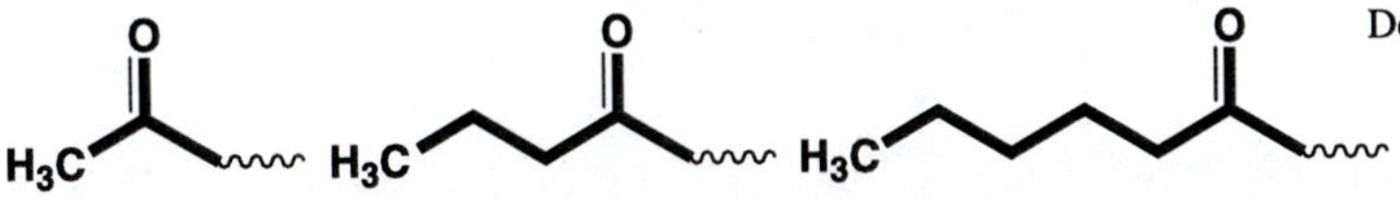 Acetyl, Butyryl and Hexanoyl	Doody et al. (2006), Gabrielson and Pack (2006)

(continued)

Table 6 (continued)

Hydrophobic moieties for PEI modifications	Reference
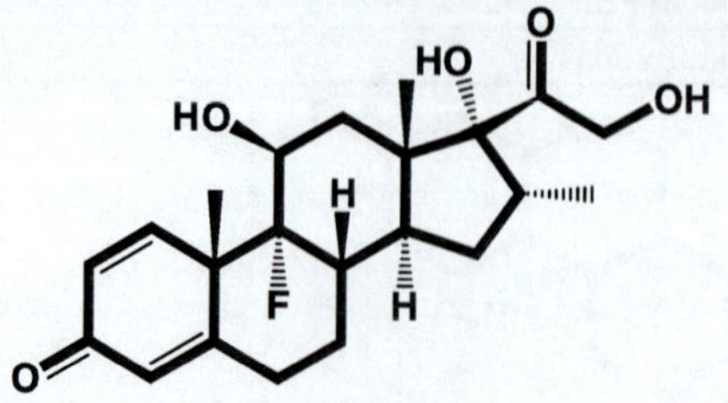 Dexamethasone	Bae et al. (2007), Kim et al. (2009a, b)
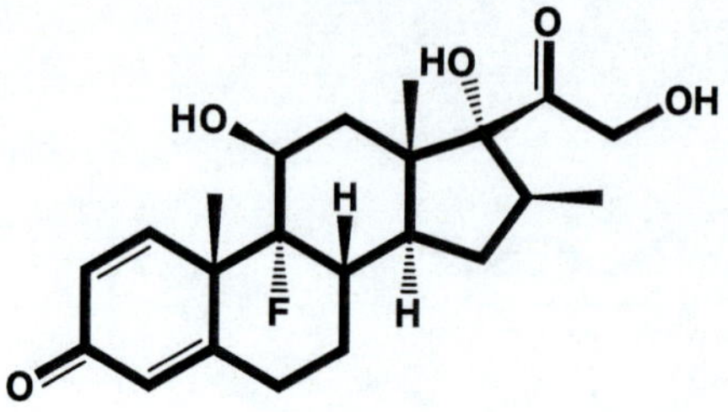 Betamethasone	Ma et al. (2009)
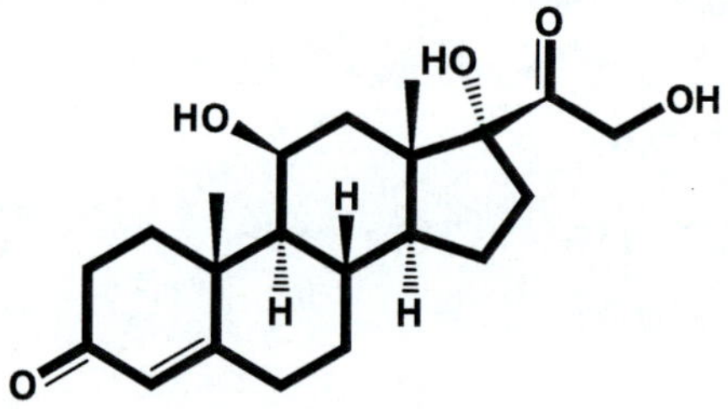 Hydrocortisone	Ma et al. (2009)
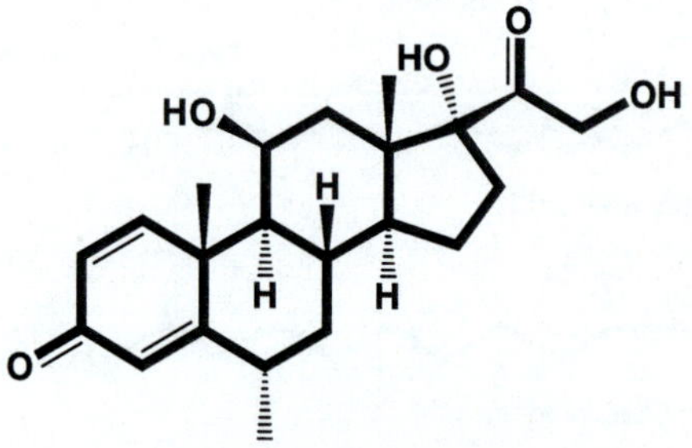 Methylprednisolone	Ma et al. (2009)
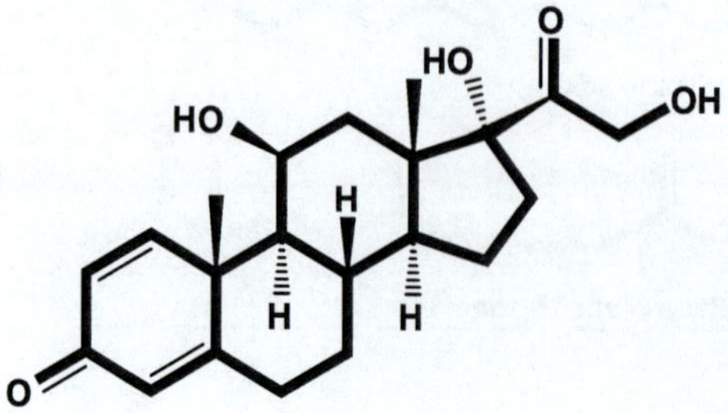 Prednisolone	Ma et al. (2009)

Modification of bPEI (1,800 Da) with cholesteryl chloroformate produced a water-soluble lipopolymer (bPEI-Chol) as the percentage of conjugated cholesterol was about 47% (Han et al. 2001). In contrast to PEI25/pDNA complexes, bPEI-Chol/pDNA polyplexes and bPEI-Chol itself were less toxic to CT-26 colon adenocarcinoma and 293 T human embryonic kidney transformed cells. Compared to 25 kDa or 1,800- Da based formulations, bPEI-Chol showed higher transfection efficiency in both CT-26 and 293 T cells. High levels of GFP expression in Jurkat cells was also observed for this type of PEI derivatives (Wang et al. 2002a). Kim et al. observed the enhanced transfection when cholesterol or myristate was conjugated onto 2 kDa PEI (Kim et al. 2001a). Increased cytotoxicity, however, was found in this case. N-Dodecylation of primary amino groups of 2,000 Da PEI yields a low toxic polycation whose transfection efficiency in the presence of serum is 400 times higher than its precursor polymer (Thomas and Klibanov 2002). This derivative was even more effective than PEI25. The same group also found that N-acylation of PEI25 with alanine nearly doubled the transfection efficiency in the presence of serum and also lowered its cytotoxicity. Acylation of PEI25 with palmitic acid (PA) created amphiphilic PEI derivatives (bPEI-PA) that could form nanoparticles or vesicles (Brownlie et al. 2004). The PEGylated bPEI-PA was 10-fold less toxic while retaining 30% of the transfection efficiency *in vitro*. After intravenous administration in a mouse model, polyplexes based on the PEGylated bPEI-PA could mediate the transgene expression of GFP in the liver. However, a recent report by Incani et al. showed negative results when they using PA modified PEI to transfer GFP-encoding pDNA to bone marrow stromal cells (Incani et al. 2007). Difference in substitution and cell line may be responsible for this discrepancy. On the other hand, modification of bPEI using oleic and stearic acid gave promising results for siRNA delivery (Alshamsan et al. 2009). The modified bPEIs showed 3-fold increased siRNA transfection in B16 cells than the parent PEI, which were superior or comparable to some of the commercially available carriers like jetPEI, Metafectene, and INTERFERin.

Small hydrophobic alkyl moieties, such as acetyl, butyryl and hexanoyl, are also able to increase the transfection capability of modified polycations. Study by Pack's group showed the increased gene delivery efficiency of acetylated PEI25 in MDA-MB-231 and C2C12 cell lines, upon acetylation of up to 43% of the primary amines with acetic anhydride (Forrest et al. 2004). They also found that transfection efficiency continued to increase (up to 58-fold in HEK293) with up to 57% acetylation of primary amines, albeit decreased as the acetylation degree further increased (Gabrielson and Pack 2006). Putnam and coworkers hydrophobically decorated PEI25 with acetyl, butyryl and hexanoyl by 1-ethyl-3-(3-dimethylaminopropyl)carbodiimide (EDAC)/N-hydroxysuccinimide (NHS) mediated condensation reactions between protonated PEI25 and acetate, butanoate, and hexanoate, respectively (Doody et al. 2006). *In vitro* transfections using HeLa, NIH/3T3 and Clone 9 cell lines showed that substitution of the primary amines generally increased transfection efficiency relative to unmodified PEI25. Increasing both the substitution degree beyond 25 mol% and hydrocarbon length, however, decreased the delivery efficacy.

It has been found that the intranuclear gene delivery is strongly limited by nuclear pore complexes (NPCs), which are proteins assembled channels that span the nuclear envelope. The presence of glucocorticoids (GC) such as dexamethasone (DEX) can lead to the dilation of NPC, which may significantly increase the efficiency of nuclear gene transfection (Shahin 2006; Dames et al. 2007). The conjugation of DEX to DNA has been demonstrated to be able to facilitate the uptake of transfected DNA into the nucleus, taking advantage of steroid receptors as shuttles (Rebuffat et al. 2001). DEX, has been therefore conjugated to PEIs and the obtained derivatives showed increased transfection efficacy (Bae et al. 2007; Kim et al. 2009a, b). By conjugating five GCs to 1,800 Da bPEI (bPEI-GC), including betamethasone, dexamethasone, hydrocortisone, methylprednisolone and prednisolone, it was found the transfection activity of bPEI-GC polymers correlated with their GC potency (Ma et al. 2009).

A delicately modulated balance between hydrophilic and hydrophobic components is crucial for designing more efficient gene vectors based on PEI. Detailed knowledge about the molecular structure of PEI derivatives is a prerequisite in order to establish clear structure-function relationships, as well as to optimize cytotoxicity and biocompatibility of resulting carriers. In general, the cytotoxicity of PEI derivatives can be significantly changed by hydrophobic or amphiphilic modifications. In addition, enhanced transfection activities can be implemented in most cases. Moreover, incorporation of hydrophobic moieties into PEI may influence the interactions of polyplexes with endo-/lysosomal membranes, resulting in a more efficient escape from these compartments.

3.3 Cyclodextrin-Containing Polymers

Due to their excellent compatibility and low immunogenicity, multiple sites for functionalizing as well as their complexation capability with a broad spectrum of substances including hydrophobic molecules and macromolecules, cyclodextrins (CDs) are promising candidates for the development of new vectors. There have been increasing interests in recent years to discover efficient delivery carriers for gene therapy base on CDs and their derivatives. Table 7 summarizes CD-containing compounds used for gene transfections.

3.3.1 Cyclodextrin Derivatives

Cationic derivatives were prepared by modifying β-CD with pyridylamino, alkylimidazole, methoxyethylamino or primary amine groups at 6-positions of the glucose units (Cryan et al. 2004). These compounds could form stable nanoparticulate complexes with DNA, while their transfection activities in COS-7 cells were dependent on the substituents and their sites. The most effective vector, heptakispyridylamino CD, produced a 4,000-fold increase in transfection level over DNA

Table 7 Cyclodextrin derivatives used for gene transfection

Chemical structure	Reference
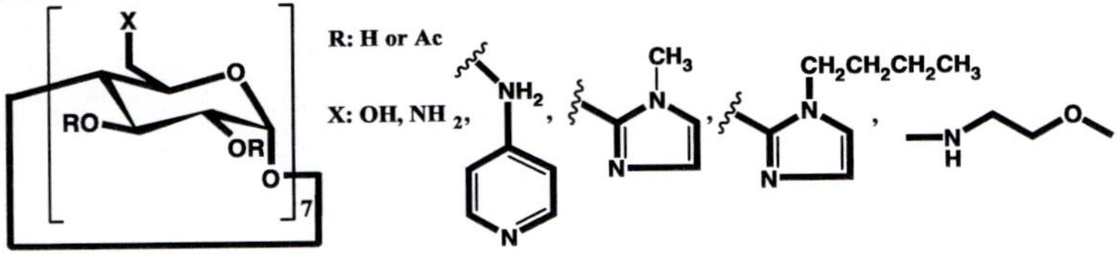	Cryan et al. (2004)
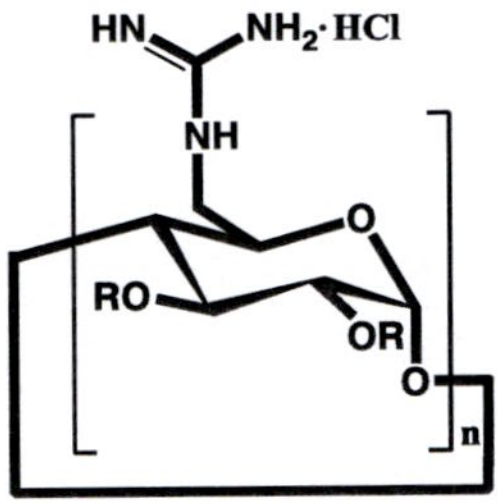 n = 6, 7, 8	Mourtzis et al. (2007, 2008)
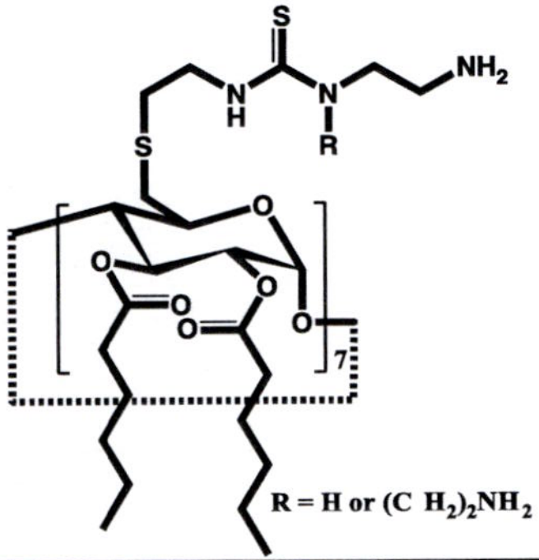	Diaz-Moscoso et al. (2008)

(continued)

Table 7 (continued)

Chemical structure	Reference
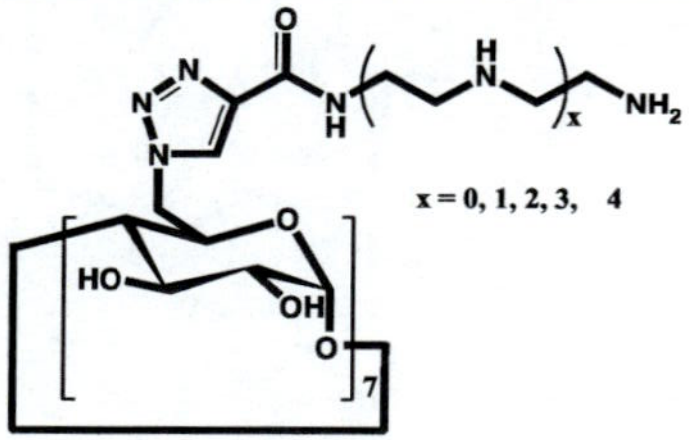	Srinivasachari et al. (2008)
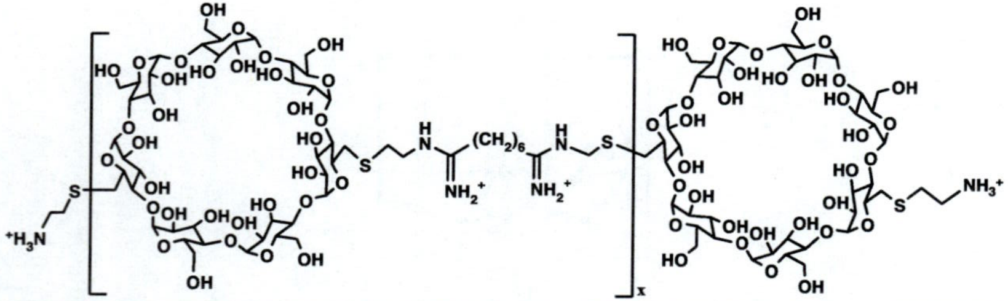	Gonzalez et al. (1999), Hwang et al. (2001), Popielarski et al. (2003), Reineke and Davis (2003a), Reineke and Davis (2003b), Mishra et al. (2004), Pun et al. (2004b), Hu-Lieskovan et al. (2005), Bartlett and Davis (2007), Heidel et al. (2007)

alone. Introducing the guanidino group at the primary side of α-, β-, and γ-CDs resulted in per(6-guanidino-6-deoxy)-cyclodextrins, which could condense DNA into nanoparticles (Mourtzis et al. 2007). However, these derivative and their analogs heptakis- and octakis(6-guanidinoalkylamino-6-deoxy)-β- and -γ-cyclodextrins could not effectively mediate the transfection of DNA *in vitro* (Mourtzis et al. 2008). On the contrary, relatively high efficient transfection and GFP expression was observed for the per-(6-aminoalkylamino-6-deoxy)-cyclodextrins in HEK 293T and HEK 283 cells. To obtain an efficient vector from β-CD, Diaz-Moscoso et al. synthesized an amphiphilic β-CD derivative conjugated with seven positively charged amino moieties and fourteen alkyl groups at primary and secondary hydroxyl positions, respectively (Diaz-Moscoso et al. 2008). This polycation with reduced cytotoxicity exhibited high efficient transfection of pDNA in BNL-CL2 cells, which was even comparable to that of PEI25. By conjugating oligoethyleneamines to the primary hydroxyl sites of β-CD via click coupling chemistry, a novel series of multivalent polycationic β-CD "click clusters" have been created (Srinivasachari et al. 2008). The cellular gene delivery experiments revealed that all these cationic β-CD derivatives could efficiently deliver Cy5-labeled pDNA and pDNA encoding the luciferase reporter gene into HeLa and H9c2 cells in Opti-MEM. The level of reporter gene expression was found to be increased with an increase in oligoethyleneamine number within the cluster arms, and derivatives armed with tetraethylenepentamine pentaethylenehexamine are the most effective and promising vectors that deserve further development. Cytotoxicity evaluation based on protein, MTT and lactate dehydrogenase assays demonstrated that all the click clusters remained nontoxic within the expected dosage range, while the positive controls, Jet PEI and Superfect, were highly cytotoxic.

3.3.2 CD-Containing Cationic Polymers

On the other hand, CDs-containing cationic polymers have been systematically examined for gene delivery in recent years. Davis and coworkers designed and synthesized a class of linear, cationic, β-CD-containing polymers (βCDPs) by copolymerizing difunctionalized β-CD monomers with other difunctionalized comonomers (Gonzalez et al. 1999). Table 7 shows a representative structure of this type of polycation. These polymers have low *in vitro* and *in vivo* toxicity. Polyplexes with size of <200 nm can be assembled by the condensation of pDNA using βCDPs. Importantly, *in vitro* transfection efficiency with βCDPs was comparable to those based on PEI (branched 25 kDa) and Lipofectamine. The transfection activity and cell toxicity are critically related to the structure of this type of carbohydrate polycations (Hwang et al. 2001), which is mainly determined by the comonomer structure, charge center type as well as the cyclodextrin type and functionalization (Reineke and Davis 2003a, b; Popielarski et al. 2003). The colloidal stability in biological fluids can be furthered increased by a surface decoration of the assembled polyplexes using PEG-Ada, which is achieved through the host-guest interactions between β-CD and Ada group(Mishra et al. 2004). Furthermore, a targeting ligand can be introduced onto the PEG shell of assemblies containing gene segments, by using a transferrin-PEG-Ada conjugate (Bartlett and Davis 2007). *In vivo* transfection experiments in both murine models and non-human primates showed that the ligand functionalized nano-assemblies can efficiently deliver the gene payload to tumor tissues, and perform their therapeutic efficacy after intravenous injections (Pun et al. 2004b; Hu-Lieskovan et al. 2005; Heidel et al. 2007). These studies also confirmed the safety of these non-viral nanoparticles for systemic administrations.

Taking advantage of the host-guest interactions between β-CD units and Ada cationic derivatives, Burckbuchler et al. were able to construct a pseudo-polycation functioning as a vector for gene delivery (Burckbuchler et al. 2008). Gel retardation, ζ-potential and surface enhanced Raman spectroscopy (SERS) experiments demonstrated the formation of polyplex assemblies by host-guest polycations and pDNA encoding a *luciferase* gene. In addition, gel retardation and small angle neutron scattering (SANS) measurements suggested the stability of assembled polyplexes was mainly related to the chemical structure of guest molecules. The polyplexes based on the polycation of Ada3 and Ada4 were relatively stable under higher NaCl concentrations. *In vitro* transfection was performed on HepG2 and KEK293 cells. The optimized polyplexes based on Ada4 exhibited a transfection efficiency with the same order as that of DOTAP, a commercialized transfection agent. Nevertheless, the former transfection was carried out in the presence of the fusogenic peptide JTS-1.

CDs based polypseudorotaxanes (CDPRs) are a family of polymers with CDs threading onto the polymer chains. Since their dependent discoveries by several groups (Ogata et al. 1976; Harada and Kamachi 1990; Wenz and Keller 1992), there have been overwhelming interests on the developments of functional polymer materials and delivery systems based on CDPRs (Harada et al. 2006; Huang and Gibson 2005; Araki and Ito 2007; Loethen et al. 2007; Li and Loh 2008; Li 2009). Nevertheless, the applications of CDPR polymers for gene delivery emerged as a new direction more recently. Yui et al. synthesized a biocleavable polyrotaxane (Ooya et al. 2006), in which dimethylaminoethyl-modified α-CDs were threaded onto a PEG chain capped with benzyloxycarbonyl tyrosine via disulfide linkages. AFM observation together with gel electrophoresis and ξ-potential measurements confirmed the formation of polyplexes as well as the successful condensation of pDNA. The cleavage of disulfide linkages under reducible conditions in cytosolic milieu facilitates the de-condensation and intracellular trafficking of pDNA cargo. *In vitro* transfection revealed that the assembled polyplexes can rapidly escape from endosome and deliver pDNA to the nucleus. On the other hand, Li et al. designed cationic CDPRs with multiple oligoethylenimine-grafted CDs threaded on a PEO-PPO-PEO chain (Fig. 2) (Li et al. 2006). These cationic polyrotaxanes with different oligoethylenimines can efficiently condense pDNA into nanoparticles, which was confirmed by agarose gel electrophoresis, particle size, and ξ-potential measurements. Importantly, these supramolecular polycations are less toxic that PEI25 for cultured L929 and HEK 293 cells. Transfection studies using luciferase as a marker gene in HEK293 showed that the efficiency of assembled polyplexes was even

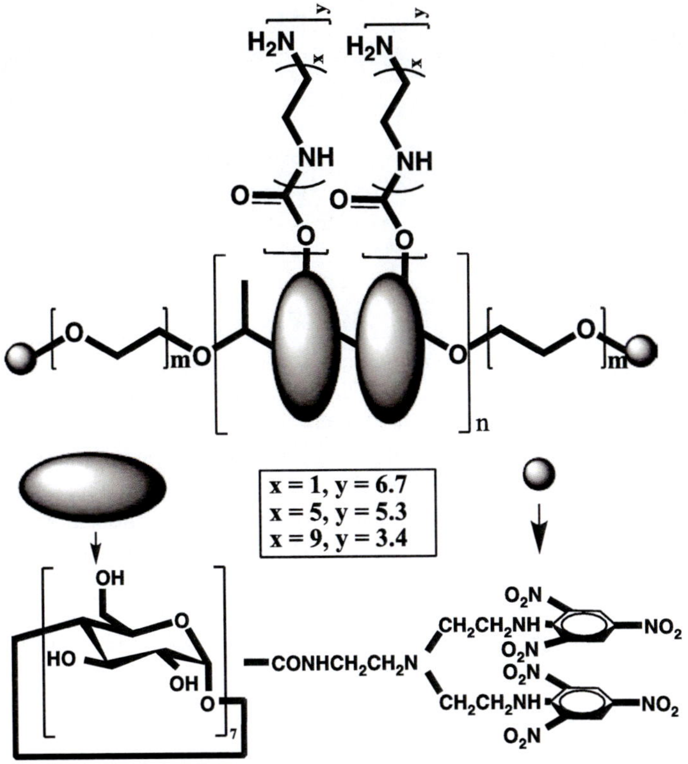

Fig. 2 Cationic polyrotaxanes with different oligoethylenimines based on β-CD and PEO-PPO-PEO triblock copolymer

comparable with those of PEI25. In addition, Liu et al. reported the condensation of calf thymus DNA by CDPRs with anthryl-modified-β-CDs threading onto polypropylene glycol (PPG) (Liu et al. 2007a). However, no cytotoxicity and transfection results have been reported for these CDPRs.

3.4 Poly[2-(Dimethylamino)Ethyl Methacrylate]

Poly(2-(dimethylamino)ethyl methacrylate) (PDMAEMA), as illustrated in Fig. 3 is a water-soluble inherent cationic polymer, which is frequently synthesized by radical polymerization of 2-(dimethylamino)ethyl methacrylate. Initial study by Hennink et al. suggested that PDMAEMA displayed higher transfection efficiency than PLL in COS-7 and OVCAR-3 cells, and its transfer activity was dependent on both molecular weight and polymer/DNA weight ratio (van de Wetering et al. 1997). Despite its relatively high *in vitro* transfection capability, the *in vivo* gene delivery

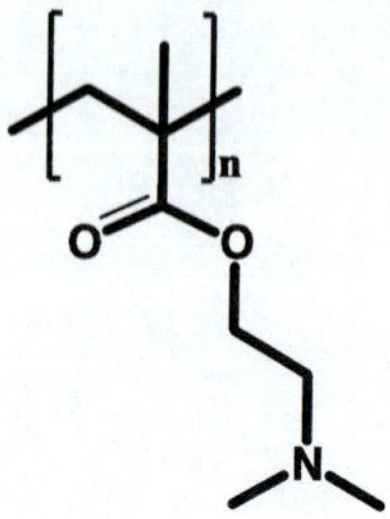

Fig. 3 Poly (2-(dimethylamino)ethyl methacrylate) (PDMAEMA)

(especially systemic administration) using PDMAEMA has been largely limited, since polyplexes based on this polymer could induce severe aggregation of erythrocytes (Verbaan et al. 2001). Attempts have been made to improve the transfection efficiency of PDMAEMA by chemical modifications. These studies concerned the incorporation of additional tertiary amino groups in each structural unit to promote the 'proton sponge' effect (Funhoff et al. 2004a), the change of ammonium groups in PDMEMA to pyridine, imidazole, and carboxylic acid functionalities to improve endosomal escape (Dubruel et al. 2003), the incorporation of guanidinium side groups to facilitate membrane penetration (Funhoff et al. 2004b). In addition, PDMEMA has been copolymerized with variety of monomers such as methyl methacrylate, N-vinyl-pyrrolidone, and ethoxytriethylene glycol methacrylate in order to reduce its cytotoxicity (van de Wetering et al. 1998, 2000). Thorough these procedures, DMAEMA based polycations with a better transfection/toxicity ratio could be reached.

3.5 *Biodegradable Polycations*

For the clinical applications of polymeric carriers for gene therapy, their long-term fate in the host has to be considered. Biocompatible and biodegradable polymers that can be degraded *in vivo* are therefore advantageous. In addition to biodegradable PEI and PLL, other degradable polymers have been synthesized and investigated for gene delivery (Fig. 4). These polymers include poly[α-(4-aminobutyl)-L-glycolic acid], poly(4-hydroxy-L-proline ester), poly(amino esters), polyphosphoesters and polyphosphoramidates, and polyphosphazenes. Depending on their structures, these polycations can be hydrolyzed within different time periods.

3.5.1 Poly[α-(4-Aminobutyl)-L-Glycolic Acid]

Poly[α-(4-aminobutyl)-L-glycolic acid] (PAGA) is a biodegradable analogue of PLL, which was synthesized by the polymerization of N_ε-cbz-L-oxylysine monomers followed by a deprotection (Fig. 4a) (Lim et al. 2000b). PAGA could effectively

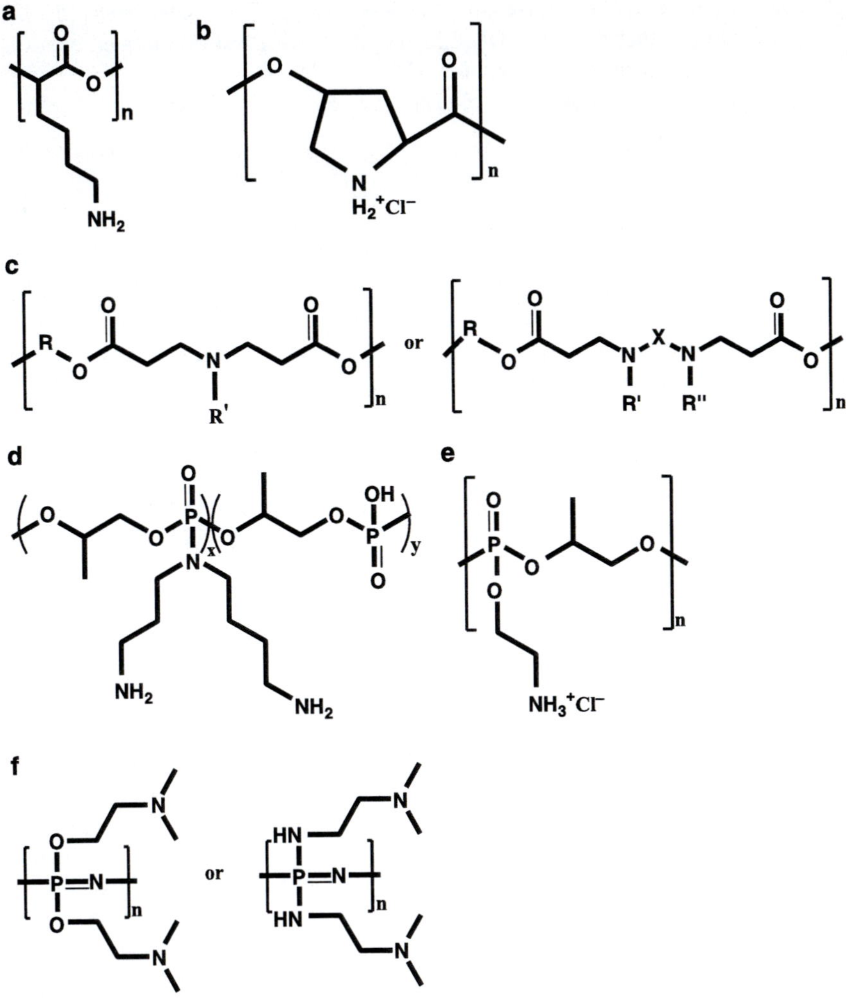

Fig. 4 Biodegradable synthetic polycations: (**a**) poly[α-(4-aminobutyl)-L-glycolic acid] (PAGA), (**b**) poly(4-hydroxy-L-proline ester) (PHP), (**c**) poly(β-amino ester)s, (**d**) poly(2-aminoethylpropylene phosphate) (PPE-EA), (**e**) polyphosphoramidate (PPA), and (**f**) cationic polyphosphazenes

condense pDNA to form nanoscaled polyplexes which were able to completely dissociated within one day of incubation at 37°C due to the degradation of PAGA (Lim et al. 2000a). PAGA showed about 2-fold higher transfection efficiency than PLL, while no measurable cytotoxicity was detected in the concentration range of experiments. *In vivo* study using nonobese diabetic (NOD) mice showed the sustained mIL-10 expression for 9 weeks post-systemic administration of PAGA-pDNA (encoding murine interleukin-10, mIL-10) nano-polyplexes (Koh et al. 2000). In addition, the prevalence of severe insulitis on 12 week-old NOD mice was evidently decreased by the intravenous injection of PAGA-DNA polyplexes (15.7%), in comparison to that of naked DNA injection (34.5%) and non-treated controls (90.9%). Combined administration of pDNAs encoding IL-4 and IL-10 with PAGA vector in NOD mice could also prevent the development of diabetes in 75% of the treated animals (Ko et al. 2001).

3.5.2 Poly(4-Hydroxy-L-Proline Ester)

Poly(4-hydroxy-L-proline) (PHP) was synthesized by the polymerization of N-cbz-4-hydroxy-L-proline followed by deprotection (Fig. 4b) (Lim et al. 1999; Putnam and Langer 1999). This polymer exhibited significantly reduced cytotoxicity as compared to PEI25 or PLL (Putnam and Langer 1999). A two-stage degradation profile was observed for PHP in aqueous solutions at a 37°C buffer condition (pH 7), and a thorough degradation took nearly 3 months. PHP-DNA polyplexes, however, showed a slower degradation profile compared to that of PHP alone. Transfection of β-galactosidase gene into CPAE cells using PHP-DNA polyplexes was successful, showing an efficiency that was higher than that of PLL with or without fetal bovine serum (Lim et al. 1999).

3.5.3 Poly(β-Amino Ester)s

To develop safe alternatives to existing polymeric vectors, Langer's group discovered a new type of degradable amine-containing polyesters, called as poly(β-amino esters) (PAEs, Fig. 4c) (Lynn and Langer 2000). These polycations possess low cytotoxicity and can be easily synthesized by the conjugate addition of a primary amine or bis(secondary amine) to a diacrylate (Lynn et al. 2001). At physiological pH, PAEs can be completely degraded to nontoxic monomers. A large library (>2,500) of PAE was discovered using the high-throughput semi-automated methodology (Anderson et al. 2003). Optimizing monomers and polymerization conditions resulted in a new PAE library (>500), part of which showed transfection activities that rival PEI25 and Lipofectamine 2000 in both COS 7 and HUVEC cell lines (Lynn et al. 2001; Green et al. 2006). *In vivo* delivery to xenografts in mice showed promising results for the local treatment of cancer(Anderson et al. 2004). Structure-function correlation based on the PAEs library suggested that molecular weight and end groups significantly affected the transfection potential of these polymers. High molecular weight and amine termination confers PAEs with high transfection efficiency (Akinc et al. 2003; Zugates et al. 2007).

By ester exchange reactions, Park's group synthesized hyperbranched poly(amino esters). These polymers displayed significantly lower cytotoxicity than PEI25. The transfection efficiency of these vectors, however, was lower than that of PEI25 (Lim et al. 2001). Other hyperbranched poly(amino esters) were also synthesized, and these polymers showed significantly improved transfer activity (Li et al. 2005; Wu et al. 2005, 2006).

In addition, network-type poly(amino esters) (n-PAE) have been prepared by Park et al. to prolong the degradation of polyplexes based on poly(amino ester). n-PAE exhibited transfection efficiency superior to that of PEI25, particularly in the presence of serum (Lim et al. 2002). Additional conjugation of n-PAE with either aminohexanoic acid or lysine gave rise to polymers with minimal toxicity and high transfection capability comparable to that of PEI25 (Kim et al. 2005a, 2006a). Furthermore, the unique feature of slow degradation of these polymers could retain the polyplexes for 5–10 days, which makes them reliable for long-term therapeutic applications (Kim et al. 2007a).

3.5.4 Polyphosphoesters and Polyphosphoramidates

Polyphosphoester and polyphosphoramidates are another type of biodegradable gene carriers, which are frequently synthesized by the ring-opening polymerization of 4-methyl-2-oxo-2-hydro-1,3,2-dioxaphospholane followed by chlorination, then substitution with either an alcohol or amine (Penczek and Pretula 1993). Their degradability is based on the hydrolytic cleavage of phosphoester bonds at physiological pH (Wang et al. 2001).

Poly(2-aminoethylpropylene phosphate) (PPE-EA, Fig. 4d) is a water soluble and biodegradable polyphosphoester that can be hydrolyzed into propylene glycol, phosphate and ethanolamine in physiological condition. Following intramuscular injection to mice, PPE-EA showed better tissue compatibility and significantly lower cytotoxicity compared to PEI25 and PLL. Stable polyplexes could be formed by PPE-EA and pDNA, while intact DNA could be released by incubation in PBS, due to the degradation of PPE-EA. In the presence of chloroquine, PPE-EA showed much higher transfection efficiency than PLL (Wang et al. 2001). Intramuscular injection of polyplexes derived from PPE-EA and p43-LacZ resulted in enhanced β-galactosidase expression in anterior tibialis muscle of Balb/c mice, as compared with naked DNA injections (Wang et al. 2002c).

Polyphosphoramidate (PPA) bearing a spermidine side chain (PPA-SP, Fig. 4e) has also been synthesized by Mao and coworkers (Wang et al. 2002b). PPA-SP could be efficiently condense pDNA and showed lower cytotoxicity than PLL and PEI25 in COS-7 cells. In the presence of chloroquine, PPA-SP mediated transgene expression was greatly enhanced, and the optimized PPA-SP/DNA polyplexes yielded gene expression levels close to PEI25/DNA complexes. To establish the structure-activity relationship, PPAs with an identical backbone, same spacer group, similar molecular weights but different charge groups containing primary to quaternary amino groups were synthesized. Consistent with the buffering capacity, PPA with ethylenediamine (PPA-EA) as charged groups gave the highest transfection efficiency in various cell lines, among all the PPAs. PPA-EA also mediated the highest transgene expression in the rat spinal cord following intrathecal injection, which was comparable to that of PEI25. These results indicated that PPA vectors with primary amino group side chains are more potent than those with secondary, tertiary or quaternary amino groups for both *in vitro* and *in vivo* gene delivery (Wang et al. 2004b).

3.5.5 Cationic Polyphosphazenes

Polyphosphazenes (PPPs) are an important family of hybrid inorganic-organic polymers with organic side groups attached onto inorganic backbone with alternative phosphorus–nitrogen connected with alternative single and double bonds. Linear polyphosphazenes were initially synthesized via ring-opening polymerization of hexachlorocyclotriphosphazene (Allcock and Kugel 1965). Hennink et al. firstly explored the use of cationic polyphosphazene polymers for gene delivery, and they synthesized PPPs with either 2-dimethylaminoethanol (DMAE) (PPP-DMAE) or 2-dimethylaminoethylamine (DMAEA) (PPP-DMAEA) as side groups (Fig. 4f) (Luten et al. 2003). These polymers exhibited slow degradation at pH 7.5 and 37°C with half-lives of 7 and 24 days for PPP-DMAE and PPP-DMAEA, respectively. *In vitro* transfection in COS-7 cells suggested that these polymers showed a transfection efficiency rivaling a well-known transfectant PDMAEMA. The toxicity of both PPP-DMAE and PPP-DMAEA was lower than PDMAEMA. Further study in tumor bearing mice suggested that PPP-DMAEA/DNA polyplexes could preferentially mediate gene expression in tumor tissue (de Wolf et al. 2005). By co-substitution of PPP with DMAEA and diaminobutane (BA), in combination with PEGylation using folate functionalized PEG, Hennink and coworkers also synthesized a polycationic vector with improved compatibility and targeting capability for *in vivo* gene delivery (Luten et al. 2008). Yang et al. have found that cationic PPP with both DMAEA and histidyl groups could effectively mediate DNA transfer, and the efficiency at a polymer/DNA ratio of 10:1 was much higher than that of PPP-DMAEA and PEI25 (Yang et al. 2008a). Also, imidazole and DMAEA co-substituted PPP showed less toxicity and superior transfection activity over PEI25 (Yang et al. 2008b). Tetra(L-lysine)-grafted poly(organophosphazene) with PEG as hydrophilic substituent has also been synthesized and its *in vitro* cytotoxicity was examined (Jun et al. 2007). However, no transfection results have been reported for this polymer.

3.6 *Dendrimers Based Vectors*

Dendrimers comprise a central core molecule which acts as the root from which a number of tree-like arms emanated in an ordered and symmetric fashion. Since their discovery in late 1970s, dendrimers have been widely used for drug delivery, gene therapy and diagnostics as well as biomedical imaging (Tomalia and Frechet 2002).

Polyamidoamine dendrimers (PAMAM) are probably the most frequently investigated dendrimers-based vectors for gene delivery (Fig. 5a). The use of PAMAM for gene transfection was originally reported by Haensler and Szoka (Haensler and Szoka 1993). The sixth-generation PAMAM was found to have the highest transfection. In addition, expression was not affected by lysomotrophic agents such as chloroquine and only mildly influenced by the presence of 10% serum in the medium. When a membrane-destabilizing peptide was covalently attached to

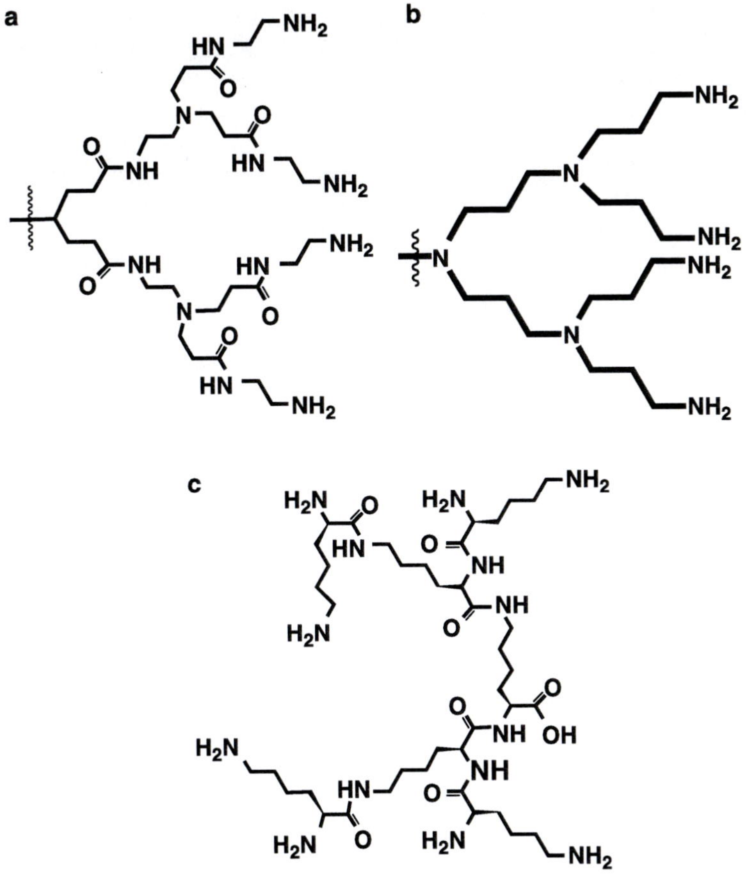

Fig. 5 Dendrimers based gene vectors: (**a**) polyamidoamine dendrimers (PAMAM), (**b**) poly(propylenimine) dendrimers, and (**c**) poly(L-lysine) dendrimers

PAMAM via a disulfide linkage, the transfection efficiency was significantly increased. *In vivo* delivery studies were also performed to substantiate the usefulness of PAMAM for gene therapy (Tanaka et al. 2000; Maruyama-Tabata et al. 2000; Rudolph et al. 2000). Whereas the good biocompatibility of PAMAM has been well demonstrated by these studies, the *in vivo* transfection efficiency of PAMAM is relatively low compared with the potent PEI25. Accordingly, efforts have been made to chemically modify PAMAM to achieve better transfer performances.

Uekama's group has systematically studied the effect of cyclodextrin modification on the transfection capacity of PAMAM dendrimers (Arima et al. 2001). All the PAMAM derivatives conjugated with α-, β-, or γ-CD showed potent luciferase gene expression. The highest transfection was observed for α-CD conjugated dendrimer (PAMAM-α-CD), which showed activity approximately 100-fold higher than that of dendrimer alone in NIH3T3 and RAW264.7 cells. This PAMAM derivative was also superior to Lipofectin. The enhanced transfection activity is considered to be resulted from the increased cellular association and the facilitated intracellular trafficking of pDNA. It was found that both PAMAM generation and the number α-CD conjugated per dendrimer affects transfection efficiency (Kihara et al. 2002, 2003). The structurally optimized dendrimer could delivery pDNA more efficiently in spleen, liver, and kidney, compared with the unmodified PAMAM (Kihara et al. 2003). PAMAM-α-CD was also effective for siRNA delivery, and it showed higher sequence-specific gene silencing effects without off-target effects that those of commercial reagents such as Lipofectamine™2000, TransFast™ and Lipofectin™ (Tsutsumi et al. 2006, 2007). Additionally, the modification of PAMAM via mannosylation with subsequent α-CD conjugation could give rise to dendrimer with improved transfer activity both *in vitro* and *in vivo* (Wada et al. 2005).

Other chemical modifications, such as internal quaternization (Lee et al. 2003), PEGylation (Kim et al. 2004a), peptide conjugation (Kim et al. 2007d) and hydrophobic tailoring (Kono et al. 2005), have been conducted to high efficient and less toxic delivery vectors based on PAMAM. Other type of dendrimer vectors such as poly(propylenimine) dendrimers (Fig. 5b) (Zinselmeyer et al. 2004), poly(L-lysine) dendrimers (Fig. 5c) (Kawano et al. 2004), phosphorus-containing dendrimers (Galliot et al. 1995), carbosilane dendrimers (Krska and Seyferth 1998) have also been developed for gene delivery applications. Detailed information can be found in recent review papers (Dufes et al. 2005; Mintzer and Simanek 2009).

3.7 *Natural Polymers*

Natural polymers including chitosan, dextran, alginate, collagen, and gelatin have been broadly studied in areas varying from pharmaceutics, gene therapy, tissue engineering, to biotechnology and bioengineering, mainly owing to their distinct advantages like the presence of multiple reactive sites, excellent biocompatibility and biodegradability.

3.7.1 Chitosan and Its Derivatives

Chitosan (CS), a linear, natural polysaccharide comprising β (1→4) linked glucosamine and N-acetyl-D-glucosamine (Fig. 6a), is obtained by the alkaline deacetylation of chitin that exists in the exoskeleton of crustaceans and insects. Due to its non-toxicity, low immunogenicity, and positive charges, CS is one of the most intensively studied non-viral naturally derived polymeric gene vectors. CS can effectively condense DNA and protect it from nuclease degradation. The transfection efficiency of CS-DNA polyplexes is mainly determined by the degree of deacetylation, molecular weight of CS, pH, serum concentration, charge ratio of

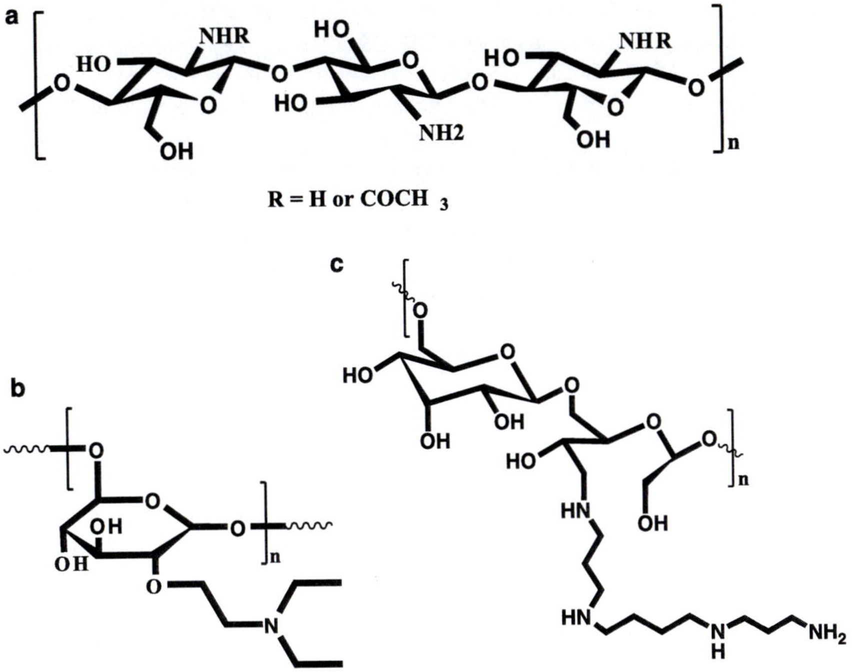

Fig. 6 Gene vectors based on natural polymers: (**a**) chitosan, (**b**) diethylaminoethyl-dextran (DEAE-dextran), and (**c**) spermine-conjugated dextran (DEX-SP)

amine (CS) to phosphate (DNA), and cell type (Ishii et al. 2001; Sato et al. 2001; Corsi et al. 2003).

The effect of CS molecular mass, plasmid concentration, N/P ratio, serum, and medium pH on the transfection efficiency of CS/pGL3-Luc(encoding luciferase) was systematically studied by Sato and coworkers. The optimized formulation by adjusting each of these factors showed the highest transfection in SOJ cells (Ishii et al. 2001). For different cell types, however, the optimized conditions were found to be different (Sato et al. 2001). Based on the study using mesenchymal stem cells, MG63 and HEK293, Corsi et al. have found that the transfection activity of CS-DNA polyplexes is cell type dependent (Corsi et al. 2003).

To improve transfection efficiency, numerous chemical modifications on CS have been carried out, which include quaternization, ligand conjugation, and hydrophobic modification. By methylation using methyl iodide in an alkaline environment, trimethylated CS oligomers (TM-CS) were prepared (Thanou et al. 2002). The quaternized chitosan oligomers were able to effectively condense DNA to form nano-polyplexes (200–500 nm). *In vitro* transfection in COS-1 and Caco-2 cells revealed that the quaternized derivatives were superior to unmodified CS. In the transfection of Caco-2 cells, TM-CS was more effective than DOTAP. The presence of fetal calf serum did not affect the transfection efficiency of the DNA/TM-CS polyplexes, whereas the transfection efficiency of DNA/DOTAP complexes was decreased. Conjugating tetragalactose to TM-CS conferred the vector with the ability of specific gene delivery hepatocytes, taking advantage of the receptor-mediated endocytosis (Murata et al. 1997). To further increase the transfection activity of TM-CS, Rojanarata et al. have synthesized a quaternized N-(4-N,N-dimethylaminobenzyl) chitosan (TM-Bz-CS), in which hydrophobic N,N-dimethylaminobenzyl was introduced. Transfection of pDNA encoding GFP in human hepatoma cell lines (Huh7 cells) showed the higher efficiency in the case of TM-Bz-CS formulated polyplexes, in comparison to those of TM-CS and CS. The improved gene transfection was due to the promoted interaction by hydrophobic N,N-dimethylaminobenzyl, and the increased water solubility by N-quaternization (Rojanarata et al. 2008). Unfortunately, quaternization of CS may increase its cytotoxicity, to some extent. This higher toxicity of TM-CS could be reduced by PEGylation (Germershaus et al. 2008).

CS has been modified by various ligands that enable specific binding to the receptors of the cells and enhance the transfection efficiency. Galactose, mannose, transferrin, and folate are frequently used ligands for modification. Kim et al. conjugated lactobionic acid bearing galactose group to CS (GA-CS) for hepatocytes specificity (Kim et al. 2004b). Cytotoxicity study in HepG2 and HeLa confirmed the extremely low cytotoxicity of GA-CS. The transfection efficiency of GA-CS/DNA complexes in HepG2, which has asialoglycoprotein receptors, was higher than that in HeLa without the receptors. This ligand-mediated increase in the transfection activity of modified CS in target cells has been also observed by other researchers (Hashimoto et al. 2006). Mannose ligand was covalently linked to CS to synthesize vectors that can target antigen presenting cells (APCs) via receptor-mediated endocytosis (Kim et al. 2006c). In addition, transferrin and folated were conjugated to CS for targeting tumor cells such as HEK293, HeLa and MCF-7 cells (Mao et al. 2001; Lee et al. 2006).

To improve the buffering capacity and interactions with cell surfaces, and reduce the aggregation of CS based polyplexes, CS has been modified with various hydrophobic moieties including deoxycholic acid, stearic acid, and alkyl chains. Lee et al. synthesized deoxycholic acid modified CS (DA-CS) by an EDC-mediated coupling reaction (Lee et al. 1998). DA-CS could self-assemble into nanoparticles of 162 nm, which could efficiently mediate DNA transfection in COS-1 cells. Precise control over the size and structure of DA-CS nano-assemblies could be implemented by depolymerizing CS with subsequent DA modification (Kim et al. 2001b). The transfection efficiency of these nano-structured assemblies in COS-1 cells was significantly influenced by their size and structure. Compared with the unmodified CS, the transfection activity of the stearic acid and alkyl modified CS was strongly enhanced (Hu et al. 2006; Liu et al. 2003).

3.7.2 Dextran Derivatives

Due to its well-demonstrated safety *in vivo*, chemically modified dextrans have been broadly investigated for drug and gene delivery. For instance, diethylaminoethyl-dextran (DEAE-dextran) has been a commonly used gene vector for decades

(Fig. 6b) (Rigby 1969). Domb's group has developed a series of oligoamines conjugated dextrans for gene delivery recently, which were synthesized by oxidizing dextran with periodate followed by reductive amination with various oligoamines (Azzam et al. 2002). In this polycation library, spermine-conjugated dextran (DEX-SP) is the most effective vector (Fig. 6c). DEX-SP showed the transfection efficiency as higher as DOTAP-based lipoplexes. High local gene expression was observed post-intramuscular injection of DEX-SP/pDNA polyplexes, while no expression was observed after intravenous injection of the same nano-formulation (Hosseinkhani et al. 2004). PEGylation of DEX-SP led to a polymer with high transfection yield in serum-rich medium. Furthermore, this dextran derivation can high efficiently mediate the gene expression in the liver of mice post-intravenous injection. Targeting studies suggested that the preferential binding to galactose receptor of liver parenchymal cells rather than the mannose receptor of liver non-parenchymal cells was responsible for the liver specific gene expression mediated by PEGylated DEX-SP. In addition, the gene transfer capability of DEX-SP *in vitro* and *in vivo* could be further enhanced by chemical modification using hydrophobic moieties such as oleic acid, fatty acid, and cholesterol (Azzam et al. 2004; Eliyahu et al. 2005).

3.7.3 Vectors Derived from Proteins

Natural proteins such as collagen and gelatin as well as their derivatives have been extensively studied for gene delivery applications. Methylated collagen could mediate a gene transfection significantly higher than that of native collagen in HEK 293 cells (Wang et al. 2004c). The transfection efficiency of this formulation was dramatically affected by the collagen/pDNA weight ratio. Different from *in vitro* result, the complexes based on native collagen showed a higher transfection level than that of methylated collagen. Gelatin and its derivatives have also been demonstrated to be effective vehicles for gene delivery. High level of gene transfection in human tracheal epithelial cells was achieved using gelatin-DNA nanoparticles (Truong-Le et al. 1999). Tabata and coworkers synthesized cationized gelatin, which could effectively deliver a pDNA (expressing siRNA for TGF-β type II receptor) to the interstitium of renal tissue, after the polyplexes were injected to the left kidney of mice via the ureter (Kushibiki et al. 2005a, b).

3.8 Biodegradable Hydrophobic Polymers

In addition to polycations, biodegradable polymers that are essentially hydrophobic can also serve as carriers for gene therapy. Due to their degradation profile at a long time scale, these polymers have been extensively employed for sustained or local delivery of genetic payload. Polyesters, such as polylactide (PLA), poly(lactide-co-glycolide) (PLGA), and poly(ortho esters) are among the most prominent materials examined for these purposes.

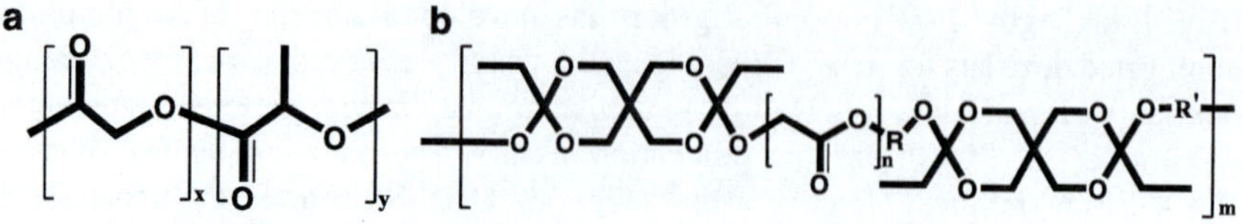

Fig. 7 Biodegradable hydrophobic polymers used for gene delivery: (**a**) poly(lactide -co-glycolide) (PLGA), and (**b**) poly(ortho esters)

3.8.1 Poly(Lactide-co-Glycolide)

PLGA has been approved by FDA for drug delivery and other biomedical applications. DNA-loaded microparticles have been extensively investigated for the induction of immune responses (Fig. 7a) (Jilek et al. 2005). The degradation half-life of PLGA can be, on one hand, modulated by its composition, while on the other regulated by molecular weight. DNA can be encapsulated into microparticles via spray-drying, w/o/w (water-in-oil-in-water) double emulsion solvent evaporation or coacervation technique. Alternatively, DNA can be adsorbed on cationic microparticles, which can be easily fabricated by introducing cationic molecules during microparticle formulation. Study by Luby et al. showed that immunization of HLA-A2/Kb transgenic mice with PLGA microparticles containing pDNA encoding human cytochrome P450 CYP1B1 (CYP1B1) elicited durable HLA-A2-restricted T cell responses that recognize naturally processed CYP1B1 peptide produced in cells endogenously expressing CYP1B1 (Luby et al. 2004). In addition, this formulation was found to be safe and well tolerated even following repeated administrations. However, hydrolysis of PLA or PLGA may lead to a low pH within microparticles that can result in DNA degradation (Fu et al. 2000). Incomplete release of DNA from PLGA microparticles is often observed (Wang et al. 1999). In addition, the release rate sometimes may be too slow to prompt a robust immune response (Kaech and Ahmed 2001).

3.8.2 Poly(Ortho Esters)

Poly(ortho esters) (POEs) are another types of biodegradable polyesters with well-controlled degradability and excellent biocompatibility (Fig. 7b). POEs were originally developed as biomaterials in the 1970s by Heller and colleagues (Heller et al. 2002). They are stable at pH >7, while degrade rapidly at reduced pH. These characteristics can be utilized to temporally and spatially control the release of DNA. To overcome the drawbacks of the polyester-based DNA delivery systems, Heller and Langer et al. have molecularly engineered POE microspheres that are non-toxic to cells (Wang et al. 2004a). These particles could protect the payload DNA from degradation, and release DNA rapidly in response to phagosomal pH post-uptake by APCc. The optimized formulation could elicit distinct primary and secondary humoral and cellular immune responses in mice, and suppress the growth of tumour cells bearing a model antigen.

3.9 *Integrating Stimuli-Responsiveness to Gene Vectors*

In biological systems, each step during gene expression is elegantly and precisely controlled by molecular machinery. Viral vectors can highly efficiently transfect host cells using their delicately assembled complex structures that have been evolved for thousands of years. These natural systems have kindled the scientist's enthusiasm to develop smart carriers by mimicking their excellent performances. To this end, stimuli-responsive vectors have been designed, and their advantages for gene transfection have been explored. Despite a small step towards the final goal, we do witness some interesting results. Responsive non-viral vectors that can regulate some structural factors and/or physiological properties in response to extracellular signals (such as pH, redox, ultrasound, light, temperature, etc.) have already showed their potential to reach site-, timing-, and duration period-specific gene expression (Dincer et al. 2005; Nagasaki and Shinkai 2007). Some stimuli-responsive moieties such as functional linkages and compounds that are employed for constructing intelligent gene vectors are briefly reviewed in the following sections.

3.9.1 pH-Sensitive Linkages

As well known, intracellular endosomes/lysosomes possess a significantly low pH, which may be considered as a useful chemical stimulus for designing environmentally sensitive gene carriers, since polyplexes taken up by the cells via endocytosis are finally localized in the endosomes and/or lysosomes. Varieties of pH-sensitive carriers have been created for gene delivery applications.

Orthoester is a functional group, in which three alkoxy groups are attached to one carbon atom. Whereas it is stable at pH > 7, orthoester can be readily hydrolyzed in mild acid conditions. As a good example, POEs has been exploited for enhanced delivery of DNA vaccines by incorporating DNA into POE microspheres (Wang et al. 2004a).

Acetal is another acid-labile group that has been broadly used for pH-responsive drug delivery (Sy et al. 2008; Bachelder et al. 2008). Murthy et al. have incorporated the acetal linkers into polymer vectors (Murthy et al. 2003). These terpolymers comprise a hydrophobic, membrane-disruptive poly(propylacrylic acid) backbone, onto which hydrophilic PEG chains have been grafted through acetal linkages. Fluorescence microscopy experiments showed that these polymers could direct endosomal escape and efficiently deliver ODNs into the cytoplasm of hepatocytes. Similar to acetal, ketal group is also acid-labile, which has been incorporated to lPEI most recently. Polyplexes based on ketalized PEI showed much higher RNA interference efficiency than those originated from unmodified PEI (Shim and Kwon 2009, 2008).

As a pH-sensitive linker, hydrazone bond is utilized to engineer drug delivery vehicles with spatially controlled release profile by pH (Bae et al. 2003). Pyridylhydrazone has been recently used by Walker et al. to conjugate PLL with PEG (Walker et al. 2005). Polyplexes formed using this polymer with pH-labile hydrazone linkage showed 100-fold higher gene expression *in vitro*, and 10 times higher gene expression in a mouse model, when compared to the stably shielded control polyplexes. For the PEI polyplexes, their peripheral decoration using amine-reactive pyridylhydrazone-based PEGs also significantly increased transaction efficiency in both *in vitro* and *in vivo* models, as compared to stably shielded polyplexes (Fella et al. 2008).

Due to its acid-labile character, imine linker, also called as Schiff base, has been used to cross-link low molecular weight PEI (1,800 Da) using glutadialdehyde (Kim et al. 2005b). *In vitro* transfection in 293 T cells and A7R5 cells showed that the transfection efficiency of the acid-labile polycations was comparable to that of PEI25. In addition, the acid-labile PEI was much less toxic than PEI25, due to the degradation of acid-labile linkage.

Alternatively, pH-dependent charge reversal moieties can be used to modulate the gene delivery performance of synthetic vectors. Maleic amide derivatives, like cis-aconitic and citraconic amide have negative charges at neutral pH, while they may degrade promptly at mildly acidic pH 5.5 to expose positively charged amines. Materials bearing these functional moieties are called masked endosomolytic agents (Wolff and Rozema 2008). Namely, when they enter the acidic environment of endosomes or lysosomes, their pH-labile bonds are broken, releasing the agent's endosomolytic capability. Polymer vectors with cis-aconitic or citraconic amide can significantly improve the transfection activity of conventional polyplexes both *in vitro* and *in vivo* as evidenced by Wolff's and Kataoka's groups (Rozema et al. 2007; Lee et al. 2008).

3.9.2 Redox-Responsive Disulfide

It is widely accepted that the cytoplasm is a highly reducing environment, where the glutathione concentration is about 50–1,000 times higher than that in the extracellular milieu. Disulfide can be effectively reduced in the intracellular environment, and therefore it can function as a responsive linkage to create redox-triggerable delivery systems. There have been considerable studies concentrating on intracellular delivery of nucleic acids based on the reductive cleavage of disulfide bonds. As aforementioned, cross-linking of LMW-PEI has been implemented using various compounds bearing disulfide. Once inside the cell, disulfide linkages will be reduced, resulting in the cleavage of disulfide-containing high molecular weight PEI into low molecular weight fragments which are less toxic. In addition, PEG and polycations can be coupled with disulfide linkage to formulate polyplexes with enhanced colloidal stability and prolonged circulation, while PEG chains will be detached after endocytosis to facilitate intracellular trafficking.

Kataoka's group have synthesized a block catiomer (PEG-SS-P[Asp(DET)]), in which PEG was linked with PAsp(DET) (Takae et al. 2008). *In vitro* transfection study showed that PEG-SS-P[Asp(DET)] based polyplex micelles displayed 1–3 orders of magnitude higher transfection efficiency than those based on a structurally similar polymer without disulfide linkages. A more rapid onset of gene expression was also observed for the former. This was attributed to much more effective endosomal escape based on the PEG detachment in endosome. The same research group has recently developed the polyplex micelle with a disulfide cross-linked core, which is characterized by the efficient release of the loaded siRNA responding to intracellular reductive conditions. As a result, this type of vector achieved 100-fold higher siRNA transfection efficacy compared with non-cross-linked polyplexes from PEG-*b*-PLL, which were less stable at physiological ionic strength (Matsumoto et al. 2009).

3.9.3 Temperature-Responsive Vectors

Temperature is one of the safest and most controllable external stimuli. Hyperthermia therapy via microwave has long been adopted for tumor therapy (Needham and Dewhirst 2001). PNIPAm is the most studied synthetic responsive polymer, which undergoes a sharp coil–globule transition in water at about 32°C, changing from a hydrophilic state to a hydrophobic state (Schild 1992). So far, most of the temperature-responsive vectors are based on PNIPAm-containing materials (Yokoyama 2002). For example Hinrichs et al. have synthesized thermosensitive copolymers of NIPAm (temperature-sensitive monomer) and DMAEMA (positively charged unit) and investigated the transfection efficiency in ovarian cancer cells (Hinrichs et al. 1999). The temperature sensitivity was found to be dependent on the polymer/pDNA ratio. However, no direct correlation between transfection efficiency and polymer LCST was found. By copolymerizing NIPAm and DMAEMA with hydrophobic butylmethacrylate, Kurisawa et al. developed a terpolymeric gene carrier that displayed temperature-controlled gene expression in COS1 cells *in vitro* (Kurisawa et al. 2000; Yokoyama et al. 2001). Other PNIPAm-bearing polycations such as PEI have also been synthesized to study the effect of temperature on the physicochemical characteristics of polyplexes and transfection activity (Twaites et al. 2004).

3.9.4 Photoresponsive Delivery Vectors

Photo-irradiation has been used to enhance the transfection efficiency by improving both the endosomal escape of exogenous DNA and the DNA release from complexes, since light is an easily controllable stimulus.

Høgest and colleagues have proposed photo-chemical internalization (PCI) for drug and gene delivery (Høgset et al. 2004). PCI is based on photochemical reactions initiated by a photosensitizer localized in endocytic vesicles such as endosomes and lysosomes, inducing the rupture of these vesicles upon light illumination. Photosensitizers, such as porphyrins and phthalocyanines, are compounds that make cells extraordinary sensitive to light irradiation. Photochemical transfection using a photosensitizer, aluminum phthalocyanine, PLL/DNA polyplexes showed a 20-fold higher transfection efficiency of GFP in human melanoma cell line upon light treatment, reaching transfection levels of about 50% of the surviving cells

(Høgset et al. 2000). The efficacy of PEI-based polyplexes can also be enhanced by PCI mediated transfection. Successful PCI-mediated gene delivery *in vivo* has been achieved by Kataoka and coworkers. To this purpose, they developed a ternary complex composed of a core polyplex containing DNA condensed with cationic peptides and enveloped in the anionic dendrimer phthalocyanine that served as photosensitizer. This ternary complex showed more than 100-fold photochemical increase in transgene expression *in vitro* with reduced photocytotoxicity. In a rat model, subconjuctival injection of the ternary complex followed by laser irradiation led to transgene expression only in the laser-irradiated site.

In addition to PCI mediated transfection, photoresponsive gene delivery can also be implemented by directly incorporating photosensitive unit into the non-viral vector. Nagasaki et al. synthesized a novel cationic lysine-modified polyazobenzene dendrimer (Lys-G2) (Nagasaki et al. 2000). Polyplexes formed by Lys-G2/pDNA showed a 50% increase in the transfection efficiency in COS1 cells upon UV irradiation, as compared with the case without UV irradiation. This was attributed to the decreased cationic density on the surface of the complex to suppress the cationic repulsion, resulting from *trans*-to-*cis* photo-isomerization.

3.9.5 Protein Kinase Responsive Vectors

Katayama et al. have developed a novel gene regulation system responding to specifically and abnormally activated intracellular enzymes in diseased cells (Katayama et al. 2002; Kawamura et al. 2005; Oishi et al. 2006a). For this purpose, a cationic polymer grafted with an oligopeptide that is a substrate for cyclic AMP-dependent protein kinase (PKA) and could regulate gene-expression in a cell-free system. PAK/DNA polyplexes showed no expression of the loaded gene in the unstimulated NIH 3T3 cells. This suggested that PAK formed a stable complex with DNA in the normal cells to totally suppress gene transfection. In contrast, significant transgene expression was observed when the PAK/DNA polyplexes were incubated with forskolin-treated cells. In this case, the activated PKA resulted in the disassociation of the polyplexes even in living cells, and therefore an efficient gene expression could be achieved.

4 Lipids as Gene Vectors

Since the landmark report by Felgner and coworkers (Felgner et al. 1987), cationic lipids or lipid-like molecules have been extensively studied for gene transfection in cultured cells, in animals and in patients enrolled in clinical trials (Dass 2004). Most of the commonly used cationic lipids are commercially available, technically simple and easy to formulate. Compared with viral vectors, lipids based vectors are less biologically hazardous. Table 8 shows the structures of cationic lipids frequently used for gene delivery.

Table 8 Various lipids employed for gene delivery

A. Neutral lipids

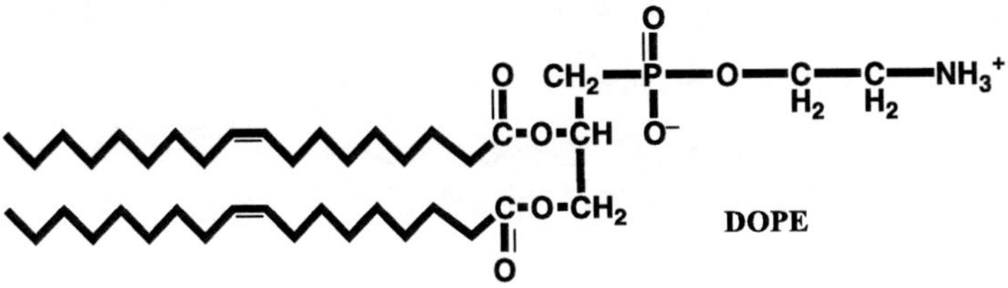

DOPE

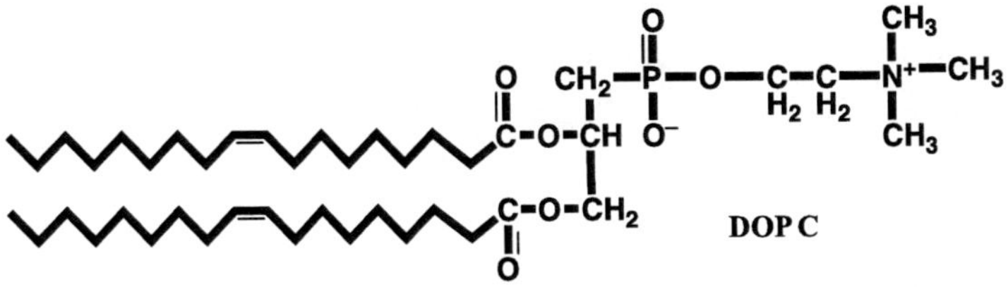

DOPC

F-PE

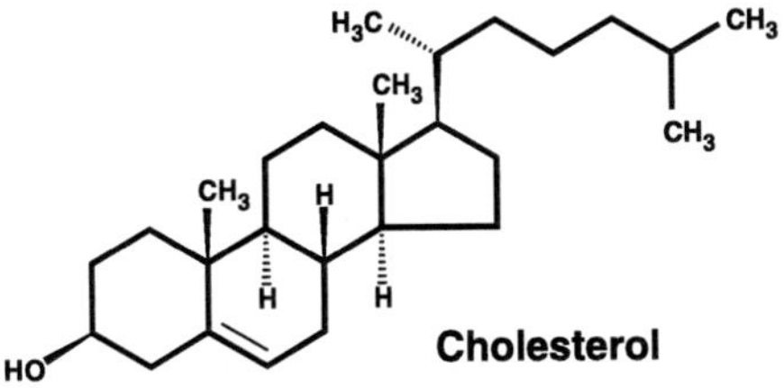

Cholesterol

B. Lipids with monovalent headgroup

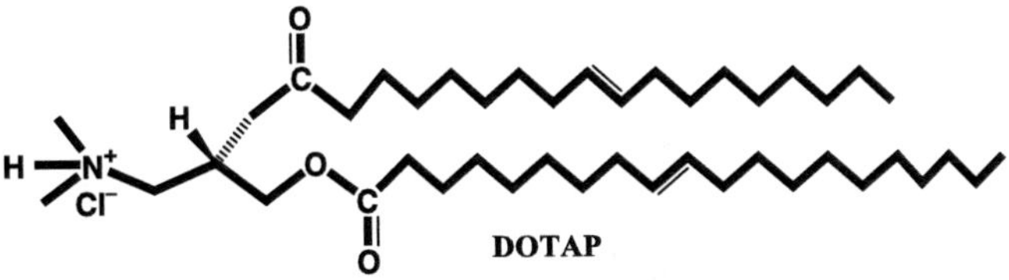

DOTAP

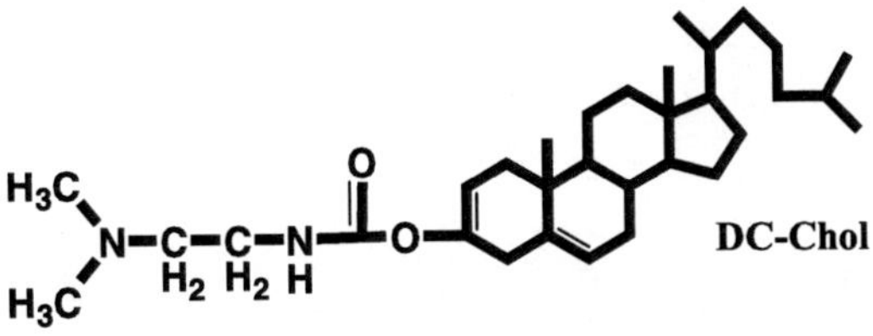

DC-Chol

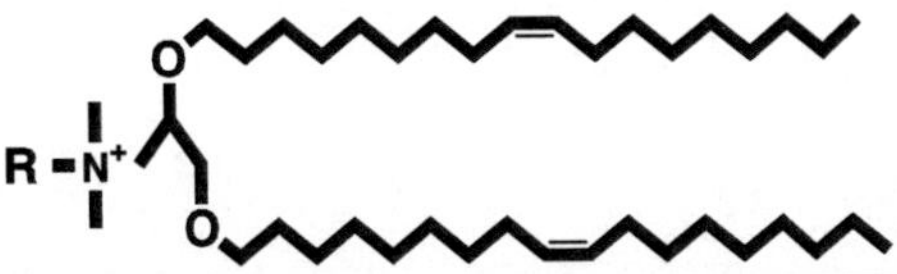

DOTMA	R = CH_3
DORIM	$HO(CH_2)_2$
DORIM-HP	$HO(CH_2)_3$
DORIM-HB	$HO(CH_2)_4$
DORIE-HPe	$HO(CH_2)_5$

(continued)

Table 8 (continued)

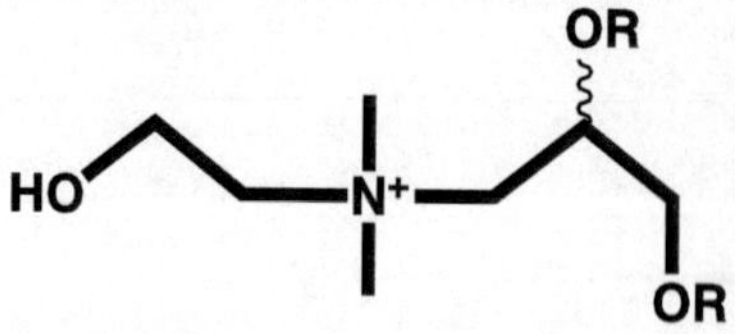

DORI	R = CO$(CH_2)_7$[z]CH=CH$(CH_2)_7CH_3$
DORIM	(CH_2)[z]CH=CH$(CH_2)_7CH_3$
DMRIM	$(CH_2)_{13}CH_3$
DPRIE	$(CH_2)_{15}CH_3$
DSRIE	$(CH_2)_{17}CH_3$

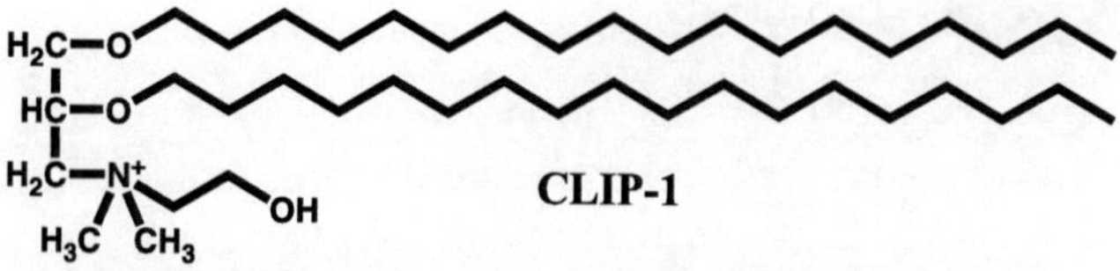

CLIP-1

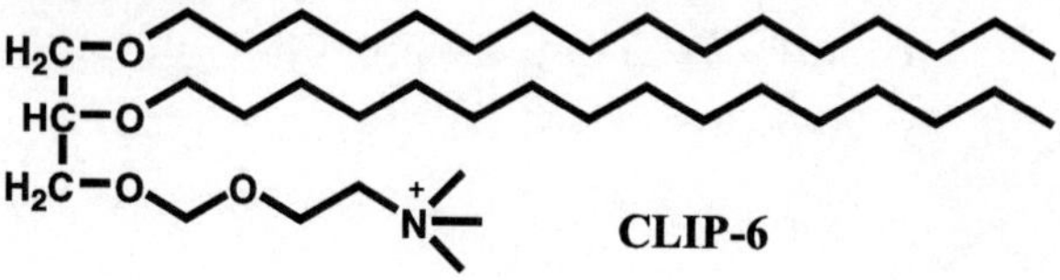

CLIP-6

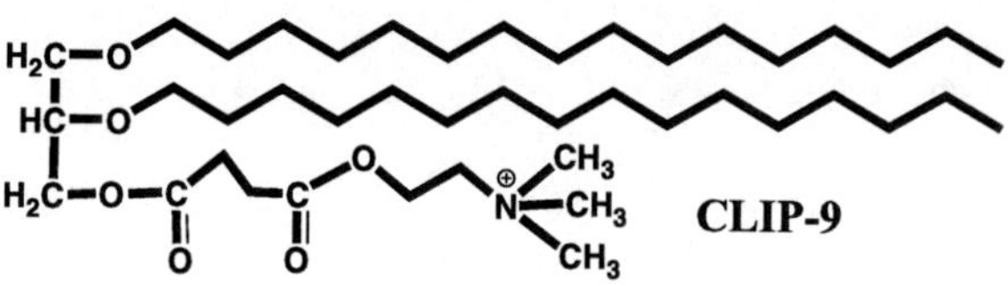

CLIP-9

C. Lipids with multivalent headgroups

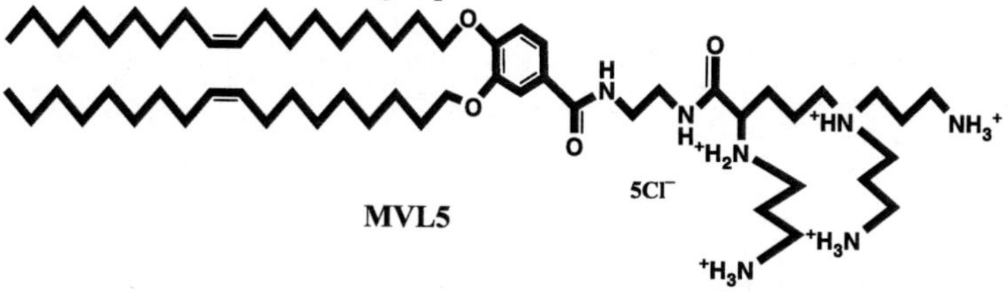

MVL5

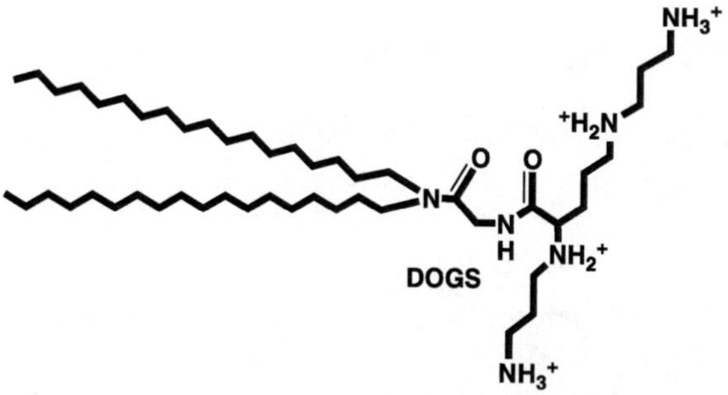

DOGS

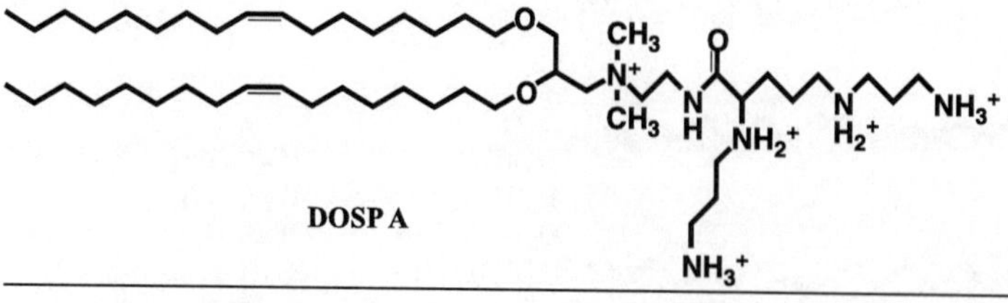

DOSP A

(continued)

Table 8 (continued)

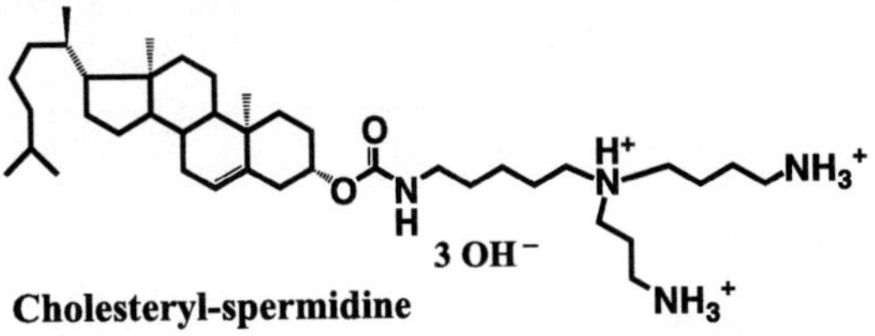

Cholesteryl-spermidine

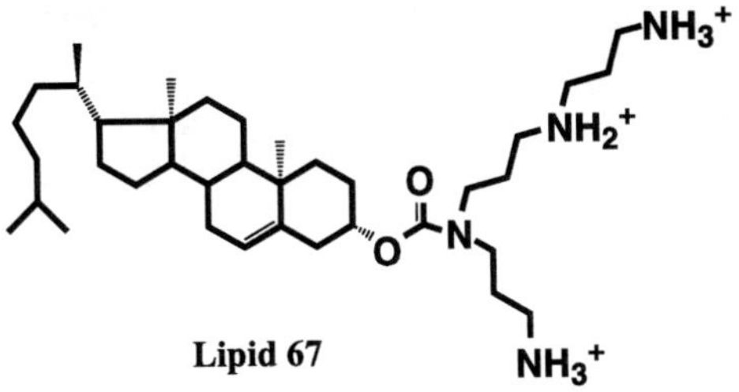

Lipid 67

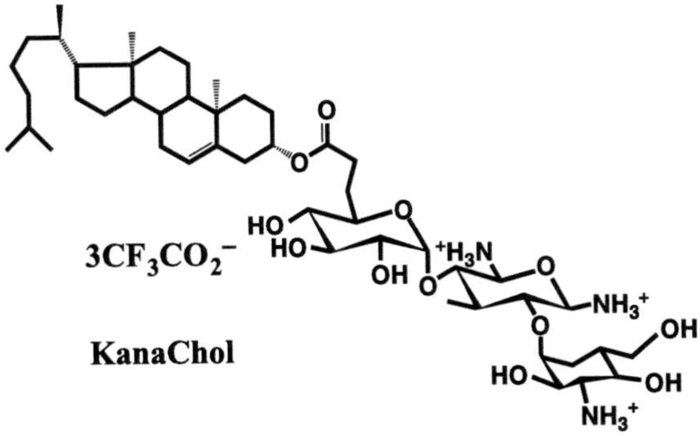

KanaChol

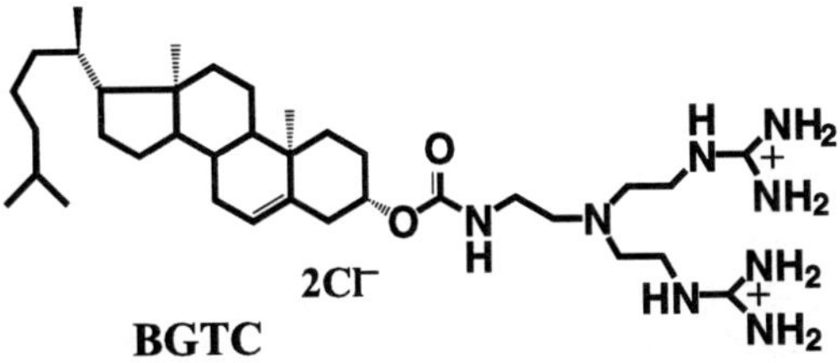

BGTC

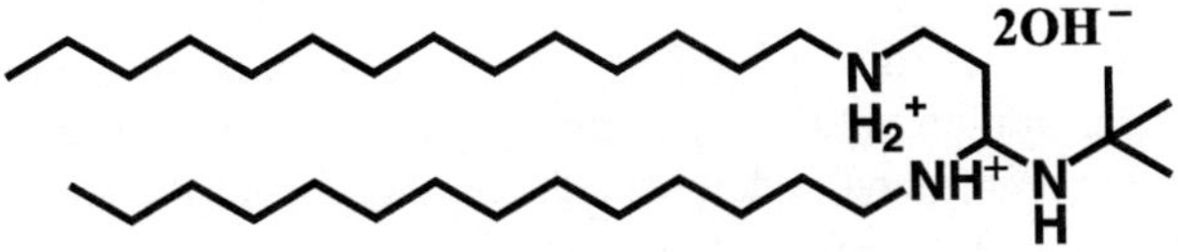

Vectamidine

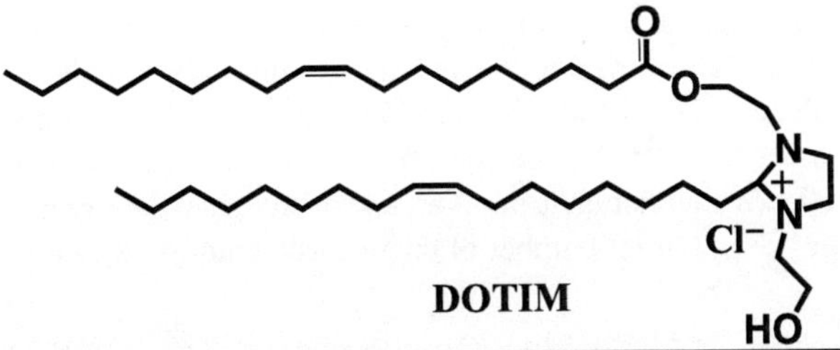

DOTIM

(continued)

Table 8 (continued)

D. Disulfide containing lipids

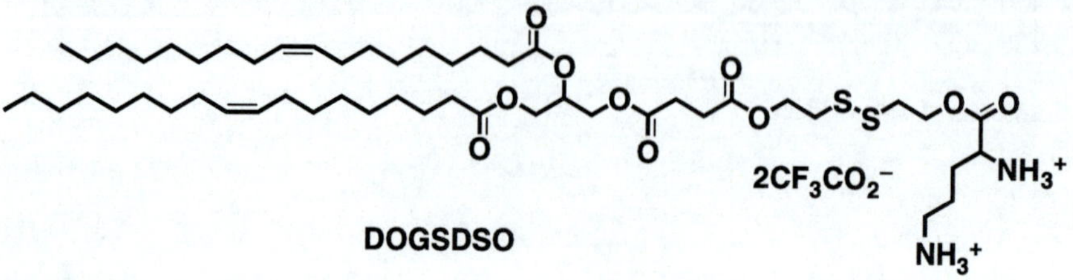

DOGSDSO

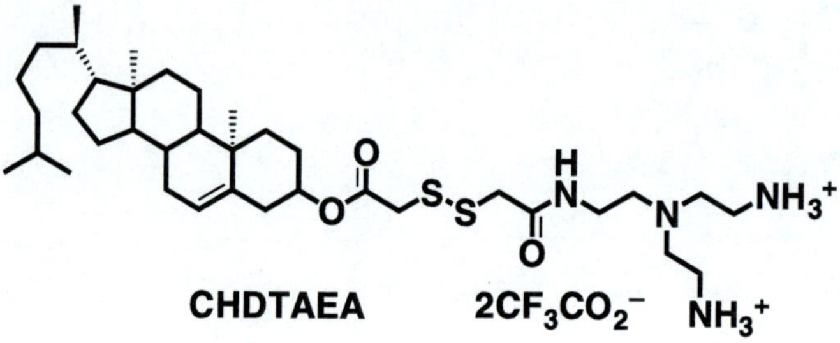

CHDTAEA

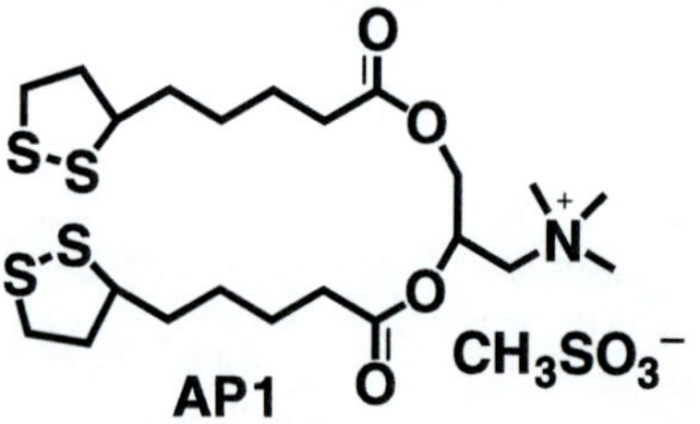

AP1

4.1 *Gene Delivery Via Vationic Liposomes*

Cationic lipids are generally formulated into cationic liposomes (CL) that can electrostatically and hydrophobically interact with anionic nucleic acids like pDNA and siRNA to form lipoplexes (Wong et al. 1996). Factors such as the lipid structure, the lipid composition of liposomes, the preparation procedure of liposomes, the kinetics of mixing for the preparation of lipoplexes, the lipid/nucleic acid ratio, and the size of nucleic acid, have been found to dominate the physicochemical characters of the resulting lipoplexes (Dass and Su 2000), which in turn may affect the transfection efficiency. Among them, the chemical structure of lipids is one of the most important factors that strongly affect both transfection activity and cytotoxicity of lipoplexes. Extensive studies on triester phosphatidylcholines (PCs) indicate a strong, systematic dependence of their transfection efficiency on the lipid hydrocarbon chain. It has been shown that transfection activity increases with the increase in chain unsaturation from 0 to 2 double bonds per lipid and decreases with increase of chain length in the range ~30–50 total number of chain carbon atoms (Koynova and Tenchov 2009). This hydrocarbon chain effect of cationic lipids is considered to be related to the lipid phase properties as well as their interactions with membrane lipids. A simple and direct correlation between phase structure and transfection activity has been proposed by some researchers. For instance, study by Zuhorn et al. suggested that the formation of hexagonal phase in lipoplexes is important for the cytoplasmic translocation of ODNs (Zuhorn et al. 2005). On the other hand, Caracciolo et al. have demonstrated that DOTAP/DOPE/cholesterol liposomes with a multilayer structure exhibit a more than 4-fold higher transfection activity in OVCAR-3 and SK-OV-3 cells, compared with DC-Chol/DOPE liposomes with a columnar inverted hexagonal phase (HII) (Caracciolo et al. 2003). Besides cationic lipids, neutral lipids, such as dioleoylphosphatidylethanolamine (DOPE) and dioleoyl phosphatidyl choline (DOPC) are required to stabilize the CL suspension, to facilitate membrane fusion and endosomal escape (Felgner et al. 1994; Zuidam and Barenholz 1998; Hafez et al. 2001). More often, the transfection activity of lipoplexes may be impaired if they are formulated without neutral lipid(s) (Felgner et al. 1994; Mui et al. 2000). It should be noted that the interactions of lipoplexes with cellular membrane may transform the original structure, rendering a new structural organization. Consequently, the transgene efficacy of lipoplexes depends not only on the formulations of CLs and their structure, but also on their interactions with cell membrane and the resulting structures (de Ilarduya et al. 2010).

Strategies employed to improve the lipoplexes-based gene delivery *in vivo* include enhancing the stability and circulation time in bloodstream, targeting to specific tissues and cells, and facilitating intracellular trafficking. To prolong the circulation time of lipoplexes, PEG-conjugated lipids can be incorporated into the CL (Fenske et al. 2002). Additionally, PEGylation of CLs may reduce the toxicity and stabilize lipoplexes *in vivo*. However, PEG-lipid substantially decreases the intracellular uptake of lipoplexes and impedes intracellular trafficking to some degree, and therefore reduces gene transfection efficiency. To partly circumvent this problem, Szoka's group synthesized an acid-labile PEG-diorthoester-lipid conjugate (POD) (Guo and Szoka 2001), in which the diorthoester linkage may be rapidly cleaved as pH decreases from 7.0 to 5.0 (pH value in the endosome). They also demonstrated that the lipoplexes with pH-sensitivity displayed better *in vitro* transfection activity than the pH-insensitive counterparts (Choi et al. 2003). Alternatively, novel lipids bearing acid-labile linkages, such as vinyl ether (Boomer et al. 2002), ketal (Zhu et al. 2002), ortho ester (Zhu et al. 2000), or acylhydrazone (Mueller et al. 1990), may be employed to construct pH-sensitive lipoplexes with improve gene transfer performance. Redox potential-sensitive lipids are another family of triggerable vectors, and disulfide is the most frequently used redox-sensitive linker. Table 8 shows the representative lipids with disulfide linkage. For example, lipoplexes based on disulfide-containing lipid DOGSDSO showed 6–15-fold or 50-fold transfection efficiency compared with those based on DOTAP or a non-cleavable analogue, respectively (Tang and Hughes 1998).

A common problem for the cationic lipids-based vectors is their toxicity, particularly in cultured cells (Dass and Choong 2006). The lipoplexes related toxicity has also been observed in animal studies and even in human patients at higher doses

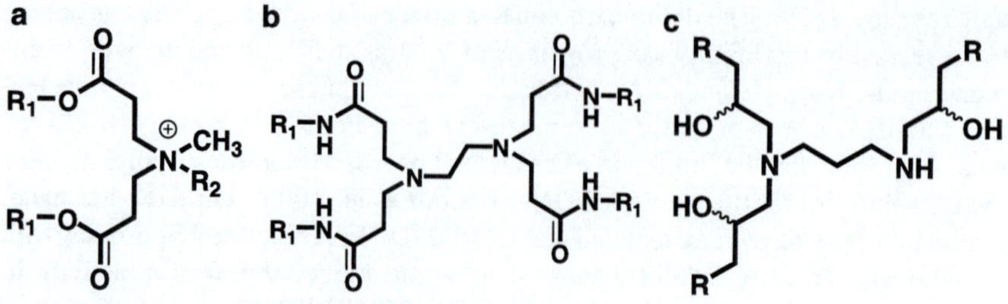

Fig. 8 The chemical structures of lipidoids

(Dass 2002; Tseng et al. 2009; Lv et al. 2006). In addition, lipoplexes exhibit an immunostimulation effect, a phenomenon that may either be harmful or beneficial. Besides toxicity, low transfection activity, limited targetability, and non-specific interactions with the blood components as well as with extracellular matrix are the major limitations of vectors based on lipoplexes (Dass 2004). Thus, efforts have been made to develop high effective and low toxic lipids.

4.2 *Lipid-Like Molecules for Gene Therapy*

More recently, Anderson et al. have developed a combinatorial library of lipid-like materials, termed lipidoids, simply through the conjugate-addition of amines to an acrylate or acrylamide (Fig. 8a, b) (Akinc et al. 2008). Over 1,200 structurally diverse lipidoids were synthesized. From this library, the authors identified lipidoids that facilitate high levels of specific silencing of endogenous gene transcripts when complexed with either siRNA or single-stranded antisense 2′-O-methyl (2′-OMe) oligoribonucleotides targeting microRNA (mRNA). The safety and efficacy of lipidoids were examined in three animal models including mice, rats and nonhuman primates. Lipidoid-based formulations containing either siRNA or anti-miR showed potent, specific and durable effects on gene expression *in vivo* in liver, lung and peritoneal macrophages. The same group has also synthesized epoxide-derived lipidoid library by efficient ring-opening reaction between epoxides and amines (Fig. 8c) (Love et al. 2010). From this library, an effective formulation has been identified that enables siRNA-directed liver gene silencing in mice at doses below 0.01 mg/kg. This formulation was also shown to specifically inhibit expression of five hepatic genes simultaneously, after a single injection. Furthermore, the potential of this formulation was validated in nonhuman primates, where high level of knockdown of the clinically relevant gene transthyretin was observed at doses as low as 0.03 mg/kg. These promising results imply that these lipid-like vectors may be developed to clinically usefully formulations for gene therapy. In addition, the common structural features of these materials may provide certain design criteria for creating high efficient transfection agents.

5 Nanomaterials as Non-viral Vectors

Nanomaterials, such as gold nanoparticles, magnetite, quantum dots, and layered double hydroxides, carbon nanotubes, and nanostructures of calcium phosphate and silica, are emerging as novel delivery carriers for a large number of therapeutics varying from small molecules drugs, proteins, to nucleic acids (Bourgeat-Lami 2002; Chowdhury and Akaike 2005; Fukumori and Ichikawa 2006). In addition to their advantages of easy production and purification, controlled shape and size, and good storage stability, these nanostructured materials have showed their special capability to break through some tissue barriers, which is unachievable by organic materials.

5.1 *Noble Metallic Nanoparticles*

Gold nanoparticles with typical size of 10–20 nm can be easily taken up by the cell (Fynan et al. 1993b; Sandhu et al. 2002; Thomas and Klibanov 2003; Jen et al. 2004). It has been recently reported by Schmid et al. that Au_{55} clusters could function as anticancer agents because of their effective interactions with DNA (Tsoli et al. 2005). The conventional method for preparing metallic nanoparticles is to reduce the corresponding metal salts in the presence of suitable protecting groups that can prevent further aggregation. For gold nanoparticles, their surface can be conveniently functionalized using citrate, amine, ODN, peptide, and antibody. Studies by Rotello and co-workers have demonstrated that 2 nm gold nanoparticles functionalized with a mixed monolayer containing alkanethilos and trimethylammonium thiols were able to bind with pDNAs and deliver them efficiently to 293T cells (Sandhu et al. 2002). These nanoconjugates showed ~8% fold higher efficiency than the commonly used PEI (60 kDa). The transfection efficiency was found to be affected by the ratio of positively charged quaternary amines to negatively charged phosphate groups on the DNA, as well as the length of alkyl chains. Furthermore, the same group has recently shown that gold nanoparticles functionalized with lysine moieties are highly efficacious for pDNA delivery, and they performed better than PLL vector by a factor of 28 (Ghosh et al. 2008). Klibanov et al. synthesized gold nanoparticles conjugated with thiol-modified PEI (2 kDa) and dodecyl-PEI (2 kDa) (Thomas and Klibanov 2003). These PEI-Au hybrid particles could deliver pDNA to COS-7 cells and show 15- and 6-fold higher transfection efficiency than PEI (2 kDa) and PEI25, respectively.

Feldheim et al. covalently conjugated various peptides to gold nanoparticles through coupling with bovine serum albumin (BSA) (Tkachenko et al. 2003, 2004). The intracellular fate of these gold-peptide nanoparticles was studied. When transfected in different cell lines, these nanomaterials showed different degrees of nuclear targeting based on the peptide and cell line studied. Only nanoparticles conjugated with the peptides that can mediate both receptor-mediated endocytosis and nuclear localization were able to enter the nucleus of these cells. ODN-loaded gold nanoparticles have been reported for gene-silencing by Mirkin's group

(Rosi et al. 2006). The researchers functionalized 13 nm gold particles with either tetrathiol or monothiol-modified antisense ODNs. These nanoparticles showed higher knockdown of EGFP gene expression when compared to antisense DNA delivered by both commercial Lipofectamine and Cytofectin. Nagasaki et al. reported gold nanoparticles complexed with PEG-*b*-PDMAEMA copolymers (Oishi et al. 2006b). These nanoplexes were then treated with either HS-siRNA or siRNA, which attached to the particle through a thiol-Au linkage or electrostatic interaction, respectively. All these nanoplexes showed the inhibition of luciferase expression in human hepatoma cells, and the HS-siRNA-based vectors showed more effective inhibition than the siRNA-based ones. Rotello et al. also reported the inhibition of the DNA transcription of T7 RNA polymerase by gold particles with terminal amine groups through Au-thiol linkage could be reversed by adding glutathione in the system (Han et al. 2005).

5.2 *Magnetic Nanoparticles*

Magnetic nanoparticles (MNPs), such as magnetite and Fe_3O_4 that have magnetic properties and other functionalities have demonstrated to hold great promise as multimodality imaging probes, and for tumor thermotherapy (Berry et al. 2003; Hautot et al. 2003; Tartaj and Serna 2003). When their surfaces were multifuntionalized by biological or drug molecules, it is possible to achieve target-specific diagnosis and therapy (Del Gaudio et al. 2005; Gupta and Gupta 2005; Neubergera et al. 2005; Dobson 2006; Sunderland et al. 2006).

Numerous synthetic methods have been developed to synthesize MNPs (Laurent et al. 2008), including co-precipitation (Jiang et al. 2004), sol–gel synthesis (Solinas et al. 2001), microemulsion technique (Deng et al. 2003), sonochemical reaction (Pol et al. 2005), hydrothermal reaction (Chen and Xu 1998), thermal decomposition (Hyeon et al. 2001), electrospray synthesis (Basaka et al. 2007), and laser pyrolysis approach (Veintemillas-Verdaguer et al. 1998).

The unfunctionalized iron oxide nanoparticles alone are cytotoxic. Therefore, a suitable coating on the surface of MNPs is necessary to improve their biocompatibility and functionality. Silica-coated magnetite nanoparticles were prepared by Bruce et al. and functionalized with amine groups (Bruce et al. 2004; del Campo et al. 2005). In these studies, selected samples were comparatively examined for their ability to adsorb, and subsequently elute, 2-deoxyguanosine-5-monophosphate (GMP) and a range of sequence defined ODNs as well as sheared salmon sperm DNA. It was found that magnetite could readily adsorb GMP via its phosphate anion, whereas silica did not, due to the electrostatic repulsion between the negatively charged surface of silica and the GMP. In comparison to commercially available silica-magnetite composite in terms of DNA adsorption and elution, the newly fabricated material was observed to perform approximately 10% more efficiently.

Plank et al. have evaluated the viability of iron oxide nanoparticles as vehicles for delivering nucleic acids into cells (Krotz et al. 2003; Plank et al. 2003a, b). They presented the concept of magnetofection and showed a strongly enhanced uptake of DNA by cells after treatment with superparamagnetic iron oxide (SPIO) particles. The SPIO particles coated with polyelectrolytes were mixed with other gene vectors, such as DNA or recombinant viruses, in a salt-containing solution. The gene vectors were supposed to collect around the magnetic particles owing to the salt-induced colloid aggregation. These tagged particles could then be directed along a magnetic field gradient to transfect appropriate cells. The researchers have shown that their 'magnetofection' method worked for *in vivo* applications. But substantial improvements are required to direct the magnetic gene vectors to specific target sites in the human body without invasive surgery.

Different cell lines show a different selectivity towards the hydrophilicity of the particle's surface when it comes to the uptake of nanoparticles. Berry et al. investigated the effect of naked and functionalized (either with dextran or albumin) iron oxide nanoparticles on the uptake by fibroblasts (Berry et al. 2003). They found that all three kinds of nanoparticles were well taken up by the cells, but only albumin-coated nanoparticles did not hinder cell proliferation, the other two particles caused cell death.

5.3 Carbon Nanotubes

Carbon nanotubes (CNTs) are considered to be first discovered by L. V. Radushkevich and V. M. Lukyanovich in 1952 (Monthioux and Kuznetsov 2006). There are two different types of CNTs: single-walled CNTs (SWCNTs) and multi-walled CNTs (MWCNTs), with diameters of a few nanometers and lengths up to 1 mm (Dicks 2006; Balani et al. 2007; Grujicica et al. 2007). Since these carbon structures are practically insoluble in aqueous environments, both noncovalent and covalent functionalization techniques have been utilized to increase their solubility (Klumpp et al. 2006). In the case of covalent functionalization, there are two methods: the oxidation of CNTs in acidic conditions to form acid-terminated structures or 1, 3-dipolar cycloaddition reactions using α-amino acid derivatives and para-formaldehyde. Noncovalent functionalization usually involves either hydrophobic or π-π stacking interactions between CNTs and surfactants, nucleic acids, peptides, polymers, or oligomers. Major efforts have been directed to improve the solubility and to reduce the toxicity of CNTs to a practical delivery system.

Bianco et al. synthesized ammonium-functionalized SWCNTs and MWCNTs by 1, 3-dipolar cycloaddition reactions to afford cationic nanoplexes (f-SWNT and f-MWNT, respectively) (Pantarotto et al. 2004). In HeLa cells, these nanoplexes were taken up by a non-endocytic route, as cellular uptake was not inhibited by sodium azide or 2, 4-dinitrophenol, which typically hinder energy-dependent cellular processes. In CHO cell lines, the f-SWNT nanoplexes showed significant β-galactosidase expression at f-SWNT/DNA ratios between 2:1 and 6:1. MWCNTs were found to bind to DNA more tightly than the single-walled analogues due to increased surface area. The same group also investigated the system that replaced the ammonium cation with a lysine group and found the DNA binding capacity was increased, because of the increased charge density of lysine group (Singh et al. 2005). Cai et al. synthesized vertically aligned CNTs by plasma-enhanced chemical vapor deposition.

These structures contain nickel particles on their tips, so that can penetrate cells through a nanotube spearing procedure (Cai et al. 2005). When these nanotubes were complexed with pDNA encoding EGFP, the nanotube spearing pathway resulted in high transfection efficiency, particularly in nondividing neuron cells.

In addition to DNA transfection, CNTs have also been used to deliver siRNA into cells for gene knockdown. Dai et al. noncovalently functionalized CNTs with phospholipid-PEG moieties containing either maleimide or amine terminal groups (Kam et al. 2005). These surface modified CNTs were reacted with thiol-terminated siRNA and yielded nanoplexes. The nanoplexes thus obtained could inhibit the gene encoding lamin A/C protein in HeLa cells. The same nanostructures also led to an efficient delivery of siRNA into human T-cells and primary cells (Liu et al. 2007b). Another example of gene knockdown using CNTs was reported by Zolk and coworkers (Krajcik et al. 2008). Oxidized CNTs were reacted with hexamethylenediamine, and then functionalized with poly(diallyldimethylammonium) chloride (PDDA). It was found that these CNT-based nanoparticles can cause 80% inhibition of the targeted genes, ERK1 and ERK2 *in vitro*, when complexed with siRNA through electrostatic interactions.

5.4 *Silica Nanoparticles*

Organically modified silica materials are normally nanocomposites and have the potential for providing unique combination of properties, which cannot be achieved by other materials. These surface-modified silica nanoparticles have many advantages as candidates for gene delivery. For instance, colloidal silica nanoparticles are inert and exhibit high biocompatibility, they are stable with respect to physical stresses comparing with liposomes (Sameti et al. 2003).

Silica nanoparticles can be prepared by suitable sol–gel processing routes (Brinker and Scherer 1990). The silanol groups on the surface allow an easy functionalization. One of the successful examples is modifying silica nanoparticles with sodium chloride reported by Chen et al. (2003). These particles had diameters of 10–100 nm and showed a transfection efficiency of about 70% exceeded than previously reported for lipoplexes mediated transfection. Lin et al. reported the first use of mesoporous silica nanoparticles for gene delivery (Radu et al. 2004). PAMAMs were covalently bound to the surface of mesoporous silica nanoparticles to promote electrostatic interactions with pDNA. These cationic mesoporous nanoplexes were successfully introduced into neural glia cells, human cervical cancer cells, and Chinese hamster ovarian (CHO) cells, and showed higher gene transfection efficiency than commercial transfection agents, such as PolyFect, SuperFect, and Metafectene. This high efficiency was attributed to the increased sedimentation. The DNA-binding capacity of cationic silica nanoparticles was first studied by Lehr and co-workers (Kneuer et al. 2000). N-(2-Aminoethyl)-3-aminopropyltrimethoxysilane and N-(6-aminohexyl)-3-aminopropyltrimethoxysilane were used to modify silica nanoparticles in their study. These nanoparticles showed significant transfection in Cos-1 cells, especially in the presence of serum and chloroquine, which was attributed to the physical stabilization of the nanoplexes and the increasing of buffering capacity, respectively. It is noteworthy that modified silica nanoparticles also exhibited promise in *in vivo* studies. Prasad et al. showed that sterotaxic injection of functionalized silica nanoplexes gave transfection efficiency that equaled or even exceeded that obtained with herpes simplex virus 1 vectors and showed less tissue damage (Bharali et al. 2005).

Luo et al. noticed that unmodified silica nanoparticles could serve as mediators for the uptake of DNA into cells by adsorbing on the cell surface (Luo et al. 2004). As a result, β-galactosidase gene expression was enhanced by up to 750%. In addition, large silica particles were found to enhance gene expression more efficiently than small ones, because of the faster settling of the larger particles onto the cell surface. The co-precipitation of other inorganic or polymeric particles together with DNA on cell surfaces also showed effective transfection, indicating that the chemical composition of the nanoparticles is not a significant factor dominating the transgene activity, which means this enhanced DNA uptake is kind of a "mechanical" effect.

5.5 *Quantum Dots*

Quantum dots (QDs) are luminescent semiconductor nanocrystals with typical diameters of a few nanometers which comprise periodic groups of II-IV (e.g. CdSe and CdTe) or III-V (e.g. InP and InAs) semiconductor materials (Alivisatos et al. 2005; Yin and Alivisatos 2005; De et al. 2008). For these nanocrystals with all three dimensions less than the Bohr exciton radius (typically a few nanometers), their energy levels are quantized, which means the spacing of their energy levels can be controlled by the crystal size (Nirmal and Brus 1999; Poznyak et al. 2005). This effect endowed QDs superior properties over other organic fluorophores, such as narrow, symmetric and size-tunable emission spectra, and broad excitation spectra. Another benefit of QDs over organic fluorophores is their stronger fluorescence (~10–100 times brighter) and higher fluorescence stability against photobleaching (~100–1,000 times more stable), which is very important to the long-term monitoring of intermolecular and intramolecular interactions in live cells and organisms (Salgueiriño-Maceira et al. 2006; Resch-Genger et al. 2008). There are different routes devised for the preparation of QDs, the predominant one is to coat a CdSe core with ZnS layer to obtain the best crystalline quality and monodispersity. The reason for introducing ZnS is to reduce toxicity by preventing the CdSe from leaching out to the surrounding solutions, and also enhances the photoluminescence yield (Pinaud et al. 2006; Norris et al. 2008; Smith et al. 2008). However, due to the fact that ZnS-coated QDs are not soluble in aqueous milieu, which is the nature of the biological environment, altering the QDs surface properties from hydrophobic to hydrophilic is quite necessary for biological applications (Rogach et al. 2000; Gaponik et al. 2002). Furthermore, the inherent toxicity of II-IV and III-V semiconductor QDs is a serious issue for *in vivo* applications. They must rely on conjugation with biological molecules such as aptamers (Chu et al. 2006), antibodies (Gao et al. 2004), peptides (Cai et al. 2006), ODNs (Zhang et al. 2005), and small molecule ligands (Lidke et al. 2004) to obtain biological affinity.

Zhang et al. developed a method to increase the biocompatibility of CdSe/ZnS QDs (Zhang et al. 2006b). The surface of CdSe/ZnS core-shell QDs was first silanized and then coated with PEG to reduce the toxicity of nanocrystals to the cells. The results showed that the gene expression of fibroblasts changed when they were exposed to silica-coated QDs. Burgess et al. showed that QDs could be covalently conjugated to pDNA for transfection studies (Srinivasan et al. 2006). Transfection studies on QD-DNA conjugates and Lipofectamine 2000 showed maximum cellular uptake at 6 h, with approximately 75% of the DNA located in the cytoplasm and 25% in the nucleus. After 10 h, approximately 2/3 of the DNA was in the nucleus of the cells, and this nuclear localization was proportional to gene expression. Ho et al. developed a QD-FRET (Förster resonance energy transfer) system to investigate the structural composition and dynamic behavior of polymeric DNA nanocomplexes intracellularly (Ho et al. 2006; Chen et al. 2008). In these studies, pDNA encoding GFP was biotinylated using PEI-psoralen-biotin, and these analogues were conjugated to streptavidin-functionalized QDs. The QD-labeled DNA was complexed with Cy5-labeled chitosan and introduced into HEK293 cells. The trafficking of the complexes within cells was monitored using confocal microscopy. At 24 h after transfection, the intact complexes were localized around the cell nucleus. At 48 h after transfection, most complexes released DNA molecules. At 72, QD-labeled DNAs could be observed within the nucleus of a cell expression GFP. This study illustrated that QD-FRET-enabled detection of nanoplexes stability combined with image-based quantification is a valuable method for investigating mechanisms involved in nanoplexes uptake and intracellular trafficking.

QD-labeling has also been utilized by Bhatia et al. for gene silencing by siRNA (Chen et al. 2005a). In this study, QDs were introduced as a traceable marker to shed light on the siRNA delivery. However, cationic transfection agents (Lipofectamine 2000, superfect, or translocation peptide) were required to promote gene transfer, indicating that the QDs worked only as a labeling moiety. Subsequently, Bhatia et al. functionalized QDs with both siRNA and the tumor-targeting F3 peptide (Derfus et al. 2007). In HeLa cells, these tumor-targeting QD-siRNA conjugates showed up to ~29% EGFP gene knockdown when cellular uptake was followed by the addition of an endosome escape agent (Lipofectamine 2000). Additionally, Tan and coworkers ameliorated the chitosan-based transfection system by doping with QDs for siRNA tracking (Tan et al. 2007). Recently, Gao's group developed both the proton-sponge-coated and amphipol-modified QDs to further enhance the delivery efficiency (Qi and Gao 2008a).

5.6 *Calcium Phosphate Nanoparticles*

Calcuim phosphates (CaP) are the inorganic components of biological hard tissues, for example, bone, teeth, and tendors. Because the calcium phosphates usually have a high biocompatibility and a good biodegradability compared to other types of nanomaterials, they are extremely attractive for gene therapy applications. Relying on the fact that divalent metal cations are easily to form ionic complexes with nucleic acids and these complexes have excellent transportability across the cell membrane via ion channel, the technique of CaP co-precipitation for *in vitro* transfection is employed as a routine laboratory procedure (Graham and van der Eb 1973).

The standard transfection method of calcium phosphate was originally established by Graham and van der Eb in 1973 (Graham and van der Eb 1973). For the preparation of the CaP carrier for transfection, calcium chloride solution with DNA and phosphate-buffered saline solution are mixed together, which results in the formation of fine precipitates (nano- and micropaticles) of calcium phosphate with DNA. There are several factors that are concerned with cell transfection efficiency of the standard calcium phosphate method, including the precipitation conditions (such as pH value, concentrations of calcium chloride and DNA, temperature, and the time between precipitation and transfection), as well as the morphology and size of the nanoparticles. Insufficiently protected nanocrystals will grow to microcrystals by Ostwald-ripening and therefore loss their transfection ability as investigated by Orrantia and colleagues (Orrantia and Chang 1990). Moreover, the transfection efficiency also strongly depends on the cell type. Quite often, the transfection reproducibility of calcium phosphate is poor.

Inspired by the observation that CaP formed nanoparticles generally possess good biocompatibility and biodegradability as well as relatively high transfection activity, a number of groups have concentrated on CaP nanoparticles for transfection (Zhang and Kataoka 2009). Jordan et al. reported the application of CaP/DNA coprecipitates to achieve efficient transfection in cultured cell lines (Jordan et al. 1996; Jordan and Wurm 2004). They found that highly efficient transient and stable transfections could be achieved by optimizing physicochemical conditions of initiation and growth of precipitate complexes, such as the concentrations of calcium and phosphate, temperature, DNA concentration, and reaction time. One of the successful examples in cancer gene therapy using DNA-loaded CaP nanoparticles (CPNP-DNA) of 23.5–34.5 nm was reported by Cai et al. in 2005 (Liu et al. 2005). The CPNP-DNA complex efficiently adhered to the cell membrane and entered the cells, leading to an increase in the amount of DNA in the cells, thus resulting in a successful enhancement in transfection efficiency. It should be noted that for these CaP-based nanosystems loaded with genetic materials, the transfection efficiency decreased with increasing incubation time for nanoparticles formation, which was attributed to the time-dependent particle growth in the presence of a protective outer layer (Fasbender et al. 1998). Furthermore, the genetic material located on the surface of the nanoparticles was very susceptible to enzymatic attacks by nuclease, resulting in the destabilization of nanoparticles and thus reduced transfection activity.

Apparently, CaP-based transfection systems alone do not meet the requirements for *in vivo* applications. Since PEGylation of CaP nanoparticles helps to minimize the nonspecific interactions between the particles and the biological species present in the biological medium, PEGylated CaP nanoparticles are expected to yield stable and biocompatible nanocarriers for nucleic acid delivery (Kakizawa and Kataoka 2002; Kakizawa et al. 2004). Kataoka and coworkers have synthesized PEG-*b*-poly(aspartic acid) (PEG-PAA) block copolymers to prepare CaP nanoparticles entrapping genetic materials (such as pDNA, ODN or siRNA) via a facile method (Kakizawa and Kataoka 2002). It was found that when the PEG-PAA concentration was above a certain value, monodispersed and stable nanoparticles could be obtained, which were essential to prevent precipitation of CaP crystals during nanoparticle formation. The PEGylated CaP particles exhibited high efficiency in encapsulating genetic materials, and also exhibited minimal cytotoxicity, in addition to considerable nuclease-resistance, and appreciable gene expression under optimized conditions. *In vitro* experiments, including cellular uptake and gene knockdown, were performed for hybrid PEG-PAA/CaP/DNA nanoparticles containing fluorescently labeled nucleic acid (Kakizawa et al. 2004). By investigating the effects of temperature and the presence of metabolic inhibitors on the uptake of the nanoparticles, it was revealed that the nanoparticles were internalized via an endocytotic pathway. Moreover, significant gene knockdown against luciferase was achieved by the hybrid nanopaticles loaded with siRNA encoding a luciferase sequence. The same group has proposed an environmentally responsive siRNA delivery system based on CaP nanostructures (Zhang et al. 2009). Block copolymers consisting of PEG and siRNA were utilized to prepare CaP nanoparticles, in which the PEG and siRNA segments were linked via a disulfide bond that can be reversibly cleaved in the presence of reducing reagents. Due to the environmental sensitivity, the PEG shell layer would be selectively cleaved off from the hybrid nanoparticles when located inside the intracellular compartments with a reducing environment, upon cellular uptake. This study clearly demonstrated that PEGylated CaP nanocarrier may be a promising candidate for *in vivo* gene delivery applications.

5.7 *Layered Double Hydroxides/Clays*

Layered double hydroxides (LDH), also known as hydrotalcite materials or anionic clays, are a family of clays which contain positively charged layers, exemplified by the natural mineral hydrotalcite $[Mg_6Al_2(OH)_{16}CO_3 \cdot 4H_2O]$. Most LDH materials can be described using the general formula $\left[M^{II}_{1-x}M^{III}_{x}(OH)_2\right]^{x+}(A^{m-})_{x/m} \cdot nH_2O\ (x = 0.2-0.4;\ n = 0.5-1)$, where M^{II} represents a divalent metal cation, M^{III} a trivalent metal cation and A^{m-} an anion (Itoh et al. 2005). LDH materials can be found in nature as minerals or readily synthesized in the laboratory (Braterman et al. 2004). Most widely used method for LDH preparation is called co-precipitation method. This procedure includes mixing an aqueous solution containing the salts of two metal ions with a base solution in the presence of the desired anion to nucleate and grow the metal hydroxide layers. Metal chlorides or nitrates are generally used (Yamaoka et al. 1989; Braterman et al. 2004). Additionally, because the anions present in the LDH interspaces are exchangeable, thus anion exchange is the second most widely used method for the synthesis of LDH hybrids (Crepaldi et al. 1999). Anion exchange is simply accomplished by stirring previously formed LDH materials in solution that contains the replacement anion species. Other methods for producing LDH materials were also developed, for instance, from metal oxides and/or hydroxides (Whilton et al. 1997; Choy et al. 2001; Gardner et al. 2001; Xu et al. 2006, 2007), precipitation via aluminate (Zikmund et al. 1996), by sol–gel techniques (Lopez et al. 1996).

LDHs have a high potential to exchange intercalated anions which allows for the direct loading of a variety of negatively charged drugs/biomolecules (such as vitamins and nucleic acids) into their interlayer galleries (Kwak et al. 2004; Zhang et al. 2006a; Xu et al. 2008; Ladewig et al. 2009). Although this intercalation decreases their positive surface charge, LDHs remain sufficiently positively charged to facilitate cellular uptake. The high anion-exchange capability of LDHs has attracted particular attention in the field of bio-hybrid nanomaterials owing to their good biocompatibility, excellent chemical stability as well as the controlled release character. Organic molecules can be released from LDHs at a rate that related to the pH value and the ionic strength of the surrounding medium (Zhang et al. 2006a).

With respect to the use of LDH nanoparticles as gene vectors, the size of nucleic acids to be delivered should be concerned. It seems that larger and sterically hindered biomolecules such as pDNA, are not completely accessible to intercalation in the interlayer galleries by means of anion exchange, because of their supercoiled structure in solution. Accordingly, most of the reports have focused on the delivery of smaller nucleic acids like antisense ODNs, siRNA, sheared genomic DNA, or PCR fragments. Choy et al. intercalated sheared genomic DNA (500–1,000 bp) in the interlayer galleries of LDHs and used these hybrids as non-viral vectors (Choy et al. 2000). Because of its negative charges, DNA could be strongly incorporated into such a layered double hydroxide, while the positive surface charges of the nanoparticles attracted them to the cell surface by electrostatic interaction. The LDH-bio-nanohybrids were then taken up by receptor-mediated endocytosis and escaped into the cytoplasm by endosome rupture. Similar phenomena were observed by Tyner et al. using PCR fragments of 800 bp length (Tyner et al. 2004). In view of their good biocompatibility and wide availability, LDHs deserve further *in vivo* studies to justify their usefulness.

6 Multifunctional Nano-Systems for Gene Delivery

Multifunctionalized nanocarriers integrated with multiple functionalities such as longevity, targetability, intracellular penetration, contrast loading, multi-therapeutics payload, and stimuli-sensitivity, have attracted great attention in recent years, owing to their great potentials in molecular pharmaceutics, medical imaging, biomedical engineering, and gene therapy (Torchilin 2006, 2009). For instance, drug delivery, cell targeting and/or multi-modal imaging can be simultaneously achieved by multifunctional inorganic nanoparticles or nanocrystals, inorganic-organic hybrid nanoparticles and polymeric assemblies (Kim et al. 2006b, 2007b; Huang et al. 2007; Li et al. 2008; Liong et al. 2008; Kim et al. 2007c, 2008; Rieter et al. 2007). Studies based on these versatile pharmaceutical nanocarriers have provided profound insights into the intracellular or *in vivo* transportation and/or distribution, and valuable information on the spatial and temporal interactions of delivery carriers with cells and tissues as well as their intracellular interactions. In addition, multifunctional nanoparticles have also been employed to implement the simultaneous gene delivery, drug delivery and imaging (Qi and Gao 2008b; Nehilla et al. 2008; Liong et al. 2008; Ghosh et al. 2008; Al-Jamal et al. 2008; Farokhzad and Langer 2009).

6.1 Multifunctional Magnetic Nanoparticles

Magnetic nanoparticles are 10–20 nm sized spherical nanocrystals with a Fe^{2+} and Fe^{3+} core surrounded by dextran or PEG chains (Lu et al. 2007). In addition to active targeting triggering using external magnetic field, their special magnetic properties make them excellent agents to label biomolecules in bioassays, as well as magnetic resonance imaging (MRI) contrast agents to trace the *in vivo* biodistribution or accumulation of delivery vehicles by high-resolution MRI.

Recently, Moore and coworkers reported the development of dual-purpose probes for *in vivo* transfer of siRNA and the simultaneous imaging of its accumulation in tumors by MRI and near-infrared *in vivo* optical imaging (NIRF) (Medarova et al. 2007). These nanoplatforms comprised magnetic nanoparticles labeled with a near-infrared dye (cyanine dye Cy5.5) and covalently linked to siRNA molecules. Additionally, these nanoparticles were peripherally decorated with a membrane translocation peptide (myristoylated polyarginine peptide) for intracellular delivery. The tumor could be imaged via T2 MR and NIR fluorescent imaging, after the intravenous injection of the nanoparticles into nude mice bearing subcutaneous human colorectal carcinoma tumors. Gene silencing was observed only when the multimodal nanoparticles-containing siRNA were administered. These studies represent the first step towards the advancement of simultaneous delivery and detection of siRNA-based therapeutic agents *in vivo*, essential for cancer therapeutic product development and optimization. Park and Cheon et al. have developed "all-in-one" nanoparticle probes for simultaneous siRNA delivery and multimodal imaging (Lee et al. 2009). To achieve this purpose, manganese-doped magnetism-engineered iron oxide (MnMEIO) nanoparticles of 15 nm in size coated with bovine serum albumin (BSA) were used as the core material, onto which RGD functionalized PEG and Cy5-dye-labeled sRNA were conjugated via the disulfide linkage. Besides the intracellular delivery of therapeutic siRNA, these multifunctional nanoprobes enable highly accurate imaging by both MRI and fluorescence imaging.

6.2 Quantum Dots as Multifunctional Vectors

Quantum dots (QDs) are colloidal fluorescent semiconductor nanocrystals of 2–10 nm. The central core of QDs composes of element combinations from groups II-VI of the periodic table (CdSe, CdTe, CdS, PbSe, ZnS and ZnSe) or III–V

(GaAs, GaN, InP and InAs), which are 'overcoated' with a layer of ZnS (Sanvicens and Marco 2008). Quantum dots are photostable and resistant to photobleaching, and they show size- and composition-tunable emission spectra and high quantum yield, as well as exceptional resistance to photo and chemical degradation. The unique optical properties of QDs make them appealing as *in vivo* and *in vitro* fluorophores in a variety of biological investigations, in which traditional fluorescent labels based on organic molecules fall short of providing long-term stability and simultaneous detection of multiple signals (Medintz et al. 2005).

Burgess et al. first showed that quantum dots could be covalently conjugated to pDNA for transfection and long-term intracellular and intranuclear tracking (Srinivasan et al. 2006). Conjugation of pDNA with phospholipid-coated QDs was achieved using a peptide nucleic acid (PNA)-N-succinimidyl-3-(2-pyridylthio) propionate linker. AFM imaging revealed that multiple QDs were attached in a cluster at the PNA-reactive site of the pDNA. These QD-DNA conjugates could express the reporter protein in Chinese hamster ovary (CHO-K1) cells with an efficiency of ca. 62%, which was comparable to the control (unconjugated) pDNA. In addition, confocal imaging could be employed to study the intracellular trafficking and distribution of QD-DNA conjugates, utilizing the inherent fluorescence possessed by QDs. Nie and Gao et al. developed multifunctional nanoparticles for siRNA delivery and imaging based on QDs and proton-sponging polymer coating (Yezhelyev et al. 2008). Transfection studies showed dramatic improvement in gene silencing efficiency by 10~20-fold, when compared with commercial transfection agents (TransIT and JetPEI) in MDA-MB-231 cells. These QD-siRNA nanoparticles are also dual-modality optical and electron-microscopy probes, allowing real-time tracking and ultrastructural localization of QDs during delivery and transfection. Most recently, siRNA conjugated QDs have been fabricated as a multifunctional nanoplatform to deliver siRNA and to elucidate the EGFRvIII-knockdown effect of PI3K signaling pathway in U87-EGFRvIII (Jung et al. 2010). Two different strategies were developed for the siRNA-QD conjugates: (1) siRNA and QDs were coupled through disulfide linkage to release the siRNA upon entering the cells by cleavage of the disulfide linkage, through enzymatic reduction or ligand exchange (e.g. glutathione); and (2) siRNA and QDs were conjugated using a robust linker (3-maleimidopropionic acid pentafluorophenyl ester) which enabled the tracking of siRNA-QDs within the cells. In addition to selective inhibition of human brain tumor cells, real-time monitoring of down-regulated signaling pathway could also be implemented.

6.3 Gold Nanorods as Multifunctional Carriers

Multifunctional nanorods offer a unique capability to combing a number of essential diagnostic, imaging, delivery and dosage properties. In addition, their properties have been designed and explored for specific biomedical applications by taking advantage of the additional degrees of freedom associated with nanorods in comparison to spherical particles (Pearce et al. 2007).

Leong et al. reported the fabrication of bifunctional Au/Ni nanorods and their application for gene delivery. The nanorods, synthesized by electrodeposition into an Al_2O_3 template with a pore diameter of 100 nm, were 100 nm in diameter and 200 nm in length with 100 nm gold segments and 100 nm nickel segments. Ni segment was modified using the 3-[(2-aminoethyl) dithio] propionic acid (AEDP) linker, through which pDNA was loaded by electrostatic forces. Rhodamine-conjugated transferrin is selectively bound to the gold segment of the nanorods for targeting. These multicomponent nanorods could mediate the effective transfection and expression of the GFP and luciferase reporter genes in human embryonic kidney (HEK293) mammalian cell line. Nanorods containing compacted plasmids showed a 4-fold increase in GFP-positive cells in comparison to naked DNA. Luciferase expression by nanorods transfection was 255 times higher than that mediated by naked DNA. Bifunctional nanorods conjugated with transferring increased GFP expression by a factor of 2 (22%) and luciferase expression by a factor of 3.4, compared with nanorods with compacted plasmids alone. The *in vivo* transfection efficacy of the nanorods was also demonstrated by preliminary studies in the skin and muscle tissues of mice. Furthermore, the cellular uptake and distribution of nanorods could be visualized by fluorescence and electronic microscopy. This approach allows the precise control of composition, size and multifunctionality of the gene delivery systems based on nanorods (Salem et al. 2003).

Studies by Huang et al. suggested that the stability of gold nanorods in biologically relevant media could be significantly enhanced by layer-by-layer depositing polyelectrolyte multilayers on their surface. These polyelectrolyte (PE)-gold nanorod assemblies (PE-GNRs) possessed a stable Arrhenius-like photothermal response, which was examined for the hyperthermic therapy of prostate cancer cells *in vitro*. PE-GNRs based on a cationic copolymer of ethyleneglycol diglycidyl ether and 3,3′-diamino-N-methyl dipropylamine showed higher transfection efficacy and lower cytotoxicity compared to those based on PEI. These results indicate that judicious engineering of biocompatible polyelectrolytes leads to multifunctional gold nanorod-based assemblies that combine a variety of simultaneous applications including optical imaging, nonviral gene delivery, and localized hyperthermia (Huang et al. 2009).

6.4 Silica Nanomaterials-Based Multifunctional Vectors

He and Minko have reported the first effort of utilizing mesoporous silica nanoparticles (MSNs) as a multifunctional system to simultaneously deliver doxorubicin (Dox) and Bcl-2-targeted siRNA into multidrug resistant A2780/AD human ovarian cancer cells for efficient chemotherapy (Chen et al. 2009). To develop this nano-platform, MSNs were modified to encapsulate Dox inside the pores to achieve minimal premature release. Then the Dox-loaded MSNs were decorated with amine-terminated PAMAM dendrimers (generation 2). The dendrimer-modified MSNs could efficiently complex with siRNAs targeted against mRNA encoding Bcl-2 protein, which is the main player for nonpump resistance. By delivering Dox and Bcl-2 siRNA simultaneously into cancer cells, the Bcl-2 siRNA could effectively silence the Bcl-2 mRNA and significantly suppress the nonpump resistance and substantially enhance the anticancer activity of Dox. It was also suggested that the Dox delivered by MSNs exhibited minimal premature release in the extracellular environment, which can greatly reduce its side effects. Furthermore, the Dox was primarily localized in perinuclear region after internalization, possibly bypassing the pump resistance and further enhancing the cytotoxicity.

Chen et al. fabricated fluorescent silica nanotubes with an inner diameter of hundreds of nanometers for gene delivery (Chen et al. 2005b). The hollow silica nanotubes were synthesized by a sol–gel reaction using an anodic aluminium oxide membrane as a template. After modification of the inner surface of the silica nanotubes with amine groups, CdSe–ZnS nanoparticles with either green or red fluorescence that were stabilized with mercaptoacetic acid, were decorated on the inner surface of the silica nanotubes through electrostatic interactions between the positively charged amine groups and negatively charged QDs. The subsequent amine-modification of the inner surface of the silica nanotubes made it possible to incorporate DNA molecules inside the fluorescent silica nanotubes. The transfection of pDNA-loaded nanotubes in COS-7 cells resulted in successful GFP expression. Additionally, the cytoplasmic existence of nanotubes was confirmed by imaging using confocal laser scanning microscopy (CLSM).

6.5 *Multifunctional Nanoparticles Based on Lipids*

While nanomaterials such as magnetic/gold nanoparticles and QDs allow the real-time tracking and ultrastructural localization of siRNA/DNA delivery systems, combining drug delivery and gene therapy in one particle has the potential to enhance the transfection efficiency or to achieve a synergistic/combined effect of drug and gene therapies (Malone et al. 1994; Zhang et al. 2001; Kishida et al. 2003; Janat-Amsbury et al. 2004).

By co-delivering inflammatory suppressors and pDNA through liposomes, Huang and coworkers have developed a non-immunostimulatory gene vector (Liu et al. 2004). To this end, many inflammatory suppressors, such as glucocorticoids, nonsteroidal anti-inflammatory drugs (NSAIDs), NF-κB inhibitor, and tetrandrine (a natural compound from an herbal medicine), could be encapsulated into the cationic liposomes for form the new lipoplexes (cationic liposome/inflammatory suppressor/DNA). The DNA and suppressor could be co-delivered to an individual immune cell where the cytokines are produced. The inflammatory suppressor will be released from the vector in the cytoplasm and interrupt the cytokine production by inhibiting NF-κB at one or multiple activation steps of the signaling pathway. Through this way, the inflammatory toxicity induced by cationic lipids and CpG motif of pDNA can be largely reduced.

To achieve an enhanced anticancer therapy, a combination of therapeutics of proapoptotic peptide D-(KLAKLAK) (2) and Bcl-2 antisense ODN G3139 was simultaneously delivered using cationic liposomes by Torchilin and coworkers (Ko et al. 2009a). For the preparation of these multifunctional liposomes, the cationic peptide D-(KLAKLAK) (2) was first compacted with the anionic ODN G3139 to form negatively charged peptide/ODN complexes, which were then incorporated into DOTAP/DOPE cationic liposomes (CL). *In vitro* treatment of mouse melanoma B16(F10) with CL containing D-(KLAKLAK)(2)/G3139 showed significantly enhanced antitumor efficacy, mediated by stimulated induction of apoptotic (caspase 3/7) activity, when compared to CL loaded with G3139 alone. Intratumoral injection of this dual-delivery formulation in 1,316(F10) mice xenograft also resulted in suppressed tumor growth associated with enhanced apoptotic activity. Accordingly, the simultaneous delivery of proapoptotic peptide D-(KLAKLAK)(2) and antisense ODN G3139 via cationic liposomes could give rise to enhanced apoptotic/antitumor efficacy, which may provide a promising tool for cancer treatment.

In addition, multifunctional nanoparticles that combine gene delivery with the ability to cross tissue and membrane barriers can function as ideal non-viral vectors for gene therapy. One of the first indications of the potential of multifunctional platforms for gene delivery is the work performed by Pardridge and coworkers (Zhang et al. 2003). With the aim to transfer exogenous genes to the entire retina, 83–14 MAb antibody was conjugated to the PEGylated liposomes. After systemic administration to rhesus monkeys, these ‘stealthy’ immunoliposomes were targeted across the blood–retinal barrier, resulting in specific gene expression in the eye by using the opsin promoter. On the other hand, targeting tumor cells or ischemic myocardium by liposomes has been achieved via conjugating TAT peptide and/or anti-myosin antibody by Torchilin’s group (Kale and Torchilin 2007; Gupta et al. 2007; Ko et al. 2009b).

Harashima and coworkers have created a multifunctional envelope-type nano device (MEND) as a non-viral gene delivery system (Kogure et al. 2008). The ideal MEND consists of a condensed DNA core and a lipid envelope structure equipped with the various functionalities such as longevity and targetability. The condensation of DNA by cationic polymers (such as PLL) allows protection of DNA from DNase, size control and improvement in packaging efficiency. To control topology, polyplexes are incorporated into lipid envelopes by hydration of a lipid film with subsequent sonication such that the DNA core and lipid envelope exist as separate structures, rather than a disordered mixture. The packaging mechanism was based on electrostatic interactions between DNA, polycations and lipids. As an example, pDNA was first condensed with a polycation such as PLL by vortexing. The kinetic control of this process is important to control the size and charge of the resulting polyplexes. In the case of PLL, small (around 100 nm) and positively charged (around 30 mV) PLL/DNA complexes were prepared at a N/P ratio of 2.4. The lipid film containing a negatively charged lipid such as cholesteryl hemisuccinate was hydrated in an aqueous solution containing PLL/DNA polyplexes. The packaging of the PLL/DNA nanoparticles into a lipid bilayer was achieved by sonication.

The diameter and zeta-potential of the MEND were around 300 nm and –40 mV, respectively. The encapsulation efficiency of the DNA could be greater than 70%. This MEND based nano-system can serve as effective vectors for the intranuclear or cytosolic delivery of a variety of therapeutics including pDNA, ODN and siRNA as well as other molecules.

6.6 Polymer Nanoparticles for Simultaneous Drug/Gene Delivery

Considerable recent studies have been focusing on the development of polymer nanoparticles, especially assembled nanostructures for dual-release of drugs and nucleic acids. The first nano-vehicle designed for the co-delivery of drugs and DNA was developed by Yang and colleagues (Wang et al. 2006). For creating the nanoparticles for dual-release by self-assembly, they synthesized a biodegradable cationic amphiphilic copolymer, consisting of cholesterol side chains and a cationic main chain. This copolymer (PMDS-Chol) was synthesized by grafting N-(2-bromoethyl) carbarmoyl onto the hydrophilic poly(N-methyldietheneamine sebacate) (PMDS) that was produced by condensation polymerization between N-methyldiethanolamine and sebacoyl chloride. Core-shell structured nanoparticles with a hydrophobic cholesterol core and a cationic shell could be spontaneously formed the self-assembly of PMDS-Chol in an aqueous solution. Hydrophobic drugs such as paclitaxel and indomethacin could be effectively incorporated into the core, while the cationic shell of the resulting drug-loaded nanoparticles could be used to bind pDNA. *In vitro* release study indicated that the DNA binding did not affect the release profiles of payload indomethacin, whereas paclitaxel release from the nanoparticles was slightly faster than that from the nanoparticle/DNA complexes. More importantly, these nanoparticles could achieve high transfection efficiency and the possibility of simultaneously delivering drugs and genes to the same cells. Enhanced transfection of pDNA or siRNA with the simultaneous delivery of paclitaxel has been demonstrated by *in vitro* and *in vivo* studies. In particular, the co-delivery of paclitaxel with an IL-12-encoded plasmid using these nano-assemblies suppressed cancer growth more efficiently than the delivery of either paclitaxel or the plasmid in a 4T1 mouse breast cancer model. Moreover, the co-delivery of paclitaxel with Bcl-2-targeted siRNA increased cytotoxicity in MDA-MB-231 human breast cancer cells. These results clearly demonstrated the synergistic/combined effect of co-delivery of a therapeutic drug and gene. Other assemblies constructed using cationic amphiphilic copolymers of various architectures have also been developed for the simultaneous delivery of various drug/gene combinations (Wang et al. 2007; Qiu and Bae 2007; Zhu et al. 2008, 2010; Cheng et al. 2009; Wiradharma et al. 2009).

To develop a multifunctional polymer nanocarrier via host-guest interactions for simultaneous drug and gene delivery, β-CDs were conjugated onto branched polyethylenimine (PEI-CD) to synthesize a host polymer (Zhang et al. 2010).

Using poly(β-benzyl L-aspartate) (PBLA) as a macromolecular guest, core-shell structured spherical nano-assemblies of 90.6 nm can be constructed. The core of assemblies composed of hydrophobic PBLA chain can accommodate hydrophobic drugs, while the cationic shell can complex with pDNA. Dexamethasone (DEX), a potent steroidal anti-inflammatory drug can be efficiently loaded into and released from the PEI-CD/PBLA nanoparticles in a sustained manner. Gel retardation combined with EtBr exclusion assays demonstrated that the pDNA could be complexed with PEI-CD/PBLA assemblies to form a condensed structure when the weight ratio was higher than 0.6. Cell culture experiments suggested that pDNA encoding red fluorescence protein can be transfected and expressed in osteoblast cells, using the polyplexes based on the assembled host-guest nanoparticles. Interestingly, introducing DEX into the polyplexes can increase the transfection efficiency as well as dramatically decrease the cytotoxicity of assemblies. The decreased cytotoxicity was considered to be related to the anti-inflammatory and immunosuppressive activity of DEX that can inhibit proinflammatory cytokines such as tumor necrosis factor α (TNF-α) and antagonize the activation of the NF-κB pathway by direct and indirect mechanisms.

7 Nonviral Therapeutics in Clinical Trials

It is clear that most of the clinically studied gene delivery vectors are based on viruses. Whereas a variety of delivery vectors have been studied for gene therapy, most of them are still in the laboratory stage. Currently, lipoplexes and polyplexes are two types of nonviral vectors that have been extensively studied for clinical trials. Below we list some examples where the nonviral vectors, mainly lipoplex and polyplex based nanomedicines have, at least in part showed their success in the therapy of diseases such as cystic fibrosis, HIV, and various cancers.

7.1 Clinical Trials of Lipoplexes Vectors

Among the various nonviral vectors, lipoplexes have been the most intensively studied carriers. Considerable amount of CL-based vectors have been evaluated in clinical trials for the therapy of diseases varying from cancer, cystic fibrosis, HIV, AAT deficiency, to cardiovascular diseases (Dass 2004; Davis and Cooper 2007; Rubanyi 2001; Edelstein et al. 2004, 2007).

In 1993, Nabel et al. reported the first clinical gene therapy of melanoma using DNA-liposome complexes (Nabel et al. 1993). Lipoplexes containing pDNA encoding a foreign major histocompatibility complex protein (HLA-B7) were locally injected into the melanoma nodule of HLA-B7-negative patients with advanced melanoma. No complications were observed during six courses of treatment in five HLA-B7-negative patients with stage IV melanoma. Three to seven days post injection, pDNA could be detected by using PCR within biopsies of treated tumor nodules, but was not found in the serum at any time. In addition, recombinant HLA-B7 protein was measured in tumor biopsy tissue in all five patients by immunochemistry, and immune responses to HLA-B7 and autologous tumors could be detected as well. No antibodies to DNA could be detected in any patient. One patient showed regression of injected nodules on two independent treatments, which was accompanied by regression at distant sites. These studies thus demonstrate the feasibility, safety, and therapeutic potential of local gene therapy in humans using lipoplexes. Local delivery of lipoplexes containing HLA-B7 in pulmonary artery of a patient with melanoma also demonstrated the well tolerance of this formulation, and no adverse respiratory, cardiac immunologic or other organ toxicities were detected (Nabel et al. 1994).

Lipid-based vectors have also been evaluated clinically in patients for the correction of cystic fibrosis (CF) defect. Lipoplexes of cystic fibrosis transmembrane conductance regulator (CFTR) cDNA and lipid 67 were delivered in the nose and the lung in CF patients in Great Britain (Alton et al. 1999). It was shown that no mRNA for CFTR was detected in epithelial samples. Modest improvement in chloride transport was detected at both sites by transepithelial potential difference measurements. However, significant inflammatory response was observed. In other trials conducted in the USA, little or no chloride transport improvement has been documented with lipoplexes based formulations, while significant inflammatory responses were still observed in these trials (Noone et al. 2000; Ruiz et al. 2001). The inflammatory toxicities have been considered to be related to the CpG of DNA. Local gene transfer by lipoplexes for the therapy of alpha1-antitrypsin (AAT) deficiency was reported by Brigham and colleagues (Brigham et al. 2000). It has been demonstrated that the delivery of a normal AAT gene by CL to the respiratory epithelium of deficient patients produces potentially therapeutic local AAT concentrations.

Currently, however, only limited success has been reached in the case of lipoplexes-based gene therapy. In addition, local delivery is the most frequently employed route in the performed clinical trials. Toxicity and efficiency are two major issues found in these trials. Further development of safe and highly efficient lipid-vectors is highly required for the successful gene therapy.

7.2 *Polymer Vectors*

In addition to lipoplexes, cationic polymers based polyplexes have also been used for gene therapy in clinical trials. Ohana et al. performed the transfection study using polyplexes based on a pDNA and jetPEI (22 kDa) in human patients with bladder carcinoma (Ohana et al. 2004). The employed pDNA encoded for the expression of diphtheria toxin A chain under the H19 promoter gene, an paternally-imprinted, oncofetal gene commonly expressed in various tumor tissue.

Nearly complete ablation of the tumor was determined by video imaging in the two human patients after treating once a week with 2 mg of pDNA-jetPEI polyplexes for a total of 9 weeks.

Fewell et al. have developed a non-viral vector (EGEN-001) consisting of a pDNA encoding the human gene for IL-12 and a synthetic polycation of PEI derivative (Kendrick et al. 2008). This delivery system is currently being investigated in phase I trial for the therapy of recurrent ovarian cancer. EGEN-001 is designed to increase the local concentration of IL-12 in the tumor microenvironment. Intraperitoneal injections of EGEN-001 resulted in an overall clinical response of 31% stable disease and 69% progressive disease after four weekly treatments. When patients were treated with EGEN-001 in combination with either docetaxel or carboplatin, all patients showed partial response by CT evaluation and reductions in CA- 125 levels following the completion of EGEN and two cycles of chemotherapy.

DermaVir, the first dendritic cell-targeting topical HIV vaccine candidate for topical immunization, is also a polyplex-based nanomedicine, that is fabricated by pDNA encoding the entire HIV genome minus the integrase gene and mannose-conjugated lPEI (Lori et al. 2005). Phase 1 clinical trial performed by Lori and colleagues showed that the DermaVir patch could significantly increase the number of HIV-specific T-cells at DNA doses varying from 0.1 to 0.8 mg. Currently, the DermaVir patch is in a phase II clinical trial. In a randomized, placebo-controlled, dose-ranging Phase II study conducted in Hamburg, Germany, that enrolled thirty-six HIV-infected individuals not yet taking antiretroviral drugs, some promising results were also obtained.

Besides PEI derivatives, PLL derivatives have been employed as nonviral vectors for gene transfection in human patients. For example, Cooper and colleagues have constructed DNA-loaded nanoparticles for the therapy of cystic fibrosis (CF) (Konstan et al. 2004). PEG-PLL was used to condense a pDNA carrying the cystic fibrosis transmembrane regulator (CFTR)-encoding gene. Twelve patients with classic CF were intranasally administered with PEG-PLL/pDNA nano-polyplexes. No serious adverse events occurred, and no events were attributed to polyplex nanoformulation. There was no association of serum or nasal washing inflammatory mediators with administration of compacted DNA. These results suggested the safety and tolerability of the employed DNA nano-therapeutics. In addition, this study also showed partial or complete restoration of CFTR chloride channel function for up to six days after treatment.

Davis's group has developed a β-CD containing polycation that can effectively mediate siRNA delivery in preclinical studies (Davis 2009). Most recently, the results obtained in Phase I clinical trial have been presented (Davis et al. 2010). The gene delivery system employed in this trial consists of synthetic nanoparticles, which were constructed by the complexation of siRNA (designed to reduce the expression of the RRM2) with a linear, cyclodextrin-based polycation (CDP). Subsequent PEGylation and targeting was implemented by host-guest interactions between β-CD and adamantyl (Ada) group, using Ada-terminated PEG and Ada/human transferrin protein (TF)-terminated PEG, respectively. Tumour biopsies from melanoma patients obtained after treatment via systemic administration show the presence of intracellularly localized nanoparticles in amounts that correlate with dose levels of the nanoparticles administered. Furthermore, a reduction was found in both the specific mRNA (M2 subunit of ribonucleotide reductase (RRM2)) and the protein (RRM2) levels when compared to pre-dosing tissue. Most importantly, the presence of an mRNA fragment was detected, which demonstrates that siRNA-mediated mRNA cleavage occurs specifically at the site predicted for an RNAi mechanism from a patient who received the highest dose of the nanoparticles. These results evidence that siRNA administered systemically to a human can produce a specific gene inhibition by an RNAi mechanism of action.

8 Concluding Remarks

Tremendous progress in the design and synthesis of nonviral vectors has been achieved in the past decade, which significantly richens the carrier library that can be selected for gene delivery. A large number of studies, however, still focused on *in vitro* transfections. Even in this case, the nonviral vectors that possess higher transfection efficiency and minimal cytotoxicity compared with commercially available transfer reagents such as Lipofectamine 2000 and PEI25, are still limited. Currently, only a small percentage of non-viral vectors have been in clinical trials, especially the extensively investigated cationic polymers. Based on the growing understanding of mechanisms underlying the gene delivery of nonviral systems, delicate design and discovery of transfer vectors that can overcome each of the extracellular and intracellular barriers is highly necessary for efficient gene therapy.

For lipoplexes and polyplexes, the establishment of the structure-activity relationships, based on a better understanding of complex structure and characters, stability in serum, and cellular uptake, intracellular translocation as well as nuclear entry, may facilitate the advances towards the rationalized and conceptualized new vectors. High throughput screening technique combined with combinatorial chemistry has already showed their advantages to rapidly identify effective vectors, and accordingly establish reliable structure-function correlations (Sunshine et al. 2009; Love et al. 2010). Besides chemical tailoring on currently studied compounds, materials with entirely new structures should be discovered to provide gene vectors with clinical significance.

Emerging delivery systems based on nanomaterials like nanocrystals and inorganic nanostructures have partly showed their benefits over organic compounds, such as ease in real-time tracing via electronic or optical imaging techniques, unique photo-, magnetic-, and thermo-responsiveness for signal-directed smart delivery, especially the capacity for penetrating some tissue barriers. Honeycomb mesoporous silica nanoparticles with 3 nm pores have been shown able to transport the payload DNA molecules into isolated plant cells and intact leaves, and achieve significant transgene expression in these organisms (Torney et al. 2007).

Alternatively, these nanostructured materials, such as quantum dots and magnetic nanoparticles may be integrated with other delivery vectors to create multi-functionalized systems armed with multimodal imaging, simultaneous imaging and gene delivery. Additional functions such as excellent stability and cell or tissue-specific targetability can be integrated. These multifunctional vectors loaded with therapeutic genes that contain functional elements which can regulate and direct the cytonuclear translocation and subsequent transgene expression of genetic payload are even more exciting directions deserving our efforts in the coming years. Simultaneous delivery of drugs and genes represents another multifunctionalizing approach, through which a synergistic therapeutic effect may be achieved. For this purpose, efficacious methods should be developed to identify appropriate drug-gene pairs. Nevertheless, despite their promising future, the study on multifunctionalized gene vectors is still in its infancy stage.

By continuing to observe and understand how naturally evolved systems have overcome all the delivery barriers via top-down approaches, we may continue to mimic these natural processes by a well-controlled top-down strategy to develop highly efficient synthetic systems, thus render nonviral vectors to be important tools for human gene therapy.

Acknowledgements We gratefully acknowledge the financial support from the National Natural Science Foundation of China (No. 21004077). JXZ would like to thank the Returnee Start-up Fund from the Third Military Medical University (2010XHG03).

References

Abbasi, M., Uludag, H., Incani, V., Hsu, C. Y. M. & Jeffery, A. (2008) Further investigation of lipid-substituted poly(l-lysine) polymers for transfection of human skin fibroblasts. *Biomacromolecules*, 9, 1618–1630.

Abbasi, M., Uludag, H., Incani, V., Olson, C., Lin, X. Y., Clements, B. A., Rutkowski, D., Ghahary, A. & Weinfeld, M. (2007) Palmitic acid-modified poly-l-lysine for non-viral delivery of plasmid DNA to skin fibroblasts. *Biomacromolecules*, 8, 1059–1063.

Adcock, I. M. & Caramori, G. (2001) Cross-talk between pro-inflammatory transcription factors and glucocorticoids. *Immunol. Cell. Biol.*, 79, 376–384.

Akinc, A., Anderson, D. G., Lynn, D. M. & Langer, R. (2003) Synthesis of poly(β-amino ester)s optimized for highly effective gene delivery. *Bioconjugate Chem.*, 14, 979–988.

Akinc, A., Zumbuehl, A., Goldberg, M., Leshchiner, E. S., Busini, V., Hossain, N., Bacallado, S. A., Nguyen, D., Fuller, J., Alvarez, R., Borodovsky, A., Borland, T., Constien, R., De Fougerolles, A., Dorkin, J. R., Jayaprakash, K. N., Jayaraman, M., John, M., Koteliansky, V., Manoharan, M., Nechev, L., Qin, J., Racie, T., Raitcheva, D., Rajeev, K. G., Sah, D. W. Y., Soutschek, J., Toudjarska, I., Vornlocher, H. P., Zimmermann, T. S., Langer, R. & Anderson, D. G. (2008) A combinatorial library of lipid-like materials for delivery of RNAi therapeutics. *Nat. Biotechnol.*, 26, 561–569.

AL-Jamal, W. T., AL-Jamal, K. T., Tian, B., Lacerda, L., Bomans, P. H., Frederik, P. M. & Kostarelos, K. (2008) Lipid-quantum dot bilayer vesicles enhance tumor cell uptake and retention in vitro and in vivo. ACS Nano, 2, 408–418.

Alivisatos, A. P., Gu, W. & Larabell, C. (2005) Quantum dots as cellular probes. Annu. Rev. Biomed. Eng., 7, 55–76.

Allcock, H. R. & Kugel, R. L. (1965) Synthesis of high polymeric alkoxy- and aryloxyphosphonitriles. *J. Am. Chem. Soc.*, 87, 4216–4217.

Alshamsan, A., Haddadi, A., Incani, V., Samuel, J., Lavasanifar, A. & Uludag, H. (2009) Formulation and delivery of siRNA by oleic acid and stearic acid modified polyethylenimine. *Mol. Pharm.*, 6, 121–133.

Alton, E. W., Stern, M., Farley, R., Jaffe, A., Chadwick, S. L., Phillips, J., Davies, J. C., Smith, S. N., Browning, J., Davies, M. G., Hodson, M. E., Durham, S. R., Li, D., Jeffery, P. K., Scallan, M., Balfour, R., Eastman, S. J., Cheng, S. H., Smith, A. E., Meeker, D. & Geddes, D. M. (1999) Cationic lipidmediated CFTR gene transfer to the lungs and nose of patients with cystic fibrosis: A double-blind placebo-controlled trial. *Lancet*, 353, 947–954.

Anderson, D. G., Lynn, D. M. & Langer, R. (2003) Semi-automated synthesis and screening of a large library of degradable cationic polymers for gene delivery. *Angew. Chem. Int. Ed.*, 42, 3153–3158.

Anderson, D. G., Peng, W., Akinc, A., Hossain, N., Kohn, A., Padera, R., Langer, R. & Sawicki, J. A. (2004) A polymer library approach to suicide gene therapy for cancer. *Proc. Natl. Acad. Sci. U. S. A.*, 101, 16028–16033.

Andre, F. & Mir, L. M. (2004) DNA electrotransfer: its principles and an updated review of its therapeutic applications. *Gene Ther.*, 11, S33-S42.

Araki, J. & Ito, K. (2007) Recent advances in the preparation of cyclodextrin-based polyrotaxanes and their applications to soft materials. *Soft Matter*, 3, 1456–1473.

Arima, H., Kihara, F., Hirayama, F. & Uekama, K. (2001) Enhancement of gene expression by polyamidoamine dendrimer conjugates with alpha-, beta-, and gama-cyclodextrins. *Bioconjugate Chem.*, 12, 476–484.

Auricchio, A., Gao, G. P., Yu, Q. C., Raper, S., Rivera, V. M., Clackson, T. & Wilson, J. M. (2002) Constitutive and regulated expression of processed insulin following in vivo hepatic gene transfer. *Gene Ther.*, 9, 963–971.

Azzam, T., Eliyahu, H., Makovitzki, A., Linial, M. & Domb, A. J. (2004) Hydrophobized dextran-spermine conjugate as potential vector for in vitro gene transfection. *J. Control. Release*, 96, 309–323.

Azzam, T., Eliyahu, H., Shapira, L., Linial, M., Barenholz, Y. & Domb, A. J. (2002) Polysaccharide-oligoamine based conjugates for gene delivery. *J. Med. Chem.*, 45, 1817–1824.

Bachelder, E. M., Beaudette, T. T., Broaders, K. E., Dashe, J. & Frechet, J. M. J. (2008) Acetal-derivatized dextran: An acid-responsive biodegradable material for therapeutic applications. *J. Am. Chem. Soc.*, 130, 10494–10495.

Bae, Y., Fukushima, S., Harada, A. & Kataoka, K. (2003) Design of environment-sensitive supramolecular assemblies for intracellular drug delivery: Polymeric micelles that are responsive to intracellular pH change. *Angew. Chem. Int. Ed.*, 42, 4640–4643.

Bae, Y. M., Choi, H., Lee, S., Kang, S. H., Kim, Y. T., Nam, K., Park, J. S., Lee, M. & Choi, J. S. (2007) Dexamethasone-conjugated low molecular weight polyethylenimine as a nucleus-targeting lipopolymer gene carrier. *Bioconjugate Chem.*, 18, 2029–2036.

Balani, K., Anderson, R., Laha, T., Andara, M., Tercero, J., Crumpler, E. & Agarwal, A. (2007) Plasma-sprayed carbon nanotube reinforced hydroxyapatite coatings and their interaction with human osteoblasts in vitro. *Biomaterials*, 28, 618–624.

Bartlett, D. W. & Davis, M. E. (2007) Physicochemical and biological characterization of targeted, nucleic acid-containing nanoparticles. *Bioconjugate Chem.*, 18, 456–468.

Basaka, S., Chen, D. R. & Biswas, P. (2007) Electrospray of ionic precursor solutions to synthesize iron oxide nanoparticles: Modified scaling law *Chem. Eng. Sci.*, 62, 1263–1268.

Belenkov, A. L., Alakhov, V. Y., Kabanov, A. V., Vinogradov, S. V., Panasci, L. C., Monia, B. P. & Chow, T. Y. (2004) Polyethyleneimine grafted with pluronic P85 enhances Ku86 antisense delivery and the ionizing radiation treatment efficacy in vivo. *Gene Ther.*, 11, 1665–1672.

Benns, J. M., Mahato, R. I. & Kim, S. W. (2002) Optimization of factors influencing the transfection efficiency of folate-PEG-folate-graftpolyethylenimine. *J. Control. Release*, 79, 255–269.

Berry, C. C., Wells, S., Charles, S. & Curtis, A. S. (2003) Dextran and albumin derivatised iron oxide nanoparticles: influence on fibroblasts in vitro. *Biomaterials*, 24, 4551–4557.

Bharali, D. J., Klejbor, I., Stachowiak, E. K., Dutta, P., Roy, I., Kaur, N., Bergey, E. J., Prasad, P. N. & Stachowiak, M. K. (2005) Organically modified silica nanoparticles: a nonviral vector for in vivo gene delivery and expression in the brain. *Proc. Natl. Acad. Sci. USA*, 102, 11539–11544.

Blessing, T., Kursa, M., Holzhauser, R., Kircheis, R. & Wagner, E. (2001) Different strategies for formation of pegylated EGF-conjugated PEI/DNA complexes for targeted gene delivery. *Bioconjugate Chem.*, 12, 529–537.

Bloomfield, V. A. (1997) DNA condensation by multivalent cations. *Biopolymers*, 44, 269–282.

Boeckle, S., Fahrmeir, J., Roedl, W., Ogris, M. & Wagner, E. (2006) Melittin analogs with high lytic activity at endosomal pH enhance transfection with purified targeted PEI polyplexes. *J. Control. Release*, 112, 240–248.

Boeckle, S., Wagner, E. & Ogris, M. (2005) C- versus N-terminally linked melittin-polyethylenimine conjugates: the site of linkage strongly influences activity of DNA polyplexes. *J. Gene Med.*, 7, 1335–1347.

Boomer, J. A., Thompson, D. H. & Sullivan, S. M. (2002) Formation of plasmid-based transfection complexes with an acid-labile cationic lipid: Characterization of in vitro and in vivo gene transfer. *Pharm. Res.*, 19, 1292–1301.

Bourgeat-Lami, E. (2002) Organic-inorganic nanostructured colloids. *J. Nanosci. Nanotechnol.*, 2, 1–24.

Boussif, O., Lezoualc'h, F., Zanta, M. A., Mergny, M. D., Scherman, D., Demeneix, B. & Behr, J. P. (1995) A versatile vector for gene and oligonucleotide transfer into cells in culture and in vivo: Polyethylenimine. *Proc. Natl. Acad. Sci. USA*, 92, 7297–7301.

Braterman, P. S., Xu, Z. P. & Yarberry, F. (2004) Layered double hydroxides (LDH). In: Auerbach, S. M., Carrado, K. A. & Dutta, P. K. (Eds.) *Handbook of layered materials.* New York, CRC press.

Bremner, K. H., Seymour, L. W., Logan, A. & Read, M. L. (2004) Factors influencing the ability of nuclear localization sequence peptides to enhance nonviral gene delivery. *Bioconjugate Chem.*, 15, 152–161.

Breyer, B., Jiang, W., Cheng, H., Zhou, L., Paul, R., Feng, T. & He, T. C. (2001) Adenoviral vector-mediated gene transfer for human gene therapy. *Curr. Gene Ther.*, 1, 149–162.

Brigham, K. L., Lane, K. B., Meyrick, B., Stecenko, A. A., Strack, S., Cannon, D. R., Caudill, M. & Canonico, A. E. (2000) Transfection of nasal mucosa with a normal alpha1-antitrypsin gene in alpha1-antitrypsin-deficient subjects: comparison with protein therapy. *Hum. Gene Ther.*, 11, 1023–1032.

Brinker, C. J. & Scherer, G. W. (1990) The physics and chemistry of sol-gel processing. *Sol-Gel Science.* Boston, Academic Press.

Bromberg, L., Alakhov, V. Y. & Hatton, T. A. (2006) Self-assembling Pluronic (R)-modified polycations in gene delivery. *Curr. Opin. Colloid. Inter.*, 11, 217–223.

Brownlie, A., Uchegbu, I. F. & Schatzlein, A. G. (2004) PEI-based vesicle-polymer hybrid gene delivery system with improved biocompatibility. *Int. J. Pharm.*, 274, 41–52.

Bruce, I. J., Taylor, J., Todd, M. J., Davies, E., Borioni, C., Sangregorio, T. & Sen, J. (2004) Synthesis, characterization and application of silica-magnetite nanocomposite. *J. Magn. Magn. Mater.*, 284, 145–160.

Brunner, S., Sauer, T., Carotta, S., Cotten, M., Saltik, M. & Wagner, E. (2000) Cell cycle dependence of gene transfer by lipoplex, polyplex and recombinant adenovirus. *Gene Ther.*, 7, 401–407.

Burckbuchler, V., Wintgens, V., Leborgne, C., Lecomte, S., Leygue, N., Scherman, D., Kichler, A. & Amiel, C. (2008) Development and characterization of new cyclodextrin polymer-based DNA delivery systems. *Bioconjugate Chem.*, 19, 2311–2320.

Cai, D., Mataraza, J. M., Qin, Z. H., Huang, Z., Huang, J., Chiles, T. C., Carnahan, D., Kempa, K. & Ren, Z. (2005) Highly efficient molecular delivery into mammalian cells using carbon nanotube spearing. *Nat. Methods*, 2, 449–454.

Cai, W., Shin, D. W., Chen, K., Gheysens, O., Cao, Q., Wang, S. X., Gambhir, S. S. & Chen, X. (2006) Peptide-labeled near-infrared quantum dots for imaging tumor vasculature in living subjects. *Nano Lett.*, 6, 669–676.

Caracciolo, G., Pozzi, D., Caminiti, R. & Congiu Castellano, A. (2003) Structural characterization of a new lipid/DNA complex showing a selective transfection efficiency in ovarian cancer cells. *Eur. Phys. J. E*, 10, 331–336.

Chen, A. A., Derfus, A. M., Khetani, S. R. & Bhatia, S. N. (2005a) Quantum dots to monitor RNAi delivery and improve gene silencing. *Nucleic Acids Res.*, 33, e190.

Chen, A. M., Zhang, M., Wei, D. G., Stueber, D., Taratula, O., Minko, T. & He, H. (2009) Mesoporous silica nanoparticles enhances the efficacy of chemotherapy in multidrug-resistant cancer cells. *Small*, 5, 2673–2677.

Chen, C. C., Liu, Y. C., Wu, C. H., Yeh, C. C., Su, M. T. & Wu, Y. C. (2005b) Preparation of fluorescent silica nanotubes and their application in gene delivery. *Adv. Mater.*, 17, 404–407.

Chen, D. & Xu, R. (1998) Hydrothermal synthesis and characterization of nanocrystallineγ-Fe2O3 particles *J. Solid State Chem.*, 137, 185–190.

Chen, H. H., Ho, Y. P., Jiang, X., Mao, H. Q., Wang, T. H. & Leong, K. W. (2008) Quantitative comparison of intracellular unpacking kinetics of polyplexes by a model constructed from quantum dot-FRET. *Mol. Ther.*, 16, 324–332.

Chen, Y., Xue, Z., Zheng, D., Xia, K., Zhao, Y., Liu, T., Long, Z. & Xia, J. (2003) Sodium chloride modified silica nanoparticles as a non-viral vector with a high efficiency of DNA transfer into cells. *Curr. Gene Ther.*, 3, 273–279.

Cheng, H., Li, Y. Y., Zeng, X., Sun, Y. X., Zhang, X. Z. & Zhuo, R. X. (2009) Protamine sulfate/poly(L-aspartic acid) polyionic complexes self-assembled via electrostatic attractions for combined delivery of drug and gene. *Biomaterials*, 30, 1246–1253.

Cheung, C. Y., Murthy, N., Stayton, P. S. & Hoffman, A. S. (2001) A pHsensitive polymer that enhances cationic lipid-mediated gene transfer. *Bioconjugate Chem.*, 12, 906–910.

Choi, J. S. & Lee, M. (2005) Effect of dexamethasone preincubation on polymer-mediated gene delivery. *Bull. Korean Chem. Soc.*, 26, 1209–1213.

Choi, J. S., Mackay, J. A. & Szoka, F. C. (2003) Low-pH-sensitive PEG-stabilized plasmid-lipid nanoparticles: preparation and characterization. *Bioconjugate Chem.*, 14, 420–429.

Chowdhury, E. H. & Akaike, T. (2005) Bio-functional inorganic materials: an attractive branch of gene-based nano-medicine delivery for 21st century. *Curr. Gene Ther.*, 5, 669–676.

Choy, J. H., Kwak, S. Y., Jeong, Y. J. & Park, J. S. (2000) Inorganic layered double hydroxides as nonviral vectors *Angew. Chem. Int. Ed.*, 39, 4041–4045.

Choy, J. H., Kwak, S. Y., Park, J. S. & Jeong, Y. J. (2001) Cellular uptake behavior of [γ-^{32}P] labeled ATP-LDH nanohybrids. *J. Mater. Chem.*, 11, 1671–1674.

Chu, T. C., Marks, J. W., Lavery, L. A., Faulkner, S., Rosenblum, M. G., Ellington, A. D. & Levy, M. (2006) Aptamer:Toxin conjugates that specifically target prostate tumor cells. *Cancer Res.*, 66, 5989–5992.

Colin, M., Maurice, M., Trugnan, G., Kornprobst, M., Harbottle, R. P., Knight, A., Cooper, R. G., Miller, A. D., Capeau, J., Coutelle, C. & Brahimi-Horn, M. C. (2000) Cell delivery, intracellular trafficking and expression of an integrin-mediated gene transfer vector in tracheal epithelial cells. *Gene Ther.*, 7, 139–152.

Corsi, K., Chellat, F., Yahia, L. & Fernandes, J. C. (2003) Mesenchymal stem cells, MG63 and HEK293 transfection using chitosan- DNA nanoparticles. *Biomaterials*, 24, 1255–1264.

Crepaldi, E. L., Pavan, P. C. & Valim, J. B. (1999) A new method of intercalation by anion exchange in layered double hydroxides *Chem. Commun.*, 2, 155–156.

Cryan, S. A., Holohan, A., Donohue, R., Darcy, R. & O'Driscoll, C. M. (2004) Cell transfection with polycationic cyclodextrin vectors. *Eur. J. Pharm. Sci.*, 21, 625–633.

Dames, P., Laner, A., Maucksch, C., Aneja, M. K. & Rudolph, C. (2007) Targeting of the glucocorticoid hormone receptor with plasmid DNA comprising glucocorticoid response elements improves nonviral gene transfer efficiency in the lungs of mice. *J. Gene Med.*, 9, 820–829.

Dass, C. R. (2002) Biochemical and biophysical characteristics of lipoplexes pertinent to solid tumour gene therapy. *Int. J. Pharm.*, 241, 1–25.

Dass, C. R. (2004) Lipoplex-mediated delivery of nucleic acids: factors affecting in vivo transfection. *J. Mol. Med.*, 82, 579–591.

Dass, C. R. & Choong, P. F. M. (2006) Selective gene delivery for cancer therapy using cationic liposomes: In vivo proof of applicability. *J. Control. Release*, 113, 155–163.

Dass, C. R. & Su, T. (2000) Delivery of lipoplexes for genotherapy of solid tumours: role of vascular endothelial cells. *J. Pharm. Pharmacol.*, 52, 1301–1317.

Dauty, E. & Verkman, A. S. (2005) Actin cytoskeleton as the principal determinant of size-dependent DNA mobility in cytoplasm: a new barrier for non-viral gene delivery. *J. Biol. Chem.*, 280, 7823–7828.

Davis, M. E. (2009) The first targeted delivery of siRNA in humans via a self-assembling, cyclodextrin polymer-based nanoparticle: From concept to clinic *Mol. Pharm.*, 6, 659–668.

Davis, M. E. & Brewster, M. E. (2004) Cyclodextrin-based pharmaceutics: Past, present and future *Nat. Rev. Drug Discov.*, 3, 1023–1035.

Davis, M. E., Zuckerman, J. E., Choi, C. H. J., Seligson, D., Tolcher, A., Alabi, C. A., Yen, Y., Heidel, J. & Ribas, A. (2010) Evidence of RNAi in humans from systemically administered siRNA via targeted nanoparticles. *Nature*, 464, 1067–1070.

Davis, P. B. & Cooper, M. J. (2007) Vectors for airway gene delivery. *AAPS J.*, 9, E11-E17.

De Ilarduya, C. T., Sun, Y. & Duzgunes, N. (2010) Gene delivery by lipoplexes and polyplexes. *Eur. J. Pharm. Sci.*, 40, 159–170.

De, M., Ghosh, P. S. & Rotello, V. M. (2008) Applications of nanoparticles in biology. *Adv. Mater.*, 20, 4225–4241.

De Wolf, H. K., Luten, J., Snel, C. J., Oussoren, C., Hennink, W. E. & Storm, G. (2005) In vivo tumor transfection mediated by polyplexes based on biodegradable poly(DMAEA)-phosphazene *J. Control. Release*, 109, 275–287.

Del Campo, A., Sen, T., Lellouchec, J. P. & Bruce, I. J. (2005) Multifunctional magnetite and silica-magnetite nanoparticles: Synthesis, surface activation and applications in life sciences *J. Magn. Magn. Mater.*, 293, 33–40.

Del Gaudio, P., Colombo, P., Colombo, G., Russo, P. & Sonvico, F. (2005) Mechanisms of formation and disintegration of alginate beads obtained by prilling. *Int. J. Pharm.*, 302, 1–9.

Demeneix, B. & Behr, J. P. (2005) Polyethylenimine (PEI) *Adv. Genetics*, 53, 217–230.

Deming, T. J. (2006) Polypeptide and polypeptide hybrid copolymer synthesis via NCA polymerization. *Adv. Polym. Sci.*, 202, 1–18.

Deng, Y., Wang, L., Yang, W., Fu, S. & Elaïssari, A. (2003) Preparation of magnetic polymeric particles via inverse microemulsion polymerization process *J. Magn. Magn. Mater.*, 257, 69–78.

Derfus, A. M., Chen, A. A., Min, D. H., Ruoslahti, E. & Bhatia, S. N. (2007) Targeted quantum dot conjugates for siRNA delivery. *Bioconjug. Chem.*, 18, 1391–1396.

Diaz-Moscoso, A., Balbuena, P., Gomez-Garcia, M., Mellet, C. O., Benito, J. M., Gourrierec, L. L., Giorgio, C. D., Vierling, P., Mazzaglia, A., Micali, N., Defaye, J. & Fernandez, J. M. G. (2008) Rational design of cationic cyclooligosaccharides as efficient gene delivery systems. *Chem. Commun.*, 2001–2003.

Dicks, A. L. (2006) The role of carbon in fuel cells *J. Power Sources*, 156, 128–141.

Diebold, S. S., Kursa, M., Wagner, E., Cotten, M. & Zenke, M. (1999) Mannose polyethylenimine conjugates for targeted DNA delivery into dendritic cells. *J. Biol. Chem.*, 274, 19087–19094.

Dincer, S., Turk, M. & Piskin, E. (2005) Intelligent polymers as nonviral vectors. *Gene Ther.*, 12, S139-S145.

Dobson, J. (2006) Magnetic nanoparticles for drug delivery. *Drug Dev. Res.*, 67, 55–60.

Doody, A. M., Korley, J. N., Dang, K. P., Zawaneh, P. N. & Putnam, D. (2006) Characterizing the structure/function parameter space of hydrocarbon-conjugated branched polyethylenimine for DNA delivery in vitro. *J. Control. Release*, 116, 227–237.

Dubruel, P., Christiaens, B., Vanloo, B., Bracke, K., Rosseneu, M., Vandekerckhove, J. & Schacht, E. (2003) Physicochemical and biological evaluation of cationic polymethacrylates as vectors for gene delivery *Eur. J. Pharm. Sci.*, 13, 211–220.

Dufes, C., Uchegbu, I. F. & Schatzlein, A. G. (2005) Dendrimers in gene delivery. *Adv. Drug Deliv. Rev.*, 57, 2177–2202.

Edelstein, M. L., Abedi, M. R. & Wixon, J. (2007) Gene therapy clinical trials worldwide to 2007 - An update. *J. Gene Med.*, 9, 833–842.

Edelstein, M. L., Abedi, M. R., Wixon, J. & Edelstein, R. M. (2004) Gene therapy clinical trials worldwide 1989–2004 - An overview. *J. Gene Med.*, 6, 597–602.

Elferink, J. G. (1985) Cytolytic effect of polylysine on rabbit polymorphonuclear leukocytes. *Inflammation*, 9, 321–331.

Eliyahu, H., Makovitzki, A., Azzam, T., Zlotkin, A., Joseph, A., Gazit, D., Barenholz, Y. & Domb, A. J. (2005) Novel dextran-spermine conjugates as transfecting agents: comparing water-soluble and micellar polymers. *Gene Ther.*, 12, 494–503.

Erbacher, P., Remy, J. S. & Behr, J. P. (1999) Gene transfer with synthetic virus-like particles via the integrin-mediated endocytosis pathway. *Gene Ther.*, 6, 138–145.

Erbacher, P., Roche, A. C., Monsigny, M. & Midoux, P. (1995) Glycosylated polylysine/DNA complexes: Gene transfer efficiency in relation with the size and the sugar substition level of glycosylated polylysines and with the plasmid size. *Bioconjugate Chem.*, 6, 401–410.

Farokhzad, O. C. & Langer, R. (2009) Impact of nanotechnology on drug delivery. *ACS Nano*, 3, 16–20.

Fasbender, A., Lee, J. H., Walters, R. W., Moninger, T. O., Zabner, J. & Welsh, M. J. (1998) Incorporation of adenovirus in calcium phosphate precipitates enhances gene transfer to airway epithelia in vitro and in vivo. *J. Clin. Invest.*, 102, 184–193.

Felgner, J. H., Kumar, R., Sridhar, A., Wheeler, C. J., Tsai, Y. J., Border, R., Ramsey, P., Martin, M. & Felgner, P. L. (1994) Enhanced gene delivery and mechanism studies with a novel series of cationic lipid formulations. *J. Biol. Chem.*, 269, 2550–2561.

Felgner, P. L., Gadek, T. R., Holm, M., Roman, R., Chan, H. W., Wenz, M., Northrop, J. P., Ringold, G. M. & Danielsen, M. (1987) Lipofection: a highly efficient, lipid mediated DNA-transfection procedure. *Proc. Natl. Acad. Sci. U. S. A.*, 84, 7413–7417.

Fella, C., Walker, G., Ogris, M. & Wagner, E. (2008) Amine-reactive pyridylhydrazone-based PEG reagents for pH-reversible PEI polyplex shielding. *Eur. J. Pharm. Sci.*, 34, 309–320.

Fenske, D. B., Maclachlan, I. & Cullis, P. R. (2002) Stabilized plasmid-lipid particles: a systemic gene therapy vector. *Method Enzymol.*, 346, 36–71.

Ferkol, T., Perales, J. C., Mularo, F. & Hanson, R. W. (1996) Receptor-mediated gene transfer into macrophages. *Proc. Natl. Acad. Sci. USA*, 93, 101–105.

Fischer, D., Von Harpe, A., Kunath, K., Petersen, H., Li, Y. & Kissel, T. (2002) Copolymers of ethylene Imine and N-(2-Hydroxyethyl)-ethylene Imine as tools to study effects of polymer structure on physicochemical and biological properties of DNA complexes. *Bioconjugate Chem.*, 13, 1124–1133.

Forrest, M. L., Gabrielson, N. & Pack, D. W. (2005) Cyclodextrin–Polyethylenimine conjugates for targeted in vitro gene delivery. *Biotechnol. Bioeng.*, 89, 416–423.

Forrest, M. L., Koerber, J. T. & Pack, D. W. (2003) A degradable polyethylenimine derivative with low toxicity for highly efficient gene delivery. *Bioconjugate Chem.*, 14, 934–940.

Forrest, M. L., Meister, G. E., Koerber, J. T. & Pack, D. W. (2004) Partial acetylation of polyethylenimine enhances in vitro gene delivery. *Pharm. Res.*, 21, 365–371.

Fu, K., Pack, D. W., Klibanov, A. M. & Langer, R. (2000) Visual evidence of acidic environment within degrading poly(lactic-co-glycolic acid) (PLGA) microspheres. *Pharm. Res.*, 17, 100–106.

Fukumori, Y. & Ichikawa, H. (2006) Nanoparticles for cancer therapy and diagnosis *Adv. Powder Technol.*, 17, 1–28.

Funhoff, A. M., Van Nostrum, C. F., Koning, G. A., Schuurmans-Nieuwenbroek, N. M. E., Crommelin, D. J. A. & Hennink, W. E. (2004a) Endosomal escape of polymeric gene delivery complexes is not always enhanced by polymers buffering at low pH *Biomacromolecules*, 5, 32–39.

Funhoff, A. M., Van Nostrum, C. F., Lok, M. C., Fretz, M. M., Crommelin, D. J. A. & Hennink, W. E. (2004b) Poly(3-guanidinopropyl methacrylate): A novel cationic polymer for gene delivery. *Bioconjugate Chem.*, 15, 1212–1220.

Fynan, E. F., Webster, R. G., Fuller, D. H., Haynes, J. R., Santoro, J. C. & Robinson, H. L. (1993a) DNA vaccines: protective immunizations by parenteral, mucosal, and gene-gun inoculations. *Proc. Natl. Acad. Sci. U. S. A.*, 90, 11478–11482.

Fynan, E. F., Webster, R. G., Fuller, D. H., Haynes, J. R., Santoro, J. C. & Robinson, H. L. (1993b) DNA vaccines: Protective immunizations by parenteral, mucosal, and gene-gun inoculations. *Proc. Natl. Acad. Sci. USA*, 90, 11478–11482.

Gabrielson, N. P. & Pack, D. W. (2006) Acetylation of polyethylenimine enhances gene delivery via weakened polymer/DNA interactions. *Biomacromolecules*, 7, 2427–2435.

Galliot, C., Prevote, D., Caminade, A. M. & Majoral, J. P. (1995) Polyaminophosphines containing dendrimers - Synthesis and characterizations. *J. Am. Chem. Soc.*, 117, 5470–5476.

Gao, X., Cui, Y., Levenson, R. M., Chung, L. W. & Nie, S. (2004) In vivo cancer targeting and imaging with semiconductor quantum dots. *Nat. Biotechnol.*, 22, 969–976.

Gaponik, N., Talapin, D. V., Rogach, A. L. H. E. K., Shevchenko, V., Kornowski, A., Eychmüller, E. & Weller, H. (2002) Thiol-capping of CdTe nanocrystals: An alternative to organometallic synthetic routes. *J. Phys. Chem. B*, 106, 7177–7185.

Gardner, E., Huntoon, K. M. & Pinnavaia, T. J. (2001) Direct synthesis of alkoxide-intercalated derivatives of hydrocalcite-like layered double hydroxides: Precursors for the formation of colloidal layered double hydroxide suspensions and transparent thin films. *Adv. Mater.*, 13, 1263–1266.

Germershaus, O., Mao, S., Sitterberg, J., Bakowsky, U. & Kissel, T. (2008) Gene delivery using chitosan, trimethyl chitosan or polyethylenglycol-graft-trimethyl chitosan block copolymers: establishment of structure-activity relationships in vitro. *J. Control. Release*, 125, 145–154.

Ghosh, P. S., Kim, C. K., Han, G., Forbes, N. S. & Rotello, V. M. (2008) Efficient gene delivery vectors by tuning the surface charge density of amino acid-functionalized gold nanoparticles. *ACS Nano*, 1, 2213–2218.

Glover, D. J., Lipps, H. J. & Jans, D. A. (2005) Towards safe, non-viral therapeutic gene expression in humans, *Nat. Rev. Genet.*, 6, 299–310.

Godbey, W. T., Wu, K. K. & Mikos, A. G. (1999) Poly(ethylenimine) and its role in gene delivery. *J. Control. Release*, 60, 149–160.

Gonçalves, C., Mennesson, E., Fuchs, R., Gorvel, J. P., Midoux, P. & Pichon, C. (2004) Macropinocytosis of polyplexes and recycling of plasmid via the clathrin-dependent pathway impair the transfection efficiency of human hepatocarcinoma cells. *Mol. Ther.*, 10, 373–385.

Gonzalez, H., Hwang, S. J. & Davis, M. E. (1999) New class of polymers for the delivery of macromolecular therapeutics. *Bioconjugate Chem.*, 10, 1068–1074.

Gosselin, M. A., Guo, W. J. & Lee, R. J. (2001) Efficient gene transfer using reversibly cross-linked low molecular weight polyethylenimine. *Bioconjugate Chem.*, 12, 989–994.

Graham, F. L. & van der Eb, A. J. (1973) A new technique for the assay of infectivity of human adenovirus 5 DNA. *Virology*, 52, 456–467.

Green, J. J., Shi, J., Chiu, E., Leshchiner, E. S., Langer, R. & Anderson, D. G. (2006) Biodegradable polymeric vectors for gene delivery to human endothelial cells. *Bioconjugate Chem.*, 17, 1162–1169.

Grujicica, M., Sun, Y. P. & Koudela, K. L. (2007) The effect of covalent functionalization of carbon nanotube reinforcements on the atomic-level mechanical properties of poly-vinyl-ester-epoxy *Appl. Surf. Sci.*, 253, 3009–3021.

Guo, X. & Szoka, F. C. (2001) Steric stabilization of fusogenic liposomes by a low-pH sensitive PEG-diortho ester-lipid conjugate. *Bioconjugate Chem.*, 12, 291–300.

Gupta, A. K. & Gupta, M. (2005) Synthesis and surface engineering of iron oxide nanoparticles for biomedical applications. *Biomaterials*, 26, 3995–4021.

Gupta, B., Levchenko, T. S. & Torchilin, V. P. (2007) TAT peptide-modified liposomes provide enhanced gene delivery to intracranial human brain tumor Xenografts in nude mice *Oncology Res.*, 16, 351–359.

Haensler, J. & Szoka, F. C. (1993) Polyamidoamine cascade polymers mediate efficient transfection of cells in culture. *Bioconjugate Chem.*, 4, 372–379.

Hafez, I. M., Maurer, N. & Cullis, P. R. (2001) On the mechanism whereby cationic lipids promote intracellular delivery of polynucleic acids. *Gene Ther.*, 8, 1188–1196.

Han, G., Chari, N. S., Verma, A., Hong, R., Martin, C. T. & Rotello, V. M. (2005) Controlled recovery of the transcription of nanoparticle-bound DNA by intracellular concentrations of glutathione. *Bioconjug. Chem.*, 16, 1356–1359.

Han, S., Mahato, R. I. & Kim, S. W. (2001) Water-soluble lipopolymer for gene delivery. *Bioconjugate Chem.*, 12, 337–345.

Harada, A., Hashidzume, A. & Takashima, Y. (2006) Cyclodextrin-based supramolecular polymers *Adv. Polym. Sci.*, 201, 1–43.

Harada, A. & Kamachi, M. (1990) Complex formation between poly(ethylene glycol) and a-cyclodextrin. *Macromolecules*, 23, 2821–2823.

Hart, S. L., Harbottle, R. P., Cooper, R., Miller, A., Willianon, R. & Coutelle, C. (1995) Gene delivery and expression mediated by an integrin-binding peptide. *Gene Ther.*, 2, 552–554.

Hashimoto, M., Morimoto, M., Saimoto, H., Shigema, Y. & Sato, T. (2006) Lactosylated chitosan for DNA delivery into hepatocytes: The effect of lactosylation on the physicochemical properties and intracellular trafficking of pDNA/chitosan complexes. *Bioconjugate Chem.*, 17, 309–316.

Hautot, D., Pankhurst, Q. A., Khan, N. & Dobson, J. (2003) Preliminary evaluation of nanoscale biogenic magnetite in Alzheimer's disease brain tissue. *Proc. Biol. Sci.*, 270, S62-S64.

Heidel, J. D., Yu, Z. P., Liu, J. Y. C., Rele, S. M., Liang, Y. C., Zeidan, R. K., Kornbrust, D. J. & Davis, M. E. (2007) Administration in non-human primates of escalating intravenous doses of targeted nanoparticles containing ribonucleotide reductase subunit M2 siRNA. *Proc. Natl. Acad. Sci. U. S. A.*, 104, 5715–5721.

Held, P. K., Olivares, E. C., Aguilar, C. P., Finegold, M., Calos, M. P. & Grompe, M. (2005) In vivo correction of murine hereditary tyrosinemia type I by phiC31 integrase-mediated gene delivery. *Mol. Ther.*, 11, 399–408.

Heller, J., Barr, J., Ng, S. Y., Schwach-Abdellaoui, K. & Gurny, R. (2002) Poly(ortho esters): synthesis, characterization, properties and uses. *Adv. Drug Deliv. Rev.*, 54, 1015–1039.

Hinrichs, W. L., Schuurmans-Nieuwenbroek, N. M., Van De Wetering, P. & Hennink, W. E. (1999) Thermosensitive polymers as carriers for DNA delivery. *J. Control. Release*, 60, 249–259.

Ho, Y. P., Chen, H. H., Leong, K. W. & Wang, T. H. (2006) Evaluating the intracellular stability and unpacking of DNA nanocomplexes by quantum dots-FRET. *J. Control. Release*, 116, 83–89.

Høgset, A., Prasmickaite, L., Selbo, P. K., Hellum, M., Engesæter, B. Ø., Bonsted, A. & Berg, K. (2004) Photochemical internalisation in drug and gene delivery. *Adv. Drug Deliv. Rev.*, 56, 95–115.

Høgset, A., Prasmickaite, L., Tjelle, T. E. & Berg, K. (2000) Photochemical transfection: a new technology for light-induced, site-directed gene delivery. *Hum. Gene Ther.*, 11, 869–880.

Hollon, T. (2000) Researchers and regulators reflect on first gene therapy death. *Nat. Med.*, 6, 6–6.

Hosseinkhani, H., Azzam, T., Tabata, Y. & Domb, A. J. (2004) Dextran-spermine polycation: an efficient nonviral vector for in vitro and in vivo gene transfection. *Gene Ther.*, 11, 194–203.

Hu-Lieskovan, S., Heidel, J. D., Bartlett, D. W., Davis, M. E. & Triche, T. J. (2005) Sequence-specific knockdown of EWS-FLI1 by targeted, nonviral delivery of small interfering RNA inhibits tumor growth in a murine model of metastatic Ewing's sarcoma. *Cancer Res.*, 65, 8984–8992.

Hu, F. Q., Zhao, M. D., Yuan, H., You, J., Du, Y. Z. & Zeng, S. (2006) A novel chitosan oligosaccharide-stearic acid micelles for gene delivery: Properties and in vitro transfection studies. *Int. J. Pharm.*, 315, 158–166.

Hu, Y., Cai, K. Y., Luo, Z. & Hu, R. (2010) Construction of polyethyleneimine-beta-cyclodextrin/pDNA multilayer structure for improved in situ gene transfection. *Adv. Eng. Mater.*, 12, B18-B25.

Huang, C. K., Lo, C. L., Chen, H. H. & Hsiue, G. H. (2007) Multifunctional micelles for cancer cell targeting, distribution imaging, and anticancer drug delivery. *Adv. Funct. Mater.*, 17, 2291–2297.

Huang, F. H. & Gibson, H. W. (2005) Polypseudorotaxanes and polyrotaxanes. *Prog. Polym. Sci.*, 30, 982–1018.

Huang, H. C., Barua, S., Kay, D. B. & Rege, K. (2009) Simultaneous enhancement of photothermal stability and gene delivery efficacy of gold nanorods using polyelectrolytes. *ACS Nano*, 3, 2941–2952.

Huang, H. L., Tang, G. P., Wang, Q. Q., Li, D., Shen, F. P., Zhou, J. & Yu, H. (2006) Two novel non-viral gene delivery vectors: low molecular weight polyethylenimine cross-linked by (2-hydroxypropyl)-b-cyclodextrin or (2-hydroxypropyl)-c-cyclodextrin. *Chem. Commun.*, 2382–2384.

Huang, H. L., Yu, H., Tang, G. P., Wang, Q. Q. & Li, J. (2010) Low molecular weight polyethylenimine cross-linked by 2-hydroxypropyl-gamma-cyclodextrin coupled to peptide targeting HER2 as a gene delivery vector. *Biomaterials*, 31, 1830–1838.

Hwang, S. J., Bellocq, N. C. & Davis, M. E. (2001) Effects of structure of alpha-cyclodextrin-containing polymers on gene delivery. *Bioconjugate Chem.*, 12, 280–290.

Hyeon, T., Lee, S. S., Park, J., Chung, Y. & Na, H. B. (2001) Synthesis of highly crystalline and monodisperse maghemite nanocrystallites without a size-selection process. *J. Am. Chem. Soc.*, 123, 12798–12801.

Incani, V., Lavasanifar, A. & Uludag, H. (2010) Lipid and hydrophobic modification of cationic carriers on route to superior gene vectors *Soft Matter*, 6, 2124–2138.

Incani, V., Lin, X. Y., Lavasanifar, A. & Uludag, H. (2009) Relationship between the extent of lipid substitution on poly(l-lysine) and the DNA delivery efficiency. *ACS Appl. Mater. Interfaces*, 1, 841–848.

Incani, V., Tunis, E., Clements, B. A., Olson, C., Kucharski, C., Lavasanifar, A. & Uludag, H. (2007) Palmitic acid substitution on cationic polymers for effective delivery of plasmid DNA to bone marrow stromal cells. *J. Biomed. Mater. Res. Part A*, 81, 493–504.

Ishii, T., Okahata, Y. & Sato, T. (2001) Mechanism of cell transfection with plasmid/chitosan complexes. *Biochim. Biophys. Acta*, 1514, 51–64.

Itaka, K., Yamauchi, K., Harada, A., Nakamura, K., Kawaguchi, H. & Kataoka, K. (2003) Polyion complex micelles from plasmid DNA and poly(ethylene glycol)–poly(l-lysine) block copolymer as serum-tolerable polyplex system: physicochemical properties of micelles relevant to gene transfection efficiency. *Biomaterials*, 24, 4495–4506.

Itoh, T., Shichi, T., Yui, T. & Takagi, K. (2005) Layered double hydroxide hybrids with dicetylphosphate. *J. Colloid Interface Sci.*, 291, 218–222.

Janat-Amsbury, M. M., Yockman, J. W., Lee, M., Kern, S., Furgeson, D. Y., Bikram, M. & Kim, S. W. (2004) Combination of local, nonviral IL12 gene therapy and systemic paclitaxel treatment in a metastatic breast cancer model. *Mol. Ther.*, 9, 829–836.

Jen, C. P., Chen, Y. H., Fan, C. S., Yeh, C. S., Lin, Y. C., Shieh, D. B., Wu, C. L., Chen, D. H. & Chou, C. H. (2004) A nonviral transfection approach in vitro: the design of a gold nanoparticle vector joint with microelectromechanical systems. *Langmuir*, 20, 1369–1374.

Jeon, E., Kim, H. D. & Kim, J. S. (2003) Pluronic-grafted poly-(l)-lysine as a new synthetic gene carrier. *J. Biomed. Mater. Res. Part A*, 66, 854–859.

Jiang, H. L., Kim, Y. K., Arote, R., Nah, J. W., Cho, M. H., Choi, Y. J., Akaike, T. & Cho, C. S. (2007) Chitosan-graft-polyethylenimine as a gene carrier. *J. Control. Release*, 117, 273–280.

Jiang, W. Q., Yang, H. C., Yang, S. Y., Horng, H. E., Hungd, J. C., Chen, C. Y. & Hong, C. Y. (2004) Preparation and properties of superparamagnetic nanoparticles with narrow size distribution and biocompatible *J. Magn. Magn. Mater.*, 283, 210–214.

Jilek, S., Merkle, H. P. & Walter, E. (2005) DNA-loaded biodegradable microparticles as vaccine delivery systems and their interaction with dendritic cells. *Adv. Drug Deliv. Rev.*, 57, 377–390.

Jordan, M., Schallhorn, A. & Wurm, F. M. (1996) Transfecting mammalian cells: optimization of critical parameters affecting calcium-phosphate precipitate formation. *Nucleic Acids Res.*, 24, 596–601.

Jordan, M. & Wurm, F. (2004) Transfection of adherent and suspended cells by calcium phosphate. *Methods*, 33, 136–143.

Jun, Y. J., Kim, J. H., Choi, S. J., Lee, H. J., Jun, M. J. & Sohn, Y. S. (2007) A tetra(L-lysine)-grafted poly(organophosphazene) for gene delivery. *Bioorg. Med. Chem. Lett.*, 17, 2975–2978.

Jung, J., Solanki, A., Memoli, K. A., Kamei, K. I., Kim, H., Drahl, M. A., Williams, L. J., Tseng, H. R. & Lee, K. (2010) Selective inhibition of human brain tumor cells through multifunctional quantum-dot-based siRNA delivery. *Angew. Chem. Int. Ed.*, 49, 103–107.

Kabanov, A. V., Lemieux, P., Vinogradov, S. V. & Alakhov, V. Y. (2002) Pluronic((R)) block copolymers: novel functional molecules for gene therapy. *Adv. Drug Deliv. Rev.*, 54, 223–233.

Kaech, S. M. & Ahmed, R. (2001) Memory CD8+ T cell differentiation: initial antigen encounter triggers a development program in naïve cells. *Nat. Immunol.*, 2, 415–422.

Kakizawa, Y., Furukawa, S. & Kataoka, K. (2004) Block copolymer-coated calcium phosphate nanoparticles sensing intracellular environment for oligodeoxynucleotide and siRNA delivery. *J. Control. Release*, 97, 345–356.

Kakizawa, Y. & Kataoka, K. (2002) Block copolymer self-assembly into monodispersive nanoparticles with hybrid core of antisense DNA and calcium phosphate. *Langmuir*, 18, 4539–4543.

Kale, A. A. & Torchilin, V. P. (2007) Enhanced transfection of tumor cells in vivo using "Smart" pH-sensitive TAT-modified pegylated liposomes. *J. Drug Target.*, 15, 538–545.

Kam, N. W., Liu, Z. & Dai, H. (2005) Functionalization of carbon nanotubes via cleavable disulfide bonds for efficient intracellular delivery of siRNA and potent gene silencing. *J. Am. Chem. Soc.*, 127, 12492–12493.

Kanayama, N., Fukushima, S., Nishiyama, N., Itaka, K., Jang, W. D., Miyata, K., Yamasaki, Y., Chung, U. I. & Kataoka, K. (2006) A PEG-based biocompatible block catiomer with high buffering capacity for the construction of polyplex micelles showing efficient gene transfer toward primary cells. *ChemMedChem*, 1, 439–444.

Kasturi, S. P., Sachaphibulkij, K. & Roy, K. (2005) Covalent conjugation of polyethyleneimine on biodegradable microparticles for delivery of plasmid DNA vaccines. *Biomaterials*, 26, 6375–6385.

Katayama, Y., Fujii, K., Ito, E., Sakakihara, S., Sonoda, T., Murata, M. & Maeda, M. (2002) Intracellular signal-responsive artificial gene regulation for novel gene delivery. *Biomacromolecules*, 3, 905–909.

Katayose, S. & Kataoka, K. (1997) Water-soluble polyion complex associates of DNA and poly(ethylene glycol)-poly(L-lysine) block copolymer. *Bioconjugate Chem.*, 8, 702–707.

Kawamura, K., Oishi, J., Kang, J. H., Kodama, K., Sonoda, T., Murata, M., Niidome, T. & Katayama, Y. (2005) Intracellular signal-responsive gene carrier for cell-specific gene expression. *Biomacromolecules*, 6, 908–913.

Kawano, T., Okuda, T., Aoyagi, H. & Niidome, T. (2004) Long circulation of intravenously administered plasmid DNA delivered with dendritic poly(L-lysine) in the blood flow. *J. Control. Release*, 99, 329–337.

Kendrick, J. E., Matthews, K. S., Straughn, J. M., Barnes, J. M. N., Fewell, J., Anwer, K. & Alvarez, R. D. (2008) A phase I trial of intraperitoneal EGEN-001, a novel IL-12 gene therapeutic, administered alone or in combination with chemotherapy in patients with recurrent ovarian cancer. *J. Clin. Oncol.*, 26, 5572–5572.

Kichler, A. (2004) Gene transfer with modified polyethylenimines. *J. Gene Med.*, 6, S3-S10.

KichleR, A., Chillon, M., Leborgne, C., Danos, O. & Frisch, B. (2002) Intranasal gene delivery with a polyethylenimine-PEG conjugate. *J. Control. Release*, 80, 379–388.

Kichler, A., Pages, J. C., Leborgne, C., Druillennec, S., Lenoir, C., Coulaud, D., Delain, E., Cam, E. L., Roques, B. P. & Danos, O. (2000) Efficient DNA transfection mediated by the C-terminal domain of human immunodeficiency virus type 1 viral protein R. *J. Virol.*, 74, 5424–5431.

Kihara, F., Arima, H., Tsutsumi, T., Hirayama, F. & Uekama, K. (2002) Effects of structure of polyamidoamine dendrimer on gene transfer efficiency of the dendrimer conjugate with α-cyclodextrin. *Bioconjugate Chem.*, 13, 1211–1219.

Kihara, F., Arima, H., Tsutsumi, T., Hirayama, F. & Uekama, K. (2003) In vitro and in vivo gene transfer by an optimized alpha-cyclodextrin conjugate with polyamidoamine dendrimer. *Bioconjugate Chem.*, 14, 342–350.

Kim, H., Kim, H. A., Bae, Y. M., Choi, J. S. & Min, M. (2009a) Dexamethasone conjugated polyethylenimine as an efficient gene carrier with an anti-apoptotic effect to cardiomyocytes. *J. Gene Med.*, 11, 515–522.

Kim, H., Kim, H. A., Choi, J. S. & Lee, M. (2009b) Delivery of hypoxia inducible heme oxygenase-1 gene using dexamethasone conjugated polyethylenimine for protection of cardiomyocytes under hypoxia. *Bull. Korean Chem. Soc.*, 30, 897–901.

Kim, H. J., Kwon, M. S., Choi, J. S., Kim, B. H., Yoon, J. K., Kim, K. & Park, J. S. (2007a) Synthesis and characterization of poly (amino ester) for slow biodegradable gene delivery vector. *Bioorg. Med. Chem.*, 15, 1708–1715.

Kim, H. J., Kwon, M. S., Choi, J. S., Yang, S. M., Yoon, J. K., Kim, K. & Park, J. S. (2006a) Highly effective and slow-biodegradable network-type cationic gene delivery polymer: Small library-like approach synthesis and characterization. *Biomaterials*, 27, 2292–2301.

Kim, J., Lee, J. E., Lee, S. H., Yu, J. H., Lee, J. H., Park, T. G. & Hyeon, T. (2008) Fabrication of a multifunctional polymer nanomedical platform for simultaneous cancertargeted imaging and magnetically guided drug delivery. *Adv. Mater.*, 20, 478–483.

Kim, J., Park, S., Lee, J. E., Jin, S. M., Lee, J. H., Lee, I. S., Yang, I., Kim, J. S., Kim, S. K., Cho, M. H. & Hyeon, T. (2006b) Designed fabrication of multifunctional magnetic gold nanoshells and their application to magnetic resonance imaging and photothermal therapy. *Angew. Chem. Int. Ed.*, 45, 7754–7758.

Kim, J. S., Rieter, W. J., Taylor, K. M. L., An, H. Y., Lin, W. L. & Lin, W. B. (2007b) Self-Assembled Hybrid Nanoparticles for Cancer-Specific Multimodal Imaging. *J. Am. Chem. Soc.*, 129, 8962–8963.

Kim, S., Choi, J. S., Jang, H. S., Suh, H. & Park, J. (2001a) Hydrophobic modification of polyethyleneimine for gene transfectants. *Bull. Korean Chem. Soc.*, 22, 1069–1075.

Kim, S., Ohulchanskyy, T. Y., Pudavar, H. E., Pandey, R. K. & Prasad, P. N. (2007c) Organically modified silica nanoparticles co-encapsulating photosensitizing drug and aggregation-enhanced two-photon absorbing fluorescent dye aggregates for two-photon photodynamic therapy. *J. Am. Chem. Soc.*, 129, 2669–2675.

Kim, T., Seo, H. J., Choi, J. S., Jang, H. S., Baek, J., Kim, K. & Park, J. S. (2004a) PAMAM-PEG-PAMAM: Novel triblock copolymer as a biocompatible and efficient gene delivery carrier. *Biomacromolecules*, 5, 2487–2492.

Kim, T. H., Nah, J. W., Cho, M. H., Park, T. G. & Cho, C. S. (2006c) Receptor mediated gene delivery into antigen presenting cells using mannosylated chitosan/DNA nanoparticles. *J. Nanosci. Nanotechnol.*, 6, 2796–2803.

Kim, T. H., Park, I. K., Nah, J. W., Choi, Y. J. & Cho, C. S. (2004b) Galactosylated chitosan/DNA nanoparticles prepared using water-soluble chitosan as a gene carrier. *Biomaterials*, 25, 3783–3792.

Kim, T. I., Baek, J. U., Yoon, J. K., Choi, J. S., Kim, K. & Park, J. S. (2007d) Synthesis and characterization of a novel arginine-grafted dendritic block copolymer for gene delivery and study of its cellular uptake pathway leading to transfection. *Bioconjugate Chem.*, 18, 309–317.

Kim, T. I., Seo, H. J., Choi, J. S., Yoon, J. K., Baek, J. U., Kim, K. & Park, J. S. (2005a) Synthesis of biodegradable cross-linked poly(beta-amino ester) for gene delivery and its modification, inducing enhanced transfection efficiency and stepwise degradation. *Bioconjugate Chem.*, 16, 1140–1148.

Kim, Y. H., Gihm, S. H., Park, C. R., Lee, K. Y., Kim, T. W., Kwon, I. C., Chung, H. & Jeong, S. Y. (2001b) Structural characteristics of size-controlled self-aggregates of deoxycholic acid-modified chitosan and their application as a DNA delivery carrier. *Bioconjugate Chem.*, 12, 932–938.

Kim, Y. H., Park, J. H., Lee, M., Kim, Y. H., Park, T. G. & Kim, S. W. (2005b) Polyethylenimine with acid-labile linkages as a biodegradable gene carrier. *J. Control. Release*, 103, 209–219.

Kim, Y. H., TewarI, M., Pajerowski, J. D., Cai, S., Sen, S., Williams, J., Sirsi, S., Lutz, G. & Discher, D. E. (2009c) Polymersome delivery of siRNA and antisense oligonucleotides. *J. Control. Release*, 134, 132–140.

Kircheis, R., Kichler, A., Wallner, G., Kursa, M., Ogris, M., Felzmann, T., Buchberger, M. & Wagner, E. (1997) Coupling of cell-binding ligands to polyethylenimine for targeted gene delivery. *Gene Ther.*, 4, 409–418.

Kircheis, R., Schuller, S., Brunner, S., Ogris, M., Heider, K. H., Zauner, W. & Wagner, E. (1999) Polycation-based DNA complexes for tumor-targeted gene delivery in vivo. *J. Gene Med.*, 1, 111–120.

Kircheis, R., Wightman, L., Schreiber, A., Robitza, B., Rossler, V., Kursa, M. & Wagner, E. (2001a) Polyethylenimine/DNA complexes shielded by transferrin target gene expression to tumors after systemic application. *Gene Ther.*, 8, 28–40.

Kircheis, R., Wightman, L. & Wagner, E. (2001b) Design and gene delivery activity of modified polyethylenimines. *Adv. Drug Deliv. Rev.*, 53, 341–358.

Kishida, T., Asada, H., Itokawa, Y., Yasutomi, K., Shin-Ya, M., Gojo, S., Cui, F. D., Ueda, Y., Yamagishi, H., Imanishi, J. & Mazda, O. (2003) Electrochemo-gene therapy of cancer: Intratumoral delivery of Interleukin-12 gene and bleomycin synergistically induced therapeutic immunity and suppressed subcutaneous and metastatic melanomas in mice. *Mol. Ther.*, 8, 738–745.

Klumpp, C., Kostarelos, K., Prato, M. & Bianco, A. (2006) Functionalized carbon nanotubes as emerging nanovectors for the delivery of therapeutics. *Biochim. Biophys. Acta.*, 1758, 404–412.

Kneuer, C., Sameti, M., Haltner, E. G., Schiestel, T., Schirra, H., Schmidt, H. & Lehr, C. M. (2000) Silica nanoparticles modified with aminosilanes as carriers for plasmid DNA. *Int. J. Pharm.*, 196, 257–261.

Ko, K. S., Lee, M., Koh, I. J. & Kim, S. W. (2001) Combined administration of plasmids encoding IL-4 and IL-10 prevents the development of autoimmune diabetes in nonobese diabetic mice. *Mol. Ther.*, 4, 313–316.

Ko, Y. T., Falcao, C. & Torchilin, V. P. (2009a) Cationic liposomes loaded with proapoptotic peptide D-(KLAKLAK)(2) and Bcl-2 antisense oligodeoxynucleotide G3139 for enhanced anticancer therapy *Mol. Pharm.*, 6, 971–977.

Ko, Y. T., Hartner, W. C., Kale, A. A. & Torchilin, V. P. (2009b) Gene delivery into ischemic myocardium by double-targeted lipoplexes with anti-myosin antibody and TAT peptide. *Gene Ther.*, 16, 52–59.

Kogure, K., Akita, H., Yamada, Y. & Harashima, H. (2008) Multifunctional envelope-type nano device (MEND) as a non-viral gene delivery system. *Adv. Drug Deliv. Rev.*, 60, 559–571.

Koh, I. J., Ko, K. S., Lee, M., Han, S., Park, J. S. & Kim, S. W. (2000) Degradable polymeric carrier for the delivery of IL-10 plasmid DNA to prevent autoimmune insulitis of NOD mice. *Gene Ther.*, 7, 2099–2104.

Kono, K., Akiyama, H., Takahashi, T., Takagishi, T. & Harada, A. (2005) Transfection activity of polyamidoamine dendrimers having hydrophobic amino acid residues in the periphery *Bioconjugate Chem.*, 16, 208–214.

Konstan, M. W., Davis, P. B., Wagener, J. S., Hilliard, K. A., Stern, R. C., Milgram, L. J. H., Kowalczyk, T. H., Hyatt, S. L., Fink, T. L., Gedeon, C. R., Oette, S. M., Payne, J. M., Muhammad, O., Ziady, A. G., Moen, R. C. & Cooper, M. J. (2004) Compacted DNA nanoparticles administered to the nasal mucosa of cystic fibrosis subjects are safe and demonstrate partial to complete cystic fibrosis transmembrane regulator reconstitution. *Hum. Gene Ther.*, 15, 1255–1269.

Kopatz, I., Remy, J. S. & Behr, J. P. (2004) A model for non-viral gene delivery: through syndecan adhesion molecules and powered by actin *J. Gene Med.*, 6, 769–776.

Kornguth, S. E. & Stahmann, M. A. (1961) Effect of polylysine on the leakage and retention of compounds by Ehrlich ascites tumor cells. *Cancer Res.*, 21, 907–912.

Koynova, R. & Tenchov, B. (2009) Cationic phospholipids: structure-transfection activity relationships. *Soft Matter*, 5, 3187–3200.

Krajcik, R., Jung, A., Hirsch, A., Neuhuber, W. & Zolk, O. (2008) Functionalization of carbon nanotubes enables non-covalent binding and intracellular delivery of small interfering RNA for efficient knock-down of genes. *Biochem. Biophys. Res. Commun.*, 369, 595–602.

Krotz, F., De Wit, C., Sohn, H. Y., Zahler, S., Gloe, T., Pohl, U. & Plank, C. (2003) Magnetofection – A highly efficient tool for antisense oligonucleotide delivery in vitro and in vivo. *Mol. Ther.*, 7, 700–710.

Krska, S. W. & Seyferth, D. (1998) Synthesis of water-soluble carbosilane dendrimers. *J. Am. Chem. Soc.*, 120, 3604–3612.

Kunath, K., Von Harpe, A., Fischer, D. & Kissel, T. (2003a) Galactose-PEI-DNA complexes for targeted gene delivery: degree of substitution affects complex size and transfection efficiency. *J. Control. Release*, 88, 159–172.

Kunath, K., Von Harpe, A., Fischer, D., Petersen, H., Bickel, U., Voigt, K. & Kissel, T. (2003b) Low-molecular-weight polyethylenimine as a non-viral vector for DNA delivery: comparison of physicochemical properties, transfection efficiency and in vivo distribution with high-molecular-weight polyethylenimine. *J. Control. Release*, 89, 113–125.

Kundahl, E., Richman, R. & Flickinger, R. A. (1981) The effect of added H1 histone and polylysine on DNA synthesis and cell division of cultured mammalian cells. *J. Cell. Physiol.*, 108, 291–298.

Kurisawa, M., Yokoyama, M. & Okano, T. (2000) Gene expression control by temperature with thermo-responsive polymeric gene carriers. *J. Control. Release*, 69, 127–137.

Kursa, M., Walker, G. F., Roessler, V., Ogris, M., Roedl, W., Kircheis, R. & Wagner, E. (2003) Novel shielded transferrin-polyethylene glycol-polyethylenimine/DNA complexes for systemic tumor-targeted gene transfer. *Bioconjugate Chem.*, 14, 222–231.

Kushibiki, T., Nagata-Nakajima, N., Sugai, M., Shimizu, A. & Tabata, Y. (2005a) Delivery of plasmid DNA expressing small interference RNA for TGF-beta type II receptor by cationized gelatin to prevent interstitial renal fibrosis. *J. Control. Release*, 105, 318–331.

Kushibiki, T., Nagata-nakajima, N., Sugai, M., Shimizu, A. & Tabata, Y. (2005b) Targeting of plasmid DNA to renal interstitial fibroblasts by cationized gelatin. *Biol. Pharm. Bull.*, 28, 2007–2010.

Kwak, S. Y., Kriven, W. M., Wallig, M. A. & Choy, J. H. (2004) Inorganic delivery vector for intravenous injection. *Biomaterials*, 25, 5995–6001.

Kyriakis, J. M. & Avruch, J. (1990) pp54 Microtubule-associated protein 2 kinase. *J. Biol. Chem.*, 265, 17353–17363.

Ladewig, K., Xu, Z. P. & Lu, G. Q. (2009) Layered double hydroxide nanoparticles in gene and drug delivery. *Expert Opin. Drug Deliv.*, 6, 907–922.

Lai, E. & Van Zanten, J. H. (2001) Monitoring DNA/poly-L-lysine polyplex formation with time-resolved multiangle laser light scattering. *Biophys. J.*, 80, 864–873.

Laurent, S., Forge, D., Port, M., Roch, A., Robic, C., Vanderelst, L. & Muller, R. N. (2008) Magnetic iron oxide nanoparticles: synthesis, stabilization, vectorization, physicochemical characterizations, and biological applications. *Chem. Rev.*, 108, 2064–2110.

Lavigne, M. D., Pennadam, S., Ellis, J., Yates, L. L., Alexander, C. & Gorecki, D. C. (2007) Enhanced gene expression through temperature profile-induced variations in molecular architecture of thermoresponsive polymer vectors *J. Gene Med.*, 9, 44–54.

Lee, D., Lockey, R. & Mohapatra, S. (2006) Folate receptor-mediated cancer cell specific gene delivery using folic acid-conjugated oligochitosans. *J. Nanosci. Nanotechnol.*, 6, 2860–2866.

Lee, H., Kim, T. H. & Park, T. G. (2002) A receptor-mediated gene delivery system using streptavidin and biotin-derivatized, pegylated epidermal growth factor. *J. Control. Release* 83, 109–119.

Lee, J. H., Lee, K., Moon, S. H., Lee, Y., Park, T. G. & Cheon, J. (2009) All-in-one target-cell-specific magnetic nanoparticles for simultaneous molecular imaging and siRNA delivery. *Angew. Chem. Int. Ed.*, 48, 4174–4179.

Lee, J. H., Lim, Y. B., Choi, J. S., Lee, Y., Kim, T. I., Kim, H. J., Yoon, J. K., Kim, K. & Park, J. S. (2003) Polyplexes assembled with internally quaternized PAMAM-OH dendrimer and plasmid DNA have a neutral surface and gene delivery potency. *Bioconjugate Chem.*, 14, 1214–1221.

Lee, K. Y., Kwon, I. C., Kim, Y. H., Jo, W. H. & Jeong, S. Y. (1998) Preparation of chitosan self-aggregates as a gene delivery system. *J. Control. Release*, 51, 213–220.

Lee, Y., Miyata, K., Oba, M., Ishii, T., Fukushima, S., Han, M., Koyama, H., Nishiyama, N. & Kataoka, K. (2008) Charge-conversion ternary polyplex with endosome disruption moiety: A technique for efficient and safe gene delivery. *Angew. Chem. Int. Ed.*, 47, 5163–5166.

Lemieux, P., Vinogradov, S. V., Gebhart, C. L., Guerin, N., Paradis, G., Nguyen, H. K., Ochietti, B., Suzdaltseva, Y. G., Bartakova, E. V., Bronich, T. K., St-Pierre, Y., Alakhov, V. Y. & Kabanov, A. V. (2000) Block and graft copolymers and NanoGel copolymer networks for DNA delivery into cell. *J. Drug Target.*, 8, 91–105.

Li, J. (2009) Cyclodextrin inclusion polymers forming hydrogels. *Adv. Polym. Sci.*, 222, 79–112.

Li, J. & Loh, X. J. (2008) Cyclodextrin-based supramolecular architectures: Syntheses, structures, and applications for drug and gene delivery. *Adv. Drug Deliv. Rev.*, 60, 1000–1017.

Li, J., Yang, C., Li, H. Z., Wang, X., Goh, S. H., Ding, J. L., Wang, D. Y. & Leong, K. W. (2006) Cationic supramolecules composed of multiple oligoethylenimine-grafted β-cyclodextrins threaded on a polymer chain for efficient gene delivery. *Adv. Mater.*, 18, 2969–2974.

Li, S., Tan, Y., Viroonchatapan, E., Pitt, B. R. & Huang, L. (2000) Targeted gene delivery to pulmonary endothelium by anti-PECAM antibody. *Am. J. Physiol. Lung Cell Mol. Physiol.*, 278, L504-511.

Li, S. D. & Huang, L. (2006) Gene therapy progress and prospects: non-viral gene therapy by systemic delivery. *Gene Ther.*, 13, 1313–1319.

Li, X., Su, Y., Chen, Q., Lin, Y., Tong, Y. & Li, Y. (2005) Synthesis and characterization of biodegradable hyperbranched poly(ester-amide)s based on natural material. *Biomacromolecules*, 6, 3181–3188.

Li, Y. Y., Cheng, H., zhang, Z. G., Wang, C., Zhu, J. L., Liang, Y., Zhang, K. L., Cheng, S. X., Zhang, X. Z. & Zhuo, R. X. (2008) Cellular internalization and in vivo tracking of thermosensitive luminescent micelles based on luminescent lanthanide chelate. *ACS Nano*, 2, 125–133.

Lidke, D. S., Nagy, P., Heintzmann, R., Arndt-Jovin, D. J., Post, J. N., Grecco, H. E., Jares-Erijman, E. A. & Jovin, T., M. (2004) Quantum dot ligands provide new insights into erbB/HER receptor-mediated signal transduction. *Nat. Biotechnol.*, 22, 198–203.

Lim, Y. B., Choi, Y. H. & Park, J. S. (1999) A self-destroying polycationic polymer: Biodegradable poly(4-hydroxy-L-proline ester). *J. Am. Chem. Soc.*, 121, 5633–5639.

Lim, Y. B., Han, S. O., Kong, H. U., Lee, Y., Park, J. S., Jeong, B. & Kim, S. W. (2000a) Biodegradable polyester, poly[alpha-(4-aminobutyl)-l-glycolic acid], as a non-toxic gene carrier,. *Pharm. Res.*, 17, 811–816.

Lim, Y. B., Kim, C. H., Kim, K., Kim, S. W. & Park, J. S. (2000b) Developement of a safe gene delivery system using biodegradable polymer, poly[alpha-(4-aminobutyl)-l-glycolic acid]. *J. Am. Chem. Soc.*, 122, 6524–6525.

Lim, Y. B., Kim, S. M., Lee, Y., Lee, W. K., Yang, T. G., Lee, M. J., Suh, H. & Park, J. S. (2001) Cationic hyperbranched poly(amino ester): A novel class of DNA condensing molecule with cationic surface, biodegradable three-dimensional structure, and tertiary amine groups in the interior. *J. Am. Chem. Soc.*, 123, 2460–2461.

Lim, Y. B., Kim, S. M., Suh, H. & Park, J. S. (2002) Biodegradable, endosome disruptive, and cationic network-type polymer as a highly efficient and nontoxic gene delivery carrier. *Bioconjugate Chem.*, 13, 952–957.

Liong, M., Lu, J., Kovochich, M., Xia, T., Ruehm, S. G., Nel, A. E., Tamanoi, F. & Zink, J. I. (2008) Multifunctional inorganic nanoparticles for imaging, targeting, and drug delivery. *ACS Nano*, 2, 889–896.

Liu, F., Shollenberger, L. M. & Huang, L. (2004) Non-immunostimulatory nonviral vectors. *FASEB J.*, 18, 1779–1781.

Liu, T., Tang, A., Zhang, G., Chen, Y., Zhang, J., Peng, S. & Cai, Z. (2005) Calcium phosphate nanoparticles as a novel nonviral vector for efficient transfection of DNA in cancer gene therapy. *Cancer Biother Radiopharm*, 20, 141–149.

Liu, W. G., Zhang, X., Sun, S. J., Sun, G. J., Yao, K. D., Liang, D. C., Guo, G. & Zhang, J. Y. (2003) N-alkylated chitosan as a potential nonviral vector for gene transfection. *Bioconjugate Chem.*, 14, 782–789.

Liu, Y., Yu, L., Chen, Y., Zhao, Y. L. & Yang, H. (2007a) Construction and DNA condensation of cyclodextrin-based polypseudorotaxanes with anthryl grafts. *J. Am. Chem. Soc.*, 129, 10656–10657.

Liu, Z., Winters, M., Holodniy, M. & Dai, H. (2007b) siRNA delivery into human T cells and primary cells with carbon-nanotube transporters. *Angew. Chem. Int. Ed.*, 46, 2023–2027.

Liu, Z. H., Zhang, Z. Y., Zhou, C. R. & Jiao, Y. P. (2010) Hydrophobic modifications of cationic polymers for gene delivery. *Prog. Polym. Sci.*, 35, 1144–1162.

Loethen, S., Kim, J. M. & Thompson, D. H. (2007) Biomedical applications of cyclodextrin based polyrotaxanes. *Polym. Rev.*, 47, 383–418.

Lopez, T., Bosch, P. & Ramos, E. (1996) Synthesis and characterization of sol-gel hydrotalcites. Strutures and texture. *Langmuir*, 12, 189–192.

Lori, F., Trocio, J., Bakare, N., Kelly, L. M. & Lisziewicz, J. (2005) DermaVir, a novel HIV immunisation technology. *Vaccine*, 23, 2030–2034.

Love, K. T., Mahon, K. P., Levins, C. G., Whitehead, K. A., Querbes, W., Dorkin, J. R., Qin, J., Cantley, W., Qin, L. L., Racie, T., Frank-Kamenetsky, M., Yip, K. N., Alvarez, R., Sah, D. W. Y., De Fougerolles, A., Fitzgerald, K., Koteliansky, V., Akinc, A., Langer, R. & Anderson, D. G. (2010) Lipid-like materials for low-dose, in vivo gene silencing. *Proc. Natl. Acad. Sci. U.S.A.*, 107, 1864–1869.

Lu, A. H., Salabas, E. L. & Schuth, F. (2007) Magnetic nanoparticles: synthesis, protection, functionalization, and application. *Angew. Chem. Int. Ed.*, 46, 1222–1244.

Lu, B., Xu, X. D., Zhang, X. Z., Cheng, S. X. & Zhuo, R. X. (2008) Low molecular weight polyethylenimine grafted N-maleated chitosan for gene delivery: Properties and in vitro transfection studies. *Biomacromolecules*, 9, 2594–2600.

Luby, T. M., Cole, G., Baker, L., Kornher, J. S., RamstedT, U. & Hedley, M. L. (2004) Repeated immunization with plasmid DNA formulated in poly(lactide-co-glycolide) microparticles is well tolerated and stimulates durable T cell responses to the tumor-associated antigen cytochrome P450 1B1. *Clin. Immun.*, 112, 45–53.

Luo, D., Han, E., Belcheva, N. & Saltzman, W. M. (2004) A self-assembled, modular DNA delivery system mediated by silica nanoparticles. *J. Control. Release*, 95, 333–341.

Luo, D. & Saltzman, W. M. (2000) Synthetic DNA delivery systems. *Nat. Biotechnol.*, 18, 33–37.

Luten, J., Van Steenbergen, M. J., Lok, M. C., De Graaff, A. M., Van Nostrum, C. F., Talsma, H. & Hennink, W. E. (2008) Degradable PEG-folate coated poly(DMAEA-co-BA)phosphazene-based polyplexes exhibit receptor-specific gene expression *Eur. J. Pharm. Sci.*, 33, 241–251.

Luten, J., Van Steenis, J. H., Van Someren, R., Kemmink, J., Schuurmans, N. M. E., Koning, G. A., Crommelin, D. J. A., Van Nostrum, C. F. & Hennink, W. E. (2003) Water-soluble biodegradable cationic polyphosphazenes for gene delivery. *J. Control. Release*, 89, 483–497.

Lv, H., Zhang, S., Wang, B., Cui, S. & Yan, J. (2006) Toxicity of cationic lipids and cationic polymers in gene delivery. *J. Control. Release*, 114, 100–109.

Lynn, D. M., Anderson, D. G., Putnam, D. & Langer, R. (2001) Accelerated discovery of synthetic transfection vectors: Parallel synthesis and screening of a degradable polymer library. *J. Am. Chem. Soc.*, 123, 8155–8156.

Lynn, D. M. & Langer, R. (2000) Degradable poly(beta-amino esters): Synthesis, characterization, and self-assembly with plasmid DNA. *J. Am. Chem. Soc.*, 122, 10761–10768.

Ma, K., Hu, M., Qi, Y., Qiu, L., Jin, Y., Yu, J. & Li, B. (2009) Structure-transfection activity relationships with glucocorticoid-polyethylenimine conjugate nuclear gene delivery systems. *Biomaterials*, 30, 3780–3789.

Malone, R. W., Hickman, M. A., Lehmann-Bruinsma, K., Sih, T. R., Walzem, R., Carlson, D. M. & Powell, J. S. (1994) Dexamethasone enhancement of gene expression after direct hepatic DNA injection. *J. Biol. Chem.*, 269, 29903–29907.

Mao, H. Q., Roy, K., Troung-Le, V. L., Janes, K. A., Lin, K. Y., Yan, W., August, J. T. & Leong, K. W. (2001) Chitosan-DNA nanoparticles as gene carriers: synthesis, characterization and transfection efficiency. *J. Control. Release*, 70, 399–421.

Maruyama-Tabata, H., Harada, Y., Matsumura, T., Satoh, E., Cui, F., Iwai, M., Kita, M., Hibi, S., Imanishi, J., Sawada, T. & Mazda, O. (2000) Effective suicide gene therapy in vivo by EBV-based plasmid vector coupled with polyamidoamine dendrimer. *Gene Ther.*, 7, 53–60.

Masago, K., Itaka, K., Nishiyama, N., Chung, U. I. & Kataoka, K. (2007) Gene delivery with biocompatible cationic polymer: Pharmacogenomic analysis on cell bioactivity. *Biomaterials*, 28, 5169–5175.

Mathiesen, I. (1999) Electropermeabilization of skeletal muscle enhances gene transfer in vivo. *Gene Ther.*, 6, 508–514.

Matsumoto, S., Christie, R. J., Nishiyama, N., Miyata, K., Ishii, A., Oba, M., Koyama, H., Yamasaki, Y. & Kataoka, K. (2009) Environment-responsive block copolymer micelles with a disulfide cross-linked core for enhanced siRNA delivery. *Biomacromolecules*, 10, 119–127.

Matteoni, R. & Kreis, T. E. (1987) Translocation and clustering of endosomes and lysosomes depends on microtubules. *J. Cell. Biol.*, 105, 1253–1265.

Medarova, Z., Pham, W., Farrar, C., Petkova, V. & Moore, A. (2007) In vivo imaging of siRNA delivery and silencing in tumors. *Nat. Med.*, 13, 372–377.

Medina-Kauwe, L. K., Xie, J. & Hamm-Alvarez, S. (2005) Intracellular trafficking of nonviral vectors. *Gene Ther.*, 12, 1734–1751.

Medintz, I. L., Uyeda, H. T., Goldman, E. R. & Mattoussi, H. (2005) Quantum dot bioconjugates for imaging, labelling and sensing. *Nat. Mater.*, 4, 435–446.

Mesika, A., Kiss, V., Brumfeld, V., Ghosh, G. & Reich, Z. (2005) Enhanced intracellular mobility and nuclear accumulation of DNA plasmids associated with a karyophilic protein. *Hum. Gene Ther.*, 16, 200–208.

Mintzer, M. A. & Simanek, E. E. (2009) Nonviral vectors for gene delivery. *Chem. Rev.*, 109, 259–302.

Mishra, S., Webster, P. & Davis, M. E. (2004) PEGylation significantly affects cellular uptake and intracellular trafficking of non-viral gene delivery particles. *Eur. J. Cell Biol.*, 83, 97–111.

Mislick, K. A. & Baldeschwieler, J. D. (1996) Evidence for the role of proteoglycans in cation-mediated gene transfer. *Proc. Natl. Acad. Sci. U. S. A.*, 93, 12349–12354.

Mislick, K. A., Baldeschwieler, J. D., Kayyem, J. F. & Meade, T. J. (1995) Transfection of folate-polysine DNA complexes: Evidence for lysosomal delivery. *Bioconjugate Chem.*, 6, 512–515.

Miyata, K., Fukushima, S., Nishiyama, N., Yamasaki, Y. & Kataoka, K. (2007) PEG-based block catiomers possessing DNA anchoring and endosomal escaping functions to form polyplex micelles with improved stability and high transfection efficacy. *J. Control. Release*, 122, 252–260.

Miyata, K., Oba, M., Kano, M. R., Fukushima, S., Vachutinsky, Y., Han, M., Koyama, H., Miyazono, K., Nishiyama, N. & Kataoka, K. (2008a) Polyplex micelles from triblock copolymers composed of tandemly aligned segments with biocompatible, endosomal escaping, and DNA-condensing functions for systemic gene delivery to pancreatic tumor tissue. *Pharm. Res.*, 25, 2924–2936.

Miyata, K., Oba, M., Nakanishi, M., Fukushima, S., Yamasaki, Y., Koyama, H., Nishiyama, N. & Kataoka, K. (2008b) Polyplexes from poly(aspartamide) bearing 1, 2-diaminoethane side chains induce pH-selective endosomal membrane destabilization with amplified transfection and negligible cytotoxicity. *J. Am. Chem. Soc.*, 130, 16287–16294.

Moffatt, S., Wiehle, S. & Cristiano, R. J. (2006) A multifunctional PEI based cationic polyplex for enhanced systemic p53-mediated gene therapy. *Gene Ther.*, 13, 1512–1523.

Monsigny, M., Rondanino, C., Duverger, E., Fajac, I. & Roche, A. C. (2004) Glyco-dependent nuclear import of glycoproteins, glycoplexes and glycosylated plasmids. *Biochim. Biophys. Acta*, 1673, 94–103.

Monthioux, M. & Kuznetsov, V. L. (2006) Who should be given the credit for the discovery of carbon nanotubes? *Carbon*, 44, 1621–1624.

Mounkes, L. C., Zhong, W., Cipres-Palacin, G., Heath, T. D. & Debs, R. J. (1998) Proteoglycans mediate cationic liposome-DNA complex-based gene delivery in vitro and in vivo. *J. Biol. Chem.*, 273, 26164–26170.

Mourtzis, N., Eliadou, K., Aggelidou, C., Sophianopoulou, V., Mavridis, I. M. & Yannakopoulou, K. (2007) Per(6-guanidino-6-deoxy)cyclodextrins: synthesis, characterisation and binding behaviour toward selected small molecules and DNA. *Org. Biomol. Chem.*, 5, 125–131.

Mourtzis, N., Paravatou, M., Mavridis, I. M., Roberts, M. L. & Yannakopoulou, K. (2008) Synthesis, characterization, and remarkable biological properties of cyclodextrins bearing guanidinoalkylamino and aminoalkylamino groups on their primary side. *Chem. Eur. J.*, 14, 4188–4200.

Mueller, B. M., Wrasildo, W. A. & Reisfeld, R. A. (1990) Antibody conjugates with morpholinodoxorubicin and acid-cleavable linkers. *Bioconjugate Chem.*, 1, 325–330.

Mui, B., Ahkong, Q. F., Chow, L. & Hope, M. J. (2000) Membrane perturbation and the mechanism of lipid-mediated transfer of DNA into cells. *Biochim. Biophys. Acta*, 1467, 281–292.

Murata, J. I., Ohya, Y. & Ouchi, T. (1997) Design of quaternary chitosan conjugate having antennary galactose residues as a gene delivery tool. *Carbohydrate Polym.*, 32, 105–109.

Murthy, N., Campbell, J., Fausto, N., Hoffman, A. S. & Stayton, P. S. (2003) Design and synthesis of pH-responsive polymeric carriers that target uptake and enhance the intracellular delivery of oligonucleotides. *J. Control. Release*, 89, 365–374.

Murthy, N., Robichaud, J. R., Tirrell, D. A., Stayton, P. S. & Hoffman, A. S. (1999) The design and synthesis of polymers for eukaryotic membrane disruption. *J. Control. Release*, 61, 137–143.

Nabel, E. G., Yang, Z., Muller, D., Chang, A. E., Gao, X., Huang, L., Cho, K. J. & Nabel, G. J. (1994) Safety and toxicity of catheter gene delivery to the pulmonary vasculature in a patient with metastatic melan. *Hum. Gene Ther.*, 5, 1089–1094.

Nabel, G. J., Nabel, E. G., Yang, Z. Y., Fox, B. A., Plautz, G. E., Gao, X., Huang, L., Shu, S., Gordon, D. & Chang, A. E. (1993) Direct gene transfer with DNA-liposome complexes in melanoma: Expression, biologic activity, and lack of toxicity in humans. *Proc. Natl. Acad. Sci. USA*, 90, 11307–11311.

Nagasaki, T., Atarashi, K., Makino, K., Noguchi, A. & Tamagaki, S. (2000) Synthesis of a novel water-soluble polyazobenzene dendrimer and photoregulation of affinity toward DNA. *Mol. Cryst. Liq. Cryst.*, 345, 227–232.

Nagasaki, T. & Shinkai, S. (2007) The concept of molecular machinery is useful for design of stimuli-responsive gene delivery systems in the mammalian cell. *J. Incl. Phenom. Macrocycl. Chem.*, 58, 205–219.

Nakanishi, M., Park, J. S., Jang, W. D., Oba, M. & Kataoka, K. (2007) Study of the quantitative aminolysis reaction of poly(b-benzyl L-aspartate) (PBLA) as a platform polymer for functionality materials. *React. Funct. Polym.*, 67, 1361–1372.

Needham, D. & Dewhirst, M. W. (2001) The development and testing of a new temperature-sensitive drug delivery system for the treatment of solid tumors. *Adv. Drug Deliv. Rev.*, 53, 285–305.

Nehilla, B. J., Allen, P. G. & Desai, T. A. (2008) Surfactant-free, drug-quantum-dot coloaded poly(lactide-co-glycolide) nanoparticles: Towards multifunctional nanoparticles. *ACS Nano*, 2, 538–544.

Neu, M., Fischer, D. & Kissel, T. (2005) Recent advances in rational gene transfer vector design based on poly(ethylene imine) and its derivatives. *J. Gene Med.*, 7, 992–1009.

Neu, M., Sitterberg, J., Bakowsky, U. & Kissel, T. (2006) Stabilized nanocarriers for plasmids based upon cross-linked poly(ethylene imine). *Biomacromolecules*, 7, 3428–3438.

Neubergera, T., Schöpfa, B., Hofmannb, H. M. H. & von Rechenberga, B. (2005) Superparamagnetic nanoparticles for biomedical applications: Possibilities and limitations of a new drug delivery system *J. Magn. Magn. Mater.*, 293, 483–496.

Nirmal, M. & Brus, L. (1999) Luminescence photophysics in semiconductor nanocrystals. *Acc. Chem. Res.*, 32, 407–414.

Noone, P. G., Hohneker, K. W., Zhou, Z., Johnson, L. G., Foy, C., Gipson, C., Jones, K., Noah, T. L., Leigh, M. W., Schwartzbach, C., Efthimiou, J., Pearlman, R., Boucher, R. C. & Knowles, M. R. (2000) Safety and biological efficacy of a lipid-CFTR complex for gene transfer in the nasal epithelium of adult patients with cystic fibrosis. *Mol. Ther.*, 1, 105–114.

Norris, D. J., Efros, A. L. & Erwin, S. C. (2008) Doped nanocrystals. *Science*, 319, 1776–1779.

O'neill, M. M., Kennedy, C. A., Barton, R. W. & Tatake, R. J. (2001) Receptormediated gene delivery to human peripheral blood mononuclear cells using anti-CD3 antibody coupled to polyethylenimine. *Gene Ther.*, 8, 362–368.

Ochietti, B., Guerin, N., Vinogradov, S. V., St-Pierre, Y., Lemieux, P., Kabanov, A. V. & Alakhov, V. Y. (2002) Altered organ accumulation of oligonucleotides using polyethyleneimine grafted with poly(ethylene oxide) or pluronic as carriers. *J. Drug Target.*, 10, 113–121.

Ogata, N., Sanui, K. & Wada, J. (1976) Novel synthesis of inclusion polyamides. *J. Polym. Sci. Polym. Lett. Ed.*, 14, 459–462.

Ogris, M., Brunner, S., Schuller, S., Kircheis, R. & Wagner, E. (1999) PEGylated DNA/transferrin-PEI complexes: reduced interaction with blood components, extended circulation in blood and potential for systemic gene delivery. *Gene Ther.*, 6, 595–605.

Ogris, M., Carlisle, R. C., Bettinger, T. & Seymour, L. W. (2001) Melittin enables efficient vesicular escape and enhanced nuclear access of nonviral gene delivery vectors. *J. Biol. Chem.*, 276, 47550–47555.

Ogris, M., Walker, G., Blessing, T., Kircheis, R., Wolschek, M. & Wagner, E. (2003) T umortargeted gene therapy: strategies for the preparation of ligand-polyethylene glycol-polyethylenimine/DNA complexes. *J. Control. Release*, 91, 173–181.

Ohana, P., Gofrit, O., Ayesh, S., Al-Sharef, W., Mizrahi, A., Birman, T., Schneider, T., Matouk, I., De Groot, N., Tavdy, E., Sidi, A. A. & Hochberg, A. (2004) Regulatory sequences of the H19 gene in DNA based therapy of bladder cancer. *Gene Ther. Mol. Biol.*, 8, 181–192.

Oishi, J., Kawamura, K., Kang, J. H., Kodama, K., Sonoda, T., Murata, M., Niidome, T. & Katayama, Y. (2006a) An intracellular kinase signal-responsive gene carrier for disordered cell-specific gene therapy. *J. Control. Release*, 110, 431–436.

Oishi, M., Nakaogami, J., Ishii, Y. & Nagasaki, Y. (2006b) Smart PEGylated gold nanoparticles for the cytoplasmic delivery of siRNA to induce enhanced gene silencing. *Chem. Lett.*, 35, 1046–1047.

Ooya, T., Choi, H. S., Yamashita, A., Yui, N., Sugaya, Y., Kano, A., Maruyama, A., Akita, H., Ito, R., Kogure, K. & Harashima, H. (2006) Biocleavable polyrotaxane-plasmid DNA polyplex for enhanced gene delivery. *J. Am. Chem. Soc.*, 128, 3852–3853.

Orrantia, E. & Chang, P. L. (1990) Intracellular distribution of DNA internalized through calcium phosphate precipitation. *Exp. Cell Res.*, 190, 170–174.

Oupicky, D., Ogris, M., Howard, K. A., Dash, P. R., Ulbrich, K. & Seymour, L. W. (2002) Importance of lateral and steric stabilization of polyelectrolyte gene delivery vectors for extended systemic circulation. *Mol. Ther.*, 5, 463–472.

Pack, D. W., Hoffman, A. S., Pun, S. & Stayton, P. S. (2005) Design and development of polymers for gene delivery. *Nat. Rev. Drug Discov.*, 4, 581–593.

Pantarotto, D., Singh, R., Mccarthy, D., Erhardt, M., Briand, J. P., Prato, M., Kostarelos, K. & Bianco, A. (2004) Functionalized carbon nanotubes for plasmid DNA gene delivery. *Angew. Chem. Int. Ed.*, 43, 5242–5246.

Pearce, M. E., Melanko, J. B. & Salem, A. K. (2007) Multifunctional nanorods for biomedical applications. *Pharm. Res.*, 24, 2335–2352.

Penczek, S. & Pretula, J. (1993) High-molecular-weight poly(alkylene phosphate)s and preparation of amphiphilic polymers thereof. *Macromolecules*, 26, 2228–2233.

Peng, Q., Hu, C., Cheng, J., Zhong, Z. L. & Zhuo, R. X. (2009) Influence of disulfide density and molecular weight on disulfide cross-linked polyethylenimine as gene vectors. *Bioconjugate Chem.*, 20, 340–346.

Peng, Q., Zhong, Z. L. & Zhuo, R. X. (2008) Disulfide cross-linked polyethylenimines (PEI) prepared via thiolation of low molecular weight PEI as highly efficient gene vectors. *Bioconjugate Chem.*, 19, 499–506.

Perales, J. C., Ferkol, T., Beegen, H., Ratnoff, O. D. & Hanson, R. W. (1994) Gene transfer in vivo: Sustained expression and regulation of genes introduced into the liver by receptor-targeted uptake. *Proc. Natl. Acad. Sci. USA*, 91, 4086–4090.

Pinaud, F., Michalet, X., Bentolila, L. A., Tsay, J. M., Doose, S., Li, J. J., Iyer, G. & Weiss, S. (2006) Advances in fluorescence imaging with quantum dot bio-probes. *Biomaterials*, 27, 1679–1687.

Plank, C., Mechtler, K., Szoka, C. P. & Wagner, E. (1996) Activation of the complement system by synthetic DNA complexes: a potential barrier for intravenous gene delivery. *Hum. Gene Ther.*, 7, 1437–1446.

Plank, C., Scherer, F., Schillinger, U., Bergemann, C. & Anton, M. (2003a) Magnetofection: Enhancing and targeting gene delivery with superparamagnetic nanoparticles and magnetic fields. *J. Lipid Res.*, 13, 29–31.

Plank, C., Schillinger, U., Scherer, F., Bergemann, C., Rémy, J. S., Krötz, F., Anton, M., Lausier, J. & Rosenecker, J. (2003b) The magnetofection method: Using magnetic force to enhance gene delivery. *Biol. Chem.*, 384, 737–747.

Pol, V. G., Wildermuth, G., Felsche, J., Gedanken, A. & Calderon-Moreno, J. (2005) Sonochemical deposition of Au nanoparticles on titania and the significant decrease in the melting point of gold. *J. Nanosci. Nanotechnol.*, 5, 975–979.

Popielarski, S. R., Mishra, S. & Davis, M. E. (2003) Structural effects of carbohydrate-containing polycations on gene delivery. 3. Cyclodextrin type and functionalization. *Bioconjugate Chem.*, 14, 672–678.

Poznyak, S. K., Talapin, D. V., Shevchenko, E. V. & Weller, H. (2005) Quantum dot chemiluminescence. *Nano Lett.*, 4, 693–698.

Pun, S. H., Bellocq, N. C., Liu, A., Jensen, G., Machemer, T., Quijano, E., Schluep, T., Wen, S., Engler, H., Heidel, J. & Davis, M. E. (2004a) Cyclodextrin-modified polyethylenimine polymers for gene delivery. *Bioconjugate Chem.*, 15, 831–840.

Pun, S. H., Tack, F., Bellocq, N. C., Cheng, J. J., Grubbs, B. H., Jensen, G. S., Davis, M. E., Brewster, M., Janicot, M., Janssens, B., Floren, W. & Bakker, A. (2004b) Targeted delivery of RNA-cleaving DNA enzyme (DNAzyme) to tumor tissue by transferrin-modified, cyclodextrin-based particles. *Cancer Biol. Ther.*, 3, 641–650.

Putnam, D. (2006) Polymers for gene delivery across length scales. *Nat. Mater.*, 5, 439–451.

Putnam, D. & Langer, R. (1999) Poly(4-hydroxy-L-proline ester): Low-temperature polycondensation and plasmid DNA complexation. *Macromolecules*, 32, 3658–3662.

Qi, L. & Gao, X. (2008a) Quantum dot-amphipol nanocomplex for intracellular delivery and real-time imaging of siRNA. *ACS Nano*, 2, 1403–1410.

Qi, L. F. & Gao, X. H. (2008b) Quantum dot-amphipol nanocomplex for intracellular delivery and real-time imaging of siRNA. *ACS Nano*, 2, 1403–1410.

Qiu, L. Y. & Bae, Y. H. (2007) Self-assembled polyethylenimine-graft-poly(e-caprolactone) micelles as potential dual carriers of genes and anticancer drugs. *Biomaterials*, 28, 4132–4142.

Radu, D. R., Lai, C. Y., Jeftinija, K., Rowe, E. W., Jeftinija, S. & Lin, V. S. (2004) A polyamidoamine dendrimer-capped mesoporous silica nanosphere-based gene transfection reagent. *J. Am. Chem. Soc.*, 126, 13216–13217.

Rebuffat, A., Bernasconi, A., Ceppi, M., Wehrli, H., Verca, S. B., Ibrahim, M., Frey, B. M., Frey, F. J. & Rusconi, S. (2001) Selective enhancement of gene transfer by steroid-mediated gene delivery. *Nat. Biotechnol.*, 19, 1155–1161.

Rebuffat, A. G., NawrockZi, A. R., Nielsen, P. E., Bernasconi, A. G., Bernal-Mendez, E., Frey, B. M. & Frey, F. J. (2002) Gene delivery by a steroid-peptide nucleic acid conjugate. *FASEB J.*, 16, 1426–1428.

Reineke, T. M. & Davis, M. E. (2003a) Structural effects of carbohydrate-containing polycations on gene delivery. 1. Carbohydrate size and Its distance from charge centers. *Bioconjugate Chem.*, 14, 247–254.

Reineke, T. M. & Davis, M. E. (2003b) Structural effects of carbohydrate-containing polycations on gene delivery. 2. charge center type. *Bioconjugate Chem.*, 14, 255–261.

Remy, J. S., Kichler, A., Mordvinov, V., Schuber, F. & Behr, J. P. (1995) Targeted gene transfer into hepatoma cells with lipopolyamine-condensed DNA particles presenting galactose ligands: a stage toward artificial viruses. *Proc. Natl. Acad. Sci. U. S. A.*, 92, 1744–1748.

Resch-Genger, U., Grabolle, M., Cavaliere-Jaricot, S., Nitschke, R. & Nann, T. (2008) Quantum dots versus organic dyes as fluorescent labels. *Nat. Methods*, 5, 763–775.

Rieter, W. J., Kim, J. S., Taylor, K. M. L., An, H. Y., Lin, W. L., Tarrant, T. & Lin, W. B. (2007) Hybrid silica nanoparticles for multimodal imaging. *Angew. Chem. Int. Ed.*, 46, 3680–3682.

Rigby, P. G. (1969) Prolongation of survival of tumour-bearing animals by transfer of 'immune' RNA with DEAE dextran. *Nature*, 221, 968–969.

Rogach, A. L., Nagesha, D., Ostrander, J. W., Giersig, M. & Kotov, N. A. (2000) "Raisin Bun"-type composite spheres of silica and semiconductor nanocrystals. *Chem. Mater.*, 12, 2676–2685.

Rojanarata, T., Petchsangsai, M., Opanasopit, P., Ngawhirunpat, T., Ruktanonchai, U., Sajomsang, W. & Tantayanon, S. (2008) Methylated N-(4-N,N-dimethylaminobenzyl) chitosan for novel effective gene carriers. *Eur. J. Pharm. Biopharm.*, 70, 207–214.

Rosette, C. & Karin, M. (1995) Cytoskeletal control of gene expression: depolymerization of microtubules activates NF-kappa B. *J. Cell. Biol.*, 128, 1111–1119.

Rosi, N. L., Giljohann, D. A., Thaxton, C. S., Lytton-Jean, A. K., Han, M. S. & Mirkin, C. A. (2006) Oligonucleotide-modified gold nanoparticles for intracellular gene regulation. *Science*, 312, 1027–1030.

Roth, C. M. & Sundaram, S. (2004) Engineering synthetic vectors for improved DNA delivery: Insights from intracellular pathways. *Annu. Rev. Biomed. Eng.*, 6, 397–426.

Rozema, D. B., Lewis, D. L., Wakefield, D. H., Wong, S. C., Klein, J. J., Roesch, P. L., Bertin, S. L., Reppen, T. W., Chu, Q., Blokhin, A. V., Hagstrom, J. E. & Wolff, J. A. (2007) Dynamic polyconjugates for targeted in vivo delivery of siRNA to hepatocytes. *Proc. Natl. Acad. Sci. USA*, 104, 12982–12987.

Rubanyi, G. M. (2001) The future of human gene therapy. *Mol. Aspects Med.*, 22, 113–142.

Rudolph, C., Lausier, J., Naundorf, S., Muller, R. H. & Rosenecker, J. (2000) In vivo gene delivery to the lung using polyethylenimine and fractured polyamidoamine dendrimers. *J. Gene Med.*, 2, 269–278.

Ruiz, F. E., Clancy, J. P., Perricone, M. A., Bebok, Z., Hong, J. S., Cheng, S. H., Meeker, D. P., Young, K. R., Schoumacher, R. A., Weatherly, M. R., Wing, L., Morris, J. E., Sindel, L., Rosenberg, M., Van Ginkel, F. W., Mcghee, J. R., Kelly, D., Lyrene, R. K. & Sorscher, E. J. (2001) A clinical infl ammatory syndrome attributable to aerosolized lipid-DNA administration in cystic fibrosis. *Hum. Gene Ther.*, 12, 751–761.

Sagara, K. & Kim, S. W. (2002) A new synthesis of galactose-poly(ethylene glycol)-polyethylenimine for gene delivery to hepatocytes. *J. Control. Release*, 79, 271–281.

Salem, A. K., Searson, P. C. & Leong, K. W. (2003) Multifunctional nanorods for gene delivery. *Nat. Mater.*, 2, 668–671.

Salgueiriño-Maceira, V., Correa-Duarte, M. A., Spasova, M., Liz-Marzán, L. M. & Farle, M. (2006) Composite silica spheres with magnetic and luminescent functionalities. *Adv. Funct. Mater.*, 16, 509–514.

Sameti, M., Bohr, G., Ravikumar, M. N., Kneuer, C., Bakowsky, U., Nacken, M., Schmid, T. H. & Lehr, C. M. (2003) Stabilisation by freeze-drying of cationically modified silica nanoparticles for gene delivery. *Int. J. Pharm.*, 266, 51–60.

Sandhu, K. K., Mcintosh, C. M., Simard, J. M., Smith, S. W. & Rotello, V. M. (2002) Gold nanoparticle-mediated transfection of mammalian cells. *Bioconjug. Chem.*, 13, 3–6.

Sanvicens, N. & Marco, M. P. (2008) Multifunctional nanoparticles- properties and prospects for their use in human medicine. *Trends Biotechnol.*, 26, 425–433.

Sato, T., Ishii, T. & Okahata, Y. (2001) In vitro gene delivery mediated by chitosan. effect of pH, serum, and molecular mass of chitosan on the transfection efficiency. *Biomaterials*, 22, 2075–2080.

Schaffer, D. V. & Lauffenburger, D. A. (1998) Optimization of cell surface binding enhances efficiency and specificity of molecular conjugate gene delivery. *J. Biol. Chem.*, 273, 28004–28009.

Schild, H. G. (1992) Poly(N-isopropylacrylamide): Experiment, theory and application. *Prog. Polym. Sci.*, 17, 163–249.

Shahin, V. (2006) The nuclear barrier is structurally and functionally highly responsive to glucocorticoids. *BioEssays*, 28, 935–942.

Shier, W. T., Dubourdieu, D. J. & Durkin, J. P. (1984) Polycations as prostaglandin synthesis inducers. Stimulation of arachidonic acid release and prostaglandin synthesis in cultured fibroblasts by poly(L-lysine) and other synthetic polycations. *Biochim. Biophys. Acta*, 793, 238–250.

Shim, M. S. & Kwon, Y. J. (2008) Controlled delivery of plasmid DNA and siRNA to intracellular targets using ketalized polyethylenimine. *Biomacromolecules*, 9, 444–455.

Shim, M. S. & Kwon, Y. J. (2009) Acid-responsive linear polyethylenimine for efficient, specific, and biocompatible siRNA delivery. *Bioconjugate Chem.*, 20, 488–499.

Simoes, S., Slepushkin, V., Pires, P., Gaspar, R., De Lima, M. C. & Duzgunes, N. (1999) Mechanisms of gene transfer mediated by lipoplexes associated with targeting ligands or pH-sensitive peptides. *Gene Ther.*, 6, 1798–1807.

Singh, M., Briones, M., Ott, G. & O'hagan, D. (2000) Cationic microparticles: A potent delivery system for DNA vaccines. *Proc. Natl. Acad. Sci. U. S. A.*, 97, 811–816.

Singh, R., Pantarotto, D., Mccarthy, D., Chaloin, O., Hoebeke, J., Partidos, C. D., Briand, J. P., Prato, M., Bianco, A. & Kostarelos, K. (2005) Binding and condensation of plasmid DNA onto functionalized carbon nanotubes: toward the construction of nanotube-based gene delivery vectors. *J. Am. Chem. Soc.*, 127, 4388–4396.

Smith, A. E. (1995) Viral vectors in gene therapy. *Annu. Rev. Microbiol.*, 49, 807–838.

Smith, A. M., Duan, H., Mohs, A. M. & Nie, S. (2008) Bioconjugated quantum dots for in vivo molecular and cellular imaging. *Adv. Drug Deliv. Rev.*, 60, 1226–1240.

Solinas, S., Piccaluga, G., Morales, M. P. & Serna, C. J. (2001) Sol-gel formation of γ-Fe2O3/SiO2 nanocomposites. *Acta Mater.*, 49, 2805–2811.

Srinivasachari, S., Fichter, K. M. & Reineke, T. M. (2008) Polycationic beta-cyclodextrin "click clusters": Monodisperse and versatile scaffolds for nucleic acid delivery. *J. Am. Chem. Soc.*, 130, 4618–4627.

Srinivasan, C., Lee, J., Papadimitrakopoulos, F., Silbart, L. K., Zhao, M. & Burgess, D. J. (2006) Labeling and intracellular tracking of functionally active plasmid DNA with semiconductor quantum dots. *Mol. Ther.*, 14, 192–201.

Suh, W., Chung, J. K., Park, S. H. & Kim, S. W. (2001) Anti-JL1 antibody-conjugated poly(L-lysine) for targeted gene delivery to leukemia T cells. *J. Control. Release*, 72, 171–178.

Sun, Y. X., Xiao, W., Cheng, S. X., Zhang, X. Z. & Zhuo, R. X. (2008a) Synthesis of (Dex-HMDI)-g-PEIs as effective and low cytotoxic nonviral gene vectors *J. Control. Release*, 128, 171–178.

Sun, Y. X., Zhang, X. Z., Cheng, H., Cheng, S. X. & Zhuo, R. X. (2008b) A low-toxic and efficient gene vector: Carboxymethyl dextran-graft-polyethylenimine. *J. Biomed. Mater. Res. Part A*, 84A, 1102–1110.

Sunderland, C. J., Steiert, M., Talmadge, J., Derfus, A. M. & Barry, S. E. (2006) Targeted nanoparticles for detecting and treating cancer. *Drug Dev. Res.*, 67, 70–93.

Sunshine, J., Green, J. J., Mahon, K. P., Yang, F., Eltoukhy, A., Nguyen, D. N., Langer, R. & Anderson, D. G. (2009) Small-molecule end-groups of linear polymer determine cell-type gene-delivery efficacy. *Adv. Mater.*, 21, 4947–4951.

Sy, J. C., Seshadri, G., Yang, S. C., Brown, M., Oh, T., Dikalov, S., Murthy, N. & Davis, M. E. (2008) Sustained release of a p38 inhibitor from non-inflammatory microspheres inhibits cardiac dysfunction. *Nat. Mater.*, 7, 863–869.

Takae, S., Miyata, K., Oba, M., Ishii, T., Nishiyama, N., Itaka, K., Yamasaki, Y., Koyama, H. & Kataoka, K. (2008) PEG-detachable polyplex micelles based on disulfide-linked block catiomers as bioresponsive nonviral gene vectors. *J. Am. Chem. Soc.*, 130, 6001–6009.

Tan, W. B., Jiang, S. & Zhang, Y. (2007) Quantum-dot based nanoparticles for targeted silencing of HER2/neu gene via RNA interference. *Biomaterials*, 28, 1565–1571.

Tanaka, S., Iwai, M., Harada, Y., Morikawa, T., Muramatsu, A., Mori, T., Okanoue, T., Kashima, K., Maruyama-Tabata, H., Hirai, H., Satoh, E., Imanishi, J. & Mazda, O. (2000) Targeted killing of carcinoembryonic antigen (CEA)-producing cholangiocarcinoma cells by polyamidoamine dendrimer-mediated transfer of an Epstein-Barr virus (EBV)-based plasmid vector carrying the CEA promoter. *Cancer Gene Ther.*, 7, 1241–1450.

Tang, F. & Hughes, J. A. (1998) Introduction of a disulfide bond into a cationic lipid enhances transgene expression of plasmid DNA. *Biochem. Biophys. Res. Commun.*, 242, 141–145.

Tang, G. P., Guo, H. Y., Alexis, F., Wang, X., Zeng, S., Lim, T. M., Ding, J., Yang, Y. Y. & Wang, S. (2006) Low molecular weight polyethylenimines linked by β-cyclodextrin for gene transfer into the nervous system. *J. Gene Med.*, 8, 736–744.

Tartaj, P. & Serna, C. J. (2003) Synthesis of monodisperse superparamagnetic Fe/silica nanospherical composites. *J. Am. Chem. Soc.*, 125, 15754–15755.

Thanou, M., Florea, B. I., Geldof, M., Junginger, H. E. & Borchard, G. (2002) Quaternized chitosan oligomers as novel gene delivery vectors in epithelial cell lines. *Biomaterials*, 23, 153–159.

Thomas, M. & Klibanov, A. M. (2002) Enhancing polyethylenimine's delivery of plasmid DNA into mammalian cells. *Proc. Natl. Acad. Sci. USA*, 99, 14640–14645.

Thomas, M. & Klibanov, A. M. (2003) Conjugation to gold nanoparticles enhances polyethylenimine's transfer of plasmid DNA into mammalian cells. *Proc. Natl. Acad. Sci. USA*, 100, 9138–9143.

Tkachenko, A. G., Xie, H., Coleman, D., Glomm, W., Ryan, J., Anderson, M. F., Franzen, S. & Feldheim, D. L. (2003) Multifunctional gold nanoparticle-peptide complexes for nuclear targeting. *J. Am. Chem. Soc.*, 125, 4700–4701.

Tkachenko, A. G., Xie, H., Liu, Y., Coleman, D., Ryan, J., Glomm, W. R., Shipton, M. K., Franzen, S. & Feldheim, D. L. (2004) Cellular trajectories of peptide-modified gold particle complexes: comparison of nuclear localization signals and peptide transduction domains. *Bioconjugate Chem.*, 15, 482–490.

Tomalia, D. A. & Frechet, J. M. J. (2002) Discovery of dendrimers and dendritic polymers: a brief historical perspective,. *J. Polym. Sci. Part A: Polym. Chem.*, 40, 2719–2728.

Torchilin, V. P. (2006) Multifunctional nanocarriers. *Adv. Drug Deliv. Rev.*, 58, 1532–1555.

Torchilin, V. P. (2009) Multifunctional and stimuli-sensitive pharmaceutical nanocarriers. *Eur. J. Pharm. Biopharm.*, 71, 431–444.

Torney, F., Trewyn, B. G., Lin, V. S. Y. & Wang, K. (2007) Mesoporous silica nanoparticles deliver DNA and chemicals into plants. *Nat. Nanotechnol.*, 2, 295–300.

Truong-Le, V. L., Walsh, S. M., Schweibert, E., Mao, H. Q., Guggino, W. B., August, J. T. & Leong, K. W. (1999) Gene transfer by DNA-Gelatin nanospheres. *Arch. Biochem. Biophys.*, 361, 47–56.

Tseng, W. C., Fang, T. Y., Su, L. Y. & Tang, C. H. (2005) Dependence of transgene expression and the relative buffering capacity of dextran-grafted polyethylenimine. *Mol. Pharm.*, 2, 224–232.

Tseng, W. C. & Jong, C. M. (2003) Improved stability of polycationic vector by dextran-grafted branched polyethylenimine *Biomacromolecules*, 4, 1277–1284.

Tseng, W. C., Tang, C. H. & Fang, T. Y. (2004) The rote of dextran conjugation in transifection mediated by dextran-grafted polyethylenimine. *J. Gene Med.*, 6, 895–905.

Tseng, Y. C., Mozumdar, S. & Huang, L. (2009) Lipid-based systemic delivery of siRNA. *Adv. Drug Deliv. Rev.*, 61, 721–731.

Tsoli, M., Kuhn, H., Brandau, W., Esche, H. & Schmid, G. (2005) Cellular uptake and toxicity of Au55 clusters. *Small*, 1, 841–844.

Tsutsumi, T., Arima, H., Hirayama, F. & Uekama, K. (2006) Potential use of dendrimer/a-cyclodextrin conjugate as a novel carrier for small interfering RNA (siRNA). *J. Inclusion Phenomena Macrocyclic Chem.*, 56, 81–84.

Tsutsumi, T., Hirayama, F., Uekama, K. & Arima, H. (2007) Evaluation of polyamidoamine dendrimer/α-cyclodextrin conjugate (generation 3, G3) as a novel carrier for small interfering RNA (siRNA). *J. Control. Release*, 119, 349–359.

Twaites, B. R., De Las Heras-Alarcón, C., Cunliffe, D., Lavigne, M., Pennadam, S., Smith, J. R., Górecki, D. C. & Alexander, C. (2004) Thermo and pH responsive polymers as gene delivery vectors:effect of polymer architecture on DNA complexation in vitro. *J. Control. Release*, 97, 551–566.

Twaites, B. R., De Las Heras Alarcon, C., Lavigne, M., Saulnier, A., Pennadam, S., Cunliffe, D., Gorecki, D. C. & Alexander, C. (2005) Thermoresponsive polymers as gene delivery vectors: Cell viability, DNA transport and transfection studies *J. Control. Release*, 108, 472–483.

Tyner, K. M., Roberson, M. S., Berghorn, K. A., Li, L., Gilmour, R. F., Batt, C. A. & Giannelis, E. P. (2004) Intercalation, delivery, and expression of the gene encoding green fluorescence protein utilizing nanobiohybrids. *J. Control. Release*, 100, 399–409.

Vaheri, A. & Pagano, J. S. (1965) Infectious poliovirus RNA: a sensitive method of assay. *Virology*, 27, 434–436.

Van de Wetering, P., Cherng, J. Y., Talsma, H., Crommelin, D. J. A. & Hennink, W. E. (1998) 2-(dimethylamino)ethyl methacrylate based (co)polymers as gene transfer agents *J. Control. Release*, 53, 145–153.

Van de Wetering, P., Cherng, J. Y., Talsma, H. & Hennink, W. E. (1997) Relation between transfection efficiency and cytotoxicity of poly(2-(dimethylamino)ethyl methacrylate)/plasmid complexes. *J. Control. Release*, 49, 59–69.

Van de Wetering, P., Schuurmans, N. M. E., Van Steenbergen, M. J., Crommelin, D. J. A. & Hennink, W. E. (2000) Copolymers of 2-(dimethylamino)ethyl methacrylate with ethoxytriethylene glycol methacrylate or N-vinyl-pyrrolidone as gene transfer agents *J. Control. Release*, 64, 193–203.

Veintemillas-Verdaguer, S., Morales, M. P. & Serna, C. J. (1998) Continuous production of γ-Fe2O3 ultrafine powders by laser pyrolysis. *Mater. lett.*, 35, 227–231.

Verbaan, F. J., Oussoren, C., Van Dam, I. M., Takakura, Y., Hashida, M., Crommelin, D. J. A., Hennink, W. E. & Storm, G. (2001) The fate of poly(2-dimethyl amino ethyl)methacrylate-based polyplexes after intravenous administration. *Int. J. Pharm.*, 214, 99–101.

Wada, K., Arima, H., Tsutsumi, T., Chihara, Y., Hattori, K., Hirayama, F. & Uekama, K. (2005) Improvement of gene delivery mediated by mannosylated dendrimer/a-cyclodextrin conjugates. *J. Control. Release*, 104, 397–413.

Wagner, E., Cotten, M., Foisner, R. & Birnstiel, M. L. (1991) Transferrin-polycation-DNA complexes: The effect of polycations on the structure of the complexes and DNA delivery to the cells. *Proc. Natl. Acad. Sci. USA*, 88, 4255–4259.

Wagner, E. & Kloeckner, J. (2006) Gene delivery using polymer therapeutics *Adv. Polym. Sci.*, 192, 135–173.

Wagner, E., Zenke, M., Cotten, M., Beug, H. & Birnstiel, M. L. (1990) Transferrin-polycation conjugates as carriers for DNA uptake into cells. *Proc. Natl. Acad. Sci. USA*, 98, 3410–3414.

Walker, G. F., Fella, C., Pelisek, J., Fahrmeir, J., Boeckle, S., Ogris, M. & Wagner, E. (2005) Toward synthetic viruses: Endosomal pH-triggered deshielding of targeted polyplexes greatly enhances gene transfer in vitro and in vivo. *Mol. Ther.*, 11, 418–425.

Wang, C., Ge, Q., Ting, D., Nguyen, D., Shen, H. R., Chen, J. Z., Elsen, H. N., Heller, J., Langer, R. & Putnam, D. (2004a) Molecularly engineered poly(ortho ester) microspheres for enhanced delivery of DNA vaccines. *Nat. Mater.*, 3, 190–196.

Wang, D., Robinson, D. R., Kwon, G. S. & Samuel, J. (1999) Encapsulation of plasmid DNA in biodegradable poly(d,l-lactic-co-glycolic acid) microspheres as a novel approach for immunogene delivery. *J. Control. Release*, 57, 9–18.

Wang, D. A., Narang, A. S., Kotb, M., Gaber, A. O., Miller, D. D., Kim, S. W. & Mahato, R. I. (2002a) Novel branched poly(ethylenimine)-cholesterol water-soluble lipopolymers for gene delivery. *Biomacromolecules*, 3, 1197–1207.

Wang, J., Gao, S. J., Zhang, P. C., Wang, S., Mao, H. Q. & Leong, K. W. (2004b) Polyphosphoramidate gene carriers: effect of charge group on gene transfer efficiency. *Gene Ther.*, 11, 1001–1010.

Wang, J., Lee, I. L., Lim, W. S., Chia, S. M., Yu, H., Leong, K. W. & Mao, H. Q. (2004c) Evaluation of collagen and methylated collagen as gene carriers. *Int. J. Pharm.*, 279, 115–126.

Wang, J., Mao, H. Q. & Leong, K. W. (2001) A novel biodegradable gene carrier based on polyphosphoester *J. Am. Chem. Soc.*, 123, 9480–9481.

Wang, J., Zhang, P. C., Lu, H. F., Ma, N., Wang, S., Mao, H. Q. & Leong, K. W. (2002b) New polyphosphoramidate with a spermidine side chain as a gene carrier. *J. Control. Release*, 83, 157–168.

Wang, J., Zhang, P. C., Mao, H. Q. & Leong, K. W. (2002c) Enhanced gene expression in mouse muscle by sustained release of plasmid DNA using PPE-EA as a carrier. *Gene Ther.*, 9, 1254–1261.

Wang, L. & Macdonald, R. C. (2004) Effects of microtubule-depolymerizing agents on the transfection of cultured vascular smooth muscle cells: enhanced expression with free drug and especially with drug-gene lipoplexes. *Mol. Ther.*, 9, 729–737.

Wang, Y., Gao, S. J., Ye, W. H., Yoon, H. S. & Yang, Y. Y. (2006) Co-delivery of drugs and DNA from cationic core-shell nanoparticles self-assembled from a biodegradable copolymer. *Nat. Mater.*, 5, 791–796.

Wang, Y., Ke, C. Y., Beh, C. W., Liu, S. Q., Goh, S. H. & Yang, Y. Y. (2007) The self-assembly of biodegradable cationic polymer micelles as vectors for gene transfection. *Biomaterials*, 28, 5358–5368.

Wen, Y., Pan, S., Luo, X., Zhang, X., Zhang, W. & Feng, M. (2009) A biodegradable low molecular weight polyethylenimine derivative as low toxicity and efficient gene vector. *Bioconjugate Chem.*, 20, 322–332.

Wenz, G. & Keller, B. (1992) Threading cyclodextrin rings on polymer chains. *Angew. Chem. Int. Ed.*, 31, 197–199.

Whilton, N. T., Vickers, P. J. & Mann, S. (1997) Bioinorganic clays: synthesis and characterization of amineo- and polyamino acid intercalated layered double hydroxides. *J. Mater. Chem.*, 7, 1623–1629.

Wiradharma, N., Tong, Y. W. & Yang, Y. Y. (2009) Self-assembled oligopeptide nanostructures for co-delivery of drug and gene with synergistic therapeutic effect *Biomaterials*, 30, 3100–3109.

Wolff, J. A., Malone, R. W., Williams, P., Chong, W., Acsadi, G., Jani, A. & Felgner, P. L. (1990) Direct gene transfer into mouse muscle in vivo. *Science*, 247, 1465–1468.

Wolff, J. A. & Rozema, D. B. (2008) Breaking the bonds: Non-viral vectors become chemically dynamic. *Mol. Ther.*, 16, 8–15.

Wong, F. M. P., Reimer, D. L. & Bally, M. B. (1996) Cationic lipid binding to DNA: characterization of complex formation. *Biochemistry*, 35, 5756–5763.

Wong, K., Sun, G. B., Zhang, X. Q., Dai, H., Liu, Y., He, C. B. & Leong, K. W. (2006) PEI-g-chitosan, a novel gene delivery system with transfection efficiency comparable to polyethylenimine in vitro and after liver administration in vivo. *Bioconjugate Chem.*, 17, 152–158.

Wong, S. Y., Pelet, J. M. & Putnam, D. (2007) Polymer systems for gene delivery – Past, present, and future. *Prog. Polym. Sci.*, 32, 799–837.

Wu, D. C., Liu, Y., Jiang, X., Chen, L., He, C. B., Goh, S. H. & Leong, K. W. (2005) Evaluation of hyperbranched poly(amino ester)s of amine constitutions similar to polyethylenimine for DNA delivery. *Biomacromolecules*, 6, 3166–3173.

Wu, D. C., Liu, Y., Jiang, X., He, C. B., Goh, S. H. & Leong, K. W. (2006) Hyperbranched Poly(amino ester)s with Different Terminal Amine Groups for DNA Delivery. *Biomacromolecules*, 7, 1879–1833.

Wu, G. Y. & Wu, C. H. (1987) Receptor-mediated in vitro gene transformation by a soluble DNA carrier system. *J. Biol. Chem.*, 262, 4429–4432.

Xu, Y. & Szoka, J. F. C. (1996) Mechanism of DNA release from cationic liposome/DNA complexes used in cell transfection. *Biochemistry*, 35, 5616–5623.

Xu, Z. P., Niebert, M., Porazik, K., Walker, T. L., Cooper, H. M., Middelberg, A. P., Gray, P. P., Bartlett, P. F. & Lu, G. Q. (2008) Subcellular compartment targeting of layered double hydroxide nanoparticles. *J. Control. Release*, 130, 86–94.

Xu, Z. P., Stevenson, G. S., Lu, C. Q., Lu, G. Q., Bartlett, P. F. & Gray, P. P. (2006) Stable suspension of layered double hydroxide nanoparticles in aqueous solution. *J. Am. Chem. Soc.*, 128, 36–37.

Xu, Z. P., Walker, T. L., Liu, K. L., Cooper, H. M., Lu, G. Q. & Bartlett, P. F. (2007) Layered double hydroxide nanoparticles as cellular delivery vectors of supercoiled plasmid DNA. *Int. J. Nanomedicine*, 2, 163–174.

Yamaoka, T., Abe, M. & Tsuji, M. (1989) Synthesis of Cu-Al hydrotalcite like compound and its ion exchange property *Mater. Res. Bulletin*, 24, 1183–1199.

Yang, N. S., Burkholder, J., Roberts, B., Martinell, B. & Mccabe, D. (1990) In vivo and in vitro gene transfer to mammalian somatic cells by particle bombardment. *Proc. Natl. Acad. Sci. U. S. A.*, 87, 9568–9572.

Yang, Y., Xu, Z., Chen, S., Gao, Y., Gu, W., Chen, L., Pei, Y. & Li, Y. (2008a) Histidylated cationic polyorganophosphazene/DNA self-assembled nanoparticles for gene delivery *Int. J. Pharm.*, 353, 277–282.

Yang, Y., Xu, Z., Jiang, J., Gao, Y., Gu, W., Chen, L., Tang, X. & Li, Y. (2008b) Poly(imidazole/DMAEA)phosphazene/DNA self-assembled nanoparticles for gene delivery: Synthesis and in vitro transfection *J. Control. Release*, 127, 273–279.

Yant, S. R., Meuse, L., Chiu, W., Ivics, Z., Izsvak, Z. & Kay, M. A. (2000) Somatic integration and long-term transgene expression in normal and haemophilic mice using a DNA transposon system. *Nat. Genet.*, 25, 35–41.

Yezhelyev, M. V., Qi, L. F., O'regan, R. M., Nie, S. M. & Gao, X. H. (2008) Proton-sponge coated quantum dots for siRNA delivery and intracellular imaging. *J. Am. Chem. Soc.*, 130, 9006–9012.

Yin, Y. & Alivisatos, A. P. (2005) Colloidal nanocrystal synthesis and the organic-inorganic interface. *Nature*, 437, 664–670.

Yokoyama, M. (2002) Gene delivery using temperatureresponsive polymeric carriers. *DDT*, 7, 426–432.

Yokoyama, M., Kurisawa, M. & Okano, T. (2001) Influential factors on temperature-controlled gene expression using thermoresponsive polymeric gene carriers. *J. Artif. Organs*, 4, 138–145.

Young, L. S., Searle, P. F., Onion, D. & Mautner, V. (2006) Viral gene therapy strategies: from basic science to clinical application. *J. Pathol.*, 208, 299–318.

Yu, J. H., Quan, J. S., Huang, J., Wang, C. Y., Sun, B., Nah, J. W., Cho, M. H. & Cho, C. S. (2009a) α,β-Poly(L-aspartate-graft-PEI): A pseudo-branched PEI as a gene carrier with low toxicity and high transfection efficiency. *Acta Biomaterialia*, 5, 2485–2494.

Yu, J. H., Quan, J. S., Kwon, J. T., Xu, C. X., Sun, B., Jiang, H. L., Nah, J. W., Kim, E. M., Jeong, H. J., Cho, M. H. & Cho, C. S. (2009b) Fabrication of a novel core-shell gene delivery system based on a brush-like polycation of α, β-poly (L-aspartate-graft-PEI). *Pharm. Res.*, 26, 2152–2163.

Zanta, M. A., Boussif, O., Adib, A. & Behr, J. P. (1997) In vitro gene delivery to hepatocytes with galactosylated polyethylenimine. *Bioconjugate Chem.*, 8, 839–844.

Zauner, W., Ogris, M. & Wagner, E. (1998) Polylysine-based transfection systems utilizing receptor-mediated delivery. *Adv. Drug Deliv. Rev.*, 30, 97–113.

Zelphati, O. & Szoka, J. F. C. (1996) Mechanism of oligonucleotide release from cationic liposomes. *Proc. Natl. Acad. Sci. U. S. A.*, 93, 11493–11498.

Zhang, C. Y., Yeh, H. C., Kuroki, M. T. & Wang, T. H. (2005) Single-quantum-dot-based DNA nanosensor. *Nat. Mater.*, 4, 826–831.

Zhang, G., Budker, V. & Wolff, J. A. (1999) High levels of foreign gene expression in hepatocytes after tail vein injections of naked plasmid DNA. *Hum. Gene Ther.*, 10, 1735–1737.

Zhang, H., Zou, K., Guo, S. H. & Duan, X. (2006a) Nanostructural drug-inorganic clay composites: Structure, thermal property and in vitro release of captopril-intercalated Mg-Al-layered double hydroxides *J. Solid State Chem.*, 179, 1792–1801.

Zhang, J. X., Sun, H. L. & Ma, P. X. (2010) Host-guest interactions mediated polymeric assemblies: Multifunctional nanoparticles for drug and gene delivery. *ACS Nano*, 4, 1049–1059.

Zhang, M. Z., Ishii, A., Nishiyama, N., Matsumoto, S., Ishii, T., Yamasaki, Y. & Kataoka, K. (2009) PEGylated calcium phosphate nanocomposites as smart environment-sensitive carriers for siRNA delivery. *Adv. Mater.*, 21, 3520–3525.

Zhang, M. Z. & Kataoka, K. (2009) Nano-structured composites based on calcium phosphate for cellular delivery of therapeutic and diagnostic agents *Nano Today*, 4, 508–517.

Zhang, T., Stilwell, J. L., Gerion, D., Ding, L., Elboudwarej, O., Cooke, P. A., Gray, J. W., Alivisatos, A. P. & Chen, F. F. (2006b) Cellular effect of high doses of silica-coated quantum dot profiled with high throughput gene expression analysis and high content cellomics measurements. *Nano Lett.*, 6, 800–808.

Zhang, X. H., Collins, L., Sawyer, G. J., Dong, X. B., Qiu, Y. & Fabre, J. W. (2001) In vivo gene delivery via portal vein and bile duct to individual lobes of the rat liver using a polylysine-based nonviral DNA vector in combination with chloroquine. *Hum. Gene Ther.*, 12, 2179–2190.

Zhang, Y., Schlachetzki, F., Li, J. Y., Boado, R. J. & Pardridge, W. M. (2003) Organ-specific gene expression in the rhesus monkey eye following intravenous non-viral gene transfer. *Mol. Vision*, 9, 465–472.

Zhu, C. H., Jung, S., Luo, S. B., Meng, F. H., Zhu, X. L., Park, T. G. & Zhong, Z. Y. (2010) Co-delivery of siRNA and paclitaxel into cancer cells by biodegradable cationic micelles based on PDMAEMA-PCL-PDMAEMA triblock copolymers. *Biomaterials*, 31, 2408–2416.

Zhu, J., Munn, R. J. & Nantz, M. H. (2000) Self-cleaving ortho ester lipids: A new class of pH-vulnerable amphiphiles. *J. Am. Chem. Soc.*, 122, 2645–2646.

Zhu, J. L., Cheng, H., Jin, Y., Cheng, S. X., Zhang, X. Z. & Zhuo, R. X. (2008) Novel polycationic micelles for drug delivery and gene transfer. *J. Mater. Chem.*, 18, 4433–4441.

Zhu, M. Z., Wu, Q. H., Zhang, G., Ren, T., Liu, D. & Guo, Q. X. (2002) Synthesis and evaluation of cationic lipids bearing cholesteryl groups for gene delivery in vitro. *Bull. Chem. Soc. Jpn.*, 75, 2207–2213.

Zikmund, M., Putyera, K. & Hrnciarova, K. (1996) A novel route for the preparation of hydrotalcite and synthesis of intercalated reversible dioxygen-carrying cobalt (II) complexes. *Chem. Pap.*, 50, 262–270.

Zinselmeyer, B. H., Mackay, S. P., Schatzlein, A. G. & Uchegbu, I. F. (2004) The lower-generation polypropylenimine dendrimers are effective gene-transfer agents. *Pharm. Res.*, 19, 960–967.

Zintchenko, A., Ogris, M. & Wagner, E. (2006) Temperature dependent gene expression induced by PNIPAM-based copolymers: Potential of hyperthermia in gene transfer *Bioconjugate Chem.*, 17, 766–772.

Zugates, G. T., Tedford, N. C., Zumbuehl, A., Jhunjhunwala, S., Kang, C. S., Griffith, L. G., Lauffenburger, D. A., Langer, R. & Anderson, D. G. (2007) Gene delivery properties of end-modified poly(β-amino ester)s. *Bioconjugate Chem.*, 18, 1887–1896.

Zuhorn, I. S., Bakowsky, U., Polushkin, E., Visser, W. H., Stuart, M. C., Engberts, J. B. & Hoekstra, D. (2005) Nonbilayer phase of lipoplex-membrane mixture determines endosomal escape of genetic cargo and transfection efficiency. *Mol. Ther.*, 11, 801–810.

Zuhorn, I. S., Kalicharan, R. & Hoekstra, D. (2002) Lipoplex-mediated transfection of mammalian cells occurs through the cholesterol-dependent clathrin-mediated pathway of endocytosis. *J. Biol. Chem.*, 277, 18021–18028.

Zuidam, N. J. & Barenholz, Y. (1998) Electrostatic and structural properties of complexes involving plasmid DNA and cationic lipids commonly used for gene delivery. *Biochim. Biophys. Acta*, 1368, 115–128.

Imaging of Nanoparticle Delivery Using Terahertz Waves

Joo-Hiuk Son, Seung Jae Oh, Jihye Choi, Jin-Suck Suh, Yong-Min Huh, and Seungjoo Haam

Abstract In this paper, the principle of terahertz molecular imaging (TMI) used for the visualization of nanoparticle delivery is presented. In addition, a brief review of terahertz (THz) technology in terms of the generation, detection, and characteristics of THz waves is given. The TMI technique entails the sensitive measurement of the change in the THz optical coefficients of water induced by the surface plasmon resonance of nanoparticles. It is shown that the measured THz response is linearly proportional to the concentration of the nanoparticles. Nanoparticle delivery to a cancerous tumor and the organs in a mouse is imaged by employing the TMI technique. This technique enables target-specific sensing of cancers as well as molecular diagnosis of drug delivery.

Keywords Cancer diagnosis • Drug delivery • Optical coefficients • Molecular imaging • Terahertz waves

1 Introduction

A number of nanoparticle probes and drugs have recently been developed for use in the areas of therapeutics and diagnostics (Lee et al. 2007, 2008; McCarthy and Weissleder 2008). It is imperative that nanoparticles be delivered to the targeted cells or organs for obtaining maximum diagnostic and therapeutic results; further, accurate monitoring of nanoparticle delivery to the target cells is important in

J.-H. Son (✉)
Department of Physics, University of Seoul, Seoul 130-743, Korea
e-mail: joohiuk@uos.ac.kr

S.J. Oh, J.-S. Suh, and Y.-M. Huh
Department of Radiology, College of Medicine, Yonsei University, Seoul 120-752, Korea

J. Choi and S. Haam
Department of Chemical and Biomolecular Engineering, Yonsei University, Seoul 120-752, Korea

A. Prokop (ed.), *Intracellular Delivery: Fundamentals and Applications*, Fundamental Biomedical Technologies 5, DOI 10.1007/978-94-007-1248-5_22,

theranostics. A number of molecular imaging techniques have been developed to visualize such biological processes at the cellular or molecular level. Among these techniques are positron emission spectroscopy (PET) (Massoud and Gambhir 2003), magnetic resonance imaging (MRI) (Weissleder et al. 2000), and optical imaging using bioluminescence or fluorescence (Moore et al. 1948, Weissleder and Ntziachristos 2003), all of which have their pros and cons. For example, PET is a highly sensitive technique but has low spatial resolution and requires the use of radioactive isotopes. MRI offers better spatial resolution than does PET but has lower sensitivity despite the fact that iron oxide or gadolinium derivatives, which are expected to increase the sensitivity of the technique, are used as contrast agents. Optical imaging techniques are relatively simple and inexpensive, but the penetration depth and spatial resolution that can be achieved are very low because of the Rayleigh scattering of optical waves by cells.

In this review, a novel molecular imaging technique based on the use of terahertz (THz) electromagnetic waves and nanoparticle probes (Oh et al. 2009) is introduced. The basics of THz technology, including the generation and detection of THz waves, are explained in Sect. 2. In addition, the THz characteristics of water are reviewed because they form the basis of TMI. In Sect. 3, the principle of TMI is presented. An example of nanoparticle delivery imaging of a cancerous tumor and the organs in a mouse is shown in Sect. 4. A brief summary is presented in Sect. 5.

2 Review of Terahertz Technology

THz waves, ranging from 0.1 to 10 THz, occupy the electromagnetic spectrum between microwave and infrared bands. This region is scientifically rich because of the rotational and vibrational energies of the materials that lie in that frequency range (see Fig. 1), but it has gained importance only in recent times because of the difficulty involved in the generation and detection of THz signals. In this section, THz technology is reviewed with respect to generation and detection techniques. In addition, the THz characteristics of water are discussed because they form the basis of the THz molecular imaging technique.

2.1 *Generation and Detection of Terahertz Waves*

The generation of electromagnetic waves in the THz frequency range is difficult because transport-type devices such as transistors cannot move electrons fast enough to generate THz signals, and transit-type devices such as lasers do not have natural gain media with a small energy separation of a only a few meV. However, two decades ago when femtosecond lasers were developed, researchers worked to combine optical pulses with semiconductors and thus generate THz electromagnetic pulses. When 800-nm laser pulses from a mode-locked Ti:sapphire laser (Spence

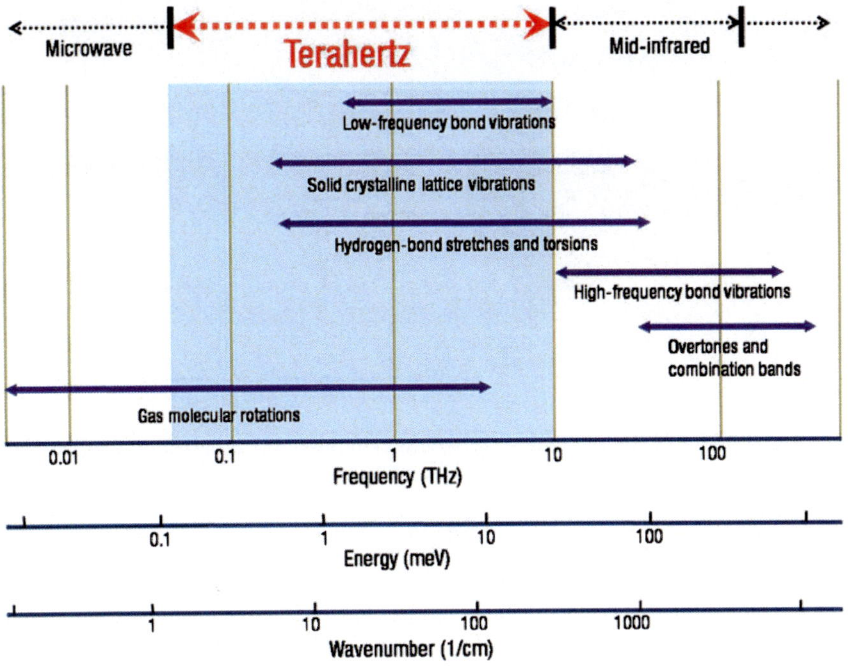

Fig. 1 Characteristic energies in electromagnetic spectrum

et al. 1991, Son et al. 1992) illuminate a semiconductor, such as semi-insulating GaAs (SI-GaAs) having a band gap of 1.45 eV, electron–hole pairs are generated and drift to the electrodes fabricated on the semiconductor wafer (see Fig. 2). The drifted charges are collected by the opposite electrodes biased with a dc voltage, which, in turn, generate a photocurrent surge in the coplanar waveguide. The increase in the photocurrent is proportional to the integration of the optical pulse, and the decrease is dependent on the semiconductor carrier lifetime. The time-varying photocurrent gives rise to the free-space radiation of electromagnetic waves with an equivalent Lienard-Wiechert potential (Schwartz 1972). This generation scheme is known as the photoconductive switching technique. Using this technique, Fattinger et al. achieved a frequency spectrum of over 1 THz by Fourier transformation of the subpicosecond electromagnetic pulse in the free space. They referred to this radiation as the THz beams (Fattinger and Grischkowsky 1988). Other generation methods using femtosecond laser interaction include techniques using a built-in surface field in semiconductors, the photo-Dember effect, and optical rectification. These techniques do not require the use of external voltage bias (Zhang and Xu 2010).

Variations in THz pulses can occur in less than a picosecond, making them impossible to measure with conventional electronic equipment such as oscilloscopes or network analyzers. The first technique to measure such high-frequency

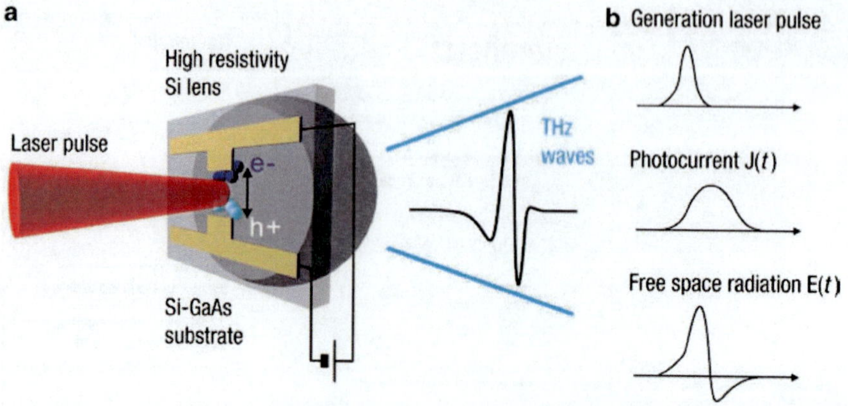

Fig. 2 Schematic of THz pulse generation. (**a**) Generation of electron–hole pairs in photoconductive switching antenna. (**b**) Waveforms of excitation laser pulse, induced photocurrent, and THz pulse in free space

signals was a photoconductive sampling technique that is the reverse of photoconductive switching. Free-space THz waves are captured by a waveguide antenna and the induced electric fields are switched on a semiconductor such as low-temperature-grown GaAs (LT-GaAs) by a femtosecond laser. The switching occurs in less than a picosecond to give a dc voltage to a lock-in amplifier because the carrier lifetime of LT-GaAs is approximately 0.5 ps (Gupta et al. 1991). The full THz waveform can be reconstructed by varying the optical delay between the generation and detection femtosecond optical pulses. The detection can also be accomplished with the electro-optic (EO) sampling technique using Pockels' effect (Valdmanis and Mourou 1986). Some examples of EO crystals that can be used for freespace detection are ZnTe, GaP, GaSe, etc. A bandwidth of over 30 THz was realized using this technique (Wu and Zhang 1997).

One example of a THz spectroscopic system is shown in Fig. 3. This system generates THz pulses from an InAs wafer by the photo-Dember effect and detects them using a photoconductive switch of LT-GaAs. It operates in a reflection mode to measure signals from an absorptive sample such as water or certain biological materials. This system can obtain spectroscopic information from 0.1 to several THz from a specific measurement point, and can also construct a spectroscopic image of a sample by raster scanning. This experimental setup was utilized to produce the measurement results shown in Sects. 3 and 4.

Because THz technology has tremendous potential for applications in scientific study, security, and medicine, many generators and detectors have been developed. Among the more notable generators are quantum cascade lasers (Kohler et al. 2002), THz parametric generators and oscillators (Kawase et al. 1996, 2001), and THz photomixers using two diode lasers (McIntosh et al. 1995). A detector fabricated with a microbolometer array can take live images of THz signals (Behnken et al. 2008). See the work by Son (2009) for a detailed review of generators and detectors.

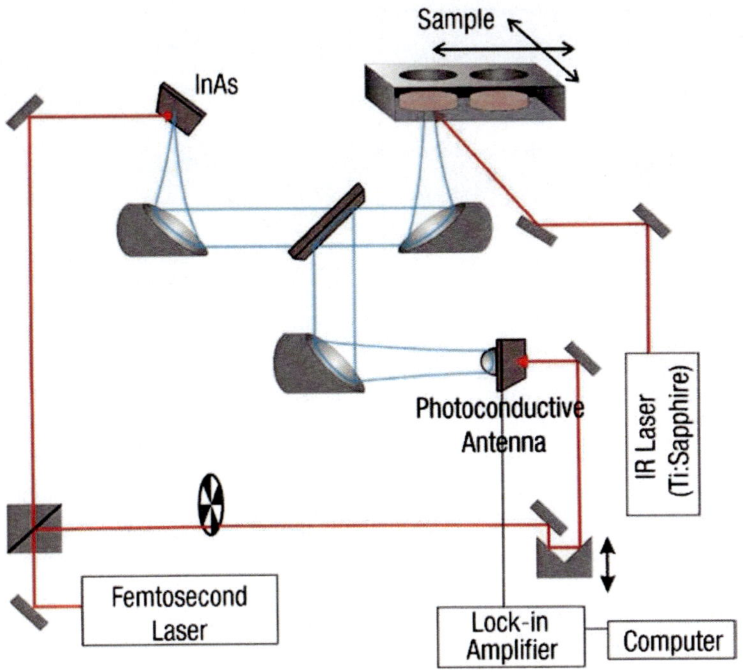

Fig. 3 Schematic of reflection-mode THz imaging setup with infrared (IR) laser for the induction of surface plasma polaritons (From Oh et al. (2009). Copyright © 2009 by Optical Society of America. Reprinted by permission of Optical Society of America)

2.2 *THz Characteristics of Water*

Water is the most abundant material in living cells and organs, and its characteristics play an important role in diagnosis based on the use of THz waves. THz waves are sensitive to the dynamics of water molecules because their rotational and vibrational energies and hydrogen bond stretching energy lie in the THz region. Another important feature is that the low energy of the THz waves prevents ionization of the water molecules. Therefore, THz waves have been utilized to study the microscopic and macroscopic characteristics of water. The microscopic dynamics between water and a solute resulted in a larger absorption of THz waves due to the coherent oscillation between the hydration water and the solute and caused the hydrogen bond rearrangement process to slow down (Heugen et al. 2006). For THz molecular imaging for nanoparticle delivery, the macroscopic characteristics are more important because the TMI technique is based on a principle that relies on the temperature-dependent property of bulk water, as will be explained in the next section. As can be seen in Fig. 4, the power absorption and the refractive index of liquid water are highly dependent on temperature because it affects bonding strength. In addition,

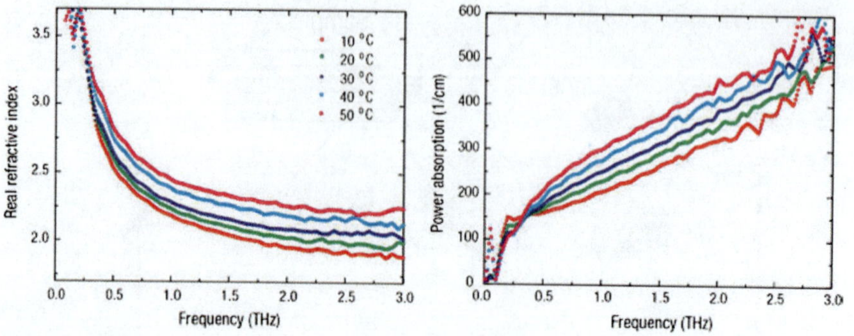

Fig. 4 Temperature-dependent power absorptions and refractive indices of water (From Son (2009). Copyright © 2009 by American Institute of Physics. Reprinted by permission of American Institute of Physics)

the molecule's motions are active in the THz frequency range (Son 2009). The power absorption becomes greater as the temperature increases, and reaches a local maximum around 6 THz due to the hydrogen-bond resonance (Zelsmann 1995). THz measurement with a Debye model fit has shown that bulk water has two relaxation dynamics with fast and slow times. The slow time of approximately 8 ps is related to the loosening of hydrogen bonds, and the fast time of 170 fs arises from the realignment of molecules under an external electric field (Oh et al. 2007).

3 Principle of Terahertz Molecular Imaging

THz molecular imaging (TMI) is achieved by monitoring the changes in water's THz optical property such as absorption and index caused by temperature variation. Cells and organs contain an abundant amount of water, and the water temperature can be controlled by nanoparticle probes (nanoprobes) delivered to the cells and organs. The gold nanorods (GNRs) shown by the transmission electron microscopy (TEM) images in Fig. 5 (Oh et al. 2009) are an example of nanoprobes. The GNRs have resonant features around 800 nm, as shown in Fig. 5d, and their resonant wavelength can be engineered by adjusting the dimension. GNRs can be targeted to specific cells or tumors by antibody conjugation, and they can be used to generate heat in the targeted location by surface-plasmon resonance under infrared (IR) irradiation at the resonant wavelength. To simulate GNRs in cells, the GNRs were immersed in water and the THz response was measured by the reflection-mode THz imaging setup shown in Fig. 3. The increased heat induced by the surface plasmon elevated the temperature of the water, and THz reflectivity increased, as shown in Fig. 6. Similar results were obtained with an A431 skin cancer cell endocytosed with GNRs. Fig. 7 shows the cancer cell images with and without GNRs. Figs 7b, c are the THz images with and without IR irradiation, respectively. The cell with GNRs subjected to IR irradiation was the only one to display higher THz reflectivity

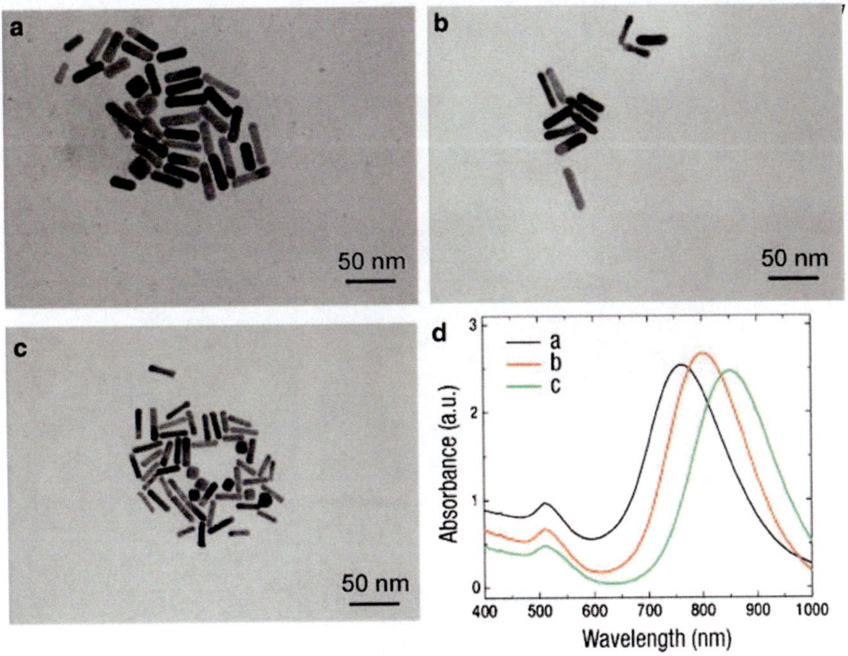

Fig. 5 TEM images of gold nanorods (GNRs) with aspect ratios of (**a**) 3.2, (**b**) 4.0, and (**c**) 4.2; (**d**) UV-visible absorption spectra of the GNRs of (**a**), (**b**), and (**c**) (From Oh et al. (2009). Copyright © 2009 by Optical Society of America. Reprinted by permission of Optical Society of America)

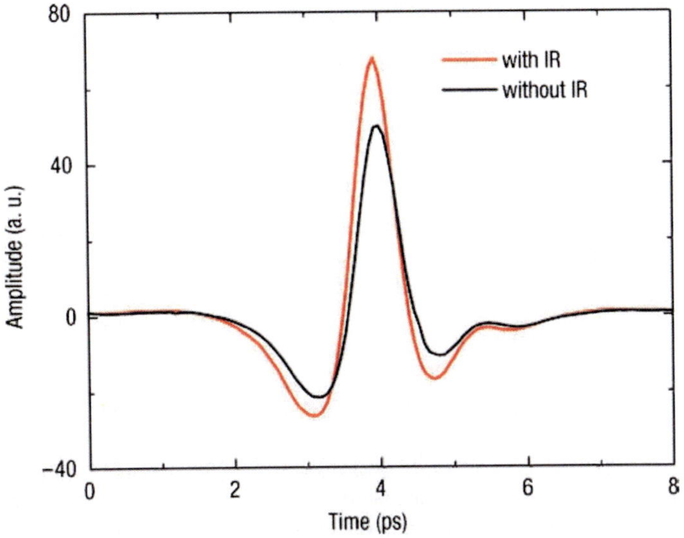

Fig. 6 Reflected THz time-domain waveforms from water with GNRs before (*black line*) and 5 minutes after their irradiation with an IR laser (*red line*). (From Oh et al. (2009). Copyright © 2009 by Optical Society of America. Reprinted by permission of Optical Society of America)

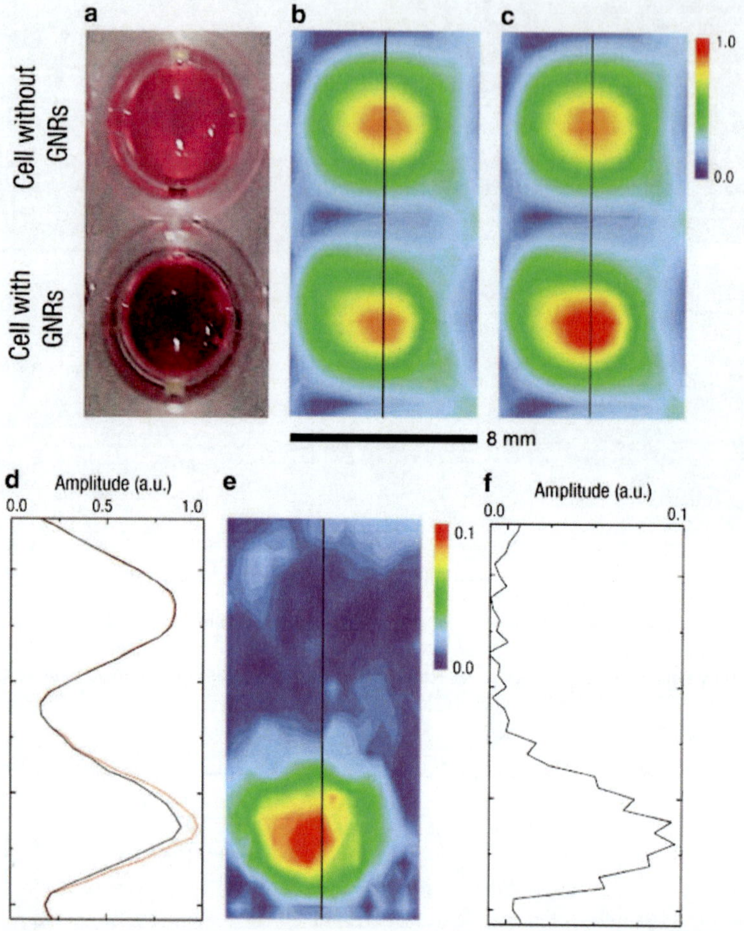

Fig. 7 Cancer cell images with and without GNRs. (**a**) Visible image; (**b**) THz image without IR irradiation; (**c**) THz image with IR irradiation; (**d**) amplitudes along the lines in (**b**) (*black*) and (**c**) (*red*); (**e**) differential image between (**b**) and (**c**); and (**f**) amplitude along the line in (**e**) (From Oh et al. (2009). Copyright © 2009 by Optical Society of America. Reprinted by permission of Optical Society of America)

resulting from the temperature rise caused by the surface-plasmon resonance. The cell without GNRs did not show any change regardless of IR irradiation. Although the increase in reflectivity was small (less than 10%), the contrast became evident when the difference was measured between Figs. 7b, c to give Fig. 7e. The differential THz image from the cell without GNRs was negligible, but that from the cell with GNRs showed clear evidence of nanoparticle distribution. The ratio between the amplitudes with and without GNRs was approximately 30. Therefore, very-high-contrast THz molecular imaging can be accomplished for the sensitive monitoring of nanoprobe delivery.

4 Imaging of Nanoparticle Delivery

Nanoparticle delivery to a cancerous tumor and the organs in a mouse was measured using the technique of THz molecular imaging with a differential modulation of the IR irradiation beam. As shown in Figs. 8a, b a male nude mouse bearing an A431 epidermoid carcinoma tumor in the proximal thigh region was prepared, and the nanoprobes were injected through the tail-vein; a similar mouse without nanoprobe injection served as a control. The nanoprobe was the GNR composite conjugated with Cetuximab (Cetuximab PEGylated gold nanorods: CET-PGNRs) for epidermal growth factor receptor (EGFR) specific tumor-cell targeting. The nanoprobes were absorbed by the liver and the spleen on their way to being delivered to the tumor. The organs were surgically removed to measure the concentration of nanoprobes delivered to specific locations by the TMI technique. As can be seen in Fig. 9a, most of the nanoprobes were captured by the liver, a small amount was captured by the spleen, and a fair amount was targeted to the cancerous tumor as the CET-PGNR was designed. Almost no nanoprobes reached the brain. Fig. 9b shows the conventional THz imaging without nanoprobe injection. It is clear that detailed quantitative imaging of nanoparticle delivery can be achieved using THz waves.

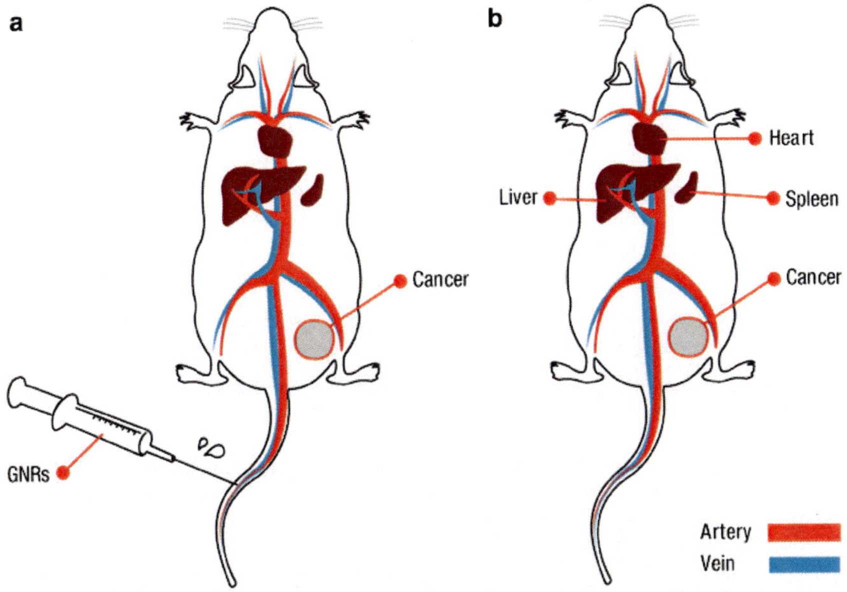

Fig. 8 Preparation of mice with A431 skin cancer tumors. (**a**) A mouse with the injection of nanoprobes; (**b**) a mouse without the nanoprobe injection to serve as a control

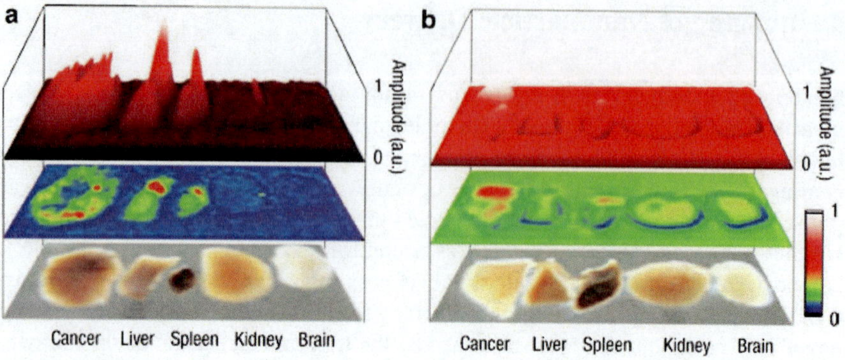

Fig. 9 THz images of cancerous tumors and organs. (**a**) From a mouse with the injection of nanoprobes; (**b**) from a mouse without the injection

5 Conclusion

The principle of TMI is presented along with a brief review of the method of generation and detection of THz waves. The use of the TMI technique enables the quantitative measurement of nanoparticle distributions and allows for the high-contrast imaging of these distributions. The effectiveness of this technique is demonstrated by considering nanoprobe delivery to a cancerous tumor and the organs in a mouse. The images show that the signal amplitudes are linearly proportional to the nanoprobe concentration (Oh et al. 2011). The TMI technique has promising applications in the target-specific sensing of cancers as well as in the molecular diagnosis of drug delivery.

Acknowledgments The authors thank Mr. Heejun Shin and Mr. Dong-Gyu Lee for their assistance in the preparation of this paper. This study was supported by a grant from the Korean Health Technology R&D Project of the Ministry for Health, Welfare and Family Affairs, Republic of Korea (A101954), and a National Research Foundation of Korea (NRF) grant funded by the Korean government (MEST) under Grant Nos. 20100020647, 20100001979, 20100015989, 20100011934, and 20090054519.

References

Behnken BN, Karunasiri G, Chamberlin DR, Robrish PR, Faist J (2008) Real-time imaging using a 2.8 THz quantum cascade laser and uncooled infrared microbolometer camera. Opt. Lett. 33: 440–442.

Fattinger Ch., Grischkowsky D (1988) Terahertz beams. Appl. Phys. Lett. 54: 490–492.

Gupta S, Frankel MY, Valdmanis JA, Whitaker JF, Mourou GA, Smith FW, Calawa AR (1991) Subpicosecond carrier lifetime in GaAs grown by molecular beam epitaxy at low temperatures. Appl. Phys. Lett. 59: 3276–3278.

Heugen U, Schwaab G, Brundermann E, Heyden M, Yu X, Leitner DM, Havenith M, (2006) Solute-induced retardation of water dynamics probed directly by terahertz spectroscopy. Proc. Natl. Acad. Sci. USA 103: 12301–12306.

Kawase K, Sato M, T. Taniuchi T, H. Ito H, (1996) Coherent tunable THz-wave generation from $LiNbO_3$ with monolithic grating coupler. Appl. Phys. Lett. 68: 2483–2485.
Kawase K, Shikata J.-I., K. Imai K, H. Ito H, (2001) Transform-limited, narrow-linewidth, terahertz-wave parametric generator. Appl. Phys. Lett. 78: 2819–2021.
Kohler R, Tredicucci A, Beltram F, Beere HE, Linfield EH, Davies AG, Ritchie DA, Iotti RC, Rossi F (2002) Terahertz semiconductor-heterostructure laser. Nature 417: 156–159.
Lee JH, Huh YM, Jun YW, Seo JW, Jang JT, Song HT, Kim S, Cho EJ, Yoon HG, Suh JS, Cheon J (2007) Artificially engineered magnetic nanoparticles for ultra-sensitive molecular imaging. Nature Med. 13: 95–99.
Lee J, Yang J, Ko H, Oh SJ, Kang J, Son JH, Lee K, Lee SW, Yoon HG, Suh JS, Huh YM, Haam S (2008) Multifunctional magnetic gold nanocomposites: Human epithelial cancer detection via magnetic resonance imaging and localized synchronous therapy. Adv. Func. Mat. 18: 258–264.
Massoud TF, Gambhir SS (2003) Molecular imaging in living subjects: Seeing fundamental biological processes in a new light. Genes & Development 17: 545–580.
McCarthy JR, Weissleder R (2008) Multifunctional magnetic nanoparticles for targeted imaging and therapy. Adv. Drug Deliv. Rev. 60: 1241–1251.
McIntosh KA, Brown ER, Nichols KB, McMahon OB, Dinatale WF, Lyszczarz TM (1995) Terahertz photomixing with diode lasers in low-temperature-grown GaAs. Appl. Phys. Lett. 67: 3844–3846.
Moore GE, Peyton WT, French LA (1948) The clinical use of fluorescein in neurosurgery. The localization of brain tumors. J. Neurosurg. 5: 392–398.
Oh SJ, Son J, Yoo O, Lee DH (2007) Terahertz characteristics of electrolytes in aqueous Luria-Bertani media. J. Appl. Phys. 102: 074702-1-5.
Oh SJ, Kang J, Maeng I, Suh JS, Huh YM, Haam S, Son JH (2009) Nanoparticle-enabled terahertz imaging for cancer diagnosis. Opt. Express 17: 3469–3475.
Oh SJ, Choi J, Maeng I, Park JY, Lee K, Huh YM, Suh JS, Haam S, Son JH (2011) Molecular imaging with terahertz waves. Opt. Express 19: 4009–4016.
Schwartz M (1972) in Principles of Electrodynamics. Dover Publications, Inc., New York.
Son J, Rudd JV, Whitaker JF (1992) Noise characterization of a self-mode-locked Ti:sapphire laser. Opt. Lett. 17: 733–735.
Son JH (2009) Terahertz electromagnetic interactions with biological matter and their applications. J. Appl. Phys. 105: 102033-1-10.
Spence DE, Kean PN, Sibbett W (1991) 60-fsec pulse generation from a self-mode-locked Ti:sapphire laser. Opt. Lett. 16: 42–44.
Valdmanis JA, Mourou GA (1986) Subpicosecond electrooptic sampling: Principles and applications. IEEE J. Quantum Electron. 22: 69–78.
Weissleder R, Moore A, Mahmood U, Bhorade R, Benveniste H, Chiocca EA, Basilion JP (2000) In vivo magnetic resonance imaging of transgene expression Nature Med. 6: 351–355.
Weissleder R, Ntziachristos V (2003) Shedding light onto live molecular targets. Nature Med. 9: 123–128.
Wu Q, Zhang XC (1997) Free-space electro-optics sampling of mid-infrared pulses. Appl. Phys. Lett. 71: 1285–1286.
Zelsmann HR (1995) Temperature dependence of the optical constants for liquid H_2O and D_2O in the far IR region. J. Mol. Struct. 350: 95–114.
Zhang XC, Xu J (2010) in Introduction to THz Wave Photonics, Springer, New York.

Calcium Phosphate and Calcium Phosphosilicate Mediated Drug Delivery and Imaging

O.A. Pinto, A. Tabaković, T.M. Goff, Y. Liu, and J.H. Adair

Abstract Nanotechnology-based imaging, drug and gene delivery systems can be used to improve current diagnostic and therapeutic approaches. Calcium phosphate (CP) and calcium phosphosilicate (CPS) based nanoparticulate systems are becoming increasingly popular as delivery vehicles for therapeutic and imaging agents due to biocompatible and biodegradable properties of the material. In this chapter, we emphasize properties of the various calcium phosphates that make these materials promising nanocarriers and we provide an overview of the architectures of different calcium phosphate systems including the nanocomposite calcium phosphosilicate nanoparticles, hybrid CP-liposome and CP-polymeric systems. The versatility of calcium phosphate based delivery systems for therapeutic and imaging applications is emphasized.

Keywords Bioimaging • Calcium phosphate • Calcium phosphosilicate • Drug delivery • Gene delivery • Nanoparticles • Targeted delivery

List of Acronyms/Abbreviations

7-OH-DPAT	7-hydroxy-2-dipropyl-aminotetralin
AAV	adeno-associated virus
ADP	adenosine diphosphate
ATP	adenosine triphosphate
ACP	amorphous calcium phosphate
APTES	aminopropyltriethoxysilane
carboxySNARF-1	carboxy-seminapthorhodafluor

O.A. Pinto, A. Tabaković, T.M. Goff, Y. Liu, and J.H. Adair (✉)
Department of Materials Science and Engineering, Penn State University,
University Park, PA 16802, USA
e-mail: jha3@psu.edu

A. Prokop (ed.), *Intracellular Delivery: Fundamentals and Applications*, Fundamental Biomedical Technologies 5, DOI 10.1007/978-94-007-1248-5_23,

CD71	Cluster of Differentiation 71 (an antibody for the transferrin receptor)
CEPA	carboxyethylphosphoric acid
CER_6	hexanoyl-ceramide
CER_{10}	decanoyl ceramide
CHM	cholesterol-bearing mannan
CHP	cholesterol-bearing pullalan
CP	calcium phosphate
CPS	calcium phosphosilicate
CPSNPs	calcium phosphosilicate nanocomposite particles
CPP	casein phosphopeptides
Cy3	Cy3 Amidite
DMSO	dimethyl sulfoxide
DNA	deoxyribonucleic acid
DOPA	1,2-dioleoylsn-glycero-3-phosphate
DOTAP	1, 2-dioleoyl-3-trimethylammonium-propane chloride salt
DPBS	Dulbecco's phosphate buffered saline
DSPE-PEG	1,2-distearoryl-sn-glycero-3-phosphoethanolamine-N-[methoxy (polyethyleneglycol-2000) ammonium salt
EPC	L-α-phosphatidylcholine
EPR	enhanced permeation and retention
FCS	fluorescence correlation spectroscopy
HA	hydroxyapatite
ICG	indocyanine green
LD50	(median) lethal dose, 50%
LCP	lipid coated calcium phosphate
LNCP	lipid-coated nano-calcium-phosphate
MPS	mononuclear phagocytic system
MTT	3-(4,5-dimethylthiazol-2-yl)-2,5-diphenyltetrazolium bromide
PBS	phosphate buffered saline
PEG	polyethylene glycol
RES	reticuloendothelial system
RNA	ribonucleic acid
RNAi	RNA interference
siRNA	small interfering RNA
ZP	zeta potential

1 Introduction

Conventional chemotherapeutic treatments have side effects caused by the systemic delivery and interaction of the drug with healthy, non-cancerous cells (Alagna 2001; Margolis 2005). Reducing or eliminating side effects caused by therapeutics and improving diagnostics by enhanced imaging agents is an area that has been under investigation for a number of years (Langer 1990; Tomlinson 1987; Leong 1988). Approaches to improve diagnosis and treatment include employing technologies such as nanotechnology-based imaging (for earlier and better detection), and targeted nanoparticulate drug and gene delivery systems for improved treatment alternatives without the systemic side effects. Moreover, nanoparticle delivery systems can deliver higher dosages of the chemotherapeutic to the targeted tissue while avoiding systemic side effects (Adair 2010).

In conventional therapy and imaging, drugs do not generally possess tunable properties that enable enhanced imaging and therapeutic effects. Nanoparticulate delivery systems, such as calcium phosphate based systems, can overcome the barriers associated with conventional treatments, yielding an improved avenue for treatment. The potential advantages of nanoparticulate systems over conventional therapies include the capability to minimize drug degradation during *in vivo* transport (by affording the drug protection from the environment), minimize side effects (by improved biocompatibility and biodegradability and better targeted delivery) and increase drug bioavailability and amount of drug delivered to the tissue(s) of interest (due to high drug loading potential and targetability of nanoparticles) (Torchilin 2006; Adair 2010).

The functionalities of nanoparticulate delivery systems provide the opportunity (1) for prolonged circulation in the bloodstream, and hence increased likelihood of accumulation at tumor sites via the enhanced permeation and retention (EPR) effect or, (2) to specifically bind to cells or tissue of interest via targets for cell surface receptors, (3) to respond to local stimuli *in vivo* (such as responses to changes in temperature and pH) and (4) to overcome the cell membrane barrier and avoid the enzyme filled lysosome where degradation occurs (Torchilin 2006). Consequently, desired characteristics of the ideal delivery system include: (1) non-toxic starting materials and degradation products, (2) small size (in the range of 10–200 nm) for improved uptake, (3) encapsulation of the active agent within the delivery system (as opposed to surface decoration or adsorption where the active agent is not protected from the environment), (4) colloidal stability of the delivery system (to prevent agglomeration *in vivo* during transport), (5) suitable clearance mechanism (to avoid side effects due to drug laden particles), (6) long clearance times (to allow adequate time for the delivery system to reach target cells and undergo endocytosis), (7) controlled release of active agent (such as a pH trigger) and (8) targetability of the delivery system (to enable delivery of particles to cells of choice) (Adair 2010).

The innate properties of calcium phosphates (such as biocompatibility and pH dependent solubility) combined with the ability to carefully engineer specific characteristics (size, surface functionalization, etc.) into the delivery system to suit a specific need or carry out a certain function make calcium phosphate based delivery systems especially useful in delivery based applications.

This chapter focuses on the potential use and efficacy of calcium phosphate based nanoparticles in biomedical applications involving imaging and therapy. An overview is presented of the inherent properties of the calcium phosphate system and practical aspects of calcium phosphate based systems that satisfy the criteria outlined earlier for an ideal delivery system. It is demonstrated that the inherent properties of calcium phosphate based systems provide advantages in the delivery and release of bio-agents (such as drugs, nucleic acids or fluorescent molecules for imaging).

The chapter draws attention to some of the different architectures of calcium phosphate based systems that have been developed and focuses on calcium phosphate based systems that are currently being investigated in the areas of gene and drug delivery and bioimaging. The importance of executing *in vivo* work to corroborate *in vitro* results is addressed, as the fate of any delivery system is dependent on the performance of the nanoparticulate system in the human body (best represented by *in vivo* experimentation).

2 Practical Aspects of Calcium Phosphate as a Delivery Vehicle

2.1 *Biocompatibility of Calcium Phosphate Nanoparticulates*

Calcium phosphate (CP) is biocompatible and biodegradable and is therefore a suitable material for biomedical applications. Calcium phosphate is naturally present in the body in bones and teeth. Additionally, both calcium and phosphate exist in the body at millimolar concentrations while intracellular concentrations are at the micromolar level (Oyane 2003; Tung 1998; Wang 2002; Guha 2009). As a biomineral, calcium phosphate biodistributes in the body with constituent ions regulated by the kidneys. The constituent ions of CP play essential roles in cell signaling, membrane regulation, programmed cell death, synaptic transmission and general cell functioning (Ghosh 1995). Phosphate is primarily concentrated in the mineral phase of bone, while smaller amounts are present as inorganic and organic compounds (such as adenosine diphosphate (ADP) and adenosine triphosphate (ATP)) in intracellular and extracellular compartments. The biocompatibility associated with calcium phosphate renders it a non-toxic material system with applications in drug delivery and imaging. The inherent non-toxicity of calcium phosphate suggests minimal side effects due to interactions of the delivery vehicle with the body.

A calcium phosphosilicate nanoparticle (CPSNP) system has been recently developed by Adair and co-workers. Among the innovations in the CPSNP system relative to pure nanoscale CP is that the silicate stabilizes the amorphous calcium phosphate (ACP) phase. Interestingly, silicon in bulk materials aids in the biomineralization process (Carlisle 1988; Dorozhkin 2010). In the aqueous precipitation of calcium phosphates, Eanes *et al.*, in a series of seminal papers beginning in 1965 demonstrate that an amorphous calcium phosphate phase almost immediately forms at reasonable supersaturations, but the high solubility of the ACP drives the dissolution-recrystallization to the more thermodynamically stable phase of hydroxyapatite (sometimes with intermediate phase formation) (Eanes 1965, 1973). The introduction of silicate into a solid solution of the ACP has been limited thus far to the confined space obtained within the reverse micelle system. The reverse micelle system used in the synthesis can be used to control the particle diameter down to small (15–40 nm) sizes (Adair 2005, 2010; Altınoğlu 2008, 2009,

2010b; Barth 2010; Kester 2008; Morgan 2008; Muddana 2009; Russin 2010). The reverse micelle synthetic route also permits the formation of well-dispersed nano-colloids that can be bio-conjugated to target molecules for specific cancers (Barth 2010). Adair and co-workers have reported in a series of manuscripts the efficacy of drugs and bioimaging agents encapsulated within the CPSNP.

2.2 pH Dependent Solubility of Calcium Phosphates

In addition to the biocompatibility and bioresorbability of calcium phosphate, the unique chemistry of the material system enables tunability of properties and hence versatility in applications. CP exists in a variety of crystalline and amorphous phases primarily dependent on the conditions under which the particles are formed and subsequent processing. ACP can be stabilized by the addition of species including Mg^{2+} and Zn^{2+} (Bertoni 1998; Bigi 1993, 1995; Rokita 1993). Additionally, the pH dependent solubility of calcium phosphates as shown in Fig. 1 underscores the

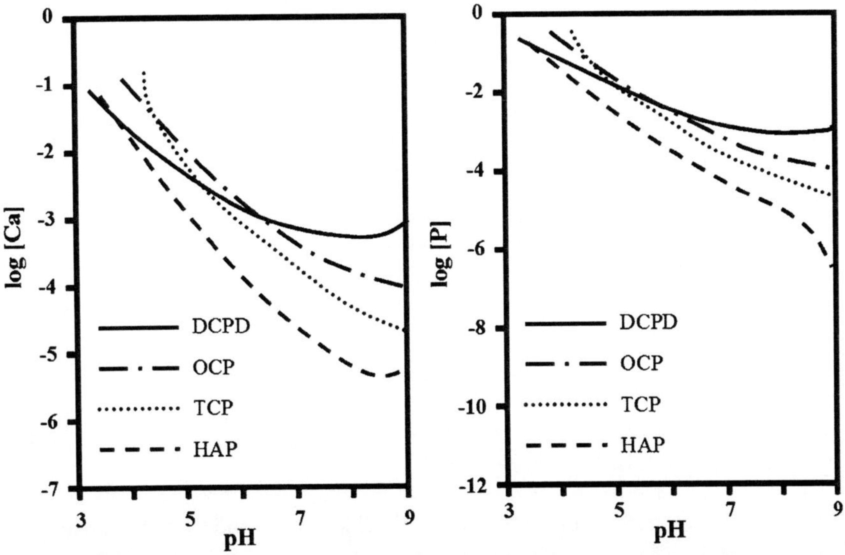

Fig. 1 pH dependent solubility behavior of calcium phosphates. As pH increases to more basic values calcium phosphate becomes increasingly insoluble. In contrast, as the pH approaches the acidic regime, calcium phosphate becomes increasingly soluble. The different phases of calcium phosphate represented above are: *DCPD* – dicalcium phosphate dihydrate, *OCP* – octacalcium phosphate, *TCP* – tricalcium phosphate and *HAP* – hydroxyapatite. The highly metastable amorphous phase and the less soluble amorphous calcium phosphosilicate phases are not shown. All phases of calcium phosphate become increasingly soluble below pH 6. This pH dependent behavior makes calcium phosphate an ideal encapsulating agent for drug and imaging agents – dissolution with release of the active agent occurs as pH decreases within the cell (Modified from Chow (1991). With permission)

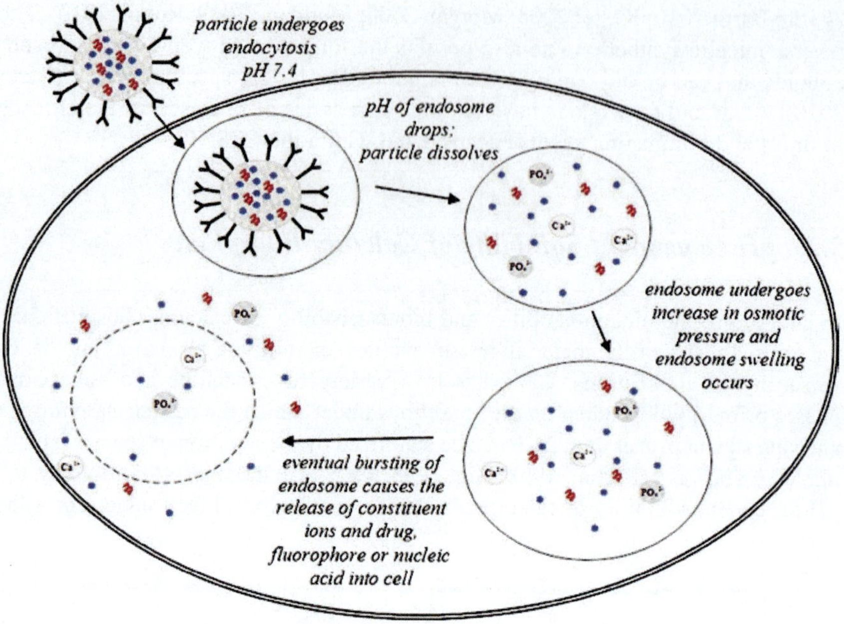

Fig. 2 Endocytosis and dissolution of calcium phosphate nanoparticle and release of active agent(s). The schematic represents the endocytosis and dissolution of a targeted calcium phosphate nanoparticle with the active agents (*blue* – fluorescent dye, *red helix* – nucleic acid) encapsulated within the nanoparticle. During endocytosis and in the early endosome, the pH is at pH 7.4 (physiological pH). As the proton pumps increase the H^+ concentration in the endosome, pH decreases dissolving the calcium phosphate or calcium phosphosilicate nanoparticles. The osmotic imbalance between the cytoplasm and the endosome causes endosomal swelling with eventual bursting, which results in the release of the active agent(s) into the cytoplasm of the cell

potential of the CP material system for controlled release (Chow 1991; Tung 1998). Calcium phosphates are sparingly soluble at pH 7.4 and at more alkaline conditions but are increasingly soluble as the pH decreases below pH 6 (where the pH in endosomes are particularly relevant, *see below*), a property that is a great advantage in drug delivery and release (Chander 1982).

In *in vitro* and *in vivo* delivery, following endocytosis (Fig. 2), nanoparticles experience a decrease in pH (pH 4.6–pH 5) while in the endosomal compartment (Alberts 2002). The pH decrease is created by osmotic shock also called the "proton sponge" effect. Vacuolar ATPase-H^+ pumps present in the membranes of endosomes regulate proton transfer from the cytoplasm into endosomal compartments (Li 2010; Alberts 2002). The influx of protons into the endosome drops the pH which induces the dissolution of the calcium phosphate. Additionally, the osmotic imbalance between the cytoplasm and the endosome causes endosomal swelling which eventually leads to endosomal rupture and subsequent release of the encapsulant into the cytoplasm of the cell (Li 2010; Alberts 2002).

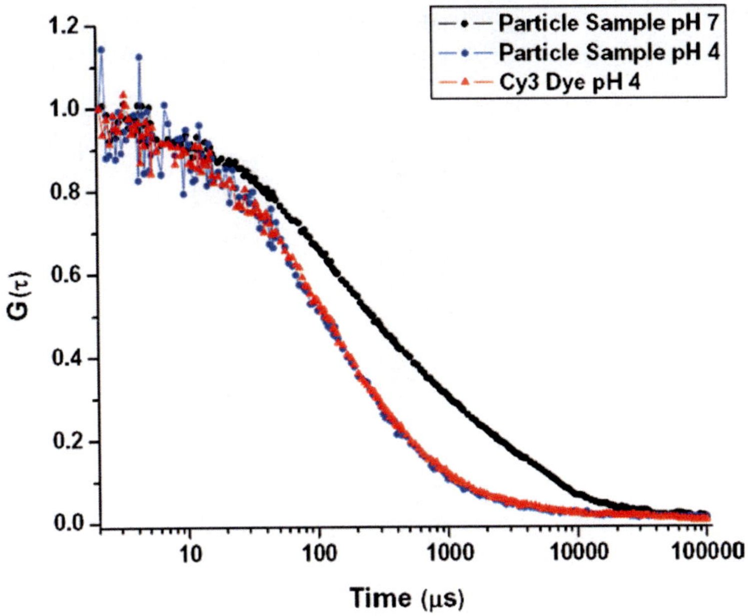

Fig. 3 Sample autocorrelation curves obtained using fluorescence correlation spectroscopy (FCS). Calcium phosphosilicate particles were characterized at pH 4 and pH 7 and free Cy3 at pH 4. Dissolved particles (pH 4) resemble behavior of free dye at pH 4 verifying calcium phosphosilicate particle dissolution and dye release as shown in Fig. 4 (Reproduced from Morgan (2008). With permission)

This pH dependent release mechanism was studied by Morgan *et al.* and Muddana *et al.* for the calcium phosphosilicate nanocomposite particles (CPSNPs) (Morgan 2008; Muddana 2009). Using fluorescence correlation spectroscopy (FCS) the solution-phase behavior of the particles was studied in DPBS (Dulbecco's phosphate buffered saline) at pH 7 and pH 4 to mimic the pH of mature endosomes (Morgan 2008; Muddana 2009). The drop in pH causes dissociation of calcium phosphate into its constituent ions and serves as a release mechanism for the payload encapsulated within the particle. As shown in Fig. 3, the sample autocorrelation curve at pH 4 resembles that of free Cy3 dye molecules, indicating particle dissolution and dye release. The change in the hydrodynamic diameter and the diffusion coefficient of the particles as a function of pH for the CPSNPs is depicted in Fig. 4 (Morgan 2008). The hydrodynamic diameter data verify that the particles at pH 4 resemble the behavior of the free dye with respect to both hydrodynamic diameter and diffusion coefficient (Morgan 2008). The pH dependent solubility paired with the biocompatibility of the material system render calcium phosphate delivery systems promising alternatives to conventional treatments (Altınoğlu 2009, 2008; Morgan 2008; Adair 2010; Li 2010).

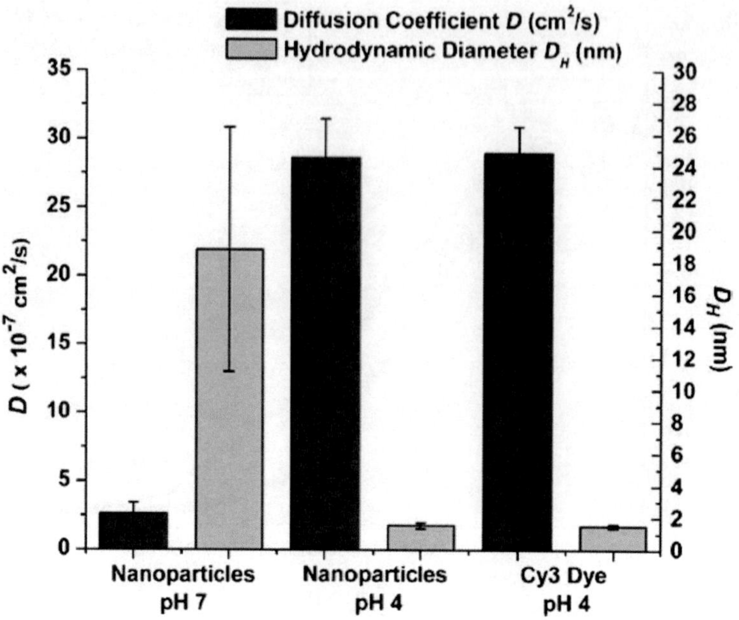

Fig. 4 Diffusion coefficient and hydrodynamic diameter of free Cy3 and Cy3-calcium phosphosilicate nanocomposite particles. Diffusion coefficient (*black*) and the hydrodynamic diameter (*gray*) of free Cy3 in Dulbecco's phosphate buffer solution (*DPBS*) at pH 4 and of Cy3 encapsulated calcium phosphosilicate nanoparticles at pH 4 and pH 7. No significant difference is detected for D (D_H) between nanoparticles sample and free Cy3 at pH 4 verifying that the encapsulated Cy3 is released via dissolution form the calcium phosphosilicate nanoparticles (Reproduced from Morgan (2008). With permission)

2.3 *Significance of Colloidal Stability in Calcium Phosphate Delivery Systems*

The earliest study of calcium phosphate as a delivery vehicle was performed over 35 years ago with use as a gene transfection agent (Graham 1973). These initial nanoparticulate systems, though promising as a transfection agent, lacked colloidal stability and size control. With the effect of colloidal stability and size becoming increasingly significant in the intracellular delivery of nanoparticles and in *in vivo* applications, greater importance is being placed on these nanoparticulate properties.

Lack of colloidal stability in a nanoparticulate system invariably leads to agglomeration (Adair 2005). Since size is a critical factor in cellular uptake, with the optimal particle size ranging between 20 and 200 nm, avoiding agglomeration is imperative (Gao 2005; Chithrani 2006; Thorek 2008). Size also affects circulation time, with larger particles easily trapped in the liver and rapidly removed from circulation while smaller particles usually remain in circulation for longer time periods (Kumar 2006; Burns 2009). Agglomerated particles (and large particles) are more likely to be sequestered by organs or membranes. Therefore, appropriate synthetic and dispersion schemes lead to better bioavailability at the tumor site while minimizing local off-target cytotoxic effects.

The three main classes of dispersion schemes are electrostatic, electrosteric and steric stabilization. For example, particles can be electrostatically dispersed with citrate (Li 2010; Morgan 2008; Kester 2008) to yield a negative zeta potential (ZP) or with aminopropyltriethoxysilane (APTES) to yield a positive ZP or sterically stabilized with a polymer such as polyethylene glycol (PEG) as shown in Fig. 5 (Morgan 2008). Other steric stabilizing molecules commonly used include dextrans, poly(vinylpyrrolidone) and polyacrylates (Kumar 2006). In addition to stabilization with synthetic molecules, a necessary criterion for successful bio-conjugation of a target molecule, such as an antigen or a polypeptide for an over-expressed receptor site on target cells, is to maintain colloidal dispersion during bio-conjugation and during the subsequent preparation of the particles for *in vitro* or *in vivo* delivery in solutions such as phosphate buffered saline (PBS) (Adair 2010; Barth 2010; Morgan 2008).

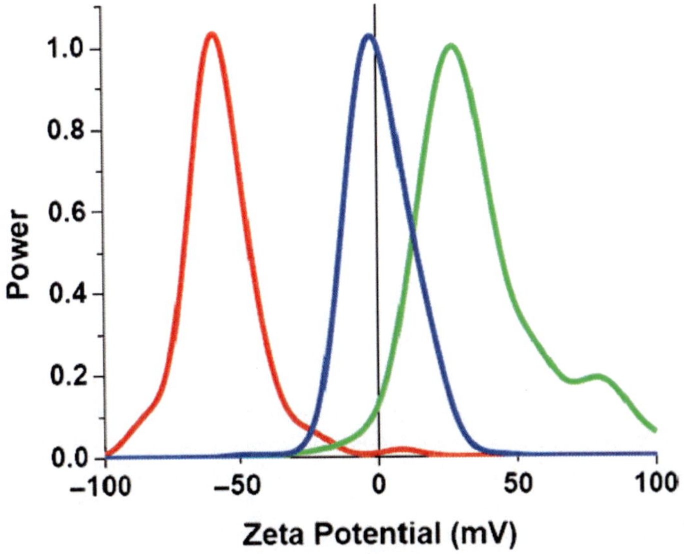

Fig. 5 The effect of the surface functionalization of calcium phosphosilicate nanocomposite particle (CPSNP) in 70:30 EtOH:H_2O on zeta potential. The COO⁻ – terminated CPSNPs (*red line*) have a mean zeta potential of −30 ± 3 mV. The $-NH_3^+$ terminated CPSNPs (*green line*) have a mean zeta potential of +24 ± 6 mV. The PEG-terminated CPSNPs (*blue line*) have a mean zeta potential of 3 ± 2 mV (Reproduced from Morgan (2008). With permission)

2.4 Significance of Surface Functionality in Calcium Phosphate Delivery Systems

The reticuloendothelial system (RES) and the mononuclear phagocytic system (MPS), identify and remove foreign particulates in the bloodstream and play a major role in drug bioavailability (Allemann 1993). A factor that affects recognition of particles by the RES and macrophages of the MPS is the particle surface chemistry. Certain surfaces are known to promote opsonization (a process that promotes detection and removal by the RES or MPS). Therefore, particle surface properties, more specifically, the hydrophobicity, hydrophilicity and charge of particles are usually modified to minimize opsonization. Hydrophobic nanoparticles are easily coated with opsonins (molecules, such as antibodies, that promote phagocytosis) and hence more recognizable by the RES and the MPS. In contrast, hydrophilic particles can resist the coating process to a greater extent and are more slowly cleared from the blood stream (Allemann 1993; Kumar 2006).

Since suppression of opsonization increases the retention of particles in the body, measures such as steric stabilization are employed to physically prevent opsonin proteins from adsorbing to particle surfaces (Owens III 2005). PEG is commonly used as a steric stabilization agent in an effort to improve dispersion of particles and to prevent opsonization. PEGylation of particles is also commonly utilized as a method to reduce renal clearance of smaller drug particles or molecules. Therefore, properly designed synthetic schemes (relevant surface functionalities) can lead to improved bioavailability (due to longer circulation times) of particles at the tumor site.

2.5 Targetability of Calcium Phosphate Nanoparticles

The two overarching schemes of drug delivery are passive and active targeting. Passive targeting takes advantage of the enhanced permeation and retention (EPR) effect present due to the leaky vasculature and poor drainage of tumors (Davis 1997; Torchilin 2000; Peer 2007). In contrast, active targeting employs cell specific targets that are bioconjugated onto the surface of nanoparticles to exclusively target the cells of interest. Targeted drug delivery is advantageous because it reduces uptake of nanoparticles by healthy, non-cancerous cells, reduces the overall drug dose necessary while increasing the dose at the target tissue and minimizing or eliminating side effects and non-specific effects. For a more comprehensive review of active and passive targeting, the reader is directed to publications by Davis, Torchilin and Peer (Davis 1997; Torchilin 2000; Peer 2007).

Examples of targeting in calcium phosphate systems include studies by Li *et al.* using lipid coated calcium phosphate (LCP) nanoparticles and Barth *et al.* using calcium phosphosilicate nanoparticles (CPSNPs) (Li 2010; Barth 2010). The study by Li *et al.* compares the gene silencing effect of untargeted LCP particles to anisamide-LCP particles. The results indicate improvement in gene silencing for targeted particles in both *in vitro* and an *in vivo* animal model. Barth *et al.* target CPSNPs with avidin-human holotransferrin and avidin-anti-CD71 antibody for breast cancer and with a ten amino acid gastrin (gastrin-10) polypeptide for pancreatic cancer via avidin-biotin and PEG-maleimide coupling, respectively. The enhanced delivery of gastrin-10 targeted CPSNPs to pancreatic cancer cells as compared to PEGylated CPSNPs and biotinylated avidin-pentagastrin CPSNPs is a promising result of the study (Barth 2010). While the methoxy-terminated PEG (passively targeted) CPSNPs display effective targeting, the anti-CD71-avidin targeted CPSNPs are the most effective for delivery to breast cancer cells (Barth 2010). However, the anti-CD71-avidin CPSNPs have non-specific binding in the stomach (Barth 2010). The off-target accumulation of the CPSNPs in the stomach is attributed to the diets of the mice which are high in biotin (which likely interacts with the avidin on the surface of the particles) and underscores an inherent limitation of the avidin-biotinylated bio-conjugation scheme for *in vivo* models (Barth 2010). The results of this study exemplify the importance of identifying appropriate target molecules and conducting both *in vitro* and *in vivo* studies to fully evaluate efficacy of the drug delivery or bioimaging nanoparticle system (Barth 2010).

3 Overview of Calcium Phosphate Systems – Synthesis and Architectures

There are various approaches in the encapsulation of therapeutic agents using calcium phosphate. Aside from the pure calcium phosphate system, the incorporation of polymers, liposomes and other organic and inorganic materials as well as various architectures to improve ease of encapsulation have been developed. The basic properties of some of these systems are addressed in this section but with new ideas and systems constantly being developed, this segment serves more as an introduction to the area rather than a detailed account.

Of great importance is to distinguish between encapsulation and surface decoration (adsorption). Table 1 summarizes a few of the calcium phosphate based delivery systems classified based on their architecture – encapsulation, surface decoration or a combination of the two architectures. As discussed earlier, the encapsulation of the active agent is a better alternative to the traditional use of calcium phosphate using surface decoration. In an encapsulation model, the active agent is protected within the delivery vehicle. Therefore, the active agent does not degrade due to interactions with other molecules in the blood stream *en route* to the cells of interest. In the surface decoration model, which has been practiced since 1973 (Graham 1973), the active agent is on the outer surface of the delivery vehicle and is susceptible to degradation while in the bloodstream. The ability to control dispersion, the EPR effect, and targeting are all compromised by the surface decoration approach for calcium phosphates. Thus, surface decoration does not readily permit *in vivo* studies, although *in vitro* evaluations may indicate

Table 1 Calcium phosphate based nanoparticulate systems

Architectures of select calcium phosphate based nanoparticulate systems			
Encapsulation			
System	Active agent	Remarks	Reference
Calcium phosphosilicate	Indocyanine green, ceramide C10, Cy3 amidite, cascade blue, fluorescein, rhodamine WT, 10-(3-sulfopropyl) acridinium betaine (SAB)	Successfully demonstrated encapsulation of a range of fluorophores with verified ability to encapsulate drugs and nucleic acids	Altınoğlu (2008), Kester (2008), Morgan (2008), Muddana (2009), Barth (2010)
Polymer core, CaP shell	β-Carotene	Demonstrates encapsulation with prospects in imaging and drug and gene delivery	Perkin (2005)
Hydroxyapatite hollow nanoparticles	Vancomycin	Hollow structure permits higher payload concentrations	Ye (2010)
Triblock co-Polymer-calcium phosphate nanoparticles	Ibuprofen	Ability to encapsulate hydrophobic drugs	Cao (2010)
Lipid coated calcium phosphate nanoparticles	Plasmid DNA	Higher DNA transfection efficiency compared to standard methods	Zhou (2010)
Calcium phosphate shell liposome core nanoparticles	carboxySNARF-1	Higher resistance to photobleaching compared to free dye	Chen (2010)
Calcium phosphate apatite doped particles	Europium	Applications in optical imaging	Al-Kattan (2010)

Surface decoration/adsorption

System	Active agent	Remarks	Reference
Calcium phosphate core nanoparticles	Oligonucleotides	Prone to degradation by nucleases	Sokolova (2007a, b)
Hydroxyapatite needle-shaped particles and plate-shaped particles	*cis*-diamminedichloroplatinum(II) (CDDP,cisplatin), di(ethylenediamineplatinum) medronate (DPM), bisphosphonate alendronate	Lacks *in vivo* work to demonstrate stability of drug under physiological conditions	Palazzo (2007)
Calcium phosphate core nanoparticles	5,10,15,20-tetrakis(4-phosphonooxyphenyl)porphine (p-TPPP)	Lacks *in vivo* work to corroborate stability of drug under physiological conditions	Ganesan (2008)

Combination structure

System	Active agent	Remarks	Reference
Calcium phosphate core nanoparticles	Oligonucleotides	Outer layer is prone to degradation while internal layer of active agent is protected	Sokolova (2007a, b)
Calcium phosphate core – polymer shell	5,10,15,20-Tetrakis(3-hydroxyphenyl)-porphyrin (mTHPP)	Outer layer is prone to degradation No improvement in efficiency over free dye Issues with toxicity of PEI	Klesing (2010)

efficacy for the surface decoration scheme. Unfortunately, a preliminary evaluation of a nanoparticle drug or bioimaging scheme limited to just *in vitro* studies can lead to a false sense of efficacy when, in fact, there is no efficacy for the critical *in vivo* tests that actually verify the true value of a nanoparticle system for eventual human health care.

3.1 Crystalline Calcium Phosphate

The most thermodynamically stable and widely used form of calcium phosphate is the crystalline phase of hydroxyapatite. As a function of time, the amorphous calcium phosphate phase will eventually transform into hydroxyapatite unless measures (such as doping) are taken to inhibit transformation (Dorozhkin 2010). Some examples of inhibitory agents are citrate (Lopez-Macipe 1999; Brecevic 1979; Johnsson 1991), phosphocitrate (Johnsson 1991), silicate (Skrtic 2002; Morgan 2008; Kester 2008; Altınoğlu 2008; Muddana 2009; Barth 2010; Altınoğlu 2010a) and casein phosphopeptides (Cross 2004). Various researchers have studied the benefits of the crystalline structure of calcium phosphate primarily for drug treatments involving bone diseases but also for other drug delivery and imaging systems.

Synthetic schemes for crystalline particles range from the simple precipitation of calcium and phosphate based on solutions such as $(NH_4)_2HPO_4$ and $Ca(NO_3)_2$ H_2O with an aqueous solution of the dopant (Mondejar 2007) to polymeric micelle templated systems (Ye 2010) and reverse-microemulsion systems (Mukesh 2009).

In terms of dispersion, for the nanoparticulate systems, the drug being delivered is the dispersant of choice for some investigators (Palazzo 2007; Kunieda 1993; Immamura 1995). Other dispersants include DNA (Mondejar 2007) and cholesterol-bearing mannan (CHM) (Yamane 2009) while no dispersant is evident in some systems (Mukesh 2009).

Crystalline calcium phosphate or hydroxyapatite (HA) has been studied as both a drug and gene delivery system (Barroug 2002, 2004; Ye 2010; Sokolova 2007b). Hollow hydroxyapatite nanospheres and nanotubes have been explored as carriers for the drug vancomycin (Ye 2010). The hollow nature of the calcium phosphate nanoparticles has the potential for higher drug loading capacity as compared to solid particles. Crystalline nanoparticles have also been investigated for their use in gene delivery (Sokolova 2006a,b, 2007a,b, 2010). The delivery vehicle model employed by Sokolova and colleagues uses the surface decoration approach and the combined surface decoration-encapsulation models. While *in vitro* work performed using their particles offer promising results, an *in vivo* model to better understand the performance of the particles in a dynamic, protein-abundant setting is lacking and should be the focus of future studies.

For bioimaging, it has been suggested that crystalline calcium phosphate has the ability to improve fluorescence properties when doped with lanthanide ions by preventing quenching (Mondejar 2007). Dynamic quenching (or quenching), a non-radiative relaxation mechanism occurs when one fluorescent molecule or ion interacts with another fluorescent molecule or ion (due to proximity or interactions). The incorporation of lanthanide ions into apatites via substitution of the calcium ions in the crystallographic lattice sites is expected to overcome the problem of quenching. The incorporation of the lanthanide ions in the appropriate lattice sites serve as fluorescent centers that act to reduce quenching, as stringent ion placement prevents interaction with other fluorescent ions.

A limitation, as suggested by Mondejar *et al.*, is that crystalline calcium phosphate usually results in larger particle sizes (due to the need for a highly crystalline structure), sometimes too large for cellular uptake. Poor size control due to the synthetic method also leads to larger particles and as explained earlier, increased susceptibility for RES and MPS mediated removal under *in vivo* conditions. Hydroxyapatite bulk implants, however, are exceptionally useful for implant treatment options (Nelson 2006) where cell uptake is irrelevant and HA exists in the form of large blocks of material. For imaging and drug delivery however, the size of particles can pose a problem and alternative structures or composite structures are sometimes preferred and are currently being developed as options to crystalline CP particles.

3.2 Polymeric – Calcium Phosphate Systems

Hybrid nanoparticles composed of calcium phosphate and polymeric materials are synthesized using polymer cross-linked micelles and hollow nanocages (Perkin 2005), nanogels (Sugawara 2006) and triblock copolymers (Cao 2010). In addition, block copolymer templates (though not incorporated in the actual particle) are used to synthesize hollow calcium phosphate nanoparticles (Tjandra 2006). The advantage of combined polymer-calcium phosphate systems is the higher loading potential of dyes (Klesing 2010) or drugs (Cao 2010) due to further loading ability in the polymer layer in addition to loading in the calcium phosphate component, and the ability to load hydrophobic drugs into the polymer layer.

Polymeric-calcium phosphate systems that are currently under investigation employ a variety of synthetic schemes. The cross-linked micelle and nanocage approach employed by Perkin *et al.* yields polymer core-amorphous calcium phosphate shell micelle particles and hollow calcium phosphate nanocages on the order of 50–70 nm in diameter. Both types of particles are assayed for the permeability of β-carotene as a function of calcium phosphate mineralization with optimal results for the unmineralized particles (Perkin 2005). The particles demonstrate potential applications as pH-responsive nanostructures for use in bioimaging and therapeutic delivery (Perkin 2005).

Hybrid polymer-calcium phosphate nanoparticles are synthesized by the precipitation of calcium phosphate in the presence of cholesterol-bearing pullalan (CHP) nanogels by the pH-gradient method (based on the pH dependent solubility of calcium phosphate) (Sugawara 2006). The assumed architecture of the hybrid particles are a composite CHP-calcium phosphate particle. The investigators report that calcium phosphate nucleation and growth first occurs within the nanogel and proceeds outward. Calcium phosphate growth is predicted to stop before the outer surface of the nanogel is reached; owing to the colloidal stability of the hybrid particles (Kakizawa 2004; Sugawara 2006). The outer surface of the hybrid particles, expected to consist of CHP polymer chains serve as a steric stabilizing agent (Sugawara 2006). Applications for the nanogel hybrid calcium phosphate particles (measuring approximately 40 nm in diameter) include protein and nucleic acid delivery as well as bone-repair material (Sugawara 2006). Other approaches to synthesize a hybrid polymer-calcium phosphate nanoparticle include the block copolymer (Kakizawa 2002) and aniomer (Kakizawa 2006) self-assembly method employed by Kakizawa *et al.* for gene therapy.

3.3 Liposome-Calcium Phosphate Systems

Liposomes have made the most progress as delivery systems for bioagents. However, liposomes are undergoing further optimization to improve *in vivo* circulation time and minimize negative effects on normal tissue while efficiently maintaining a high level of accumulation and sustained drug release to target sites (Hong 1999). The advantage of liposomal systems is the potential to encapsulate hydrophilic molecules in the interior cavity while hydrophobic molecules can be stored within the phospholipid membrane (Zhang 2009).

Synthetic methods for the formation of calcium phosphate-liposome composite systems include the precipitation of calcium followed by phosphate on the exterior of the liposome forming a CP shell with a liposome/solvent core forming particles 45–100 nm in diameter (Schmidt 2002). Another technique involves the coprecipitation of supersaturated solutions containing calcium and phosphate on 1,2-dioleoylsn-glycero-3-phosphate (DOPA) liposomes. In this method, carboxyethylphosphoric acid (CEPA) is used as a surface capping agent, preventing additional precipitation of CP and thus controlling size to around 100–200 nm (Schmidt 2004). The calcium phosphate shell is intended to mitigate the leaky nature of liposomes by supporting the liposome and protecting its contents from degradation while in the bloodstream.

Similarly, Chen *et al.* encapsulated carboxy-seminapthorhodafluor (carboxySNARF-1) in liposomes with a calcium phosphate shell (Chen 2010). In this approach, carboxySNARF-1 was prepared in a phosphate buffered solution which was used in the synthesis of the L-α-phosphatidylcholine (EPC) liposomes. To form the calcium phosphate shell on the surface, the carboxySNARF-1 EPC liposomes were added to a calcium chloride solution which initiates the precipitation of calcium phosphate with the already present phosphate. Again, CEPA was used as a capping agent to control particle size (Chen 2010). The ability to encapsulate carboxy-SNARF-1 (a pH sensitive dye) has applications in sensing and imaging.

In contrast to the previous systems in which a liposomal core is surrounded by a calcium phosphate shell, Zhou *et al.* synthesized nanoparticles consisting of a core of calcium phosphate surrounded by a lipid layer (Zhou 2010). The lipid-coated nano-calcium-phosphate (LNCP) particles were synthesized via ethanol injection: the lipid components (dimethyldioctadecylammonium bromide/cholesterol/d-α-Tocopheryl Polyethylene Glycol-1000 Succinate) dissolved in ethanol were injected into a solution containing a mixture of calcium acetate and monosodium phosphate, sonicated and then dialyzed. The LNCP particles (on the order of 100s of nanometers) demonstrate improved DNA transfection in a mammalian cell line as compared to the liposomal control (Zhou 2010). The calcium phosphate-liposomal systems have use in fluorophore (Chen 2010) and drug delivery (Schmidt 2002, 2004) with applications in gene delivery (Zhou 2010).

3.4 Layer by Layer Calcium Phosphates

Another approach to preparing calcium phosphate nanoparticles is the layer by layer or multi-shell method in which molecules of interest are surface decorated on a calcium phosphate core as opposed to encapsulated within the particle. In this technique, calcium and phosphate solutions are coprecipitated and then stabilized with DNA (Welzel 2004; Sokolova 2006a, b, 2007a) or oligonucleotides (Sokolova 2007b, 2010).

The layer by layer method, with modifications, is also used to create double and triple-shell layered nanoparticles (Sokolova 2006a, b) as shown in Fig. 6. For the double-shell approach, a layer of calcium phosphate is precipitated on to the

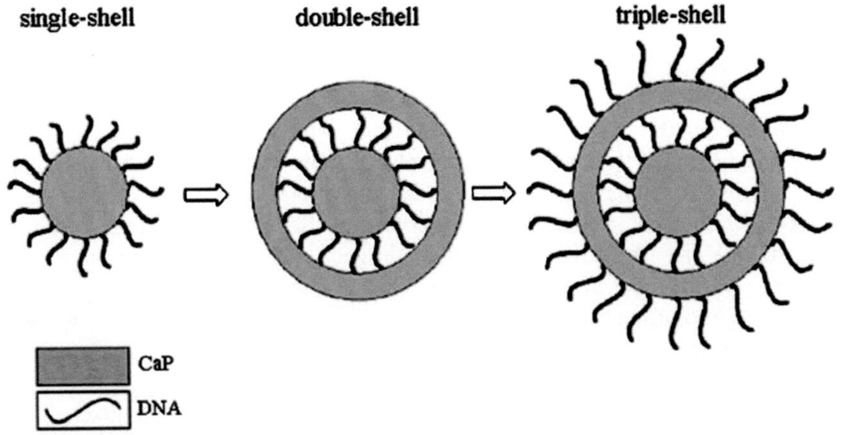

Fig. 6 Schematic representation of the three different shell structures of calcium phosphate nanoparticles synthesized by Sokolova *et al.* The single-shell consists of a core of calcium phosphate surrounded by a layer of DNA as a stabilizing agent. The double-shell structure has an additional layer of calcium phosphate on top of the DNA layer of the single-shell method. The triple-shell structure consists of the initial single-shell structure with an additional layer of calcium phosphate stabilized with DNA (Reproduced from Sokolova (2006a, b). With permission)

single-shell, DNA stabilized nanoparticles. The problem with the double-shell approach is the lack of colloidal stability due to the absence of electrostatic or steric stabilizers (Sokolova 2006a). The triple-shell architecture consists of a layer of DNA or oligonucleotide on the double-shell nanoparticle. The additional shell is an improvement to the original single-shell and double-shell method. The challenge associated with the single shell method is the lack of protection of the DNA or oligonucleotides from enzymatic attack, resulting in low transfection and knock-down efficiency while the double-shell lacks colloidal stability (Sokolova 2006a). By creating another shell of calcium phosphate and DNA, the interior DNA is now encapsulated and protected from the outside environment, and thus available for transfection. The outer layer of DNA stabilizes the particle but at the expense of increased size.

A potential problem that is yet to be addressed in the layer by layer system is the lack of colloidal stability of the particles in the presence of serum proteins under in vivo conditions. Initial results indicate the instability of the DNA stabilized particles in the presence of serum containing media is due to the adsorption of proteins on the surface which cause agglomeration (Sokolova 2006b, 2007a, b, 2010).

3.5 Amorphous Calcium Phosphate

Amorphous calcium phosphate (ACP) is another class of calcium phosphates with use as a nanocarrier. ACP is the phase of calcium phosphate that first precipitates from aqueous solution. Unless stabilized by dopants or stored under dry conditions, transformation to the thermodynamically stable hydroxyapatite occurs (Dorozhkin 2010). The biological relevance of amorphous calcium phosphates is addressed in detail by a recent review published by Dorozhkin (Dorozhkin 2010).

The amorphous phase of calcium phosphate is selected as the material system of choice by a handful of investigators. Cross *et al.* used an amorphous calcium fluoride phosphate stabilized with casein phosphopeptides (CPP) as calcium, phosphate and fluoride ion carriers for the remineralization of teeth (Cross 2004). The amorphous particles were synthesized by the coprecipitation of calcium, phosphate and fluoride solutions with CPP (the stabilization agent) with the resulting nanoparticle diameters on the order of 4–5 nm (Cross 2004).

Another promising amorphous calcium phosphate system developed by Adair and coworkers includes the synthesis of calcium phosphosilicate nanoparticles (CPSNPs) in double reverse microemulsions (Altınoğlu 2008; Morgan 2008; Kester 2008; Muddana 2009; Russin 2010; Barth 2010). The investigators utilize silicate to stabilize the amorphous phase. Citrate is employed as a dispersant during the synthetic process. The core CPSNPs are 20 nm in diameter but can range in hydrodynamic diameter between 20 and 200 nm depending on the surface functionalization scheme (Barth 2010). The CPSNPs encapsulate therapeutics (Kester 2008) and a wide variety of fluorescent dyes (Altınoğlu 2008; Morgan 2008) for imaging and photodynamic therapy (Altınoğlu 2010b; Barth 2010). Additionally, the efficacy of the particles has been demonstrated under *in vitro* (Kester 2008; Morgan 2008) and *in vivo* (Barth 2010; Altınoğlu 2008, 2009, 2010b) conditions.

4 Applications of Calcium Phosphate Delivery Systems

4.1 Nucleic Acid Delivery Using Calcium Phosphate Systems

RNA interference (RNAi) is an exploratory treatment option that is of great interest in the development of personalized treatment options particularly for cancer treatment. RNAi involves the delivery of small interfering RNA (siRNA) to target cells with the intention of down-regulating (or up-regulating in some instances) gene activity and hence protein synthesis. Although promising as a treatment option, the primary limitation of this technique is the susceptibility of the siRNA to degrade by nucleases present in the blood and tissue in systemic delivery (Nguyen 2008). While viral vectors (primarily the adeno-associated viruses (AAVs) and adenoviruses) have been used experimentally as a delivery vehicle, severe toxicity (hepatoxicity and immunogenicity) associated with these agents have raised safety concerns and hence prevented their use in humans (Grimm 2006).

The earliest studies on calcium phosphate for the transfection of DNA date back to 1973 (Graham 1973). However, though promising, this initial approach had problems with dispersion, size control and low transfection efficiency. Most calcium phosphate delivery systems being developed today have taken into consideration the importance of nucleic acid encapsulation, delivery vehicle colloidal stability and size, and address it in their synthetic schemes.

The biocompatibility and bioresorbability of calcium phosphates make this material an advantageous gene delivery vehicle. Syntheses of nucleic acid-calcium phosphate complexes take advantage of the affinity of the positively charged calcium ions for the negatively charged phosphate backbone of RNA and DNA. The electrostatic interaction of the RNA/DNA with the calcium ion stabilizes the nucleic acid/CP complex which can consequently be carried across the cell membrane via ion-channel mediated endocytosis (Truong-Le 1999; Roy 2003).

The calcium-nucleic acid complex is a non-toxic alternative to the viral delivery systems (Kulkarni 2006). Investigators that pursue calcium phosphate as a gene delivery agent have employed synthetic methods such as reverse microemulsions (Liu 2005; Roy 2003; Bisht 2005; Li 2010), liposomal coated calcium phosphate nanoparticles by the ethanol injection method (Zhou 2010), and the coprecipitation of calcium and phosphate solutions with oligonucleotides for single shell and multishell calcium phosphate nanoparticles (Zhang 2010; Sokolova 2006a,b, 2007b).

Since particle size and stability as well as nucleic acid stability are important factors affecting intracellular delivery, investigators are addressing these issues when designing delivery systems for nucleic acids. In an effort to protect the genetic material from *in vivo* degradation by nucleases, the coprecipitation of calcium and phosphate with the negatively charged nucleic acid was investigated (Welzel 2004).

The resulting particulate system is a calcium phosphate particle with some nucleic acid on the interior of the particle, but the majority on the exterior (surface decoration). The problem with this scheme is the lack of protection of the nucleic acid from enzymatic attack.

Calcium phosphate-DNA particles in the 10–20 nm size range are synthesized using the coprecipitation method but degradation of the DNA is evident (Welzel 2004). In addition, the use of DNA as a dispersant proves unsuccessful as agglomeration tends to occur in serum-containing media resulting in decreased bioavailability. Improvements to this scheme (single-shell method) are made by the triple-shell approach. Agglomeration is avoided for the triple-shell method and DNA transfection efficiency is comparable to that of commercially available Polyfect (Sokolova 2006b, 2007a,b, 2010) but only when serum proteins are absent. Agglomeration takes place in the presence of proteins. Thus, the major limitation of the triple layer method is the likelihood of agglomeration in an *in vivo* model where serum proteins naturally exist in abundance.

Another example of siRNA encapsulation with calcium phosphate was summarized by Li *et al.* (Li 2010). The investigators successfully showed the encapsulation of siRNA in lipid coated calcium phosphate (LCP) nanoparticles. Calcium phosphate core nanoparticles with siRNA are synthesized via a double reverse microemulsion method, similar to the process employed by Morgan *et al.* with sodium citrate as the dispersant (Morgan 2008). The LCP nanoparticles are synthesized by mixing the siRNA-calcium phosphate particles obtained via the microemulsion method with a 1:1 molar ratio of 1, 2-dioleoyl-3-trimethylammonium-propane chloride salt (DOTAP)/cholesterol liposomes. The investigators demonstrated successful *in vitro* and *in vivo* gene silencing with the siRNA-LCP nanoparticles (Li 2010). Additionally, the investigators successfully demonstrated the use of targeted particles using 1,2-distearoryl-sn-glycero-3-phosphoethanolamine-N-[methoxy(polyethyleneglycol-2000) ammonium salt (DSPE-PEG)-anisamide in a nude mouse model with human lung cancer H-460 cell xenografts with a 50% down regulation effect (Li 2010). The study also illustrated the effectiveness of a targeted nanoparticle delivery system over an untargeted system.

4.2 Drug Delivery Using Calcium Phosphate Systems

Advances in drug delivery systems have the ability to: increase drug stability and carrier capacity, incorporate both hydrophilic and hydrophobic drugs, tune carrier properties to suit route of administration (oral, inhalation, ophthalmic, etc.), and control drug delivery and release (prolonged release) properties. An adequate system can not only improve the bioavailability of the drug but also reduce dosing concentration and frequency (Adair 2010).

Among the many desired characteristics of a nanoparticle drug delivery vehicle, some of the most important are: protection of encapsulated therapeutic from inactivation during transport, small size (10–200 nm), colloidal stability in physiological conditions, non-toxicity, and targetability to specific tissues in the body (Adair 2010). Such drug delivery vehicles are especially desirable in chemotherapy, since various barriers are encountered at the cellular and tumor levels as well as *en route* to the diseased tissue when trying to deliver a chemotherapeutic substance (Serpe 2006). Calcium phosphate nanoparticles have not been studied in drug delivery as extensively as in gene delivery applications, but studies thus far have demonstrated the potential efficacy of these nanoparticles as drug delivery vehicles. Most importantly, improvements in synthetic approaches for calcium phosphate and calcium phosphosilicate nanoparticles have allowed researchers to develop delivery carriers which meet the desired criteria for an efficacious drug delivery system. This is especially promising for *in vivo* drug delivery applications, where, for example, the protection of encapsulants from inactivation is vital since delivery of therapeutics adsorbed to the nanoparticle surface may suffer from some of the same drawbacks as conventional drug delivery approaches (Adair 2010).

In vitro and *in vivo* drug delivery studies with calcium phosphate based composite nanoparticle systems have included delivery of insulin, cisplatin, methotrexate, 7-hydroxy-2-dipropyl-aminotetralin (7-OH-DPAT), hexanoyl-ceramide (Cer_6) and decanoyl-ceramide (Cer_{10}) (Barroug 2002, 2004; Chu 2002; Kester 2008; Morgan 2008; Mukesh 2009; Ramachandran 2009). Most of these studies have fallen short in demonstrating the desired characteristics of a drug delivery vehicle for calcium phosphate nanoparticles, especially in terms of size and colloidal stability. Furthermore, most studies have focused on using calcium phosphate nanoparticles where the drug entity is adsorbed to the surface (i.e., surface decoration) rather than protected within the interior of the nanoparticle via encapsulation. The most recent study on insulin focused on demonstrating the use of calcium phosphate nanocomposites with enteric coatings for oral delivery as a way to overcome the drawbacks of traditional delivery via subcutaneous injections. Calcium phosphate nanoparticles were coated with PEG and Eudragit 100 copolymer, and *in vitro* delivery of insulin was assessed over time in phosphate buffered saline (PBS) solution (Ramachandran 2009). The nanoparticles did not exhibit cytotoxicity as confirmed via an *in vitro* MTT (3-(4,5-dimethylthiazol-2-yl)-2,5-diphenyltetrazolium bromide) assay and the released insulin molecules maintained the same size and conformation as before delivery (Ramachandran 2009). Cytotoxicity and release studies were also conducted with apatite nanoparticles containing the chemotherapeutic cisplatin, where drug induced cytotoxicity was assessed in K8 mouse osteosarcoma cells (Barroug 2002, 2004). However, these studies were limited to *in vitro* delivery, falling short of demonstrating *in vivo* efficacy. Similarly, the chemotherapeutic drug methotrexate was loaded into the surfaces of calcium phosphate nanoparticles, where the drug was both an encapsulant and the dispersing agent. The nanoparticles did not exhibit long term colloidal stability, as shown by measured particle size increase over a period of 90 days at 2–8°C and room temperature (Mukesh 2009). The ocular hypotensive agent 7-OH-DPAT in a calcium phosphate nanoparticle formulation was tested *in vivo* in rabbits, and exhibited improved efficacy when compared to the free 7-OH-DPAT (Chu 2002). While this drug study demonstrated potential for calcium phosphate composite nanoparticles as *in vivo* drug delivery vehicles, the reported synthetic approach yielded a range of particle sizes, most likely due to a lack of colloidal stability for this system.

As microemulsion synthetic routes are known to produce nanoparticles with controlled morphologies and particle sizes, a reverse microemulsion route was utilized to encapsulate Cer_6 and Cer_{10}, experimental hydrophobic therapeutic molecules, inside a calcium phosphosilicate composite nanoparticle (Kester 2008; Morgan 2008). Along with the drugs, the nanoparticles were also doped with fluorescein and rhodamine WT dyes, and used to illustrate simultaneous drug delivery and imaging capabilities. An *in vitro* study demonstrated that Cer_6-fluorescein calcium phosphate composite nanoparticles induced 80% growth inhibition in human vascular smooth muscle cells at a 0.2 μM ceramide concentration, which was reported to be 25-fold less than when administering ceramide in dimethyl sulfoxide (DMSO), while the cells exhibited no apparent morphological changes as a result of nanoparticle cytotoxicity (Fig. 7) (Morgan 2008). More importantly, in *in vitro* studies with Cer_{10}-rhodamine WT calcium phosphosilicate nanoparticles in UACC 903 melanoma cells it was shown that melanoma cell survival was reduced to less than 5% at 5 μM concentrations of delivered therapeutic (Fig. 8) (Kester 2008). In similar studies with Cer_{10}-CPSNPs in both drug sensitive and drug resistant breast cancer cell lines, the nanoparticle formulations exhibited significantly greater efficacy when compared to Cer_{10}-DMSO controls (Kester 2008). Unpublished *in vivo* results for the nude mouse model did not indicate any chronic or acute toxicity for control calcium phosphosilicate formulations. Based

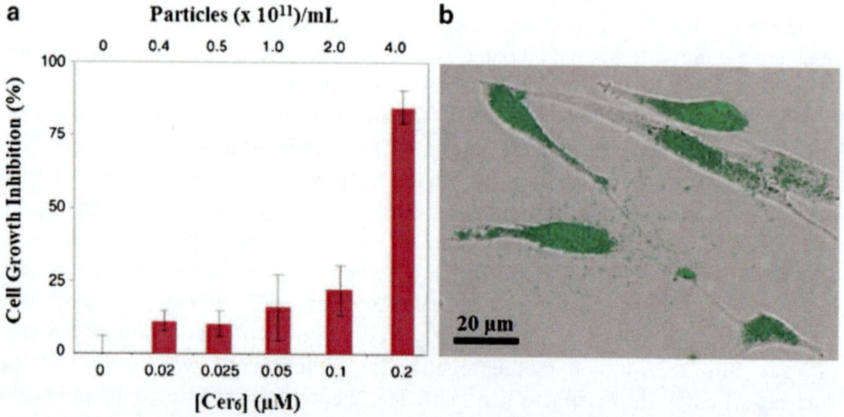

Fig. 7 Effect of experimental chemotherapeutic, ceramide C6, encapsulated in calcium phosphosilicate nanocomposite particles on cell growth. The growth of vascular smooth muscle cells is increasingly inhibited as a function of increasing hexanoyl-ceramide (Cer6) concentration delivered via calcium phosphosilicate nanoparticles (**a**) and the cells show no apparent nanoparticle cytotoxicity induced morphology changes (**b**) (Reproduced from Morgan (2008). With permission)

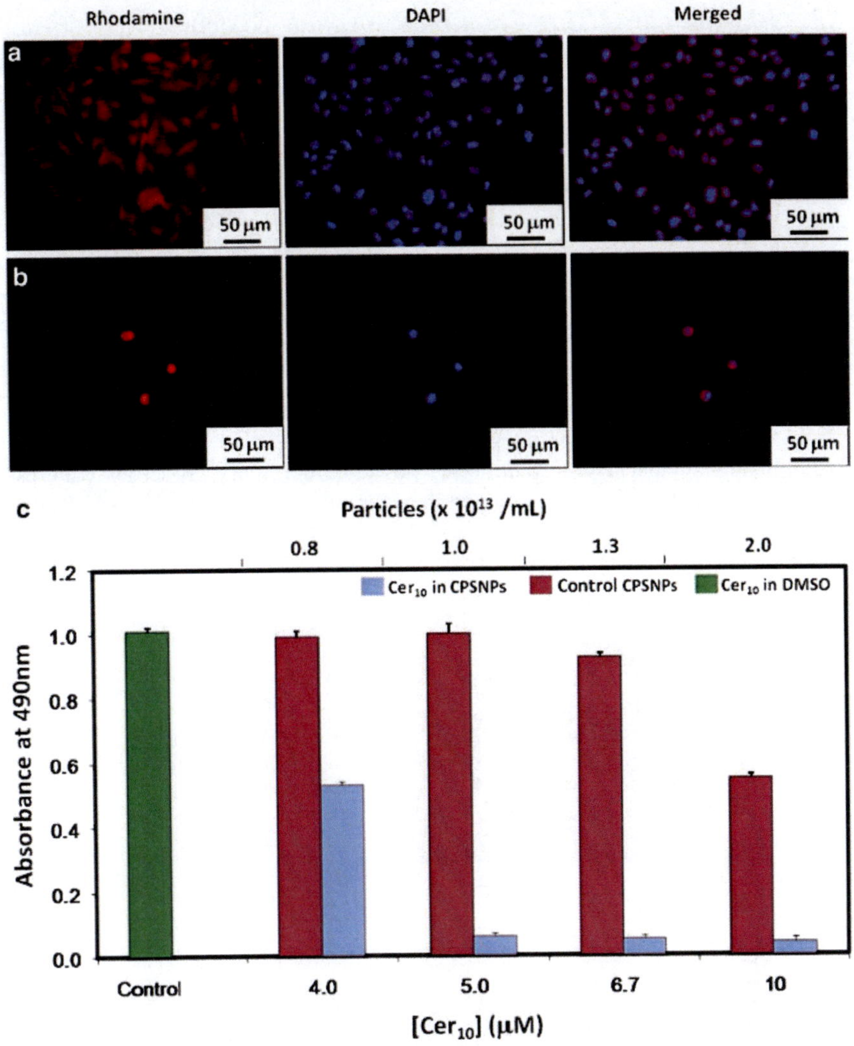

Fig. 8 Effect of Ceramide C10 on UACC 933 melanoma cells. The cell death of UACC 933 melanoma cells is induced (**b**) by treatment with decanoyl-ceramide (Cer_{10}) rhodamine WT-calcium phosphosilicate nanoparticles (CPSNPs) when compared to (**a**) the treatment with control calcium phosphate nanoparticles. (**c**) An MTS cytotoxicity assay demonstrates dosage dependent cell death for Cer_{10} CPSNPs in UACC 903 cells (Reproduced from Kester (2008). With permission)

on these results, CPSNPs could be further investigated *in vivo* for the delivery of other hydrophobic chemotherapeutics, especially using similar synthetic approaches where the drug is shielded from degradation via encapsulation inside the nanoparticles.

4.3 Bio-Imaging Using Calcium Phosphate Systems

Nanoparticulate delivery systems have the advantage of overcoming many of the drawbacks associated with conventional contrast agents (organic dyes) such as poor photostability, low quantum yield and instability in *in vitro* and *in vivo* applications. An array of nanoparticulate material systems are employed for this application with great emphasis on quantum dots and silica and gold nanoparticles.

With the incorporation of organic fluorophores and lanthanide ions, calcium phosphate based nanoparticles have found wide use in imaging applications (Doat 2003, 2004; Lebugle 2006; Mondejar 2007; Ganesan 2008; Morgan 2008; Kester 2008; Altınoğlu 2008; Barth 2010; Al-Kattan 2010). Organic fluorophores that are encapsulated in the calcium phosphosilicate particles span the optical spectrum (Fig. 9) and permit analysis of cells using fluorescent microscopy and imaging of whole mice in the near-infrared (Altınoğlu 2008; Barth 2010).

Lanthanides, such as europium (Eu) possess fluorescent properties (fluorescence emission at 620 nm) when doped into calcium phosphate crystals (Al-Kattan

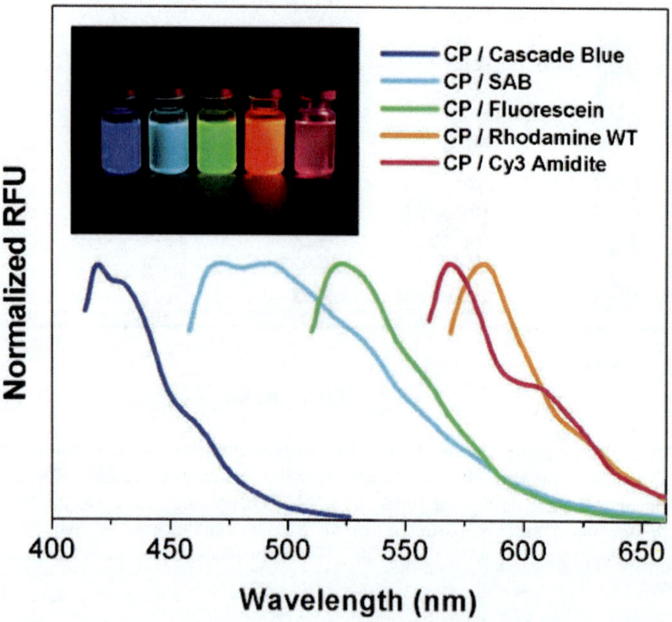

Fig. 9 Fluorescence spectra of organic fluorophores encapsulated in calcium phosphosilicate nanoparticles. Inset shows the five fluorescent nanocolloids of CPSNP illuminated with a UV lamp. From left to right, cascade blue, SAB, fluorescein, rhodamine WT, Cy3 Amidite (Reproduced from Morgan (2008). With permission)

2010; Doat 2004; Lebugle 2006; Mondejar 2007). Europium ions have an atomic radii similar to calcium ions and will substitute for calcium in a calcium phosphate lattice (Doat 2004). Eu-doped calcium deficient apatites form when the particles precipitate in the presence of Eu (Mondejar 2007; Doat 2004; Lebugle 2006; Al-Kattan 2010). Unfortunately, efficient fluorescent behavior requires well crystallized apatites and with increased crystallinity the size of the europium doped particles are greater than 100 nm in diameter; sacrificing the small size required for efficient endocytosis (Mondejar 2007; Al-Kattan 2010).

The europium ion content in the lanthanide-doped apatite particles is cause for some concern in terms of toxicity for use in biological systems (Doat 2004; Mondejar 2007). *In vitro* studies show cells experience morphological changes when exposed to the particles, however no toxicity is reported (Mondejar 2007). It should be noted that these particles were developed due to the relative low toxicity as compared to semiconductor nanoparticles and the Eu^{3+} ion has an LD50 dosage of 5,000 mg/kg in rats (Doat 2003).

Calcium phosphosilicate nanoparticles have been shown to be biocompatible both *in vitro* and *in vivo* (Kester 2008; Morgan 2008; Altınoğlu 2008; Barth 2010). The calcium phosphosilicate particles additionally have been shown to not interfere with calcium channels in even very sensitive cells such as adult rat stellate ganglion neurons in cell culture (Kester 2008).

Association of organic fluorescent molecules with calcium phosphate nanoparticles is accomplished through both surface adsorption and encapsulation during synthesis. Some synthetic schemes use the fluorophore or DNA as the dispersant by adsorbing it onto the surface of the particle (Sokolova 2007b; Ganesan 2008). Calcium phosphates that are dispersed with a fluorescent porphyrin are well dispersed spheres but have a large diameter of 250 nm (Ganesan 2008). Using DNA as the dispersant permits transfection of fluorescent nanoparticles into cells but the particles are poorly dispersed with sizes of agglomerates ranging from tens to hundreds of nanometers (Sokolova 2007a).

In contrast, fluorescent calcium phosphosilicate nanoparticles restrain the fluorophore within the rigid, inorganic matrix giving the molecule improved optical performance. The encapsulated molecules show increased quantum efficiency and resistance to solvatochromic effects (Altınoğlu 2008; Morgan 2008; Russin 2010). The increase in quantum efficiency is due to the matrix shielding effect, preventing conformational changes which lead to non-radiative decay (Altınoğlu 2008). Calcium phosphosilicate particles containing Cy3 are used to investigate the route by which particles are taken up into cells (Morgan 2008). As shown in Fig. 10 particles enter the cell through endocytosis followed by particle dissolution in the late endosome (Morgan 2008).

Using fluorescent molecules to image living animals requires the penetration of tissue by excited and emitted photons. At near infrared wavelengths, absorption and scattering from tissue components is minimal (Altınoğlu 2010b) presenting a window of opportunity for enhanced imaging. Calcium phosphosilicate encapsulated indocyanine green (ICG-CPSNPs) nanoparticles excite and emit within this

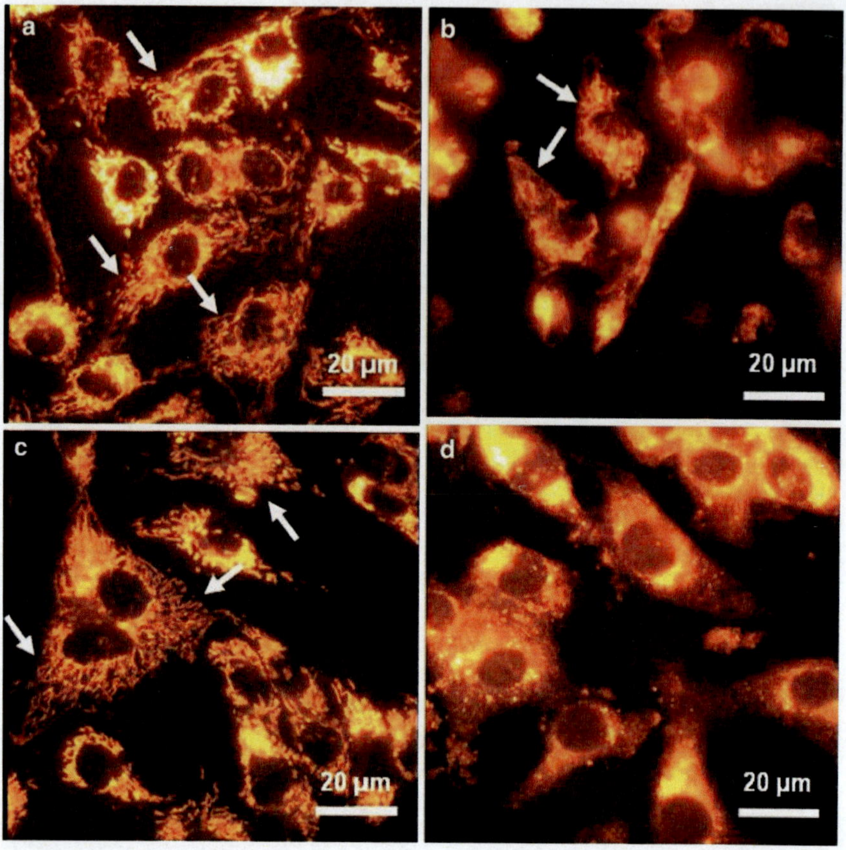

Fig. 10 Bovine aortic endothelial cells treated with free Cy3 or Cy3 encapsulated in calcium phosphosilicate nanocomposite particles dosed with cytochalasin-D to inhibit endosome formation. Bovine aortic endothelial cells were stained using Cy3 dye, either as free dye (**a**, **b**) or as dye encapsulated in calcium phosphosilicate nanoparticles (**c**, **d**). Cell in panels (**b**, **d**) have been treated with cytochalasin-D to inhibit formation of endosomes. Arrows indicate staining of organelles which does not occur for cells treated with cytochalasin-D and incubated with Cy3 indicating that endosomal capture with pH changes as shown in Fig. 2 are required to dissolve the calcium phosphosilicate particles to release the encapsulated Cy3 (Reproduced from Morgan (2008). With permission)

window and fluoresce bright enough for deep tissue imaging (Altınoğlu 2008). *In vivo* images of these particles show successful targeting to both breast and pancreatic cancers in nude mice (Altınoğlu 2008; Barth 2010). With a PEGylated surface, ICG-CPSNPs circulate for over 96 h (Fig. 11) and accumulate in breast tumors via the enhanced permeation and retention effect (Altınoğlu 2008; Morgan 2008). In addition, covalently bound antibodies on the PEG surface, target specific cancers, such as pancreatic cancer and show increased localization

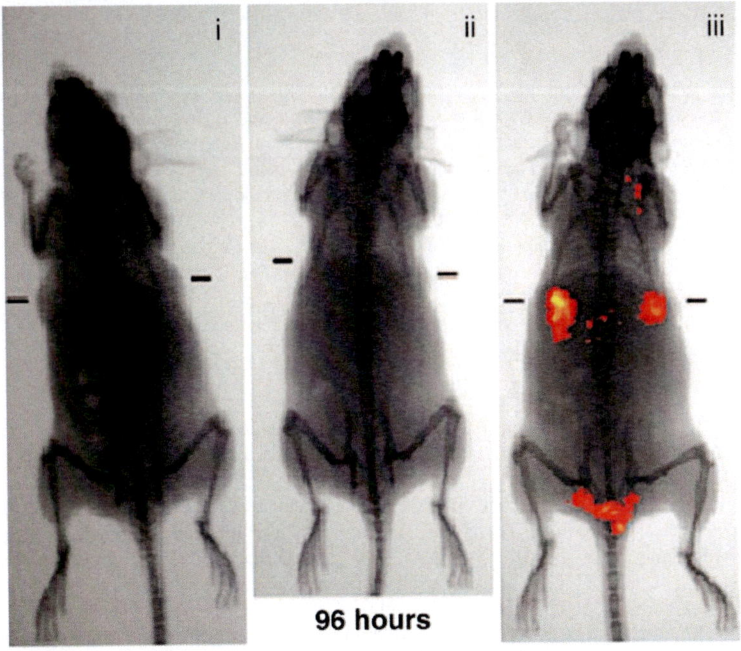

Fig. 11 Near infrared imaging of nude mice with indocyanine green. Near infrared imaging of nude mice 96 h after injection with **(i)** carboxylate terminated calcium phosphosilicate particles without encapsulated dye (i.e., Ghost CPSNP), **(ii)** ICG free dye, **(iii)** PEGylated calcium phosphosilicate particles with encapsulated ICG. The near infra-red emitting nanoparticles accumulate via enhanced permeation and retention in the xenografted breast cancer tumors (Reproduced from Altınoğlu (2008). With permission)

at the tumor site; (Barth 2010) a promising result in the step towards improved diagnostic tools, particularly for the early detection and diagnosis of cancers such as pancreatic cancer.

5 Conclusions

Based on the progress made thus far and the continued use of calcium phosphate based delivery systems over the past 37 years, it is apparent that the calcium phosphate material system has great potential as the delivery vehicle of choice. With inherent properties such as biocompatibility, non-toxicity, and pH dependent solubility paired with improved synthetic routes yielding colloidally stable and targeted nanoparticles, calcium phosphate based systems are bound to enter clinical trials for imaging and therapeutic applications (in addition to the current bone treatment clinical trials) in the near future.

References

Adair JH, Kumar, R., Antolino, N., Szepesi, C.J., Kimel, R.A., Rouse, S.M. "Colloidal Lessons Learned for Dispersion of Nanosize Particulate Suspensions" In: World Academy of Ceramics Conference Proceedings, Faenza, Italy, 2005. Lessons in Nanotechnology from Traditional and Advanced Ceramics. Techna Group S.R.L, pp 93–145

Adair JH, Parette, M.P., Altınoğlu, E.İ., Kester, M. (2010) Nanoparticulate Alternatives for Drug Delivery. ACS Nano 4 (9):4967–4970

Al-Kattan A, Dufour, P., Dexpert-Ghys, J., Drouet, C. (2010) Preparation and Physicochemical Characteristics of Luminescent Apatite-Based Colloids. The Journal of Physical Chemistry C 114:2918–2924

Alagna M (2001) Everything you need to know about chemotherapy. 1st edn. The Rosen Publishing Group, New York

Alberts B, Johnson, A., Lewis, J., Raff, M., Roberts, K., Walter, P. (2002) Molecular Biology of the Cell. 4th edn. Garland Science, New York

Allemann E, Gurny, R., Doelker, E. (1993) Drug-Loaded Nanoparticles - Preparation Methods and Drug Targeting Issues. European Journal of Pharmaceutics and Biopharmaceutics 39 (5):173–191

Altınoğlu E.İ (2010a) Indocyanine Green-Encapsulating Calcium Phosphosilicate Nanoparticles:Bifunctional Theranostic Vectors for Near Infrared Diagnosic Imaging and Photodynamic Therapy. Doctorate Thesis, Pennsylvania State University, University Park

Altınoğlu E.İ, Adair, J.H. (2009) Calcium phosphate nanocomposite particles: a safer and more effective alternative to conventional chemotherapy? Future Oncology 5 (3):279–281

Altınoğlu E.İ, Adair, J.H. (2010b) Near infrared imaging with nanoparticles. Nanomedicine and Nanobiotechnology 2 (5):461–477

Altınoğlu E.İ, Russin, T.J., Kaiser, J.M., Barth, B.M., Eklund, P.C., Kester, M., Adair, J.H. (2008) Near-Infrared Emitting Fluorophore-Doped Calcium Phosphate Nanoparticles for *In Vivo* Imaging of Human Breast Cancer. ACS Nano 2 (10):2075–2085

Barroug A, Glimcher, M.J. (2002) Hydroxyapatite crystals as a local delivery system for cisplatin: adsorption and release of cisplatin *in vitro*. Journal of Orthopaedic Research 20 (2):274–280

Barroug A, Kuhn, L.T., Gerstenfeld, L.C., Glimcher, M.J. (2004) Interactions of cisplatin with calcium phosphate nanoparticles: *in vitro* controlled adsorption and release. Journal of Orthopaedic Research 22:703–708

Barth BM, Sharma, R., Altınoğlu, E.İ., Morgan, T.T., Shanmugavelandy, S.S., Kaiser, J.M., McGovern, C., Matters, G.L., Smith, J.P., Kester, M., Adair, J.H. (2010) Bioconjugation of calcium phosphosilicate composite nanoparticles for selective targeting of human breast and pancreatic cancers *in vivo*. ACS nano 4:1279–1287

Bertoni E, Bigi, A., Cojazzi, G. Gandolfi, M., Panzavolta, S., Roveri, N. (1998) Nanocrystals of magnesium and fluoride substituted hydroxyapatite. Journal of Inorganic Biochemistry 72 (1–2):29–35

Bigi A, Falini, G., Foresti, E., Ripamonti, A., Gazzano, M., Roveri, N. (1993) Magnesium influence on hydroxyapatite crystallization. Journal of Inorganic Biochemistry 49 (1):69–78

Bigi A, Foresti, E., Gandolfi, M., Gazzano, M., Roveri, N. (1995) Inhibiting effect of zinc on hydroxylapatite crystallization. Journal of Inorganic Biochemistry 58 (1):49–58

Bisht S, Bhakta, G., Mitra, S., Maitra, A. (2005) pDNA loaded calcium phosphate nanoparticles: highly efficient non-viral vector for gene delivery. International Journal of Pharmaceutics 288:157–168

Brecevic LJ, Furedi-Milhofer, H. (1979) Precipitation of Calcium Phosphates from Electrolyte Solutions V. The Influence of Citrate Ions. Calcified Tissue Research 28:131–136

Burns AA, Vider, J., Ow, H., Herz, E., Penate-Medina, O., Baumgart, M., Larson, S.M., Wiesner, U., Bradbury, M. (2009) Fluorescent Silica Nanoparticles with Efficient Urinary Excretion for Nanomedicine. Nano Letters 9 (1):442–448

Cao S-W, Zhu, Y.-J., Wu, J., Wang, K.-W., Tang, Q.-L. (2010) Preparation and Sustained-Release Property of Triblock Copolymer/Calcium Phosphate Nanocomposite as Nanocarrier for Hydrophobic Drug. Nanoscale Research Letters 5:781–785

Carlisle EM (1988) Silicon as a trace nutrient. Science of The Total Environment 73 (1–2): 95–106

Chander S, Fuerstenau, D.W. (1982) On the dissolution and interfacial properties of hydroxyapatite. Colloids and Surfaces 4 (2):101–120

Chen YC, Ostafin, A., Mizukami, H. (2010) Synthesis and characterization of pH sensitive carboxySNARF-1 nanoreactors. Nanotechnology 21 (21)

Chithrani BD, Ghazani, A.A., Chan, W.C.W. (2006) Determining the Size and Shape Dependence of Gold Nanoparticle Uptake into Mammalian Cells. Nano Letters 6 (4):662–668

Chow LC (1991) Development of self-setting calcium phosphate cements. Journal of the Ceramic Society of Japan 99 (10):954–964

Chu T-C, He, Q., Potter, D.E. (2002) Biodegradable Calcium Phosphate Nanoparticles as a New Vehicle for Delivery of a Potential Ocular Hypotensive Agent. Journal of Ocular Pharmacology and Therapeutics 18 (6):507–514

Cross KJ, Huq, N.L., Stanton, D.P., Sum. M., Reynolds, E.C. (2004) NMR studies of a novel calcium, phosphate and fluoride delivery vehicle-α_{S1}-casein (59–79) by stabilized amorphour calcium fluoride phosphate nanocomplexes. Biomaterials 25:5061–5069

Davis SS (1997) Biomedical applications of nanotechnology - implications for drug targeting and gene therapy. Trends in Biotechnology 15:217–224

Doat A, Fanjul, M., Pellé, F., Hollande, E., Lebugle, A. (2003) Europium-doped bioapatite: a new photostable biological probe, internalizable by human cells. Biomaterials 24 (19):3365–3371

Doat A, Pellé, F., Gardant, N., Lebugle, A. (2004) Synthesis of luminescent bioapatite nanoparticles for utilization as a biological probe. Journal of Solid State Chemistry 177 (4–5):1179–1187

Dorozhkin SV (2010) Amorphous calcium (ortho)phosphates. Acta Biomaterialia 6 (12): 4457–4475

Eanes ED, Gillessen, I.H., Posner, A.S. (1965) Intermediate States in the Precipitation of Hydroxyapatite. Nature 208 (5008):365–367

Eanes ED, Termine, J.D., Nylen, M.U. (1973) An electron microscopic study of the formation of amorphous calcium phosphate and its transformation to crystalline apatite. Calcified Tissue Research 12 (1):143–158

Ganesan K, Kovtun, A., Neumann, S., Heumann, R., Epple, M. (2008) Calcium phosphate nanoparticles: colloidally stabilized and made fluorescent by a phosphate-functionalized porphyrin. Journal of Materials Chemistry 18:3655–3661

Gao H, Shi, W., Freund, L.B. (2005) Mechanics of receptor-mediated endocytosis. Proceedings of the National Academy of Science 102 (27):9469–9474

Ghosh A, Greenberg, M.E. (1995) Calcium signaling in neurons: molecular mechanisms and cellular consequences. Science 268:239–247

Graham FL, van der Eb, A.J. (1973) A new technique for the assay of infectivity of human adenovirus 5 DNA. Virology 52 (2):456–467

Grimm D, Streetz, K.L., Jopling, C.L., Storm, T.A., Pandey, K., Davis, C.R., Marion, P., Salazar, F., Kay, M.A. (2006) Fatality in mice due to oversaturation of cellular microRNA/short hairpin RNA pathways. Nature 441:537–541

Guha AK, Singh, S., Kumaresan, R., Sayar, S., Sinha, A. (2009) Mesenchymal cell response to nanosized biphasic calcium phosphate composites. Colloids and Surfaces B: Biointerfaces 73:146–151

Hong RL, Huang, C.J., Tseng, Y.L., Pang, V.F., Chen, S.T., Liu, J.J., Chang, F.H. (1999) Direct comparison of liposomal doxorubicin with or without polyethylene glycol coating in C-26 tumor-bearing mice: is surface coating with polyethylene glycol beneficial? Clinical Cancer Research 5 (11):3645–3652

Immamura M, Seki, T., Kunieda, K., Nakatani, S., Inque, K., Nakano, T., Harada, K. (1995) Hydroxyapatite beads containing Doxorubicin Hydrochloride (DOX) and Buthionine

Sulfoximine (BSO) - A new anticancer drug for local treatment with multiple drugs. Oncology Reports 2:33–36

Johnsson M, Richardson, C.F., Sallis, J.D., Nancollas, G.H. (1991) Adsorption and Mineralization Effects of Citrate and Phosphocitrate on Hydroxyapatite. Calcified Tissue Research 49:134–137

Kakizawa Y, Furukawa, S., Ishii, A., Kataoka, K. (2006) Organic-inorganic hybrid-nanocarrier for siRNA constructing through the self-assembly of calcium phosphate and PEG-based block aniomer. Journal of Controlled Release 111:368–370

Kakizawa Y, Furukawa, S., Kataoka, K. (2004) Block copolymer-coated calcium phosphate nanoparticles sensing intracellular environment for oligodeoxynucleotide and siRNA delivery. Journal of Controlled Release 97:345–356

Kakizawa Y, Kataoka, K. (2002) Block Copolymer Self-Assembly into Monodispersive Nanoparticles with Hybrid Core of Antisense DNA and Calcium Phosphate. Langmuir 18 (12):4539–4543

Kester M, Heakal, Y., Fox, T., Sharma, A., Robertson, G., Morgan, T.T., Altinoglu, E.I., Tabakovic, A., Parette, M.R., Rouse, S.M., Ruiz-Velasco, V., Adair, J.H. (2008) Calcium Phosphate Nanocomposite Particles for *In Vitro* Imaging and Encapsulated Chemotherapeutic Drug Delivery to Cancer Cells. Nano Letters 8 (12):4116–4121

Klesing J, Wiehe, A., Gitter, B., Grafe, S., Epple, M. (2010) Positively charged calcium phosphate/polymer nanoparticles for photodynamic therapy. Journal of Materials Science-Materials in Medicine 21:887–892

Kulkarni VI, Shenoy, V.S., Dodiya, S.S., Rajyaguru, T.H., Murthy, R.R. (2006) Role of calcium in gene delivery. Expert Opinion on Drug Delivery 3 (2):235–245

Kumar C (ed) (2006) Nanomaterials for Cancer Therapy, vol 6. Nanotechnologies for the Life Sciences. Wiley-VCH, Weinheim

Kunieda K, Seki, T., Nakatami, S., Wakabayashi, M., Shiro, T., Inoue, K., Sougawa, M., Kimura, R., Harada, K. (1993) Implantation treatment method of slow release anticancer doxorubicin containing hydroxyapatite (DOX-HAP) complex. A basic study of a new treatment of hepatic cancer. British Journal of Cancer 67:668–673

Langer R (1990) New methods of drug delivery. Science 249 (4976):1527–1533

Lebugle A, Pellé, F., Charvillat, C., Rousselot, I., Chane-Ching, J.Y. (2006) Colloidal and monocrystalline Ln3+ doped apatite calcium phosphate as biocompatible fluorescent probes. Chemical Communications 6:606–608

Leong KW, Langer, R. (1988) Polymeric controlled drug delivery. Advanced Drug Delivery Reviews 1 (3):199–233

Li J, Chen, Y.-C., Tseng, Y.-C., Mozumdar, S, Huang, L. (2010) Biodegradable calcium phosphate nanoparticle with lipid coating for systemic siRNA delivery. Journal of Controlled Release 142:416–421

Liu T, Tang, A., Zhang, G., Chen, Y., Zhang, J., Peng, S., Cai, Z. (2005) Calcium Phosphate Nanoparticles as a Novel Nonviral Vector for Efficient Transfection of DNA in Cancer Gene Therapy. Cancer Biotherapy & Radiopharmaceuticals 20 (2):141–149

Lopez-Macipe A, Gomez-Morales, J., Rodriguez-Clemente, R. (1999) Nanosized Hydroxyapatite Precipitation from Homogeneous Calcium/Citrate/Phosphate Solutions Using Microwave and Conventional Heating. Advanced Materials 10 (1):49–53

Margolis S (ed) (2005) The Johns Hopkins Medical Guide to Health After 50. Black Dog & Leventhal Publishers, Inc., New York

Mondejar SP, Kovtun, A., Epple, M. (2007) Lanthanide-doped calcium phosphate nanoparticles with high internal crystallinity and with a shell of DNA as fluorescent probes in cell experiments. Journal of Materials Chemistry 17:4153–4159

Morgan TT, Muddana, H.S., Altınoğlu, E.İ., Rouse, S.M., Tabakovic, A., Tabouillot, T., Russin, T.J., Shanmugavelandy, S.S., Butler, P.J., Eklund, P.C., Yun, J.K., Kester, M., Adair, J.H. (2008) Encapsulation of Organic Molecules in Calcium Phosphate Nanocomposite Particles for Intracellular Imaging and Drug Delivery. Nano Letters 8 (12):4108–4115

Muddana HS, Morgan, T.T., Adair, J.H., Butler, P.J. (2009) Photophysics of Cy3-encapsulated calcium phosphate nanoparticles. Nano Letters 9 (4):1559–1566
Mukesh U, Kulkarni, V., Tushur, R., Murthy, R.S.R. (2009) Methotrexate Loaded Self Stabilized Calcium Phosphate Nanoparticles: A Novel Inorganic Carrier for Intracellular Drug Delivery. Journal of Biomedical Nanotechnology 5:99–105
Nelson M, Balasundaram, G., Webster, T.J. (2006) Increased osteoblast adhesion on nanoparticulate crystalline hydroxyapatite functionalized with KRSR. International Journal of Nanomedicine 1 (3):339–349
Nguyen T, Menocal, E.M., Harborth, J., Fruehauf, J.H. (2008) RNAi therapeutics: An update on delivery. Current Opinion in Molecular Therapeutics 10 (2):158–167
Owens III DE, Peppas, N.A. (2005) Opsonization, biodistribution and pharmacokinetics of polymeric nanoparticles. International Journal of Pharmaceutics 307 (1):93–102
Oyane A, Kim, H.-M., Furuya, T., Kokubo, T., Miyazaki, T., Nakamura, T. (2003) Preparation and assessment of revised simulated body fluids. Journal of Biomedical Materials Research Part A 65A (2):188–195
Palazzo B, Iafisco, M., Laforgia, M., Margiotta, N., Natile, G., Bianchi, C.L., Walsh. D., Mann, S., Roveri, N. (2007) Biomimetic Hydroxyapatite-Drug Nanocrystals as Potential Bone Substitutes with Antitumor Drug Delivery Properties. Advanced Functional Materials 17: 2180–2188
Peer D, Karp, J.M., Hong, S., Farokhzad, O.C., Margalit, R., Langer, R. (2007) Nanocarriers as an emerging platform for cancer therapy. Nature Nanotechnology 2:751–760
Perkin KK, Turner, J.L., Wooley, K.L., Mann, S. (2005) Fabrication of Hybrid Nanocapsules by Calcium Phosphate Mineralization of Shell Cross-Linked Polymer Micelles and Nanocages. Nano Letters 5 (7):1457–1461
Ramachandran R, Paul, W., Sharma, C.P. (2009) Synthesis and Characterization of PEGylated Calcium Phosphate Nanoparticles for Oral Insulin Delivery. Journal of Biomedical Materials Research B: Applied Biomaterials 88 (1):41–48
Rokita E, Hermes, C., Nolting, H.-F., Ryczek, J. (1993) Substitution of calcium by strontium within selected calcium phosphates. Journal of Crystal Growth 130 (3–4):543–552
Roy I, Mitra, S., Maitra, A., Mozumdar, S. (2003) Calcium phosphate nanoparticles as novel non-viral vectors for targeted gene delivery. International Journal of Pharmaceutics 250:25–33
Russin TJ, Altinoğlu, E.İ., Adair, J.H., Eklund, P.C. (2010) Measuring the fluorescent quantum efficiency of indocyanine green encapsulated in nanocomposite particulates. Journal of Physics: Condensed Matter 22:334217
Schmidt H, Gray, B.L., Wingert, P.A., Ostafin, A.E. (2004) Assembly of Aqueous-Cored Calcium Phosphate Nanoparticles for Drug Delivery. Chemistry of Materials 16:4942–4947
Schmidt H, Ostafin, A. (2002) Liposome directed growth of calcium phosphate nanoshells. Advanced Materials 14 (7):532–535
Serpe L (2006) Conventional Chemotherapeutic Drug Nanoparticles for Cancer Treatment. In: Nanomaterials for Cancer Therapy. vol 6. Wiley-VCH, Weinheim, pp 1–39
Skrtic D, Antonucci, J.M., Eanes, E.D., Brunworth, R.T. (2002) Silica- and zirconia-hybridized amorphous calcium phosphate: Effect on transformation to hydroxyapatite. Journal of Biomedical Materials Research 59 (4):597–604. doi:10.1002/jbm.10017
Sokolova V, Knuschke, T., Kovtun, A., Buer, J., Epple, M., Westendorf, A.M. (2010) The use of calcium phosphate nanoparticles encapsulating Toll-like receptor ligands and the antigen hemagglutinin to induce dendritic cell maturation and T cell activation. Biomaterials 21: 5627–5633
Sokolova V, Kovtun, A., Heumann, R. Epple, M. (2007a) Tracking the pathway of calcium phosphate/DNA nanoparticles during cell transfection by incorporation of red-fluorescing tetramethylrhodamine isothiocyanate – bovine serum albumin into these nanoparticles. Journal of Biological Inorganic Chemistry 12 (2):174–179
Sokolova V, Kovtun, A., Prymak, O., Meyer-Zaika, W., Kubareva, E.A., Romanova, E.A., Oretskaya, T.S. Heumann, R., Epple, M. (2007b) Functionalisation of calcium phosphate

nanoparticles by oligonucleotides and their application for gene silencing. Journal of Materials Chemistry 17:721–727
Sokolova V, Prymak, O., Meyer-Zaika, W., Colfen, H., Rehage, H., Shukla, A., Epple, M. (2006a) Synthesis and characterization of DNA-functionalized calcium phosphate nanoparticles. Materialwissenschaft und Werkstofftechnik 37 (6):441–445
Sokolova V, Radtke, I., Heumann, R., Epple, M. (2006b) Effective transfection of cells with multi-shell calcium phosphate-DNA nanoparticles. Biomaterials 27 (16):3147–3153
Sugawara A, Yamane, S., Akiyoshi, K. (2006) Nanogel-templated mineralization: Polymer-calcium phosphate hybrid nanomaterials. Macromolecular Rapid Communications 27 (6):441–446
Thorek DLJ, Tsourkas, A. (2008) Size, charge and concentration dependent uptake of iron oxide particles by non-phagocytic cells. Biomaterials 29 (26):3583–3590
Tjandra W, Ravi, P., Yao, J., Tam, K.C. (2006) Synthesis of hollow spherical calcium phosphate nanoparticles using polymeric nanotemplates. Nanotechnology 17:5988–5994
Tomlinson E (1987) Theory and practice of site-specific drug delivery. Advanced Drug Delivery Reviews 1 (2):87–198
Torchilin VP (2000) Drug Targeting. European Journal of Pharmaceutical Sciences 11 (S2):S81–S91
Torchilin VP (ed) (2006) Nanoparticulates as Drug Carriers. 1st Ed. edn. Imperial College Press, London
Truong-Le VL, Walsh, S.M., Schweibert, E., Mao, H.-Q., Guggino, W.B., August, J.T., Leong, K.W. (1999) Gene Transfer by DNA–Gelatin Nanospheres. Archives of Biochemistry and Biophysics 361 (1):47–56
Tung MS (1998) Calcium Phosphates: Structure, Composition, Solubility, and Stability. In: Amjad Z (ed) Calcium Phosphates in Biological and Industrial Systems. Kluwer Academic Publishers, Boston, pp 1–19
Wang S, McDonnell, E.H., Sedor, F.A., Toffaletti, J.G. (2002) pH Effects on Measurements of Ionized Calcium and Ionized Magnesium in Blood. Archives of Pathology & Laboratory Medicine 126 (8):947–950
Welzel T, Radtke, I., Meyer-Zaika, W., Heumann, R., Epple, M. (2004) Transfection of cells with custom-made calcium phosphate nanoparticles coated with DNA. Journal of Materials Chemistry 14 (14):2213–2217
Yamane S, Sugawara, A., Watanabe, A., Akiyoshi, K. (2009) Hydrid Nanoapatite by Polysaccharide Nanogel-templated Mineralization. Journal of Bioactive and Compatible Polymers 24:151–168
Ye F, Guo, H., Zhang, H., He, X. (2010) Polymeric micelle-templated synthesis of hydroxyapatite hollow nanoparticles for a drug delivery system. Acta Biomaterialia 6:2212–2218
Zhang M, Kataoka, K. (2009) Nano-structured composites based on calcium phosphate for cellular delivery of therapeutic and diagnostic agents. Nano Today 4:508–517
Zhang X, Kovtun, A.. Mendoza-Palomares, C., Oulad-Abdelghani, M., Fioretti, F., Rinckenbach, S., Mainard, D., Epple, M., Benkirane-Jessel, N. (2010) SiRNA-loaded multi-shell nanoparticles incorporated into a multilayered film as a reservoir for gene silencing. Biomaterials 31:6013–6018
Zhou C, Yu, B., Yang, X., Huo, T., Lee, L.J., Barth, R.F., Lee, R.J. (2010) Lipid-coated nano-calcium-phosphate (LNCP) for gene delivery. International Journal of Pharmaceutics 392:201–208

Intracellular Bacteria and Protozoa

Maria Jose Morilla and Eder Lilia Romero

Abstract Selecting antimicrobials as a function of their chemical structure is a classical paradigm that was challenged by the advent of nanomedicines. Drugs carried in nanoparticles can be delivered into intracellular compartments in cells from selected tissues, independently of free drug's pharmacokinetics and pharmacodynamics. Inflammation associated to most infections favors the passive targeting of intravenous or topical nanomedicines against those accessible from blood circulation or close to the skin surface, infected macrophages. As a result treatments become more selective and less toxic. Up to date however preclinical development of nanomedicines is limited to increase the targeting to infected cells. After cell uptake, drugs carried in nanoparticles are delivered to the endolysosomal pathway. The use of nanoparticles that could provide drug delivery to cell cytoplasm – site of residence of most of the infectious targets excepting leishmania-remains unexplored.

In this chapter we will survey the use of nano or microparticles as antimicrobials carriers against experimental tuberculosis, salmonellosis, tularemia, chlamydiasis, malaria, leishmaniasis, Chagas's disease and toxoplasmosis. The inhalation of anti tuberculose drugs in nano and microparticles directly targets alveolar macrophages, reduces the number of administrations and provides surfactant material to atelectatic lungs. Inhalation has been the only succeeding via of administration for nanomedicines against chlamydiasis. The same as against salmonellosis and tularemia, antimycobacterial drugs loaded in intravenously administered nanoparticles effectively target extrapulmonary infected macrophages in the disseminated form of the diseases. Intravenously nanoparticles did not succeed in targeting infected erythrocytes, but were effective against infected hepatocytes in malaria. On the other hand, targets in visceral leishmaniasis are treated with intravenous nanomedicines, but the cutaneous and muco-cutaneous clinical forms need improved delivery strategies.

M.J. Morilla and E.L. Romero (✉)
Programa de Nanomedicinas, Departamento de Ciencia y Tecnologia,
Universidad Nacional de Quilmes, Roque Saenz Peña 180,
Bernal 1876, Buenos Aires, Argentina
e-mail: elromero@unq.edu.ar

A. Prokop (ed.), *Intracellular Delivery: Fundamentals and Applications*, Fundamental Biomedical Technologies 5, DOI 10.1007/978-94-007-1248-5_24,

Chagas's diseases and toxoplasmosis, two diseases with intracytoplasmic targets within non phagocytic cells in tissues where inflammation is almost absent remain as unsurpassed challenges for nanomedical approaches.

Keywords Chlamydiasis • Intracellular drug delivery • Intracellular bacteria • Intracellular protozoa

1 Nanomedicines in Drug Delivery Against Intracellular Bacteria and Protozoa

The branch of Nanotechnology that employs nano-objects (namely metal and oxides nanoparticles, polymeric and lipid nanoparticles, dendrimers, micelles, liposomes and other vesicles under 200–300 nm, that will be referred in general as "nanoparticles") (ISO 2008), such as therapeutic drug carriers diagnosis agents, nanodevices and nano-scaffolds for tissue engineering in the medical field, is known as Nanomedicine (European Science Foundation 2005). A drug loaded into a nano-object is regarded as a "nanomedicine". In this way, drug's pharmacokinetics and pharmacodynamics become independent from it's chemical structure, to depend on the structure of the nano-object. On the other hand, eucariotic cells recognize and capturate nano-objects. If properly designed, nano-objects can cross anatomical and phenomenological barriers such as the gastrointestinal mucus, the skin and the blood brain barrier (Alonso 2004). Therefore, carried drugs can be targeted delivered to cells in spite of their poor bioavailability, lability in circulation, poor cell penetration and/or non selective distribution in their free form (Tan et al. 2010). From the advent of peguilated liposomes, such attractive goals are responsible for the exponential increase in the number of articles dealing with preclinical applications of nanomedicine that initiated circa 25 years ago (Rannard and Owen 2009). Preclinical research has also fostered the entering of nanomedicines in new clinical trials (Emerich and Thanos 2006; Sakamoto et al. 2010).

In order to succeed, a delivery strategy mediated by nanomedicines neccesitates of a suitable anatomo-pathological environment or a target site having an adequate tissue/cell activity. A clear example are the oncologic therapies, where the EPR (enhanced permeation and retention) effect enables the circulating nanoparticles to extravasate and accumulate in the neighborhood of target cells (passive targeting), diminishing the toxicity of treatments to solid tumors (Stern et al. 2010). Passive targeting mediated by EPR effect, plus a structural design that allows its cellular uptake by caveolin mediated endocytosis, is responsible for the improved therapeutic efficacy of products such as Abraxane (paclitaxel in albumin nanoparticles) (Petrelli et al. 2010). In vaccinology on the other hand, the phagocytic activity of antigen presenting cells is used to favour the uptake, processing and presentation of antigens loaded in nanoparticles. In pre-clinics, different types of nanoparticles are used as human and veterinary vaccine adjuvants (Peek et al. 2008).

Nanomedical pharmacotherapy is classically administered by intravenous (i.v.) route. To avoid opsonization and aggregation while in circulation, the surface of intravenously injected nanoparticles is pegylated (a coverage made of a ~50 Å thick polyethylene glycol (PEG) hydrophilic corona) (Dos Santos et al. 2007). In this way, the early capture of circulating nanoparticles by cells from the retículo endothelial system (RES) in liver, spleen and bone marrow is diminished. Pegylated, long circulating nanoparticles have increased chances of extravasation in EPR sites, where diseased cells can be passively targeted (Fang et al. 2010). In the last 10 years, new nano-objects have been designed in order to make them suitable for administration by alternative vias to parenteral, such as mucosal and topical (Antosova et al. 2009; Chadwick et al. 2010; Csaba et al. 2009). Moreover, improved knowledge has been gained on the relationship between structure (size, shape, surface) of the nano-objects and their rate of cell capture and subsequent intracellular pathways (Decuzzi et al. 2010; Ruenraroengsak et al. 2010). In this manner, it is in theory possible achieve intracellular targeted drug delivery (Sahay et al. 2010). Classical examples are the pH-sensitive liposomes that after cell uptake by clathrin mediated endocytosis or phagocytosis, experience a phase transition triggered by the low pH and massively deliver their payload to the cytosol (Torchilin 2009). Overall nanomedicines solve problems derived from incorrect drug targeting/selectivity. Other problems such as poor gastrointestinal absorption and bioavailability and transient or low maximal drug concentration in plasma, can be solved by other approaches. For instance, micro or nanonization of drugs to increase absorption or inclusion in matrices (not necessarily nano-objects) as drug depots for sustained release.

As seen in Table 1, some intracellular bacteria survive and multiply within intracytoplasmic vacuola, such as *Mycobacteria* and *Salmonella spp*. Other, such as *Lysteria, Rickketsia* and *Shigella* escape from vacuola and cross the cytoplasm to invade adjacent cells where reinitiate the infection cycle. Certain intracellular protozoa survive within acid intracellular compartments such as the *Leishmania spp* or simply perdurate inside the cytoplasm such as *Trypanosoma cruzi.* Other protozoa

Table 1 Intracellular bacteria and protozoa (Kumar and Valdivia 2009; Maurin and Raoult 1997)

Bacteria	Cell type infected	Subcellular localization
Afilia felis	Macrophage, vascular endothelial cells	Phagosome
Anaplasma phagocytophilum	Neutrophil	Vacuole
Brucella abortus	Macrophage	Vacuole
Bartonella bacilliformis	Erythrocyte/endotelial cell	Vacuole
Burkholderia pseudomallei	Macrophage	Vacuole
Calymmatobacterium granulomatosis	Monocyte	Vacuole
Chlamydia trachomatis	Mucosal epithelial cell	Vacuole
Coxiella burnetti	Macrophage	Phagolysosome without alkalinization
Ehrlichia canis	Neutrophile/monocyte	Phagosome
Franciscella tularensis	Macrophage	Phagosome
Legionella pneumophila	Macrophage	Vacuole
Lysteria monocytogenes	Macrophage/hepatocyte	Cytoplasm
Mycobacterium tuberculosis,M. avium	Macrophage	Vacuole
Mycobacterium leprae	Macrophage/schwann cell	Vacuole

(continued)

Table 1 (continued)

Bacteria	Cell type infected	Subcellular localization
Rickettsia rickettsii, R. prowazekii, R. shigella	Endotelial cell	cytoplasm/nucleus
Salmonella enterica	Intestinal epithelial cell/macrophage	Vacuole
S. typhimurium	Macrophages	Vacuole
Shigella flexneri	Epithelial cell	Cytoplasm
Staphylococus aureus	Human endothelial cells, bovine epithelial cells and murine fibroblasts	Phagolysosome without alkalinization
Yersinia enterocolitica	Intestinal epithelial cell	Vacuole
Y. pestis	Macrophage	Phagolysosome without alkalinization
Protozoa		
Plasmodium spp.	Erythrocyte/hepatocyte	Parasitophorous vacuole
Leishmania spp.	Macrophage	Phagolysosome
Trypanosoma cruzi	Macrophage, smooth and striated muscle cells and fibroblast	Cytoplasm
Toxoplasma gondii	All nucleated cells	Parasitophorous vacuole

live inside a parasitophorous vacuola (PV) like *Plasmodium spp* and *Toxoplasma gondii.* Besides tissue accessibility, the host cell/PV/parasite and food vacuola or endoplasmic reticulum membranes are the barrieres interposed to the access of antiparasitic agents against *Plasmodium spp* and *T. gondii*. Host cell, phagolysosome and parasite membranes are the barriers interposed by *Leishmania spp,* whereas host cell and parasite membranes are interposed by *T. cruzi*.

Clearly, the intracellular location of potozoaria and bacteria is a true challenge for the delivery of antimicrobial agents. As seen, a series of structural/phenomenological barriers impairs the antimicrobials to achieve therapeutic intracellular concentrations during the required time period. Besides, a potential drug may be too toxic, leading to a maximum tolerated dose well below what is necessary for efficient eradication of the infection (Mehta 1996). Pharmacokinetic and pharmacodynamic parameters of antimicrobials are dependent of their chemical structure (Tulkens 1991; van den Broek 1989).

To be used against intracellular target, a given antimicrobial has to satisfy many selection criteria for optimal activity including entry, retention, subcellular distribution, expression of activity in the infected compartments, and susceptibility of microorganism at the infection site with regard to the growth phase.

The ratio of the intracellular concentration to the extracellular concentration (*Ci*:*Ce*) is the parameter used to classify the ability for intraphagocytic accumulation of antibiotics. According to this, antibiotics can be poor cell penetrating (*Ci:Ce* <1, penicyllin G), antibiotics that penetrate but do not concentrate (*Ci:Ce* ~1 such as aminoglycosides, rifampin, sulfonamides, chloramphenicol, tetracyclines) and antibiotics that concentrate within cells (*Ci:Ce* >1, macrolides and azoles) (Asanuma et al. 1983;

Labro 1996; Maurin and Raoult 1997; Tulkens 1991). Poor penetration into the cells and decreased activity intracellularly are the major reasons for the limited activity of most antibiotics (penicillins, cephalosporins and aminoglycosides) in intracellular infections. An equivalent classification for antiprotozoan agents is not available. Macrolides such as erythromycin and tilmicosin have been shown to be 75-fold more concentrated in lung macrophages compared with parenchymal cells (Blais and Chamberland 1994). This differential accumulation of macrolides within phagocytes and other cells involved in immunity explains in part the excellent clinical effect of these antibiotics against organisms for which they possess a moderate minimum inhibitory concentration (MIC) value (Smith 1988) as well as high activity against phagocytosed intracellular gram-negative bacteria, compared with their weaker activity against extracellular bacteria (Rakita et al. 1994). While macrolides generally distribute evenly in the cell cytosol and phagosomes/lysosomes, azolides concentrate in phagosomes. The accumulation of these antibiotics within host cells is partly owing to protonation of the weak base macrolides intracellularly, which enhances their retention (Labro 1996). Such localized concentrations are likely to increase the activity of these antibiotics against bacteria located in the same compartment. The enhanced concentration of macrolides and azolides in phagosomes could therefore retard the development of resistant strains of intraphagosomal bacteria, such as *Chlamydia spp.* The intracellular location of tetracyclines, however, remains poorly defined. The various members of the macrolide and azolide group have a relatively wide range of MIC values for any particular intracellular organism, possibly reflecting subtle variations in the intracellular location of the antibiotic (McOrist 2000).

Most antibiotics are more effective at neutral or basic pHs, although rifampin, pyrazinamide, and to a lesser extent the tetracyclines, which more active at acidic pH, are exceptions (Maurin and Raoult 1997). Weak base antibiotics such as the aminoglycosides and the macrolides are concentrated within lysosomes by a pH-dependent mechanism. Although these antibiotics may be concentrated several folds within cells, their localization within the lysosomal compartment and partial inactivation of their activity by acidic pH represent major disadvantages (Rakita et al. 1994). Intracellular location and activity as a function of pH of antibiotics are shown in Table 2.

Table 2 Mode of entry within eukaryotic cells, subcellular localization and pH activity of antibiotics (Maurin and Raoult 1997)

		Localization		Antibiotic activity at pH		
Antibiotics	Mode of entry	Cytoplasm	Lysosomes	Basic	Neutral	Acidic
Betalactams	Diffusion	+		+++	++	+
Aminoglicosides	Pinocytosis		+++	+++	+	−
Tetracyclines	Diffusion	++	?	++	+++	+
Chloramphenicol	Diffusion	++	?	+++	+	−
Rifampin	Diffusion	++	++	++	++	+++
Erythromycin	Transport	+	+++	+++	+	−
Clindamycin	Transport	+	+++	+++	+	−
Fluoroquinolones	Unknown	++	++	+++	+	−

An additional difficulty with classical antibiotic therapy is that many intracellular bacteria are quiescent or dormant. These bacteria are present in a reversible state and can persist for extended periods without division (Kaprelyants et al. 1993) under a viable but non-culturable state. So the drug targets are downregulated and the cellular permeability is altered dramatically, influencing their susceptibility to antibacterial agents (Gilbert et al. 1990).

On the other hand, protozoan pathogens (eukaryotic unicellular organism) like plasmodium, leishmania, trypanosoma and toxoplasma have tremendous negative impact on human health and prosperity. These protozoans are obligate intracellular parasites. The intracellular habitat make possible to avoid detection and destruction by host antibodies. As yet, there are no safe vaccines for any of these parasites, leaving drug treatments as the major strategy for control. Available drugs are compromised by low efficacy, high toxicity, and wide spread resistance.

Overall, performing a strategy for passive or active drug targeting by means of nanoparticles is far more challenging than developing a strategy for sustained drug delivery. Tissue location of infected host cells and intracellular ubication of the microorganism, play a protagonic role in rationale drug targeting strategies. Non selective, widely distributed, toxic or short half live antimicrobials can be suitable candidates for targeted delivery. Toxic antimicrobials that do not accumulate inside the cells (Ebert 2004) such as aminoglycosides and antibiotics that do not enter infected intracellular compartments such as quinolonas should be candidates to targeted delivery instead of sustained release (Hand et al. 1987; Prokesch and Hand 1982). Poor penetrating aminoglycosides that the same as macrolides inactivate at low pH are candidate to early escape from endo-phagosomal pathway and delivery against cytoplasmic microorganism. An early escape strategy is also suitable against for agents against cytoplasmic protozoa such as *T. cruzi* and *T. gondii*. The phagosomal via naturally culminates in delivering phagocytosed material against *Leishmania spp.* Specific targeting and intracellular delivery on erythrocytes and hepatocytes are needed in against *Plasmodium spp.*

In this chapter we will review selected preclinical approaches that made use of nanomedicines for intracellular delivery of antibiotic and antiprotozoal agents.

2 Mycobacterium spp

Mycobacterium tuberculosis is a facultative intracellular pathogen, nonmotile nonencapsulated rod shaped Gram positive aerobe (growing most successfully in tissues with high oxygen content such as lungs) or facultative anaerobe (Godreuil et al. 2007).

M. tuberculosis is the main cause of tuberculosis, a deadly infection (Kaufmann and McMichael 2005), considered one of the main challenges in public health (Gaspar et al. 2008b) Approximately two billion people (~30% of the global population) are currently infected (The union 2008). After HIV/AIDS, tuberculosis is the second most deadly infectious disease (Frieden et al. 2003), every year 9.2 million people develop the disease and the annual mortality rate is 1.7 million people (WHO 2008). Tuberculosis is endemic in developing countries and resurged in developed countries (Jain et al. 2008). Moreover, a high prevalence of HIV/tuberculosis co-infected patients has spurred the WHO to declare a global sanitary emergency since 1993 (WHO 1993).

Tuberculosis is considered a disease with an *interhuman transmission*. Tuberculous bacilli are spread out by infected patients coughing, sneezing or speaking and they can be inhaled by another individual in close contact (Godreuil et al. 2007). Upon entering the respiratory system *M. tuberculosis* are nonspecifically phagocytosed by alveolar macrophages (Smith 2003). The majority of these bacilli are destroyed or inhibited by macrophages that process the bacterial antigens and present them to T lymphocytes. The bacilli that are not destroyed by alveolar macrophages multiply, in what is know as a second or symbiotic stage that occurs between 7 and 21 days after infection. Bacilli are released after cellular lysis and can thus infect other circulating macrophages and continue their multiplication. Extracellular bacilli attract macrophages from the bloodstream. At the end of this stage, a huge number of macrophages and bacilli are concentrated at the level of early pulmonary lesions (Dannenberg 1989). Within 2–10 weeks after infection, the immune system is usually able to halt the multiplication of the tubercle bacilli, preventing further spread and leading to the formation of a granuloma. At this point the person has tuberculosis infection. Tuberculosis disease results when the immune system is not able to halt the multiplication of tubercle bacilli. Three months after infection, these bacilli can spread through the lymphatic channels to regional lymph nodes and then through the bloodstream to more distant tissues and organs from the apices of the lungs, kidneys, genitalia and spongy bone to the brain. Tuberculosis disease presents as pulmonary tuberculosis (80%), extrapulmonary tuberculosis (20%) or a combination of the two. Acute tuberculosis meningitis or disseminated tuberculosis can sometimes result in death. The release of the bacteria to the pleura 3–7 months after infection results in pleurisy (Banu et al. 2004). Persons who are infected with *M. tuberculosis* but who do not have tuberculosis disease cannot spread the infection to other people. Tuberculosis infection in a person who does not have tuberculosis disease is not considered a case of tuberculosis and is often referred to as latent tuberculosis infection (Godreuil et al. 2007). Only 6–10% of HIV-negative patients develop the disease and, in most of the cases, because of the reactivation of a preexisting infection. In contrast, 50–60% of HIV-infected patients have chances to show reactivation during lifetime. Also, a fast progression toward the disease is found; higher mortalities in earlier stages of the disease have been found for HIV/ tuberculosis co-infected patients (Schluger 2005).

The *M. avium–M. intracellulare* complexes (MAC) are aerobic Gram positive bacilli, responsible for several different syndromes. MAC is the main cause of complications in immunocompromised patients or with underlying lung disease. MAC is the most common cause of disseminated infection by nontuberculous mycobacteria in patients with HIV. Less commonly, pulmonary disease in nonimmunocompromised persons is a result of infection with MAC. In children, the most common syndrome is cervical lymphadenitis (Centers for Disease Control and Prevention). Unlike tuberculosis, a MAC infection cannot be passed from one person to another.

MAC can invade and translocate across the mucosal epithelium. The bacteria subsequently infect the resting macrophages in the lamina propria and spread in the submucosal tissue; they are then carried to the local lymph nodes by lymphatics. In immunocompromised hosts, they are subsequently spread hematogenously to the liver, spleen, bone marrow, and other sites.

Mycobacterium spp replicates within alveolar macrophages (Cosma et al. 2003). The Mycobacteria pathogen vacuole (MPV) arrests at the early endosomal (EEA1, Rab5-positive) stage (Philips 2008). The MPV retains the ability to interact with early and recycling endosomes, presumably to acquire nutrients delivered by endosomal recycling pathways. Despite the arrest in MPV maturation, *Mycobacterium spp* can be delivered to phagolysosomes in macrophages after induction of autophagy with rapamycin treatment or in response to IFNγ treatment (Gutierrez et al. 2004; Hope et al. 2004). As with the salmonella containing vacuola (SCV), ubiquitination of bacterial products and autophagy are potent host defenses against establishment of the MPV. Not surprisingly, *Mycobacterium spp* have acquired mechanisms to minimize the induction of autophagy (Kumar and Valdivia 2009).

2.1 Conventional Treatments

If well there are drugs that successfully act against tuberculosis, the complexity of treatment (de Cock and Wilkinson 1995; Gelband 2000), the adverse side effects (Myers 2005) and acquisition of drug resistance (Mitchison 2005; Toit et al. 2006) are impediments to eradicate this infection.

Long treatment (6–8 months of daily intake) stems from the presence of non-metabolizing bacteria that are not killed by the antibiotics and of pathogens in stationary phase or proliferating at extremely low rates in old lesions or within fibrotic or calcified sites (Rook and Hernandez-Pando 1996). The prolonged pharmacotherapy and the pill burden can hamper patient lifestyle and thus low compliance and adherence to administration schedules remain the main reasons for therapeutic failure and contribute to the development of multi-drug resistant strains (Dye 2006). In developing countries and rural areas, additional difficulties to complete the treatment are the difficult supervision of the regimens and the prohibitive costs of the drugs involved (Gupta et al. 2001).

The most effective pharmacotherapy against tuberculosis is comprised of a multi-drug combination of isoniazid (INH), pyrazinamide (PYZ) and rifampicin (RIF). INH is a synthetic prodrug that inhibits the synthesis of mycolic acid, required for the mycobacterial cell wall. It eliminates only active (growing) bacteria. Since the bacteria may exist in a resting (nongrowing) state for long periods, therapy with INH must be continued for a long time (usually 6–12 months) (National Institutes of Health). PYZ is a bacteriostatic/bacteriocide prodrug that transforms in pyrazinoic acid. RIF is a bactericidal semisynthetic drug of the rifamycin group, that inhibits DNA-dependent RNA polymerase by binding its β-subunit, thus preventing transcription to RNA and subsequent translation to proteins. RIF is active against microorganisms of the genus Mycobacterium, including *M. tuberculosis*, *M. kansasii*, *M. marinum*, MAC and *M. leprae*. The limited bioavailability of RIF is a main clinical issue (Gohel and Sarvaiya 2007).

During the initial intensive stage (2 months), these three agents are administered together with ethambutol (ETB), a bacteriostatic drug against actively growing tuberculosis bacilli that works by obstructing the formation of cell wall and is responsible for permanent visual loss (Lim 2006; WHO 2003). The second phase (4 months) comprises exclusively RIF and INH. These four drugs together with the aminoglycoside streptomycin constitute the first-line therapy. INH, PYZ, RIF and streptomycin have been shown to induce hepatotoxicity when used alone or in combination (Davidson and Quec 1992).

RIF is the first line drug currently used for treatment of latent *M. tuberculosis* infection in adults. But a number of side effects like lack of appetite, nausea, hepatotoxicity, fever, chill, allergic rashes, itching and immunological disturbances, patient non-compliance in long term therapy limit its use (Laura et al. 2000).

MAC is intrinsically resistant to many antibiotics and antituberculosis drugs (Bermudez 1994; Leitzke et al. 1998) but is fairly susceptible to macrolides (clarithromycin, azithromycin), rifamycins (RIF, rifabutin), ETB, clofazimine, fluoroquinolones, amikacin and streptomycin (Koirala 2010).

2.2 *Preclinical Delivery Systems*

Delivery systems against *Mycobacteria spp* were by far the most varied, outnumbering those used against other bacterial infections. In first place, the performance of oral polymeric nanoparticles, which are not absorbed but that increased drug bioavailability was extensively explored. Their use led to a reduced number of administrations and increased accumulation of drug in tissues. On the other hand, i.v. nanoparticles which are captured by fixed accessible or circulating macrophages, allowed a targeted delivery of drugs to infected macrophages. Once phagocitosed, drugs within nanoparticles probably followed the same intracellular pathway and unless being carried by pH-sensitive nanoparticles, drugs were delivered inside phagosomes. Alveolar macrophages were refractary to the uptake of i.v. nanoparticles and so the disease burden was less diminished than in liver and spleen. Only pulmonary administration allowed an effective targeting to alveolar macrophages and therefore the frequency of administrations could be pronouncedly reduced as compared to the other vias.

In the recent review of Sosnik et al. (2010) the strategies based in the use of nanoparticles for sustained or targeted delivery of antibiotics against tuberculosis are extensively discussed. In this chapter will only be included works showing biodistribution of nanoparticles and or preclinical therapeutic efficacy against disease models.

2.2.1 Oral and Parenteral Administration

Biodistribution and Therapeutic Efficacy of Non Liposomal Nanoparticles

Between 2003 and 2005 there were published a series of articles where first line drugs were loaded into poly(lactic-*co*-glycolic acid) (PLGA) nanoparticles that were orally administered to different healthy or infected animal models. In an early article RIF, INH and PYZ loaded in 190–290 nm PLGA nanoparticles administered to mice, resulted in drugs detection in plasma for up to 9 days while therapeutic concentrations in tissues were maintained for 9–11 days. Five oral doses every 10 days of drug-loaded PLGA nanoparticles effectively eliminated the pathogen from the different organs (Pandey et al. 2003a). In a next step, the same drug-loaded PLGA nanoparticles were administered to guinea pig infected with aerosolized *M. tuberculosis*, at PLG-NP 250 mg/kg, RIF 12 mg/kg, INH 10 mg/kg and PYZ 25 mg/kg of body weight (bw), in five oral doses every 10 days. It was found that the bacterial count and lung histopathology were reduced to the same extent than free drugs daily administered by 6 weeks (Johnson et al. 2005).

A further work showed that RIF, INH, PYZ and ETB loaded in 190–290 nm PLGA nanoparticles orally administered to mice i.v. infected with 1.5×10^7 bacilli of *M. tuberculosis* at RIF 10 mg/kg, INH 25 mg/kg, PYZ 150 mg/kg and ETB 100 mg/kg bw, rendered therapeutic levels maintained for 5–8 and 9 days in plasma and brain, respectively. Additionally five oral doses every 10 days of drug-loaded PLGA nanoparticles eliminated the bacteria in the meninges (Pandey and Khuller 2006).

Azoles have the potential to replace INH and RIF in murine tuberculosis and are effective against latent tuberculosis (Ahmad et al. 2006a), while quinolones may be used to shorten the duration of chemotherapy (Duncan and Barry 2004). Therefore, the antimicrobial activity of azoles and quinolones loaded in PLGA nanoparticles has also been determined. It a recent work, econazole (ECZ, an antifungal imidazole) and moxifloxacin (MOX, a fourth-generation synthetic fluoroquinolone) loaded in 220 nm PLGA nanoparticles were orally administered to mice at MOX 8 mg/kg and ECZ 3.3 mg/kg bw (Ahmad et al. 2008). A single dose of drug-loaded PLGA nanoparticles rendered therapeutic drug concentrations in plasma for up to 5 (ECZ) or 4 days (MOX), whilst in lungs, liver and spleen it was up to 6 days. In comparison, free drugs were cleared from the same organs within 12–24 h. In *M. tuberculosis*-infected mice, eight oral doses of drug-loaded PLGA nanoparticles administered weekly were found to be equipotent to 56 doses of MOX daily administered or 112 doses of ECZ administered twice daily. Furthermore, the combination of MOX plus ECZ proved to be significantly efficacious compared with individual drugs. Addition of RIF to this combination resulted in total bacterial clearance from the organs of mice in 8 weeks.

More recently, similar results were obtained with alginate nanoparticles. In a first approach, RIF, INH, PYZ and ETB were loaded in 235 nm alginate nanoparticles and orally administered to mice at two dose levels (Ahmad et al. 2006b). While free drugs were cleared from circulation after 12–24 h after administration and were detectable in spleen, liver and lung only until day 1, when loaded in alginate nanoparticles were observed in plasma up to 7, 9, 11 and 11 days after administration for ETB, RIF, INH and PYZ, respectively, and in tissues until day 15. In a second approach, eight oral doses of ECZ at 3.3 mg/kg bw loaded in 230 nm alginate nanoparticles or 112 doses of free ECZ administered twice daily, reduced bacterial burden by more than 90% in the lungs and spleen of mice i.v. infected with 10^5–10^7 bacilli of *M. tuberculosis*. ECZ (free or loaded in nanoparticles) could replace RIF and INH during chemotherapy of murine tuberculosis. Alginate nanoparticles reduced the dosing frequency of azoles and antitubercular drugs by 15-fold (Ahmad et al. 2007).

In a different approach, solid lipid nanoparticles (SLN) were loaded with RIF, INH and PYZ and administered to mice. A single oral dose resulted in drug concentrations detectable after 3 h and for up to 8 days in therapeutic concentrations in plasma and lungs, and for 10 days in liver and spleen. Free drugs on the other hand, were cleared within 1–2 days. Also in *M. tuberculosis* infected mice, no tubercle bacilli was detected in the lungs and spleen after five oral doses of drug-loaded SLN administered every 10 days; whereas 46 daily doses of oral free drugs were necessary to reach the same therapeutic effect (Pandey et al. 2005).

The performance of experimental drugs loaded in cyclodextrins has also been explored, as the case of PA-824 (a poorly aqueous soluble three-substituted nitroimidazopyran), which is highly specific for the *M. tuberculosis* complex (*M. avium*, *M. smegmatis*, *M. chelonae*, and *M. fortuitum*) and showed only modest or no activity against other mycobacteria. PA-824 was included in hydroxypropyl-βcyclodextrin /lecithin and orally administered to mice infected with a low dose of aerosolized *M. tuberculosis* Erdman (a short-course mouse infection model). PA-824 at 100 mg/kg in cyclodextrin/lecithin was as active as free MOX at 100 mg/kg and INH at 25 mg/kg and was slightly more active than RIF at 20 mg/kg, administered in nine daily consecutive doses. Long-term treatment with PA-824 at 100 mg/kg in cyclodextrin/lecithin administered in 5 days per week along 12 weeks reduced the bacterial load below 500 colony forming units (CFU) in lungs and spleen. No significant differences in activity between PA-824 and the other single drug treatments tested at INH 25 mg/kg, RIF 10 mg/kg, gatifloxacin 100 mg/kg and MOX 100 mg/kg bw, could be observed (Lenaerts et al. 2005).

An active targeting strategy that did not lead to superior results as compared to non targeted carriers, was explored by employing RIF, INH and PYZ loaded in 350–400 nm PLGA nanoparticles surface grafted with lectins recognizable by glycosylated structures in the gut and the lung mucosa (Sharma et al. 2004). Drug-loaded PLGA nanoparticles were orally or aerosol administered to guinea pigs at RIF 12 mg/kg, INH 10 mg/kg and PYZ 25 mg/kg bw. Plasma half life times of the different drugs loaded into these surface-modified nanoparticles were extended from 4 to 9 days (uncoated) to 6–14 days upon oral or inhalation administration, respectively. In addition, total bacterial clearance was achieved in lungs, liver and spleen after only three orally or aerosolized doses (one every 14 days) on guinea pigs infected intramuscularly with 1.5×10^5 bacilli of *M. tuberculosis*.

The i.v. via has been barely explored. In a recent work for instance, RIF loaded in 265 nm gelatin nanoparticles was administered at 10 mg/kg bw to Balb/c mice infected with aerosolized 1×10^5 *M. tuberculosis* (Saraogi et al. 2010). After a 4-week treatment period, RIF-loaded gelatine nanoparticles depicted almost two fold reduction in mycobacteria CFU in lungs and spleen when compared with untreated animals. A significant reduction in the number of bacilli recovered both from the lungs and spleen as compared with of free RIF given orally was also produced. The i.v. administration of RIF gelatin nanoparticles allowed to reduce 28 doses (oral RIF daily) to ten doses (i.v. administered every 3 days). Gelatin nanoparticles offer a number of benefits because of their biocompatibility, biodegradability, low antigenicity, low cost and ease of their use as parenteral formulations (Coester et al. 2000; Kaur et al. 2008; Verma et al. 2005). Though similar results were obtained with other drugs and carrier systems such as PLGA nanoparticles (Pandey et al. 2003a, b) liposomes (Gaspar et al. 2008a, b) as well as with ligand-appended synthetic polymer based nanoparticles (Sharma et al. 2004), the need for large amounts of the polymer, biodegradability, cost restriction of the polymer, the hazard of residual organic solvents and a lower drug stability are some of the critical factors that shift the choice of formulation towards biodegradable nanoparticles such as gelatin nanoparticles.

Two additional works explored the performance of alternative drugs loaded in different formulations. In the first of them, clofazimine (CLF), a riminophenazine drug with activity against disseminated *M. avium* nanosuspensions were prepared. CLF accumulate within macrophages, where mycobacteria multiply but possess poor solubility, being unsuitable for i.v. administration in patients with drug malabsorption. It is doubtful whether oral administration can lead to intracellular concentrations high enough to kill the bacteria; additionally showed toxic side affects. Nanosuspensions (100–1,000 nm) of nanocrystalline CLF i.v. administered to C57BL/6 mice infected with *M. avium* at 20 mg/kg bw (Peters et al. 2000), resulted in CLF acummulation in spleen, liver and lung (81.4, 72.5 and 35.0 mg/kg tissue, respectively) higher that the MIC. Nanocrystallyne CLF was as effective as liposomal CLF in reducing bacterial loads in liver, spleen and lungs. Preparation of drug in form of nanosuspensions was shown to be a more cost effective and technically simpler alternative, particularly for poorly soluble drugs, and to yield to a physically more stable product than liposome dispersions. In the second one, MOX, loaded in non biodegradable poly (butyl-2-cyanoacrylate) nanoparticles and i.v. administered to mice infected with *M. tuberculosis* resulted in a significant decrease in mycobacteria counts in lungs (Shipulo et al. 2008).

Two further works explored the intraperitoneal (i.p.) administration. In 2003, the quinolone norfloxacin was bound (through an α(N–C) bond sensitive to lysosomal proteases, allowing release in the lysosomes (Peters et al. 2000)), to a dextran polymer (a conjugate large enough to impair renal clearance) bound to mannose and administered to mice infected with *M. bovis*. The number of CFUs in lungs was decreased, whereas the free drug was ineffective (Roseeuw et al. 2003). More recently, RIF loaded in niosomes i.p. administered to mice rendered greater accumulation of RIF in thoracic lymph nodes (46.2% of the administered dose) than free drug (Jain et al. 2006). Other administration routes such as i.v. rendered poor RIF accumulation in torax, while intratracheally 128,585 and 885 lung/plasma ratios for niosomes and free drug, respectively were achieved in rats. This represented a 145-fold increase in the accumulation of RIF-loaded niosomes in lungs as compared to the free drug (Mullaicharam and Murthy 2004).

Sustained release formulations help guard against the acquisition of drug resistance, as there is less chance for missed doses leading to suboptimal drug concentrations in the blood (Kisich et al. 2007). Recently, the performance of a subcutaneous (s.c.) depot was surveyed. A single administration of RIF, INH and PYZ loaded in 186–290 nm PLGA nanoparticles at RIF 12 mg/kg, INH 10 mg/kg and PYZ 25 mg/kg bw, maintained plasma, lungs and spleen drugs concentrations for more than 1 month, while free drugs were cleared from circulation after 12–24 h. When administered to mice i.v. infected with 1×10^5 bacilli of *M. tuberculosis*, there were undetectable bacterial counts in the different organs (Pandey and Khuller 2004).

Biodistribution and Therapeutic Efficacy of Liposomes

Intravenous liposomes carrying aminoglycosides showed to be effective antimycobacterial agents. Aminoglycosides are potent antibiotics with bactericidal activity against Gram-negative and Gram-positive organisms as well as a series of intracellular pathogens. Aminoglycosides have post-antibiotic effect, synergy with β-lactam antibiotics, rapid concentration dependent bactericidal effect and low cost as well as low frequency of resistance to them for the treatment of serious bacterial infections (Begg and Barclay 1995; Lacy et al. 1998; Lortholary et al. 1995; Poole 2005; Zembower et al. 1998). However, they require parenteral administration and their high dosage often results in serious toxicity including ototoxicity and nephrotoxicity (Kumana and Yuen 1994; Nagai and Takano 2004). Therefore, administration of aminoglycosides are only limited to patients with severe bacterial infections, especially in immunocompromised patients with mycobacterial infections (Cometta and Glauser 1996; Maertens and Boogaerts 1998; Schiffelers et al. 2001).

One of the limitations of liposomes-based aminoglycoside antibiotics is their low encapsulation efficiency, which results in formulations with low drug-to lipid ratio. This, in turn, should require high amount of liposomal formulations to achieve desirable therapeutic dose, a feature that defeat the purpose (Mugabe et al. 2006). However, for nephrotoxic drugs such as aminoglycosides an advantage of liposome entrapment is their decreased renal excretion. Free aminoglycosides are usually rapidly cleared from the blood by glomerular filtration, exposing renal tubules to potentially toxic concentrations. An additional advantage is that upon liposomal encapsulation aminoglycosides can be delivered to macropages, and used against intracellular microorganisms that invade these cells.

In the first work, amikasine (AMK) loaded in multilamellar vesicles (MLV) (egg phosphatydilcholine, eggPC) i.v. administered at 40 mg/kg increased the delivery of AMK to the liver and spleen and showed higher and more prolonged levels in serum (up to 24 h after administration) than free AMK that showed a rapid decline in serum (below the level of detection at 5 h). Daily administered for 5 days on beige mouse acute infection model (10^7 MAC i.v. incoculated) at 40 mg/kg bw, liposomal AMK reduced the viable cell count in the liver and spleen compared with free AMK. At 110 mg/kg bw twice a week for 3 weeks, liposomal AMK and free AMK reduced the viable cell counts in the liver, spleen and lung compared with counts in the control group. Liposomal AMK was more active than the free drug for organisms in the liver and spleen but not for organisms in the lung. The ability of both free and liposomal amikacin to reduce the viable count of organisms per organ was directly related to the interval between infection and the initiation of therapy. The earlier therapy was started, the greater the reduction in bacterial load (Cynamon et al. 1989).

The very high levels of AMK activity in the spleen and liver as well as sustained levels in the serum, suggested a saturation of the RES to some extent with the lipid doses used. On the other hand, treatment with empty liposomes resulted in increased numbers of organisms in the spleens and livers of mice. Since distribution to nonhepatic RES tissues (such as the spleen, bone marrow, and lung) and sustained levels in serum would appear to be desirable in the treatment of MAC, it may be necessary to carefully balance the advantages and potential disadvantages of high lipid doses to achieve the optimum therapy of this disease.

In a second approach, AMK and the quinolone ciprofloxacin (CIP) were loaded in oligolamelar liposomes and either a single dose of 600 mg of free or liposomal AMK or 1,000 mg of free or liposomal CIP were administered to C57BL/6 mice (Leitzke et al. 1998). At 24 h postinjection, only liposomal AMK resulted in a sustained accumulation of drug in liver and spleen (59.1 +/– 7.5 and 194.0 +/– 51.0 mg/kg, respectively), while CIP concentrations had fallen below the threshold of detection. These results indicated a fast re-distribution of liposomal CIP from the target tissues and excluded its further application. Upon administering 2,000 mg of liposomal AMK, a concentration exceeding 8 mg/kg was found in liver and spleen at the end of 28 days. In contrast, the same dose of free AMK dissapaired within 2–9 h. Different to liver and spleen, levels of AMK in lungs were not increased neither sustained. Liposomal encapsulation of AMK, but not of CIP, resulted in high and sustained drug levels in infected tissues, exceeding the MIC for *M. avium* for at least 28 days. The therapeutic efficacy of 2,000 mg of liposomal AMK administered once-weekly and once-monthly (6 consecutive weeks or months) on C57BL/6 mice i.v. infected with 10^5 CFU of the virulent *M. avium* strain TMC 724 was determined. This model of infection leads to the death of even immunocompetent mice 120–150 days after infection. Thus, a once-weekly treatment schedule proved to be as effective as previously reported regimens of twice weekly or even daily injections of other liposomal AMK (Bermudez et al. 1990; Cynamon et al. 1989; Duzgunes et al. 1988; Ehlers et al. 1996). Monthly treatment led to a significantly prolonged survival time of the treated animals, which all survived the observation period of 7 months, whereas all control infected animals died approximately 4 months after time of infection. This benefit was also observed when treatment was initiated at an advanced stage of infection. Again, overall treatment efficacy in the lungs was poor.

In a further work, the aminoglycoside gentamicin (GEN) was loaded in MLV (eggPC) and administered to murine model of disseminated MAC infection at

20 mg/kg of bw either daily for 5 consecutive days or twice weekly for 3 weeks (Klemens et al. 1990). Liposomal GEN significantly reduced viable cell counts in the spleen and liver compared with the free GEN. Consecutive and intermittent treatments produced a 1.5-log and a 2.5-log reduction in cell counts in spleen and liver, respectively. None of the regimens resulted in sterilization of spleen and liver and the bactericidal activity of liposomal GEN was not present in lungs.

A further assay employed CLF, used for leprosy, although it is also used in cases of tuberculosis resistant to other drugs (Sepkowitz et al. 1995), and it shows activity against MAC. This drug, in addition to causing many side effects, it has not been therapeutically beneficial for tuberculosis. Liposomal encapsulation of CLF allowed its parenteral administration and resulted more effective in treatment of MAC infection than free CLF i.v. or orally administered to beige mice (Kansal et al. 1997; Mehta 1996). The maximum tolerated dose of free CLF was 5 mg/kg bw, which was not enough to cause a significant reduction in the number of viable *M. tuberculosis* bacilli (Mehta 1996). Even this concentration of free drug cannot be administered i.v. to patients because of its lipophilicity, the presence of organic solvents and crystallization in the aqueous phase. An equivalent concentration of liposomal CLF did not improve the treatment outcome; however, encapsulation of CLF in liposomes reduced toxicity and allowed i.v. administration as well as administration of higher doses, enhancing therapeutic efficacy as observed with MAC (Kansal et al. 1997; Mehta 1996).

After the liposomal CLF activity against experimental tuberculosis was proven, its performance against acute and chronic infection was determined. CLF was loaded in MLV (dipalmitoyl phosphatidylcholine (DPPC): dipalmitoyl phosphatidylglycerol (DPPG), 7:3 M ratio) and administered to infected mice at 50 mg/kg bw. It was found that the antibacterial activity was more pronounced in the chronic-infection model. This was attributed to the combined effect of the drug, the acquired specific immunity, and granuloma formation. Overall, the treatment was more effective in liver and spleen than in lungs, similar to previous studies with MAC infection (Kansal et al. 1997; Mehta 1996). Once again, the lungs responded poorly to liposomal CLF therapy when the animals were infected with greater numbers of bacteria. In the lungs, a gradual decrease in CFU was observed, though a rebound was found 2 months post-treatment. Regardless of the apparently total clearance of the bacilli following a second treatment with the liposomal CLF, a similar phenomenon was observed after 2 months (Adams et al. 1999).

A recent study reinforced the effectivity of liposomal antibiotics in the intermittent treatment of tuberculosis. PYZ was loaded in liposomes (DPPC: cholesterol (chol), 7: 2 M ratio) and administered to mice infected with *M. tuberculosis* (El-Ridy et al. 2007). Liposomal PYZ caused a significant reduction in bacterial counts (CFU/g) in lungs, 10, 20 and 30 days after the last treatment dose. Histopathological examination of mice lungs showed highest severity of infection in empty liposomes>liposomal PYZ>free PYZ 6 days/week. The results indicate high therapeutic efficacy of liposomal PYZ injected twice weekly in treatment of *M. tuberculosis* in mice.

In a second study, the performance of liposomal Rifabutin (RFB a rifamycin used for patients taking drugs (particularly antiretroviral drugs) that have unacceptable interactions with RIF), was explored. Its action is similar to RIF, but it has been associated with uveitis when used with clarithromycin or fluconazole. Different RFB liposomal formulations were developed and characterised and their *in vivo* profile was compared with free RFB following i.v. administration to mice (Gaspar et al. 2008a, b). Liposomal RFB rendered higher concentration in liver, spleen and lungs 24 h post administration compared with free RFB. The concentration of RFB in these organs was dependent on the rigidity of liposomal matrix. Liposomal RFB matrix prepared with DPPC and DPPG was the most effective and was selected for biological evaluation in a mouse model of disseminated tuberculosis. Compared with mice treated with free RFB, mice treated with rigid liposomal RFB exhibited lower bacterial loads in spleen (5.53 $\log_{10}$ vs. 5.18 $\log_{10}$) and liver (5.79 $\log_{10}$ vs. 5.41 $\log_{10}$). In the lungs, the level of pathology was lower in mice treated with liposomal RFB. These results suggest that liposomal RFB was a promising approach for the treatment of extrapulmonary tuberculosis in human immunodeficiency virus co-infected patients.

Finally, attempting to increase the delivery to alveolar macrophages, targeted liposomes were developed. *O*-steroyl amylopectin (*O*-SAP) is a recognizement signal for the uptake of pegylated liposomes by lung macrophages (Deol and Khuller 1997). Biodistribution of lung-specific pegylated MLV (PC): chol, dicetylphosphate (DCP): monosialogangliosides: distearylphosphatidylethanolamine-poly(ethylene glycol) 2000 (DSPE-PEG 2000) were studied in mice. Findings showed a pronounced increase from 5.1% with conventional liposomes to 31% with pegylated liposomes containing *O*-SAP in lungs accumulation 30 min after administration. Moreover, pretreatment of both healthy and infected animals with conventional liposomes (1 h before the administration of modified liposomes) saturated the RES and the uptake level in the lungs rose to approximately 40% for the targeted pegylated liposomes, after 30 min. On the contrary, these liposomes showed 30–50% reduced uptake and accumulation in liver and spleen. Also, the biodistribution in the different organs was similar in both animal groups showing 40% accumulation in lungs of normal and tuberculous mice.

Simmilarly targeted pegylated MLV (eggPC: chol: DCP: DSPE-PEG 2000, 2:1.5:0.2:0.2 M ratio, with *O*-SAP included at a ratio to eggPC of 1:3 (wt/wt)) were loaded with INH or RIF and administered into mice infected with a low dose (1.5×10^4 CFU/mouse) of *M. tuberculosis* twice a week for 6 weeks. Liposomal drugs at and below therapeutic concentrations (INH 12 and 4 mg/kg, RIF 10 and 3 mg/kg bw) were more effective than free drugs against tuberculosis, as evaluated on the basis of CFUs detected, organomegaly and histopathology. Furthermore, liposomal drugs had marginal hepatotoxicities as determined from the levels of total bilirubin and hepatic enzymes in serum. The elimination of mycobacteria from the liver and spleen was also higher with liposomal drugs than with free drugs (Deol et al. 1997).

2.2.2 Pulmonary Administration

Pulmonary drug delivery is a non-invasive technique and offers a large surface area for solute transport, fast drug uptake and improved drug bioavailability. It is, therefore, not surprising that oral inhalation therapy is considered not only as the preferred mode of administration of drugs to the lungs (regional drug delivery), but also as a potential pathway to the blood stream (Patton and Byron 2007). However, many challenges in formulation development (Rogueda 2005) and little understanding of the interactions between the therapeutic molecules and the respiratory surfaces have significantly constrained advances in pulmonary drug delivery strategies (Sakagami 2006).

Alveolar macrophages remove inhaled particulate matter from the air spaces and red blood cells from the septum by phagocytosis and endocytosis, kill intracellular infecting bacteria, produce pro-inflammatory cytokines, migrate to local lymphoid tissue, and presentate antigen and activate T cells. In contrast to antigen presenting cells from lymphoid tissue, the lung alveolar macrophages and dentritic cells are biased towards an immunosuppressive response, and need strong inflammatory inducing signals via toll like receptors, complement, or Fc receptors to overcome this. Alveolar macrophages are found in the connective tissue of the septum and in the air space of the alveoli, lying on the surface of the alveolar cells type II and type I, immersed within alveolar fluid which is covered by a surfactant layer. However, compared to Kupffer cells (macrophages lining the endotelial liver vasculature) alveolar macrophages are not readily accessible for liposomes or other nanoparticles in systemic circulation. Because of this, the inhalatory via is the only way of administration that enables a direct contact between particulate drug carriers and alveolar macrophages.

There are two objectives of targeting alveolar macrophages using inhaled therapies. First, such targeting can deliver extremely large amounts of the anti-mycobacterial drugs to the macrophage cytosol, even potentially sufficient to overcome drug resistance. Second, uptake of particles by macrophages has the potential to activate infected macrophages. A recent paper on RIF-loaded PLGA microparticles reported that infected macrophages possess an enhanced phagocytic activity with respect to uninfected cells. This is a very interesting finding since it demonstrates that phagocytic activity, during infection, is not only maintained but even increased (Hirota et al. 2008).

Inhalatory techniques can be classified according to the mechanism of aerosolization (Misra et al. 2010) in: (i) nebulization is the generation of a mist of droplets through the use of compressed air or oxygen and consists of dispersing solid or phase-separated drug delivery systems into droplets suspended in a small amount of medium; (ii) pressurized metered-dose inhaler (pMDI or MDI), the most familiar technique of administering medication to the airways and lungs. Such a device uses an aerosol propellant and a dose-metering valve to regulate the egress of defined amounts of medication to the respiratory tract; (iii) dry powder inhalers (DPI) represent an option that lies mid-way between nebulization of a medicament in an external aqueous or gaseous phase, and its delivery under positive pressure such as MDI. DPI relies on the indrawn breath of the patient to pull in a dry powder. The aerosol is generated by turbulence-creating devices actuated by the airstream of indrawn breath (Onoue et al. 2009). A technique usually employed in experimental studies on tuberculosis, accomplishes pulmonary delivery by intratracheal (i.t.) administration which bypasses the nose and throat, making it easier to precisely quantify the delivered dose. This is in contrast to MDI or DPI inhalers and/or nebulizers in animal studies, which filter the aerosol through the nasopharynx, are prone to inter subject variability and are not capable of targeting a desired, predetermined area of the lungs.

A limited number of substances are approved by drug regulatory agencies worldwide for use in inhalation therapy. Excipients may be categorized according to the type of inhalation technique. For nebulization solutions, excipients are typically the same as used in injectables or large-volume parenterals for pH and osmolarity control, preservation or solubilization and wettability. MDI require the use of propellant. Currently, hydrofluoroalkanes are used as propellants, since the elimination of chlorofluorocarbons. Lipids, phospholipids or surfactants are often present in the MDI formulations. DPI often avoids excipients if the shape and size of dry particles is sufficient for ensuring respirability of the formulation. In some cases, sugars, amino acids or soluble polymers are used as particle shapers. These excipients are employed for constructing respirable particles with techniques such as spray drying. Sustained-release polymers, such as poly lactic acid (PLA), PLGA or other polyesters, polyacrylates, albumin, chitosan or alginates are studied for extending the lung residence time of compounds. The goal is to increase efficacy and allow safer low doses to be effective. For highly potent anti-tuberculosis agents (e.g. siRNA), when the dose is very low, lactose may be used as a 'carrier' for constructing ordered mixtures. It is important to stress that lactose intended for use in pulmonary delivery is a large-sized powder used for flow and packing improvement. During aerosolization, micronized drug particles are detached and lactose particles are deposited in the mouth (Misra et al. 2010). Hypoallergenic materials such as lactose, mannitol, sodium citrate dihydrate, glycine, lysine, leucine, etc. are generally used in inhalation products. Such material is 'generally regarded as safe' and is not known to modify immune response to the inhalant in the lung (Chimote and Banerjee 2009).

To date, most studies have focused on developments of nanoparticles administered by DPIs and nebulizers and not by MDI (Azarmi et al. 2008; Bharatwaj et al. 2010). MDIs are of great technological relevance as they are the least expensive oral inhalation devices. Moreover, they are portable, easy to use and have high compliance, making MDIs the most widely used devices for administration of therapeutics to the pulmonary tract (Laube 2005). Compared to DPIs and nebulizers, however, there is yet an additional challenge in the development of polymeric nanoparticles-based MDIs: it has been shown to be difficult to stabilize particle dispersions in the low dielectric hydrofluoroalkane propellants (Rogueda 2005). The particles for MDIs thus need not only to be within the desired size range, but must also have an appropriate surface chemistry-one that is well-solvated by the propellant so as to prevent particle aggregation, which may otherwise negatively impact the aerosol characteristics of the formulation (Bharatwaj et al. 2010; Rogueda 2005).

Pulmonary delivery of nanoparticles intended for macrophage targeting should be easily recognized and phagocytosed by infected alveolar macrophages.

Nanoparticles should possess appropriate physical dimensions and surface characteristics. Thus, although there is *in vitro* evidence of activation of human macrophages with particles as large as 90 mm, the size amenable to phagocytosis is 1–10 μm, with an optimum of 3 μm (Hirota et al. 2007). Similarly, the surface characteristics are important for both phagocytosis and macrophage activation. Phagocytosis of particles made up of poly(styrene), an extremely hydrophobic material, is less efficient than that of PLGA particles of equivalent size and shape but of lower hydrophobicity. However, extremely hydrophobic biodegradable polymer, high molecular weight PLA nanoparticles, still report efficient phagocytosis (Misra et al. 2010; Muttil et al. 2007; Sharma et al. 2001). This may be due to the fact that their formulations contain an unusually high drug payload, which could modify the surface characteristics of the particles by virtue of high surface density of drug molecules. It was demonstrated the advantage of particles with an elliptical geometry for macrophage uptake (Hickey et al. 1996). However, surface and shape can also affect lung deposition of inhaled therapies (Misra et al. 2010). A stringent aerosol particle size requirement represents another important challenge. The delivery efficiency of many traditional aerosol formulations is low due to the fact that a large fraction of the particles have aerosol sizes that are too large (>~5 μm), and are thus retained in the mouth and throat to reach in the digestive tract rather than lungs. In contrast, aerosol particles that are too small (<0.4 μm, which is within the range of typical polymeric nanoparticles described in the literature) are also expected to have very low deposition efficiency (Rogueda and Traini 2007). The 'fine particle fraction', defined as % particulate matter below a threshold size, can be used to predict lung deposition (Hickey et al. 1996). In the airways, the humidity is very high, at almost 100%. Therefore, the drug carrier has to resist wetting and subsequent aggregation in the respiratory tract. It can thus be appreciated that the twin objectives of lung deposition and macrophage targeting will depend strongly on size, shape and surface properties of drug carrier systems. Particles with an aerodynamic diameter between 1 and 5 μm are deposited in the lower lungs, with larger particles remaining in the throat and smaller particles being exhaled (Cryan et al. 2007).

Medication using DPI can result in high local levels of drug in epithelial lining fluid of the airways and lower respiratory tract. Drugs administered topically to the lungs, via aerosol are attractive in that they achieve higher levels in the lungs with fewer systemic side effects.

There are several issues associated with the delivery of polymeric nanoparticles to the lungs. Biocompatibility and clearance are two very important considerations (Davis et al. 2008).

Biodistribution and Therapeutic Efficacy of Inhaled/Instilled Liposomes

Upon pulmonary administration, liposomes are preferentially taken up by alveolar macrophages instead of type I or II alveolar cells. This was determined in an experiment where small (161 nm) and large (1,455 nm) liposomes (DPPC: DPPG:

phosphatidylinositol (PI): phosphatidylethanolamine (PE): chol, 55:21:8:2:6:8 weight ratio) were i.t. administered to rats. It was found that alveolar macrophages internalize more than three times as many liposomes as alveolar type II cells, whereas only 30% of the alveolar type II cells ingest liposomes vs. 70% of the alveolar macrophages. Hence, only a small subpopulation of alveolar type II cells is able to internalize liposomes (Poelma et al. 2002).

RIF was loaded in re-dispersible freeze-dried vesicles (soyPC). RIF-liposome aerosols were generated using an efficient high-output continuous-flow nebulizer, driven by a compressor. Liposomes showed a good stability during nebulization and were able to retain more than 65% of the incorporated RIF. The effect of liposome encapsulation on lung levels of RIF following aerosol inhalation was determined in rats. The *in vitro* intracellular activity of liposomal RIF against MAC residing in macrophage-like J774 cells was also evaluated. Results indicated that liposomes are able to inhibit the growth of MAC in infected macrophages and to reach the lower airways in rats (Zaru et al. 2009).

Independently of their internalization by alveolar macrophages upon pulmonary administration, liposomes can behave as surfactant suppliers because phospholipids can interac with endogenous lung surfactant to reduce alveolar atelectasis induced by mycobacteria (Chimote and Banerjee 2008). For instance, RIF, INH and ETB were loaded in 2 μm liposomes (DPPC). *In vitro* at 37°C, these liposomes had better surfactant function with quicker reduction of surface tension on adsorption (32.71 +/− 0.65 mN/m) than empty liposomes (44.67 +/− 0.57 mN/m) and maintained 100% airway patency in a capillary surfactometer. Sustained release of the drugs from liposomes was observed over 24 h. *In vitro* alveolar deposition efficiency using the twin impinger showed 12.06% +/− 1.87% of INH, 43.30% +/− 0.87% of RIF and 22.07% +/− 2.02% of ETB deposited in the alveolar chamber upon nebulization for a minute using a jet nebulizer. The formulation was biocompatible and stable with physicochemical properties being retained even after storage for a month at 4°C (Chimote and Banerjee 2009, 2010).

Biodistribution and Therapeutic Efficacy of Targeted Inhaled/Instilled Liposomes

Alveolar macrophages posses sugar receptors such as lectins that enable increasing their targeting by sugar grafted liposomes upon pulmonary administration (Irache et al. 2008; Lane et al. 1998; Wijagkanalan et al. 2008).

For instance, RIF loaded in aerosolized micrometric liposomes were designed to target the alveolar macrophages, by anchoring maleylated bovine serum albumin (MBSA, a ligand that is recognized by macrophage scavenger receptors) or *O*-SAP to the surface of eggPC: chol liposomes (Vyas et al. 2004). Liposomes were formulated in chlorofluorocarbons propellants, packed in a pressurized container and administered to rats. Biodistribution showed that both liposomes, independently of the modification, led to higher concentrations in lungs and lower concentrations in the plasma when compared to the free RIF. For example, after 30 min, lung and serum accumulation of RIF encapsulated in MBSA and *O*-SAP liposomes was 61.5% and 11.3%, and 65.1% and 9.2% of the initial dose, respectively. Contrary to this, free RIF showed 39.1% and 29.8% accumulation in lungs and plasma, respectively. 10.8% of *O*-SAP liposomes were retained in the lungs after 24 h, as compared to 8.2% with MBSA modified ones; this showed the absence of significant differences in the *in vivo* accumulation of these two formulations.

Lately, it was shown that mannose-coated liposomes (1,000 nm) were more pronouncedly captured than unmodified liposomes (100–2,000 nm) by rat alveolar macrophages and a greater accumulation of mannosylated-liposomes was found in the lungs after pulmonary administration to rats (Chono et al. 2007). Accumulation in alveolar cells indicated the higher uptake of mannosylated-liposomes (15–17-folds) by rat alveolar macrophages over alveolar epithelial type II cells.

CIP is a synthetic second generation bactericidal fluoroquinolone antibiotic which inhibits the activity of bacterial DNA gyrase, resulting in the degradation of bacterial DNA by exonuclease activity. Access to the target site is a major determinant of antibacterial activity of CIP, the outer membrane being the major permeability barrier in Gram-negative bacteria. CIP has broad-spectrum efficacy against a wide variety of bacteria, including *Staphylococcus aureus*, streptococci, *Pseudomonas aeruginosa*, *Klebsiella pneumoniae*, *M. tuberculosis* and *MAC*. CIP has been shown to have a superior ability to penetrate most tissues compared to other antibiotics (Nix et al. 1991; Kuhlmann et al. 1998), accumulates in macrophages (Easmon and Crane 1985a) and neutrophils (Easmon and Crane 1985b) and is bactericidal in low pH environment (Rastogi and Blom Potar 1990). These attributes contribute partly to CIP being the drug of choice for the treatment of infectious diseases caused by intracellular pathogens. Furthermore, CIP, when orally or i.v. administered, is known to reach such organs as liver, spleen, lungs and lymph nodes (Hanan et al. 2000), which are important infection sites for intracellular bacteria. However, CIP does not preferentially accumulate well at these tissues and may therefore not reach high sustain therapeutic levels at these sites. Despite its enormous success, there are some factors which limit CIP's clinical utility, such as its poor solubility at physiological pH, bitter taste in solution and rapid renal clearance. For example, in order to administer a typical ~0.5-g i.v. dose, the drug must first be diluted to <2 mg/ml and infused slowly to avoid precipitation at the site of injection (Webb et al. 1998). In order to increase CIP delivery to alveolar macrophages upon pulmonary administration, CIP was loaded in 1,000 nm mannose-modified liposomes (hydrogenated soybean PC: dioleoyl PC: chol: DCP: 4-aminophenyl-a-D-mannopyranoside, 3.5:3.5:2:1:1 M ratio) and its performance *in vivo* after inhalation was evaluated (Chono et al. 2008). Mannosylated liposomal CIP led to a sharp increase in the uptake by alveolar macrophages from approximately 12–22% of dose/mg of cell protein. In addition, pharmacokinetic studies in rats indicated 1.5- and 2.4-fold greater AUC and maximum concentration values in alveolar macrophages, respectively, with the modified liposomes. Also, plasma levels were reduced by encapsulation as compared to free CIP. Finally, a preferential uptake of the liposomes by alveolar macrophages was achieved for the mannose-coated liposomes as compared to plain liposomes.

Biodistribution and Therapeutic Efficacy of Non Liposomal Inhaled/Instilled Nanoparticles

Between 2001 and 2009, the performance of RIF alone or together with other antibiotics, loaded in microspheres and PLGA nanoparticles and administered by nebulisation or instillation, was investigated.

For instance, RIF was loaded in 2.76–3.45 μm PLGA microspheres and insufflated or nebulizated directly to the lungs at 1–1.7 mg/kg in guinea pig infected with aerosolized low dose (2×10^5 CFU/ml) of virulent H37Rv strain of *M. tuberculosis.* There was a dose-effect relationship between insufflated RIF- loaded PLGA microspheres and burden of bacteria in lungs. In addition, guinea pigs treated with RIF- loaded PLGA microspheres had a significantly smaller number of viable bacteria, reduced inflammation and lung damage than lactose, saline control, empty PLGA microspheres or free RIF treated animals (Suárez et al. 2001). A further assay employed RIF, INH and PYZ loaded in mass median aerodynamic diameter of 1.88 ± 0.11 μm PLGA microspheres, and administered by nebulization at RIF 12 mg/kg, INH 10 mg/kg and PYZ 25 mg/kg bw to guinea pigs. A single nebulization resulted in sustained therapeutic drug levels in plasma for 6–8 days and in the lungs for up to 11 days (Pandey et al. 2003b). The elimination half-life and mean residence time of the drugs were significantly prolonged compared to oral free drugs, resulting in an enhanced relative bioavailability for encapsulated drugs (12.7-, 32.8- and 14.7-fold for RIF, INH and PYZ, respectively). The absolute bioavailability (compared to i.v. administration) was also increased by 6.5-, 19.1- and 13.4-fold for RIF, INH and PYZ, respectively. On nebulization of microspheres containing drugs to *M. tuberculosis* infected guinea pigs at every tenth day, no tubercle bacilli could be detected in the lung after five doses of treatment whereas 46 daily doses of orally administered drug were required to obtain an equivalent therapeutic benefit. Hence, these PLGA microspheres could be used for intermitent treatments.

In another approach, RIF was loaded in 3.5 μm PLGA microspheres that were i.t. administered to rats i.t. infected with 2×10^3 CFU with of *M. tuberculosis* Kurono, at 0.4 and 0.04 mg/kg bw (Yoshida et al. 2006). These RIF- loaded PLGA microspheres were superior to free RIF for killing of intracellular bacilli and preventing granuloma formation in some lobes.

Finally, RIF loaded in 213 nm PLGA nanoparticles were incorporated into mannitol microspheres in one single step by means of a four-fluid nozzle spray drier. Encapsulation of the nanoparticles in mannitol improved the *in vivo* uptake of the drug by alveolar macrophages in rat lungs as compared to RIF-loaded in PLGA nanoparticles and mannitol microparticles alone (Ohashi et al. 2009).

Microparticles other than PLGA were also employed in pulmonary delivery, with simmilar successs. For instance, RIF and INH were loaded in ~5 μm PLA microparticles by spray-drying. Microparticles showed satisfactory aerosol characteristics (median mass aerodynamic diameter 3.57 μm; geometric standard deviation 1.41 μm; fine particle fraction$_{< 4.6 \mu m} = 78.91 \pm 8.4\%$) and were administered to mice using an in-house inhalation apparatus or by i.t. instillation. About 10 mg microparticles were placed in the in-house inhalation apparatus and the animals inhaled the powder during 1930s. Microparticles targeted macrophages and not epithelial cells on inhalation. Drug concentrations in macrophages were ~20 times higher when microparticles were inhaled rather than free drug solutions. Microparticles were thus suitable for enhanced targeted drug delivery to lung macrophages (Muttil et al. 2007).

In the following study, a comparison between PLGA and alginate microparticles was established. RIF, INH and PYZ were loaded in mass median aerodynamic diameter 1.1 μm alginate microparticles and administered by nebulization (Ahmad et al. 2005). Drug concentrations were detected in plasma after 3 h, which was faster than PLGA microparticles. Also, INH, RIF and PYZ were detectable up to 14, 10 and 14 days, respectively, as opposed to the fast clearance of the free drugs after 12–24 h. Nebulization of drugs-loaded alginate nanoparticles (one dose biweekly, three doses) over the course of 45 days was as effective in clearing the lungs and the spleen of *M. tuberculosis*-infected guinea pigs as 45 oral daily doses of the free drugs. Therefore, alginate microparticles could also be suitable for intermitent treatments.

Different cyclodextrins (CDs) were used to elaborate pharmaceutical formulations of hydrophobic drugs for the inhalation route. The short-term toxicity of such formulations administered by inhalation to C57BL/6 mice was tested (Evrard et al. 2004). Hydroxypropyl-β-CD (HP-β-CD), γ-CD, as well as randomly methylated β-CD (RAMEB) aqueous solutions can undergo aerosolization; the resulting droplet-size ranges were compatible with pulmonary deposition. *In vivo*, it was demonstrated that short-term exposure to inhaled HP-β-CD, γ-CD and RAMEB solutions are non-toxic after assessing bronchoalveolar lavage, lung and kidney histology, bronchial responsiveness to methacholine and blood urea. The only change noted was a slight increase in lymphocyte count in the bronchoalveolar lavage after HP-β-CD and γ-CD inhalation. Therefore, CDs are useful in significantly enhancing the solubility of apolar drugs with a view to inhalation therapy although an increase in lymphocyte counts in the bronchoalveolar lavage after CDs inhalations needs further investigations.

3 Salmonella spp

Salmonella enterica are facultative intracellular, motile Gram negative anaerobes bacteria that cause a variety of diseases in humans and animal hosts. Salmonella serovar falls into two general categories: those that cause enteric (typhoid) fever in humans (*S. typhi* and *S. parathyphi*) and those that do not (*S. enteritidis* and *S. typhimurium*). Enteric fever is a systemic illness characterized by a high sustained fever, abdominal pain and weakness. Millions of cases of enteric fever are reported annually throughout the world and without antimicrobial therapy the mortality rate is 15% (Hook 1990). Nontyphoidal serovars of *S. typhimurium* produce an acute gastroenteritis, characterized by intestinal pain and non bloody diarrhea, which is a serious public health problem in developing countries. Nontyphoidal salmonella are responsible for more deaths in USA than any other foodborne pathogen, with a mortality rate of nearly 1% and an estimated total number of annual cases of one to three millions. In infants, in the elderly, in immunocompromised or in response to certain serovar, nontyphoidal salmonella can also result in bacteriemia and establishment of secondary focal infections (meningithis, septic arthritis, or pneumonia) (McCormick 2004).

Most infections with *Salmonella* are contracted through contaminated food or water (Groisman 1997) and although it is very rare, direct person to person transmission can occur. Once ingested, most organisms are killed by the low pH of the stomach, but those that persist target the colon and small intestine (distal illeum) as a portal entry to the host (Galán et al. 1992). In mice, the primary sites of *Salmonella* invasion are the ileal Peyer's patches and possibly also the cecal lymphoid patches (Carter and Collins 1974). The follicle-associated epithelium overlying these gut-associated lymphoid tissues includes the specialized antigen-sampling M cells, which are a major site of invasion by diverse pathogens (Giannasca and Neutra 1993). Interaction of salmonellae with intestinal M cells is accompanied by loss of cell surface microvilli and remodelling of the apical membrane to form structures resembling membrane ruffles, and massive uptake of extracellular fluid in the form of macropinosomes (Clark et al. 1996; Finlay 1994; García del Portillo and Finlay 1994). Bacterial invasion correlates with the appearance of transient *Salmonella*-surface appendage-like structures termed "invasomes" (Ginocchio et al. 1994), formation of filamentous endocytic vacuoles that contain lysosomal membrane glycoproteins (García del Portillo et al. 1993) and the intracellularly replicating bacteria.

Salmonella spp. pass through the intestinal epithelial barrier to gain access to mesenteric lymph nodes from which point the bacteria are eventually transported to the bloodstream and more distant sites of infection such as the liver and spleen (Hook 1990). Once colonized, the bacteria may then potentially disseminate to the lungs, gallbladder, kidneys or central nervous system. The nontyphoid species of *Salmonella* tend to produce a more localized response where they elicit an acute inflammatory response through dendritic cells that manifests itself as fever, diarrhea and abdominal cramping. This inflammatory host response can actually benefit the intestinal pathogens and contribute to the nature and severity of the infection by establishing a competitive advantage against the native flora (Owens and Warren 2010). However, the *typhi* serotype can develop the more invasive disease resulting in bacteremia. The severity of disease is related to the serotype, number of organisms, and host factors (Finlay 1994; Owens and Warren 2010).

Soon after entry, the Salmonella-containing vacuole (SCV) displays markers of early endosomes. The SCV recruits Rab7 and a subset of lysosomal markers (LAMP1 and vATPase) by selective fusion with late endosomes and/or lysosomes. A small portion of intracellular Salmonella exits the SCV and resides in the host cytoplasm. These bacteria are targeted for ubiquitination (Perrin et al. 2004) and subsequently cleared by autophagy (Birmingham et al. 2006).

3.1 Conventional Treatment

Antibiotic treatment for nontyphoidal salmonellosis is not recommended because it may prolong the illness (Grassl and Finlay 2008; McCormick 2004; Owens and Warren 2010).

Antimicrobial therapy is indicated for patients with typhoid fever and should be considered on a case-by-case basis to include patients with severe symptoms (Wiström et al. 1992). Antibiotics are currently indicated for infants up to 2 months of age, elderly, immunocompromised, those with a history of sickle cell disease or prosthetic grafts, or patients who have extraintestinal findings. Treatment of those at-risk patients should last 2–5 days or until the patient is afebrile (McCormick 2004).

Traditional first-line antibiotic medications include ampicillin, chloramphenicol, and trimethoprim-sulfamethoxazole. *Salmonella* antibiotic resistance is a global concern that includes multi-drug resistant strains (Weinberger and Keller 2005). Resistance to these first-line antibiotics defines multi-drug resistance in *S. enterica* (Crump and Mintz 2010). *Salmonella* infections are commonly treated with fluoroquinolones such as CIP Despite the increase in CIP resistance in typhoid and paratyphoid, it is still considered the drug of choice by many physicians. However, in the case of treatment failures, a third-generation cephalosporin such as ceftriaxone and macrolide are good alternatives (Owens and Warren 2010; Threlfall et al. 2008).

3.2 Preclinical Delivery Systems

3.2.1 Biodistribution and Therapeutic Efficacy of Parenteral Liposomes

Early work was focused in the encapsulation of aminoglycosides in multilamellar vesicles. In one of them, GEN was loaded in ~1 μm oligolamellar liposomes (partially hydrogenated eggPC: eggPG: chol: alpha-tocopherol) and i.v. administered to mice orally infected with *S. dublin*. Of ten mice, eight survived after a single i.v. injection of 2 mg of liposomal GEN /kg compared with 0 of 10 treated with the same amount of free GEN. All mice survived after treatment with a single i.v. or i.p. injection of 20 mg of liposomal GEN/ kg, whereas that dose of free drug was completely ineffective and caused neuromuscular paralysis when injected rapidly i.v. In mice treated with liposomal GEN, there was a steady decrease in the number of salmonellae in spleens for 2 weeks after treatment. High concentrations of GEN were present in the spleen for at least 10 days after treatment. Although GEN was not detected in the mesenteric lymph nodes of mice treated with liposomal GEN, bacterial counts in the nodes also decreased over time. Small numbers of bacteria remained viable in the mesenteric lymph nodes and Peyer's patches but not in the spleens of mice treated with 20–80 mg/kg. Mice treated with doses of liposomal

GEN as high as 80 mg/kg showed only a transient increase in blood urea nitrogen and no rise in serum creatinine. These results confirm that liposomal GEN is less toxic in mice than is free GEN and is extremely effective-therapy for disseminated Salmonella infections in mice (Fierer et al. 1990).

In a simmilar approach, GEN was loaded in 1.2–10.0 μm MLV (eggPC) and i.v. administered to rats and mice at 20 mg/kg. The half-life of liposomal GEN in serum was substantially prolonged (mean half-lives in serum of free GEN in mice and rats were 1.0 and 0.6 h, respectively, whereas liposomal GEN had apparent mean half-lives of 3.8 h in mice and 4.0 h in rats) and the apparent rate of elimination and clearance of active drug from the blood was dose dependent. As the total dose increased, the rate of elimination and systemic clearance decreased (the apparent half-life in serum of liposomal GEN increased as the dose Increased). This was consistent with a saturable process (such as phagocytosis by fixed or circulating cells) as a primary mechanism in the removal of drug from the blood. Liposome encapsulation resulted in higher and more prolonged activity in organs rich in RES cells (especially spleen and liver). In a model of murine salmonellosis where infection is by the i.v. route with *S. typhimurium*, the initial septicemia is followed by localization of the organism within the major RES organs, liposomal GEN greatly enhanced survival when given as a single dose (10 mg/kg) at 1 or 2 days after infection as well as up to 7 days before infection. The mechanism of the enhanced therapeutic effect of liposomal GEN in this model was probably related to targeting of the antibiotic to the organ, cellular, and possibly subcellular site of bacterial residence (Swenson et al. 1990).

Two further works determined the biodistribution and therapeutic efficacy of liposomal fluoroquinolone CIP. For instance, CIP was loaded in MLV (DPPC: DPPG: chol, 4.1:0.9:4 M ratio) and i.v. injected in BALB/c mice infected per os with the fatal *S. dublin* at 20 mg/kg. Untreated, the infection begins in Peyer's patches, spreads to the regional lymph nodes, and then disseminates to the liver and spleen, leading to bacteremia and death 10–14 days after infection. A single administration of liposomal CIP was ten times more effective than a single sc injection of free CIP at preventing mortality and the result was nearly identical to the one achieved with comparable doses of free CIP given twice daily for 5 days. Treatment with liposomal CIP produced dose-dependent decreases in bacterial counts in spleen, stool, and Peyer's patches, indicating that the drug had distributed to all areas of inflammation, not just to the major reticuloendothelial system organs. Although liposomal CIP was cleared rapidly from the blood, drug persisted in the liver and spleen for at least 48 h after administration of a dose of liposomal CIP. The improved survival may have been due to the persistence of CIP in the liver and spleen for at least 2 days after a single injection of 20 mg/kg. This had the effect of prolonging therapy well beyond the time when the drug had disappeared from the circulation. In contrast, at 8 h after administration of the same dose of free CIP, there was no measurable drug in the spleen (Magallanes et al. 1993). In general, a drug that is actively retained inside a liposome by a transmembrane ion gradient leaks more slowly than a drug that has been passively encapsulated (Hope and Wong 1995; Cullis et al. 1997). CIP pharmacokinetics, accumulation at infection site, and antibacterial efficacy are significantly altered both by the liposome size (Hwang 1987; Zou et al. 1995) and by the rate of drug leakage. In the following study, the zwitterionic molecule CIP was actively loaded of into three different unilamellar liposomes (DPPC: chol; distearoylphosphatidylcholine (DSPC): chol or sphingomyelin (SM): chol) with a transmembrane gradient of methylammonium sulfate. Drug pharmacokinetics after i.v., i.p. and i.t. administration and antibacterial efficacy of liposomal CIP at 20 mg/kg against an *S. typhimurium* infected mice was determined. Intravenous liposomal CIP increased the circulation lifetime of CIP by >15-fold. Increased circulation lifetimes were associated with enhanced delivery of the drug to the liver, spleen, kidneys, and lungs of mice. The retention of CIP in liposomes in the circulation decreased in the sequence SM: chol > DSPC: chol > DPPC: chol. Encapsulation of CIP also conferred significant increases in the longevity of the drug in the plasma after i.p. administration and in the lungs after i.t. administration in comparison to free CIP. At 20 mg/kg SM: chol liposomal CIP resulted in 10^3- to 10^4-fold fewer viable bacteria in the liver and spleens of infected mice than was observed for animals treated with free CIP (Webb et al. 1998).

Finally, GEN was loaded in pH-sensitive liposomes (dioleoylphosphatidylethanolamine (DOPE): N-succinyl–DOPE: chol: Peg-ceramide, 35:30:30:5% mol) especially formulated against pathogens residing in the cytosol. The pharmacokinetics and biodistribution of the free and liposomal GEN were examined in mice bearing a systemic *S. enterica* serovar Typhimurium infection. Encapsulation of GEN in pH-sensitive liposomes significantly increased the concentration of GEN in plasma compared to those of free GEN, precluded accumulation in kidneys and redirected the antibiotic to the liver and spleen. While the accumulation of drugs in the liver and spleen is well known for liposomal formulations, the rate of pH-sensitve liposomal GEN accumulation in these organs was substantially faster than that for neutrally charged DPPC-chol liposomes. Furthermore, the levels of accumulation of drug in the infected liver and spleen were increased by 153- and 437-fold, respectively, as a result of liposomal encapsulation. The increased accumulation of GEN in the liver and spleen affected by liposomal delivery was associated with 10^4-fold greater antibacterial activity than that associated with free GEN in a murine salmonellosis model. However, the antibacterial effects observed with the pH-sensitive liposomes were also achieved with two different nonfusogenic control formulations of GEN. Hence, the therapeutic benefits that were achieved by fusogenic carrier-mediated intracellular delivery of GEN were outweighed by those benefits that arise from the substantial increases in the AUC for the antibiotic at the infection site for both fusogenic and nonfusogenic liposomes (Cordeiro et al. 2000; Kumana and Yuen 1994). It should be noted that the numbers of bacteria that survived in the spleens of mice treated with liposomal gentamicin were similar to those that were present at the start of gentamicin therapy (Richter-Dahlfors et al. 1997). Therefore, it cannot be excluded that the liposomal formulations may be exerting cytostatic rather than cytotoxic effects. However, since the infection is exclusively present intracellularly, these data demonstrate that the use of these liposomal formulations enhanced the intracellular delivery of the drug to increase either the cytotoxic or the cytostatic activities of the drug compared to those achieved with free GEN.

3.2.2 Biodistribution and Therapeutic Efficacy of Nanoparticles

Early work employed ampicillin, that was loaded to ~190 nm polyisohexylcyanoacrylate (PIHCA) nanoparticles and administered to C57BL/6 mice experimentally infected with *S. typhimurium* C5 (Fattal et al. 1989). All control mice and all those treated with nonloaded nanoparticles died within 10 days of infection. By contrast, all mice treated with a single injection of 0.8 mg of nanoparticle-bound ampicillin survived. With free ampicillin, a similar curative effect required three doses of 32 mg each. Lower doses delayed but did not reduce mortality. The therapeutic index of ampicillin, calculated on the basis of mouse mortality, was therefore increased by 120-fold when it was bound to nanoparticles, probably due to its delivery to liver and spleen, the primary foci of infection in the experimental model that was used. However, neither the liver nor the spleen was sterilized in any of the surviving mice, even 60 days after injection.

More recently, azithromycin was loaded into PLGA nanoparticles and its antibacterial activity *in vitro* was examined against *S. typhi*. The results showed that physicochemical properties were affected by drug to polymer ratio. Zeta potential of the nanoparticles was fairly negative. *In vitro* release study showed two phases: an initial burst for 4 h followed by a very slow release pattern during a period of 24 h. The results of antimicrobial activity test showed that the nanoparticles were more effective than pure azithromycin against *S. typhi* with the nanoparticles showing equal antibacterial effect at 1/8 concentration of the free drug. In conclusion, the azithromycin-loaded PLGA nanoparticles showed appropriate physicochemical and improved antimicrobial properties which can be useful for oral administration (Mohammadi et al. 2010).

Lately, a series of non convencional micro and nanoparticles were employed as carriers for aminoglycosides. Most of the assays were proofs of principle and their performance against typhoid fever was not determined.

For instance, the encapsulation of antibiotics in carrier erythrocytes is able to give rise to a sustained release of the antibiotic, since the plasma half-life and the area under the curve of the antibiotic are increased, together with possible RES targeting when the loaded erythrocytes are phagocytosed (Gening and Manuilov 1991; Gutiérrez Millán et al. 2005). AMK was loaded into glutaraldehyde-treated erythrocytes as a delivery system and i.v. administered to rats at 2.7 +/– 1 mg/kg to determine pharmacokinetics and tissue distribution (Gutiérrez Millán et al. 2008). AMK is a polar drug and, once entrapped inside the erythrocyte, it does not undergo processes of diffusion across the cell membrane owing to its polar nature. Thus, release must occur through a process of cell lysis, as happens with other aminoglycosides.

In another approach, GEN cores were prepared by complexing the antibiotic with polyacrylate anions that were further incorporated into nanostructures made of having either amphiphilic N1 (containing poly(ethylene oxide), poly(propylene oxide) and polyacrylate) or hydrophilic N2 (containing polyacrylate) surfaces and incubated with cells (Ranjan et al. 2009; Tian et al. 2007). Flow cytometry and confocal microscopy studies demonstrated higher rate of uptake for amphiphilic surfaces. The majority of N1 were localized in the cytoplasm, whereas N2 colocalized with the endosomes/lysosomes. Colocalization was not observed between nanostructures and intracellular *Salmonella* bacteria. However, significant *in vitro* reductions in bacterial counts (0.44 $\log_{10}$) were observed after incubation with N1, suggesting that the surface property of the nanostructure influences intracellular bacterial clearance. In view of these results, GEN loaded in the same core-shell nanostructures (nonionic amphiphilic shells N1 and ionic cores) and employed for therapy against intracellular pathogens, including *Salmonella* (Ranjan et al. 2010). The intracellular pathway of GEN-N1 nanostructures was determined by flow cytometry and confocal microscopy, finding that their uptake into J774A.1 macrophages proceeded mainly by fluid-phase endocytosis and clathrin-mediated pathways. The nanostructures were nontoxic in vitro at doses of 50–250 μg/ml, and they significantly reduced the amounts of intracellular *Salmonella* (0.53 log).

Finally, GEN was uniformly loaded in a porous (1.7–3.3 μm pore diameter) biodegradable silica xerogel matrix for the prolonged release for treatment of Salmonella infection in a mouse model. Three ip doses administered to mice of porous silica loaded with GEN reduced the CFU of *S. thyphimurium* in livers of infected mice by 0.48 log compared to 0.13 log with free drug (Munusamy et al. 2009).

4 Francisella

F. tularensis is a highly virulent and contagious, facultative intracellular bacterium. It causes the zoonotic disease tularemia in a large number of mammals and can be a potentially fatal human disease if untreated. *F. tularensis* is a potential biological warfare/bioterrorism agent, and medical and public health management of the infectious disease caused by this bacterial agent is important (Dennis et al. 2001). The bacteria are spread via multiple transmission routes including arthropod bites, physical contact with infected animal tissues, contaminated water and inhalation of aerosolized organisms. Inhalation of as few as 10 CFU are sufficient to initiate lung colonization and the subsequent development of pulmonary tularemia, which is the most lethal form of the disease exhibiting mortality rates as high as 60% (Fuller et al. 2009).

Infection by *F. tularensis* involves the RES system, dendritic and alveolar epithelial cells and leads to bacterial growth within the lungs, liver and spleen (Fortier et al. 1991). Following uptake or invasion of a host cell wild type, *F. tularensis* cells escape the phagosome and replicate within the cytoplasm of infected cells (Clemens et al. 2004; Fortier et al. 1995; Golovliov et al. 2003; Santic et al. 2005). Following a period of multiplication where relatively high numbers of bacteria are produced when compared to other pathogens such as *Salmonella*, intracellular *F. tularensis* induces apoptosis leading to death of the infected macrophage and release of the bacteria into the extracellular environment (Lai et al. 2001). Extracellular bacteria are subsequently thought to infect naive macrophages and bacterial multiplication begins again.

Initially, bacteria colocalized with the late endosomal/lysosomal markers LAMPs, but not with cathepsin D. Moreover, the phagosomes containing

F. tularensis bacteria were not significantly acidified (Clemens et al. 2004). *F. tularensis* bacteria are capable to alter the maturation of the phagosome, as evidenced by the exclusion of cathepsin D, lysosomal tracers, and the lack of acidification, and then, by an unknown mechanism, disrupt the phagosomal membrane.

4.1 Conventional Treatments

Tularemia in humans can be treated with antibiotics, including streptomycin, gentamicin and chloramphenicol, but even with prolonged daily therapy, relapse and failure rates can range from 0% to 33% (Enderlin et al. 1994).

4.2 Preclinical Delivery Systems

CIP was encapsulated in 100 nm large unilamellar vesicles (LUV) by remote loading using an ammonium sulfate gradient (Wong et al. 2003). Biodistribution of liposomal CIP upon i.v. administration or aerosol inhalation in mice were determined using ^{14}C-CIP. Intravenous administration of liposomal CIP resulted in higher serum levels of drug in serum, as well as increased drug retention in lungs, liver and spleen, compared to that of free drug. Liposomal CIP has been shown to be stable when aerosolized with no measurable liposome disruption (Finlay and Wong 1998). Aerosol administration of liposomal CIP by jet nebulization resulted in significantly higher drug levels and prolonged drug retention in the lower respiratory tract compared to the free drug. Aerosol inhalation of liposome-encapsulated CIP, given either prophylactically or therapeutically, provided complete protection to BALB/c mice against a pulmonary lethal infection of *F. tularensis*. In contrast, CIP given in its free form was ineffective. The aerosol delivery of liposomal CIP provided very effective prophylaxis and post exposure treatment of pulmonary *F. tularensis* infection in mice, while aerosolized free CIP, provided little or no protection. The enhanced therapeutic efficacy can be partly attributable to increased lung retention of CIP provided by the sustained release from the liposomes, as well as to enhanced intracellular delivery of CIP by liposomes. Previous results showed the total eradication of intracellular bacteria from the lungs, spleen and liver (Conley et al. 1997), which serve as the primary infection sites for these agents.

5 Chlamydia

Chlamydophila pneumoniae, C. psittaci and *C. trachomatis* are small, gram-negative, obligate intracellular organisms pathogenic to humans. The three species can cause pneumonia in humans. *C. pneumoniae* is responsible for respiratory infections such as sinusitis, bronchitis, and 6–12% of community-acquired pneumonias (Grayston et al. 1989, 1993), as well as chronic, persistent and often asymptomatic infections (Hammerschlag et al. 1992) Persistent *C. pneumoniae* infection has been associated with chronic human diseases, including atherosclerosis, arthritis and asthma (Cook et al. 1998; Danesh et al. 2000).

All *Chlamydiae* undergo an unusual biphasic developmental cycle, which is of great relevance when considering the design of novel drug delivery strategies (Whittum-Hudson and Hudson 2005). After the attachment of the infectious form of the organism to the outer membrane of susceptible host cells, the organism subsequently produces cytoplasmic inclusions known as elementary body. These elementary bodies reside within a vacuole termed an inclusion and subsequently develop into the non-infectious, metabolically active form called the reticulate body. Reinfection occurs after reticular bodies dedifferentiate back to elementary bodies and are released back into the tissue milieu by host cell lysis. The three species can cause systemic disease by hematogenous spread. Respiratory secretions transmit *C. pneumoniae* from human to human.

5.1 Conventional Treatments

Treatment is empirical. Doxycycline is the drug of choice except in children younger than 9 years and in pregnant women. If symptoms persist, a second course with a different class of antibiotics is usually effective. Telithromycin is the first antibiotic in a new class called ketolides and is approved for *C. pneumoniae* pneumonia by the US Food and Drug Administration. Fluoroquinolones, including levofloxacin and moxifloxacin, also have some activity, although less than that of tetracyclines or macrolides (Yuji 2010).

5.2 Preclinical Delivery Systems

Core-shell polymeric microparticles consisting of water soluble, hydrofluoroalkane-philic biodegradable copolymer of chitosan, and PLA shell and a core of PLGA nanoparticles were employed for pulmonary delivery by MDI. The PLGA nanoparticles were entrapped within these micron-sized particles whose shell is designed to be well-solvated by the propellant, and thus to provide adequate physical stability to the formulation, and consequently good aerosol characteristics. Dispersions of the core-shell particles in hydrofluoroalkane propellant revealed enhanced physical stability compared to PLGA nanoparticles alone, and more importantly, excellent aerosol characteristics as determined by inertial impaction studies. The shell is soluble in water and breaks down releasing the core. These microparticles were specially designed to target region in the upper airways instead of the alveolar region. In the upper airways, microparticles break down, the core is released and penetrates mucus; its uptake by underlying epithelium is enhanced. The core-shell particles thus formed provide for an opportunity to efficiently deliver nanoparticles to the lungs, by enhancing their dispersibility in the propellant, and by providing the appropriate aerodynamic size so as to enhance the aerosol characteristics of the corresponding MDI (fine particle fraction of 72% with a mass median aerodynamic diameter of 2.3 mm). *In vitro* results reveal that nanoparticles from such core-shell formulations can be readily internalized by Calu-3 (airway epithelial) cells infected with *C. pneumoniae*, and more importantly, they can gain access to chlamydial inclusions (Bharatwaj et al. 2010; Lai et al. 2007).

6 Plasmodium spp

Malaria is responsible for >300 million cases and more than one million deaths each year (Gardella et al. 2008) and is caused by four species of *Plasmodium* (*P. falciparum, P. vivax, P. malariae and P. ovale)* that enter the humans by the bites of female mosquito of *Anopheles* genus. *P. falciparum* infection (80% of malaria cases) cause severe malaria; neurological symptoms occur and may result in death if treatment is not promptly instituted, especially in acute primary infections (Whitty and Sanderson 1999). The pathogenesis of cerebral malaria is multifactorial, being caused by parasite sequestration and blockage of blood flow in small vessels of the brain that includes clogging, sequestration, rosette formation, release of cytokines, cerebral oedema, increased intracranial hypertension, etc. (Coltel et al. 2004; Pouvelle et al. 2007). *P. vivax* and *P. ovale* infection cause chronic malaria. The disease can relapse months or years after exposure, due to the presence of dormant liver stage parasites, named hypnozoites. Reactivation has been reported for up to 30 years after the initial infection in humans. *P. malariae* produces long-lasting infections and if untreated can persist asymptomatically in the human host for years, even a lifetime (Krettli et al. 2001).

The inoculated sporozoites migrate to the liver and invade the hepatocytes by mechanisms still not entirely understood (Krettli and Miller 2001). A sporozoite generates about 30,000 new invasive forms of the parasites (merozoites) in 9–10 days in the case of *P. falciparum* and 10,000 in 6–8 days in *P. vivax*. Released merozoites enter erythrocytes and remain relatively inactive metabolically for 10–15 h (the ring stage). The parasite then undergoes a rapid phase of growth over the next 25 h, forming the trophozoite stage, during which the parasite digests the majority of the haemoglobin and grows to fill >50% of the volume of the host cell. At the end of the trophozoite stage the parasite divides several times (the schizont stage) before the host cell lyses (some 48 h after invasion) to release the newly formed merozoites that continue the cycle. Some merozoites turn into male and female gametocytes that also invade erythrocytes. This concerted cell lysis triggers the classical malaria symptoms that are mainly headache, periodically recurrent high fever (every 48–72 h), myalgia, anemia, hepato- and splenomegaly (Greenwood et al. 2008; WHO 2006).

Plasmodium invades cells by pulling itself into the cell by an actin-myosin motor that is present in their own cytoskeleton. Then a vacuola is formed by the host plasma membrane. Parasites resist phagosome-lysosome fusion by exclusion of host-cell integral membrane proteins that regulates the endocytic/phagocytic pathway from these vacuola, whereas parasite's proteins are incorporated. This result in the formation of a unique PV enclosed by a semi-permeable membrane with channels or pores with a size exclusion limit of <1,300–1,400 Da (Desai et al. 1993; Schwab et al. 1994; Desai and Rosenberg 1997; Werner-Meier and Entzeroth 1997). It is important to note that following infection by *Plasmodium*; membrane channels called 'new permeability pathways' (NPPs) appear on host plasma membrane that allow the passage of solutes to the intracellular space (Biagini et al. 2005). Macromolecules such as dextran, A protein and IgG2a (Pouvelle et al. 1991) and also latex nanoparticles of 50–80 nm have access to parasite through these channels (Goodyer et al. 1997).

Additionally the PV membrane of *Plasmodium* extends into the host cell cytoplasm and forms a network of tubulovesicular membranes (TVM). The TVM acts as a molecular sieve which forms a junction with the erythrocyte membrane which allows import of specific nutrients such as adenosine, glutamate and orotic acid to the parasite (Lauer et al. 1997).

The erythrocytic stages (trophozoites, schizontes and gametocytes) living inside PV and the exoerythrocytic stages (intra-hepatocytes hypnozoites) present major structural and phenomenological barriers that antimalarial drugs have to overcome.

6.1 Conventional Treatments

Therapy depends on *Plasmodium* specie, severity of infection, previous exposition to infection, clinic status of patient and area where the infection was acquired and its drug-resistance status. Most drugs are active against the parasite forms in the blood and include: chloroquine, atovaquone-proguanil, artemether-lumefantrine, mefloquine, quinine, quinidine, doxycycline (used in combination with quinine), clindamycin (used in combination with quinine) and artesunate (Santos-Magalhães and Furtado Mosqueira 2010). The main drawbacks of conventional malaria chemotherapy are the development of multiple drug resistance and the non-specific targeting to cells free from intracellular parasites, resulting in high dose requirements and subsequent intolerable toxicity.

The four-aminoquinolone chloroquine phosphate was until recently the most widely used anti-malarial. It is accumulated in the food vacuoles (pH 5.2–5.8) and acts by inhibiting the hemozoin biocrystallization, thus facilitating an aggregation of cytotoxic heme. The main advantages of chloroquine therapy are the fast action in blood parasite stages, low toxicity, good bioavailability from oral dosage form, water solubility, high volume of distribution in the body and low cost. The main disadvantage is the widespread resistance due to a very long terminal elimination half-time (1–2 months).

The eight-aminoquinolone primaquine is orally used in the treatment (prophylactic or therapeutic) of all types of malaria infection. It remains as the only drug that destroys late hepatic stages and latent tissue forms of *P. vivax* and *P. ovale* and it is effective against gametocytes in only one dose. The main disadvantages are the dose-limiting severe toxicities and side-effects (hemolytic anemia, methemoglobinemia and hemeolysis in patients with G6PD deficiency) and limited oral availability.

The alkaloid quinine is used as blood schizonticidal and weak gametocide against *P. vivax* and *P. malariae.* Quinine is less effective and more toxic as a blood schizonticidal agent than chloroquine but it is especially useful in areas where high level of resistance to chloroquine, mefloquine, and sulfa drug combinations with pyrimethamine.

Artemsinin is a sesquiterpene lactone isolated from the plant *Artemisia annua,* with an endoperoxide bridge, that is is thought to be responsible for their action and is used against all forms of multi-drug resistant *P. falciparum* under the control of WHO. The semi-synthetic artemisinin derivatives artesunate (hemisuccinate), artemether (methyl ether) and arteether (ethyl ether) are easier to use than the parent compound and are converted rapidly once in the body to the active compound dihydroartemesinin. However the therapeutic potential of derivatives is considerably hampered due to: (i) short half-life usually between 3 h and 5 h; (ii) poor aqueous solubility, and thus low oral bioavailability (40%); (iii) risk of degradation in acidic conditions; (iv) associated risk of toxicity and (v) the i.m. injections currently available in the market are associated with low patient compliance and inconsistent assimilation (Akinlolu and Shokunbi 2010). The mainstay of cerebral malaria therapy is either quinine or artemisinin.

Halofrantine (phenanthrene methanol) is used as blood schizonticide effective against all plasmodium parasites. Cytotoxic complexes are formed with ferritoporphyrin XI that cause plasmodial membrane damage. Despite being effective against drug resistant parasites, halofantrine is not commonly used due to its high cost, variable bioavailability and potentially high levels of cardiotoxicity.

6.2 Preclinical Delivery Systems

6.2.1 Non-targeted Systems

Recently Santos-Magalhães and Furtado Mosqueira have thoroughly reviewed the latest preclinical nanomedical developments against malaria (Santos-Magalhães and Furtado Mosqueira 2010). Most of the approaches however, were based on changing antimalarial's pharmacokinetics by drugs loading in liposomes, nanocapsules and lipid nanoparticles. A minoritary fraction of articles were focused on targeting to infected erythrocytes or hepatocytes and subsequent cytoplasmic delivery.

The main findings can be grouped according to the delivery strategy:

i. *Modified biodistribution.* The first works in the 1980s, found that primaquine loaded in neutral or anionic MLV (PC: PS: chol, 4:1:5 M ratio) was increasedly accumulated in liver and spleen, while its concentration was decreased in other organs (lungs, kidney, heart and brain), upon i.v. administration. However, the decreased toxicity of primaquine was not followed by an increase of anti-malaria activity. This was due to the preferential capture of high sized MLV by Kupffer cells in the liver and spleen macrophages instead and not by infected hepatocytes (Smith et al. 1983; Pirson et al. 1982).

ii. *Increased half life upon i.v. administration.* The first work at the end of the 1990s showed that i.v. administered arteether loaded in MLV (DPPC: dibehenoylphosphatidylcholine: chol, 1:1:2 M ratio) increased its half life in circulation and also upon oral administration the bioavailability was increased in rabbits (Bayomi et al. 1998). Further work employed nanocapsules, for instance made of copolymer (PLA-PEG) polymeric wall nanocapsules and medium chain triglycerides core, to load halofrantine. These nanocapsules showed simmilar or higher antimalarial activity than that of free drug, after a single dose to mice severely infected with *P. berghei* (Mosqueira et al. 2004). Halofantrine in nanocapsules had also six folds higher plasma AUC than free drug (up to 70 h). It also allowed a faster control of parasitemia in the absence of toxic effects up to a maximal dose of 100 mg/kg, while toxicity of free halofrantine dissolved in PEG-dimethylacetamide was shown at lower concentration. Halofantrine loaded in poly-ε-caprolactone showed better results than PLA-PEG nanocapsules at a dose of 1 mg/kg in *P. berghei* infected mice and a significant reduction in halofantrine cardiotoxicity, with a reduction of the QT interval prolongation of ECG in rats (Leite et al. 2007). On the other hand, quinine loaded in poly-ε-caprolactone and polysorbate 80 nanocapsules showed increased efficacy against *P. berghei* infected rats compared to the free drug. However, pharmacokinetic parameters of quinine in nanocapsules were similar to those of free quinine. It was observed that nanoencapsulation doubled the drug penetration into red blood cells of *P. berghei*-infected rats, justifying the improvement in quinine efficacy when nanoencapsulated (Haas et al. 2009). Recently, artemether loaded in solid lipid nanoparticles (consisting of a solid matrix made of a blend of solid lipid trimyrstin (triglyceride of C14) with the liquid soybean oil) at 5 mg/kg body weight, significantly prolonged the survival period of *P. berghei* infected mice in comparison to the untreated group, and to those treated with artemether solution and marketed formulation. The three treated groups resulted in an initial decrease of parasitemia; parasitemia however only reappeared in the groups treated with marketed formulation and artemether solution. This result was attributed to the bioavailability of the drug when formulated as lipid nanoparticles, provided by a burst release followed by a sustained release. In this way, the fast metabolism of artemether could be minimized reducing thereafter the production of high toxicity levels of dihydroartemisinin (Aditya et al. 2010a).

iii. *Sustained drug release.* Chloroquine loaded in anionic MLV (DPPC/DPPG) led to a dozen-fold increase of maximum permissible dose, as compared to free drug and a 100% efficacy on chloroquine-resistant *P. berghei* infected mice upon i.p. administration (Peeters et al. 1989a, b). On the other hand, multiple s.c. administration of desferroxamine B, (DFO, an iron chelate) loaded in 300–500 nm liposomes at 800 mg DFO/kg/day at days −1 and 0, reduced parasitemia and long-term survival of mice infected with *P. vinckei* (Postma et al. 1998). Finally, artemether loaded in MLV (EggPC:chol, 4:3 M ratio) caused 100% cure by clearing the recrudescent parasitaemia of mice infected with the virulent rodent malaria parasite *P. chabaudi chabaudi* upon i.p. administration (Chimanuka et al. 2002).

Recently, nanostructured lipid carriers (NLC) have been explored as antimalarial drug depots. In a first approach, artemether loaded in 60 nm NLC made of glyceryl dilaurate (solid lipid), Capmul® MCM (liquid lipid), tween 80 and Solutol HS 15 were i.p. administered at 0.15 g (equivalent to the marketed i.m. formulation) to mice infected with *P. berghei*. Artemether NLC was responsible for an increased survival as compared with the same dose of i.m. commercial formulation (Joshi et al. 2008a). In the second work, curcuminoids (hidrofobic, thermolabile and chemically sensitive molecules) were loaded in NLC made of trimyristin, tristerin and glyceryl monostearate as solid lipids and medium chain triglyceride as liquid lipid. Curcuminoids NLC increased the survival period of mice as compared to control and free curcuminoids treated groups after 3 i.p. doses in 3 consecutive days. Increased bioavailability of drug in its native form was due to sustained release from the lipid nanoparticles depository, which prevented the fast metabolism of curcuminoids (Aditya et al. 2010b).

iv. *Increased bioavailability upon oral administration*. Primaquine loaded in Miglyol nanoemulsions increased the AUC by 1.3 folds and produced higher drug levels in the liver as compared to free primaquine. Primaquine nanoemulsions also suppressed parasitemia (~93%) and extended survival of mice infected with *P. berghei* at doses 25% lower than free drug (Singh and Vingkar 2008). Artemether loaded in a self-microemulsifying drug delivery system (Mandawgade et al. 2008) and solid microemulsions (Joshi et al. 2008b) produced an improvement in antimalarial activity after oral administration to mice infected with *P. berghei* as compared to i.m. administration of commercial formulation.

v. Increased brain delivery. The transferring (Tf) receptor is highly polarized at the blood-brain barrier and is localized only on the apical membrane of endothelial cells. The evidence supports an iron transport model in which iron-loaded Tf is taken up by receptor-mediated endocytosis at the luminal membrane of brain capillaries. The iron then dissociates from Tf in endosomal compartments and is transcytosed by unknown mechanisms, while the Tf is retroendocytosed (Roberts et al. 1993). On these bases, Tf-conjugated SLN were investigated for their ability to deliver quinine dihydrochloride to the brain, for the management of cerebral malaria (Gupta et al. 2007). Quinine dihydrochloride was loaded in (108–126 nm) Tf grafted to SLN (hydrogenated SoyPC, triolein, chol, DSPE). Upon i.v. administration, conjugation of SLNs with Tf significantly enhanced the brain uptake of quinine. A higher dose percentage of quinine was recovered from the brain following administration of Tf-SLNs compared with unconjugated SLNs or drug solution. However, not the presence of quinine in brain parenchima neither the antimalarial activity was determined.

6.2.2 Targeted Delivery to Erythrocytes

The first studies from the end of 1980s and early 1990s showed that chloroquine loaded in surface grafted anti-erythrocyte antibodies liposomes markedly increased its efficacy against both chloroquine-susceptible and chloroquine-resistant *P. berghei* infections in mice (Agrawal et al. 1987; Chandra et al. 1991). Later, a more elaborated approach employed F(ab´)$_2$ fragments of a mouse monoclonal antibody (MAb F10), raised against cell membranes isolated from *P. berghei*-infected mouse erythrocytes were covalently attached to the liposome surface (60 nm, egg PC: chol: gangliosides, 20:20: 4 M ratio) (Owais et al. 1995). Double radiolabelled liposomes by attaching ^{125}I-labelled F(ab)$_2$ fragments to the liposome surface, and entrapping [^{14}C]inulin in the aqueous compartment were i.v. administered to healthy and *P. berghei*-infected BALB/c mice, and its distribution was determined. MAb F10-liposomes did not significantly bind to the erythrocytes in healthy animals, but it readily recognized these cells in *P. berghei*-infected mice. MAb F10-liposomes containing chloroquine given once at 5.0 mg/kg effectively controlled not only the chloroquine-susceptible but also the resistant *P. berghei* infections. On the other hand upon two doses on days 4 and 6 post-infection, 75–90% of the animals survived the chloroquine-resistant *P. berghei* infections and no parasites were detectable in their blood even on day 30 post-treatment. Only 40–50% survived when liposomes were conjugated to an antibody non specifically targeted against erythrocytes, that recognized infected erythrocytes.

These studies were discontinued, owed probably to the structural complexity of the carrier construct that made difficult it's further scaling up.

6.2.3 Targeted Delivery to Hepatocytes

More recently, two different strategies could successfully target hepatocytes. The first of them was based on the use of reconstituted artificial chylomicron emulsion made of commercially available lipids (Dierling and Cui 2005). Lipoproteins are endogenous particles that transport lipids through the blood to various cell types, where they are recognised and taken up via specific receptors. For example, the remnant receptor and the asialoglycoprotein receptor (ASGPr), which are uniquely localised on hepatocytes, recognise chylomicrons and lactosylated high density lipopoteins (HDL), respectively. Being endogenous, lipoproteins are biodegradable, do not trigger immune reactions, and are not recognised by the reticuloendothelial system. Primaquine loaded into a chylomicron emulsion, was significantly retained by the emulsion and was more stable than free primaquine in the presence of serum. This lipophilized-primaquine was i.v. administered to mice and showed an enhanced accumulation in liver, as compared to free primaquine. The antimalarial activity was not determined.

The second strategy employed LUV grafted to sporozoite recognizement peptides. The liver specificity of this construct is attributed to two major surface proteins on the sporozoite, circumsporozoite protein (CSP) and thrombospondin-related anonymous protein (Ménard 2000; Mota and Rodriguez 2002). The CSP contains two conserved amino acid sequences (termed regions I and II) that are found in the various species that infect mammalian hosts. Recently, a peptide containing the conserved region I amino acids (KLKQP) plus the basic amino acid domain upstream from region I has been shown to bind strongly to affinity columns of heparin and heparan sulfate (Ancsin and Kisilevsky 2004). It is proposed that this 19-amino-acid sequence of the CSP of *P. falciparum* is at least partially responsible for Plasmodium targeting *in vivo* by binding to the highly sulfated heparan sulfate proteoglycans (HSPGs) found in liver (Ancsin and Kisilevsky 2004; Lyon et al. 1994). These HSPGs are located primarily on the basolateral side of the hepatocytes and within the space of Disse (Frevert et al. 1993; Pradel et al. 2002). These 19 aa peptide were attached to the distal end of a lipid PEG conjugate, and incorporated to 100 nm LUV (Longmuir et al. 2006). Due to the tendency of this targeting peptide to induce rapid aggregation when the liposomes were mixed with serum, a relatively high mole fractions of PEG5000 without peptide (10% mol PEG5000 plus 4% mol PEG3400-peptide) had to be added as well as employing long-chain C22:1 fatty acid as a lipid ingredient (82% mol). Fluorescently labelled liposomes (Bodipy-TR-X) or liposomes containing gold-particle-labeled phosphatidylethanolamine were i.v. administered to healthy mice. Fluorescence and electron microscopy demonstrated that the liposomes were accumulated both by non-parenchymal cells (endothelial and Kupffer cells) and hepatocytes, with the majority of the liposomal material associated with hepatocytes. Further works showed that liver cells uptake was more than 600-fold higher than uptake by those in the heart, and more than 200-fold higher than uptake by lung or kidney cells. Effective targeting to liver *in vivo* was successful after repeated (up to three) administrations to the host at 14-day intervals (Haynes et al. 2008). Intracellular localization of drugs as well as its antimalarial activity remained to be determined.

7 Leishmania spp

Leishmaniasis is caused by flagellated promastigotes of the *Leishmania* genus that is injected in the skin by sandflies. *Leishmania spp* infect 12–14 million people worldwide (Cunningham 2002). Depending on the leishmania genus, different clinical manifestations are produced: ulcerative skin lesions (cutaneous leishmaniasis, caused by *L. amazonensis* , *L. mexicana, L. guyanensis, L. panamensis, L. peruviana, L. venezuelensis and L. pifanoi* in America, *L. major, L. tropica and L. aethiopica* in Europe), destructive mucosal inflammation (muco-cutaneous caused by *L. braziliensis* in America), and disseminated visceral infection (kala-azar caused by *L. donovani* in India and Bangladesh, *L. infantum* in rest of Asia, Africa and Europe and *L. chagasi* in America) (Desjeux 2004; Herwaldt 1999). Visceral leishmaniasis is mortal if untreated; cutaneous leishmaniasis heals by reepithelisation with scarring, but it is treated to avoid disfiguring lesions; mucosal leishmaniasis can produce potentially life-threatening inflammatory disease and must be treated (Garcia and Bruckner 1997; Murray et al. 2005).

Once injected into the skin, the parasites invade the local phagocytic host cells, including neutrophils. Promastigotes, which are the infective form are acid sensitive and inhibit phagosome-endosome fusion through unique lipophosphoglycan molecules on their surface (Descoteaux and Turco 1999; Miao et al. 1995). Promastigotes transform into amastigotes and after that, the inhibition is lifted. Amastigotes, that are metabolically most active at acidic environment, depend on the harsh environment of the phagolysosome of resident macrophages where they survive and multiply (Joshi et al. 1993). Dissemination then occurs locally and distant macrophages are infected. In visceral leishmaniasis liver, spleen and bone marrow macrophages are colonized, in cutaneous and mucocutaneous leishmaniasis skin macrophages and dendritic cells including Langerhans cells are colonized.

The amastigotes living inside the macrophage's phagolysosomes located in different anatomical areas present major structural and phenomenological barriers that antileishmanial drugs have to overcome.

7.1 *Conventional Treatments*

Pentavalent antimonials (Sb^v) (meglumine antimoniate and sodium stibogluconate) were introduced in 1945 and remain effective treatment for some forms leishmaniasis, but the requirement for up to 28 days of parenteral administration, the variable efficacy against visceral and cutaneous leishmaniasis and the emergence of significant resistance are all factors limiting the drug's usefulness.

Diamidine pentamidine is used as a second-line drug when Sb^v have proved ineffective although it is highly toxic. Amphotericin B (AmB) (Fungizone, colloidal dispersion of amphotericin and deoxycholate), is highly effective for the treatment of Sb^v resistant *L. donovani* visceral leishmaniasis and cases of muco-cutaneous leishmaniasis that have not responded to Sb^v, but it highly toxic and needs slow parenteral infusion over 4 h. The development of lipid-associated amphotericine formulations has reduced toxicity an extended plasma half-life in comparison to the parent drug. Liposomal AmB (AmBisome) has proved to be effective and has been approved by the Food and Drug Administration, but high cost have limited its use (Croft and Coombs 2003).

7.2 *Preclinical Delivery Systems*

7.2.1 Visceral Leishmaniasis

From the middle ´1970s, a series of articles addressed the use of antileishmanial drugs loaded in liposomes for selective delivery to the liver and spleen macrophages infected with visceral leishmaniasis. Early applications started with liposomal pentavalent antimony (Sb^v), that was found to be 700 times more active than free drug (Alving et al. 1978; New et al. 1978), while MLV containing AmB were 170–750 and 60 times more active as active as Sb^v in *L. donovani* infected hamsters and monkeys, respectively (Berman et al. 1986). Once AmB in unilamellar liposomes was marketed, it was rapidly realized that AmBisome eliminated liver parasites to the same extent but faster than conventional treatment with Sb^v (Gradoni et al. 1993). AmBisome was effective in cases of Sb^v unresponsiveness in *L. infantum* and *L. donovani* foci, both in immunocompetent and in immunosuppressed patients, as well as being less toxic than other amphotericine lipid formulations such as Amphocil (mixed micelles of amphotericine and surfactant) (Minodier et al. 2003). A comparison of the efficacy and pharmacology of amphotericine formulations in treatment of leishmaniasis has been reviewed in (Barratt and Legrand 2005) and for recommendations on AmBisome doses and uses see Bern et al. (2006). Besides, several other drugs like inosine analogs, atovaquone, harmine and miltefosine, improved their performance compared with free drugs when loaded in plain liposomes and i.v. administered against experimental visceral leishmaniasis (see Romero and Morilla 2008).

Other strategies relied on the use of grafted liposomes in order to increase targeting to Kupffer cells. For instance, mannose-grafted liposomes that specifically recognised by receptors for terminal mannose residua on Kuppfer cells (Ghosh and Bacchawat 1980); tuftsin-bearing liposomes that specifically binds to macrophages and potentates their natural killer activity against pathogens (Agrawal et al. 2002; Guru et al. 1989) and IgG grafted-liposomes that are recognized by the Fc receptor on macrophages (Dasgupta et al. 2000), have shown improved activity over the free drugs on experimental visceral leishmaniasis, due to the increased macrophage uptake.

Niosomes, emulsions and micro/nano polymeric particles increased the effectivity of carried drugs upon parenteral administration to different diseased animal models (see Romero and Morilla 2008).

Since polymer conjugates are taken into cells by endocytosis and then trafficked through endosomes to lysosomes (Duncan 2003), antileishmanial drugs conjugated were employed as egntes for lysosomotropic delivery. The PV has many similarities to late endosomes/lysosomes and multiple vacuole trafficking pathways can intersect with *Leishmania* PV, therefore it is likely that the polymer–drug conjugates can also be trafficked to this compartment. Quinolone was linked to a biocompatible hydrophilic lysosomotropic polymer (*N*-(2-hydroxypropyl)methacrylamide, HPMA) bearing a 5% mole mannose targeting moiety. The mannosegrafted polymers were designed on the premise that one of the routes of phagocytosis of *leishmania* is dependent on the interaction between the mannose-containing lipopolysaccharides on the parasite cell surface and the macrophage mannose receptors. Therefore, by mimicking the invasion process (mannose-dependent receptor-mediated endocytosis) the systems can maximise the potential of the drug to destroy the parasite at the site where it resides. These conjugates showed higher *in vivo* anti leishmanial activity than those with lower mannose content upon i.v. administration (Nan et al. 2004).

Later, AmB was attached to poly(HPMA)–polymer through a degradable GlyPheLeuGly linker (Nicoletti et al. 2009). The copolymers had an IC_{50} of 0.03 μg/ml against *L. donovani* amastigotes in murine peritoneal exudate macrophages and in murine bone marrow-derived macrophages and an IC_{50} 0.57 μg/ml against *L. donovani* in differentiated THP-1 cells. This activity was comparable with free AmB (IC_{50}: 0.03–0.07 μg/ml and 0.24–0.42 μg/ml against *L. donovani* amastigotes in mice derived macrophages and amastigotes in THP-1 cells, respectively) and Fungizone (IC_{50}: 0.04–0.07 μg/ml against amastigotes in mice derived macrophages). Conjugates also showed potent *in vivo* activity with ca. 50% inhibition of parasite burden at 1 mg/kg body weight upon i.v. administration. Recently, the poly(HPMA)-AmB copolymer was conjugated with alendronic acid (Nicoletti et al. 2010). However, no advantage of the combinatorial conjugates was observed as compared to single-drug poly(HPMA)–AmB conjugates. Furthermore, there was no significant activity for the poly(HPMA)–alendronic acid conjugate within the concentration range tested.

A single attempt of using less invasive treatments administered by oral via employed nanosuspensiones made of AmB in an aqueous solution of Tween 80, Pluronics F68 and sodium cholate, but only reduced 29% the liver parasite load on experimental visceral leishmaniasis (Kayser et al. 2003).

7.2.2 Cutaneous Leishmaniasis

Intravenously administered liposomal Sb^{v} (New and Chance 1980; New et al. 1981) and AmBisome (Yardley and Croft 1997) were more effective than free drugs. However, when administered by i.p. or s.c. vias liposomal treatments failed against experimental cutaneous leishmaniasis (*L. major*).

The success of treatment for early stages of cutaneous leishmaniasis depends on how physically accessible the infected macrophages in the stratum spinosum are to the drug delivery system. Circulating nanoparticles can became close to the *stratum spinosum* only from local extravasation. The higher IC_{50} values (25 mg/kg) for cutaneous leishmaniasis than for visceral leishmaniasis (0.3 mg/kg) of i.v. AmBisome reflect the difficulty for particulate material to target the infected cells (Yardley and Croft 2000). Although clinical applications of i.v. AmBisome against cutaneous leishmaniasis have been reported, up to date there are no available data on optimum dosage regimen and extention of treatment (del Rosal Rabes et al. 2010; Wortmann et al. 2010). Succesful short term treatments (5 days at 3 mg/kg/day plus a reinforcement dose at the tenth day) (total dose: 18 mg/kg) in adults have been reported (Rapp et al. 2003; Solomon et al. 2007). Other authors recommend total doses of 40 mg/kg (Brown et al. 2005). However, the need for i.v. administration together with its high cost may limit the use of AmBisome.

AmB loaded in apolipoprotein-stabilised phospholipids bilayer disk complexes, completely cleared parasites, with no lesions remaining on experimental cutaneous leishmaniasis upon i.p. administration (Nelson et al. 2006). Apolipoproteins modified by acetylation, oxidization and others serve as ligands for class A scavenger receptor on macrophages. It was proposed that the small size (8–20 nm) of these nanodisks was responsible for their recognisement by macrophages.

In the search for less invasive vias of administration, oral administration of meglumine antimoniate-β-cyclodextrin complexes led to smaller skin lesions on experimental cutaneous leishmaniasis (Demicheli et al. 2004; Frézard et al. 2008).

Equally less invasive are the topical treatments, but to date no satisfactory results have been achieved by this route. A few attempts were accomplished in order to improve drug absorption. The use of permeation enhancers like ethanol significantly reduced the lesion size on experimental cutaneous leishmaniasis (*L. major*) upon topically applied dispersion of Amphocil and Abelcet (amphotericin intercalated with sheets of phospholipid) in solution containing 5–25% ethanol, over animals treated with dispersed Fungizone (Frankenburg et al. 1998). On the other hand, topical AmB (Amphocil in 5% ethanol) has been successful in treatment of *L. major* infected patients in Israel (Vardy et al. 2001; Zvulunov et al. 2003).

A 15% ointment of the highly hydrophilic antibiotic paromomycin, associated to the permeation enhancer methylbenzethonium chloride (12%) (marketed in Israel as Leshcutan), is relatively effective for treatment (*L major, L. tropica, L. mexicana and L. panamensis*) (El-On et al. 1986, 1992), but local side-effects are observed frequently due to the permeation enhancer (Arana et al. 2001; Armijos et al. 2004; Minodier and Parola 2007). A second formulation combined paromomycin with urea and while non-irritating its efficacy remains largely undistinguished, with reported cure rates little better than placebo in Iran (47% vs. 44%) and Tunisia (27% vs. 18%) (Asilian et al. 1995; Ben Salah et al. 1995; Iraji and Sadeghinia 2005). New cream formulation (WR279,396) containing paromomycin and gentamicin, showed high cure rates (94% vs. 71% placebo) in treating patiens with *L. major* cutaneous leishmaniasis with only mild local irritation (Ben Salah et al. 2009).

Recently, the effect of topical liposomes containing paromomycin sulfate was evaluated in *L. major* infected mice (Jaafari et al. 2009). Mice that received ~500 nm liposomes (SoyPc: chol: propylenglicol: vitamin E, 15: 27: 0.3 wt.%), twice a day for 4 weeks showed a significantly smaller lesion size in the mice in the treated groups than in the mice in the control group. Eight weeks after the beginning of the treatment, the mice treated with liposomes were completely cured. The spleen parasite burden was significantly lower in mice treated with liposomes than in mice treated with PBS or control liposomes. Using a simmilar approach (Carneiro et al. 2010) showed that 500–300 nm fluid liposomes containing paromomycin enhanced *in vitro* drug permeation across stripped skin and improved the *in vivo* antileishmanial activity in *L. major* infected mice.

Topical administration was also addressed by our research group, by determining the *in vitro* leishmanicidal activity of sunlight triggered photodynamic ultradeformable liposomes (Montanari et al. 2010). A hydrophobic Zn phthalocyanine with 20% anti-promastigote activity and 20% anti-amastigote activity against *L. braziliensis* after 15 min sunlight irradiation (15 J/cm^2), hadd 100% and 80% anti-promastigote and anti-amastigote activity respectively, at the same light dose when loaded in ultradeformable liposomes. In the absence of host cells toxicity, ultradeformable liposomes also showed non-photodynamic leishmanicidal activity. Confocal laser scanning microscopy of cryosectioned human skin mounted in non-occlusive Saarbrücken Penetration Model, showed that upon transcutaneous administration Zn phthalocyanine penetrated nearly ten folds deeper when incorporated in ultradeformable liposomes that if loaded in conventional liposomes. Quantitative determination confirmed that when loaded in ultradeformable liposomes phthalocyanine penetrated homogeneously in the *stratum corneum,* carrying seven folds higher amount eight folds deeper than conventional liposomes. It is envisioned that the multiple leishmanicidal effects of ultradeformable liposomes contaning phthalocyanine could play a synergistic role in prophylaxis or therapeutic at the first stages of the infection.

8 Trypanosoma Cruzi

The causative agent of Chagas disease is the flagellate protozoan *Trypanosoma cruzi* which is transmitted to the human by hematophagous reduviidae bugs. *T. cruzi* infect 10–12 million people in endemic areas of Latin America with 15,000 deaths each day (Clayton 2010). Transmission is associated with the faeces of triatomine bugs (>80%), blood transfusion or organ transplant (~15%) (Leiby et al. 2002), congenital transmission (the parasite crosses placenta) (4%), ingestion of contaminated food (<1%) and laboratory accidents (<1%) (Strosberg et al. 2007). After entering a mammal through a skin wound or a mucosal membrane, trypomastigote forms of *T. cruzi* invade many different cell types. In the cytoplasm, it is transformed in *amastigotes* that multiply by means of binary fission producing cell lysis, and releasing new *trypomastigotes* into the blood stream that can invade any nucleated cell to begin a new reproductive cycle. After the generally asymptomatic acute phase (characterized by detectable parasitemia of *trypomastigotes*) that lasts from a few weeks to several months, the infection is well controlled by the host immune response. Nevertheless, without specific treatment during the acute phase, the infection progresses to a long chronic asymptomatic period, this may last for years. It is estimated that in one-third of those infected develops clinical symptoms of different degrees of severity, which include cardiomyopathy, heart failure and digestive-tract abnormalies such megacolon and megaoesophagous, will appear in what is known as symptomatic chronic Chagas disease (Andrade et al. 1996; Sosa Estani et al. 1998). This irreversible structural damage to the heart, the oesophagus and the colon, with severe disorders of nerve conduction in these organs are caused by the intracellular *amastigotes.*

In contrast to Leishmania, *T. cruzi* parasites are not only taken up by phagocytic cells via classic phagocytosis, but can also actively invade cells by induced phagocytosis (Nogueira and Cohn 1976). *T. cruzi* induce the recruitment of lysosomes by a unique calcium-regulated microtubule-mediated pathway. The binding of *T. cruzi* to cells causes a transient increase of intracellular calcium, which induces actin disruption and the movilization of lysosomes along microtubules towards the site of parasite attachment, where the lysosomes fuse with the plasma membrane and contribute to the formation of the PV (Rodriguez et al. 1996; Schenkman et al. 1988). Desialylation of the PV by *T. cruzi* trans-sialidase then facilitates the lysis of the membrane by a parasite-secreted pore forming toxin (Tc-TOX), which is most active at acidic pH and leads to liberation of the parasite into the cytosol (Andrew 1994).

Because of the short and asymptomatic acute phase, the cytoplasmatic amastigotes presents the major structural and phenomenological barriers that antichagasic drugs have to overcome.

8.1 *Conventional Treatments*

Two drugs that were introduced for clinical use in the late 1960s and early 1970s are commonly indicated for the treatment of Chagas disease: benznidazole (*N*-benzyl-2-nitroimidazole acetamide) and nifurtimox (3-methyl-4-[(5-nitrofurfurylidene) amino] thiomorpholine-1, 1-dioxide). Benznidazole at 5 mg/kg/day 30–60 days acts via reductive stress and involves covalent modification of macromolecules by nitroreduction intermediates Nifurtimox at 8–10 mg/kg/day 90 days acts via the reduction of the nitro group to unstable nitroanion radicals, which react to produce highly toxic, reduced oxygen metabolites (superoxide anion, hydrogen peroxide). *T. cruzi* is deficient in detoxification mechanisms for oxygen metabolites and is more sensitive to oxidative stress than vertebrate cells (Rodrigues Coura and de Castro 2002). Theses drugs have significant activity in the acute (up to 80% of parasitological cures in treated patients, defined as a negative result for all parasitological and serological tests) and early chronic phases (up to 60% cures) (de Andrade et al. 1996; Sosa Estani et al. 1998; 1999).

The main disadvantages of these drugs are: (i) low effectiveness in the chronic phase of the disease (<20%), that is the most frequent clinical presentation (Cançado 1999); (ii) regional variations in efficacy due to naturally resistant *T.cruzi* strains (Murta et al. 1998); (iii) high rate of patient noncompliance due to drug side effects; (iv) long period of treatment (30–60 days), and dose-dependent toxicity; (v) need for monitoring under specialized medical supervision; (vi) nifurtimox has been only intermittently produced since 1997 by Bayer while benznidazole, Roche has transferred rights and technology to Brazilian government in 2007, making Pharmaceutical Laboratory of Pernambuco the drug sole manufacture in world (with help of DNDi).

8.2 *Preclinical Delivery Systems*

Recently we have extensively reviewed the state-of-the-art in the development of nanotechnological strategies against Chagas disease (Romero and Morilla 2010).

First strategies were only tested *in vitro*. Stearylamine liposomes showed activity against epimastigotes (only found in bugs), trypomastigotes and axenic amastigotes; these liposomes however, were not assayed against amastigotes in infected cells, the most relevant target of the disease (Yoshihara et al. 1987). The same occurred with alkylpolycyanoacrylate nanospheres containing nifurtimox (Gonzalez-Martin et al. 1998; Sánchez et al. 2002) and allopurinol (Gonzalez-Martín et al. 2000), that showed some activity against epimastigotes and intracellular amastigotes, although empty nanospheres showed also trypanocidal activity.

The activities of four AmB formulations, Fungizone, AmBisome, Amphocil and Abelcet, were compared *in vitro* and *in vivo* on an acute murine model infected with the Y and the Tulahuen strains (Yardley and Croft 1999). When tested against trypomastigotes of the Y strain upon 24 h incubation, Fungizone showed the highest activity (LD_{50} 0.3 μg/ml), followed by Amphocil (LD_{50} between 1 and 0.5 μg/ml). Abelcet and AmBisome did not show activity up to a 30 μg/ml. When tested against amastigotes infecting murine peritoneal macrophages, Fungizone and Amphocil showed the highest activities (IC_{50} between 0.027 and 0.028 μg/ml), followed by AmBisome, Abelcet and benznidazole, with IC_{50} of 0.19, 1.2 and 1.43 μg/ml, respectively. On amastigote-infected Vero cells, Fungizone, Amphocil, AmBisome, Abelcet and benznidazole IC_{50} increased to 2, 4.1, 3.6, 2.3 and 4.2 μg/ml, respectively.

The effect of multiple and single doses of amphotericine B doxycholate and liposomal amphotericine (AmBisome) on Balb/c mice infected with Y and Tulahuen strain was also evaluated. On *T. cruzi* Y strain-infected mice, the administration of six doses of Fungizone (0.5 mg/kg, maximum tolerated dose) or AmBisome (12.5 mg/kg) enabled 4 of 5 and 5 of 5 mice to survive until the end of the experiment (60 days post-infection). The effect of AmBisome was not curative since trypomastigotes were observed in the tail blood of treated mice for 3 weeks post-treatment. Untreated control animals died 18 days post-infection. On *T. cruzi* Tulahuen strain–infected mice, the lowest dose of AmBisome (6.25 mg/kg) produced survival of all animals until 60 days post-infection, meanwhile untreated controls survived until 13 days post-infection.

Used as a single dose at 25 mg /kg AmBisome induced the survival of 5 of 5 mice, whereas Abelcet and Amphocil and five consecutive oral doses of benznidazole at 45 mg/kg only induced the survival of 3/5 at the end of 60 days (lower doses failed).

In general terms, *in vivo* AmBisome was more effective and less toxic than other lipid amphotericine formulations, despite of a delayed elimination of blood parasites when compared to benznidazole (3 vs. 1 week). A crucial difference between the use of AmBisome against leishmaniasis (a parasite that colonizes the phago-lysosomal system in cells of the RES) and against Chagas is that in the latter, even if properly biodistributed, the intracellular pathway followed by amphotericine could not be appropriate. It is possible that once inside the target cell, the hydrophobic nature of the amphotericine impaired its diffusion from the endo/phago-lisosomal confinement.

To minimize opsonization, RES uptake and to act as a long circulating drug depot, different azoles (itraconazole, ketokonazole and the fourth generation bis-triazole D0870) were loaded in PEG-co-PLA nanospheres. Nanoparticles were daily i.v. administered to swiss albino mice infected with a nitroimidazoles/nitrofurans-susceptible and a partially resistant strain, for 30 days (Molina et al. 2001). D0870 in nanospheres (3 mg/kg) produced similar percentages of cure (90%) than free D0870 (30 oral doses at 5 mg/kg) (80%) on the susceptible strain. In the case of the resistant strain, D0870 containing nanospheres (60% cure) were more effective than itraconazole- and ketokonazole-nanospheres (0% cure) and 30 daily oral doses of benznidazole (100 mg/kg) (47%). However, 20 oral doses of free D0870 (5 mg/kg day) in the same animal model produced 100% cure (Molina et al. 2000).

Aiming to increase the efficacy of benznidazole as trypanocidal agent by modifying its pharmacokinetics and biodistribution, our research group developed a MLV formulation of benznidazole (hydrogenated soyPC: chol: distearoyl-phosphatidylglycerol, 2:2:1 M ratio) (Morilla et al. 2002). When i.v. administered in rats as a 0.2 mg/kg, a three-fold higher benznidazole accumulation in the liver than the achieved with the same dose of the free drug was found. However, liposomal benznidazole (i.v., 0.4 mg/kg, twice-a-week from day 5 to day 22 post infection) did not decrease parasitemia levels in mice infected with the RA strain. These results indicated that the relationship between the increased liver delivery and the therapeutic effect of liposomal benznidazol was not that simple. The hydrophobic benznidazole remained associated to the liposomal bilayer that kept trapped within the endo-lysosomal pathway, instead of being released to the cell cytoplasm (Morilla et al. 2004).

Acid media trigger a phase transition in pH sensitive liposomes, from bilayer at neutral or alcaline media, to inverted hexagonal phase II when the pH drops below 5–6. The hexagonal phase II is responsible for the fusion of liposome with the endo-lysosomal bilayer; as a result, the aqueous content of pH sensitive liposomes is released to cell cytoplasm. The hydrophobic nature of benznidazole made it unsuitable for delivery from pH-sensitive liposomes, and had to be replaced by a hydrophilic compound with trypanocydal activity. Hence, pH-sensitive liposomes (DOPE: cholesteryl hemisuccinate, 6:4, mol:mol, ~400 nm) containing the hydrophilic etanidazole were prepared (Morilla et al. 2005). This formulation ensured a fast and massive delivery of etanidazole into the cytosol of murine J774 macrophages. Liposomes containing etanidazole were phagocytosed by both uninfected and *T. cruzi*-infected macrophages and at 200 μg/ml showed 72% anti-amastigote activity in J774 cells after 2 h. Contrary to this, the same dose of free etanidazole rendered 0% activity. I.v. administration of etanidazole loaded in pH-sensitive liposomes (0.56 mg/kg, starting 5 day post infection, 3 days-a week over 3 weeks) resulted in a significant decrease in parasitemia (days 12, 19, 21 and 23) of Balb/c mice infected with 50 trypomastigotes of the RA strain. Administration of a 180-fold higher dose of free etanidazole failed to reduce the number trypomastigotes in blood.

Remarkably, previous *in vitro* determinations of free etanidazole trypanocidal activity showed that on RA trypomastigotes, the LD_{50} of etanidazole was 18 μM (8.2-fold less active than benznidazole). In the case of amastigote-infected Vero cells, IC_{50} for etanidazole was 40 μM (23.5-fold less active than benznidazole). Finally, on amastigotes infecting J774 cells, 15-fold lower activity than benznidazole was found (Petray et al. 2004). According to the classical interpretation, these *in vitro* data should rule out the etanidazole as trypanocidal agent. However, we showed that this poor performance could be rescued without modifying its chemical structure just by encapsulating the drug in pH-sensitive liposomes. The example of etanidazole, that *in vivo* proved to be less toxic than benznidazole, could be extensive to other trypanocidal drugs. Mice infected with 10^2 trypomastigotes (Tulahuen strain) and treated with liposomal etanidazole at 3.2 mg/mouse led to the complete elimination of parasitemia. The administration of an equivalent dose of empty liposomes (293 μg liposomal lipid/mouse) or free etanidazole had no effect on parasitemia (unpublished results). The etanidazole liposomal therapy however, remains to be tested against higher infecting inoculums from different *T. cruzi* strains, both in the acute and the chronic stages of the disease.

9 Toxoplasma Gondii

Toxoplasmosis is primarily acquired through the ingestion of food or water contaminated with oocysts or undercooked meat infected with tissue cysts of the parasite *Toxoplasma gondii* (Dubey 2004). It can also be transmitted from mother to fetus, after mother acquisition for the first time during gestation, or by organ transplantation. After ingestion, bradyzoites released from the cysts, or sporozoites released from oocysts enter intestinal tissues and convert into taquyzoites. During the usually asymptomatic acute phase, tachyzoites multiply inside the parasitophorous vacuola in the epithelial cells from the intestine. Tachyzoites propagate to mesenteric ganglia and invade macrophages and nucleate cells by lymphohaematogenous via. Nearly after 10–14 days of acquisition, the immune system from the host controllates the infection. In this point the parasite differentiates to a metabolically quiescent bradyzoite and are found in cyst more prevalent in neural and muscular tisúes, including the brain, eyes, and skeletal and cardiac muscles. Cysts grow and remain inside the cells while bradyzoites multiply by endodyogeny (Duval and Leport 2001). Brain cysts are spherical, rarely beyond 70 μm radii, while in muscle are found ~100 μm length elongated cysts. Cyst wall produced by modifications in the PV is elastic and thin lacking of glycogen and other polysaccharides and encloses hundreds of bradyzoites. These cysts remain dormant in the tissue until they can become reactivated. The host's innate immune system acts against these cysts to prevent their reactivation in normal, healthy conditions, although it's unclear by which mechanism this is managed. Evidence suggests that cysts periodically are disrupted and the released bradizoites are destroyed by host's immune system (Djurkovic-Djakovic et al. 2000). However, the immune systems of immunosuppressed patients may fail to keep these cysts from reactiving, allowing a full-blown reinfection to occur. The absence of antibodies in the brain makes it more likely for cysts in the central nervous system (CNS) to persist and not being cleared as the protozoa in other tissues might have been. As a result, a strong reinfection in an immunocompromised state is often centralized in the CNS and can lead to the severe pathogenesis associated with toxoplasmosis called toxoplasma encephalitis that is lethal if untreated.

Congenital infections are more likely to be symptomatic and can be severe, 5–10% causing death, 8–10% severe brain and eye damage, 10–13% moderate-severe visual impairments, 58–72% being asymptomatic at birth with some later development of retino-choroiditis or mental impairments.

T. gondii invades cells by pulling itself into the cell by mechanism simmilar than *Plasmodium* and lives in a simmilar PV.

Because of the short term and absence of clinical symptoms of the acute phase, the main disease targets are the cysts in brain and muscle, both offering impaired access to the medication.

9.1 Conventional Treatments

Due to the self-limited and benign nature of the disease, immunocompetent adults do not receive treatment during the chronic phase of toxoplasmosis. Only are treated those patients with reactivation by immunosuppression, with ocular lesions or after acquisition by congenital transmission.

Firts line treatment is a sinergic combination of pyrimethamine and sulfadiazine (a 4-amino-N-2-pirimidil-benzen-sulfonamida). Pyrimethamine interferes with folic acid synthesis by inhibiting the enzyme dihydrofolate reductase (in mammals and parasites) thus preventing the biosynthesis of purines and pyrimidines, thereby halting the processes of DNA synthesis, cell division and reproduction. Sulfonamides inhibit the dihydropteroate synthetase, a parasitic enzyme that participates in folic acid synthesis from para-aminobenzoic acid. Hence, sulfonamides work synergistically with pyrimethamine by blocking a different enzyme needed for folic acid synthesis. Pyrimethamine cause bone marrow supresión (and folic acid has to be co-administered), hematologic toxicity, and/or life-threatening allergic reactions in up to 50% of cases (Katlama et al. 1996).

Clindamycine can substitute sulfadiazine in patient that do not tolerate sulfa drugs. Spiramycin is recommended for prophylactic use during pregnancy.

These drugs however are not effective against cysts in brain and muscle (Couvreur et al. 1991; Djurkovic-Djakovic et al. 2000).

9.2 Preclinical Delivery Systems

9.2.1 Targeted Delivery to Tachyzoites

Two approaches employed sulfadiazine as a single drug in the treatment of experimental infection:

In the first of them, arginine olygomers linked to immunomodulators and antitumoral compounds have been shown to translocate across different epithelial barriers, including skin, lungs and eyes (Mitchell et al. 2000; Rothbard et al. 2000; Wender et al. 2000). To make use of this property, short arginine olygomers were linked by a hydrolizable ester linkage to triclosan, an inhibitor of the key enzyme of fatty acid synthesis enoyl-ACP reductase from *T. gondii* (Samuel et al. 2003). Short arginine (seven or eight arginines) peptides were found to cross all membranes and to enter each life cycle stage and compartment (intracellular and extracellular tachyzoites and extracellular and encysted bradyzoites). The triclosan conjugate completely inhibited the in vitro growth of intracellular *T gondii* tachyzoites after 48 h incubation. Starting the day of infection, and after four ip consecutive doses, the conjugate showed comparable activity to sulfadiazyne at reducing parasite burden in peritoneal cavity, on an ip infected mice model.

The second work employed, sulfadiazine was non covalently loaded in cationic G4 (DG4) or anionic G4.5 (DG4.5) poly(amidoamine) (PAMAM) dendrimers, that were used as sulfadiazine carriers (Prieto et al. 2006). Dendrimeric (DG4.5) sulfadiazine was no toxic on Vero and J774 cells up to 33 μM DG4,5, wile dendrimeric (DG4) sulfadiazine was cytotoxic on the two cell lines above 3.3 μM DG4. The anti toxoplasmic activity of dendrimeric sulfadiazine and void dendrimers was tested on Vero cells infected with tachizoytes of RH strain upon 4 h incubation. Dendrimeric (G4) sulfadiazine showed the highest anti toxoplasmic activity, since a concentration of 0.03 μM produced a 60% decrease of the in vitro infection. Dendrimeric (G4,5) sulfadiazine on the other hand, required a 1,000 folds higher dendrimer concentration to render twice lower activity. Free sulfadiazine showed no antitoxoplasmic activity. The combination of surfacial activity and endosomolytic effect could be at least in part responsible for the antiparasitic effect of dendrimeric (DG4) sulfadiazine at the ultra-low concentration of 0.03 μM. Dendrimeric (DG4) sulfadiazine on the other hand, contained 1×10^{-3} mM sulfadiazine (250 ng/ml), being this concentration 1×10^{9} lower than *in vitro* IC_{50} for sulfadiazine, 2.5 g/ml, determined upon 72 h on infected MRC5 fibroblasts (Duval and Leport 2001).

More recently, gold nanorods were functionalized with anti-*T. gondii* antibody against the major antigen on the surface of *T. gondii* tachyzoite for photothermal therapy (Pissuwan et al. 2007, 2009). Tachyzoites that had antibody-nanorod conjugates specifically bound to them were found to be strongly effected by laser irradiation. The percentage of cell death rose from 28 at the lowest power density (~0.7 W/m^2) to 83 at the highest power density (~51 W/m^2). In contrast, a control sample of untreated gold-antibody conjugate tachyzoites that had been irradiated contained an average of 15% dead tachyzoites, irrespective of the laser power applied. This was comparable to the sample of control tachyzoites that had not been irradiated at all (which contained 15% dead tachyzoites). The specificity of the photothermal effect was obtained by incubating U937 cells with anti-*T. gondii*-gold nanorod conjugates. In this case, the number of dead U937 cells was not different from non-irradiated cells even at the highest power density.

9.2.2 Targeted Delivery to Bradyzoites

Atovaquone is a highly lipophilic hydroxynaphthoquinone that has potent *in vitro* activity against *T. gondii*, but it is poorly absorbed after oral administration and shows poor therapeutic efficacy against toxoplasma encephalitis. I.v. injection of an atovaquone solution is not a feasible alternative to oral administration because of the poor solubility of this compound in the solvent mixtures acceptable for i.v. administration. Atovaquone nanosuspensions (either of 0.3% Tween 80, 0.3% Poloxamer 188 and 0.05% sodium cholate or of 1.0% Tween 80, 280 nm) for i.v. administration were prepared (Schöler et al. 2001). At concentrations higher than 1 μg/ml nanosuspensions did not exert cytotoxicity and was as effective as free atovaquone against *T. gondii* in freshly isolated peritoneal macrophages.

A murine model of toxoplasma encephalitis that closely mimics reactivated toxoplasmosis in immunocompromised hosts was used. Mice have a targeted mutation in the gene encoding the interferon consensus sequence binding protein. Mice were treated with sulfadiazine in drinking water for 3 weeks beginning 2 days after infection. Treatment of infected mice with sulfadiazine resulted in latent infections involving the development of brain cysts. Following discontinuation of sulfadiazine, remarkable histological changes were observed in brains. At day 8 brains showed inflammatory changes associated with parasitophorous vacuoles and/or free parasitic antigen, consistent with toxoplasma encephalitis, characterized by infiltration of mononuclear cells around vessels and, to a lesser extent, in meninges, and mice died by day 14.

Two days after discontinuation of sulfadiazine, nanosuspension were i.v. administered (every other day from day 2 until day 16), at doses of 10 mg/kg of body weight 50% of animals survived until day 24 (the end of the study period). On the other hand, mice orally treated with free atovaquone (100 mg/kg every other day) died within 22 days. Brains of mice treated with nanosuspensions showed neither inflammatory foci nor foci of parasitophorous vacuoles and/or parasite antigen. Oral treatment with atovaquone suspension at 100 mg/kg resulted in serum, lung and brain drug levels comparable to those measured after i.v. treatment with 10 mg /kg. The amount of atovaquone in liver was nearly six folds higher after i.v. administration of nanosuspensions than after oral administration of free atovaquone.

References

Adams LB, Sinha I, Franzblau SG, Krahenbuhl JL, Mehta RT (1999) Effective treatment of acute and chronic murine tuberculosis with liposome-encapsulated clofazimine. Antimicrob Agents Chemother 43: 1638–1643

Aditya NP, Patankar S, Madhusudhan B, Murthy RSR, Souto EB (2010 a). Arthemeter-loaded lipid nanoparticles produced by modified thin-film hydration: Pharmacokinetics, toxicological and *in vivo* anti-malarial activity. Eur J Pharm Sci 40: 448–455

Aditya PN, Tiyaboonchai W, Patankar S, Madhusudhan B, Souto EB (2010 b) Curcuminoids-loaded lipid nanoparticles: Novel approach towards malaria treatment. Colloids Surf B Biointerfaces 81: 263–273

Agrawal AK, Singhal A, Gupta CM (1987) Functional drug targeting to erythrocytes *in vivo* using antibody bearing liposomes as drug vehicles. Biochem Biophys Res Commun 148: 357–361

Agrawal AK, Agrawal A, Pal A, Guru PY, Gupta CM (2002) Superior chemotherapeutic efficacy of amphotericin B in tuftsin–bearing liposomes against *Leishmania donovani* infection in hamsters. J Drug Target 10: 41–45

Ahmad Z, Sharma S, Khuller GK (2005) Inhalable alginate nanoparticles as antitubercular drug carriers against experimental tuberculosis. Int J Antimicrob Agents 26: 298–303

Ahmad Z, Sharma S, Khuller GK (2006 a) Azole antifungals as novel chemotherapeutic agents against murine tuberculosis. FEMS Microbiol Lett 261: 181–186

Ahmad Z, Pandey R, Sharma S, Khuller GK (2006 b) Pharmacokinetic and pharmacodynamic behavior of antitubercular drugs encapsulated in alginate nanoparticles at two doses. Int J Antimicrob Agents 27: 409–416

Ahmad Z, Sharma S, Khuller GK (2007) Chemotherapeutic evaluation of alginate nanoparticle-encapsulated azole antifungal and antitubercular drugs against murine tuberculosis. Nanomedicine 3: 239–243

Ahmad Z, Pandey R, Sharma S, Khuller GK (2008) Novel chemotherapy for tuberculosis: chemotherapeutic potential of econazole- and moxifloxacin-loaded PLG nanoparticles. Int J Antimicrob Agents 31: 142–146

Akinlolu AA, Shokunbi MT (2010) Neurotoxic effects of 25 mg/kg/bodyweight of artemether on the histology of the trapezoid nuclei and behavioural functions in adult male Wistar rats. Acta Histochem 112: 193–198.

Alonso MJ (2004) Nanomedicines for overcoming biological barriers. Biomed Pharmacother 58: 168–172

Alving CR, Steck EA, Chapman WL, Waits VB, Hendricks LD, Swartz GM Jr, Hanson WL (1978) Therapy of leishmaniasis: superior efficacies of liposome-encapsulated drugs. Proc Natl Acad Sci USA 75: 2959–2963

Ancsin JB, Kisilevsky R (2004) A binding site for highly sulfated heparan sulfate is identified in the N terminus of the circumsporozoite protein: significance for malarial sporozoite attachment to hepatocytes. J Biol Chem 279: 21824–21832

Andrew NW (1994) From lysosomes into the cytosol: the intracellular pathway of Trypanosoma cruzi. Braz J Med Biol Res 27: 471–475

Antosova Z, Mackova M, Kral V, Macek T (2009) Therapeutic application of peptides and proteins: parenteral forever?. Trends Biotechnol 27: 628–634

Arana BA, Mendoza CE, Rizzo NR, Kroeger A (2001) Randomized, controlled, double-blind trial of topical treatment of cutaneous leishmaniasis with paromomycin plus methylbenzethonium chloride ointment in Guatemala. Am J Trop Med Hyg 65: 466–470

Armijos RX, Weigel MM, Calvopina M, Mancheno M, Rodriguez R (2004) Comparison of the effectiveness of two topical paromomycin treatments versus meglumine antimoniate for New World cutaneous leishmaniasis. Acta Trop 91: 153–160

Asanuma K, Kenji H, Shimazak S (1983) A study concerning the distribution of OTC and CTC in lung and blood of pigs. J Anim Drugs 6: 1–6

Asilian A, Jalayer T, Whitworth JA, Ghasemi RL, Nilforooshzadeh M, Olliaro P (1995) A randomized, placebo-controlled trial of a two-week regimen of aminosidine (paromomycin) ointment for treatment of cutaneous leishmaniasis in Iran. Am J Trop Med Hyg 53: 648–651

Azarmi S, Roa WH, Lobenberg R (2008) Targeted delivery of nanoparticles for the treatment of lung diseases. Adv Drug Deliver Rev 60: 863–875

Banu S, Gordon SV, Palmer S, Islam MR, Ahmed S, Alam KM, Cole ST, Brosch R (2004) Genotypic analysis of *Mycobacterium tuberculosis* in Bangladesh and prevalence of the Beijing strain. J Clin Microbiol 42(2): 674–82

Barratt B, Legrand P (2005) Comparison of the effi cacy and pharmacology of formulations of amphotericin B used in treatment of leishmaniasis. Curr Opin Infect Dis 18: 527–530

Bayomi MA, Al-Angary AA, Al-Meshal MA, Al-Dardiri MM (1998) *In vivo* evaluation of arteether liposomes. Int J Pharm 175: 1–7

Begg EJ, Barclay ML (1995) Aminoglycosides-50 years on. British J Clin Pharmacol 39: 597–603

Ben Salah A, Zakraoui H, Zaatour A, Ftaiti A, Zaafouri B, Garraoui A, Olliaro PL, Dellagi K, Ben Ismail R (1995) A randomized, placebo-controlled trial in Tunisia treating cutaneous leishmaniasis with paromomycin ointment. Am J Trop Med Hyg 53: 162–166

Ben Salah A, Buffet PA, Morizot G, Ben Massoud N, Zâatour A, Ben Alaya N, Haj Hamida NB, El Ahmadi Z, Downs MT, Smith PL, Dellagi K, Grögl M (2009) WR279,396, a third generation aminoglycoside ointment for the treatment of *Leishmania major* cutaneous leishmaniasis: a phase 2, randomized, double blind, placebo controlled study. PLoS Negl Trop Dis 3(5): e432

Berman JD, Hanson WL, Chapman WL, Alving CR, Lopez-Berestein G (1986) Antileishmanial activity of liposome-encapsulated amphotericin B in hamsters and monkeys. Antimicrob Agents Chemother 30: 847–851

Bermudez LE (1994) Use of liposome preparation to treat mycobacterial infections. Immunobiology 191 (4–5): 578–583

Bermudez LE, Yau-Young AO, Lin J-P, Cogger J, Young LS (1990) Treatment of disseminated *Mycobacterium avium* complex infection of beige mice with liposome-encapsulated aminoglycosides. J Infect Dis 161: 1262–1268

Bern C, Adler-Moore J, Berenguer J, Boelaert M, den Boer M, Davidson RN, Figueras C, Gradoni L, Kafetzis DA, Ritmeijer K, Rosenthal E, Royce C, Russo R, Sundar S, Alvar J (2006) Liposomal amphotericin B for the treatment of visceral leishmaniasis. Clin Infect Dis 43: 917–924

Bharatwaj B, Wu L, Whittum-Hudson JA, da Rocha SRP (2010) The potential for the noninvasive delivery of polymeric nanocarriers using propellant-based inhalers in the treatment of Chlamydial respiratory infections. Biomaterials 31: 7376–7385

Biagini GA, Ward SA, Bray PG (2005) Malaria parasite transporters as a drug-delivery strategy. Trends Parasitol 21: 299–301

Birmingham CL, Smith AC, Bakowski MA, Yoshimori T, Brumell JH (2006) Autophagy controls Salmonella infection in response to damage to the Salmonella-containing vacuole. J Biol Chem 281: 11374–11383

Blais J, Chamberland S (1994) Intracellular accumulation of tilmicosin in primary swine alveolar macrophages. Proc Int Pig Vet Soc 13: 331

Brown M, Noursadeghi M, Boyle J, Davidson RN (2005) Successful liposomal amphotericin B treatment of *Leishmania braziliensis* cutaneous leishmaniasis. Br J Dermatol 153: 203–205

Cançado JR (1999) Criteria of Chagas disease cure. Mem Inst Oswaldo Cruz 94: 331–336

Carneiro G, Santos DC, Oliveira MC, Fernandes AP, Ferreira LS, Ramaldes GA, Nunan EA, Ferreira LA (2010) Topical delivery and *in vivo* antileishmanial activity of paromomycin-loaded liposomes for treatment of cutaneous leishmaniasis. J Liposome Res 20(1):16–23.

Carter PB, Collins FM (1974) The route of enteric infection in normal mice. J Exp Med 139: 1189–1203

Centers for Disease Control and Prevention, CDC. http://www.cdc.gov/ncidod/dbmd/diseaseinfo/mycobacteriumavium_t.htm

Chadwick S, Kriegel C, Amiji M (2010) Nanotechnology solutions for mucosal immunization. Adv Drug Del Rev 62: 394–407

Chandra SA, Agrawal K, Gupta CM (1991) Chloroquine delivery to erythrocytes in *Plasmodium berghei*-infected mice using antibody bearing liposomes as drug vehicles. J Biosci 16: 37–144

Chimanuka B, Gabriëls M, Detaevernier MR, Plaizier-Vercammen JA (2002) Preparation of beta-artemether liposomes, their HPLC-UV evaluation and relevance for clearing recrudescent parasitaemia in *Plasmodium chabaudi* malaria-infected mice. J Pharm Biomed 28: 13–22

Chimote G, Banerjee R (2008) Effect of mycobacterial lipids on surface properties of Curosurf: implications for lung surfactant dysfunction in tuberculosis. Respir Physiol Neurobiol 162: 73–9

Chimote G, Banerjee R (2009) Evaluation of antitubercular drug-loaded surfactants as inhalable drug-delivery systems for pulmonary tuberculosis. J Biomed Mater Res A 89: 281–292

Chimote G, Banerjee R (2010) *In vitro* evaluation of inhalable isoniazid-loaded surfactant liposomes as an adjunct therapy in pulmonary tuberculosis. JBiomed Mater Res B Appl Biomater 94(1): 1–10

Chono S, Tanino T, Seki T, Morimoto K (2007) Uptake characteristics of liposomes by rat alveolar macrophages: influence of particle size and surface mannose modification. J Pharm Pharmacol 59: 75–80

Chono S, Tanino T, Seki T, Morimoto K (2008) Efficient drug targeting to rat alveolar macrophages by pulmonary administration of ciprofloxacin incorporated into mannosylated liposomes for treatment of respiratory intracellular parasitic infections. J Control Release 127: 50–58

Clark A, Reed KA, Lodge J, Stephen J, Hirst BH, Jepson MA (1996) Invasion of murine intestinal M cells by *Salmonella typhimurium inv* mutants severely deficient for invasion of cultured cells. Infect Immun 64: 4363–4368

Clayton J (2010) Chagas disease. Nature 465: S4–S5

Clemens DL, Lee BY, Horwitz MA (2004) Virulent and avirulent strains of *Francisella tularensis* prevent acidification and maturation of their phagosomes and escape into the cytoplasm in human macrophages. Infect Immun 72: 3204–3217

Coester CJ, Langer K, Van-Briesen H, Kreuter J (2000) Gelatin nanoparticles by two-step desolvation - a new preparation method, surface modifications and cell uptake. J Microencapsul 17: 187–193

Coltel N, Combes V, Hunt NH, Grau GE (2004) Cerebral malaria – a neurovascular pathology with many riddles still to be solved. Curr Neurovasc Res 1: 91–110

Cometta A, Glauser MP (1996) The use of aminoglycosides in neutropenic patients. SCh Med Wochensch S, 76: 21 S–7 S

Conley J, Yang H, Wilson T, Blasetti K, Di Ninno V, Schnell G, Wong JP (1997) Aerosol delivery of liposome-encapsulated ciprofloxacin: aerosol characterization and efficacy against *Francsella tularensis* infection in mice. Antimicrob Agents Chemother 41: 1288–1292

Cook PJ, Davies P, Tunnicliffe W, Ayeres JG, Honeybourne D, Wise R (1998) Chlamydia pneumoniae and asthma. Thorax 53: 254–259

Cordeiro C, Wiseman DJ, Lutwyche P, Uh M, Evans JC, Finlay BB, Webb MS (2000) Antibacterial efficacy of gentamicin encapsulated in pH-sensitive liposomes against an *in vivo Salmonella enterica* serovar typhimurium intracellular infection model. Antimicrob Agents Chemother 44 (3): 533–539

Cosma CL, Sherman DR, Ramakrishnan L (2003) The secret lives of pathogenic mycobacteria. Annu Rev Microbiol 57: 641–676

Couvreur J, Thulliez P, Daffos F, Aufrant C, Bompard Y, Gesquière A, Desmonts G (1991) Fetal toxoplasmosis. *In utero* treatment with pyrimethamine sulfamides. Arch Fr Pediatr 48(6): 397–403

Croft SL, Coombs GH (2003) Leishmaniasis-current chemotherapy and recent advances in the search for novel drugs. Trends Parasitol 19: 502–507

Crump JA, Mintz ED (2010) Global trends in typhoid and paratyphoid Fever. Clin Infect Dis 50: 241–246

Cryan S-A, Sivadas N, Garcia-Contreras L (2007) *In vivo* animal models for drug delivery across the lung mucosal barrier. Adv Drug Del Rev 59: 1133–1151

Csaba N, Garcia-Fuentes M, Alonso MJ (2009) Nanoparticles for nasal vaccination. Adv Drug Del Rev 61: 140–157

Cullis PR, Hope MJ, Bally MB, Madden TD, Mayer LD, Fenske DB (1997) Influence of pH gradients on the transbilayer transport of drugs, lipids, peptides and metal ions into large unilamellar vesicles. Biochim Biophys Acta 1331: 187–211

Cunningham AC (2002) Parasitic adaptive mechanisms in infection by *Leishmania*. Exp Mol Pathol 72: 132–141

Cynamon MH, Swenson CE, Palmer GS, Ginsberg RS (1989) Liposome-encapsulated-amikacin therapy of *Mycobacterium avium* complex infection in beige mice. Antimicrob Agents Chemother 33:1179–1183

Danesh J, Whincup P, Walker M, Lennon L, Thomson A, Appleby P, Wong Y, Bernardes-Silva M, Ward M (2000) Chlamydia pneumoniae IgG titres and coronary heart disease: prospective study and meta-analysis. BMJ 321: 208–213

Dannenberg AMJr (1989) Immune mechanisms in the pathogenesis of pulmonary tuberculosis. Rev Infect Dis 11(Suppl 2): S369–378

Dasgupta D, Chakraborty P, Basu MK (2000) Ligation of Fc receptor of macrophages stimulates protein kinase C and anti-leishmanial activity. Mol Cell Biochem 209: 1–8

Davidson PT, Quec LH (1992) Drug treatment of tuberculosis- 1992. Drugs 43: 651–673

Davis ME, Chen Z, Shin DM (2008) Nanoparticle therapeutics: an emerging treatment modality for cancer. Nat Rev Drug Discov 7: 771–782

de Andrade AL, Zicker F, de Oliveira RM, Almeida Silva S, Luquetti A, Travassos LR, Almeida IC, de Andrade SS, de Andrade JG, Martelli CM (1996) Randomised trial of efficacy of benznidazole in treatment of early *Trypanosoma cruzi* infection. Lancet 348: 1407–1413

de Cock KM, Wilkinson D (1995) Tuberculosis control in resource-poor countries: alternative approaches in the era of HIV. Lancet 346: 675–677

Decuzzi P, Godin B, Tanaka T, Lee S-Y, Chiappini C, Liu X, Ferrari M (2010) Size and shape effects in the biodistribution of intravascularly injected particles. J Control Release 141: 320–327

del Rosal Rabes T, Baquero-Artigao F, Gómez Fernández C, García Miguel MJ, de Lucas Laguna R (2010) Cartas del editor. Tratamientodelaleishmaniasiscutánea con anfotericinaB liposomal. An Pediatr (Barc) 73(2):101–111

Demicheli C, Ochoa R, Da Silva JBB, Falcão CA, Rossi-Bergmann B, de Melo AL, Sinisterra RD, Frézard F (2004) Oral delivery of meglumine antimoniate-cyclodextrin complex for treatment of leishmaniasis. Antimicrob Agents Chemother 48: 100–103

Dennis DT, Inglesby TV, Henderson DA, Bartlett JG, Ascher MS, Eitzen E, Fine AD, Friedlander AM, Hauer J, Layton M, Lillibridge SR, McDade JE, Osterholm MT, O'Toole T, Parker G, Perl TM, Russell PK, Tonat K (2001) Tularemia as a biological weapon: medical and public health management. JAMA 285: 2763–2773

Deol P, Khuller GK (1997) Lung specific liposomes: stability, biodistribution and toxicity of liposomal antitubercular drugs in mice. Biochem Biophys Acta 1334: 161–172

Deol P, Khuller GK, Joshi K (1997) Therapeutic efficacies of isoniazid and rifampin encapsulated in lung-specific stealth liposomes against Mycobacterium tuberculosis infection induced in mice. Antimicrob Agents Chemother 41(6): 1211–1214

Desai SA, Krogstad DJ, McCleskey EW (1993) A nutrient-permeable channel on the intraerythrocytic malaria parasite. Nature 362: 643–646

Desai SA, Rosenberg RL (1997) Pore size of the malaria parasite's nutrient channel. Proc Natl Acad Sci USA 94: 2045–2049

Descoteaux A, Turco SJ (1999) Glycoconjugates in *Leishmania* infectivity. Biochim Biophys Acta 1455: 341–352

Desjeux P (2004) Leishmaniasis: current situation and new perspectives. Comp Immunol Microbiol Infect Dis 27: 305–318

Dierling AM, Cui Z (2005) Targeting primaquine into liver using chylomicron emulsions for potential vivax malaria therapy. Int J Pharm 303: 143–152.

Djurkovic-Djakovic O, Romand S, Nobrè R, Couvreur J, Thulliez P (2000) Serologic rebounds after one-year-long treatment for congenital toxoplasmosis. Pediatr Infect Dis J 19(1): 81–83.

Dos Santos N, Allen C, Doppen A-M, Anantha M, Cox KAK, Gallagher RC, Karlsson G, Edwards K, Kenner G, Samuels L, Webb MS, Bally MB (2007) Influence of poly(ethylene glycol) grafting density and polymer length on liposomes: Relating plasma circulation lifetimes to protein binding. Biochim Biophys Acta 1768: 1367–1377

Dubey JP (2004) Toxoplasmosis – a waterborne zoonosis. Vet Parasitol 126: 57–72

Duncan R (2003) The dawning era of polymer therapeutics. Nat Rev Drug Disc 2: 347–360

Duncan K, Barry CE III (2004) Prospects for new antitubercular drugs. Curr Opinion Microbiol 7: 460–465

Duval X, Leport C (2001) Toxoplasmosis in AIDS. Curr Treatment Options Infect Dis 3: 113–128

Duzgunes N, Perumal VK, Kesavalu L, Goldstein JA, Debs RJ, Gangadharam PRJ (1988) Enhanced effect of liposome-encapsulated amikacin on *Mycobacterium avium-M. intracellulare* complex infection in beige mice. Antimicrob Agents Chemother 32: 1404–1411

Dye C (2006) Global epidemiology of tuberculosis. Lancet 367: 938–940

Easmon CSF, Crane JP (1985 a) Uptake of ciprofloxacin by macrophages. J Clin Pathol 38: 442– 444

Easmon CSF, Crane JP (1985 b) Uptake of ciprofloxacin by human neutrophils. J Antimicrob Chemother 16: 67– 73

Ebert SC (2004) Application of pharmacokinetics and pharmacodynamics to antibiotic selection. PharmD, BCPS 29: 244–252

Ehlers S, Bucke W, Leitzke S, Smith D, Fortmann L, Hänsch H, Hahn H, Bancroft G, Müller R (1996) Liposomal amikacin for treatment of experimental *Mycobacterium avium* infections mimicking clinically relevant disease. Int J Med Microbiol Virol Parasitol Infect Dis 284: 218– 231

El-On J, Livshin R, Even-Paz Z, Hamburger D, Weinrauch L (1986) Topical treatment of cutaneous leishmaniasis. J Invest Dermatol 87: 284–288

El-On J, Halevy S, Grunwald MH, Weinrauch L (1992) Topical treatment of Old World cutaneous leishmaniasis caused by *Leishmania major*: a double-blind control study. J Am Acad Dermatol 27: 227–231

El-Ridy MS, Mostafa DM, Shehab A, Nasr EA, El-Alim SA (2007) Biological evaluation of pyrazinamide liposomes for treatment of *Mycobacterium tuberculosis*. Int J Pharm 330: 82–88

Emerich DF, Thanos CG (2006) The pinpoint promise of nanoparticle-based drug delivery and molecular diagnosis. Biomol Eng 23: 171–184

Enderlin G, Morales L, Jacobs RF, Cross LT (1994) Streptomycin and alternative agents for the treatment of tularaemia; review of literature. Clin Infect Dis 19: 42– 47

European Science Foundation (2005) Nanomedicine. An ESF-European Medical Research Councils (EMRC) Forward Look report. http://www.nanopharmaceuticals.org/files/nanomedicine.pdf

Evrard B, Bertholet P, Gueders M, Flament MP, Piel G, Delattre L, Gayot A, Leterme P, Foidart JM, Cataldo D (2004) Cyclodextrins as a potential carrier in drug nebulization. J Control Release 96: 403–410

Fang J, Nakamura H, Maeda H (2010) The EPR effect: Unique features of tumor blood vessels for drug delivery, factors involved, and limitations and augmentation of the effect. Adv Drug Del Rev [Epub ahead of print]

Fattal E, Youssef M, Couvreur P, Andremont A (1989) Treatment of experimental Salmonellosis in mice with ampicillin-bound nanoparticles. Antimicrob Agents Chemother 33: 1540–1543

Fierer J, Hatlen L, Lin JP, Estrella D, Mihalko P, Yau-Young A (1990) Successful treatment using gentamicin liposomes of Salmonella Dublin infections in mice. Antimicrob Agents Chemother 34 (2): 343–348

Finlay BB (1994) Molecular and cellular mechanisms of *Salmonella* pathogenesis. Curr Top Microbiol 192: 163–185

Finlay WH, Wong JP (1998) Regional lung deposition of nebulized liposome-encapsulated ciprofloxacin. Int J Pharm 167: 121–127

Fortier AH, Slayter MV, Ziemba R, Meltzer MS, Nacy CA (1991) Live vaccine strain of *Francisella tularensis*: infection and immunity in mice. Infect Immun 59: 2922–2928

Fortier AH, Leiby DA, Narayanan RB, Asafoadjei E, Crawford RM, Nacy CA, Meltzer MS, (1995) Growth of *Francisella tularensis* LVS in macrophages: the acidic intracellular compartment provides essential iron required for growth. Infect Immun 63: 1478–1483

Frankenburg S, Glick D, Klaus S, Barenholz Y (1998) Efficacious topical treatment for murine cutaneous leishmaniasis with ethanolic formulations of amphotericin B. Antimicrob Agents Chemother 42: 3092–3096

Frevert U, Sinnis P, Cerami PC, Shreffler W, Takacs B, Nussenzweig V (1993) Malaria circumsporozoite protein binds to heparan sulfate proteoglycans associated with the surface membrane of hepatocytes. J Exp Med 177: 1287–1298

Frézard F, Martins PS, Bahia AP, Le Moyec L, de Melo AL, Pimenta AM, Salerno M, da Silva JB, Demicheli C (2008) Enhanced oral delivery of antimony from meglumine antimoniate/beta-cyclodextrin nanoassemblies. Int J Pharm 347: 102–108

Frieden TR, Sterling TR, Munsiff SS, Watt CJ, Dye C (2003) Tuberculosis (Seminar). Lancet 362: 887–899

Fuller JR, Kijek TM, Taft-Benz S, Kawula TH (2009) Environmental and intracellular regulation of *Francisella tularensis ripA*. BMC Microbiology 9:216

Galán JE, Pace J, Hayman MJ (1992) Involvement of the epidermal growth factor receptor in the invasion of the epithelial cells by *Salmonella typhimurium*. Nature 357: 588–589

García del Portillo F, Finlay BB (1994) *Salmonella* invasion of nonphagocytic cells induces formation of macropinosomes in the host cell. Infect Immun 62: 4641–4645

García del Portillo F, Zwick MB, Leung KY, Finlay BB (1993) *Salmonella* induces the formation of filamentous structures containing lysosomal membrane glycoproteins in epithelial cells. Proc Natl Acad Sci USA 90: 10544–10548

Garcia LS, Bruckner DA (1997) Diagnostic medical parasitology, 3 rd edn. ASM Press, Washington

Gardella F, Assi S, Simon F, Bogreau H, Eggelte T, Ba F, Foumane V, Henry MC, Kientega PT, Basco L, Trape JF, Lalou R, Martelloni M, Desbordes M, Baragatti M, Briolant S, Almeras L, Pradines B, Fusai T, Rogier C (2008) Antimalarial drug use in general populations of tropical Africa. Malaria J 7: 124

Gaspar MM, Cruz A, Fraga AG, Castro AG, Cruz MEM, Pedrosa J (2008 a) Developments on drug delivery systems for the treatment of mycobacterial infections. Curr Top Med Chem 8: 579–591

Gaspar MM, Cruz A, Penha AF, Reymao J, Sousa AC, Eleuterio CV, Domingues SA, Fraga AG, Longatto Filho A, Cruz MEM, Pedrosa J (2008 b) Rifabutin encapsulated in liposomes exhibits increased therapeutic activity in a model of disseminated tuberculosis. Int J Antimicrob Agents 31: 37–45

Gelband H (2000) Regimens of less than six months for treating tuberculosis. Cochrane Database Syst Rev 2: CD001362

Gening TP, Manuilov KK (1991) Pharmacokinetics of an antibiotic administered in cell carriers. Antibiot Khimioter 36: 19–20

Ghosh P, Bacchawat BK (1980) Grafting of different glycosides on the surface of liposomes and its effect on their tissue distribution of ^{125}I-labelled gamma globulin encapsulated in liposomes. Biochem Biophys Acta 632: 562–572

Giannasca PJ, Neutra MR (1993) Interactions of microorganisms with intestinal M cells: mucosal invasion and induction of secretory immunity. Infect Agents Dis 2: 242–248

Gilbert P, Collier PJ, Brown MRW (1990) Influence of growth rate on susceptibility to antimicrobial agents: biofilms, cell cycle, dormancy and stringent response. Antimicrob Agents Chemother 34: 1865–1868

Ginocchio C, Olmsted SB, Wells CL, Galán JE (1994) Contact with epithelial cells induces the formation of surface appendages on *Salmonella typhimurium*. Cell 76: 717–724

Godreuil S, Tazi L, Bañuls A-L (2007) Chapter 1: Pulmonary Tuberculosis and *Mycobacterium Tuberculosis*: Modern Molecular Epidemiology and Perspectives. In: Tibayrenc M (ed) Encyclopedia of Infectious Diseases: Modern Methodologies, John Wiley & Sons Inc, New York

Gohel MC, Sarvaiya KG (2007) A novel solid dosage form of rifampicin and isoniazid with improved functionality. AAPS PharmSciTech 8 Article 68

Golovliov I, Baranov V, Krocova Z, Kovarova H, Sjöstedt A (2003) An attenuated strain of the facultative intracellular bacterium *Francisella tularensis* can escape the phagosome of monocytic cells. Infect Immun 71: 5940–5950

Gonzalez-Martin G, Merino I, Rodriguez-Cabezas MN, Torres M, Nuñez R, Osuna A (1998) Characterization and trypanocidal activity of nifurtimox-containing and empty nanoparticles of polyethylcyanoacrylates. J Pharm Pharmacol 50: 29–35

Gonzalez-Martín G, Figueroa C, Merino I, Osuna A (2000) Allopurinol encapsulated in polycyanoacrylate nanoparticles as potential lysosomatropic carrier: preparation and trypanocidal activity. Eur J Pharm Biopharm 49: 137–142

Goodyer ID, Pouvelle B, Schneider TG, Trelka DP, Taraschi TF (1997) Characterization of macromolecular transport pathways in malaria-infected erythrocytes. Mol Biochem Parasitol 87: 13–28

Gradoni L, Davidson RN, Orsini S, Betto P, Giambenedetti M (1993) Activity of liposomal amphotericin B (AmBisome) against *Leishmania infantum* and tissue distribution in mice. J Drug Target 1: 311–316

Grassl GA, Finlay BB (2008) Pathogenesis of enteric *Salmonella* infections. Curr Opin Gastroenterol 24: 22–26

Grayston JT, Kuo CC, Campbell LA, Wang SP (1989) *Chlamydia pneumoniae* sp. nov. for *Chlamydia* sp. strain TWAR. Int J Syst Bacteriol 39: 88–90

Grayston JT, Aldous MB, Easton A, Wang SP, Kuo CC, Campbell LA, Altman J (1993) Evidence that Chlamydia pneumoniae causes pneumonia and bronchitis. J Infect Dis 168: 1231–1235

Greenwood BM, Fidock DA, Kyle DE, Kappe SHI, Alonso PL, Collins FH, Duffy PE (2008) Malaria: Progress, perils, and prospects for eradication. J Clin Invest 118: 1266–1276

Groisman EA (1997) How *Salmonella* became a pathogen. Trends Microbiol 5: 343–349

Gupta R, Kim JY, Espinal MA, Caudron J-M, Pecoul B, Farmer PE, Raviglione MC (2001) Responding to market failures in tuberculosis control. Science 293: 1049–1051

Gupta Y, Jain A, Jain SK (2007) Transferrin-conjugated solid lipid nanoparticles for enhanced delivery of quinine dihydrochloride to the brain. J Pharm Pharmacol 59: 935–940

Guru PY, Agrawal AK, Singha UK, Singhal A, Gupta CM (1989) Drug targeting in *Leishmania donovani* infections using tuftsin-bearing liposomes as drug vehicles. FEBS Lett 245: 204–208

Gutierrez MG, Master SS, Singh SB, Taylor GA, Colombo MI, Deretic V (2004) Autophagy is a defense mechanism inhibiting BCG and *Mycobacterium tuberculosis* survival in infected macrophages. Cell 119: 753–766

Gutiérrez Millán C, Castañeda AZ, López FG Marinero ML, Lanao JM, Arévalo M (2005) Encapsulation and *in vitro* evaluation of amikacin-loaded erythrocytes. Drug Deliv 12: 409–416

Gutiérrez Millán C, Castaneda AZ, López FG, Marinero ML, Lanao JM (2008) Pharmacokinetics and biodistribution of amikacin encapsulated in carrier erythrocytes. J Antimicrobial Chemother 61: 375–381

Haas SE, Bettoni CC, de Oliveira LK, Guterres SS, Dalla Costa T (2009) Nanoencapsulation increases quinine antimalarial efficacy against *Plasmodium berghei in vivo. Int J Antimicrob Agents* 34: 156–161

Hammerschlag MR, Chirgwing K, Roblin PM, Gelling M, Dumornay W, Mandel L, Smith P, Schachter J (1992) Persistent infection with Chlamydia pneumoniae following acute respiratory illness. Clin Infect Dis 14: 178–182

Hanan MS, Riad EM, el-Khouly NA (2000) Antibacterial efficacy and pharmacokinetic studies of ciprofloxacin on *Pasteurella multocida* infected rabbits. Dtsch Tierarztl Wochenschr 107: 151–155

Hand WL, King-Thompson N, Holman JW (1987) Entry of roxithromycin (RU 965), imipenem, cefotaxime, trimethoprim, and metronidazole into human polymorphonuclear leukocytes, Antimicrob Agents Chemother 31(10): 1553–1557

Haynes SM, Longmuir KJ, Robertson RT, Baratta JL, Waring AJ (2008) Liposomal polyethyleneglycol and polyethyleneglycol-peptide combinations for active targeting to liver *in vivo.* Drug Deliv 15: 207–217

Herwaldt BL (1999) Leishmaniasis. Lancet 354: 1191–1199

Hickey AJ, Martonen TB, Yang Y (1996) Theoretical relationship of lung deposition to the fine particle fraction of inhalation aerosols. Pharm Acta Helv 71: 185–190

Hirota K, Hasegawa T, Hinata H, Ito F, Inagawa H, Kochi C, Soma G, Makino K, Terada H (2007) Optimum conditions for efficient phagocytosis of rifampicin-loaded PLGA microspheres by alveolar macrophages. J Control Release 119: 69–76

Hirota K, Tomoda K, Inagawa H, Kohchi C, Soma G, Makino K, Terada H (2008) Stimulation of phagocytic activity of alveolar macrophages toward artificial microspheres by infection with mycobacteria. Pharm Res 25: 1420–1430

Hook EW (1990) *Salmonella* species (including typhoid fever). In: Mandell GL, Douglas RG Jr (eds) Principles and practice of infectiuos disease, 3 rd edn. Churchill Livingstonc, New York

Hope MJ, Wong KF (1995) Liposomal formulation of ciprofloxacin. In: Shek PN (ed) Liposomes in biomedical applications. Harwood Academic Publishers, Germany

Hope JC, Thom ML, McCormick PA, Howard CJ (2004) Interaction of antigen presenting cells with mycobacteria. Vet Immunol Immunopathol 100: 187–195

Hwang KJ (1987) Liposome pharmacokinetics. In: Ostro MJ (ed) Liposomes: from biophysics to therapeutics. Marcel Dekker, Inc., New York

Irache JM, Salman HH, Gamazo C, Espuelas S (2008) Mannose-targeted systems for the delivery of therapeutics. Expert Opin Drug Deliv 5: 703–724

Iraji F, Sadeghinia A (2005) Efficacy of paromomycin ointment in the treatment of cutaneous leishmaniasis: results of a double-blind, randomized trial in Isfahan, Iran. Ann Trop Med Parasitol 99: 3–9

ISO/TS 27687:2008 nanotechnologies – terminology and definitions for nano-objects – nanoparticle, nanofibre and nanoplate

Jaafari MR, Bavarsad N, Bazzaz BSF, Samiei A, Soroush D, Ghorbani S, Heravi MML, Khamesipour A (2009) Effect of topical liposomes containing paromomycin sulfate in the course of *Leishmania major* infection in susceptible BALB/c mice. Antimicrob Agents Chemother 53: 2259–2265

Jain CP, Vyas SP, Dixit VK (2006) Niosomal system for delivery of rifampicin to lymphatics. Indian J Pharm Sci 68: 575–578

Jain SK, Lamichhane G, Nimmagadda S, Pomper MG, Bishai WR (2008) Antibiotic treatment of tuberculosis: old problems, newsolutions. Microbe 3: 285–292

Johnson CM, Pandey R, Sharma S, Khuller GK, Basaraba RJ, Orme IM, Lenaerts AJ, (2005) Oral therapy using nanoparticle-encapsulated antituberculosis drugs in guinea pigs infected with Mycobacterium tuberculosis. Antimicrob Agents Chemother 49 (10): 4335–4338

Joshi M, Dwyer DM, Nakhasi HL (1993) Cloning and characterization of differentially expressed genes from *in vitro*-grown 'amastigotes' of *Leishmania donovani*. Mol Biochem Parasitol 58: 345–354

Joshi M, Pathak S, Sharma S, Patravale V (2008 a) Design and *in vivo* pharmacodynamic evaluation of nanostructured lipid carriers for parenteral delivery of artemether: Nanoject. Int J Pharm 364: 119–126

Joshi M, Pathak S, Sharma S, Patravale V (2008 b) Solid microemulsion preconcentrate (NanOsorb) of artemether for effective treatment of malaria. Int J Pharm 362: 172–178

Kansal RG, Gómez-Flores R, Sinha I, Mehta RT (1997) Therapeutic efficacy of liposomal clofazimine against Mycobacterium avium complex in mice depends on size of initial inoculum and duration of infection. Antimicrob Agents Chemother 41: 17–23

Kaprelyants AS, Gottschal JC, Kell DB (1993) Dormancy in nonsporulating bacteria. FEMS Microbiol Reviews 104: 271– 286

Katlama C, De Wit S, O'Doherty E, Van Glabeke M, Clumeck N (1996) Pyrimethamine-clindamycin versus pyrimethamine-sulfadiazine as acute and long-term therapy for toxoplasmic encephalitis in patients with AIDS. Clin Infect Dis 22: 268–275

Kaufmann SH, McMichael AJ (2005) Annulling a dangerous liaison: vaccination strategies against AIDS and tuberculosis. Nat Med 11: S33–S44

Kaur A, Jain S, Tiwary AK (2008). Mannan-coated gelatin nanoparticles for sustained and targeted delivery of didanosine: *in vitro* and *in vivo* evaluation. Acta Pharm 58: 61–74

Kayser O, Olbrich C, Yardley V, Kiderlen AF, Croft SL (2003) Formulation of amphotericin B as nanosuspension for oral administration. Int J Pharm 254: 73–75

Kisich KO, Gelperina S, Higgins MP, Wilson S, Shipulo E, Oganesyan E, Heifets L (2007) Encapsulation of moxifloxacin within poly(butyl cyanoacrylate) nanoparticles enhances efficacy against intracellular *Mycobacterium tuberculosis*. Int J Pharm 345: 154–162

Klemens SP, Cynamon MH, Swenson CE, Ginsberg RS (1990) Liposome-encapsulated-gentamicin therapy of *Mycobacterium avium* complex infection in beige mice. Antimicrob Agents Chemother 34: 967–970

Koirala J (2010) *Mycobacterium Avium*–Intracellulare. http://emedicine.medscape.com/article/222664-overview

Krettli AU, Miller LH (2001) Malaria: A sporozoite runs through it. Curr Biol 11: R409–R412

Krettli AU, Andrade-Neto VF, Brandão MGL, Ferrari WMS (2001) The search for new antimalarial drugs from plants used to treat fever and malaria or plants randomly selected: A review. Mem I Oswaldo Cruz 96: 1033–1042

Kuhlmann J, Schaefer H.G., Beermann D. (1998) Clinical pharmacology. In: Kuhlmann J, Dalhoff A, Zeiler HJ (eds) Quinolone Antibacterials. Handbook of Experimental Pharmacology 127, Springer-Verlag, Berlin

Kumana CR, Yuen KY (1994) Parenteral aminoglycoside therapy. Selection, administration and monitoring. Drugs 47: 902–913

Kumar Y, Valdivia RH (2009) Leading a sheltered life: intracellular pathogens and maintenance of vacuolar compartments. Cell Host & Microbe 5: 593–601

Labro MT (1996) Intracellular bioactivity of macrolides. Clin Microbiol Infect 1: S24–S30

Lacy MK, Nicolau DP, Nightingale CH, Quintiliani R (1998) The pharmacodynamics of aminoglycosides. Clinical Infectious Diseases 27: 23–7

Lai XH, Golovliov I, Sjostedt A (2001) *Francisella tularensis* induces cytopathogenicity and apoptosis in murine macrophages via a mechanism that requires intracellular bacterial multiplication. Infect Immun 69: 4691–4694

Lai SK, O'Hanlon DE, Harrold S, Man ST, Wang YY, Cone R, Hanes J (2007) Rapid transport of large polymeric nanoparticles in fresh undiluted human mucus. Proc Natl Acad Sci USA 104: 1482–1487

Lane KB, Egan B, Vick S, Abdolrasulnia R, Shepherd VL (1998) Characterization of a rat alveolar macrophage cell line that expresses a functional mannose receptor. J Leukoc Biol 64: 345–350

Lauer SA, Rathod PK, Ghori N, Haldar K (1997) A membrane network for nutrient import in red cells infected with the malaria parasite. Science 276: 1122–1125

Laube BL (2005) The expanding role of aerosols in systemic drug delivery, gene delivery and vaccination. Respir Care 50: 1161–1176

Laura JVP, Kerstin JW, Vito, R (2000) Accumulation of rifampicin by *Mycobacterium aurum*, *Mycobacterium smegmatis* and *Mycobacterium tuberculosis*. J Antimicrob Chemother 45: 159–165

Leiby DA, Herron JRM, Read EJ, Lenes BA, Stumpf RJ (2002) *Trypanosoma cruzi* in Los Angeles and Miami blood donors: impact of evolving donor demographics on seroprevalence and implications for transfusion transmission. Transfusion 42: 549–555

Leite EA, Grabe-Guimarães A, Guimarães HN, Machado-Coelho GLL, Barratt G, Mosqueira VCF (2007) Cardiotoxicity reduction induced by halofantrine entrapped in nanocapsule devices. Life Sci 80: 1327–1334

Lenaerts AJ, Gruppo V, Marietta KS, Johnson CM, Driscoll DK, Tompkins NM, Rose JD, Reynolds RC, Orme IM (2005) Preclinical testing of the nitroimidazopyran PA-824 for activity against Mycobacterium tuberculosis in a series of *in vitro* and *in vivo* models. Antimicrob Agents Chemother 49: 2294–2301

Leitzke S, Bucke W, Borner K, Muller R, Hahn H, Ehlers S (1998) Rationale for and efficacy of prolonged-interval treatment using liposome-encapsulated amikacin in experimental *Mycobacterium avium* infection. Antimicrob Agents Chemother 42: 459–461

Lim SA 2006 Ethambutol-associated Optic Neuropathy. Ann Acad Med Singapore 35: 274–278

Longmuir KJ, Robertson RT, Haynes SM, Baratta JL, Waring AJ (2006) Effective targeting of liposomes to liver and hepatocytes *in vivo* by incorporation of a Plasmodium amino acid sequence. Pharm Res 23: 759–769

Lortholary O, Tod M, Cohen Y, Petitjean O (1995) Aminoglycosides. Medical Clinics of North America 79: 761–87

Lyon M, Deakin JA, Gallagher JT (1994) Liver heparin sulfate structure. A novel molecular design. J Biol Chem 269: 11208–11215

Maertens J, Boogaerts MA (1998) Anti-infective strategies in neutropenic patients. Acta Clinica Belgica 53: 168–77

Maurin M, Raoult D (1997) Intracellular organisms. Int J Antimicrob Agents 9: 61–70

Magallanes M, Dijkstra J, Fiererl J (1993) Liposome-Incorporated ciprofloxacin in treatment of murine Salmonellosis. Antimicrob Agents Chemother 37: 2293–2297

Mandawgade SD, Sharma S, Pathak S, Patravale VB (2008) Development of SMEDDS using natural lipophile: Application to beta-Artemether delivery. Int J Pharm 362: 179–183

McCormick BA (2004) Chapter 1: Invasion mechanisms of *Salmonella*. In: Lamont RJ (ed) Advances in molecular and cell microbiology. Bacterial invasion of host cells. Cambridge University Press, Cambridge

McOrist S (2000) Obligate intracellular bacteria and antibiotic resistance. Trends Microbiol 8: 483

Mehta RT (1996) Liposome encapsulation of clofazimine reduces toxicity *in vitro* and *in vivo* and improves therapeutic efficacy in the beige mouse model of disseminated *Mycobacterium avium-M. intracellulare* complex infection. Antimicrob Agents Chemother 40: 1893–1902

Ménard R (2000) The journey of the malaria sporozoite through its hosts: two parasite proteins lead the way. Microbes Infect 2: 633–642

Miao L, Stafford A, Nir S, Turco SJ, Flanagan TD, Epand RM (1995) Potent inhibition of viral fusion by the lipophosphoglycan of *Leishmania donovani*. Biochemistry 34: 4676–4683

Minodier P, Parola P (2007) Cutaneous leishmaniasis treatment. Travel Med Infect Dis 5(3): 150–158

Minodier P, Retornaz K, Horelt A, Garnier JM (2003) Liposomal amphotericin B in the treatment of visceral leishmaniasis in immunocompetent patients. Fundam Clin Pharmacol 17: 183–188

Misra A, Hickey AJ, Rossi C, Borchard G, Terada H, Makino K, Fourie PB, Colombo P (2010) Inhaled drug therapy for treatment of tuberculosis. Tuberculosis [Epub ahead of print]

Mitchell DJ, Kim DT, Steinman L, Fathman CG, Rothbard JB (2000) Polyarginine enters cells more efficiently than other polycationic homopolymers. J Peptide Res 56, 318–325

Mitchison DA (2005) Drug resistance in tuberculosis. Eur Respir J 25: 376–379

Mohammadi G, Valizadeh H, Barzegar-Jalali M, Lotfipour F, Adibkia K, Milani M, Azhdarzadeh M, Kiafar F, Nokhodchi A (2010) Development of azithromycin–PLGA nanoparticles: Physicochemical characterization and antibacterial effect against *Salmonella typhi*. Colloids Surf B Biointerfaces 80: 34–39

Molina J, Brener Z, Romanha A, Urbina JA (2000) *In vivo* activity of the bis-triazole D0870 against drug-susceptible and drug-resistant strains of the protozoan parasite *Trypanosoma cruzi*. J Antimicrob Chemother 46: 137–140

Molina J, Urbina JA, Gref R, Brener Z, Maciel Rodrigues JJ (2001) Cure of experimental Chagas's disease by the bis-triazole D0870 incorporated into stealth polyethyleneglycol-polylactide nanospheres. J Antimicrob Chemother 47: 101–104

Montanari J, Maidana C, Esteva MI, Salomon C, Morilla MJ, Romero EL (2010) Sunlight triggered photodynamic ultradeformable liposomes against *Leishmania braziliensis* are also leishmanicidal in the dark. J Control Release 147: 368–376

Morilla MJ, Benavidez P, Lopez MO, Bakas L, Romero EL (2002) Development and *in vitro* characterisation of a benznidazole liposomal formulation. Int J Pharm 249: 89–99

Morilla MJ, Montanari J, Frank F, Malchiodi E, Corral R, Petray P, Romero EL (2005) Etanidazole in pH-sensitive liposomes: design, characterization and *in vitro/in vivo* anti-*Trypanosoma cruzi* activity. J Control Release 103: 599–607

Morilla MJ, Montanari JA, Prieto MJ, Lopez MO, Petray PB, Romero EL (2004) Intravenous liposomal benznidazole as trypanocidal agent: increasing drug delivery to liver is not enough. Int J Pharm 278: 311–318

Mosqueira VCF, Loiseau PM, Bories C, Legrand P, Devissaguet JP, Barratt G (2004) Efficacy and pharmacokinetics of intravenous nanocapsule formulations of halofantrine in *Plasmodium berghei*-infected mice. Antimicrob Agents Chemother 48: 1222–1228.

Mota MM, Rodriguez A (2002) Invasion of mammalian host cells by Plasmodium sporozoites. Bioessays 24: 149–156

Mugabe C, Azghani AO, Omri A (2006) Preparation and characterization of dehydration–rehydration vesicles loaded with aminoglycoside and macrolide antibiotics. Int J Pharm 307: 244–250

Mullaicharam AR, Murthy RSR (2004) Lung accumulation of niosome-entrapped rifampicin following intravenous and intratracheal administration in the rat. J Drug Deliv Sci Technol 14: 99–104

Munusamy P, Seleem MN, Alqublan H, Tyler R Jr, Sriranganathan N, Pickrell G (2009) Targeted drug delivery using silica xerogel systems to treat diseases due to intracellular pathogens. Mater Sci Eng C 29: 2313–2318

Murray HW, Berman JD, Davies RR, Saravia NG (2005) Advances in Leishmaniasis. Lancet 366: 1561–1577

Murta S, Gazzinelli R, Brener Z, Romanha A (1998) Molecular characterization of susceptible and naturally resistant strains of *Trypanosoma cruzi* to benznidazole and nifurtimox. Mol Biochem Parasitol 93: 203–214

Muttil P, Kaur J, Kumar K, Yadav AB, Sharma R, Misra A (2007) Inhalable microparticles containing large payload of anti-tuberculosis drugs. Eur J Pharm Sci 32: 140–150

Myers JP (2005) New recommendations for the treatment of tuberculosis. Curr Opin Infect Dis 18: 133–140

Nagai J, Takano M 2004 Molecular aspects of renal handling of aminoglycosides and strategies for preventing the nephrotoxicity. Drug Metab Pharmacokinet 19(3): 159–70

Nan A, Croft SL, Yardley V, Ghandehari H (2004) Targetable water-soluble polymer–drug conjugates for the treatment of visceral leishmaniasis. J Control Release 94: 115–27

Nelson KG, Bishop JV, Ryan RO, Titus R (2006) Nanodisk-associated amphotericin B clears leishmanis major cutaneous infection in susceptible Balb/c mice. Antimicrob Agents Chemother 50: 1238–1244

New RRC, Chance ML, Thomas SC, Peters W (1978) Antileishmanial activity of antimonials entrapped in liposomes. Nature 272: 55–56

New RRC, Chance ML (1980) Treatment of experimental cutaneous leishmaniasis by liposome-entrapped Pentostam. Acta Trop 37: 253–256

New RRC, Chance ML, Heath S (1981) The treatment of experimental cutaneous leishmaniasis with liposome-entrapped Pentostam. Parasitology 83: 519–527

Nicoletti S, Seifert K, Gilbert IH (2009) N-(2-hydroxypropyl)methacrylamide-amphotericin B (HPMA-AmB) copolymer conjugates as antileishmanial agents. Int J Antimicrob Agents 33: 441–448

Nicoletti S, Seifert K, Gilbert IH (2010) Water-soluble polymer–drug conjugates for combination chemotherapy against visceral leishmaniasis. Bioorg Med Chem 18: 2559–2565

Nix DE, Goodwin D, Peloquin CA, Rotella DL, Schentag JJ (1991) Antibiotic tissue penetration and its relevance: models of tissue penetration and their meaning. Antimicrob Agents Chemother 35: 1947–1950

Nogueira N, Cohn Z (1976) *Trypanosoma cruzi*: mechanism of entry and intracellular fate in mammalian cells. J Exp Med 143: 1402–1420

Ohashi K, Kabasawa T, Ozeki T, Okada H (2009) One-step preparation of rifampicin/ poly(lactic-co-glycolic acid) nanoparticle-containing mannitol microspheres using a four-fluid nozzle spray drier for inhalation therapy of tuberculosis. J Control Release 135: 19–24

Onoue S, Misaka S, Kawabata Y, Yamada S (2009) New treatments for chronic obstructive pulmonary disease and viable formulation/device options for inhalation therapy. Expert Opin Drug Deliv 6: 793–811

Owais M, Varshney GC, Choudhury A, Chandra S, Gupta CM (1995) Chloroquine encapsulated in malaria-infected erythrocyte-specific antibody-bearing liposomes effectively controls chloroquine-resistant *Plasmodium berghei* infections in mice. Antimicrob Agents Chemother 39: 180–184

Owens MD, Warren DA (2010) Salmonella Infection. http://emedicine.medscape.com/article/785774-overview. Updated: May 20, 2010

Pandey R, Khuller GK (2004) Subcutaneous nanoparticle-based antitubercular chemotherapy in an experimental model. J Antimicrob Chemother 54: 266–268

Pandey R, Sharma s, Khuller GK (2005) Oral solid lipid nanoparticle based antitubercular chemotherapy. Tuberculosis 85: 415–420

Pandey R, Khuller GK (2006) Oral nanoparticle-based antituberculosis drug delivery to the brain in an experimental model. J Antimicrob Chemother 57: 1146–1152

Pandey R, Zahoor A, Sharma S, Khuller GK (2003 a) Nanoparticle encapsulated antitubercular drugs as a potential oral drug delivery system against murine tuberculosis. Tuberculosis 83: 373–378

Pandey R, Sharma A, Zahoor A, Sharma S, Khuller GK, Prasad B (2003 b) Poly (DL-lactide-co-glycolide) nanoparticle-based inhalable sustained drug delivery system for experimental tuberculosis. J Antimicrob Chemother 52: 981–986

Patton JS, Byron PR (2007) Inhaling medicines: delivering drugs to the body through lungs. Nat Rev Drug Discov 6: 67–74

Peek LJ, Middaugh CR, Berkland C (2008) Nanotechnology in vaccine delivery. Adv Drug Del Rev 60: 915–928

Peeters PAM, Brunink BG, Eling WMC, Crommelin DJA (1989 a) Therapeutic effect of chloroquine (CQ)-containing immunoliposomes in rats infected with *Plasmodium berghei* parasitized mouse red blood cells: Comparison with combinations of antibodies and CQ or liposomal CQ. Biochem Biophys Acta Biomembranes 981: 269–276

Peeters PAM, De Leest K, Eling WMC, Crommelin DJA (1989 b) Chloroquine blood levels after administration of the liposome-encapsulated drug in relation to therapy of murine malaria. Pharm Res 6: 787–793

Perrin AJ, Jiang X, Birmingham CL, So NSY, Brumell JH (2004) Recognition of bacteria in the cytosol of mammalian cells by the ubiquitin system. Curr Biol 14: 806–811

Peters K, Leitzke S, Diederichs JE, Borner K, Hahn H, Müller RH, Ehlers S (2000) Preparation of a clofazimine nanosuspension for intravenous use and evaluation of its therapeutic efficacy in murine *Mycobacterium avium* infection. J Antimicrob Chemother 45: 77–83

Petray PB, Morilla MJ, Corral RS, Romero EL (2004) *In vitro* activity of etanidazole against the protozoan parasite *Trypanosoma cruzi*. Mem Inst Oswaldo Cruz 99: 233–235

Petrelli F, Borgonovo K, Barni S (2010) Targeted delivery for breast cancer therapy: the history of nanoparticle-albumin-bound paclitaxel. Expert Opin Pharmacother 11(8): 1413–32

Philips JA (2008) Mycobacterial manipulation of vacuolar sorting. Cell Microbiol 10: 2408–2415

Pirson P, Steiger R, Trouet A (1982) The disposition of free and liposomally encapsulated antimalarial primaquine in mice. Biochem Pharmacol 31: 3501–3507

Pissuwan D, Valenzuela SM, Miller CM, Cortie MB (2007) A golden bullet? Selective targeting of *Toxoplasma gondii* tachyzoites using antibody-functionalized gold nanorods. Nano Letters 7: 3808–3812

Pissuwan D, Valenzuela SM, Miller CM, Killingsworth MC, Cortie MB (2009) Destruction and control of *Toxoplasma gondii* tachyzoites using gold nanosphere/antibody conjugates. Small 5(9): 1030–1034

Poelma DL, Zimmermann LJI, Scholten HH, Lachmann B, Iwaarden van JF (2002) *In vivo* and *in vitro* uptake of surfactant lipids by alveolar type II cells and macrophages. Am J Physiol Lung Cell Mol Physiol 283: L648–L654

Poole K (2005) Efflux mediated antimicrobial resistance. J Antimicrob Chemother 56: 20–51

Postma NS, Hermsen CC, Zuidema J, Eling WMC (1998) *Plasmodium vinckei*: Optimization of desferrioxamine B delivery in the treatment of murine malaria. Exp Parasitol 89: 323–330

Pouvelle B, Spiegel R, Hsiao L, Howard RJ, Morris RL, Thomas AP, Taraschi TF (1991) Direct access to serum macromolecules by intraerythrocytic malaria parasites. Nature 353: 73–75

Pouvelle B, Matarazzo V, Jurzynski C, Nemeth J, Ramharter M, Rougon G, Gysin J (2007) Neural cell adhesion molecule, a new cytoadhesion receptor for *Plasmodium falciparum*-infected erythrocytes capable of aggregation. Infect Immun 75: 3516–3522

Pradel G, Garapaty S, Frevert U (2002) Proteoglycans mediate malaria sporozoite targeting to the liver. Mol Microbiol 45: 637–651

Prieto MJ, Bacigalupe D, Pardini O, Amalvy JI, Venturini C, Morilla MJ, Romero EL (2006) Nanomolar cationic dendrimeric sulfadiazine as potential antitoxoplasmic agent. Int J Pharm 326: 160–168

Prokesch RC, Hand WL (1982) Antibiotic entry into human polymorphonuclear leukocytes. Antimicrob Agents Chemother 21: 373–380

Rakita RM, Jacques-Palaz K, Murray BE (1994) Intracellular activity of azithromycin against bacterial enteric pathogens. Antimicrob Agents Chemother 38: 1915–1921

Ranjan A, Pothayee N, Seleem MN, Sriranganathan N, Kasimanickam R, Makris M, Riffle JS (2009) *In Vitro* trafficking and efficacy of core-shell nanostructures for treating intracellular *Salmonella* Infections. Antimicrob Agents Chemother 53: 3985–3988

Ranjan A, Pothayee N, Vadala TP, Seleem MN, Restis E, Sriranganathan N, Riffle JS, Kasimanickam R (2010) Efficacy of amphiphilic core-shell nanostructures encapsulating gentamicin in an *in vitro Salmonella* and *Listeria* intracellular infection model. Antimicrob Agents Chemother 54: 3524–3526

Rannard S, Owen A (2009) Nanomedicine: Not a case of "One size fits all". Nano Today 4: 382–384

Rapp C, Imbert P, Darie H, Simon F, Gros P, Debord T, Roué R (2003) Traitement par amphotéricine B liposomale d´une leishmaniose cutanée contractée à Djibouti et résistante à l´antimoniate de méglumine [Liposomal amphotericin B treatment of cutaneous leishmaniasis contracted in Djibouti and resistant to meglumine antimoniate]. Bull Soc Pathol Exot 96: 209–211

Rastogi N, Blom Potar MC (1990) Intracellular bactericidal activity of ciprofloxacin and ofloxacin against *Mycobacterium tuberculosis* H37Rv multiplying in the J-775 macrophage cell line. Zbl Bakt 273: 195–199

Richter-Dahlfors A, Buchan AMJ, Finlay BB (1997) Murine salmonellosis studies by confocal microscopy: *Salmonella typhimurium* resides intracellularly inside macrophages and exerts a cytotoxic effect on phagocytes *in vivo*. J Exp Med 186: 569–580

Roberts RL, Fine RE, Sandra A (1993) Receptor-mediated endocytosis of transferrin at the blood-brain barrier. J Cell Sci 104: 521–532

Rodriguez A, Samoff E, Rioult MG, Chung A, Andrews NW (1996) Host cell invasion by trypanosomes requires lysosomes and microtubule/kinesin-mediated transport. J Cell Biol 134: 349–362

Rodrigues Coura J, de Castro SL (2002) A critical review on Chagas disease chemotherapy. Mem Inst Oswaldo Cruz 97: 3–24

Rogueda PG (2005) Novel hydrofluoroalkane suspension formulations for respiratory drug delivery. Expert Opin Drug Deliv 2: 625–638

Rogueda PG, Traini D (2007) The nanoscale in pulmonary delivery. Part 1: deposition, fate, toxicology and effects. Expert Opin Drug Del 4: 595–606

Romero EL, Morilla MJ (2008) Drug delivery systems against leishmaniasis? Still an open question. Expert Opin Drug Deliv 5(7): 805–823

Romero EL, Morilla MJ (2010) Nanotechnological approaches against Chagas disease. Adv Drug Del Rev 62: 576–88

Rook GAW, Hernandez-Pando R (1996) The pathogenesis of tuberculosis. Annu Rev Microbiol 50: 259–284

Roseeuw E, Coessens V, Balazuc AM, Lagranderie M, Chavarot P, Pessina A, Neri MG, Schacht E, Marchal G, Domurado D (2003) Synthesis, degradation, and antimicrobial properties of targeted macromolecular prodrugs of norfloxacin. Antimicrob Agents Chemother 47: 3435–3441

Rothbard JB, Garlington S, Lin Q, Kirschberg T, Kreider E, McGrane PL, Wender PA, Khavari PA (2000) Conjugation of arginine oligomers to cyclosporin A facilitates topical delivery and inhibition of inflammation.. Nat Med 6: 1253–1257

Ruenraroengsak P, Cook JM, Florence AT (2010) Nanosystem drug targeting: Facing up to complex realities. J Control Release 141: 265–276

Sahay G, Gautam V, Luxenhofer R, Kabanov AV (2010) The utilization of pathogen-like cellular trafficking by single chain block copolymer. Biomaterials 31: 1757–1764

Sakagami M (2006) *In vivo, in vitro* and *ex vivo* models to assess pulmonary absorption and disposition of inhaled therapeutics for systemic delivery. Adv Drug Deliver Rev 58: 1030–1060

Sakamoto JH, van de Ven AL, Godin B, Blanco E, Serda RE, Grattoni A, Ziemys A, Bouamrani A, Hu T, Ranganathan SI, De Rosa E, Martinez JO, Smid CA, Buchanan RM, Lee S-Y, Srinivasan S, Landry M, Meyn A, Tasciotti E, Liu X, Decuzzi P, Ferrari M (2010) Enabling individualized therapy through nanotechnology. Pharmacol Res 62: 57–89

Samuel BU, Hearn B, Mack D, Wender P, Rothbard J, Kirisits MJ, Mui E, Wernimont S, Roberts CW, Muench SP, Rice DW, Prigge ST, Law AB, McLeod R (2003) Delivery of antimicrobials into parasites. Proc Natl Acad Sci USA 24: 14281–14286

Sánchez G, Cuellar D, Zulantay I, Gajardo M, González-Martin G (2002) Cytotoxicity and trypanocidal activity of nifurtimox encapsulated in ethylcyanoacrylate nanoparticles. Biol Res 35: 39–45

Santic]M, Molmeret M, Abu Kwaik Y (2005) Modulation of biogenesis of the *Francisella tularensis* subsp. *novicida*-containing phagosome in quiescent human macrophages and its maturation into a phagolysosome upon activation by IFN-gamma. Cell Microbiol 7: 957– 967

Santos-Magalhães NS, Furtado Mosqueira VC (2010) Nanotechnology applied to the treatment of malaria. Adv Drug Del Rev 62: 560–575

Saraogi GK, Gupta P, Gupta UD, Jain NK, Agrawal GP (2010) Gelatin nanocarriers as potential vectors for effective management of Tuberculosis. Int J Pharm 385: 143–149

Schenkman S, Andrews NW, Nussenzweig V, Robbins ES (1988) *Trypanosoma cruzi* invade a mammalian epithelial cell in a polarized manner. Cell 55: 157–165

Schiffelers R, Storm G, Bakker-Woudenberg I (2001) Liposomes-encapsulated aminoglycosides in pre-clinical and clinical studies. J Antimicrob Chemother 48: 333–344

Schluger NW (2005) The pathogenesis of tuberculosis. The first one hundred (and twenty-three) years. Am J Respir Cell Mol Biol 32: 251–256

Schöler N, Krause K, Kayser O, Müller RH, Borner K, Hahn H, Liesenfeld O (2001) Atovaquone Nanosuspensions show excellent therapeutic effect in a new murine model of reactivated toxoplasmosis. Antimocrob Agents Chemother 45: 1771–1779

Schwab JC, Beckers CJ, Joiner KA (1994) The parasitophorous vacuole membrane surrounding intracellular *Toxoplasma gondii* functions as a molecular sieve. Proc Natl Acad Sci USA 91: 509–513

Sepkowitz KA, Raffalli J, Riley L, Kiehn TE, Armstrong D (1995) Tuberculosis in the AIDS era. Clin Microbiol Rev 8: 180–199

Sharma R, Saxena D, Dwivedi AK, Misra A (2001) Inhalable microparticles containing drug combinations to target alveolar macrophages for treatment of pulmonary tuberculosis. Pharm Res 18: 1405–1410

Sharma A, Sharma S, Khuller GK (2004) Lectin functionalised poly(lactide-coglycolide) nanoparticles as oral/aerosolized antitubercular drug carriers for treatment of tuberculosis. J Antimicrob Chemother 54: 761–766

Shipulo EV, Lyubimov II, Maksimenko OO, Vanchugova LV, Oganesyan EA, Sveshnikov PG, Biketov SF, Severin ES, Heifets LB, Gelperina SE (2008) Development of a nanosomal formulation of moxifloxacin based on poly (butyl-2-cyanoacrylate). Pharm Chem J 42: 145–149

Singh KK, Vingkar SK (2008) Formulation, antimalarial activity and biodistribution of oral lipid nanoemulsion of primaquine. Int J Pharm 347: 136–143

Smith CR (1988) The spiramycin paradox. J Antimicrob Chemother 22 (Suppl. B): 1141–1444

Smith I (2003) Mycobacterium tuberculosis pathogenesis and molecular determinants of virulence. Clin Microbiol Rev 16: 463–496
Smith JE, Pirson P, Sinden RE (1983) Studies on the kinetics of uptake and distribution of free and liposome-entrapped primaquine, and of sporozoites by isolated perfused rat liver. Ann Trop Med Parasitol 77: 379–386
Solomon M, Baum S, Barzilai A, Scope A, Trau H, Schwartz E (2007) Liposomal amphotericin B in comparison to sodium stibogluconate for cutaneous infection due to *Leishmania braziliensis*. J Am Acad Dermatol 56: 612–616
Sosa Estani S, Segura EL (1999) Treatment of *Trypanosoma cruzi* infection in the undetermined phase. Experience and current guidelines in Argentina. Mem Inst Oswaldo Cruz 94: 363–365
Sosa Estani S, Segura EL, Ruiz AM, Velazquez E, Porcel BM, Yampotis C (1998) Efficacy of chemotherapy with benznidazole in children in the indeterminate phase of Chagas disease. Am J Trop Med Hyg 59: 526–529
Sosnik A, Carcaboso ÁM, Glisoni RJ, Moretton MA, Chiappetta DA (2010) New old challenges in tuberculosis: Potentially effective nanotechnologies in drug delivery. Adv Drug Del Rev 62: 547–559
Stern ST, Hall JB, Yu LL, Wood LJ, Paciotti GF, Tamarkin L, Long SE, McNeil SE (2010) Translational considerations for cancer nanomedicine. J Control Release 146: 164–174
Strosberg AM, Barrio K, Stinger VH, Tashker J, Wilbur JC, Wilson L, Woo K (2007) Chagas disease: a Latin American nemesis. Report Institute for OneWorld Health
Suárez S, O'Hara P, Kazantseva M, Newcomer CE, Hopfer R, McMurray DN, Hickey AJ (2001) Respirable PLGA microspheres containing rifampicin for the treatment of tuberculosis: screening in an infectious disease model. Pharm Res 18 (9): 1315–1319
Swenson CE, Stewart KA, Hammett JL, Fitzsimmons WE, Ginsberg RS (1990) Pharmacokinetics and *in vivo* activity of liposome-encapsulated gentamicin. Antimicrob Agents Chemother 34: 235–240
Tan ML, Choong PFM, Dass CR (2010) Recent developments in liposomes, microparticles and nanoparticles for protein and peptide drug delivery. Peptides 31: 184–193
The Union (2008) Tuberculosis. The International Union Against Tuberculosis and Lung Disease. http://www.theunion.org/tuberculosis/tuberculosis.html. Accessed February 2008
Threlfall EJ, de Pinna E, Day M, Lawrence J, Jones J (2008) Alternatives to ciprofloxacin use for enteric Fever, United kingdom. Emerg Infect Dis 14: 860–861
Tian Y, Bromberg L, Lin SN, Hatton TA, Tam KC (2007) Complexation and release of doxorubicin from its complexes with pluronic P85-b-poly(acrylic acid) block copolymers. J Control Release 121:137–145
Toit LC, Pillay V, Danckwerts MP (2006) Tuberculosis chemotherapy: current drug delivery approaches. Resp Res 7: 1–18
Torchilin V (2009) Intracellular delivery of protein and peptide therapeutics. Drug Discov Today Technol 5: e95-e103
Tulkens PM (1991) Intracellular distribution and activity of antibiotics. Eur J Clin Microbiol Infect Dis 10: 100–106
van den Broek PJ (1989) Antimicrobial drugs, microorganisms and phagocytes. Rev Infect Dis 11: 213–245
Vardy D, Barenholz Y, Naftoliev N, Klaus S, Gilead L, Frankenburg S (2001) Efficacious topical treatment for human cutaneous leishmaniasis with ethanolic lipid amphotericin B. Trans R Soc Trop Med Hyg 95(2): 184–186
Verma AK, Sachin K, Saxena A, Bohidar HB (2005) Release kinetics from bio-polymeric nanoparticles encapsulating protein synthesis inhibitor-cycloheximide, for possible therapeutic applications. Curr Pharm Biotechnol 6: 121–130
Vyas SP, Kannan ME, Jain S, Mishra V, Singh P (2004) Design of liposomal aerosols for improved delivery of rifampicin to alveolar macrophages. Int J Pharm 269: 37–49
Webb MS, Boman NL, Wiseman DJ, Saxon D, Sutton K, Wong KF, Logan P, Hope MJ (1998) Antibacterial efficacy against an in vivo Salmonella typhimurium infection model and

pharmacokinetics of a liposomal ciprofloxacin formulation. Antimicrob Agents Chemother 42(1): 45–52

Weinberger M, Keller N (2005) Recent trends in the epidemiology of non-typhoid Salmonella and antimicrobial resistance: the Israeli experience and worldwide review. Curr Opin Infect Dis 18(6): 513–521

Wender PA, Mitchell DJ, Pattabiraman K, Pelkey ET, Steinman L, Rothbard JB (2000) The design, synthesis, and evaluation of molecules that enable or enhance cellular uptake: peptoid molecular transporters. Proc Natl Acad Sci USA 97: 13003–13008

Werner-Meier R, Entzeroth R (1997) Diffusion of microinjected markers across the parasitophorous vacuole membrane in cells infected with *Eimeria nieschulzi* (Coccidia, Apicomplexa). Parasitol Res 83: 611–613

Whittum-Hudson JA, Hudson AP (2005) Human chlamydial infections: persistence, prevalence and outlook for the future. Nat Sci Soc 13: 371–382

Whitty CJM, Sanderson F (1999) New therapies and changing patterns of treatment for malaria. Curr Opin Infect Dis 12: 579–584

Wijagkanalan W, Kawakami S, Takenaga M, Igarashi R, Yamashita F, Hashida M (2008) Efficient targeting to alveolar macrophages by intratracheal administration of mannosylated liposomes in rats. J Control Release 125: 121–130

Wiström J, Jertborn M, Ekwall E, Norlin K, Söderquist B, Strömberg A, Lundholm R, Hogevik H, Lagergren L, Englund G, *et al.* (1992) Empiric treatment of acute diarrheal disease with norfloxacin. A randomized, placebo-controlled study. Swedish Study Group. Ann Intern Med 117(3): 202–208

Wong JP, Yang H, Blasetti KL, Schnell G, Conley J, Schofield LN (2003) Liposome delivery of ciprofloxacin against intracellular *Francisella tularensis* infection. J Control Release 92: 265– 273

World Health Organization (WHO) (1993) declares tuberculosis a global emergency. Soc. Prevent. Med. 38: 251–252

World Health Organization (WHO) (2003) Treatment of tuberculosis: guidelines for national programmes 3rd Geneva

World Health Organization (WHO) (2006) Guidelines for the Treatment of Malaria http://www.whqlibdoc.who.int/publications /2006/9241546948_eng.pdf, accessed on 20th March 2009

World Health Organization (WHO) (2008) Global tuberculosis control: surveillance, planning, financing: report http://www.who.int/tb/publications/global_report/2008/pdf/fullreport.pdf. Accessed February 2008

Wortmann G, Zapor M, Ressner R, Fraser S, Hartzell J, Pierson J, Weintrob A, Magill A (2010) Lipsosomal amphotericin B for treatment of cutaneous leishmaniasis. Am J Trop Med Hyg 83(5): 1028–1033

Yardley V, Croft SL (1997) Activity of liposomal amphotericin B against experimental cutaneous leishmaniasis. Antimicrob Agents Chemother 41: 752–756

Yardley V, Croft SL (1999) *In vitro* and *in vivo* activity of amphotericin B-lipid formulations against experimental *Trypanosoma cruzi* infections. Am J Trop Med Hyg 61: 193–197

Yardley V, Croft SL (2000) A comparison of the activities of three amphotericin B lipid formulations against experimental visceral and cutaneous leishmaniasis. Int J Antimicrob Agents 13: 243–248

Yoshida A, Matumoto M, Hshizume H, Oba Y, Tomishige T, Inagawa H, Kohchi C, Hino M, Ito F, Tomoda K, Nakajima T, Makino K, Terada H, Hori H, Soma G (2006) Selective delivery of rifampicin incorporated into poly (DL-lactic-co-glycolic) acid microspheres after phagocytotic uptake by alveolar macrophages, and the killing effect against intracellular *Mycobacterium bovis* Calmettee Guérin. Microbes Infect 8: 2484–2491

Yoshihara E, Tachaban H, Nakae T (1987) Trypanocidal activity of the stearylamine - bearing liposome. Life Sci 40: 2153–2159

Yuji O (2010) Chlamydial Pneumonias. http://emedicine.medscape.com/article/297351-overview.

Zaru M, Sinico C, De Logu A, Caddeo C, Lai F, Manca ML, Fadda AM (2009) Rifampicin-loaded liposomes for the passive targeting to alveolar macrophages: *in vitro* and *in vivo* evaluation. J Liposome Res 19(1): 68–76

Zembower TR, Noskin G A, Postelnick MJ, Nguyen C, Peterson LR (1998) The utility of aminolgycosides in an era of emerging drug resistance. Int J Antimicrob Agents 10: 95–105

Zou YY, Ling YH, Reddy S, Priebe W, Perez-Soler R (1995) Effect of vesicle size and lipid composition on the *in vivo* tumor selectivity and toxicity of the non-cross-resistant anthracycline Anamycin incorporated in liposomes. Int J Cancer 61: 666–671

Zvulunov A, Cagnano E, Frankenburg S, Barenholz Y, Vardy D (2003) Topical treatment of persistent cutaneous leishmaniasis with ethanolic lipid amphotericin B. Pediatr Infect Dis J 22(6): 567–569

Gene Therapy in Bone Regeneration: A Summary of Delivery Approaches for Effective Therapies

Laura Rose, Ross Fitzsimmons, Tarek El-Bialy, and Hasan Uludağ

Abstract Injuries involving bones are a significant burden on the health care system. While biological therapies for stimulation of bone formation and repair have advanced to clinical practise, gene delivery towards the same goal is actively explored to provide more cost-effective and efficacious therapies. The viral carriers have proven to be effective for delivery of genes involved in bone tissue regeneration, but the safety concerns associated with viral carriers prevent their widespread clinical use. Instead, a better understanding of the viral mechanisms of gene delivery is guiding the design of effective non-viral gene carriers. This chapter will review the critical features of viral carriers that make them effective in gene delivery to human cells. Intracellular events in trafficking of viral vectors are linked to viral structural features and use of functional peptide motifs for mimicking viral delivery have been summarized. Non-viral approaches to gene delivery in the context of bone regeneration and repair have been reviewed, with emphasis on the challenges facing such a gene delivery. Therapeutic outcomes obtained from animal studies were critically summarized and the range of promising agents for gene delivery has been reviewed.

Keywords Bone regeneration • Gene delivery • Bone morphogenetic proteins • Viral carriers • Non-viral carriers • Biomaterials

---

L. Rose , R. Fitzsimmons, T. El-Bialy, and H. Uludağ
Department of Biomedical Engineering, Faculty of Medicine & Dentistry and Engineering, University of Alberta, Alberta, Canada
e-mail: lrose@ualberta.ca; rossf@ualberta.ca

H. Uludağ (✉)
Department of Chemical & Materials Engineering, Faculty of Engineering, University of Alberta, Alberta, Canada
and
Faculty of Pharmacy & Pharmaceutical Sciences, University of Alberta, Alberta, Canada
e-mail: Huludag@ualberta.ca

T. El-Bialy
Department of Dentistry, Faculty of Medicine & Dentistry, University of Alberta, Alberta, Canada
e-mail: telbialy@ualberta.ca

A. Prokop (ed.), *Intracellular Delivery: Fundamentals and Applications*, Fundamental Biomedical Technologies 5, DOI 10.1007/978-94-007-1248-5_25,

Abbreviations

AV	Adenovirus
AAV	Adeno-Associated Virus
BMP	Bone Morphogenetic Protein
FGF	Fibroblast Growth Factor
HA	Hydroxyapatite
LIPUS	Low Intensity Pulsed Ultrasound
NLS	Nuclear Localization Signal
NPC	Nuclear Pore Complex
OP	Osteogenic Protein
PEI	Polyethyleneimine
PLGA	Poly(lactic-glycolic acid)
PLL	Poly-*L*-lysine
RGD	Arginine-Glycine-Aspartic Acid
VEGF	Vascular Endothelial Growth Factor

1 Introduction

Musculoskeletal conditions affect one in four Americans, and the combined direct and indirect costs of bone and joint health are estimated to be nearly $850 billion (AAOS 2010). There are approximately 20.2 million musculoskeletal injuries in the United States annually and up to 20% of fractures are associated with impaired healing (Verettas et al. 2002). Approximately 2.2 million bone grafts are performed worldwide each year (Giannoudis et al. 2005), which places a huge burden on medical health care systems: spinal fusions alone cost approximately $20 billion annually (Porter et al. 2009). The complex biology of bone healing undoubtedly plays a role in assurance of successful outcome as well as the complications typically seen. Fracture healing, as a prototypical bone healing model, is generally divided into four stages (Marsh and Li 1999): hematoma formation and inflammation, soft callus phase, maturation to a hard callus, and remodelling. Following the injury, platelets at the injury site are activated and produce growth factors to recruit inflammatory cells to the injury site. Cells involved in repair, such as fibroblasts, pre-osteoblasts and mesenchymal stem cells, are also recruited. New connective tissue induced after the injury forms a soft callus, along with a robust angiogenic network. This callus eventually hardens, either via intramembranous ossification (i.e., osteoblast formation directly from mesenchymal stem cells) or endochondral ossification with a cartilage intermediate. Because the initially deposited woven bone does not have enough mechanical strength, it is remodelled into stronger laminar bone through the dual efforts of osteoblasts and osteoclasts. This complex series of well-coordinated events is orchestrated by a myriad of growth factors produced by cells involved in the healing process. Growth factors produced during fracture healing include platelet derived growth factor (Bourque et al. 1993), transforming growth factor-β (Andrew et al. 1993), insulin-like growth factor, vascular endothelial growth factor (Uchida et al. 2003) and fibroblast growth factors (Scmid et al. 2009). Failure during the sequence of events leads to either a hypertrophic or atrophic non-union. Hypertrophic non-unions have a robust blood supply and the capacity to fully heal. In contrast, atrophic non-unions have poor vascularisation and are incapable of biological response needed for successful healing. In the case of atrophic non-union, *de novo* bone regeneration is required. Thus, the clinical challenge is to address these failures of biology and to intervene with therapeutics that can restore and/or stimulate natural healing response.

2 Current Treatment of Non-unions and Use of Bioactive Agents for Bone Regeneration

Autologous bone grafts remain the gold standard for non-union bone defects, where tissue from another site, usually the iliac crest, is harvested to repair the defect. Assuming the quantity and quality of bone harvested are sufficient, pain, second-site morbidity and risks associated with a second surgical site present obstacles to be overcome. Allografts are used as a viable alternative to autografts; both the US Food and Drug Administration and American Associated of Tissue Banks works to develop standardized procedures for harvest, sterilization and storage of tissue suitable for grafting (Cook and Cook 2009). The quality of allografts varies greatly; extensive processing helps to decrease the likelihood of disease transmission, but it also reduces the osteogenic potential of the graft while increasing the variation in their bone induction capability. However, the risk of disease transmission remains high despite extensive graft processing, screening and sterilization. Synthetic biomaterials, such as hydroxyapatite (HA) or collagen-based bone substitutes, are another option as bone implants, but they are only osteoconductive and therefore not ideal for large defects where bone ingrowth from the surrounding tissue needs to be extensive. HA and other calcium phosphate cements generally lack appropriate mechanical properties for bone, as their brittleness leaves them prone to fracture very easily. Demineralized bone matrix, decalcified cortical bone, has demonstrated to be successful in bone grafting but their success is sporadic and not always guaranteed, which has been attributed to the differences in processing of the bone tissues. Although thoroughly processed, demineralised bone matrix still contains certain growth factors that make this biomaterial osteoinductive (i.e., capable of inducing de novo bone from the surrounding/infiltrating mesenchymal stem cells). One group of the proteins responsible for the osteoinductivity of demineralised bone matrix is the bone morphogenetic proteins (BMP), first discovered by Marshall Urist (1956). This group of proteins is best known for its ability to induce ectopic bone formation in the absence of any bone tissue at the implant site. At lower femtomolar concentrations, BMP-2 can act as a chemotactic agent attracting cells (Feidler et al. 2002); however, at high concentrations, BMPs act as a morphogen, causing stem cells to differentiate to an osteogenic fate. There are several BMP-based therapies approved for orthopaedic applications, where BMP-2 and BMP-7 (also known as Osteogenic Protein-1 or OP-1) are the bioactive ingredients.

The early recognition of the ability of BMPs to induce de novo bone, combined with better realization of the importance of BMPs in fracture-healing have led to the development of growth factor-based therapies that incorporate recombinant human proteins. A biomaterial scaffold, that provides a three-dimensional space for bone deposition and is intended to mimic an extracellular matrix, has been an indispensable part of these therapies. INFUSE™ uses recombinant human bone morphogenetic protein-2 (rhBMP-2) loaded in collagen sponges. It was first FDA approved for spine and oral/maxillofacial applications (McKay et al. 2007) and then for orthopaedic trauma (Govender et al. 2002). When evaluated in surgery for tibial trauma, patients receiving recombinant human BMP-2 (rhBMP-2) therapy had faster wound-healing, reduced incidence of secondary intervention, as well as reduced infection risk (Govender et al. 2002). BMP-7 (OP-1) has been used in a similar fashion for the treatment of tibial non-unions (Friedlaender et al. 2001) with similar results. Patients receiving rhOP-1 in bone-derived bovine collagen (i.e., demineralised bone matrix from bovine bones) had success rates comparable to those receiving an autograft. The concentration rhBMP-2 used for implantation was ~1.5 mg/cc (McKay et al. 2007), while we estimate the concentration of rhOP-1 to be 0.9 mg/cc (Friedlaender et al. 2001). These are exceedingly high growth factor concentrations, considering that the endogenous levels of the BMPs are at expected to be at ng/mL to μg/mL levels. Despite the success of these protein-based therapies, the expense associated with milligram quantities of protein greatly limits their wide-spread use (especially in developing countries) and presents an opportunity for improvement for wider application of these therapeutics. A recent study has also cast a shadow on the expected benefits of BMP protein therapeutics (Cahill et al. 2009). In this retrospective study, patients receiving BMPs for spinal fusion (anterior cervical fusion) were associated with a higher rate of complication and higher inpatient hospital charges.

BMPs are arguably the most studied growth factor for bone regeneration, but several other proteins have been investigated for similar clinical use. In particular, growth factors that stimulate new blood vessel formation have shown great promise. Angiogenesis is tightly intertwined with bone regeneration, as a good blood supply is imperative for growth of new bone tissue (Carano and Filvaroff 2003). BMPs alone can indirectly stimulate angiogenesis in the early stages of bone induction. Treatment of osteoblasts with BMPs causes them to produce the vascular endothelial growth factor (VEGF; Deckers et al. 2002). VEGF stimulates angiogenesis, and delivery of this growth factor alone is useful for bone healing, although not as effectives as the BMPs as a stand alone therapeutics in animal models. Regeneration is impaired in the absence of VEGF, and exogenous VEGF accelerates the healing process (Street et al. 2002). In addition to its role in cell proliferation and migration, fibroblast growth factors (FGFs) are also involved in angiogenesis (Montesano et al. 1986). Basic FGF (bFGF) stimulates bone formation (Nakamura et al. 1995) and facilitates healing of bone defects (Hong et al. 2010). Both VEGF and bFGF are being pursued for bone regeneration, and pre-clinical studies have demonstrated their synergistic effect in BMP-induced bone repair.

In addition to bioactive proteins, low intensity pulsed ultrasound (LIPUS) has been employed for enhancing bone fracture healing, both in animals and in human

(Heckman et al. 1994). Daily application of LIPUS can stimulate mandibular bone growth in baboon monkeys (El-Bialy et al. 2006) and in growing human patients with craniofacial syndrome (Hemifacial Microsomia) (El-Bialy et al. 2005). LIPUS application for 20 min/day was shown to enhance bone healing at bone surgical lengthening (Distraction osteogenesis) in humans (El-Mowafi and Mohsen 2005). It has been suggested that the stimulatory effect of LIPUS on bone fracture healing and new bone formation is due to the stimulation of angiogenesis (Lu et al. 2008; Doan et al. 1999), enhancement of BMP expression (Suzuki et al. 2009), and Type I and X collagen expression (Scheven et al. 2009). Moreover, LIPUS stimulates other bone-related and calcification genes in cell culture, via intracellular signalling pathways affecting many osteogenesis related genes (Zhou et al. 2008).

3 Gene-Based Approaches for Bone Regeneration and Associated Challenges

Employing gene-based agents, rather than bioactive proteins, is an alternative for effective bone regeneration. Rather than mass-producing a bioactive protein by recombinant technology, one employs the genes coding for bioactive proteins and relies on *in situ* production of the proteins for stimulation of bone regeneration (Fig. 1). The most common approach for gene-based therapy involves *ex vivo* transfection of cells with therapeutic genes for administration into a patient. This approach typically requires harvesting bone marrow aspirate from a patient, and expanding the plastic-adherent population of cells (Fig. 2a). Although other sources, such as adipose tissue, might be more convenient, bone marrow aspirates are likely to provide more osteogenic cells critical for regeneration. Following expansion, cultures are transfected, and the transformed cells purified from the unmodified cells. The cells now expressing the desired growth factor would be administered to the patient, allowing *in vivo* production of the growth factor by the host's own cells. Since the host's own cells are grafted, there is no risk of immune response and the *ex vivo* modification gives more control over the transfection procedure to ensure that cells are properly expressing the therapeutic gene. However, this method is not without its road-blocks, particularly in translating such technology to a clinical setting. To begin, the costs associated with a separate surgery to collect marrow aspirate and subsequent cell culture required would be significant. Several millions of cells are estimated to be required for such cell-based therapies, meaning the original sample would have to be expanded over several weeks. If the cell-based therapy is to be used to deal with trauma, this lengthy culture period would mean a delay in treatment, usually unacceptable in a clinical. A lengthy expansion time may also result in changes in the desired characteristics of the bone marrow stromal cells. Ideally, a population of mesenchymal stem cells would be best for cell-based bone regeneration, as they have multi-lineage potential and thus able to differentiate into chondrocytes and osteoblasts. The cells at the end

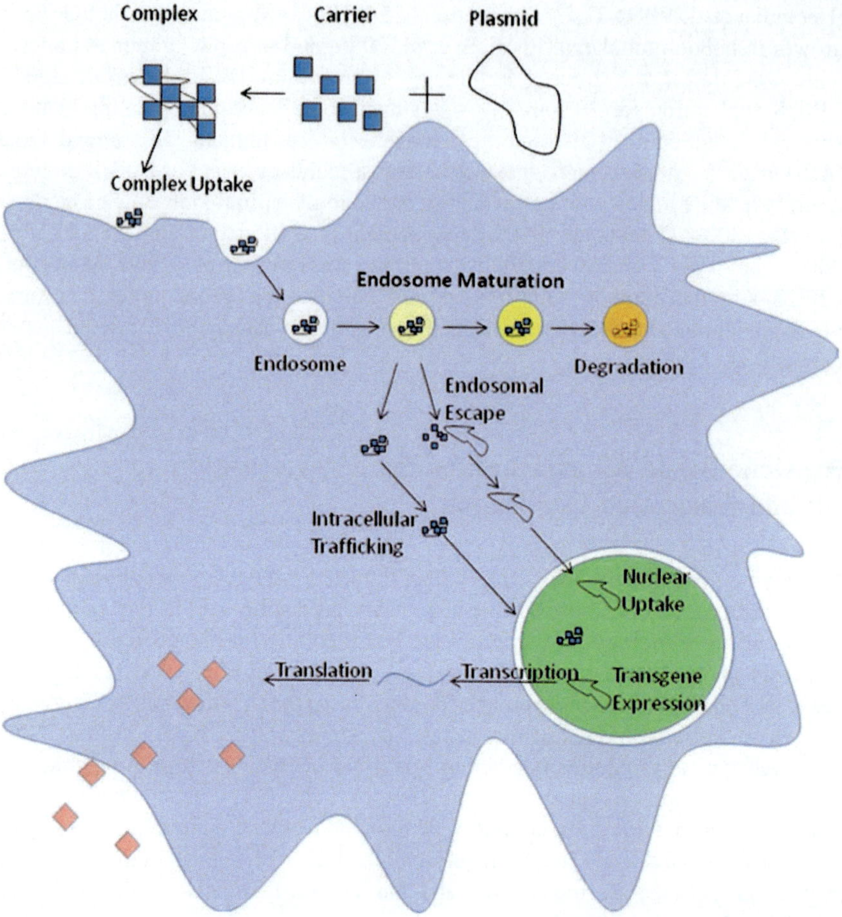

Fig. 1 Scheme for effective gene delivery. Plasmid DNA coding for a protein is mixed with carriers to form a complex for cellular uptake by endocytosis. The newly formed endosome begins to mature, and the complex must escape before lysosome formation in order to avoid degradation. Following endosomal escape, the complex is expected to target to nucleus for the plasmid to be transcribed. The messenger RNA is then translated in the cytoplasm to synthesize the coded protein

of a lengthy culture period may not possess the same multi-lineage potential as fresh bone marrow cells.

An alternative to *ex-vivo* gene transfer is the direct non-viral gene delivery that offers some advantages with respect to cost and ease-of-use. Direct gene therapy can be employed immediately after the trauma, and without the issues associated with cell harvesting and expansion. Many groups are exploring virus-based direct gene therapy, but a handful of groups are also investigating non-viral carriers due to safety concerns associated with viruses. Many of the viral therapies are less effective in immunocompotent animals (Kang et al. 2004; Li et al. 2003a, b; Alden et al. 1999), and require immuno-suppression for a therapeutic effect (Kaihara et al. 2004), making them difficult to translate to a clinical setting. The non-viral approach aims to delivery carrier/DNA complexes locally, so that the patient cells at the repair site become transfected and produce the growth factor (Fig. 2b). Biomaterial scaffolds provide an easy vehicle with which to deliver the complexes, making the administration of the complexes facile to a repair site. Cells invade the scaffold and become transfected after taking up the resident complexes. Growth factor for which the plasmid codes is produced locally, facilitating bone regeneration and healing. Although much effort has been devoted to non-viral gene delivery, the majority of such systems have been tested *in vitro* only and *in vitro* performances do not correlate well with *in vivo* studies (Wang et al. 1998) due to unrealistic choice of experimental conditions *in vitro*. The immune system is thought to

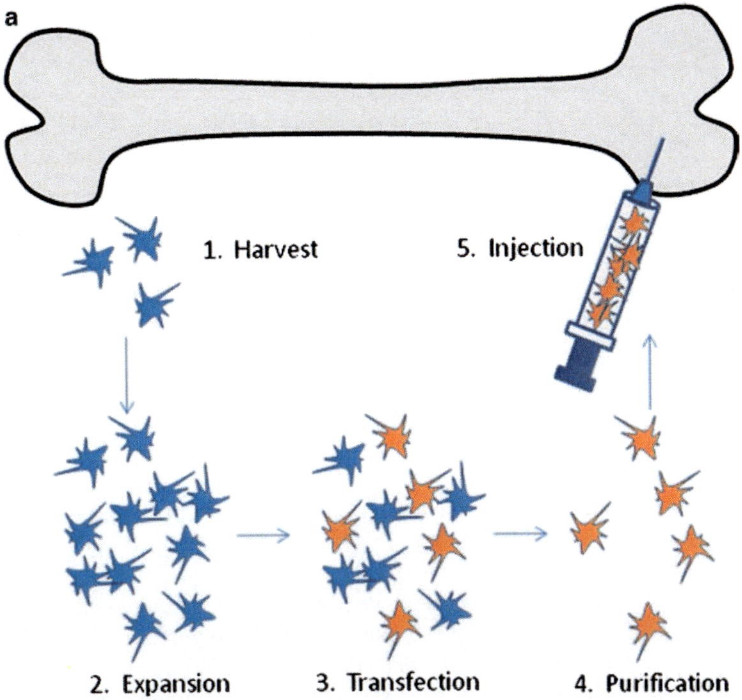

Fig. 2 (**a**) Ex-vivo modification of cells for grafting into a host. The patient's cells are harvested from bone marrow aspirate (or other sites), expanded in culture, and then transfected with the growth factor gene of interest. Cells that are positive for growth factor are selectively expanded or purified from unmodified cells, and administered to the patient. (**b**) In vivo approach to gene delivery. A scaffold is loaded with complexes containing plasmid DNA for a growth factor expression, and administered into a bone defect. The patient's cell invade the scaffold, internalize the complexes and expresses the desired growth factor, facilitating healing and bone regeneration

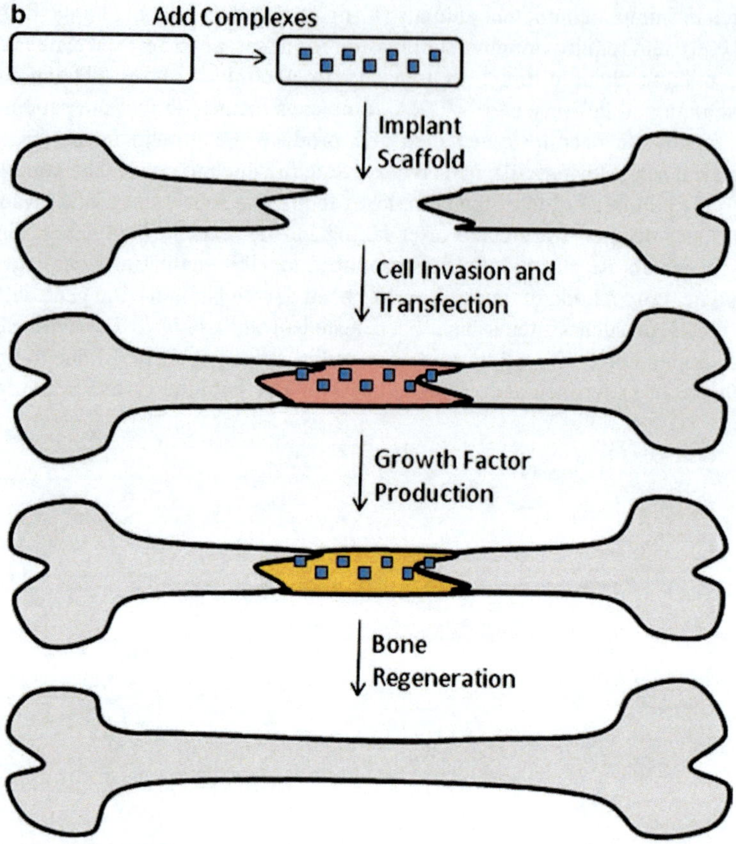

Fig. 2 (continued)

be partly responsible for this by clearing plasmid particles prematurely. Extended gene expression is required to attain a biological response, so the plasmid DNA complexes must be taken up by the cells over a sustained period of time. This means the complexes must retain their stability in a high serum environment, in addition to avoiding clearance by the immune system. A scaffold is beneficial to prolong the presence of plasmid/carrier complexes and additionally provides an attachment site for incoming cells and extracellular matrix production (De LaPorte and Shea 2007). Although *in vivo* transfection is possible with naked DNA, a large and economically-unsustainable amount of plasmid DNA is required in the absence of a carrier. The latter is involved in compacting DNA molecules into a suitable size for cellular uptake, lowering the dose required for effective cellular modification. Gene carriers can also help protect the unmethylated plasmid DNA from premature clearance by the immune system (Krieg 2002).

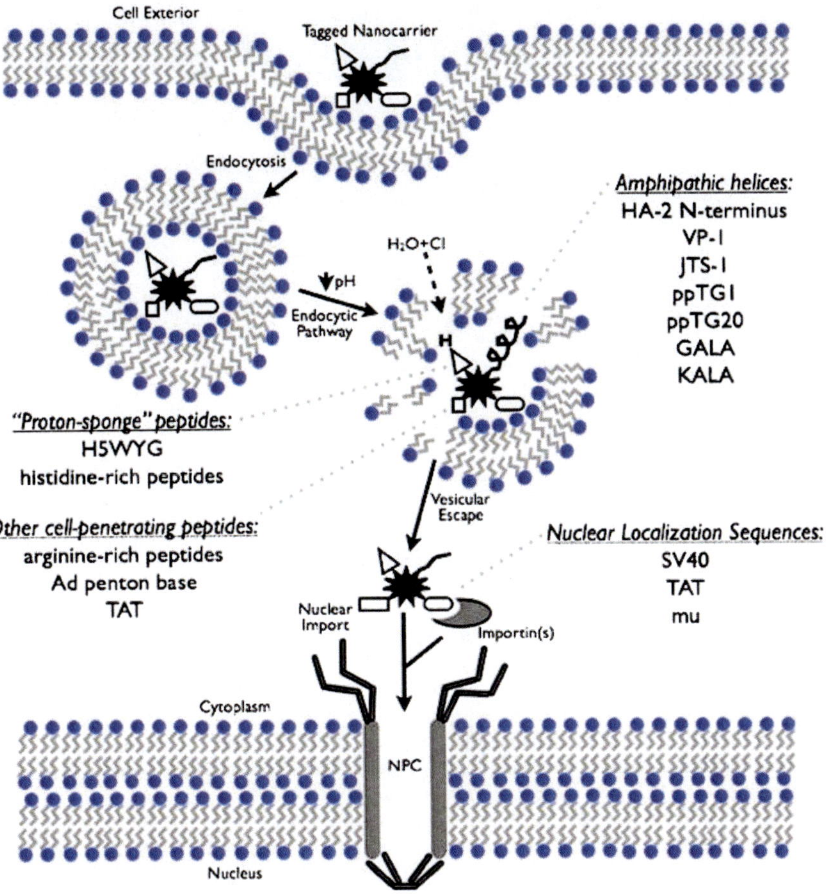

Fig. 3 Viral peptide-mediated trafficking of DNA intracellularly. The association of carriers with peptides derived from viral proteins can facilitate the navigation of DNA past cellular barriers to the nucleus. Peptides which facilitate endocytic escape include those that form membrane-destabilizing amphipathic helices upon acidification, proton-sponge peptides which disrupt endosomes by osmotic effects, and those that penetrate membranes by alternate (or unknown) means. Nuclear targeting can be accomplished by utilizing nuclear localization sequences that recruit importin proteins which guide the complexes through the nuclear pore complex

We previously reviewed gene delivery efforts for bone regeneration (Varkey et al. 2004) and now provide an update on non-viral (Fig. 4) and viral-mediated (Table 1) gene delivery in bone regeneration. Below, we further review the gene delivery characteristics of select groups of viruses and emphasize the features that are critical for their functioning and have been employed to design viral-mimetics based on synthetic carriers. The considerations for non-viral approaches for gene delivery are highlighted and current efforts on gene-based bone regeneration are presented. Given that the effective gene therapies rely on appropriate intracellular

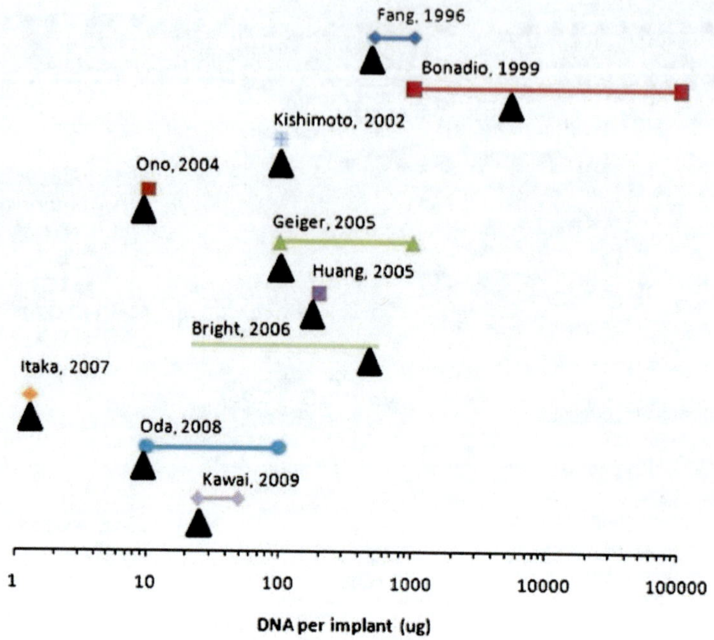

Fig. 4 (**a**) Summary of non-viral gene delivery studies intended for bone regeneration. The amount or the range of implanted DNA (μg/implant) reported in the published reports are shown in the horizontal axis. Arrows indicate the DNA dose at which a bone formation was detected. (**b**) Experimental details of non-viral gene delivery studies intended for bone regeneration. Abbreviations: BMP: bone morphogenetic protein, hPTH1-34: 34-peptide segment of human parathyroid hormone 1, VEGF: vascular endothelial growth factor, Runx2: runt-related transcription factor 2 (Cbfa1), caALK6: constitutively active form of activin receptor-like kinase 6

Table 1 Summary of virus-mediated in vivo gene delivery for osteogenic induction

Gene	Site	Carrier	Comments
BMP2	Ectopic	AV	Bone observed in athymic nude rats, none in immunocompetent rats (Kang et al. 2004; Li et al. 2003a)
			Immunosuppressant required for osteoinduction (Kaihara et al. 2004)
			No osteoinduction with BMP-2 AV injection in immunocompetent rats; BMP-2 AV delivered in collagen led to bone formation (Sonobe et al. 2004)
			Immunosuppression was required for bone formation (Okubo et al. 2000, 2001a, b)
			Endochondral bone formation was observed in athymic nude rats, however an attenuated response was observed in immunocompetent rats (Alden et al. 1999)
			Bone formation observed in muscle of immunocompetent rats only with ischemic degeneration. No osteoinduction was observed in the absence of muscle grafting (Gonda et al. 2000)

(continued)

Table 1 (continued)

Gene	Site	Carrier	Comments
		AAV	BMP-2 AAV delivery to muscle led to ectopic bone formation in immunocompetent mice (Chen et al. 2003, 2004)
			Recombinant AAV expressing BMP-2 delivered on hydroxyapatite led to bone formation in immunocompetent rats (Nasu et al. 2009)
			A tetracycline-sensitive promoter controlled bone formation in mice, with no bone formation observed when the promoter was not activated by drug administration (Gafni et al. 2004)
	Orthotopic	AV	New bone formation observed in distraction osteogenesis models (Ashinoff et al. 2004), critical sized mandibular (Alden et al. 2000), nasal defects (Lindsey 2001), osteoporotic fracture models (Egermann et al. 2006a), metatarsal defect (Ishihara et al. 2008), rib defect (Ishihara et al. 2010), and femoral defect (Betz et al. 2007a)
			When injected, BMP-2 expressing AV led to modest bone formation in calvarial defect; robust bone formation was observed when AV were loaded on a gelatin scaffold (Hu et al. 2007)
			Healing of iliac crest defects was retarded following injections of BMP-2 AV in sheep (Egermann et al. 2006b)
			Improved healing of critical size defect was observed when BMP-2 expression was delayed (Betz et al. 2007b)
	Orthotopic	AAV	AAV-BMP2 injections led to accelerated healing in equine osteotomy model (Ishihara et al. 2008)
BMP4	Ectopic	AV	Bone formation was observed in athymic nude rats following delivery of a BMP-4 expressing AV (Chen et al. 2002; Jane et al. 2002; Kang et al. 2004)
			BMP-4 expressing AV led to bone formation in athymic nude rats, but not in immunocompetent rats (Li et al. 2003a, b)
	Ectopic	AAV	An injection of AAV expressing BMP-4 led to bone formation in hindlimb muscle of immunocompetent rats (Luk et al. 2003)
	Orthotopic	RV	A BMP-4 hybrid RV administered to periosteum fracture in immunocompetent rats led to increased callus size, enhancing healing (Rundle et al. 2003)
BMP6	Ectopic	AV	Bone induction observed in athymic nude rats and in immunocompetent rats (Kang et al. 2004; Li et al. 2003b; Jane et al. 2002)
BMP7	Ectopic	AV	Bone induction observed in athymic nude rats, none in immunocompetent rats (Kang et al. 2004; Li et al. 2003a)
	Orthotopic	AV	Enhanced alveolar bone and osseointegration of dental implant was observed with BMP-7 expressing AV (Dunn et al. 2005)
BMP9	Ectopic	AV	Bone induction in athymic nude rats and in immunocompetent rats(Kang et al. 2004; Li et al. 2003a, b, c; Varady et al. 2001)
	Orthotopic	AV	Bony healing observed in critical-sized mandibular defects and model of spinal arthrodesis in athymic nude mice (Alden et al. 2000; Helm et al. 2000)
TGFβ	Orthotopic	AV	Injection of TGF-β AV led in increased epiphyseal thickness (Mehara et al. 1999)

(continued)

Table 1 (continued)

Gene	Site	Carrier	Comments
VEGF	Orthotopic	AV	VEGF-A expressing AV injected into the muscle surrounding a bone defect resulted in enhanced periosteal cartilage and femoral mineral content, and faster endochondral phase (Tarkka et al. 2003)
PDGF	Orthotopic	AV	Periodontal lesions filled with collagen containing PDGF expressing AV led to increased enhanced alveolar bone regeneration (Jin et al. 2004; Chang et al. 2010)
LMP1	Orthotopic	RV	LMP-1 led to improved bony union of fracture in rats (Strohbach et al. 2008)
Cbfa1	Orthotopic	AV	Cbfa1 expressing adenovirus led to robust bone regeneration in an idiopathic osteonecrosis model (Sakai et al. 2008)
Nell1	Orthotopic	AV	Adenoviral Nell-1 delivery in demineralised bone matrix led to better spinal fusion rates than BMP-2 and –7 (Lu et al. 2007)

Abbreviations: *AV* adenovirus, *AAV* adeno-associated virus, *RV* retrovirus, *BMP* bone morphogenetic protein, *TGFβ* transforming growth factor-β, *VEGF* vascular endothelial growth factor, *PDGF* platelet derived growth factor, *LMP1* surface located membrance protein 1, *Cbfa1* core binding factor alpha1 subunit protein (Runx2), *Nell1* NEL-like 1
Modified from Varkey et al.

trafficking of DNA molecules, intracellular processing of gene complexes were emphasized where appropriate and we highlighted the technological approaches to this end.

3.1 *Delivery Considerations for Growth Factors and Gene Expression Systems*

For a therapeutic effect, cells in a tissue must be exposed to protein growth factors for a prolonged period of time. In animal models utilizing rhBMP-2 implants, the longer the duration of growth factor retention in implants, the more potent was its ability to induce bone formation (Uludag et al. 2000, 2001). The concentration of the growth factor is equally important for healing, as too little may be insufficient for bone induction/regeneration while too much may be detrimental to healing process due to excess swelling or down-regulation of specific intracellular mediators. A local injection of growth factor is marred by the short half-life of proteins *in vivo*; growth factors in particular are cleared from the local site quickly, and can be easily denatured and degraded, leading to a loss of the protein's function. Current clinical delivery systems for BMP-2 and OP-1 navigate this barrier by delivering large amounts of protein in an implantable format. Even with a fast clearance, sufficient protein remains locally to keep a therapeutic concentration for a period of time sufficient to allow bone regeneration. The milligrams of protein growth factor required for efficacy make this type of therapy very expensive, and may show less than ideal pharmacokinetics. Since BMPs act via cell-surface receptors (Rosen 2009), intracellular delivery and trafficking is not a challenge in the case of growth factor delivery, as long as the proteins are freely available in the extracellular matrix to the target cells.

Gene delivery provides a potential alternative to address some of the shortfalls of protein delivery. Compared to bolus protein delivery, gene delivery scaffolds can be designed in such a way that DNA released over several weeks can to be taken up by cells gradually (Shea et al. 1999). The burst release kinetics associated with a bolus protein injection are also absent in a gene delivery system. The kinetics of growth factor expression after gene delivery may better mimic natural growth factor production during fracture healing, since a relatively constant amount of protein is expected to be produced by the cells and the level of the produced protein could be lower. Since *in situ* expressed growth factor concentrations are closer to physiological levels, there may potentially be less bone ectopic to the site that must be remodelled. Similar to protein delivery, however, gene delivery is not without its challenges. Unlike protein therapies, the therapeutic genes not only need to enter the cell but also undergo the appropriate intracellular trafficking and must be delivered to the nucleus for transcription. A carrier usually facilitates cellular uptake of the protein-coding expression vector, and forms a complex with the DNA containing the gene of interest, or is an attenuated virus harbouring the gene of interest. Viruses are generally considered the most effective agent for gene expression but their safety profile is worrisome. The risks of inflammation, immune response and insertion mutagenesis make the use of viruses justifiable in only life-threatening diseases (such as cancer) and not in wide-spread clinical applications. However, their high transfection efficiency offers insight to the barriers facing non-viral delivery.

3.2 *Molecular Features Making Viral Gene Delivery Effective*

Fostered by millions of years of evolution, viruses are masters of intracellular delivery of nucleic acid. Their extraordinary efficiency permits effective gene delivery to a multitude of cell types while providing an effective model of gene delivery for emulation with non-viral carriers. The general paradigm for engineering viruses into useful experimental reagents and therapeutics requires disabling their replication and pathogenicity while forcing them to express a desired genetic material. The usual method for this involves separating the viral genome into two (or more) DNA molecules (Verma and Weitzman 2005); one (*the vector construct*) that carries the desired foreign sequences and any *cis*-acting viral elements necessary for transcription and packaging the foreign DNA into viral particles, and the other (*the packaging construct*, which is either a plasmid or a genomically-inserted cassette of a packaging cell line) that encodes viral structural proteins and any other proteins essential for the production of viral particles. Genes that are deemed non-essential or associated with pathogenicity are excised to maximize the space available for foreign sequences, as well as to prevent deleterious effects on infected cells in the event of unintended recombination between constructs (Kay et al. 2001). In this way, the engineered viral particles cannot replicate post-infection due to the absence of the packaging construct and cause the detrimental effects of viral replication, cell lysis, immunoreaction, and rampant infection.

Viral vectors often utilized include members of the retrovirus family (lentivirus, oncoretroviruses, spumaviruses), adenovirus, adeno-associated virus, and herpesvirus (Kay et al. 2001; Verma and Weitzman 2005). While retroviral vectors are particularly easy to engineer, more complicated vectors such as adenovirus pose more of a challenge. The retroviral genome can essentially be swapped out of its flanking LTRs (long terminal repeats) in exchange for foreign sequence to make the vector construct and retroviral particles can easily be manufactured in packaging cell lines (Sinn et al. 2005). Viruses such as adenovirus, however, require elaborate recombination strategies and helper-dependent systems to produce functional particles free of pathogenic and replication-associated genes (Segura et al. 2008; Howarth et al. 2010; Alba et al. 2005). The complex genetic engineering and optimization of viral gene combinations to create safe and functional constructs are two hurdles associated with viruses that non-viral gene delivery largely circumvents. That said, the necessary engineering of viral carriers is rewarded with exceptional transduction efficiencies due to their ability to efficiently navigate their genome past numerous cellular barriers. Cellular barriers for effective viral nucleic acid delivery include passage through cell membrane (either by direct fusion or vesicular escape) and trafficking through the cytoplasm and the nuclear envelope. As outlined below, each virus has acquired highly specialized features (due to evolution and human modifications) allowing them to overcome these barriers and successfully express a gene sequence.

3.2.1 Retrovirus

Retroviral infection begins with nonspecific binding to cell surface followed by specific attachment between the virus's envelope glycoproteins (Env proteins) and their target receptors (Pizzato et al. 1999; Haynes et al. 2003). Following this binding, the membranes of both the cell and enveloped-virus fuse mediating the entry of the retroviral nucleocapsid into the cytoplasm (Harrison 2005). The tropism of retroviral vectors has been greatly expanded by exchanging the wild-type attachment-protein genes with those from related viruses (in the packaging cell line or packaging construct), a technique known as pseudotyping (Cronin et al. 2005). Pseudotyping is most commonly performed with the Env glycoprotein of vesicular stomatitis virus (VSV-G), which bestows an extremely broad tropism (Burns et al. 1993). Following the release of the nucleocapsid into the cytoplasm, the viral RNA genome is reverse-transcribed into dsDNA (with the exception of Foamy Virus particles which can reverse-transcribe their genome and deliver a DNA genome) (Sinn et al. 2005; Deyle et al. 2010). Prior to inserting into the host's genome, the preintegration complex can either be actively imported through the nuclear envelope by cellular proteins (in the case of lentiviruses, like HIV) or remain cytoplasmic until the nuclear membrane is disrupted in mitosis (De Rijck et al. 2007; Roe et al. 1993; Lewis and Emerman 1994). With the exception of lentiviruses, this inability to pass through the nuclear envelope is a major bottleneck to most retroviral vectors and hence generally limits their use to cells with high mitotic activity (Miller et al. 1990).

3.2.2 Adenovirus

Cellular internalization is initiated by most adenoviral serotypes through binding of their fibre knobs to the Coxsackie-adenovirus receptor (CAR) (Bergelson et al. 1997). Subsequent binding of their penton base proteins to alpha-V integrins predominantly results in clathrin-mediated endocytosis (Wickham et al. 1993; Meier and Greber 2004). Expanding the tropism of non-enveloped viruses through pseudotyping is slightly more challenging than their enveloped counterparts due to the fact that foreign proteins engineered into the virus must be structurally compatible with the native capsid proteins and cannot simply be inserted into a lipid bilayer (Waehler et al. 2007). That said, the tropism of individual adenoviral strains have been successfully expanded by replacing fibre proteins with homologous proteins from related serotypes, unrelated viruses and even with single-chain variable fragment antibodies to confer specific targeting (Kanerva et al. 2002; Mercier and Campbell 2004; Hedley et al. 2005). Once endocytosed, it is generally accepted that the decrease in pH along the endocytic pathway induces a conformational change of the penton base proteins causing membrane disruption and permitting escape to the cytosol (Blumenthal et al. 1986; Wang et al. 2000). The virion is then transported to the nucleus in a dynein/dynactin-dependant manner where the dsDNA genome is actively imported through the nuclear pore complex (NPC) (Kelkar et al. 2004; Saphire et al. 2000). Nuclear uptake is an elaborate process whereby NPC docking is mediated by viral hexamer proteins followed by the entry of the genome and the DNA-condensing proteins mu and VII through the NPC (Trotman et al. 2001; Wodrich et al. 2006). The host proteins Nup214, histone H1, and hsc70 have also been implicated in adenovirus docking, capsid disintegration and DNA uptake (Trotman et al. 2001; Saphire et al. 2000). This nuclear targeting is a considerable feat considering that the commonly used serotypes Ad2 and Ad5 have a ~36 Kb genome (virion size, 70–100 nm), which is much larger than the genome of most non-viral plasmid-based systems (Davidson et al. 2003; McConnell and Imperiale 2004; Kennedy and Parks 2009). Generally, the adenoviral genome then remains episomal in the nucleus (not integrating into the host genome) (Harui et al. 1999). The episomal preference of adenoviral vectors, and their ability to actively import into the nucleus, make them both safer and more efficient compared to most retroviral strains. That said, adenoviral vectors carry the disadvantages of occasional integration, transient transgene expression, and immunological reactivity resulting from prior exposure (Bangari and Mittal 2006). Some studies have focused on using less common serotypes, than the typically used Ad2 and Ad5, in order to decrease the possibility of an immune reaction (Loser et al. 2000).

3.2.3 Adeno-Associated Virus

Unlike adenovirus, AAV carries the distinct advantage of not being associated with any pathogenic conditions and is generally considered to produce a less severe immune response (Kay et al. 2001; Chirmule et al. 1999). Unlike adenoviruses, AAV serotypes used in gene delivery studies bind to a diverse array of cellular receptors naturally resulting in a variety of preferential cell-types. For example, the commonly used serotype AAV2 utilizes heparan sulfate proteoglycan and the coreceptors FGFR1 and integrin $\alpha_V\beta_5$ (Summerford and Samulski 1998; Summerford et al. 1999; Qing et al. 1999). The AAV4 uses sialic acid, as does AAV5, though with a different carbohydrate linkage and an interaction with PDGF receptors (Kaludov et al. 2001; Di Pasquale et al. 2003). The precise mechanism by which AAV escapes the endocytic pathway after clathrin-mediated endocytosis has not been fully elucidated. Endosomal acidification is required, but unlike adenovirus, the intact virion appears to enter the nucleus without the loss its capsid coat; this is debateable, however, and may be a serotype specific phenomenon (Bartlett et al. 2000; Wu et al. 2006). The ability of the entire capsid to pass through the NPC is likely a result of its small size of ~26 nm, as molecules <40–60 kD may passively diffuse through the NPC (Harbison et al. 2008; Weis 2003). Following viral uncoating in the nucleus, the ssDNA load may be converted to dsDNA and then remain episomal like adenovirus, or integrate into the host's genome if the viral *rep* gene is present (McCarty et al. 2004; Dyall et al. 1999). Much safer compared to retroviruses, AAV usually integrates at a highly specific location (AAVS1) in the human genome on chromosome 19, hence having a lower probability of oncogenic mutation (Giraud et al. 1994). AAV does carry disadvantages, however, most notably its limited payload of ~5 Kb and the necessity of infecting packaging cells with helper viruses (such as adenovirus or herpesvirus) due to its inability to replicate itself (Grieger and Samulski 2005; Wu et al. 2006).

3.2.4 Herpesvirus

In stark contrast to AAV, Herpes simplex virus type I can hold over 30 Kb of foreign sequence (Pechan et al. 1999). HSV infection is more complicated than the fore-mentioned vectors in that cellular entry requires many protein interactions. Cell binding is mediated by viral proteins gB and gC interacting with heparan sulfate, while gD binds to herpesvirus entry mediators HveA and HveC (Laquerre et al. 1998). Membrane fusion between the cell and the enveloped virion seems to require the presence of gB, gD, and gH/gL (Fink et al. 2000). The tropism of HSV may be expanded by genetically altering the glycoproteins gB, gD, and more successfully gC (Laquerre et al. 1998; Anderson et al. 2000; Grandi et al. 2004, 2010). Following active uptake into the nucleus, wild-type HSV may either follow a lytic cycle initiating transcription or (as in the case of engineered constructs) exist in a latent episomal state (Jackson and DeLuca 2003; Fink et al. 2000). The transcriptional persistence of the foreign sequences largely depends on what specific class of vector was used: a recombinant HSV-1 vector or a HSV-derived amplicon vector. A rHSV-1 vector is replication deficient and can be produced in large quantities but its transgene expression is transient; additionally, it still contains large regions of the wild-type genome and carries the risk of expressing toxic and immunogenic proteins (Verma and Weitzman 2005). HSV-derived amplicon vectors only carry essential *cis*-acting sequences along with a transgene that can be expressed long-term, but their production requires a replicating helper virus or the use of cosmids or artificial chromosomes expressing the essential structural proteins (Smith et al. 1995; Stavropoulos and Strathdee 1998; Kong et al. 1999).

The appropriate choice of a viral vector for a clinical application will ultimately depend on the nature of the condition being treated in terms of target cell-type, time-frame of protein expression, size of the transgene needed, and the degree of risk appropriate for the condition. Beyond their direct utility, viral vectors provide useful models of nucleic acid delivery and their effective mechanisms may be exploited to create safer non-viral vectors.

3.3 Mimicking Viral Mechanisms for the Purpose of Non-viral Gene Delivery

Non-viral carriers share general properties with viral particles, such as the nano-scale size and a capacity to penetrate the plasma membrane. With non-specific carriers (i.e., in the absence of specific ligands), plasmid particle uptake appears to mostly follow clathrin-mediated endocytosis, however some particles may enter the cell via other forms of endocytosis. The physiochemical properties of particles appear to influence whether clathrin- or caveolae-mediated endocytosis is the major uptake method. For example, PLGA nanospheres alone were shown to be mainly taken up via clathrin-mediated endocytosis, however, surface coating of such nanospheres with polysorbate 80 led to equivalent uptake with clathrin- and caveola-mediated endocytosis (Tahara et al. 2010). In an effort to increase particle uptake, many groups have focused on modifying gene carriers with specific peptides that display cell affinity. The expectation is that increased cell membrane binding will lead to increased complex uptake, ultimately improving the transfection efficiency. This is generally accomplished through incorporating short viral peptides into the non-viral system permitting binding to and penetration through the plasma membrane, and targeting to and through the nuclear envelope (Fig. 3).

3.3.1 Cellular Uptake and Vesicular Escape

The internalization of carrier/DNA complexes by clathrin-mediated endocytosis or other mechanisms (such as caveolae-mediated endocytosis, clathrin/caveole-independent endocytosis and macropinocytosis) must be followed by vesicular escape to allow nuclear uptake. The fate of an endosome is dependent on the payload, and can vary from maturation and intracellular trafficking to lysosome formation. The latter is not desirable for nucleic acid payloads, as the acidic environment in lysosome may damage the plasmid DNA. However, a moderate decrease in pH can be used as a trigger to disrupt the endosome and allow the plasmid DNA to be released into the cytoplasm. The escape of one important class of cationic polymers, i.e., PEI, has been attributed to the 'proton sponge' hypothesis (Boussif et al. 1995); excess protons during endosome acidification are absorbed by the amine groups on PEI, which causes an excess influx of counterions and water, leading to the endosome swelling, burst, and eventual release of complexes into the cytoplasm. Although this mechanism has been generally accepted for PEI-mediated delivery, exceptions to this mechanism were also noted (Hoekstra et al. 2007).

A number of peptides have been utilized to disrupt vesicular membranes to facilitate the necessary escape. One commonly used motif is a pH-sensitive amphipathic helix, which has aspartic and/or glutamic residues concentrated on one side of the helix. At neutral pH, the peptide has no secondary structure, but as the pH decreases along the endocytic pathway and the acidic residues become protonated, alpha-helix formation results that can lead to multimerization and membrane destabilization (Wagner 1999). The fusogenic N-terminus of hemagglutinin HA-2 from influenza virus is one example and has been used in both poly-L-lysine (PLL) and liposomal systems to successfully increase endosomal escape (Wagner et al. 1992; Subramanian et al. 2002). Endosomal escape has been also increased with the synthetic peptide H_5WYG (GLFHAIAHFIHGGWHGLIHGWYG) which is a derivative of the HA-2 N-terminus (GLFGAIAGFIEGGWTGMIDGWYG) with five histidines residues replacing the native amino acids (Pichon et al. 2001). The proton sponge effect is thought to impart an endosomolytic property to H_5WYG due to histidine protonation. Another molecule employing membrane-destabilizing helix formation is the VP-1 protein from rhinovirus and the synthetic peptides JTS-1, ppTG1, ppTG20, GALA, and KALA which all mimic the same viral model of helix-mediated endosomal escape (Zauner et al. 1995; Gottschalk et al. 1996; Rittner et al. 2002; Futaki et al. 2005; Wyman et al. 1997).

The arginine-rich TAT peptide from the lentivirus HIV has also been used to facilitate vesicular escape. This peptide has been utilized in both polymer and liposomal systems to increase transfection (Xiong et al. 2010; Rudolph et al. 2003; Torchilin et al. 2001). A number of linear and branched arginine-rich synthetic peptides inspired by the TAT peptide have been developed that can also act as cell-penetrating peptides (Futaki et al. 2001, 2002; Nakamura et al. 2007). While TAT and similar synthetic peptides have been used in a variety of non-viral systems, both their preferential endocytic route (likely a combination of direct fusion, clathrin-mediated endocytosis and macropinocytois) and their mechanism of traversing membranes remain ill-defined (Futaki 2006; Futaki et al. 2007). It appears that their ability to bind to cell membrane plays a functional role as much as their ability to facilitate endosomal escape.

Another notable viral peptide utilized by nanocarriers is the penton base protein of adenovirus serotype 5 (Ad5) (Rentsendorj et al. 2004). As mentioned before, this peptide permits binding to alpha-V integrins mediating cellular internalization and causes membrane destabilization at low pH permitting endocytic escape. While penton base complexes have been shown to follow multiple endocytic routes (as with native adenovirus), it is thought that the complexes that contribute to nuclear delivery follow a clathrin-mediated endocytic route (Rentsendorj et al. 2006). Also similar to native adenovirus, penton base proteins seem to be capable of utilizing microtubules and dynein to traffic to the perinuclear region, potentially making them useful for not only vesicular escape but for increasing nuclear uptake as well (Rentsendorj et al. 2006). The RGD peptide sequences found in integrin-binding adenoviral proteins have been also incorporated into synthetic carriers; however, this approach met with mixed success. Linear RGD modification of PEI was shown to be beneficial for transgene expression in a melanoma cell line, which appeared to be due to increased complex attachment to the cell surface (Kunath et al. 2003). However, our own group found that modifying PEI (branched, 25 kDa) with RGD did not increase polymer binding to bone marrow stromal cells (Clements et al. 2006). Attaching RGD to non-cationic polymers (i.e., polymers unable to interact with anionic cell surface), on the other hand, were shown to be beneficial in increasing nucleic acid delivery (Xiong et al. 2010). As much as the ligand, the polymeric carrier used for complex formation seems to be critical for the ultimate success of this strategy.

Unlike peptide-mediated specific uptake mechanisms, it is possible to take a non-specific approach to enhance cellular crossing of the complexes by imparting a hydrophobic character to the carriers. This could be achieved by lipid substitution (Incani et al. 2009). Modifying cationic polymers, such as PEI (Neamnark et al. 2009) and PLL (Abbasi et al. 2007), with various lipids resulted in increased transfection efficiency, in parallel with increased uptake of polymer/plasmid complexes.

3.3.2 Nuclear Import

While many carriers can successfully achieve cytoplasmic localization, nuclear targeting remains a major rate-limiting step. This is especially true in cells with low mitotic activity where carriers face the barrier of an intact nuclear envelope. Entry into the nucleus is regulated by NPC and, although small segments of DNA (1 Kb) can enter the nucleus unassisted (Hagstrom et al. 1997), DNA larger than 2 Kb mostly remains in the cytoplasm. The NPC is generally regarded to be permeable to molecules up to 39 nm (Pante and Kann 2002). Considering that the plasmid DNAs for typical therapeutic genes are several Kbs in size and form complexes >100 nm in size with carriers, passage through the NPC is a major challenge. Movement from the cytoplasm to the nucleus can be a passive diffusion process, or an active process based on the properties of the plasmid carrier. Some polymeric carriers, such as PEI and PLL, appear to have the intrinsic affinity to nucleus (Farrell et al. 2007), which lipid-based systems lack (Pollard et al. 1998). Cationic lipids appear to be dependent on cell mitosis, and dissolution of the nuclear membrane, for efficient nuclear delivery of plasmid DNA and transgene expression

(Brunner et al. 2000). This 'passive' uptake seems to be operational with polymeric carriers as well, since our recent studies with lipid-substituted polymers also indicate a strong correlation between the localization of the particles in the nuclear periphery and the ability of a cell to express a transgene (Hsu et al. 2010).

Most viral vectors are far more effective at nuclear delivery and hence integrating their targeting peptides into non-viral systems has become a common approach to overcoming this bottleneck. Studies using nuclear import signals have overwhelmingly utilized the Large T-antigen NLS (nuclear localization sequence) of SV40 (Simian Virus 40). This monopartite NLS binds importin-α, which itself has a bipartite NLS that can bind to importin-β resulting in the import of the entire complex through the nuclear pore (Fontes et al. 2000). A number of approaches have been undertaken to facilitate nuclear DNA delivery using the SV40 NLS. Plasmid DNA has been modified with the peptide by simple electrostatic interactions, peptide nucleic acid coupling, streptavidin-biotin systems, and covalent linking (Aronsohn and Hughes 1998; Branden et al. 1999; Ludtke et al. 1999; Sebestyen et al. 1998). The peptide can also be associated with polymer or lipid-based carriers instead of being linked to the DNA directly (Talsma et al. 2006; Chan et al. 2000; Kurihara et al. 2009). While eclipsed by the use of SV40 NLS, other viral peptides can facilitate nuclear targeting of non-viral carriers as well. The HIV-derived TAT peptide, mentioned above for its membrane-destabilizing properties, can also serve as a nuclear-localizing agent when complexed with DNA (Rudolph et al. 2003). Similarly, the mu peptide from the core complex of adenovirus was initially used to condense DNA but was later found to have nuclear localizing properties when used in liposomal formulations (Murray et al. 2001; Tagawa et al. 2002; Keller et al. 2003).

Although nuclear transcription of exogenous DNA is not generally thought to be rate-limiting, the choice of sequences surrounding the transgene can significantly influence expression. Plasmids with viral sequences have had poorer expression compared to non-viral promoters; following robust expression, a steady decline in transgene expression was observed *in vitro* and *in vivo* (Aviles et al. 2010). In addition to this steady decline, the immune response to foreign material must be considered. Unmethylated CpG plasmid DNA is recognized by Toll-like Receptor 9 (Hemmi et al. 2000), and can lead to cytokine production, inflammation and other undesirable immune responses.

4 Current State of Affairs on Non-viral Gene Delivery for Bone Regeneration

Local injection of plasmid, rather than systemic injection, is favoured for bone regeneration because of undesired effects should the recombinant protein be expressed ectopically. Injections of the therapeutic agents are advantageous with respect to their minimum invasiveness compared to surgical implantation. The latter approach, however, enables one to employ biomaterial scaffolds whose beneficial effects were articulated before. Whereas the use of DNA-condensing carriers seems to be paramount for effective gene expression, *in vivo* attempts for bone regeneration have occasionally relied on plasmid DNA delivery with scaffold alone without the appropriate carriers. Figure 4 summarizes the reported attempts at non-viral gene delivery for bone regeneration.

4.1 Plasmid DNA Delivery Without Carriers

One can avoid the use of carriers and rely on electroporation for cellular uptake of genes; the feasibility of this approach was shown in a rat subcutaneous model where radiopague bone was obtained after the delivery of 25–50 μg BMP-2/-7 plasmid to a rat ectopic site (Kawai et al. 2009). However, this physical method to disrupt plasma membranes may leave too much tissue trauma, as dystrophic calcification has been seen in controls (Kishimoto et al. 2002) when 100 μg of BMP-4 plasmid was delivered ectopically to rat muscle using electroporation. Additionally, it may be difficult to control to where the plasmid DNA is delivered, especially in intraosseous applications where the bone is well-surrounded of muscle and soft tissue. Plasmid DNA encoding for OP-1 was also delivered by a collagen scaffold without a specific carrier (Bright et al. 2006). Histological bone formation was evident with the use of the collagen carrier but no radio-opaque bone formation was observed even with 500 μg of plasmid, indicating less than efficacious bone induction from a clinical perspective. No bone formation was detected histologically in that study when naked plasmid DNA was delivered, supporting the role of collagen in localizing or retaining the plasmid DNA. It was not known if collagen condensed the plasmid DNA into a nanoparticle for cell uptake, or simply retained the plasmid DNA in its native configuration at the site for a prolonged time.

More success was seen with a collagen scaffold delivering a plasmid coding for human parathyroid hormone 1–34 (PTH1-34) in a canine tibia critical defect model (Bonadio et al. 1999). The plasmid was added to a collagen solution, which was frozen and lyophilized as a unitary device. A 25% increase in bone mass was seen with 40 mg of plasmid DNA after 4 weeks, and the majority of the bone gap was filled at 6 weeks with 100 mg of plasmid. While this model shows good promise for clinical utility, the high amount of plasmid (100 mg) required for healing this tibial defect leaves room for improvement. Another study investigated a combination of therapeutic genes delivered by a collagen sponge to a rat critical defect femur model (Fang et al. 1996). Only fibrous tissue was seen in control groups; however the delivery of BMP-4 plasmid resulting in new bone forming to bridge the defect at 9 weeks and sufficient new bone had formed by 18 weeks to remove external fixation supporting the defect. Co-delivery of a BMP-4 and PTH1-34 expression plasmid was found to be superior to BMP-4 alone. The combination of BMP-4 and PTH1-34 led to accelerated healing, where defect was bridged at 4 weeks, and external fixation could be removed at 12 weeks, but even this favourable configuration still required 0.5–1 mg of plasmid DNA. Since a group looking at bone healing with PTH1-34 alone was not included in this study, the healing time following PTH1-34 expression alone is not known.

Delivery of 0.1 mg or 1 mg of VEGF expressing plasmid in a collagen sponge was investigated to a rabbit critical-sized radial defect (Geiger et al. 2005). VEGF delivery resulted in bone formation at 6 and 12 weeks compared to controls, where no defect bridging was observed. In addition to bone formation, delivery of VEGF plasmid led to increased vasculature at 6 and 12 weeks compared to controls. Although an increase in vasculature was seen from 0.1 mg to 1 mg of VEGF plasmid, this did not appear to correspond to increased regeneration, as no difference was seen with respect to the volume of new bone.

These studies demonstrate the feasibility of plasmid delivery for bone regeneration; however, the large amount of plasmid DNA required for a measurable biological response led groups to investigate gene carriers to facilitate delivery, which may increase the efficiency of gene delivery. In addition, these results highlight the importance of gene selection for bone regeneration studies of gene treatment success. This is evident when comparing delivery of BMP-4 (Fang et al. 1996) and VEGF (Geiger et al. 2005) genes in a collagen sponge for healing of rat femur and rabbit radial defects. Despite potential differences between the healing and regenerative potential of each of the defect sites, only 0.1 mg of VEGF plasmid was required to bridge the defect at 6 weeks, whereas 0.5 mg of BMP-4 plasmid was required to bridge the gap at 9 weeks. Given the importance of sufficient blood supply for fracture healing and regeneration, a growth factor able to induce angiogenesis may be more potent.

Recent studies have also shown that ultrasound can also be used to introduce genes into different cell types in the absence of a gene carrier (Li et al. 2009). This process, also known as sonoporation, creates pores in the cell membrane and facilitates uptake of genes both *in vitro* and *in vivo*. Using a BMP-2 expression system, *de novo* induction of bone was demonstrated at intramuscular sites (rodent) using microbubble-enhanced transcutaneous sonoporation (Osawa et al. 2009). No carrier was used in that study since muscle tissue is unique in picking up DNA molecules. BMP-9 expression was also shown to lead to bone induction with a similar approach (Sheyn et al. 2008). The nature of the microbubbles used to enhance the sonoporation process was shown to be critical in optimizing gene expression at muscular sites (Kodama et al. 2010), although a therapeutic effect was not explored in the latter study. It remains to be seen if sonoporation could be effectively used for gene expression in deep tissues, so that an effective bone repair could be achieved.

4.2 *Plasmid DNA Delivery with Synthetic Carriers*

The success of branched 25 kDa PEI for *in vitro* gene delivery led to its investigation for *in vivo* applications. A poly(lactic-co-glycolic acid) (PLGA) scaffold was used to deliver 0.2 mg of BMP-4 plasmid condensed with PEI to a rat skull critical defect (Huang et al. 2005). At 8 weeks, defects with scaffolds containing the PEI-condensed plasmid showed bone formation around the edges of the defect, compared to naked (uncondensed) plasmid and the empty scaffolds which showed minimal regeneration. The amount of 25 kDa PEI contained in each implant scaffold is unknown, and there is no discussion on the possible toxicity of the PEI delivered in the implant. A previous study from our lab found that small amounts of PEI (16–32 μg) interfered with bone formation at an ectopic site following delivery of BMP-2 albumin nanoparticles (Zhang et al. 2009). Given that 0.2 mg of plasmid DNA was delivered, more than 32 μg of PEI is expected to be required to condense the plasmid DNA. Thus, some cytotoxicity is expected in these implants. Although a ~4.5-fold increase in bone formation was detected over no plasmid and naked plasmid controls, the majority of the defect still was not healed after 15 weeks with PEI-condensed plasmid. When naked plasmid DNA or the scaffold alone was delivered, bone formation was only seen along the periphery, with no differences between the two groups, suggesting differences in either cell penetration into the scaffold or the regenerative capacity at each of the defect sites. The delivery of 0.5 mg of BMP-4 in collagen led to bridging of a femur defect by 9 weeks (Fang et al. 1996), while here 0.2 mg of BMP-4 delivered in a PLGA scaffold was not able to show significant bone formation over a blank PLGA scaffold in a skull defect.

The HA, the main non-organic component of bone, and other calcium phosphates are commonly used scaffolds in bone regeneration, so that it is natural to employ them for plasmid DNA delivery as well. An HA fiber scaffold loaded with BMP-2 plasmid condensed in a CaP solution was implanted subcutaneously in rats (Oda et al. 2009). At 4 weeks, radiopague volume of the implants containing 50 or 100 μg of BMP-2 plasmid was higher than the HA scaffold alone. However, the HA fibers themselves contained radiopaque regions detected by micro-CT as well, suggesting that the scaffold itself may induce calcification even in the absence of therapeutic gene expression. By week 12, implants containing 10, 50 or 100 μg of BMP-2 plasmid all had higher bone volume than the empty scaffold. The mineral content of scaffolds containing 50 μg of BMP-2 plasmid maintained high levels from week 4 to 12, and although there was no significant increase, there was no decrease in mineral content, as was observed in other groups.

A similar study employed porous HA to deliver 10 μg of BMP-2 plasmid via cationic liposomes, using the commercially available SuperFect™ transfection reagent. The scaffolds were used to fill a rabbit critical-sized cranial defect (Ono et al. 2004). After 3 weeks, groups with the BMP-2 gene in liposomes showed new bone formation on the bottom side of the scaffold. At 9 weeks, new bone tissue had penetrated the lower half of the HA scaffold. Bone formation was also observed in the absence of the HA scaffold when the BMP-2 plasmid liposomes were administered directly to the site, and closed the defect by week 6. In total, 40 μg of the liposome was administered to the site, an amount the authors have chosen based on the desire to avoid cytotoxicity.

Finally, one group used employed as little as 1.3 μg of plasmid DNA (the lowest amount known to the authors) and showed successful bone regeneration in a mouse skull defect (Itaka et al. 2007). A block polymer consisting of polyethyleneglycol, aspartate and diethylenetriamine was used to deliver the plasmid DNA. Two genes were delivered to the defect site: runt-related transcription factor 2 (Runx2) and constitutively active form of activin receptor-like kinase 6 (caALK6), both intracellular mediators involved in osteogenic differentiation. This is a unique choice of genes since they are not extra-cellularly acting proteins (whose target is usually stem/responsive cells at the site), but rather intracellularly active species that will have the power to transform the expressing cells, rather than by acting on the neighbouring cells. The resulting complexes were mixed in a calcium phosphate paste, and after 4 weeks of implantation, bone formation was observed histologically covering the implant. Bone formation was quantified by summation of the regenerated area to that of the original defect area, and was found to be more than 50%. Similar results were not seen with 22 kDa linear PEI or Fugene 6™, both of which are successful transfection reagents *in vitro*. Such results highlight a major issue in developing non-viral vectors for bone regeneration. While PEI and many commercially available reagents work very well *in vitro*, their gene delivery success often does not translate in vivo.

5 Concluding Comments and Future Challenges

Large strides have been made in non-viral gene delivery for bone regeneration, but many challenges remain. It must be stated that it is difficult to compare outcomes from different groups, as each study uses different locations, species and genes. Additionally, each group uses different measures of success, some of which correspond to better clinical relevance. In some studies, bone observed by histological assessment did not correspond to a desirable radiopaque bone formation. Other studies use a completely healed defect as the standard for success, which is more clinically relevant, whereas others simply used an increase in bone volume or histological detection of calcified tissue, regardless of whether the gap had fully healed. More standardized approaches will help better elucidate the relative effectiveness of a proposed approach. Significant challenges remain on non-viral gene carrier development, whose efficiency currently lags behind that of viral vectors. A better safety profile makes non-viral carriers ideal over viruses, however their transfection efficiency must improve in order to achieve robust bone formation required for future clinical use. Understanding the biology of both viral and non-viral gene delivery can lead to the development of non-viral carriers with specific design features to overcome intracellular barriers. Given the progress in the field of non-viral vectors, further research is warranted to develop improved non-viral gene delivery systems for osteogenic applications.

References

Abbasi M, Uludag H, Incani V, Olson C, Lin X, Clements BA, Rutkowski D, Ghahary A, Weinfeld M. *Palmitic acid-modified poly-L-lysine for non-viral delivery of plasmid DNA to skin fibroblasts*. Biomacromolecules 2007;8(4):1059–63.

Alba R, Bosch A, Chillon M. Gutless adenovirus: *Last-generation adenovirus for gene therapy*. Gene Ther. 2005;12:S18-27.

Alden T, Pittman D, Hankins F, Beres E, Engh J, Das S, Hudson S, Kerns K, Kallmes D, Helm G. *In vivo endochondral bone formation using a bone morphogenetic protein 2 adenoviral vector.* Hum Gene Ther 1999;10(13):2245–53.

Alden T, Beres E, Laurent J, Engh J, Das S, London S, Jane J, Hudson S, Helm G. *The use of bone morphogenetic protein gene therapy in craniofacial bone repair.* J Craniofac Surg 2000;11(1): 24–30.

American Association of Orthopaedic Surgeons. *The Burden of Muscoluskeletal Diseases in the United States.* In: AAOS Website (Available at: http://boneandjointburden.org/) Cited August 3 2010.

Anderson D, Laquerre S, Goins W, Cohen J, Glorioso J. *Pseudotyping of glycoprotein D-deficient herpes simplex virus type 1 with vesicular stomatitis virus glycoprotein G enables mutant virus attachment and entry.* J Virol. 2000;74(5):2481.

Andrew J, Hoyland J, Andrew S, Freemont A, Marsh D. *Demonstration of TGF-beta 1 mRNA by in situ hybridization in normal human fracture healing.* Calcif Tissue Int 1993;52(2):74–78.

Aronsohn A, Hughes J. *Nuclear localization signal peptides enhance cationic liposome-mediated gene therapy.* J Drug Target. 1998;5(3):163–9.

Ashinoff R, Cetrulo C, Galiano R, Dobryansky M, Bhatt K, Ceradini D, Michaels J, McCarthy J, Gurtner G. *Bone morphogenic protein-2 gene therapy for mandibular distraction osteogenesis.* Ann Plast Surg 2004;52(6):585–90.

Aviles M, Lin CH, Zelivyankaya M, Graham J, Boehler R, Messersmith P, Shea L. *The contribution of plasmid design and release to in vivo gene expression following delivery from cationic polymer modified scaffolds.* Biomaterials 2010;31:1140–1147.

Bangari D, Mittal S. *Current strategies and future directions for eluding adenoviral vector immunity.* Current Gene Therapy 2006;6(2):215.

Bartlett J, Wilcher R, Samulski R. *Infectious entry pathway of adeno-associated virus and adeno-associated virus vectors.* J Virol. 2000;74(6):2777.

Bergelson J, Cunningham J, Droguett G, Kurt-Jones E, Krithivas A, Hong J, et al. *Isolation of a common receptor for coxsackie B viruses and adenoviruses 2 and 5.* Science. 1997;275(5304): 1320.

Betz V, Betz O, Glatt V, Gerstenfeld L, Einhorn T, Bouxsein M, Vrahas M, Evans C. *Healing of segmental bone defects by direct percutaneous gene delivery: effect of vector dose.* Hum Gene Ther 2007a;18(10):907–15.

Betz O, Betz V, Nazarian A, Egermann M, Gerstenfeld L, Einhorn T, Vrahas M, Bouxsein M, Evans C. *Delayed administration of adenoviral BMP-2 vector improves the formation of bone in osseous defects.* Gene Therapy 2007b; 14:1039–1044.

Blumenthal R, Seth P, Willingham M, Pastan I. *pH-dependent lysis of liposomes by adenovirus.* Biochemistry. 1986;25:2231–2237.

Bonadio J, Smiley E, Patil P, Goldstein S. *Localized, direct plasmid gene delivery in vivo: prolonged therapy results in reproducible tissue regeneration.* Nature Medicine 1999;5(7): 753–759.

Bourque W, Gross M, Hall B. *Expression of four growth factors during fracture repair.* Int J Dev Biol 1993;37:573–579.

Boussif O, Lezoualch F, Zanta MA, Mergny MD, Scherman D, Demeneix B, Behr JP. *A versatile vector for gene and oligonucleotide transfer into cells in culture and in vivo: polyethylenimine.* Proc Natl Acad Sci USA 1995;92:7297–7301.

Branden L, Mohamed A, Smith C. *A peptide nucleic acid-nuclear localization signal fusion that mediates nuclear transport of DNA.* Nat Biotechnol. 1999;17:784–7.

Bright C, Park YS, Sieber AN, Kostuik JP, Leong KW. *In vivo evaluation of plasmid DNA encoding OP-1 protein for spine fusion.* Spine 2006;31:2163.

Brunner S, Sauer T, Carotta S, Cotton M, Saltik M, Wagner E. *Cell cycle dependence of gene transfer by lipoplex, polyplex and recombinant adenovirus.* Gene Therapy 2000;7:401–407.

Burns JC, Friedmann T, Driever W, Burrascano M, Yee JK. *Vesicular stomatitis virus G glycoprotein pseudotyped retroviral vectors: concentration to very high titer and efficient gene transfer into mammalian and nonmammalian cells.* 1993;90:8033–8037.

Cahill K, Chi J, Day A, Claus E. *Prevalence, complications, and hospitals charges associated with use of Bone Morphogenetic Proteins in spinal fusion procedures.* JAMA 2009;302(1):58–66.

Carano R, Filvaroff E. *Angiogenesis and bone repair.* Drug Delivery Today 2003;8(21): 980–989.

Chan C, Senden T, Jans D. *Supramolecular structure and nuclear targeting efficiency determine the enhancement of transfection by modified polylysines.* Gene Ther. 2000;7(19):1690.

Chang P, Seol Y, Cirelly J, Pellegrini G, Jin Q, Franco L, Goldstein S, Chandler L, Sosnowski B, Giannobile. *PDGF-B gene therapy accelerates bone engineering and oral implant osseointegration.* Gene Ther 2010;17(1):95–104.

Chen Y, Cheung K, Kung H, Leong J, Lu W, Luk K. *In vivo new bone formation by direct transfer of adenoviral-mediated bone morphogenetic protein-4 gene.* Biochem Biophys Res Commun 2002;298(1):121–127.

Chen Y, Luk K, Cheung K, Xu R, Lin M, Lu W, Leong J, Kung H. *Gene therapy for new bone formation using adeno-associated viral bone morphogenetic protein-2 vectors.* Gene Therapy 2003;10:1345–1353.

Chen Y, Luk K, Cheung K, Lu A, An X, Ng S, Lin M, Kung H. *Combination of adeno-associated virus and adenovirus vectors expressing bone morphogenetic protein-2 produces enhanced osteogenic activity in immunocompetent rats.* Biochem Biophys Res Commun 2004;317(3):675–81.

Chirmule N, Propert K, Magosin S, Qian Y, Qian R, Wilson J. *Immune responses to adenovirus and adeno-associated virus in humans.* Gene Ther. 1999;6(9):1574.

Clements BA, Bai J, Kucharski C, Farrell LL, Lavasanifar A, Ritchie B, Ghahary A, Uludag H. *RGD conjugation to polyethyleneimine does not improve DNA delivery to bone marrow stromal cells.* Biomacromolecules 2006;7:1481–1488.

Cook E, Cook J. *Bone graft substitutes and allografts for reconstruction of the foot and ankle.* Clin Podiatr Med Surg 2009;26:589–605.

Cronin J, Zhang X, Reiser J. *Altering the tropism of lentiviral vectors through pseudotyping.* Current gene therapy. 2005;5(4):387–398.

Davidson A, Benko M, Harrach B. *Genetic content and evolution of adenovirus.* J Gen. Virol. 2003;84:2895–2908.

De LaPorte L, Shea L. *Matrices and scaffolds for DNA delivery in tissue engineering.* Adv Drug Deliv Rev 2007;59:292–307.

De Rijck J, Vandekerckhove L, Christ F, Debyser P. *Lentiviral nuclear import: a complex interplay between virus and host.* Bioessays. 2007;29(5):441–451.

Deckers M, Bezooijen R, Horst G, Hoogendam J, Bent C, Papapoulos S, Lowik C. *Bone morphogenetic proteins stimulate angiogenesis through osteoblast-derived vascular endothelial growth factor A.* Endocrinology 2002;143(4):1545–1553.

Deyle D, Li Y, Olson E, Russell D. *Nonintegrating foamy virus vectors.* J Virol. 2010;84(18): 9341.

Di Pasquale G, B Davidson, C Stein, I Martins. *Identification of PDGFR as a receptor for AAV-5 transduction.* Nat. Med. 2003;9(10):1306–1312

Doan N, Reher P, Meghji S, Harris M. *In vitro effects of therapeutic ultrasound on cell proliferation, protein synthesis, and cytokine production by human fibroblasts, osteoblasts, and monocytes.* J Oral Maxillofac Surg. 1999;57(4):409–419.

Dunn C, Jin Q, Taba M, Franceschi R, Bruce Rutherford R, Giannnobile W. *BMP gene delivery for alveolar bone engineering at dental implant defects.* Mol Ther 2005;11(2):294–9

Dyall J, Szabo P, Berns K. *Adeno-associated virus (AAV) site-specific integration: Formation of AAV-AAVS1 junctions in an in vitro system.* Proc Natl Acad Sci U S A. 1999;96(22):12849.

Egermann M, Baltzer A, Adamaszek S, Evans C, Robbins P, Schneider E, Lill C. *Direct adenoviral transfer of bone morphogenetic protein-2 cDNA enhances fracture healing in osteoporotic sheep.* Hum Gene Ther 2006a;17(5):507–17.

Egermann M, Lill C, Griesbeck K, Evans C, Robbins P, Schneider E, Baltzer A. *Effect of BMP-2 gene transfer on bone healing in sheep.* Gene Ther 2006b;13(17):1290–9.

El-Bialy TH, Hassan A, Albaghdadi T, Fouad HA, Maimani AR. *Growth modification of the mandible using ultrasound in monkeys: A preliminary report.* Am. J. Orthod. Dentofac. Orthop. 2006;130:435 e7-435e14

El-Bialy T, Alyamani A, Albaghdadi T, Hassan A. Major PW. *Treatment of hemifacial microsomia without surgery: an evidence-based approach.* Proceeding of the 6th International Congress, 6, World Federation of Orthodontists, 8 Sep 2005.

El-Mowafi H, Mohsen M. *The effect of low-intensity pulsed ultrasound on callus maturation in tibial distraction osteogenesis.* Int. Orthop. 2005;29(2):121–124.

Fang J, Zhu YY, Smiley E, Bonadio J, Rouleau J, Goldstein S, McCauley L, Davidson B, Roessler B. *Stimulation of new bone formation by direct transfer of osteogenic plasmid genes.* Proc Natl Acad Sci USA 1996;93:5753.

Farrell L-L, Pepin J, Kucharski C, Lin X, Xu Z, Uludag H. *A comparison of the effectiveness of cationic polymers poly-L-lysine (PLL) and polyethylenimine (PEI) for non-viral delivery of plasmid DNA to bone marrow stromal cells.* Eur J Pharmaceut Biopharmaceut 2007;65: 388–397.

Feidler J, Roderer G, Gunther K, Brenner R. *BMP-2, BMP-4 and PDGF-bb stimulate chemotactic migration of primary human mesenchymal progenitor cells.* J Cell Biochem 2002;87(3): 305–12.

Fink D, DeLuca N, Yamada M, Wolfe D, Glorioso J. *Design and application of HSV vectors for neuroprotection.* Gene Ther. 2000;7(2):115–9.

Fontes M, Teh T, Kobe B. *Structural basis of recognition of monopartite and bipartite nuclear localization sequences by mammalian importin-alpha 1.* J Mol Biol. 2000;297(5):1183–94.

Friedlaender G, Perry C, Cole JD, Cook S, Cierny G, Muscheler G, Zych G, Calhoun J, LaForte A, Yin S. *Osteogenic protein-1 (bone morphogenetic protein-7) in the treatment of tibial non-unions.* J Bone Joint Surg Am 2001;83-A Suppl 1:S151-158

Futaki S, Suzuki T, Ohashi W, Yagami T, Tanaka S, Ueda K, et al. *Arginine-rich peptides.* J Biol Chem. 2001;276(8):5836.

Futaki S, Nakase I, Suzuki T, Youjun Z, Sugiura Y. *Translocation of branched-chain arginine peptides through cell membranes: Flexibility in the spatial disposition of positive charges in membrane-permeable peptides.* Biochemistry. 2002;41(25):7925–30.

Futaki S, Masui Y, Nakase I, Sugiura Y, Nakamura T, Kogure K, et al. *Unique features of a pH-sensitive fusogenic peptide that improves the transfection efficiency of cationic liposomes.* J Gene Med. 2005;7(11):1450–8.

Futaki S. *Oligoarginine vectors for intracellular delivery: Design and cellular-uptake mechanisms.* Peptide Science. 2006;84(3):241–9.

Futaki S, Nakase I, Tadokoro A, Takeuchi T, Jones A. *Arginine-rich peptides and their internalization mechanisms.* Biochem Soc Trans. 2007;35:784–7.

Gafni Y, Pelled G, Zilberman Y, Turgeman G, Apparailly F, Yotvat H, Galun E, Gazit Z, Jorgensen C, Gazit D. *Gene therapy platform for bone regeneration using an exogenously regulated, AAV-2-based gene expression system.* Mol Ther 2004;9(4):587–95.

Geiger F, Betram H, Berger I, Lorenz H, Wall O, Eckhardt C, Simank HG, Richter W. *Vascular endothelial growth factor gene-activated matrix (VEGF165-GAM) enhances osteogenesis and engiogenesis in large segmental bone defects.* J Bone Miner Res 2005;20:2028.

Gerstenfeld L, Cullinane D, Barnes G, Graves D, Einhorn T. *Fracture healing as a post-natal development process: molecular, spatial, and temporal aspects of its regulation.* J Cell Biochem 2003;88:873–884.

Giannoudis P, Dinopoulos H, Tsiridis E. *Bone substitutes: an update.* Injury 2005;36 S:S20-27.

Giraud C, Winocour E, Berns K. *Site-specific integration by adeno-associated virus is directed by a cellular DNA sequence.* Proc Natl Acad Sci. 1994;91(21):10039.

Gonda K, Nakaoka T, Yoshimura K, Otawara-Hamamoto Y, Harrii K. *Heterotopic ossification of degenerating rat skeletal muscle induced by adeno-virus mediated transfer of bone morphogenetic protein-2 gene.* J Bone Miner Res 2000;15(6):1056–65.

Gottschalk S, Sparrow JT, Hauer J, Mims MP, Leland FE, Woo SL, Smith LC. *A novel DNA-peptide complex for efficient gene transfer and expression in mammalian cells.* Gene Ther. 1996;3(5):448–57.

Govender S, Csimma C, Genant HK, Valentin-Opran A. *Recombinant human bone morphogenetic protein-2 for treatment of open tibial fractures.* J Bone Joint Surg Am 2002;84(12): 2123–2134.

Grandi P, Wang S, Schuback D, Krasnykh V, Spear M, Curiel D, et al. *HSV-1 virions engineered for specific binding to cell surface receptors.* Molecular Therapy. 2004;9(3):419–27.

Grandi P, Fernandez J, Szentirmai O, Carter R. *Targeting HSV-1 virions for specific binding to epidermal growth factor receptor-vIII-bearing tumor cells.* Cancer Gene. 2010;17:655–663.

Grieger J, Samulski R. *Packaging capacity of adeno-associated virus serotypes: Impact of larger genomes on infectivity and postentry steps.* J Virol. 2005;79(15):9933.

Hagstrom J, Ludtke J, Bassik M, Sebestyen M, Adam S, Wolff J. *Nuclear import of DNA in digitonin-permeabilized cells.* J Cell Sci 1997;110:2323–2331.

Heckman JD, Ryaby JB, McCabe J, Frey JJ, Kilcoyne RF. *Acceleration of tibial fracture – healing by non-invasive, low-intensity pulsed ultrasound.* J. Bone Joint Surg. 1994;76: 26–34.

Harbison C, Chiorini J, Parrish C. *The parvovirus capsid odyssey: from the cell surface to the nucleus.* Trends in Microbiology. 2008;16(5):208–214.

Harrison S. *Mechanism of membrane fusion by viral envelope proteins.* Adv Virus Res. 2005;64:231–261.

Harui A, Suzuki S, Kochanek S, Mitani K. *Frequency and stability of chromosomal integration of adenovirus vectors.* J Virol. 1999;73(7):6141.

Haynes C, Erlwein O, Schnierle BS. *Modified envelope glycoproteins to retarget retroviral vectors.* Curr Gene Ther. 2003;3(5):405–10.

Hedley S, Maur Ad, Hohn S, Escher D, Barberis A. *An adenovirus vector with a chimeric fiber incorporating stabilized single chain antibody achieves targeted gene delivery.* Gene Ther. 2005;13:88–94.

Helm G, Alden T, Beres E, Hudson S, Das S, Engh J, Pittman D, Kern K, Kallmes D. *Use of bone morphogenetic protein-9 gene therapy to induce spinal arthrodesis in the rodent.* J Neurosurg 2000;92(2 S):191–6.

Hemmi H, Takeuchi O, Kawai T, Kaisho T, Sat S, Sanjo H, Matsumoto M, Hoshino K, Wagner H, Takeda K, Akira S. *A Toll-like receptor recognizes bacterial DNA.* Nature 2000;408: 740–745.

Hoekstra D, Rejman J, Wasungu L, Shi F, Zuhorn I. *Gene delivery by cationic lipids: in and out of an endosome.* Biochem Soc Trans 2007;35(1):68–71.

Hong KS, Kim EC, Bang SH, Chung CH, Lee YI, Hyun JK, Lee HH, Jang JH, Kim TI, Kim HW. *Bone regeneration by bioactive hybrid membrance containing FGF2 within rat calvarium.* J Biomed Mat Res 2010; 94A(4):1187–1194.

Howarth J, Lee Y, Uney J. *Using viral vectors as gene transfer tools (cell biology and toxicology special issue: ETCS-UK 1 day meeting on genetic manipulation of cells).* Cell Biol Toxicol. 2010;26(1):1–20.

Hsu C, Hendzel M, Uludag H. *Improved transfection efficiency of an aliphatic lipid substituted 2 kDa polyethylenimine is attributed to enhanced nuclear association and uptake in rat bone marrow stromal cells.* J Gene Med 2010, in press.

Hu W, Wang Z, Hollister S, Krebsbach P. *Localized viral vector delivery to enhance in situ regenerative gene therapy.* Gene Ther 2007;12(11):891–901.

Huang YC, Simmons C, Kaigler D, Rice KG, Mooney DJ. *Bone regeneration in a rat cranial defect with delivery of PEI-condensed plasmid DNA encoding for bone morphogenetic protein-4 (BMP-4).* Gene Therapy 2005; 12:418.

Incani V, Lin X, Lavansanifar A, Uludag H. *Relationship between the extent of lipid substitution on poly(L-lysine) and the DNA delivery efficiency.* ACS Appl Mater Interfaces 2009;1(4):841–8.

Ishihara A, Shields K, Litsky A, Mattoon J, Weisbrode S, Bartlett J, Bertone A. *Osteogenic gene regulation and relative acceleration of healing by adenoviral-mediated transfer of human BMP-2 or −6 in equine osteotomy and ostectomy models.* J Orthop Res 2008;26(6):764–71.

Ishihara A, Zekas L, Weisbrode S, Bertone A. *Comparative efficacy of dermal fibroblast-mediated and direct adenoviral bone morphogenetic protein-2 gene therapy for bone regeneration in an equine rib model.* Gene Ther 2010;17(6):733–44.

Itaka K, Ohba S, Miyata K, Kawaguchi H, Nakamura K, Takato T, Chung U, Kataoka K. *Bone regeneration by regulated in vivo gene transfer using biocompatible polyplex nanomicelles.* Molecular Therapy 2007;15(9):1655.

Jackson S, DeLuca N. *Relationship of herpes simplex virus genome configuration to productive and persistent infections.* Proc Natl Acad Sci. 2003;100(13):7871.

Jane J, Dunford B, Kron A, Pittman D, Sasaki T, Li J, Li H, Alden T, Dayoub H, Hankins G, Kallmes D, Helm G. *Ectopic osteogenesis using adenoviral bone morphogenetic protein (BMP)-4 and BMP-6 gene transfer.* Mol Ther 2002;6(4):464–70.

Jin Q, Anusaksathien O, Webb S, Printz M, Giannobile W. *Engineering of tooth-supporting structures by delivery of PDGF gene therapy vectors.* Mol Ther 2004;9:519–526.

Kaihara S, Dessho K, Okubo Y, Sonobe J, Kawai M, IIzuka T. *Simple and effective osteoinductive gene therapy by local injection of a bone morphogenetic protein-2-expressing recombinant adenoviral vector and FK506 in rats.* Gene Therapy 2004;11:439–447.

Kaludov N, Brown K, Walters R, Zabner J. *Adeno-associated virus serotype 4 (AAV4) and AAV5 both require sialic acid binding for hemagglutination and efficient transduction but differ in sialic acid linkage specificity.* J Virol. 2001;6884–6893.

Kang Q, Sun M, Cheng H, Peng Y, Montag A, Deyrup A, Jiang W, Luu H, Luo J, Szatkowski J, Vanichakarn P, Park J, Li Y, Haydon R, He T. *Characterization of the distinct orthotopic bone formting activity of 14 BMPs using recombinant adenovirus-mediated gene delivery.* Gene Ther 2004;11(17):1312–20.

Kanerva A, Wang M, Bauerschmitz G, Lam J, Desmond R, Bhoola S, et al. *Gene transfer to ovarian cancer versus normal tissues with fiber-modified adenoviruses.* Molecular therapy. 2002;5(6):695–704.

Kawai M, Maruyama H, Bessho K, Yamamoto H, Miyazaki J, Yamamoto T. *Simple strategy for bone regeneration with a BMP-2/7 gene expression cassette vector.* Biochem Biophys Res Commun 2009;390(3):1012–7.

Kay M, Glorioso J, Naldini L. *Viral vectors for gene therapy: The art of turning infectious agents into vehicles of therapeutics.* Nat Med. 2001;7(1):33–40.

Kelkar S, Pfister K, Crystal R, Leopold P. *Cytoplasmic dynein mediates adenovirus binding to microtubules.* J Virol. 2004;78(18):10122.

Keller M, Harbottle R, Perouzel E, Colin M, Shah I, Rahim A, et al. *Nuclear localisation sequence templated nonviral gene delivery vectors: Investigation of intracellular trafficking events of LMD and LD vector systems.* Chembiochem. 2003;4(4):286–98.

Kennedy M, Parks R. *Adenovirus virion stability and the viral genome: size matters.* Molecular Therapy. 2009;17(10): 1664–1666.

Kishimoto K, Watanabe Y, Nakamura H, Kokubun S. *Ectopic bone formation by electroporatic transfer of bone morphogenetic protein-4 gene.* Bone 2002;31(2):340–347.

Kodama T, Aoi A, Watanabe Y, Horie S, Kodama M, Li L, Chen R, Teramoto N, Morikawa H, Mori S, Fukumoto M. *Evaluation of transfection efficiency in skeletal muscle using nano/microbubbles and ultrasound.* Ultrasound Med Biol. 2010;36(7):1196–205.

Kong Y, Yang T, Geller A. *An efficient in vivo recombination cloning procedure for modifying and combining HSV-1 cosmids.* J Virol Methods. 1999;80(2):129–36.

Krieg A. *CpG Motifs in bacterial DNA and their immune effects.* Annu Rev Immunol 2002;20: 709–760.

Kunath K, Merdan T, Hegener O, Haberlein H. *Integrin targeting using RGD-PEI conjugates for in vitro gene transfer.* J Gene Med 2003;5:588–599.

Kurihara D, Akita H, Kudo A, Masuda T, Futaki S, Harashima H. *Effect of polyethyleneglycol spacer on the binding properties of nuclear localization signal-modified liposomes to isolated nucleus.* Biol Pharmacet. Bull. 2009;32(7):1303–6.

Laquerre S, Anderson D, Stolz D. *Recombinant herpes simplex virus type 1 engineered for targeted binding to erythropoietin receptor-bearing cells.* J Virol. 1998:9683–9697.

Lewis P, Emerman M. *Passage through mitosis is required for oncoretroviruses but not for the human immunodeficiency virus.* J Virol. 1994;68(1):510–516.

Li J, Li H, Sasaki T, Holman D, Beres B, Dumont J, Pittman D, Hankins G, Helm G. *Osteogenic potential of five different recombinant human bone morphogenetic protein adenoviral vectors in rat.* Gene Therapy 2003a;10:1735–1743.

Li J, Dunford B, Holman D, Beres B, Pittman D, Hankins G, Helm G. *Rat strain differences in the ectopic osteogenic potential of recombinant human BMP adenoviruses.* Mol Ther 2003b; 8(5):822–9.

Li J, Hankins G, Kao C, Li H, Kammauff J, Helm G. *Osteogenesis in rats induced by a novel recombinant helper-dependent bone morphogenetic protein-9 (BMP-9) adenovirus.* J Gene Med 2003c; 5(9):748–56.

Li YS, Davidson E, Reid CN, McHale AP. *Optimising ultrasound-mediated gene transfer (sonoporation) in vitro and prolonged expression of a transgene in vivo: potential applications for gene therapy of cancer.* Cancer Lett. 2009;273(1):62–69.

Lindsey W. *Osseous tissue engineering with gene therapy for facial bone reconstruction.* Laryngoscope 2001;111(7):1128–36.

Loser P, Hillgenberg M, Arnold W, Both G, Hofmann C. *Ovine adenovirus vectors mediate efficient gene transfer to skeletal muscle.* Gene Ther. 2000;7(17):1491–8.

Lu S, Zhang X, Soo C, Hsu T, Napoli A, Aghaloo T, Wu B, Tsou P, Ting K, Wang J. *The osteoinductive properties of Nell-1 in a rat spinal fusion model.* Spine J 2007;7(1):50–60.

Lu H, Qin L, Cheung W, Lee K, Wong W, Leung K. *Low-intensity pulsed ultrasound accelerated bone-tendon junction healing through regulation of vascular endothelial growth factor expression and cartilage formation.* Ultrasound Med Biol. 2008;34(8):1248–1260.

Ludtke J, Zhang G, Sebestyen M, Wolff J. *A nuclear localization signal can enhance both the nuclear transport and expression of 1 kb DNA.* J Cell Sci. 1999;112:2033.

Luk K, Chen Y, Cheung K, Kung H, Lu W, Leong J. *Adeno-associated virus-mediated bone morphogenetic protein-4 gene therapy for in vivo bone formation.* Biochem Biophys Res Commun 2003;308(3):636–45.

Marsh D, Li G. *The biology of fracture healing: optimising outcome.* British Medical Bulletin 1999;55(4):856–869.

McCarty D, Jr SY, Samulski R. *Integration of adeno-associated virus (AAV) and recombinant AAV vectors.* Genetics. 2004;38.

McConnell M, Imperiale M. *Biology of adenovirus and its use as a vector for gene therapy.* Human Gene Ther. 2004;15:1022–1033.

McKay W, Peckham S, Badura J. *A comprehensive review of recombinant human bone morphogenetic protein-2 (INFUSE® Bone Graft).* International Orthopaedics 2007;31: 729–734.

Mehara B, Saadeh P, Steinbrech D, Dudziak M, Spector J, Greenwald J, Gittes G, Longaker M. *Adenovirus-mediated gene therapy of osteoblasts in vitro and in vivo.* J Bone Miner Res 1999;14(8):1290–301.

Meier O, Greber U. *Adenovirus endocytosis.* J Gene Med. 2004;6(S1):S152-63.

Mercier G, Campbell J. *A chimeric adenovirus vector encoding reovirus attachment protein Sigma1 targets cells expressing junctional adhesion molecule 1.* Proc Natl Acad Sci. 2004;101(16):6188–6193.

Miller D, Adam M, Miller A. *Gene transfer by retrovirus vectors occurs only in cells that are actively replicating at the time of infection.* Mol Cell Biol. 1990;10(8):4239–4242.

Montesano R, Vassalli JD, Baird A, Guillemin R, Orci L. *Basic fibroblast growth factor induces angiogenesis in vitro.* Proc Natl Acad Sci USA 1986; 83:7297–7301.

Murray K, Etheridge C, Shah S, Matthews D, Russell W, Gurling H, et al. *Enhanced cationic liposome-mediated transfection using the DNA-binding peptide mu from the adenovirus core.* Gene Ther. 2001;8(6):453–60.

Nakamura T, Hanada K, Tamura M, Shibanushi T, Nigi H, Tagawa M, Fukumoto S, Matsumoto T. *Stimulation of endosteal bone formation by systemic injections of recombinant basic fibroblast growth factor in rats.* Endocrinology 1995;136:1276–1284.

Nakamura Y, Kogure K, Futaki S, Harashima H. *Octaarginine-modified multifunctional envelope-type nano device for siRNA.* J Controlled Release. 2007;119(3):360–7.

Nasu T, Ito H, Tsutsumi R, Kitaori T, Takemoto M, Schwarz E, Nakamura T. *Biological activation of bone-related biomaterials by recombinant adeno-associated virus vector.* J Orthop Res 2009;27(9):1162–8.

Neamnark A, Suwantong O, Bahadur R, Hsu C, Supaphol P, Uludag H. *Aliphatic lipid substitution on 2 kDa polyethylenimine improves plasmid delivery and transgene expression.* Mol Pharm 2009;6(6):1798–815.

Oda M, Kuroda S, Kondo H, Kasugai S. *Hydroxyapatite fiber material with BMP-2 gene induces ectopic bone formation.* J Biomed Mat Res Part B 2009;90(1):101.

Okubo Y, Bessho K, Fujimura K, Iizuka T, Miyatake S. *Osteoinduction by bone morphogenetic protein-2 via adenoviral vector under transient immunosuppression.* Biochem Biophys Res Commun 2000;267(1):382–7.

Okubo Y, Bessho K, Fujimura K, Kaihara S, Iizuka T, Miyatake S. *The time course study of osteoinduction by bone morphogenetic protein-2 via adenoviral vector.* Life Sci 2001a;70:325–336.

Okubo Y, Bessho K, Fujimura K, Iizuka T, Miyatake S. *In vitro and in vivo studies of a bone morphogenetic protein-2 expressing adenoviral vector.* J Bone Joint Surg Am 2001b;83-A:S99-104.

Ono I, Yamashita T, Jin HY, Ito Y, Hamada H, Akasaka Y, Nakasu M, Ogawa T, Jombow K. *Combination of porous hydroxyapatite and cationic liposomes as a vector for BMP-2 gene therapy.* Biomaterials 2004;25:4709–4718.

Osawa K, Okubo Y, Nakao K, Koyama N, Bessho K. *Osteoinduction by microbubble-enhanced transcutaneous sonoporation of human bone morphogenetic protein-2.* J Gene Med. 2009;11(7):633–41.

Pante N, Kann M. *Nuclear pore complex is able to transport macromolecules with diameters of ~39 nm.* Mol Biol Cell 2002;13:425–34.

Pechan P, Herrlinger U, Aghi M, Jacobs A, Breakefield X. *Combined HSV-1 recombinant and amplicon piggyback vectors: Replication-competent and defective forms, and therapeutic efficacy for experimental gliomas.* J Gene Med. 1999;1(3):176–85.

Pichon C, Goncalves C, Midoux P. *Histidine-rich peptides and polymers for nucleic acids delivery.* Adv Drug Deliv Rev. 2001;53(1):75–94.

Pizzato M, Marlow S, Blair E, Takeuchi Y. *Initial binding of murine leukemia virus particles to cells does not require specific env-receptor interaction.* J Virol. 1999;73(10):8599.

Pollard H, Remys JS, LoussouarnG, Demolombe S, Behr JP. *Polyethylenimine but not cationic lipids promotes transgene delivery to the nucleus in mammalian cells.* J Biol Chem1998; 273(13):7507–7511.

Porter J, Ruckh T, Popat K. *Bone tissue engineering: a review in bone biomimetics and drug delivery strategies.* Biotechnol Prog 2009;25(6):1539–1560.

Qing K, Mah C, Hansen J, Zhou S, Dwarki V. *Human fibroblast growth factor receptor 1 is a co-receptor for infection by adeno-associated virus 2.* Nat Med. 1999;5(1):71–77.

Rentsendorj A, Agadjanian H, Chen X, Cirivello M, MacVeigh M, Kedes L, et al. *The Ad5 fiber mediates nonviral gene transfer in the absence of the whole virus, utilizing a novel cell entry pathway.* Gene Ther. 2004;12(3):225–37.

Rentsendorj A, Xie J, MacVeigh M, Agadjanian H, Bass S, Kim D, et al. *Typical and atypical trafficking pathways of Ad5 penton base recombinant protein: Implications for gene transfer.* Gene Ther. 2006;13(10):821–36.

Rittner K, Benavente A, Bompard-Sorlet A, Heitz F. *New basic membrane-destabilizing peptides for plasmid-based gene delivery in vitro and in vivo.* Molecular Therapy. 2002;5(2):104–114.

Roe T, Reynolds T, Yu G, Brown P. *Integration of murine leukemia virus DNA depends on mitosis.* EMBO J. 1993;12(5):2099.

Rosen V. *BMP2 signaling in bone development and repair.* Cytokine Growth Factor Rev 2009;20(5–6):475–80.

Rudolph C, Plank C, Lausier J, Schillinger U, Muller R, Rosenecker J. *Oligomers of the arginine-rich motif of the HIV-1 TAT protein are capable of transferring plasmid DNA into cells.* J Biol Chem. 2003;278(13):11411.

Rundle C, Miyakoshi N, Kasukawa Y, Chen S, Sheng M, Wergedal J, Lau K, Baylink D. *In vivo bone formation in fracture repair induced by direct retroviral-based gene therapy with bone morphogenetic protein-4*. Bone 2003;32(6):591–601.

Sakai S, Tamura M, Mishima H, Kojima H, Uemura T. *Bone regeneration induced by adenoviral vectors carrying til-1/Cbfa1 genes implanted with biodegradable porous materials in animal models of osteonecrosis of the femoral head.* J Tissue Eng Regen Med 2008;2(2–3):164–7.

Saphire A, Guan T, Schirmer E, Nemerow G, Gerace L. *Nuclear import of adenovirus DNA in vitro involves the nuclear protein import pathway and hsc70.* J Biol Chem. 2000;275(6): 4298.

Scheven BA, Man J, Millard JL, Cooper PR, Lea SC, Walmsley AD et al. *VEGF and odontoblast-like cells: stimulation by low frequency ultrasound.* Arch Oral Biol 2009;54:185–191.

Scmid G, Kobayashi C, Sandell L, Ornitz D. *Fibroblast growth factor expression during skeletal fracture healing in mice.* Dev Dyn 2009; 238(3):766–774.

Sebestyen M, Ludtke J, Bassik M, Zhang G, Budker V, Lukhtanov E, et al. *DNA vector chemistry: The covalent attachment of signal peptides to plasmid DNA.* Nat Biotechnol. 1998;16(1):80–5.

Segura MM, Alba R, Bosch A, Chillón M. *Advances in helper-dependent adenoviral vector research.* Curr Gene Ther. 2008;8(4):222–35.

Shea L, Smiley E, Bonadio J, Mooney D. *DNA delivery from polymer matrices for tissue engineering*. Nature Biotechnology 1999;17:551–554.

Sheyn D, Kimelman-Bleich N, Pelled G, Zilberman Y, Gazit D, Gazit Z.: *Ultrasound-based non-viral gene delivery induces bone formation in vivo*. Gene Ther. 2008;15(4):257–66.

Sinn P, Sauter S, McCray P. *Gene therapy progress and prospects: Development of improved lentiviral and retroviral vectors--design, biosafety, and production.* Gene Ther. 2005;12(14):1089–98.

Smith R, Geller A, Escudero K, Wilcox C. *Long-term expression in sensory neurons in tissue culture from herpes simplex virus type 1 (HSV-1) promoters in an HSV-1-derived vector.* J Virol. 1995;69(8):4593.

Sonobe J, OkuboY, Kaihara S, Miyatake S, Bessho K. *Osteoinduction by bone morphogenetic protein 2-expressing adenoviral vector: application of biomaterial to mask the host immune response.* Hum Gene Ther 2004;15:659–668.

Stavropoulos T, Strathdee C. *An enhanced packaging system for helper-dependent herpes simplex virus vectors.* J Virol. 1998;72(9):7137.

Street J, Bao M, deGuzman L, Bunting S, Peale F, Ferrara N, Stelnmetz H, Hoeffel J, Cleland J, Daugherty A, van Bruggen N, Redmond H, Carano R, Filvaroff E. *Vascular endothelial growth factor stimulates bone repair by promoting angiogenesis and bone turnover.* PNAS 2002;99(15):9656–9661.

Strohbach C, Rundle C, Wergedal J, Chen S, Linkhart T, Lau K, Strong D. *LMP-1 retroviral gene therapy influences osteoblast differentiation and fracture repair: a preliminary study.* Calcif Tissue Int 2008;83(3):202–11.

Subramanian A, Ma H, Dahl K, Zhu J, Diamond S. *Adenovirus or HA-2 fusogenic peptide-assisted lipofection increases cytoplasmic levels of plasmid in nondividing endothelium with little enhancement of transgene expression.* J Gene Med. 2002;4(1):75–83.

Summerford C, Samulski R. *Membrane-associated heparan sulfate proteoglycan is a receptor for adeno-associated virus type 2 virions.* J Virol. 1998;72(2):1438–1445.

Summerford C, Bartlett J, Samulski R. *Alpha-V Beta5 integrin: A co-receptor for adeno-associated virus type 2 infection.* Nat Med. 1999;5(1):78–82.

Suzuki A, Takayama T, Suzuki N, Kojima T, Ota N, Asano S, Ito K. *Daily low-intensity pulsed ultrasound stimulates production of bone morphogenetic protein in ROS 17/2.8 cells.* J Oral Sci. 2009;51(1):29–36.

Tagawa T, Manvell M, Brown N, Keller M, Perouzel E, Murray K, et al. *Characterisation of LMD virus-like nanoparticles self-assembled from cationic liposomes, adenovirus core peptide mu and plasmid DNA.* Gene Ther. 2002;9(9):564–76.

Tahara K, Yamamoto H, Kawashima Y. *Cellular uptake mechanisms and intracellular distribution of polysorbate 80-modified poly(D,L-lactide-co-glycolide) nanospheres for gene delivery.* Eup J Pharm Biopharm 2010;75:218–224.

Talsma S, Babensee J, Murthy N, Williams I. *Development and in vitro validation of a targeted delivery vehicle for DNA vaccines.* J Controlled Release. 2006;112(2):271–9.

Tarkka T, Sipola A, Jamsa T, Soini Y, Yla-Herttuala S, Tuukkanen J, Hautala T. *Adenoviral VEGF-A gene transfer induces angiogenesis and promotes bone formation in healing osseous tissues.* J Gene Med 2003;5(7):560–6.

Torchilin V, Rammohan R, Weissig V, Levchenko T. *TAT peptide on the surface of liposomes affords their efficient intracellular delivery even at low temperature and in the presence of metabolic inhibitors.* Proc Natl Acad Sci. 2001;98(15):8786.

Trotman L, Mosberger N, Fornerod M, Stidwill R, Greber U. *Import of adenovirus DNA involves the nuclear pore complex receptor CAN/Nup214 and histone H1.* Nat. Cell Biol. 2001;3:1092–1100.

Uchida S, Sakai A, Kudo H, Otomo H, Watanuki M, Tanaka M, Nagashima M, Nakamura T. *Vascular endothelial growth factor is expressed along with its receptors during the healing process of bone and bone marrow after drill-hole injury in rats.* Bone 2003;32:491–501.

Uludag H, D'Augusta D, Golden J, Li J, Timony G, Riedel R, Wozney J. *Implantation of recombinant human bone morphogenetic proteins with biomaterial carriers: a correlation between protein pharmacokinetics and osteoinduction in the rat ectopic model.* J Biomed Mater Res 2000;50(2):227–238.

Uludag H, Gao T, Porter T, Friess W, Wozney J. *Delivery system for BMPs: factors contributing to protein retention at an application site.* J Bone Joint Surg Am 2001;83-A Suppl 1:S128–135.

Urist M. *Bone formation by autoinduction.* Science 1956;150:893–899.

Varady P, Li J, Cunningham M, Beres E, Das S, Engh J, Alden T, Pittman D, Kerns K, Kallmes D, Helm G. *Morphologic analysis of BMP-9 gene therapy-induced osteogenesis.* Hum Gene Ther 2001;12(6):697–710.

Varkey M, Gittens S, Uludag H. *Growth factor delivery for bone tissue repair: an update.* Expert Opin Drug Deliv 2004;1(1):19–36.

Verettas D, Galanis B, Kazakos K, Hatziyiannakis A, Kotsios E. *Fractures of the proximal part of the femur in patients under 50 years of age.* Injury 2002;33:41–51.

Verma I, Weitzman M. Gene therapy: *Twenty-first century medicine.* Annu. Rev. Biochem. 2005;74:711–38.

Waehler R, Russell S, Curiel D. *Engineering targeted viral vectors for gene therapy.* Nature Reviews Genetics. 2007;8(8):573–87.

Wagner E, Plank C, Zatloukal K, Cotten M, Birnstiel M. *Influenza virus hemagglutinin HA-2 N-terminal fusogenic peptides augment gene transfer by transferrin-polylysine-DNA complexes: Toward a synthetic virus-like gene-transfer vehicle.* Proc Natl Acad Sci. 1992;89(17):7934.

Wagner E. *Application of membrane-active peptides for nonviral gene delivery.* Adv Drug Deliv Rev. 1999;38(3):279–89.

Wang J, Guo X, Xu Y, Barron L, Szoka F. *Synthesis and characterization of long chain alkyl acyl carnitine esters. Potentially biodegradable cationic lipids for use in gene delivery.* J Med Chem 1998;41(13):2207–15.

Wang K, Guan T, Cheresh D. *Regulation of adenovirus membrane penetration by the cytoplasmic tail of integrin beta 5.* J Virol. 2000;74(6):2731–2739.

Weis K. *Regulating access to the genome: nucleocytoplasmic transport throughout the cell cycle.* Cell. 2003;112(4):441–451.

Wickham T, Mathias P, Cheresh D, Nemerow R. *Integrins $\alpha_v\beta_3$ and $\alpha_v\beta_5$ promote adenovirus internalization but not virus attachment.* Cell. 1993;73(2):309–319.

Wodrich H, Cassany A, D'Angelo M, Guan T, Nemerow G, Gerace L. *Adenovirus core protein pVII is translocated into the nucleus by multiple import receptor pathways.* J Virol 2006;80:9608–9618.

Wu Z, Asokan A, Samulski R. *Adeno-associated virus serotypes: Vector toolkit for human gene therapy.* Molecular therapy. 2006;14(3):316–27.

Wyman T, Nicol F, Zelphati O, Scaria P, Plank C, Jr FS. *Design, synthesis, and characterization of a cationic peptide that binds to nucleic acids and permeabilizes bilayers.* Biochemistry. 1997;36(10):3008–17.

Xiong X, Uludag H, Lavasanifar A. *Virus-mimetic polymeric micelles for targeted siRNA delivery.* Biomaterials. 2010;31(22):5886–5893.
Zauner W, Blaas D, Kuechler E, Wagner E. *Rhinovirus-mediated endosomal release of transfection complexes.* J Virol. 1995;69(2):1085–1092.
Zhang S, Doschak M, Uludag H. *Pharmacokinetics and bone formation by BMP-2 entrapped in polyethylenimine-coated albumin nanoparticle.* Biomaterials 2009;30:5143–5155.
Zhou S, Bachem MG, Seufferlein T, Li Y, Gross HJ, Schmelz A. *Low intensity pulsed ultrasound accelerates macrophage phagocytosis by a pathway that requires actin polymerization, Rho, and Src/MAPKs activity.* Cell Signal. 2008;20(4):695–704.

Nanoinformatics: Developing Advanced Informatics Applications for Nanomedicine

Victor Maojo, Miguel García-Remesal, Diana de la Iglesia, José Crespo, David Pérez-Rey, Stefano Chiesa, Martin Fritts, and Casimir A. Kulikowski

Abstract In this chapter we introduce a new informatics field called "Nanoinformatics" based on our own research and on consensus work carried out under the auspices of the US National Cancer Institute (NCI) and the European Commission (EC). Nanoinformatics can be defined as the "use of informatics techniques for analyzing and processing information about the structure and physicochemical characteristics of nanoparticles, their environments, and applications". Its goal is to use information to accelerate research in nanomedicine. We summarize various areas of research and applications, such as ontologies and semantic search, text mining, imaging, standards, together with an example of the research our group has performed in this field on text mining for extracting information about nanotoxicity from the literature.

Keywords Nanomedicine • Biomedical Informatics • Nanoinformatics • Nanoparticles • Nanoresources • Nanotoxicity • Data Integration • Ontologies • Semantic search • Modeling and Simulation

Abbreviations

BI	Bioinformatics
BMI	Biomedical Informatics
EC	European Commission

V. Maojo (✉), M. García-Remesal, D. de la Iglesia, J. Crespo, D. Pérez-Rey, and S. Chiesa
Departamento Inteligencia Artificial, Facultad de Informatica, Biomedical Informatics Group, Universidad Politécnica de Madrid, Campus de Montegancedo, Madrid, Spain
e-mail: vmaojo@fi.upm.es

M. Fritts
Nanotechnology Characterization Laboratory and SAIC-Frederick of Computer Science, National Cancer Institute, US National Institutes of Health, Baltimore, MD, USA

C.A. Kulikowski
Department of Computer Science, Rutgers, The State University of New Jersey, Newark, NJ, USA

A. Prokop (ed.), *Intracellular Delivery: Fundamentals and Applications*, Fundamental Biomedical Technologies 5, DOI 10.1007/978-94-007-1248-5_26,

HIS	Hospital Information Systems
IMIA	International Medical Informatics Association
MI	Medical Informatics
NCI	National Cancer Institute
NIH	National Institutes of Health
NIOSH	National Institute for Occupational Safety and Health
NPO	NanoParticle Ontology
NSF	National Science Foundation
NT	Nanomedicine Taxonomy
ONAMI	Oregon Nanoscience and Microtechnologies Institute
PMID	PubMed Identifier
QM/MM	Quantum Mechanics and Molecular Mechanics
TM	Text Mining
UMLS	Unified Medical Language System

1 Introduction

Computers applications in biomedicine started appearing only a few years after the first commercial computers came onto the market in the 1950s. Some physicians and biologists soon realized that computers were suited scientific and professional problem-solving, going beyond their first uses in financial and resource management. In about a decade many medical applications, involving computerized medical records, Hospital Information Systems (HIS), decision support systems, radiology and departmental information systems had been developed. In biology, cell and chromosome analysis was followed by work on DNA and RNA sequencing, databases for storing genetic information, programs for molecular visualization and simulation and much else besides.

Over the past 15 years, different informatics disciplines addressing various biomedical subfields have led to various kinds of synergies emerging between them. From different areas, originally labeled "medical informatics", "bioinformatics", "systems biology", "public health informatics", "computational biology", "clinical informatics" and others a new broad and challenging field emerged in 1998–2000 (Altman 1998; Maojo and Kulikowski 2003), called "Biomedical Informatics" (BMI). Scientific recognition has come from the many informatics challenges resulting from the very large, publicly available datasets generated as result of the Human Genome and other—omics projects. Now, a new area termed "Nanoinformatics" involves the analysis of information-related issues arising in nanomedicine and nanotechnology, and the research and development of new methods and tools at the biomedical nano-level. At the present time of writing, only six papers can be retrieved in Medline by searching the term "nanoinformatics"—including papers published in major journals (Maojo et al. 2010); (De la Calle et al. 2009; De la Iglesia et al. 2011)—but a larger set of papers can be found searching with the terms "Nanotechnology

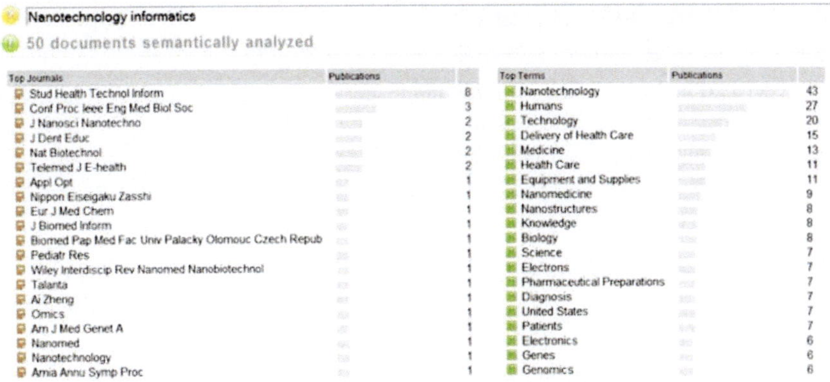
Nanotechnology informatics

50 documents semantically analyzed

Top Journals	Publications	Top Terms	Publications
Stud Health Technol Inform	8	Nanotechnology	43
Conf Proc Ieee Eng Med Biol Soc	3	Humans	27
J Nanosci Nanotechno	2	Technology	20
J Dent Educ	2	Delivery of Health Care	15
Nat Biotechnol	2	Medicine	13
Telemed J E-health	2	Health Care	11
Appl Opt	1	Equipment and Supplies	11
Nippon Eiseigaku Zasshi	1	Nanomedicine	9
Eur J Med Chem	1	Nanostructures	8
J Biomed Inform	1	Knowledge	8
Biomed Pap Med Fac Univ Palacky Olomouc Czech Repub	1	Biology	8
Pediatr Res	1	Science	7
Wiley Interdiscip Rev Nanomed Nanobiotechnol	1	Electrons	7
Talanta	1	Pharmaceutical Preparations	7
Ai Zheng	1	Diagnosis	7
Omics	1	United States	7
Am J Med Genet A	1	Patients	7
Nanomed	1	Electronics	6
Nanotechnology	1	Genes	6
Amia Annu Symp Proc	1	Genomics	6

Fig. 1 An analysis of the papers published and available in Medline, indexed with the term "Nanotechnology informatics". On the left, a summary of the publication sources; on the right, top keyword terms included in those papers. Search carried out by using GoPubMed (http://www.gopubmed.org/)

Informatics". Figure 1 shows an analysis, carried out with GoPubMed (http://www.gopubmed.org/), of different results from such a search, including the year of publication and the areas covered. However, we use the term "Nanoinformatics" here, since we believe that it might result in greater long-term acceptance. In fact, the three main conferences held or planned at the time of writing—in 2007, 2008 and 2010—included the term "Nanoinformatics" in their titles.

Nanoinformatics refers "to the use of informatics techniques for analyzing and processing information about the structure and physicochemical characteristics of nanoparticles, their environments, and applications" (Maojo et al. 2010). It incorporates, under its umbrella, concepts from areas such as informatics (computer science and information technologies), nanotechnology, medicine and other traditional disciplines such as biology, chemistry and physics. Two preliminary White Papers, written by researchers with support from the NCI (Baker et al. 2009) and the EC (http://www.action-grid.eu), are already available for reference. The authors actively participated in writing the second one. These and other pioneering initiatives have helped launch Nanoinformatics as a novel discipline designed to accelerate the different research and development directions in nanomedicine and propose new challenges and a roadmap for the field.

In the following we give a brief overview of a discipline which is still at an early stage, barely beyond its inception. First, we summarize various nanomedical topics where informatics methods and tools can be effectively applied. Second, we list and discuss a range of emerging and challenging Nanoinformatics topics. Finally, we describe text-mining based research on Nanoinformatics developed by the authors—a first in the literature.

2 Topics in Nanomedicine

As described in this volume, Nanomedicine deals with a wide range of topics, some of which specifically relate to potential applications of Nanoinformatics, as illustrated by the examples that follow (Jain 2008).

2.1 *Delivery Mechanisms*

Nanoparticles can be used as carriers to deliver specific substances (e.g., genes, drugs, etc.) to target various specific parts of an organism (e.g., organs, cells, molecules). Various particles can be used; each of them has its own special characteristics—like dendrimers, liposomes, buckyballs, etc. Nanoparticles can be designed to release drugs in a controlled manner if certain conditions are met. Computerized simulations can be carried out "in silico" to design their structure and anticipate their properties. This area emphasizes controlled release which implies, for instance, a sensing system to start and stop release or a detection system.

2.2 *Implantable Devices*

Nanodevices—such as nanorobotos—can be designed to obtain an almost non-invasive continuous monitoring for some diseases. These nanodevices integrate different modules to detect variations on standard biological parameters or, if necessary, to release a specific drug to reach a specified target, reducing expected toxicity. Research in this field presents great challenges and opportunities for personalizing the practice of medicine. This area will require tailoring the device to a patient's individual response by monitoring a portion of her gene expression, proteome and/or metabolome.

2.3 *Diagnosis*

Nanoparticles can interact directly with molecules in living organisms. This makes them specific and effective tools for diagnosing some pathological conditions and diseases. When an engineered nanoparticle detects a specific molecule, it can bind to it and "highlight" it by using, for instance, quantum dots. Decision support is a very active area of research in BMI which extends naturally to nanomedicine. Examples might include, for instance, decision support in surgery to help in visualization. This area of diagnosis would supplement information on disease and pathology with actively correlating known biomarkers to aid a doctor in, for instance, traversing a vastly more complicated decision tree based on the reported patient data.

2.4 Therapy

Nanotechnology allows for therapies that are less invasive for the patient and usually require lower dosage levels than conventional drugs. By improving therapeutic efficacy they can help reduce the toxic effect of drugs. And, nanoparticles can be designed to have their own therapeutic properties including those which are anti-bacterial (Thomas et al. 2009) or antimicrobial (Chamundeeswari et al. 2009). To test their efficacy and potential secondary effects new models of nano-clinical trials must be designed. Similar efforts have been already carried out to design clinical-genomic trials (Anguita et al. 2008). New models of computerized clinical practice guidelines can be adapted to nanomedicine. This area, therapy, would also have to bring in antimicrobial and antibacterial information—as well as help to design trials that can identify and compensate for individual reactions to these pathogens and help generate the needed clinical guidelines.

3 Materials

Studies of the biological interactions between specific nanostructures (such as dendrimers) and cell physiology are needed to help in the design of new nanomedical techniques. New biocompatible materials can be also used in implants for replacement therapies and in regenerative medicine.

4 Nanoinformatics: Areas of Research and Development

Given its novelty, Nanoinformatics is a potential discipline which has yet to be fully defined. Computers have been used for many aspects of nanotechnological research, but the field still needs a better specification of goals, subfields, topics and educational requirements, which could help lead to more concrete agendas for research and development. We can consider nanoinformatics to be the ideal way to integrate various areas of research and translation in a reasonable timeframe. Nanotechnology offers the hope of developing and testing these systems now so that we develop a rational method of expanding them as our knowledge grows.

Even at this early stage we can suggest some specific nanoinformatics areas of investigation as follows:

4.1 Ontologies and Semantic Search

Biomedical ontologies describe concepts or classes—for instance, organisms, organs, cells, molecules, etc.—with their associated properties and semantic relationships. They have contributed to the development of various significant approaches towards systematizing information and knowledge, particularly in the biomedical field (Smith et al. 2007). Ontologies are currently a fundamental technology for facilitating system interoperability and information search, retrieval and extraction.

Structuring information in the area of nanomedicine is essential for current research, so ontology design becomes a fundamental challenge for anchoring data and knowledge representations. In this regard, two taxonomies or ontologies have already been designed. These are the Nanomedicine Taxonomy (NT) (Gordon and Sagman 2003) and the Nanoparticle Ontology (NPO) (Thomas et al. 2010). In addition, there are proposals to develop domain ontologies and taxonomies in the area of cancer nanotechnology, with support from the NCI and other organizations. The NCI White paper (Baker et al. 2009) described a set of initiatives, including a ontology for discovery of new nanomaterials, a functional ontology, the Nanotech Index Ontology, an atlas of nanotechnology or BiomedGT, among others.

4.2 Nanoparticle Characterization

Nanotechnology requires a lengthy development pipeline from nanomaterial synthesis to physical characterization, to *in vitro* and *in vivo* experimentation, and to clinical applications of all these materials. How to integrate all the information that will be involved in this becomes a key challenge. Informatics methods and tools are necessary to complement wet bench research, and involves much empirical heterogeneous that must be collected and analyzed. As suggested by Fritts (2009), this should be one of the central challenges for Nanoinformatics, suggesting that applying previous results obtained in the area of BMI becomes a natural and most efficient option.

4.3 Data Integration and Exchange

Establishing an inventory of nanoresources will become a fundamental prerequisite for accessing the different materials, software tools, databases, registries, etc. In Nanomedicine, making this information easily accessible for biomedical professionals and practitioners will facilitate developing personalized therapies and reducing potential nanotoxic side-effects.

The authors have been particularly focused on carrying out research on system interoperability in BMI areas. These included syntactic and semantic-focused systems which used ontologies as a conceptual reference to achieve database integration (Alonso-Calvo et al. 2007). Over the next few years data integration might evolve from its original clinical and/or genomic foci to address multilevel integration down to the nano level. From a computational perspective, data integration at the nano level poses even more difficult informational challenges that must be addressed—parallel to those faced at higher scales by biomedical informatics. These include, for instance, developing central repositories of nanoparticle-related information, new nano-ontologies or creating new standards for storing information. NanoHub or caNanolab are interesting examples of extended repositories of different information and tools.

4.4 Linking Nano-Information to Computerized Medical Records

Since the creation of computerized medical records, standardization of their informational contents has been a key issue in their development. Since the turn of the century, a new challenge has been the introduction of genomic information. In this context, linking genotypic and phenotypic information is still a topic of research. For nanomedicine, toxicity will play a critical role, which must be introduced into computerized medical records. There are already various databases of toxic effects such as NIOSH's (National Institute for Occupational Safety and Health, http://www.cdc.gov/niosh/) and ONAMI's toxicity analysis (Oregon Nanoscience and Microtechnologies Institute, http://www.greennano.org). Researchers can address new nanoinformatics methods to model and simulate nanotoxicity processes, linking them with actual patient data from computerized medical records to predict human response. In Europe and the USA some recent initiatives—like ACGT (Anguita et al. 2008) and caBIG (McCusker et al. 2009)—have already carried out research to develop new approaches to data acquistion, exchange, modeling and simulation of nano-related issues such as drug delivery. They use Grid-based infrastructures to facilitate high performance computing of complex processes like simulation. They propose to optimize clinico-genomic trials, enhancing at the same time other issues like testing drug efficacy and preventive care. By using advanced computational methods, researchers can reduce the time needed to develop and evaluate drugs—nanoparticles, in nanomedicine—, which introduce different requirements, leading to "nano-clinical" trials—from the lab to clinical practice.

4.5 Nanoinformatics Education

Training in nano-related topics has been, until now, usually restricted to programmes mostly focused in the chemical and physical fields. In the future nanoinformatics training should be expanded to other areas—e.g. medicine, engineering, computer science—to deal with numerous additional issues. In this regard, members of the International Medical Informatics Association (IMIA) have developed new guidelines for nano-related education (Mantas et al. 2010). New academic programmes can be designed or new sections of established programmes—e.g. in BMI or medicine—can be modified by introducing nano-related topics and areas.

Education in the new field of Nanoinformatics reminds one of the original problems faced by BMI years ago (early1990s), when traditional programs had to be redesigned. In addition, Nanoinformatics relies heavily on work in fields such as physics and chemistry, involving new modes of imaging or pharmacodynamics, which go well beyond traditional research and curricular efforts in BMI, biomedicine or nanotechnology. As we have already proposed before (Maojo et al. 2010), future professionals with expertise in nanoinformatics may have an important role in future research frameworks as information brokers, linking professionals with very different backgrounds and expertise who might, otherwise, experience difficulties understanding each other. Informatics methods and tools could help understand and manage new concepts needed (fruit fly vs. other kind of fly), by means of different educational systems created in the area.

4.6 Ethical Issues

Nanotechnology and nanomedicine have important associated ethical issues, like nanotoxicity—both animal and environmental—, these included not just ethical issues, but also cultural, media, entertainment, philosophical and even religious aspects that have been previously proposed elsewhere (http://www.action-grid.eu). In this regard, nano-clinical trials and research must account for these issues, while informatics tools can contribute to managing the required information and help following the specific ethical considerations, rules and laws that regulate these issues in the different countries.

4.7 Imaging Informatics

Imaging techniques—particularly, in biomedicine—based on nanotechnological research can be less invasive and more precise. For example, quantum dots generate long-duration signals that can improve the quality of images with respect to traditional contrast agents. Monitoring diseases in living organisms should improve the efficacy of therapies, allowing doctors to design better case-specific approaches. So what is needed are novel informatics methods to link tissue banks to images of tissues and histology results, imaging annotation from radiology, computerized medical records containing large amounts of images, can improve imaging in concrete medical cases.

4.8 Simulation and Modeling

In the context of nanomedicine, molecular modeling and simulation techniques have been proposed as key methodologies for bionanotechnology research. Quantum mechanics, molecular modeling and molecular simulations techniques provide an adequate basis for characterizing diverse nanosytems' physicochemical properties. In this subfield, several simulations of nanoparticles and other nanosystems suggest the suitability of these computer-based techniques to explore and understand basic chemical and biological properties of nanoparticles and nanomaterials (Khurana et al. 2006). These tools can be used not only to model basic nanoparticle properties, but also to gain insight into specific interactions with biological systems. The application of computer-intensive methods like hybrid quantum mechanics and molecular mechanics (QM/MM) simulations, augmented with algorithms that efficiently sample the vast conformational space, will facilitate the characterization of a large number of nanoparticle properties, obtaining additional information about the biological phenomena that explains nanoparticle recognition by physiological systems (Archakov and Ivanov 2007). In this regard, this type of approach needs several different layers of structural modelling and collaborative sharing to accelerate both research and translation.

Molecular modeling and simulations facilitate exploring interactions at different scales. Through these simulations, one can study the fundamental physicochemical properties of nanoparticles. High performance computing infrastructures—like Grid, Cloud Computing or dedicated supercomputers facilitate this work considerably.

4.9 Terminologies and Standards

Various challenging issues arise in trying to understand standard syntax, semantics, and the pragmatics of design, development and practical use in the field of different application-focused and generic clinical terminologies. In this regard, standards and terminologies like HL7, LOINC, SNOMED, DICOM and others will have to be extended to include nano-related information that address the requirements of this new field. In BMI, for instance, the Unified Medical Language System (UMLS) was able to incorporate genomic information (like Gene Ontology), going beyond the original, medically-focused information. In this regard, standards like HL7 have introduced new approaches to incorporate genomic models and information in its recent releases. Similarly, nanomedicine might lead to new approaches and future versions of current standards. Important examples are the ISA-TAB standards and their extension to nano in Nano-TAB.

4.10 Translational Nanoinformatics and Biomedicine

As we have earlier stated (Kulikowski and Kulikowski 2009), translational bioinformatics aims to translate basic research in the -omics areas into new knowledge that can provide, for instance, personalized medical diagnosis and therapy for patients, taking into account the complexity of genetic and environmental interactions in human diseases. In a similar way, Nanomedicine requires novel insights beyond the current technology of informatics which is typically focused on collecting, representing and linking information and managing various aspects of both systemic and semantic heterogeneity. In this regard, translational nanoinformatics can extend the vision of translating basic research into clinical applications down to the nano level. A wide number of applications can be envisioned along these lines.

4.11 Networks of International Researchers, Projects and Labs

Collaborative research and the development of tools, standards, databases, and publications to support it have shown the feasibility of linking people with related objectives to work together towards a common target. Recently, new opportunities for funding have arisen, and European groups can now be funded by the US National Institutes of Health (NIH) and National Science Foundation (NSF) and vice versa. This new environment will facilitate collaborative research and exchange, as happened earlier in joint initiatives like the Human Genome Project. In the nano fields, such international collaborations are likely to be fundamental in exploiting complementary strengths for advancing research.

4.12 Text Mining for Nanomedical Research

Text Mining (TM) techniques have been extensively used in Medical Informatics (MI) and Bioinformatics (BI) research for different tasks involving the extraction of data and knowledge from textual sources. These tasks include, among others, (i) identifying and extracting named entities such as genes and/or proteins (Chang et al. 2004; Mika and Rost 2004; Settles 2005; Tanabe et al. 2005; Torii et al. 2009) or sequences of nucleic acids and proteins (Mika and Rost 2004; Wren et al. 2005; Aerts et al. 2008; García-Remesal et al. 2010a), (ii) supporting the semantic annotation of textual documents by matching document terms to concepts belonging to controlled vocabularies and ontologies (Aronson 2001), (iii) building ontologies and concept networks from textual resources (García-Remesal et al. 2007), (iv) identifying and extracting relationships between different concepts (Rindflesch and Fiszman 2003; Lussier et al. 2006; García-Remesal et al. 2010b), and (v) automatically populating electronic health records and other biomedical databases with information extracted from texts (Meystre et al. 2008; de la Calle et al. 2009, García-Remesal et al. 2010c).

From an informatics perspective, we believe that previous TM research carried out within MI and BI can be adapted and reused to create novel methods, tools and data resources to facilitate nanomedicine research. For instance, we have developed an informatics tool to identify and extract the toxic effects of nanoparticles reported in the literature. Our tool was built by adapting and modifying the methods and source code of ABNER (Settles 2005), a tool for recognizing mention of genes and proteins in scientific papers. Once the text has been analyzed by our tool and the targets have been detected, the tool highlights the recognized entities with different color codes, each denoting a different category. The extracted data can be useful, among other applications, to create and automatically populate a database of toxic effects of nanoparticles, or to semantically annotate and search the manuscripts based on the different entities that appear in them.

We have recently also developed an automated text indexing and retrieval engine that links scientific articles to mention of concepts from the Nanoparticle Ontology (Thomas et al. 2010) and the Foundational Model of Anatomy (Rosse and Mejino 2003). The indexing engine was built following a dictionary-based approach, similar to that adopted in (Garten and Altman 2009). Using the search engine, users can search for papers reporting which nanoparticles are more suitable for delivering a certain drug to a given anatomical location, or on the other hand, identifying which nanoparticles are toxic to a specific anatomical part. Figure 2 shows the interface of the search engine after executing a query for retrieving all documents in the database that were semantically annotated with the nanoparticle "Paclitaxel" (concept

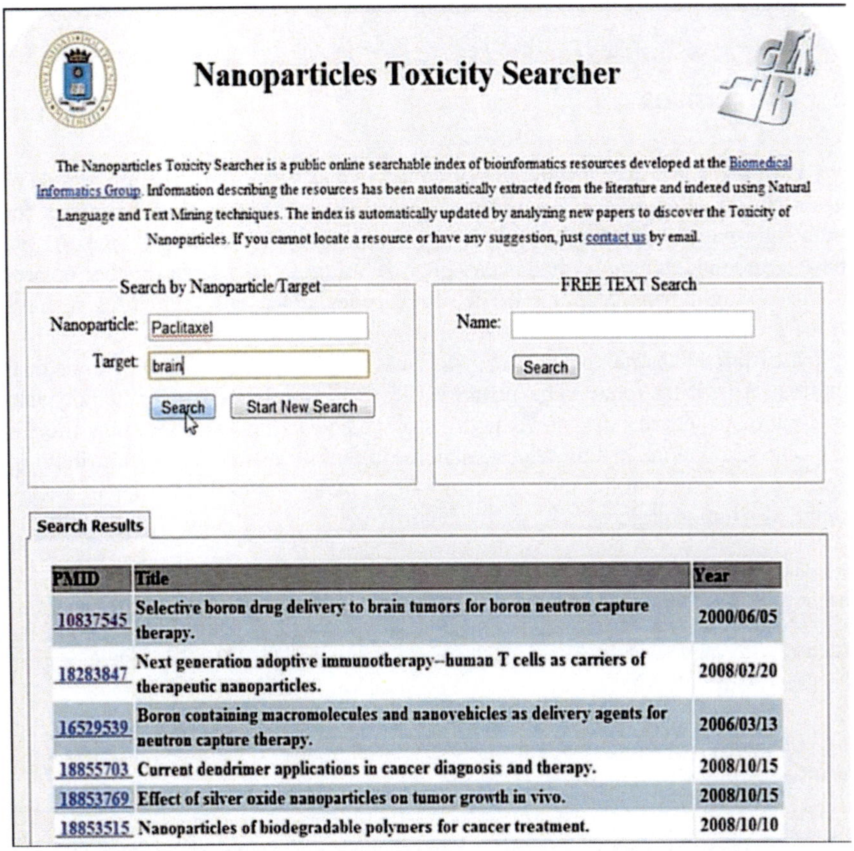

Fig. 2 Screenshot of the search engine linking manuscripts to mentions of nanoparticles and anatomical targets belonging to the Nanoparticle Ontology and to the Foundational Model of Anatomy, respectively

belonging to the Nanoparticle Ontology) and the anatomical location "Brain" (term belonging to the Foundational Model of Anatomy). Results show the PMIDs (PubMed Identifiers) of the articles matching the query together with other relevant information such as the title of the paper and the date of publication. To complete this approach a collaborative effort will be needed—for instance, to improve searching, annotations, integration, etc. This is an area facing many challenges.

Another nanomedicine-related text mining project we are currently working on is the development of an inventory of existing nanoparticles/nanodevices. The inventory is automatically populated by extracting relevant information regarding nanoparticles from scientific papers reporting recent advances in nanotechnology and nanomedicine. The extracted information includes, for instance, the names of the nanoparticles, their nano-scale size-dependent properties (e.g. quantum confinement, surface plasmon resonance, superparamagnetism, etc.), their morphology (e.g. nanospheres, nanoreefs, nanoboxes, etc.), their synthesis methods (e.g. attrition, pyrolysis, etc.), their intended functionality (e.g. drug delivery, tissue repair, etc.), and their potential toxicity (both medical and environmental).

5 Conclusions

We have presented here a summary of concepts and suggested structure related to a new informatics field of application, which presents significant challenges for both informaticians and specialists in nanotechnology and nanomedicine. As we have mentioned, nanoinformatics can provide solutions to a large number of problems arising in nanomedicine. In the past decades, physicians and biologists have faced problems in their research and practice areas which do not differ too much—at least from an abstract perspective—to what nanotechnologists and nanomedical experts are facing today. The primary information-related issues include data integration, standards, decision support, storage and access to information, modeling and simulation. In this regard, nanoinformatics can decisively contribute to accelerate research in the different nano areas, as has happened earlier in a wide range of—omics projects.

Acknowledgments The present work has been funded, in part, by the ACTION-Grid support action (FP7-ICT-2007-2-224176), the Spanish Ministry of Science and Innovation (FIS/AES PS09/00069 and COMBIOMED-RETICS), the Ibero-NBIC network (CYTED 209RT0366), the European Commission through the ACGT integrated project (FP6-2005-IST-026996), and the Comunidad de Madrid, Spain.

References

Aerts S, Haeussler M, van Vooren S, Griffith OL, Hulpiau P, Jones SJ, Montgomery SB, Bergman CM; Open Regulatory Annotation Consortium (2008). Text-mining assisted regulatory annotation. Genome Biol, 9(2): R31.

Alonso-Calvo R, Maojo V, Billhardt H, Martin-Sanchez F, García-Remesal M, Pérez-Rey D. An agent- and ontology-based system for integrating public gene, protein, and disease databases. J Biomed Inform. 2007 Feb;40(1):17–29.

Altman RB. Bioinformatics in support of molecular medicine (1998). Proc AMIA Symp. 1998:53–61.

Anguita A, Martín L, Crespo J, Tsiknakis M (2008). An ontology-based method to solve query identifier heterogeneity in post-genomic clinical trials. Stud Health Technol Inform 136:3–8.

Archakov AI, Ivanov YD (2007). Analytical nanobiotechnology for medicine diagnostics. Molecular Biosystems 3, 336–342.

Aronson AR (2001). Effective mapping of biomedical text to the UMLS Metathesaurus: the metamap program. Proc AMIA Symp 2001, 17–21.

Baker NA, Fritts M, Guccione S, Paik DS, Pappu RV, Patri A, Rubin D, Shaw SY, Thomas DG. Nanotechnology Informatics White Paper. caBIG, National Cancer Institute. February 2009. Non-peer Reviewed Manuscript.

Chamundeeswari M, Liji Sobhana SS, Jacob JP, Kumar MG, Devi MP, Sastry TP, Mandal AB (2009). Preparation, characterization and evaluation of biopolymeric gold nanocomposite with antimicrobial activity. Biotechnol Appl Biochem. 2009 Nov 20. [Epub ahead of print].

Chang JT, Schütze H, Altman RB (2004). GAPSCORE: finding gene and protein names one word at a time. Bioinformatics, 20(2):216–25.

De la Calle G, García-Remesal M, Chiesa S, De la Iglesia D, Maojo V (2009). BIRI: a new approach for automatically discovering and indexing available public bioinformatics resources from the literature. BMC Bioinformatics, 10: 320.

De la Iglesia D, Maojo V, Chiesa S, Martin-Sanchez F, Kern J, Potamias G, Crespo J, García-Remesal M, Keuchkerian S, Kulikowski C, Mitchell JA (2011). International efforts in nanoinformatics research applied to nanomedicine. Methods Inf Med. 50(1):84–95.

Fritts, M. Personal Communication. 2009.

García-Remesal M, Maojo V, Crespo J, Billhardt H (2007). Logical schema acquisition from text-based sources for structured and non-structured biomedical sources integration. Proc AMIA Symp 2007, 259–263.

García-Remesal M, Cuevas A, López-Alonso V, López-Campos G, de la Calle G, de la Iglesia D, Pérez-Rey D, Crespo J, Maojo V (2010a). A method for automatically extracting infectious disease-related primers and probes from the literature.BMC Bioinformatics, 11:410

García-Remesal M, Maojo V, Billhardt H, Crespo J (2010b). Integration of relational and textual biomedical sources.A pilot experiment using a semi-automated method for logical schema acquisition. Methods Inf Med, 49(4): 337–348

García-Remesal M, Cuevas A, Pérez-Rey D, Martín L, Anguita A, de la Iglesia D, de la Calle G, Crespo J, Maojo V (2010c). PubDNA Finder: a web database linking full-text articles to sequences of nucleic acids. Bioinformatics 2010; doi: 10.1093/bioinformatics/btq520 (epub ahead of print).

Garten Y, Altman RB (2009). Pharmspresso: a text mining tool for extraction of pharmacogenomic concepts and relationships from full text. BMC Bioinformatics 10 Suppl 2.

Gordon N, Sagman U (2003). Nanomedicine Taxonomy. Canadian Institute of Health Research & Canadian NanoBusiness Alliance.

Jain KK. The Handbook of Nanomedicine. 1st ed. Totowa, New Jersey: Humana; 2008.

Khurana E, Nielsen SO, Ensing B. Klein ML. (2006). Self-assembling cyclic peptides: molecular dynamics studies of dimers in polar and nonpolar solvents. J Phys Chem B 110, 18965–18972

Kulikowski CA, Kulikowski CW. Biomedical and health informatics in translational medicine. Methods Inf Med. 2009;48(1):4–10.

Lussier YA, Borlawski T, Rappaport D, Liu Y and Friedman C (2006). PhenoGO: assigning phenotypic context to Gene Ontology annotations with natural language processing. Pac Symp Bio, pp 64–75.

Mantas J, Ammenwerth E, Demiris G, Hasman A, Haux R, Hersh W, Hovenga E, Lun KC, Marin H, Martin-Sanchez F, Wright G; IMIA Recommendations on Education Task Force.

Recommendations of the International Medical Informatics Association(IMIA) on Education in Biomedical and Health Informatics. First Revision. Methods Inf Med. 2010 Jan 7; 49(2):105–120.

Maojo V, Kulikowski CA (2003). Bioinformatics and medical informatics: collaborations on the road to genomic medicine? Journal of the American Medical Informatics Association Nov-Dec 2003;10(6):515–22.

Maojo V, Martin-Sanchez F, Kulikowski C, Rodriguez-Paton A, Fritts M (2010). Nanoinformatics and DNA-based computing: catalyzing nanomedicine. Pediatr Res. 2010 May;67(5):481–9.

McCusker JP, Phillips JA, Beltrán AG, Finkelstein A, Krauthammer M. (2009). Semantic web data warehousing for caGrid. BMC Bioinformatics 10 Suppl 10:S2

Meystre SM, Savova GK, Kipper-Schuler KC, Hurdle JF (2008). Extracting Information from Textual Documents in the Electronic Health Record: a Review of Recent Research.Yearb Med Inform. 2008, 128–44.

Mika S, Rost B (2004).NLProt: extracting protein names and sequences from papers. Nucleic Acids Res, 32(suppl 2): W364-W367

Rindflesch TC, Fiszman M (2003). The interaction of do- main knowledge and linguistic structure in natural language processing: interpreting hypernymic propositions in biomedical text. J Biomed Inform, 36:462–477.

Rosse C, Mejino JVL (2003). A reference ontology for biomedical informatics: the Foundational Model of Anatomy. J Biomed Inform. 36:478–500.

Settles B (2005). ABNER: an open source tool for automatically tagging genes, proteins, and other entity names in text. Bioinformatics, 21(14):3191–3192.

Smith B, Ashburner M, Rosse C, Bard J, Bug W, Ceusters W, Goldberg LJ, Eilbeck K, Ireland A, Mungall CJ, OBI Consortium, Leontis N, Rocca-Serra P, Ruttenberg A, Sansone SA, Scheuermann RH, Shah N, Whetzel PL, Lewis S (2007). The OBO Foundry: coordinated evolution of ontologies to support biomedical data integration. Nat Biotechnol 25:1251–1255

Tanabe L, Xie N, Thom LH, Matten W, Wilbur WJ (2005). GENETAG: a tagged corpus for gene/protein named entity recognition. BMC Bioinformatics, 6(Suppl 1): S3.

Thomas V, Yallapu MM, Sreedhar B, Bajpai SK (2009). Fabrication, characterization of chitosan/nanosilver film and its potential antibacterial application. J Biomater Sci Polym Ed. 20(14):2129–2144.

Thomas DG, Pappu RV, Baker NA (2010). NanoParticle Ontology for cancer nanotechnology research. J Biomed Inform. 2010 Mar 6. doi:10.1016/j.jbi.2010.03. 001 DOI:dx.doi.org [Epub preprint]

Torii M, Hu Z, Wu CH, Liu H (2009). BioTagger-GM: A gene/protein name recognition system. J Am Med Inform Assn, 16(2):247–255.

Wren JD, Hildebrand WH, Chandrasekaran S, Melcher U (2005). Markov model recognition and classification of DNA/protein sequences within large text databases. Bioinformatics 21(21): 4046–4053.

Index

A. Prokop (ed.), *Intracellular Delivery: Fundamentals and Applications*, Fundamental Biomedical Technologies 5, DOI 10.1007/978-94-007-1248-5,

Q

R

S

T